D1606628

Interdisciplinary Applied Mathematics

Volume 18

Problems in engineering, computational science, and the physical and biological sciences are using increasingly sophisticated mathematical techniques. Thus, the bridge between the mathematical sciences and other disciplines is heavily traveled. The correspondingly increased dialog between the disciplines has led to the establishment of the series: *Interdisciplinary Applied Mathematics*.

The purpose of this series is to meet the current and future needs for the interaction between various science and technology areas on the one hand and mathematics on the other. This is done, firstly, by encouraging the ways that mathematics may be applied in traditional areas, as well as point towards new and innovative areas of applications; and secondly, by encouraging other scientific disciplines to engage in a dialog with mathematicians outlining their problems to both access new methods and suggest innovative developments within mathematics itself.

The series will consist of monographs and high-level texts from researchers working on the interplay between mathematics and other fields of science and technology.

Interdisciplinary Applied Mathematics

Volumes published are listed at the end of the book.

Springer

New York
Berlin
Heidelberg
Hong Kong
London
Milan
Paris
Tokyo

J.D. Murray

Mathematical Biology

II: Spatial Models and Biomedical Applications

Third Edition

With 298 Illustrations

J.D. Murray, FRS
Emeritus Professor
University of Oxford *and*
University of Washington
Box 352420
Department of Applied Mathematics
Seattle, WA 98195-2420
USA

Mathematics Subject Classification (2000): 92B05, 92-01, 92C05, 92D30, 34Cxx

Library of Congress Cataloging-in-Publication Data
Murray, J.D. (James Dickson)
Mathematical biology. II: Spatial models and biomedical applications / J.D. Murray.—3rd ed.
p. cm.—(Interdisciplinary applied mathematics)
Rev. ed. of: Mathematical biology. 2nd ed. c1993.
Includes bibliographical references (p.).
ISBN 0-387-95228-4 (alk. paper)
1. Biology—Mathematical models. I. Murray, J.D. (James Dickson) Mathematical biology. II. Title. III. Series.
QH323.5 .M88 ~~2001b~~ 2002 v.2
570′.1′5118—dc21 2001020447

ISBN 0-387-95228-4 Printed on acid-free paper.

Printed in the United States of America.

9 8 7 6 5 4 3 2 1 SPIN 10792366

www.springer-ny.com

Springer-Verlag New York Berlin Heidelberg
A member of BertelsmannSpringer Science+Business Media GmbH

To my wife Sheila, whom I married more than forty years ago and lived happily ever after, and to our children Mark and Sarah

... que se él fuera de su consejo al tiempo de la general criación del mundo, i de lo que en él se encierra, i se hallá ra con él, se huvieran producido *i formado algunas cosas mejor que fueran hechas, i otras ni se hicieran, u se enmendaran i corrigieran.*

Alphonso X (Alphonso the Wise), 1221–1284
King of Castile and Leon (attributed)

If the Lord Almighty had consulted me before embarking on creation I should have recommended something simpler.

Preface to the Third Edition

In the thirteen years since the first edition of this book appeared the growth of mathematical biology and the diversity of applications has been astonishing. Its establishment as a distinct discipline is no longer in question. One pragmatic indication is the increasing number of advertised positions in academia, medicine and industry around the world; another is the burgeoning membership of societies. People working in the field now number in the thousands. Mathematical modelling is being applied in every major discipline in the biomedical sciences. A very different application, and surprisingly successful, is in psychology such as modelling various human interactions, escalation to date rape and predicting divorce.

The field has become so large that, inevitably, specialised areas have developed which are, in effect, separate disciplines such as biofluid mechanics, theoretical ecology and so on. It is relevant therefore to ask why I felt there was a case for a new edition of a book called simply *Mathematical Biology*. It is unrealistic to think that a single book could cover even a significant part of each subdiscipline and this new edition certainly does not even try to do this. I feel, however, that there is still justification for a book which can demonstrate to the uninitiated some of the exciting problems that arise in biology and give some indication of the wide spectrum of topics that modelling can address.

In many areas the basics are more or less unchanged but the developments during the past thirteen years have made it impossible to give as comprehensive a picture of the current approaches in and the state of the field as was possible in the late 1980s. Even then important areas were not included such as stochastic modelling, biofluid mechanics and others. Accordingly in this new edition only some of the basic modelling concepts are discussed—such as in ecology and to a lesser extent epidemiology—but references are provided for further reading. In other areas recent advances are discussed together with some new applications of modelling such as in marital interaction (Volume I), growth of cancer tumours (Volume II), temperature-dependent sex determination (Volume I) and wolf territoriality (Volume II). There have been many new and fascinating developments that I would have liked to include but practical space limitations made it impossible and necessitated difficult choices. I have tried to give some idea of the diversity of new developments but the choice is inevitably prejudiced.

As to general approach, if anything it is even more practical in that more emphasis is given to the close connection many of the models have with experiment, clinical data and in estimating real parameter values. In several of the chapters it is not yet possible to relate the mathematical models to specific experiments or even biological entities. Nevertheless such an approach has spawned numerous experiments based as much on the modelling approach as on the actual mechanism studied. Some of the more mathematical parts in which the biological connection was less immediate have been excised while others that have been kept have a mathematical and technical pedagogical aim but all within the context of their application to biomedical problems. I feel even more strongly about the philosophy of mathematical modelling espoused in the original preface as regards what constitutes good mathematical biology. One of the most exciting aspects regarding the new chapters has been their genuine interdisciplinary collaborative character. Mathematical or theoretical biology is unquestionably an interdisciplinary science *par excellence*.

The unifying aim of theoretical modelling and experimental investigation in the biomedical sciences is the elucidation of the underlying biological processes that result in a particular observed phenomenon, whether it is pattern formation in development, the dynamics of interacting populations in epidemiology, neuronal connectivity and information processing, the growth of tumours, marital interaction and so on. I must stress, however, that mathematical descriptions of biological phenomena are not biological explanations. The principal use of any theory is in its predictions and, even though different models might be able to create similar spatiotemporal behaviours, they are mainly distinguished by the different experiments they suggest and, of course, how closely they relate to the real biology. There are numerous examples in the book.

Why use mathematics to study something as intrinsically complicated and ill understood as development, angiogenesis, wound healing, interacting population dynamics, regulatory networks, marital interaction and so on? We suggest that mathematics, rather theoretical modelling, must be used if we ever hope to genuinely and realistically convert an understanding of the underlying mechanisms into a predictive science. Mathematics is required to bridge the gap between the level on which most of our knowledge is accumulating (in developmental biology it is cellular and below) and the macroscopic level of the patterns we see. In wound healing and scar formation, for example, a mathematical approach lets us explore the logic of the repair process. Even if the mechanisms were well understood (and they certainly are far from it at this stage) mathematics would be required to explore the consequences of manipulating the various parameters associated with any particular scenario. In the case of such things as wound healing and cancer growth—and now in angiogensesis with its relation to possible cancer therapy—the number of options that are fast becoming available to wound and cancer managers will become overwhelming unless we can find a way to simulate particular treatment protocols before applying them in practice. The latter has been already of use in understanding the efficacy of various treatment scenarios with brain tumours (glioblastomas) and new two step regimes for skin cancer.

The aim in all these applications is not to derive a mathematical model that takes into account every single process because, even if this were possible, the resulting model would yield little or no insight on the crucial interactions within the system. Rather the goal is to develop models which capture the essence of various interactions allowing their outcome to be more fully understood. As more data emerge from the biological system, the models become more sophisticated and the mathematics increasingly challenging.

In development (by way of example) it is true that we are a long way from being able to reliably simulate actual biological development, in spite of the plethora of models and theory that abound. Key processes are generally still poorly understood. Despite these limitations, I feel that exploring the logic of pattern formation is worthwhile, or rather essential, even in our present state of knowledge. It allows us to take a hypothetical mechanism and examine its consequences in the form of a mathematical model, make predictions and suggest experiments that would verify or invalidate the model; even the latter casts light on the biology. The very process of constructing a mathematical model can be useful in its own right. Not only must we commit to a particular mechanism, but we are also forced to consider what is truly essential to the process, the central players (variables) and mechanisms by which they evolve. We are thus involved in constructing frameworks on which we can hang our understanding. The model equations, the mathematical analysis and the numerical simulations that follow serve to reveal quantitatively as well as qualitatively the consequences of that logical structure.

This new edition is published in two volumes. Volume I is an introduction to the field; the mathematics mainly involves ordinary differential equations but with some basic partial differential equation models and is suitable for undergraduate and graduate courses at different levels. Volume II requires more knowledge of partial differential equations and is more suitable for graduate courses and reference.

I would like to acknowledge the encouragement and generosity of the many people who have written to me (including a prison inmate in New England) since the appearance of the first edition of this book, many of whom took the trouble to send me details of errors, misprints, suggestions for extending some of the models, suggesting collaborations and so on. Their input has resulted in many successful interdisciplinary research projects several of which are discussed in this new edition. I would like to thank my colleagues Mark Kot and Hong Qian, many of my former students, in particular Patricia Burgess, Julian Cook, Tracé Jackson, Mark Lewis, Philip Maini, Patrick Nelson, Jonathan Sherratt, Kristin Swanson and Rebecca Tyson for their advice or careful reading of parts of the manuscript. I would also like to thank my former secretary Erik Hinkle for the care, thoughtfulness and dedication with which he put much of the manuscript into LaTeX and his general help in tracking down numerous obscure references and material.

I am very grateful to Professor John Gottman of the Psychology Department at the University of Washington, a world leader in the clinical study of marital and family interactions, with whom I have had the good fortune to collaborate for nearly ten years. Without his infectious enthusiasm, strong belief in the use of mathematical modelling, perseverance in the face of my initial scepticism and his practical insight into human interactions I would never have become involved in developing with him a general theory of marital interaction. I would also like to acknowledge my debt to Professor Ellworth C. Alvord, Jr., Head of Neuropathology in the University of Washington with whom I have collaborated for the past seven years on the modelling of the growth and control of brain tumours. As to my general, and I hope practical, approach to modelling I am most indebted to Professor George F. Carrier who had the major influence on me when I went to Harvard on first coming to the U.S.A. in 1956. His astonishing insight and ability to extract the key elements from a complex problem and incorporate them into a realistic and informative model is a talent I have tried to acquire throughout my career. Finally, although it is not possible to thank by name all of my past students, postdoctorals, numerous collaborators and colleagues around the world who have encouraged me in this field, I am certainly very much in their debt.

Looking back on my involvement with mathematics and the biomedical sciences over the past nearly thirty years my major regret is that I did not start working in the field years earlier.

Bainbridge Island, Washington *J.D. Murray*
January 2002

Preface to the First Edition

Mathematics has always benefited from its involvement with developing sciences. Each successive interaction revitalises and enhances the field. Biomedical science is clearly the premier science of the foreseeable future. For the continuing health of their subject, mathematicians must become involved with biology. With the example of how mathematics has benefited from and influenced physics, it is clear that if mathematicians do not become involved in the biosciences they will simply not be a part of what are likely to be the most important and exciting scientific discoveries of all time.

Mathematical biology is a fast-growing, well-recognised, albeit not clearly defined, subject and is, to my mind, the most exciting modern application of mathematics. The increasing use of mathematics in biology is inevitable as biology becomes more quantitative. The complexity of the biological sciences makes interdisciplinary involvement essential. For the mathematician, biology opens up new and exciting branches, while for the biologist, mathematical modelling offers another research tool commensurate with a new powerful laboratory technique but *only* if used appropriately and its limitations recognised. However, the use of esoteric mathematics arrogantly applied to biological problems by mathematicians who know little about the real biology, together with unsubstantiated claims as to how important such theories are, do little to promote the interdisciplinary involvement which is so essential.

Mathematical biology research, to be useful and interesting, must be relevant *biologically*. The best models show how a process works and then predict what may follow. If these are not already obvious to the biologists *and* the predictions turn out to be right, then you will have the biologists' attention. Suggestions as to what the governing mechanisms are may evolve from this. *Genuine* interdisciplinary research and the use of models can produce exciting results, many of which are described in this book.

No previous knowledge of biology is assumed of the reader. With each topic discussed I give a brief description of the biological background sufficient to understand the models studied. Although stochastic models are important, to keep the book within reasonable bounds, I deal exclusively with deterministic models. The book provides a toolkit of modelling techniques with numerous examples drawn from population ecology, reaction kinetics, biological oscillators, developmental biology, evolution, epidemiology and other areas.

The emphasis throughout the book is on the practical application of mathematical models in helping to unravel the underlying mechanisms involved in the biological processes. The book also illustrates some of the pitfalls of indiscriminate, naive or uninformed use of models. I hope the reader will acquire a practical and realistic view of biological modelling and the mathematical techniques needed to get approximate quantitative solutions and will thereby realise the importance of relating the models and results to the real biological problems under study. If the use of a model stimulates experiments—even if the model is subsequently shown to be wrong—then it has been successful. Models can provide biological insight and be very useful in summarising, interpreting and interpolating real data. I hope the reader will also learn that (certainly at this stage) there is usually no 'right' model: producing similar temporal or spatial patterns to those experimentally observed is only a first step and does not imply the model mechanism is the one which applies. Mathematical descriptions are *not* explanations. Mathematics can never provide the complete solution to a biological problem on its own. Modern biology is certainly not at the stage where it is appropriate for mathematicians to try to construct comprehensive theories. A close collaboration with biologists is needed for realism, stimulation and help in modifying the model mechanisms to reflect the biology more accurately.

Although this book is titled *mathematical biology* it is not, and could not be, a definitive all-encompassing text. The immense breadth of the field necessitates a restricted choice of topics. Some of the models have been deliberately kept simple for pedagogical purposes. The exclusion of a particular topic—population genetics, for example—in no way reflects my view as to its importance. However, I hope the range of topics discussed will show how exciting intercollaborative research can be and how significant a role mathematics can play. The main purpose of the book is to present some of the basic and, to a large extent, generally accepted theoretical frameworks for a variety of biological models. The material presented does not purport to be the latest developments in the various fields, many of which are constantly expanding. The already lengthy list of references is by no means exhaustive and I apologise for the exclusion of many that should be included in a definitive list.

With the specimen models discussed and the philosophy which pervades the book, the reader should be in a position to tackle the modelling of genuinely practical problems with realism. From a *mathematical* point of view, the art of good modelling relies on: (i) a sound understanding and appreciation of the biological problem; (ii) a realistic mathematical representation of the important biological phenomena; (iii) finding useful solutions, preferably quantitative; and what is crucially important; (iv) a biological interpretation of the mathematical results in terms of insights and predictions. The mathematics is dictated by the biology and not vice versa. Sometimes the mathematics can be very simple. Useful mathematical biology research is not judged by mathematical standards but by different and no less demanding ones.

The book is suitable for physical science courses at various levels. The level of mathematics needed in collaborative biomedical research varies from the very simple to the sophisticated. Selected chapters have been used for applied mathematics courses in the University of Oxford at the final-year undergraduate and first-year graduate levels. In the U.S.A. the material has also been used for courses for students from the second-year undergraduate level through graduate level. It is also accessible to the more theoretically oriented bioscientists who have some knowledge of calculus and differential equations.

I would like to express my gratitude to the many colleagues around the world who have, over the past few years, commented on various chapters of the manuscript, made valuable suggestions and kindly provided me with photographs. I would particularly like to thank Drs. Philip Maini, David Lane, and Diana Woodward and my present graduate students who read various drafts with such care, specifically Daniel Bentil, Meghan Burke, David Crawford, Michael Jenkins, Mark Lewis, Gwen Littlewort, Mary Myerscough, Katherine Rogers and Louisa Shaw.

Oxford *J.D. Murray*
January 1989

Table of Contents

CONTENTS, VOLUME II

Table of Contents (*continued*)

CONTENTS, VOLUME I

J.D. Murray: ***Mathematical Biology, I: An Introduction***

1. Multi-Species Waves and Practical Applications

1.1 Intuitive Expectations

In Volume 1 we saw that if we allowed spatial dispersal in the single reactant or species, travelling wavefront solutions were possible. Such solutions effected a smooth transition between two steady states of the space independent system. For example, in the case of the Fisher–Kolmogoroff equation (13.4), Volume I, wavefront solutions joined the steady state $u = 0$ to the one at $u = 1$ as shown in the evolution to a propagating wave in Figure 13.1, Volume I. In Section 13.5, Volume I, where we considered a model for the spatial spread of the spruce budworm, we saw how such travelling wave solutions could be found to join any two steady states of the spatially independent dynamics. In this and the next few chapters, we shall consider systems where several species—cells, reactants, populations, bacteria and so on—are involved, concentrating, but not exclusively, on reaction diffusion chemotaxis mechanisms, of the type derived in Sections 11.2 and 11.4, Volume I. In the case of reaction diffusion systems (11.18), Volume I, we have

$$\frac{\partial \mathbf{u}}{\partial t} = \mathbf{f}(\mathbf{u}) + D\nabla^2 \mathbf{u}, \tag{1.1}$$

where $\mathbf{u}$ is the vector of reactants, $\mathbf{f}$ the nonlinear reaction kinetics and D the matrix of diffusivities, taken here to be constant.

Before analysing such systems let us try to get some intuitive idea of what kind of solutions we might expect to find. As we shall see, a very rich spectrum of solutions it turns out to be. Because of the analytical difficulties and algebraic complexities that can be involved in the study of nonlinear systems of reaction diffusion chemotaxis equations, an intuitive approach can often be the key to getting started and to what might be expected. In keeping with the philosophy in this book such intuition is a crucial element in the modelling and analytical processes. We should add the usual cautionary caveat, that it is mainly stable travelling wave solutions that are of principal interest, but not always. The study of the stability of such solutions is not usually at all simple, particularly in two or more space dimensions, and in many cases has still not yet been done.

Consider first a single reactant model in one space dimension x, with multiple steady states, such as we discussed in Section 13.5, Volume I, where there are 3 steady states u_i, $i = 1, 2, 3$ of which u_1 and u_3 are stable in the spatially homogeneous situation. Suppose that initially u is at one steady state, $u = u_1$ say, for all x. Now suppose we suddenly change u to u_3 in $x < 0$. With u_3 dominant the effect of diffusion is to initiate a travelling wavefront, which propagates into the $u = u_1$ region and so eventually $u = u_3$ everywhere. As we saw, the inclusion of diffusion effects in this situation resulted in a smooth travelling wavefront solution for the reaction diffusion equation. In the case of a multi-species system, where **f** has several steady states, we should reasonably expect similar travelling wave solutions that join steady states. Although mathematically a spectrum of solutions may exist we are, of course, only interested here in nonnegative solutions. Such multi-species wavefront solutions are usually more difficult to determine analytically but the essential concepts involved are more or less the same, although there are some interesting differences. One of these can arise with interacting predator–prey models with spatial dispersal by diffusion. Here the travelling front is like a wave of pursuit by the predator and of evasion by the prey: we discuss one such case in Section 1.2. In Section 1.5 we consider a model for travelling wavefronts in the Belousov–Zhabotinskii reaction and compare the analytical results with experiment. We also consider practical examples of competition waves associated with the spatial spread of genetically engineered organisms and another with the red and grey squirrel.

In the case of a single reactant or population we saw in Chapter 13, Volume I that limit cycle periodic solutions are not possible, unless there are delay effects, which we do not consider here. With multi-reactant kinetics or interacting species, however, as we saw in Chapter 3, Volume I we can have stable periodic limit cycle solutions which bifurcate from a stable steady state as a parameter, γ say, increases through a critical γ_c. Let us now suppose we have such reaction kinetics in our reaction diffusion system (1.1) and that initially $\gamma > \gamma_c$ for all x; that is, the system is oscillating. If we now locally perturb the oscillation for a short time in a small spatial domain, say, $0 < |x| \leq \varepsilon \ll 1$, then the oscillation there will be at a different phase from the surrounding medium. We then have a kind of localised 'pacemaker' and the effect of diffusion is to try to smooth out the differences between this pacemaker and the surrounding medium. As we noted above, a sudden change in u can initiate a propagating wave. So, in this case as u regularly changes in the small circular domain relative to the outside domain, it is like regularly initiating a travelling wave from the pacemaker. In our reaction diffusion situation we would thus expect a travelling *wave train* of concentration differences moving through the medium. We discuss such wave train solutions in Section 1.7.

It is possible to have chaotic oscillations when three or more equations are involved, as we noted in Chapter 3, Volume I, and indeed with only a single *delay* equation in Chapter 1, Volume I. There is thus the possibility of quite complicated wave phenomena if we introduce, say, a small chaotic oscillating region in an otherwise regular oscillation. These more complicated wave solutions can occur with only one space dimension. In two or three space dimensions the solution behaviour can become quite baroque. Interestingly, chaotic behaviour can occur without a chaotic pacemaker; see Figure 1.23 in Section 1.9.

Suppose we now consider two space dimensions. If we have a small circular domain, which is oscillating at a different frequency from the surrounding medium, we should expect a travelling wave train of concentric circles propagating out from the pacemaker centre; they are often referred to as *target patterns* for obvious reasons. Such waves were originally found experimentally by Zaikin and Zhabotinskii (1970) in the Belousov–Zhabotinskii reaction: Figure 1.1(a) is an example. Tyson and Fife (1980)

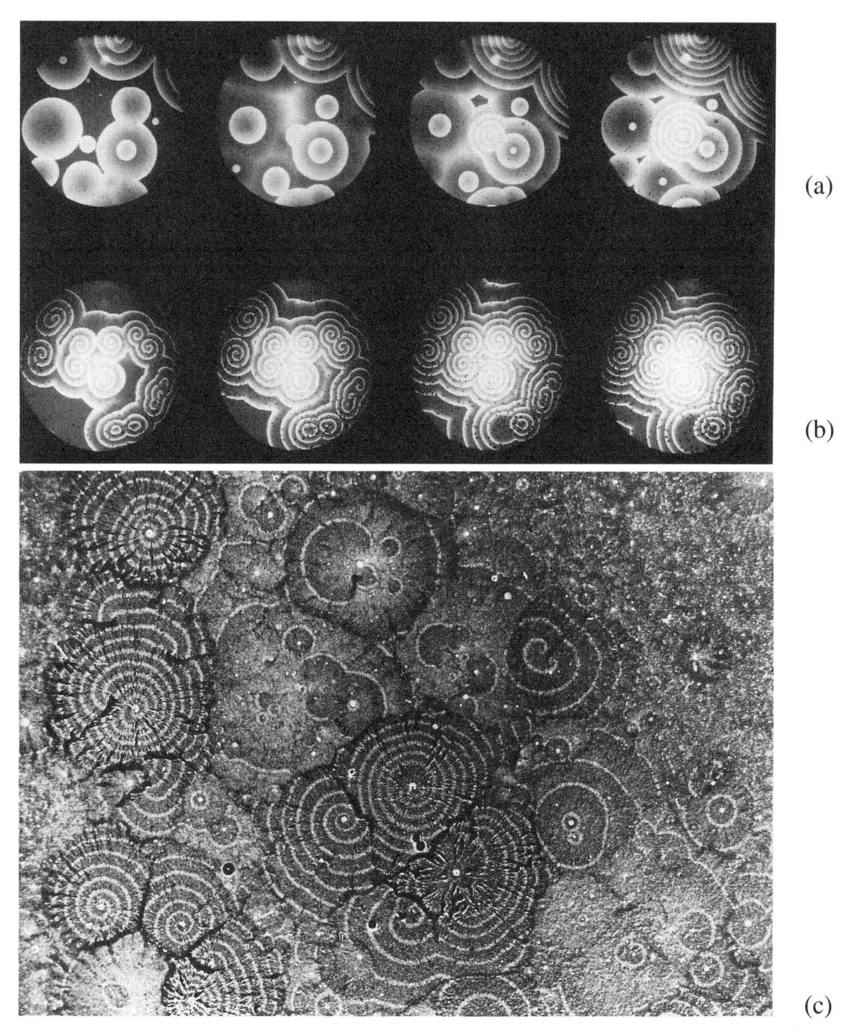

Figure 1.1. (**a**) Target patterns (circular waves) generated by pacemaker nuclei in the Belousov–Zhabotinskii reaction. The photographs are about 1 min apart. (**b**) Spiral waves, initiated by gently stirring the reagent. The spirals rotate with a period of about 2 min. (Reproduced with permission of A. T. Winfree) (**c**) In the slime mould *Dictyostelium*, the cells (amoebae) at a certain state in their group development, emit a periodic signal of the chemical, cyclic AMP, which is a chemoattractant for the cells. Certain pacemaker cells initiate target-like and spiral waves. The light and dark bands arise from the different optical properties between moving and stationary amoebae. The cells look bright when moving and dark when stationary. (Courtesy of P. C. Newell from Newell 1983)

discuss target patterns in the Field–Noyes model for the Belousov–Zhabotinskii reaction, which we considered in detail in Chapter 8. Their analytical methods can also be applied to other systems.

We can think of an oscillator as a pacemaker which continuously moves round a circular ring. If we carry this analogy over to reaction diffusion systems, as the 'pacemaker' moves round a small core ring it continuously creates a wave, which propagates out into the surrounding domain, from each point on the circle. This would produce, not target patterns, but spiral waves with the 'core' the limit cycle pacemaker. Once again these have been found in the Belousov–Zhabotinskii reaction; see Figure 1.1(b) and, for example, Winfree (1974), Müller et al. (1985) and Agladze and Krinskii (1982). See also the dramatic experimental examples in Figures 1.16 to 1.20 in Section 1.8 on spiral waves. Kuramoto and Koga (1981) and Agladze and Krinskii (1982), for example, demonstrate the onset of chaotic wave patterns; see Figure 1.23 below. If we consider such waves in three space dimensions the topological structure is remarkable; each part of the basic 'two-dimensional' spiral is itself a spiral; see, for example, Winfree (1974), Welsh et al., (1983) for photographs of actual three-dimensional waves and Winfree and Strogatz (1984) and Winfree (2000) for a discussion of the topological aspects. Much work (analytical and numerical) on spherical waves has also been done by Mimura and his colleagues; see, for example, Yagisita et al. (1998) and earlier references there.

Such target patterns and spiral waves are common in biology. Spiral waves, in particular, are of considerable practical importance in a variety of medical situations, particularly in cardiology and neurobiology. We touch on some of these aspects below. A particularly good biological example is provided by the slime mould *Dictyostelium discoideum* (Newell 1983) and illustrated in Figure 1.1(c); see also Figure 1.18.

Suppose we now consider the reaction diffusion situation in which the reaction kinetics has a single stable steady state but which, if perturbed enough, can exhibit a threshold behaviour, such as we discussed in Section 3.8, Volume I, and also in Section 7.5; the latter is the FitzHugh–Nagumo (FHN) model for the propagation of Hodgkin–Huxley nerve action potentials. Suppose initially the spatial domain is everywhere at the stable steady state and we perturb a small region so that the perturbation locally initiates a threshold behaviour. Although eventually the perturbation will disappear it will undergo a large excursion in phase space before doing so. So, for a time the situation will appear to be like that described above in which there are two quite different states which, because of the diffusion, try to initiate a travelling wavefront. The effect of a threshold capability is thus to provide a basis for a travelling pulse wave. We discuss these threshold waves in Section 1.6.

When waves are transversely coupled it is possible to analyse a basic excitable model system, as was done by Gáspár et al. (1991). They show, among other things, how interacting circular waves can give rise to spiral waves and how complex planar wave patterns can evolve. Petrov et al. (1994) also examined a model reaction diffusion system with cubic autocatalysis and investigated such things as wave reflection and wave slitting. Pascual (1993) demonstrated numerically that certain standard predator–prey models that diffuse along a spatial gradient can exhibit temporal chaos at a fixed point in space and presented evidence for a quasiperiodic route to it as the diffusion increase. Sherratt et al. (1995) studied a caricature of a predator–prey system in one space dimension and demonstrated that chaos can arise in the wake of an invasion wave. The appearance of seemingly chaotic behaviour used to be considered an artifact of the numerical scheme used to study the wave propogation. Merkin et al. (1996) also investigated wave-induced chaos using a two-species model with cubic reaction terms. Epstein and Showalter (1996) gave an interesting overview of the complexity in oscillations, wave pattern and chaos that are possible with nonlinear chemical dynamics.

The collection of articles edited by Maini (1995) shows how ubiquitous and diverse spatiotemporal wave phenomena are in the biomedical disciplines with examples from wound healing, tumour growth, embryology, individual movement in populations, cell–cell interaction and others.

Travelling waves also exist, for certain parameter domains, in model chemotaxis mechanisms such as proposed for the slime mould *Dictyostelium* (cf. Section 11.4, Volume I); see, for example, Keller and Segel (1971) and Keller and Odell (1975). More complex bacterial waves which leave behind a pseudosteady state spatial pattern have been described by Tyson et al. (1998, 1999) some of which will be discussed in detail in Chapter 5.

It is clear that the variety of spatial wave phenomena in multi-species reaction diffusion chemotaxis mechanisms is very much richer than in single species models. If we allow chaotic pacemakers, delay kinetics and so on, the spectrum of phenomena is even wider. Although there have been many studies, only a few of which we have just mentioned, many practical wave problems have still to be studied, and will, no doubt, generate dramatic and new spatiotemporal phenomena of relevance. It is clear that here we can only consider a few which we shall now study in more detail. Later in Chapter 13 we shall see another case study involving rabies when we discuss the spatial spread of epidemics.

1.2 Waves of Pursuit and Evasion in Predator–Prey Systems

If predators and their prey are spatially distributed it is obvious that there will be temporal spatial variations in the populations as the predators move to catch the prey and the prey move to evade the predators. Travelling bands have been observed in oceanic plankton, a small marine organism (Wyatt 1973), animal migration, fungi and vegetation (for example, Lefever and Lejeune, 1997 and Lejeune and Tlidi, 1999) to mention only a few. They are also fairly common, for example, in the movement of primitive organisms invading a source of nutrient. We discuss in some detail in Chapter 5 some of the models and the complex spatial wave and spatial phenomena exhibited by specific bacteria in response to chemotactic cues. In this section we consider, mainly for illustration of the analytical technique, a simple predator–prey system with diffusion and show how travelling wavefront solutions occur. The specific model we study is a modified Lotka–Volterra system (see Section 3.1, Volume I) with logistic growth of the prey and with both predator and prey dispersing by diffusion. Dunbar (1983, 1984) discussed this model in detail. The model mechanism we consider is

$$\begin{aligned}\frac{\partial U}{\partial t} &= AU\left(1-\frac{U}{K}\right) - BUV + D_1\nabla^2 U,\\ \frac{\partial V}{\partial t} &= CUV - DV + D_2\nabla^2 V,\end{aligned} \tag{1.2}$$

where U is the prey, V is the predator, A, B, C, D and K, the prey carrying capacity, are positive constants and D_1 and D_2 are the diffusion coefficients. We nondimensionalise the system by setting

$$u = \frac{U}{K}, \quad v = \frac{BV}{A}, \quad t^* = At, \quad x^* = x\left(\frac{A}{D_2}\right)^{1/2},$$
$$D^* = \frac{D_1}{D_2}, \quad a = \frac{CK}{A}, \quad b = \frac{D}{CK}.$$

We consider only the one-dimensional problem, so (1.2) become, on dropping the asterisks for notational simplicity,

$$\begin{aligned}\frac{\partial u}{\partial t} &= u(1 - u - v) + D\frac{\partial^2 u}{\partial x^2},\\ \frac{\partial v}{\partial t} &= av(u - b) + \frac{\partial^2 v}{\partial x^2},\end{aligned} \tag{1.3}$$

and, of course, we are only interested in non-negative solutions.

The analysis of the spatially independent system is a direct application of the procedure in Chapter 3, Volume I; it is simply a phase plane analysis. There are three steady states (i) $(0, 0)$; (ii) $(1, 0)$, that is, no predator and the prey at its carrying capacity; and (iii) $(b, 1 - b)$, that is, coexistence of both species if $b < 1$, which henceforth we assume to be the case. It is left as a revision exercise to show that both $(0, 0)$ and $(1, 0)$ are unstable and $(b, 1 - b)$ is a stable node if $4a \leq b/(1 - b)$, and a stable spiral if $4a > b/(1 - b)$. In fact in the positive (u, v) quadrant it is a globally stable steady state since (1.3), with $\partial/\partial x \equiv 0$, has a Lyapunov function given by

$$L(u, v) = a\left[u - b - b\ln\left(\frac{u}{b}\right)\right] + \left[v - 1 + b - (1 - b)\ln\left(\frac{v}{1 - b}\right)\right].$$

That is, $L(b, 1 - b) = 0$, $L(u, v)$ is positive for all other (u, v) in the positive quadrant and $dL/dt < 0$ (see, for example, Jordan and Smith 1999 for a readable exposition of Lyapunov functions and their use). Recall, from Section 3.1, Volume I that in the simplest Lotka–Volterra system, namely, (1.2) without the prey saturation term, the nonzero coexistence steady state was only neutrally stable and so was of no use practically. The modified system (1.2) is more realistic.

Let us now look for constant shape travelling wavefront solutions of (1.3) by setting

$$u(x, t) = U(z), \quad v(x, t) = V(z), \quad z = x + ct, \tag{1.4}$$

in the usual way (see Chapter 13, Volume I) where c is the positive wavespeed which has to be determined. If solutions of the type (1.4) exist they represent travelling waves moving to the left in the x-plane. Substitution of these forms into (1.3) gives the ordinary differential equation system

$$\begin{aligned}cU' &= U(1 - U - V) + DU'',\\ cV' &= aV(U - b) + V'',\end{aligned} \tag{1.5}$$

where the prime denotes differentiation with respect to z.

The analysis of (1.5) involves the study of a four-dimensional phase space. Here we consider a simpler case, namely, that in which the diffusion, D_1, of the prey is very much smaller than that of the predator, namely D_2, and so to a first approximation we take $D(= D_1/D_2) = 0$. This would be the equivalent of thinking of a plankton–herbivore system in which only the herbivores were capable of moving. We might reasonably expect the qualitative behaviour of the solutions of the system with $D \neq 0$ to be more or less similar to those with $D = 0$ and this is indeed the case (Dunbar 1984). With $D = 0$ in (1.5) we write the system as a set of first-order ordinary equations, namely,

$$U' = \frac{U(1-U-V)}{c}, \quad V' = W, \quad W' = cW - aV(U-b). \tag{1.6}$$

In the (U, V, W) phase space there are two unstable steady states $(0, 0, 0)$ and $(1, 0, 0)$, and one stable one $(b, 1-b, 0)$; we are, as noted above, only interested in the case $b < 1$. From the experience gained from the analysis of Fisher–Kolmogoroff equation, discussed in detail in Section 13.2, Volume I, there is thus the possibility of a travelling wave solution from $(1, 0, 0)$ to $(b, 1-b, 0)$ and from $(0, 0, 0)$ to $(b, 1-b, 0)$. So we should look for solutions $(U(z), V(z))$ of (1.6) with the boundary conditions

$$U(-\infty) = 1, \quad V(-\infty) = 0, \quad U(\infty) = b, \quad V(\infty) = 1-b \tag{1.7}$$

and

$$U(-\infty) = 0, \quad V(-\infty) = 0, \quad U(\infty) = b, \quad V(\infty) = 1-b. \tag{1.8}$$

We consider here only the boundary value problem (1.6) with (1.7). First linearise the system about the singular point $(1, 0, 0)$, that is, the steady state $u = 1, v = 0$, and determine the eigenvalues λ in the usual way as described in detail in Chapter 3, Volume I. They are given by the roots of

$$\begin{vmatrix} -\lambda - \dfrac{1}{c} & -\dfrac{1}{c} & 0 \\ 0 & -\lambda & 1 \\ 0 & -a(1-b) & c-\lambda \end{vmatrix} = 0,$$

namely,

$$\lambda_1 = -\frac{1}{c}, \quad \lambda_2, \lambda_3 = \frac{1}{2}\{c \pm [c^2 - 4a(1-b)]^{1/2}\}. \tag{1.9}$$

Thus there is an unstable manifold defined by the eigenvectors associated with the eigenvalues λ_2 and λ_3 which are positive for all $c > 0$. Further, $(1, 0, 0)$ is unstable in an oscillatory manner if $c^2 < 4a(1-b)$. So, the only possibility for a travelling wavefront solution to exist with non-negative U and V is if

$$c \geq [4a(1-b)]^{1/2}, \quad b < 1. \tag{1.10}$$

With c satisfying this condition, a realistic solution, with a lower bound on the wave speed, may exist which tends to $u = 1$ and $v = 0$ as $z \to -\infty$. This is reminiscent of the travelling wavefront solutions described in Chapter 13, Volume I.

The solutions here, however, can be qualitatively different from those in Chapter 13, Volume I, as we see by considering the approach of (U, V) to the steady state $(b, 1-b)$. Linearising (1.6) about the singular point $(b, 1-b, 0)$ the eigenvalues λ are given by

$$\begin{vmatrix} -\lambda - \frac{b}{c} & -\frac{b}{c} & 0 \\ 0 & -\lambda & 1 \\ -a(1-b) & 0 & c-\lambda \end{vmatrix} = 0$$

and so are the roots of the cubic characteristic polynomial

$$p(\lambda) \equiv \lambda^3 - \lambda^2\left(c - \frac{b}{c}\right) - \lambda b - \frac{ab(1-b)}{c} = 0. \tag{1.11}$$

To see how the solutions of this polynomial behave as the parameters vary we consider the plot of $p(\lambda)$ for real λ and see where it crosses $p(\lambda) = 0$. Differentiating $p(\lambda)$, the local maximum and minimum are at

$$\lambda_M, \lambda_m = \frac{1}{3}\left[\left(c - \frac{b}{c}\right) \pm \left[\left(c - \frac{b}{c}\right)^2 + 3b\right]^{1/2}\right]$$

and are independent of a. For $a = 0$ the roots of (1.11) are

$$\lambda = 0, \quad \lambda_1, \lambda_2 = \frac{1}{2}\left[\left(c - \frac{b}{c}\right) \pm \left[\left(c - \frac{b}{c}\right)^2 + 4b\right]^{1/2}\right]$$

as illustrated in Figure 1.2. We can now see how the roots vary with a. From (1.11), as a increases from zero the effect is simply to subtract $ab(1-b)/c$ everywhere from

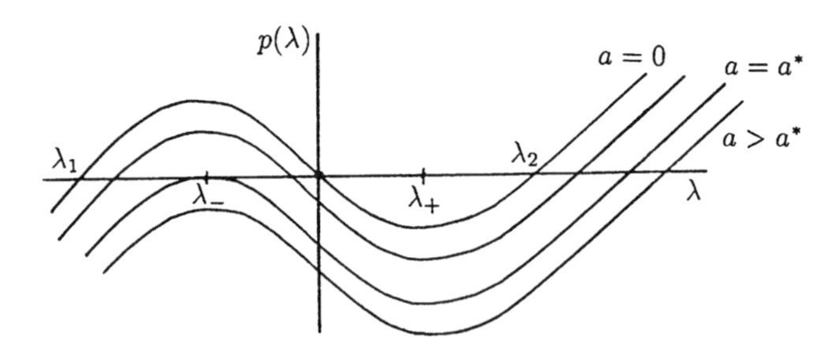

Figure 1.2. The characteristic polynomial $p(\lambda)$ from (1.11) as a function of λ as a varies. There is a critical value a^* such that for $a > a^*$ there is only one real positive root and two complex ones with negative real parts.

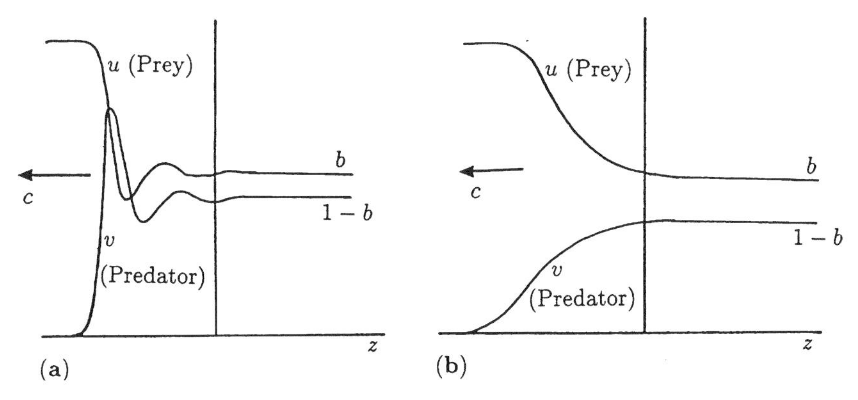

Figure 1.3. Typical examples of the two types of waves of pursuit given by wavefront solutions of the predator (v)–prey (u) system (1.3) with negligible dispersal of the prey. The waves move to the left with speed c. (**a**) Oscillatory approach to the steady state $(b, 1-b)$, when $a > a^*$. (**b**) Monotonic approach of (u, v) to $(b, 1-b)$ when $a \leq a^*$.

the $p(\lambda; a = 0)$ curve. Since the local extrema are independent of a, we then have the situation illustrated in the figure. For $0 < a < a^*$ there are 2 negative roots and one positive one. For $a = a^*$ the negative roots are equal while for $a > a^*$ the negative roots become complex with negative real parts. This latter result is certainly the case for a just greater than a^* by continuity arguments. The determination of a^* can be carried out analytically. The same conclusions can be derived using the Routh–Hurwitz conditions (see Appendix B, Volume I) but here if we use them it is intuitively less clear.

The existence of a critical a^* means that, for $a > a^*$, the wavefront solutions (U, V) of (1.6) with boundary conditions (1.7) approach the steady state $(b, 1-b)$ in an *oscillatory* manner while for $a < a^*$ they are monotonic. Figure 1.3 illustrates the two types of solution behaviour.

The full predator–prey system (1.3), in which both the predator and prey diffuse, also gives rise to travelling wavefront solutions which can display oscillatory behaviour (Dunbar 1983, 1984). The proof of existence of these waves involves a careful analysis of the phase plane system to show that there is a trajectory, lying in the positive quadrant, which joins the relevant singular points. These waves are sometimes described as 'waves of pursuit and evasion' even though there is little evidence of prey evasion in the solutions in Figure 1.3, since other than quietly reproducing, the prey simply wait to be consumed.

Convective Predator–Prey Pursuit and Evasion Models

A totally different kind of 'pursuit and evasion' predator–prey system is one in which the prey try to evade the predators and the predators try to catch the prey only if they interact. This results in a basically different kind of spatial interaction. Here, by way of illustration, we briefly describe one possible model, in its one-dimensional form. Let us suppose that the prey (u) and predator (v) can move with speeds c_1 and c_2, respectively, that diffusion plays a negligible role in the dispersal of the populations and that each population obeys its own dynamics with its own steady state or states. Refer now to

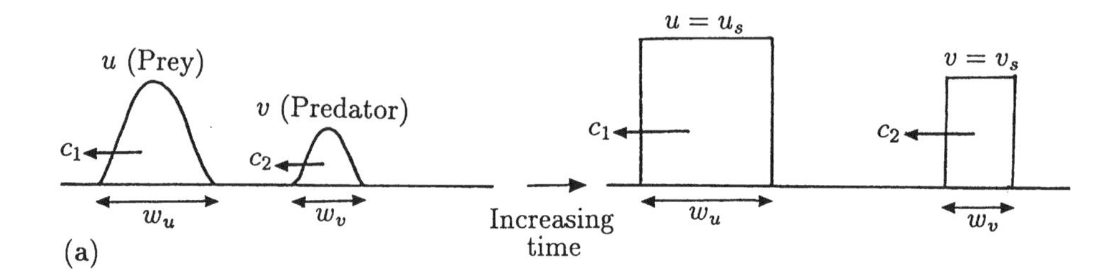

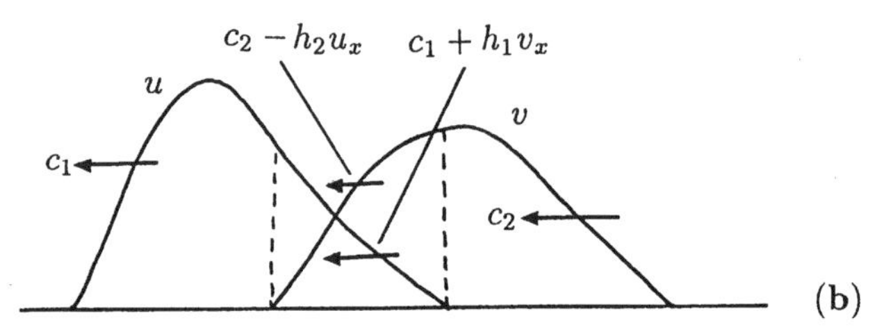

Figure 1.4. (**a**) The prey and predator populations are spatially separate and each satisfies its own dynamics: they do not interact and simply move at their own undisturbed speed c_1 and c_2. Each population grows until it is at the steady state (u_s, v_s) determined by its individual dynamics. Note that there is no dispersion so the spatial width of the 'waves' w_u and w_v remain fixed. (**b**) When the two populations overlap, the prey put on an extra burst of speed $h_1 v_x$, $h_1 > 0$ to try and get away from the predators while the predators put on an extra spurt of speed, namely, $-h_2 u_x$, $h_2 > 0$, to pursue them: the motivation for these terms is discussed in the text.

Figure 1.4 and consider first Figure 1.4(a). Here the populations do not interact and, since there is no diffusive spatial dispersal, the population at any given spatial position simply grows or decays until the whole region is at that population's steady state. The dynamic situation is then as in Figure 1.4(a) with both populations simply moving at their undisturbed speeds c_1 and c_2 and without spatial dispersion, so the width of the bands remains fixed as u and v tend to their steady states. Now suppose that when the predators overtake the prey, the prey try to evade the predators by moving away from them with an extra burst of speed proportional to the predator gradient. In other words, if the overlap is as in Figure 1.4(b), the prey try to move away from the increasing number of predators. By the same token the predators try to move further into the prey and so move in the direction of increasing prey. At a basic, but nontrivial, level we can model this situation by writing the conservation equations (see Chapter 11, Volume I) to include convective effects as

$$u_t - [(c_1 + h_1 v_x)u]_x = f(u, v), \tag{1.12}$$

$$v_t - [(c_2 - h_2 u_x)v]_x = g(v, u), \tag{1.13}$$

where f and g represent the population dynamics and h_1 and h_2 are the positive parameters associated with the retreat and pursuit of the prey and predator as a consequence of the interaction. These are conservation laws for u and v so the terms on the left-hand sides of the equations must be in divergence form. We now motivate the various terms in the equations.

The interaction terms f and g are whatever predator–prey situation we are considering. Typically $f(u, 0)$ represents the prey dynamics where the population simply grows or decays to a nonzero steady state. The effect of the predators is to reduce the size of the prey's steady state, so $f(u, 0) > f(u, v > 0)$. By the same token the steady state generated by $g(v, u \neq 0)$ is larger than that produced by $g(v, 0)$.

To see what is going on physically with the convective terms, suppose, in (1.12), $h_1 = 0$. Then

$$u_t - c_1 u_x = f(u, v),$$

which simply represents the prey dynamics in a travelling frame moving with speed c_1. We see this if we use $z = x + c_1 t$ and t as the independent variables in which case the equation simply becomes $u_t = f(u, v)$. If $c_2 = c_1$, the predator equation, with $h_2 = 0$, becomes $v_t = g(v, u)$. Thus we have travelling waves of changing populations until they have reached their steady states as in Figure 1.4(a), after which they become travelling (top hat) waves of constant shape.

Consider now the more complex case where h_1 and h_2 are positive and $c_1 \neq c_2$. Referring to the overlap region in Figure 1.4(b), the effect in (1.12) of the $h_1 v_x$ term, positive because $v_x > 0$, is to increase locally the speed of the wave of the prey to the left. The effect of $-h_2 u_x$, positive because $u_x < 0$, is to increase the local convection of the predator. The intricate nature of interaction depends on the form of the solutions, specifically u_x and v_x, the relative size of the parameters c_1, c_2, h_1 and h_2 and the interaction dynamics. Because the equations are nonlinear through the convection terms (as well as the dynamics) the possibility exists of shock solutions in which u and v undergo discontinuous jumps; see, for example, Murray (1968, 1970, 1973) and, for a reaction diffusion example, Section 13.5 in Chapter 13 (Volume 1).

Before leaving this topic it is interesting to write the model system (1.12), (1.13) in a different form. Carrying out the differentiation of the left-hand sides, the equation system becomes

$$\begin{aligned} u_t - [(c_1 + h_1 v_x)]u_x &= f(u, v) + h_1 u v_{xx}, \\ v_t - [(c_2 - h_2 u_x)]v_x &= g(v, u) - h_2 v u_{xx}. \end{aligned} \tag{1.14}$$

In this form we see that the h_1 and h_2 terms on the right-hand sides represent *cross diffusion*, one positive and the other negative. Cross diffusion, which, of course, is only of relevance in multi-species models was defined in Section 11.2, Volume I: it occurs when the diffusion matrix is not strictly diagonal. It is a diffusion-type term in the equation for one species which involves another species. For example, in the u-equation, $h_1 u v_{xx}$ is like a diffusion term in v, with 'diffusion' coefficient $h_1 u$. Typically a cross diffusion would be a term $\partial(D v_x)/\partial x$ in the u-equation. The above is an example where cross diffusion arises in a practical modelling problem—it is not common.

The mathematical analysis of systems like (1.12)–(1.14) is a challenging one which is largely undeveloped. Some analytical work has been done by Hasimoto (1974), Yoshikawa and Yamaguti (1974), who investigated the situation in which $h_1 = h_2 = 0$ and

Murray and Cohen (1983), who studied the system with h_1 and h_2 nonzero. Hasimoto (1974) obtained analytical solutions to the system (1.12) and (1.13), where $h_1 = h_2 = 0$ and with the special forms $f(u, v) = l_1 uv$, $g(u, v) = l_2 uv$, where l_1 and l_2 are constants. He showed how blow-up can occur in certain circumstances. Interesting new solution behaviour is likely for general systems of the type (1.12)–(1.14).

Two-dimensional problems involving convective pursuit and evasion are of ecological significance and are particularly challenging; they have not been investigated. For example, in the first edition of this book, it was hypothesized that it would be very interesting to try and model a predator–prey situation in which species territory is specifically involved. With the wolf–moose predator–prey situation in Canada we suggested that it should be possible to build into a model the effect of wolf territory boundaries to see if the territorial 'no man's land' provides a partial safe haven for the prey. The intuitive reasoning for this speculation is that there is less tendency for the wolves to stray into the neighbouring territory. There seems to be some evidence that moose do travel along wolf territory boundaries. A study along these lines has been done and will be discussed in detail in Chapter 14.

A related class of wave phenomena occurs when convection is coupled with kinetics, such as occurs in biochemical ion exchange in fixed columns. The case of a single-reaction kinetics equation coupled to the convection process, was investigated in detail by Goldstein and Murray (1959). Interesting shock wave solutions evolve from smooth initial data. The mathematical techniques developed there are of direct relevance to the above problems. When several ion exchanges are occurring at the same time in this convective situation we then have chromatography, a powerful analytical technique in biochemistry.

1.3 Competition Model for the Spatial Spread of the Grey Squirrel in Britain

Introduction and Some Facts

About the beginning of the 20th century North American grey squirrels (*Sciurus carolinensis*) were released from various sites in Britain, the most important of which was in the southeast. Since then the grey squirrel has successfully spread through much of Britain as far north as the Scottish Lowlands and at the same time the indigenous red squirrel *Sciurus vulgaris* has disappeared from these localities.

Lloyd (1983) noted that the influx of the grey squirrel into areas previously occupied by the red squirrel usually coincided with a decline and subsequent disappearance of the red squirrel after only a few years of overlap in distribution.

The squirrel distribution records in Britain seem to indicate a definite negative effect of the greys on the reds (Williamson 1996). MacKinnon (1978) gave some reasons why competition would be the most likely among three hypotheses which had been made (Reynolds 1985), namely, competition with the grey squirrel, environmental changes that reduced red squirrel populations independent of the grey squirrel and diseases, such as 'squirrel flu' passed on to the red squirrels. These are not mutually exclusive of course.

Prior to the introduction of the grey, the red squirrel had evolved without any interspecific competition and so selection favoured modest levels of reproduction with low numerical wastage. The grey squirrel, on the other hand, evolved within the context of strong interspecific competition with the American red squirrel and fox squirrel and so selection favoured overbreeding. Both red and grey squirrels can breed twice a year but the smaller red squirrels rarely have more than two or three offspring per litter, whereas grey squirrels frequently have litters of four or five (Barkalow 1967).

In North America the red and grey squirrels occupy separate niches that rarely overlap: the grey favour mixed hardwood forests while the red favour northern conifer forests. On the other hand, in Britain the native red squirrel must have evolved, in the absence of the grey squirrel, in such a way that it adapted to live in hardwood forests as well as coniferous forests. Work by Holm (1987) also tends to support the hypothesis that grey squirrels may be at a competitive advantage in deciduous woodland areas where the native red squirrel has mostly been replaced by the grey. Also the North American grey squirrel is a large robust squirrel, with roughly twice the body weight of the red squirrel. In separate habitats the two squirrel species show similar social organisation, feeding and ranging ecology but within the same habitat we would expect even greater similarity in their exploitation of resources, and so it seems inevitable that two species of such close similarity could not coexist in sharing the same resources.

In summary it seems reasonable to assume that an interaction between the two species, probably largely through indirect competition for resources, but also with some direct interaction, for example, chasing, has acted in favour of the grey squirrel to drive off the red squirrel mostly from deciduous forests in Britain. Okubo et al. (1989) investigated this displacement of the red squirrel by the grey squirrel and, based on the above, proposed and studied a competiton model. It is their work we follow in this section. They also used the model to simulate the random introduction of grey squirrels into red squirrel areas to show how colonisation might spread. They compared the results of the modelling with the available data.

Competition Model System

Denote by $S_1(\mathbf{X}, T)$ and $S_2(\mathbf{X}, T)$ the population densities at position $\mathbf{X}$ and time T of grey and red squirrels respectively. Assuming that they compete for the same food resources, a possible model is the modified competition Lotka–Volterra system with diffusion, (cf. Chapter 5, Volume I), namely,

$$\begin{aligned}\frac{\partial S_1}{\partial T} &= D_1\nabla^2 S_1 + a_1 S_1(1 - b_1 S_1 - c_1 S_2),\\ \frac{\partial S_2}{\partial T} &= D_2\nabla^2 S_2 + a_2 S_2(1 - b_2 S_2 - c_2 S_1),\end{aligned} \tag{1.15}$$

where, for $i = 1, 2$, a_i are net birth rates, $1/b_i$ are carrying capacities, c_i are competition coefficients and D_i are diffusion coefficients, all non-negative. The interaction (kinetics) terms simply represent logistic growth with competition. For the reasons discussed above we assume that the greys outcompete the reds so

$$b_2 > c_1, \quad c_2 > b_1. \tag{1.16}$$

We now want to investigate the possibility of travelling waves of invasion of grey squirrels which drive out the reds. We first nondimensionalise the model system by setting

$$\begin{aligned}\theta_i &= b_i S_i, i = 1, 2, t = a_1 T, \mathbf{x} = (a_1/D_1)^{1/2}\mathbf{X},\\ \gamma_1 &= c_1/b_2, \gamma_2 = c_2/b_1, \kappa = D_2/D_1, \alpha = a_1/a_2\end{aligned} \tag{1.17}$$

and (1.15) becomes

$$\begin{aligned}\frac{\partial \theta_1}{\partial t} &= \nabla^2\theta_1 + \theta_1(1 - \theta_1 - \gamma_1\theta_2),\\ \frac{\partial \theta_2}{\partial t} &= \kappa\nabla^2\theta_1 + \alpha\theta_2(1 - \theta_2 - \gamma_2\theta_1).\end{aligned} \tag{1.18}$$

Because of (1.16),

$$\gamma_1 < 1, \quad \gamma_2 > 1. \tag{1.19}$$

In the absence of diffusion we analysed this specific competition model system (1.18) in detail in Chapter 5, Volume I. It has three homogeneous steady states which, in the absence of diffusion, by a standard phase plane analysis, are $(0, 0)$ an unstable node, $(1, 0)$ a stable node and $(0, 1)$ a saddle point. So, with the inclusion of diffusion, by the now usual procedure, there is the possibility of a solution trajectory from $(0, 1)$ to $(1, 0)$ and a travelling wave joining these critical points. This corresponds to the ecological situation where the grey squirrels (θ_1) outcompete the reds (θ_2) to extinction: it comes into the category of competitive exclusion (cf. Chapter 5, Volume I).

In one space dimension $\mathbf{x} = x$ we look for travelling wave solutions to (1.18) of the form

$$\theta_i = \theta_i(z), i = 1, 2, \quad z = x - ct, c > 0, \tag{1.20}$$

where c is the wavespeed. $\theta_1(z)$ and $\theta_2(z)$ represent wave solutions of constant shape travelling with velocity c in the positive x-direction. With this, equations (1.18) become

$$\begin{aligned}\frac{d^2\theta_1}{dz^2} + c\frac{d\theta_1}{dz} + \theta_1(1 - \theta_1 - \gamma_1\theta_2) &= 0,\\ \kappa\frac{d^2\theta_2}{dz^2} + c\frac{d\theta_2}{dz} + \alpha\theta_2(1 - \theta_2 - \gamma_2\theta_1) &= 0,\end{aligned} \tag{1.21}$$

subject to the boundary conditions

$$\theta_1 = 1,\ \theta_2 = 0,\ \text{at } z = -\infty, \quad \theta_1 = 0, \theta_2 = 1,\ \text{at } z = \infty. \tag{1.22}$$

That is, asymptotically the grey (θ_1) squirrels drive out the red (θ_2) squirrels as the wave propagates with speed c, which we still have to determine.

Hosono (1988) investigated the existence of travelling waves for the system (1.18) with (1.19) and (1.22) under certain conditions on the values of the parameters. In general, the system of ordinary differential equations (1.18) cannot be solved analytically. However, in the special case where $\kappa = \alpha = 1$, $\gamma_1 + \gamma_2 = 2$ we can get some analytical results. We add the two equations in (1.21) to get

$$\frac{d^2\theta}{dz^2} + c\frac{d\theta}{dz} + \theta(1-\theta) = 0, \quad \theta = \theta_1 + \theta_2, \tag{1.23}$$

which is the well-known Fisher–Kolmogoroff equation discussed in depth in Chapter 13, Volume I which we know has travelling wave solutions with appropriate boundary conditions at $\pm\infty$. However, the boundary conditions here are different to those for the classical Fisher–Kolmogoroff equation: they are, from (1.22),

$$\theta = 1, \quad \text{at} \quad z = \pm\infty \tag{1.24}$$

which suggest that for all z,

$$\theta = 1 \Rightarrow \theta_1 + \theta_2 = 1. \tag{1.25}$$

Substituting this into the first of (1.21) we get

$$\frac{d^2\theta_1}{dz^2} + c\frac{d\theta_1}{dz} + (1-\gamma_1)\theta_1(1-\theta_1) = 0, \tag{1.26}$$

which is again the Fisher–Kolmogoroff equation for θ_1 with boundary conditions (1.22). From the results on the wave speed we deduce that the wavefront speed for the grey squirrels will be greater than or equal to the minimum Fisher–Kolmogoroff wave speed for (1.26); that is,

$$c \geq c_{\text{min}} = 2(1-\gamma_1)^{1/2}, \quad \gamma_1 < 1. \tag{1.27}$$

Similarly, from the second of (1.21) with (1.25) the equation for θ_2 is

$$\frac{d^2\theta_2}{dz^2} + c\frac{d\theta_2}{dz} + (\gamma_2 - 1)\theta_2(1-\theta_2) = 0, \tag{1.28}$$

with boundary conditions (1.22). This gives the result that for the red squirrels

$$c \geq c_{\text{min}} = 2(\gamma_2 - 1)^{1/2}, \quad \gamma_2 > 1. \tag{1.29}$$

Since $\gamma_1 + \gamma_2 = 2$ (and remember too that $\kappa = \alpha = 1$) these two minimum wavespeeds are equal. In terms of dimensional quantities, we thus get the dimensional minimum wavespeed, C_{min}, as

$$C_{\text{min}} = 2\left[a_1 D_1 \left(1 - \frac{c_1}{b_2}\right)\right]^{1/2}. \tag{1.30}$$

Parameter Estimation

We must now relate the analysis to the real world competition situation that obtains in Britain. The travelling wavespeeds depend upon the parameters in the model system (1.15) so we need estimates for the parameters in order to compare the theoretical wavespeed with available data. As reiterated many times, this is a crucial aspect of realistic modelling.

Let us first consider the intrinsic net growth rates a_1 and a_2. Okubo et al. (1989) used a modified Leslie matrix described in detail by Williamson and Brown (1986). In principle the estimates should be those at zero population density, but demographic data usually refer to populations near their equilibrium density. Three components are considered in estimating the intrinsic net growth rate, specifically the sex ratio, the birth rate and the death rate. The sex ratio is taken to be one to one. Determining the birth and death rates, however, is not easy. It depends on such things as litter size and frequency and their dependence on age, age distribution, where and when the data are collected, food source levels and life expectancy; the paper by Okubo et al. (1989) shows what is involved. After a careful analysis of the numerous, sometimes conflicting, sources they estimated the intrinsic birth rate for the grey squirrels as $a_1 = 0.82$/year with a stable age distribution of nearly three young to one adult and for the red squirrels, $a_2 = 0.61$/year with an age distribution of just over two young for each adult.

Determining estimates for the carrying capacities $1/b_1$ and $1/b_2$ involves a similar detailed examination of the available literature which Okubo et al. (1989) also did. They suggested values for the carrying capacities of $1/b_1 = 10$/hectare and $1/b_2 = 0.75$/hectare respectively for the grey and red squirrels.

Unfortunately there is no quantitative information on the competition coefficients c_1 and c_2. In the model, however, only the ratios $c_1/b_2 = \gamma_1$ and $c_2/b_1 = \gamma_2$ are needed to estimate the minimum speed of the travelling waves. As far as the speed of propagation of the grey squirrel is concerned, we only need an estimate of γ_1: recall that $0 < \gamma_1 < 1$. Since γ_1 appears in the expression of the minimum wavespeed in the term $(1 - \gamma_1)^{1/2}$, the speed is not very sensitive to the value of γ_1 if it is small, in fact unless it is larger than around 0.6. We expect that the competition coefficient c_1, that is, red against grey, should have a small value. So, this, together with the smallness of the carrying capacity b_2^{-1}, it is reasonable to assume that the value of γ_1 is close to zero, so that the minimum speed of the travelling wave of the grey squirrel, C_{min}, is approximately given from (1.30) by $2(D_1 a_1)^{1/2}$. In the numerical simulations carried out by Okubo et al. (1989) they used several different values for the γs since the analysis we carried out above was for special values which allowed us to do some analysis.

Let us now consider the diffusion coefficients, D_1 and D_2. These are crucial parameters in wave propagation and notoriously difficult to estimate. (The same problem of diffusion estimation comes up again later in the book when we discuss the spatial spread of rabies in a fox population, bacterial patterns and tumour cells in the brain.) Direct observation of dispersal is difficult and usually short term. The reported values for movement vary widely. There is also the movement between woodlands.

For grey squirrels, a maximum for a one-dimensional diffusion coefficient of 1.25 km^2/yr, and for a two-dimensional diffusion coefficient of 0.63 km^2/yr, was derived based on individual movement. However, this may not correspond to the squirrels'

Table 1.1. Two-dimensional diffusion coefficients for the grey squirrel as a function of the distance l km between woodland areas. The minimum wavespeed $C_{\min}$ km/year $= 2(a_1 D_1)^{1/2}$ with $a_1 = 0.82$/year. (From Okubo et al. 1989)

l(km)	1	2	5	10	15	20
D_1(km^2/year)	0.179	0.714	4.46	17.9	40.2	71.4
$C_{\min}$(km/year)	0.77	1.53	3.82	7.66	11.5	15.3

movement between woodlands. If the annual dispersal in the grey squirrel takes place primarily between woodlands rather than within a woodland, then the values of these diffusion coefficients should be too small to be considered representative. Okubo et al. (1989) speculated that it might not be unreasonable to expect a diffusion coefficient for grey squirrels of the order of 10–20 km^2/year rather than of the order of 1 km^2/year. They gave a heuristic argument, which we now give, to support these much larger values for the diffusion coefficient.

Consider a patch of woodlands, each having an equal area of A hectare (ha) with four neighbours and separated from each other by a distance l. Suppose a woodland is filled with grey squirrels and the carrying capacity for grey squirrels is 10/ha. This implies that the woodland carries $n = 10A$ individuals. With the intrinsic growth rate $a_1 = 0.82$/yr, the woodland will contain $22.7A$ animals (since $e^{0.82} = 2.27$) in the following year of which $12.7A$ individuals have to disperse. Assuming the animals disperse into the nearest neighbouring woodlands, $12.7/4A = 3.175A$ individuals will arrive at a neighbouring woodland. This woodland will then be filled with grey squirrels in $\tau = 1.4$ years ($10A = 3.175Ae^{0.82\tau}$), after which another dispersal will occur. In other words, the grey squirrels make dispersal to the nearest neighbouring woodlands, on average, every 1.4 years. Thus, a two-dimensional diffusion coefficient for the grey squirrel is estimated as

$$D_1 = l^2/(4 \times 1.4) \text{ km}^2/\text{year} = l^2/5.6 \text{ km}^2/\text{year}. \tag{1.31}$$

Table 1.1 gives the calculated diffusion coefficient, D_1, using (1.31) as a function l. Williamson and Brown (1986) estimated the speed of dispersal of grey squirrels to be 7.7 km/year. If we take this value we then get, from the table, a value of $D_1 = 17.9$ km^2/year. So, a mean separation between neighbouring woodlands of 10 km, which is reasonable, would give the minimum speed of travelling waves that agrees well with the data.

Comparison of the Theoretical Rate of Spread with the Data

One of the best sources of information on the spread of the grey squirrel in Britain is given by Reynolds (1985), who studied it in detail in East Anglia during the period 1960 to 1981. Colonization of East Anglia by the grey squirrel has been comparatively recent. In 1959 no grey squirrels were found and red squirrels were still present more or less throughout the county of Norfolk both in 1959 and at the later survey in 1971. However, by 1971 the grey squirrel was also recorded over about half the area of Norfolk.

Reynolds (1985) constructed a series of maps showing the annual distribution of the grey and red squirrels for the period of 1960 to 1981 using grids of 5×5 km squares. Based on these maps Williamson and Brown (1986) calculated the rate of spread of the grey squirrel during the period 1965–1981 and obtained a rate of spread of between 5 and 10 km per year; the mean rate of spread of the grey squirrel was calculated to be 7.7 km/year, the value mentioned above which, if we use the Fisher–Kolmogoroff minimum wavespeed gives an estimated value of the diffusion coefficient for the grey squirrels of approximately $D_1 = 17.9\ \mathrm{km}^2/\mathrm{year}$. So, there is a certain data justification for the heuristic estimation of the diffusion ceofficient we have just given.

Solutions to the dimensionless model system (1.18) have to be done numerically if we use values for the γ's other than those satisfying $\gamma_1 + \gamma_2 = 2$. In one dimension the waves are qualitatively as we would expect from the boundary conditions and the form of the equations, even with unrelated values of γ. For the grey squirrels there will be a wave of advance qualitatively similar to a typical Fisher–Kolmogoroff wave with a corresponding wave of retreat (almost a mirror image in fact) for the red squirrels; these are shown in Okubo et al. (1989). Colonization is, of course, two-dimensional where analytical solutions of propagating wavelike fronts are not available—it is a very hard problem. In the special case of a radially symmetric distribution of grey squirrels, the velocity of the invasive wave is less than in the corresponding one-dimensional case (because of the term $(1/r)(\partial\theta_1/\partial r)$ and the equivalent for θ_2 in the Lapalacian). Numerical solutions, however, can be found relatively easily. We started with an initial small scattered distribution of grey squirrels in a predominantly red population. These small areas of grey squirrels moved outward, coalesced with other areas of greys and eventually drove out the red population completely. Figure 1.5 shows a typical numerical solution with a specimen set of parameter values.

The basic models of population spread via diffusion and growth, such as with the Fisher–Kolmogoroff model, start with an initial seed which spreads out radially eventually becoming effectively a one-dimensional wave because the $(1/r)(\partial\theta_1/\partial r)$ term tends to zero as $r \to \infty$. The same holds with the model we have discussed here, although the competition wave of advance is slower, which is not surprising since the effective birth rate of the grey squirrels is less than a simple logistic growth. There are numerous maps (references are given in Okubo et al. 1989) of the advance of the grey squirrel and retreat of the red squirrel in Britain dating back to 1930. The behaviour exhibited in Figure 1.5 is a fair representation of the major patterns seen. The parameter values used were based on the detailed survey of Reynolds (1985). The parameters and hence the course of the competition, however, inevitably vary with the climate, density of trees and their type and so on. It seems that the broad features of the displacement of the red squirrels by the grey is captured in this simple competition model and is a practical example of the principle of competitive exclusion discussed in Chapter 5, Volume I.

1.4 Spread of Genetically Engineered Organisms

There is a rapidly increasing use of recombinant DNA technology to modify plants (and animals) to perform special agricultural functions. However, there is an increasing con-

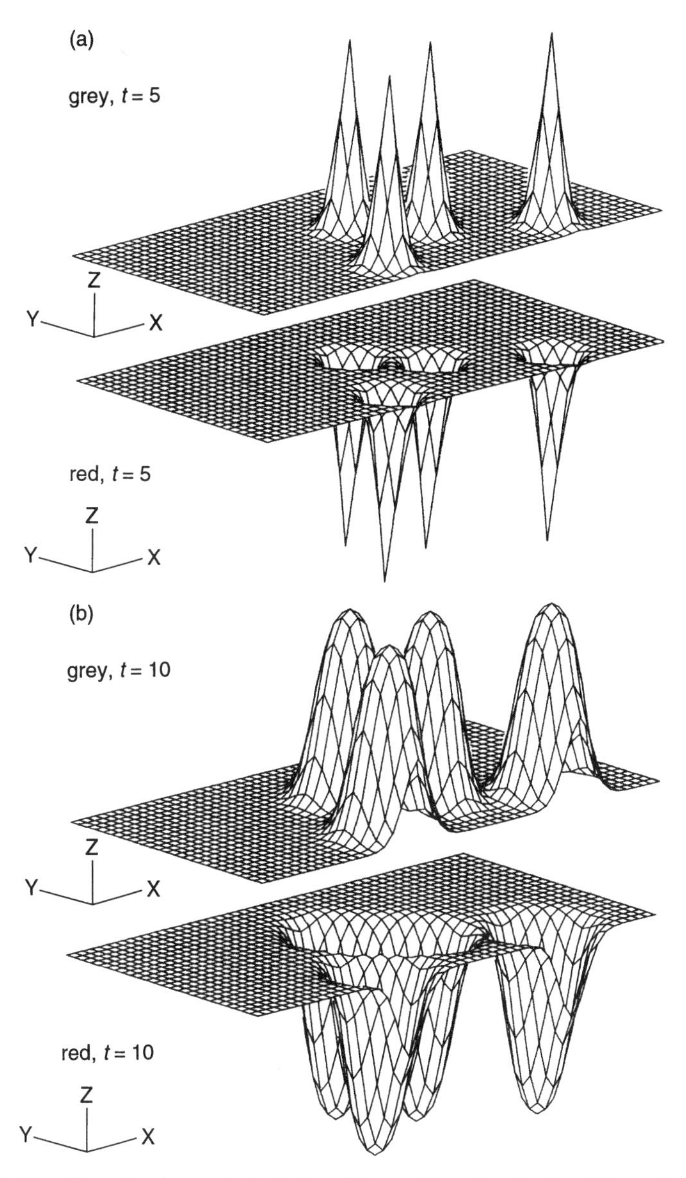

Figure 1.5a,b. Two-dimensional numerical solution of the nondimensionalised model equations (1.18) on a 4.9×2.4 rectangle with zero flux boundary conditions. The initial distribution consists of red squirrels at unit normalised density, seeded with small pockets of grey squirrels of density 0.1 at points (1.9,0.4),(3.9,0.4), (2.9,0.9) and (2.4,1.4). (**a**) Surface plot of the solution at time $t = 5$; the base density of greys is 0.0 and of the reds 1.0. Solutions at subsequent times: (**b**) $t = 10$, (**c**) $t = 20$, (**d**) $t = 30$. As the system evolves the greys begin to increase in density and spread outwards while the reds recede. Eventually the greys drive out the reds. Parameter values: $\gamma_1 = 0.2$, $\gamma_2 = 1.5$, $\alpha = 0.82/0.61$, $\kappa = 1 (D_1 = D_2 = 0.001)$. (From Okubo et al. 1989)

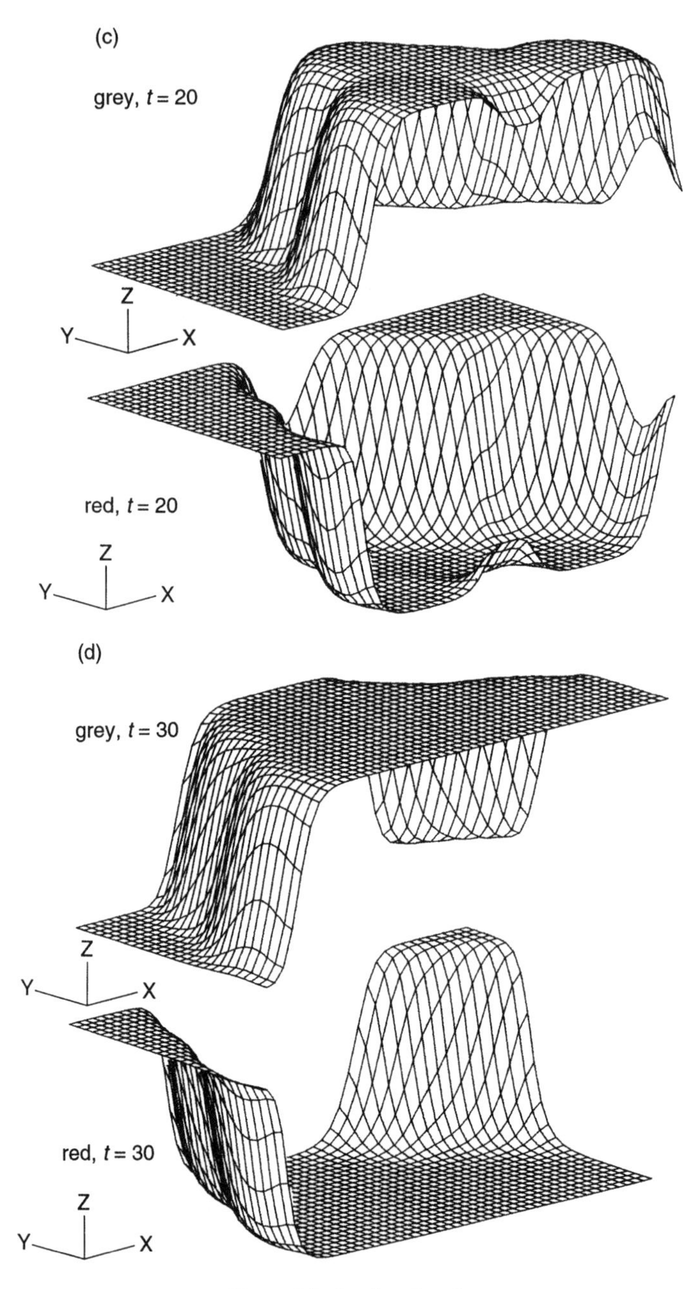

Figure 1.5c,d. (continued)

cern about its possible disruption of the ecosystem and even the climatic system caused by the release of such genetically engineered organisms. Studies of the spatiotemporal dynamics of genetically engineered organisms in the natural environment are clearly increasingly important. Scientists have certainly not reached a consensus regarding the risks or containment of genetically engineered organisms. In the case of plants the initial timescale is short compared with genetically engineered trees. For example fruit trees have been modified to kill pests who land on the leaves. Means are also being studied to have trees clean up polluted ground. Hybrid plants, of course, have been widely used for a very long time but without the direct genetically designed input. In the case of the even more controversial genetically modified (and also cloned) animals there are other serious risks. Their use, associated with animal development for human transplants, poses different epidemiological problems. Whatever the protesting Luddites say, or do, genetic manipuation of both plants and animals (including humans) is here to stay.

One of the main concerns regarding the release of engineered organisms is how far and how rapidly they are likely to spread, under different ecological scenarios and management plans. An unbiased assessment of the risks associated with releasing such organisms should lead to strategies for the effective containment of an outbreak. There is still little reliable quantitative information for estimating spread rates and analysing possible containment strategies. Some initial work along these lines, however, was carried out by Cruywagen et al. (1996).

Genetically engineered microbes are especially amenable to mathematical analyses because they continuously reproduce, lack complex behaviours and exhibit population dynamics well described by simple models. One example of such a microbe is *Pseudomonas syringae* (ice minus bacteria), which can reduce frost damage to crops by occupying crop foliage to the exclusion of *Pseudomonas syringae* strains that do cause frost damage (Lindow 1987).

In this section we develop a model to obtain quantitative results on the spatiotemporal spread of genetically engineered organisms in a spatially heterogeneous environment; we follow the work of Cruywagen et al. (1996). We get information regarding the risk of outbreak of an engineered population from its release site in terms of its dispersal and growth rates as well as those of a competing species. The nature of the environment plays a key role in the spread of the organisms. We focus specifically on whether containment can be guaranteed by the use of geographical barriers, for example, water, a different crop or just barren land.

For the basic model we start with a system of two competing and diffusing species, namely, the one we used in the last section for the spatial spread of the grey squirrel in Britain. As we saw this model provides an explanation as to why the externally introduced grey squirrel invaded at the cost of the indigenous red squirrel which was driven to extinction in areas of competition.

Most invasion models deal with invasion as travelling waves propagating in a homogeneous environment. However, because of variations in the environment (natural or made), this is almost never the case. Not only is spatial heterogeneity one of the most obvious features in the natural world, it is likely to be one of the more important factors influencing population dynamics.

A first analysis of propagating frontal waves in a heterogeneous unbounded habitat was carried out by Shigesada et al. (1986) (see also the book by Shigesada and Kawasaki 1997) for the Fisher–Kolmogoroff equation which describes a single species with logistic population growth and dispersal. Here we again use the Lotka–Volterra competition model with diffusion to model the population dynamics of natural microbes and competing engineered microbes. However, we modify this model to account for a spatially heterogeneous environment by assuming a periodically varying domain consisting of good and bad patches. The good patches signify the favourable regions in which the microbes are released, while the bad patches model the unfavourable barriers for inhibiting the spread of the microbes. We are particularly interested in the invasion and containment conditions for the genetically engineered population.

Although the motivation for this discussion is to determine the conditions for the spread of genetically engineered organisms, the models and analysis also apply to the introduction of other exotic species where containment, or in some cases deliberate propagation, is the goal.

Let $E(x, t)$ denote the engineered microbes and $N(x, t)$ the unmodified microbes. Here we consider only the one-dimensional situation. We use classical Lotka–Volterra dynamics to describe competition between our engineered and natural microbes and allow key model parameters to vary spatially to reflect habitat heterogeneity. So, we model the dynamics of the system by

$$\frac{\partial E}{\partial t} = \frac{\partial}{\partial x}\left(D(x)\frac{\partial E}{\partial x}\right) + r_E E[G(x) - a_E E - b_E N], \tag{1.32}$$

$$\frac{\partial N}{\partial t} = \frac{\partial}{\partial x}\left(d(x)\frac{\partial N}{\partial x}\right) + r_N N[g(x) - a_N N - b_N E], \tag{1.33}$$

where $D(x)$ are $d(x)$ are the space-dependent diffusion coefficients and r_E and r_N are the intrinsic growth rates of the organisms. These are scaled so that the maximum values of the functions $G(x)$ and $g(x)$, which quantify the respective carrying capacities, are unity. The positive parameters a_E and a_N measure the effects of *intra*specific competition, while b_E and b_N are the *inter*specific competition coefficients.

In this section we model the environmental heterogeneity by considering the dispersal and carrying capacities $D(x)$, $d(x)$, $G(x)$ and $g(x)$ to be spatially periodic. We assume that l is the periodicity of the environmental variation and so define

$$D(x) = D(x + l), d(x) = d(x + l), G(x) = G(x + l), g(x) = g(x + l). \tag{1.34}$$

Initially we assume there are no engineered microbes, that is, $E(x, 0) \equiv 0$, so the natural microbes $N(x, 0)$ satisfy the equation

$$\frac{\partial}{\partial x}\left(d(x)\frac{\partial N}{\partial x}\right) + r_N N(g(x) - a_N N) = 0. \tag{1.35}$$

The engineered organisms are then introduced at a release site, which we take as the origin. This initial distribution in $E(x, t)$ is represented by the initial conditions

$$E(x,0) = \begin{cases} H > 0 & \text{if } |x| \leq x_c \\ 0 & \text{if } |x| > x_c, \end{cases} \tag{1.36}$$

where H is a positive constant.

To bring in the idea of favourable and unfavourable patches we consider the environment consists of two kinds of homogeneous patches, say, Patch 1 of length l_1, the favourable patch, and Patch 2 of length l_2, the unfavourable patch, connected alternately along the x-axis such that $l = l_1 + l_2$. In the unfavourable patches the diffusion and carrying capacity of the organisms are less than in the favourable patches. This could occur because the unfavourable patch is a hostile environment that either limits a population or interferes with its dispersal. Correspondingly, the functions $D(x)$, $d(x)$, $G(x)$ and $g(x)$ are periodic functions of x. In Patch 1, where $ml < x < ml + l_1$ for $m = 0, 1, 2, \ldots$,

$$D(x) = D_1 > 0, d(x) = d_1 > 0, \quad G(x) = 1, g(x) = 1, \tag{1.37}$$

and in Patch 2, where $ml - l_2 < x < ml$ for $m = 0, 1, 2, \ldots$,

$$D(x) = D_2 > 0, d(x) = d_2 > 0, G(x) = G_2, g(x) = g_2. \tag{1.38}$$

Since Patch 1 is favourable,

$$D_1 \geq D_2, d_1 \geq d_2; 1 \geq G_2, 1 \geq g_2. \tag{1.39}$$

Figure 1.6 shows an example of how the diffusion of the engineered microbes could vary in space.

At the boundaries between the patches, say, $x = x_i$, with

$$x_{2m} = ml, x_{2m+1} = ml + l_1 \quad \text{for } m = 0, \pm 1, \pm 2, \ldots, \tag{1.40}$$

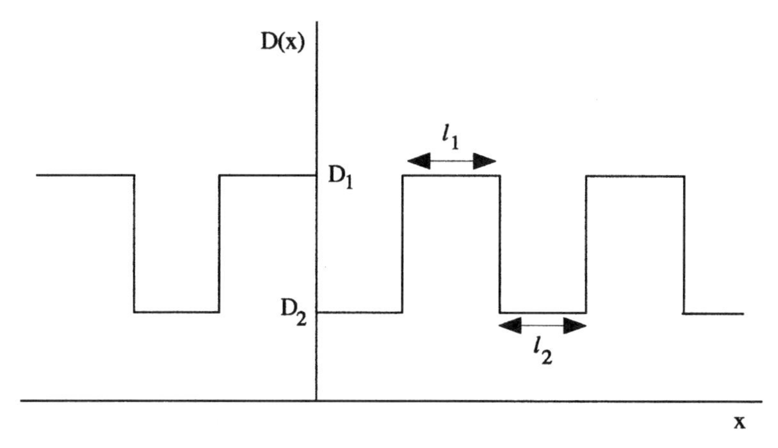

Figure 1.6. The spatial pattern in the diffusion coefficient of the genetically engineered microbes, $D(x)$, in the periodic environment. There are two patch types with a higher diffusion in the favourable patch, Patch 1, of length l_1, than in the unfavourable patch, Patch 2, of length l_2.

the population densities and fluxes are continuous so

$$\lim_{x\to x_i^+} E(x,t) = \lim_{x\to x_i^-} E(x,t),$$

$$\lim_{x\to x_i^+} N(x,t) = \lim_{x\to x_i^-} N(x,t),$$

$$\lim_{x\to x_i^+} D(x)\frac{\partial E(x,t)}{\partial x} = \lim_{x\to x_i^-} D(x)\frac{\partial E(x,t)}{\partial x},$$

$$\lim_{x\to x_i^+} d(x)\frac{\partial N(x,t)}{\partial x} = \lim_{x\to x_i^-} d(x)\frac{\partial N(x,t)}{\partial x}.$$

The mathematical problem is now defined. The key questions we want to answer are: (i) Under which conditions will the engineered organisms invade successfully when rare? and (ii) If invasion succeeds, will the engineered species drive the natural population to invader-dominant or will a coexistent state be reached? Here we follow Shigesada et al. (1986) and consider the problem on an infinite domain and assume that the diffusion and carrying capacities vary among the different patch types. We focus on the stability of the system to invasions initiated by a very small number of engineered organisms. Mathematically this means we can use a linear analysis with spatiotemporal perturbations about the steady state solutions.

Nondimensionalisation

We nondimensionalise equations by introducing

$$\begin{aligned} &e = a_E E, n = a_N N, t^* = r_E t, x^* = x\left(\frac{r_E}{D_1}\right)^{1/2} \\ &d^*(x) = \frac{d(x)}{D_1}, D^*(x) = \frac{D(x)}{D_1}, r = \frac{r_N}{r_E}, \gamma_e = \frac{b_E}{a_N}, \gamma_N = \frac{b_N}{a_E}, \\ &l^* = l\left(\frac{r_E}{D_1}\right)^{1/2}, l_1^* = l_1\left(\frac{r_E}{D_1}\right)^{1/2}, l_2^* = l_2\left(\frac{r_E}{D_1}\right)^{1/2}, \end{aligned} \tag{1.41}$$

and the nondimensional model equations, where we have dropped the asterisks for algebraic convenience, become

$$\frac{\partial e}{\partial t} = \frac{\partial}{\partial x}\left(D(x)\frac{\partial e}{\partial x}\right) + e[G(x) - e - \gamma_e n], \tag{1.42}$$

$$\frac{\partial n}{\partial t} = \frac{\partial}{\partial x}\left(d(x)\frac{\partial n}{\partial x}\right) + rn[g(x) - n - \gamma_n e], \tag{1.43}$$

where

$$D(x) = \begin{cases} 1 & \text{if } ml < x < ml + l_1 \\ D_2 & \text{if } ml - l_2 < x < ml, \end{cases}$$
$$d(x) = \begin{cases} d_1 & \text{if } ml < x < ml + l_1 \\ d_2 & \text{if } ml - l_2 < x < ml, \end{cases} \tag{1.44}$$

and the functions $G(x)$ and $g(x)$ are as before (see (1.37)–(1.39)).

At the boundaries between the patches, $x = x_i$, where $x_i = ml$ for $i = 2m$ and $x_i = ml + l_1$ for $i = 2m + 1 (m = 0, 1, 2, \ldots)$ the nondimensional conditions are now

$$\lim_{x \to x_i^+} e(x,t) = \lim_{x \to x_i^-} e(x,t), \qquad \lim_{x \to x_i^+} n(x,t) = \lim_{x \to x_i^-} n(x,t), \tag{1.45}$$

and

$$\lim_{x \to x_i^+} D(x) \frac{\partial e(x,t)}{\partial x} = \lim_{x \to x_i^-} D(x) \frac{\partial e(x,t)}{\partial x},$$
$$\lim_{x \to x_i^+} d(x) \frac{\partial n(x,t)}{\partial x} = \lim_{x \to x_i^-} d(x) \frac{\partial n(x,t)}{\partial x}, \tag{1.46}$$

for all integers i.

No Patchiness and Conditions for Containment

In the case when the whole domain is favourable the unfavourable patch has zero length, $l_2 = 0$. So $D(x) = 1, d(x) = d_1, G(x) = 1$ and $g(x) = 1$ everywhere. This results in the Lotka–Volterra competition model with diffusion that we considered in the last section.

The initial steady state reduces to $e_1 = 0, n_1 = 1$, which below we refer to as the native-dominant steady state. There are two other relevant steady states: the invader-dominant steady state, where the engineered organisms have driven the natural organisms to invader-dominant, that is, $e_2 = 1, n_2 = 0$, and the coexistence steady state given by

$$e_3 = \frac{\gamma_n - 1}{\gamma_n \gamma_e - 1}, \quad n_e = \frac{\gamma_e - 1}{\gamma_n \gamma_e - 1}. \tag{1.47}$$

The latter is only relevant, of course, if it is positive, which means that either $\gamma_e < 1$ and $\gamma_n < 1$, that is, weak interspecific competition for both species, or $\gamma_e > 1$ and $\gamma_n > 1$, which represents strong interspecific competition for both species. The trivial zero steady state is of no significance here. With these specific competition interactions there are no other steady state solutions in a Turing sense, that is, with zero flux boundary conditions.

As with the red and grey squirrel competition we know it is possible to have travelling wave solutions connecting the native-dominant steady state (e_1, n_1), to the existence steady state, (e_2, n_2), or the invader-dominant steady state, (e_3, n_3). Such solutions here correspond to waves of microbial invasion, either driving the natural species to invader-dominant or to a new, but lower, steady state.

A usual linear stability analysis about the initial native-dominant steady state, (e_1, n_1) determines under which conditions invasion succeeds. By looking for solutions of the form $e^{ikx+\lambda t}$ in the linearised system we get the dispersion relationship (this is left as an exercise)

$$\lambda(k^2) = \frac{1}{2}\left[-b(k^2) \pm \sqrt{b^2(k^2) - 4c(k^2)}\right], \tag{1.48}$$

where

$$\begin{aligned} b(k^2) &= k^2(d_1 + 1) + (\gamma_e - 1) + r, \\ c(k^2) &= d_1 k^4 + [\gamma_e - 1 + r]k^2 + r(\gamma_e - 1). \end{aligned} \tag{1.49}$$

The native-dominant steady state is linearly unstable if there exists a k^2 so that $\lambda(k^2) > 0$. From the dispersion relationship we can see that if $\gamma_e > 1$ then the initial steady state will always be linearly stable, since $b(k^2)$ and $c(k^2)$ are always positive. However if $\gamma_e < 1$ then there are values of k for which the steady state is unstable and the invasion of the engineered species, e, will succeed.

If we now linearize about the other steady states we can also determine their stability in a similar way (another exercise). The invader-dominant steady state, (e_2, n_2), is stable if $\gamma_n > 1$, and unstable if $\gamma_n < 1$. The coexistence steady state (e_3, n_3) is stable if $\gamma_n < 1$ and $\gamma_e < 1$, and unstable if $\gamma_n > 1$ and $\gamma_e > 1$. If $\gamma_e < 1 < \gamma_n$ or $\gamma_n < 1 < \gamma_e$ the coexistence steady state is no longer relevant, since either e_3 or n_3 in (1.47) becomes negative. The trivial steady state is always linearly unstable since in the absence of an indigenous species either the natural strain, the engineered strain, or both, would invade.

Since we have already considered the travelling wave of invasion in the last section in this situation we do not repeat it here. In summary, if the native-dominant steady state is unstable, then if $\gamma_e < 1$ a travelling wave connecting the native-dominant steady state to the invader-dominant steady state results, but only if $\gamma_n > 1$. On the other hand, a travelling wave connecting the native-dominant steady state to the coexistence steady state results only if $\gamma_n < 1$. The numerical solutions for the case when $\gamma_e < 1$ and $\gamma_n < 1$ are similar to those in Figure 1.5.

The requirement, $\gamma_e < 1$, for the native-dominant steady state to be unstable, implies, in terms of our original dimensional variables, defined in (1.41), that the interspecific competitive effect of the natural organisms, n, on the engineered species, e, is dominated by the intraspecific competition of the natural species.

If $\gamma_n > 1$ the natural species is driven to invader-dominant and, in terms of the original dimensional parameters, this happens when increases in density of the engineered species reduce the population growth of the natural species more than they reduce their own population's growth rate. When $\gamma_n < 1$ the situation is just reversed; again refer to (1.41).

If, on the other hand, the native-dominant state is stable, $\gamma_e > 1$, we can, simultaneously, have the invader-dominant steady state stable if $\gamma_n > 1$. These are also the conditions for the coexistence steady state to be unstable. In this case the stability of the native-dominant steady state depends on the initial conditions (1.36). If H represents a small perturbation about $e = 0$ then the native-dominant steady state remains the final steady state solution. However, Cruywagen et al. (1996) found from numerical experimentation that for very large perturbations corresponding to a very large initial release of e, a travelling wave solution results and the invader-dominant steady state becomes the final solution. So, containment can only be guaranteed for all initial release strategies if $\gamma_e > 1$ and $\gamma_n < 1$.

If we consider the whole domain as *unfavourable*, instead of favourable, by setting $l_1 = 0$ instead of $l_2 = 0$, then we obtain analogous results. In this situation, however, the nonzero steady states are now different. The native-dominant steady state is $e_1 = 0, n_1 = g_2$, the invader-dominant steady state is $e_2 = G_2, n_2 = 0$, while the coexistence steady state is

$$e_3 = \frac{\gamma_n G_2 - g_2}{\gamma_e \gamma_n - 1}, \quad n_3 = \frac{\gamma_e g_2 - G_2}{\gamma_e \gamma_n - 1}. \tag{1.50}$$

The stability conditions are now determined from whether γ_e and γ_n are respectively larger or smaller than G_2/g_2 and $\gamma_n < g_2/G_2$: the coexistence steady state is stable, and all other steady states are unstable.

Spatially Varying Diffusion

As above we again carry out a linear stability analysis about the various steady states but now consider spatial variations in the diffusion, that is, when patchiness affects the diffusion functions. Here we only investigate how spatially varying diffusion coefficients affect the ability of the engineered species to invade.

Conditions for Invasion

The initial native-dominant steady state is (e_1, n_1), where again $e_1 = 0$ and, depending on the function $g(x)$, either $n_1 = 1$ or n_1 is a periodic function of x with period related to the length of the patches.

As a first case, let us assume, however, that $g(x) = 1$ so that n_1 is then independent of x: this simplifies the problem considerably. Cruywagen et al. (1996) consider the much more involved problem when $g(x)$ is a periodic function of x.

To determine the stability of the initial native-dominant steady state we linearise about $(e_1, n_1) = (0, 1)$ to obtain

$$\frac{\partial e}{\partial t} = \frac{\partial}{\partial x}\left(D(x)\frac{\partial e}{\partial x}\right) + e[G(x) - \gamma_e], \tag{1.51}$$

$$\frac{\partial n}{\partial t} = \frac{\partial}{\partial x}\left(d(x)\frac{\partial n}{\partial x}\right) + r[-n - \gamma_n e], \tag{1.52}$$

where e and n now represent small perturbations from the steady state (n_1, e_1) $(|e| \ll 1, |n| \ll 1)$.

Here we can determine the stability of this system by looking at just the equation for the engineered species, (1.51), since it is independent of n. This reduces the linear stability problem to analysing

$$\frac{\partial e}{\partial t} = \frac{\partial^2 e}{\partial x^2} + e[1 - \gamma_e] \qquad \text{in Patch 1,} \tag{1.53}$$

$$\frac{\partial e}{\partial t} = D_2 \frac{\partial^2 e}{\partial x^2} + e[G_2 - \gamma_e] \quad \text{in Patch 2.} \tag{1.54}$$

Substituting $e(x, t) = e^{-\lambda t} f(x)$ into these equations gives the characteristic equation of the form

$$\frac{d}{dx}\left(D(x)\frac{df}{dx}\right) + [G(x) - \gamma_e + \lambda] f = 0, \tag{1.55}$$

which is known as Hill's equation. Here, according to our definition, $G(x) - \gamma_e$ and $D(x)$ are both periodic functions of period l. There is a well-established theory on the solution behaviour of Hill's equation in numerous ordinary differential equation books.

It is known from the theory of Hill's equation with periodic coefficients that there exists a monotonically increasing infinite sequence of real eigenvalues λ,

$$-\infty < \lambda_0 < \tilde{\lambda}_1 \leqslant \tilde{\lambda}_2 < \lambda_1 \leqslant \lambda_2 \leqslant \tilde{\lambda}_3 \leqslant \tilde{\lambda}_4 < \cdots, \tag{1.56}$$

associated with (1.55), for which it has nonzero solutions. The solutions are of period l if and only if $\lambda = \lambda_i$, and of period $2l$ if and only if $\lambda = \tilde{\lambda}_i$. Furthermore, the solution associated with $\lambda = \lambda_0$ has no zeros and is globally unstable (in the spatial sense) in that $f \to \infty$ as $|x| \to \infty$; this is discussed in detail by Shigesada et al. (1986). For the detailed theory see, for example, the book by Coddington and Levinson (1972).

So, the stability of the native dominant steady state of the partial differential equation system (1.53) and (1.54) is determined by the sign of λ_0. If $\lambda_0 < 0$ the trivial solution $e_0 = 0$ of (1.55) is dynamically unstable and if $\lambda_0 > 0$ it is dynamically stable. Cruywagen et al. (1996) obtained a bound on λ_0, which then gives the containment conditions we require, and which we now derive.

By defining the function

$$Q(x) = G(x) - \frac{l_1 + G_2 l_2}{l}, \tag{1.57}$$

we can write (1.55) in the form

$$\frac{d}{dx}\left(D(x)\frac{df}{dx}\right) + \left[Q(x) + \frac{l_1 + G_2 l_2}{l} - \gamma_e + \lambda\right] f = 0. \tag{1.58}$$

As a preliminary we have to derive a result associated with the equation

$$\frac{d}{dx}\left(D(x)\frac{du}{dx}\right) + [\sigma + Q(x)] u = 0, \tag{1.59}$$

where $D(x)$ and $Q(x)$ are periodic with period l and $D(x) > 0$. From the above quoted result on Hill's equation we know that the periodic solution $u(x) = u_0(x)$ of period l corresponding to the smallest eigenvalue $\sigma = \sigma_0$ has no zeros. We can assume that $u_0(x) > 0$ for all x, and then define the integrating factor $h(x)$ as

$$h(x) = \frac{d}{dx} \ln u_0(x). \tag{1.60}$$

So, $h(x)$ is periodic of period l and is a solution of

$$\frac{d}{dx}\left[D(x)h(x)\right] + D(x)h^2(x) = -\sigma_0 - Q(x). \tag{1.61}$$

If we now integrate this over one period of length l we get

$$\int_{\zeta l}^{(\zeta+1)l} D(x)h^2(x)\,dx = -l\sigma_0, \quad \text{for real } \zeta, \tag{1.62}$$

since $D(x)$, $h(x)$ and $Q(x)$ are periodic. So, $\sigma_0 = 0$ if the integral over $h^2(x)$ is zero; otherwise $\sigma_0 < 0$ because $D(x) > 0$.

Now, since

$$\int_{\zeta l}^{(\zeta+1)l} Q(x)\,dx = 0 \quad \text{for arbitrary real } \zeta, \tag{1.63}$$

and comparing (1.58) and (1.59) and using the result we have just derived,

$$\frac{l_1 + G_2 l_2}{l} - \gamma_e + \lambda_0 < 0. \tag{1.64}$$

So, we now have the following sufficient condition for $\lambda_0 < 0$, and hence for the system (1.53) and (1.54), to be unstable,

$$(1 - \gamma_e)l_1 \geq (\gamma_e - G_2)l_2. \tag{1.65}$$

There are now three relevant cases to consider. Remember that $G_2 < 1$.

In the case when $\gamma_e > 1 > G_2$ the native-dominant steady state is stable in both the favourable and unfavourable patches if they are considered in isolation. Refer again to the above detailed discussion of the stability conditions for either the favourable or unfavourable patches (note that $g_2 = 1$ here). Although it seems reasonable, we cannot conclude from (1.65) that the native-dominant steady state will be stable for the full problem, since this is only a sufficient condition for instability.

On the other hand, when $1 > G_2 > \gamma_e$ the native-dominant steady state is unstable in both patches if they are considered in isolation. Not only that, as expected, it follows from (1.65) that the native-dominant steady state is also unstable for the problem on the full domain considered here. So, if the carrying capacity for the engineered microbes in the unfavourable patch, Patch 2 (reflected by G_2), exceeds their loss due to the interspecific competitive effect of the natural organisms (reflected by γ_e), the engineered organisms always invade.

However, if $1 > \gamma_e > G_2$, the native-dominant steady state is unstable in the favourable patch but stable in the unfavourable patch when considered in isolation. Which of these patches dominates the actual stability of the native-dominant steady state depends on the relative sizes of these patches, as can be seen from inequality (1.65). By increasing the favourable patch length, l_1, and/or decreasing the unfavourable patch length, l_2, the native-dominant steady state will become unstable so that invasion does occur. So, the condition (1.65) is not a necessary condition for instability and so does not provide exact conditions for ensuring the stability of the native-dominant steady state.

We start by deriving separable solutions for (1.53) and (1.54) for each of the two types of patches. Since we expect periodic solutions this suggests that we use Fourier series expansions to find the solutions.

In Patch 1 we get, after a little algebra, the solution

$$e(x,t) = \sum_{i=0}^{\infty} A_i e^{-\gamma_i t} \cos\left[\left(x - \frac{l_1}{2} - ml\right)\sqrt{1 - \gamma_e + \lambda_i}\right], \tag{1.66}$$

while in Patch 2 we have

$$e(x,t) = \sum_{i=0}^{\infty} B_i e^{-\lambda_i t} \cos\left[\left(x + \frac{l_2}{2} - (m+1)l\right)\sqrt{\frac{G_2 - \gamma_e + \lambda_i}{D_2}}\right], \tag{1.67}$$

with A_i and B_i constants.

Applying the continuity conditions (1.45) and (1.46) the following series of equalities must hold

$$\begin{aligned} &\sqrt{1 - \gamma_e + \lambda_i} \tan\left(\frac{l_1}{2}\sqrt{1 - \gamma_e + \lambda_i}\right) \\ &= -D_2\sqrt{\frac{G_2 - \gamma_e + \lambda_i}{D_2}} \tan\left(\frac{l_2}{2}\sqrt{\frac{G_2 - \gamma_e + \lambda_i}{D_2}}\right), \end{aligned} \tag{1.68}$$

for $i = 0, 1, 2, \ldots$. If the expressions inside the square roots become negative we have to use the identities

$$\tan iz = i \tanh z, \qquad \arctan iz = i \,\mathrm{arctanh}\, z. \tag{1.69}$$

We are, of course, interested in the sign of the smallest eigenvalue, $\lambda = \lambda_0$, which satisfies the above equality. It is easy to show that λ_0 is negative if and only if the expressions $1 - \gamma_e + \lambda_0$ and $G_2 - \gamma_e + \lambda_0$ appearing under the square roots have opposite signs. Since, by definition $G_2 < 1$, this can occur only if $\gamma_e < 1$. So, since this is the case in the problem with spatially uniform coefficients, a necessary condition for the native-dominant steady state to be unstable, thereby letting engineered microbes invade, is that the competitive effect b_E of the natural species on the engineered species is smaller than the intraspecific competition effect, a_N, of the natural species; refer to the dimensionless forms in (1.41).

If $G_2 \geq \gamma_e$ then λ_0 is negative and invasion will succeed regardless of the other parameters and the patch sizes, as we discussed above. However, if $G_2 < \gamma_e < 1$, then depending on the various parameter values, λ_0 can be either negative or positive. We now consider this case, in which the native-dominant steady state is unstable in the favourable patch but stable in the unfavourable patch, in further detail. As we have seen above, the relative sizes of the patches now become important.

At the critical value, $\lambda_0 = 0$, the following holds,

$$\sqrt{1-\gamma_e}\tan\left[\frac{l_1}{2}\sqrt{1-\gamma_e}\right] = D_2\sqrt{\frac{\gamma_e - G_2}{D_2}}\tanh\left[\frac{l_2}{2}\sqrt{\frac{\gamma_e - G_2}{D_2}}\right], \tag{1.70}$$

from which we determine the critical length, l_1^*, of Patch 1 as

$$l_1^* = \frac{2}{\sqrt{1-\gamma_e}}\arctan\left[\sqrt{\frac{D_2(\gamma_e - G_2)}{1-\gamma_3}}\tanh\left\{\frac{l_2}{2}\sqrt{\frac{\gamma_e - G_2}{D_2}}\right\}\right]. \tag{1.71}$$

For $l_1 < l_1^*$ the native-dominant steady state would be stable, since λ_0 would be positive, while for $l_1 > l_1^*$, λ_0 would be negative and the native-dominant steady state unstable. So, as we showed earlier in this section, invasion will succeed if the favourable patch is large enough compared to the unfavourable patch.

Note that as l_2 tends to infinity the boundary curve approaches the asymptote

$$\lim_{l_2\to\infty} l_1(l_2) = l_1^c = \frac{2}{\sqrt{1-\gamma_e}}\arctan\sqrt{\frac{D_2(\gamma_e - G_2)}{1-\gamma_e}}. \tag{1.72}$$

So, invasion will always succeed, regardless of the unfavourable patch size, if $l_1 \geq l_1^c$. Furthermore, since

$$l_l^c > l_l^m = \frac{2\arctan\infty}{\sqrt{1-\gamma_e}} = \frac{\pi}{\sqrt{1-\gamma_e}}, \tag{1.73}$$

invasion will succeed regardless of the values of l_2, $G_2(<\gamma_e)$ and D_2 if $l \geq l_1^m$. The stability region in terms of l_1 and l_2, for the case when $G_2 < \gamma_e$, is shown in Figure 1.7(a).

In a similar way we can draw a stability curve for γ_e versus l_2. We have shown that if $\gamma_e < G_2$ invasion will always succeed independent of the length of Patch 2 (l_2). However, as γ_e increases beyond G_2 a stability curve appears from infinity at some critical value $\gamma_e = \gamma_e^c$. The asymptote of the curve, γ_e^c, can be obtained from the following nonlinear relationship

$$\frac{D_2(\gamma_e^c - G_2)}{1-\gamma_e^c} = \tan^2\left(\frac{l_1}{2}\sqrt{1-\gamma_e^c}\right). \tag{1.74}$$

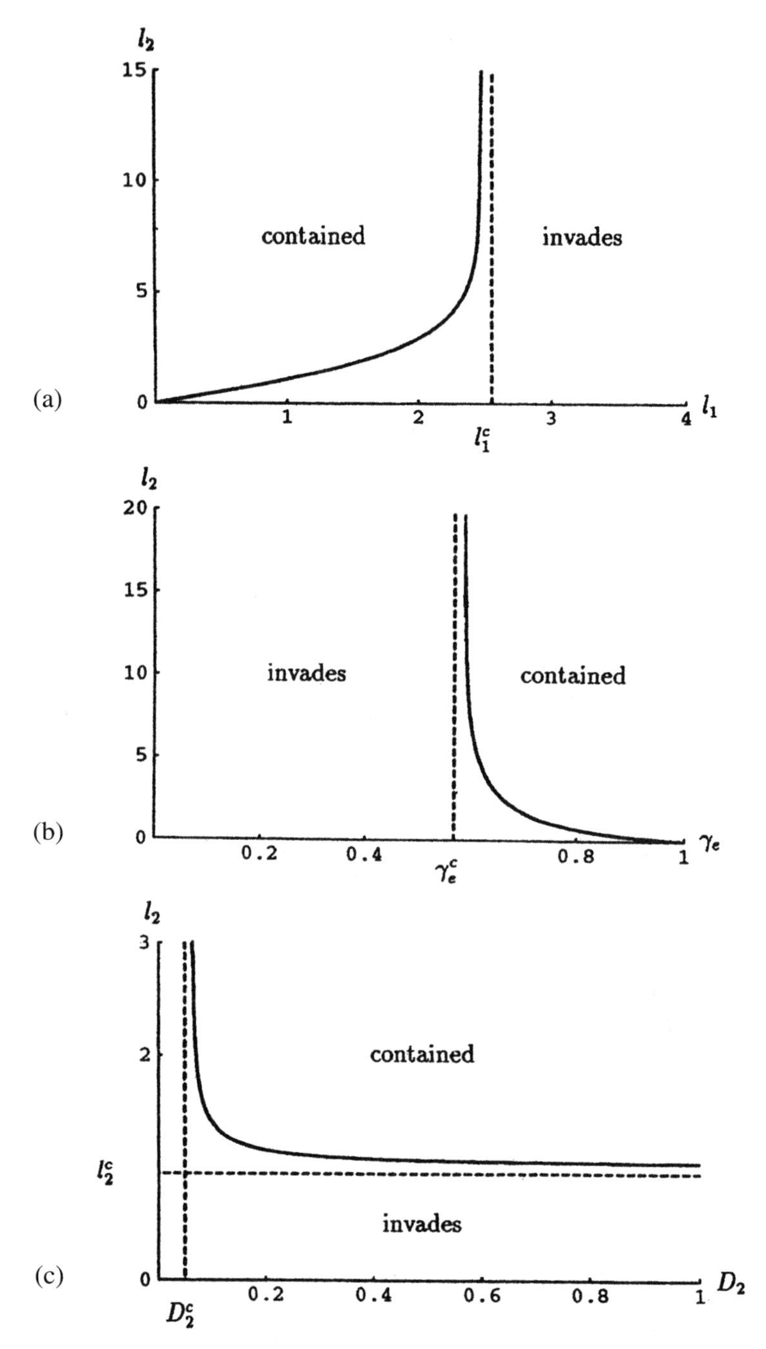

Figure 1.7. The stability diagram for the native-dominant steady state, obtained from (1.70), with spatially periodic diffusion coefficients and a spatially periodic carrying capacity for the engineered population. The boundary curves are indicated by the solid line, while the asymptotes are indicated by the dotted lines. (**a**) The (l_1, l_2) plane for $D_2 = 0.5$, $\gamma_e = 0.75$ and $G_2 = 0.5$. (**b**) The (γ_e, l_2) plane for $D_2 = 0.5$, $l_1 = 1.0$ and $G_2 = 0.5$. (**c**) The (D_2, l_2) plane for $\gamma_e = 0.75$, $l_1 = 1.0$ and $G_2 = 0.5$. The algebraic expressions for the asymptotes are given in the text. (From Cruywagen et al. 1996)

This stability region is shown in Figure 1.7(b). Note here that, as l_1 increases towards l_1^c, the stability curve would appear for increasingly larger values of γ_e^c, while for $l_1 \geq l_1^c$, the stability curve would not appear at all.

The properties of the stability graph of G_2 versus l_2 is similar to that of γ_e versus l_2. For $G_2 > \gamma_e$ invasion is successful, however, because the value of G_2 decreases beyond γ_e; a stability curve appears from infinity at the asymptote

$$G_2^c = \gamma_e + \frac{\gamma_e - 1}{D_2} \tan^2 \left(\frac{l_1}{2} \sqrt{1 - \gamma_\epsilon} \right). \tag{1.75}$$

The diffusion rate of the engineered species in the unfavourable patch, Patch 2, also plays an important role in determining the stability of the native-dominant steady state if $l_1 < l_1^*$. For sufficiently small values of D_2 invasion succeeds regardless of the value of l_2. Biologically this implies that the diffusion is so small in the unfavourable patch that the effect on the favourable patch is minimal. However, as D_2 increases, a stability curve appears from infinity at the critical asymptotic value $D_2 = D_2^c$, with

$$D_2^c = \frac{1 - \gamma_e}{\gamma_e - G_2} \tan^2 \left(\frac{l_1}{2} \sqrt{1 - \gamma_e} \right). \tag{1.76}$$

On the other hand, as $D_2 \to \infty$ the stability curve approaches an asymptote at l_2^c. Since, from (1.70),

$$l_2 = 2 \sqrt{\frac{D_2}{\gamma_e - G_2}} \operatorname{arc\,tanh} \left[\sqrt{\frac{1 - \gamma_e}{D_2(\gamma_e - G_2)}} \tan \left(\frac{l_1}{2} \sqrt{1 - \gamma_e} \right) \right] \tag{1.77}$$

and, after applying L'Hôpital's rule, we get

$$\lim_{D_2 \to \infty} l_2(D_2) = l_2^c = \frac{2\sqrt{1 - \gamma_e}}{\gamma_e - G_2} \tan \left(\frac{l_1}{2} \sqrt{1 - \gamma_e} \right). \tag{1.78}$$

So for $l_2 < l_2^c$ the engineered microbes invade for any diffusion rate. Figure 1.7(c) shows the graph of the D_2 versus l_2 stability curve.

We can conclude that for $\gamma_e < 1$ necessary conditions for containment are $G_2 < \gamma_e$, and

$$l_1 < l_1^c, \quad l_2 > l_2^c, \quad \gamma_e > \gamma_e^c, \quad G_2 < G_2^c, \quad D_2 > D_2^c. \tag{1.79}$$

These inequalities show that containment can be ensured either by decreasing l_1 or G_2, or by increasing γ_e, D_2 or l_2. Recall from above that $\gamma_e < 1$ implies that the native-dominant steady state is unstable in the favourable patch, while, on the other hand, $G_2 < \gamma_e$ implies that it is stable in the unfavourable patch. The simplest strategy for ensuring the stability of the native-dominant steady state is, however, to have $\gamma_e > 1$ as we discussed above.

It is important to note here that linear stability has been discussed for the patchy domain as a whole. Even though invasion succeeds somewhere in the whole domain, it might be the case that, depending on initial conditions, invasion is only local and, in effect, contained in a certain patch. This could be the case when $1 > \gamma_e > G_2$. Although the native-dominant steady state can be stable for the full problem it is locally unstable in the favourable patch. Local invasion in a favourable patch, resulting from small nonzero initial perturbations for e in that patch, might thus occur in some cases. Cruywagen et al. (1996) considered such an example.

We have just obtained the stability conditions for the native-dominant steady state, (e_1, n_1). There are three other possible steady states, the zero steady state (e_0, n_0), the invader-dominant steady state (e_2, n_2), and the coexistence steady state (e_3, n_3). We can determine whether, and under what conditions, the solution would evolve into any of these steady states. Naturally in these cases, travelling wave solutions, connecting the unstable native-dominant steady state to non-native-dominant steady states, are to be expected. Cruywagen et al. (1996) examined these other steady states in a similar way to the above and showed that $\gamma_e > 1$ and $\gamma_n < 1$ is the only strategy for containment that is safe for any initial microbe release density. Comparing this result with the one we obtained above we see that varying diffusion does not affect the stability conditions.

Cruywagen et al. (1996) also considered the effect of a varying carrying capacity by considering $G(x)$ and $g(x)$ to be spatially periodic which means that all three nonzero steady states have spatially periodic solutions. Determining their stability is difficult so they used perturbation techniques to obtain approximate stability conditions.

Although the model of Cruywagen et al. (1996) is really too simple to be realistic, it identifies several key scenarios for the effect of a heterogeneous environment on invasion in competitive systems. In spite of the relative simplicity the detailed analysis is complicated and involves considerable bookkeeping. Cruywagen et al. (1996) produced quite a complex summary table of approximate predictions of coexistence, extinction and invasion for genetically engineered microbes competing with a wild-type strain in terms of the various parameters and forms of the diffusion and variable carrying capacities, including patch sizes. First, although invasion is less likely to occur for large unfavourable 'moats' surrounding an interior suitable habitat island, once the interior island gets sufficiently large, no size of the surrounding hostile region can prevent an invasion. In general, a decrease in the relative size of patch type in which the native species dominates increases the chances that the exotic species can invade. Perhaps the most interesting scenario involves a situation in which an invasion can succeed locally (within one patch) but fail globally. It is in this situation where results are not intuitively clear and the mathematical model is particularly helpful.

Further work on models that include convective transport as well as a possible distinction of sedentary and mobile classes has been done by Lewis et al. (1996); see also Lewis and Schmitz (1996). This work takes into account the fact that microbes enter different mobile and immobile compartments, such as the roots of plants, various hosts, ground water or wind. The paper by Lewis (1997) and the book of articles edited by Tilman and Kareiva (1997) are of particular relevance to the spatial spread of invading species. Finally, any study and extension of the basic model considered should be associated with field studies to estimate the interspecific competition parameters to ensure realistic field predictions (Kareiva 1990).

1.5 Travelling Fronts in the Belousov–Zhabotinskii Reaction

One of the reasons for studying chemical waves in the Belousov–Zhabotinskii reaction is that it is now just one of many such reactions which exhibit similar wave and pattern formation phenomena. These reactions are used as paradigms for biological pattern formation systems and as such have initiated numerous experiments in embryonic development and greatly enhanced our understanding of some of the complex wave behaviour in the heart and other organs in the body. So, although the study of these chemical waves is interesting in its own right, we discuss them here with the pedagogical aim of their use in furthering our understanding of biological pattern formation mechanisms.

The waves in Figure 1.1(a) are travelling bands of chemical concentrations in the Belousov–Zhabotinskii reaction; they are generated by localised pacemakers. In this section we derive and analyse a model for the propagating *front* of such a wave. Far from the centre the wave is essentially plane, so we consider here the one-dimensional problem and follow in part the analysis of Murray (1976). The reason for investigating this specific problem is the assumption that the speed of the wavefront depends primarily on the concentrations of the key chemicals, bromous acid ($HBrO_2$) and the bromide ion (Br^-) denoted respectively by x and y. Refer to Chapter 8, Volume I, specifically Section 8.1, for the details of the model reaction kinetics. This section can, however, be read independently by starting with the reaction scheme in (1.80) below. We assume that these reactants diffuse with diffusion coefficient D. We believe that the wavefront is dominated by process I of the reaction, namely, the sequence of reactions which (i) reduces the bromide concentration to a small value, (ii) increases the bromous acid to its maximum concentration and in which (iii) the cerium ion catalyst is in the Ce^{3+} state. Since the concentration of Ce^{4+} was denoted by z in Section 8.1, Volume I, the last assumption implies that $z = 0$. The simplified reaction sequence, from equation (8.2), Volume I without the cerium reaction and with $z = 0$, is then

$$
\begin{aligned}
&A + Y \overset{k_1}{\to} X + P, \quad X + Y \overset{k_2}{\to} 2P, \\
&A + X \overset{k_3}{\to} 2X, \quad 2X \overset{k_4}{\to} P + A,
\end{aligned}
\tag{1.80}
$$

where X and Y denote the bromous acid and bromide ion respectively and the k's are rate constants. P (the compound HBrO) does not appear in our analysis and the concentration $A(BrO^{3-})$ is constant.

Applying the Law of Mass Action (see Chapter 6, Volume I) to this scheme, using lowercase letters for concentrations, and including diffusion of X and Y, we get

$$
\begin{aligned}
\frac{\partial x}{\partial t} &= k_1 ay - k_2 xy + k_3 a_x - k_4 x^2 + D\frac{\partial^2 x}{\partial s^2}, \\
\frac{\partial y}{\partial t} &= -k_1 ay - k_2 xy + D\frac{\partial^2 y}{\partial s^2},
\end{aligned}
\tag{1.81}
$$

where s is the space variable. An appropriate nondimensionalisation here is

$$u = \frac{k_4 x}{k_3 a}, \quad v = \frac{k_2 y}{k_3 a r}, \quad s^* = \left(\frac{k_3 a}{D}\right)^{1/2} s,$$
$$t^* = k_3 a t, \quad L = \frac{k_1 k_4}{k_2 k_3}, \quad M = \frac{k_1}{k_3}, \quad b = \frac{k_2}{k_4}, \tag{1.82}$$

where r is a parameter which reflects the fact that the bromide ion concentration far ahead of the wavefront can be varied experimentally. With these, (1.81) becomes, on omitting the asterisks for notational simplicity,

$$\frac{\partial u}{\partial t} = Lrv + u(1 - u - rv) + \frac{\partial^2 u}{\partial s^2}$$
$$\frac{\partial v}{\partial t} = -Mv - buv + \frac{\partial^2 v}{\partial s^2}. \tag{1.83}$$

Using the estimated values for the various rate constants and parameters from Chapter 8, Volume I, equations (8.4), we find

$$L \approx M = O(10^{-4}), \quad b = O(1).$$

The parameter r can be varied from about 5 to 50.

With the nondimensionalisation (1.82) the realistic steady states are

$$u = v = 0; \quad u = 1, v = 0, \tag{1.84}$$

so we expect u and v to be $O(1)$-bounded. So, to a first approximation, since $L \ll 1$ and $M \ll 1$ in (1.83), we may neglect these terms and thus arrive at a model for the leading edge of travelling waves in the Belousov–Zhabotinskii reaction, namely,

$$\frac{\partial u}{\partial t} = u(1 - u - rv) + \frac{\partial^2 u}{\partial s^2}$$
$$\frac{\partial v}{\partial t} = -buv + \frac{\partial^2 v}{\partial s^2}, \tag{1.85}$$

where r and b are positive parameters of $O(1)$. Note that this model approximation introduces a new steady state $(0, S)$, where $S > 0$ can take any value. This is because this is only a model for the front, not the whole wave pulse, on either side of which $v \to 0$.

Let us now look for travelling wavefront solutions of (1.85) where the wave moves from a region of high bromous acid concentration to one of low bromous acid concentration as it reduces the level of the bromide ion. With (1.84) we therefore look for waves with boundary conditions

$$u(-\infty, t) = 0, \quad v(-\infty, t) = 1, \quad u(\infty, t) = 1, \quad v(\infty, t) = 0 \tag{1.86}$$

and the wave moves to the left.

Before looking for travelling wave solutions we should note that there are some special cases which reduce the problem to a Fisher–Kolmogoroff equation. Setting

$$v = \frac{1-b}{r}(1-u), \quad b \neq 1, \quad r \neq 0 \tag{1.87}$$

the system (1.85) reduces to

$$\frac{\partial u}{\partial t} = bu(1-u) + \frac{\partial^2 u}{\partial s^2},$$

the Fisher–Kolmogoroff equation (13.4) in Volume I, which has travelling monotonic wavefront solutions going from $u = 0$ to $u = 1$ which travel at speeds $c \geq 2\sqrt{b}$. Since we are only concerned here with nonnegative u and v, we must have $b < 1$ in (1.87). If we take the initial condition

$$u(s, 0) \sim O(\exp[-\beta s]) \quad \text{as} \quad s \to \infty$$

we saw in Chapter 13, Volume I that the asymptotic speed of the resulting travelling wavefront is

$$c = \begin{cases} \beta + \dfrac{b}{\beta}, & 0 < \beta \leq \sqrt{b} \\ 2\sqrt{b}, & \beta > \sqrt{b}. \end{cases} \tag{1.88}$$

The wavefront solutions given by the Fisher–Kolmogoroff wave with v as in (1.87) are not, however, of practical relevance unless $1 - b = r$ since we require u and v to satisfy the boundary conditions (1.86), where $v = 0$ when $u = 1$ and $v = 1$ when $u = 0$. The appropriate Fisher–Kolmogoroff solution with suitable initial conditions, namely,

$$u(s, 0) = \begin{cases} 0 \\ h(s) \\ 1 \end{cases} \text{for} \quad \begin{cases} s < s_1 \\ s_1 < s < s_2 \\ s_2 < s \end{cases}, \tag{1.89}$$

where $h(s)$ is a positive monotonic continuous function with $h(s_1) = 0$ and $h(s_2) = 1$, then has wavespeed $c = 2\sqrt{b} = 2\sqrt{1-r}$ from (1.88) and $v = 1 - u$. Necessarily $0 < r \leq 1$.

We can further exploit the results for the Fisher–Kolmogoroff equation by using the maximum principle for parabolic equations. Let $u_f(s, t)$ denote the unique solution of

$$\begin{gathered} \frac{\partial u_f}{\partial t} = u_f(1 - u_f) + \frac{\partial^2 u_f}{\partial s^2} \\ u_f(-\infty, t) = 0, \quad u_f(\infty, t) = 1 \end{gathered} \tag{1.90}$$

with initial conditions (1.89). The asymptotic travelling wavefront solution has speed $c = 2$. Now write

$$w(s,t) = u(s,t) - u_f(s,t)$$

and let $u(s,t)$ have the same initial conditions as u_f, namely, (1.89). Subtracting equation (1.90) from the equation for u given by (1.85) and using the definition of w in the last equation we get

$$w_{ss} - w_t + [1 - (u + u_f)]w = ruv.$$

We are restricting our solutions to $0 \leq u \leq 1$ and, since $0 \leq u_f \leq 1$, we have $[1 - (u + u_f)] \leq 1$ and so we cannot use the usual maximum principle immediately. If we set $W = w \exp[-Kt]$, where $K > 0$ is a finite constant, the last equation becomes

$$W_{ss} - W_t + [1 - (u + u_f) - K]W = ruve^{-Kt} \geq 0.$$

Choosing $K > 1$ we then have $[1 - (u + u_f) - K] < 0$ and the maximum principle can now be used on the W-equation. It says that W, and hence w, has its maximum at $t = 0$ or at $s = \infty$. But $w_{\max} = (u - u_f)_{\max} = 0$ at $t = 0$ and at $s = \pm\infty$ so we have the result

$$u(s,t) \leq u_f(s,t) \quad \text{for all} \quad s, \quad t > 0.$$

This says that the solution for u of (1.85) is at all points less than or equal to the Fisher–Kolmogoroff solution u_f which evolves from initial conditions (1.89). So, if the solutions of (1.85) have travelling wave solutions with boundary conditions (1.86) and equivalent initial conditions to (1.89), then their wavespeeds c must be bounded by the Fisher–Kolmogoroff speed and so we have the upper bound $c(r, b) \leq 2$ for all values of the parameters r and b. Intuitively we would expect any such travelling wave solution of (1.85) to have speed $c \leq 2$ since with $uv \geq 0$ the term $-ruv$ in the first of (1.85) is like a sink term in addition to the kinetics $u(1 - u)$. This inhibits the growth of u at any point as compared with the Fisher–Kolmogoroff wave solution so we would expect u and its speed to be bounded above by the Fisher–Kolmogoroff solution.

Various limiting values for the wavespeed c, as a function of r and b, can be derived from the equation system (1.85). Care, however, has to be taken in their derivation because of nonuniform limiting situations; these will be pointed out at the appropriate places.

If $b = 0$, the equation for v from (1.85) becomes the basic diffusion equation $v_t = v_{ss}$ which cannot have wave solutions. This means that neither can the first of (1.85) for u, since a wave solution requires u and v to have the same speed of propagation. This suggests that the limit $c(b \to 0, r) = 0$ for $r > 0$. If $b \to \infty$, (1.85) says that $v = 0$ (we exclude the trivial solution $u = 0$) in which case $c(b \to \infty, r) = 2$ for all $r \geq 0$. Now if $r = 0$, the u and v equations are uncoupled with the u-equation being the basic Fisher–Kolmogoroff equation (1.90) which, with initial conditions (1.89), has wavefront solutions with $c = 2$: this means that the relevant v-solution also has speed 2. This gives the limiting case $c(b, r \to 0) = 2$ for $b > 0$. If $r \to \infty$ then $u = 0$ or $v = 0$, either of which implies that there is no wave solution, so $c(b, r \to \infty) = 0$. Note the nonuniform limiting situation with this case: the limit $r \to \infty$ with $v \neq 0$ is not the same as the situation with $v = 0$ and then letting $r \to \infty$. In the latter, u is governed by the Fisher–Kolmogoroff equation and r is irrelevant. As we said above, however, we are here concerned with travelling waves in which neither u nor v are identically zero. With that in mind we then have in summary

$$\begin{aligned} c(0, r) &= 0, \quad r > 0; \qquad c(\infty, r) = 2, \quad r \geq 0 \\ c(b, 0) &= 2, \quad b > 0; \qquad c(b, \infty) = 0, \quad b \geq 0. \end{aligned} \tag{1.91}$$

The first of these does not give the whole story for small b as we see below.

The travelling wavefront problem for the system (1.85), on using the travelling wave transformation

$$u(s, t) = f(z), \quad v(s, t) = g(z), \quad z = s + ct,$$

and the boundary conditions (1.86), become

$$\begin{aligned} &f'' - cf' + f(1 - f - rg) = 0, \quad g'' - cg' - bfg = 0 \\ &f(\infty) = g(-\infty) = 1, \quad f(-\infty) = g(\infty) = 0. \end{aligned} \tag{1.92}$$

Using various bounds and estimation techniques for monotonic solutions of (1.92) with $f \geq 0$ and $g \geq 0$, Murray (1976) obtained the general bounds on c in terms of the parameters r and b given by

$$\left[\left(r^2 + \frac{2b}{3}\right)^{1/2} - r\right] [2(b + 2r)]^{-1/2} \leq c \leq 2. \tag{1.93}$$

The system (1.85), with initial and boundary conditions (1.86) and (1.89), were solved numerically (Murray 1976) and some of the results are shown in Figure 1.8. Note, in Figure 1.8(b), the region bounded by $b = 0$, $c^2 = 4b$ and $c = 2$ within which nonnegative solutions do not exist. The limit curve $c^2 = 4b$ is obtained using the special

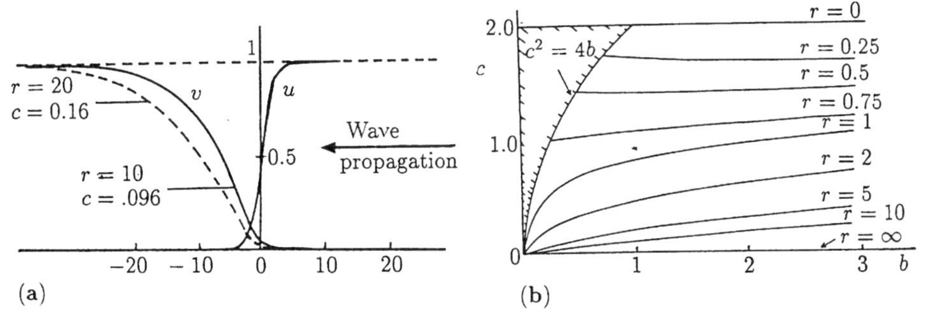

Figure 1.8. (**a**) Typical computed wavefront solution of the Belousov–Zhabotinskii model system (1.85) for u and v for $b = 1.25$ and two values of the upstream bromide parameter r. The u-curves for both values of r are effectively indistinguishable. (**b**) Wavespeed c of wavefront solutions as a function of b for various r. (From Murray 1976)

case solution in which $v = 1 - u$ and $b = 1 - r$, $r < 1$. A fuller numerical study of the model system (1.85) was carried out by Manoranjan and Mitchell (1983).

Let us now return to the experimental situation. From Figure 1.8(a), if we keep b fixed, the effect of increasing r (which with the nondimensionalisation (1.82) is the equivalent of increasing the upstream bromide ion (Br^-) concentration) is to flatten the v-curve. That is, the wavefront becomes less sharp. On the other hand, for a fixed r and increasing b, the front becomes sharper. Although it is imprecise, we can get some estimate of the actual width of the wavefront from the width, ω say, of the computed wavefront solution. In dimensional terms this is ω_D where, from (1.82),

$$\omega_D = \left(\frac{D}{k_3 a}\right)^{1/2} \omega \approx 4.5 \times 10^{-4} \omega \text{ cm},$$

where we have taken $D \approx 2 \times 10^{-5}$ cm^{2}s^{-1}, a typical value for reasonably small molecules such as we are concerned with here, and $k_3 a \approx 10^2$s^{-1} obtained from the parameter values in (8.4) in Chapter 8. From Figure 1.8(a), ω is around 10 which then gives ω_D of the order of 10^{-3} cm. This is of the order found experimentally; the front is very thin.

Another practical prediction, from Figure 1.8(b), is that for b larger than about 2, we see that for a fixed r the wavespeed is fairly independent of b. Computations for values of b up to about 50 confirm this observation. This has also been observed experimentally.

From the nondimensionalisation (1.82) the dimensional wavespeed, c_D say, is given by

$$c_D = (k_3 a D)^{1/2} c(r, b),$$

where r is a measure of the upstream bromide concentration and $b = k_2/k_4$. From the parameter estimates given in Chapter 8, Volume I, equations (8.4), we get $b \approx 1$. Assigning r is not very easy and values of 5–50 are reasonable experimentally. With r of $O(10)$ and b about 1 we get the dimensionless wavespeed from Figure 1.8(b) to be $O(10^{-1})$; the precise value for c can be calculated from the model system. With the values for D and $k_3 a$ above we thus get c_D to be $O(4.5 \times 10^{-3}$ cm s$^{-1})$ or $O(2.7 \times 10^{-1}$ cm min$^{-1})$, which is again in the experimental range observed. In view of the reasonable quantitative comparison with experiment and the results derived here from a model which mimics the propagation of the wavefront, we suggest that the speed of propagation of Belousov–Zhabotinskii wavefronts is mainly determined by the leading edge and not the trailing edge.

Finally, in relation to the speed of propagation of a reaction diffusion wavefront compared to simple diffusion we get the time for a wavefront to move 1 cm as $O(10/2.7$ min), that is, about 4 minutes as compared to the diffusional time which is $O(1 \text{ cm}^2/D)$, namely, $O(5 \times 10^4$ s) or about 850 minutes. So, as a means of transmitting information via a change in chemical concentration, we can safely say that reaction diffusion waves are orders of magnitude faster than pure diffusion, if the distances involved are other than very small. Later we discuss in detail the problem of pattern formation in embryological contexts where distances of interest are of the order of cell diameters, so diffusion is again a relevant mechanism for conveying information. However, as we shall see, it is not the only possible mechanism in embryological contexts.

The model system (1.85) has been studied by several authors. Gibbs (1980), for example, proved the existence and monotonicity of the travelling waves. An interesting formulation of such travelling wave phenomena as Stefan problems, together with a singular perturbation analysis of the Stefan problem associated with the Murray model (1.85) in which the parameters r and b are both large, is given by Ortoleva and Schmidt (1985).

Showalter and his colleagues have made major contributions to this area of travelling fronts, crucially associated with experiments, many involving the Belousov–Zhabotinskii reaction. For example, Merkin et al. (1996) looked at wave-induced chaos based on a simple cubic autocatalytic model involving chemical feedback. Their model involves a cubic autocatalytic feedback represented by

$$A + 2B \rightarrow 3B, \text{ reaction rate} = k_1 ab^2,$$

coupled with a decay step given by

$$B \rightarrow C, \text{ reaction rate} = k_2 b,$$

where a and b are the concentrations of A and B. The setup consists of a gel reaction zone on one side of which there is a reservoir kept at a constant a_0 with which it can exchange a and on the other side, kept at b_0 with which it can exchange b. The reaction diffusion system is then given by

$$\begin{aligned}
\frac{\partial a}{\partial t} &= D_A \nabla^2 a + k_f(a_0 - a) - k_1 ab^2, \\
\frac{\partial b}{\partial t} &= D_B \nabla^2 b + k_f(b_0 - b) + k_1 ab^2 - k_2 b,
\end{aligned}$$

where k_f is a constant parameter associated with the inflow and outflow of the reactants and the D's are their respective diffusion coefficients. Although these equations do not look very different, except in reaction details, to many that have been studied in the past they exhibit an interesting spectrum of solutions. Merkin et al. (1996) show, for example, by linear, travelling wave and numerical (in one and two dimensions) analyses, that there are ranges of the dimensionless parameters $\mu = k_1 a_0^2/k_f$ and $\phi = (k_f + k_2)/k_f$ for which there can be travelling fronts, travelling pulses, Hopf bifurcations and interestingly chaotic behaviour, the latter being induced behind a travelling wave.

1.6 Waves in Excitable Media

One of the most widely studied systems with excitable behaviour is neural communication by nerve cells via electrical signalling. We discussed the important Hodgkin–Huxley model in Chapter 7, Volume I, and derived a mathematical caricature, the Fitzhugh–Nagumo equations (FHN). Here we first consider, simply by way of example, the spatiotemporal FHN model and demonstrate the existence of travelling pulses which only propagate if a certain threshold perturbation is exceeded. By a pulse here, we mean a wave which represents an excursion from a steady state and back to it—like a solitary wave on water; see, for example, Figure 1.10. We shall consider models with kinetics

$$u_t = f(u, v), \quad v_t = g(u, v)$$

for a specific class of f and g. The approach we shall describe is quite general and applies to a wide class of qualitative models of excitable media whose null clines are qualitatively similar to those in Figure 1.9(a). This section can be read without reference to the actual physiological situation if the equation system (1.94) below is simply considered as a specific model example for an excitable medium.

As we saw in Section 7.5, Volume I, without any spatial variation, that is, the space-clamped situation, the FHN equations exhibited a threshold behaviour in time as illustrated in Figure 7.12, Volume I (refer also to Section 3.8, Volume I). The FHN system without any applied current ($I_a = 0$), but where we allow spatial 'diffusion' in the transmembrane potential and with a slight change in notation for consistency in this chapter, is

$$\begin{aligned} \frac{\partial u}{\partial t} &= f(u) - v + D\frac{\partial^2 u}{\partial x^2}, \quad \frac{\partial v}{\partial t} = bu - \gamma v, \\ f(u) &= u(a - u)(u - 1). \end{aligned} \tag{1.94}$$

Here u is directly related to the membrane potential (V in Section 7.5, Volume I) and v plays the role of several variables associated with terms in the contribution to the membrane current from sodium, potassium and other ions. The 'diffusion' coefficient D is associated with the axial current in the axon and, referring to the conservation of current equation (7.38) in Section 7.5, Volume I, the spatial variation in the potential V gives a contribution $(d/4r_i)V_{xx}$ on the right-hand side, where r_i is the resistivity and d is the axon diameter. The parameters $0 < a < 1$, b and γ are all positive. The null clines of the 'kinetics' in the (u, v) plane are shown in Figure 1.9(b).

We want to demonstrate in this section how travelling wave solutions arise for reaction diffusion systems with excitable kinetics. There are several important physiological applications in addition to that for the propagation of nerve action potentials modelled with the FHN model. One such important application is concerned with the waves which arise in muscle tissue, particularly heart muscle: in their two- and three-dimensional context these excitable waves are intimately related to the problem of atrial flutter and fibrillation (see, for example, Winfree 1983a,b). Another example is the reverberating cortical depression waves in the brain cortex (Shibata and Bureš 1974). Two- and three-dimensional excitable waves can also arise in the Belousov–Zhabotinskii reaction and others. We shall come back to these applications below. The system (1.94) has been studied in some detail and the following is only a very small sample from the long and ever-increasing list of references. The review by Rinzel (1981) specifically discussed models in neurobiology. Rinzel and Keller (1973) considered the piecewise linear caricature of (1.94) where $\gamma = 0$ and obtained analytical results for travelling pulses and

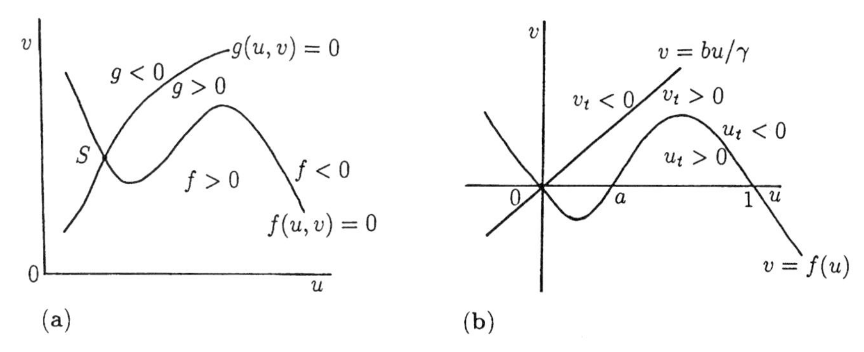

Figure 1.9. (**a**) Typical null clines for excitable kinetics. The kinetics here have only one steady state, S, which is globally stable but excitable. (**b**) Null clines for the excitable Fitzhugh–Nagumo system (1.94): the origin is the single steady state.

periodic wavetrains: the method of analysis here has more general applicability and is often the only way to investigate such nonlinear problems analytically. The caricature form when $f(u)$ is replaced by the piecewise linear approximation $f(u) = H(u-a)-u$ where H denotes the Heaviside function ($H(x) = 0$ if $x < 0$, $H(x) = 1$ if $x > 0$) has been studied by McKean (1970) while Feroe (1982) looked at the stability of multiple pulse solutions of this caricature. Ikeda et al. (1986) considered the Hodgkin–Huxley system and demonstrated the instability of certain slow wave solutions. The situation when b and γ in (1.94) are such that $v = bu/\gamma$ intersects the u-null cline to give three steady states was studied by Rinzel and Terman (1982). General discussions and reviews of waves in excitable media have been given, for example, by Keener (1980), Zykov (1988) and Tyson and Keener (1988) and of periodic bursting phenomena in excitable membranes by Carpenter (1979). The book by Keener and Sneyd (1998) also discusses the phenomenon; see other references there.

Travelling wave solutions of (1.94), in which u and v are functions only of the travelling coordinate variable $z = x - ct$, satisfy the travelling coordinate form of (1.94), namely,

$$Du'' + cu' + f(u) - v = 0, \quad cv' + bu - \gamma v = 0, \quad z = x - ct, \tag{1.95}$$

where the prime denotes differentiation with respect to z and the wavespeed c is to be determined. The boundary conditions corresponding to a solitary pulse are

$$u \to 0, \quad u' \to 0, \quad v \to 0 \quad \text{as} \quad |z| \to \infty \tag{1.96}$$

and the pulse is typically as illustrated in Figure 1.10(a). The corresponding phase trajectory in the (u, v) plane is schematically as in Figure 1.10(b).

Initial conditions play a critical role in the existence of travelling pulses. Intuitively we can see why. Suppose we have a spatial domain with (u, v) initially at the zero rest state and we perturb it by a local rise in u over a small domain, keeping $v = 0$, as in Figure 1.11(a). If the perturbation has a maximum u less than the threshold u_A in Figure 1.10(b) (and Figure 1.11(c)) then the kinetics cause u to return to the origin and

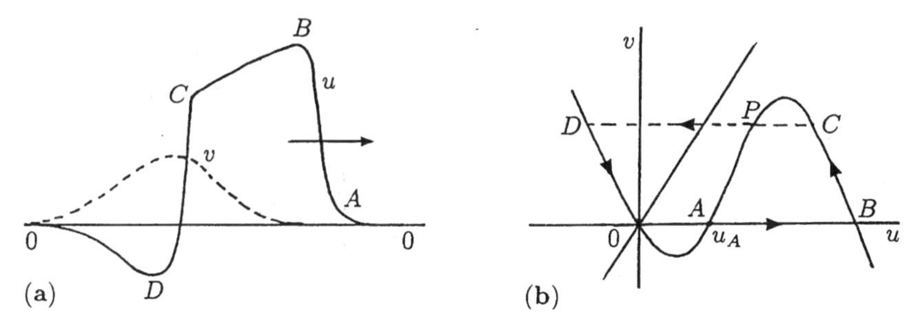

Figure 1.10. (**a**) Typical travelling pulse, or solitary wave, solution for the excitable system (1.94). (**b**) Corresponding phase trajectory in the (u, v) plane. Note the threshold characteristic of the null clines. A perturbation from the origin to a value $u < u_A$ will simply return to the origin with u always less than u_A. A perturbation to $u > u_A$ initiates a large excursion qualitatively like $ABCD0$. The position of C is obtained from the analysis as explained in the text.

the spatial perturbation simply dies out. On the other hand if the perturbation is larger than the threshold u_A then the kinetics initiate a large excursion in both u and v as shown by $0BCD0$ in Figure 1.11(c). When a wave is initiated the trailing edge is represented in the phase plane by CD. Whereas it is intuitively clear that the leading edge should be at $0B$ the positioning of CD is not so obvious. We now consider this important aspect of travelling pulses.

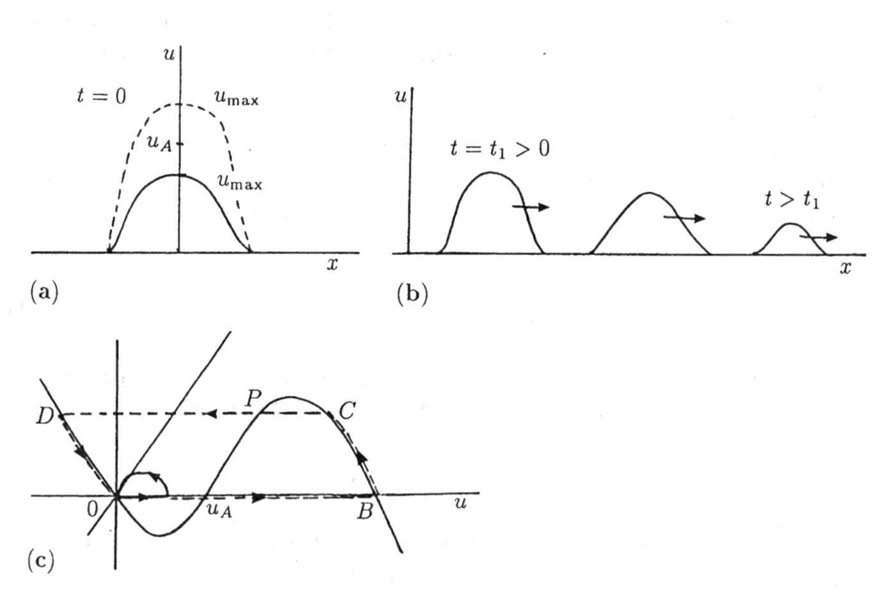

Figure 1.11. (**a**) The perturbation given by the solid line has $u_{max} < u_A$, where u_A is the threshold value in Figure 1.10(b). The solution is then simply a decaying transient such as illustrated in (**b**). With the dashed line as initial conditions the maximum value of u is larger than the threshold u_A and this initiates a travelling pulse such as in Figure 1.10(a). (**c**) Typical phase trajectories for u and v depending on whether the initial u is greater or less than u_A. The positioning of the trailing edge part of the trajectory, CD, is discussed in the text.

It is analytically easier to see what is going on if we consider (1.94) with b and γ small, so we write

$$b = \varepsilon L, \quad \gamma = \varepsilon M, \quad 0 < \varepsilon \ll 1$$

and (1.94) becomes

$$u_t = Du_{xx} + f(u) - v, \quad v_t = \varepsilon(Lu - Mv). \tag{1.97}$$

Now refer back to Figure 1.10(a) and consider the leading front $0AB$. In the limiting situation $\varepsilon \to 0$ the last equation says that $v \approx$ constant and from Figures 1.10(a) and 1.10(b), this constant is zero. The u-equation in (1.97) then becomes

$$u_t = Du_{xx} + f(u), \quad f(u) = u(a - u)(u - 1), \tag{1.98}$$

where $f(u)$ is sketched as a function of u in Figure 1.10(b). It has three steady states $u = 0$, $u = a$ and $u = 1$. In the absence of diffusion (1.98) implies that $u = 0$ and $u = 1$ are linearly stable and $u = a$ is unstable. We can thus have a travelling wave solution which joins $u = 0$ to $u = 1$. Equation (1.98) is a specific example of the one studied in Section 13.5, Volume I, specifically equation (13.73), which has an exact analytical solution (13.78) with a unique wavespeed given by (13.77). For the wave solution here we thus get

$$u = u(z), \quad z = x - ct; \quad c = \left(\frac{D}{2}\right)^{1/2}(1 - 2a), \tag{1.99}$$

and so the wavespeed is positive only if $a < 1/2$. This is the same condition we get from the sign determination given by (13.70) in Volume I, which, for (1.98), is

$$c \gtreqless 0 \quad \text{if} \quad \int_0^1 f(u)du \gtreqless 0.$$

Referring now to Figure 1.10(b), the area bounded by $0A$ and the curve $v = f(u)$ is less than the area enclosed by AB and the curve $v = f(u)$ so $c > 0$; carrying out the integration gives $c > 0$ for all $a < 1/2$.

We arrived at the equation system for the wave pulse front by neglecting the ε-terms in (1.97). This gave us the contribution to the pulse corresponding to $0AB$ in Figure 1.10(b). Along BC, v changes. From (1.97) a change in v will take a long time, $O(1/\varepsilon)$ in fact, since $v_t = O(\varepsilon)$. To get this part of the solution we would have to carry out a singular perturbation analysis (see, for example, Keener 1980), the upshot of which gives a slow transition period where u does not change much but v does. This is the part of the pulse designated BC in Figures 1.10(a) and 1.10(b).

The crucial question immediately arises as to where the next fast transition takes place, in other words where C is on the phase trajectory. Remember we are investigating the existence of pulse solutions which travel without change of shape. For this to be so the wavespeed of the trailing edge, namely, the speed of a wavefront solution that goes from C to D via P in Figure 1.10(a) (and Figure 1.11(c)), has to be the same as that for the leading edge $0AB$. On this part of the trajectory $v \approx v_C$ and the equation for the trailing edge wavefront from (1.97) is then given by

$$u_t = Du_{xx} + f(u) - v_C. \tag{1.100}$$

A travelling wavefront solution of this equation has to have

$$u = u(z), \quad z = x - ct; \qquad u(-\infty) = u_D, \quad u(\infty) = u_C.$$

The analytical solution and its unique wavespeed are again given in terms of u_C, u_P and u_D by the analysis in Section 13.5, Volume I. It gives the wavespeed as

$$c = \left(\frac{D}{2}\right)^{1/2} (u_C - 2u_P + u_D). \tag{1.101}$$

From the expression for $f(u)$ in (1.98), the roots u_C, u_D and u_P of $f(u) = v_C$ are determined in terms of v_C. The wavespeed c in (1.101) is then $c(v_C)$, a function of v_C. We now determine the value of v_C by requiring this $c(v_C)$ to be equal to the previously calculated wavespeed for the pulse front, namely, $c = (D/2)^{1/2}(1 - 2a)$ from (1.99). In principle it is possible to determine it in this way since the expression for v_C is the solution of a polynomial.

To complete the analytical determination of the wave pulse we now have to consider the part of the solution and the phase trajectory $D0$ in Figures 1.10 and 1.11(c). As for the part BC, during this stage v again changes by $O(1)$ in a time $O(1/\varepsilon)$. This is referred to as the *refractory phase* of the phenomenon. Figure 1.12 shows a computed example for the system (1.94) where the cubic $f(u)$ is approximated by the piecewise linear expression $f(u) = H(u - a) - u$.

Threshold waves are also obtained for more general excitable media models. To highlight the analytical concepts let us consider the two-species system in which one of the reactions is fast. To facilitate the analysis we consider the reaction diffusion system

$$\varepsilon u_t = \varepsilon^2 D_1 u_{xx} + f(u, v), \quad v_t = \varepsilon^2 D_2 v_{xx} + g(u, v), \tag{1.102}$$

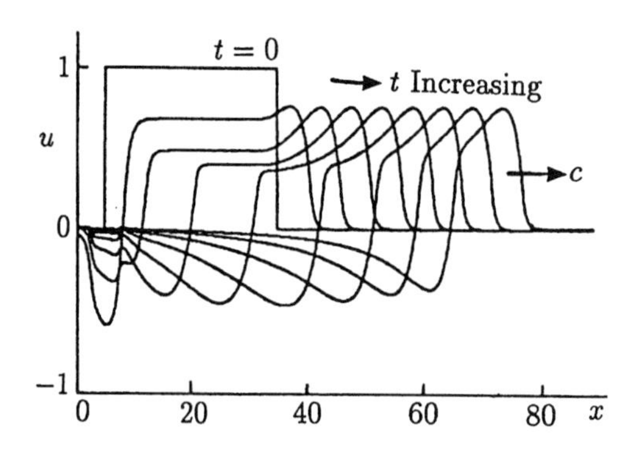

Figure 1.12. Development of a travelling wave pulse solution from square initial data for the excitable system (1.97) with $f(u) = H(u - a) - u$ with $a = 0.25$, $D = 1$, $\varepsilon = 0.1$, $L = 1$, $M = 0.5$. The wave is moving to the right. (Redrawn from Rinzel and Terman 1982 with permission of J. Rinzel)

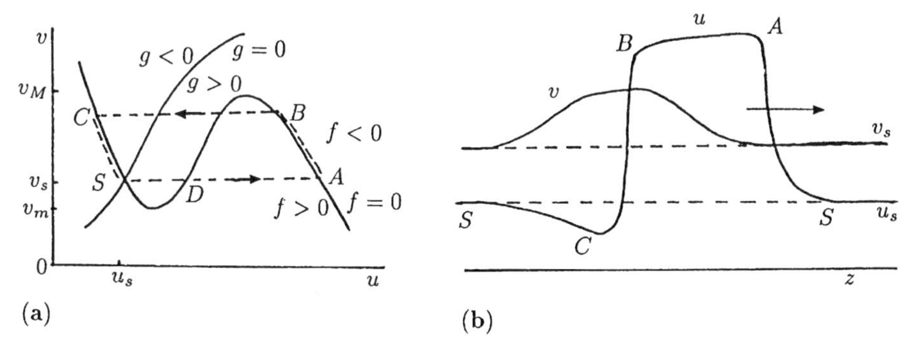

Figure 1.13. (**a**) Schematic null clines $f(u, v) = 0$, $g(u, v) = 0$ for the excitable system (1.102). The travelling pulse solution corresponds to the dashed trajectory. (**b**) Typical pulse solutions for u and v.

where $0 < \varepsilon \ll 1$ and the kinetics f and g have null clines like those in Figure 1.13(a), and we exploit the fact that ε is small. The key qualitative shape for $f(u, v) = 0$ is a cubic. This form is typical of many reactions where activation and inhibition are involved (cf. Section 6.6 and Section 6.7, Volume I).

The system (1.102) is excitable in the absence of diffusion. In Section 3.8, Volume I the description of a threshold mechanism was rather vague. We talked there of a system where the reactants underwent a large excursion in the phase plane if the perturbation was of the appropriate kind and of sufficient size. A better, and much more precise, definition is that a mechanism is excitable if a stimulus of sufficient size can initiate a travelling pulse which will propagate through the medium.

For $0 < \varepsilon \ll 1$, the $O(1)$ form of (1.102) is $f(u, v) = 0$ which we assume can be solved to give u as a multivalued function of v. From Figure 1.13(a) we see that for all given $v_m < v < v_M$ there are three solutions for u of $f(u, v) = 0$: they are the intersections of the line $v =$ constant with the null cline $f(u, v) = 0$. In an analogous way to the above discussion of the FHN system we can thus have a wavefront type solution joining S to A and a trailing wavefront from B to C with slow transitions in between. The time for u to change from its value at S to that at A is fast. This is what $f(u, v) = 0$ means since (referring to Figure 1.13(a)) if there is a perturbation to a value u to the right of D, u goes to the value at A instantaneously since u moves so that $f(u, v)$ is again zero. It takes, in fact, a time $O(\varepsilon)$. On the other hand it takes a relatively long time, $O(1)$, to traverse the AB and CS parts of the curve, while BC is covered again in $O(\varepsilon)$.

The analytical investigation of the pulse solution is quite involved and the detailed analysis of this general case has been given by Keener (1980). Here we consider by way of illustration how to go about carrying out the analysis for the leading front: that is, we consider the transition from S to A in Figures 1.13(a) and 1.13(b). The transition takes place quickly, in a time $O(\varepsilon)$, and spatially it is a sharp front of thickness $O(\varepsilon)$; see Figure 1.13(b). These scales are indicated by a singular perturbation appraisal of equations (1.102). This suggests that we introduce new independent variables by the transformations

$$\tau = \frac{t}{\varepsilon}, \quad \xi = \frac{x - x_T}{\varepsilon\sqrt{D_1}}, \tag{1.103}$$

where x_T is the position of the transition front, which we do not need at this stage or level of analysis. All the introduction of x_T does is to make the leading edge of the wave pulse at the origin $\xi = 0$ in the ξ-plane. Substitution into (1.102) and letting $\varepsilon \to 0$, keeping τ and ξ fixed in the usual singular perturbation way (see Murray 1984 or Kevorkian and Cole 1996) gives the $O(1)$ system as

$$u_\tau = u_{\xi\xi} + f(u, v), \quad v_\tau = 0. \tag{1.104}$$

So, considering the line SA, the second equation is simply $v = v_S$ and we then have to solve

$$u_\tau = u_{\xi\xi} + f(u, v_S). \tag{1.105}$$

This is just a scalar equation for u in which f has three steady states, namely u_S, u_D and u_A and is qualitatively the same as those studied in the last chapter. It is essentially the same as equation (13.62) which was discussed in detail in Section 13.5, Volume I. In the absence of diffusion the steady states at S, D and A are respectively stable, unstable and stable. We have already shown in Section 13.5, Volume I how a travelling monotonic wave solution with a unique wavespeed exists which can join u_S and u_A as in Figure 1.13(b).

The complete solution requires determining the wavespeed and the other parts of the pulse, namely, AB, BC and CS, making sure that they all join up consistently. It is an interesting singular perturbation analysis. This was carried out by Keener (1980), who also presented numerical solutions as well as an analysis of threshold waves in two space dimensions.

Another type of threshold wave of practical interest occurs when the null clines $f(u, v) = 0$, $g(u, v) = 0$ intersect such as in Figure 1.14(a). That is there are 3 steady states. With the scaling as in (1.102) we now have the sharp front from A to D and the slower DC transition essentially the same as above. We can get the sign of the wavespeed in the same way as described in Section 13.5, Volume I. Now there is a tail

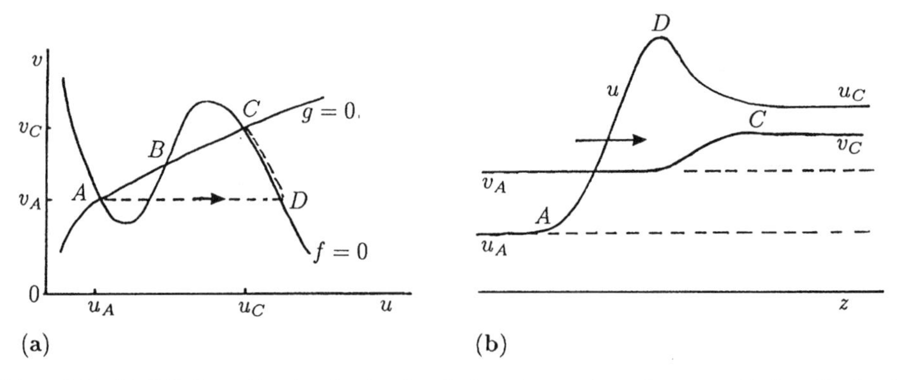

Figure 1.14. (**a**) Null clines $f(u, v) = 0$, $g(u, v) = 0$ where there are 3 steady states: with these kinetics A and C are linearly stable while B is unstable. (**b**) Typical travelling wave effecting a transition from A to C if the initial perturbation from A is sufficiently large and ε is small in (1.102).

to the wave since C is a linearly stable steady state. Figure 1.14(b) is a typical example of such threshold front waves. There is also the possibility of a transition wave obtained by perturbing the uniform steady state at C thus effecting a transition to A. Rinzel and Terman (1982) studied such waves in the FHN context.

Threshold waves exist for quite a varied spectrum of real world systems—any in fact which can exhibit threshold kinetics. For example, Britton and Murray (1979) studied them in a class of substrate inhibition oscillators (see also the book by Britton 1986). New wave phenomena and their potential use is described in an article (Steinbock et al. 1996) from Showalter's group which describes the new concept of chemical wave logic gates in excitable systems and (like most of this group's articles) the ideas and analysis are backed up by original and illuminating experiments.

1.7 Travelling Wave Trains in Reaction Diffusion Systems with Oscillatory Kinetics

Wavetrain solutions for general reaction diffusion systems with limit cycle kinetics have been widely studied; the mathematical papers by Kopell and Howard (1973) and Howard and Kopell (1977) are seminal. Several review articles in the book edited by Field and Burger (1985) are apposite to this section; other references will be given at appropriate places below.

The general evolution system we shall be concerned with is (1.1), which, for our purposes we restrict to one spatial dimension and for algebraic simplicity we incorporate the diffusion coefficient in a new scaled space variable $x \to x/D^{1/2}$. The equation system is then

$$\frac{\partial \mathbf{u}}{\partial t} = \mathbf{f}(\mathbf{u}) + \frac{\partial^2 \mathbf{u}}{\partial x^2}. \tag{1.106}$$

We assume that the spatially homogeneous system

$$\frac{d\mathbf{u}}{dt} = \mathbf{f}(\mathbf{u}; \gamma), \tag{1.107}$$

where γ is a bifurcation parameter, has a stable steady state for $\gamma < \gamma_c$ and, via a Hopf bifurcation (see, for example, Strogatz 1994), evolves to a stable limit cycle solution for $\gamma > \gamma_c$; that is, for $\gamma = \gamma_c + \varepsilon$, where $0 < \varepsilon \ll 1$, a small amplitude limit cycle solution exists and is stable.

Travelling plane wavetrain solutions are of the form

$$\mathbf{u}(x, t) = \mathbf{U}(z), \quad z = \sigma t - kx, \tag{1.108}$$

where $\mathbf{U}$ is a 2π-periodic function of z, the 'phase.' Here $\sigma > 0$ is the frequency and k the wavenumber; the wavelength $w = 2\pi/k$. The wave travels with speed $c = \sigma/k$. This form is only a slight variant of the general travelling waveform used in Chapter 13, Volume I and above and can be reduced to that form by rescaling the time. Substituting

(1.108) into (1.106) gives the following system of ordinary differential equations for **U**,

$$k^2\mathbf{U}'' - \sigma\mathbf{U}' + \mathbf{f}(\mathbf{U}) = 0, \tag{1.109}$$

where prime denotes differentiation with respect to z. We want to find σ and k so that the last equation has a 2π-periodic solution for **U**.

Rather than consider the general situation (see Kopell and Howard 1973 and the comments below) it is instructive and algebraically simpler to discuss, by way of demonstration, the analysis of the λ–ω model system described in Section 7.4, Volume I and given by equations (7.30). Later we shall relate it to general reaction diffusion systems which can arise from real biological situations. This two-reactant model mechanism, for (u, v) say, is

$$\frac{\partial}{\partial t}\begin{pmatrix} u \\ v \end{pmatrix} = \begin{pmatrix} \lambda(r) & -\omega(r) \\ \omega(r) & \lambda(r) \end{pmatrix}\begin{pmatrix} u \\ v \end{pmatrix} + \frac{\partial^2}{\partial x^2}\begin{pmatrix} u \\ v \end{pmatrix}, \quad \text{where} \quad r^2 = u^2 + v^2. \tag{1.110}$$

Here $\omega(r)$ and $\lambda(r)$ are real functions of r. If r_0 is an isolated zero of $\lambda(r)$ for some $r_0 > 0$, $\lambda'(r_0) < 0$ and $\omega(r_0) \neq 0$, then the spatially homogeneous system, that is, with $\partial^2/\partial x^2 = 0$, has a limit cycle solution (see Section 7.4, Volume I and (1.113) below).

It is convenient to change variables from (u, v) to polar variables (r, θ), where θ is the phase, defined by

$$u = r\cos\theta, \quad v = r\sin\theta \tag{1.111}$$

with which (1.110) becomes

$$\begin{aligned} r_t &= r\lambda(r) + r_{xx} - r\theta_x^2, \\ \theta_t &= \omega(r) + \frac{1}{r^2}(r^2\theta_x)_x. \end{aligned} \tag{1.112}$$

If $r_0 > 0$ exists and $\lambda'(r_0) < 0$ the asymptotically stable limit cycle solution of the kinetics is given immediately by

$$r = r_0, \quad \theta = \theta_0 + \omega(r_0)t, \tag{1.113}$$

where θ_0 is some arbitrary phase. Substituting into (1.111) gives the limit cycle solutions u and v as

$$u = r_0\cos[\omega(r_0)t + \theta_0], \quad v = r_0\sin[\omega(r_0)t + \theta_0], \tag{1.114}$$

which have frequency $\omega(r_0)$ and amplitude r_0.

Suppose we look for travelling plane wave solutions of the type (1.108) in the polar form

$$r = \alpha, \quad \theta = \sigma t - kx. \tag{1.115}$$

Substituting into (1.112) we get the necessary and sufficient conditions for these to be travelling wave solutions as

$$\sigma = \omega(\alpha), \quad k^2 = \lambda(\alpha). \tag{1.116}$$

So, with α the convenient parameter, there is a one-parameter family of travelling wave-train solutions of (1.110) given by

$$u = \alpha \cos\left[\omega(\alpha)t - x\lambda^{1/2}(\alpha)\right], \quad v = \alpha \sin\left[\omega(\alpha)t - x\lambda^{1/2}(\alpha)\right]. \tag{1.117}$$

The wavespeed is given by

$$c = \frac{\sigma}{k} = \frac{\omega(\alpha)}{\lambda^{1/2}(\alpha)}. \tag{1.118}$$

If $r = \alpha \to r_0$, that is, there is a limit cycle solution of the λ–ω dynamics, the wave number of the plane waves tends to zero. This suggests that we should look for travelling plane wavetrain solutions near the limit cycle. Kopell and Howard (1973) showed how to do this in general. Here we consider a specific simple, but nontrivial, example where $\lambda(r)$ and $\omega(r)$ are such that the kinetics satisfy the Hopf requirements of (1.107), and on which we can carry out the analysis simply, so as to derive travelling wavetrain solutions; we mainly follow the analysis of Ermentrout (1981).

Suppose

$$\omega(r) \equiv 1, \quad \lambda(r) = \gamma - r^2. \tag{1.119}$$

The dynamics in (1.110) then has $u = v = 0$ as a steady state which is stable for $\gamma < 0$ and unstable for $\gamma > 0$. $\gamma = 0$ is the bifurcation value γ_c such that at $\gamma = 0$ the eigenvalues of the linearization about $u = v = 0$ are $\pm i$. This is a standard Hopf bifurcation requirement (see Strogatz 1994) so we expect small amplitude limit cycle solutions for small positive γ; that is, $\gamma = \gamma_c + \varepsilon$ with $0 < \varepsilon \ll 1$. With the above general solutions (1.117), since $\lambda = 0$ when $r = \sqrt{\gamma}$, these limit cycle solutions are given by

$$u_\gamma(t) = \sqrt{\gamma}\cos t, \quad v_\gamma(t) = \sqrt{\gamma}\sin t, \quad \gamma > 0 \tag{1.120}$$

and in polar variables by

$$r_0 = \sqrt{\gamma}, \quad \theta = t + \theta_0, \tag{1.121}$$

where θ_0 is some arbitrary phase which we can take to be zero.

Now consider the reaction diffusion system (1.112) with λ and ω from (1.119). On substituting travelling plane wave solutions of the form

$$r = r_0, \quad \theta = \sigma t - kx$$

we find, as expected from (1.116),

$$\sigma = 1, \quad k^2 = \gamma - r_0^2, \quad 0 < r_0 < \sqrt{\gamma},$$

which result in the small amplitude travelling wavetrain solutions

$$\begin{aligned} u &= r_0 \cos(t - x[\gamma - r_0^2]^{1/2}), \\ v &= r_0 \sin(t - x[\gamma - r_0^2]^{1/2}). \end{aligned} \tag{1.122}$$

Figure 1.15 illustrates these solutions, which have amplitude $r_0 < \sqrt{\gamma}$ and wavelength $L = 2\pi/(\gamma - r_0^2)^{1/2}$.

Such travelling wavetrains are really only of relevance, for example, to the target patterns in Figure 1.1, if they are stable. Linear stability for this particular system can be done but in general it is far from trivial. It is a rare example where we can carry out the analysis fairly easily.

The effect of diffusion on reaction kinetics which exhibit periodic behaviour is to generate travelling periodic wavetrain solutions. The specific nonlinearity in the above λ–ω example, namely, $\lambda(r) = \gamma - r^2$, is typical of a Hopf bifurcation problem. It seems likely that reaction diffusion mechanisms where the reaction kinetics alone exhibit periodic limit cycle behaviour via a Hopf bifurcation will also generate periodic wavetrain solutions. To show this it suffices to demonstrate that general reaction diffusion systems with this property are similar to λ–ω systems in the vicinity of the Hopf bifurcation.

Consider the two-species system

$$u_t = F(u, v; \gamma) + D\nabla^2 u, \quad v_t = G(u, v; \gamma) + D\nabla^2 v, \tag{1.123}$$

where F and G are the reaction kinetics. For algebraic simplicity suppose (1.123) has a steady state at $u = v = 0$ and the diffusionless ($D = 0$) system exhibits a Hopf bifurcation to a limit cycle at the bifurcation value γ_c. Now consider u and v as the perturbations about the zero steady state and write

$$T = \begin{pmatrix} u \\ v \end{pmatrix}, \quad M = \begin{pmatrix} F_u & F_v \\ G_u & G_v \end{pmatrix}_{u=v=0}, \quad P = \begin{pmatrix} D & 0 \\ 0 & D \end{pmatrix}.$$

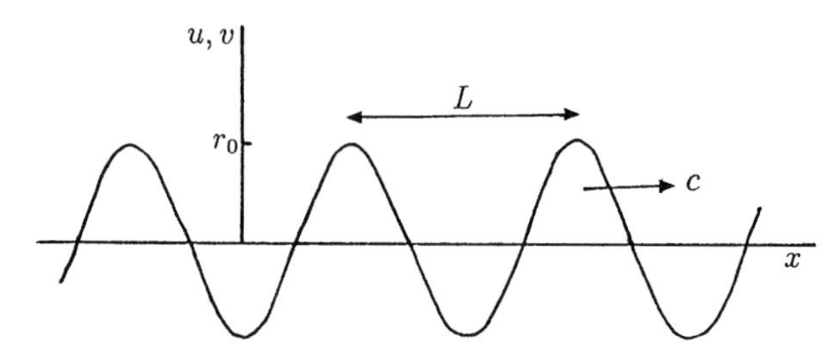

Figure 1.15. Small amplitude travelling wave solution for the $\lambda - \omega$ system (1.110) with λ and ω given by (1.119). The wavespeed $c = \sigma/k = [\gamma - r_0^2]^{-1/2}$ and the wavelength $L = 2\pi[\gamma - r_0^2]^{-1/2}$ depend on the amplitude r_0; $0 < r_0 < \sqrt{\gamma}$.

The terms in M are functions of the bifurcation parameter γ. The linearised form of (1.123) is then

$$T_t = MT + P\nabla^2 T \tag{1.124}$$

and the full system (1.123) can be written as

$$T_t = MT + P\nabla^2 T + H, \quad H = \begin{pmatrix} f(u, v, \gamma) \\ g(u, v, \gamma) \end{pmatrix}, \tag{1.125}$$

where f and g are the nonlinear contributions in u and v to F and G in the vicinity of $u = v = 0$.

Since the kinetics undergo a Hopf bifurcation at $\gamma = \gamma_c$, the eigenvalues, σ say, of the matrix M are such that $\operatorname{Re}\sigma(\gamma) < 0$ for $\gamma < \gamma_c$, $\operatorname{Re}\sigma(\gamma_c) = 0$, $\operatorname{Im}\sigma(\gamma_c) \neq 0$ and $\operatorname{Re}\sigma(\gamma) > 0$ for $\gamma > \gamma_c$. So, at $\gamma = \gamma_c$,

$$\operatorname{Tr} M = 0, \quad \det M > 0 \quad \Rightarrow \quad \sigma(\gamma_c) = \pm i(\det M)^{1/2}. \tag{1.126}$$

Introduce the constant matrix N and the nonconstant matrix W by

$$T = NW \quad \Rightarrow \quad W_t = N^{-1}MNW + N^{-1}PN\nabla^2 W + N^{-1}H \tag{1.127}$$

from (1.125). Now choose N such that

$$N^{-1}MN = \begin{pmatrix} 0 & -k \\ k & 0 \end{pmatrix} \quad \text{at} \quad \gamma = \gamma_c \quad \Rightarrow \quad k^2 = \det M]_{\gamma=\gamma_c}.$$

In the transformed system (1.127) we now have the coefficients in the linearised matrix

$$N^{-1}MN = \begin{pmatrix} \alpha(\gamma) & -\beta(\gamma) \\ \beta(\gamma) & \delta(\gamma) \end{pmatrix}, \tag{1.128}$$

where

$$\alpha(\gamma_c) = 0 = \delta(\gamma_c), \quad \beta(\gamma_c) \neq 0. \tag{1.129}$$

That is, the general system (1.123) with a Hopf bifurcation at the steady state can be transformed to a form in which, near the bifurcation γ_c, it has a λ–ω form (cf. (1.110)). This result is of some importance since analysis valid for λ–ω systems can be carried over in many situations to real reaction diffusion systems. This result is not restricted to equal diffusion coefficients for u and v as was shown by Duffy et al. (1980) who discussed the implications for spiral waves, a subject we discuss in the next section.

It is perhaps to be expected that if we have travelling wavetrains associated with oscillatory kinetics we should see even more complex wave phenomena from kinetics which exhibit period doubling and chaos. The type of chaos found by Merkin et al. (1996), and briefly described above, is not of this period-doubling type.

1.8 Spiral Waves

Rotating spiral waves occur naturally in a wide variety of biological, physiological and chemical contexts. A class which has been extensively studied is those which arise in the Belousov–Zhabotinskii reaction. Relatively, it is a very much simpler system than those which arise in physiology where we do not know the detailed mechanism involved unlike the Belousov–Zhabotinskii mechanism. Experimental work on spiral solutions in this reaction has been done by many people such as Winfree (1974), one of the major figures in their early study and subsequent application of the concepts to cardiac problems, Krinskii et al. (1986) and by Müller et al. (1985, 1986, 1987). The latter's novel experimental technique, using light absorption, highlights actual concentration levels quantitatively. Figure 1.16 as well as Figures 1.19 and 1.20, show some experimentally observed spiral waves in the Belousov–Zhabotinskii reaction; refer also to Figure 1.1(b). Although the spirals in these figures are symmetric, this is by no means the only pattern form; see, for example, Winfree (1974) in particular, and Müller et al. (1986), who exhibit dramatic examples of complex spiral patterns. A lot of work has gone into the mathematical study of spiral waves and in particular the diffusion version of the FKN model system. Keener and Tyson (1986) analysed spiral waves in excitable reaction diffusion systems with general excitable kinetics. They applied their technique to the FKN model with diffusion and the results are in good agreement with experiment. Although in a different context, see Figure 1.18 for other examples of nonsymmetric, as well as symmetric, spirals. There are numerous examples of spiral waves in the book by Keener and Sneyd (1998). General discussions of spiral waves have been given, for example, by Keener (1986), who presents a geometric theory, Zykov (1988) in his book on wave processes in excitable media and in the book by Grindrod (1996).

Much novel and seminal work on chemical spiral waves has been carried out by Showalter and his colleagues. In the paper by Amemiya et al. (1996), for example, they use a Field–Körös–Noyes (FKN) model system for the Belousov–Zhabotinskii reaction (similar to that studied in Chapter 8, Volume I) which exhibits excitability kinetics, to investigate three-dimensional spiral waves and carry out related experiments to back up their analysis; see other references there.

There are many other important occurrences of spiral waves. Brain tissue can exhibit electrochemical waves of 'spreading depression' which spread through the cortex

Figure 1.16. Spiral waves in a thin (1 mm) layer of an excitable Belousov–Zhabotinskii reaction. The section shown is 9 mm square. (Courtesy of T. Plesser from Müller et al. 1986)

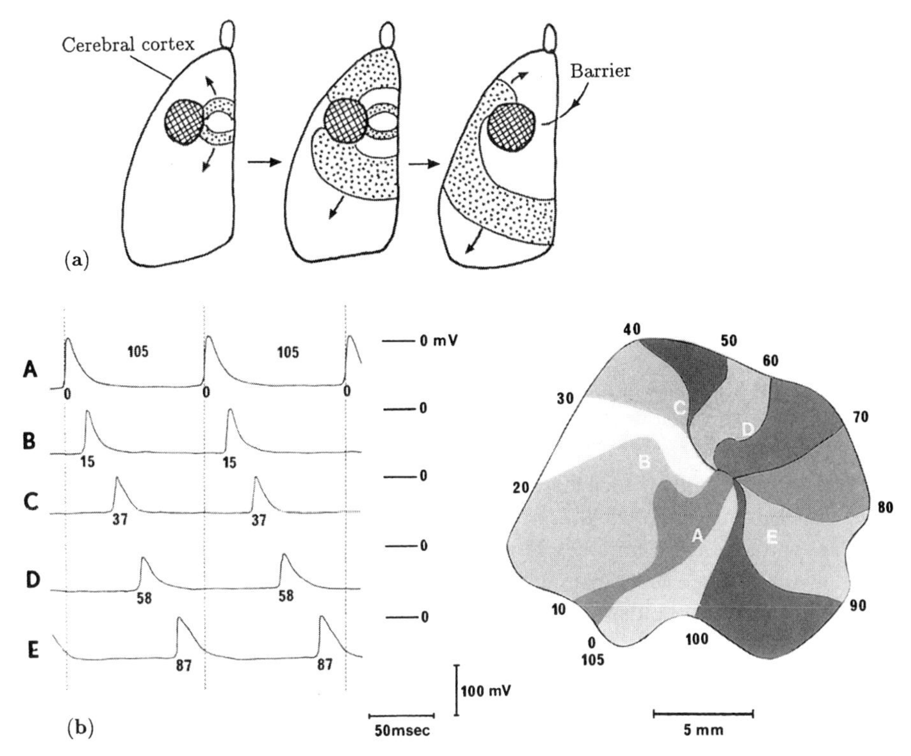

Figure 1.17. (**a**) Evolution of spiralling reverberating waves of cortical spreading depression about a lesion (a thermal coagulation barrier) in the right hemisphere of a rat cerebral cortex. The waves were initiated chemically. The shaded regions have different potential from the rest of the tissue. (After Shibata and Bureš 1974) (**b**) Rotating spiral waves experimentally induced in rabbit heart (left atria) muscle: the numbers represent milliseconds. Each region was traversed in 10 msec with the lettering corresponding to the points in the heart muscle on the right. The right also shows the isochronic lines, that is, lines where the potential is the same during passage of the wave. (Reproduced from Allessie et al. 1977, courtesy of M.A. Allessie and the American Heart Association, Inc.)

of the brain. These waves are characterised by a depolarisation of the neuronal membrane and decreased neural activity. Shibata and Bureš (1972, 1974) studied this phenomenon experimentally and demonstrated the existence of spiral waves which rotate about a lesion in the brain tissue from the cortex of a rat. Figure 1.17(a) schematically shows the wave behaviour they observed. Keener and Sneyd (1998) discuss wave motion in general and in particular the types of wave propagation found in the Hodgkin–Huxley equations and their caricature system, the FitzHugh–Nagumo equations. They also describe cardiac rhythmicity and wave propagation and calcium waves; some of the wave phenomena discussed are quite different to those covered in this book, such as wave curvature effects. A general procedure, based on an eikonal approach, for including curvature effects, particularly on curved surfaces is given by Grindrod et al. (1991).

There are new phenomena and new applications of reaction diffusion models and spiral waves that are continuing to be discovered. A good place to start is with papers by

Winfree and the references given in them. For example, Winfree et al. (1996) relate travelling waves to aspects of movement in heart muscle and nerves which are 'excitable.' They obtain complex periodic travelling waves which resemble scrolls radiating from vortex rings which are organising centres. Winfree (1994b) uses a generic excitable reaction diffusion system and shows that the general configuration of these vortex lines is a turbulent tangle. The generic system he used is

$$\frac{\partial u}{\partial t} = \nabla^2 u + \left[u - \frac{u^3}{3} - v \right] / \varepsilon,$$
$$\frac{\partial v}{\partial t} = \nabla^2 v + \varepsilon \left(u + \beta - \frac{v}{2} \right). \tag{1.130}$$

When death results from a disruption of the coordinated contractions of heart muscle fibres, the cause is often due to fibrillation. In a fibrillating heart, small regions undergo contractions essentially independent of each other. The heart looks, as noted before, like a handful of squirming worms — it is a quivering mass of tissue. If this disruption lasts for more than a few minutes death usually results. Krinskii (1978) and Krinskii et al. (1986), for example, discussed spiral waves in mathematical models of cardiac arrhythmias. Winfree (1983a,b) considered the possible application to sudden cardiac death. He suggested there that the precursor to fibrillation is the appearance of rotating waves of electrical impulses. Figure 1.17(b) illustrates such waves induced in rabbit heart tissue by Allessie et al. (1977). These authors (Allessie et al. 1973, 1976, Smeets et al. 1986) also carried out an extensive experimental programme on rotating wave propagation in heart muscle.

Winfree (1994a, 1995) put forward the interesting hypothesis that sudden cardiac death could involve three-dimensional rotors (spiral-type waves) of electrical activity which suddenly become unstable when the heart thickness exceeds some critical value; see there references to other articles in this general area. He has made extensive studies of the complex wave phenomena in muscle tissue over the past 20 years. For example, electrical aspects, activation fronts, anisotropy, and so on are important in cardiac physiology and have been discussed in detail by Winfree (1997) within a reaction diffusion context. He also discusses the roles of electrical potential diffusion, electrical turbulence, activation front curvature and anisotropy in heart muscle and their possible role in cardiac failure. His work suggests possible scenarios for remedial therapy for serious cardiac failures such as ventricular fibrillation. With his series of papers modelling muscle tissue activity in the heart Winfree has greatly enhanced our understanding of sudden cardiac death and changed in a major way previously held (medical) beliefs.

The spirals that arise in signalling patterns of the slime mould *Dictyostelium discoideum* are equally dramatic as seen in Figure 1.18. A model for these, based on an experimentally motivated kinetics scheme, was proposed by Tyson et al. (1989a,b).

It is really important (and generally) that, although the similarity between Figures 1.1(b) and 1.18 is striking, one must not be tempted to assume that the model for the Belousov–Zhabotinskii reaction is then an appropriate model for the slime mould patterns—the mechanisms are very different. Although producing the right kind of pat-

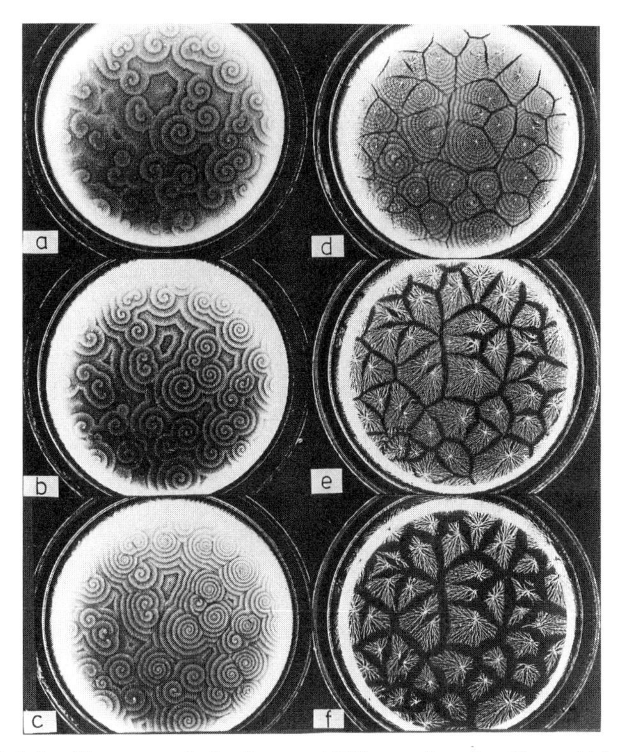

Figure 1.18. Spiral signalling patterns in the slime mould *Dictyostelium discoideum* which show the increasing chemoattractant (cyclic AMP) signalling. The photographs are taken about 10 min apart, and each shows about 5×10^7 amoebae. The Petri dish is 50 mm in diameter. The amoebae move periodically and the light and dark bands which show up under dark-field illumination arise from the differences in optical properties between moving and stationary amoebae. The cells are bright when moving and dark when stationary. The patterns eventually lead to the formation of bacterial territories. (Courtesy of P. C. Newell from Newell 1983)

terns is an important and essential aspect of successful modelling, understanding the basic mechanism is the ultimate objective.

The possible existence of large-scale spirals in interacting population situations does not seem to have been considered with a view to practical applications, but, given the reaction diffusion character of the models, they certainly exist in theory.

From a mathematical point of view, what do we mean by a spiral wave? In the case of the Belousov–Zhabotinskii reaction, for example, it is a rotating, time periodic, spatial structure of reactant concentrations; see Figures 1.17 and 1.20. At a fixed time a snapshot shows a typical spiral pattern. A movie of the process shows the whole spiral pattern moving like a rotating clock spring. Figure 1.19 shows such a snapshot and a superposition of the patterns taken at fixed time intervals. The sharp wavefronts are contours of constant concentration, that is, isoconcentration lines.

Consider now a spiral wave rotating around its centre. If you stand at a fixed position in the medium it seems locally as though a periodic wavetrain is passing you by since every time the spiral turns a wavefront moves past you.

As we saw in Chapter 9, Volume I, the state or concentration of a reactant can be described by a function of its phase, ϕ. It is clearly appropriate to use polar coordinates

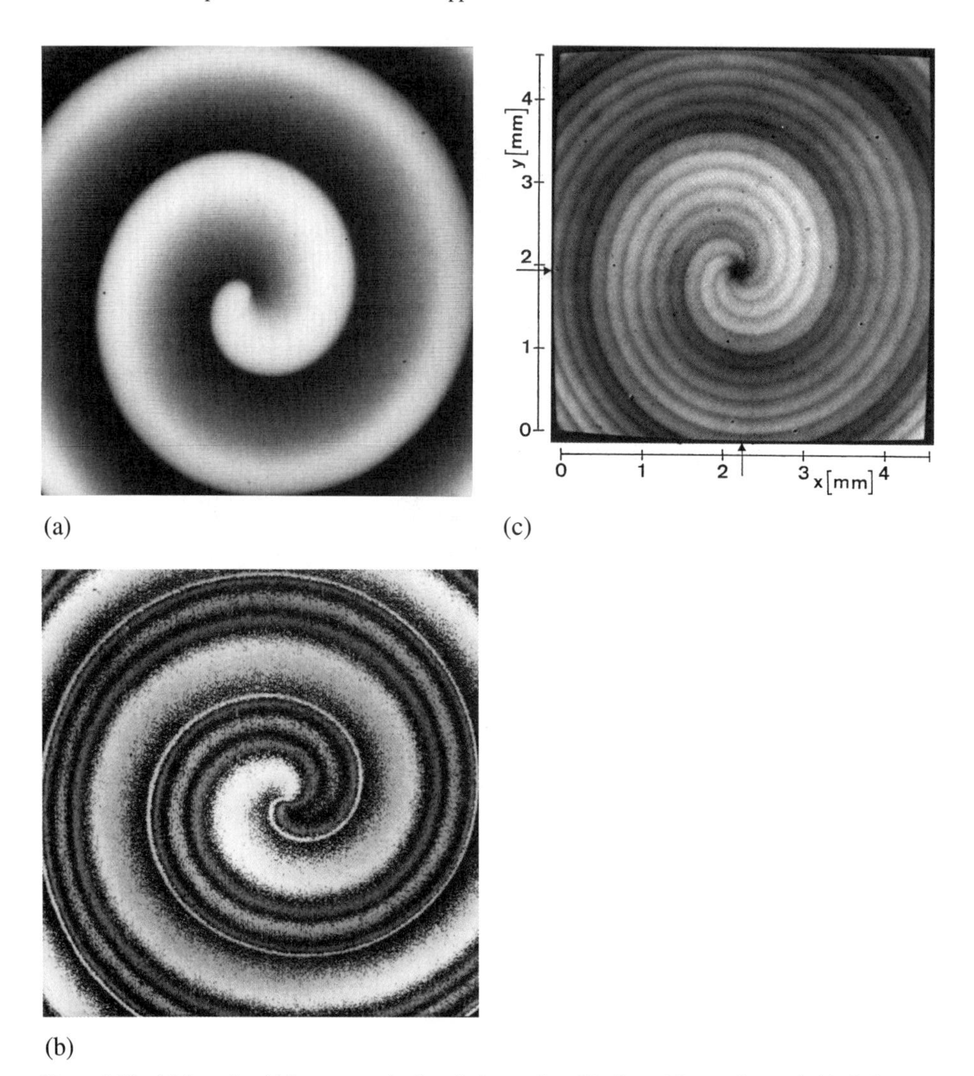

Figure 1.19. (**a**) Snapshot (4.5 mm square) of a spiral wave in a thin (1 mm) layer of an excitable Belousov–Zhabotinskii reagent. The grey-scale image is a measure of the level of transmitted light intensity (7 intensity levels were measured), which in turn corresponds to isoconcentration lines of one of the reactants. (**b**) The grey-scale highlights the geometric details of the isoconcentration lines of one of the reactants in the reaction. (**c**) Superposition of snapshots (4.5 mm square) taken at three-second intervals, including the one in (**a**). The series covers approximately one complete revolution of the spiral. Here six light intensity levels were measured. Note the small core region. (From Müller et al. 1985 courtesy of T. Plesser and the American Association for the Advancement of Science: Copyright 1985 AAAS)

r and θ when discussing spiral waves. A simple rotating spiral is described by a periodic function of the phase ϕ with

$$\phi = \Omega t \pm m\theta + \psi(r), \tag{1.131}$$

where Ω is the frequency, m is the number of arms on the spiral and $\psi(r)$ is a function which describes the type of spiral. The $\pm$ in the $m\theta$ term determines the sense of rotation. Figure 1.20 shows examples of 1-armed and 3-armed spirals including an experimental example of the latter. Suppose, for example, we set $\phi = 0$ and look at the steady state situation; we get a simple geometric description of a spiral from (1.131): a 1-armed spiral, for example, is given by $\theta = \psi(r)$. Specific $\psi(r)$ are

$$\theta = ar, \quad \theta = a \ln r \tag{1.132}$$

with $a > 0$; these are respectively Archimedian and logarithmic spirals. For a spiral about a central core the corresponding forms are

$$\theta = a(r - r_0), \quad \theta = a \ln (r - r_0). \tag{1.133}$$

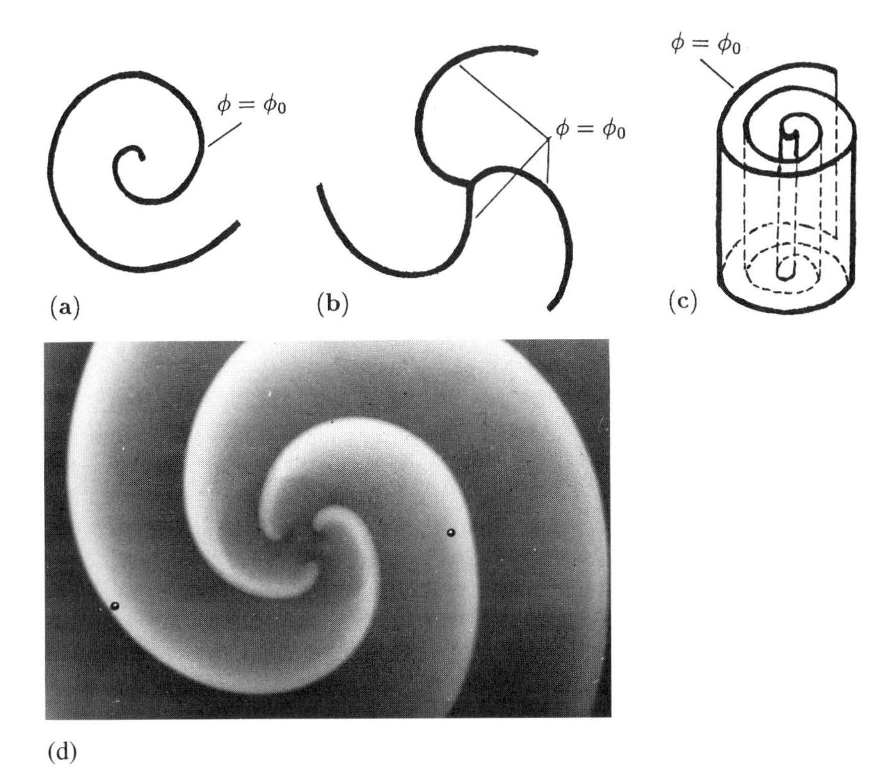

Figure 1.20. (**a**) Typical 1-armed Archimedian spiral. The actual spiral line is a line of constant phase ϕ, that is, a line of constant concentration. (**b**) Typical 3-armed spiral. (**c**) Three-dimensional spiral. These have a scroll-like quality and have been demonstrated experimentally by Welsh et al. (1983) with the Belousov–Zhabotinskii reaction. (**d**) Experimentally demonstrated 3-armed spiral in the Belousov–Zhabotinskii reaction. (From Agladze and Krinskii 1982 courtesy of V. Krinskii)

Figure 1.20(a) is a typical Archimedian spiral with Figure 1.20(b) an example with $m = 3$.

A mathematical description of a spiral configuration in a reactant, u say, could then be expressed by

$$u(r, \theta, t) = F(\phi), \tag{1.134}$$

where $F(\phi)$ is a 2π-periodic function of the phase ϕ given by (1.131). If t is fixed we get a snapshot of a spiral, the form of which puts certain constraints on $\psi(r)$ in (1.131); ar and $a \ln r$ in (1.132) are but two simple cases. A mixed type, for example, has $\psi(r) = ar + b \ln r$ with a and b constants. In (1.134) with ϕ as in (1.131), if we fix r and t and circle around the centre we have m-fold symmetry where m is the number of arms; an example with $m = 3$ is shown in Figure 1.20(b) and in Figure 1.20(d), one obtained experimentally by Agladze and Krinskii (1982). If we fix r and θ, that is, we stay at a fixed point, we see a succession of wavefronts as we described above. If a wavefront passes at $t = t_0$ with say, $\phi = \phi_0$, the next wave passes by at time $t = t_0 + 2\pi/\Omega$ which is when $\phi = \phi_0 + 2\pi$.

If we look at a snapshot of a spiral and move out from the centre along a ray we see intuitively that there is a wavelength associated with the spiral; it varies however as we move out from the centre. If one wavefront is at r_1 and the next, moving out, is at r_2, we can define the wavelength λ by

$$\lambda = r_2 - r_1, \quad \theta(r_2) = \theta(r_1) + 2\pi.$$

From (1.131), with t fixed, we have, along the curve $\phi =$ constant,

$$\phi_\theta + \phi_r \left[\frac{dr}{d\theta}\right]_{\phi=\text{constant}} = 0$$

and so, if, to be specific we take $-m\theta$ in (1.131),

$$\left[\frac{dr}{d\theta}\right]_{\phi=\text{constant}} = -\frac{\phi_\theta}{\phi_r} = \frac{m}{\psi'(r)}.$$

The wavelength $\lambda(r)$ is now given by

$$\lambda(r) = \int_{\theta(r)}^{\theta(r)+2\pi} \left[\frac{dr}{d\theta}\right] d\theta = \int_{\theta(r)}^{\theta(r)+2\pi} \left[\frac{m}{\psi'(r(\theta))}\right] d\theta,$$

where r, as a function of θ, is given by (1.131) with $t =$constant and $\phi =$ constant which we can take to be zero. For an Archimedian spiral $r = \theta/a$, so $\psi' = a$ and the wavelength is $\lambda = m/a$.

The pitch of the spiral is defined by

$$\left[\frac{dr}{d\theta}\right]_{\phi=\text{constant}} = \frac{m}{\psi'(r)}$$

Figure 1.21. Evolution of spiral waves for the two-dimensional FitzHugh–Nagumo model mechanism (1.94), namely, $u_t = u(a-u)(1-u) - v + D\nabla^2 u$, $v_t = bu - \gamma v$ for excitable nerve action potentials. Parameter values: $D = 2 \times 10^{-6}$, $a = 0.25$, $b = 10^{-3}$, $\gamma = 3 \times 10^{-3}$. The dark regions are where $u \geq a$; that is, u is in the excited state. (From Tsujikawa et al. 1989 courtesy of M. Mimura)

which, for an Archimedian spiral where $\psi'(r) = a$, gives a constant pitch m/a, while for a logarithmic spiral gives the pitch as mr/a since $\psi'(r) = a/r$. For large r, the pitch of the latter is large, that is, loosely wound, while for small r the pitch is small; that is, the spiral is tightly wound.

Before discussing the analytic solutions of a specific reaction diffusion system we should note some numerical studies on the birth of spiral waves carried out by Krinskii et al. (1986) and Tsujikawa et al. (1989). The latter considered the FitzHugh–Nagumo excitable mechanism (1.94) and investigated numerically the propagation of a wave of excitation of finite spatial extent. Figure 1.21 shows a time sequence of the travelling wave of excitation and shows the evolution of spiral waves; similar evolution figures were obtained by Krinskii et al. (1986) who discussed the evolution of spiral waves in some detail. The evolution patterns in this figure are similar to developing spirals observed experimentally in the Belousov–Zhabotinskii reaction.

1.9 Spiral Wave Solutions of λ–ω Reaction Diffusion Systems

Numerous authors have investigated spiral wave solutions of general reaction diffusion models, such as Cohen et al. (1978), Duffy et al. (1980), Kopell and Howard (1981) and Mikhailov and Krinskii (1983). The papers by Keener and Tyson (1986), dealing with the Belousov–Zhabotinskii reaction, and Tyson et al. (1989a,b) with the slime mould *Dictyostelium*, are specific examples. The analysis is usually quite involved with much use being made of asymptotic methods. The λ–ω system, which exhibits wavetrain solutions as we saw above in Section 1.7, has been used as a model system because of the relative algebraic simplicity of the analysis. Spiral wave solutions of λ–ω systems have been investigated, for example, by Greenberg (1981), Hagan (1982), Kuramoto and Koga (1981) and Koga (1982). The list of references is fairly extensive; other relevant references are given in these papers and the book by Keener and Sneyd (1998). In this section we develop some solutions for the λ–ω system.

The λ–ω reaction diffusion mechanism for two reactants is

$$\frac{\partial}{\partial t}\begin{pmatrix} u \\ v \end{pmatrix} = \begin{pmatrix} \lambda(a) & -\omega(a) \\ \omega(a) & \lambda(a) \end{pmatrix}\begin{pmatrix} u \\ v \end{pmatrix} + D\nabla^2\begin{pmatrix} u \\ v \end{pmatrix}, \quad A^2 = u^2 + v^2, \tag{1.135}$$

where $\omega(a)$ and $\lambda(a)$ are real functions of A. (The change of notation from (1.110) is so that we can use r as the usual polar coordinate.) We assume the kinetics sustain limit cycle oscillations; this puts the usual constraints on λ and ω; namely, if A_0 is an isolated zero of $\lambda(a)$ for some $A_0 > 0$ and $\lambda'(A_0) < 0$ and $\omega(A_0) \neq 0$, then the spatially homogeneous system, that is, with $D = 0$, has a stable limit cycle solution $u^2 + v^2 = A_0$ with cycle frequency $\omega(A_0)$ (see Section 7.4, Volume I).

Setting $w = u + iv$, (1.135) becomes the single complex equation

$$w_t = (\lambda + i\omega)w + D\nabla^2 w. \tag{1.136}$$

The form of this equation suggests setting

$$w = A \exp[i\phi], \tag{1.137}$$

where A is the amplitude of w and ϕ its phase. Substituting this into (1.136) and equating real and imaginary parts gives the following equation system for A and ϕ,

$$\begin{aligned} A_t &= A\lambda(a) - DA|\nabla\phi|^2 + D\nabla^2 A, \\ \phi_t &= \omega(a) + 2A^{-1}D(\nabla A \cdot \nabla\phi) + D\nabla^2\phi, \end{aligned} \tag{1.138}$$

which is the polar form of the λ–ω system. Polar coordinates r and θ are the appropriate ones to use for spiral waves. Motivated by (1.131) and the discussion in the last section, we look for solutions of the form

$$A = A(r), \quad \phi = \Omega t + m\theta + \psi(r), \tag{1.139}$$

where Ω is the unknown frequency and m the number of spiral arms. Substituting these into (1.138) gives the ordinary differential equations for A and ψ:

$$\begin{aligned} &DA'' + \frac{D}{r}A' + A\left[\lambda(a) - D\psi'^2 - \frac{Dm^2}{r^2}\right] = 0, \\ &D\psi'' + D\left(\frac{1}{r} + \frac{2A'}{A}\right)\psi' = \Omega - \omega(a), \end{aligned} \tag{1.140}$$

where the prime denotes differentiation with respect to r. On multiplying the second equation by rA^2, integration gives

$$\psi'(r) = \frac{1}{DrA^2(r)} \int_0^r sA^2(s)[\Omega - \omega(A(s))]ds. \tag{1.141}$$

The forms of (1.140) and (1.141) are convenient for analysis and are the equations which have been the basis for many of the papers on spiral waves of λ–ω systems, using asymptotic methods, fixed point theorems, phase space analysis and so on.

Before analysing (1.140) we have to decide on suitable boundary conditions. We want the solutions to be regular at the origin and bounded as $r \to \infty$. The former, together with the form of the equations for A and ψ', thus requires

$$A(0) = 0, \quad \psi'(0) = 0. \tag{1.142}$$

If $A \to A_\infty$ as $r \to \infty$ we have, from (1.141),

$$\begin{aligned}\psi'(r) &\sim \frac{1}{DrA_\infty^2} \int_0^r sA_\infty^2 [\Omega - \omega(A_\infty)]\, ds \\ &= \frac{[\Omega - \omega(A_\infty)]r}{2D} \quad \text{as} \quad r \to \infty\end{aligned}$$

and so ψ' is bounded only if $\Omega = \omega(A_\infty)$. The first of (1.140) determines $\psi'(\infty)$ as $[\lambda(A_\infty)/D]^{1/2}$. We thus have the dispersion relation

$$\psi'(\infty) = \left[\frac{\lambda(A_\infty)}{D}\right]^{1/2}, \quad \Omega = \omega(A_\infty) \tag{1.143}$$

which shows how the amplitude at infinity determines the frequency Ω.

Near $r = 0$, set

$$A(r) \sim r^c \sum_{n=0}^{\infty} a_n r^n, \quad \text{as} \quad r \to 0,$$

where $a_0 \neq 0$, substitute into the first of (1.140) and equate powers of r in the usual way. The coefficient of lowest order, namely, r^{c-2}, set equal to zero gives

$$c(c-1) + c - m^2 = 0 \quad \Rightarrow \quad c = \pm m.$$

For $A(r)$ to be nonsingular as $r \to 0$ we must choose $c = m$, with which

$$A(r) \sim a_0 r^m, \quad \text{as} \quad r \to 0,$$

where a_0 is an undetermined nonzero constant. The mathematical problem is to determine a_0 and Ω so that $A(r)$ and $\psi'(r)$ remain bounded as $r \to \infty$. From (1.137), (1.139) and the last equation, we get the behaviour of u and v near $r = 0$ as

$$\begin{pmatrix} u \\ v \end{pmatrix} \propto \begin{pmatrix} r^m \cos[\Omega t + m\theta + \psi(0)] \\ r^m \sin[\Omega t + m\theta + \psi(0)] \end{pmatrix}. \tag{1.144}$$

Koga (1982) studied phase singularities and multi-armed spirals, analytically and numerically, for the λ–ω system with

$$\lambda(a) = 1 - A^2, \quad \omega(a) = -\beta A^2, \tag{1.145}$$

where $\beta > 0$. Figure 1.22 shows his computed solutions for 1-armed and 2-armed spirals.

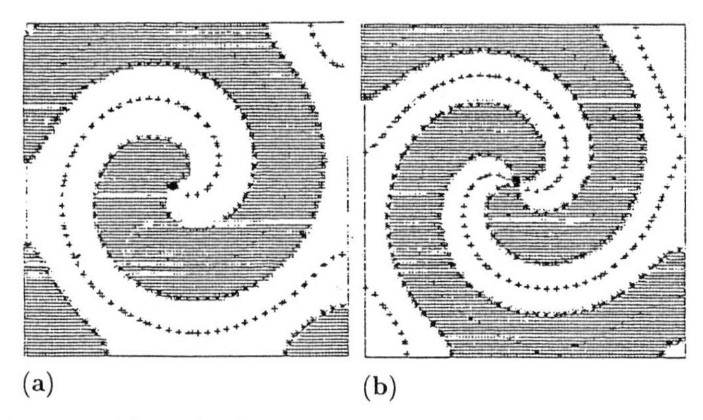

Figure 1.22. Computed (**a**) 1-armed and (**b**) 2-armed spiral wave solutions of the λ–ω system (1.135) with the λ and ω given by (1.145) with $\beta = 1$. Zero flux boundary conditions were taken on the square boundary. The shaded region is where $u > 0$. (From Koga 1982, courtesy of S. Koga)

The basic starting point to look for solutions is the assumption of the functional form for u and v given by

$$\begin{pmatrix} u \\ v \end{pmatrix} = \begin{pmatrix} A(r)\cos[\Omega t + m\theta + \psi(r)] \\ A(r)\sin[\Omega t + m\theta + \psi(r)] \end{pmatrix}. \tag{1.146}$$

With $A(r)$ a constant and $\psi(r) \propto \ln r$ these represent rotating spiral waves as we have shown. Cohen et al. (1978) proved that for a class of $\lambda(a)$ and $\omega(a)$ the system (1.136) has rotating spiral waves of the form (1.146) which satisfy boundary conditions which asymptote to Archimedian and logarithmic spirals; that is, $\psi \sim cr$ and $\psi \sim c \ln r$ as $r \to \infty$. Duffy et al. (1980) showed how to reduce a general reaction diffusion system with limit cycle kinetics and *unequal* diffusion coefficients for u and v, to the case analysed by Cohen et al. (1978).

Kuramoto and Koga (1981) studied numerically the specific λ–ω system where

$$\lambda(a) = \varepsilon - aA^2, \quad \omega(a) = c - bA^2,$$

where $\varepsilon > 0$ and $a > 0$. With these the system (1.136) becomes

$$w_t = (\varepsilon + ic)w - (a + ib)|\, w\,|^2 w + D\nabla^2 w.$$

We can remove the c-term by setting $w \to we^{ict}$ (algebraically the same as setting $c = 0$) and then rescale w, t and the space coordinates according to

$$w \to \left(\frac{a}{\varepsilon}\right)^{1/2} w, \quad t \to \varepsilon t, \quad \mathbf{r} \to \left(\frac{\varepsilon}{D}\right)^{1/2} \mathbf{r}$$

to get the simpler form

$$w_t = w - (1 + i\beta)|\, w\,|^2 w + \nabla^2 w, \tag{1.147}$$

where $\beta = b/a$. The space-independent form of the last equation has a limit cycle solution $w = \exp(-i\beta t)$.

Kuramoto and Koga (1981) numerically investigated the spiral wave solutions of (1.147) as $|\beta|$ varies. They found that for small $|\beta|$ a steadily rotating spiral wave developed, like that in Figure 1.22(a) and of the form (1.144). As $|\beta|$ was increased these spiral waves became unstable and appeared to become chaotic for larger $|\beta|$. Figure 1.23 shows the results for $|\beta| = 3.5$.

Kuramoto and Koga (1981) suggest that 'phaseless' points, or black holes, such as we discussed in Chapter 9, Volume I, start to appear and cause the chaotic instabilities. Comparing (1.147) with (1.136) we have $\lambda = 1 - A^2$, $\omega = -\beta A^2$ and so β is a measure of how strong the local limit cycle frequency depends on the amplitude A. Since A varies with the spatial coordinate r we have a situation akin to an array of coupled, appropriately synchronized, oscillators. As $|\beta|$ increases the variation in the oscillators increases. Since stable rotating waves require a certain synchrony, increasing the variation in the local 'oscillators' tends to disrupt the synchrony giving rise to phaseless points and hence chaos. Chaos or turbulence in wavefronts in reaction diffusion mechanisms has been considered in detail by Kuramoto (1980); see other references there to this interesting problem of spatial chaos.

To conclude this section let us look at the 1-dimensional analogue of a spiral wave, namely, a pulse which is emitted from the core, situated at the origin, periodically and on alternating sides of the core. If the pulses were emitted symmetrically then we would

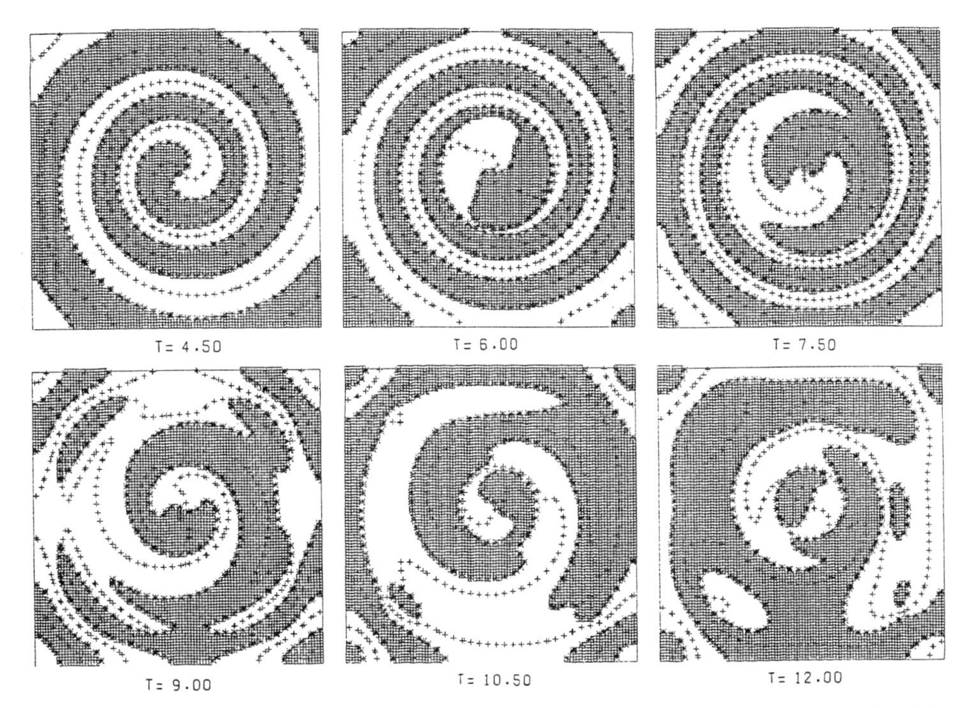

Figure 1.23. Temporal development (time T) of chaotic patterns for the $\lambda - \omega$ system (1.147) for $\beta = 3.5$ and zero flux boundary conditions. (From Kuramoto and Koga 1981, courtesy of Y. Kuramoto)

have the analogue of target patterns. Let us consider (1.138) with $\nabla^2 = \partial^2/\partial x^2$, $\lambda(a) = 1 - A^2$ and $\omega(a) = qA^2$. Now set

$$x \to \frac{x}{D^{1/2}}, \quad A = A(x), \quad \phi = \Omega t + \psi(x)$$

to get as the equations for A and ψ,

$$\begin{aligned} A_{xx} + A(1 - A^2 - \psi_x^2) &= 0, \\ \psi_{xx} + \frac{2A_x\psi_x}{A} &= \Omega - qA^2. \end{aligned} \tag{1.148}$$

Boundary conditions are

$$\begin{aligned} &A(x) \sim a_0 x \quad \text{as} \quad x \to 0, \quad \psi_x(0) = 0, \\ &A(x), \ \psi_x(x) \quad \text{bounded as} \quad x \to \infty. \end{aligned}$$

The problem boils down to finding a_0 and Ω as functions of q so that the solution of the initial value problem to the time-dependent equations is bounded. One such solution is

$$\begin{aligned} A(x) &= \left(\frac{\Omega}{q}\right)^{1/2} \tanh(x/\sqrt{2}), \\ \psi_x(x) &= \left(1 - \frac{\Omega}{q}\right)^{1/2} \tanh(x/\sqrt{2}), \qquad \Omega^2 + \frac{9}{2q}\Omega - \frac{9}{2} = 0, \end{aligned} \tag{1.149}$$

as can be verified. These solutions are generated periodically at the origin, alternatively on either side.

The stability of travelling waves and particularly spiral waves can often be quite difficult to demonstrate analytically; the paper by Feroe (1982) on the stability of excitable FHN waves amply illustrates this. However, some stability results can be obtained, without long and complicated analysis, in the case of the wavetrain solutions of the λ–ω system. In general analytical determination of the stability of spiral waves is still far from complete although numerical evidence, suggests that many are indeed stable. As briefly mentioned above, more recently Yagisita et al. (1998) have investigated spiral waves on a sphere in an excitable reaction diffusion system. They show, among other things, that the spiral tip rotates. They consider the propagation in both a homogeneous and inhomogeneous medium.

Biological waves exist which are solutions of model mechanisms other than reaction diffusion systems. For example, several of the mechanochemical models for generating pattern and form, which we discuss later in Chapter 6, also sustain travelling wave solutions. Waves which lay down a spatial pattern after passage are also of considerable importance as we shall also see in later chapters. In concluding this chapter, perhaps we should reiterate how important wave phenomena are in biology. Although this is clear just from the material in this chapter they are perhaps even more important in tissue communication during the process of embryological development. Generation of steady state spatial pattern and form is a topic of equal importance and will be discussed at length in subsequent chapters.

Exercises

1. Consider the modified Lotka–Volterra predator–prey system in which the predator disperses via diffusion much faster than the prey; the dimensionless equations are

$$\frac{\partial u}{\partial t} = u(1-u-v), \quad \frac{\partial v}{\partial t} = av(u-b) + \frac{\partial^2 v}{\partial x^2},$$

where $a > 0$, $0 < b < 1$ and u and v represent the predator and prey respectively. Investigate the existence of realistic travelling wavefront solutions of speed c in terms of the travelling wave variable $x+ct$, in which the wavefront joins the steady states $u = v = 0$ and $u = b$, $v = 1-b$. Show that if c satisfies $0 < c < [4a(1-b)]^{1/2}$ such wave solutions cannot exist whereas they can if $c \geq [4a(1-b)]^{1/2}$. Further show that there is a value a^* such that for $a > a^*$ (u, v) tend to $(b, 1-b)$ exponentially in a damped oscillatory way for large $x+ct$.

2. Consider the modified Lotka–Volterra predator–prey system

$$\frac{\partial U}{\partial t} = AU\left(1-\frac{U}{K}\right) - BUV + D_1 U_{xx},$$

$$\frac{\partial V}{\partial t} = CUV - DV + D_2 V_{xx},$$

where U and V are respectively the predator and prey densities, A, B, C, D and K, the prey carrying capacity, are positive constants and D_1 and D_2 are the diffusion coefficients. If the dispersal of the predator is slow compared with that of the prey, show that an appropriate nondimensionalisation to a first approximation for $D_2/D_1 \approx 0$ results in the system

$$\frac{\partial u}{\partial t} = u(1-u-v) + \frac{\partial^2 u}{\partial x^2}, \quad \frac{\partial v}{\partial t} = av(u-b).$$

Investigate the possible existence of travelling wavefront solutions.

3. Quadratic and cubic autocatalytic reaction steps in a chemical reaction can be represented by

$$A + B \rightarrow 2B, \quad \text{reaction rate} = k_q ab,$$

$$A + 2B \rightarrow 3B, \quad \text{reaction rate} = k_c ab^2,$$

where a and b are the concentrations of A and B and the k's are the rate constants. With equal diffusion coefficients D for A and B these give rise to reaction diffusion systems of the form

$$\frac{\partial a}{\partial t} = D \nabla^2 a - k_c ab^2 - k_q ab,$$

$$\frac{\partial b}{\partial t} = D \nabla^2 b + +k_c ab^2 + k_q ab.$$

First nondimensionalise the system and then show that it can be reduced to the study of a single scalar equation with polynomial reaction terms.

Consider in turn the situations when there is only a cubic autocatalysis, that is, $k_q = 0, k_c \neq 0$, and then when there is only a quadratic autocatalysis term, that is, when $k_q \neq 0, k_c = 0$. In the one-dimensional situation investigate possible travelling waves in terms of the travelling wave coordinate $z = x - ct$ where c is the wavespeed.

4. A primitive predator–prey system is governed by the model equations

$$\frac{\partial u}{\partial t} = -uv + D\frac{\partial^2 u}{\partial x^2}, \quad \frac{\partial v}{\partial t} = uv + \lambda D\frac{\partial^2 v}{\partial x^2}.$$

Investigate the possible existence of realistic travelling wavefront solutions in which $\lambda > 0$ and $u(-\infty, t) = v(\infty, t) = 0$, $u(\infty, t) = v(-\infty, t) = K$, a positive constant. Note any special cases. What type of situation might this system model?

5. Travelling bands of microorganisms, chemotactically directed, move into a food source, consuming it as they go. A model for this is given by

$$b_t = \frac{\partial}{\partial x}\left[Db_x - \frac{b\chi}{a}a_x\right], \quad a_t = -kb,$$

where $b(x, t)$ and $a(x, t)$ are the bacteria and nutrient respectively and D, χ and k are positive constants. Look for travelling wave solutions, as functions of $z = x-ct$, where c is the wavespeed, with the boundary conditions $b \to 0$ as $|z| \to \infty$, $a \to 0$ as $z \to -\infty$, $a \to 1$ as $z \to \infty$. Hence show that $b(z)$ and $a(z)$ satisfy

$$b' = \frac{b}{cD}\left[\frac{kb\chi}{a} - c^2\right], \quad a' = \frac{kb}{c},$$

where the prime denotes differentiation with respect to z, and then obtain a relationship between $b(z)$ and $a(z)$.

In the special case where $\chi = 2D$ show that

$$a(z) = \left[1 + Ke^{-cz/D}\right]^{-1}, \quad b(z) = \frac{c^2}{kD}e^{-cz/D}\left[1 + Ke^{-cz/D}\right]^{-2},$$

where K is an arbitrary positive constant which is equivalent to a linear translation; it may be set to 1. Sketch the wave solutions and explain what is happening biologically.

6. Consider the two species reduction model (1.85) for the wavefront spatial variation in the Belousov–Zhabotinskii reaction in which the parameters $r = b \gg 1$. That is, consider the system

$$u_t = -\frac{uv}{\varepsilon} + u(1-u) + u_{xx}, \quad v_t = -\frac{uv}{\varepsilon} + v_{xx}, \quad 0 < \varepsilon \ll 1.$$

By looking for travelling wave solutions in powers of ε,

$$\begin{pmatrix} u \\ v \end{pmatrix} = \sum_{n=0} \begin{pmatrix} u_n(z) \\ v_n(z) \end{pmatrix} \varepsilon^n, \quad z = x - ct$$

show, by going to $O(\varepsilon)$, that the equations governing u_0, v_0 are

$$u_0 v_0 = 0, \quad u_0'' + cu_0' + u_0(1-u_0) = v_0'' + cv_0'.$$

Hence deduce that the wave problem is split into two parts, one in which $u_0 \neq 0$, $v_0 = 0$ and the other in which $u_0 = 0$, $v_0 \neq 0$. Suppose that $z = 0$ is the point where the transition takes place. Sketch the form of the solutions you expect.

In the domain where $v_0 = 0$ the equation for u_0 is the Fisher–Kolmogoroff equation, but with different boundary conditions. We must have $u(-\infty) = 1$ and $u_0(0) = 0$. Although $u_0 = 0$ for all $z \geq 0$, in general $u_0'(0)$ would not be zero. This would in turn result in an inconsistency since $u_0'(0) \neq 0$ implies there is a flux of u_0 into $z > 0$ which violates the restriction $u_0 = 0$ for all $z \geq 0$. This is called a *Stefan problem*. To be physically consistent we must augment the boundary conditions to ensure that there is no flux of u_0 into the region $z > 0$. So the complete formulation of the u_0 problem is the Fisher–Kolmogoroff equation plus the boundary conditions $u_0(0) = u_0'(0) = 0$, $u(-\infty) = 1$. Does this modify your sketch of the wave? Do you think such a wave moves faster or slower than the Fisher–Kolmogoroff wave? Give mathematical and physical reasons for your answer. [This asymptotic form, with $b \neq r$, has been studied in detail by Schmidt and Ortoleva (1980), who formulated the problem in the way described here and obtained analytical results for the wave characteristics.]

7. The piecewise linear model for the FHN model given by (1.94) is

$$u_t = f(u) - v + Du_{xx}, \quad v_t = bu - \gamma v, \qquad f(u) = H(u-a) - u,$$

where H is the Heaviside function defined by $H(s) = 0$ for $s < 0$, $H(s) = 1$ for $s > 0$, and $0 < a < 1$, b, D and γ are all positive constants. If b and γ are small and the only rest state is $u = v = 0$, investigate qualitatively the existence, form and speed of a travelling solitary pulse.

8. The piecewise linear FHN caricature can be written in the form

$$u_t = f(u) - v + Du_{xx}, \quad v_t = \varepsilon(u - \gamma v), \qquad f(u) = H(u-a) - u,$$

where H is the Heaviside function defined by $H(s) = 0$ for $s < 0$, $H(s) = 1$ for $s > 0$, and $0 < a < 1$, ε, D and γ are all positive constants. Sketch the null clines and determine the condition on the parameters a and γ such that three steady states exist.

When $\varepsilon \ll 1$ and a and γ are in the parameter domain such that three steady states exist, investigate the existence of a threshold travelling wavefront from the zero rest state to the other stable rest state if the domain is originally at the zero rest state. Determine the unique wavespeed of such a front to $O(1)$ for ε small. Sketch the wave solution and make any relevant remarks about the qualitative size and form of the initial conditions which could give rise to such a travelling front. If initially u and v are everywhere at the non-zero steady state discuss the possibility of a wave from it to the zero rest state.

[Problems 6 and 7 have been investigated in depth analytically by Rinzel and Terman (1982).]

2. Spatial Pattern Formation with Reaction Diffusion Systems

2.1 Role of Pattern in Biology

Embryology is that part of biology which is concerned with the formation and development of the embryo from fertilization until birth. Development of the embryo is a sequential process and follows a ground plan, which is usually laid down very early in gestation. In humans, for example, it is set up roughly by the 5th week. There are many books on the subject; the one by Slack (1983) is a readable account of the early stages of development from egg to embryo. Morphogenesis, the part of embryology with which we are mainly concerned, is the development of pattern and form. How the developmental ground plan is established is unknown as are the mechanisms which produce the spatial patterning necessary for specifying the various organs.[1]

The following chapters and most of this one will be devoted to mechanisms which can generate spatial pattern and form, and which have been proposed as possible pattern formation processes in a variety of developmental situations. Wave phenomena create spatial patterns, of course, but these are spatio-temporal patterns. Here we shall be concerned with the formation of steady state spatially heterogeneous spatial patterns. In this chapter we introduce and analyse reaction diffusion pattern formation mechanisms mainly with developmental biology in mind. Section 2.7, however, is concerned with an ecological aspect of pattern formation, which suggests a possible strategy for pest control—the mathematical analysis is different but directly relevant to many embryological situations.

The questions we would like to answer, or, more realistically, get any enlightenment about, are legion. For example, are there any general patterning principles that are shared by bacteria, which can form complex patterns and wolf packs when they mark

[1]Professor Jean-Pierre Aubin in his book on *Mutational and Morphological Analysis* notes that the adjective *morphological* is due to Goethe (1749–1832). Goethe spent a lot of time thinking and writing about biology (the discipline 'biology' dates from 1802). Aubin goes on to describe Goethe's theory of plant evolution that hypothesizes that most plants come from one archetypal plant. Except for Geoffroy St. Hilaire, a distinguished early 19th century French biologist who wrote extensively about teratologies, or monsters (see Chapter 7), it was not taken seriously. Goethe was very bitter about it and complained about the difficulties in trying to work in a discipline other than one's own: it is still a problem. Goethe's work had some reevaluation in the 20th century, primarily by historians of science (see, for example, Lenoir 1984, Brady 1984 and other references there).

out territory? Spatial patterns are ubiquitous in the biomedical sciences and understanding how they are formed is without question one of the major fundamental scientific challenges. In the rest of the book we shall study a variety of pattern formation mechanisms which generate pattern in a variety of diverse areas.

Cell division starts after fertilisation. When sufficient cell division has taken place in a developing embryo the key problem is how the homogeneous mass of cells are spatially organised so that the sequential process of development can progress. Cells differentiate, in a biological sense, according to where they are in the spatial organisation. They also move around in the embryo. This latter phenomenon is an important element in morphogenesis and has given rise to a new approach to the generation of pattern and form discussed in some detail in Chapter 6.

It is impossible not to be fascinated and enthralled with the wealth, diversity and beauty of pattern in biology. Figure 2.1 shows only four examples. How such patterns, and millions of others, were laid down is still unknown although considerable progress has been made in several different fronts such as in the early patterning in the embryo of the fruit fly, spatial patterning in slime moulds and bacterial patterns discussed in Chapter 5. The patterning problems posed by only the few patterns in Figure 2.1 are quite diverse.

As a footnote to Figure 2.1(c), note the antennae on the moth. These antennae very effectively collect molecules of the chemical odorant, called a pheromone called bombykol, which is exuded by the female to attract the male. In the case of the silk moth the male, which cannot fly, can detect the pheromone from the female as far away as a kilometre and can move up the concentration gradient towards the female. The filtering efficiency of such antennae, which collect, and in effect count, the molecules, poses a very different and interesting mathematical biology patterning problem to those discussed in this book, namely, how such a filter antenna should be designed to be most efficient. This specific problem—an interesting fluid mechanics and diffusion one—was discussed in detail by Murray (1977).

The fundamental importance of pattern and form in biology is self-evident. Our understanding is such that whatever pattern we observe in the animal world, it is almost certain that the process that produced it is still unknown. Pattern formation studies have often been criticized for their lack of inclusion of genes in the models. But then the criticism can be levelled at any modelling abstraction of a complex system to a relatively simple one. It should be remembered that the generation of pattern and form, particularly in development, is usually a long way from the level of the genome. Of course genes play crucial roles and the mechanisms must be genetically controlled; the genes, however, themselves cannot create the pattern. They only provide a blueprint or recipe, for the pattern generation. Many of the evolving patterns could hardly have been anticipated solely by genetic information. Another of the major problems in biology is how genetic information is physically translated into the necessary pattern and form. Much of the research in developmental biology, both experimental and theoretical, is devoted to trying to determine the underlying mechanisms which generate pattern and form in early development. The detailed discussion in these next few chapters discusses some of the mechanisms which have been proposed and gives an indication of the role of mathematical modelling in trying to unravel the underlying mechanisms involved in morphogenesis.

(a)

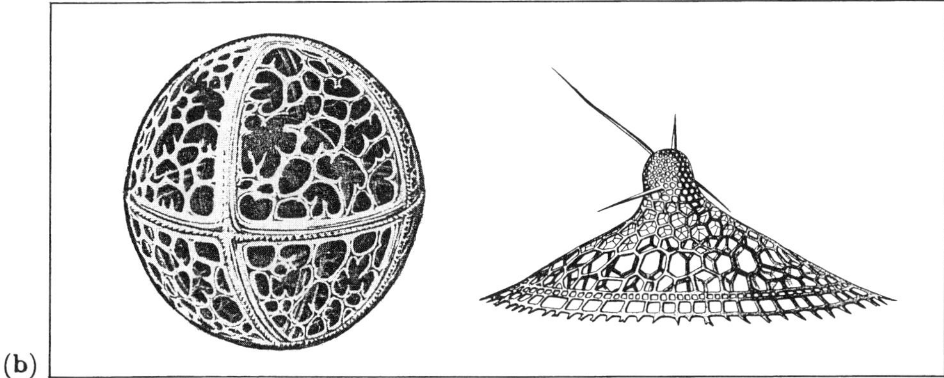

(b)

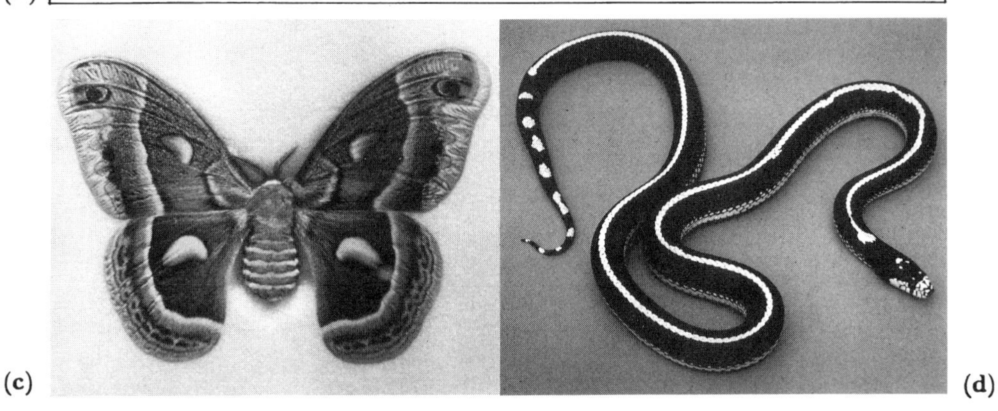

(c) (d)

Figure 2.1. (**a**) Leopard (*Panthera pardus*) in the Serengeti National Park, Tanzania. Note the individual spot structure. (Photograph courtesy of Hans Kruuk) (**b**) Radiolarians (*Trissocyclus spaeridium* and *Eucecryphalus genbouri*). These are small marine organisms—protozoa—of the order of a millimeter across. (After Haekel 1862, 1887) The structural architecture of radiolarians is amazingly diverse (see, for example, the plate reproductions of some of Haeckel's drawings in the Dover Archive Series, Haeckel 1974, but see also the historical aside on Haeckel in Section 6.1). (**c**) Moth (*Hyalophora cecropia*). As well as the wing patterns note the stripe pattern on the body and the structure of the antennae. (**d**) California king snake. Sometimes the pattern consists of crossbands rather than a backstripe. (Photograph courtesy of Lloyd Lemke)

A phenomenological concept of pattern formation and differentiation called *positional information* was proposed by Wolpert (1969, see the reviews in 1971, 1981). He suggested that cells are preprogrammed to react to a chemical (or morphogen) concentration and differentiate accordingly, into different kinds of cells such as cartilage cells. The general introductory paper by Wolpert (1977) gives a very clear and nontechnical description of development of pattern and form in animals and the concepts and application of his positional information scenario. Although it is a phenomenological approach, with no actual mechanism involved it has given rise to an immense number of illuminating experimental studies, many associated with the development of the limb cartilage patterning in chick embryos and feather patterns on other bird embryos, such as the quail and guinea fowl (see, for example, Richardson et al. 1991 and references there). A literature search of positional information in development will produce an enormous number of references. Although it is a simple and attractive concept, which has resulted in significant advances in our knowledge of certain aspects of development, it is not a mechanism.

The chemical prepattern viewpoint of embryogenesis separates the process of development into several steps; the essential first step is the creation of a morphogen concentration spatial pattern. The name 'morphogen' is used for such a chemical because it effects morphogenesis. The notion of positional information relies on a chemical pre-specification so that the cell can read out its position in the coordinates of chemical concentration, and differentiate, undergo appropriate cell shape change, or migrate accordingly. So, once the prepattern is established, morphogenesis is a slave process. Positional information is not dependent on the specific mechanism which sets up the spatial prepattern of morphogen concentration. This chapter is concerned with reaction diffusion models as the possible mechanisms for generating biological pattern. The basic chemical theory or reaction diffusion theory of morphogenesis was put forward in the classical paper by Turing (1952). Reaction diffusion theory, which now has a vast literature, is a field of research in its own right.

With the complexity of animal forms the concept of positional information necessarily implies a very sophisticated interpretation of the 'morphogen map' by the cell. This need not pose any problem when we recall how immensely complex a cell is whether or not it is positional information or simply a cell responding in some way to small differences in chemical concentration. The scale of pattern that can be formed by reaction diffusion can be very small as seen in the experimental patterns shown in Figure 2.11. A very rough idea of cell complexity is given by comparing the weight per bit of information of the cell's DNA molecule (deoxyribonucleic acid) of around 10^{-22}, to that of, say, imaging by an electron beam of around 10^{-10} or of a magnetic tape of about 10^{-5}. The most sophisticated and compact computer chip is simply not in the same class as a cell.

An important point arising from theoretical models is that any pattern contains its own history. Consider the following simple engineering analogy (Murray et al. 1998) of our role in trying to understand a biological process. It is one thing to suggest that a bridge requires a thousand tons of steel, that any less will result in too weak a structure, and any more will result in excessive rigidity. It is quite another matter to instruct the workers on how best to put the pieces together. In morphogenesis, for example, it is conceivable that the cells involved in tissue formation and deformation have enough expertise that given the right set of ingredients and initial instructions they could be persuaded to construct whatever element one wants. This is the hope of many who are searching for a full and predictive understanding. However, it seems very likely that the global effect of all this sophisticated cellular activity would be critically sensitive to the sequence of events occurring during development. As scientists we should concern ourselves with how to take advantage of the limited opportunities we have for communicating with the workforce so as to direct experiment towards an acceptable end-product.

None of the individual models that have been suggested for any biological patterning process, and not even all of them put together, could be considered a complete model. In the case of some of the widely studied problems (such as patterning in the developing limb bud), each model has shed light on different aspects of the process and we can now say what the important conceptual elements have to be in a complete model. These studies have served to highlight where our knowledge is deficient and to suggest directions in which fruitful experimentation might lead us. Indeed, a critical test of these theoretical constructs is in their impact on the experimental community.[2]

To conclude this section it must be stressed again that mathematical descriptions, including phenomenological descriptions, of patterning scenarios are *not* explanations. This is generally accepted, but often forgotten.

2.2 Reaction Diffusion (Turing) Mechanisms

Turing (1952) suggested that, under certain conditions, chemicals can react and diffuse in such a way as to produce steady state heterogeneous spatial patterns of chemical or morphogen concentration. In Chapter 11, Volume I we derived the governing equations for reaction diffusion mechanisms, namely, (11.16), which we consider here in the form:

$$\frac{\partial \mathbf{c}}{\partial t} = \mathbf{f}(\mathbf{c}) + D\nabla^2 \mathbf{c}, \tag{2.1}$$

where $\mathbf{c}$ is the vector of morphogen concentrations, $\mathbf{f}$ represents the reaction kinetics and D is the diagonal matrix of positive constant diffusion coefficients. This chapter is mainly concerned with models for two chemical species, $A(\mathbf{r}, t)$ and $B(\mathbf{r}, t)$ say. The equation system is then of the form

$$\begin{aligned} \frac{\partial A}{\partial t} &= F(A, B) + D_A \nabla^2 A, \\ \frac{\partial B}{\partial t} &= G(A, B) + D_B \nabla^2 B, \end{aligned} \tag{2.2}$$

where F and G are the kinetics, which will always be nonlinear.

[2]In the case of the mechanical theory of pattern formation discussed later, after some discussion, Lewis Wolpert (a friend and colleague of many years) who did not believe in the mechanical theory of pattern formation, designed some experiments specifically to disprove the theory. Although the experiments did not in fact do so, he discovered something else about the biological process he was studying—patterning in the chick limb bud. The impact of the theory was biologically illuminating even if the motivation was not verification. As he freely admits, he would not have done these specific experiments had he not been stimulated (or rather provoked) to do so by the theory.

Turing's (1952) idea is a simple but profound one. He said that, if in the absence of diffusion (effectively $D_A = D_B = 0$), A and B tend to a linearly stable uniform steady state then, under certain conditions, which we shall derive, spatially inhomogeneous patterns can evolve by *diffusion driven instability* if $D_A \neq D_B$. Diffusion is usually considered a *stabilising* process which is why this was such a novel concept. To see intuitively how diffusion can be destablising consider the following, albeit unrealistic, but informative analogy.

Consider a field of dry grass in which there is a large number of grasshoppers which can generate a lot of moisture by sweating if they get warm. Now suppose the grass is set alight at some point and a flame front starts to propagate. We can think of the grasshopper as an inhibitor and the fire as an activator. If there were no moisture to quench the flames the fire would simply spread over the whole field which would result in a uniform charred area. Suppose, however, that when the grasshoppers get warm enough they can generate enough moisture to dampen the grass so that when the flames reach such a pre-moistened area the grass will not burn. The scenario for spatial pattern is then as follows. The fire starts to spread—it is one of the 'reactants,' the activator, with a 'diffusion' coefficient D_F say. When the grasshoppers, the inhibitor 'reactant,' ahead of the flame front feel it coming they move quickly well ahead of it; that is, they have a 'diffusion' coefficient, D_G say, which is much larger than D_F. The grasshoppers then sweat profusely and generate enough moisture to prevent the fire spreading into the moistened area. In this way the charred area is restricted to a finite domain which depends on the 'diffusion' coefficients of the reactants—fire and grasshoppers—and various 'reaction' parameters. If, instead of a single initial fire, there were a random scattering of them we can see how this process would result in a final spatially heterogeneous steady state distribution of charred and uncharred regions in the field and a spatial distribution of grasshoppers, since around each fire the above scenario would take place. If the grasshoppers and flame front 'diffused' at the same speed no such spatial pattern could evolve. It is clear how to construct other analogies; other examples are given below in Section 2.3 and another in the *Scientific American* article by Murray (1988).

In the following section we describe the process in terms of reacting and diffusing morphogens and derive the necessary conditions on the reaction kinetics and diffusion coefficients. We also derive the type of spatial patterns we might expect. Here we briefly record for subsequent use two particularly simple hypothetical systems and one experimentally realised example, which are capable of satisfying Turing's conditions for a pattern formation system. There are now many other systems which have been used in studies of spatial patterning. These have varying degrees of experimental plausibility. With the extensive discussion of the Belousov–Zhabotinskii reaction in Chapter 8, Volume I and the last chapter we should particularly note it. Even though many other real reaction systems have been found it is still the major experimental system.

The simplest system is the Schnakenberg (1979) reaction discussed in Chapter 7, Volume I which, with reference to the system form (2.2), has kinetics

$$\begin{aligned} F(A, B) &= k_1 - k_2 A + k_3 A^2 B, \\ G(A, B) &= k_4 - k_3 A^2 B, \end{aligned} \tag{2.3}$$

where the k's are the positive rate constants. Here A is created autocatalytically by the k_3A^2B term in $F(A, B)$. This is one of the prototype reaction diffusion systems. Another is the influential activator–inhibitor mechanism suggested by Gierer and Meinhardt (1972) and widely studied and used since then. Their system was discussed in Chapter 6, Volume I and is

$$F(A, B) = k_1 - k_2A + \frac{k_3A^2}{B}, \quad G(A, B) = k_4A^2 - k_5B, \tag{2.4}$$

where here A is the activator and B the inhibitor. The k_3A^2/B term is again autocatalytic. Koch and Meinhardt (1994) review the applications of the Gierer–Meinhardt reaction diffusion system to biological pattern formation of complex structures. They give an extensive bibliography of applications of this specific model and its variations.

The real empirical substrate-inhibition system studied experimentally by Thomas (1975) and also described in detail in Chapter 6, Volume I, has

$$\begin{aligned} F(A, B) &= k_1 - k_2A - H(A, B), \quad G(A, B) = k_3 - k_4B - H(A, B), \\ H(A, B) &= \frac{k_5AB}{k_6 + k_7A + k_8A^2}. \end{aligned} \tag{2.5}$$

Here A and B are respectively the concentrations of the substrate oxygen and the enzyme uricase. The substrate inhibition is evident in the H-term via k_8A^2. Since the H-terms are negative they contribute to reducing A and B; the rate of reduction is inhibited for large enough A. Reaction diffusion systems based on the Field–Körös–Noyes (FKN) model kinetics (cf. Chapter 8, Volume I) is a particularly important example because of its potential for experimental verification of the theory; references are given at the appropriate places below.

Before commenting on the types of reaction kinetics capable of generating pattern we must nondimensionalise the systems given by (2.2) with reaction kinetics from such as (2.3) to (2.5). By way of example we carry out the details here for (2.2) with F and G given by (2.3) because of its algebraic simplicity and our detailed analysis of it in Chapter 7, Volume I. Introduce L as a typical length scale and set

$$\begin{aligned} u &= A\left(\frac{k_3}{k_2}\right)^{1/2}, \quad v = B\left(\frac{k_3}{k_2}\right)^{1/2}, \quad t^* = \frac{D_At}{L^2}, \quad \mathbf{x}^* = \frac{\mathbf{x}}{L}, \\ d &= \frac{D_B}{D_A}, \quad a = \frac{k_1}{k_2}\left(\frac{k_3}{k_2}\right)^{1/2}, \quad b = \frac{k_4}{k_2}\left(\frac{k_3}{k_2}\right)^{1/2}, \quad \gamma = \frac{L^2k_2}{D_A}. \end{aligned} \tag{2.6}$$

The dimensionless reaction diffusion system becomes, on dropping the asterisks for algebraic convenience,

$$\begin{aligned} u_t &= \gamma(a - u + u^2v) + \nabla^2u = \gamma f(u, v) + \nabla^2u, \\ v_t &= \gamma(b - u^2v) + d\nabla^2v = \gamma g(u, v) + d\nabla^2v, \end{aligned} \tag{2.7}$$

where f and g are defined by these equations. We could incorporate γ into new length and timescales by setting $\gamma^{1/2}\mathbf{r}$ and γt for $\mathbf{r}$ and t respectively. This is equivalent to defining the length scale L such that $\gamma = 1$; that is, $L = (D_A/k_2)^{1/2}$. We retain the specific form (2.7) for reasons which become clear shortly as well as for the analysis in the next section and for the applications in the following chapters.

An appropriate nondimensionalisation of the reaction kinetics (2.4) and (2.5) give (see Exercise 1)

$$\begin{aligned} f(u,v) &= a - bu + \frac{u^2}{v}, \quad g(u,v) = u^2 - v, \\ f(u,v) &= a - u - h(u,v), \quad g(u,v) = \alpha(b - v) - h(u,v), \\ h(u,v) &= \frac{\rho u v}{1 + u + K u^2}, \end{aligned} \tag{2.8}$$

where a, b, α, ρ and K are positive parameters. If we include activator inhibition in the activator–inhibitor system in the first of these we have, for f and g,

$$f(u,v) = a - bu + \frac{u^2}{v(1 + ku^2)}, \quad g(u,v) = u^2 - v, \tag{2.9}$$

where k is a measure of the inhibition; see also Section 6.7 in Chapter 6 in Volume I. Murray (1982) discussed each of these systems in detail and drew conclusions as to their relative merits as pattern generators; he presented a systematic analytical method for studying any two-species reaction diffusion system. For most pattern formation (analytical) illustrations with reaction diffusion mechanisms the simplest, namely, (2.7), turned out to be the most robust of those considered and, fortunately, the easiest to study.

All such reaction diffusion systems can be nondimensionalised and scaled to take the general form

$$u_t = \gamma f(u,v) + \nabla^2 u, \quad v_t = \gamma g(u,v) + d\nabla^2 v, \tag{2.10}$$

where d is the ratio of diffusion coefficients and γ can have any of the following interpretations.

(i) $\gamma^{1/2}$ is proportional to the *linear* size of the spatial domain in one dimension. In two dimensions γ is proportional to the area. This meaning is particularly important as we shall see later in Section 2.5 and in Chapter 3.

(ii) γ represents the relative strength of the reaction terms. This means, for example, that an increase in γ may represent an increase in activity of some rate-limiting step in the reaction sequence.

(iii) An increase in γ can also be thought of as equivalent to a decrease in the diffusion coefficient ratio d.

Particular advantages of this general form are: (a) the dimensionless parameters γ and d admit a wider biological interpretation than do the dimensional parameters and (b) when we consider the domains in parameter space where particular spatial patterns appear, the results can be conveniently displayed in (γ, d) space. This aspect was exploited by Arcuri and Murray (1986).

Whether or not the systems (2.2) are capable of generating Turing-type spatial patterns crucially depends on the reaction kinetics f and g, and the values of γ and d. The detailed form of the null clines provides essential initial information. Figure 2.2 illustrates typical null clines for f and g defined by (2.7)–(2.9).

In spite of their different chemical motivation and derivation all of these kinetics are equivalent to some activation–inhibition interpretation and when coupled with unequal diffusion of the reactants, are capable of generating spatial patterns. The spatial activation–inhibition concept was discussed in detail in Section 11.5 in Chapter 11 in Volume I, and arose from an integral equation formulation: refer to equation (11.41) there. As we shall see in the next section the crucial aspect of the kinetics regarding pattern generation is incorporated in the form of the null clines and how they intersect in the vicinity of the steady state. There are two broad types illustrated in the last fig-

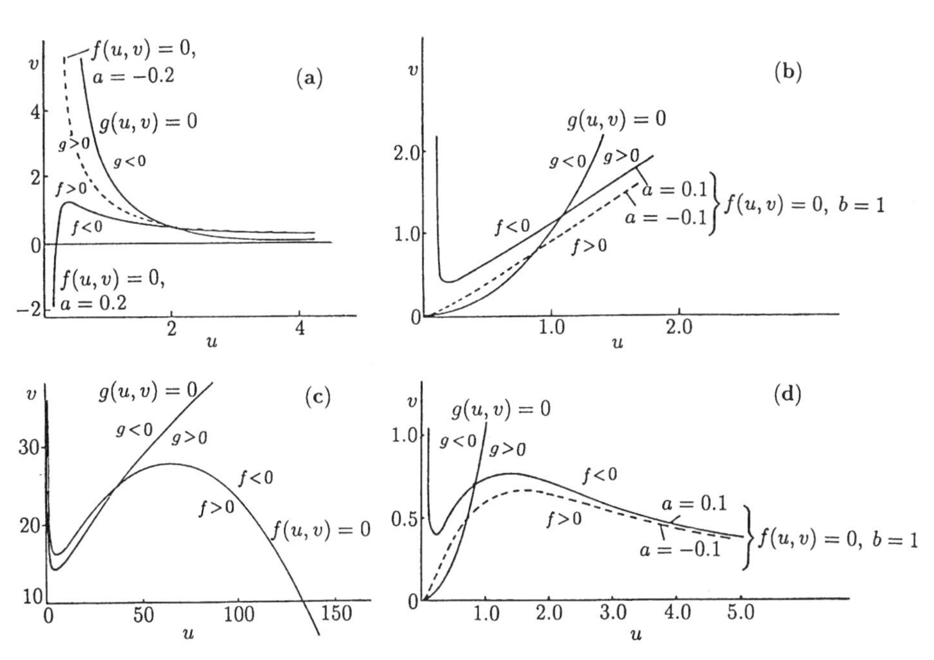

Figure 2.2. Null clines $f(u, v) = 0$, $g(u, v) = 0$: (**a**) The dimensionless Schnakenberg (1979) kinetics (2.7) with $a = 0.2$ and $b = 2.0$ with the dashed curve, where $a = -0.2$ and which is typical of the situation when $a < 0$. (**b**) The dimensionless Gierer and Meinhardt (1972) system with $a = \pm 0.1$, $b = 1$ and no activator inhibition. (**c**) The empirical Thomas (1975) system defined by (2.8) with parameter values $a = 150$, $b = 100$, $\alpha = 1.5$, $\rho = 13$, $K = 0.05$. (**d**) The kinetics in (2.9) with $a > 0$, $b > 0$ and $k > 0$, which implies activator inhibition; the dashed curve has $a < 0$.

ure. The steady state neighbourhood of the null clines in Figures 2.2(b), (c) and (d) are similar and represent one class, while that in Figure 2.2(a) is the other.

We should note here that there are other important classes of null clines which we do not consider, such as those in which there is more than one positive steady state; we discussed such kinetics in Chapter 7, Volume I for example. Reaction diffusion systems with such kinetics can generate even more complex spatial patterns: initial conditions here are particularly important. We also do not discuss here systems in which the diffusion coefficients are space-dependent and concentration- or population-dependent; these are important in ecological contexts (recall the discussion in Chapter 1 on the spread of genetically engineered organisms). We briefly considered density-dependent diffusion cases in Chapter 11, Volume I. Later in the book we discuss an important application in which the diffusion coefficient is space-dependent when we model the spread of brain tumours in anatomically realistic brains.

It is often useful and intuitively helpful in model building to express the mechanism's kinetics in schematic terms with some convention to indicate autocatalysis, activation, inhibition, degradation and unequal diffusion. If we do this, by way of illustration, with the activator–inhibitor kinetics given by the first of (2.8) in (2.10) we can adopt the convention shown in Figure 2.3(a).

The effect of different diffusion coefficients, here with $d > 1$, is to illustrate the prototype spatial concept of local activation and lateral inhibition illustrated in Figures 2.3(b) and 2.4(b). The general concept was introduced before in Chapter 11, Volume I. It is this generic spatial behaviour which is necessary for spatial patterning: the grasshoppers and the fire analogy is an obvious example with the fire the local activation and the grasshoppers providing the long range inhibition. It is intuitively clear that the diffusion coefficient of the inhibitor must be larger than that of the activator.

The concept of local activation and lateral inhibition is quite old going back at least to Ernst Mach in 1885 with his Mach bands. This is a visual illusion which occurs when dark and light bands are juxtaposed. Figure 2.4 is a schematic illustration of what happens together with an example of the Hermann illusion which is based on it.

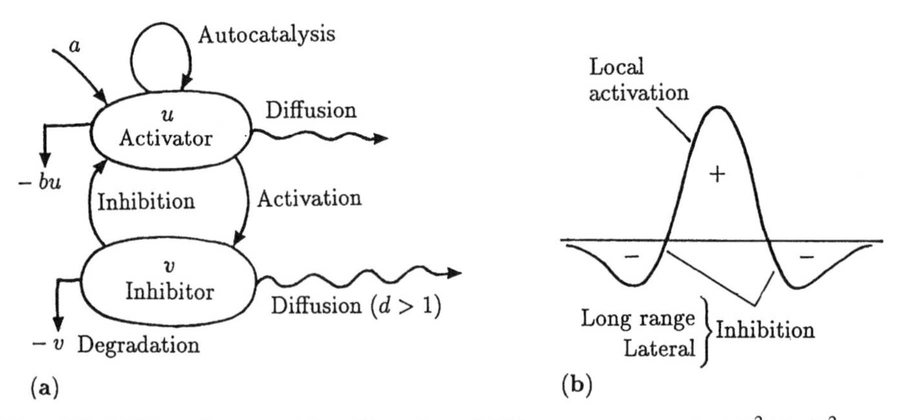

Figure 2.3. (**a**) Schematic representation of the activator–inhibitor system $u_t = a - bu + (u^2/v) + \nabla^2 u$, $v_t = u^2 - v + d\nabla^2 v$. (**b**) Spatial representation of local activation and long range inhibition.

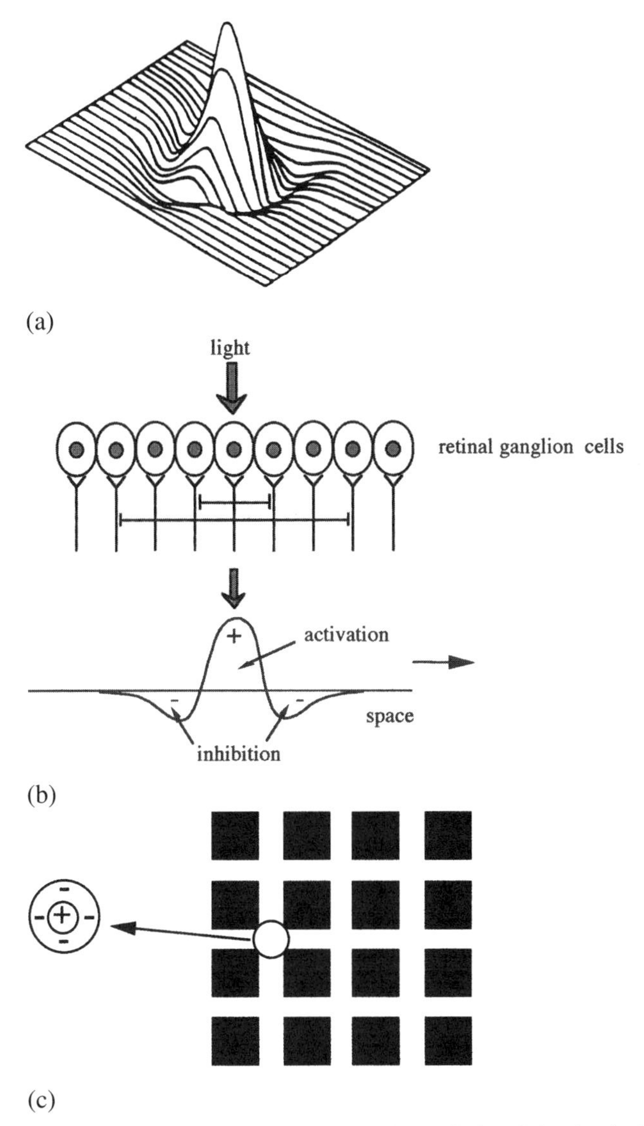

Figure 2.4. (**a**) If a light is shone on an array of retinal ganglion cells there is local activation of the cells in the immediate neighbourhood of the light with lateral inhibition of the cells farther away from the light source. The result is a landscape of local activation and lateral inhibition as illustrated in (**b**). (**c**) This illustrates the Hermann illusion. Here cells have more illumination in their inhibitory surrounding regions than cells in other white regions and so are more strongly inhibited and appear darker.

2.3 General Conditions for Diffusion-Driven Instability: Linear Stability Analysis and Evolution of Spatial Pattern

A reaction diffusion system exhibits diffusion-driven instability, sometimes called Turing instability, if the homogeneous steady state is stable to small perturbations in the absence of diffusion but unstable to small *spatial* perturbations when diffusion is present. The concept of instability in biology is often in the context of ecology, where a uniform steady state becomes unstable to small perturbations and the populations typically exhibit some temporal oscillatory behaviour. The instability we are concerned with here is of a quite different kind. The main process driving the spatially inhomogeneous instability is diffusion: the mechanism determines the spatial pattern that evolves. How the pattern or mode is selected is an important aspect of the analysis, a topic we discuss in this (and later) chapters.

We derive here the necessary and sufficient conditions for diffusion-driven instability of the steady state and the initiation of spatial pattern for the general system (2.10). To formulate the problem mathematically we require boundary and initial conditions. These we take to be zero flux boundary conditions and given initial conditions. The mathematical problem is then defined by

$$\begin{gathered} u_t = \gamma f(u,v) + \nabla^2 u, \quad v_t = \gamma g(u,v) + d\nabla^2 v, \\ (\mathbf{n}\cdot\nabla)\begin{pmatrix} u \\ v \end{pmatrix} = 0, \quad \mathbf{r} \text{ on } \partial B; \quad u(\mathbf{r},0),\ v(\mathbf{r},0) \text{ given,} \end{gathered} \tag{2.11}$$

where ∂B is the closed boundary of the reaction diffusion domain B and $\mathbf{n}$ is the unit outward normal to ∂B. There are several reasons for choosing zero flux boundary conditions. The major one is that we are interested in self-organisation of pattern; zero flux conditions imply no external input. If we imposed fixed boundary conditions on u and v the spatial patterning could be a direct consequence of the boundary conditions as we shall see in the ecological problem below in Section 2.7. In Section 2.4 we carry out the analysis for a specific one- and two-dimensional situation with the kinetics given by (2.7).

The relevant homogeneous steady state (u_0, v_0) of (2.11) is the positive solution of

$$f(u,v) = 0, \quad g(u,v) = 0. \tag{2.12}$$

Since we are concerned with *diffusion-driven* instability we are interested in linear instability of this steady state that is solely *spatially* dependent. So, in the absence of any spatial variation the homogeneous steady state must be linearly stable: we first determine the conditions for this to hold. These were derived in Chapter 3, Volume I but as a reminder and for notational completeness we briefly rederive them here.

With no spatial variation u and v satisfy

$$u_t = \gamma f(u,v), \quad v_t = \gamma g(u,v). \tag{2.13}$$

Linearising about the steady state (u_0, v_0) in exactly the same way as we did in Chapter 3, Volume I, we set

$$\mathbf{w} = \begin{pmatrix} u - u_0 \\ v - v_0 \end{pmatrix} \tag{2.14}$$

and (2.13) becomes, for $|\,\mathbf{w}\,|$ small,

$$\mathbf{w}_t = \gamma A \mathbf{w}, \quad A = \begin{pmatrix} f_u & f_v \\ g_u & g_v \end{pmatrix}_{u_0, v_0}, \tag{2.15}$$

where A is the stability matrix. From now on we take the partial derivatives of f and g to be evaluated at the steady state unless stated otherwise. We now look for solutions in the form

$$\mathbf{w} \propto e^{\lambda t}, \tag{2.16}$$

where λ is the eigenvalue. The steady state $\mathbf{w} = 0$ is linearly stable if $\mathrm{Re}\,\lambda < 0$ since in this case the perturbation $\mathbf{w} \to 0$ as $t \to \infty$. Substitution of (2.16) into (2.15) determines the eigenvalues λ as the solutions of

$$\begin{aligned} |\,\gamma A - \lambda I\,| = \begin{vmatrix} \gamma f_u - \lambda & \gamma f_v \\ \gamma g_u & \gamma g_v - \lambda \end{vmatrix} = 0 \\ \Rightarrow \quad \lambda^2 - \gamma(f_u + g_v)\lambda + \gamma^2(f_u g_v - f_v g_u) = 0, \end{aligned} \tag{2.17}$$

so

$$\lambda_1, \lambda_2 = \frac{1}{2}\gamma\left[(f_u + g_v) \pm \left\{(f_u + g_v)^2 - 4(f_u g_v - f_v g_u)\right\}^{1/2}\right]. \tag{2.18}$$

Linear stability, that is, $\mathrm{Re}\,\lambda < 0$, is guaranteed if

$$\mathrm{tr}\,A = f_u + g_v < 0, \quad |\,A\,| = f_u g_v - f_v g_u > 0. \tag{2.19}$$

Since (u_0, v_0) are functions of the parameters of the kinetics, these inequalities thus impose certain constraints on the parameters. Note that for all cases in Figure 2.2 in the neighbourhood of the steady state, $f_u > 0$, $g_v < 0$, and for Figure 2.2(a) $f_v > 0$, $g_u < 0$ while for Figure 2.2(b) to (d) $f_v < 0$, $g_u > 0$. So $\mathrm{tr}\,A$ and $|\,A\,|$ could be positive or negative: here we are only concerned with the conditions and parameter ranges which satisfy (2.19).

Now consider the full reaction diffusion system (2.11) and again linearise about the steady state, which with (2.14) is $\mathbf{w} = 0$, to get

$$\mathbf{w}_t = \gamma A \mathbf{w} + D\nabla^2 \mathbf{w}, \quad D = \begin{pmatrix} 1 & 0 \\ 0 & d \end{pmatrix}. \tag{2.20}$$

To solve this system of equations subject to the boundary conditions (2.11) we first define $\mathbf{W}(\mathbf{r})$ to be the time-independent solution of the spatial eigenvalue problem defined by

$$\nabla^2\mathbf{W} + k^2\mathbf{W} = 0, \quad (\mathbf{n}\cdot\nabla)\mathbf{W} = 0 \quad \text{for} \quad \mathbf{r} \text{ on } \partial B, \tag{2.21}$$

where k is the eigenvalue. For example, if the domain is one-dimensional, say, $0 \leq x \leq a$, $\mathbf{W} \propto \cos(n\pi x/a)$ where n is an integer; this satisfies zero flux conditions at $x = 0$ and $x = a$. The eigenvalue in this case is $k = n\pi/a$. So, $1/k = a/n\pi$ is a measure of the wavelike pattern: the eigenvalue k is called the *wavenumber* and $1/k$ is proportional to the wavelength ω; $\omega = 2\pi/k = 2a/n$ in this example. From now on we shall refer to k in this context as the wavenumber. With finite domains there is a discrete set of possible wavenumbers since n is an integer.

Let $\mathbf{W}_k(\mathbf{r})$ be the eigenfunction corresponding to the wavenumber k. Each eigenfunction $\mathbf{W}_k$ satisfies zero flux boundary conditions. Because the problem is linear we now look for solutions $\mathbf{w}(\mathbf{r}, t)$ of (2.20) in the form

$$\mathbf{w}(\mathbf{r}, t) = \sum_k c_k e^{\lambda t}\mathbf{W}_k(\mathbf{r}), \tag{2.22}$$

where the constants c_k are determined by a Fourier expansion of the initial conditions in terms of $\mathbf{W}_k(\mathbf{r})$. λ is the eigenvalue which determines temporal growth. Substituting this form into (2.20) with (2.21) and cancelling $e^{\lambda t}$, we get, for each k,

$$\begin{aligned}\lambda\mathbf{W}_k &= \gamma A\mathbf{W}_k + D\nabla^2\mathbf{W}_k\\ &= \gamma A\mathbf{W}_k - Dk^2\mathbf{W}_k.\end{aligned}$$

We require nontrivial solutions for $\mathbf{W}_k$ so the λ are determined by the roots of the characteristic polynomial

$$|\,\lambda I - \gamma A + Dk^2\,| = 0.$$

Evaluating the determinant with A and D from (2.15) and (2.20) we get the eigenvalues $\lambda(k)$ as functions of the wavenumber k as the roots of

$$\begin{aligned}&\lambda^2 + \lambda[k^2(1+d) - \gamma(f_u + g_v)] + h(k^2) = 0,\\ &h(k^2) = dk^4 - \gamma(df_u + g_v)k^2 + \gamma^2|\,A\,|.\end{aligned} \tag{2.23}$$

The steady state (u_0, v_0) is linearly stable if both solutions of (2.23) have $\operatorname{Re}\lambda < 0$. We have already imposed the constraints that the steady state is stable in the absence of any spatial effects; that is, $\operatorname{Re}\lambda(k^2 = 0) < 0$. The quadratic (2.23) in this case is (2.17) and the requirement that $\operatorname{Re}\lambda < 0$ gave conditions (2.19). For the steady state to be unstable to *spatial* disturbances we require $\operatorname{Re}\lambda(k) > 0$ for some $k \neq 0$. This can happen if either the coefficient of λ in (2.23) is negative, or if $h(k^2) < 0$ for some $k \neq 0$. Since $(f_u + g_v) < 0$ from conditions (2.19) and $k^2(1+d) > 0$ for all $k \neq 0$ the coefficient of λ, namely,

$$[k^2(1+d) - \gamma(f_u + g_v)] > 0,$$

so the only way $\operatorname{Re}\lambda(k^2)$ can be positive is if $h(k^2) < 0$ for some k. This is immediately clear from the solutions of (2.23), namely,

$$2\lambda = -[k^2(1+d) - \gamma(f_u + g_v)] \pm \{[k^2(1+d) - \gamma(f_u + g_v)]^2 - 4h(k^2)\}^{1/2}.$$

Since we required the determinant $|A| > 0$ from (2.19) the only possibility for $h(k^2)$ in (2.23) to be negative is if $(df_u + g_v) > 0$. Since $(f_u + g_v) < 0$ from (2.19) this implies that $d \neq 1$ and f_u and g_v must have opposite signs. So, a further requirement to those in (2.19) is

$$df_u + g_v > 0 \quad \Rightarrow \quad d \neq 1. \tag{2.24}$$

With the reaction kinetics giving the null clines in Figure 2.2 we noted that $f_u > 0$ and $g_v < 0$, so the first condition in (2.19) and the last inequality (2.24) require that the diffusion coefficient ratio $d > 1$. For example, in terms of the activator–inhibitor mechanism (2.8) this means that the inhibitor must diffuse faster than the activator as we noted above.

The inequality (2.24) is necessary but not sufficient for $\operatorname{Re}\lambda > 0$. For $h(k^2)$ to be negative for some nonzero k, the minimum $h_{\min}$ must be negative. From (2.23), elementary differentiation with respect to k^2 shows that

$$h_{\min} = \gamma^2\left[|A| - \frac{(df_u + g_v)^2}{4d}\right], \quad k^2 = k_m^2 = \gamma\frac{df_u + g_v}{2d}. \tag{2.25}$$

Thus the condition that $h(k^2) < 0$ for some $k^2 \neq 0$ is

$$\frac{(df_u + g_v)^2}{4d} > |A|. \tag{2.26}$$

At bifurcation, when $h_{\min} = 0$, we require $|A| = (df_u + g_v)^2/4d$ and so for fixed kinetics parameters this defines a critical diffusion coefficient ratio $d_c(> 1)$ as the appropriate root of

$$d_c^2 f_u^2 + 2(2f_v g_u - f_u g_v)d_c + g_v^2 = 0. \tag{2.27}$$

The critical wavenumber k_c is then given (using (2.26)) by

$$k_c^2 = \gamma\frac{d_c f_u + g_v}{2d_c} = \gamma\left[\frac{|A|}{d_c}\right]^{1/2} = \gamma\left[\frac{f_u g_v - f_v g_u}{d_c}\right]^{1/2}. \tag{2.28}$$

Figure 2.5(a) shows how $h(k^2)$ varies as a function of k^2 for various d.

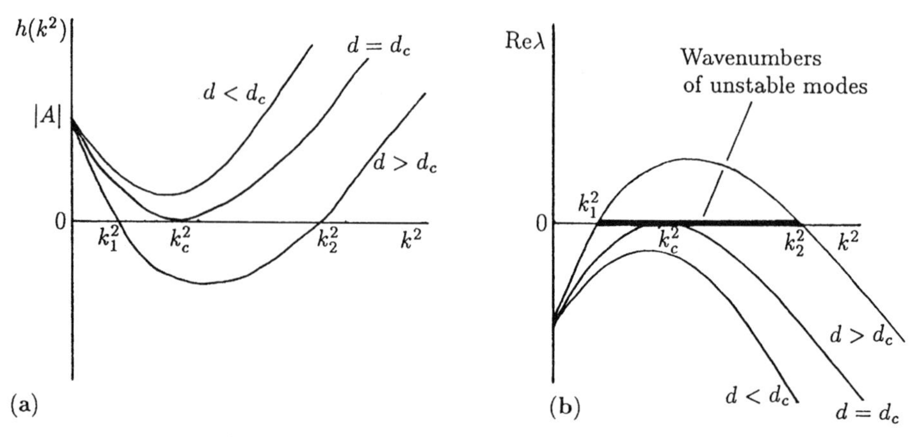

Figure 2.5. (**a**) Plot of $h(k^2)$ defined by (2.23) for typical kinetics illustrated in Figure 2.2. When the diffusion coefficient ratio d increases beyond the critical value d_c, $h(k^2)$ becomes negative for a finite range of $k^2 > 0$. (**b**) Plot of the largest of the eigenvalues $\lambda(k^2)$ from (2.23) as a function of k^2. When $d > d_c$ there is a range of wavenumbers $k_1^2 < k^2 < k_2^2$ which are linearly unstable.

Whenever $h(k^2) < 0$, (2.23) has a solution λ which is positive for the same range of wavenumbers that make $h < 0$. From (2.23) with $d > d_c$ the range of unstable wavenumbers $k_1^2 < k^2 < k_2^2$ is obtained from the zeros k_1^2 and k_2^2 of $h(k^2) = 0$ as

$$\begin{aligned} k_1^2 &= \frac{\gamma}{2d}\left[(df_u + g_v) - \{(df_u + g_v)^2 - 4d|\,A\,|\}^{1/2}\right] < k^2 \\ &< \frac{\gamma}{2d}\left[(df_u + g_v) + \{(df_u + g_v)^2 - 4d|\,A\,|\}^{1/2}\right] = k_2^2. \end{aligned} \tag{2.29}$$

Figure 2.5(b) plots a typical $\lambda(k^2)$ against k^2. The expression $\lambda = \lambda(k^2)$ is called a *dispersion relation*. We discuss the importance and use of dispersion relations in more detail in the next two sections. Note that, within the unstable range, $\text{Re}\,\lambda(k^2) > 0$ has a maximum for the wavenumber k_m obtained from (2.25) with $d > d_c$. This implies that there is a fastest growing mode in the summation (2.22) for **w**; this is an attribute we now exploit.

If we consider the solution **w** given by (2.22), the dominant contributions as t increases are those modes for which $\text{Re}\,\lambda(k^2) > 0$ since all other modes tend to zero exponentially. From Figure 2.5, or analytically from (2.29), we determine the range, $k_1^2 < k^2 < k_2^2$, where $h(k^2) < 0$, and hence $\text{Re}\,\lambda(k^2) > 0$, and so from (2.22)

$$\mathbf{w}(\mathbf{r}, t) \sim \sum_{k_1}^{k_2} c_k e^{\lambda(k^2)t}\mathbf{W}_k(\mathbf{r}) \quad \text{for large } t. \tag{2.30}$$

An analysis and graph of the dispersion relation are thus extremely informative in that they immediately say which eigenfunctions, that is, which spatial patterns, are linearly unstable and grow exponentially with time. We must keep in mind that, with finite domain eigenvalue problems, the wavenumbers are discrete and so only certain k in the range (2.29) are of relevance; we discuss the implications of this later.

The key assumption, and what in fact happens, is that these linearly unstable eigenfunctions in (2.30) which are growing exponentially with time will eventually be bounded by the nonlinear terms in the reaction diffusion equations and an ultimate steady state spatially inhomogeneous solution will emerge. A key element in this assumption is the existence of a confined set (or bounding domain) for the kinetics (see Chapter 3, Volume I). We would intuitively expect that if a confined set exists for the kinetics, the same set would also contain the solutions when diffusion is included. This is indeed the case and can be rigorously proved; see Smoller (1983). So, part of the analysis of a specific mechanism involves the demonstration of a confined set within the *positive* quadrant. A general nonlinear analysis for the evolution to the finite amplitude steady state spatial patterns is still lacking but singular perturbation analyses for d near the bifurcation value d_c have been carried out and a nonuniform spatially heterogeneous solution is indeed obtained (see, for example, Lara-Ochoa and Murray 1983, Zhu and Murray 1995). Singular perturbation analyses can be done near any of the critical parameters near bifurcation. There have now been many spatially inhomogeneous solutions evaluated numerically using a variety of specific reaction diffusion mechanisms; the numerical methods are now quite standard. The results presented in the next chapter illustrate some of the richness of pattern which can be generated.

To recap, we have now obtained conditions for the generation of spatial patterns by *two*-species reaction diffusion mechanisms of the form (2.11). For convenience we reproduce them here. Remembering that all derivatives are evaluated at the steady state (u_0, v_0), they are, from (2.19), (2.24) and (2.26),

$$\begin{aligned} f_u + g_v < 0, \quad & f_u g_v - f_v g_u > 0, \\ d f_u + g_v > 0, \quad & (d f_u + g_v)^2 - 4d(f_u g_v - f_v g_u) > 0. \end{aligned} \tag{2.31}$$

The derivatives f_u and g_v must be of opposite sign: with the reaction kinetics exhibited in Figure 2.2, $f_u > 0$, $g_v < 0$ so the first and third of (2.31) imply that the ratio of diffusion coefficients $d > 1$.

There are two possibilities for the cross-terms f_v and g_u since the only restriction is that $f_v g_u < 0$. So, we must have $f_v < 0$ and $g_u > 0$ or the other way round. These correspond to qualitatively different reactions. The two cases are illustrated schematically in Figure 2.6. Recall that the reactant which promotes growth in one is the activator and the other the inhibitor. In the case illustrated in Figure 2.6(a), u is the activator, which is also self-activating, while the inhibitor, v, inhibits not only u, but also itself. For pattern formation to take place the inhibitor must diffuse more quickly than the activator. In the case illustrated in Figure 2.6(b), v is the activator but is still self-inhibiting and diffuses more quickly. There is another difference between the two cases. The pattern grows along the unstable manifold associated with the positive eigenvalue. In Figure 2.6(a) this means that the two species are at high or low density in the same region as the pattern grows as in Figure 2.6(c); in case Figure 2.6(b) u is at a high density where v is low, and vice versa as in Figure 2.6(d). The qualitative features of the phase plane (just for the reaction terms) in the vicinity of the steady state are shown in Figure 2.6(e) and (f) for the two cases. The fact that the patterns are either in or out of phase has fundamental implications for biological applications.

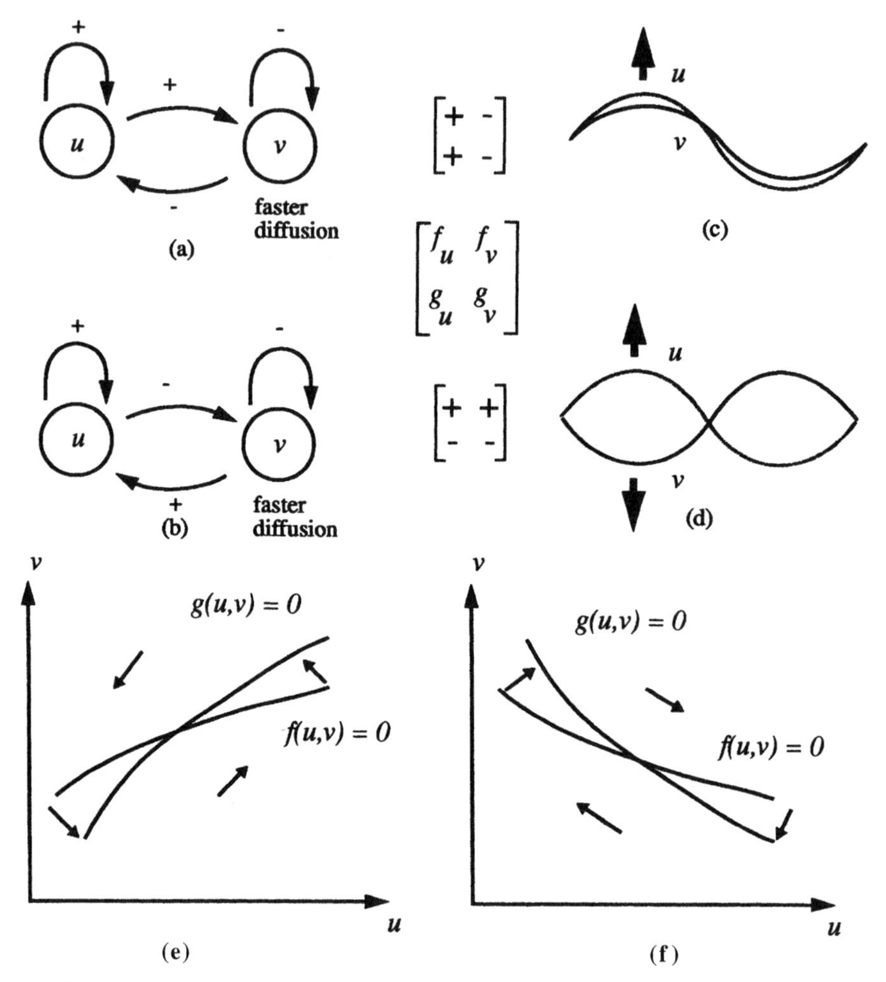

Figure 2.6. Schematic illustration of the two qualitatively different cases of diffusion driven instability. (**a**) self-activating u also activates v, which inhibits both reactants. The resulting initially growing pattern is shown in (**c**). (**b**) Here the self-activating u inhibits v but is itself activated by v with the resulting pattern illustrated in (**d**). The matrices give the signs of f_u, f_v, g_u, g_v evaluated at the steady state. (**e**) and (**f**) The reaction phase planes near the steady state. The arrows indicate the direction of change due to reaction (in the absence of diffusion). Case (**e**) corresponds to the interactions illustrated in (**a**) and (**c**) while that in (**f**) corresponds to the interactions illustrated in (**b**) and (**d**).

To get an intuitive feel for these two cases let us consider two different ecological predator–prey scenarios. In the first, that is, Figure 2.6(e), let u and v represent the prey and predator respectively. At high predator density prey numbers are reduced but at low densities their number is increased. Near the steady state the prey benefit from each other in that an increase in number is temporarily amplified. Predators decrease in numbers if the predator-to-prey ratio is high, but otherwise increase. Another example, from parasitology, is if v is a parasite dispersing via a motile host while u is a more sedentary host that is severely affected by the parasite. In these the interaction near the steady state is as in Figure 2.6(a) with the local null clines and qualitative growth as in Figure 2.6(e).

A necessary condition for diffusion-driven instability in this predator–prey situation is that the predators disperse faster than the prey. In this case the patterns form as in Figure 2.6(c). Let us suppose there is a region of increased prey density. Without diffusion this would be damped out since the predators would temporarily increase and then drop back towards the steady state. However, with the predators diffusing it is possible that the local increase in predators (due to an increase in the prey) partially disperses and so is not strong enough to push the prey population back towards equilibrium. When predators disperse they lower the prey density in the neighbourhood. It is therefore possible to end up with clumps of high prey and predator populations interspersed with areas in which both densities are low. In the parasite analogy clumping of the sedentary prey (the host) coincides with areas of high parasite density. Hosts can also exist at high levels because the parasites continue to disperse into the nearby 'dead zone' in which there are few of this type of host. The scale on which patterning takes place depends on the ratio of the diffusion coefficients d.

Now consider the second type of interaction illustrated in Figures 2.6(b), (d) and (f). Again with a predator–prey situation let u now be the predator and v the prey. In this case the predators are 'autocatalytic' since when densities are close to the steady state, an increase in predator density is temporarily amplified, a not uncommon situation. For example, increased predator densities could improve hunting or reproductive efficiency. Another difference between this case and the first one is that it is now the prey that disperse at a faster rate.

Suppose again that there is a high prey density area. Without diffusion the predator numbers would increase and eventually make both populations return to the steady state. However, it could happen that the predators grow and reduce the prey population to a level below the steady state value (the temporary increase in prey is enough to prompt the autocatalytic growth of predators to kick in). This would result in a net flux of prey from neighbouring regions which in turn would cause the predator density to drop in those regions (as autocatalysis works in the other direction) thereby letting the prey populations grow above their steady state value. A pattern could become established in which areas of low predator/high prey supply with extra prey in those areas in which there are few prey and large numbers of predators. In effect, autocatalytic predators benefit both from being at a high density locally and also because nearby there are regions containing few predators which thus supply them with a constant extra flux of prey. Prey continue to flow towards regions of high predation because of the random nature of diffusion.

If the conditions (2.31) are satisfied there is a scale (γ)-dependent range of patterns, with wavenumbers defined by (2.29), which are linearly unstable. The spatial patterns which initially grow (exponentially) are those eigenfunctions $\mathbf{W}_k(\mathbf{r})$ with wavenumbers between k_1 and k_2 determined by (2.29), namely, those in (2.30). Note that the scale parameter γ plays a crucial role in these expressions, a point we consider further in the next section. Generally we would expect the kinetics and diffusion coefficients to be fixed. In the case of embryogenesis the natural variable parameter is then γ which reflects the size of the embryo or rather the embryonic domain (such as a developing limb bud) we are considering.

Diffusion-Driven Instability in Infinite Domains: Continuous Spectrum of Eigenvalues

In a finite domain the possible wavenumbers k and corresponding spatial wavelengths of allowable patterns are discrete and depend in part on the boundary conditions. In developmental biology the size of the embryo during the period of spatial patterning is often sufficiently large, relative to the pattern to be formed, that the 'boundaries' cannot play a major role in isolating specific wavelengths, as, for example, in the generation of patterns of hair, scale and feather primordia discussed in later chapters. Thus, for practical purposes the pattern formation domain is effectively infinite. Here we describe how to determine the spectrum of unstable eigenvalues for an infinite domain—it is easier than for a finite domain.

We start with the linearised system (2.20) and look for solutions in the form

$$\mathbf{w}(\mathbf{r}, t) \propto \exp[\lambda t + i\mathbf{k} \cdot \mathbf{r}],$$

where $\mathbf{k}$ is the wave vector with magnitude $k = |\,\mathbf{k}\,|$. Substitution into (2.20) again gives

$$|\,\lambda I - \gamma A + Dk^2\,| = 0$$

and so the dispersion relation giving λ in terms of the wavenumbers k is again given by (2.23). The range of eigenvalues for which $\operatorname{Re}\lambda(k^2) > 0$ is again given by (2.29). The crucial difference between the situation here and that for a finite domain is that there is always a spatial pattern if, in (2.29), $0 < k_1^2 < k_2^2$ since we are not restricted to a discrete class of k^2 defined by the eigenvalue problem (2.21). So, at bifurcation when k_c^2, given by (2.28), is linearly unstable the mechanism will evolve to a spatial pattern with the critical wavelength $\omega_c = 2\pi/k_c$. Thus the wavelength with the maximum exponential growth in Figure 2.5(b) will be the pattern which generally emerges at least in one dimension: it is not always the case and depends on the number of unstable modes and initial conditions. In the next chapter on biological applications we shall see that the difference between a finite domain and an effectively infinite one has important biological implications: finite domains put considerable restrictions on the allowable patterns.

2.4 Detailed Analysis of Pattern Initiation in a Reaction Diffusion Mechanism

Here we consider, by way of example, a specific two-species reaction diffusion system and carry out the detailed analysis. We lay the groundwork in this section for the subsequent applications to real biological pattern formation problems. We calculate the eigenfunctions, obtain the specific conditions on the parameters necessary to initiate spatial patterns and determine the wavenumbers and wavelengths of the spatial disturbances which initially grow exponentially.

We study the simplest reaction diffusion mechanism (2.7), first in one space dimension; namely,

$$
\begin{aligned}
u_t &= \gamma f(u, v) + u_{xx} = \gamma(a - u + u^2 v) + u_{xx}, \\
v_t &= \gamma g(u, v) + d v_{xx} = \gamma(b - u^2 v) + d v_{xx}.
\end{aligned}
\tag{2.32}
$$

The kinetics null clines $f = 0$ and $g = 0$ are illustrated in Figure 2.2(a). The uniform positive steady state (u_0, v_0) is

$$
u_0 = a + b, \quad v_0 = \frac{b}{(a+b)^2}, \quad b > 0, \quad a + b > 0 \tag{2.33}
$$

and, at the steady state,

$$
\begin{aligned}
&f_u = \frac{b-a}{a+b}, \quad f_v = (a+b)^2 > 0, \quad g_u = \frac{-2b}{a+b} < 0, \\
&g_v = -(a+b)^2 < 0, \quad f_u g_v - f_v g_u = (a+b)^2 > 0.
\end{aligned}
\tag{2.34}
$$

Since f_u and g_v must have opposite signs we must have $b > a$. With these expressions, conditions (2.31) require

$$
\begin{aligned}
&f_u + g_v < 0 \quad \Rightarrow \quad 0 < b - a < (a+b)^3, \\
&f_u g_v - f_v g_u > 0 \quad \Rightarrow \quad (a+b)^2 > 0, \\
&d f_u + g_v > 0 \quad \Rightarrow \quad d(b-a) > (a+b)^3, \\
&(d f_u + g_v)^2 - 4d(f_u g_v - f_v g_u) > 0 \\
&\Rightarrow \quad [d(b-a) - (a+b)^3]^2 > 4d(a+b)^4.
\end{aligned}
\tag{2.35}
$$

These inequalities define a domain in (a, b, d) parameter space, called the pattern formation space (or Turing space), within which the mechanism is unstable to certain spatial disturbances of given wavenumbers k, which we now determine.

Consider the related eigenvalue problem (2.21) and let us choose the domain to be $x \in (0, p)$ with $p > 0$. We then have

$$
\mathbf{W}_{xx} + k^2 \mathbf{W} = 0, \quad \mathbf{W}_x = 0 \text{ for } x = 0, p \tag{2.36}
$$

the solutions of which are

$$
\mathbf{W}_n(x) = \mathbf{A}_n \cos(n\pi x/p), \quad n = \pm 1, \pm 2, \ldots, \tag{2.37}
$$

where the $\mathbf{A}_n$ are arbitrary constants. The eigenvalues are the *discrete* wavenumbers $k = n\pi/p$. Whenever (2.34) are satisfied and there is a range of wavenumbers $k = n\pi/p$ lying within the bounds defined by (2.29), then the corresponding eigenfunctions $\mathbf{W}_n$ are linearly unstable. Thus the eigenfunctions (2.37) with wavelengths $\omega = 2\pi/k = 2p/n$ are the ones which initially grow with time like $\exp\{\lambda([n\pi/p]^2)t\}$. The band of wavenumbers from (2.29), with (2.34), is given by

$$\gamma L(a,b,d) = k_1^2 < k^2 = \left(\frac{n\pi}{p}\right)^2 < k_2^2 = \gamma M(a,b,d)$$

$$L = \frac{[d(b-a)-(a+b)^3] - \{[d(b-a)-(a+b)^3]^2 - 4d(a+b)^4\}^{1/2}}{2d(a+b)}, \tag{2.38}$$

$$M = \frac{[d(b-a)-(a+b)^3] + \{[d(b-a)-(a+b)^3]^2 - 4d(a+b)^4\}^{1/2}}{2d(a+b)}.$$

In terms of the wavelength $\omega = 2\pi/k$, the range of unstable modes $\mathbf{W}_n$ have wavelengths bounded by ω_1 and ω_2, where

$$\frac{4\pi^2}{\gamma L(a,b,d)} = \omega_1^2 > \omega^2 = \left(\frac{2p}{n}\right)^2 > \omega_2^2 = \frac{4\pi^2}{\gamma M(a,b,d)}. \tag{2.39}$$

Note in (2.38) the importance of scale, quantified by γ. The smallest wavenumber is π/p; that is, $n = 1$. For fixed parameters a, b and d, if γ is sufficiently small (2.38) says that there is *no allowable* k in the range, and hence no mode $\mathbf{W}_n$ in (2.37), which can be driven unstable. This means that all modes in the solution $\mathbf{w}$ in (2.30) tend to zero exponentially and the steady state is stable. We discuss this important role of scale in more detail below.

From (2.30) the spatially heterogeneous solution which emerges is the sum of the unstable modes, namely,

$$\mathbf{w}(x,t) \sim \sum_{n_1}^{n_2} \mathbf{C}_n \exp\left[\lambda\left(\frac{n^2\pi^2}{p^2}\right)t\right] \cos\frac{n\pi x}{p}, \tag{2.40}$$

where λ is given by the positive solution of the quadratic (2.23) with the derivatives from (2.34), n_1 is the smallest integer greater than or equal to pk_1/π, n_2 the largest integer less than or equal to pk_2/π and $\mathbf{C}_n$ are constants which are determined by a Fourier series analysis of the initial conditions for $\mathbf{w}$. Initial conditions in any biological context involve a certain stochasticity and so it is inevitable that the Fourier spectrum will contain the whole range of Fourier modes; that is, the $\mathbf{C}_n$ are nonzero. We can therefore assume at this stage that γ is sufficiently large to ensure that allowable wavenumbers exist in the unstable range of k. Before discussing the possible patterns which emerge let us first obtain the corresponding two-dimensional result.

Consider the two-dimensional domain defined by $0 < x < p$, $0 < y < q$ whose rectangular boundary we denote by ∂B. The spatial eigenvalue problem in place of that in (2.36) is now

$$\nabla^2\mathbf{W} + k^2\mathbf{W} = 0, \quad (\mathbf{n}\cdot\nabla)\mathbf{W} = 0 \quad \text{for} \quad (x,y) \text{ on } \partial B \tag{2.41}$$

the eigenfunctions of which are

$$\mathbf{W}_{p,q}(x,y) = \mathbf{C}_{n,m}\cos\frac{n\pi x}{p}\cos\frac{m\pi y}{q}, \quad k^2 = \pi^2\left(\frac{n^2}{p^2} + \frac{m^2}{q^2}\right), \tag{2.42}$$

where n and m are integers. The two-dimensional modes $\mathbf{W}_k(x, y)$ which are linearly unstable are those with wavenumbers k, defined by the last equation, lying within the unstable band of wavenumbers defined in terms of a, b and d by (2.38). We again assume that γ is sufficiently large so that the range of unstable wavenumbers contains at least one possible mode. Now the unstable spatially patterned solution is given by (2.30) with (2.42) as

$$\begin{aligned} \mathbf{w}(x, y, t) &\sim \sum_{n,m} \mathbf{C}_{n,m} e^{\lambda(k^2)t} \cos \frac{n\pi x}{p} \cos \frac{m\pi y}{q}, \\ \gamma L(a, b, d) = k_1^2 < k^2 &= \pi^2 \left(\frac{n^2}{p^2} + \frac{m^2}{q^2} \right) < k_2^2 = \gamma M(a, b, d), \end{aligned} \tag{2.43}$$

where the summation is over all pairs (n, m) which satisfy the inequality, L and M are defined by (2.38) as before and $\lambda(k^2)$ is again the positive solution of (2.23) with the expressions for the derivatives of f and g given by (2.34). As t increases a spatial pattern evolves which is initially made up of the modes in (2.43).

Now consider the type of spatial patterns we might expect from the unstable solutions in (2.40) and (2.43). Suppose first that the domain size, as measured by γ, is such that the range of unstable wavenumbers in (2.38) admits only the wavenumber $n = 1$: the corresponding dispersion relation for λ in terms of the wavelengths $\omega = 2p/n$ is illustrated in Figure 2.7(a). The only unstable mode, from (2.37) is then $\cos(\pi x/p)$ and the growing instability is given by (2.40) as

$$\mathbf{w}(x, t) \sim \mathbf{C}_1 \exp\left[\lambda\left(\frac{\pi^2}{p^2}\right)t\right] \cos \frac{\pi x}{p},$$

where λ is the positive root of the quadratic (2.23) with f_u, f_v, g_u and g_v from (2.34) and with $k^2 = (\pi/p)^2$. Here all other modes decay exponentially with time. We can only determine the $\mathbf{C}_1$ from initial conditions. To get an intuitive understanding for what is going on, let us simply take $\mathbf{C}_1$ as $(\varepsilon, \varepsilon)$ for some small positive ε and consider the morphogen u; that is, from the last equation and the definition of $\mathbf{w}$ from (2.14),

$$u(x, t) \sim u_0 + \varepsilon \exp\left[\lambda\left(\frac{\pi^2}{p^2}\right)t\right] \cos \frac{\pi x}{p}. \tag{2.44}$$

This unstable mode, which is the dominant solution which emerges as t increases, is illustrated in Figure 2.7(b). In other words, this is the pattern predicted by the dispersion relation in Figure 2.7(a).

Clearly if the exponentially growing solution were valid for all time it would imply $u \to \infty$ as $t \to \infty$. For the mechanism (2.32), the kinetics has a confined set, within the positive quadrant, which bounds the solution. So the solution in the last equation must be bounded and lie in the positive quadrant. We hypothesise that this growing solution eventually settles down to a spatial pattern which is similar to the single cosine mode shown in Figure 2.7(b). As mentioned before, singular perturbation analyses in the

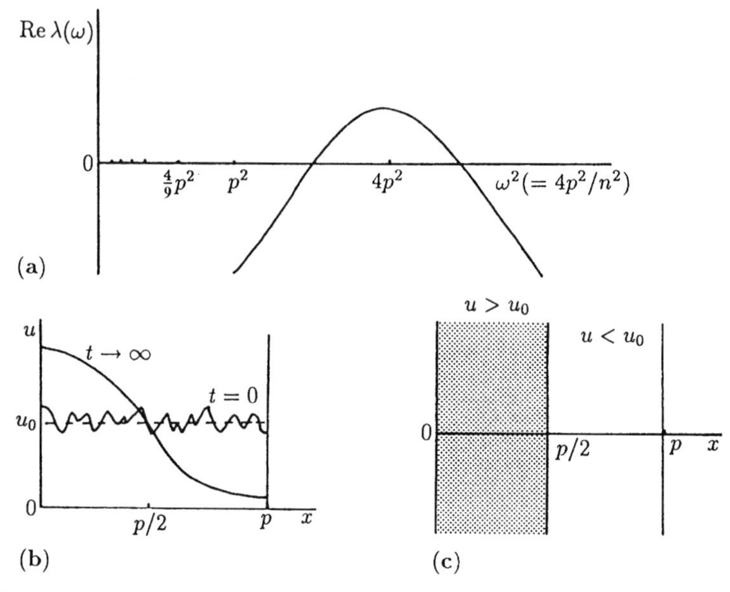

Figure 2.7. (**a**) Typical dispersion relation for the growth factor Re λ as a function of the wavelength ω obtained from a linearization about the steady state. The only mode which is linearly unstable has $n = 1$; all other modes have Re $\lambda < 0$. (**b**) The temporally growing linear mode which eventually evolves from random initial conditions into a finite amplitude spatial pattern such as shown in (**c**), where the shaded area corresponds to a concentration higher than the steady state u_0 and the unshaded area to a concentration lower than the steady state value.

vicinity of the bifurcation in one of the parameters, for example, near the critical domain size for γ, such that a single wavenumber is just unstable, or when the critical diffusion coefficient ratio is near d_c, bear this out as do the many numerical simulations of the full nonlinear equations. Figure 2.7(c) is a useful way of presenting spatial patterned results for reaction diffusion mechanisms—the shaded region represents a concentration above the steady state value while the unshaded region represents concentrations below the steady state value. As we shall see, this simple way of presenting the results is very useful in the application of chemical prepattern theory to patterning problems in developmental biology, where it is postulated that cells differentiate when one of the morphogen concentrations is above (or below) some threshold level.

Let us now suppose that the domain size is doubled, say. With the definition of γ chosen to represent scale this is equivalent to multiplying the original γ by 4 since in the one-dimensional situation $\sqrt{\gamma}$ is proportional to size, that is, the length here, of the domain. This means that the dispersion relation and the unstable range are simply moved along the k^2-axis or along the ω^2-axis. Suppose the original $\gamma = \gamma_1$. The inequalities (2.38) determine the unstable modes as those with wavelengths $\omega(= 2\pi/k)$ determined by (2.39); namely,

$$\frac{4\pi^2}{\gamma_1 L(a,b,d)} > \omega^2 > \frac{4\pi^2}{\gamma_1 M(a,b,d)}. \tag{2.45}$$

Let this be the case illustrated in Figure 2.7(a) and which gives rise to the pattern in Figure 2.7(c). Now let the domain double in size. We consider exactly the same domain as in Figure 2.7 but with an increased γ to $4\gamma_1$. This is equivalent to having the same γ_1 but with a domain 4 times that in Figure 2.7. We choose the former means of representing a change in scale. The equivalent dispersion relation is now illustrated in Figure 2.8(a)—it is just the original one of Figure 2.7(a) moved along so that the wavelength of the excited or unstable mode now has $\omega = p$; that is, $n = 2$. The equivalent spatial pattern is then as in Figure 2.8(b). As we shall see in the applications chapters which follow, it is a

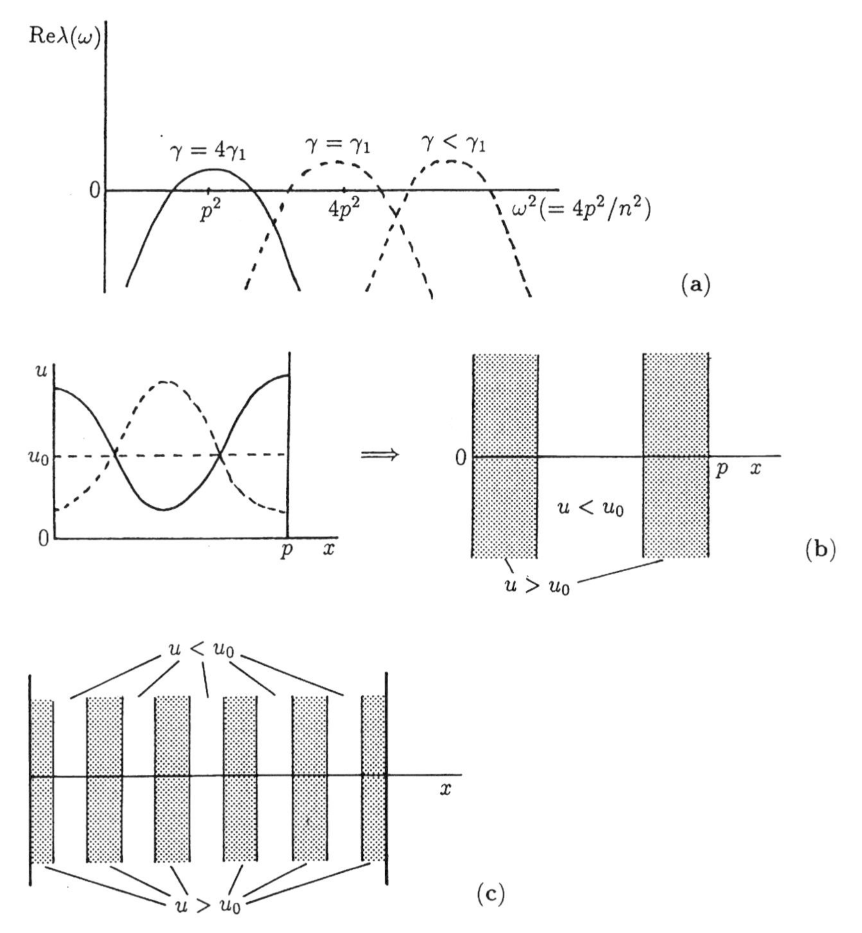

Figure 2.8. (**a**) Dispersion relation Re λ as a function of the wavelength ω when the single mode with $n = 2$ is unstable for a domain size $4\gamma_1$; the dashed curves are those with $\gamma = \gamma_1$ and $\gamma < \gamma_c < \gamma_1$, where γ_c is the critical bifurcation scale value of the domain that will not admit any heterogeneous pattern. (**b**) The spatial pattern in the morphogen u predicted by the dispersion relation in (**a**). The dashed line, the mirror image about $u = u_0$, is also an allowable form of this solution. The initial conditions determine which pattern is obtained. (**c**) The spatial pattern obtained when the domain is sufficiently large to fit in the number of unstable modes equivalent to $n = 10$: the shaded regions represent morphogen levels $u > u_0$, the uniform steady state.

particularly convenient way, when presenting spatial patterned solutions, to incorporate scale solely via a change in γ.

We can thus see with this example how the patterning process works as regards domain size. There is a basic wavelength picked out by the analysis for a given $\gamma = \gamma_1$, in this example that with $n = 1$. As the domain grows it eventually can incorporate the pattern with $n = 2$ and progressively higher modes the larger the domain, as shown in Figure 2.8(c). In the same way if the domain is sufficiently small there is clearly a $\gamma = \gamma_c$ such that the dispersion relation, now moved to the right in Figure 2.8(a), will not even admit the wavelength with $n = 1$. In this case no mode is unstable and so no spatial pattern can be generated. The concept of a critical domain size for the existence of spatial pattern is an important one both in developmental biology, and in spatially dependent ecological models as we show later.

Note in Figure 2.8(b) the two possible solutions for the same parameters and zero flux boundary conditions. Which of these is obtained depends on the bias in the initial conditions. Their existence poses certain conceptual difficulties from a developmental biology point of view within the context of positional information. If cells differentiate when the morphogen concentration is larger than some threshold then the differentiated cell pattern is obviously different for each of the two possible solutions. Development, however, is a sequential process and carries with it its own history so, a previous stage generally cues the next. In the context of reaction diffusion models this implies a bias in the initial conditions towards one of the patterns.

Now consider the two-dimensional problem with a dispersion relation such that the unstable modes are given by (2.43). Here the situation is not so straightforward since for a given γ, representing the *scale*, the actual modes which are unstable now depend on the domain *geometry* as measured by the length p and the width q. Referring to (2.43), first note that if the width is sufficiently small, that is, q is small enough, even the first mode with $m = 1$ lies outside the unstable range. The problem is then equivalent to the above one-dimensional situation. As the width increases, that is, q increases, genuine two-dimensional modes with $n \neq 0$ and $m \neq 0$ become unstable since

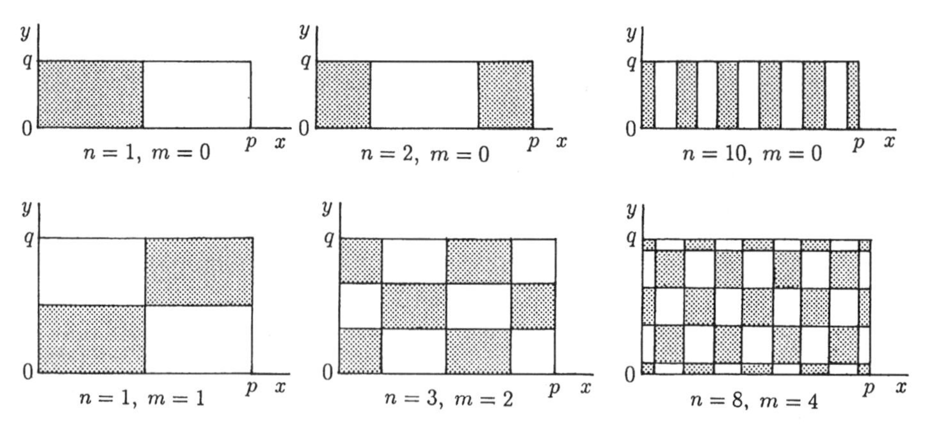

Figure 2.9. Typical two-dimensional spatial patterns indicated by the linearly unstable solution (2.43) when various wavenumbers are in the unstable range. The shaded regions are where $u > u_0$, the uniform steady state.

$\pi^2(n^2/p^2 + m^2/q^2)$ lies in the range of unstable wavenumbers. Figure 2.9 illustrates typical temporally growing spatial patterns indicated by (2.43) with various nonzero n and m.

Regular Planar Tesselation Patterns

The linear patterns illustrated in the last figure arise from the simplest two-dimensional eigenfunctions of (2.41). Less simple domains require the solutions of

$$\nabla^2\psi + k^2\psi = 0, \quad (\mathbf{n}\cdot\nabla)\psi = 0 \quad \text{for} \quad \mathbf{r} \text{ on } \partial B. \tag{2.46}$$

Except for simple geometries the analysis quickly becomes quite complicated. Even for circular domains the eigenvalues have to be determined numerically. There are, however, some elementary solutions for symmetric domains which tesselate the plane, namely, squares, hexagons, rhombi and, by subdivision, triangles; these were found by Christopherson (1940). In other words we can cover the complete plane with, for example, regular hexagonal tiles. (The basic symmetry group of regular polygons are hexagons, squares and rhombi, with, of course, triangles, which are subunits of these.) Hexagonal patterns, as we shall see, are common in many real developmental situations—feather distribution on the skin of birds is just one example (just look at the skin of a plucked chicken). Refer also to Figure 2.11 below where a variety of experimentally obtained patterns is shown. Thus we want solutions ψ where the unit cell, with zero flux conditions on its boundary, is one of the regular tesselations which can cover the plane. That is, we want solutions which are *cell* periodic; here the word 'cell' is, of course, meant as the unit of tesselation.

The solution of (2.46) for a hexagon is

$$\begin{aligned}\psi(x,y) &= \frac{\cos k\left(\frac{\sqrt{3}y}{2} + \frac{x}{2}\right) + \cos k\left(\frac{\sqrt{3}y}{2} - \frac{x}{2}\right) + \cos kx}{3} \\ &= \frac{\cos\left\{kr\sin\left(\theta + \frac{\pi}{6}\right)\right\} + \cos\left\{kr\sin\left(\theta - \frac{\pi}{6}\right)\right\} + \cos\left\{kr\sin\left(\theta - \frac{\pi}{2}\right)\right\}}{3}.\end{aligned} \tag{2.47}$$

From (2.46), a linear equation, ψ is independent to the extent of multiplication by an arbitrary constant: the form chosen here makes $\psi = 1$ at the origin. This solution satisfies zero flux boundary conditions on the hexagonal symmetry boundaries if $k = n\pi, n = \pm1, \pm2, \ldots$. Figure 2.10(a) shows the type of pattern the solution can generate.

The polar coordinate form shows the invariance to hexagonal rotation, that is, invariance to rotation by $\pi/3$, as it must. That is,

$$\psi(r,\theta) = \psi\left(r, \theta + \frac{\pi}{3}\right) = H\psi(r,\theta) = \psi(r,\theta),$$

where H is the hexagonal rotation operator.

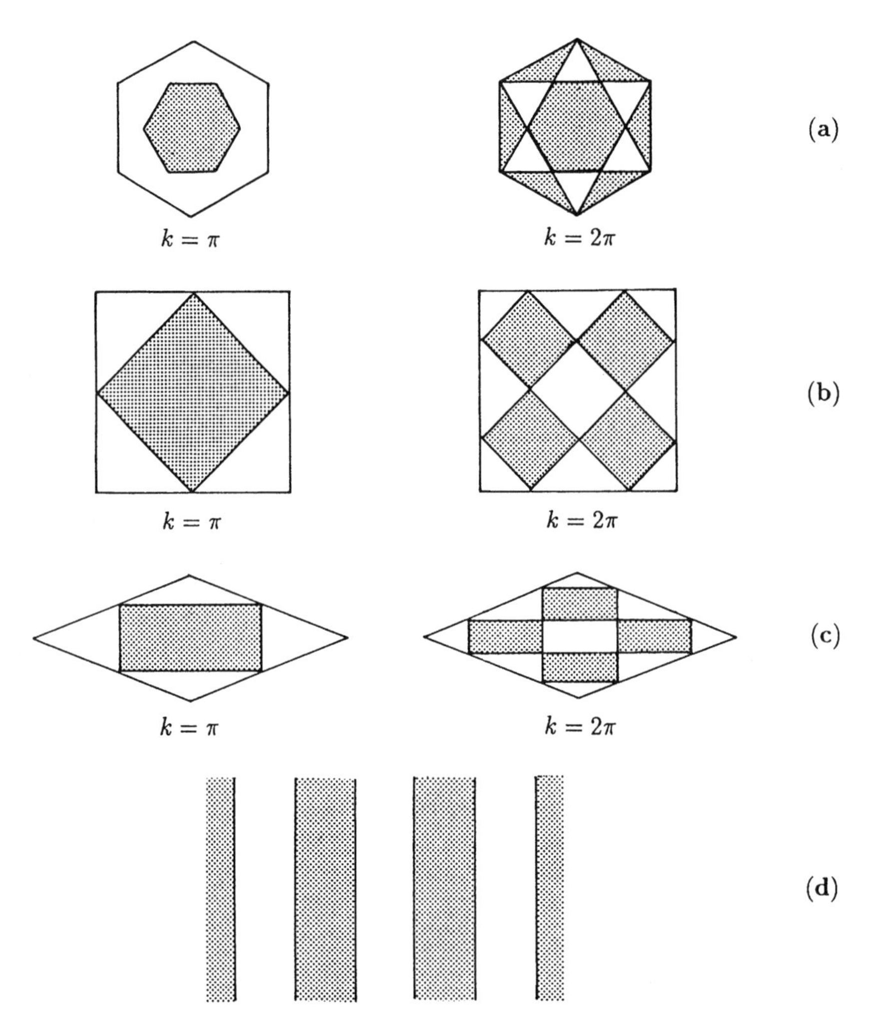

Figure 2.10. (**a**) Patterns which are obtained with the solution (2.47) with $k = \pi$ and $k = 2\pi$. The shaded region is where $\psi > 0$ and the unshaded region where $\psi < 0$. (**b**) Patterns generated by the solution (2.48) for a square tesselation with $k = \pi$ and $k = 2\pi$. (**c**) Rhombic patterns from (2.49) with $k = \pi$ and $k = 2\pi$. (**d**) One-dimensional roll patterns from (2.50).

The solution for the square is

$$\begin{aligned}\psi(x, y) &= \frac{\cos kx + \cos ky}{2} \\ &= \frac{\cos(kr\cos\theta) + \cos(kr\sin\theta)}{2},\end{aligned} \tag{2.48}$$

where $k = \pm 1, \pm 2 \ldots$ and $\psi(0, 0) = 1$. This solution is square rotationally invariant since

$$\psi(r,\theta) = \psi\left(r, \theta + \frac{\pi}{2}\right) = S\psi(r,\theta) = \psi(r,\theta),$$

where S is the square rotational operator. Typical patterns are illustrated in Figure 2.10(b).

The solution for the rhombus is

$$\begin{aligned}\psi(x,y) &= \frac{\cos kx + \cos\{k(x\cos\phi + y\sin\phi)\}}{2} \\ &= \frac{\cos\{kr\cos\theta\} + \cos\{kr\cos(\theta - \phi)\}}{2},\end{aligned} \tag{2.49}$$

where ϕ is the rhombus angle and again $k = \pm 1, \pm 2, \ldots$. This solution is invariant under a rhombic rotation; that is,

$$\psi(r,\theta;\phi) = \psi(r,\theta+\pi;\phi) = R\psi(r,\theta;\phi),$$

where R is the rhombic rotation operator. Illustrative patterns are shown in Figure 2.10(c).

A further cell periodic solution is the one-dimensional version of the square; that is, there is only variation in x. The solutions here are of the form

$$\psi(x,y) = \cos kx, \quad k = n\pi, \quad n = \pm 1, \pm 2, \ldots \tag{2.50}$$

and represent rolls with patterns as in Figure 2.10(d). These, of course, are simply the one-dimensional solutions (2.37).

When the full nonlinear equations are solved numerically with initial conditions taken to be small random perturbations about the steady state, linear theory turns out to be a good predictor of the ultimate steady state in the one-dimensional situation, particularly if the unstable modes have large wavelengths, that is, small wavenumbers. With larger wavenumbers the predictions are less reliable—and even more so with two-dimensional structures. Since the equations we have studied are linear and invariant when multiplied by a constant, we can have equivalent solutions which are simply mirror images in the line $u = u_0$; refer to Figure 2.8(b). Thus the pattern that evolves depends on the initial conditions and the final pattern tends to be the one closest to the initial conditions. There is, in a sense, a basin of attraction for the spatial patterns as regards the initial conditions. Once again near bifurcation situations singular perturbation analysis indicates nonlinear patterns closely related to the linear predictions. In general, however, away from the bifurcation boundaries linear predictions are much less reliable; see the computed patterns exhibited in the next chapter. Except for the simplest patterns, we should really use linear theory for two and three dimensions only as a guide to the wealth of patterns which can be generated by pattern formation mechanisms. Linear theory does, however, determine parameter ranges for pattern generation.

Figure 2.10 shows a selection of regular patterns that can be formed by reaction diffusion equations based on linear theory. Mathematically (and experimentally of course) a key question is which of these will be formed from given initial conditions. If one pattern is formed, variation of which parameters will effect changing to another? To determine which of the various possible patterns—hexagons, rhombi, squares or rolls—will be stable we have to go beyond linear theory and carry out a weakly nonlinear analysis; that is, the parameters are such that they are close to the bifurcation boundary from homogeneity to heterogeneity. When we do such a nonlinear analysis we can determine the conditions on the parameters for stability of these steady state spatially heterogeneous solutions. This has been done for reaction diffusion equations by Ermentrout (1991) and Nagorcka and Mooney (1992) using a multi-scale singular perturbation analysis. Other pattern formation mechanisms, namely, cell-chemotaxis and mechanical mechanisms for pattern formation were studied by Zhu and Murray (1995). The latter compare chemotaxis systems and their patterning potential with reaction diffusion systems. Zhu and Murray (1995) were particularly interested in determining the parameter spaces which give rise to stable stripes, spots, squares and hexagons and their spatial characteristics such as wavelength and so on.

In the case of spots they could also determine which of the tessalation spot arrangement patterns would be stable. They compared the robustness and sensitivity of different models and confirmed the results with extensive numerical simulations of the equations. The analytical technique is well established but the details are fairly complex. Zhu and Murray (1995) show from their numerical study of the equations how the transition takes place from stripes to spots and then to hexagonal patterns and the converse pathway how hexagons become unstable and eventually end up in stripes. The hexagons in effect become elongated and rhombic in character with the spots lining up in lines and eventually fusing; it makes intuitive sense. The analytical procedure near bifurcation is referred to as weakly nonlinear stability analysis, an extensive review of which is given by Wollkind et al. (1994). Generally the form of the interaction kinetics plays a major role in what patterns are obtained. Cubic interactions tend to favour stripes while quadratic interactions tend to produce spots. When different boundary conditions (other than zero flux ones) are used the patterns obtained can be very different and less predictable. Barrio et al. (1999) investigated the effect of these and the role of the nonlinearities in the patterns obtained. From extensive numerical simulations they sugggest that such reaction diffusion mechanisms could play a role in some of the complex patterns observed on fish.

The patterns we have discussed up to now have mainly been regular in the sense that they are stripes, spots, hexagonal patterns and so on. Reaction diffusion systems can generate an enormous range of irregular patterns as we shall see in the following chapter where we discuss a few practical examples. The recent article by Meinhardt (2000: see also other references there) discusses complex patterns and in particular the application of reaction diffusion mechanisms to patterns of gene activation, a subject not treated in this book. He also reviews other important applications not treated here such as branching structures in plant morphology.

In the analyses of the pattern formation potential of reaction diffusion systems here the reactants, or morphogens, must have different diffusion coefficients. In many developmental situations there are often preferred directions in which the diffusion of the same morphogen may have different values in different directions; that is, the diffusion is anisotropic. Although we do not discuss it here, this can have, as we would expect, a marked affect on the patterns formed in a Turing instability of the uniform state (see Exercise 10).

It has been known for a long time, from the 1970's in fact, from many numerical studies that reaction diffusion systems can produce steady state finite amplitude spatial patterns. It is only in the last 10 years, however, that such steady state patterns, sometimes called Turing structures or Turing patterns, have been found experimentally. The experimental breakthrough started in 1989; see Ouyang et al. (1990, 1993), Castets et al. (1990), Ouyang and Swinney (1991), Gunaratne et al. (1994), De Kepper et al. (1994) and other references in these articles. The last two are good reviews to get an overall picture of some of these developments. The latter also describe the complex structures which are obtained when the Turing structures interact with travelling waves; they can be highly complex with such phenomena as spatiotemporal intermittency and spot splitting to form more complex patterns and so on. Ouyang and Swinney (1991) experimentally demonstrate the transition from a uniform state to hexagonal and eventually striped patterns; the transition is similar to that found by Zhu and Murray (1995) for both reaction diffusion and cell-chemotaxis pattern formation mechanisms. Since these early experimental studies, Turing patterns have been found with several quite different reaction systems; the details of the chemistry and experimental arrangements are given in detail in the papers. Figure 2.11 shows chemical Turing patterns obtained experimentally with a chlorite-iodide-malonic acid reaction diffusion system from Gunaratne et al. (1994). Note the small size of the domain and the accurately defined wavelength of the patterns which vary from 0.11 mm to 0.18 mm; these are certainly in the range we would expect of many morphogenetic situations and clearly demonstrate the potential for fine-scale delineation of pattern with reaction diffusion mechanisms and, from the theoretical studies of Zhu and Murray (1994) with other pattern generators. That they are of morphogenetic scale, in the case of the developing chick limb, at the time of the patterning associated with cartilage formation the width of the limb bud is of the order of 2 mm (see the discussion on a limb bud patterning scenario in Chapter 6). Wollkind and Stephenson (2000a,b) give a thorough and comprehensive discussion of the various transitions between patterns, including the black eye pattern shown in Figure 2.11. They specifically study the chlorite-iodide-malonic acid reaction system which was used in the experiments and importantly compare their results with experiment. They also relate these transitions between symmetry breaking structures in the chemical system to similar ones in quite different scientific contexts.

The application of reaction diffusion pattern generation to specific developmental biology problems is often within the context of a prepattern theory whereby cells differentiate according to the level of the morphogen concentration in which they find themselves. If the spatial patterns is quite distinct, as described above or with relatively large gradients, less sensitive tuning is required of the cells in order to carry out their assigned roles than if the pattern variation or the concentration gradients are small. It is perhaps useful therefore to try to get a quantitative measure of spatial heterogeneity, which is meaningful biologically, so as to compare different mechanisms. Another biologically relevant method will be discussed in the next section.

Berding (1987) introduced a 'heterogeneity' function for the spatial patterns generated by reaction diffusion systems with zero flux boundary conditions. Suppose the general mechanism (2.10), in one space variable, is diffusionally unstable and the solutions evolve to spatially inhomogeneous steady state solutions $U(x)$ and $V(x)$ as $t \to \infty$. With the definition of γ in (2.6) proportional to the square of the domain length we can

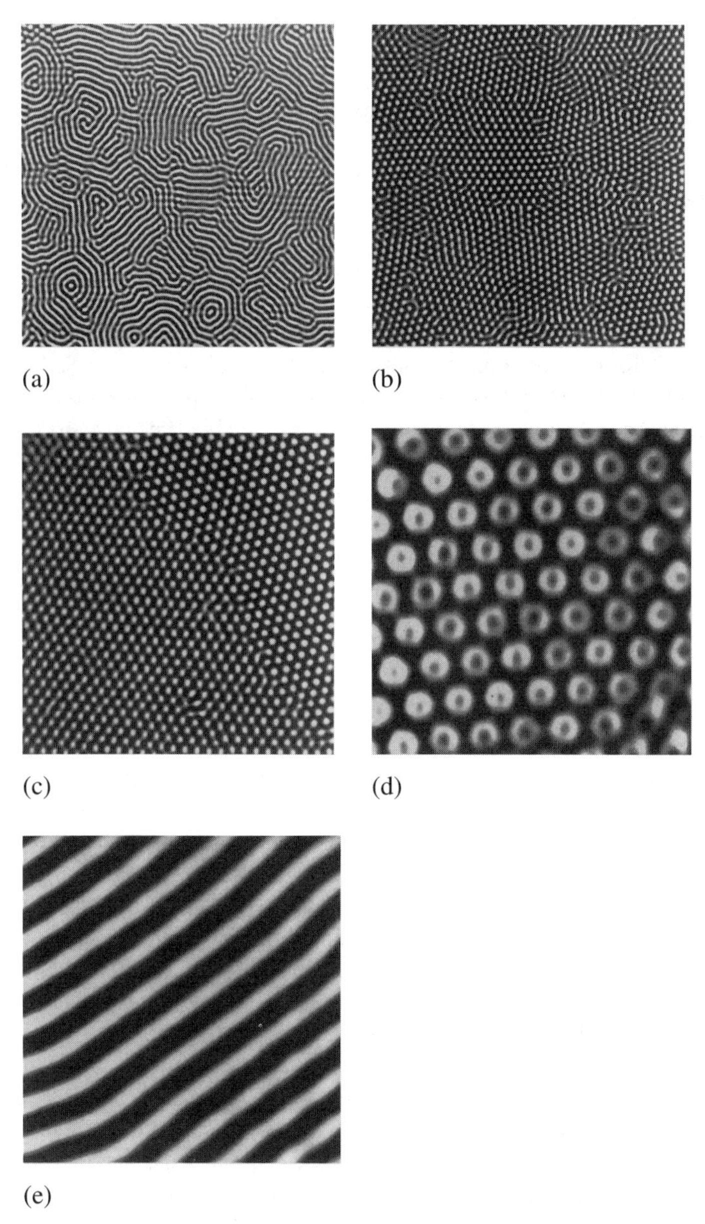

Figure 2.11. Chemical patterns obtained with the reaction diffusion system with the chlorite-iodide-malonic reaction from Gunaratne et al. (1994). The domain size is 6 mm $\times$ 6 mm. (**a**) Multiple domains with stripes with wavelength 0.11 mm. (**b**) This shows multiple domains of hexagonal patterns with different orientations. Here the wavelength is 0.12 mm. (**c**) This again shows hexagonal patterns with a single boundary separating the hexagonal lattices with different orientations: the wavelength here is 0.18 mm. (**d**) This shows a fully developed complex black eye pattern: the domain size is 1.6 mm $\times$ 1.6 mm. (**e**) When the hexagonal pattern in (**d**) becomes unstable it deforms into rhombic structures and the spots line up eventually becoming the striped pattern (in a similar way to the transition patterns in Zhu and Murray 1995): the domain is again 1.6 mm $\times$ 1.6 mm. The experimental details are given in Gunaratne et al. (1994). (Photographs reproduced courtesy of Harry Swinney)

measure domain size by γ and hence take x to be in $(0, 1)$. Then (U, V) satisfy the dimensionless equations

$$\begin{aligned} &U'' + \gamma f(U, V) = 0, \quad dV'' + \gamma g(U, V) = 0, \\ &U'(0) = U'(1) = V'(0) = V'(1) = 0. \end{aligned} \tag{2.51}$$

The non-negative heterogeneity function is defined by

$$H = \int_0^1 (U'^2 + V'^2)\, dx \geq 0, \tag{2.52}$$

which depends only on the parameters of the system and the domain scale γ. H is an 'energy function.' If we now integrate by parts, using the zero flux boundary conditions in (2.51),

$$H = -\int_0^1 (UU'' + VV'')\, dx$$

which, on using (2.51) for U'' and V'', becomes

$$H = \frac{\gamma}{d} \int_0^1 [dUf(U, V) + Vg(U, V)]\, dx. \tag{2.53}$$

If there is no spatial patterning, U and V are simply the uniform steady state solutions of $f(U, V) = g(U, V) = 0$ and so $H = 0$, as also follows, of course, from the definition (2.52).

From (2.53) we see how the scale parameter and diffusion coefficient ratio appear in the definition of heterogeneity. For example, suppose the domain is such that it sustains a single wave for $\gamma = \gamma_1$, in dimensional terms a domain length $L = L_1$ say. If we then double the domain size to $2L_1$ we can fit in two waves and so, intuitively from (2.52), H must increase as there is more heterogeneity. Since $\gamma \propto L^2$, H from (2.53) is simply quadrupled. From an embryological point of view, for example, this means that as the embryo grows we expect more and more structure. An example of this increase in structure in a growing domain is illustrated in Figure 2.18 below. Berding (1987) discusses particular applications and compares specific reaction diffusion mechanisms as regards their potential for heterogeneity.

2.5 Dispersion Relation, Turing Space, Scale and Geometry Effects in Pattern Formation Models

We first note some general properties about the dispersion relation and then exploit it further with the specific case we analysed in the last section. The formation of spatial patterns by any morphogenetic model is principally a nonlinear phenomenon. However, as we noted, a good indication of the patterns in one dimension can be obtained by a simple linear analysis. For spatial patterns to form, we saw that two conditions must hold simultaneously. First, the spatially uniform state must be stable to small perturbations, that is, all $\lambda(k^2)$ in (2.22) have $\text{Re}\,\lambda(k^2 = 0) < 0$, and second, only patterns of a certain spatial extent, that is, patterns within a definite range of wavelengths k, can begin to grow, with $\text{Re}\,\lambda(k^2 \neq 0) > 0$. These conditions are encapsulated in the dispersion relation in either the (λ, k^2) or (λ, ω^2) forms such as in Figure 2.5(b) and Figure 2.7(a). The latter, for example, also says that if the spatial pattern of the disturbances has k^2 large, that is, very small wavelength disturbances, the steady state is again linearly stable. A dispersion relation therefore immediately gives the initial rate of growth or decay of patterns of various sizes. Dispersion relations are obtained from the general evolution equations of the pattern formation mechanism. A general and nontechnical biologically oriented discussion of pattern formation models is given by Oster and Murray (1989).

Since the solutions to the linear eigenfunction equations such as (2.36) are simply sines and cosines, the 'size' of various spatial patterns is measured by the wavelength of the trigonometric functions; for example, $\cos(n\pi x/p)$ has a wavelength $\omega = 2p/n$. So, the search for growing spatial patterns comes down to seeing how many sine or cosine waves can 'fit' into a domain of a given size. The two-dimensional situation is similar, but with more flexibility as to how they fit together.

A very important use of the dispersion relation is that it shows immediately whether patterns can grow, and if so, what the sizes of the patterns are. The curves in Figure 2.5(b) and Figure 2.7(a) are the prototype—no frills—dispersion relation for generating spatial patterns. We show later that other forms are possible and imply different pattern formation scenarios. However, these are less common and much is still not known as to which patterns will evolve from them. The mechanochemical models discussed in detail in Chapter 6 can in fact generate a surprisingly rich spectrum of dispersion relations (see Murray and Oster 1984) most of which cannot be generated by two- or three-species reaction diffusion models.

The prototype dispersion relation has the two essential characteristics mentioned above: (i) the spatially featureless state ($k = 0, \omega = \infty$) is stable; that is, the growth rate of very large wavelength waves is negative, and (ii) there is a small band, or window, of wavelengths which can grow (that is, a finite band of unstable 'modes,' $\cos(n\pi x/L)$, for a finite number of integers n in the case of a finite domain). Of these growing modes, one grows fastest, the one closest to the peak of the dispersion curve. This mode, k_m say, is the solution of

$$\frac{\partial \lambda}{\partial k^2} = 0 \quad \Rightarrow \quad \max[\text{Re}\,\lambda] = \text{Re}\,\lambda(k_m^2).$$

Strictly k_m may not be an allowable mode in a finite domain situation. In this case it is the possible mode closest to the analytically determined k_m.

Thus the dispersion curve shows that while the spatially homogeneous state is stable, the system will amplify patterns of a particular spatial extent, should they be excited by random fluctuations, which are always present in a biological system, or by cues from earlier patterns in development. Generally, one of the model parameters is 'tuned' until the dispersion curve achieves the qualitative shape shown. For example, in Figure 2.5(b) if the diffusion ratio d is less than the critical d_c, $\text{Re}\,\lambda < 0$ for all k^2. As d increases, the curve rises until $d = d_c$ after which it pushes its head above the axis at some wavenumber k_c, that is, wavelength $\omega_c = 2\pi/k_c$, whereupon a cosine wave of that wavelength can start to grow, assuming it is an allowable eigenfunction. This critical wavenumber is given by (2.28) and, with $d = d_c$ from (2.27), we thus have the critical wavelength

$$\omega_c = \frac{2\pi}{k_c} = 2\pi \left\{ \frac{d_c}{\gamma^2(f_u g_v - f_v g_u)} \right\}^{1/4}. \tag{2.54}$$

With the illustrative example (2.32) there are 4 dimensionless parameters: a and b, the kinetics parameters, d, the ratio of diffusion coefficients and γ, the scale parameter. We concentrated on how the dispersion relation varied with d and showed how a bifurcation value d_c existed when the homogeneous steady state became unstable, with the pattern 'size' determined by k_c or ω_c given by the last equation. It is very useful to know the parameter space, involving all the parameters, wherein pattern forms and how we move into this pattern forming domain by varying whatever parameter we choose, or indeed when we vary more than one parameter. Clearly the more parameters there are the more complicated is this corresponding parameter or Turing space. We now determine (analytically) the parameter space for the model (2.32) by extending the parametric method we described in Chapter 3, Volume I, for determining the space in which oscillatory solutions were possible. The technique was developed and applied to several reaction diffusion models by Murray (1982); it is a general procedure applicable to other pattern formation mechanisms. Numerical procedures can also be used such as those developed by Zhu and Murray (1995)

The conditions on the parameters a, b and d for the mechanism (2.32) to generate spatial patterns, if the domain is sufficiently large, are given by (2.35) with γ coming into the picture via the possible unstable modes determined by (2.38). Even though the inequalities (2.35) are probably the simplest realistic set we could have in any reaction diffusion mechanism they are still algebraically quite messy to deal with. With other than extremely simple kinetics it is not possible to carry out a similar analysis analytically. So let us start with the representation of the steady state used in Section 7.4, Volume I, namely, (7.24), with u_0 as the nonnegative parametric variable; that is, v_0 and b are given in terms of a and u_0 from (7.24) (Volume I), or (2.33) above, as

$$v_0 = \frac{u_0 - a}{u_0^2}, \quad b = u_0 - a. \tag{2.55}$$

The inequalities (2.35), which define the conditions on the parameters for spatial patterns to grow, involve, on using the last expressions,

$$\begin{aligned} f_u &= -1 + 2u_0v_0 = 1 - \frac{2a}{u_0}, \quad f_v = u_0^2, \\ g_u &= -2u_0v_0 = -\frac{2(u_0 - a)}{u_0}, \quad g_v = -u_0^2. \end{aligned} \tag{2.56}$$

We now express the conditions for diffusion-driven instability given by (2.31) as inequalities in terms of the parameter u_0; these define boundary curves for domains in parameter space. With the first,

$$\begin{aligned}
f_u + g_v < 0 \quad &\Rightarrow \quad 1 - \frac{2a}{u_0} - u_0^2 < 0 \\
&\Rightarrow \quad a > \frac{u_0(1 - u_0^2)}{2}, \qquad b = u_0 - a > \frac{u_0(1 + u_0^2)}{2}, \\
&\Rightarrow \quad b = \frac{u_0(1 + u_0^2)}{2}
\end{aligned} \tag{2.57}$$

as the boundary where, since we are interested in the boundary curve, the $b = u_0 - a$ comes from the steady state definition (2.55) and where we replace a by its expression from the inequality involving only u_0 and a. These define a domain parametrically in (a, b) space as we let u_0 take all positive values; if the inequality is replaced by an equality sign, (2.57) define the boundary curve parametrically. We now do this with each of the conditions in (2.31).

The second condition of (2.31), using (2.56), requires

$$f_u g_v - f_v g_u > 0 \quad \Rightarrow \quad u_0^2 > 0, \tag{2.58}$$

which is automatically satisfied. The third condition requires

$$\begin{aligned}
df_u + g_v > 0 \Rightarrow \quad & a < \frac{u_0(d - u_0^2)}{2d}, \qquad b = u_0 - a > \frac{u_0(d + u_0^2)}{2d}, \\
\Rightarrow \quad & b = \frac{u_0(d + u_0^2)}{2d}
\end{aligned} \tag{2.59}$$

as the boundary curve.

The fourth condition in (2.31) is a little more complicated. Here

$$\begin{aligned}
&(df_u + g_v)^2 - 4d(f_u g_v - f_v g_u) > 0 \\
&\Rightarrow \quad [u_0(d - u_0^2) - 2da]^2 - 4du_0^4 > 0 \\
&\Rightarrow \quad 4a^2d^2 - 4adu_0(d - u_0^2) + [u_0^2(d - u_0^2)^2 - 4u_0^4 d] > 0
\end{aligned}$$

which, on factorising the left-hand side, implies

$$a < \frac{u_0}{2}\left(1 - \frac{2u_0}{\sqrt{d}} - \frac{u_0^2}{d}\right) \quad \text{or} \quad a > \frac{u_0}{2}\left(1 + \frac{2u_0}{\sqrt{d}} - \frac{u_0^2}{d}\right).$$

Thus this inequality results in *two* boundary curves, namely,

$$\begin{aligned}
a &= \frac{1}{2}u_0\left(1 - \frac{2u_0}{\sqrt{d}} - \frac{u_0^2}{d}\right), \quad b = u_0 - a = \frac{1}{2}u_0\left(1 + \frac{2u_0}{\sqrt{d}} + \frac{u_0^2}{d}\right), \\
a &= \frac{1}{2}u_0\left(1 + \frac{2u_0}{\sqrt{d}} - \frac{u_0^2}{d}\right), \quad b = u_0 - a = \frac{1}{2}u_0\left(1 - \frac{2u_0}{\sqrt{d}} + \frac{u_0^2}{d}\right).
\end{aligned} \tag{2.60}$$

The curves, and the enclosed domains, defined parametrically by (2.57)–(2.60), define the parameter space or *Turing space* (see Murray 1982), where the steady state can be diffusionally driven unstable and hence create spatial patterns. As we noted in Section 2.4 the first and third conditions in (2.35) require f_u and g_v to have opposite signs which require $b > a$ and hence $d > 1$.

It is now a straightforward plotting exercise to obtain the curves defined by (2.57)–(2.60); we simply let u_0 take on a range of positive values and calculate the corresponding a and b for a given d. In general, with inequalities (2.57)–(2.60), five curves are involved in defining the boundaries. Here, as is often the case, several are redundant in that they are covered by one of the others. For example, in the first of (2.60),

$$a < \frac{1}{2}u_0\left(1 - \frac{2u_0}{\sqrt{d}} - \frac{u_0^2}{d}\right) < \frac{1}{2}u_0\left(1 - \frac{u_0^2}{d}\right)$$

since we are considering $u_0 > 0$, so (2.59) is automatically satisfied if we satisfy the first condition in (2.60). Also, since $d > 1$,

$$\frac{1}{2}u_0\left(1 - \frac{u_0^2}{d}\right) > \frac{1}{2}u_0\left(1 - u_0^2\right)$$

so the curve defined by (2.57) lies below the curve defined by (2.59); the former is a lower limiting boundary curve, so a suitable domain is defined if we use the first of (2.60). Furthermore, since

$$\frac{1}{2}u_0\left(1 - \frac{u_0^2}{d}\right) < \frac{1}{2}u_0\left(1 + \frac{2u_0}{\sqrt{d}} - \frac{u_0^2}{d}\right)$$

there can be no domain satisfying (2.59) and the second curve in (2.60).

Finally, therefore, for this mechanism we need only two parametric curves, namely, those defined by (2.57) and the first of (2.60), and the Turing space is determined by

$$\begin{aligned} &a > \frac{1}{2}u_0(1 - u_0^2), \quad b = \frac{1}{2}u_0(1 + u_0^2), \\ &a < \frac{1}{2}u_0\left(1 - \frac{2u_0}{\sqrt{d}} - \frac{u_0^2}{d}\right), \quad b = \frac{1}{2}u_0\left(1 + \frac{2u_0}{\sqrt{d}} + \frac{u_0^2}{d}\right). \end{aligned} \tag{2.61}$$

We know that when $d = 1$ there is no Turing space; that, is there is no domain where spatial patterns can be generated. The curves defined by (2.61) with $d = 1$ contradict each other and hence no Turing space exists. Now let d take on values greater than 1. For a critical d, d_c say, a Turing space starts to grow for $d > d_c$. Specifically $d = d_c = 3 + 2\sqrt{2}$, calculated from (2.61) by determining the d such that both curves give $a = 0$ at $b = 1$ and at this value the two inequalities are no longer contradictory. The space is defined, in fact, by two surfaces in (a, b, d) space. Figure 2.12 shows the cross-

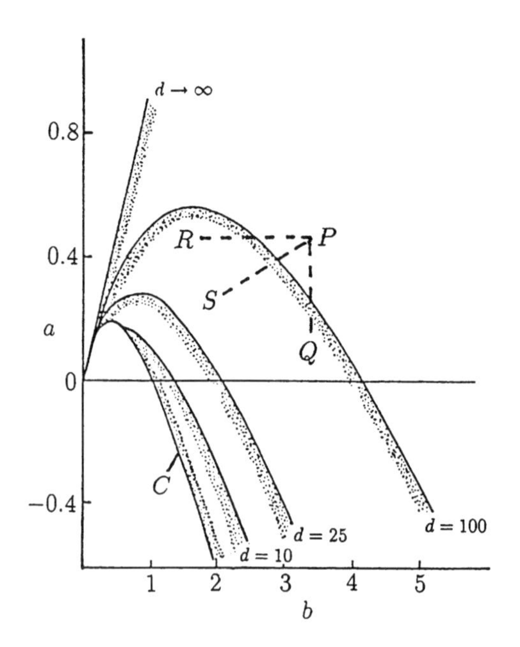

Figure 2.12. Turing space for (2.32), that is, the parameter space where spatial patterns can be generated by the reaction diffusion mechanism (2.32). For example, if $d = 25$ any values for a and b lying within the domain bounded by the curves marked C (that is, $d = 1$) and $d = 25$ will result in diffusion-driven instability. Spatial pattern will evolve if the domain (quantified by γ) is sufficiently large for allowable k^2, defined by (2.38) and (2.43).

sectional regions in (a, b) parameter space where the mechanism (2.32) can generate spatial patterns.

Even if a and b, for a given $d > 1$, lie within the Turing space this does not guarantee that the mechanism will generate spatial patterns, because scale and geometry play a major role. Depending on the size of γ and the actual spatial domain in which the mechanism operates, the unstable eigenfunctions, or modes, may not be allowable solutions. It is here that the detailed form of the dispersion relation comes in again. To be specific consider the one-dimensional finite domain problem defined by (2.36). The eigenvalues, that is, the wavenumbers, $k = n\pi/p,\ n = \pm 1, \pm 2 \ldots$ are *discrete*. So, referring to Figure 2.5(b), unless the dispersion relation includes in its range of unstable modes at least one of these discrete values no structure can develop. We must therefore superimpose on the Turing space in Figure 2.12 another axis representing the scale parameter γ. If γ is included in the parameters of the Turing space it is not necessarily simply connected since, if the dispersion relation, as γ varies, does not include an allowable eigenfunction in its unstable modes, no pattern evolves. Let us consider this aspect and examine the dispersion relation in more detail.

The Turing space involves only dimensionless parameters which are appropriate groupings of the dimensional parameters of the model. The parameters a, b and d in the last figure are, from (2.6),

$$a = \frac{k_1}{k_2}\left(\frac{k_3}{k_2}\right)^{1/2}, \quad b = \frac{k_4}{k_2}\left(\frac{k_3}{k_2}\right)^{1/2}, \quad d = \frac{D_B}{D_A}.$$

Suppose, for example, $d = 100$ and a and b have values associated with P in Figure 2.12; that is, the mechanism is not in a pattern formation mode. There is no unique way to move into the pattern formation domain; we could decrease either a or b so that we arrive at Q or R respectively. In dimensional terms we can reduce a, for example, by appropriately changing k_1, k_2 or k_3—or all of them. Varying other than k_1 will also affect b, so we have to keep track of b as well. If we only varied k_2 the path in the Turing space is qualitatively like that from P to S. If d can vary, which means either of D_A or D_B can vary, we can envelope P in the pattern formation region by simply increasing d. Interpreting the results from a biological point of view, therefore, we see that it is the *orchestration of several effects which produce pattern*, not just one, since we can move into the pattern formation regime by varying one of several parameters. Clearly we can arrive at a specific point in the space by one of several paths. The concept of equivalent effects, via parameter variation, producing the same pattern is an important one in the interpretation and design of relevant experiments associated with any model. It is not a widely appreciated concept in biology. We shall discuss some important biological applications of the practical use of dimensionless groupings in subsequent chapters.

To recap briefly, the dispersion relation for the general reaction diffusion system (2.10) is given by the root $\lambda(k^2)$ of (2.23) with the larger real part. The key to the existence of unstable spatial modes is whether or not the function

$$h(k^2) = dk^4 - \gamma(df_u + g_v)k^2 + \gamma^2(f_u g_v - f_v g_u) \tag{2.62}$$

is negative for a range of $k^2 \neq 0$; see Figure 2.5(a). Remember that the f and g derivatives are evaluated at the steady state (u_0, v_0) where $f(u_0, v_0) = g(u_0, v_0) = 0$, so $h(k^2)$ is a quadratic in k^2 whose coefficients are functions only of the parameters of the kinetics, d the diffusion coefficient ratio and the scale parameter γ. The minimum, $h_{\min}$, at $k = k_m$ corresponds to the λ with the maximum $\operatorname{Re}\lambda$ and hence the mode with the largest growth factor $\exp[\lambda(k_m^2)t]$. From (2.25), or simply from the last equation, $h_{\min}$ is given by

$$\begin{aligned} h_{\min} = h(k_m^2) &= -\frac{1}{4}\gamma^2\left[df_u^2 + \frac{g_v^2}{d} - 2(f_u g_v - 2f_v g_u)\right], \\ k_m^2 &= \gamma\frac{df_u + g_v}{2d}. \end{aligned} \tag{2.63}$$

The bifurcation between spatially stable and unstable modes is when $h_{\min} = 0$. When this holds there is a critical wavenumber k_c which, from (2.28) or again simply derived from (2.62), is when the parameters are such that

$$(df_u + g_v)^2 = 4d(f_u g_v - f_v g_u) \quad \Rightarrow \quad k_c^2 = \gamma\frac{df_u + g_v}{2d}. \tag{2.64}$$

As the parameters move around the Turing space we can achieve the required equality, the first of the equations in (2.64), by letting one or other of the parameters pass through its bifurcation values, all other parameters being kept fixed. In the last section, and in Figure 2.5(b), for example, we chose d as the parameter to vary and for given a and b we evaluated the bifurcation value d_c. In this situation, just at bifurcation, that is, when $h_{\min}(k_c^2) = 0$, a single spatial pattern with wavenumber k_c is driven unstable, or excited, for $d = d_c + \varepsilon$, where $0 < \varepsilon \ll 1$. This critical wavenumber from (2.64) is proportional to $\sqrt{\gamma}$ and so we can vary which spatial pattern is initiated by varying γ. This is called *mode selection* and is important in applications as we shall see later.

In the case of finite domains we can isolate a specific mode to be excited, or driven unstable, by choosing the width of the band of unstable wavenumbers to be narrow and centred round the desired mode. Let us take the parameters in the kinetics to be fixed and let $d = d_c + \varepsilon$, $0 < \varepsilon \ll 1$. We then get from (2.64) the appropriate γ for a specified k as

$$\gamma \approx \frac{2d_c k^2}{d_c f_u + g_v}, \tag{2.65}$$

where the kinetics parameters at bifurcation, sometimes called the *marginal kinetics state*, satisfy the first of (2.64). So, by varying γ we can isolate whatever mode we wish to be excited. Figure 2.13(a) shows a typical situation. Arcuri and Murray (1986) have carried out an extensive Turing space analysis for the much more complex Thomas (1975) mechanism in such a case. Note in Figure 2.13(a) that as γ increases $h_{\min}$ becomes more negative, as is indicated by (2.63).

Suppose now we keep γ and the kinetics parameters fixed, and let d increase from its bifurcation value d_c. From (2.63) $h_{\min} \sim -(d/4)(\gamma f_u)^2$ for d large and so $\lambda \to \infty$ with d. The width of the band of unstable modes has wavenumbers bounded by the zeros k_1 and k_2 of $h(k^2)$ in (2.62). These are given by (2.29), or immediately from (2.62) as

$$\begin{aligned} k_1^2 &= \frac{\gamma}{4d}\left[(df_u + g_v) - \{(df_u + g_v)^2 - 4d(f_u g_v - f_v g_u)\}^{1/2}\right], \\ k_2^2 &= \frac{\gamma}{4d}\left[(df_u + g_v) + \{(df_u + g_v)^2 - 4d(f_u g_v - f_v g_u)\}^{1/2}\right], \end{aligned} \tag{2.66}$$

from which we get

$$k_1^2 \sim 0, \quad k_2^2 \sim \gamma f_u \quad \text{as} \quad d \to \infty. \tag{2.67}$$

So, for a fixed scale there is an upper limit for the unstable mode wavenumber and hence a lower limit for the possible wavelengths of the spatial patterns. Figure 2.13(b) illustrates a typical case for the Thomas (1975) system given by (2.8).

With all kinetics parameters fixed, each parameter pair (d, γ) defines a unique parabola $h(k^2)$ in (2.62), which in turn specifies a set of unstable modes. We can thus consider the (d, γ) plane to be divided into regions where specific modes or a group of modes are diffusively unstable. When there are several unstable modes, because of the form of the dispersion relation, such as in Figure 2.5(b), there is clearly a mode with the largest growth rate since there is a maximum $\operatorname{Re}\lambda$ for some k_m^2 say. From (2.23), the positive eigenvalue $\lambda_+(k^2)$ is given by

$$2\lambda_+(k^2) = \gamma(f_u + g_v) - k^2(1+d) + \{[\gamma(f_u + g_v) - k^2(1+d)]^2 - 4h(k^2)\}^{1/2}$$

which has a maximum for the wavenumber k_m given by

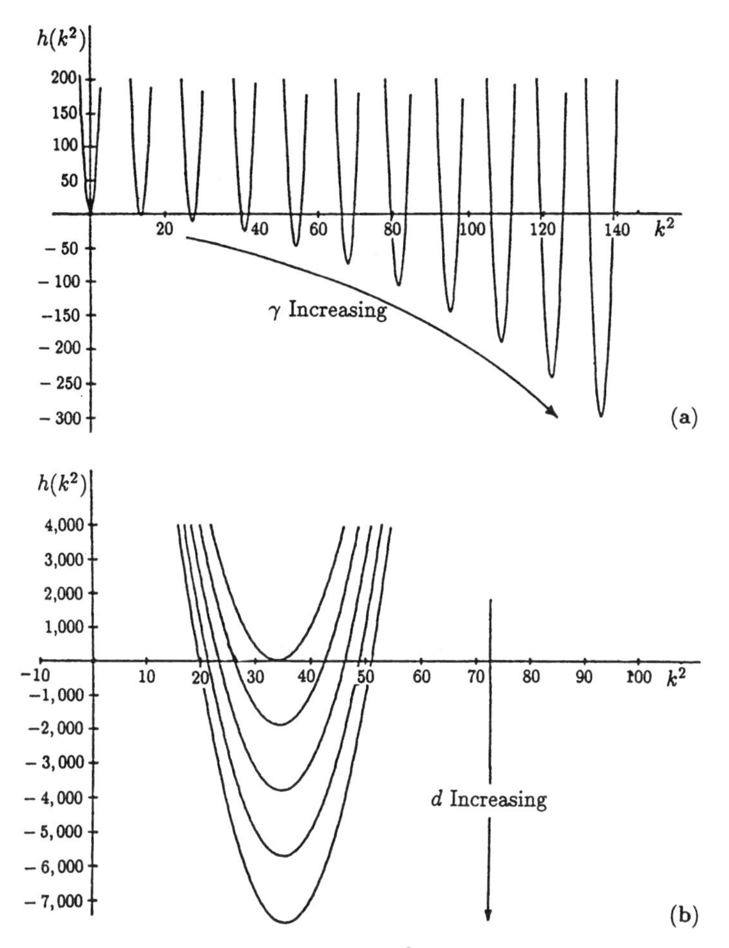

Figure 2.13. (**a**) Isolation of unstable modes (that is, $h(k^2) < 0$ in (2.23)) by setting the diffusion ratio $d = d_c + \varepsilon$, $0 < \varepsilon \ll 1$ and varying the scale γ for the Thomas (1975) kinetics (2.8) with $a = 150$, $b = 100$, $\alpha = 1.5$, $\rho = 13$, $K = 0.05$, $d = 27.03$: the critical $d_c = 27.02$. (**b**) The effect of increasing d with all other parameters fixed as in (**a**). As $d \to \infty$ the range of unstable modes is bounded by $k^2 = 0$ and $k^2 = \gamma f_u$.

$$k^2 = k_m^2 = \frac{\gamma}{d-1}\left\{(d+1)\left[-\frac{f_v g_u}{d}\right]^{1/2} - f_u + g_v\right\}. \tag{2.68}$$

As we have noted the prediction is that the fastest growing k_m-mode will be that which dominates and hence will be the mode which evolves into the steady state nonlinear pattern. This is only a reasonable prediction for the lower modes. The reason is that with the higher modes the interaction caused by the nonlinearities is more complex than when only the simpler modes are linearly unstable. Thus using (2.68) we can map the regions in (d, γ) space where a specific mode, and hence pattern, will evolve; see Arcuri and Murray (1986). Figures 2.14(a) and (b) show the mappings for the Thomas (1975)

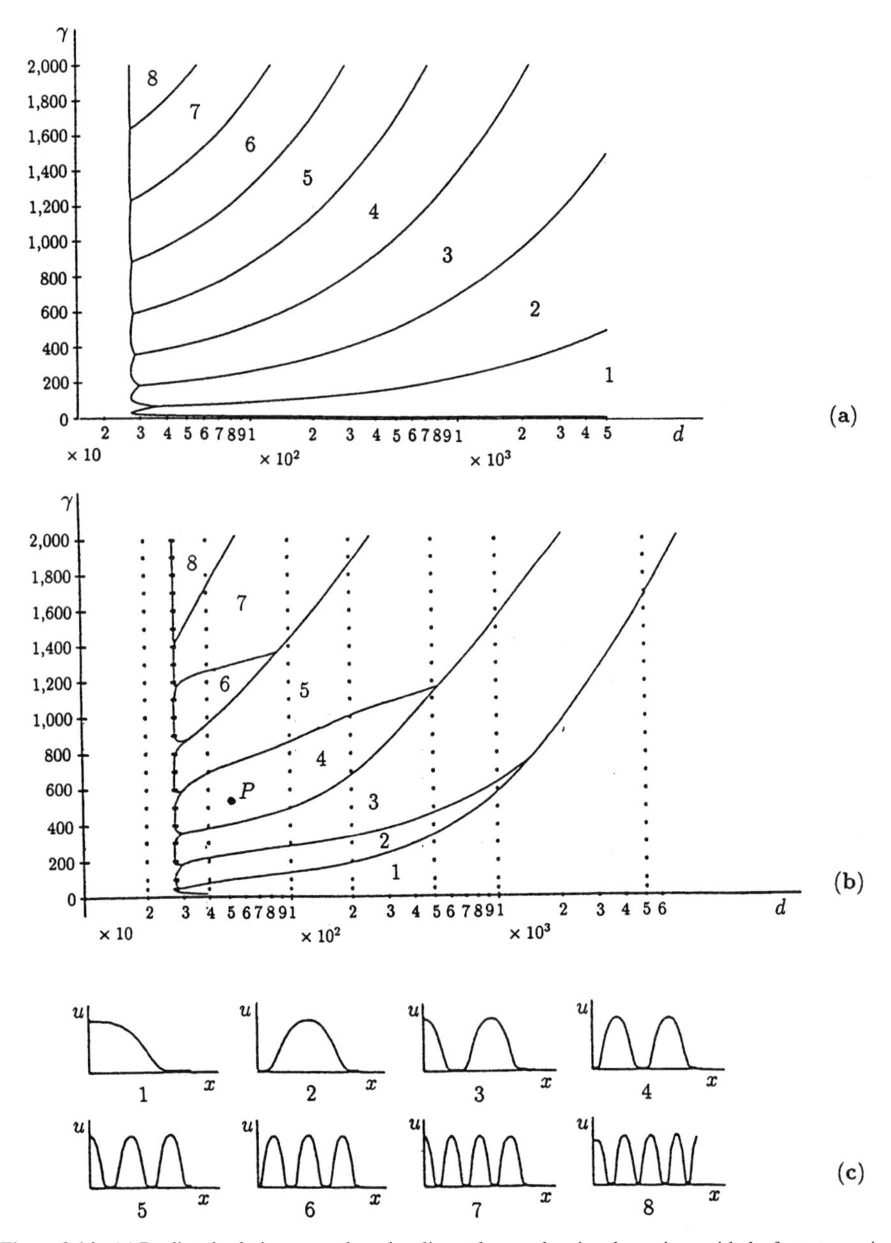

Figure 2.14. (**a**) Predicted solution space, based on linear theory, showing the regions with the fastest growing modes for the Thomas (1975) system (2.8) with parameter values as in Figure 2.13 and zero flux boundary conditions. (**b**) A typical space as evaluated from the numerical simulation of the full nonlinear Thomas system (2.8) with the same parameter values and zero flux boundary conditions as in (**a**). Each (γ, d) point marked with a period represents a specific simulation of the full nonlinear system. (**c**) The corresponding spatial concentration patterns obtained with parameters d and γ in the regions indicated in (**b**). Spatial patterns can be visualised by setting a threshold u^* and shading for $u > u^*$. The first two morphogen distributions, for example, correspond to the first two patterns in Figure 2.9. (From Arcuri and Murray 1986)

system in one space dimension calculated from the linear theory and the full nonlinear system, while Figure 2.14(c) shows the corresponding spatial morphogen patterns indicated by Figure 2.14(b).

An important use of such parameter spaces is the measure of the robustness of the mechanism to random parameter variation. With Figure 2.14(b), for example, suppose the biological conditions result in a (d, γ) parameter pair giving P, say, in the region which evolves to the 4-mode. A key property of any model is how sensitive it is to the inevitable random perturbations which exist in the real world. From Figure 2.14(b) we see what leeway there is if a 4-mode pattern is required in the developmental sequence. This (d, γ) space is but one of the relevant spaces to consider, of course, since any mechanism involves other parameters. So, in assessing robustness, or model sensitivity, we must also take into account the size and shape of the Turing space which involves all of the kinetics parameters. Probably (d, γ) spaces will not be too different qualitatively from one reaction diffusion system to another. What certainly is different, however, is the size and shape of the Turing space, and it is this space which provides another useful criterion for comparing relevant robustness of models. Murray (1982) studied this specific problem and compared various specific reaction diffusion mechanisms with this in mind. He came to certain conclusions as to the more robust mechanisms: both the Thomas (1975) and Schnakenberg (1979) systems, given respectively by (2.7) and (2.8), have relatively large Turing spaces, whereas that of the activator–inhibitor model of Gierer and Meinhardt (1972), given by (2.9) is quite small and implies a considerable sensitivity of pattern to small parameter variation. In the next chapter on specific pattern formation problems in biology we touch on other important aspects of model relevance which are implied by the form of the dispersion relation and the nondimensionalisation used.

The parameter spaces designating areas for specific patterns were all obtained with initial conditions taken to be random perturbations about the uniform steady state. Even in the low modes the polarity can be definitively influenced by biased initial conditions. We can, for example, create a single hump pattern with a single maximum in the centre of the domain or with a single minimum in the centre; see Figure 2.8(b). So even though specific modes can be isolated, initial conditions can strongly influence the polarity. When several of the modes are excitable, and one is naturally dominant from the dispersion relation, we can still influence the ultimate pattern by appropriate initial conditions. If the initial conditions include a mode within the unstable band and whose amplitude is sufficiently large, then this mode can persist through the nonlinear region to dominate the other unstable modes and the final pattern qualitatively often has roughly that wavelength. We discuss this in more detail in the next section. These facts also have highly relevant implications for biological applications.

2.6 Mode Selection and the Dispersion Relation

Consider a typical no frills or simplest dispersion relation giving the growth factor λ as a function of the wavenumber or wavelength ω such as shown in Figure 2.5(b) where a band of wavenumbers is linearly unstable. Let us also suppose the domain is finite so that the spectrum of eigenvalues is discrete. In the last section we saw how geometry and scale played key roles in determining the particular pattern predicted from linear theory, and this was borne out by numerical simulation of the nonlinear system; see also the results presented in the next chapter. We pointed out that initial conditions can play a role in determining, for example, the polarity of a pattern or whether a specific pattern will emerge. If the initial conditions consist of small random perturbations about the uniform steady state then the likely pattern to evolve is that with the largest linear growth. In many developmental problems, however, the trigger for pattern initiation is scale; there are several examples in the following chapter, Chapter 4 in particular, but also in later chapters. In other developmental situations a perturbation from the uniform steady state is initiated at one end of the spatial domain and the spatial pattern develops from there, eventually spreading throughout the whole domain. The specific pattern that evolves for a given mechanism therefore can depend critically on how the instability is initiated. In this section we investigate this further so as to suggest what patterns will evolve from which initial conditions, for given dispersion relations, as key parameters pass through bifurcation values. The problem of which pattern will evolve, namely, mode selection, is a constantly recurring one. The following discussion, although motivated by reaction diffusion pattern generators, is quite general and applies to any pattern formation model which produces a similar type of dispersion relation.

Consider a basic dispersion relation $\lambda(\omega^2)$ where the wavelength $\omega = 2\pi/k$ with k the wavenumber, such as in Figure 2.15(a). Now take a one-dimensional domain and consider in turn the three possible ways of initiating pattern as shown in Figures 2.15(b), (c) and (d).

Consider first the case in Figure 2.15(b). Here the initial perturbation has all modes present in its expansion in terms of the eigenfunctions and so all modes in the unstable band of wavelengths in Figure 2.15(a) are stimulated. The mode with the maximum λ, ω_2, is the one with the fastest growth and it ultimately dominates. The steady state inhomogeneous pattern that persists is then that with wavelength ω_2.

In Figure 2.15(c) we envisage the domain to be growing at a rate that is slow compared with the time to generate spatial pattern. Later in this section we describe a caricature system where growth is not small. In Chapter 4 we go into the important effect of growth on pattern in more detail; the interaction is crucial there. For small $L(t)$ the domain is such that it cannot contain any wave with wavelengths in the unstable band. When it reaches L_c, the critical domain size for pattern, it can sustain the smallest wavelength pattern, namely, that with wavelength ω_1. In the time it takes $L(t)$ to grow sufficiently to allow growth of the next wavenumber, that with wavelength ω_1 is sufficiently established to dominate the nonlinear stage. So the final pattern that emerges is that with the base wavelength ω_1.

Travelling Wave Initiation of Pattern

Consider the situation, as in Figure 2.15(d), where the pattern is initiated at one end of the domain; what happens here is more subtle. We expect the final pattern to have a wavelength somewhere within the unstable band predicted by the dispersion relation. To see how to calculate the wavenumber in general let us start with an infinite one-dimensional domain and a general linear system

$$\mathcal{J}\mathbf{w} = 0, \quad \mathbf{w}(x,t) \propto \exp(ikx + \lambda t) \quad \Rightarrow \quad \lambda = \lambda(k), \tag{2.69}$$

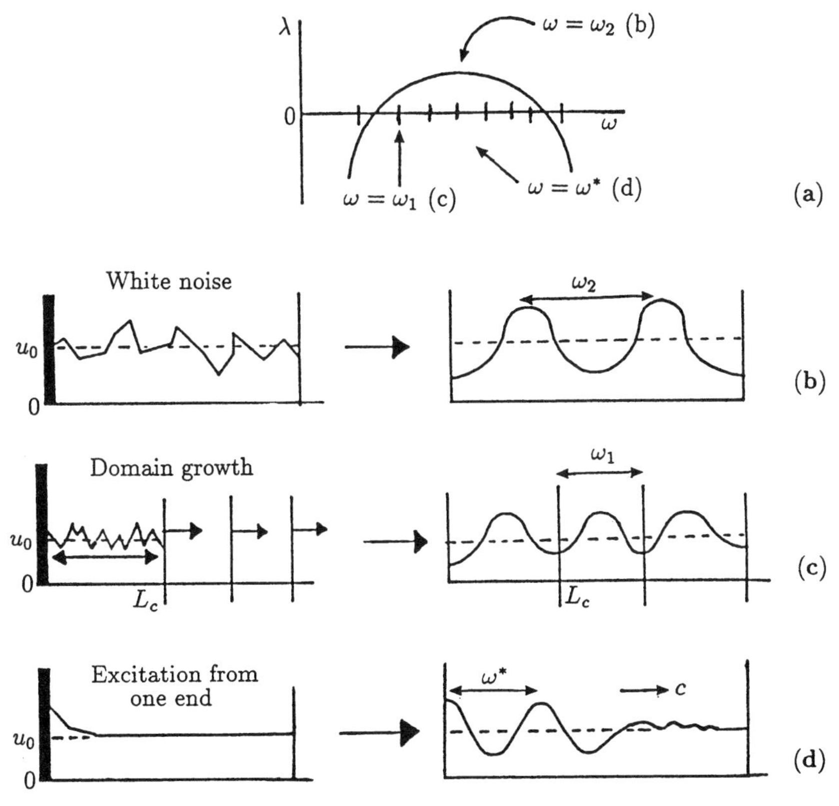

Figure 2.15. (**a**) Typical basic dispersion relation giving the growth coefficient λ as a function of the wavelength ω of the spatial pattern. (**b**) Here the initial disturbance is a random perturbation (white noise) about the uniform steady state u_0. The pattern which evolves corresponds to ω_2 in (**a**), the mode with the largest growth rate. (**c**) Pattern evolution in a growing domain. The first unstable mode to be excited, ω_1, remains dominant. (**d**) Here the initial disturbance is at one end and it lays down a pattern as the disturbance moves through the domain. The pattern which evolves has a wavelength ω^* somewhere within the band of unstable wavelengths.

where $\mathcal{J}$ is a linear operator such as that associated with the linear form of reaction diffusion equations. For (2.20), for example, $\mathcal{J} = (\partial/\partial t) - \gamma A - D\nabla^2)$ and the dispersion relation $\lambda(k)$ is like that in Figure 2.5(b) or in the last figure, Figure 2.16(a), with ω replaced by k; in other words the classic form. The general solution $\mathbf{w}$ of the linear system in (2.69) is

$$\mathbf{w}(x, t) = \int \mathbf{A}(k) \exp[ikx + \lambda(k)t]\, dk, \tag{2.70}$$

where the $\mathbf{A}(k)$ are determined by a Fourier transform of the initial conditions $\mathbf{w}(x, 0)$. Since we are concerned with the final structure and not the transients we do not need to evaluate $\mathbf{A}(k)$ here.

Suppose the initial conditions $\mathbf{w}(x, 0)$ are confined to a small finite domain around $x = 0$ and the pattern propagates out from this region. We are interested in the wavelike generation of pattern as shown in the second figure in Figure 2.15(d). This means that we should look at the form of the solution well away from the origin. In other words, we should focus our attention on the asymptotic form of the solution for x and t large but such that x/t is $O(1)$, which means we move with a velocity $c = x/t$ and so are in the vicinity of the 'front,' roughly where the arrow (with c at the end) is in the second figure in Figure 2.15(d). We write (2.70) in the form

$$\mathbf{w}(x,t) = \int \mathbf{A}(k)\exp[\sigma(k)t]\,dk, \quad \sigma(k) = ikc + \lambda(k), \quad c = \frac{x}{t}. \tag{2.71}$$

The asymptotic evaluation of this integral for $t \to \infty$ is given by analytically continuing the integrand into the complex k-plane and using the method of steepest descents (see Murray's 1984 book, Chapter 3) which gives

$$\mathbf{w}(x,t) \sim \mathbf{J}(k_0)\left[\frac{2\pi}{t|\,\sigma''(k_0)\,|}\right]^{1/2} \exp\{t[ik_0c + \lambda(k_0)]\},$$

where $\mathbf{J}$ is a constant and k_0 (now complex) is given by

$$\sigma'(k_0) = ic + \lambda'(k_0) = 0. \tag{2.72}$$

The asymptotic form of the solution is thus

$$\mathbf{w}(x,t) \sim \frac{\mathbf{K}}{t^{1/2}} \exp\{t[ick_0 + \lambda(k_0)]\}, \tag{2.73}$$

where $\mathbf{K}$ is a constant.

For large t the wave 'front' is roughly the point between the pattern forming tail and the leading edge which initiates the disturbances, that is, where $\mathbf{w}$ neither grows nor decays. This is thus the point where

$$\text{Re}\,[ick_0 + \lambda(k_0)] \approx 0. \tag{2.74}$$

At the 'front' the wavenumber is $\text{Re}\,k_0$ and the solution frequency of oscillation ω is

$$\omega = \text{Im}\,[ick_0 + \lambda(k_0)].$$

Denote by k^* the wavenumber of the pattern laid down behind the 'front.' We now assume there is conservation of nodes across the 'front' which implies

$$k^*c = \omega = \text{Im}\,[ick_0 + \lambda(k_0)]. \tag{2.75}$$

The three equations (2.72), (2.74) and (2.75) now determine k_0 and the quantities we are interested in, namely, c and k^*, which are respectively the speed at which the pattern is laid down and the steady state pattern wavenumber. Because of the complex variables this is not as simple as it might appear. Myerscough and Murray (1992), for the case of a cell-chemotaxis system (see also Chapter 4), used the technique and developed a caricature dispersion relation to solve the three equations analytically. They compared their analytical results with the exact numerical simulations; the comparison was good and, of course, qualitatively useful since analytical results were obtained. This technique has also been used (numerically) by Dee and Langer (1983) for a reaction diffusion mechanism.

Dynamics of Pattern Formation in Growing Domains

The time evolution of patterns in growing domains can be quite complex, particularly if the domain growth is comparable with the generation time of the spatial pattern and there are two or more space dimensions. The form of the dispersion relation as the scale γ increases can have highly pertinent biological implications as we shall see in Chapter 6 when we consider, as one example, cartilage formation in the developing limb. Here we introduce the phenomenon and discuss some of the implications of two specific classes of dispersion relation behaviour as γ increases with time.

The form of reaction diffusion equations in growing domains has to be derived carefully and this is done in Chapter 4 which is mainly concerned with patterning problems and the effect of growth on the patterns formed. Here we consider only a caricature form to demonstrate certain time-dependent effects of growing domains. With a simplified caricature we should not expect to capture all of the possible sequential spatial patterns and this is indeed the case. Crampin et al. (1999), in a comprehensive analytical and numerical examination of reaction diffusion systems in growing one-dimensional domains, categorise the sequential patterns which evolve. They consider different growth forms. They use a self-similarity argument to predict frequency doubling in the case of exponential growth and show how growth may be a mechanism for increasing pattern robustness. Kulesa et al. (1996a,b) (see also Chapter 4) show that the sequential positioning of teeth primordia (precursors of teeth) is intimately related to jaw growth, which is experimentally determined. The correct sequence of primordial appearance depends crucially on the interaction of the pattern formation process and domain growth dynamics. A somewhat different approach to including domain change was used by Murray and Myerscough (1992) in their study of snake patterns which we discuss in some detail in Chapter 4: they looked at bifurcations from solutions of the steady state equations (cell-chemotaxis equations in this case).

In Figure 2.8(a) we saw that as the scale γ increased the dispersion curve was moved along the axis where it successively excited modes with smaller wavelengths. Figure 2.16(a) is a repeat example of this behaviour. Figure 2.16(b) is another possible behaviour of a dispersion relation as the scale γ increases. They imply different pattern generation scenarios for growing domains.

Consider first the situation in Figure 2.16(a). Here for $\gamma = \gamma_1$ the mode with wavelength ω_1 is excited and starts to grow. As the domain increases we see that for $\gamma = \gamma_2$ no mode lies within the unstable band and so the pattern decays to the spatially uniform steady state. With further increase in scale, to $\gamma = \gamma_3$ say, we see that a pattern with wavelength ω_2 is created. So the pattern formation is effectively a discrete process with successively more structure created as γ increases but with each increase in structure interspersed with a regime of spatial homogeneity. Figure 2.17(a) illustrates the sequence of events as γ increases in the way we have just described.

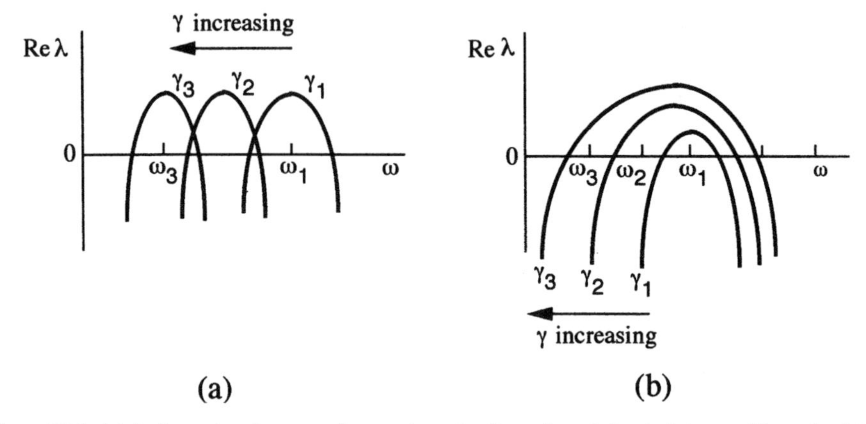

Figure 2.16. (**a**) As the scale γ increases from γ_1 to γ_3 the dispersion relation isolates specific modes interspersed with gaps during which no pattern can form. (**b**) Here as γ increases the number of unstable modes increases: the mode with maximum growth varies with γ. Unstable modes exist for all $\gamma \geq \gamma_1$.

Consider now the behaviour implied by the dispersion relation dependence on scale implied by Figure 2.16(b). Here the effect of scale is simply to increase the band of unstable modes. The dominant mode changes with γ so there is a continuous evolution from one mode, dominant for $\gamma = \gamma_1$ say, to another mode as it becomes dominant for $\gamma = \gamma_3$ say. This dynamic development of pattern is illustrated in Figure 2.17(b). We shall see later in Section 6.6 how the implications of Figures 2.16 and 2.17 have a direct bearing on how cartilage patterns form in the developing limb.

When comparing different models with experiments it is not always possible to choose a given time as regards pattern generation to carry out the experiments since it is not generally known exactly when the pattern formation takes place in embryogenesis.

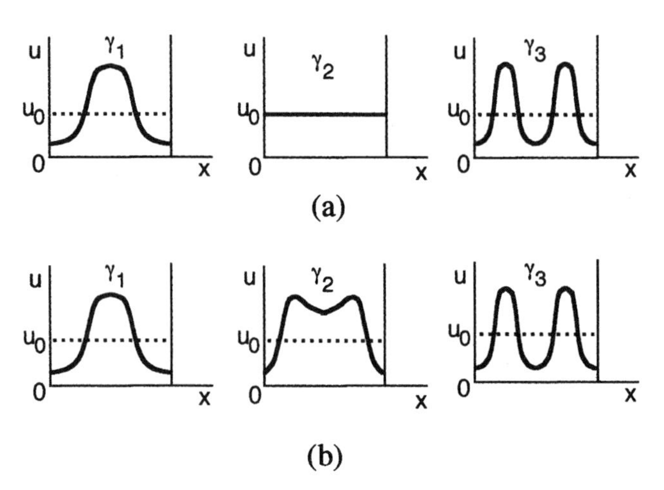

Figure 2.17. (**a**) Development of spatial patterns with a dispersion relation dependence on scale, via γ, as shown in Figure 2.16(a). (**b**) Sequential development of pattern as γ varies according to the dispersion relation in Figure 2.16(b).

When it is possible, then similarity of pattern is a necessary first step in comparison with theory. When it is not possible, the dynamic form of the pattern can be important and can be the key step in deciding which mechanism is the more appropriate. We shall recall these comments later in Chapters 4 and 6.

A computed example of dynamic pattern formation as the scale γ is increased is shown in Figure 2.18.

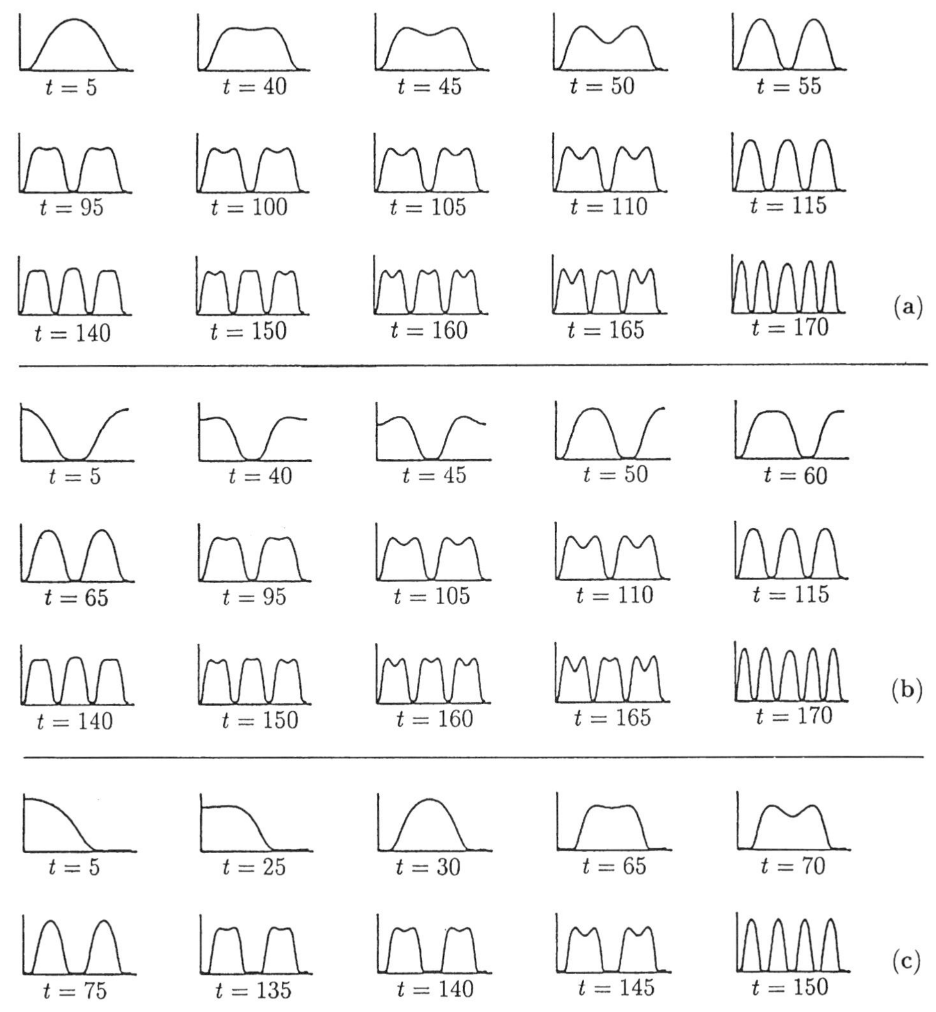

Figure 2.18. Sequence of one-dimensional spatial patterns obtained numerically with the mechanism (2.10) and kinetics given by the second of (2.8). Zero flux boundary conditions were used and the domain growth is incorporated in the scale parameter $\gamma(t) = s + 0.1t^2$ with s fixed. This gives a caricature for the reaction diffusion system for a linear rate of domain growth since $\gamma \propto (length)^2$. Parameter values for the kinetics are as in Figure 2.13(a) except for d. (**a**) and (**b**) have $d = 30$ ($d_c \approx 27$), $s = 100$ and two different sets of initial random perturbations. Note how the two sets of patterns converge as time t increases. (**c**) has $d = 60$, $s = 50$. As d increases more modes are missed in the pattern sequence and there is a distinct tendency towards frequency doubling. (After Arcuri and Murray 1986)

In these simulations the mechanism's pattern generation time is smaller than a representative growth time since the sequence of patterns clearly forms before breaking up to initiate the subsequent pattern. This is an example of a dispersion relation behaviour like that in Figure 2.16(b); that is, there is no regime of spatial homogeneity. The tendency to period doubling indicated by Figure 2.18(c) is interesting and as yet unexplained. Arcuri and Murray (1986) consider this and other aspects of pattern formation in growing domains. It must be kept in mind that the latter study is a caricature reaction diffusion system in a growing domain; refer to Chapter 4 for the exact formulation with exponential domain growth and Crampin et al. (1999) for a comprehensive discussion.

2.7 Pattern Generation with Single-Species Models: Spatial Heterogeneity with the Spruce Budworm Model

We saw above that if the domain size is not large enough, that is, γ is too small, reaction diffusion models with zero flux boundary conditions cannot generate spatial patterns. Zero flux conditions imply that the reaction diffusion domain is isolated from the external environment. We now consider different boundary conditions which take into account the influence of the region exterior to the reaction diffusion domain. To be specific, consider the single reaction diffusion equation in the form

$$u_t = f(u) + D\nabla^2 u, \tag{2.76}$$

and think of the model in an ecological setting; that is, u denotes the population density of a species. Here $f(u)$ is the species' dynamics and so we assume $f(0) = 0$, $f'(0) \neq 0$, $f(u_i) = 0$ for $i = 1$ if there is only one (positive) steady state or $i = 1, 2, 3$ if there are three. Later we shall consider the population dynamics $f(u)$ to be those of the spruce budworm, which we studied in detail in Chapter 1, Volume I, Section 1.2 and which has three steady states as in Figure 1.5(b) (Volume I). The diffusion coefficient D is a measure of the dispersal efficiency of the relevant species.

We consider in the first instance the one-dimensional problem for a domain $x \in (0, L)$, the exterior of which is completely hostile to the species. This means that on the domain boundaries $u = 0$. The mathematical problem we consider is then

$$\begin{aligned} &u_t = f(u) + Du_{xx}, \\ &u(0,t) = 0 = u(L,t), \quad u(x,0) = u_0(x), \\ &f(0) = 0, \quad f'(0) > 0, \quad f(u_2) = 0, \quad f'(u_2) > 0, \\ &f(u_i) = 0, \quad f'(u_i) < 0, \quad i = 1, 3, \end{aligned} \tag{2.77}$$

where u_0 is the initial population distribution. The question we want to answer is whether or not such a model can sustain spatial patterns.

In the spatially homogeneous situation $u = 0$ and $u = u_2$ are unstable and u_1 and u_3 are stable steady states. In the absence of diffusion the dynamics imply that u tends to one or other of the stable steady states and which it is depends on the initial conditions. In the spatial situation, therefore, we would expect $u(x,t)$ to try to grow from $u = 0$ except at the boundaries. Because $u_x \neq 0$ at the boundaries the effect of diffusion implies that there is a flux of u out of the domain $(0, L)$. So, for u small there are two competing effects, the growth from the dynamics and the loss from the boundaries. As a first step we examine the linear problem obtained by linearising about $u = 0$. The relevant formulation is, from (2.77),

$$\begin{aligned} u_t &= f'(0)u + Du_{xx}, \\ u(0,t) &= u(L,t) = 0, \quad u(x,0) = u_0(x). \end{aligned} \tag{2.78}$$

We look for solutions in the form

$$u(x,t) = \sum_n a_n e^{\lambda t} \sin(n\pi x/L),$$

which by inspection satisfy the boundary conditions at $x = 0, L$. Substitution of this into (2.78) and equating coefficients of $\sin(n\pi x/L)$ determines λ as $\lambda = [f'(0) - D(n\pi/L)^2]$ and so the solution is given by

$$u(x,t) = \sum_n a_n \exp\left\{\left[f'(0) - D\left(\frac{n\pi}{L}\right)^2\right]t\right\} \sin\frac{n\pi x}{L}, \tag{2.79}$$

where the a_n are determined by a Fourier series expansion of the initial conditions $u_0(x)$. We do not need the a_n in this analysis. From (2.79) we see that the dominant mode in the expression for u is that with the largest λ, namely, that with $n = 1$, since

$$\exp\left[f'(0) - D\left(\frac{n\pi}{L}\right)^2\right]t < \exp\left[f'(0) - D\left(\frac{\pi}{L}\right)^2\right]t, \quad \text{for all} \quad n \geq 2.$$

So, if the dominant mode tends to zero as $t \to \infty$, so then do all the rest. We thus get as our condition for the linear stability of $u = 0$,

$$f'(0) - D\left(\frac{\pi}{L}\right)^2 < 0 \quad \Rightarrow \quad L < L_c = \pi\left[\frac{D}{f'(0)}\right]^{1/2}. \tag{2.80}$$

In dimensional terms D has units cm^2s^{-1} (or $\text{km}^2\text{yr}^{-1}$ or whatever scale we are interested in) and $f'(0)$ has units s^{-1} since it is the linear birth rate (for u small $f(u) \approx f'(0)u$) which together give L_c in centimetres. Thus if the domain size L is less than the critical size L_c, $u \to 0$ as $t \to \infty$ and no spatial structure evolves. The larger the diffusion coefficient the larger is the critical domain size; this is in keeping with the observation that as D increases so also does the flux out of the region.

The scenario for spatial structure in a growing domain is that as the domain grows and L just passes the bifurcation length L_c, $u = 0$ becomes unstable and the first mode

$$a_1 \exp\left[f'(0) - D\left(\frac{\pi}{L}\right)^2 t\right] \sin\frac{\pi x}{L}$$

starts to grow with time. Eventually the nonlinear effects come into play and $u(x,t)$ tends to a steady state spatially inhomogeneous solution $U(x)$, which, from (2.77), is determined by

$$DU'' + f(U) = 0, \quad U(0) = U(L) = 0, \tag{2.81}$$

where the prime denotes differentiation with respect to x. Because $f(U)$ is nonlinear we cannot, in general, get an explicit solution for U.

From the spatial symmetry in (2.77) and (2.81)—setting $x \to -x$ leaves the equations unchanged—we expect the solutions to be symmetric in x about the midpoint $x = L/2$. Since $u = 0$ at the boundaries we assume the midpoint is the maximum, u_m say, where $U' = 0$; it is helpful now to refer to Figure 2.19(a). If we multiply (2.81) by U' and integrate with respect to x from 0 to L we get

$$\frac{1}{2}DU'^2 + F(U) = F(u_m), \quad F(U) = \int_0^U f(s)\,ds \tag{2.82}$$

since $U = u_m$ when $U' = 0$. It is convenient to change the origin to $L/2$ so that $U'(0) = 0$ and $U(0) = u_m$; that is, set $x \to x - L/2$. Then

$$\left(\frac{D}{2}\right)^{1/2} \frac{dU}{dx} = [F(u_m) - F(U)]^{1/2}$$

which integrates to give

$$|\,x\,| = \left(\frac{D}{2}\right)^{1/2} \int_{U(x)}^{u_m} [F(u_m) - F(w)]^{-1/2} dw, \tag{2.83}$$

which gives the solution $U(x)$ implicitly; typical solutions are illustrated schematically in Figure 2.19(b). The boundary conditions $u = 0$ at $x = \pm L/2$ and the last equation give

$$L = (2D)^{1/2} \int_0^{u_m} [F(u_m) - F(w)]^{-1/2} dw \quad \Rightarrow \quad u_m = u_m(L). \tag{2.84}$$

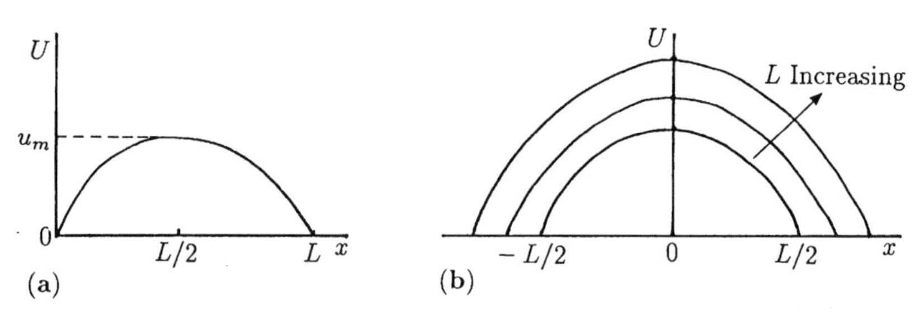

Figure 2.19. (**a**) Typical steady state pattern in the population u governed by (2.77) when the domain length $L > L_c$, the critical size for instability in the zero steady state. Note the symmetry about $L/2$. (**b**) Schematic steady state solution with the origin at the symmetry point where $u = u_m$ and $u_x = 0$.

We thus obtain, implicitly, u_m as a function of L. The actual determination of the dependence of u_m on L has to be carried out numerically. Note the singularity in the integrand when $w = u_m$, but because of the square root it is integrable. Typically u_m increases with L as illustrated in Figure 2.19(b).

Spatial Patterning of the Spruce Budworm

Now consider the model for the spruce budworm, the dynamics for which we derived in Chapter 1, Volume I. Here, using (1.8) for $f(u)$, (2.77) becomes

$$u_t = ru\left(1 - \frac{u}{q}\right) - \frac{u^2}{1+u^2} + Du_{xx} = f(u) + Du_{xx}, \tag{2.85}$$

where the positive parameters r and q relate to the dimensionless quantities associated with the dimensional parameters in the model defined by (1.7) in Chapter 1, Volume I; q is proportional to the carrying capacity and r is directly proportional to the linear birth rate and inversely proportional to the intensity of predation. The population dynamics $f(u)$ is sketched in Figure 2.20(a) when the parameters are in the parameter domain giving three positive steady states u_1, u_2 and u_3, the first and third being linearly stable and the second unstable. With $F(u)$ defined by (2.82) and substituted into (2.84) we have u_m as a function of the domain size L. This was evaluated numerically by Ludwig et al. (1979) the form of which is shown in Figure 2.20(b); there is another critical length, L_0 say, such that for $L > L_0$ more than one solution exists. We analyse this phenomenon below.

From an ecological viewpoint we would like to know the critical domain size L_0 when the maximum population can be in the outbreak regime; that is, $u_m > u_2$ in Figure 2.20(a). This is determined from numerical integration of (2.84) and is shown in Figure 2.20(b). When $L > L_0$ we see from Figure 2.20(b) that there are three possible solutions with different u_m. The ones with u in the refuge and outbreak regimes are

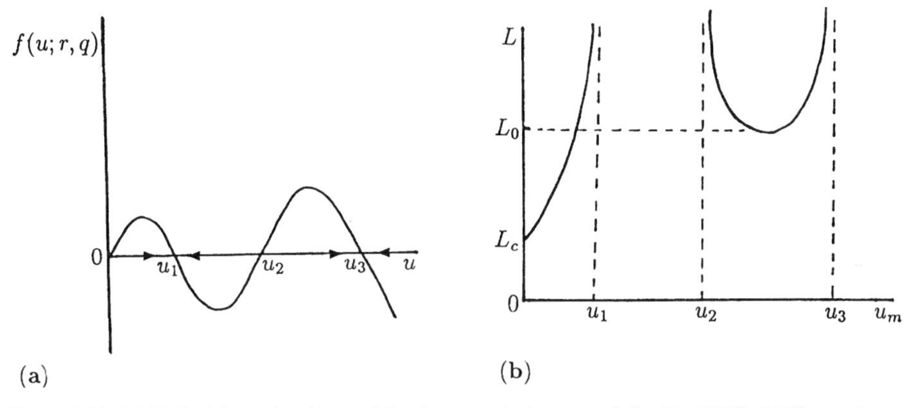

Figure 2.20. (**a**) Typical dynamics $f(u; r, q)$ for the spruce budworm as defined by (2.85). (**b**) The maximum population u_m as a function of the domain size L. For $u_m < u_1$ the population is in the refuge range, whereas $u_m > u_2$ for $L > L_0$, which is in the outbreak regime.

stable and the other, the middle one, is unstable. Which solution is obtained depends on the initial conditions. Later we shall consider possible ecological uses of this model in the control of the budworm. Before doing so we describe a useful technique for determining approximate values for L_0 analytically.

Analytical Method for Determining Critical Domain Sizes and Maximum Populations

The numerical evaluation of $u_m(L)$ when there are three possible u_m's for a given L is not completely trivial. Since the critical domain size L_0, which sustains an outbreak, is one of the important and useful quantities we require for practical applications, we now derive an ad hoc analytical method for obtaining it by exploiting an idea described by Lions (1982).

The steady state problem is defined by (2.81). Let us rescale the problem so that the domain is $x \in (0, 1)$ by setting $x \to x/L$ so that the equivalent $U(x)$ is now determined from

$$DU'' + L^2 f(U) = 0, \quad U(0) = U(1) = 0. \tag{2.86}$$

From Figure 2.19 the solution looks qualitatively like a sine. With the rescaling so that $x \in (0, 1)$ the solution is thus qualitatively like $\sin(\pi x)$. This means that $U'' \approx -\pi^2 U$ and so the last equation implies

$$-D\pi^2 U + L^2 f(U) \approx 0 \quad \Rightarrow \quad \frac{D\pi^2 U}{L^2} \approx f(U). \tag{2.87}$$

We are interested in the value of L such that the last equation has three roots for U: this corresponds to the situation in Figure 2.20(b) when $L > L_0$. Thus all we need do to determine an approximate L_0 is simply to plot the last equation as in Figure 2.21 and determine the value L such that three solutions exist.

For a fixed dispersal coefficient D we see how the solutions U vary with L. As L increases from $L \approx 0$ the first critical L, L_c, is given when the straight line $D\pi^2 U/L^2$ intersects $f(U)$, that is, when $D\pi^2/L^2 = f'(0)$, as given by (2.80). As L increases further we can determine the critical L_0 when $D\pi^2 U/L_0^2$ is tangent to the curve $f(U)$, at P in Figure 2.21. It is simply a matter of determining L which gives a double positive root of

$$\frac{D\pi^2 U}{L^2} = f(U).$$

It is left as an exercise (Exercise 7) to determine L_0 as a function of r, q and D when $f(U)$ is given by (2.85). For any given L the procedure also determines, approximately, the maximum U. From Figure 2.21 we clearly obtain by this procedure a figure similar to that in Figure 2.20(b). This simple procedure is quite general for determining critical domain sizes, both for structure bifurcating from the zero steady state and for domains which can sustain larger populations arising from population dynamics with multiple positive steady states.

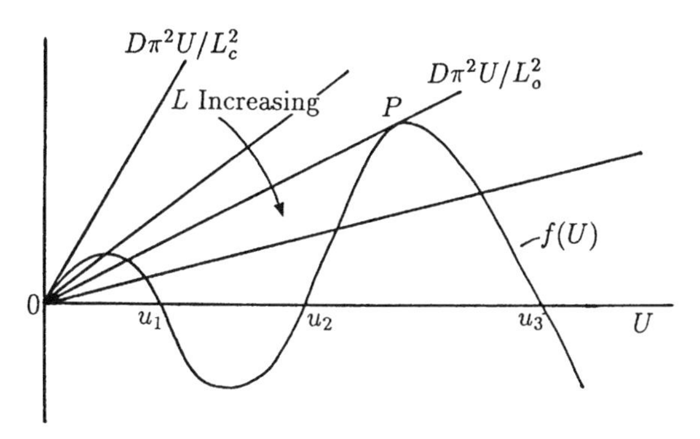

Figure 2.21. Approximate analytic procedure for determining the critical domain sizes L_c and L_0 which can sustain respectively a refuge and an outbreak in the species population where the dynamics is described by $f(U)$. L_c is the value of L when $D\pi^2U/L^2$ is tangent to $f(U)$ at $U = 0$. L_0 is given by the value of L when $D\pi U/L^2$ is just tangent to $F(U)$ at P.

2.8 Spatial Patterns in Scalar Population Interaction Diffusion Equations with Convection: Ecological Control Strategies

In practical applications of such models the domains of interest are usually two-dimensional and so we must consider (2.76). Also, with insect pests in mind, the exterior region is not generally completely hostile, so $u = 0$ on the boundaries is too restrictive a condition. Here we briefly consider a one- and two-dimensional problem in which the exterior domain is not completely hostile and there is a prevailing wind. This is common in many insect dispersal situations and can modify the spatial distribution of the population in a major way.

Suppose, for algebraic simplicity, that the two-dimensional domain is a rectangular region B defined by $0 \le x \le a$, $0 \le y \le b$ having area A. The completely hostile problem is then given by

$$u_t = f(u) + D\left(\frac{\partial^2 u}{\partial x^2} + \frac{\partial^2 u}{\partial y^2}\right), \qquad \text{(2.88)}$$
$$u = 0 \quad \text{for} \quad (x, y) \text{ on } \partial B.$$

Following the same procedure as in the last section for u small we get the solution of the linearised problem to be

$$u(x, y, t) = \sum_{n,m} a_{mn} \exp\left\{\left[f'(0) - D\pi^2\left(\frac{n^2}{a^2} + \frac{m^2}{b^2}\right)\right]t\right\} \sin\frac{n\pi x}{a} \sin\frac{m\pi y}{b}. \qquad \text{(2.89)}$$

So the critical domain size, which involves both a and b, is given by any combination of a and b such that

$$\frac{a^2b^2}{a^2+b^2} = \frac{D\pi^2}{f'(0)}.$$

Since

$$a^2 + b^2 > 2ab = 2A \quad \Rightarrow \quad \frac{a^2b^2}{a^2+b^2} < \frac{A}{2}$$

we get an inequality estimate for spatial patterning to exist, namely,

$$A > \frac{2D\pi^2}{f'(0)}. \tag{2.90}$$

Estimates for general two-dimensional domains were obtained by Murray and Sperb (1983). Clearly the mathematical problem is that of finding the smallest eigenvalue for the spatial domain considered.

In all the scalar models considered above the spatial patterns obtained have only a single maximum. With completely hostile boundary conditions these are the only type of patterns that can be generated. With two-species reaction diffusion systems, however, we saw that more diverse patterns could be generated. It is natural to ask whether there are ways in which similar multi-peak patterns could be obtained with single-species models in a one-dimensional context. We now show how such patterns could occur.

Suppose there is a constant prevailing wind $\mathbf{w}$ which contributes a convective flux $(\mathbf{w} \cdot \nabla)u$ to the conservation equation for the population $u(\mathbf{r}, t)$. Also suppose that the exterior environment is not completely hostile in which case appropriate boundary conditions are

$$(\mathbf{n} \cdot \nabla)u + hu = 0, \quad \mathbf{r} \text{ on } \partial B, \tag{2.91}$$

where $\mathbf{n}$ is the unit normal to the domain boundary ∂B. The parameter h is a measure of the hostility: $h = \infty$ implies a completely hostile exterior, whereas $h = 0$ implies a closed environment, that is, zero flux boundaries. We briefly consider the latter case later. The mathematical problem is thus

$$u_t + (\mathbf{w} \cdot \nabla)u = f(u) + D\nabla^2 u, \tag{2.92}$$

with boundary conditions (2.91) and given initial distribution $u(\mathbf{r}, 0)$. Here we consider the one-dimensional problem and follow the analysis of Murray and Sperb (1983), who also deal with the two-dimensional analogue and more general aspects of such problems.

The problem we briefly consider is the one-dimensional system which defines the steady state spatially inhomogeneous solutions $U(x)$. From (2.91) and (2.92), since

$$(\mathbf{w} \cdot \nabla)u = w_1 u_x,$$
$$(\mathbf{n} \cdot \nabla)u + hu = 0 \quad \Rightarrow \quad u_x + hu = 0, \ x = L; \quad u_x - hu = 0, \ x = 0,$$

where w_1 is the x-component of the wind $\mathbf{w}$, the mathematical problem for $U(x)$ is

$$\begin{aligned} &DU'' - w_1U' + f(U) = 0, \\ &U'(0) - hU(0) = 0, \quad U'(L) + hU(L) = 0. \end{aligned} \tag{2.93}$$

We study the problem using phase plane analysis by setting

$$U' = V, \quad DV' = w_1V - f(U) \quad \Rightarrow \quad \frac{dV}{dU} = \frac{w_1V - f(U)}{DV}, \tag{2.94}$$

and we look for phase plane trajectories which, from the boundary conditions in (2.93), join any point on one of the following lines to any point on the other line,

$$V = hU, \quad V = -hU. \tag{2.95}$$

The phase plane situation is illustrated in Figures 2.22(a) and (b) as we now show.

Refer first to Figure 2.22(a). From (2.94) we get the sign of dV/dU at any point (U, V). On the curve $V = f(U)/w_1$, $dV/dU = 0$ with dV/dU positive and negative when (U, V) lies respectively above (if $V > 0$) and below it. So, if we start on the boundary line $V = hU$ at say, P, the trajectory will qualitatively be like T_1

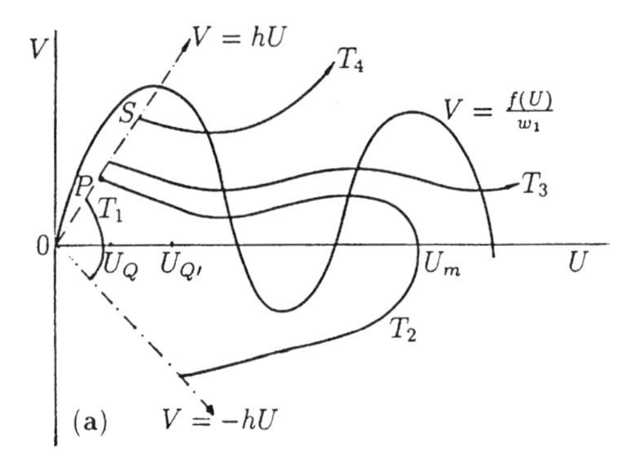

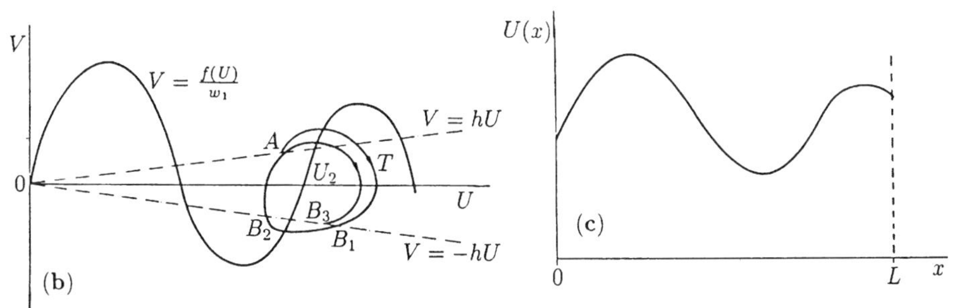

Figure 2.22. (**a**) With h sufficiently large the possible trajectories from $V = hU$ to $V = -hU$ admit solution trajectories like T_1 and T_2 with only a single maximum U_m. (**b**) For small enough h it is possible to have more complex patterns as indicated by the specimen trajectory T. (**c**) A typical solution $U(x)$ for a phase trajectory like T in (**b**).

since $dV/dU < 0$ everywhere on it. If we start at S, say, although the trajectory starts with $dV/dU < 0$, it intersects the $dV/dU = 0$ line and passes through to the region where $dV/dU > 0$ and so the trajectory turns up. The trajectories T_2, T_3 and T_4 are all possible scenarios depending on the parameters and where the solution trajectory starts. T_3 and T_4 are not solution trajectories satisfying (2.94) since they do not terminate on the boundary curve $V = -hU$. T_1 and T_2 are allowable solution paths and each has a single maximum U where the trajectory crosses the $V = 0$ axis.

We now have to relate the corresponding domain length L to these solution trajectories. To be specific let us focus on the trajectory T_2. Denote the part of the solution with $V > 0$ by $V^+(U)$ and that with $V < 0$ by $V^-(U)$. If we now integrate the first equation in (2.94) from U_Q to U'_Q, that is, the U-values at either end of the T_2 trajectory, we get the corresponding length of the domain for the solution represented by T_2 as

$$L = \int_{U_Q}^{U_m} [V^+(U)]^{-1}\, dU + \int_{U_m}^{U'_Q} [V^-(U)]^{-1}\, dU. \tag{2.96}$$

So, for each allowable solution trajectory we can obtain the corresponding size of the solution domain. The qualitative form of the solution $U(x)$ as a function of x can be deduced from the phase trajectory since we know U and U' everywhere on it and from the last equation we can calculate the domain size. With the situation represented by Figure 2.22(a) there can only be a single maximum in $U(x)$. Because of the wind convection term, however, there is no longer the solution symmetry of the solutions as in the last Section 2.7.

Now suppose the exterior hostility decreases, that is, h in (2.95) decreases, so that the boundary lines are now as illustrated in Figure 2.22(b). Proceeding in the same way as for the solution trajectories in Figure 2.22(a) we see that it is possible for a solution to exist corresponding to the trajectory T. On sketching the corresponding solution $U(x)$ we see that here there are two maxima in the domain: see Figure 2.22(c). In this situation, however, we are in fact patching several possible solutions together. Referring to Figure 2.22(b) we see that a possible solution is represented by that part of the trajectory T from A to B_1. It has a single maximum and a domain length L_1 given by the equivalent of (2.96). So if we restrict the domain size to be L_1 this is the relevant solution. However, if we allow a larger L the continuation from B_1 to B_2 is now possible and so the trajectory AB_1B_2 corresponds to a solution of (2.93). Increasing L further we can include the rest of the trajectory to B_3. It is thus possible to have multi-humped solutions if the domain is large enough. The length L corresponding to the solution path T is obtained in exactly the same way as above, using the equivalent of (2.96).

So, for small enough values of h it is possible to have more and more structure as the trajectory winds round the point u_2 in the (U, V) phase plane. For such solutions to exist, of course, it is essential that $w_1 \neq 0$. If $w_1 = 0$ the solutions are symmetric about the U-axis and so no spiral solutions are possible. Thus, a prevailing wind is essential for complex patterning. It also affects the critical domain size for patterns to exist. General results and further analysis are given by Murray and Sperb (1983).

An Insect Pest Control Strategy

Consider now the problem of insect pest control. The forest budworm problem is very much a two-dimensional spatial problem. As we pointed out in Chapter 1, Volume I, Section 1.2, a good control strategy would be to maintain the population at a refuge level. As we also showed in Section 1.2, Volume I it would be strategically advantageous if the dynamics parameters r and q in (2.85) could be changed so that only a single positive steady state exists. This is not really ecologically feasible. With the more realistic spatial problem, however, we have a further possible means of keeping the pest levels within the refuge range by ensuring that their spatial domains are of a size that does not permit populations in the outbreak regime. The arguments go through for two-dimensional domains, but for illustrative purposes let us consider first the one-dimensional situation.

Refer to Figure 2.20(b). If the spatial region were divided up into regions with size $L < L_0$, that is, so that the maximum u_m was always less than u_1, the refuge population level, we would have achieved our goal. So, a possible strategy is to spray the region in strips so that the non-sprayed regions impose an effective $L < L_0$ as in Figure 2.23(a): the solid vertical lines separating the sprayed regions are the boundaries to a completely hostile exterior.

Of course it is not practical to destroy all pests that stray out of the unsprayed region, so a more realistic model is that with boundary conditions (2.91) where some insects can survive outside the untreated domain. The key mathematical problem to be solved then is the determination of the critical width of the insect 'break' L_b. This must be such that the contributions from neighbouring untreated areas do not contribute a sufficient number of insects, which diffuse through the break, to initiate an outbreak in the neighbouring patches even though $L < L_0$, the critical size in isolation. A qualitative population distribution would typically be as shown by the dashed line in Figure 2.23(a).

The two-dimensional analogue is clear but the solution of the optimisation problem is more complicated. First the critical domain A_0 which can sustain an insect pest

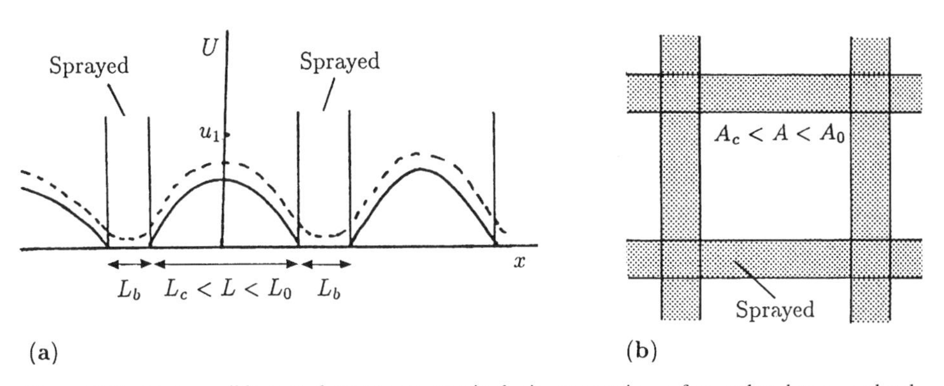

Figure 2.23. (**a**) A possible control strategy to contain the insect pest in a refuge rather than an outbreak environment. Strips—insect 'breaks'—are sprayed to maintain an effective domain size $L < L_0$, the critical size for an outbreak. The broken line is a more typical situation in practice. (**b**) Equivalent two-dimensional analogue where $A > A_c$ is a typical domain which can sustain a pest refuge population but which is not sufficient to sustain an outbreak; that is, $A < A_0$.

outbreak has to be determined for boundary conditions (2.91). Then the width of the sprayed strips has to be determined. It is not a trivial problem to solve, but certainly a possible one. A preliminary investigation of these problems was carried out by Ben-Yu et al. (1986).

Although we have concentrated on the budworm problem the techniques and control strategies are equally applicable to other insect pests. The field of insect dispersal presents some very important ecological problems, such as the control of killer bees now sweeping up through the western United States (see, for example, Taylor 1977) and locust plagues in Africa. Levin (see, for example, 1981a,b) has made realistic and practical studies of these and other problems associated with spatially heterogeneous ecological models. The concept of a break control strategy to prevent the spatial spread of a disease epidemic will be discussed in some detail later in Chapter 13 when we discuss the spatial spread of rabies; it is now in use.

In an interesting ecological application of diffusion-driven instability Hastings et al. (1997) investigated an outbreak of the western tussock moth which had been hypothesised to be the result of a predator–prey interaction between the moths and parasitoids. The model they analyse qualitatively is a two-species system in which the prey do not move. We saw in Chapter 13, Section 13.7, Volume I how the effect of having a percentage of the prey sessile gave rise to counterintuitive results. Hastings et al. (1997) also obtain counterintuitive results by considering a quite general system with typical predator–prey interactions in which the prey does not diffuse. Their new analysis is very simple but highly illuminating. Their approach is reminiscent of that in Section 1.6 on excitable waves for determining where the trajectory for the wave in the phase plane is closed, thereby determining the wavespeed among other things, and revolves around the existence of three possible steady states at each spatial position, two of which are stable, with the possibility of a jump, or rather a steep singular region joining them. They then show that, counterintuitively, the spatial distribution of the prey will have its highest density at the edge of the outbreak domain. This phenomenon has been observed in western tussock moth outbreaks. The role of theory in predicting counterintuitive behaviours and subsequent experimental or observational confirmation is particularly important when, as is frequently the case, the plethora of observational facts is bewildering rather than illuminating. The article by Kareiva (1990) is particularly relevant to this relation of theory to data.

2.9 Nonexistence of Spatial Patterns in Reaction Diffusion Systems: General and Particular Results

The scalar one-dimensional reaction diffusion system with zero flux is typically of the form

$$\begin{aligned} u_t &= f(u) + u_{xx}, \quad x \in (0,1), \quad t > 0 \\ u_x(0,t) &= u_x(1,t) = 0, \quad t > 0. \end{aligned} \tag{2.97}$$

Intuitively the only stable solution is the spatially homogeneous one $u = u_0$, the steady state solution of $f(u) = 0$: if there is more than one stable solution of $f(u) = 0$

which will obtain depends on the initial conditions. It can be proved that any spatially nonuniform steady state solution is unstable (the analysis is given in the first edition of this book: it involves estimating eigenvalues). This result does *not* carry over completely to scalar equations in more than one space dimension as has been shown by Matano (1979) in the case where $f(u)$ has two linearly stable steady states. The spatial patterns that can be obtained, however, depend on specific domain boundaries, non-convex to be specific. For example, a dumbbell shaped domain with a sufficiently narrow neck is an example. The pattern depends on the difficulty of diffusionally transporting enough flux of material through the neck to effect a change from one steady state to another so as to achieve homogeneity.

We saw in Sections 2.3 and 2.4 how reaction diffusion systems with zero flux boundary conditions could generate a rich spectrum of spatial patterns if the parameters and kinetics satisfied appropriate conditions: crucially the diffusion coefficients of the reactants had to be different. Here we show that for general multi-species systems patterning can be destroyed if the diffusion is sufficiently large. This is intuitively what we might expect, but it is not obvious if the diffusion coefficients are unequal. This we now prove. The analysis, as we show, gives another condition involving the kinetic relaxation time of the mechanism which is certainly not immediately obvious.

Before discussing the multi-species multi-dimensional theory it is pedagogically helpful to consider first the general one-dimensional two-species reaction diffusion system

$$\begin{aligned} u_t &= f(u, v) + D_1 u_{xx}, \\ v_t &= g(u, v) + D_2 v_{xx} \end{aligned} \tag{2.98}$$

with zero flux boundary conditions and initial conditions

$$\begin{aligned} &u_x(0, t) = u_x(1, t) = v_x(0, t) = v_x(1, t) = 0 \\ &u(x, 0) = u_0(x), \\ &v(x, 0) = v_0(x), \end{aligned} \tag{2.99}$$

where $u_0'(x)$ and $v_0'(x)$ are zero on $x = 0, 1$. Define an energy integral E by

$$E(t) = \frac{1}{2}\int_0^1 (u_x^2 + v_x^2)\, dx. \tag{2.100}$$

This is, except for the $1/2$, the heterogeneity function introduced in (2.52). Differentiate E with respect to t to get

$$\frac{dE}{dt} = \int_0^1 (u_x u_{xt} + v_x v_{xt})\, dx$$

and substitute from (2.98), on differentiating with respect to x, to get, on integrating by parts,

$$
\begin{aligned}
\frac{dE}{dt} &= \int_0^1 [u_x(D_1u_{xx})_x + u_x(f_u u_x + f_v v_x) \\
&\quad + v_x(D_2v_{xx})_x + v_x(g_u u_x + g_v v_x)]\, dx, \\
&= [u_x D_1 u_{xx} + v_x D_2 v_{xx}]_0^1 - \int_0^1 (D_1 u_{xx}^2 + D_2 v_{xx}^2)\, dx \\
&\quad + \int_0^1 \left[f_u u_x^2 + g_v v_x^2 + (f_v + g_u) u_x v_x \right] dx.
\end{aligned}
$$

Because of the zero flux conditions the integrated terms are zero.

Now define the quantities d and m by

$$
d = \min(D_1, D_2), \quad m = \max_{u,v} \left(f_u^2 + f_v^2 + g_u^2 + g_v^2 \right)^{1/2}, \tag{2.101}
$$

where $\max_{u,v}$ means the maximum for u and v taking all possible solution values. If we want we could define m by some norm involving the derivatives of f and g; it is not crucial for our result. From the equation for dE/dt, with these definitions, we then have

$$
\begin{aligned}
\frac{dE}{dt} &\le -d \int_0^1 (u_{xx}^2 + v_{xx}^2)\, dx + 4m \int_0^1 (u_x^2 + v_x^2)\, dx \\
&\le (4m - 2\pi^2 d) E,
\end{aligned} \tag{2.102}
$$

where we have used the result

$$
\int_0^1 u_{xx}^2\, dx \ge \pi^2 \int_0^1 u_x^2\, dx \tag{2.103}
$$

with a similar inequality for v; see Appendix A for a derivation of (2.103).

From the inequality (2.102) we now see that if the minimum diffusion coefficient d, from (2.101), is large enough so that $(4m - 2\pi^2 d) < 0$ then $dE/dt < 0$, which implies that $E \to 0$ as $t \to \infty$ since $E(t) \ge 0$. This implies, with the definition of E from (2.100), that $u_x \to 0$ and $v_x \to 0$ which implies spatial homogeneity in the solutions u and v as $t \to \infty$. The result is not precise since there are many appropriate choices for m; (2.101) is just one example. The purpose of the result is simply to show that it is possible for diffusion to dampen *all* spatial heterogeneities. We comment briefly on the biological implication of this result below.

We now prove the analogous result for general reaction diffusion systems. Consider

$$
\mathbf{u}_t = \mathbf{f}(\mathbf{u}) + D\nabla^2 \mathbf{u}, \tag{2.104}
$$

where $\mathbf{u}$, with components u_i, $i = 1, 2, \ldots, n$, is the vector of concentrations or populations, and D is a diagonal matrix of the positive diffusion coefficients D_i, $i = 1, 2, \ldots, n$ and $\mathbf{f}$ is the nonlinear kinetics. The results we prove are also valid for a diffusion matrix with certain cross-diffusion terms, but for simplicity here we only deal with (2.104). Zero flux boundary and initial conditions for $\mathbf{u}$ are

$$(\mathbf{n}\cdot\nabla)\mathbf{u}=0 \quad \mathbf{r} \text{ on } \partial B, \quad \mathbf{u}(\mathbf{r},0)=\mathbf{u}_0(\mathbf{r}), \tag{2.105}$$

where $\mathbf{n}$ is the unit outward normal to ∂B, the boundary of the domain B. As before we assume that all solutions $\mathbf{u}$ are bounded for all $t \geq 0$. Practically this is effectively assured if a confined set exists for the reaction kinetics.

We now generalise the previous analysis; it helps to refer to the equivalent steps in the above. Define the energy $E(t)$ by

$$E(t)=\frac{1}{2}\int_B \|\nabla\mathbf{u}\|^2\, d\mathbf{r}, \tag{2.106}$$

where the norm

$$\|\nabla\mathbf{u}\|^2=\sum_{i=1}^{n}|\nabla u_i|^2.$$

Let d be the smallest eigenvalue of the matrix D, which in the case of a diagonal matrix is simply the smallest diffusion coefficient of all the species. Now define

$$m=\max_{\mathbf{u}}\|\nabla_{\mathbf{u}}\mathbf{f}(\mathbf{u})\|, \tag{2.107}$$

where $\mathbf{u}$ takes on all possible solution values and $\nabla_{\mathbf{u}}$ is the gradient operator with respect to $\mathbf{u}$.

Differentiating $E(t)$ in (2.106), using integration by parts, the boundary conditions (2.105) and the original system (2.104) we get, with $\langle\mathbf{a},\mathbf{b}\rangle$ denoting the inner product of $\mathbf{a}$ and $\mathbf{b}$,

$$\begin{aligned}\frac{dE}{dt}&=\int_B\langle\nabla\mathbf{u},\nabla\mathbf{u}_t\rangle\, d\mathbf{r}\\&=\int_B\langle\nabla\mathbf{u},\nabla D\nabla^2\mathbf{u}\rangle\, d\mathbf{r}+\int_B\langle\nabla\mathbf{u},\nabla\mathbf{f}\rangle\, d\mathbf{r}\\&=\int_{\partial B}\langle\nabla\mathbf{u},D\nabla^2\mathbf{u}\rangle\, d\mathbf{r}-\int_B\langle\nabla^2\mathbf{u},D\nabla^2\mathbf{u}\rangle\, d\mathbf{r}+\int_B\langle\nabla\mathbf{u},\nabla_{\mathbf{u}}\mathbf{f}\cdot\nabla\mathbf{u}\rangle\, d\mathbf{r}\\&\leq -d\int_B|\nabla^2\mathbf{u}|^2\, d\mathbf{r}+mE.\end{aligned} \tag{2.108}$$

In Appendix A we show that when $(\mathbf{n}\cdot\nabla)\mathbf{u}=0$ on ∂B,

$$\int_B|\nabla^2\mathbf{u}|^2\, d\mathbf{r}\geq\mu\int_B\|\nabla\mathbf{u}\|^2\, d\mathbf{r}, \tag{2.109}$$

where μ is the least positive eigenvalue of

$$\nabla^2\phi+\mu\phi=0, \quad (\mathbf{n}\cdot\nabla)\phi=0 \quad \mathbf{r} \text{ on } \partial B,$$

where ϕ is a scalar. Using the result (2.109) in (2.108) we get

$$\frac{dE}{dt} \leq (m - 2\mu d)E \quad \Rightarrow \quad \lim_{t\to\infty} E(t) = 0 \quad \text{if} \quad m < 2\mu d \tag{2.110}$$

and so, once again, if the smallest diffusion coefficient is large enough this implies that $\nabla \mathbf{u} \to 0$ and so all spatial patterns tend to zero as $t \to \infty$.

Othmer (1977) has pointed out that the parameter m defined by (2.101) and (2.107) is a measure of the sensitivity of the reaction rates to changes in $\mathbf{u}$ since $1/m$ is the shortest kinetic relaxation time of the mechanism. On the other hand $1/(2\mu d)$ is a measure of the longest diffusion time. So the result (2.110), which is $1/m > 1/(2\mu d)$, then implies that if the shortest relaxation time for the kinetics is greater than the longest diffusion time then all spatial patterning will die out as $t \to \infty$. The mechanism will then be governed solely by kinetics dynamics. Remember that the solution of the latter can include limit cycle oscillations.

Suppose we consider the one-dimensional situation with a typical embryological domain of interest, say, $L = O$ (1 mm). With $d = O(10^{-6}\text{cm}^2\text{s}^{-1})$ the result (2.110) then implies that homogeneity will result if the shortest relaxation time of the kinetics $1/m > L^2/(2\pi^2 d)$, that is, a time of $O(500\text{ s})$.

Consider the general system (2.104) rescaled so that the length scale is 1 and the diffusion coefficients are scaled relative to D_1 say. Now return to the formulation used earlier, in (2.10), for instance, in which the scale γ appears with the kinetics in the form $\gamma\mathbf{f}$. The effect of this on the condition (2.110) now produces $\gamma m - 2\mu < 0$ as the stability requirement. We immediately see from this form that there is a critical γ, proportional to the domain area, which in one dimension is $(\text{length})^2$, below which no structure can exist. This is of course a similar result to the one we found in Sections 2.3 and 2.4.

We should reiterate that the results here give qualitative bounds and not estimates for the various parameters associated with the model mechanisms. The evaluation of an appropriate m is not easy. In Sections 2.3 and 2.4 we derived specific quantitative relations between the parameters, when the kinetics were of a particular class, to give spatially structured solutions. The general results in this section, however, apply to all types of kinetics, oscillatory or otherwise, as long as the solutions are bounded.

In this chapter we have dealt primarily with reaction or population interaction kinetics which, in the absence of diffusion, do not exhibit oscillatory behaviour in the restricted regions of parameter space which we have considered. We may ask what kind of spatial structure can be obtained when oscillatory kinetics is coupled with diffusion. We saw in Chapter 1 that such a combination could give rise to travelling wavetrains when the domain is infinite. If the domain is finite we could anticipate a kind of regular sloshing around within the domain which is a reflection of the existence of spatially and temporally unstable modes. This can in fact occur but it is not always so. One case to point is the classical Lotka–Volterra system with equal diffusion coefficients for the species. Murray (1975) showed that in a finite domain all spatial heterogeneities must die out (see Exercise 11).

There are now several pattern formation mechanisms, other than reaction-diffusion-chemotaxis systems. One of the best critical and thorough reviews on models for self-organisation in development is by Wittenberg (1993). He describes the models in detail and compares and critically reviews several of the diverse mechanisms including reaction-diffusion-chemotaxis systems, mechanochemical mechanisms and cellular automaton models.

In the next chapter we shall discuss several specific practical biological pattern formation problems. In later chapters we shall describe other mechanisms which can generate spatial patterns. An important system which has been widely studied is the reaction-diffusion-chemotaxis mechanism for generating aggregation patterns in bacteria and also for slime mould amoebae, one model for which we derived in Chapter 11, Volume I, Section 11.4. Using exactly the same kind of analysis we discussed above for diffusion-driven instability we can show how spatial patterns can arise in these model equations and the conditions on the parameters under which this will happen (see Exercise 9). As mentioned above these chemotaxis systems are becoming increasingly important with the upsurge in interest in bacterial patterns and is the reason for including Chapter 5 below. We discuss other quite different applications of cell-chemotaxis mechanisms in Chapter 4 when we consider the effect of growing domains on patterning, such as the complex patterning observed on snakes.

Exercises

1. Determine the appropriate nondimensionalisation for the reaction kinetics in (2.4) and (2.5) which result in the forms (2.8).

2. An activator–inhibitor reaction diffusion system in dimensionless form is given by

$$u_t = \frac{u^2}{v} - bu + u_{xx}, \quad v_t = u^2 - v + dv_{xx},$$

where b and d are positive constants. Which is the activator and which the inhibitor? Determine the positive steady states and show, by an examination of the eigenvalues in a linear stability analysis of the diffusionless situation, that the reaction kinetics cannot exhibit oscillatory solutions if $b < 1$.

Determine the conditions for the steady state to be driven unstable by diffusion. Show that the parameter domain for diffusion-driven instability is given by $0 < b < 1$, $db > 3 + 2\sqrt{2}$ and sketch the (b, d) parameter space in which diffusion-driven instability occurs. Further show that at the bifurcation to such an instability the critical wavenumber k_c is given by $k_c^2 = (1 + \sqrt{2})/d$.

3. An activator–inhibitor reaction diffusion system with activator inhibition is modelled by

$$\begin{aligned} u_t &= a - bu + \frac{u^2}{v(1 + Ku^2)} + u_{xx}, \\ v_t &= u^2 - v + dv_{xx}, \end{aligned}$$

where K is a measure of the inhibition and a, b and d are constants. Sketch the null clines for positive b, various $K > 0$ and positive or negative a.

Show that the (a, b) Turing (parameter) space for diffusion-driven instability is defined parametrically by

$$a = bu_0 - (1 + Ku_0^2)^2$$

combined with

$$b > 2[u(1 + Ku_0^2)]^{-1} - 1, \quad b > 0, \quad b > 2[u(1 + Ku_0^2)]^{-2} - \frac{1}{d},$$
$$b < 2[u(1 + Ku_0^2)]^{-2} - 2\sqrt{2}[du(1 + Ku_0^2)]^{-1/2} + \frac{1}{d},$$

where the parameter u_0 takes on all values in the range $(0, \infty)$. Sketch the Turing space for (i) $K = 0$ and (ii) $K \neq 0$ for various d (Murray 1982).

4. Determine the relevant axisymmetric eigenfunctions $\mathbf{W}$ and eigenvalues k^2 for the circular domain bounded by R defined by

$$\nabla^2 \mathbf{W} + k^2 \mathbf{W} = 0, \quad \frac{d\mathbf{W}}{dr} = 0 \text{ on } r = R.$$

Given that the linearly unstable range of wavenumbers k^2 for the reaction diffusion mechanism (2.7) is given by

$$\gamma L(a, b, d) < k^2 < \gamma M(a, b, d),$$

where L and M are defined by (2.38), determine the critical radius R_c of the domain below which no spatial pattern can be generated. For R just greater than R_c sketch the spatial pattern you would expect to evolve.

5. Consider the reaction diffusion mechanism given by

$$u_t = \gamma \left(\frac{u^2}{v} - bu \right) + u_{xx}, \quad v_t = \gamma (u^2 - v) + dv_{xx},$$

where γ, b and d are positive constants. For the domain $0 \leq x \leq 1$ with zero flux conditions determine the dispersion relation $\lambda(k^2)$ as a function of the wavenumbers k of small spatial perturbations about the uniform steady state. Is it possible with this mechanism to isolate successive modes by judicious variation of the parameters? Is there a bound on the excitable modes as $d \to \infty$ with b and γ fixed?

6. Suppose fishing is regulated within a zone H km from a country's shore (taken to be a straight line) but outside of this zone overfishing is so excessive that the population is effectively zero. Assume that the fish reproduce logistically, disperse by diffusion and within the zone are harvested with an effort E. Justify the following model for the fish population $u(x, t)$.

$$u_t = ru\left(1 - \frac{u}{K}\right) - EU + Du_{xx},$$
$$u = 0 \text{ on } x = H, \quad u_x = 0 \text{ on } x = 0,$$

where r, K, $E(< r)$ and D are positive constants.

If the fish stock is not to collapse show that the fishing zone H must be greater than $\frac{\pi}{2}[D/(r-E)]^{1/2}$ km. Briefly discuss any ecological implications.

7. Use the approximation method described in Section 2.7 to determine analytically the critical length L_0 as function of r, q and D such that an outbreak can exist in the spruce budworm population model

$$u_t = ru\left(1 - \frac{u}{q}\right) - \frac{u^2}{1+u^2} + Du_{xx}, \quad u = 0 \text{ on } x = 0, 1.$$

Determine the maximum population u_m when $L = L_0$.

8. Consider the Lotka–Volterra predator–prey system (see Chapter 3, Volume I, Section 3.1) with diffusion given by

$$u_t = u(1-v) + Du_{xx}, \quad v_t = av(u-1) + Dv_{xx}$$

in the domain $0 \le x \le 1$ with zero flux boundary conditions. By multiplying the first equation by $a(u-1)$ and the second by $(v-1)$ show that

$$S_t = DS_{xx} - D\sigma^2,$$
$$S = au + v - \ln(u^a v), \quad \sigma^2 = a\left(\frac{u_x}{u}\right)^2 + \left(\frac{v_x}{v}\right)^2 \ge 0.$$

Determine the minimum S for all u and v. Show that necessarily $\sigma \to 0$ as $t \to \infty$ by supposing σ^2 tends to a nonzero bound, the consequences of which are not possible. Hence deduce that no spatial patterns can be generated by this model in a finite domain with zero flux boundary conditions.

(This result can also be obtained rigorously, using maximum principles; the detailed analysis is given by Murray (1975).)

9. The amoebae of the slime mould *Dictyostelium discoideum*, with density $n(x,t)$, secrete a chemical attractant, cyclic-AMP, and spatial aggregations of amoebae start to form. One of the models for this process (and discussed in Section 11.4, Volume I) gives rise to the system of equations, which in their one-dimensional form, are

$$n_t = D_n n_{xx} - \chi(na_x)_x, \quad a_t = hn - ka + D_a a_{xx},$$

where a is the attractant concentration and h, k, χ and the diffusion coefficients D_n and D_a are all positive constants. Nondimensionalise the system.

Consider (i) a finite domain with zero flux boundary conditions and (ii) an infinite domain. Examine the linear stability about the steady state (which introduces a further parameter here), derive the dispersion relation and discuss the role of the various parameter groupings. Hence obtain the conditions on the parameters and domain size for the mechanism to initiate spatially heterogeneous solutions.

Experimentally the chemotactic parameter χ increases during the life cycle of the slime mould. Using χ as the bifurcation parameter determine the critical wavelength when the system bifurcates to spatially structured solutions in an infinite domain. In the finite domain situation examine the bifurcating instability as the domain is increased.

Briefly describe the physical processes operating and explain intuitively how spatial aggregation takes place.

10. Consider the dimensionless reaction anisotropic diffusion system

$$\frac{\partial u}{\partial t} = \gamma f(u, v) + d_1 \frac{\partial^2 u}{\partial x^2} + d_2 \frac{\partial^2 u}{\partial y^2},$$

$$\frac{\partial v}{\partial t} = \gamma g(u, v) + d_3 \frac{\partial^2 u}{\partial x^2} + d_4 \frac{\partial^2 u}{\partial y^2}.$$

In the absence of diffusion the steady state $\mathbf{u} = (u_0, v_0)$ is stable. By carrying out a linear analysis about the steady state by looking for solutions in the form $\mathbf{u} - \mathbf{u_0} \propto e^{\lambda t + i(k_x x + k_y y)}$, where k_x and k_y are the wavenumbers, show that if

$$H(k_x^2, k_y^2) = d_1 d_3 k_x^4 + p_1 k_x^2 k_y^2 + d_2 d_4 k_y^4 - \gamma p_2 k_x^2 - \gamma k_y^2 p_3 + \gamma^2 (f_u g_v - f_v g_u),$$

where

$$p_1 = d_1 d_4 + d_2 d_3, \quad p_2 = d_3 f_u + d_1 g_v, \quad p_3 = d_4 f_u + d_2 g_v$$

is such that $H < 0$ for some $k_x^2 k_y^2 \neq 0$ then the system can be diffusionally unstable to spatial perturbations. The maximum linear growth is given by the values k_x^2 and k_y^2 which give the minimum of $H(k_x^2, k_y^2)$. Show that the minimum of H is given by

$$\begin{bmatrix} k_x^2 \\ k_y^2 \end{bmatrix} = -\gamma \frac{(d_1 d_4 - d_2 d_3)}{(d_1 d_4 - d_2 d_3)^2} \begin{bmatrix} -d_4 f_u + d_2 g_v \\ -d_1 g_v + d_3 f_u \end{bmatrix}.$$

For a spatial pattern to evolve we need real values of k_x, k_y which requires, from the last equation, that

$$(d_1 d_4 - d_2 d_3) \begin{bmatrix} -d_4 f_u + d_2 g_v \\ -d_1 g_v + d_3 f_u \end{bmatrix} < \begin{bmatrix} 0 \\ 0 \end{bmatrix}.$$

By considering the two cases $(d_1 d_4 - d_2 d_3) < 0$ and $(d_1 d_4 - d_2 d_3) > 0$ show that the minimum of H does not lie in the first quadrant of the $k_x^2 - k_y^2$ plane and that diffusion-driven instability will first occur, for increasing ratios d_3/d_1, d_4/d_2 on one of the axial boundaries of the positive quadrant.

By setting k_x^2, k_y^2 equal to 0 in turn in the expression for H show that the conditions for diffusion-driven instability are

$$\begin{aligned} d_4 f_u + d_2 g_v > 0, \quad & d_3 f_u + d_1 g_v > 0 \\ (d_4 f_u - d_2 g_v)^2 + 4 d_2 d_4 f_v g_u > 0, \quad & (d_3 f_u - d_1 g_v)^2 + 4 d_1 d_3 f_v g_u > 0 \end{aligned}$$

so for it to occur $(d_3/d_1) > d_c$ and/or $(d_4/d_2) > d_c$ where

$$d_c = -\frac{1}{f_u^2}\left[(2 f_v g_u - f_u g_v) + [(2 f_v g_u - f_u g_v)^2 - f_u^2 g_v^2]^{1/2}\right].$$

Now consider the rectangular domain $0 < x < a, 0 < y < b$ with zero flux boundary conditions with a, b constants with a sufficiently greater than b so that the domain is a relatively thin rectangular domain. Show that it is possible to have the first unstable mode 2 bifurcation result in a striped pattern along the rectangle if the diffusion coefficient ratio in one direction exceeds the critical ratio. (Such a result is what we would expect intuitively since if only one ratio, $d_4/d_2 > d_c$, then the diffusion ratio in the x-direction is less than the critical ratio and we would expect spatial variation only in the y-direction, hence a striped pattern along the rectangle. A nonlinear analysis of this problem shows that such a pattern is stable. It further shows that if both ratios exceed the critical ratio a stable modulated (wavy) stripe pattern solution can be obtained along the rectangle.)

11. Suppose that a two-species reaction diffusion mechanism in u and v generates steady state spatial patterns $U(x)$, $V(x)$ in a one-dimensional domain of size L with zero flux boundary conditions $u_x = v_x = 0$ at both boundaries $x = 0$ and $x = L$. Consider the heterogeneity functions defined by

$$H_G(w) = \frac{1}{L}\int_0^L w_x\, dx, \quad H_S(w) = \frac{1}{L}\int_0^L [w_x - H_G(w)]^2\, dx.$$

Biologically the first of these simply measures the gradient while the second measures the deviation from the simple gradient. Show that the heterogeneity or energy integral

$$H(t) = \frac{1}{L}\int_0^L (U'^2 + V'^2)\, dx = [H_G(U)]^2 + [H_G(V)]^2 + H_S(U) + H_S(V).$$

(Berding 1987)

12. Show that the reaction diffusion mechanism

$$\mathbf{u}_t = \mathbf{f}(\mathbf{u}) + D\nabla^2\mathbf{u},$$

where the concentration vector $\mathbf{u}$ has n components, D is a diagonal diffusion matrix with elements d_i, $i = 1, 2, \ldots, n$ and $\mathbf{f}$ is the nonlinear kinetics, linearises about a positive steady state to

$$\mathbf{w}_t = A\mathbf{w} + D\nabla^2\mathbf{w},$$

where A is the Jacobian matrix of $\mathbf{f}$ at the steady state.

Let k be the eigenvalue of the problem defined by

$$\nabla^2\mathbf{W} + k^2\mathbf{W} = 0, \quad (\mathbf{n}\cdot\nabla)\mathbf{W} = 0 \quad \mathbf{r} \text{ on } \partial B.$$

On setting $\mathbf{w} \propto \exp[\lambda t + i\mathbf{k}\cdot\mathbf{r})$ show that the dispersion relation $\lambda(k^2)$ is given by the solutions of the characteristic polynomial

$$P(\lambda) = |\, A - k^2 D - \lambda I \,| = 0.$$

Denote the eigenvalues of $P(\lambda)$, with and without diffusion, by λ_i^+ and λ_i^- respectively. Diffusion-driven instability occurs if $\mathrm{Re}\,\lambda_i^- < 0$, $i = 1, 2, \ldots, n$ and at least one $\mathrm{Re}\,\lambda_i^+ > 0$ for some $k^2 \neq 0$.

From matrix algebra there exists a transformation T such that

$$|\, A - \lambda I \,| = |\, T^{-1}(A - \lambda I)T \,| = \prod_{i=1}^{n}(\lambda_i^- - \lambda).$$

Use this result and the fact that $\mathrm{Re}\,\lambda_i^- < 0$ to show that if $d_i = d$ for all i then $\mathrm{Re}\,\lambda_i^+ < 0$ for all i and hence that a necessary condition for diffusion-driven instability is that at least one diffusion coefficient is different from the rest.

13. The linearisation of a reaction diffusion mechanism about a positive steady state is

$$\mathbf{w}_t = A\mathbf{w} + D\nabla^2\mathbf{w},$$

where A is the Jacobian matrix of the reaction kinetics evaluated at the steady state.

If the matrix $A + A^T$, where T denotes the transpose, is stable this means that all of its eigenvalues λ are real and negative. Show that $\mathbf{w}\cdot A\mathbf{w} < -\delta\mathbf{w}\cdot\mathbf{w}$, for some $\delta > 0$.

[Hint: By considering $\mathbf{w}_t = A\mathbf{w}$ first show that $(\mathbf{w}^2)_t = 2\mathbf{w}A\mathbf{w}$. Then show that $\mathbf{w}_t^T\cdot\mathbf{w} = \mathbf{w}^T A^T\mathbf{w}$ and $\mathbf{w}^T\cdot\mathbf{w}_t = \mathbf{w}^T A\mathbf{w}$ to obtain $(\mathbf{w}^2)_t = \mathbf{w}^T(A + A^T)\mathbf{w}$. Thus deduce that $\mathbf{w}A\mathbf{w} = (1/2)\mathbf{w}^T(A^T + A)\mathbf{w} < -\delta\mathbf{w}\cdot\mathbf{w}$ for some $\delta > 0$.]

Let k^2 be the eigenvalues of the eigenvalue problem

$$\nabla^2\mathbf{w} + k^2\mathbf{w} = 0.$$

By considering dE/dt, where

$$E(t) = \int_B \mathbf{w}\cdot\mathbf{w}\,d\mathbf{r}$$

with B the spatial domain, show that $\mathbf{w}^2 \to 0$ as $t \to \infty$ and hence that such reaction diffusion systems cannot generate spatial patterns if the Jacobian matrix is of this particular form.

3. Animal Coat Patterns and Other Practical Applications of Reaction Diffusion Mechanisms

In this chapter we discuss some real biological pattern formation problems and show how the modelling discussed above, particularly that in the last chapter, has been applied. As an applications chapter of theories developed earlier, it contains considerably more biology than mathematics. Since all models for spatial pattern generation are necessarily nonlinear, practical applications require numerical solutions since useful analytical solutions are not available, nor likely to be. A preliminary linear analysis, however, is always useful, generally a necessity in fact. In each of the applications the biological modelling is discussed in detail. Most of the finite amplitude patterns reproduced are numerical solutions of the model equations and are applied directly to the specific biological situation.

In Section 3.1 we show how the pattern of animal coat markings such as on the zebra, leopard and so on, could be generated using a reaction diffusion mechanism. In the other sections we describe other pattern formation problems, namely, butterfly wing patterns in Section 3.3, and patterns which presage hairs in whorls during regeneration in *Acetabularia*, an important marine alga, in Section 3.4.

Reaction diffusion theory has now been applied from a patterning point of view to a large number of biological situations. For example, Kauffman et al. (1978) presented one of the first practical applications to the early segmentation of the embryo of the fruit fly *Drosophila melanogaster*. With the greatly increased understanding of the early stages of development the model is not valid; it was nevertheless useful at the time. There have been many more recent reaction diffusion models proposed for early insect development and how they could interact with gene expression; see especially the articles by Hunding and Engelhardt (1995), Meinhardt (1999, 2000) and Hunding (1999) and references given there. There have also been several very good reviews, such as those by Maini (1997, 1999). The early paper by Bunow et al. (1980) is specifically related to some of the material in this chapter; they discuss pattern sensitivity among other things. The book by Meinhardt (1982) has many examples based on activation–inhibition type of reaction diffusion models. The applications cover a wide spectrum of real biological problems; see the collection of articles in the books edited by Othmer et al. (1993), Chaplain et al. (1999) and Maini and Othmer (2000).

A key difficulty with the application of Turing's (1952) theory of morphogenesis is the identification of the morphogens and this has been a major obstacle to its acceptance as one of the essential processes in development; it still is. The fact that certain chemicals are essential for development does not necessarily mean they are morphogens. Identification of their role in the *patterning* process is necessary for this. It is partially for this reason that we discuss in Section 3.4 hair initiation in *Acetabularia*, where calcium is proposed as an example of a real morphogen. Theoretical and experimental evidence is given to back up the hypothesis. Even now, nearly 50 years after Turing's classic paper, the evidence for reaction diffusion mechanisms' role in development is still largely circumstantial. However, the large amount of circumstantial evidence is such that their importance is gaining acceptance as a necessary element in development. There is no question about its acceptance in ecology of course. The fact that the evidence is mainly circumstantial by no means implies that they have not been responsible for many major advances in our understanding of many developmental processes. Numerous case studies are described in detail in the rest of the book and in references to other studies. It is not possible to give other than a flavour of many successful (in the sense of having had a positive effect on our understanding) applications of the theory.

3.1 Mammalian Coat Patterns—'How the Leopard Got Its Spots'

Mammals exhibit a rich and varied spectrum of coat patterns; Figure 3.1 shows some typical markings. The beautifully illustrated (all drawings) multi-volume (seven) series of books, *East African Mammals*, produced since 1971 by Jonathan Kingdon (see, for example, the volume on carnivores, 1978, and large mammals, 1979) give, among other things, the most comprehensive and accurate survey of the wealth and variety of animal coat patterns. The book by Portmann (1952) has some interesting observations (some quite wild) on animal forms and patterns. However, as with almost all biological pattern generation problems (and repeated like some mantra in this book) the mechanism involved has not yet been determined. Murray (1980, 1981a,b) studied this particular pattern formation problem in some depth and it is mainly this work we discuss here: see also the general article in *Scientific American* by Murray (1988). Among other things he suggested that a *single* mechanism could be responsible for generating practically all of the common patterns observed. Murray's theory is based on a chemical concentration hypothesis by Searle (1968) who was one of the first to mention the potential of a Turing mechanism.[1] It had been more or less ignored up to then and would not be

[1] In the search for the earliest reference to a model for animal coat patterns, I think it is the one in the Book of Genesis in the Old Testament. It is a typical tale of exploitation, deception, scheming, fornication, perseverance, lack of trust and revenge. It is in the story of Laban, Laban's daughters Leah and Rachel, and Jacob who worked for seven years on Laban's promise that he could marry Rachel. Laban forced him to marry Leah first. Laban kept reneging on his bargains. Finally, Laban and Jacob agreed on what animals Jacob should have for all his labours. He said Jacob could take all the spotted and speckled cattle as part of his wages. Jacob decided to slant the count in his favour. The exact reference (Genesis, Ch. 30, Verses 38–39 in the King James Version of the Bible) to Jacob's mechanism as to how pattern comes about is the following.

'And he (Jacob) set the rods which he had pilled before the flocks in the gutters in the watering troughs when the flocks came to drink, that they would conceive when they came to drink. And the flocks conceived before the rods, and brought forth cattle ringstraked, speckled and spotted.'

These are the well-known Jacob sheep, of course, which are spotted, and so beloved by those wishing to play farmer. As a theory it might be a little difficult to get published in a reputable journal today.

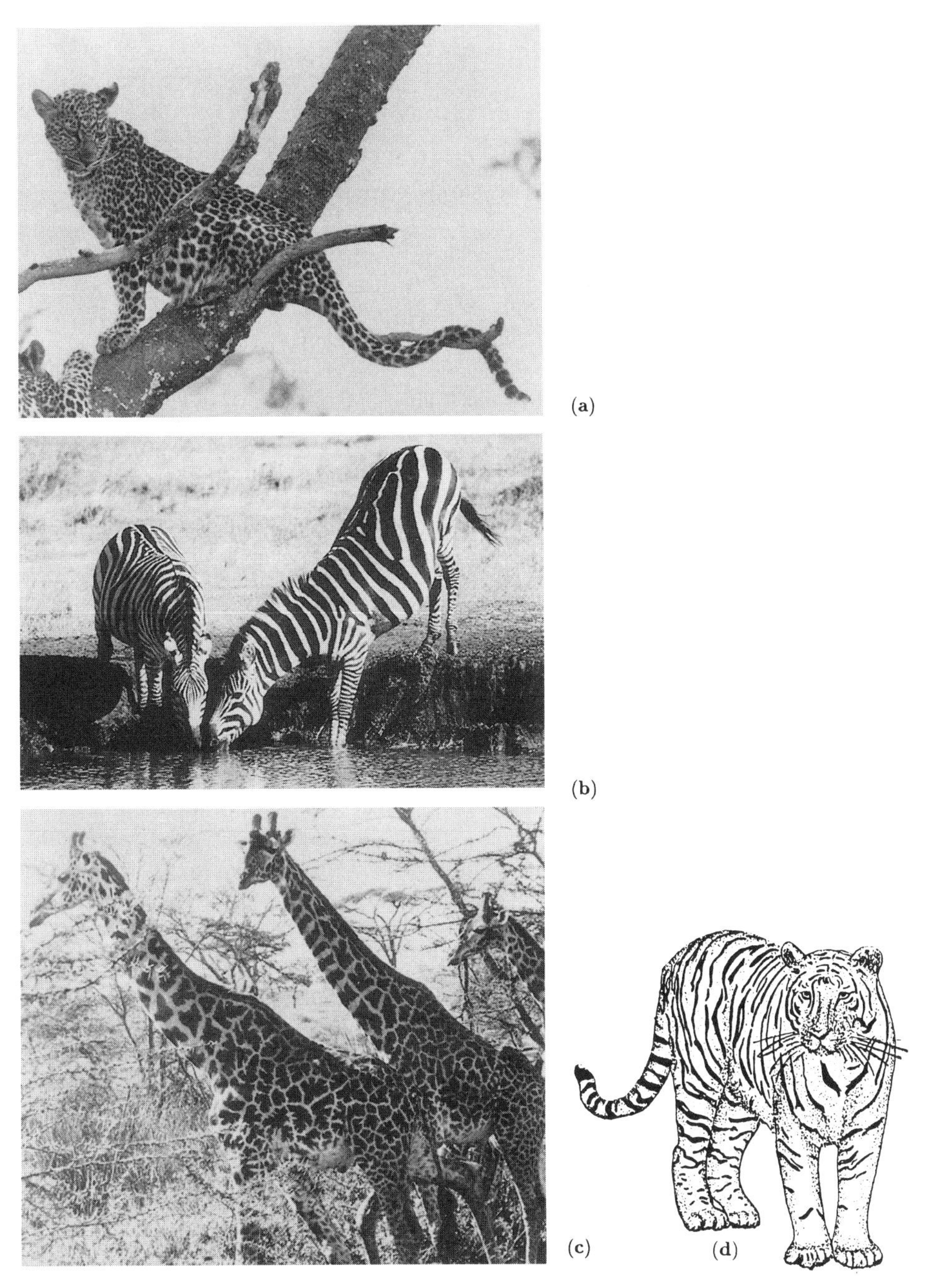

Figure 3.1. Typical animal coat markings on the (**a**) leopard (*Panthera pardus*); (**b**) zebra (*Equus grevyi*); (**c**) giraffe (*Giraffa camelopardis tippelskirchi*); (Photographs courtesy of Dr. Hans Kruuk) (**d**) tiger (*Felis tigris*).

picked up seriously again until the 1970's from which time it has become a veritable industry. Murray (1980, 1981) took a reaction diffusion system, which could be diffusively driven unstable, as the possible mechanism responsible for laying down most of the spacing patterns; these are the morphogen prepatterns for the animal coat markings. The fundamental assumption is that subsequent differentiation of the cells to produce melanin simply reflects the spatial pattern of morphogen concentration.

In the last chapter we showed how such reaction diffusion mechanisms can generate spatial patterns. In this section we (i) present results of numerical simulations of a specific reaction diffusion system with geometries relevant to the zoological problem, (ii) compare the patterns with those observed in many animals and finally (iii) highlight the circumstantial evidence to substantiate the hypothesis that a single mechanism is all that is possibly required. Bard (1981) and Young (1984) also investigated animal coat patterns from a reaction diffusion point of view. Cocho et al. (1987) proposed a quite different model based on cell–cell interaction and energy considerations; it is essentially a cellular automata approach. Savic (1995) used a mechanochemical model (see Chapter 6) based on a prepattern of polarized cells in the epithelium, the surface of the skin. Several of the subsequent models are based on a computer graphics approach such as the one by Walter et al. (1998) who also review the general area and recent contributions. Their 'clonal mosaic model' is based on cell–cell interactions (with specific rules) and involves cell division. They specifically use it to generate the spot and stripe patterns on giraffes and the large cats. The patterns they obtain are remarkably similar in detail to those found on specific animals. They relate their model to recent experiments on pigment cells. Their model and graphics procedure can be used on complex surfaces and include growth. It is certainly now possible to generate more or less most of the observed patterns. In many ways there are enough models for generating animal coat paterns. Progress now really depends on the experiments that are suggested by the models and how the results can be interpreted from a modelling point of view. The work on the patterns on the alligator and on their teeth patterning discussed in depth in Chapter 4 are case studies where some progress has been made along these lines.

Although the development of the colour pattern on the integument, that is, the skin, of mammals occurs towards the end of embryogenesis, we suggest that it reflects an underlying prepattern that is laid down much earlier. In mammals the prepattern is formed in the early stages of embryonic development—in the first few weeks of gestation. In the case of the zebra, for example, this is around 21–35 days; the gestation period is about 360 days. In the case of alligator stripes it is about halfway through the gestation period (around 65 days); see Chapter 4.

To create the colour patterns certain genetically determined cells, called melanoblasts, migrate over the surface of the embryo and become specialised pigment cells, called melanocytes, which lie in the basal layer of the epidermis. Hair colour comes from the melanocytes generating melanin, within the hair follicle, which then passes into the hair. The book by Prota (1992) discusses the whole processs of melanogenesis. As a result of graft experiments, it is generally agreed that whether or not a melanocyte produces melanin depends on the presence of a chemical although we still do not yet know what it is. In this way the observed coat colour pattern reflects an underlying chemical prepattern, to which the melanocytes are reacting to produce melanin.

For any pattern formation mechanism to be applicable the scale of the actual size of the patterns has to be large compared to the cell diameter. For example, the number of cells in a leopard spot, which, at the time of laying down the pattern, is probably of the order of 0.5 mm, that is, of the order of 100 cells. Since we do not know what reaction diffusion mechanism is involved, and since all such systems are effectively mathematically equivalent as we saw in the previous chapter, all we need at this stage is a specific system to study numerically. Solely for illustrative purposes Murray (1980, 1981a) chose the Thomas (1975) system, the kinetics of which is given in (2.5) in the last chapter: it was chosen because it is a real experimental system with parameters associated with real kinetics. There are now several others equally reasonable to use, but until we know what the patterning mechanism is it suffices for our study. The nondimensional system is given by (2.7) with (2.8), namely,

$$\frac{\partial u}{\partial t} = \gamma f(u, v) + \nabla^2 u, \quad \frac{\partial v}{\partial t} = \gamma g(u, v) + d\nabla^2 v$$
$$f(u, v) = a - u - h(u, v), \quad g(u, v) = \alpha(b - v) - h(u, v) \tag{3.1}$$
$$h(u, v) = \frac{\rho u v}{1 + u + K u^2}.$$

Here a, b, α, ρ and K are positive parameters. The ratio of diffusion coefficients, d, must be such that $d > 1$ for diffusion-driven instability to be possible. Recall from the last chapter that the scale factor γ is a measure of the domain size.

With the integument of the mammalian embryo in mind the domain is a closed surface and appropriate conditions for the simulations are periodic boundary conditions with relevant initial conditions—random perturbations about a steady state. We envisage the process of pattern formation to be activated at a specific time in development, which implies that the reaction diffusion domain size and geometry is prescribed. The initiation switch could, for example, be a wave progressing over the surface of the embryo which effects the bifurcation parameter in the mechanism which in turn activates diffusion-driven instability. What initiates the pattern formation process, and how it is initiated, are not the problems we address here. We consider only the pattern formation potential of the mechanism, to see whether or not the evidence for such a system is borne out, when we compare the patterns generated by the mechanism and observe animal coat markings with similar geometrical constraints to those in the embryo.

In Sections 2.4 and 2.5 in the last chapter we saw how crucially important the scale and geometry of the reaction diffusion domain were in determining the actual spatial patterns of morphogen concentration which start to grow when the parameters are in the Turing space where the system is diffusionally unstable. Refer also to Figures 2.8 and 2.9. (It will be helpful in the following to have the analysis and discussions in Sections 2.1–2.6 in mind.)

To investigate the effects of geometry and scale on the type of spatial patterns generated by the full nonlinear system (3.1) we chose for numerical simulation a series of two-dimensional domains which reflect the geometric constraints of an embryo's integument.

Let us first consider the typical markings found on the tails and legs of animals, which we can represent as tapering cylinders, the surface of which is the reaction diffusion domain. From the analysis in Sections 2.4 and 2.5 when the mechanism undergoes diffusion-driven instability, linear theory gives the range of unstable modes, k^2, in terms of the parameters of the model system: in two space dimensions with the domain defined by $0 < x < p$, $0 < y < q$, these are given by (2.43) as

$$\gamma L = k_1^2 < k^2 = \pi^2 \left(\frac{n^2}{p^2} + \frac{m^2}{q^2} \right) < k_2^2 = \gamma M, \tag{3.2}$$

where L and M are functions only of the kinetics parameters of the reaction diffusion mechanism. With zero flux boundary conditions, the solution of the linear problem involves exponentially growing modes about the uniform steady state, and is given by (2.43) as

$$\sum_{n,m} \mathbf{C}_{n,m} \exp[\lambda(k^2)t] \cos \frac{n\pi x}{p} \cos \frac{m\pi y}{q}, \quad \text{where } k^2 = \pi^2 \left(\frac{n^2}{p^2} + \frac{m^2}{q^2} \right), \tag{3.3}$$

where the $\mathbf{C}$ are constants which are obtained from a Fourier series of the initial conditions (they are not needed here) and summation is over all pairs (n, m) satisfying (3.2).

Now consider the surface of a tapering cylinder of length s with $0 \le z \le s$ and with circumferential variable q. The linear eigenvalue problem equivalent to that in (2.41) requires the solutions $\mathbf{W}(\theta, z; r)$ of

$$\nabla^2 \mathbf{W} + k^2 \mathbf{W} = 0, \tag{3.4}$$

with zero flux conditions at $z = 0$ and $z = s$ and periodicity in θ. Since we are only concerned here with the surface of the tapering cylinder as the domain, the radius of the cone, r, at any point is essentially a 'parameter' which reflects the thickness of the cylinder at a given z. The equivalent solution to (3.3) is

$$\sum_{n,m} \mathbf{C}_{n,m} \exp[\lambda(k^2)t] \cos(n\theta) \cos \frac{m\pi z}{s}, \quad \text{where} \quad k^2 = \frac{n^2}{r^2} + \frac{m^2\pi^2}{s^2}, \tag{3.5}$$

where the summation is over all pairs (n, m) satisfying the equivalent of (3.2); namely,

$$\gamma L = k_1^2 < k^2 < \frac{n^2}{r^2} + \frac{m^2\pi^2}{s^2} < k_2^2 = \gamma M. \tag{3.6}$$

Note that r appears here as a parameter.

Now consider the implications as regards the linearly growing spatial patterns, which we know for simple patterns usually predict the finite amplitude spatial patterns which are ultimately obtained. If the tapering cylinder is everywhere very thin this means r is small. This in turn implies that the first circumferential mode with $n = 1$, and all others with $n > 1$, in (3.5) lie outside the unstable range defined by (3.6). In this case the unstable modes involve only z-variations. In other words it is equivalent to the one-dimensional situation with only one-dimensional (stripe) patterns as in Figure 2.8;

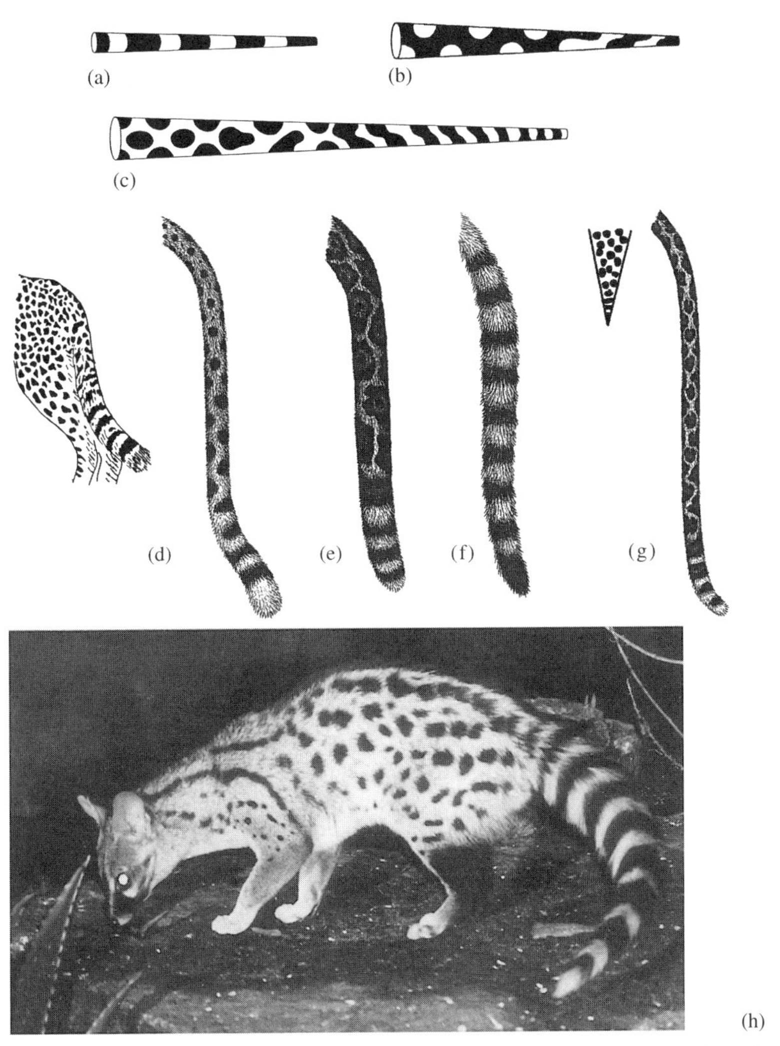

Figure 3.2. Computed solutions of the nonlinear reaction diffusion system (3.1) with zero flux boundary conditions and initial conditions taken as random perturbations about the steady state. The dark regions represent concentrations in the morphogen u above the steady state u_s. Parameter values $\alpha = 1.5$, $K = 0.1$, $\rho = 18.5$, $a = 92$, $b = 64$ (these imply a steady state $u_s = 10$, $v_s = 9$), $d = 10$. With the same geometry, in (**a**) the scale factor $\gamma = 9$ and in (**b**) $\gamma = 15$. Note how the pattern bifurcates to more complex patterns as γ increases. In (**c**) the scale factor $\gamma = 25$ and a longer domain is used to illustrate clearly the spot-to-strip transition: here the dark regions have $u < u_s$. (**d**) Typical tail markings from an adult cheetah (*Acinonyx jubatis*). (**e**) Typical adult jaguar (*Panthera onca*) tail pattern. (**f**) Prenatal (but just prenatal) tail markings in a male genet (*Genetta genetta*). (After Murray 1981a,b) (**g**) Typical markings on the tail of an adult leopard. Note how far down the tail the spots are with only a few stripes near the tip. See also the photograph in Figure 3.1(a) where the leopard's tail is conveniently draped so as to demonstrate this trait clearly. The prenatal leopard tail is very much shorter and shows why the adult pattern is as shown. (**h**) A common genet (*Genetta genetta*) showing the distinctly striped tail emerging from a spotted body. (Photograph courtesy of Dr. Hans Kruuk)

see also Figure 3.2(a). If, however, r is large enough near one end so that $n \neq 0$ is in the unstable range defined by (3.6), θ-variations appear. We thus have the situation in which there is a gradation from a two-dimensional pattern in z and θ at the thick end to the one-dimensional pattern at the thin end. Figure 3.2 shows some numerically computed solutions (using a finite element procedure) of (3.1) for various sizes of a tapering domain. In Figures 3.2(a) and (b) the only difference is the scale parameter γ, which is the bifurcation parameter here. In fact, in all the numerical simulations of the nonlinear system (3.1) reproduced in Figures 3.2 to 3.4, the mechanism parameters were kept fixed and *only* the scale and geometry varied. Although the bifurcation parameter is scale (γ), geometry plays a crucial role.

The tail patterns illustrated in Figure 3.2 are typical of many spotted animals, particularly the cats (*Felidae*). The cheetah, jaguar and genet are good examples of this pattern behaviour. In the case of the leopard (*Panthera pardus*) the spots almost reach the tip of the tail, whereas with the cheetah there is always a distinct striped part and the genet has a totally striped tail. This is consistent with the embryonic tail structure of these animals around the time we suppose the pattern formation mechanism is operative. The genet embryo tail has a remarkably uniform diameter which is relatively quite thin; in the photograph of the fully grown genet in Figure 3.2(h) the hair is typically fluffed up. The prenatal leopard tail sketched in Figure 3.2(g) is sharply tapered and relatively short; the adult leopard tail (see Figures 3.1(a) and 3.2(g)) is long but it has the same number of vertebrae. Thus the fact that the spots go almost to the tip is consistent with a rapid taper, with stripes, often incomplete, only appearing at the tail tip, if at all. This postnatal stretching is also reflected in the larger spots farther down the tail as compared to those near the base or the body generally; refer to Figures 3.1(a) and 3.2(g).

Consider now the typical striping on the zebra as in Figures 3.1(b) and 3.3(a) and (b). From the simulations reproduced in Figure 3.2(a) we see that reaction diffusion mechanisms can generate stripes easily. Zebra striping was investigated in detail by Bard (1977) who argued that the pattern was laid down around the 3rd to 5th week through gestation. He did not discuss any actual patterning mechanism but from the results in this section this is not a problem. The different species of zebra have different stripe patterns and he suggested that the stripes were therefore laid down at different times in gestation. Figure 3.3 shows the hypothesised patterning on embryos at different stages in gestation and schematically shows the effect of growth.

By noting the number of adult stripes, and how they had been distorted by growth if laid down as a regular stripe array, Bard (1977) deduced that the distance between the stripes when they were laid down was about 0.4 mm. He also deduced the time in gestation when they were created. Figure 3.3(c), with pattern distortion with growth as shown in Figure 3.3(d) and (e), is consistent with the stripe pattern on the zebra *Equus burchelli* as in Figure 3.3(b); see also the photograph in Figure 3.1(b). Grevy's zebra, *Equus grevyi*, in Figure 3.3(a) has many more stripes and these are laid down later in gestation, around 5 weeks, as in Figure 3.3(e) where again they are taken to be 0.4-mm apart.

If we now look at the scapular stripes on the foreleg of zebras as illustrated in Figure 3.4(a), we have to consider an actual pattern formation mechanism as was done by Murray (1980, 1981a,b). Here we see that the mathematical problem is that of the junction between a linear striped domain joined at right angles to another striped domain;

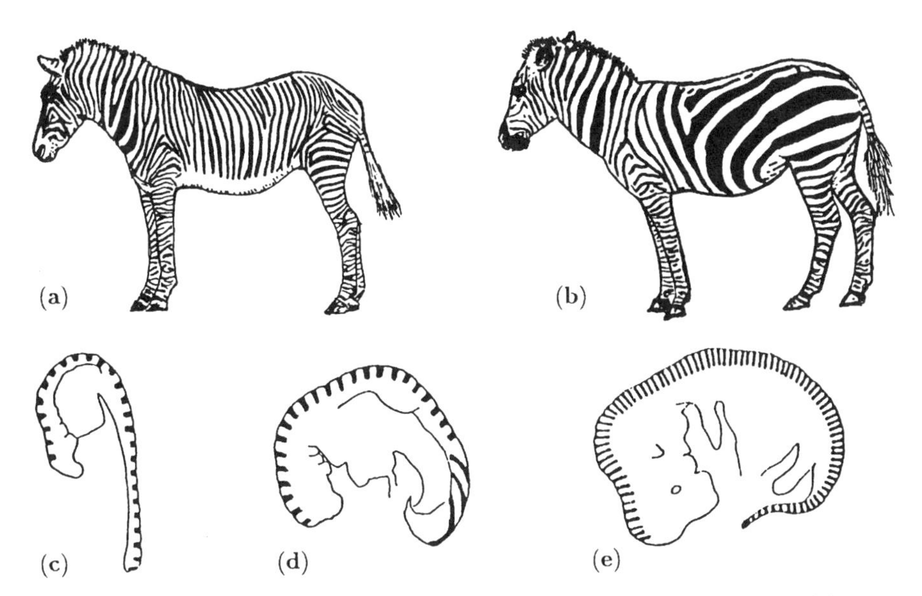

Figure 3.3. Typical zebra patterns: (**a**) *Equus grevyi*; (**b**) *Equus burchelli*. Proposed strip pattern 0.4 mm apart superimposed on two zebra embryos: (**c**) 12 day embryo; (**d**) The effect of 3–4 days of the pattern in (**c**). (**e**) A similar stripe pattern laid down on a 5 week old embryo. ((**c**)–(**e**) redrawn after Bard 1977)

Figure 3.4(b) is the pattern predicted by the reaction diffusion mechanism for such a domain. The experimentally obtained pattern displayed below in Figure 3.8(e) confirms this mathematical prediction.

The markings on zebras are extremely variable yet remain within a general stripe theme. Animals which are almost completely black with lines of white spots as well as those almost completely white have been seen (see, for example, Kingdon 1979). We come back to the question of pattern abnormalities below in Section 3.2 on coat marking teratologies.

If we now consider the usual markings on the tiger (*Felis tigris*) as in Figure 3.1(d) we can see how its stripe pattern could be formed by analogy to the zebra. The gestation period for the tiger is around 105 days. We anticipate the pattern to be laid down quite early on, within the first few weeks, and that the mechanism generates a regularly spaced

Figure 3.4. (**a**) Typical examples of scapular stripes on the foreleg of zebra (*Equus zebra zebra*). (**b**) Predicted spatial pattern from the reaction diffusion mechanism: see also Figure 3.8(e). (After Murray, 1980, 1981a)

stripe pattern at that time. Similar remarks to those made for the zebra regarding growth deformation on the stripe pattern equally apply to tiger stripes. Many tigers show similar distortions in the adult animal.

Let us now consider the giraffe, which is one of the largest animals that still exhibits a spotted pattern. Figure 3.5(a) is a sketch of a giraffe embryo 35 to 45 days old; it already has a clearly recognizable giraffe shape, even though the gestation period is about 457 days. The prepattern for the giraffe coat pattern has almost certainly been laid down by this time. Figure 3.5(b) is a sketch of typical neck markings on the reticulated giraffe. Figures 3.5(c)–(e) are tracings, on approximately the same scale, of trunk spots from the major giraffe species. Figure 3.5(f) shows a typical pattern computed from the mechanism (3.1) with the same kinetics parameter values as for Figure 3.2.

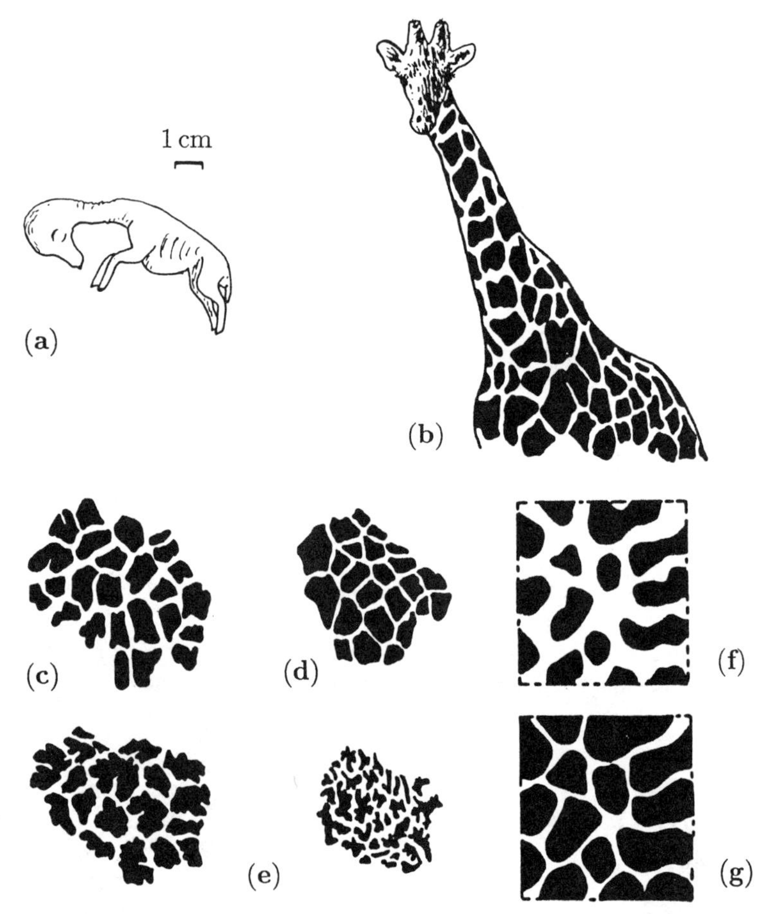

Figure 3.5. (**a**) Giraffe (*Giraffa camelopardis*): 35–45-day embryo. (**b**) Typical neck spots on the reticulated giraffe (*Giraffa camelopardis reticulata*). (**c**)–(**e**) Tracings (after Dagg 1968) of trunk spots (to the same scale) of giraffe, *Giraffa camelopardis* (**c**) *rothschildi*, (**d**) *reticulata*, (**e**) *tippelskirchi*. (**f**) Spatial patterns obtained from the model mechanism (3.1) with kinetics parameter values as in Figure 3.2. (**g**) Spatial pattern obtained when a lower threshold than in (**f**) is considered to initiate melanogenesis in the same simulations which gave (**f**). (From Murray 1981b, 1988)

We arbitrarily chose the homogeneous steady state as the threshold for melanocytes to produce melanin, represented by the dark regions in the figure. It is possible, of course, that the threshold which triggers melanogenesis is either lower or larger than the homogeneous steady state. For example, if we choose a lower threshold we get a different pattern: Figure 3.5(g) is an example in which we chose a lower threshold in the simulations which gave Figure 3.5(f). This produces larger areas of melanin. We can thus see how the markings on different species of giraffe could be achieved simply, if the melanocytes are programmed to react to a lower morphogen concentration. The giraffe photographs in Murray (1988) illustrate this particularly clearly.

The dramatic effect of scale is clearly demonstrated in Figure 3.6 where only the scale varies from one picture to the other—as indicated by the different values for γ. It is not suggested that this is necessarily the typical shape of the integument at the time of prepattern formation; it is only a nontrivial specimen shape to illustrate the results and highlight the striking effect of scale on the patterns generated. If the domain size (γ) is too small, then no spatial pattern can be generated. We discussed this in detail in the last chapter, but it is clear from the range of unstable modes, m and n in (3.6), for example. With a small enough domain, that is, small enough γ, even the lowest nonzero m and n lie outside the unstable range. This implies that in general very small animals can be expected to be uniform in colour; most of them are. As the size increases, γ

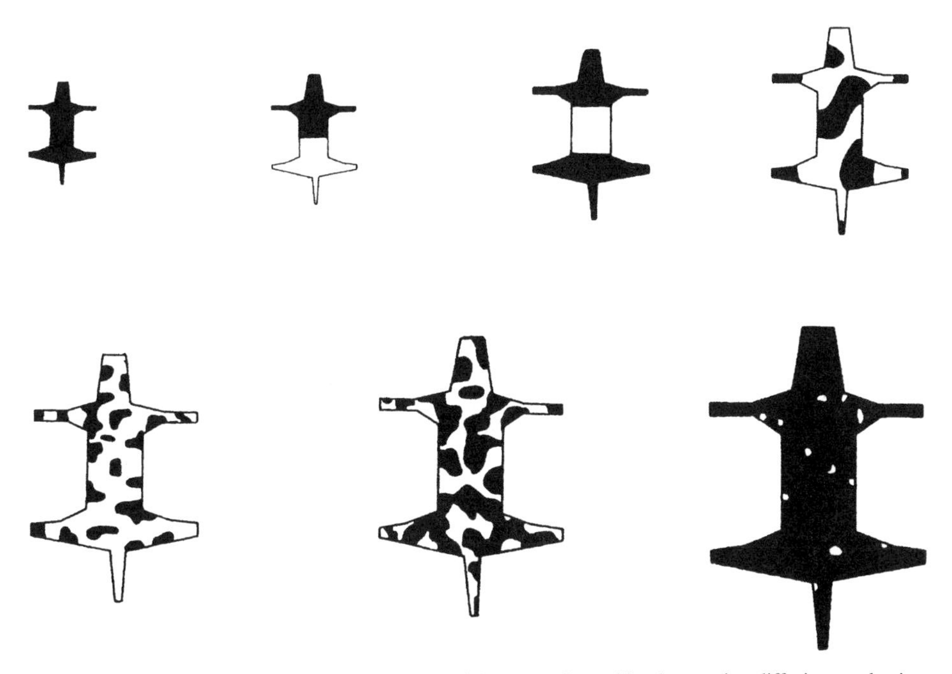

Figure 3.6. Effect of body surface scale on the spatial patterns formed by the reaction diffusion mechanism (3.1) with parameter values $\alpha = 1.5$, $K = 0.125$, $\rho = 13$, $a = 103$, $b = 77$ (steady state $u_s = 23$, $v_s = 24$), $d = 7$. Domain dimension is related directly to γ. From top to bottom, left to right, the γ-values are: $\gamma < 0.1$; $\gamma = 0.5$; $\gamma = 25$; $\gamma = 250$; $\gamma = 1250$; $\gamma = 3000$; $\gamma = 5000$. The same size shape was used for all simulations. The variable size here is for illustrative purposes. (From Murray 1988 based on Murray 1980, 1981a)

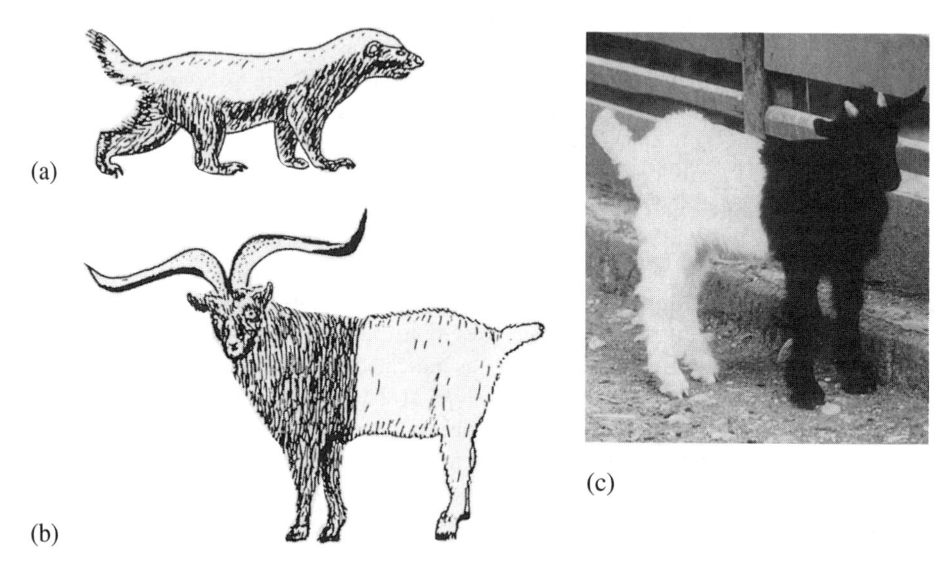

Figure 3.7. Examples of the simplest coat patterns found in animals: (**a**) ratel or honey badger (*Mellivora capensis*). (**b**) Adult Valais goat (*Capra aegragrus hircus*). (After Herán 1976). (**c**) Young Valais goat. (Photograph courtesy of Avi Baron and Paul Munro)

passes through a series of bifurcation values and different spatial patterns are generated. However, for very large domains as in the last figure in Figure 3.6, the morphogen concentration distribution is again almost uniform: the structure is very fine. This might appear, at first sight, somewhat puzzling. It is due to the fact that for large domains, large γ, the linearly unstable solutions derived from (3.3) have the equivalent of large m and n, which implies a very fine scale pattern; so small, in fact, that essentially no pattern can be seen. This suggests that most very large animals, such as elephants, should be almost uniform in colour, as indeed most are.

Consider now the first bifurcation from a uniform coat pattern as implied by the second figure in Figure 3.6: Figures 3.7(a)–(c) are sketches and a photograph of two striking examples of the half-black, half-white pattern, namely, the ratel, or honey badger, and the Valais goat. The next bifurcation for a longer and still quite thin embryo at pattern formation is elegantly illustrated in Figure 3.7(d) below which relates to the third pattern in Figure 3.6. Figure 3.7(e) below is another example of this latter pattern in the Belted Galloway cows, common in South Scotland (where I grew up) where they are known as 'belties.'

In all the numerical simulations with patterns other than the simplest, such as the first three in Figure 3.6, the final patterns were dependent on the initial conditions. However, for a given set of parameters, geometry and scale, the patterns for all initial conditions are *qualitatively* similar. From the point of view of the applicability of such mechanisms for generating animal coat patterns, this dependence on initial conditions is a very positive attribute of such models. The reason is that the initial random conditions for each animal are unique to that animal and hence so is its coat pattern, but each lies within its own general class. So, all leopards have a spotted pattern yet each has a unique

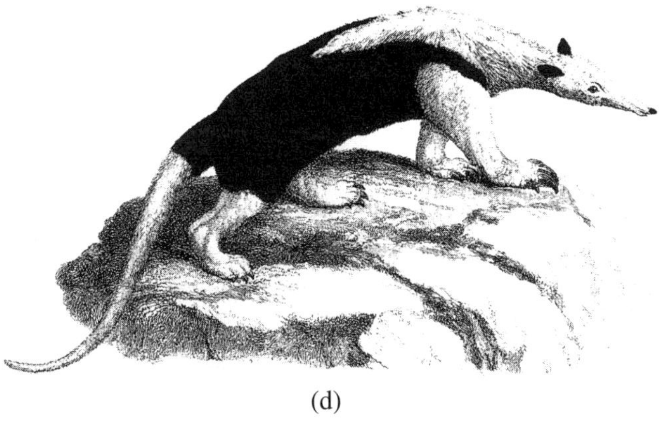

(d)

(e)

Figure 3.7. (continued) (**d**) The next pattern bifurcation is dramatically and elegantly illustrated in an early 19th century print of the anteater (*Tamandua tetradactyl*) and in (**e**) of Galloway belted cows in South Scotland.(Photograph by Allan Wright, Castle Douglas, Scotland and reproduced with permission)

distribution of spots. On tigers and zebras, for example, the stripe patterns can be quite diverse while still adhering to a general theme.

Although we have considered only a few specific coat markings (see Murray 1981a, b, 1988 for further discussion) we see that there is a striking similarity between the patterns that the model generates and those found on a wide variety of animals. Even with the restrictions we imposed on the parameters for our simulations the wealth of possible patterns is remarkable. The patterns depend strongly on the geometry and scale of the reaction domain although later growth may distort the initial pattern.

To summarise we hypothesise that almost all animal coat patterns can be generated by a single mechanism. Any reaction diffusion mechanism capable of generating diffusion-driven spatial patterns would be a plausible model. The pattern which evolves is determined at the time the mechanism is activated since this relates directly to the geometry and scale of the embryo's integument. The time of the activation wave (such as that illustrated in Figure 2.15(d)), or activation switch, is inherited. With most small animals with short gestation periods we would expect uniformity in color, which is generally the case. For larger surface integument at the time of activation, the first bifurcation produces patterns where animals can be half black and half white; see Figure 3.7. For progressively larger domains at activation more and more pattern structures emerge, with a progression through certain anteaters, zebras, on to the large cats and so on. The simpler patterns are remarkably stable; that is, they are quite insensitive to conditions at the time the mechanism is activated. At the upper end of the size scale we have more variability within a class as in the close spotted giraffes. As mentioned we expect very large animals to be uniform in colour again, which indeed is generally the case, with elephants, rhinoceri and hippopotami being typical examples.

As mentioned above, we expect the time of activation of the mechanism to be inherited and so, at least in animals where the pattern is important for survival, pattern formation is initiated when the embryo is a given size. Of course, the conditions on the embryo's surface at the time of activation naturally exhibit a certain randomness which produces patterns which depend uniquely on the initial conditions, the geometry and the scale. A very important aspect of this type of mechanism is that, for a given geometry and scale, the patterns found for a variety of random initial conditions are qualitatively similar. For example, with a spotted pattern it is essentially only the distribution of spots which varies. The resultant individuality is important for both kin and group recognition. Where the pattern is of little importance to the animal's survival, as with domestic cats, the activation time need not be so carefully controlled and so pattern polymorphism, or variation, is much greater.

It is an appealing idea that a single mechanism could generate all of the observed mammalian coat patterns. Reaction diffusion models, cell-chemotaxis models and the powerful mechanochemical models discussed later in Chapter 6 have many of the attributes such a pattern formation mechanism must have. The latter in fact have a pattern generation potential even richer than that of reaction diffusion mechanisms. The considerable circumstantial evidence which comes from comparing the patterns generated by the model mechanism with specific animal pattern features is encouraging. The fact that many general and specific features of mammalian coat patterns can be explained by this simple theory does not, of course, mean that it is correct, but so far they have not all been explained satisfactorily by any other general theory. The above results nevertheless support a single all-encompassing mechanism for pattern formation for animal coat markings.

As an interesting mathematical footnote, the initial stages of spatial pattern formation by reaction diffusion mechanisms (when departures from uniformity are small) poses the same type of mathematical eigenvalue problem as that describing the vibration of thin plates or drum surfaces. The vibrational modes are also governed by (3.4) except that **W** now represents the amplitude of the vibration. So, we can highlight experimentally how crucial geometry and scale are to the patterns by examining analogous vibrating drum surfaces. If the size of the surface is too small it will simply not sustain vibrations; that is, the disturbances simply die out very quickly. Thus a minimum size is required before we can excite any sustainable vibration. If we consider a domain similar to that in Figure 3.6 for the drum surface, which in our model is the reaction diffusion domain, we get a set of increasingly complicated modes of possible vibration as we increase the size.

Although it is not easy to use the same boundary conditions for the vibrations that we used in the reaction diffusion simulations, the general features of the patterns exhibited must be qualitatively similar from mathematical considerations. The equivalent of γ in the vibrating plate problem is the frequency of the forcing vibration. So, if a pattern

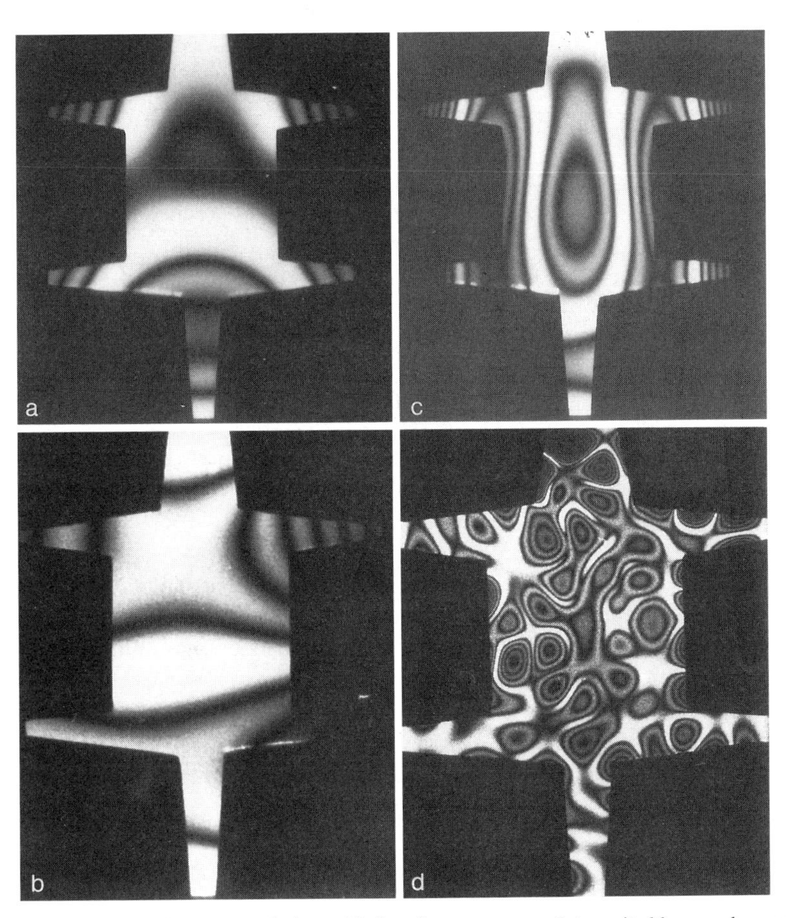

Figure 3.8. Sequence of time-average holographic interferograms on a plate excited by sound waves of increasing frequency from (**a**) to (**d**). The vibrational patterns are broadly in line with the patterns shown in Figure 3.6. Increasing frequency is equivalent to vibration at a constant frequency and increasing the plate size. (**e**) This shows vibrations very similar to the predicted pattern in Figure 3.4(b) while in (**f**) the spot-to-stripe transition in a tapering geometry is clearly demonstrated. (From Xu et al. 1983)

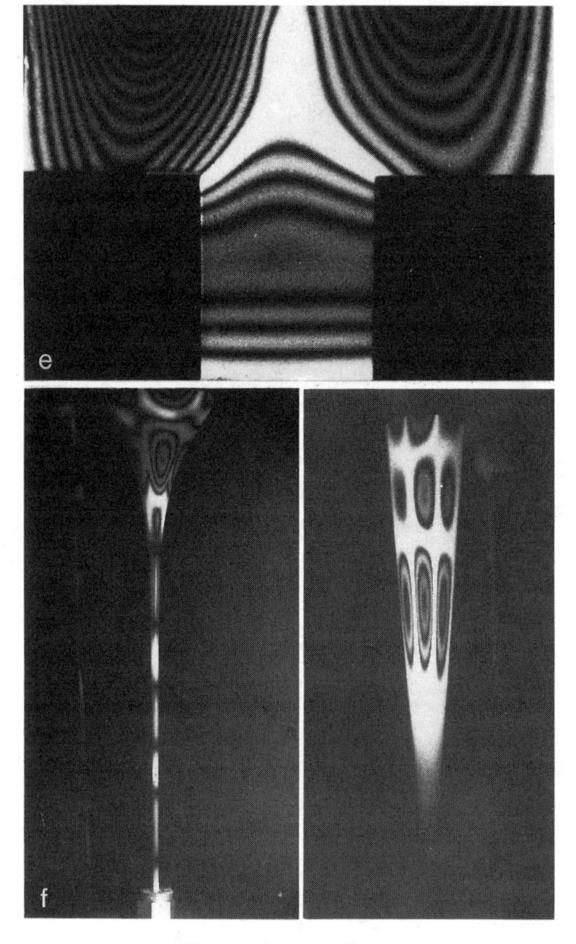

Figure 3.8. (continued)

forms on a plate vibrating at a given frequency then the pattern formed on a larger but similar plate is the same as on the original plate vibrated at a proportionally larger frequency. According to linear vibration theory, a doubling of the plate size, for example, is equivalent to keeping the original plate size and doubling the frequency. These experiments were carried out for geometries similar to those in Figures 3.2(c), 3.4 and 3.6 and the results are shown in Figure 3.8 (see Xu et al. 1983 for further details).

3.2 Teratologies: Examples of Animal Coat Pattern Abnormalities

The model we have discussed above offers possible explanations for various pattern anomalies on some animals. Under certain circumstances a change in the value of one of the parameters can result in a very marked change in the pattern obtained. An early activation, for example, in a zebra would result in an all black animal. A delay in activating the mechanism would give rise to spots on the underlying black field. Examples of both of these have been observed and recorded, for example, by Kingdon (1979).

Whether a parameter change affects the pattern markedly depends on how close the parameter value is to a bifurcation value (recall Figure 2.14 and the discussion in Section 2.5). The fact that a small change in a parameter near a bifurcation boundary can result in relatively large changes in pattern has important implications for evolutionary theory as we show later in Chapter 7.

Disruption of the usual patterning mechanism can clearly be effected by a change in the timing of the pattern formation mechanism or of any of the parameters involved in the process. There are many such examples of coat pattern teratologies. For example, a delay in the pattern formation mechanism in the case of the zebra, say, would result in the animal being more spotted than striped since the domain is larger at the time of pattern generation. Such pattern aberrations in the past have frequently given rise to a 'new' species. Since we do not know how most patterns are formed in development it is not surprising that the discovery of some unusual coat pattern could spawn a 'new' species. This was the case with a cheetah trapped in 1926 in the Umvukwe area in Zimbabwe (Rhodesia as it then was) and which was reported to the Natural History Museum in London and a photograph of it published in the English magazine *The Field* in 1927. Figure 3.9(a) is from a drawing of it from Pocock (1927; see also Ewer 1973). Pocock, a distinguished biologist of the time, was so convinced that it was a new species of cheetah (*Acinonyx*) based primarily on the coat pattern but also on other aspects of the anatomy that he was convinced were different to the normal cheetah (claw length etc.), that he declared it a new species and called it *Acinonyx rex*. With our knowledge of how only a small variation in the timing of the mechanism which generates the spatial pattern on animal coats could result in major variations, it seems likely that this is what happened in this case. Figure 3.9(b) is a recent photograph, from South Africa, of a similar kind of coat pattern abnormality on a cheetah. Interestingly Pocock comments that it is unusual that there are so few other examples of this new species, which of course, adds support to the theory that it is only a slight change in the patterning mechanism which is responsible for the pattern aberration. There are, in fact, few cases of relatively stable polymorphic forms.

It is interesting that the abnormal coat patterns in both animals in Figure 3.9 are quite similar with stripes down the back and spots appearing towards the belly. We can envisage how such a pattern could have arisen. If the above mechanism for generating coat patterns obtains (or an equivalent one which is geometry and scale-dependent in a similar way) then the mechanism was probably activated at an earlier time in gestation when the embryo was considerably smaller than the size at the normal activation time. The pattern in such circumstances would be less complex and the possibility of stripe formation would be more likely. Later in Chapter 6 we shall see that many patterns, such as the precursors of hair on the back of animals, spread out laterally from the dorsal midline. This is also probably the scenario in more complex animal coat pattern formation. In this way if the time for the mechanism to form pattern is comparable in the normal and abnormal animals, complex (spot-like) patterns will form where the domain becomes sufficiently large to sustain them and become more like the usual spots on a cheetah. In both cases however, the spots are less distinct and much larger, also as we would expect.

Figure 3.9. (**a**) A drawing of the coat pattern abnormality found on a cheetah in Zimbabwe in 1926. It was originally thought to be a hybrid between a cheetah and a leopard but then it was decided that it was a new species and was called *Acynonix rex*. (From Pocock 1927) (**b**) A recent photograph of such a cheetah in the Kruger National Park in South Africa. (Photograph by Anthony Bannister, reproduced with permission of ABPL Image Library, Parklands, South Africa)

There are some other interesting examples of mechanism disruption in animal coat pattern formation. Figure 3.10 is a photograph of a zebra which is almost totally black, which is clearly the default colour and so zebras are therefore black animals with white stripes rather than white animals with black stripes.[2]

[2]In a small not very serious survey I have taken over a number of years, when an audience is asked whether they think a zebra is a black animal with white stripes or the converse, with surprising unanimity black people say it is a black animal with white stripes with white people saying the opposite.

Figure 3.10. A photograph of a coat pattern teratology in the case of a zebra *Equus burchelli* taken in the Kruger National Park in South Africa.

In this case since the default colour of the zebra is black a possible explanation is that the mechanism could have been activated at the usual time in gestation but its time for pattern formation was curtailed thus giving rise to poorly formed and very thin stripes. The zebra seemed in all other respects normal except that it clearly knew it was different and was somewhat of a social outcast, spending most of its time on the edge of the group.

Another example of pattern disruption is that of the zebra-like striped sheep as shown in Figure 3.11 that appeared in a flock of sheep in Australia. The default colour of sheep used to be black but they were bred out of the flocks because the white fleece was more desirable.[3] Here the mechanism of pattern formation is, in a sense, to make the sheep white in which case the embryo at the time the 'pattern' is laid down is small so that a uniform colour is obtained. If the embryo is larger when the mechanism operates it produces spatial patterns in the form of stripes.

In relation to the default colour in sheep being black, in an article in *Nature* in 1880, Charles Darwin noted: 'the appearance of dark-coloured or piebald sheep is due to a reversion to the primeval colouring of the species'—a tendency 'most difficult to eradicate, and quickly to gain in strength if there is no selection.' He went on to quote

[3] I was told by an Australian friend and colleague, who went to school with the man in the photograph, that there was enormous interest by people wanting to buy the fleece of this striped sheep. Apparently the owner has been trying to reproduce it by various mating strategies but, so far, without success. It would be interesting to see if a 'Dolly'-like clone would produce a similar but not, of course, exactly the same coat pattern.

Figure 3.11. A photograph of a striped coat pattern teratology on a sheep in Australia. (Photograph reproduced with permission of *The Canberra Times*, Australia)

from a letter from a Mr. Sanderson which referred to the declining percentage of spotted or black sheep due to the Australian woolgrowers' selection which certainly speeds up evolutionary development: 'In the early days before fences were erected and when shepherds had charge of very large flocks (occasionally 4000 and 5000) it was important to have a few sheep easily noticed amongst the rest; and hence the value of a certain number of black or partly black sheep, so that coloured lambs were then carefully preserved. It was easy to count ten or a dozen such sheep in a flock, and when one was missing it was pretty safe to conclude that a good many had strayed with it, so that the shepherd really kept count of his flock by counting the speckled sheep. As fences were erected the flocks were made smaller and the necessity for having spotted sheep passed away.'

Perhaps all we can say, at this stage, about all of these mutations is that the mechanism of coat pattern formation has been disrupted. The tightness in the timing of activation of the pattern formation mechanism is important in animals in the wild since their survival is intimately tied up with their visual markings. The mechanism and its genetic control are important hereditary traits, aberrations from which are generally less successful, such as in the case of the black zebra in Figure 3.10. Where it is not important we would expect considerably more variation. This is the case with markings on domestic cats and dogs, for example.

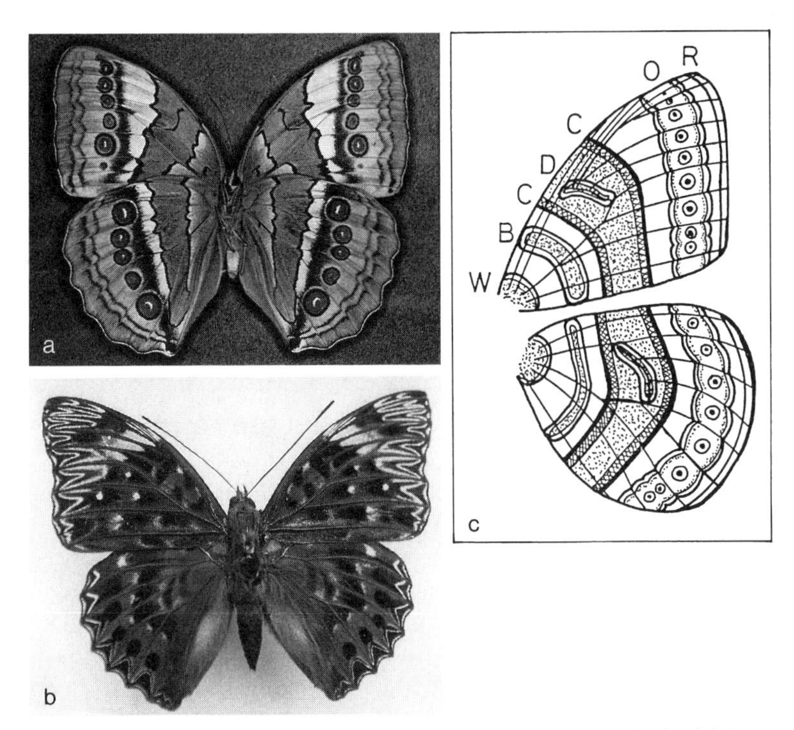

Figure 3.12. Examples of the varied and complex patterns on butterfly wings: (**a**) *Stichophthalma camadeva*. (Photograph courtesy of H.F. Nijhout) (**b**) *Dichorragia nesimachus*. (**c**) Basic groundplan of the pattern elements in the forewing and backwings in the Nymphalids. (After Schwanwitsch 1924, Suffert 1927) The letters denote: marginal bands (R), border ocelli (O), central symmetry bands (C), distal spots (D), basal bands (B), wing root bands (W). The butterfly in (**a**) exhibits almost all of the basic pattern elements. The arrowhead patterns in (**b**) pose a particular challenge to any pattern formation mechanism.

3.3 A Pattern Formation Mechanism for Butterfly Wing Patterns

The variety of different patterns, as well as their spectacular colouring, on butterfly and moth wings is astonishing. Figures 3.12(a) and (b) show but two examples; see also Figure 3.22 below. There are close to a million different types of butterflies and moths.[4] The study of butterfly wing colours and patterns has a long history, often carried out by gifted amateur scientists, particularly in the 19th century. In the 20th century there was a burgeoning of scientific activity. A review of the major elements in lepidopteran wing

[4]With such a vast number of types it is not surprising that there is a vast number of interesting aspects about their evolved social behaviour and reproduction. Andersson et al. (2000), in a recent study describe a fascinating aspect of sexual cooperation in the pierid butterfly (*Pieris napi*) the implications of which would be interesting to study from a modelling point of view. Traditional selection theory implies different selection pressures on the male and female which give rise to sexual conflict. In these butterflies, however, there is a remarkable cooperation between the male and female associated with mating. After mating they both share a common interest in reducing harassment of the female by other males wanting to mate. Andersson et al. (2000) found that the male, at mating, transfers a volatile anti-aphrodisiac which the female then emits when courted by other males; it very quickly makes them lose interest. This anti-aphrodisiac is so strong that males even avoid virgin females on which the aphrodisiac has been applied artificially.

patterns is given by Nijhout (1978; see also 1985a, 1991) and French (1999). Sekimura et al. (1998) review wing pattern formation from a different point of view, namely, at the scale level. Although the spectrum of different wing patterns is at first sight bewildering, Schwanwitsch (1924) and Suffert (1927) showed that in the case of the Nymphalids there are relatively few pattern elements; see, for example, the review by Nijhout (1978). Figure 3.12(c) shows the basic groundplan for the wing patterns in the Nymphalids; each pattern has a specific name. In this section we discuss a possible model mechanism for generating some of these regularly recurring patterns and compare the results with specific butterfly patterns and experiments.

Broadly speaking two types of butterfly wing patterns have been studied, namely, the gross colour patterns we discuss in this chapter, and the spacing patterns of cells on the wings. These patterns are on two different spatial scales. In the former cell interaction takes place over large distances while in the latter it is on the length scale of the cells. Also in the latter, precursors of these scale cells spread in a monolayer throughout the epidermis and migrate into rows approximately parallel to the body axis about 50 μ apart. Sekimura et al. (1999) developed a model for generating these parallel rows of cells on lepidopteran wings based on differential origin-dependent cell adhesion. Among other things they showed that biologically realistic cell adhesive properties were sufficient to generate these rows and, in particular, in the right orientation.

As with the development of the coat patterns on mammals the patterns on the wings of lepidoptera (butterflies and moths) appear towards the end of morphogenesis but they reflect an underlying prepattern that was laid down much earlier. The prepattern in lepidoptera is probably laid down during the early pupal stage or in some cases it perhaps starts just before (Nijhout 1980a).

Here we describe and analyse a possible model mechanism for wing patterns proposed by Murray (1981b). We apply it to various experiments concerned with the effect on wing patterns of cautery at the pupal stage in the case of the 'determination stream hypothesis' (Kuhn and von Engelhardt 1933), and on transplant results associated with the growth of ocelli or eyespots (Nijhout 1980a) all of which will be described below. As in the last section, a major feature of the model is the crucial dependence of the pattern on the geometry and scale of the wing when the pattern is laid down. Although the diversity of wing patterns might indicate that several mechanisms are required, among other things we shall show here how seemingly different patterns can be generated by the same mechanism.

As just mentioned, the formation of wing pattern can be made up by a combination of relatively few pattern elements. Of these, the *central symmetry patterns* (refer to Figure 3.12(c)) are common, particularly so in moth wings, and roughly consist of mirror image patterns about a central anterior–posterior axis across the middle of the wing (see, for example, Figure 3.12(a)). They were studied extensively by Kuhn and von Engelhardt (1933) in an attempt to understand the pattern formation on the wings of the small moth *Ephestia kuhniella*. They proposed a phenomenological model in which a 'determination stream' emanates from sources at the anterior and posterior edges of the wing and progresses as a wave across the wing to produce anterior–posterior bands of pigment; see Figure 3.13(b). They carried out microcautery experiments on the pupal wing and their results were consistent with their phenomenological hypothesis. Work by Henke (1943) on 'spreading fields' in *Lymantria dispar* (see Figure 3.15(g)) also sup-

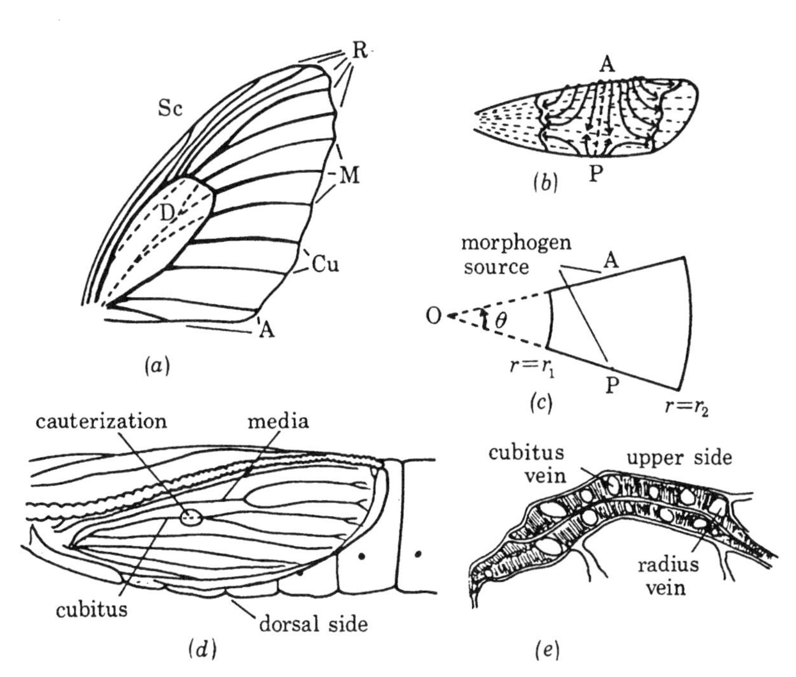

Figure 3.13. (**a**) Forewing of a generalised lepidopteran with the basic venation nomenclature: A, anal; Cu cubitus; M, media; R, radius; Sc, subcostal; D, distal. The regions between veins are wing cells. Dotted lines represent veins that exists at the pupal state, but later atrophy. (**b**) Hypothesised 'determinations stream' for central symmetry pattern formation (after Kuhn and von Engelhardt 1933). (**c**) Idealised pupal wing with A and P the anterior and posterior sources of the determination stream (morphogen). (**d**) Schematic representation of the right pupa wing approximately 6-12 hours old (after Kuhn and von Engelhardt 1933). (**e**) Schematic cross section through the wing vertically through the cauterised region, showing the upper and lower epithelia and veins. (After Kuhn and von Engelhardt 1933)

ports this hypothesis. The results from the model mechanism discussed in this section will also be related to his experiments. The model relies on scale-forming stem cells in the epithelium reacting to underlying patterns laid down during the pupal or just prepupal stage. Goldschmidt (1920) suggested that primary patterns may be laid down before the pattern is seen; this seems to be borne out by more recent experimental studies.

Eyespots or ocelli are important elements in many butterfly wings; see the examples in Figure 3.12(a). Nijhout (1980a,b) presents evidence, from experiments on the Nymphalid butterfly *Precis coenia*, that the foci of the eyespots are the influencing factors in their pattern formation. Carroll et al. (1994) and French and Brakefield (1995) also discuss eyespot development with the latter paper dealing with the focal signal. The foci generate a morphogen, the level of which activates a colour-specific enzyme. Colour production, that is, melanogenesis, in *Precis coenia* involves melanins which are not all produced at the same time (Nijhout 1980b). In another survey Sibatani (1981) proposes an alternative model based on the existence of an underlying prepattern and suggests that the ocellus-forming process involves several interacting variables. These two models are not necessarily mutually exclusive since a 'positional information' (Wolpert 1969) model relies on cells reacting in a specified manner to the concentration level of some morphogen.

The cautery work of Kuhn and von Engelhardt (1933) suggests that there are at least two mechanisms in the pattern formation in *Ephestia kuhniella*, since different effects are obtained depending on the time after pupation at which the cauterization occurs. There are possibly several independent pattern-formation systems operating, as was suggested by Schwanwitsch (1924) and Suffert (1927). However, the same mechanism, such as that discussed here, could simply be operating at different times, which could imply different parameter values and different geometries and scale to produce quite different patterns. It would also not be unreasonable to postulate that the number of melanins present indicates the minimum number of mechanisms, or separate runs of the same mechanism.

Although the main reason for studying wing pattern in lepidoptera is to try to understand their formation with a view to finding a pattern generation mechanism (or mechanisms), another is to show evidence for the existence of diffusion fields greater than about 100 cells (about 0.5 mm), which is about the maximum found so far. With butterfly wing patterns, fields of $O(5\text{ mm})$ seem to exist. From a modelling point of view an interesting aspect is that the evolution of pattern looks essentially two-dimensional, so we must again consider the roles of both geometry and scale. As in the last section we shall see that seemingly different patterns can be generated with the same mechanism simply by its activation at different times on different geometries and on different scales.

Model Mechanism: Diffusing Morphogen—Gene Activation System

We first briefly discuss central symmetry patterns (see Figure 3.12(c)) since it is the experimental work on these which motivates the model mechanism. These crossbands of pigment generally run from the anterior to the posterior of the wings and are possibly the most prevalent patterns. Dislocation of these bands along wing cells, namely, regions bounded by veins and a wing edge, can give rise to a remarkably wide variety of patterns (see, for example, Nijhout 1978, 1991); Figure 3.12(a) displays a good example. Figure 3.13(a) is a diagram of the forewing of a generalised lepidopteran, and illustrates typical venation including those in the distal cell(D) where the veins later atrophy and effectively disappear.

Khun and von Engelhardt (1933) carried out a series of experiments, using microcautery at the pupal stage, to try to see how central symmetry patterns arose on the forewing of the moth *Ephestia khuniella*. Some of their results are illustrated in Figures 3.15(a)–(c). They seem consistent with a 'determination stream' or wave emanating from sources on the anterior and posterior edges of the wing, namely, at A and P in Figure 3.13(b). The front of this wave is associated with the position of the crossbands of the central symmetry system. The work of Schwartz (1962) on another moth tends to confirm the existence of such a determination wave for central symmetry systems. Here we develop a possible mechanism, which we suggest operates just after pupation, for generating this specific pattern (as well as others) and we compare the results with experiment.

We assume that there are sources of a morphogen, with concentration S, situated at A and P on the anterior and posterior edges of the wing, which for simplicity (not necessity) in the numerical calculations is idealised as shown in Figure 3.13(c) to be a circular sector of angle θ bounded by radii r_1 and r_2. At a given time in the pupal stage, which we assume to be genetically determined, a given amount of morphogen S_0 is released and it diffuses across the wing. The wing has an upper and lower epithelial surface layer of cells and vein distribution such as illustrated in Figures 3.13(d) and (e). The pattern on the upper and lower sides of the wing are determined independently. As the morphogen diffuses we assume it is degraded via first-order kinetics. The diffusion field is the wing surface and so we have zero flux boundary conditions for the morphogen at the wing edges. The governing equation for the morphogen concentration $S(r, \theta, t)$ is then

$$\frac{\partial S}{\partial t} = D\left(\frac{\partial^2 S}{\partial r^2} + \frac{1}{r}\frac{\partial S}{\partial r} + \frac{1}{r^2}\frac{\partial^2 S}{\partial \theta^2}\right) - KS, \tag{3.7}$$

where $D(\mathrm{cm}^2\mathrm{s}^{-1})$ is the diffusion coefficient and $K(\mathrm{s}^{-1})$ the degradation rate constant.

As S diffuses across the wing surface, suppose the cells react in response to the local morphogen level, and a gene G is activated by S to produce a product g. We assume that the kinetics of the gene product exhibits a biochemical switch behaviour such as we discussed in Chapter 1, Volume I and in more detail in Chapter 6, Volume I (note specifically Exercise 3): see also Figures 3.14(a) and (b). Such a mechanism can effect a permanent change in the gene product level as we show. (A model, with similar

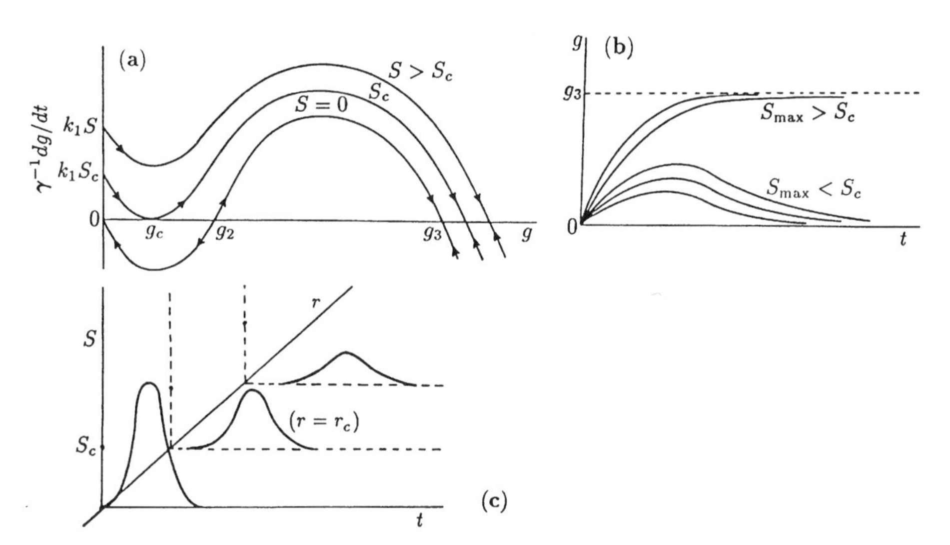

Figure 3.14. (**a**) Biochemical switch mechanism with typical bistable kinetics such as from (3.12). The graph shows $\gamma^{-1}dg/dt$ against g for appropriate k_1, k_2, k_3 and several values of S. The critical S_c is defined as having two stable steady states for $S < S_c$ and one, like $g = g_3$, for $S > S_c$. (**b**) Schematic behaviour of g as a function of t from (3.12) for various pulses of S which increase from $S = 0$ to a maximum $S_{\max}$ and then decrease to $S = 0$ again. The lowest curve is for the pulse with the smallest $S_{\max}$. The final stage of g, for large time, changes discontinuously from $g = 0$ to $g = g_3$ if $S_{\max}$ passes through a critical threshold $S_{th}(> S_c)$. (**c**) Schematic solution, (3.16) below, for the morphogen S as a function of position r measured from the release point of the morphogen.

kinetics, in which the morphogen activates a colour-specific enzyme that depends on the local morphogen level is another possible mechanism.) There are now several such biochemically plausible switch mechanisms (see, for example, Edelstein 1972). It is not important at this stage which switch mechanism we use for the gene product g, but, to be specific we use a standard cubic-like form

$$\frac{dg}{dt} = K_1 S + \frac{K_2 g^2}{K_4 + g^2} - K_3 g, \tag{3.8}$$

where the K's are positive parameters. Here g is activated linearly by the morphogen S, by its own product in a nonlinear positive feedback way and linearly degraded proportional to itself. $g(t; r, \theta)$ is a function of position through S. It is easy to construct other examples: a simple polynomial with three roots would suffice for the last two terms in Figure 3.8.

The model involves S_0 of morphogen released on the wing boundaries at A and P as in Figure 3.13(c) in the idealised wing geometry we consider. The morphogen satisfies (3.7) within the domain defined by

$$r_1 \leq r \leq r_2, \quad 0 \leq \theta \leq \theta_0, \tag{3.9}$$

and S satisfies zero flux boundary conditions. S_0 is released from $A(r = r_A, \theta = \theta_0)$ and $P(r = r_P, \theta = 0)$ as delta functions at $t = 0$: initially $S = 0$ everywhere. We take the gene product to be initially zero; that is, $g(0; r, \theta) = 0$. The appropriate boundary and initial conditions for the mathematical problem are then

$$\begin{aligned} &S(r, \theta, 0) = 0, \quad r_1 < r < r_2, \quad 0 < \theta < \theta_0, \\ &S(r_P, 0, t) = S_0 \delta(t) \quad S(r_A, \theta_0, t) = S_0 \delta(t), \\ &\frac{\partial S}{\partial r} = 0, \quad 0 \leq \theta \leq \theta_0, \quad r = r_1, \quad r = r_2, \\ &\frac{\partial S}{\partial \theta} = 0, \quad r_1 < r < r_2, \quad \theta = 0, \quad \theta = \theta_0, \\ &g(0; r, \theta) = 0, \end{aligned} \tag{3.10}$$

where $\delta(t)$ is the Dirac delta function. Equations (3.7) and (3.8) with (3.10) uniquely determine S and g for all $t > 0$.

As always, it is useful to introduce nondimensional quantities to isolate the key parameter groupings and indicate the relative importance of different terms in the equations. Let L(cm) be a standard reference length and a(cm), for example, $r_2 - r_1$, a relevant length of interest in the wing. Introduce dimensionless quantities by

$$\begin{aligned} &\gamma = \left(\frac{a}{L}\right)^2, \quad S^* = \frac{S}{S_0}, \quad r^* = \frac{r}{a}, \quad t^* = \frac{D}{a^2} t, \\ &k = \frac{KL^2}{D}, \quad k_1 = \frac{K_1 S_0 L^2}{D\sqrt{K_4}}, \quad k_2 = \frac{K_2 L^2}{D\sqrt{K_4}}, \quad k_3 = \frac{K_3 L^2}{D}, \quad g^* = \frac{g}{\sqrt{K_4}} \end{aligned} \tag{3.11}$$

and the model system (3.7), (3.8) with (3.10) becomes, on dropping the asterisks for notational simplicity,

$$\begin{aligned}\frac{\partial S}{\partial t} &= \frac{\partial^2 S}{\partial r^2} + \frac{1}{r}\frac{\partial S}{\partial r} + \frac{1}{r^2}\frac{\partial^2 S}{\partial \theta^2} - \gamma k S,\\ \frac{dg}{dt} &= \gamma\left(k_1 S + \frac{k_2 g^2}{1+g^2} - k_3 g\right) = \gamma f(g; S).\end{aligned} \tag{3.12}$$

The last equation defines $f(g; S)$.

Recall, from the last section, the reason for introducing the scale parameter γ, namely, the convenience in making scale changes easier. If our 'standard' wing has $a = L$, that is, $\gamma = 1$, then for the *same* parameters a similar wing but twice the size has $a = 2L$, that is, $\gamma = 4$, but it can be represented diagrammatically by the same-sized figure as for $\gamma = 1$. (Recall Figure 3.6 which exploited this aspect.)

The initial and boundary conditions (3.10) in nondimensional form are algebraically the same except that now

$$S(r_P, 0, t) = \delta(t) \quad S(r_A, \theta_0, t) = \delta(t). \tag{3.13}$$

The switch and threshold nature of the gene kinetics mechanism in the second of (3.12) can be seen by considering the schematic graph of $\gamma^{-1}dg/dt$ as a function of g, as in Figure 3.14(a), for various constant values for S and appropriate k's.

To determine a range of k's in (3.12) so that the kinetics exhibit a switch mechanism, consider first $f(g; 0)$ from (3.12): refer also to the last figure. We simply require $\gamma^{-1}dg/dt = f(g; 0)$ to have 2 positive steady states; that is, the solutions of

$$f(g; 0) = 0 \quad \Rightarrow \quad g = 0, \quad g_1, g_2 = \frac{k_2 \pm (k_2^2 - 4k_3^2)^{1/2}}{2k_2} \tag{3.14}$$

must all be real. This is the case if $k_2 > 2k_3$. For $S > 0$ the curve of $f(g; S)$ is simply moved up and, for small enough S, there are 3 steady states, two of which are stable. The S-shape plot of $\gamma^{-1}dg/dt$ against g is typical of a switch mechanism.

Now suppose that at a given time, say, $t = 0$, $g = 0$ everywhere and a pulse of morphogen S is released. It activates the gene product since with $S > 0$, $dg/dt > 0$ and so, at each position, g increases with time typically as in Figure 3.14(b), which are curves for g as a function of time for a given S. If S never reaches the critical threshold $S_{th}(> Sc)$, then as S decreases again to zero after a long time so does g. However, if S exceeds Sc for a sufficient time, then g can increase sufficiently so that it tends, eventually, to the local steady state equivalent to g_3, thus effecting a switch from $g = 0$ to $g = g_3$. That S must reach a threshold is intuitively clear. What is also clear is that the detailed kinetics in the second of (3.12) are not critical as long as they exhibit the threshold characteristics illustrated in Figure 3.14.

Even the linear problem for S posed by (3.12) with the relevant boundary and initial conditions (3.10) is not easily solved analytically in a usable form. We know S qualitatively looks like that in Figure 3.14(c) with S reaching a different maximum at each r. Also for a given $S(r, t)$ the solution for g has to be found numerically. The critical threshold S_{th} which effects the switch is also not trivial to determine analytically; Kath and Murray (1986) present a singular perturbation solution to this problem for fast switch kinetics. Later, when we consider eyespot formation and what are called dependent patterns, we shall derive some approximate analytical results. For central symmetry patterns, however, we shall rely on numerical simulations of the model system.

Intuitively we can see how the mechanism (3.12), in which a finite amount of morphogen S_0 is released from A and P in Figure 3.13(c), can generate a spatial pattern in gene product (or colour-specific enzyme). The morphogen pulse diffuses and decays as it spreads across the wing surface, and as it does so it activates the gene G to produce g. If over a region of the wing $S > S_{th}$, then g increases sufficiently from $g = 0$ to move towards g_3 so that when S finally decreases g continues to move towards g_3 rather than returning to $g = 0$. The growth in g, governed by (3.12), is not instantaneous and so the critical S_{th} is larger than S_c in Figure 3.14(a). The coupling of the two processes, diffusion and gene transcription, in effect introduces a time lag. Thus as the pulse of morphogen diffuses across the wing as a quasi-wave (see Figure 3.14(c)), it generates a domain of permanently nonzero values of g, namely, g_3, until along some curve on the wing, S has decreased sufficiently ($S < S_{th}$) so that g returns to $g = 0$ rather than continuing to increase to $g = g_3$. We are interested in determining the switched-on or activated domain; Kath and Murray (1985) also did this for fast switch kinetics. We now apply the mechanism to several specific pattern elements.

Central Symmetry Patterns; Scale and Geometry Effects; Comparison with Experiments

We first consider how the model may apply to central symmetry patterns and specifically to the experiments of Kuhn and von Engelhardt (1933). We assume that the morphogen S emanates from morphogen sources at A and P on the wing edges, as in Figure 3.13(c). The morphogen 'wave' progresses and decays as it moves across the wing until the morphogen level S is reduced to the critical concentration S_{th}, below which the gene-activation kinetics cannot generate a permanent nonzero product level as described above. Now relate the spatial boundary between the two steady state gene product levels, the threshold front, with the determination front of Kuhn and von Engelhardt (1933). The cells, which manifest the ultimate pigment distribution, are considered to react differentially in the vicinity of this threshold front. The idea that cells react differentially at marked boundaries of morphogen concentrations was also suggested by Meinhardt (1986) in his model for the early segmentation in the fruit fly *Drosophila* embryo.

We require the ultimate steady state solution for the gene product g which requires the solution of the full nonlinear time- and space-dependent problem (3.12) with (3.13) and the dimensionless form of (3.10). The numerical results below were obtained using a finite difference scheme. The main parameters that can be varied are the k's and γ. The qualitative behaviour of the pattern formation mechanism and the critical roles played by the geometry and scale can best be highlighted by choosing an appropriate set of values for the parameters and keeping them fixed for *all* of the calculations. The results are shown in Figures 3.12 to 3.15. The parameter values did not have to be carefully selected. In all of the simulations the same amount of morphogen was released at the sources A and P in the middle of the anterior and posterior wing edges.

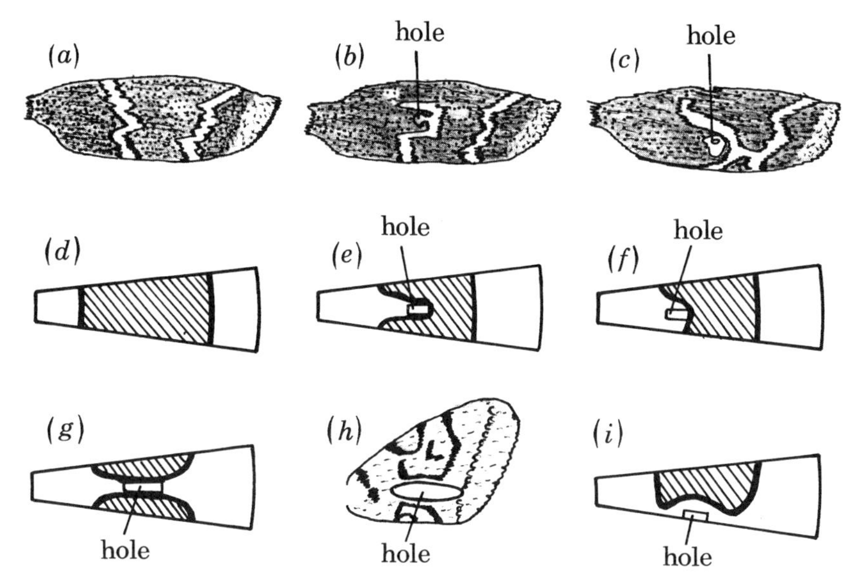

Figure 3.15. Effect of cauterization on the central symmetry pattern. (**a**)–(**c**) are from the experimental results of Kuhn and von Engelhardt (1933) on the moth *Ephestia kuhniella* during the first day after the pupation: (**a**) normal wing, (**b**) and (**c**) cauterised wings with the hole as indicated. (**d**) Idealised model normal wing in which the 'determination stream' has come from morphogen sources at A and P on the anterior and posterior wing edges: the hatched region represents a steady state nonzero gene product. (**e**), (**f**), (**g**) and (**i**) are computed solutions from the model mechanism with cauterised holes as indicated. (**h**) Effect of cauterization, during the first day after pupation, on the cross-bands of the forewing of *Lymantria dispar*. (After Henke 1943) Simulation 'experiments' on the model for comparison; the correspondence is (**a**)–(**d**). (**b**)–(**e**), (**c**)–(**f**), (**g**)–(**h**). If cauterization removes the determination stream's source of morphogen at the posterior edge, the pattern predicted by the model is as shown in (**i**). Parameter values used in the calculation for (3.12) for the idealised wing in Figure 3.13(c) for all of (**d**)–(**g**), and (**i**): $k_1 = 1.0 = k_3$, $k_2 = 2.1$, $k = 0.1$, $\gamma = 160$ and unit sources of S at $\theta = 0$ and $\theta = 0.25$ radians, $r_1 = 1$, $r_2 = 3$. (From Murray 1981b)

Consider first the experiments on the moth *Ephestia kuhniella*: its wing is, in fact, quite small, the actual size is about that of the nail on one's little finger. Figure 3.15(a) illustrates a normal wing with typical markings while Figures 3.15(b) and (c) show the results of thermal microcautery (Kuhn and von Engelhardt 1933). Figure 3.15(d) is the idealised normal wing: the shaded region is the residual nonzero gene product left behind by the determination wave of morphogen.

When a hole, corresponding to thermal cautery, is inserted in the idealised wing we assume that the morphogen level in the hole is zero. That is, we set $S = 0$ on the hole boundary on the assumption that any morphogen which diffuses into the hole is destroyed. The numerical results corresponding to the geometry of the experiments are shown in Figures 3.15(e) and (f), which relate respectively to the experimental results in Figures 3.15(b) and (c). Figure 3.15(g) is another example with a larger cauterization, while Figure 3.15(h) is of a comparably cauterised wing of *Lymantria dispar* (Henke 1943). Figure 3.15(i) is the model's prediction if cauterization removes the source of morphogen at the posterior edge of the wing. No such experiments appear to have been done to establish where the sources of the determination stream are.

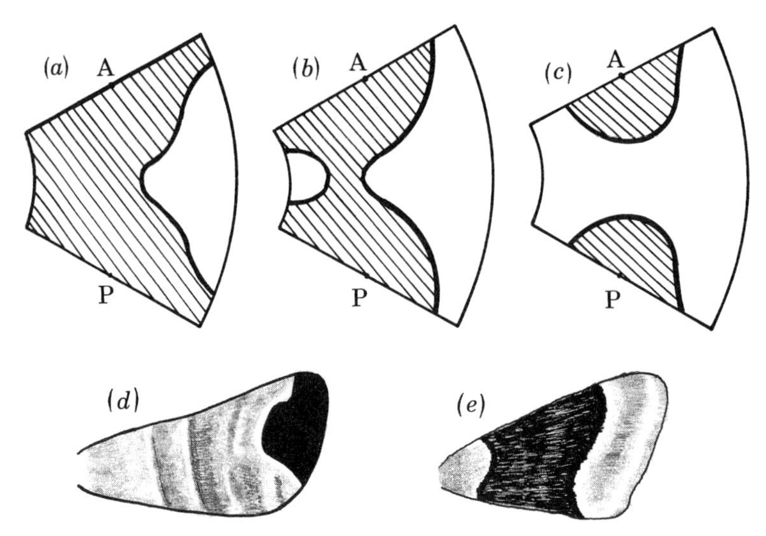

Figure 3.16. Some effects of scale on spatial patterns generated by the mechanism (3.12) when morphogen is released from sources at A and P. **(a)–(c)** have the same set of parameter values as in Figure 3.15, namely $k_1 = 1.0 = k_3$, $k_2 = 2.1$, $k = 0.1$, with the wing defined by $r_1 = 1$, $r_2 = 3$, $\theta = 1.0$ radians, for different domain sizes: **(a)** $\gamma = 2$; **(b)** $\gamma = 6$; **(c)** $\gamma = 40$. A wing with $\gamma = \gamma_2 (> \gamma_1)$ has linear dimensions $\sqrt{\gamma_2/\gamma_1}$ larger than that with $\gamma = \gamma_1$. The shaded region has a nonzero gene product. **(d)** *Psodos coracina* and **(e)** *Clostera curtula* are examples of fairly common patterns on moth wings. (From Murray 1981b)

Let us now consider the effect of geometry and scale. Even with such a simple model the variety of patterns that can be generated is quite impressive. For the same values for the kinetics parameters k, k_1, k_2 and k_3 in (3.12), Figures 3.16(a)–(c) illustrate, for a *fixed geometry*, some of the effects of *scale* on the spatial patterns. These, of course, are qualitatively as we would expect intuitively. As mentioned above, central symmetry patterns are particularly common in moth wings. Figures 3.16(d) and (e) show just two such examples, namely, the chocolate chip (*Psodos coracina*) and black mountain (*Clostera curtula*) moths respectively; compare these with Figures 3.16(a) and (b).

The effect of geometry is also important and again we can intuitively predict its general effect with this model when the morphogen is released at the same points on the wing edges. The patterns illustrated in Figure 3.17 were obtained by simply varying the angle subtended by the wing edges.

The comparison between the experimental results and the results from solving the model system's equations for appropriate domains is encouraging. The solutions generate a region where the morphogen has effected a switch from a zero to a nonzero steady state for the gene product g. If the process were repeated with a different gene product and with a slightly smaller morphogen release it is clear that we could generate a single sharp band of differentiated cells.

Dependent Patterns

Consider now dependent patterns, which are also very common, in which pigment is restricted to the vicinity of the veins. The pattern depends on the position in the wing

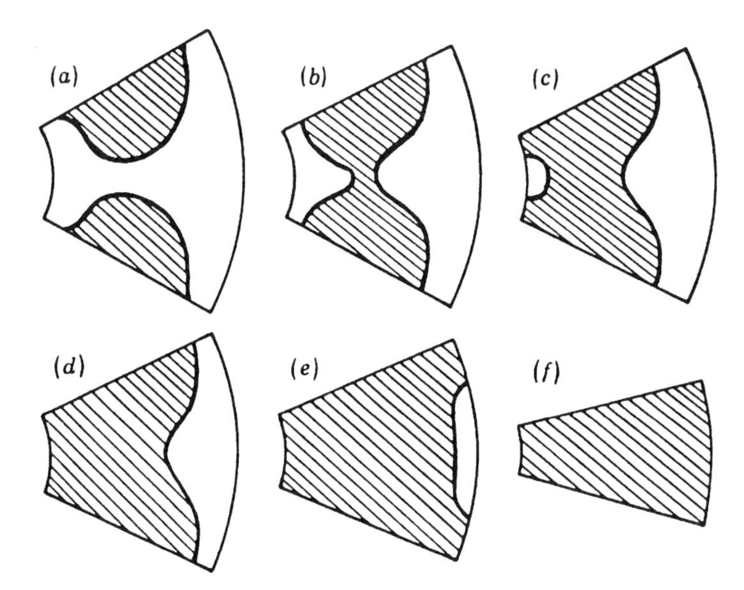

Figure 3.17. Simple effects of geometry on the spatial patterns. The morphogen is released in the same way as for Figure 3.16 with the same k-parameter values and a fixed scale parameter $\gamma = 10$ with $r_1 = 1, r_2 = 3$ taken for all simulations of (3.12). Geometry is changed by simply varying the angle, in radians, of the sector: **(a)** $\theta = 1.0$; **(b)** $\theta = 0.975$; **(c)** $\theta = 0.95$; **(d)** $\theta = 0.9$; **(e)** $\theta = 0.8$; and **(f)** $\theta = 0.5$.

of the veins, hence the name for these patterns. Here we consider the morphogen to be released from the boundary veins of the wing cells and so a nonzero gene product g is created near the veins and it is this pattern which is reflected by the pigment-generating cells. If we consider the wing cell to be modelled by the sector of a circle, just as the wing in the situations discussed above, we now have the morphogen released all along the cell boundaries except the outer edge. In this case if we consider the wing cell to be very long so that the problem is quasi-one-dimensional we can derive, given S_{th}, an analytical expression for the width of the gene product spatial pattern.

Consider the one-dimensional problem in which a given amount of morphogen is released at $x = 0$. The idealised mathematical problem is defined by

$$\begin{gathered}\frac{\partial S}{\partial t} = \frac{\partial^2 S}{\partial x^2} - \gamma k S, \\ S(x, 0) = 0, x > 0, \quad S(0, t) = \delta(t), \quad S(\infty, t) = 0,\end{gathered} \tag{3.15}$$

with solution

$$S(x, t) = \frac{1}{2(\pi t)^{1/2}} \exp\left[-\gamma k t - \frac{x^2}{4t}\right], \quad t > 0. \tag{3.16}$$

This is qualitatively like that sketched in Figure 3.14(c). For a given x the maximum S, $S_{\max}$ say, is given at time t_m where

$$\frac{\partial S}{\partial t} = 0 \quad \Rightarrow \quad t_m = \frac{-1 + \{1 + 4\gamma k x^2\}^{1/2}}{4\gamma k} \tag{3.17}$$

which on substitution in (3.16) gives

$$S_{\max}(x) = \left[\frac{\gamma k}{\pi(z-1)}\right]^{1/2} \exp\left(-\frac{z}{2}\right), \quad \text{where} \quad z = (1 + 4\gamma k x^2)^{1/2}. \tag{3.18}$$

Now, from the kinetics mechanism (3.12), $S_{\max} = S_{th}$ is the level which effects a switch from $g = 0$ to $g = g_3$ in Figure 3.14(a). Substituting in (3.8) we can calculate the distance x_{th} from the vein where $g = g_3$ and hence, in our model, the domain of a specific pigmentation. Thus x_{th} is the solution of

$$S_{th} = \left[\frac{\gamma k}{\pi(z_{th}-1)}\right]^{1/2} \exp\left(-\frac{z_{th}}{2}\right), \quad x_{th} = \left(\frac{z_{th}^2 - 1}{4\gamma k}\right)^{1/2}. \tag{3.19}$$

An alternative form of the equation for x_{th} is

$$z_{th} + \ln\left[\frac{S_{th}^2 \pi(z_{th}-1)}{\gamma k}\right] = 0, \quad x_{th} = \left(\frac{z_{th}^2 - 1}{4\gamma k}\right)^{1/2}. \tag{3.20}$$

In dimensional terms, from (3.11), the critical distance x_{th}(cm) is thus given in terms of the morphogen pulse strength S_0, the diffusion coefficient $D(\text{cm}^2\text{s}^{-1})$ and the rate constant $K\,(\text{s}^{-1})$. The role of the kinetics parameters in (3.12) comes into the determination of the critical S_{th}. The analytical evaluation of S_{th} when the gene-activation kinetics is very fast is given by Kath and Murray (1986): it is a nontrivial singular pertubation problem.

Let us now consider dependent patterns with the mechanism operating in a domain which is an idealised wing cell with a given amount of morphogen released in a pulse from the bounding veins: refer to Figure 3.18. From the above analysis we expect a width of the nonzero gene product g on either side of the vein. As in the central symmetry patterns we expect geometry and scale to play major roles in the final pattern obtained for given values of the model parameters, and we can now intuitively predict the qualitative behaviour of the solutions.

Now apply the model mechanism (3.12) to the idealised wing cell with a given amount, ρ, of morphogen released per unit length of the bounding veins (refer also to Figure 3.13(a)). With the same parameter values as for Figures 3.16 and 3.17 the equations were again solved using a finite difference scheme with zero flux conditions along the boundaries after a pulse of morphogen had been released from the three edges representing the veins. Figures 3.18(a) and (d) show examples of the computed solutions with Figures 3.18(b) and (e) the approximate resulting wing patterns generated on a full wing. The role of geometry in the patterns is as we would now expect. Figures 3.15(c) and (f) are specific, but typical, examples of the forewing of *Troides hy-*

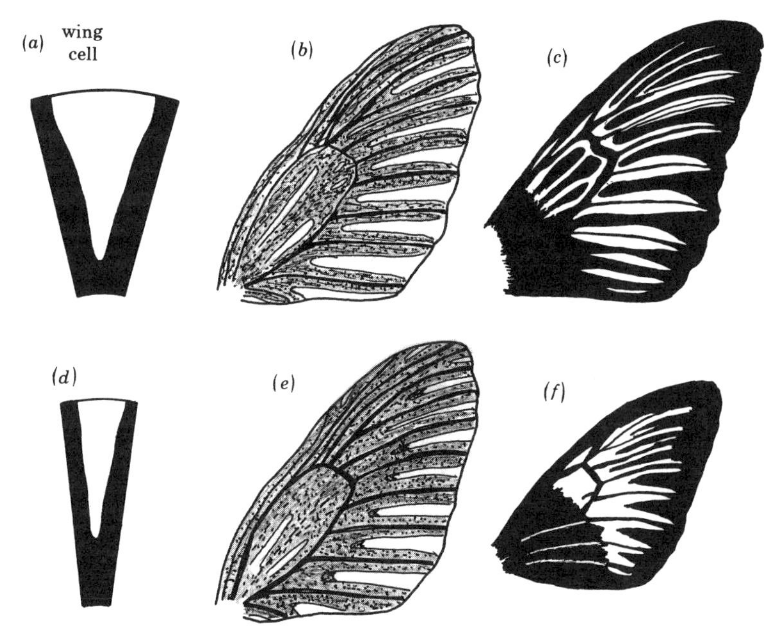

Figure 3.18. Examples of dependent patterns. (**a**), (**d**) computed patterns from the mechanism (3.12) for a wing cell with parameter values: $k = 0.1$, $k_1 = 1.0 = k_3$, $k_2 = 2.1$, $\gamma = 250$ with the morphogen source strength ρ in the anterior and posterior veins; (**a**) $\rho = 0.075$, (**d**) $\rho = 0.015$, and $\rho = 0$ on the cross veins. (**b**) and (**e**) Schematic predicted pattern from the wing cell patterns in (**a**) and (**d**) applied to the generalised wing of Figure 3.13(a): shaded regions have a nonzero gene product g. (**c**) and (**f**) Specific examples of dependent patterns on the forewing of two Papilionidae: (**c**) *Troides hypolitus*, (**f**) *Troides haliphron*.

politus and *Troides haliphron* respectively. Such dependent patterns are quite common in the Papilionidae.

Now consider scale effects. Figures 3.19(a) and (b) directly illustrate these schematically when the veins are approximately parallel. Figures 3.19(c) and (d) show examples of the forewings of *Troides prattorum* and *Iterus zalmoxis*. The distance from the vein of the pigmented pattern depends in a nonlinear way on the parameters and the amount of morphogen released. If these values are fixed, the distance from the vein is *independent* of scale. That is, the mechanism shows that the *intravenous* strips between pigmented regions vary according to how large the wing cell is. This is in agreement with the observations of Schwanwitsch (1924) and the results in Figure 3.19 exemplify this.

These results are also consistent with the observation of Schwanwitsch (1924) on Nymphalids and certain other families. He noted that although the width of intravenous stripes (in our model the region between the veins where $g = 0$) is species-dependent, the pigmented regions in the vicinity of the veins are the same size. In several species the patterns observed in the distal cell (D in Figure 3.13(a)) reflect the existence of the veins that subsequently atrophy; see Figures 3.18(c) and 3.19(c) of the forewing of the female *Troides prattorum*.

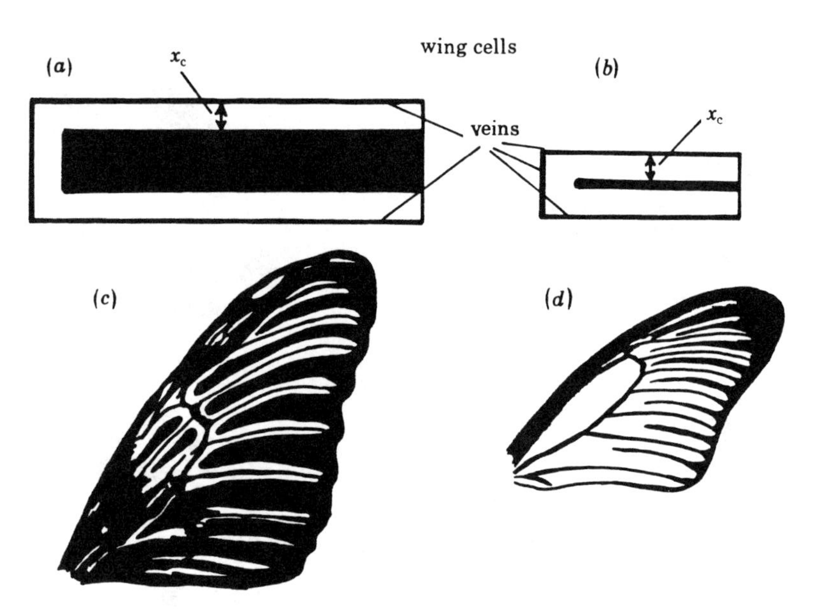

Figure 3.19. (**a**) and (**b**) Idealised wing cells based on the analytical solution (3.20) for the switched-on domain. These clearly illustrate the effect of scale. The pattern (unshaded in this case) width is fixed for given parameter values. (**c**) and (**d**) Examples of dependent patterns from two Papilionidae: (**c**) *Troides prattorum*; (**d**) *Iterus zalmoxis*.

Eyespot or Ocelli Patterns

Eyespot patterns are very common; see, for example, Figure 3.12(a). Brakefield and French (1995; see other references there) investigated the response to epidermal damage on eyespot development, while Carroll et al. (1994) suggest that a gradient controls gene expression and their association with eyespot determination in butterfly wings. Nijhout (1980b) performed transplant experiments on the buckeye butterfly (*Precis coenia*) wherein he moved an incipient eyespot from one position on the wing to another where normally an eyespot does not form. The result was that an eyespot formed at the new position. This suggests that there is possibly a source of some morphogen at the eyespot centre from which the morphogen diffuses outwards and activates the cells to produce the circular patterns observed. So, once again it seems reasonable to investigate the application of the above mechanism to some of these results.

We assume that the eyespot centre emits a pulse of morphogen in exactly the same way as for the central symmetry patterns. The idealised mathematical problem for the morphogen from the first of (3.12) in plane axisymmetric polar coordinates, with initial and boundary conditions from (3.10) and (3.13), is

$$\frac{\partial S}{\partial t} = \frac{\partial^2 S}{\partial r^2} + r^{-1}\frac{\partial S}{\partial r} - \gamma k S,$$
$$S(r,0) = 0,\, r > 0, \quad S(0,t) = \delta(t), \quad S(\infty,t) = 0. \tag{3.21}$$

The solution is

$$S(r,t) = \frac{1}{4\pi t}\exp\left[-\gamma k t - \frac{r^2}{4t}\right], \quad t > 0, \tag{3.22}$$

which is like the function of time and space sketched in Figure 3.14(c).

As before we can calculate the size of the gene-activated region given the critical threshold concentration $S_{th}(> S_c)$ which effects the transition from $g = 0$ to $g = g_3$ in Figure 3.14(a). In exactly the same way as we used to obtain (3.20) the maximum S_{max} for a given r from (3.22) is

$$S_{max} = \frac{\gamma k}{2\pi(z-1)}e^{-z}, \quad z = (1+\gamma k r^2)^{1/2}$$

and so the radius of the activated domain, r_{th} say, is given by the last equation with $S_{max} = S_{th}$, as

$$z_{th} + \ln\left[\frac{2\pi(z_{th}-1)S_{th}}{\gamma k}\right] = 0, \quad z_{th} = (1+\gamma k r_{th}^2)^{1/2}. \tag{3.23}$$

Many eyespots have several concentric ring bands of colour. With the above experience we now know what the patterns will be when we solve the system (3.12) with conditions that pertain to this eyespot situation. If, instead, we let the mechanism run twice with slightly different amounts of morphogen released we shall obtain two separate domains which overlap. Figure 3.20(a) shows the numerical result of such a case together with the predicted pattern in Figure 3.20(b) if an eyespot is situated in each distal wing cell; an example of a specific butterfly which exhibits this result is shown in Figure 3.20(c).

The simple model proposed in this section can clearly generate some of the major pattern elements observed on lepidopteran wings. As we keep reiterating in this book, this is not sufficient to say that such a mechanism is that which necessarily occurs. The evidence from comparison with experiment is, however, suggestive of a diffusion based model. From the material discussed in detail in Chapter 2 we could also generate such patterns by appropriately manipulating a reaction diffusion system capable of diffusion-driven pattern generation. What is required at this stage, if such a model mechanism is indeed that which operates, is an estimation of parameter values and how they might be varied under controlled experimental conditions. We thus consider how the model might apply quantitatively to the experimental results of Nijhout (1980a) who measured the diameter of a growing eyespot as a function of time; the results are reproduced in Figure 3.21.

Let us now relate the model analysis to the experiments. The solution for $S(r,t)$ is given by (3.22). We want the value of r such that $S = S_{th}$. Denote this by R; it is a function of t. From (3.22) with $S = S_{th}$ we have

$$S_{th} = \frac{1}{4\pi t}\exp\left[-\gamma k t - \frac{R^2}{(4t)}\right], \quad t > 0$$

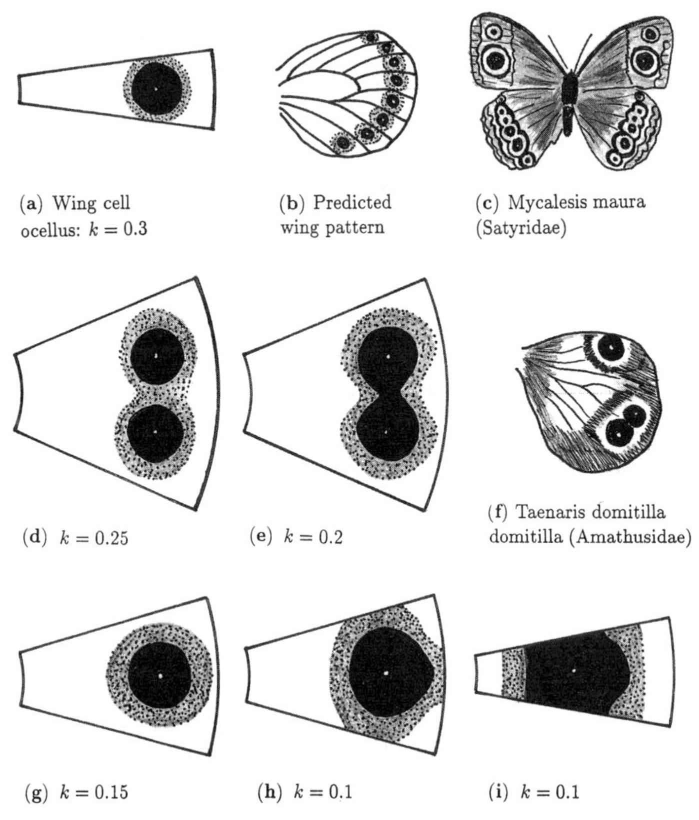

Figure 3.20. (**a**) A patterned eyespot generated within a wing cell by two emissions of morphogen each with its own gene product. With the same parameter values, the dark region had less morphogen injected than that with the shaded domain. The k-parameter values are as in Figure 3.12 to Figure 3.16 except for $k = 0.3$. (**b**) The predicted overall wing pattern if an eyespot was situated in each wing cell with (**c**) a typical example (*Mycalesis maura*). (**d**) and (**e**) illustrate the effect of different degradation constants k which result in coalescing eyespots with (**f**) an actual example (*Taenaris domitilla*). (**g**)–(**i**) demonstrate the effect of different geometries.

which gives

$$R^2(t) = -4t[\gamma kt + \ln(4\pi t S_{th})]. \tag{3.24}$$

For comparison with the experiments we require the diameter $d(= 2R)$ in dimensional terms. We consider a single eyespot with the standard length a in the nondimensionalisation (3.11) to be the diameter of the control in the experiment. Since we are interested in the growth of the eyespot to its normal size this means that $L = a$ and hence $\gamma = 1$. Thus the time varying diameter $d(t)$ in Figure 3.21 is simply $2R(t)$ which, on using (3.11) and (3.24) gives

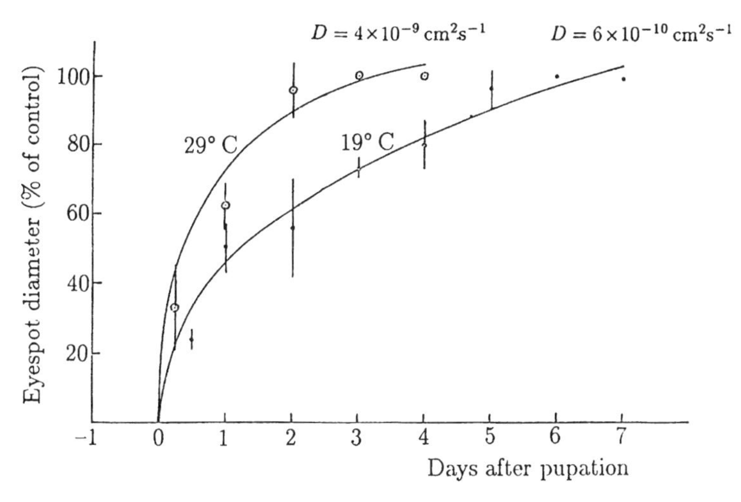

Figure 3.21. The diameter of a growing eyespot in the buckeye butterfly (*Precis coenia*) as a function of time after pupation. The experiments by Nijhout (1980a) were carried out at two different temperatures, 19°C and 29°C. The continuous curves are best fits from the analytical expression (3.25), which is derived from the simple morphogen diffusion model, (3.21).

$$\begin{aligned} d^2(t) &= -16Dt\left\{Kt + \ln\left[\frac{4\pi S_{th} t D}{S_0 a^2}\right]\right\} \\ &= -16Dt[Kt + \ln t + C], \end{aligned} \tag{3.25}$$

where $C = \ln[4\pi S_{th} D/(S_0 a^2)]$ is simply a constant and $D(\text{cm}^2\text{s}^{-1})$ and $t(\text{sec})$ are now dimensional. Note from (3.25) that

$$d(t) \sim O([t \ln t]^{1/2}) \quad \text{as} \quad t \to 0. \tag{3.26}$$

The maximum diameter d_m is obtained in the same way as above for the gene-activated domain size for dependent patterns, specifically (3.20). Here it is given by d_m, the solution of

$$z + \ln\left(\frac{z-1}{2K}\right) + C = 0, \quad z = \left(1 + \frac{Kd_m^2}{4D}\right)^{1/2}. \tag{3.27}$$

If we now use (3.25) and the experimental points from Figure 3.21, we can determine D, k and C from a best fit analysis. From the point of view of experimental manipulation it is difficult to predict any variation in the degradation constant K since we do not know what the morphogen is. There is, however, some information as to how diffusion coefficients vary with temperature. Thus, the parameter whose value we can deduce, and which we can potentially use at this stage, is the diffusion coefficient D. From the experimental results in Figure 3.21 and the best fit with (3.25), we obtained values of $D = 4 \times 10^{-9}\text{cm}^2\text{s}^{-1}$ at 29°C and $D = 6 \times 10^{-10}\text{cm}^2\text{s}^{-1}$ at 19°C. Although we cannot independently measure the diffusion coefficient of a morphogen that we cannot yet identify, the order of magnitude of these values and how D varies with temperature are not unreasonable.

From (3.25) we obtain the velocity of spread, $v(t)$, of the eyespot as $dd(t)/dt$, and so

$$v(t) = -2D\frac{\{[Kt + \ln t + C] + t(K + \frac{1}{t})\}}{\{-Dt[Kt + \ln t + C]\}^{1/2}} \tag{3.28}$$

from which we deduce that

$$v(t) \sim 2\left(-D\frac{\ln t}{t}\right)^{1/2} \quad \text{as} \quad t \to 0. \tag{3.29}$$

Nijhout (1980a) found the average wavespeed to be 0.27 mm/day at 29°C and 0.12 mm/day at 19°C. With the best fit values of the parameters, (3.28) gives the velocity of spread as a function of t with (3.29) showing how quick the initial growth rate is. With the diffusion coefficient estimates deduced above, the ratio of the initial wavespeeds from (3.29) at 29°C and 19°C is $(D_{29^\circ}/D_{19^\circ})^{1/2} = (4 \times 10^{-9}/6 \times 10^{-10})^{1/2} \approx 2.58$. This compares favorably with the ratio of the average wavespeeds found experimentally, namely, $0.27/0.12 = 2.25$. This adds to the evidence for such a diffusion controlled pattern formation mechanism as above.[5]

It is known that temperature affects colour patterns in animals; the colour change that accompanies seasons is just one example. Etchberger et al. (1993) investigated (experimentally) the effect of temperature, carbon dioxide, oxygen and maternal influences on the pigmentation on turtles (specifically *Trachemys scripta elegans*). They obtained quantitative results and showed that carbon dioxide levels, for example, had a greater impact than temperature on the hatchling patterns. They argued that the effects are not on developmental time but on the actual pattern formation mechanisms. The problem seems ripe for some interesting modelling.

If mechanisms such as we have discussed in this section are those which operate, the dimension of the diffusion field of pattern formation is of the order of several millimetres. This is much larger than those found in other embryonic situations. One reason for assuming they do not occur is that development of pattern via diffusion would, in general, take too long if distances were larger than a millimetre; over this time enough growth and development would take place to imply considerable sensitivity in pattern formation. In pupal wings, however, this is not so, since pattern can develop over a period of days during which the scale and geometry vary little. With the experience

[5]Temperature has a marked effect on wing patterns in general. An amateur entomologist who lived near us in Oxford regularly incubated butterfly pupae found in the woods. He was interested in their patterns after subjecting them to cold shocks (he simply put them in the fridge for various periods) and developed a remarkable intuition for what would happen as a consequence. Some years ago he found a pupa lying on the ground the morning after a freak and very short snowfall in May: the pupa was lying so that the underside was not touched by the snow. He incubated it and the butterfly (a fratillary) had one normal wing pattern with the other highly irregular. He was astonished at the reaction he received when he exhibited it at a meeting of amateur entomologists in London. Instead of fascination as expected there was general anger because it was not possible for them to acquire such an example.

from the last few chapters, the original misgiving is no longer valid since if we include reaction diffusion mechanisms, not only can complex patterns be formed but biochemical messages can also be transmitted very much faster than pure diffusion. A final point regarding eyespots is that the positioning of the centres can easily be achieved with reaction diffusion models and the emission of the morphogen triggered by a wave sweeping over the wing or by nerve activation or a genetic switch; we discuss neural models in Chapter 12.

It is most likely that several independent mechanisms are operating, possibly at different stages, to produce the diverse patterns on butterfly wings (Schwanwitsch 1924, Suffert 1927). It is reasonable to assume, as a first modelling step, that the number of mechanisms is the same as the number of melanins present. In the case of the Nymphalid *Precis coenia* there are four differently coloured melanins (Nijhout 1980b).

With the relatively few pattern elements (in comparison with the vast and varied number of patterns that exist) in Suffert's (1927) groundplan, it seems worthwhile to explore further the scope of pattern formation possibilities of plausible biochemical diffusion models such as discussed here. Figure 3.22, however, shows a few more of

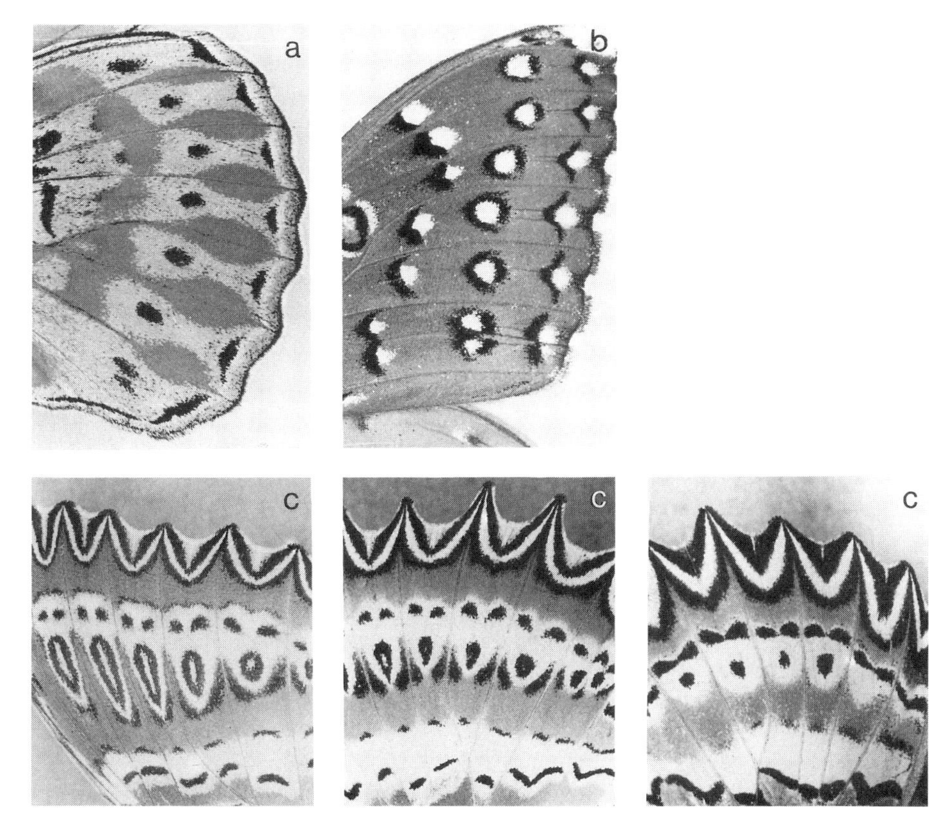

Figure 3.22. Photographs of examples of butterfly wing patterns the formation of which has not yet been modelled. (**a**) *Crenidomimas cocordiae*; (**b**) *Hamanumida daedalus*; (**c**) three examples from the genus *Cethosia*. Note here the topological similarity of these three patterns from the elongated pattern on the left to the much flatter form on the right. (Photographs courtesy of Professor H.F. Nijhout)

the complex wing patterns which have yet to be generated with such reaction diffusion mechanisms. By introducing anisotropy in the diffusion (that is, diffusion depends on the direction) it is possible to generate further patterns which are observed, such as the arrowhead in Figure 3.12(b) and in the last figure (cf. Exercise 10 in Chapter 2).

Perhaps we should turn the pattern formation question around and ask: 'What patterns *cannot* be formed by such simple mechanisms?' Some of the patterns on fish and snakes fall into this category. In the next chapter we discuss some of these and the kind of model modification sufficient to generate them. As a pattern generation problem, butterfly wing patterns seem particularly appropriate to study since it appears that pattern in the wings is developed comparatively late in development and interesting transplant experiments (Nijhout 1980a) and cautery-induced colour patterns (Nijhout 1985b) are feasible, as are the colour pattern modifications induced by temperature shocks (see, for example, Nijhout 1984).

3.4 Modelling Hair Patterns in a Whorl in *Acetabularia*

The green marine alga *Acetabularia*, a giant unicellular organism (see the beautiful photograph in Figure 3.23) is a fascinating plant which constitutes a link in the marine food chain (Bonotto 1985). The feature of particular interest to us here is its highly efficient self-regenerative properties which allow for laboratory controlled regulation

Figure 3.23. The marine algae *Acetabularia ryukyuensis*. (Photograph courtesy of Dr. I. Shihira-Ishikawa)

of its growth. *Acetabularia* has been the subject of several meetings; see, for example, the proceedings edited by Bonotto et al. (1985). In this section we describe a model, proposed by Goodwin et al. (1985), for the mechanism which controls the periodic hair spacing in the whorl of a regenerating head of *Acetabularia*. Experimental evidence is presented, not only to corroborate the analytical quantitative results of the mechanism, but more importantly to suggest that the initiation of hairs is controlled by calcium, possibly one of the elusive morphogens in a developmental situation. Fuller biological details are given in Goodwin et al. (1984).

The alga consists of a narrow stalk around 4–5 cm long on the top of which is a round cap about 1 cm across; see Figure 3.24(a). The stalk is a thin cylindrical shell of cytoplasm. After amputation free calcium, Ca^{2+}, plays a crucial role in the regeneration of the periodic distribution of the whorl hairs and eventually the cap. There are various stages in regeneration as schematically shown in Figures 3.24(b)–(d). After amputation there is an extension of the stalk, then a tip flattening and finally the formation of a whorl. Further extension of the stalk can take place with formation of other whorls. Figure 3.24(e) is a schematic cross-section of the stalk at the growth region and is the relevant spatial domain in our model.

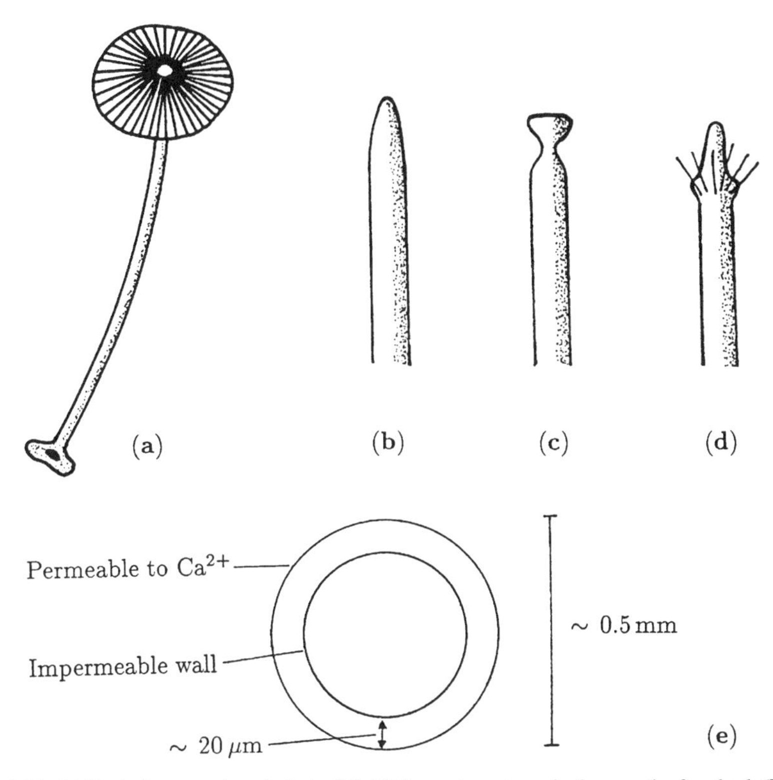

Figure 3.24. (**a**) Typical mature *Acetabularia*. (**b**)–(**d**) the various stages in the growth of a whorl: (**b**) extension, (**c**) flattening of the tip, (**d**) formation of the whorl. (**e**) Transverse cross-section of the growth region of the stalk: note the typical dimensions.

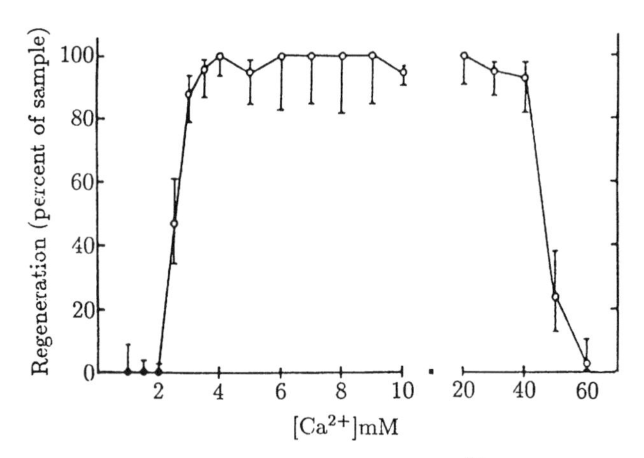

Figure 3.25. Experimental results on the effect of external calcium Ca^{2+} in the surrounding medium in the formation of whorls in *Acetabularia* after amputation. Note the lower and upper limits below and above which no regeneration occurs. (From Goodwin et al. 1984)

The model we develop and the mechanism we analyse is specifically concerned with the spatial pattern which cues the periodic distribution of hairs on a whorl. Experiments (see Goodwin et al. 1984) show that there are definite limits to the concentration of Ca^{2+} in the external medium, within which whorl formation will take place. Figure 3.25 shows the experimental results; below about 2 mM external calcium and above about 60 mM, whorls do not form. The normal value in artificial sea water is 10 mM Ca^{2+}. With about 5 mM only one whorl is produced after which the cap forms.

The experimental results suggest that the rate of movement of calcium from the external medium through the outer wall of the plant is intimately involved in growth determination and the initiation of a whorl of hairs. It is for this reason that calcium is proposed as a true morphogen in *Acetabularia*. If it is indeed a morphogen then it should play a role in the distribution of hairs or rather the mean distance between them, the wavelength of hair distribution. Experiments were conducted to determine the effect of the external free calcium concentration on the hair wavelength; see Figure 3.25 as well as Figure 3.28. Analysis of the model mechanism, which we discuss below, also corroborates the spacing hypothesis.

Let us consider some of the evidence for a reaction diffusion mechanism. First recall from Chapter 2 that if we have a spatial structure generated by a Turing-type reaction diffusion system the number of structures is not scale invariant. For example, if we have a given one-dimensional domain with several waves in morphogen concentration, a domain twice the size will have twice the number of waves as long as the parameters are kept fixed. This is an intrinsic characteristic of the spatial properties of reaction diffusion models of the kind discussed in the last chapter. We should perhaps note here the model reaction diffusion mechanisms proposed by Pate and Othmer (1984) which *are* scale invariant with regard to pattern formation. This problem of size invariance has also been addressed, for example, by Hunding and Sørensen (1988). Any model can be made to display size adaption if the parameters vary appropriately.

There is considerable variability in the number of hairs in a whorl; they vary from about 5 to 35. Experiments show that for plants maintained under the same conditions the hair spacing, w say, is almost constant and that the number of hairs is proportional to the radius of the stalk. The mechanism thus regulates the hair *spacing* irrespective of the size of the plant. This relation between scale and pattern number is a property of reaction diffusion systems as we demonstrated in the last chapter. In fact it is a property of other pattern formation mechanisms which involve the space variables in a similar way as we show later in Chapter 6 when we discuss mechanochemical pattern generators, which are quite different; so such a property is by no means conclusive evidence.

Harrison et al. (1981) showed that the spacing, w, of hairs depends on the ambient temperature, T, according to $\ln w \propto 1/T$. This Arrhenius-type of temperature variation suggests a chemical reaction kinetics factor, again in keeping with a reaction diffusion theory. In other words the spacing depends on the kinetics parameters.

The model we now develop is for the generation of the spatial distribution of a morphogen, identified with calcium, which is reflected in the spatial distribution of the hairs in the whorl. We assume initiation is governed by the overall reactions of two species u and v, the latter considered to be the concentration of Ca^{2+}, with u the other morphogen, as yet unknown. The spatial domain we consider is the annular cross-section of the stalk as illustrated in Figure 3.24(e). The available evidence is not sufficient for us to suggest any specific reaction kinetics for the reaction diffusion system so we choose the simplest two-species mechanism, the Schnackenberg (1979) system we considered in some detail in Chapter 7, Volume I, specifically the dimensionless form (2.10). It is

$$u_t = \gamma(a - u + vu^2) + \nabla^2 u = \gamma f(u, v) + \nabla^2 u, \tag{3.30}$$

$$v_t = \gamma(b - vu^2) + d\nabla^2 v = \gamma g(u, v) + d\nabla^2 v, \tag{3.31}$$

which define $f(u, v)$ and $g(u, v)$ and where a, b, γ and d are positive parameters. With the annular domain, u and v are functions of r, θ and t with the domain defined by

$$R_i \leq r \leq R_0, \quad 0 \leq \theta < 2\pi, \tag{3.32}$$

where R_i and R_0 are the dimensionless inside and outside radii of the annulus respectively, and the Laplacian

$$\nabla^2 = \frac{\partial^2}{\partial r^2} + \frac{1}{r}\frac{\partial}{\partial r} + \frac{1}{r^2}\frac{\partial^2}{\partial \theta^2}. \tag{3.33}$$

The scale parameter γ is proportional to R_i^2 here.

We introduce further nondimensional quantities and redefine the already dimensionless variable r by

$$r^* = \frac{r}{R_i}, \quad \delta = \frac{R_0}{R_i}, \quad R^2 = R_i^2\gamma \tag{3.34}$$

and the system (3.30)–(3.32) becomes, on dropping the asterisks for notational convenience,

$$u_t = R^2(a - u + vu^2) + \frac{\partial^2 u}{\partial r^2} + \frac{1}{r}\frac{\partial u}{\partial r} + \frac{1}{r^2}\frac{\partial^2 u}{\partial \theta^2}, \tag{3.35}$$

$$v_t = R^2(b - vu^2) + d\left(\frac{\partial^2 v}{\partial r^2} + \frac{1}{r}\frac{\partial v}{\partial r} + \frac{1}{r^2}\frac{\partial^2 v}{\partial \theta^2}\right), \tag{3.36}$$

with the reaction diffusion domain now given by

$$1 \le r \le \delta, \quad 0 \le \theta < 2\pi. \tag{3.37}$$

Biologically the inner wall of the stalk is impermeable to calcium so we assume zero flux conditions for both u and v on $r = 1$. There is a net flux of calcium into the annulus. However, the intracellular concentration level of calcium is $O(10^{-4}\ \text{mM})$ compared with the external level of 1 mM to 100 mM. Thus the influx of calcium is essentially independent of the internal concentration. The spatial dimensions of the annulus give values for δ of about 1.05 to 1.1 which implies that it is sufficiently thin for the geometry to be considered quasi-one-dimensional. We can thus reflect the inward flux of calcium by the source term b in the v (that is, calcium, equation (3.35)). We can then take zero flux conditions at the outer boundary $r = \delta$ as well as on $r = 1$. We are thus concerned with the system (3.35) and (3.36) in the domain (3.37) with boundary conditions

$$u_r = v_r = 0 \quad \text{on} \quad r = 1, \delta. \tag{3.38}$$

In the last chapter we discussed in detail the diffusion-driven spatial patterns generated by such reaction diffusion mechanisms and obtained the various conditions the parameters must satisfy. Here we only give a brief sketch of the analysis, which in principle is the same. We consider small perturbations about the uniform steady state (u_0, v_0) of (3.35) and (3.36), namely,

$$u_0 = a + b, \quad v_0 = \frac{b}{(a+b)^2}, \tag{3.39}$$

by setting

$$\mathbf{w} = \begin{pmatrix} u - u_0 \\ v - v_0 \end{pmatrix} \propto \psi(r, \theta)e^{\lambda t}, \tag{3.40}$$

where $\psi(r, \theta)$ is an eigenfunction of the Laplacian on the annular domain (3.37) with zero flux boundary conditions (3.38). That is,

$$\nabla^2 \psi + k_2 \psi = 0, \quad (\mathbf{n} \cdot \nabla)\psi = 0 \quad \text{on} \quad r = 1, \delta, \tag{3.41}$$

where the possible k are the wavenumber eigenvalues which we must determine. In the usual way of Chapter 2 we are interested in wavenumbers k such that $\operatorname{Re} \lambda(k^2) > 0$. The only difference between the analysis here and that in the last chapter is the different analysis required for the eigenvalue problem.

The Eigenvalue Problem

Since the dimensions of the relevant annular region in *Acetabularia* implies $\delta \sim 1$ we can neglect the r-variation as a first approximation and the eigenvalue problem is one-dimensional and periodic in θ with $\psi = \psi(\theta)$. So in (3.41) the r-variation is ignored ($r \sim 1$) and the eigenvalue problem becomes

$$\frac{d^2\psi}{d\theta^2} + k^2\psi = 0, \quad \psi(0) = \psi(2\pi), \quad \psi'(0) = \psi'(2\pi) \tag{3.42}$$

which has solutions

$$k = n, \quad \psi(\theta) = a_n \sin n\theta + b_n \cos n\theta \quad \text{for integers } n \geq 1, \tag{3.43}$$

where the a_n and b_n are constants. The exact problem from (3.41) is

$$\frac{\partial^2\psi}{\partial r^2} + \frac{1}{r}\frac{\partial\psi}{\partial r} + \frac{1}{r^2}\frac{\partial^2\psi}{\partial\theta^2} + k^2\psi = 0 \tag{3.44}$$

with

$$\begin{aligned} &\psi_r(1,\theta) = \psi_r(\delta,\theta) = 0 \\ &\psi(r,0) = \psi(r,2\pi), \quad \psi_\theta(r,0) = \psi_\theta(r,2\pi). \end{aligned} \tag{3.45}$$

We solve (3.44) by separation of variables by setting

$$\psi(r,\theta) = R_n(r)(a_n \sin n\theta + b_n \cos n\theta) \tag{3.46}$$

which on substituting into (3.44) gives

$$R_n'' + r^{-1}R_n' + \left(k^2 - \frac{n^2}{r^2}\right)R_n = 0, \quad R_n'(1) = R_n'(\delta) = 0. \tag{3.47}$$

The solution is

$$R_n(r) = J_n(k_n r)Y_n'(k_n) - J_n'(k_n)Y_n(k_n r), \tag{3.48}$$

where the J_n and Y_n are the nth order Bessel functions and the eigenvalues $k^2 = k_n^2$ are determined by the boundary conditions. The form (3.48) automatically satisfies the first of (3.45) while the second requires

$$J_n(k_n\delta)Y_n'(k_n) - J_n'(k_n)Y_n(k_n\delta) = 0. \tag{3.49}$$

For each n in the last equation there is an infinity of solutions k_n^j, $j = 1, 2, \ldots$. These values have been evaluated numerically by Bridge and Angrist (1962). We know, of course, that as $\delta \to 1$ the problem becomes one-dimensional and the eigenvalues

$k \to n$, so we expect $k_n^j(\delta) \to n$ as $\delta \to 1$. (In fact, this can be shown analytically by setting $\delta = 1 + \varepsilon$ in (3.49) and carrying out a little asymptotic analysis as $\varepsilon \to 0$.) This completes our discussion of the eigenvalue problem.

In the last chapter, specifically Section 2.5, we discussed the role of the dispersion relation in pattern creation and obtained the Turing space of the parameters wherein spatial perturbations of specific wavenumbers about the uniform steady state (u_0, v_0) in (3.39) could be driven unstable. That is, in (3.40), $\operatorname{Re}\lambda(k^2) > 0$ for a range of wavenumbers. The range of wavenumbers is obtained from the general expressions in (2.66) in Chapter 2, Section 2.5; there is a slight change in notation, with R^2 for γ used in (3.35) and (3.36). With the notation here the range is given by

$$
\begin{aligned}
&K_1^2 < k^2 < K_2^2 \\
&K_2^2, K_1^2 = R^2 \frac{(df_u + g_v) \pm \{(df_u + g_v)^2 - 4d(f_u g_v - f_v g_u)\}^{1/2}}{2d},
\end{aligned}
\tag{3.50}
$$

where here, and in the rest of this section, the derivatives are evaluated at the steady state (u_0, v_0) given by (3.39). In the quasi-one-dimensional situation with eigenvalue problem (3.42), the eigenvalues k are simply the positive integers $n \geq 1$. From the last equation and (3.34) the range of spatial patterns which are linearly unstable is thus proportional to the radius of the annulus R_i.

For each eigenvalue k satisfying (3.50) there is a corresponding $\operatorname{Re}\lambda(k^2) > 0$ and among all these (discrete) k's there is one which gives a maximum $\operatorname{Re}\lambda = \operatorname{Re}\lambda_M = \operatorname{Re}\lambda(k_M^2)$. k_M is again obtained as in Section 2.5, specifically equation (2.68) which in the notation here gives

$$
\begin{aligned}
k_M^2 &= \frac{R^2}{d-1}\left\{(d+1)\left[\frac{-f_v g_u}{d}\right]^{1/2} - f_u + g_v\right\} \\
&= \frac{R^2}{d-1}\left\{-\frac{b-a}{b+a} - (b+a)^2 + (d+1)\left[\frac{2b(b+a)}{d}\right]^{1/2}\right\}
\end{aligned}
\tag{3.51}
$$

on evaluating the derivatives at (u_0, v_0) from (3.39) (or simply getting them from (2.34)).

As we also discussed in Chapter 2, at least in a one-dimensional situation the fastest growing mode is a good indicator of the ultimate finite amplitude steady state spatial pattern. That is, the pattern wavelength w in the quasi one-dimensional situation is given by the dimensionless length $w = 2\pi/k_M$, with k_M from (3.51). If we now choose the basic length to be the radius r_i of the annulus then in dimensional terms from (3.34) we see that the dimensional wavenumber $k_{Md} = k_M/r_i$ and so the dimensional wavelength

$$
w_d = r_i w = \frac{r_i 2\pi}{r_i k_{Md}} = \frac{2\pi}{k_{Md}},
$$

which is independent of the radius r_i. Thus, in our model, hair *spacing* is *independent* of the stalk radius. With the experience from Chapter 2 this, of course, is exactly as we should expect.

For qualitative comparison with the experimental results in Figure 3.25 we must consider the effect on the pattern formed as the external calcium concentration varies, that is, as b varies. The major experimental facts (see Goodwin et al. 1984) are: (i) There is a range of external calcium concentrations within which whorls will form. That is, if b is too high or too low no hairs are initiated. (ii) Within this range, the hair spacing decreases as the calcium concentration increases, quickly at first but then becoming more gradual. (iii) The amplitude of the pattern decreases to zero as the concentration of Ca^{2+} approaches the upper and lower limits. We now want to derive relevant quantities from the model to compare with these basic experimental facts.

We must derive some analytical measure of the amplitude of the pattern which is formed by the mechanism. In practical terms only a finite amount of time is available to generate required patterns. In reaction diffusion models the steady state pattern is obtained, from a mathematical viewpoint, only as $t \to \infty$. Linear theory, however, provides information on the fastest growing mode which generally dominates the patterning, thus giving a good prediction of the final qualitative picture of steady state morphogen concentrations. It is quite likely, if a morphogen theory obtains, that differentiation to initiate a hair takes place when the morphogen level reaches some threshold value. So, it is reasonable to suppose that the maximum linear growth rate $\text{Re}\,\lambda(k_M^2)$ gives some indication of the actual morphogen amplitude observed—certainly if $\lambda_M = 0$ the amplitude must be zero. We thus use $\text{Re}\,\lambda(k_M^2)$ as our amplitude measure which we get by substituting k_M from (3.51) into the expression for λ (the larger of the two solutions of (2.23)), namely,

$$2\lambda_M = \gamma(f_u + g_v) - \frac{k_M^2}{d+1} + \left\{\left[\gamma(f_u + g_v) - \frac{k_M^2}{d+1}\right]^2 - 4h(k_M^2)\right\}^{1/2},$$

$$h(k_M^2) = dk_M^4 - \gamma(df_u + g_v)k_M^2 + \gamma(f_u g_v - f_v g_u).$$

With the kinetics from (3.35) and (3.36) and k_M from (3.51), a little tedious algebra gives the maximum growth rate as

$$\lambda_M = \lambda(k_M^2) = \frac{1}{d-1}\left\{d\frac{b-a}{b+a} + (b+a)^2 - [2bd(b+a)]^{1/2}\right\}. \tag{3.52}$$

Consider now the (a, b) Turing space for the system (3.30) and (3.31) given in Figure 2.12 for various values of the diffusion ratio d. We reproduce one of the curves for reference in Figure 3.26(a) and relate the parameter b to the external calcium concentration. Referring to Figure 3.26(a) if we consider a fixed a, $a_1 (> a_m)$ say, then, as we increase the calcium concentration b from zero, we see that no pattern is formed until it reaches the lower threshold value $b_{\min}$. Further increase in b moves the parameters into the parameter space for pattern formation. Figure 3.26(b) shows a typical computed pattern obtained numerically in the quasi-one-dimensional situation. This is the case when the stalk wall is sufficiently thin; that is, $\delta \approx 1$ in (3.37), and r-variations in (3.35) and (3.36) can be neglected. Figure 3.26(c) shows the corresponding pattern on the annulus where the shaded region is above a concentration threshold. We assume that when this happens a hair is initiated. If the annular region is wider, that is, δ is larger

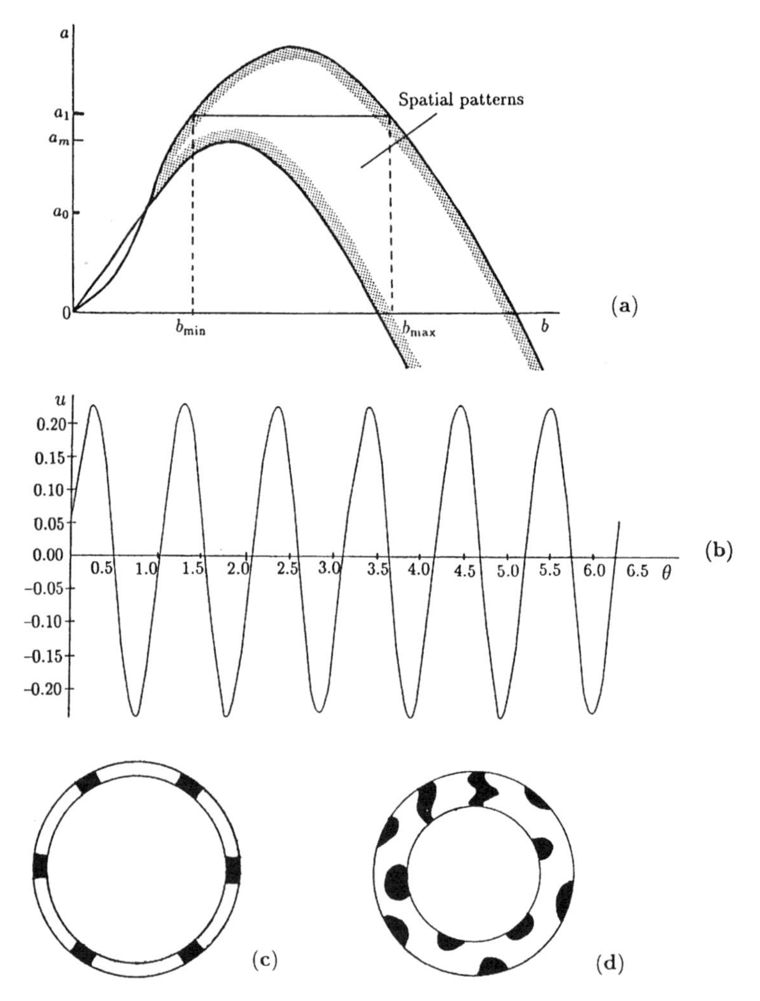

Figure 3.26. (**a**) Typical Turing space for the mechanism (3.30) and (3.31) for a fixed d. Spatial patterns can be generated when a and b lie within the region indicated. (**b**) Computed solution structure for u graphed relative to the steady state from (3.35)–(3.37) as a function of θ in the quasi-one-dimensional situation where $\delta \approx 1$ (that is, we set $\partial/\partial r \equiv 0$): parameter values $a = 0.1$, $b = 0.9$, $d = 9$, $R = 3.45$ (steady state $u_0 = 1.9$, $v_0 \approx 0.25$). (**c**) The equivalent pattern on the stalk: shaded regions represent high concentration levels of Ca^{2+} above a given threshold. (**d**) As the width of the annular region increases the pattern generated becomes more two-dimensional and less regular.

(approximately $\delta > 1.2$) so that r-variations have to be considered, the spatial pattern generated takes on a more two-dimensional aspect, an example of which is shown in Figure 3.26(d); here $\delta = 1.5$.

As b increases beyond b_{max} the parameters move out of the Turing space and the mechanism can no longer create a spatial pattern. This is in keeping with the experimental fact (i) above and illustrated in the quantitative experimental results in Figure 3.25. Note from Figure 3.26(a) that this qualitative behaviour only happens if the fixed a is

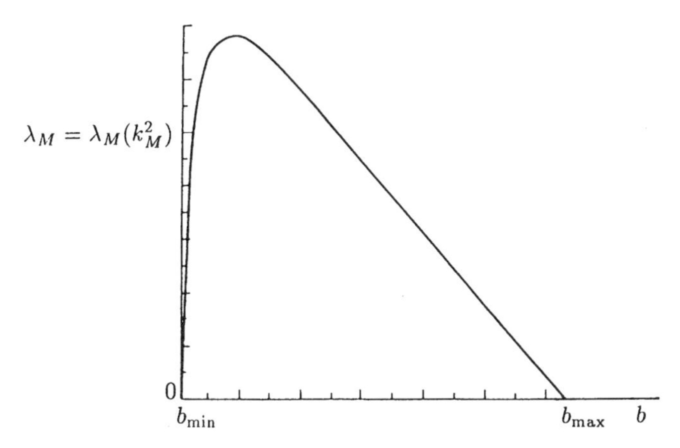

Figure 3.27. Typical computed maximal growth λ_M from (3.52) as a function of b when $a(> a_m)$ and b lie within the parameter range giving spatial structure in Figure 3.26(a). λ_M relates directly to the amplitude of the steady state standing wave in calcium Ca^{2+}. Compare this form with the experimental results shown in Figure 3.25: $b_{\min}$ and $b_{\max}$ are equivalent to the 2 mM and 60 mM points respectively.

greater than a_m and is not too large. For a fixed $a_0 < a < a_m$ we see that as b increases from zero there are two separate domains where pattern can be generated.

In Figure 3.26(a), if $a = a_1$, for example, the maximum linear growth rate λ_M is zero at $b = b_{\min}$ and $b = b_{\max}$: this can be derived analytically from (3.52). For $b_{\min} < b < b_{\max}$ the growth rate $\lambda_M > 0$. Using the analytical expression for the maximal growth rate λ_M in (3.52) we computed its variation with b as b took increasing values from $b_{\min}$ to $b_{\max}$. As discussed above we relate the maximal growth rate with the amplitude of the resulting pattern; Figure 3.27 displays the results.

Since the experiments also measure the effect on the wavelength w of varying the external Ca^{2+} concentration we also examine the predicted behaviour of w from the above analysis as b varies. For fixed a and d, (3.51) gives the dependence of k_M on b and hence of the pattern wavelength $w = 2\pi/k_M$. We find from (3.51), with appropriate $a(> a_m)$, that the wavelength decreases with b as we move through the pattern formation region in Figure 3.26(a). Figure 3.28 illustrates the computed behaviour using the dimensional wavelength obtained from k_M in (3.51) as compared with the experimental results from Goodwin et al. (1984). The parameters a, d and R were fitted to give a best fit, but nevertheless the quantitative comparison is reasonable when b is varied.

The material presented here is an example of how a model mechanism and an experimental programme can be directly related and developed together. The hypothesis that calcium could be one of the morphogens in a reaction diffusion system was explored and a specific mechanism suggested which satisfied some of the required conditions dictated by experiment, such as a window of external calcium concentration where hair patterns could be formed. Certainly not all reaction diffusion mechanisms exhibit this behaviour.

We chose a simple two-species mechanism which incorporated key biological facts and identified one of the morphogens with calcium. The question arises as to what the other morphogen, u, could be. One candidate proposed was cyclic-AMP (cAMP)

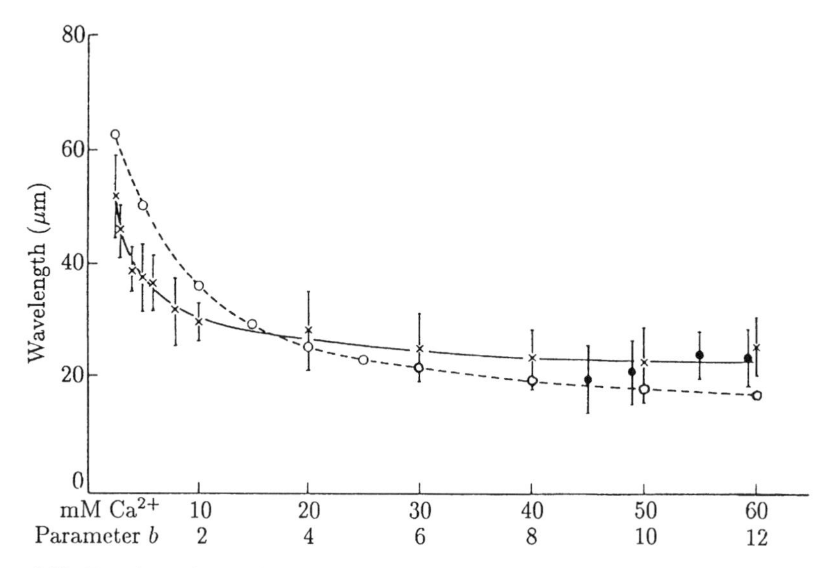

Figure 3.28. Experimental (x) and theoretical (o) results for the variation of the hair spacing wavelength (distance in microns (μm) between hairs) on a regenerating whorl of *Acetabularia* as a function of the external calcium concentration. This relates to the parameter b in the model mechanism (3.35) and (3.36). The bars show standard deviations from the average distance between hairs on groups of plants and not on individual hairs. The solid (•) period denotes where whorls were formed but, with extensive gaps where hairs failed to form, although the mean hair spacing where they formed normally was the same as in plants with complete whorls, for that calcium concentration. (From Goodwin et al. 1984)

which is important in cellular metabolism. cAMP induces the release of calcium from mitochondria while calcium inhibits cAMP production. However, the conditions for spatial structure require $d > 1$ which means cAMP has to diffuse faster than calcium which, with cAMP's larger molecular weight, is not the case. Another candidate is the proton H^+ since there is some evidence of a close connection between calcium and the proton pump activity and pH in the morphogenesis of *Acetabularia*. With the present state of knowledge, however, the identity of either morphogen must still be speculative.

The formation of a spatial pattern in calcium concentration is viewed as the prepattern for hair initiation. Actual hair growth with its mechanical deformation of the plant is a subsequent process which uses and reflects the prepattern. It is possible that calcium is directly coupled to the mechanical properties of the cytoplasm, the shell of the stalk. Such a coupling could be incorporated into the mechanochemical theory of morphogenesis discussed in detail later in Chapter 6. In fact the mechanisms proposed there do not need a prepattern prior to hair initiation; the whole process takes place simultaneously.

The Need for More Complex Processes

Although in the last chapter we briefly touched on the effect of domain growth on the patterns generated by reaction diffusion models, it is clear that pattern formation does not always take place on a static domain. Not only that, zero flux boundary conditions are not always appropriate. These aspects require us to consider pattern formation mechanisms on growing domains and the effect of different boundary conditions on the ultimate patterns formed by a mechanism. Other than the fact that we still do not know what the patterning mechanisms and the morphogens really are in development, one of the more important reasons for studying more complex systems theoretically is that there are a large number of patterns which do not seem to be able to be formed by reaction diffusion systems other than by unrealistic tinkering with the parameters and so on at different stages in the pattern formation process. The 13-lined chipmunk with its lines of spots running longitudinally along the back is an animal coat pattern example. This specific pattern was obtained by Aragón et al. (1998) in their detailed study of fish patterns. In the next chapter we consider the effect of growing domains and some of the naturally occurring complex patterns not discussed here.

4. Pattern Formation on Growing Domains: Alligators and Snakes

In Chapter 4, Volume I we discussed how sex is determined in the alligator—in the crocodilia in general—and the possible critical role it has played in their astonishingly long survival. They also have other remarkable attributes, associated, for example, with their metabolism, unique physiology and, of course, their awesome predation skills,[1] all of which contribute to their survival. In this chapter we discuss spatial patterning problems specifically associated with the alligator, *Alligator mississippiensis*. Unlike the patterning problems we have already discussed those we consider here involve different aspects of embryonic growth and the effect it has on the actual patterning process. We study two practical patterning problems for which we have excellent experimental data, namely, the stripe pattern on the skin and the spatial patterning of teeth primordia (the precursors of teeth). We shall see that growth plays a crucial role in the developmental process.

The alligator embryo is a particularly convenient (except from the alligator's point of view) embryo to study and manipulate since it does not suffer from the inaccessibility that mammalian embryos pose during their development. Development takes place in the egg external to the adult. Extensive studies of the eggshell and the embryonic membranes (Ferguson 1981a,b,c, 1985) have resulted in the development of techniques for the semi-shell-less culture and manipulation of the embryo over the whole incubation period. Because of the egg composition, particularly the calcium kinetics, alligator embryos develop quite normally in a shell-less culture. Not only that, another characteristic of the crocodilia is that they exhibit certain mammal-like characteristics which are important when studying craniofacial development in humans (Ferguson, 1981c,d); we come back to this point below when we discuss the spatial patterning of teeth primordia. The experimental data described in this chapter have all come from direct observation of the embryo at different developmental stages and of the effects of surgical manipulation, in the case of the teeth studies, by cutting windows in the eggshell and eggshell membrane. It is because of this accessibility that we have been able to relate the theory to experiment so effectively. With the first topic, namely, the stripe patterning on alligators, the theory suggested specific experiments which confirmed the theoretical hypothesis.

---

[1]Food intake for 7 crocodiles for a month is about 100 lbs compared to 95 lbs for a large dog. They create high temperatures for fever treatment. Perhaps less easy to verify is Edward Topsell's description of how some people deal with crocodiles: in his *Historie of Foure-footed Beastes* (1607) he writes, 'according to some people you could chase away a crocodile by closing the left eye and staring at it fixedly with the right.'

In the second, that of the highly regular sequential spatial patterning of teeth primordia, the theory, developed from extant knowledge of the developmental biology, makes predictions as to the outcome of a series of experiments which may help us to further understand the underlying biological pattern formation mechanism.

4.1 Stripe Pattern Formation in the Alligator: Experiments

As has been reiterated many times in this book in regard to spatial pattern formation we simply do not know what the actual mechanism is in the developmental process. Although we study several possible pattern generating mechanisms such as reaction diffusion, chemotaxis diffusion and mechanical systems the experimental evidence for a specific mechanism is still lacking. One of the major drawbacks is that in many situations we do not know *when* in development the pattern generating mechanism is operative; we only observe the results. In the case of the striping on the zebra discussed in Chapter 3 we deduced that the mechanism was probably operative at a specific time in development determined by counting the stripes and measuring their size and number relative to the size of the embryo at that time. However, firm experimental verification was lacking due to the paucity of data on developing zebra embryos. Because of the relative ease of embryonic manipulation and reliability of growth data obtainable with alligator embryos Murray et al. (1990) decided to study the stripe patterning on the alligator to try and determine the time of initiation of the patterning mechanism and to quantify the effect of size on the stripe pigmentation pattern. It is in part their work that we describe. We then use the results to show how the theory (Murray 1989) suggested specific experiments (Deeming and Ferguson 1989a) to resolve a recurring question in development, namely the role of genetics in pattern determination, as it applies to the stripe pattern on alligators. An interesting result is that we show that genetics does *not* play a role in the detailed stripe formation as had been often stated.

Experimental Results of the Effect of Incubation Temperature on the Number of Stripes on A. Mississippiensis

Hatchling alligators are dark brown/black (a result of melanin production) with a series of white stripes down their dorsal side from the nape of the neck to tail tip as typically shown in Figure 4.1. Individual hatchlings exhibit variation in both the intensity of the dark regions of the body and in the number of stripes along the body and tail (Deeming and Ferguson 1989a). Less regular 'shadow stripes' often exist on the body sides; these approximately interdigitate with the principal stripe patterning as seen in Figure 4.1; we discuss these shadow stripes later. We know from Chapter 4 that incubation temperature determines sex in alligators with females from eggs incubated at lower temperatures and males at higher temperatures. Females at 30°C are generally paler and have fewer white stripes than males, incubated from eggs incubated at 33°C (Deeming and Ferguson 1990). During development the pattern is first apparent at stage 23 of development which corresponds approximately to 41–45 days of incubation; see the detailed facts in the seminal review by Ferguson (1985). The gestation period is around 65 to 70 days.

In alligators an incubation temperature of 33°C gives rapid differentiation and growth of the embryo compared to 30°C. For any given day of incubation, or stage

Figure 4.1. Typical stripe pattern on an alligator hatchling. There are generally 8 white stripes from the nape of the head to the rump with 12 from the rump to tail tip. Note the shadow stripes on the trunk towards the ventral side and their position relative to the main stripes. This embryo is at stage 25 which is around 51–60 days of gestation. Ruler is in millimetres. (Photograph courtesy of Professor M.W.J. Ferguson)

of development, prior to hatching, embryos at the higher temperature are heavier (Ferguson 1985; Deeming and Ferguson 1989b). As pointed out in Chapter 4, Volume I the optimal temperature for both females and males is in the region of 32°C.

In the first three sections we examine the link between embryo size and the pigmentation pattern exhibited by individual hatchlings and address the important question of *when* in gestation the actual pattern generation mechanism is operative. If we are ever to discover a real biological pattern formation mechanism from experiment it is clearly essential to know when (and of course where) during development to look for it: it is too late after we see the pattern. Although there is essentially no mathematics in this part of the chapter we use mathematical modelling concepts which result in real verifiable (and already verified) biological implications.

Murray et al. (1990) counted the number of white stripes along the dorsal (top) side of alligator hatchlings, from the nape of the neck to the tip of the tail; there are no stripes on the head. The number of stripes on the body (nape to rump) and on the tail (rump to tail tip) was recorded together with the colour of the tail tip. The total length of the animal, the nape–rump length and rump–tail tip length were also measured to the nearest 0.1 mm at various times during development. Hatchlings from two incubation temperatures, 30°C and 33°C (which resulted respectively in 100% female and 100% male hatchlings) were examined (these were identical animals to those examined by Deeming and Ferguson 1989a).

To investigate the effects of sex on pigmentation pattern (specifically the number of stripes), hatchlings from eggs of a pulsed 'shift twice' experiment (Deeming and Ferguson 1988) were analysed. In these 'shift twice' experiments eggs were incubated at 33°C, except for days 7 to 14 when they were incubated at 30°C. This incubation treatment produced 23 male and 5 female hatchlings despite the male-inducing temperature of 33°C for the rest of the incubation (Deeming and Ferguson 1988). The fact that there were some females is that sometime during this time period and with this temperature the sex was determined to be female.

In a second experiment various measurements of embryos, at 30°C and 33°C, were taken from days 10 to 50 of incubation. These included total length of the animal, nape–rump length and rump–tail tip length (Deeming and Ferguson 1989b). The embryos were assigned a stage of development (Ferguson 1985). Regression estimates were calculated for embryo growth at the two temperatures. Morphometric measurements were also taken for a third group of embryos incubated for 32, 36, 40, 44, 48 and 52 days. These embryos were also assigned a stage of development and were examined using a dissecting microscope for the earliest macroscopic indication of pigmentation pattern.

Samples of skin containing both black and white stripes were removed from the tails of alligator embryos at stage 28, that is, well after the stripe pattern was evident, and methods for demonstrating neural and melanistic cells were applied to them.

Temperature clearly affected the pigmentation pattern of hatchling alligators (see Table 1 in Murray et al. 1990). There was a higher number of stripes on animals incubated at 33°C compared to those incubated at 30°C. Those animals with a white tip to their tail had, on average, one more stripe than those with a black tip at both temperatures. Generally there are 8 stripes on the body and 12 on the tail. There was no significant effect of temperature upon hatchling length nor was there any direct relationship between hatchling size and the number of stripes. The number of stripes was *not sex linked*: male hatchlings from eggs incubated at 33°C (30°C between 7 to 14 days) had a mean number of stripes of 19.96 ($+/-$ 1.15) whereas females from the same treatment had 20.00 ($+/-$ 0.71) stripes.

Regression estimates for the relationship between time and the length of the tail and the nape–rump length for embryos incubated at 30°C and 33°C are shown in the following section in Figures 4.3 and 4.4 respectively. For ease of comparison the growth at the two temperatures is shown in Figure 4.5. Embryos at the higher temperature grew more rapidly. The time at which a ratio of nape–rump to rump–tail tip equalled $8/12$, as predicted from the growth curve data, was influenced by temperature: at 30°C this occurred at 46.5 days and at 33°C at 36.5 days. These times of incubation are similar to those recorded for stage 23 (Ferguson 1985). Total length of the embryos was also related to temperature (Figure 4.6(a)). A regression analysis showed that despite being longer at any given time, embryos at 33°C were equivalent to embryos at 30°C at any given stage (Figure 4.6(b)).

The time at which pigmentation was apparent in embryos occurred much earlier at 33°C (day 36 compared with day 44 at 30°C). The pattern was also more apparent on the body of the embryo before that on the tail.

Melanocytes (melanin producing cells) were found to be present in the basal layer of the epidermis of the alligator embryos although the distribution of both cells and melanin was different in white and black stripes. In black stripes melanocytes were abundant and there was a high concentration of melanin. In white stripes melanocytes were present but rare and although they produced some melanin it was limited to the cells and their immediate environment. There was a very sharp demarcation between the black and white stripe regions of the skin, that is, those with and without melanin; see Murray et al. (1990) for experimental details and data.

4.2 Modelling Concepts: Determining the Time of Stripe Formation

Although we do not know the actual mechanism the basic concept of essentially all pattern generation models can be couched intuitively in terms of short range activation and long range inhibition as we point out in Chapter 2. There we discussed in detail reaction-diffusion-chemotaxis mechanisms and later in the book we introduce other pattern generator systems, such as neural network models and the Murray–Oster mechanochemical theory of biological pattern formation. Any of them at this stage of our knowledge of the detailed biological processes involved could be a candidate mechanism for generating the stripes on the alligator. However, from the experimental evidence from skin histological sections given in Murray et al. (1990) and briefly described above there is some justification in taking (although really by way of example) a cell-chemotaxis-diffusion sytem in which the cells create their own chemoattractant. How such a mechanism generates spatial heterogeneity was discussed in detail in Chapter 2 and is dealt with in much more detail in Chapter 5. When the aggregative effects (chemotaxis) are greater than the dispersal effects (diffusion), pattern evolves. Of particular relevance here, however, is the sequential laying down of a simple stripe pattern by a travelling wave as given in Chapter 2, Section 2.6. The reason for this is that from observation the pattern of stripes on the alligator embryo appear first at the nape of the neck and progress down the body in a wavelike manner.

The ability of a model mechanism to generate a specific pattern is no indication as to its relevance to the biological problem under study. However, different models usually suggest different experiments. A major drawback in checking or trying out any theory, is that experimentalists generally do not know when the actual patterning takes place. In the case of the alligator the actual stripe pattern, as in Figure 4.1, becomes visually evident around 40 days through the gestation period of approximately 70 days. It is almost certain that the pattern is laid down much earlier. Thus, before it is possible to determine what mechanism is actually producing the pattern it is clearly essential to know when in gestation the mechanism is active.

Pigment deposition may depend on the number of cells present in a region. Three possible ideas can explain white stripes: either (i) melanocytes are absent, or (ii) all melanocytes produce melanin to the same extent but the concentration of cells in an area is too low for the region to look dark, or (iii) the formation of melanin by cells is dependent on cell number; for example, a threshold in cell number within a certain area has to be reached before melanin is produced. In alligator embryos, white stripes appear to be due to a low number of melanocytes in white stripes which do not produce melanin in large quantities (Murray et al. 1990).

Development generally begins at the anterior end of the embryo; the extent of differentiation at the head is always much greater than at the tail. Thus in the process of melanin deposition it is the head which shows the first significant signs of pigmentation. The pattern of white stripes is first seen on the body and gradually moves down the tail.

Whether or not the skin develops a pigmented patch depends on whether pigment cells produce melanin in sufficient quantity. The specimen cell-chemotaxis-diffusion model can certainly produce a series of stripes, or spots, depending on the geometry and scale, in which the cell concentration is high. The idea then is that at the time the mechanism is activated to produce pattern the embryo is long and thin, essentially a one-dimensional domain, and so, if the mechanism starts at one end, the nape of the neck, stripes are laid down in a sequential manner as illustrated in Figure 4.2(a) via a wave-like pattern generator which then appears as shown in Figure 4.2(b). This waveform is a one-dimensional solution (Myerscough amd Murray 1992) of a basic cell-chemotaxis-diffusion system (in dimensionless form) given in the legend of Figure 4.2(a) with the parameter values given in the figure. We discuss the model in more detail later in the chapter. Here n denotes the cells, which diffuse with coefficient D (relative to the diffusion coefficient of the chemoattractant) and which produce their own chemoattractant,

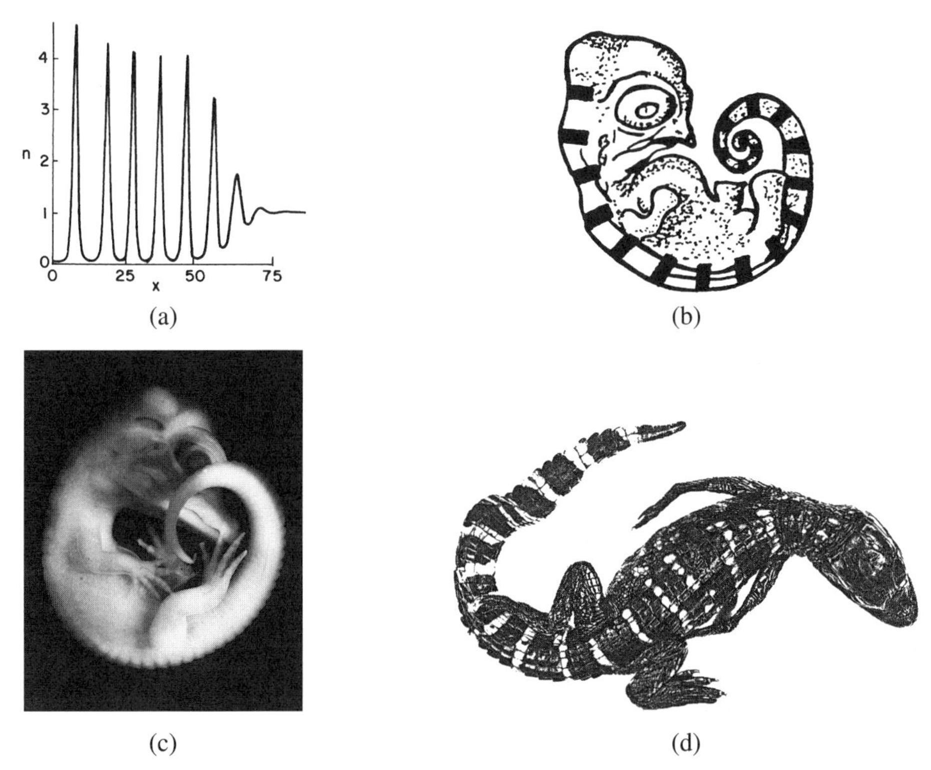

Figure 4.2. **(a)** Typical evolution of a steady state pattern in cell density, n, obtained from the cell-chemotaxis model system of equations given by (4.31) below, namely $\partial n/\partial t = D(\partial^2 n/\partial x^2) - \alpha(\partial/\partial x)(n(\partial c/\partial x)), \partial c/\partial t = (\partial^2 c/\partial x^2) + (n/n+1) - c$, where c is the chemoattractant, with parameter values $D = 0.25, \alpha b/\mu = 2$ solved with a small initial $t = 0$ perturbation at $x = 0$ about the uniform steady state $n = 1$, $c = 0.5$. The pattern forming wavefront moves to the right leaving behind the regular wavelike pattern: $x = 0$ corresponds to the nape of the neck. (From Murray et al. 1990) **(b)** Proposed evenly spaced stripe pattern of cell density on the embryo at the time the mechanism generates the pattern. **(c)** Embryo of *Alligator mississippiensis* at stage 21 (about 31 days in gestation). This is around the time when the pattern is laid down. **(d)** Alligator hatchling showing regular striping. (Photographs courtesy of Professor M.W. J. Ferguson)

c, which also diffuses. The a-term is the chemotaxis contribution which induces the cells to move up a concentration gradient in c. The cells produce c via the b-term and c degrades with first-order kinetics. The initial conditions consisted of a uniform steady state cell distribution which was perturbed by a small increase in cell density at one end, namely, the nape; the wavelike steady state pattern then moves down the back to the tip of the tail creating a regular heterogeneous stripe pattern in cell density. With the experimental evidence described above regarding the histological sections we could argue that all of the cells involved are melanocytes but they exhibit dark stripes only because of the high density of cells. This is still mainly supposition, however.

Irrespective of the actual mechanism, we hypothesise, with some justification, that the time it takes the mechanism to generate patterns—the order of a few hours in part because it is a waveformed pattern—is small compared with the time for significant embryonic growth. This being the case the mechanism will produce a regular evenly spaced pattern on the back of the alligator as illustrated in Figure 4.1.

We can go no further with the biological problem, namely, when in development the pattern was created without recourse to experimental data. With the above hypothesis, as regards the time required by the mechanism to generate the patterns, all we require is the time during embryonic development when the head to rump length and the rump to tail tip length is in the appropriate ratio consistent with the number of stripes on each as determined from the adult forms. With surprising regularity this ratio is 8:12. Such growth data for embryos during gestation is rare for any species. Fortunately it is available in the case of the alligator (Deeming and Ferguson 1990) and presented in

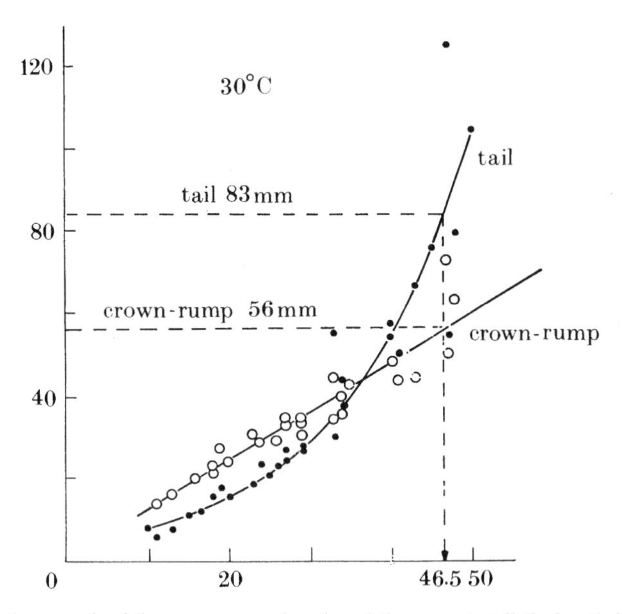

Figure 4.3. Relative growth of the crown–rump length and the rump to tail tip length in the embryo of *A. mississippiensis* at 30°C incubation temperature, which gives 100% females. Time gives days from egg laying. (From Murray et al. 1990)

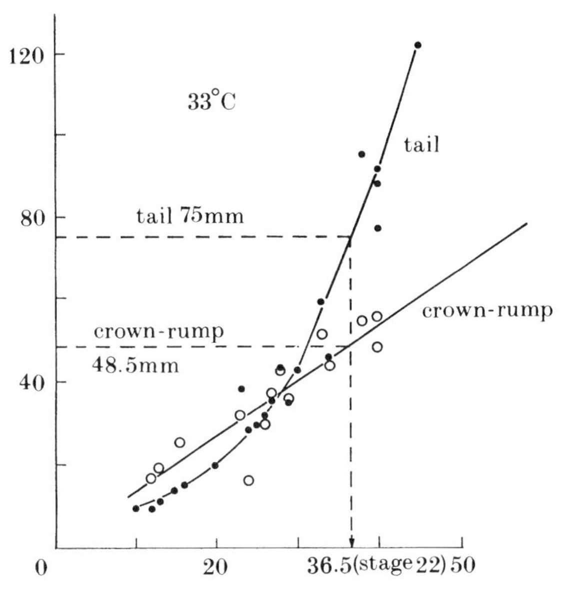

Figure 4.4. Relative growth of the crown–rump length and the rump to tail tip length in the embryo of *A. mississippiensis* at 33°C incubation temperatures, which results in 100% males. Time gives days from egg laying. (From Murray et al. 1990)

a different form by Murray et al. (1990); Figures 4.3 and 4.4 show the relative growth data needed for our model.

From the results in Figures 4.3 and 4.4 we can immediately read off when in gestation the ratio of the head–rump to rump–tail tip lengths is 8:12; namely, around day

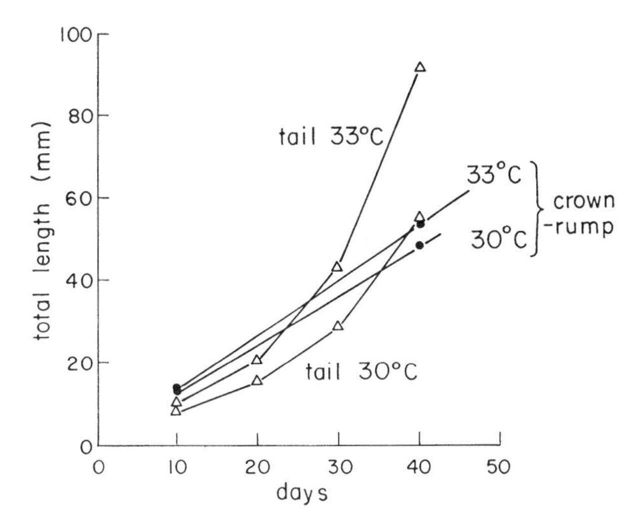

Figure 4.5. Comparison of the crown–rump length and the rump to tail tip length in the embryos of *A. mississippiensis* at two different incubation temperatures, 30°C and 33°C as a function of time from egg laying. (From Murray et al. 1990)

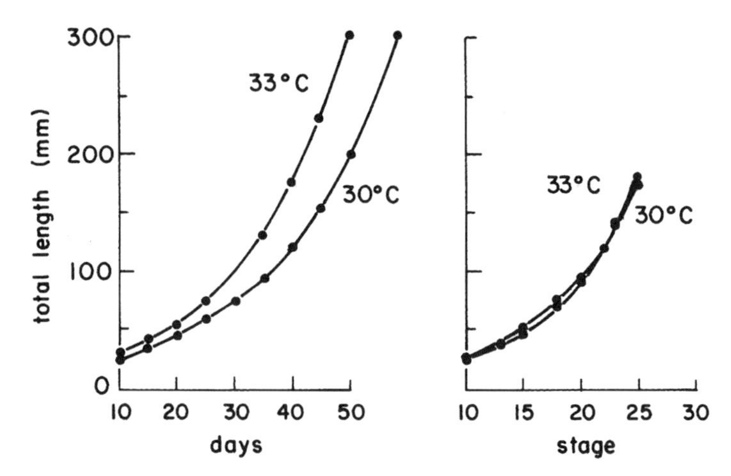

Figure 4.6. **(a)** Total length of the embryo of *A. mississippiensis* for two different temperatures, 30°C and 33°C, as a function of time from egg laying. **(b)** The length of the embryo as a function of developmental stage at 30°C and 33°C.

46.5 in the case of embryos incubated at 30°C while for embryos incubated at 33°C it is around day 35.5. From an experimental viewpoint this provides an estimate for the actual time when the mechanism is probably operative and hence focuses, much more sharply, the experimental search for the actual mechanism. Since development progresses sequentially, with different mechanisms operating at different times, it is essential to have some firm idea when in development to look. In the case of the alligator focusing attention on the period just before the pattern is visible is much too late.

4.3 Stripes and Shadow Stripes on the Alligator

As noted above the pigmentation first starts in the head region and then proceeds towards the tail. Thus, the first requirement of the mechanism is to be able to generate a travelling wave of stripe patterning from the head. Numerical simulation of the cell-chemotaxis mathematical model above with appropriate parameter values, can generate such a sequential laying down of a regular stripe pattern; Figure 4.2(a) shows a typical case with $x = 0$ corresponding to the head region and the pattern proceeding towards the tail, that is, in the positive x-direction.

In Figure 4.2(a), the peaked stripe pattern is in cell density; there is a qualitatively similar one for the chemoattractant concentration. The observed white stripe we associate with the troughs in cell density; that is, there are insufficient melanocytes to produce a pigmented area of any significance. This is in line with the observation (Table 1 in Murray et al. 1990) that when more stripes can be accommodated the tail tip is often white. A corollary of this hypothesis is that the default colour of the embryo is the dark pigmented form in that if the pattern formation mechanism were not activated there would be a uniform cell density over the integument all of which could become competent to produce melanin, probably lighter than that found between the white stripes in the normal situation. This lack of pattern would result in a melanistic form which is a very rare occurrence (Ferguson 1985).

If we hypothesise that the time it takes the mechanism to generate pattern is small compared with the time for significant embryonic growth, then the mechanism would produce a regular evenly spaced pattern on the back of the alligator as illustrated schematically in Figures 4.2(b) and in 4.7(a). On the other hand if the mechanism operates over a period of several days during which competency of the cells to generate pattern slowly decreases the striping can be less regular. We suggest that these patterns are the prepatterns for the observed pigmentation. The actual wavelength, that is, the distance between the stripes, is determined by the parameters. What is striking about these specific wavelike patterns in Figure 4.2(a) is their sharpness, which is in keeping with those found on *A. mississippiensis*. Other mechanisms mentioned above can also produce similar sequential pattern formation but it is less easy to obtain such sharpness in the peaks.

It is reasonable to assume that the mechanism is activated during a specific stage in development. If the parameter values remain fixed, the number of stripes depends principally on the length of the embryo at that stage in development. A given length of the embryo can accommodate a specific number of wavelengths, so the longer the embryo is when the mechanism is operative the greater the number of stripes. Thus the parameter values and size of the embryo determine the number of stripes.

Let us now relate these mathematical results to the experimental results in the previous sections. The first point to recall is that a higher temperature accelerates the growth of the embryo and the time to reach a given stage. So at the higher temperature the embryo is larger when the mechanism is activated. Although temperature can also affect the parameter values in a pattern generator, we do not expect this to be very significant with only a 3°C difference; we come back to this point below.

Once the cell pattern has been formed, such as in Figure 4.2 and Figure 4.7(a), a certain time must elapse for the pigmentation process and hence for the stripes to become visible. Thus the key time in development for stripe patterning must be before

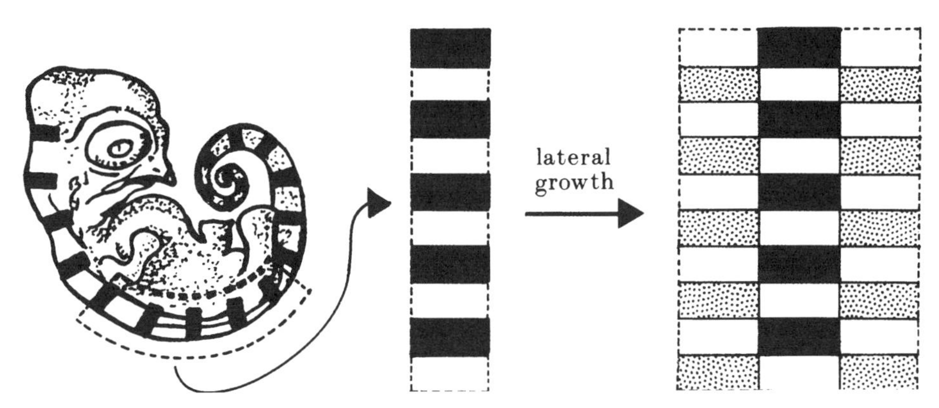

Figure 4.7. **(a)** Proposed evenly spaced stripe pattern of cell density on the embryo at the time the mechanism generates the pattern which we associate with the cell density stripes in Figure 4.2(a). **(b)** The appearance of 'shadow stripes' (cf. Figure 4.1) predicted by the model as a consequence of embryonic lateral growth in the trunk during the pattern formation process. (From Murray et al. 1990)

the stripes are visible. There is approximately an eight-day difference between embryos incubated at the two different temperatures. As also noted above, at a time prior to the pattern becoming discernible, embryos were significantly larger when incubated at 33°C than those at 30°C. Thus it is in keeping with the above patterning process that males (eggs incubated at 33°C) should have more stripes than females (30°C incubation) and that it is simply length and *not* sex which determines the number of stripes. As already mentioned the number of genes is limited and to use them up, for example, in prescribing the precise number and position of stripes on male and female alligator embryos would be an unbelievably inefficient and unnecessary use.

The theoretical prediction and resulting experimental verification described here clearly show that the number of stripes is just a question of embryonic length and size when the mechanism is activated. It is a simply a direct application of the results derived in Chapter 2 where we showed that the number of stripes in a one-dimensional domain was directly related to the length of the domain. In other words if, for example, two stripes can be formed on a domain by a typical pattern formation mechanism, whether reaction-diffusion, reaction-chemotaxis or mechanical, then four stripes will be formed if the domain is twice as long.

At the time of pattern formation a stripe (defined as from the front of a white stripe to the front of the next white stripe) must occur, on the trunk and tail, on average every L mm where

$$L = \frac{\text{tail length} + \text{nape to rump length (at pattern initiation)}}{\text{number of stripes at hatching}}.$$

Although the size of the stripes at hatching is not uniform, being wider towards the tail tip, this can be accounted for simply by the very different growth rates of the trunk and tail during embryogenesis. On the basis that the pattern generation by the mechanism is relatively fast compared to embryonic growth, and that the mechanism is activated just prior to stripe visibility, the data in Table 2 of Murray et al. (1990), from which Figures 4.3 to 4.6 were obtained, show that for embryos incubated at

$$\begin{aligned} &33^\circ\text{C: Day 32} \quad L = (56.25 + 39.50)/20.35 = 4.71 \text{ mm}, \\ &30^\circ\text{C: Day 40} \quad L = (53.60 + 37.20)/18.55 = 4.89 \text{ mm}. \end{aligned}$$

The extra 7.75 mm in length, 33°C versus 30°C, at this stage would allow for a mean of 1.64 (7.75/4.71) more stripes on male embryos. This is qualitatively in line with the observations on alligator hatchlings reported in Murray et al. (1990).

Note also that the size of the stripe, at formation, on the male is slightly smaller than on the female. Although this small size variation may not be significant it could be due to the temperature effect on the model parameters.

Some effects of temperature on pigment pattern, albeit in a very different situation, were reported by Nijhout (1980a) for the spread of an eyespot on the wing of the butterfly *Precis coenia*: the temperature difference in his experiments was 10°C. In Chapter 3 we investigated this temperature effect using a reaction diffusion model and saw that the size variation in the wing spot could be accounted for by parameter variations which were not inconsistent with those to be expected in diffusion coefficients and reaction rate constants of possible biochemicals with such a temperature change. The results of Harrison et al. (1981) on the effect of temperature on hair spacing in regenerating *Acetabularia*, which we also discussed in Chapter 3 tend to lend substance to this hypothesis: they also used a reaction diffusion model.

Shadow Stripes

We assumed that the time to generate the stripe pattern is fast compared to the time for significant embryonic growth. If we relax this assumption it implies that the patterning is taking place on a growing domain. The implication is that the cells remain competent to secrete chemoattractant, albeit with a decreasing efficiency, but nevertheless with the ability to create patterns for longer. In this case it means that the size of the integument wherein the mechanism acts is both longer and, on the trunk particularly, wider. The mathematical model implies that the type of pattern which will appear when a long thin domain becomes slightly wider, after the main stripes have been initiated, consists of less distinct 'shadow stripes' positioned between the principal stripes: see Figures 4.1 and 4.7(b). This is as we would expect from our knowledge of the effect of scale and geometry on pattern (Chapter 2). A further consequence of increasing the time of pattern formation is that since the pattern is initated from the head the density of the stripes towards the tail is less sharp in colour definition, a feature which is also often observed. Thus the mechanism, or even just knowledge of how pattern is formed by these pattern generators, can offer explanations for the more complex patterning typical of *A. mississippiensis*.

As we commented before, similar patterns can be generated by any of the models mentioned above, so the conclusions are independent of the actual mechanism involved, as long as it can form the pattern relatively quickly. So, the reaction diffusion and mechanochemical models are also candidate mechanisms. As also mentioned what distinguishes one mechanism from another are the experiments each suggests. With mechanisms which directly involve real biological quantities, such as those used here and the Murray–Oster mechanochemical models, actual cells are directly involved. It is considerably easier to manipulate embryonic cell density (a key parameter here) than it is to manipulate unknown chemicals in a reaction diffusion theory. In general it is easier to disprove theories which involve cells and tissue directly but in the process of doing so we generally increase our understanding of the patterning processes. One example of this comes from the development of cartilage patterns in the vertebrate limb discussed in detail later in the book.

In summary we have shown that the incubation temperature of *A. mississippiensis* significantly affects the number of stripes on the dorsal side of the hatchling with incubation at 30°C resulting in fewer stripes than at 33°C. Although we used a pattern formation mechanism based on the idea that cells secrete a chemoattractant to which the cells react the evidence for it is certainly not conclusive. All we require of a mechanism is that it can generate patterns in a similar way and relate similarly to the constraints of geometry and scale. The time at which the mechanism, which establishes the pattern of stripes, acts at 33°C is advanced compared to 30°C. Although the stage of development at which the mechanism is invoked is also earlier at the higher temperature, the time and developmental stage are combined in such a way that when the mechanism is activated the embryo at 33°C is longer than at 30°C and hence more stripes are laid down. The pattern of stripes is only related to the sex of the animal through temperature; the pattern is not specifically sex linked. All of the results we have found here are independent of the detailed pattern formation mechanism. This case study is another example where the modelling suggested the explanation as to why there is a larger number of stripes on male alligators as compared with female ones. The experiments described above were developed to investigate the model's hypothesis, namely, that it is simply length and not sex as such, which accounts for the difference. The theoretical predictions were confirmed.

Stripe Formation on the Juvenile Angelfish (Pomacanthus)

Another example of what might appear to start out as shadow stripes arises with growing juvenile angelfish (*Pomacanthus*), an example of which is shown in Figure 4.8. The small fish initially has three stripes but as it grows it adds to this number by inserting new stripes in between those already there: it is a gradual process which repeats itself until the fish is fully grown. However, unlike the shadow stripes on the alligator those on this fish are different in that the interdigitating stripes first appear as faint very narrow stripes unlike the same size stripes that would appear if it were simply an increase in domain space as the fish grew; this is the explanation put forward by Kondo and Asai (1995). There have been various attempts at tinkering with reaction diffusion models but none were able to explain the distinct character of the observed sequential patterning. Painter et al. (1999) obtained growth rates from extant experiments and concluded that all subepidermal tissue cell types undergo mitosis, including pigment cells. They proposed a model which included an equation for diffusional and chemotactic cell movement in which the chemotaxis is modulated by morphogens from an independnt reaction diffusion system. They show that it is the chemotaxis which produces the slow growth of the new stripes. The sequential striping they obtained, in two dimensions,

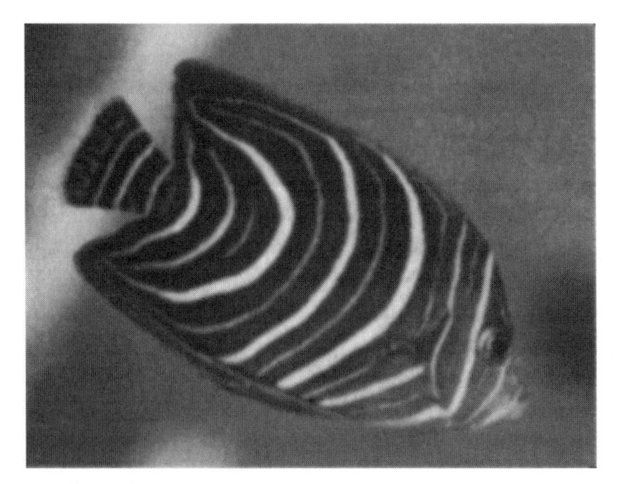

Figure 4.8. Typical stripe pattern on the angelfish, *Pomacanthus circulatis* at age 12 months. The number of stripes go from 3 at 2 months to 6 at 6 months to 12 at 12 months. (Photograph from the National Aquarium, Washington D.C. and reproduced with permission)

quantitatively mimics that observed on the angelfish. The patterning challenges on fish pose different problems to those on animal coats and snakes (discussed below). Aragón et al. (1998) investigated the role of boundary conditions, domain growth and the coupling of reaction diffusion models on the patterns that can be formed. They compared their results with the pigmentation patterns on several marine fish.

An interesting article by Denton and Rowe (1998) suggests a very practical use for the stripes on the backs of mackerel (*Scomber scombrus L.*) other than just for camouflage (or being lost in the crowd) as usually thought. They argue that the stripes are used for precise signalling information about the fish's movement to neighbours in the school. There is a thin layer of reflecting platelets which overlies the central parts of the light and dark stripes on each side of the dorsal surfaces of the fish. Denton and Rowe (1998) show how these reflecting platelets and the distribution of the body stripes can greatly facilitate information communication. When the fish changes its orientation or its velocity with respect to its neighbours there is a change in the patterns of brightness on the dorsal surfaces. They show convincingly that very slight changes in the roll, yaw and pitch immediately produce marked shifts in the observed patterns and they suggest that this is a major role for the stripe arrangement. It is well known that schools of fish coordinate their movement extremely quickly and this could be the means of doing it.

4.4 Spatial Patterning of Teeth Primordia in the Alligator: Background and Relevance

We have already given several reasons in Volume I for studying the crocodilia in general when discussing their remarkably long survivorship. In the previous sections in this chapter we saw how the study of stripe patterns on the alligator increased our understanding of the biology. In the case of teeth there are numerous reasons, other than pedagogical ones, for studying the development of dentition in *A. mississippiensis*.

The development of teeth primordia in the vertebrate jaw of the alligator (crocodilia in general) is another example in which a highly regular spatial pattern is formed in a dynamic way. Cells are coordinated to build each tooth primordium, the precursor of the actual tooth, which fits into a precise spatial and temporal sequence of teeth primordia. This process takes place as the jaw is dynamically growing and the interaction between growth and pattern formation is crucial in determining the final spacing and order of appearance of each primordium. The final pattern of teeth primordia form the foundation for the functional dentition.

The evolution of dentition has not come without occasional problems. Congenital malformations such as cleft lip and palate affect children worldwide and can cause difficulties in feeding, breathing and speech.[2] Even now little is known about the details

[2]Harelip and cleft palate medical problems have been treated (in a fashion) for a long time. Galen (130–200 AD) mentions them while James Cooke (1614–1688) a dentist in Warwick, England, a French dentist Le Monier in 1764 and others actually carried out surgery to try and correct the problems: the results were usually disastrous. Sometimes it involved putting wire ligatures through the jaw to try and correct a cleft palate while others used lead supports such as were used for George Washington's false teeth.

of biological events in craniofacial development[3] in part hampered by the difficulty in studying mammalian embryonic development *in vivo*. Although the alligator is a reptile it has certain mammalian-like characteristics with regard to its teeth and jaws and this has allowed experimental observations of certain biological processes to be made. The ability to perform non-invasive as well as invasive *in vivo* studies on the alligator *A. mississippiensis* lets us look closely at the various stages of embryonic development, where palate development and tooth initiation occur simultaneously in the embryonic jaw. This makes it possible to start investigating the biological mechanisms by which these normal and abnormal developmental processes occur. The investigations of the developmental patterning of teeth primordia complement palate studies and represent an important aspect of craniofacial development. The study of dentition development and palate closure in alligators could help in determining the developmental process in humans or at least getting clues as to how to prevent such birth deformities and predicting the effects of prenatal repair of cleft palates in humans (Ferguson 1981c, 1994). With the benefits of fetal wound repair as compared with adult wound repair (both of which we discuss in later chapters on wound healing) the benefits could be very important. At the basic level therefore their intensive study is easily justified.

Figure 4.9 shows the relative similarity (as compared with other mammalian jaws) of alligator and human jaws and their dentition. There are other similarities and of course many differences. They both have secondary palates, one row of teeth and similar palate structure but humans are biphydont (two sets during their life) and have three types of teeth while the alligator is polyphydont (many teeth sets during their life span) with one type of teeth but different sizes.

Experimental studies by Westergaard and Ferguson (1986, 1987, 1990) detailing the initiation and spatial patterning of teeth primordia provide a database from which experimental observations and hypotheses can be incorporated into a theoretical modelling framework. In this chapter we address two fundamental questions on the initiation and patterning of teeth primordia. First—the perennial question—what are the mechanisms involved in the initiation of an individual primordium? Second, how is the precise spatial distribution of these teeth primordia determined? Using the available biological data, we construct a model mechanism for teeth primordia initiation, in this case a reaction diffusion one but fundamentally different to those we have so far studied. In Section 4.8 we give the results of extensive simulations and compare them with the experimental data. We then use the model to predict possible experimental outcomes which may help to guide further experiment. By gaining a better understanding of possible mechanisms involved in teeth primordia initiation and how the mechanism actually achieves this we hope to provide clues to the formation of the human craniofacial birth defects mentioned above.

---

[3]In 1920 the U.S. War Department published some interesting statistics in a report 'Defects Found in Drafted Men.' The extensive report lists, among others, the number of draftees with cleft palates and harelips. They list the relative frequency of each defect state by state. Surprisingly the incidence in Vermont and Maine, approximately 1.5 per 1000 men, was about three times the U.S. average and half as much again as the third state in frequency. This was around the height of the eugenics fad and the report was widely studied and used by those eugenicists seeking confirmation for their loony ideas.

Human

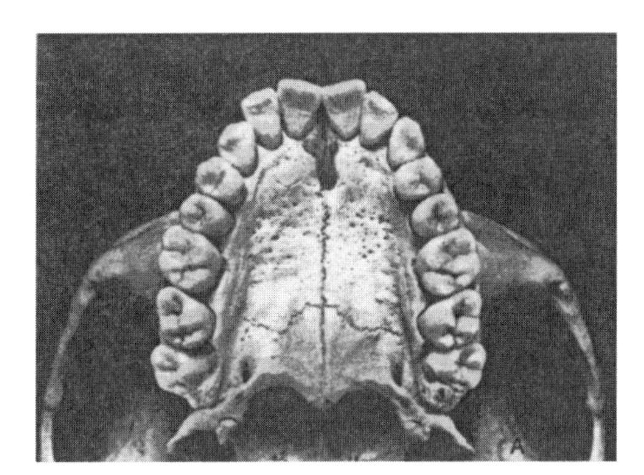

Alligator

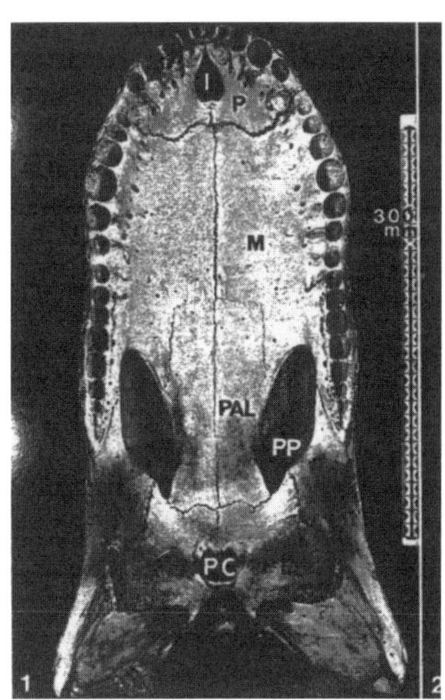

Figure 4.9. Visual comparison between human jaw dentition and that of a 13-foot (4-metre) alligator. (From Ferguson 1981d)

4.5 Biology of Tooth Initiation

Vertebrate teeth vary in size and shape yet all pass through similar stages of development. In the vertebrate jaw, there are two primary cell layers: the epithelium, which is arranged in sheets, and the underlying mesenchyme, a conglomerate of motile cells, connective tissue and collagen. Figure 4.10 schematically shows the early events in their initiation. The first sign of developing structure of the tooth organ is the tooth primordium. The tooth primordium first becomes evident in the formation of a placode, which is a localised thickening of the oral epithelium. Through a series of complex epithelial–mesenchymal (dermal) interactions which occur while the jaw is growing, these clumps of epithelial cells invaginate into the mesenchyme and cause a local aggregation of mesenchymal cells (a papilla), forming a tooth bud. In some vertebrates, early primordia degenerate into the mesenchyme and are reabsorbed or shed, while for others even early primordia develop into functioning teeth. Subsequent teeth primordia form in a similar manner and, as is the case in the alligator, in a highly stable and precise spatial and temporal sequence, and continue the formation of the set of teeth. A similar process takes place in feather germ initiation (see Chapter 6). Here we are interested in the spatial patterning of the placodes as in Figure 4.10(f).

This brings up the subject of tissue interaction and its modelling which we discuss in Chapter 6. Such models are directly related to tooth formation and are necessarily complex since the pattern generators in each tissue are coupled and the coupled sys-

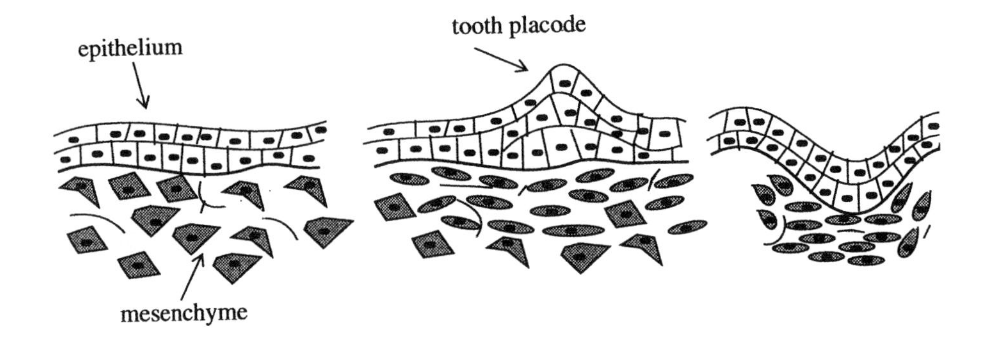

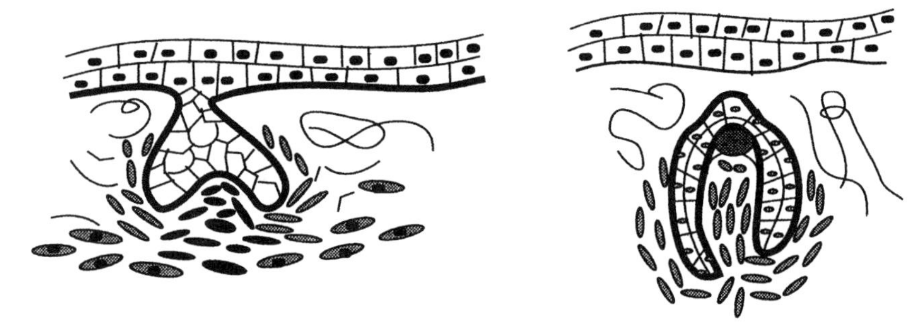

Figure 4.10. Schematic scenario for the early developmental events associated with teeth initiation. (**a**) Cell layer structure prior to primordium initiation. (**b**) Condensation in the epithelial layer marks the primordium position. (**c**) Invagination into the mesenchyme. (**d**) The dental papilla in the mesenchyme. (**e**) Differentiation of the cells to form dentin and enamel. (From Kulesa 1995)

tem has be investigated as a unit. Relevant to epidermal–dermal/mesenchyme interaction and the creation of placodes and papillae (see Figure 4.10), Cruywagen (1992) and Cruywagen and Murray (1992) present a tissue interaction model and various caricature models of the full system and carry out some analysis; Cruywagen (1992) also gives a thorough exposition of the problem of tissue interaction. A study of tooth formation through the papillae stage will require a tissue interaction model which will certainly be very complicated since it involves two complex pattern formation mechanisms interacting.

Since tooth initiation and the formation of the palate are both embryonic events, experimental investigations of the precise details have been hindered by the inaccessibility of *in vivo* observation. To understand the mechanisms involved in both processes, it is necessary to have a detailed study of the stages in development. This requires the experimentalist to have observation capabilities throughout incubation and a means of surgical manipulation of the embryo all of which are available when the embryogenesis of an animal develops in an external egg. As we mentioned, the experimentalist can carry out surgical manipulation and detailed observation throughout the whole incubation period in the case of the alligator (by cutting windows in the eggshell and its membrane).

These aspects were fully used in the embryological investigation of the reproductive biology of the crocodilia (Ferguson 1981d), the structure and composition of the eggshell and embryonic membranes (Ferguson 1981a), the mechanisms of stripe patterning discussed above in this chapter and, more relevant to teeth, in craniofacial studies of palate and tooth formation (Ferguson 1981c,d, 1988). The crocodilians possess numerous morphological features which are not characteristic of reptiles in general as pointed out above and which make them a useful model for comparison to human dentition development. Of all the crocodilians, the alligator possesses the most mammal-like snout and secondary palate (Ferguson 1981b,c,d).

Spatial and Temporal Sequence of Teeth Primordia

Westergaard and Ferguson (1986, 1987, 1990) experimentally investigated the precise spatial and temporal initiation sequence of teeth primordia in the alligator *A. mississippiensis* during development. The first tooth primordium, called the dental determinant, forms in the anterior part of the lower jaw, but it is not the most anterior tooth to form. Tooth initiation spreads from the dental determinant both forwards and backwards in the jaw. Interstitial primordia form where space is available and closer to the more mature of the two neighbours. The precise spacing and order of appearance is shown schematically in Figure 4.11, in a real alligator embryo in Figure 4.13(a) and in adults in Figure 4.18.

A major conclusion from the experimental studies of Westergaard and Ferguson (1986, 1987, 1990) is that teeth primordia initiation is directly related to jaw growth and as such necessitates modelling dynamic pattern formation on a growing domain.

From Westergaard and Ferguson (1986), the early teeth primordia form forward and backward in the jaw, with interstitial teeth forming in growing spaces between the ear-

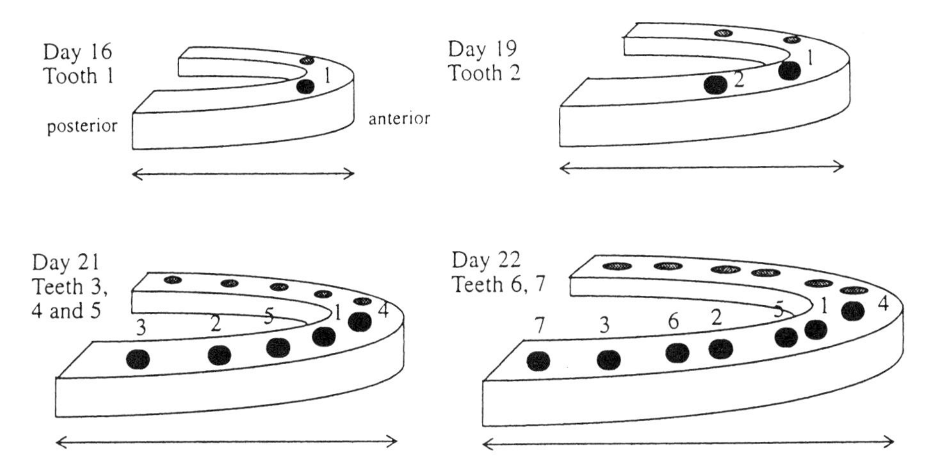

Figure 4.11. Order and time of appearance of the first seven teeth in the lower jaw of *A. mississippiensis*. (Derived from Westergaard and Ferguson 1986) The jaw is about 0.6 mm when Tooth 1 forms and about 3.8 mm by the time Tooth 4 is formed.

lier primordia. This link between the number of early teeth primordia and jaw growth led Kulesa and his coworkers (Kulesa 1995, Kulesa et al. 1993, 1996a,b, Kulesa and Murray 1995, Murray and Kulesa 1996) to investigate this relationship. If we count the number of teeth which are formed as of each developmental day, we find an unmistakable exponential relationship for the early development (Figure 4.12(a)) and a Gompertz-like growth in the number of teeth primordia over the entire course of incubation (Figure 4.12(b)). This evolution from an exponential to a Gompertz growth is a common feature of cell growth and has been modelled and explained by Murray and Frenzen (1986). From this experimental evidence we assume that the early jaw domain must be growing exponentially at a constant rate for the developmental period with which we are concerned. We incorporate this in the formulation of the mathematical model in the following section.

In the early stages of teeth primordia initiation in the jaw quadrant, the first 7 primordia form in an alternating sequence. However, the eighth primordium starts a non-alternating series which then reverts back to an alternating sequence with tooth primordium 11. Subsequent primordia formation is complicated by reabsorption of the previous teeth primordia; tooth 9a forms in the location where the first primordium was reabsorbed. The spatial sequence of the first several teeth primordia shown in Figures 4.12 and 4.13 is one of the major modelling challenges. During the full 65–70 day incubation period, approximately 19 early teeth primordia (reabsorptive group) are reabsorbed or shed without becoming functional. Seven teeth (transitional group) function for a short period (less than two weeks) or are sometimes reabsorbed or shed without becoming functional. The functional group, comprised of 36 teeth, are initiated during embryonic life and function for longer periods. From the experimental data (Westergaard and Ferguson 1987) for the lower jaw, after the initiation of the seventeenth primordium, the many remaining placodes form approximately near the spaces where older primordia have been reabsorbed. That is, jaw growth seems to slow and become less crucial to the initiation of new primordia. Thus, the role for the early primordia seems to be to set down a marker for initiation of future primordia.

The initial localised condensation of cells which mark a tooth placode occurs in the epithelium but the precise signalling mechanism for initiation is not known. Studies of signalling in tooth initiation have focused on the local occurrence of epidermal growth factor and its receptors (Thesleff and Partanen 1987, Partanen et al. 1985, Kronmiller et al. 1991), tissue interactions (Mina and Kollar 1987) and the local expression of homeobox genes (MacKenzie et al. 1991, 1992). In work on mice, Thesleff and Partanen (1987) showed that epidermal growth factor caused proliferation of dental epithelium. Kronmiller et al. (1991) then demonstrated the necessity for the presence of this epidermal growth factor during tooth initiation by showing that initiation did not occur when epidermal growth factor was chemically blocked.

More recent experimental investigations have focused on finding the molecular mechanisms involved in teeth primordia initiation and formation; for a brief review see Ferguson (1994). The presence of certain homeobox genes, namely, Msx1 and Msx2, have been identified as being expressed in the local region of tooth formation (MacKenzie et al. 1991, 1992) and these expressions are a result of epithelial–mesenchymal interactions (Jowett et al. 1993, Vainio et al. 1993). The complete formation of dentition then is a series of processes which are coordinated by signalling and the physical

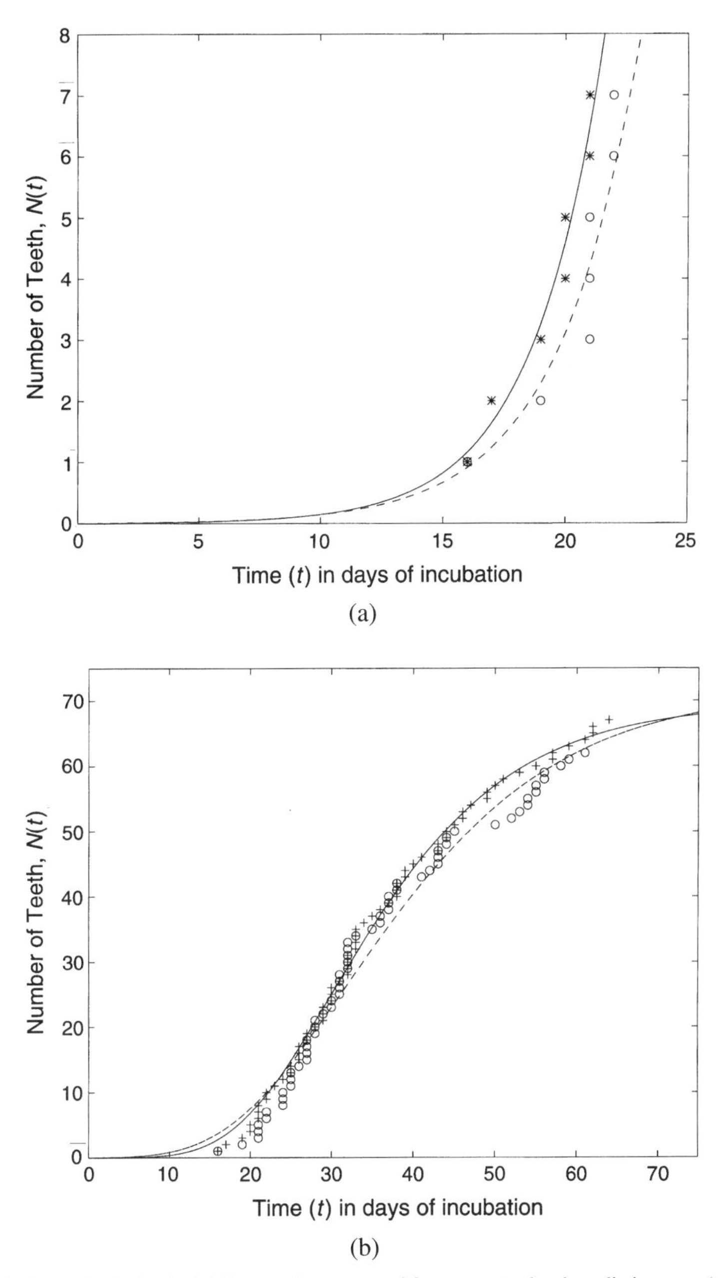

Figure 4.12. *A. mississippiensis*. (**a**) Temporal sequence of first seven teeth primordia in upper ($*$) vs. lower ($\circ$) jaws (Derived from Westergaard and Ferguson 1986, 1990) Data have been fit with exponential curve: $N(t) = N_0 \exp(rt)$. Lower jaw (dashed): $N_0 = 0.0066$, $r = 0.3077$/day. Upper jaw (solid): $N_0 = 0.0047$, $r = 0.3442$/day. (**b**) Number of teeth primordia $N(t)$ vs. incubation time t (days) for upper ($+$) and lower ($\circ$) jaws during the whole gestation period. (Derived from Westergaard and Ferguson 1986, 1987, 1990) Each data set has been fit with a Gompertz curve: $N(t) = N_1 \exp[-N_2 \exp(-rt)]$. Lower jaw (dashed): $N_1 = 71.8$, $N_2 = 8.9$, $r = 0.068$/day. Upper jaw (solid): $N_1 = 69.6$, $N_2 = 12.0$, $r = 0.082$/day. (From Kulesa and Murray 1995)

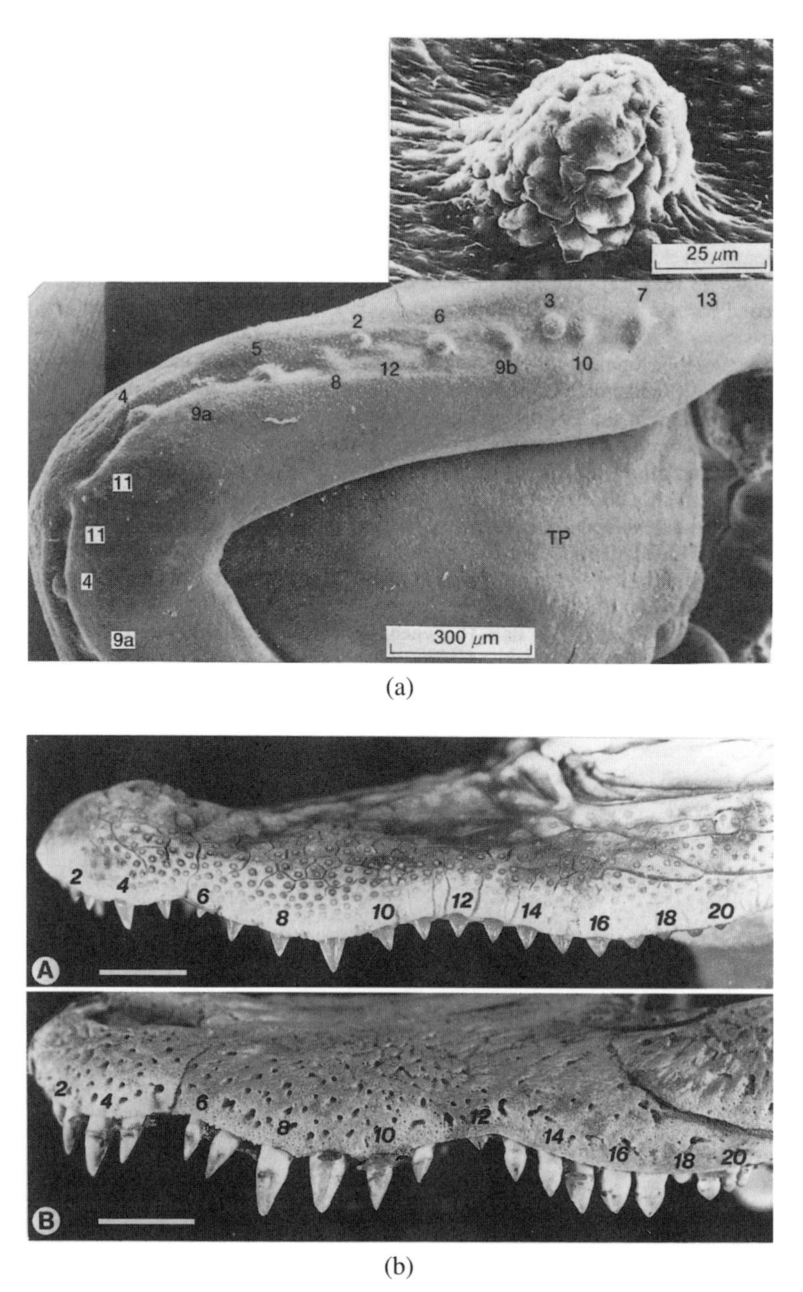

Figure 4.13. **(a)** The lower right half jaw of *A. mississippiensis* on day 26 showing the tooth primordia. The insert shows what a tooth germ (the most prominent) looks like at this time. (From Westergaard and Ferguson 1986) **(b)** Upper and lower jaws of young (A) and old (B) adult alligators. The numbers indicate tooth positions and order. In (A) the dentition is the same 20 functional teeth as at hatching: bar(A) = 5 mm, bar(B) = 20 mm. (From Westergaard and Ferguson 1990. Photographs courtesy of Professor M.W.J. Ferguson)

interaction of tissue. A critical and thorough review of the various findings, models and hypotheses, some dating back to 1894, has been given by Kulesa (1995).

4.6 Modelling Tooth Primordium Initiation: Background

The early embryological investigations of dentition development in reptiles (Röse 1894, Woerdeman 1919, 1921) form the basis for descriptive models of tooth formation, which in general fall into either a prepattern or dynamic model category. Prepattern models, which have been discussed at length in Chapter 2, are characterised by morphogens and rely on the concept of positional information (Wolpert 1969). To briefly recap, reaction and diffusion of a morphogen through a tissue creates a concentration heterogeneous landscape of the morphogen to which the cells react and differentiate accordingly. In the tooth initiation process, prepattern models hypothesise that a primordium is formed when cells respond to differences in the morphogen concentration and so the challenge is to determine the mechanism which produces the appropriate spatial pattern in morphogen concentration. The usual morphogen prepattern models are not dynamic, in the sense that once the pattern is laid down it cannot be changed by the dynamic behavior, for example, domain growth, of the system.

Woerdeman's investigations (1919, 1921) were with the reptiles, *Gongylus ocellatus* and *Crocodylus porosus*, and his data showed that the first tooth primordium formed was the most anterior (front) one in the jaw. Subsequent primordia form all along the jaw. Using these data, Edmund (1960a,b) proposed that the tooth initiation process was as if a wavelike stimulus passed along the jaw from front to back. When the wave passed over a predetermined or prepatterned tooth site, a tooth placode was initiated. The stimulus resulted from a chemical transmitter at the front of the jaw. For each passage of the hypothetical initiating pulse, a row of teeth was formed. This series of waves, called a 'Zahnreihe' (from the German 'Zahn', meaning tooth and 'reihe' meaning row) was adopted by Edmund (1960a,b) from the work of Woerdeman (1919) who first described the phenomenon. The Zahnreihe 'theory' was accepted by most investigators at the time until Osborn (1970, 1971) demonstrated, firstly, that there was no known pattern of tooth initiation in embryos which fit into the Zahnreihe theory, and secondly, that Edmund (1960) had actually rearranged Woerdeman's data (1921) to fit his theories. Although the Zahnreihe theory has been shown to be incompatible with the experimental data, it is still widely thought that a chemical wavelike impulse stimulates the initiation of teeth primordia.

In contrast to traditional prepatterning, dynamic models describe the development of pattern as self-generating, that is, a dynamic process which takes place and depends crucially on the dynamic growth of the system. We saw in the above section on shadow stripes on the alligator how growth played a crucial role. Teeth primordia may continue to be initiated by morphogen gradients, but with the pattern formation mechanism responding to the underlying growth of the system. The concept of a dynamic patterning mechanism has led to the formation of two models: the clone model (Osborn 1978) and the mechanochemical model (Sneyd et al. 1993).

Osborn (1978) was the first to try and incorporate the dynamics of jaw growth into a descriptive model for tooth initiation. His clone model postulates that tooth primordia are initiated from one or more clones of neural crest cells, situated in the mesenchyme, just under the epithelium. The pattern of teeth primordia forms as a result of the dynamic growth of the clone. In his model, a clone source has edges, called progress zones, which may expand in either direction, anterior or posterior along the jaw. As the clone expands, cell divisions give rise to competent tissue setting up a gradient in cell age. Space becomes available, at the margins of the clone, produced by successive cell generations. New primordia are initiated when there is sufficient mature tissue and available space within the expanding clone. Based on experimental evidence (Osborn 1971) that new teeth primordia tend to form closer to the older of two neighbors, Osborn (1978) suggested that each new primordium generates a zone which inhibits the initiation of further local primordia. In the model, the total number of primordia formed in the jaw depends on the size to which the clone grows and the sizes of the inhibitory zones.

The mechanochemical model by Sneyd et al. (1993) describes how the mechanical movements of cells and related tissue could create the structure and form of the tooth primordium. The model incorporates both mechanochemical and reaction diffusion mechanisms. The initiation of the dental determinant (the first tooth primordium) is presented as a process controlled by jaw growth, the age structure of the epithelial cells in the developing jaw, and the age-dependent production of cellular adhesion molecules or CAMs (for example, Chuong and Edelman 1985, Obrink 1986) in the dental epithelium. Their numerical simulations predict the initiation of the dental determinant occurs in the proper spatial position only if CAM concentrations rise quickly in the neighborhood of the forming placode and immediately prior to its development. Thus, unless CAM production in the jaw anterior increases on a timescale much faster than that of jaw growth, an incorrect pattern will form. This crucial conclusion, arrived at by simulation of the model, points out the danger of extrapolating results from a model on fixed domains to pattern formation on growing domains.

There are serious difficulties with all of these theories. The series of detailed investigations on the embryonic development of the dentition of the lower and upper jaws of *A. mississippiensis* (Westergaard and Ferguson 1986, 1987, 1990) provide accurate initiation data from day 1 to 75. The definitive sequences of initiation and replacement were derived and the development of individual teeth followed through the 65-day incubation period. These experimental studies were the first to be detailed enough to elucidate the highly precise dentition patterning during embryonic development and highlight the inadequacies, both qualitative and quantitative, of all of the above models. From a modelling point of view, the principal findings suggest that initiation of teeth is intimately related to: (1) jaw growth, (2) the distance between existing teeth and (3) the size and developmental maturity of the latter.

The experimental results confirm the inadequacies of the Zahnreihe theory and also led to a rejection (Westergaard and Ferguson 1986) of the clone model (Osborn 1978), based on the criticism that new teeth do not develop in the sequence suggested by the growing clone. Osborn (1993) suggested that the teeth, in a jaw quadrant of an alligator embryo, may develop from multiple clones; however, the initial positions and growth dynamics of each clone are unspecified. The mechanochemical model (Sneyd et al. 1993), although it produces the dental determinant in the correct position with a restriction, is primarily useful in showing the necessity of including the growth of the jaw domain. Their model also does not address the sequence of subsequent primordia. Westergaard and Ferguson (1986, 1987, 1990) make it very clear that the spatial pattern of teeth primordia is not laid down at one time, but is dynamically developing as the embryonic jaw is growing. Clearly, whatever the pattern forming mechanism for tooth initiation, it must at least be capable of reproducing the spatial *and* temporal sequence of the first few teeth primordia in the alligator from the experimental data. This is not a trivial challenge for a theoretical model. Based on the biological data, we develop and analyse in the next four sections such a dynamic model mechanism, which crucially includes jaw growth, for the initiation and spatial positioning of teeth primordia and meets the criterion as to the sequential appearance of the primordia in the *A. mississippiensis*.

4.7 Model Mechanism for Alligator Teeth Patterning

The embryonic dentition development in the alligator offers a model system to study many facets of tooth formation. As we have said we only focus here on the teeth primordia initiation process. We use the known biological facts as a guide in constructing a mathematical model mechanism which describes the initiation and spatial patterning of the teeth primordia. The preliminary goal is to reproduce the observed spatial pattern of the first seven teeth primordia in the jaw of *Alligator mississippiensis*; seven because after this these teeth begin to be reabsorbed (but in a systematic way as we described above).

We begin by discussing how the biological data let us make certain quantitative assumptions which are part of the model. The seminal experimental work of Westergaard and Ferguson (1986, 1987, 1990) forms the basis for the development of the model mechanism. The recent biological investigations and experimental studies in mice supplement the biological database. Although some experimental results have yet to be shown in alligators, we reasonably assume certain similar characteristics. We attempt to incorporate as much of the known biological data as possible.

Of crucial importance for the model mechanism, is the incorporation of the physical growth of the jaw based on the known biological data. The initiation of the first seven teeth primordia occurs during the first one-third of the incubation period of the alligator. As noted, the number of teeth primordia seems to follow an exponential (Figure 4.12(a)) relationship for the first several primordia and a Gompertz-type growth (Figure 4.12(b)) for the full set of primordia during incubation.

To begin with we have to quantitatively characterise the growth of the jaw and then incorporate this into the model system of equations. We then describe how these chemical based model equations attempt to capture the physical process by which the biological mechanisms initiate a tooth primordium and the subsequent spatial patterning of teeth primordia. Since we have considerable experience with such chemically based systems, namely, reaction diffusion systems from earlier chapters, we shall only describe briefly how the system forms the heterogeneous spatial patterns which we hypothesise give rise to the cell condensations (the placodes) in the epithelium which mark the tooth initiation sites.

Model Assumptions

Although we make certain biological assumptions, the goal of the model is to capture the essential components of the biological mechanism and as much of the known biological data as possible. From the experimental work of Westergaard and Ferguson (1986), a comparison of tooth initiation sequences and positions between left and right sides of the jaw for both the same and other specimens of *A. mississippiensis* show no evidence of significant differences. So, we assume a symmetry in the initiation processes between the left and right sides of the jaw. The region in the jaw on which the primordia form is very thin compared with its length as is evident from Figure 4.13(a) so we consider that the teeth primordia form along a one-dimensional row. We construct a line from the posterior (back) of the jaw to the anterior (front) (Figure 4.14). This posterior–anterior one-dimensional axis is further justified by the experimental results (Westergaard and Ferguson 1986, 1987, 1990) which show that tooth initiation sites have very little lateral shift from an imaginary line drawn from posterior to anterior along the jaw epithelium.

Biologically, it is unknown how the signal to start tooth initiation is switched on. It is believed that this signal is controlled by neural crest cells in the mesenchyme which somehow send a message to the epithelium to start condensing.

The cell condensations in the epithelium mark the sites of teeth primordia initiation. Since this initiating source is also unknown, we make the assumption that there is a source of chemical at the posterior end of the jaw which starts the initiation process. How the source is switched on is unspecified, which is in keeping with what we know about the biology. The role of the chemical source diminishes quickly after the first tooth primordium is formed.

The experimental identification of certain components involved in tooth initiation and formation revealed epidermal growth factor, bone morphogenetic protein (BMP-4) and certain homeobox genes (Msx1 and Msx2). This suggests a chemical mechanism for the initiation of a primordium, where certain chemical concentrations stimulate an area of the epithelium to form a placode. So, we consider a reaction diffusion system but with some very different features from those in earlier chapters.

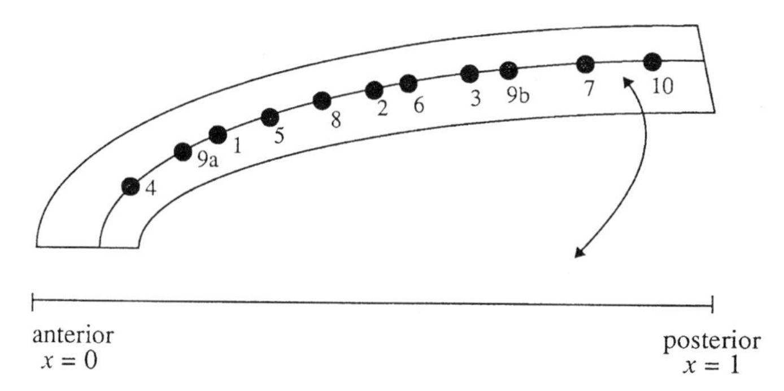

Figure 4.14. One-dimensional line approximating the anterior–posterior jaw axis along which teeth primordia form.

Model Equations

The aim then is to show that the proposed class of mechanisms for the initiation of the teeth primordia, which we now construct, following Kulesa (1995), Kulesa and Murray (1995), Murray and Kulesa (1996), and Kulesa et al. (1996a,b), are sufficient to explain the pattern of tooth sites in *A. mississippiensis*. To construct a dynamic pattern formation system, we must incorporate the physical growth of the jaw into a system capable of forming pattern. To do this, we combine the aspects of a static pattern formation mechanism, which is mediated by a control chemical, with the physical growth of the jaw domain. The result is a dynamic patterning mechanism.

Experimental evidence requires that the pattern arises dynamically as a consequence of jaw growth and not as the result of a prepattern of tooth initiation sites and so is a crucial element in the mechanism. From the experimental evidence of exponential jaw growth (Figure 4.12(a)), we assume that the jaw length, $L = L(t)$, grows at a constant strain rate, r, according to

$$\frac{dL}{dt} = rL \Rightarrow L(t) = L_0 e^{rt} \tag{4.1}$$

where, with the nondimensionalisation we use, $L_0 = 1$. The experimental data let us obtain good estimates for the parameters. Basically the growing domain dilutes the chemical concentrations.

Consider the scalar reaction diffusion equation

$$c_t = Dc_{\xi\xi} + \gamma f(c) \tag{4.2}$$

on a growing domain where D, the diffusion coefficient, and γ, the scale factor, are constants and $f(c)$ is a reaction term, a function of the concentration c. Let

$$s = \text{quantity of reactant in length } l \Rightarrow c = s/l.$$

Then, in the time interval

$$(t, t + \Delta t), \quad l \to l + \Delta l, \quad c = \frac{s}{l} \to \frac{s + \Delta s}{l + \Delta l}$$

which implies that change in the concentration

$$\Delta c = \frac{s + \Delta s}{l + \Delta l} - \frac{s}{l} = \frac{s + lf(c)\Delta t}{l + rl\Delta t} - \frac{s}{l} = \gamma f(c)\Delta t - rc\Delta t,$$

which implies that

$$\lim_{\Delta t \to 0} \frac{\Delta c}{\Delta t} = \gamma f(c) - rc$$

and so (4.2) becomes

$$c_t = Dc_{\xi\xi} + \gamma f(c) - rc \tag{4.3}$$

The dilution term is the $-rc$.

The domain grows at an exponential rate. For ease of numerical simulation of the model it is easiest to change the variable so that the equations are on a fixed domain. Here we set

$$x = \xi e^{-rt} \Rightarrow c_t = De^{-2rt}c_{xx} + \gamma f(c) - rc, \quad x\varepsilon[x, L], L \text{ fixed.} \tag{4.4}$$

The reaction diffusion equation (4.3) on a growing domain becomes a non-autonomous reaction diffusion equation in a fixed domain with a diffusion coefficient decreasing exponentially with time.

For the chemical patterning mechanism, we take, by way of example, a basic dimensionless reaction diffusion system, namely, (2.32) from Chapter 2, which we studied in some depth in Sections 2.4 and 2.5. We modify it, guided by the above discussion of the biology, by combining it with an equation for a chemical c, which controls the substrate u, in a simple inhibitory way to get

$$\left.\begin{aligned} \frac{\partial u}{\partial t} &= \gamma\left[hc - u + u^2 v\right] + \frac{\partial^2 u}{\partial x^2} \\ \frac{\partial v}{\partial t} &= \gamma\left[b - u^2 v\right] + d\frac{\partial^2 v}{\partial x^2} \\ \frac{\partial c}{\partial t} &= -\delta c + p\frac{\partial^2 c}{\partial x^2} \end{aligned}\right\}. \tag{4.5}$$

Here $u(x, t)$ and $v(x, t)$ represent the respective concentrations of a substrate and an activator with γ the usual scale factor (see Chapter 3), b and h constants, δ the assumed first-order kinetics decay of c, d the usual ratio of the activator's diffusion coefficient to that of the substrate and p the ratio of diffusion coefficient of c to that of the substrate u.

We assume the source of u is controlled by c, an inhibitor related to the epidermal growth factor, EGF. That is, we assume the existence of an inhibitory substance whose concentration decreases as the concentration of EGF increases and vice-versa. It comes from the source at the posterior end of the jaw.

If we now carry out the scale transformation in (4.3) and a nondimensionalisation to make the domain fixed to be 1 the resulting nondimensionalised equations for the substrate, u, the activator, v, and the inhibitor, c, are, on a scaled domain $0 \le x \le 1$:

$$\frac{\partial u}{\partial t} = \gamma\left[hc - u + u^2 v\right] - ru + \left(e^{-2rt}\right)\frac{\partial^2 u}{\partial x^2}, \tag{4.6}$$

$$\frac{\partial v}{\partial t} = \gamma\left[b - u^2 v\right] - rv + d\left(e^{-2rt}\right)\frac{\partial^2 v}{\partial x^2}, \tag{4.7}$$

$$\frac{\partial c}{\partial t} = -\delta c - rc + p\left(e^{-2rt}\right)\frac{\partial^2 c}{\partial x^2}. \tag{4.8}$$

Each equation (4.6)–(4.8) has a dilution term, due to jaw growth, and a time-dependent diffusion coefficient which arises from the coordinate transformation to the scaled domain in x. These equations govern the variables on the growing domain of the jaw and were studied in depth by Kulesa (1995). Relevant boundary conditions are

$$u_x(0,t) = u_x(1,t) = 0 = v_x(0,t) = v_x(1,t), \tag{4.9}$$

$$c_x(1,t) = 0, \quad c(0,t) = c_0(t), \tag{4.10}$$

where $c_0(t)$ is a decreasing function corresponding to a source term at the posterior end as discussed above. The condition (4.9) implies zero flux of u and v at either end of the domain, while for c there is zero flux only at the anterior end (4.10). Recall that we have taken only half of the jaw to scale to 1 which means that at the anterior end it is the symmetry condition which gives the condition at $x = 1$ in (4.10).

From the analyses in Chapter 2, specifically Sections 2.4 and 2.5, for a range of parameter values and a domain size larger than some minimum, the reaction diffusion system given by the first two equations of (4.5) with c a constant, is capable of producing steady state spatial patterns in u and v. By varying one or more of the parameters in these equations the system can select a stable heterogeneous state or a specific regular spatial pattern. When the inhibitor, c, in the first of (4.6) is above a threshold value, pattern formation in u and v is inhibited. For c below this threshold, the pattern formation mechanism is switched on, via diffusion-driven instability and a spatial pattern forms in u and v when the subthreshold portion of the domain is large enough. The interplay between the parameters and the domain scale is important in determining the specific pattern formed with certain spaces giving rise to specific patterns; refer specifically to Figure 4.21 below where the parameter spaces are given specifically for the hump-like patterns we require here. The analysis there is for hc a constant parameter (equivalent to the 'a' there) and for a fixed domain size. The situation we consider here has a variable domain and c varies in space and time. It is not easy to carry out an equivalent analysis in such a case. We can however get an intuitive feel for what is going to happen as we show below.

The representation of the tooth primordium becoming a source of inhibitor $c(x,t)$ is strongly suggested by the biology. The experimental studies (Westergaard and Ferguson 1986, 1987, 1990, Osborn 1971) led us to postulate a zone around a newly formed tooth which inhibits subsequent teeth primordia from forming in the local region. Westergaard and Ferguson (1986) noted that if a new primordium was forming in between older primordia, the new primordium would form closer to the older neighbour. Mathematically, we characterise this zone of inhibition by allowing each new tooth primordium to become a source of chemical growth factor, whose role is to inhibit tooth primordium formation in the local region of a newly formed primordium. We now define a new tooth site where the substrate, $u(x,t)$, crosses a threshold on a subdomain of $0 \le x \le 1$. This turns on a new tooth source, $\bar{c}_i$, of c at this site, which simulates a zone of inhibition. The concentration of this tooth source we model according to logistic growth,

$$\frac{d}{dt}\bar{c}_i = k_1 \bar{c}_i \left(1 - \frac{\bar{c}_i}{k_2}\right), \tag{4.11}$$

where the constants $k_1, k_2 > 0$ and $c(x_i, t) = \bar{c}_i(t)$ for $t \geq t_i$, where the subscript i refers to tooth i.

Mathematically these new tooth sources of inhibitor can be thought of as delta functions in both time, which 'turn on' the source when u crosses the threshold, and space which 'turn on' at the position where u crosses the threshold. We define 'turning on' of a tooth source as the rise in the inhibitory chemical c at the tooth site according to (4.11). These additional tooth sources can be added to the right-hand side of (4.8) in the form

$$\sum_{i=1} \delta(x - x_i) H(u - u_{\text{threshold}}) F(t - t_i) \quad i = i\text{th tooth}, \tag{4.12}$$

where $\delta(x - x_i)$ is the usual Dirac delta function and $H(u - u_{\text{threshold}})$ the step function defined by

$$\delta(x) = 0 \text{ if } x \neq 0 \text{ where } \int_{-\infty}^{\infty} \delta(x)\, dx = 1,\ H(u) = \begin{cases} 1 & \text{if u} > 0 \\ 0 & \text{if u} < 0 \end{cases}, \tag{4.13}$$

but with the added constraint that $H(u - u_{th}) = 1$ at any stage it remains 1 even if u decreases below u_{th}. This ensures the new tooth source stays on after it starts. The funtion $F(t)$ is the right-hand side of (4.11).

How the Mechanism Works

To get an intuitive idea of how the patterning process(4.6)–(4.8) works, let us consider, to begin with, the basic system (4.5) where we consider $c = a$ a constant and a domain of length $1 - L_c$ where L_c is determined as we describe below.

$$\left.\begin{aligned} \frac{\partial u}{\partial t} &= \gamma \left[ha - u + u^2 v \right] + \frac{\partial^2 u}{\partial x^2} \\ \frac{\partial v}{\partial t} &= \gamma \left[b - u^2 v \right] + d \frac{\partial^2 v}{\partial x^2} \end{aligned}\right\}. \tag{4.14}$$

We know from Section 2.4 that with γ, b, d and ha in the appropriate parameter domain there is a minimum length below which no pattern can form and above which it can with the mode depending on the domain length (and the other parameters, of course). We also saw that there is a parameter space (with the dimension equal to the number of parameters) in which patterns can form. It is exactly the same with this model system except that we cannot determine it analytically as we did in Chapter 2. Examples of two-dimensional cross-sections of the real parameter space are reproduced in Figure 4.21 below.

Consider now Figure 4.15(a) which schematically shows a hypothetical parameter space for some group of parameters plotted against another group of parameters and that there is a subspace in which mode 1-like and another in which mode 2-like patterns will start to grow in a large enough domain. Now consider Figure 4.15(b) which illustrates a typical solution of the third equation of (4.5) with a source of c at $x = 0$ and zero flux at $x = 1$. Now as c decreases (because of the $-\delta c$ term in the equation) there will be

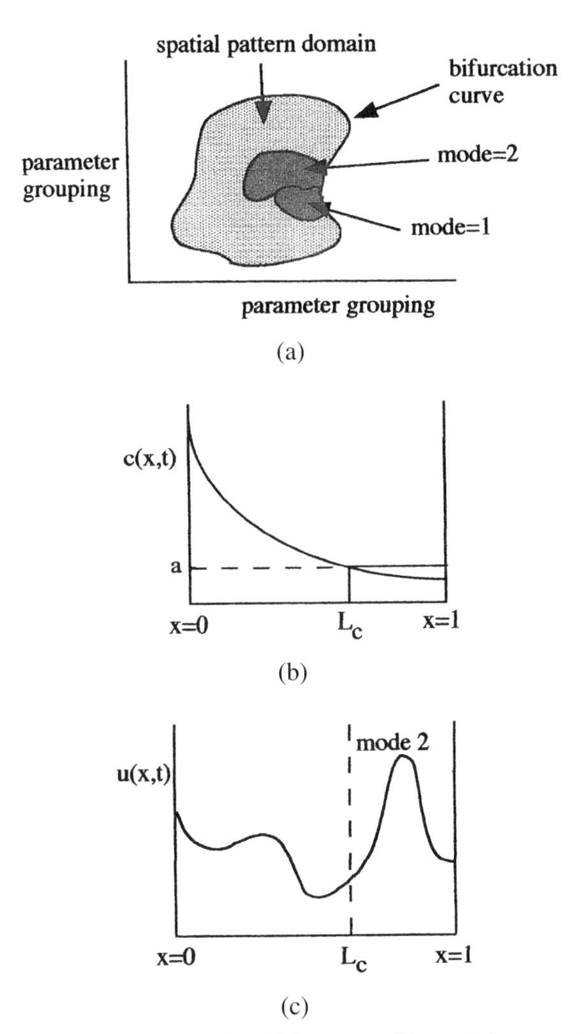

Figure 4.15. **(a)** Schematic parameter space for which patterns will grow if the parameters lie in the appropriate space and specific dominant patterns if they lie in their respective subspaces. **(b)** Typical qualitative solution c of the third equation of the sytem (4.5) with a source at $x = 0$. **(c)** Schematic mode 2 solution of the full system (4.5) for some critical L_C.

a value of L_c such that the average c (or rather, approximately the average) for $L_c < x < 1$ is such that the system (4.14) can generate a spatial pattern, specifically a mode 2-like pattern as in Figure 4.15(b) if the parameters are in the appropriate subspace. With $c = a$, a constant, we can certainly calculate what this critical L_c will be for the mode 2 pattern to be formed with the methods of Chapter 2. With $c(x, t)$ a solution of its own reaction diffusion equation (the third of (4.5)) it is considerably less easy since the full system must be solved. Nevertheless it is intuitively clear for the full system (4.5) that for some length L_c a mode 2-like solution as in Figure 4.15(b) will start to form and that u will increase. It is again intuitively clear that a somewhat similar patterning scenario will occur for the model system (4.6)–(4.8) in a growing domain and that a critical L_c exists for a mode 2-like pattern to start to form in $L_c < x < 1$. All of these behaviours are confirmed by the numerical simulation of the full system as we see below.

Let us now return to the full model set of equations (4.6)–(4.8) with (4.9) and (4.10) on the fixed domain $0 < x < 1$; recall that it is a fixed domain because of the scale transformation. We assume there is an initial source of epidermal growth factor, c, at the posterior end of the jaw ($x = 0$). This chemical diffuses through the jaw epithelium, is degraded and diluted by growth according to (4.8). As the jaw grows, c decreases further towards the anterior end $x = 1$ until it crosses below the critical threshold on a sufficiently large subdomain $L_c < x < 1$ to drive the substrate–activator system unstable (through diffusion-driven instability more or less in the usual way). The specific mode that starts to grow depends on the parameters. We choose parameter values such that the single hump (mode 2) spatial pattern is the first unstable mode. So, when the subdomain, on which c is below the threshold, has grown large enough, a single mode spatial pattern in u and v will start to grow like the mode 2 pattern in Figure 4.15(b). Eventually, the substrate concentration, u, crosses an upper threshold which triggers initiation of a placode (tooth primordium) fixing the spatial position of tooth 1: this is the dental determinant; see Figure 4.16(a).

As mentioned, the experimental evidence (Westergaard and Ferguson 1986) suggests that the dental determinant and each subsequent tooth primordium become a source of inhibitor thus simulating an inhibition zone. So, in our model, when u grows above a certain threshold, we make the location of the peak in u a source of c, the inhibitory substance. Mathematically, this is equivalent to an internal boundary condition at each tooth

$$c(x_i, t) = \bar{c}_i(t) \tag{4.15}$$

for $x = x_i$ and $t > t_i$, where $\bar{c}_i(t)$ is the solution of (4.11). So, with the appearance of the dental determinant there are now two sources of inhibitor, one at the posterior end of the jaw and the other at the dental determinant position x_1. Now, as the jaw grows, c eventually drops below the critical threshold in the region between the two sources and another hump-like pattern in u starts to appear in the posterior end of the jaw. The second primordium forms in the region where u again crosses the patterning threshold, and the tooth that is initiated becomes another source of c as illustrated in Figure 4.16(b). In this way, tooth development proceeds: $c(x, t)$ dips below a threshold, causing a local pattern to form when the domain size is large enough. In forming the pattern, u crosses a threshold, and creates a source of c, hence another tooth primordium is created. Subsequent primordia appear in an analogous manner. Based on the tooth formation scenario in Figure 4.10 the placode, which we assume forms where u crosses the threshold, induces cell aggregation, the papilla, in the mesenchyme. The exact order of which comes first, the placode or the papilla is still not generally agreed. At this stage in our modelling we do not address this question; we consider it in Chapter 6 when discussing the mechanical theory of pattern formation.

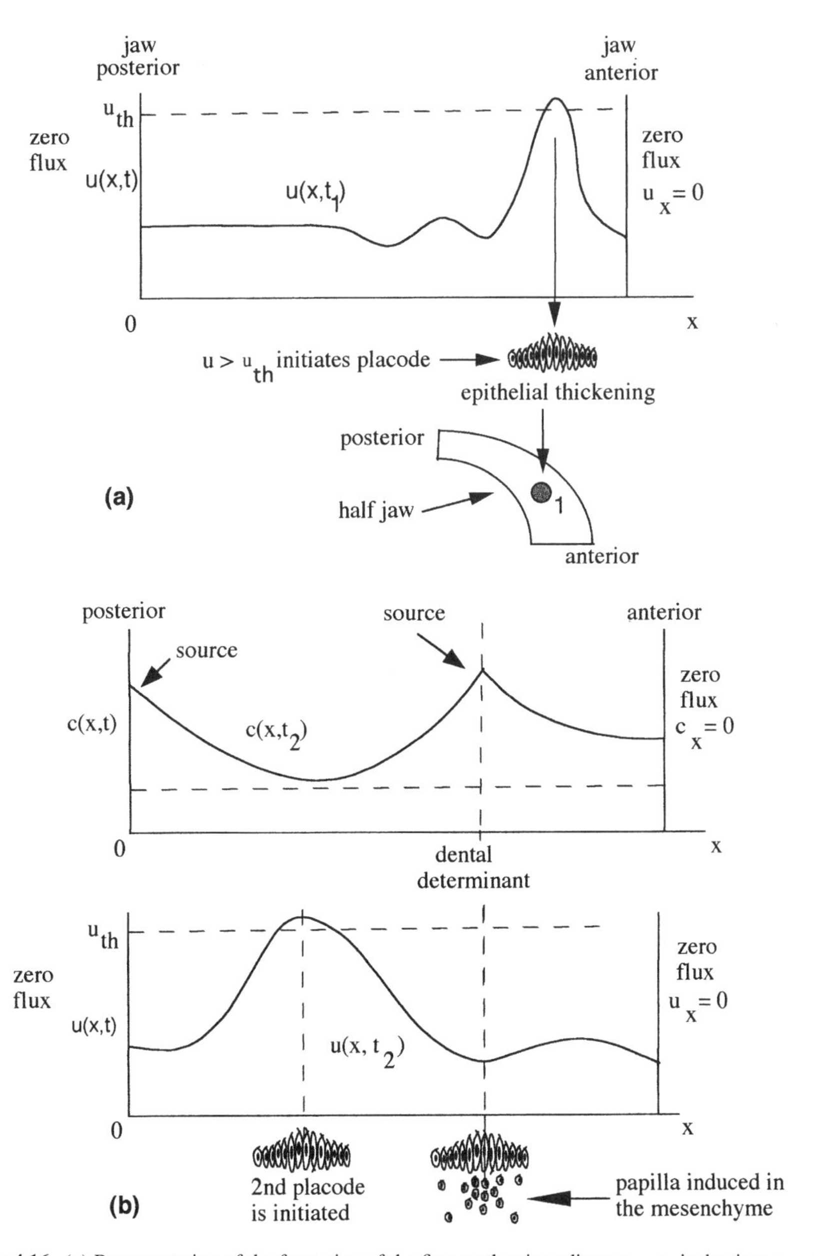

Figure 4.16. (**a**) Representation of the formation of the first tooth primordium: $t = t_1$ is the time $u = u_{th}$. The jaw length at this stage is approximately 0.6 mm. (**b**) Representation of the formation of the second tooth primordium: $t = t_2$ is the time when $u = u_{th}$ again. According to the developmental scenario described above in Figure 4.10 the placode, initiated by the substrate u, and the cell aggregation in the mesenchyme, the papilla, appear at the same place.

It is always helpful, however complicated the equation system, to try and get an intuitive feel of what the solutions will or could look like.

4.8 Results and Comparison with Experimental Data

With the above heuristic scenario we are now ready to simulate the complete model mechanism and solve for the spatial and temporal sequence of the first seven teeth primordia; full details of the numerical analysis and discussion of the results are given by Kulesa (1995) and these and others in Kulesa and Murray (1995), Murray and Kulesa (1996) and Kulesa et al. (1995, 1996a,b). We first have to estimate a set of model parameters. Some we can obtain from experiment but others must be determined to ensure the appropriate spatial pattern in u. We can estimate the crucial growth rate parameter r from Westergaard and Ferguson (1986, 1990). We can get real biological estimates for the diffusion coefficient p since it is related to the epidermal growth factor. Since the degradation constant δ is also related to epidermal growth factor we also have estimates for it. For the others, we have to use the equations and the ultimate pattern we require. This was done using a straightforward linear analysis of a simplified form (4.6) and (4.7), namely the system (4.14). We do not repeat the analysis since it is exactly the same as that carried out in detail in Chapter 2, Sections 2.4 and 2.5 where we derived the patterning space for this simplified system. We simply chose the set for the appropriate mode 2-like pattern in u to be initated when the system was diffusion-driven unstable. There is a systematic, and simple to use, logical numerical procedure for determining the parameter set for specific patterns in pattern generator mechanisms in general (Bentil and Murray 1991) when it is not possible to do it analytically which is more the norm. The boundary conditions are given by (4.9) and (4.10) and initial conditions:

$$u(x,0) = u_0 = h + b, \quad v(x,0) = v_0 = \frac{b}{(u_0)^2}, \quad c(x,0) = a\exp(-kx), \qquad (4.16)$$

where a and k are positive parameters. A representative continuous function of time for the source of inhibitor at the anterior end of the jaw, (4.16), was taken as

$$c(0,t) = c_o(t) = -m\tanh(t-f)/g + j, \qquad (4.17)$$

where m, f, g and j are constant parameters: this gives a smooth step function form for the initial switching on of the patterning process. A small random spatial perturbation was then introduced to u_0 and v_0 and the full nonlinear model system (4.6)–(4.8) solved numerically for the spatial and temporal sequence of the first seven teeth primordia of the lower jaw. A finite difference scheme based on the Crank–Nicholson method was used with $\Delta x = 0.01$, $\Delta t = (\Delta x)^2$. All parameter values are given in the legend to Figure 4.19 where the results are compared with experiment.

Figure 4.17 shows the numerical simulation for u as far as the time of formation for the first five teeth together with the solution for the inhibitory chemical c.

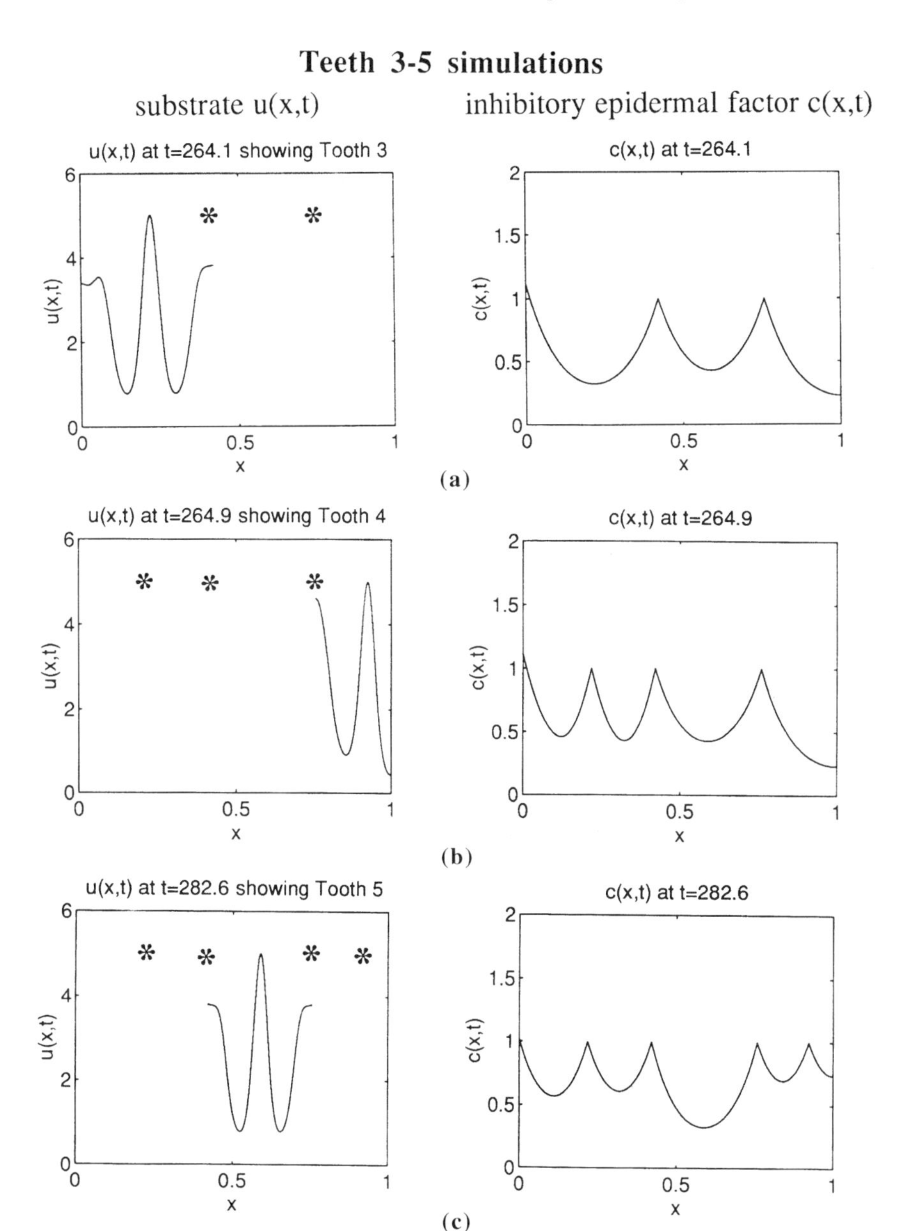

Figure 4.17. **(a–c)** The computed concentration profiles for the substrate u and the inhibitor c for times up to tooth formation time t_i where i denotes the ith tooth. Tooth 1 is the dental determinant. The spatial positioning of teeth primordia is given by x_i, the positions where u reaches the threshold u_{th}. Solutions are given on [0, 1] but the actual domain size at each time is given by $[0, \exp rt]$. The parameter values for all of the simulations are given in the legend of Figure 4.19.

In Figure 4.17(a) the time t_1 is the time u reaches the threshold u_{th} which initiates a tooth primordium at x_1 and this switches on a source of the inhibitor c at this position; at x_1 there is then a zero flux barrier. At this stage a new simulation, with the u, v and c distributions from the first calculation, was started in the two regions $0 < x < x_1$ and $x_1 < x < 1$ including the new source of c at x_1. This simulation was carried through until c again decreased sufficiently over a large enough region and the spatial pattern in u again reached u_{th} at time t_2 at some place in one of the domains at a position denoted x_2 thereby fixing the position of the next tooth, tooth 2. The next simulation was then carried out in a similar way but now in the three regions $0 < x < x_2$ and $x_2 < x < x_1$ and $x_1 < x < 1$ and the position was determined where u next crossed the threshold u_{th} at time t_3 at position x_3. The sequence of simulations was then carried out to determine the position and time for each tooth primordium position. All of the results shown in Figure 4.17 are plotted on the domain $[0, 1]$ but the actual domain size is $[0, \exp rt_1]$ where r is the growth rate parameter of the jaw.

We can represent the results in Figure 4.17 in a different way in Figure 4.18 to make it clearer where and when the teeth primordia are determined to form as the jaw grows. In Figure 4.18(a) we show the three-dimensional evolution of the substrate concentration u as a function of distance along the jaw (normalised to 1) and of time in days. The dimensionless time in the simulations is related to the actual time in days via the estimation of the expontial growth rate parameter r from the experimental data reproduced graphically in Figure 4.12: values are given in Figure 4.19. Figure 4.18(b) plots the time of incubation in days against the length of the simulated jaw with the order of teeth primordia superimposed.

The experimental data (Westergaard and Ferguson 1990) give the sequence of teeth appearance for both the upper and lower jaw of *A. mississippiensis* and it was from these data that Figure 4.12 was obtained. From Figure 4.12(a) we see that the upper jaw varies slightly for the first several primordia, namely primordia 6 and 7 are in different spatial locations. When compared to the lower jaw data, it can also be seen (Figure 4.12(a)) that the upper jaw grows at a slightly increased rate. For the simulations described we used exactly the same parameter set for both the upper and lower jaw except for the slightly higher growth rate for the upper jaw, specifically $r = 0.34/\text{day}$ as compared with $r = 0.31/\text{day}$. The experimental data also let us directly relate simulation time T to real time t as described in the figure legend.

The numerical simulations, with the parameter sets, for the sequence of teeth in both the upper and lower jaw are presented in Figure 4.19 together with the experimental data for comparison. The model mechanism reproduces the correct spatial and temporal sequence for the first 8 teeth in the lower jaw and the correct sequence for the first 6 teeth in the upper jaw. The results are encouraging.

Kulesa et al. (1996a) also relate the numerical results with the teeth primordia appearance on the actual jaw; again the agreement is very good. The spatial positions from the experimental data were obtained from a Cartesian xy-coordinate system attached to Figure 4.13(a). A line in this figure lying along the teeth primordia, represented by a parabola $f(x)$, was fit to a curve of the experimental data (($f(x) = b_1x^2 + b_2x + b_3$; $b_1 = -0.256$, $b_2 = 0$, $b_3 = 7.28$). The numerical spatial positions were then related by using the length l, of $f(x)$ from $x = 0$ to $x = x^*$ (where $f(x^*) = 0$), as the nondimensional length of the anterior–posterior axis in their simulations.

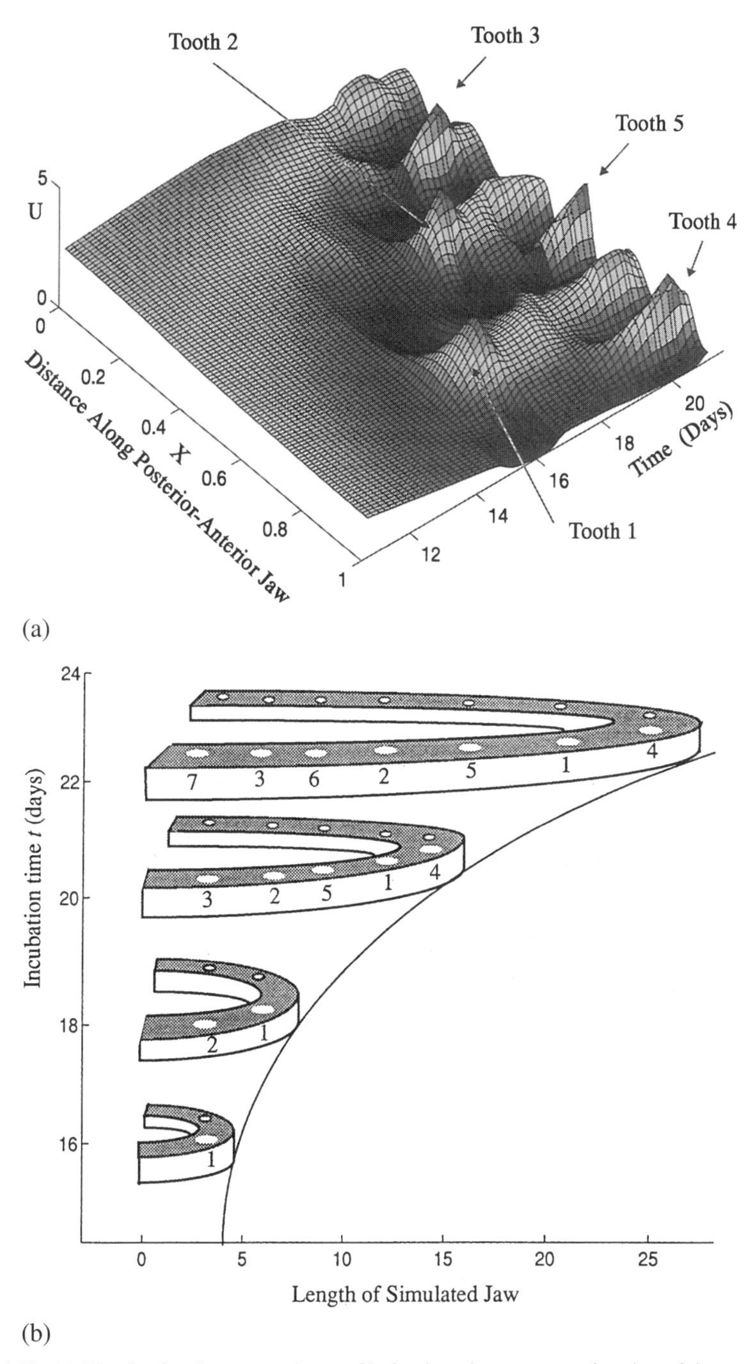

Figure 4.18. (**a**) The simulated concentration profile for the substrate u as a function of time and distance along the jaw. The positioning of the teeth is determined by the positions in the jaw where u reaches a threshold value. (**b**) The jaw length and appearance of the tooth primordia as they relate to the incubation time in days. All the parameter values are given in the legend of Figure 4.19. (From Kulesa 1995)

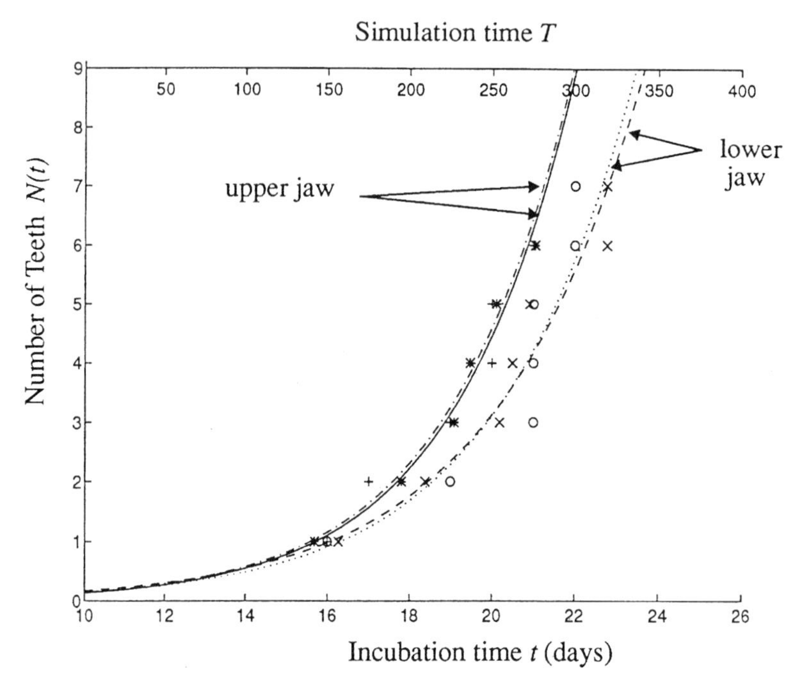

Figure 4.19. Comparison of the numerical versus experimental data for the teeth primordia initiation sequence in both the upper and lower half-jaws of *A. mississippiensis*. Upper jaw: '$*$' denotes the numerical data with a solid line: $N(t) = N_3 \exp(r_3 t)$; $N_3 = 0.0042$, $r_3 = 0.35$/day. The dash-dot line ($\cdot - \cdot - \cdot$) and '+' denote the experimental data: $N(t) = N_4 \exp(r_4 t)$; $N_4 = 0.0047$, $r_4 = 0.34$/day. Lower jaw: '$\times$' denotes the numerical data with a dashed line ($---$): $N(t) = N_1 \exp(r_1 t)$; $N_1 = 0.012$, $r_1 = 0.28$/day. The dotted line ($\cdot \cdot \cdot$) and 'o' denote the experimental data: $N(t) = N_2 \exp(r_2 t)$; $N_2 = 0.0066$, $r_2 = 0.31$/day. Time t (days) was scaled to time T (simulation) using $T = a_1 t + a_2$; $a_1 = 27.06$ and $a_2 = -286.6$. The model parameters are (refer to (4.6)–(4.11), (4.16) and (4.17)): $h = 1$, $b = 1$, $d = 150$, $r = 0.01$, $p = 0.5$, $\gamma = 40$, $\sigma = 0.2$, $k_1 = 0.3$, $k_2 = 1.0$, $a = 2.21$, $k = 0.9$ and $c_0(t) = c(0, t) = -m \tanh[(t - f)/g] + j$; with $m = 0.65$, $f = 200$, $g = 34$, $j = 1.5$. (From Kulesa et al. 1996a)

4.9 Prediction Experiments

One of the primary benefits of, and justifications for, constructing a theoretical model mechanism directly related to a real biological problem is the ability to use it as a predictive tool. Numerical simulation of the model can determine whether the model mechanism is capable of reproducing the observed experimental data. Also of importance is the ability to numerically simulate possible experiment scenarios. These virtual experiments can be conducted computationally and therefore can cover a much wider range of scenarios in a much shorter time than physical experiment. The simulations can also allow for changes in parameters and conditions which are possible, sometimes difficult and sometimes impossible, to reproduce in the laboratory. Results from the model predictions may help to identify key biological processes which are critical to the behaviour of the model mechanism and may serve to guide further illuminating experiments. Another reason, of course, for carrying out these virtual experiments is to test some of the model hypotheses.

Since we have seen that the model can reproduce the major experimental data it is reasonable to consider a group of intervention experiments which we simulate numerically to arrive at predicted outcomes of the teeth initiation sequence. We have identified several different types of feasible laboratory experiments which could be performed on the embryonic jaw of the alligator. We describe the removal of tooth placodes and replacement with sections of oral epithelium and the transplantation of tooth placodes. We also insert physical barriers in the jaw epithelium which affect the diffusion processes. Although the model and simulation procedures we have described in this chapter can be used to mimic several different scenarios, we only show here some of the more interesting prediction results of the simulated primordia initiation sequence. For reference the normal sequence is illustrated in Figure 4.20(a) for both the upper and lower jaw of *A. mississippiensis*.

Primordia Removal and Replacement with Oral Tissue

Let us consider removing the first tooth primordium immediately after it forms (Murray and Kulesa 1996). This is a test of the existence of an inhibitory growth factor at the posterior end. Experimentally, we assume that when the cell condensation forms to mark the first primordium site, the placode is removed and replaced with another section of oral epithelium. In the numerical simulations of the model system (4.6)–(4.8), when $u(x, t)$ crosses the threshold to mark the first tooth site, we do not add a new tooth source. Instead, we continue the simulation as if the tooth primordium 1, the dental determinant, had been removed. With the given parameter set used in Figure 4.19, this experiment was simulated. The prediction from the model shows that a new tooth primordium forms in the same area in which the old primordium was removed: in other words the old tooth reforms. The inhibitor, $c(x, t)$, is still below the threshold on this subdomain where the tooth primordium has been removed. The removal and insertion of oral tissue merely delays the subsequent primordium from forming at this site. A similar prediction is made if tooth primordium 2 is removed and new tissue is replaced. The remainder of the teeth primordia initiation sequence remains unchanged in both cases. The same holds for the removal of any tooth primordium: the primordium reforms in the same place. This prediction is what we would expect intuitively.

Primordia Transplantation

Murray and Kulesa (1996) next considered a series of experiments in which a tooth primordium is transplanted to various sections of the oral epithelium. One of the results is shown in Figure 4.20 where the dental determinant was transplanted to $x = 0.9$. We assume experimentally that when a tooth primordium forms it may be transplanted and replaced with a section of oral tissue. We began the simulated experiments with the first tooth primordium to test the importance of the dental determinant.

Once the first tooth primordium forms, we numerically simulate the transplantation of this placode to different spatial locations of the jaw domain, $0 \leq x \leq 1$ both anterior and posterior to its original position. We simulated transplanting the dental determinant to the positions $x = 0.25$, 0.5 and 0.9 of the jaw domain $0 \leq x \leq 1$. The case of trans-

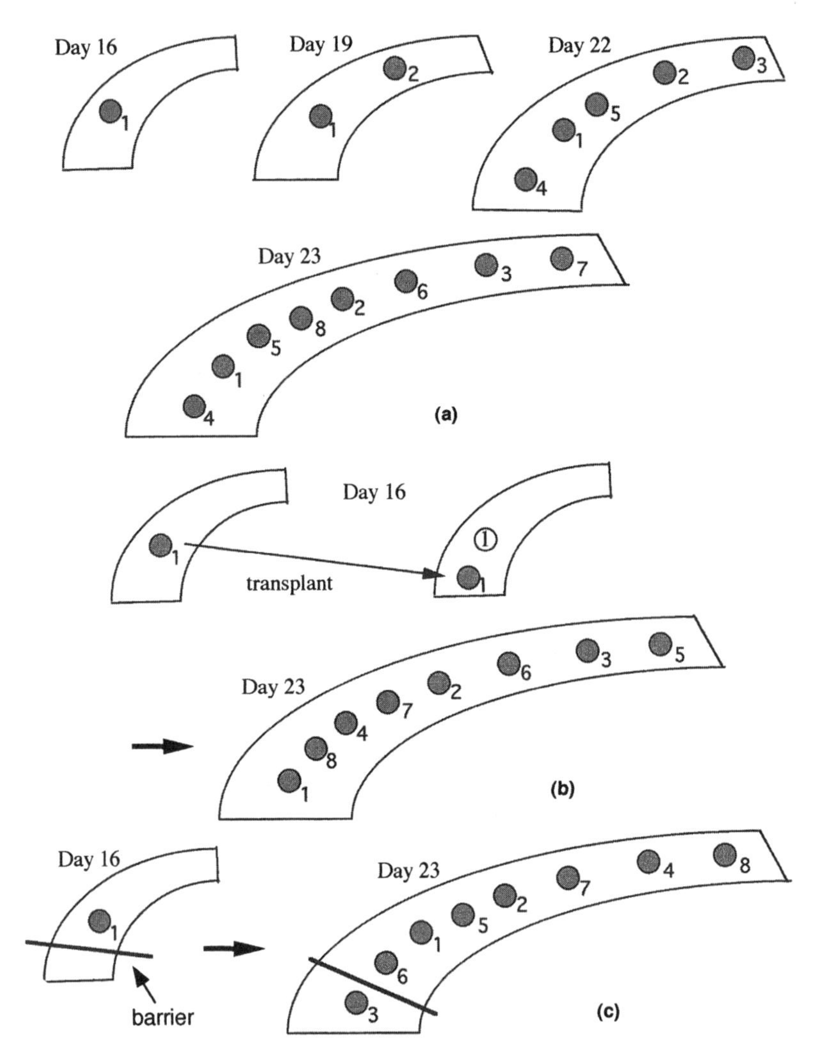

Figure 4.20. (**a**) Normal sequence of appearance of the first eight teeth primordia in the alligator *A. mississippiensis*: counting from $x = 0$, the posterior end of the jaw, the order is 7-3-6-2-8-5-1-4. The size of the jaw at Day 16 is approximately 0.6 mm while by the time Tooth 4 has formed it is approximately 3.8 mm. (**b**) Tooth 1 is transplanted to $x = 0.9$, that is, anterior to its original position and the subsequent ordering is changed from the normal as shown to 5-3-6-2-7-4-8-1. When Tooth 1 is transplanted to position $x = 0.5$ the altered sequence is 5-3-6-1-4-2-7 while for $x = 0.25$ it is 1-3-5-2-4. (**c**) Effect of a barrier inserted at $x = 0.9$: the altered sequence is 8-4-7-2-5-1-6-3. When a barrier is inserted at 0.25 the sequence becomes 4-2-3-1-5 while a barrier at 0.5 results in the order 5-2-3-1-4. (From Murray and Kulesa 1996)

planting to $x = 0.9$ is shown in Figure 4.20(b). In all cases, the numerical simulations predict that the initiation sequence is altered, again as we should expect. What is less clear is how it alters the subsequent sequence. The altered sequences when Tooth 1 was transplanted to $x = 0.25$ and $x = 0.5$ are given in the legend of Figure 4.20(b).

Transplantation of the first tooth primordium to $x = 0.9$ is perhaps the most revealing. The numerical simulations predict that after the dental determinant is transplanted, the subsequent seven teeth primordia all form posteriorly. That is, the domain between the transplanted placode and the anterior end of the jaw domain, $0.9 \leq x \leq 1$, does not grow large enough to support a tooth primordium until at least the next seven primordia are formed as illustrated in Figure 4.20(b): now Tooth 4 forms posterior to Tooth 1. The dental determinant is clearly important. The effect of transplanting the dental determinant to positions posterior to the usual position is given in the legend of Figure 4.20(b). Other numerical prediction experiments can be simulated for the transplantation of different teeth primordia, with the most dramatic changes predicted by the transplants situated at the extreme ends of the jaw domain.

Barriers in the Jaw

Possibly the simplest way of affecting the initiation sequence may be the placing of physical barriers in the jaw epithelium. Although barriers are physically invasive the experimental interference of inserting the barrier into the jaw epithelium may be minimised. These barriers create a deterrent to chemical diffusion which in the model mechanism plays a crucial role.

Here we placed a barrier at various locations of the domain, $0 \leq x \leq 1$, using the initiation of the dental determinant as a reference time of when to insert the barrier. Both the time and spatial location at which these barriers are placed is again important. We numerically simulated the placing of a barrier at the positions $x = 0.25, 0.5$ and 0.9 of the jaw domain. Mathematically, the barrier is represented by an internal zero flux boundary condition at the tooth site,

$$u_x(x_b, t) = v_x(x_b, t) = c_x(x_b, t) = 0 \quad t > t_b, \tag{4.18}$$

where $x = x_b$ is the position of the barrier and $t = t_b$ is the time of insertion. The most striking prediction result is again obtained when the near-end (posterior) regions of the jaw domain are initially affected. The insertion of a physical barrier at $x = 0.9$ blocks the effect of the tooth 1 source on the anterior part of the domain, $0.9 \leq x \leq 1$. This causes the premature initiation of a tooth primordium (in this case, tooth primordium 3) since the gradient of the epidermal growth factor, $c(x, t)$, crosses the lower threshold more rapidly on $0.9 \leq x \leq 1$ as shown in Figure 4.20(c). The results for the first few teeth when a barrier is inserted at 0.25 and 0.5 are given in the legend of Figure 4.20(c).

Robustness

Before concluding the discussion on teeth formation the question of robustness of the pattern formation and its timing must be addressed. Kulesa (1995) investigated this problem in depth and calculated a selection of parameter subspaces for the full model mechanism (4.6)–(4.8)) with a wide range of parameter values. The parameter set in the equations consists of h, b, δ, p, d and γ and so the spatial patterning spatial domain is a 6-dimensional space. In Chapter 2, Section 2.5 we examined in detail the three dimensional parameter space, or Turing space, of what are the first two equations of (4.5) with $c = a$, a constant. Kulesa's (1995) initial aim was to see if there was a parameter set which gave the correct initiation sequence for the first seven teeth. Figure 4.21(a) shows a two-dimensional cross-section in the $\gamma - d$ plane for the dominant spatial modes, n, which form in u and v: $n = 2$ corresponds to a single hump as in Figure 4.15.This model system is clearly quite robust in that the parameter domain for the first seven teeth is not small. Robustness relates directly to the small random variability in parameters that generally occur in real developent. If a parameter set is within the 7-range, for example, in Figure 4.21(b), then small variations in p and δ will still give the correct spacing, unless of course the set is near a bifurcation boundary. It is probably correct to say that if the parameter space for any spatial pattern formation mechanism is very small it is highly unlikely to continue to give the same type of pattern with the inevitable small variations in parameters in the real world.

4.10 Concluding Remarks on Alligator Tooth Spatial Patterning

The two problems associated with the alligator *A. mississippiensis* that we have discussed in this chapter are in marked contrast to each other from a modelling point of view but each highlights what are some basic principles about realistic biological modelling. The discussion on the timing of stripe formation (and the formation of shadow stripes) did not include any specific model other than a diffusion chemotaxis one for illustrative purposes. It was the modelling concepts rather than the actual model that turned out to be useful for the experimentalist. The mathematical analysis, such as it is, is quite simple. Of course it would be interesting and useful if a specific mechanism, with reasonable biological justification, could be proposed or one on which it was possible to carry out biologically possible virtual experiments to try and test its plausibility. On the other hand in the sections on teeth formation the mechanism proposed was based on known biological facts about the formation of teeth primordia. It is more like what is generally needed and not surprisingly is very much more complex.

It is perhaps helpful to recap what the more complex modelling exercise with regard to teeth has achieved and mention some of the other results. Based on the known biology, we developed a quantitative model mechanism based on a dynamic reaction diffusion system, which crucially incorporates jaw growth. We showed that the model mechanism is capable of reproducing the spatial and temporal sequence of the first seven teeth primordia in the jaw of the alligator and that the results compare well with experiment.

The robustness of the model to variations in the model parameters was tested by Kulesa (1995) who showed that there was a significant region in parameter space where we could maintain the correct sequence of the first seven teeth primordia. The patterning mechanism is very robust to changes in parameters in that for a wide range of the patterning parameters, γ, b and d, we were able to maintain a single hump spatial pattern in u and v and get the right order of primordial positions. Since this portion of the mechanism was based on a previously analysed robust system this result was not unexpected.

The numerical simulations verified the experimental hypothesis that jaw growth is crucial to the development of the precise spatial and temporal sequence of the teeth primordia. The variation of the growth rate of the jaw, r, was the most sensitive parameter

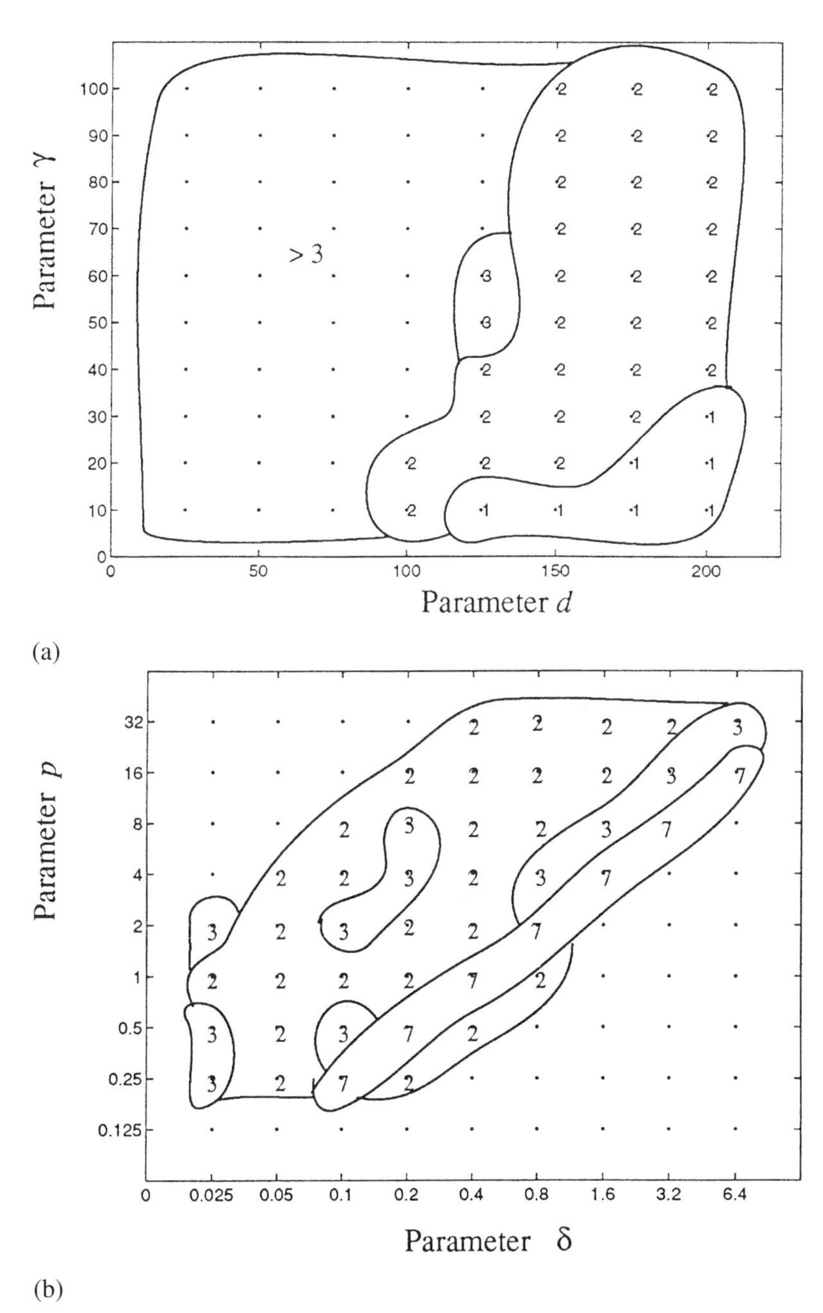

Figure 4.21. **(a)** Two-dimensional $\gamma - d$ parameter space for (4.6)–(4.11) for specific nodes, n, where $n = 2$ is a single hump. γ is the scale parameter and d the ratio of the diffusion coefficient of the substrate u to that of the activator v. The other parameter values are as in Figure 4.19 except for $b = 2.0$ which is a representative b within the range of b considered. Each dot in the figure is a simulation with these parameter values. **(b)** A representative $p - \delta$ parameter space, where p and δ are the diffusion coefficient ratio of the inhibitor c to that of the activator v and the first-order decay rate of c for $r = 0.01$ (approximately in the middle of the r-range investigated) and with the other parameters as in Figure 4.19. The numbers are number of teeth primordia which appear in the correct order. (From Kulesa 1995)

in obtaining the correct teeth primordia initiation sequence. In the upper jaw simulations, the same model parameters were used: all that was required was an increase in r—which came from experimental data—to obtain the correct upper jaw sequence, which is different from the lower jaw.

The model construction involved the incorporation of the physical growth of the jaw domain into a substrate–activator patterning mechanism. This patterning mechanism was mediated by an inhibitor, $c(x, t)$ and Kulesa (1995) showed that the patterning mechanism alone, made up of a substrate, activator and inhibitor system, in the absence of jaw growth, could not produce, without the inhibitor, the correct teeth primordia initiation sequence. He also showed that the substrate and activator system alone, including jaw growth, did not produce the correct initiation sequence. We concluded that each component, namely, jaw growth and the substrate–activator–inhibitor patterning mechanism, are necessary for the tooth initiation sequence. Only if all components are included in a full model mechanism is it capable of producing the first seven teeth primordia in the precise spatial *and* temporal sequence.

From the prediction experiments we showed that transplanting a primordium can significantly alter the initiation sequence. Teeth primordia transplanted to the ends of the jaw domain cause the most dramatic changes in the spatial ordering of the initiation sequence. We also investigated the insertion of zero flux physical barriers in the jaw epithelium and showed how these alter the initiation sequence. Barriers placed at ends of the jaw domain cause the more significant changes in the spatial ordering of the teeth primordia.

With more experiments, the mechanisms of teeth primordia initiation may become more precisely defined, incorporating actual chemical components and more precise growth data. Molecular level experimental results linking specific gene expression to biological events in tooth initiation provide further information for the theoretical modelling. New experiments would help to form a more detailed teeth initiation model mechanism, incorporating tissue interaction, which could use the qualitative results gained from the modelling here. The analysis of such a model would be very complicated.

Of course we cannot say that the actual biological pattern formation mechanism for teeth primordia, or even a realistic caricature of it, is of this reaction diffusion genre. What we can say, though, is that it is highly likely that growth plays a continuous and essential role in the pattern formed and that each tooth position is almost certainly not preassigned. The final arbiter of its biological relevance, however, must lie with further detailed molecular level experiments. We can also say that the interaction between experimental investigation and theoretical modelling in both topics discussed here resulted in a better understanding of the underlying biology.

4.11 Pigmentation Pattern Formation on Snakes

Biological Background

Snakes (order: squamata)—reptiles and amphibians in general, in fact—are numerous and highly diverse in their morphology and physiology. Snakes and lizards exhibit a particularly rich variety of patterns many of which are specific to snakes. The fascinating, and visually beautiful, book by Greene (2000) is a good place to start. He discusses their evolution, diversity, conservation, biology, venoms, social behaviour and so on. Another very good book, by Klauber (1998), is more specific and is essentially an encyclopedia on rattlesnakes.

Even within the same species there is often extreme pattern polymorphism. The common California king snake *Lampropeltis getulus californiae* is a very good example (Zweifel 1981). Pattern anomalies often occur even on an individual snake as illustrated in Figures 4.22(e) and (f). A browse through any field guide shows not only straightforward pattern elements such as lateral and longitudinal stripes and simple spots but

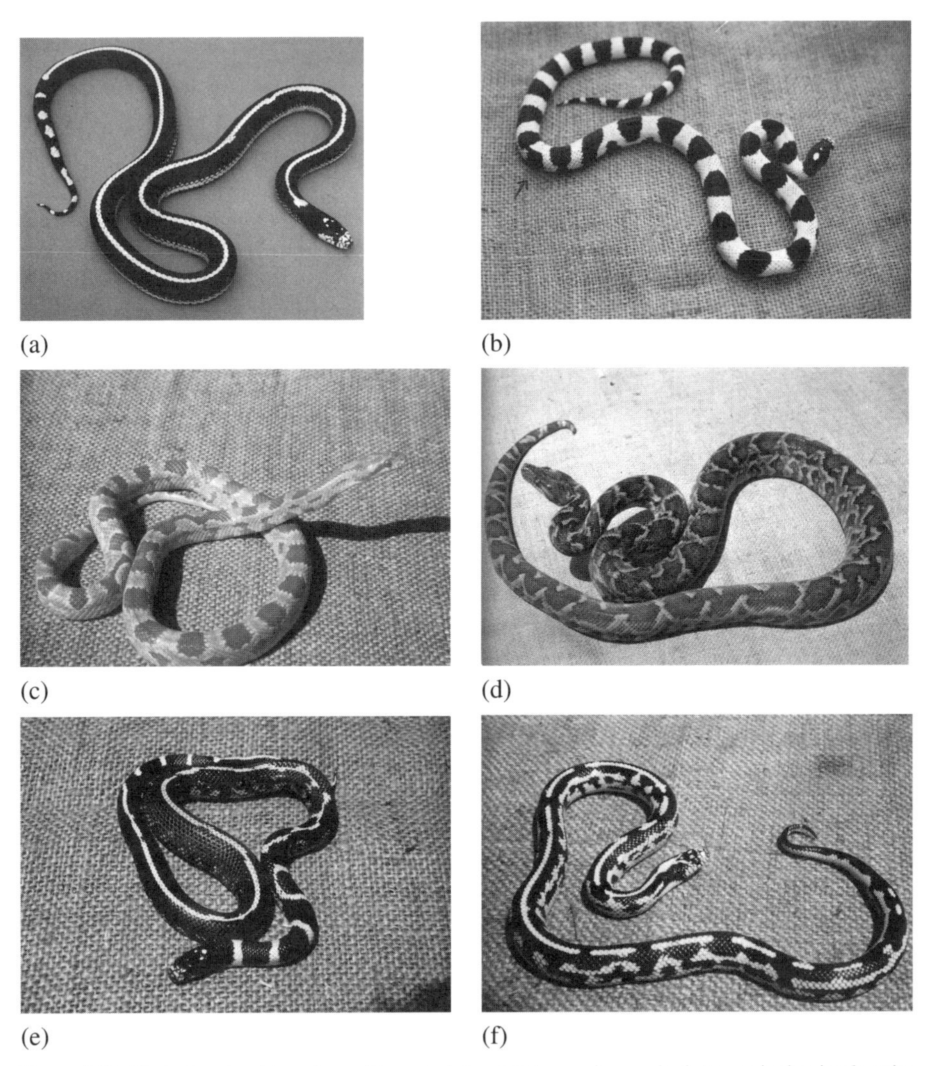

Figure 4.22. Typical snake skin patterns. Those in (**a**) and (**b**) are quite regular but even in the simple stripe pattern in (**a**) there is some aberration. In (**e**) and (**f**) there is a mixture of the basic stripe patterns. (Photographs are courtesy of Lloyd Lemke)

also a wide range of patterns based on various complex pattern elements which are not found on other animals. Figure 4.22 shows a few snake patterns, both regular and irregular. We saw in the case of butterfly wing patterns discussed in Chapter 3 that the seemingly complex patterns can be generated by a relatively small number of pattern elements. On the other hand, many of the common snake patterns do not seem to fall into any of the usual classes of patterns which can be generated by the usual reaction diffusion models unless they are modified by different boundary conditions, growing or changing domains, spatially varying parameters and so on. In the case of spatially varying parameters it is difficult to relate the results from these models to the biology.

The skin of reptiles is the largest organ in their bodies and poses many interesting developmental problems (see, for example, the review by Maderson 1985). The skin essentially consists of an external epidermis with an underlying dermis. Although we do not know the pattern formation mechanisms we do know that the pattern is fixed in the dermis. The basic skin pigment pattern remains the same after the periodic replacement of the epidermis—the well-known skin shedding exhibited by snakes and lizards. Pigment cell precursors, called chromatoblasts, migrate from the neural crest during development and more or less distribute themselves uniformly in the dermal skin. As with animal coat markings, whether or not the skin develops a pigmented patch depends on whether presumptive pigment cells produce pigment or remain quiescent. Interactions between these precursor cells and possibly directed movement may result in pigmented and unpigmented cells gathering in different regions to produce stripes or spots (Bagnara and Hadley 1973) as we supposed in the above discussion on alligator stripes. It is not known when chromatoblasts become committed to producing pigment. From evidence from the studies on alligator development cells may be able to produce pigment long before it is actually seen. Cells which are committed to pigmentation can also divide for some time.

Experimental studies of pigment development and the migration of pigment cell precursors have been largely confined to amphibians, mammals and birds although there is a large body of work on the crocodilia as we described above. However, little has been done specifically on skin patterns on reptiles except for the alligator work described earlier. An underlying assumption there and here is that the basic processes of migration, division and differentiation will be the same in snakes and other reptiles as in other animals. Relatively little research has been done on snake embryology (a partial list is given by Hubert 1985; see also the rest of the edited volume of this series on reptilia in which this paper appears), ecology and evolutionary biology (the above-mentioned book by Greene (2000) is the most definitive). As we have noted, of course, extensive embryological studies have been carried out on alligators (Ferguson 1985, Deeming and Ferguson 1989a; see also earlier references in these papers).

Hubert and Dufaure (1968) mapped the development of the asp viper (*Vipera aspis*). Pigmentation was first observed on the scales of the body, when the embryo was about 106 mm long, and extended to the head as development proceeded. The pattern is almost certainly laid down earlier in development than when it first becomes visible as was the case with the alligator. Zehr's (1962) observations of the development of the common garter snake (*Thamnophis sirtalis sirtalis*) suggest a similar developmental process. He noted that when the pigmentation pattern first appears it is not well formed but becomes more defined as development proceeds. (A similar progressive development of final pattern occurs on many butterfly wing patterns.) Treadwell (1962) noted that in embryos of the bullsnake (*Pituophis melanoleucus sayi*) three rows of spots appear on the sides of the embryo at 29 days and blotches appear on the dorsal midline at 31 days. The timing of developmental events in snakes should be regarded with caution since the rate of development is significantly affected by the temperature of incubation of the eggs or the body temperature of the mother in live-bearing species. In the case of the asp viper, for example, gestation periods from 90 to 110 days have been observed. In the case of the alligator we were able to get an estimate of when in development the stripe pattern was laid down; it was well before the pattern was visible. The cell-chemotaxis-diffusion mechanism considered there is the same as the one we use here. Recall that histological sections of the skin of the alligator embryo showed that there were melanocytes present in the white regions between the dark stripes but these did not appear to produce melanin. We suggested that one possible explanation for the lack of melanogenesis by these cells is that a threshold density of melanocytes could well be necessary before melanogenesis can take place. We noted that this could be responsible for the lighter shadow stripes observed on alligator bodies; these shadow stripes lie towards the ventral side of the body and lie between the distinct darker stripes on the dorsal side. Similar interdigitating stripes are found on certain fish patterns, specifically angelfish (*Pomacanthus*) a problem studied by Aragón et al. (1998), Painter et al. (1999) and Painter (2000). There are thus two potentially important implications relevant to snake integumental patterns which arise from the alligator studies. One is that potentially quite different patterns can be generated on an embryo when significant growth occurs during the patterning process. The other is that presumptive pigment cell patterns may be generated some time before they start to produce pigment and the pattern becomes discernable.

As in the above Sections 4.1–4.4 we are interested here in the patterns which can be formed by the mechanism when the integumental domain is growing during the patterning process. We find that the spatially heterogeneous solutions can be quite different from, and considerably more complex than those obtained by patterning mechanisms in a fixed domain. It is likely that many of the pattern forming mechanisms involved in embryogenesis are operative on a timescale commensurate with embryonic growth. Maini and his coworkers (for example, Aragón et al. 1998, Painter et al. 1999, Crampin et al. 1999) made a particular study of the interplay between growth and pattern formation in the case of reaction diffusion systems.

The surprising novelty and complexity of new patterns which are generated by this basic system as a consequence of domain growth are likely to occur with all other pattern forming mechanisms.

At this stage, because of the paucity of morphoglogical data on snake embryology it is not possible to suggest when in development of the snake embryo the mechanism is operative. Murray and Myerscough's (1991) purpose was to suggest how some of the diverse complex skin patterns on snakes could be generated. This is the usual necessary first requirement for any potential mechanism.

4.12 Cell-Chemotaxis Model Mechanism

The model we consider here involves actual cell movement. Pattern formation models which directly involve cells are potentially more amenable to related experimental investigation. There is also some experimental justification from the evidence on pigement cell density variation observed in histological sections which we described above. Also, Le Douarin (1982) speculated that chemotaxis may be a factor in the migration of pigment cells into the skin. Heuristically we can see how chemotaxis could well be responsible for the rounding up and sharpening of spots and stripes. In the model, we propose that chromatoblasts both respond to and produce their own chemoattractant. Such a mechanism can promote localisation of differentiated cells in certain regions of the skin which we associate with the observed patterns on the snake integument. The cells, as well as responding chemotactically, are assumed to diffuse. It is the interaction of the cell mitosis, diffusion and chemotaxis which can result in spatial heterogeneity. The relatively simple mechanism we propose is

$$\frac{\partial n}{\partial t} = D_n \nabla^2 n - \alpha \nabla \cdot (n \nabla c) v - rn(N - n), \tag{4.19}$$

$$\frac{\partial c}{\partial t} = D_c \nabla^2 c + \frac{Sn}{\beta + n} - \gamma c, \tag{4.20}$$

where n and c denote the cell and chemoattractant densities respectively; D_n and D_c are their diffusion coefficients. We have taken a simple logistic growth form for the cell mitotic rate with constant linear mitotic rate r and initial uniform cell density N. The chemotaxis parameter α is a measure of the strength of the chemotaxis effect. The parameters S and γ are measures, respectively, of the maximum secretion rate of the chemicals by the cells and how quickly the chemoattractant is naturally degraded; β is the equivalent Michaelis constant associated with the chemoattractant production. This is the specific model discussed by Oster and Murray (1989) in relation to developmental constraints. In spite of its relative simplicity it can display remarkably complex spatial pattern evolution, particularly when varying chemotactic response and growth are allowed to take place during the pattern formation process. We first nondimensionalise the system by setting

$$\begin{aligned} \mathbf{x}^* &= [\gamma/(D_c s)]^{1/2}\mathbf{x}, \quad t^* = \gamma t/s, \quad n^* = n/\beta, \quad c^* = \gamma c/S, \\ N^* &= N/\beta, \quad D^* = D_n/D_c, \quad \alpha^* = \alpha S/(\gamma D_c), \quad r^* = r\beta/\gamma, \end{aligned} \tag{4.21}$$

where s is a scale factor. We can think of $s = 1$ as the basic integument size, carry out the simulations on a fixed domain size and then increase s to simulate larger integuments. We used this procedure in the last chapter. With (4.21) the nondimensional equations become, on omitting the asterisks for notational simplicity,

$$\begin{aligned} \frac{\partial n}{\partial t} &= D\nabla^2 n - \alpha \nabla.(n\nabla c) + srn(N - n), \\ \frac{\partial c}{\partial t} &= \nabla^2 c + s\left[\frac{n}{1+n} - c\right]. \end{aligned} \tag{4.22}$$

The numerical simulations of these equations (including growth) were carried out on a simple rectangular domain in which length is considerably longer than the width, with zero flux of cells and chemoattractant on the boundaries. The detailed numerical simulations and complex bifurcating pattern sequences which can occur as the parameters vary are given in Winters et al. (1990), Myerscough et al. (1990) and Maini et al. (1991).

The reason we consider a long rectangular domain is that the skin patterns are probably laid down at a stage when the embryo is already distinctly snake-like; that is, it is already long and more or less cylindrical, even if it is in a coiled state. Details of the embryo of the asp viper (*Vipera aspis*), for example, are given by Hubert and Defaure (1968) and Hubert (1985). Although it would be more realistic to study the model mechanism on the surface of a coiled cylindrical domain the numerical simulation difficulties were already considerable even on a plane domain. Here we are mainly concerned with the variety of patterns that the mechanism can generate so we consider the cylindrical snake integument laid out on a plane. The main features of the patterns on an equivalent cylindrical surface will be similar. We could, of course, equally well have taken periodic boundary conditions. Equations (4.22) have one positive homogeneous steady state

$$n_0 = N, \quad c_0 = \frac{N}{1+N}. \tag{4.23}$$

We now linearise about the steady state in the usual way by setting

$$n = N + u, \quad c = c_0 + v, \quad |u| \ll 1, |v| \ll 1$$

to obtain the linear system

$$\begin{aligned} \frac{\partial u}{\partial t} &= D\nabla^2 u - \alpha N \nabla^2 v - srNu \\ \frac{\partial v}{\partial t} &= \nabla^2 v + s\left[\frac{u}{(1+N)^2} - v\right] \end{aligned} \tag{4.24}$$

with boundary conditions on the boundary ∂D, of the domain D,

$$\mathbf{n}.\nabla u = \mathbf{n}.\nabla v = 0, \quad \mathbf{x}\varepsilon\partial D, \tag{4.25}$$

where $\mathbf{n}$ is the unit outward normal to the boundary of D. Again (as usual) we look for solutions in the form

$$\begin{bmatrix} u \\ v \end{bmatrix} \propto \exp[i\mathbf{k}\cdot\mathbf{x} + \lambda t]$$

which on subsituting into (4.24) gives the characteristic polynomial for the dispersion relation $\lambda(k^2)$ as a function of the wavenumber $k = |\mathbf{k}|$:

$$\begin{aligned} &\lambda^2 + [(D+1)k^2 + rsN + s]\lambda \\ &\quad + [Dk^4 + \{rsN + Ds - (sN\alpha)/(1+N)^2\}k^2 + rNs^2] = 0. \end{aligned} \tag{4.26}$$

The solution with wave vector **k** now has to satisfy the boundary conditions (4.25). We follow exactly the same procedure described in detail in Section 2.3 in Chapter 2 to determine the conditions for spatial instabilities.

With a two-dimensional domain with sides L_x and L_y, we consider the wavevector $\mathbf{k} = k_x, k_y$, where $k_x = m\pi/L, k_y = l\pi/L$ with m and l integers. These forms come from the zero flux boundary conditions and the linear eigenfunctions $\cos(m\pi x/L_x \cos(l\pi y)/L_y$ (refer to Section 2.3). So, on this rectangular domain the values of k^2 which will generate a pattern are those where $\lambda(k^2) > 0$, where

$$\mathbf{k}.\mathbf{k} = k^2 = \pi^2\left(\frac{m^2}{L_x^2} + \frac{l^2}{L_y^2}\right). \tag{4.27}$$

We can choose parameters D, a, s, r and N to isolate only one unstable wavevector. This mode selection is simply a way to force a particular pattern to grow initially. The wavevector for the isolated mode occurs when $\lambda(k^2) = 0$, that is, when k^2 satisfies

$$Dk^4 + \left[srN + Ds - \frac{sN\alpha}{(1+N)^2}\right]k^2 + s^2rN = 0.$$

We require that this equation has only one solution for k^2, so we further impose the condition for equal roots; namely,

$$\left[srN + Ds - \frac{sN\alpha}{(1+N)^2}\right]^2 - 4Ds^2rN = 0. \tag{4.28}$$

The modulus of the critical wavevector is then given by

$$k_c^2 = [s^2rN/D]^{1/2}. \tag{4.29}$$

By choosing D, s, r and N appropriately, we can find a k^2 from (4.27) which satisfies (4.29), and then solve for α from (4.28) (we take the larger root for α so it is positive). This determines the point in the (N, D, r, s, α) parameter space where mode (4.29) is isolated. So, solutions of the equations with appropriate parameters are spatially heterogeneous.

Note that if we decrease r or N, the critical wavenumber decreases, and so the spacing of the pattern increases, or if decreased enough, the pattern disappears altogether. This is a prediction of the model which could, in principle, be tested experimentally, such as experimental manipulations of the base cell density. How the bifurcation to spatially structured solutions can be influenced by experimentally varying parameters of the system (specifically the cell density) in the case of limb development is discussed in detail in Chapter 7. There the theory is remarkably accurate in its predictions.

We now suggest that this cell-chemotaxis mechanism (4.19) and (4.20) is a candidate mechanism for generating the patterns found on snake skins and that the observed patterns reflect an underlying spatial pattern in cell density. As mentioned, we do not know when in development the pattern is formed nor how long it takes to develop as compared with significant growth in the embryo integument, that is, the relevant spatial domain for the equations. The size of the domain is, of course, a significant parameter as shown in the analysis in Chapter 3. Patterns can start to evolve as one of several parameters pass through bifurcation values which make the uniform steady state unstable. In the next section we show some of the patterns that the equation system can produce and relate them to specific snake patterns.

The simulations were carried out in a totally different way to those obtained for animal coat patterns and butterfly wing patterns. We assumed that the rate of cell differentiation and the development of the snake embryo are such that the chemotactic system has come to a steady state, or nearly so, by the time the pigmentation patterns become fixed. So Murray and Myerscough (1991) solved (4.24) at steady state; that is,

$$\begin{aligned} D\nabla^2 n - \alpha\nabla.(n\nabla c) + srn(N-n) &= 0, \\ \nabla^2 c + s\left[\frac{n}{1+n} - c\right] &= 0. \end{aligned} \tag{4.30}$$

These equations were solved numerically on a long two-dimensional rectangular domain using the package ENTWIFE as various parameters were varied; the procedure is described in some detail in Winters et al. (1990). It is a procedure for following bifurcating solutions as parameters vary: here we are mainly concerned with the biological implications and relevance of the solutions. The package lets one follow different bifurcation paths which branch off from different steady state solutions. Some analytical work on finite amplitude steady state solutions of (4.30), with $s = 0$, in one space dimension was done by Grindrod et al. (1989).

4.13 Simple and Complex Snake Pattern Elements

Murray and Myerscough (1991) showed that by altering the values of the mitotic rate r and chemotactic parameters α they were able to generate a wide range of different stripe patterns as well as regular spot patterns. Some examples of basic lateral striping are shown in Figure 4.23(a). Lateral banding is a common pattern element in snakes.[4] One example, the bandy-bandy (*Vermicella annulata*) is illustrated in Figure 4.23(b). Other examples include the coral snakes, *Micrurus* species, the banded krait (*Bungarus fasciatus*) and the ringed version of *Lampropeltis getulus californiae* (Figure 4.23(d)).

Lateral striping is what we would now come to expect from the discussion in Chapter 3. What is now more interesting is that the model can also produce longitudinal stripes parallel to the long edges of the domain as shown in Figure 4.24(a). This type of striping occurs, for example, in the ribbon snake (*Thamnophis sauritus sauritus*) as in Figure 4.24(b), in the garter snake (*Thamnophis sirtalis sirtalis*) and the four-lined snake (*Elaphe quatuorlineata*). What was initially surprising was that for some parameter sets relatively small changes in domain size or parameter values were sufficient to change the lateral stripes produced by this model to longitudinal stripes and vice versa: the pat-

[4]The highly venomous sea snakes typically have these types of stripes. An interesting unrelated feature of them is that they can come up from very deep depths to the sea surface very quickly but surprisingly do not get the bends. It is an interesting physiological feature that is still not understood.

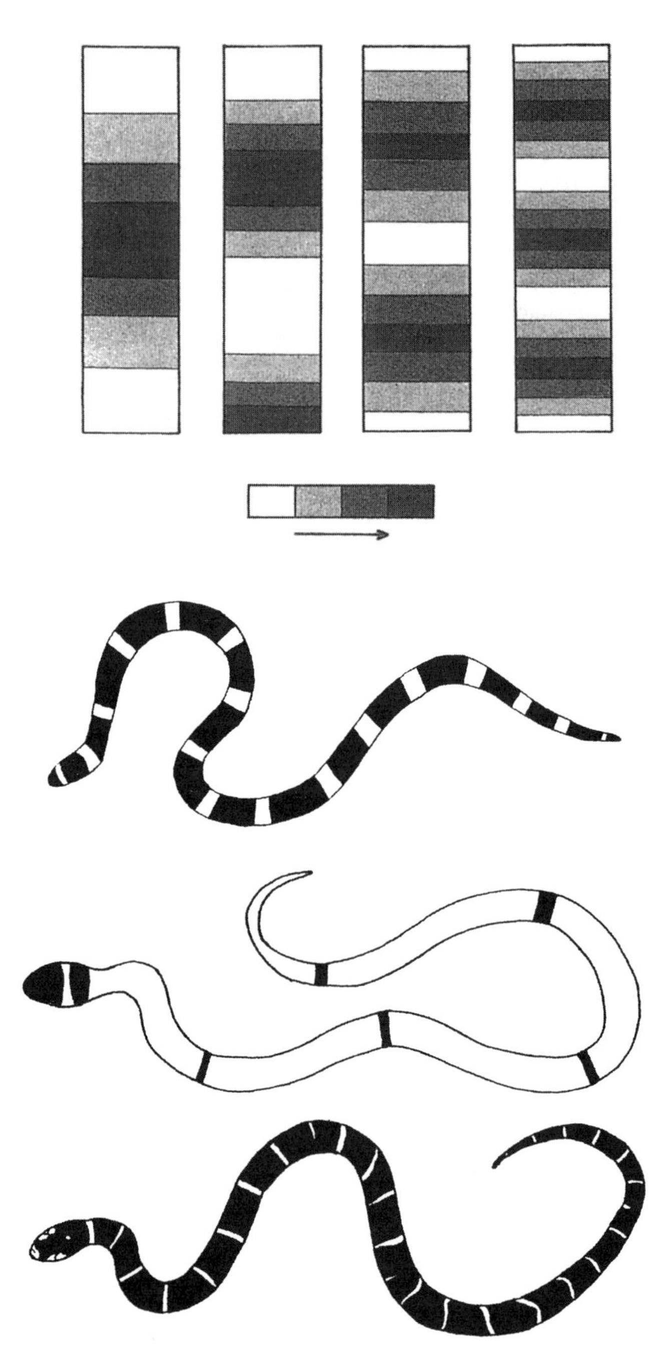

Figure 4.23. (**a**) Computed examples of simple lateral stripe patterns: the arrow denotes increasing cell density. Parameter values in (4.30) vary for each example except for $D = 0.25, N = s = 1$: for the first $r = 1.52, a = 12.31$, the second $r = 1.52, \alpha = 13.4$, the third $r = 24.4, \alpha = 118.68$ and the fourth $r = 1.52, \alpha = 29.61$. (**b**) Lateral stripes on the striped snake *Vermicella annulata*. (**c**) An example of sparse narrow stripes on the snake *Pseudonaja modesta*. (**d**) Laterally striped *Lampropeltis getulus californiae*. (From Murray and Myerscough 1991)

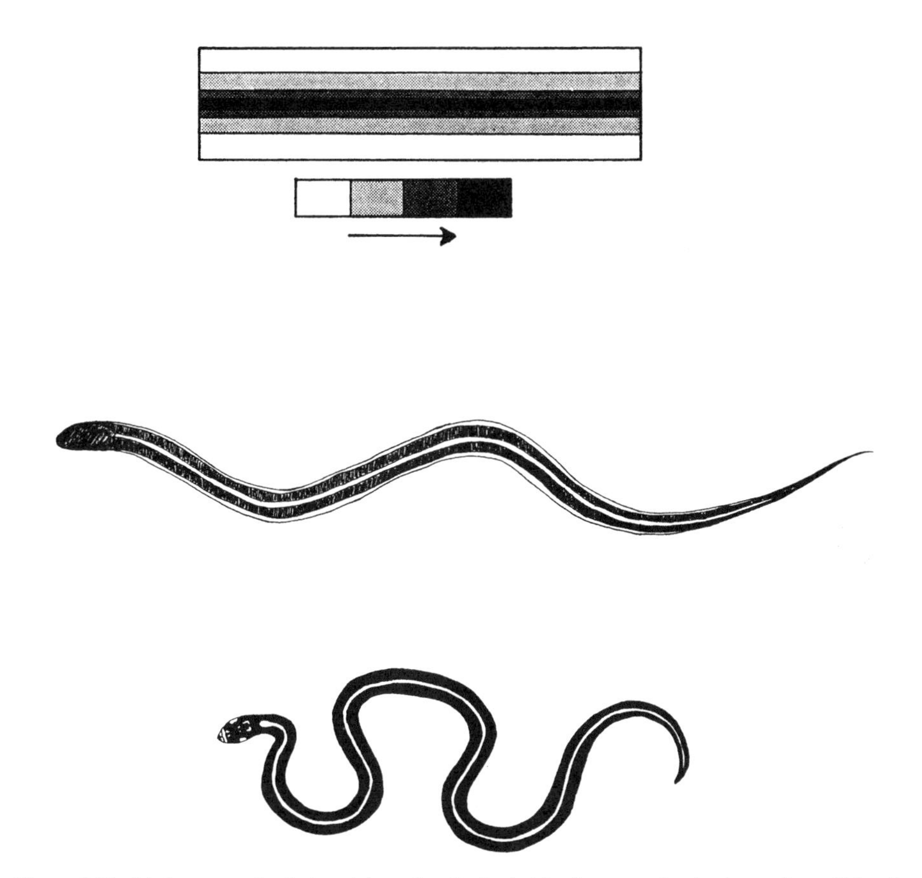

Figure 4.24. (**a**) A computed solution giving a longitudinal strip: the arrow denotes increasing cell density. Parameter values for equations (4.30) are: $D = 0.25$, $N = s = 1$, $r = 389.6$, $\alpha = 1782$. (**b**) The snake *Thamnophis sauritus sauritus* generally exhibits longitudinal striping. (**c**) Longitudinally striped California king snake *Lampropeltis getulus californiae*. (From Murray and Myerscough 1991)

terns are certainly not robust. This has important biological implications in that both lateral and longitudinal stripes can occur on different individuals of the same species. This is, for example, the case with the California king snake (*Lampropeltis getulus californiae*), which can be either laterally (Figure 4.23(d)) or longitudinally striped as in Figure 4.24(c); see also the photograph in Figure 4.22(a).

By an appropriate choice of parameter values we can also generate regular spot and blotch patterned solutions to equations (4.30) some of which are shown in Figure 4.25. Regular spots form part of the pattern on many snakes. For example, the Cape mountain adder (*Bitis atropos atropos*) displays an alternating semicircular pattern similar to those generated by this model.

Whether spots or stripes form in this chemotactic model depends on initial conditions, domain shape and size and on the values of the parameters α, D, r and N. Murray and Myerscough (1991) only considered changes in the chemotactic parameter α but other parameters can also be used to illustrate the argument. For any particular

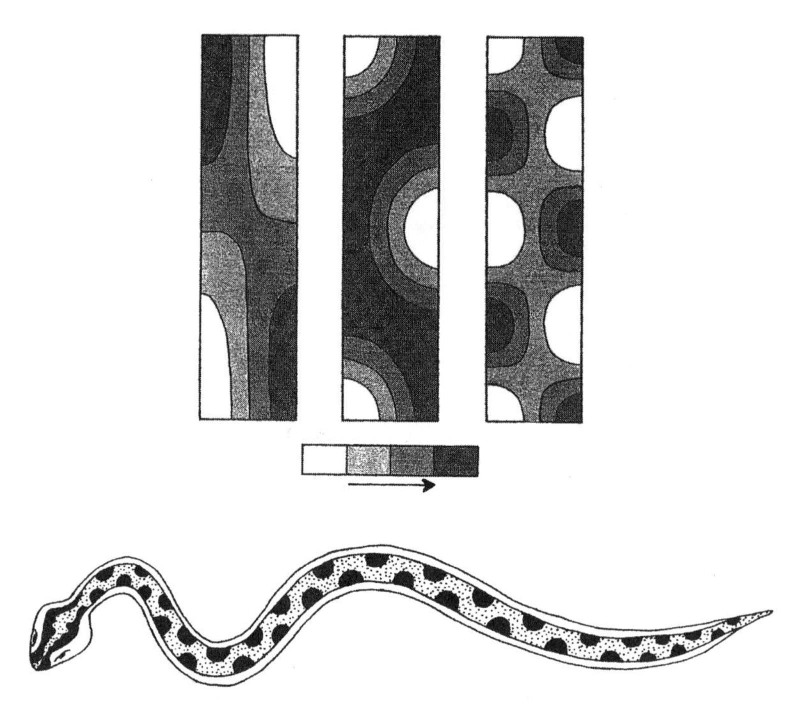

Figure 4.25. (**a**) Solutions of the system (4.30) which give regular spot patterns: the arrow denotes increasing cell density. Parameter values $D = 0.25$, $N = s = 1$, with the following for the three patterns respectively: $r = 28.22, \alpha = 135.16$; $r = 28.05, \alpha = 135$; $r = 1.52, \alpha = 27.06$. (**b**) The regular spot pattern on the leopard snake *Bitis atropos atropos*. (From Murray and Myerscough 1991)

initial conditions and domain size, stripes are most likely to form when the chemotactic response α is low (see, for example, the bifurcation diagrams in Maini et al. 1991). From the nondimensionalisation (4.39) this also corresponds to slow production or rapid diffusion or decay of chemoattractant in the dimensional problem. We can get some intuitive idea of why this is the case. When the chemotactic response is weak, cells must be in large regions of high cell density to produce enough chemoattractant to form a sufficiently steep gradient of chemoattractant concentration. This steep gradient is needed to recruit enough cells to the cluster to balance the logistic loss. For steep gradients in chemoattractant concentration to exist, a region of low cell density, where chemoattractant production is low, must be near the region of high cell density. Thus stripes, consisting of many cells but with regions with few cells nearby, will be preferred to spots when α is low. When the cells' chemotactic response is faster, gradients in chemoattractant concentration do not need to be so steep. Hence fewer cells can produce enough chemoattractant to recruit other cells to the resulting cluster and these clusters are more likely to be spots.

There is experimental justification for considering changing the chemotaxis factor, α, from the work on bacterial patterns, discussed in depth in Chapter 5. Here we have taken the simplest form of the chemotaxis term in the model equation. In the case of bacterial patterns the chemotaxis changes significantly during the patterning process since it depends on the level of the chemoattractant. If chemotaxis is indeed implicated in snake pattern formation it is quite possible that the chemotactic response of the cells will change during the patterning process.

Once steady state patterning has been formed, increasing α may either define the existing pattern more sharply or, in some cases, lead to qualitative changes, as we describe below. No cases were found where stripes split into spots although there were some instances where stripes split into two (see also Maini et al. 1991 and Myerscough et al. 1990): this happened in the absence of any mitosis. For an established stripe pattern, an increase in chemotactic response, α, gave very sharp, well-defined bands. Such isolated narrow bands of pigment cells are regularly observed in the ringed brown snake (*Pseudonaja modesta*) as illustrated in Figure 4.23(c). For a reaction diffusion mechanism to produce such sharp banding the threshold for pigment production would have to be very finely tuned. In contrast, with a chemotaxis mechanism all that is required is that α/D be large. In the case of a spot pattern increasing α usually leads to sharper focusing of the cluster although qualitative changes were observed in one case which we now discuss.

The patterning potential of this seemingly simple model system (albeit nonlinear) is not restricted to simple elements such as stripes and regular spots. It is not easy, however, to predict the type of complex patterns which can be obtained. For example, if we take a regular spot pattern and then solve the equations as the chemotactic component of the cells' motion, α, increases but without changing domain shape, the pattern shifts and changes its type. For sufficiently large α we get a pattern of pairs of spots as shown in Figure 4.26(a). Such patterns occur, for example, in the leopard snake (*Elaphe situla*) as in Figure 4.26(b). This species also exhibits a phase with single spots instead of paired spots.

A crucial aspect in the development of pattern could be the changing integument domain as a result of growth during the patterning process. It is because of the growth aspect that we discuss snake patterns in this chapter. We found, as we should now expect, that changing the shape of the domain also produces more complex patterns. Simple longitudinal growth of a laterally striped domain leads to the formation of additional stripes between established stripes just as we saw in the case of alligator skin patterns discussed above. Two examples of patterns obtained when lateral growth of the domain occurred are shown in Figures 4.27 and 4.28. In the first case, Figure 4.27(a), lateral growth causes the asymmetric spot pattern to become symmetric as the aggregates of pigmented cells move into the centre of the domain. This type of centred spot pattern is very common. Examples include the corn snake (*Elaphe guttata*) illustrated in Figure 4.27(b) and various *Vipera* species. Starting from a slightly different spot pattern lateral growth can produce diamond patterns as in Figure 4.28(a)(i). If the domain then becomes slightly narrower a wavy stripe pattern is generated as in Figure 4.28(a)(ii). Diamond patterns are a characteristic feature of many rattlesnakes, such as the eastern diamondback rattlesnake (*Crotalus adamanteus*) illustrated in Figure 4.28(b). The horseshoe snake (*Coluber hippocrepis*) also shows this type of diamond patterning. Near the tail where the body is narrower the diamond pattern may change to a wavy stripe as expected from the mathematical analysis, a feature pointed out in Chapter 3 as an example of a developmental constraint. These results suggest that growth of the domain during the laying down of pigment patterns probably has a crucial role to play in the ultimate pattern that develops.

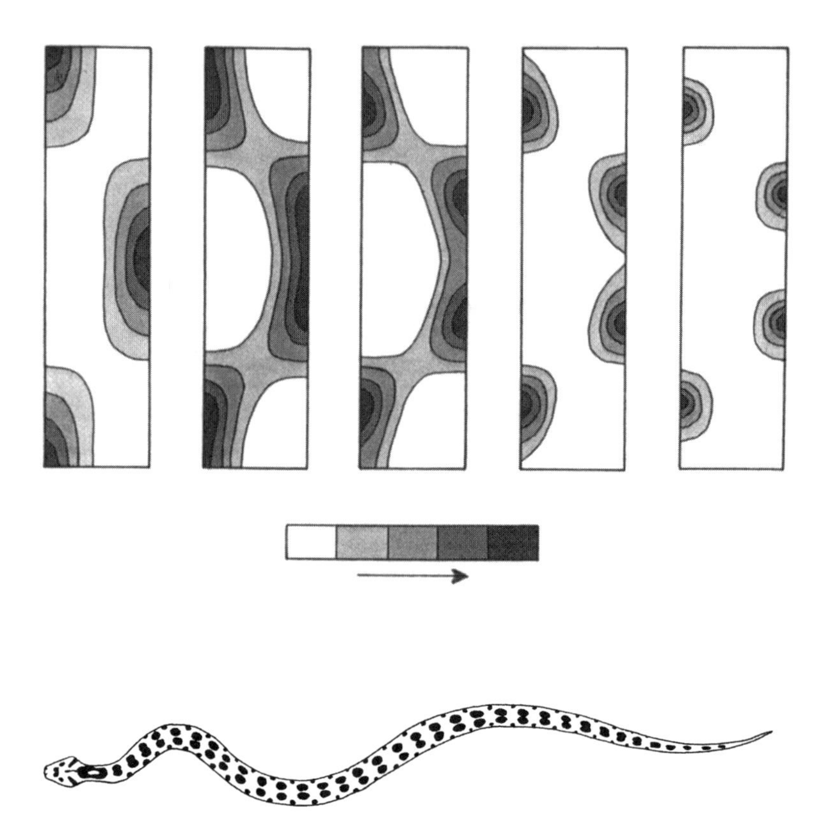

Figure 4.26. (**a**) The changing pattern as the chemotactic parameter α is increased giving a paired spot pattern: the arrow denotes increasing cell density. Parameter values for (4.30): $D = 0.25$, $N = s = 1$, $r = 1.52$ with α increasing from $\alpha = 19.92$ to 63.43. (**b**) The snake *Elaphe situla* showing paired spots. (From Murray and Myerscough 1991)

Chemotaxis was chosen as a plausible initial mechanism in view of its importance in other developing systems. In Chapter 5 on bacterial patterns there is absolutely no question as to its direct biological relevance. Further evidence for a mechanism which directly involves cells comes from the experimental work reported in Murray et al. (1990) on the alligator and touched on earlier in the chapter.

Although our experience from earlier chapters would lead us to expect a variety of complex patterns Murray and Myerscough (1991) also found other unexpected patterns, such as the paired spot pattern and the wavy stripe and diamond patterns often observed on snakes.

In many snakes and lizards the body pattern continues to the end of the tail with little alteration, even where the tail is sharply tapered. One example is shown in Figure 4.27(b) which contrasts with many mammalian patterns where if the tail pattern is mainly spots these almost always change to lateral stripes as the domain tapers. It is possible that the aggregative effect of strong chemotaxis means that spots may be able to form even on tapering domains although further numerical calculations are needed to confirm this. Ultimately, of course, we expect all spots to become stripes if the domain

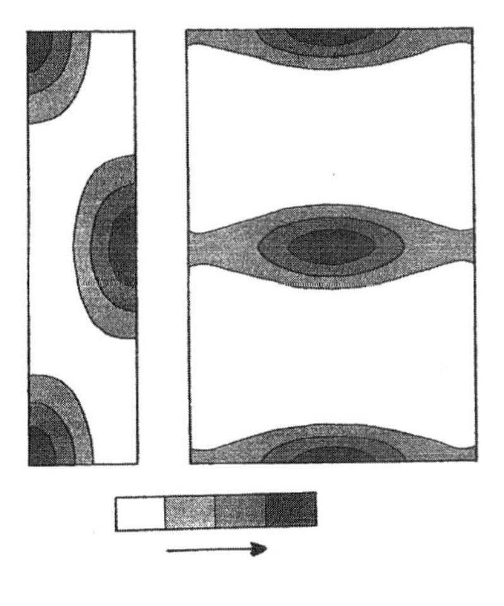

Figure 4.27. (**a**) The effect of lateral growth of the domain can result in centred spots: the arrow denotes increasing cell density. Here the domain width is from left to right (i) 1 unit; (ii) 2.7 units. Parameter values for (4.30) are $D = 0.25$, $N = s = 1$, $r = 1.52$, $\alpha = 20.5$. (**b**) The snake *Elaphe guttata* typically has centred spot patterns. (From Murray and Myerscough 1991)

is thin enough during the pattern formation process. Another point, however, is that the taper in most snakes is considerably more gradual than is found on those animals, such as the cheetah and leopard, where spots degenerate into stripes just toward the tail tip.

The phenomenological similarities between actual snake patterns and the complex patterns produced by our chemotactic model is encouraging and provides motivation for further theoretical investigations and also for further experiments to investigate the possible roles of chemotaxis, cell density and domain size in pigment pattern formation

This cell-chemotaxis model mechanism for pigment pattern formation is, of course, like previous models, speculative. Unlike straight reaction diffusion models, however, it explicitly includes cell motility and cell–cell interaction through a chemical mediator for which there is considerable biological evidence. We have shown that this relatively simple model is capable of generating many of the observed simple and complex pigment markings found on a variety of snakes.

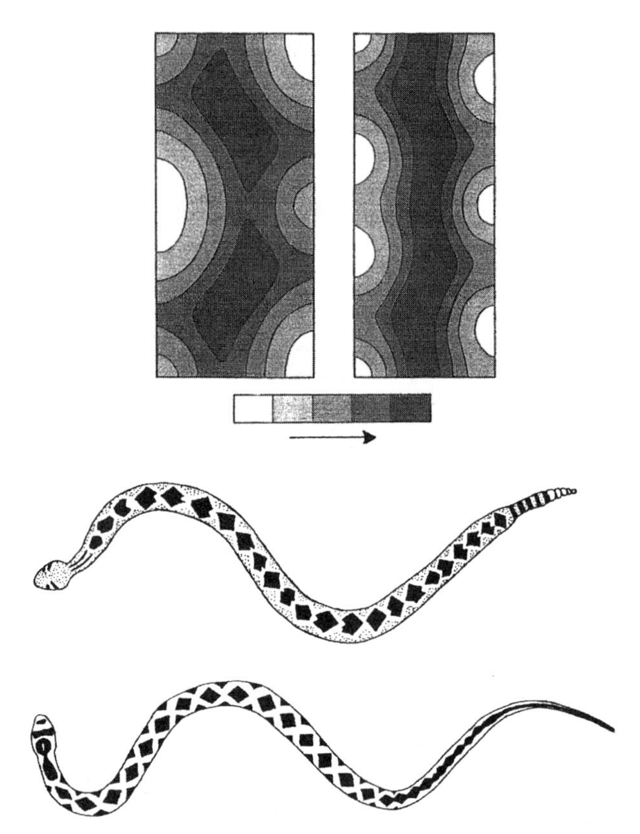

Figure 4.28. (**a**) In this case lateral growth of the domain gives diamond patterns and wavy stripes: the arrow denotes increasing cell density. Here the domain width is: (i) 1.84 units; (ii) 1.74 units. Parameter values for (4.30) are: $D = 0.25, N = s = 1, r = 38.05, \alpha = 177.7$. (**b**) Examples of diamond patterns on snakes: (i) *Crotalus adamanteus*; (ii) *Coluber hippocrepis*. Note in (ii) the result of tapering on the pattern. (From Murray and Myerscough 1991)

4.14 Propagating Pattern Generation with the Cell-Chemotaxis System

In the case of the alligator stripes we noted in Section 4.2 that the stripe pattern appears to start at the head and move progressively down the back to the tail. We showed in Section 4.12 that with parameters in the cell-chemotaxis system (4.30) can generate a variety of spatial patterns if the parameters are in the appropriate parameter space. Patterns such as given in the above figures can arise from specific initial conditions. We can see intuitively that if we initiated the perturbation about the steady state at one edge of the domain it would initiate a spatial patterned solution in that region which would move progressively along the domain. Myerscough and Murray (1992) studied this specific problem in the situation in which there is no cell mitosis, that is, $r = 0$ in (4.22), and the system is then given by

$$\begin{aligned}\frac{\partial n}{\partial t} &= D\nabla^2 n - \alpha\nabla.(n\nabla c),\\ \frac{\partial c}{\partial t} &= \nabla^2 c + s\left[\frac{n}{1+n} - c\right].\end{aligned} \tag{4.31}$$

They found that if the parameters are in the space which gives rise to a regular striped pattern then an inital spatial perturbation at one end gives rise to a travelling wave which leaves behind a steady state spatial pattern of regular stripes. Figure 4.2 shows a typical evolution of such a pattern for the system (4.30). Using the technique described in Section 2.6 in Chapter 2 we can obtain analytical estimates for the pattern wavelength and speed of spread by using asymptotic techniques. We follow in part the work of Myerscough and Murray (1992) and, in passing, recap the method which we discussed in general terms in Section 2.6 but with more specific details.

We consider the one-dimensional problem and look at the behaviour of the propagating disturbance close to its leading edge where the pattern is just beginning to form. At the leading edge the amplitude of the disturbance is small and we suggest that the behaviour of the pattern is governed by the linearised form of the equations, with the linear dispersion relation given by (4.24) with $r = 0$, namely,

$$\begin{aligned}\frac{\partial u}{\partial t} &= D\nabla^2 u - \alpha N\nabla^2 v\\ \frac{\partial v}{\partial t} &= \nabla^2 v + s\left[\frac{u}{(1+N)^2} - v\right].\end{aligned} \tag{4.32}$$

This has solutions of the form given by

$$\begin{bmatrix} u \\ v \end{bmatrix} \propto \exp[i\mathbf{k}.\mathbf{x} + \lambda t], \tag{4.33}$$

where the dispersion relation $\lambda(k^2)$ as a function of the wavenumber k is given by the positive solution of

$$\lambda^2 + [(D+1)k^2 + s]\lambda + \left[Dk^2(k^2+s) - \frac{sN\alpha}{(1+N)^2}\right]k^2 = 0. \tag{4.34}$$

We can use this linear approximation to write the solutions to (4.32) at the leading edge as the integral of Fourier modes

$$n(x,t) = \int_{-\infty}^{\infty} A(k)\exp[ikx + \lambda(k^2)t]\,dk, \tag{4.35}$$

where $\lambda(k^2)$ is the dispersion relation given by (4.34).

We are interested in the solution for large time and far from the initial disturbance. As we follow the leading edge of the disturbance as it propagates over the domain, (4.35) remains valid and we rewrite it as

$$n(Vt, t) = \int_{-\infty}^{\infty} A(k) \exp[tg(k)]\, dk, \tag{4.36}$$

where $A(k^2)$ is obtained from initial conditions for the linear system (which does not concern us here) and $g(k) = ikV + \lambda(k^2)$ where V is the finite 'speed of propagation' and is the speed of travel of the leading edge. This implies that x and t are of the same order of magnitude and that the front of the pattern is sufficiently far from the initial point of disturbance for v to be constant. We now use the method of steepest descents (see, for example, the book by Murray 1984) to evaluate this integral asymptotically for large t and hence large x with $V = 0(1)$ to obtain

$$n \approx \frac{F(k^*)}{\sqrt{t}} \exp\left[t\left\{ik^*V + \lambda(k^{*2})\right\}\right], \tag{4.37}$$

where F is a function of k which is not of importance in the following analysis. Here k^* is a saddle point of $g(k)$ in the complex k-plane chosen so that where $Rl[ik^*V + \lambda(k^{*2})] > Rl[ikV + \lambda(k^2)]|_{k\epsilon K}$, where K is the set of all k in the plane other than k^*.

Saddle points of $g(k)$ are solutions of

$$\frac{dg}{dk} = iV + \frac{d\lambda}{dk} = 0. \tag{4.38}$$

From the numerical solutions, typically like those shown in Figure 4.2, we see that the envelope of the pattern is essentially of constant shape far from the initial perturbation. So, we therefore impose a restriction on $g(k)$ of the form

$$Rl(g(k^*)) = Rl[ik^*V + \lambda(k^{*2})] = 0. \tag{4.39}$$

This is a marginal stability hypothesis, which together with (4.38) is sufficient for us to be able to solve (in principle) for the wavenumber, k^*, and the velocity of the patterning wave, V.

If we consider the pattern in the moving frame of the pattern envelope we see (refer to Figure 4.2) an oscillating pattern which starts at the leading edge of the envelope. The frequency of oscillation of this pattern is given by

$$\Omega = Im(g(k)). \tag{4.40}$$

This is the frequency with which nodes are created at the front of the envelope. If peaks are conserved, that is, peaks do not coalesce, then Ω is also the frequency of oscillation far from the leading edge in the moving frame of the envelope. Let k' be the wave number of the pattern far behind the leading edge. We can then write $\Omega = k'V$ and so

$$k' = Rl(k^*) + Im\frac{\lambda(k^{*2})}{V} \tag{4.41}$$

which gives the wavenumber for the final steady state pattern.

This method depends, of course, on the linear behaviour of the system near the leading edge of the disturbance. That is, we assume that the equations are only weakly nonlinear in the vicinity of the leading edge in which case we can write the solutions in the form of (4.35) as the integral over Fourier modes. However, this weakly nonlinear assumption will not necessarily hold for the nonlinear system (4.31) when solutions are far from the homogeneous steady state. The method also implicitly assumes that the behaviour of the solution is governed by events at the leading edge which may not be strictly true. It is unlikely, therefore, that this approach will give good quantitative agreement with values of the wavelength $w = 2\pi/k'$ and V from the computed solutions. We may expect, however, a reasonable qualitative agreement which may be used to predict V and w.

Analysing the Wave Pattern

Applying this method to the system (4.31) with the dispersion relation $\lambda = \lambda(k^2)$ given by (4.34) is not easy. To solve for k and V we have to substitute for $\lambda(k^2)$ into (4.38) and (4.39). The equations which result are very complicated because of the form of the dispersion relation given by (4.34) and we simply cannot extract the necessary information analytically. To circumvent this problem and still gain some insight into pattern initiation via wave propagation Myerscough and Murray (1992) considered a caricature dispersion relation

$$\lambda(k^2) = \Gamma k^2(k_0^2 - k^2), \quad \Gamma > 0$$

on which they did a best fit (choosing Γ and k_0^2) to the exact dispersion relation given by (4.34). The parameters Γ and k_0^2 govern the steepness of the parabolic approximation and its intercept on the k^2-axis respectively. The caricature is qualitatively similar to that given by (4.34) in the crucial region where λ is positive. Myerscough and Murray (1992) carried out the detailed analysis (which involves a steepest descent analysis) and compared their approximate results with those obtained from a numerical solution of the full nonlinear system (4.31). The comparison was not very close, but importantly such an anlalytical procedure provides qualitative dependence on the parameters. With only two parameters we cannot expect an especially close fit and some error arose from the looseness of the fit regardless of how it was done. What is interesting, however, is that Myerscough and Murray (1992) showed that the error is not due to the fitting of the caricature dispersion relation but rather to the linearisation process.

The approximate method implicitly assumes that the speed of spread and the wavelength of the final pattern is determined by events at the leading edge where linear theory is applicable. These comparison results suggest that while events at the leading edge are important in determining w and V, events behind the leading edge, in the nonlinear regime, have a significant effect in the case of a chemotaxis system. Myerscough and Murray (1992) suggested how this could come about.

The developing peaks in cell density behind the leading edge generate peaks in chemoattractant concentration. This concentration gradient then begins to act on the cells on the leading edge side of the peak inhibiting them from moving forward into the next peak which is developing at the leading edge. This means that cells which, in a linear regime might have joined the newly formed peak, either join the previously formed cluster of cells or are recruited more slowly to the peak at the leading edge. The first alternative explains why the solution wavelength is longer than the wavelength found by the analytical method. The second explains why the solution speed of spread is slower than found analytically. These effects, caused by cell recruitment by chemotaxis, do not occur in reaction diffusion systems. Nevertheless this analytical method when applied to the highly nonlinear problem posed by (4.31) gives a useful estimate of the speed of propagation of the disturbance and the wavelength of the final pattern provided peaks do not coalesce behind the leading edge. The only information needed to derive the estimates is a reasonable approximation to the dispersion relation of the linear system.

5. Bacterial Patterns and Chemotaxis

5.1 Background and Experimental Results

There is an obvious case for studying bacteria. For example, bacteria are responsible for a large number of diseases and they are responsible for most of the recycling that takes place. Their use in other areas is clearly going to increase as our understanding of their complex biology becomes clearer. Here we are interested in how a global pattern in bacterial populations can arise from local interactions. Under a variety of experimental conditions numerous strains of bacteria aggregate to form stable (or rather temporarily stable) macroscopic patterns of surprising complexity but with remarkable regularity. It is not easy to explain how these patterns are formed solely by experiment, but they can, however, be explained for the most part, with the use of mathematical models based on the known biology. It is not possible to discuss all the patterns which have been studied experimentally so we concentrate on the collection of diverse patterns observed by Berg and his colleagues (see Budrene and Berg 1991, 1995 and earlier references there) in the bacteria *Escherichia coli* (*E. coli*) and *Salmonella typhimurium* (*S. typhimurium*).

E. coli and *S. typhimurium* are common bacteria: for example, *E. coli* is abundant in the human intestine and *S. typhimurium* can occur in incompletely cooked poultry and meat. These bacteria are motile and they move by propelling themselves by means of long hairlike flagella (Berg 1983). Berg also mapped the movement of a bacterium over time, and found that the organism's motion approximates a random walk and so the usual Fickian diffusion can be used to describe their random motility; their diffusion coefficient has been measured experimentally.

A key property of many bacteria is that in the presence of certain chemicals they move preferentially towards higher concentration of the chemical, when it is a chemoattractant, or towards a lower concentration when it is a repellent. The sensitivity to such gradients often depends on the concentration levels. In the modelling below we shall be concerned with chemotaxis and the sensitivity issue. The basic concept of chemotaxis (and diffusion) was discussed in Chapter 11, Volume I, Section 11.4 and in more detail in the last chapter. Basically, whether or not a pattern will form depends on the appropriate interplay between the bacterial populations and the chemical kinetics and the competition between diffusive dispersal and chemotactic aggregation.

Budrene and Berg (1991, 1995) carried out a series of experimental studies on the patterns that can be formed by *S. typhimurium* and *E. coli*. They showed that a bacterial colony can form interesting and remarkably regular patterns when they feed on, or are exposed to, intermediates of the tricarboxylic acid (TCA) cycle especially succinate and fumarate. They used two experimental methods which resulted in three pattern forming mechanisms. The bacteria are placed in a liquid medium in one procedure, and on a semi-solid substrate (0.24% water agar) in the other. They found one mechanism for pattern formation in the liquid medium, and two in the semi-solid medium. In all of the experiments, the bacteria are known to secrete aspartate, a potent chemoattractant.

Liquid Experiments with E. coli and S. typhimurium

These experiments produce relatively simple patterns which appear quickly, on the order of minutes, and last about half an hour before disappearing permanently. Two types of patterns are observed and are selected according to the initial conditions. The simplest patterns are produced when the liquid medium contains a uniform distribution of bacteria and a small amount of the TCA cycle intermediate. The bacteria collect in aggregates of roughly the same size over the entire surface of the liquid, although the pattern often starts in one general area and spreads from there (Figure 5.1(a)).

In the second type of liquid experiment, the initial density of bacteria is uniform, and the TCA cycle intermediate is added locally to a particular spot, referred to as the 'origin.' Subsequently, the bacteria are seen to form aggregates which occur on a ring centred about the origin, and in a random arrangement inside the ring (Figure 5.1(b)).

Importantly, in these liquid experiments, the patterns are generated on a timescale which is less than the time required for bacterial reproduction and so proliferation does not contribute to the pattern formation process. Also, the bacteria are not chemotactic to any of the chemicals initially placed in the medium, including the stimulant. The experimentalists (H.C. Berg, personal communication 1994) also confirmed that fluid dynamic effects are not responsible for the observed patterns.

Semi-Solid Experiments with E. coli and S. typhimurium

The most interesting patterns are observed in the semi-solid experiments and in particular with *E. coli*. For these experiments, a high density inoculum of bacteria is placed on a petri dish which contains a uniform distribution of stimulant in the semi-solid medium, namely, 0.24% water agar. Here the stimulant also acts as the main food source for the bacteria, and so the concentrations are much larger than in the liquid experiments. After two or three days, the population of bacteria has gone through 25–40 generations during which time it spreads out from the inoculum, eventually covering the entire surface of the dish with a stationary pattern of high density aggregates separated by regions of near zero cell density. Some typical final patterns are shown in Figures 5.2 and 5.3. The *S. typhimurium* patterns are concentric rings and are either continuous or spotted while *E. coli* patterns are more complex, involving a greater degree of positional symmetry between individual aggregates. An enormous variety of patterns has been observed, the most common being sunflower type spirals, radial stripes, radial spots and chevrons. Using time-lapse videography of the experiments reveals the very different kinematics by which *S. typhimurium* and *E. coli* form their patterns.

The simple *S. typhimurium* patterns (Figure 5.2) begin with a very low density bacterial lawn spreading out from the initial inoculum. Some time later, a high density ring of bacteria appears at some radius less than the radius of the lawn and after another

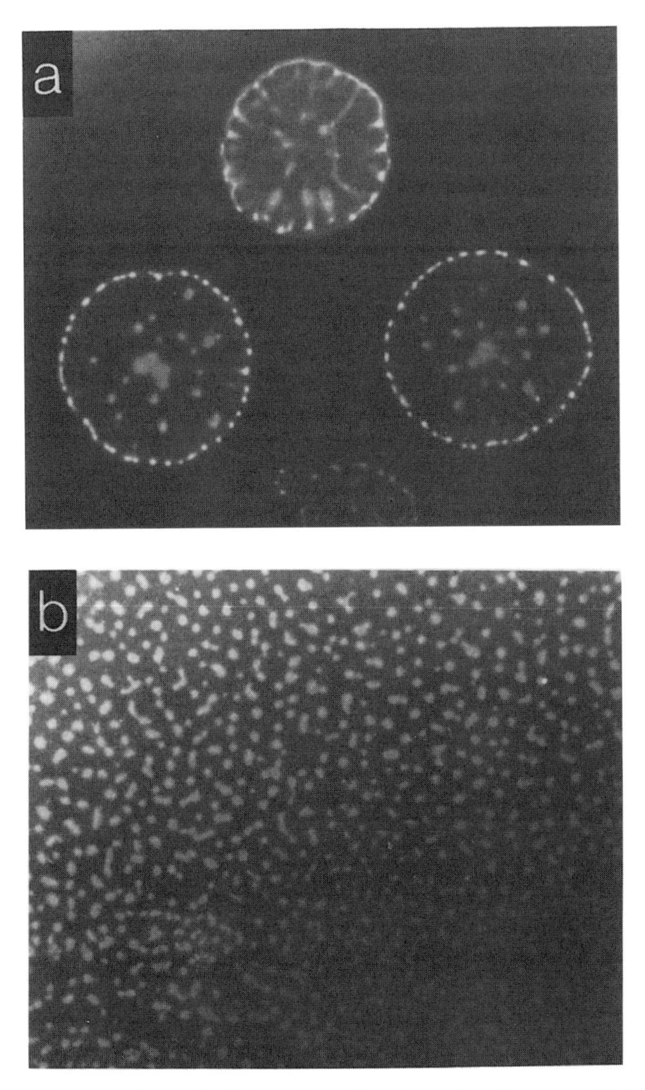

Figure 5.1. Bacterial patterns formed by *Salmonella typhimurium* in liquid medium. (**a**) A small amount of TCA was added to a uniform distribution of bacteria. (**b**) A small amount of TCA was added locally to a uniform distribution of bacteria. (Unpublished results of H. Berg and E.O. Budrene; photographs were kindly provided by courtesy of Dr. Howard Berg and Dr. Elena Budrene and reproduced with their permission.)

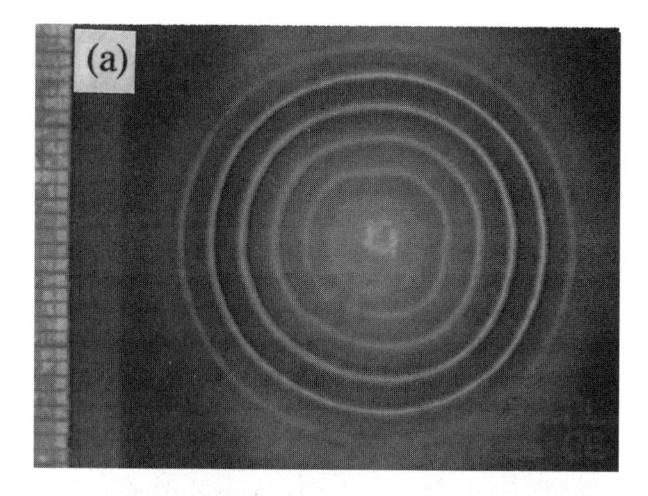

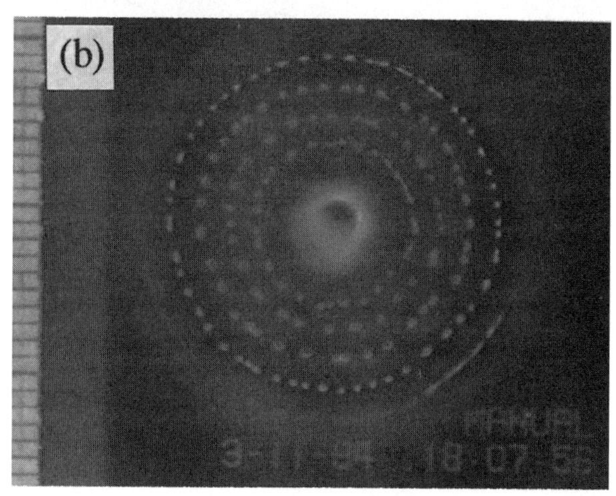

Figure 5.2. Typical *S. typhimurium* patterns obtained in semi-solid medium and visualized by scattered light. Experiments were carried out by Howard Berg and Elena Budrene using the techniques described in Budrene and Berg (1991). About 10^4 bacteria were inoculated at the centre of the dish containing 10 ml of soft agar in succinate. In (**a**) the rings remain more or less intact while in (**b**) they break up as described in the text: the time from inoculation in (**a**) is 48 hours and in (**b**), 70 hours. The 1-mm grid on the left of each figure gives an indication of the scale of the patterns. (From Woodward et al. 1995 where more experimental details and results are provided)

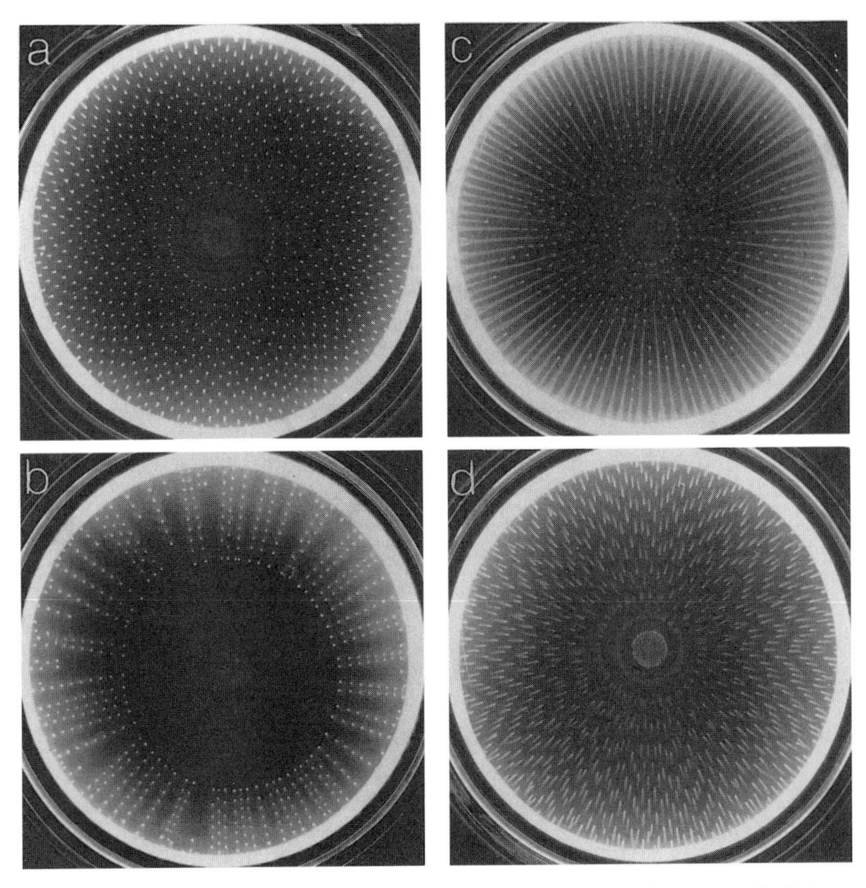

Figure 5.3. *E. coli* patterns obtained in semi-solid medium. Note the highly regular patterns. The light regions represent high density of bacteria. The different patterns (**a**)–(**d**) are discussed in the text at several places. (From Budrene and Berg (1991); photographs courtesy of Dr. Howard Berg and reproduced with permission)

time interval, when the lawn has expanded further, a second high density bacterial ring appears at some radius larger than that of the first ring. The rings, once they are formed, are stationary. The rings may remain continuous as in Figure 5.2(a) or break up into a ring of spots as in Figure 5.2(b). The high density aggregates of bacteria in one ring have no obvious positional relation to the aggregates in the two neighbouring rings.

The more dramatic patterns exhibited by *E. coli*, such as those shown in Figure 5.3, have definite positional relationships between radially and angularly neighbouring aggregates. These relationships seem to be the result of existing aggregates inducing the formation of subsequent ones (Budrene and Berg 1995). Instead of an initial bacterial lawn, a swarm ring of highly active motile bacteria forms and expands outwards from the initial inoculum. The bacterial density in the swarm ring increases until the ring becomes unstable and some percentage of the bacteria are left behind as aggregates. These aggregates remain bright and full of vigorously motile bacteria for a short time, but then dissolve as the bacteria rejoin the swarm ring. Left behind in the aggregate's original location is a clump of bacteria which, for some unknown reason, are non-motile: it is these non-motile bacteria that are the markers of the pattern.

It appears that one, or both, of the speed of the swarm ring and the time at which the dissolution of aggregates occurs, are key elements in the formation of any one pattern. If the dissolution happens quickly, the aggregates appear to be pulled along by the swarm ring, and the non-motile bacteria are left behind as a radial streak as in Figures 5.3(c) and (d). On the other hand if the dissolution happens a little less quickly, the cells from the dissolved aggregate rejoin the swarm ring and induce the formation of aggregates at the rejoining locations and this results in a radial spot pattern. If the dissolution happens even more slowly, the swarm ring becomes unstable before the bacteria from the aggregates have time to rejoin the ring. The ring then tends to form aggregates in between the locations where aggregates already exist which results in a sunflower spiral type of pattern.

Remember that, just as in the liquid experiments, none of the chemicals placed in the petri dish is a chemoattractant. The timescale of these patterns, however, is long enough to accommodate several generations of *E. coli*, and so proliferation is important here. Consumption of stimulant is also non-negligible, especially in the swarm ring patterns.

Since none of the substrates used are chemoattractants the patterns of *E. coli* in Figure 5.3 cannot be explained by some external chemoattractants. Chemoattractants, however, play a major role since the bacteria themselves produce a potent chemoattractant, namely, aspartate (Budrene and Berg 1991). Up to this time it had been assumed that the phenomenon of chemotaxis existed in *E. coli* and *S. typhimurium* only to guide the bacteria towards a food source. It was only in these experiments of Budrene and Berg (1991, 1995) that evidence was found that the bacteria can produce and secrete a chemoattractant as a signalling mechanism. This is reminiscent of the slime mould *Dictyostelium discoideum* where the cells produce the chemical cyclic AMP as a chemotactic aggregative signalling mechanism. In our modelling therefore we focus primarily on the processes of diffusion and chemotaxis towards an endogenously produced chemoattractant and how they interact to produce the bacterial patterns.

We should remember that all of these patterns are formed on a two-dimensional domain of the petri dish. With the wide variety of patterns possible it is clear that if the medium were three-dimensional the pattern complexity would be even greater. It would be quite an experimental challenge to photograph them. Here we model only the two-dimensional patterns and show that the models, which reflect the biology, can create the experimentally observed patterns.

What is clear is that chemotaxis phenomena can give rise to complex and varied geometric patterns. How these complex geometries form from interactions between individual bacteria is not easy to determine intuitively from experiments alone.[1] In cases such as this, when biological intuition seems unable to provide an adequate explanation, mathematical modelling can play an important, even crucial, role. To understand the patterns, many questions must be answered, often associated with the fine details of the biological assumptions and parameter estimates. For example, are diffusion and

[1]It was for this reason that Howard Berg initially got in touch with me. A consequence of the first joint modelling attempts (Woodward et al. 1995) generated informative and interesting experimental and biological questions.

chemotaxis towards an endogenously produced chemoattractant sufficient to explain the formation of these patterns? What is the quantitative role of the chemoattractant stimulant and how quickly must it be produced? What other patterns are possible and what key elements in the experiments should be changed to get them? A review of the biology, the modelling and the numerical schemes used to simulate the model equations is given by Tyson (1996).

Most of the material we consider in detail is based on a series of theoretical studies of these specific patterns (Woodward et al. 1995, Tyson 1996, Murray et al. 1998, Tyson et al. 1999). They proposed mathematical models which closely mimic the known biology. Woodward et al. (1995)—a collaborative work with the experimentalists Drs. Berg and Budrene—considered the less complex patterns formed by *S. typhimurium* and proposed an explanation for the observed self-organization of the bacteria. Ben-Jacob and his colleagues (Ben-Jacob et al. 1995, 2000; see other references there) have also studied a variety of bacteria both theoretically and experimentally: many of the patterns they obtain are also highly complex and dramatic. The patterns depend, of course, on the parameter values and experimental conditions. In nature, however, bacteria have to deal with a variety of conditions, both hostile and friendly. To accommodate such environmental factors bacteria have developed strategies for dealing with such conditions. These strategies involve cooperative communication and this affects the type of patterns they form. Ben-Jacob (1997; see other references there) has investigated the effect of possible communication processes, such as chemotactic feedback. The consequences of including such cooperativity in the model chemotactic systems is, as would be expected, that the spectrum of pattern complexity is even greater. The analytical and extensive numerical studies of Mimura and his colleagues (see, for example, Mimura and Tsujikawa 1996, Matsushita et al. 1998, 1999, Mimura et al. 2000 and other references there) are particularly important in highlighting some of the complex solution behaviour reaction diffusion chemotaxis systems can exhibit. For example, Mimura et al. (2000) classify the various pattern classes (five of them) and suggest that, with one exception, the morphological diversity can be generated by reaction diffusion models. Mimura and Tsujikawa (1996) considered a diffusion-chemotaxis with population growth and in the situation of small diffusion and chemotaxis they derived an equation for the time evolution of the aggregating pattern. In this chapter we discuss specific bacterial patterns obtained with *S. typhimurium* and *E. coli* and, very briefly, those exhibited by *Bacillus subtilis* which are quite different.

These bacterial patterns are far more elaborate than those observed when chemotactic strains grow on media containing nutrients that are attractants (for example, Agladze et al. 1993). They also differ from the travelling waves of aggregating cells of the slime mould *Dictyostelium discoideum* in that the structures formed by *E. coli* and *S. typhimurium*, for example, are only temporally stable.

The spatial pattern potential of chemotaxis has been exploited in a variety of different biological contexts. Mathematical models involving chemotaxis (along with reaction diffusion models (Chapters 2 and 3) and mechanochemical models (Chapters 6 and 7) are simply part of the general area of integrodifferential equation models for the development of spatial patterns. The basic Keller–Segel continuum mechanism for pattern formation in the slime mould *Dictyostelium discoideum* was proposed by Keller and Segel (1970) and was discussed in Chapter 11, Volume I, Section 11.4. A discrete, more biologically based, model (as a consequence of the new biological insights found since then) for the aggregation with appropriate cell signalling is given by Dallon and Othmer (1997). Othmer and Schaap (1998) give an extensive and thorough review of oscillatory cyclic AMP signalling in the development of this slime mould. Since the pioneering work of Keller and Segel (1970, 1971), a considerable amount of modelling effort has been expended on these patterns such as the work on bacteria by Ben-Jacob et al. (1995) who had thresholding behaviour in aspartate production and a cell-secreted waste field in their model. They obtained spatial patterns resembling some of the experimentally observed *E. coli* patterns. Brenner et al. (1998) performed a one-dimensional analysis of a model mechanism for swarm ring formation of *E. coli* patterns in a semi-solid medium. They studied the relative importance of the terms in their equations from the point of view of pattern formation and obtained some analytical results: for example, they derived an expression for the number of clumps in a given domain in terms of the model parameters.

Chemotaxis plays an important role in a wide range of practical phenomena such as in wound healing (see Chapter 10), cancer growth (see Chapter 11) and leukocytes moving in response to bacterial inflammation (for example, Lauffenburger and Kennedy 1983 and Alt and Lauffenburger 1987). Until recently, relatively little work had been done where cell populations are not constant; one exception was the travelling wave model of Kennedy and Aris (1980) where the bacteria reproduce and die as well as migrate. It appears that the presence of chemotaxis (or haptotaxis, a similar guidance phenomenon for cells in the mechanical theory of pattern formation discussed in Chapter 6) results in a wider variety of patterns than only reaction and diffusion, for example. Of course, when significant growth occurs during the patterning process the spectrum of patterns is even wider.

It is possibly pertinent to note here that the specific patterns formed by many bacteria depend sensitively on the parameters and on the conditions that obtain in the experiments, including the initial conditions. As such a potential practical application of bacterial patterning is as a quantitative measurement of pollution.

5.2 Model Mechanism for *E. coli* in the Semi-Solid Experiments

Basically we want to construct the biological mechanisms which govern the bacterial pattern formation processes in the experiments of Budrene and Berg (1991, 1995). We first consider the model for the semi-solid medium experiment with *E. coli*. The key players in the experiment seem to be the bacteria (of course), the chemoattractant (aspartate) and the stimulant (succinate or fumarate) so we consider the three variables: the cell density, n, the chemoattractant concentration, c, and the stimulant concentration, s. The bacteria diffuse, move chemotactically up gradients of the chemoattractant, proliferate and become non-motile. The non-motile cells can be thought of as dead, for the purpose of the model. The chemoattractant diffuses, and is produced and ingested by the bacteria while the stimulant diffuses and is consumed by the bacteria. To begin with it is usually helpful to simply write down a word equation for what you think is going on.[2] From the description of the biological processes described in the last section we suggest the following model consisting of three conservation equations of the form

$$\boxed{\begin{array}{c}\text{rate of change}\\ \text{of cell density, } n\end{array}} = \boxed{\begin{array}{c}\text{diffusion}\\ \text{of } n\end{array}} + \boxed{\begin{array}{c}\text{chemotaxis}\\ \text{of } n \text{ to } c\end{array}} + \boxed{\begin{array}{c}\text{proliferation}\\ \text{(growth and death)}\\ \text{of } n\end{array}} \tag{5.1}$$

$$\boxed{\begin{array}{c}\text{rate of change of}\\ \text{chemoattractant}\\ \text{concentration, } c\end{array}} = \boxed{\begin{array}{c}\text{diffusion}\\ \text{of } c\end{array}} + \boxed{\begin{array}{c}\text{production}\\ \text{of } c \text{ by } n\end{array}} - \boxed{\begin{array}{c}\text{uptake}\\ \text{of } c \text{ by } n\end{array}} \tag{5.2}$$

$$\boxed{\begin{array}{c}\text{rate of change}\\ \text{of stimulant}\\ \text{concentration, } s\end{array}} = \boxed{\begin{array}{c}\text{diffusion}\\ \text{of } s\end{array}} - \boxed{\begin{array}{c}\text{uptake}\\ \text{of } s \text{ by } n\end{array}} \tag{5.3}$$

The crucial part of the modelling is how to quantify the individual terms in these equations. A full discussion of the modelling is given by Tyson (1996) who also estimates the various parameters which appear from an extensive literature survey. It is mainly her work together with that in Tyson et al. (1999) and Murray et al. (1998) that we discuss here.

Diffusion

The diffusion terms for the chemoattractant and the stimulant in (5.2) and (5.3) are straightforward. These chemicals diffuse according to simple Fickian diffusion with diffusion coefficients D_c and D_s respectively with estimates (H.C. Berg, personal communication 1993)

$$D_c \approx D_s \approx 9 \times 10^{-6}\ \text{cm}^2\ \text{s}^{-1}.$$

All the parameter estimates are gathered together in Table 5.1 below.

We assume that the bacteria as a population also diffuse in a Fickian manner. The estimation of their diffusion coefficient, D_n, however, is less straightforward. In his book, Berg (1983) gives an expression for the diffusion constant of bacteria derived from the individual motion of the cells and comes up with an estimate of $D_n \approx 2 \times 10^{-6}\text{cm}^2\text{s}^{-1}$. Phillips et al. (1994) catalogued values of the diffusion coefficient reported in the literature over the previous 10 years: the values range from $D_n \approx 1.9\,(\pm 0.9) \times 10^{-4}\ \text{cm}^2\ \text{s}^{-1}$ at the upper end to $D_n \approx$ 1–10 $\times 10^{-7}\ \text{cm}^2\ \text{s}^{-1}$ at the lower end. The most recent measurements, however, fall in the range $D_n \approx$ 1–3 $\times 10^{-6}\ \text{cm}^2\ \text{s}^{-1}$, which corresponds with the theoretical value determined by Berg (1983).

In deciding on the appropriate diffusion coefficient for the model, we must keep in mind the number of space dimensions for movement which were available to the bacteria when the measurements were made. In the experiments we model, the depth of the liquid or agar mixture in the petri dish is of the order of 1.8 mm. When restricted

[2]It is essential when discussing with experimentalists who have little experience of mathematical modelling.

to motion through 10-mm capillary arrays, *E. coli* diffuse with a diffusion coefficient of 5.2×10^{-6} cm^2 s^{-1}, while through 50-mm capillary arrays, the diffusion coefficient was found to be 2.6×10^{-6} cm^2 s^{-1} (Berg and Turner, 1990). As we would expect, diffusion through the larger capillaries is slower.

Estimating diffusion coefficients of cells or bacteria, or really anything other than chemicals, is always a problem (see also Chapters 9, 10, 11, 13, and 14). A variety of theoretical approaches has been used to estimate diffusion coefficients such as that by Sherratt et al. (1993b) in their study of eukaryotic cell movement. Ford and Lauffenburger (1991) and Sherratt (1994) developed models beginning with receptor level kinetics in their analysis to determine the diffusion coefficient of bacteria.

For our purposes, it is not necessary to include such receptor level detail since the experimentalists believe that the absolute chemoattractant concentration does not affect the diffusion coefficient of the bacteria. So, we assume the cells diffuse with a constant diffusion coefficient with a value $D_n = 2\text{–}4 \times 10^{-6}\text{cm}^2\ \text{s}^{-1}$.

Chemotaxis

Chemotaxis, as we saw in Chapter 11, Volume I, involves the directed movement of organisms up a concentration gradient, and so is like a negative diffusion. However, whereas diffusion of cells depends only on their density gradient, chemotaxis depends on the interaction between the cells, the chemoattractant and the chemoattractant gradient. The general form of the chemotaxis term in the conservation equations is the divergence of the chemotactic flux:

$$\nabla \cdot \mathbf{J_c} = \nabla \cdot [\chi(n, c)\nabla c], \tag{5.4}$$

where $\mathbf{J_c}$ is the chemotactic flux, $\chi(n, c)$ is the chemotaxis response function, as yet unknown, with n and c the cell density and chemoattractant concentration respectively. A lot of research has been directed to finding a biologically accurate expression for the chemotaxis function $\chi(n, c)$. Ford and Lauffenburger (1991) reviewed the main types of functions tried. As in the case of the diffusion term, forms for $\chi(n, c)$ have been proposed either by working up from a microscopic description of cell behaviour, or by curve fitting to macroscopic results from population experiments. A synthesis of the various approaches suggests that the macroscopic form of Lapidus and Schiller (1976), namely,

$$\chi(n, c) = \frac{n}{(k + c)^2},$$

where k is a parameter is a good one. This seems to give the best results when compared to experimental data, in particular the experiments by Dahlquist et al. (1972) which were designed specifically to pinpoint the functional form of the chemotactic response. Interestingly, the inclusion of all of the receptor level complexity by other researchers did not give any significant improvement over the results of Lapidus and Schiller (1976). The main advantage, a significant one of course, of receptor models is that the parameters can be directly applied to experimentally observable physicochemical properties of the bacteria. It is not necessary, however, to include such detail in a population study. For the modelling and analyses here we are primarily interested in describing the behaviour of the populations of *E. coli* and *S. typhimurium* as a whole, so a macroscopically derived chemotaxis coefficient is most appropriate. Based on the above form, we choose (Woodward et al. 1995)

$$\chi(n, c) = \frac{k_1 n}{(k_2 + c)^2}. \tag{5.5}$$

The parameters k_1 and k_2 can be determined from the experimental results of Dahlquist et al. (1972) and give $k_2 = 5 \times 10^{-6}$ M and $k_1 = 3.9 \times 10^{-9}$ M cm^{2}s^{-1}.

Cell Proliferation

The proliferation term involves both growth and death of the bacteria. From Budrene and Berg (1995) cells grow at a constant rate that is affected by the availability of succinate. In the semi-solid experiments the stimulant, succinate or fumarate, is the main carbon source (nutrient) for the bacteria whereas in the liquid experiments, nutrient is provided in other forms and is not limiting. We thus assume a proliferation term of the form

$$\text{cell growth and death} = k_3 n \left(k_4 \frac{s^2}{k_9 + s^2} - n \right), \tag{5.6}$$

where the k's are parameters.

Intuitively this form is a reasonable one to take: it looks like logistic growth with a carrying capacity which depends on the availability of nutrient, s. When the bacterial density is below the carrying capacity, the expression (5.6) is positive and the population of bacteria increases. When n is larger than the carrying capacity, the expression is negative and there is a net decrease in population density. Implied in this form is the assumption that the death rate per cell is proportional to n; another possibility, but less plausible perhaps, is simply a constant death rate per cell.

Production and Consumption of Chemoattractant and Stimulant

The model contains one production term (production of chemoattractant), and two consumption terms (uptake of chemoattractant and of stimulant). Due to lack of available data, we have to rely on intuition to decide what are reasonable forms for the production and uptake of chemoattractant.

For the nutrient consumption, we expect that nutrient will disappear from the medium at a rate proportional to that at which cells are appearing. Since the linear birth rate of the cells is taken to be $k_3 k_4 s^2/(k_9 + s^2)$ this suggests the following form for the consumption of nutrient by the cells,

$$\text{nutrient consumption} = k_8 n \frac{s^2}{k_9 + s^2}, \tag{5.7}$$

where the k's are parameters. The consumption form has a sigmoid-like characteristic.

Chemoattractant consumption by the cells could have a similar sigmoidal character. However, the chemical is not necessary for growth and so there is likely very little created during the experiment. So, we simply assume that if a cell comes in contact with an aspartate molecule, it ingests it, which thus suggests a chemoattractant consumption, with parameter k_7, of the form

$$\text{chemoattractant consumption} = k_7 nc. \tag{5.8}$$

The chemoattractant production term has also not been measured in any great detail. We simply know (H.C. Berg, personal communication 1993) that the amount of chemoattractant produced increases with nutrient concentration and probably saturates over time which suggests a saturating function, of which there are many possible forms. To be specific we choose

$$\text{chemoattractant production} = k_5 s \frac{n^2}{k_6 + n^2}, \tag{5.9}$$

where k_5 and k_6 are other parameters. An alternative nonsaturating possibility which is also plausible is

$$\text{chemoattractant production} = k_5 s n^2. \tag{5.10}$$

In fact both these forms give rise to the required patterns, so further experimentation is needed to distinguish between them, or to come up with some other function. The critical characteristic is the behaviour when n is small since there the derivative of the production function must be positive.

Mathematical Model for Bacterial Pattern Formation in a Semi-Solid Medium

Let us now put these various functional forms into the model word equation system (5.1)–(5.3) which becomes:

$$\frac{\partial n}{\partial t} = D_n \nabla^2 n - \nabla \left[\frac{k_1 n}{(k_2 + c)^2} \nabla c \right] + k_3 n \left(\frac{k_4 s^2}{k_9 + s^2} - n \right) \tag{5.11}$$

$$\frac{\partial c}{\partial t} = D_c \nabla^2 c + k_5 s \frac{n^2}{k_6 + n^2} - k_7 nc \tag{5.12}$$

$$\frac{\partial s}{\partial t} = D_s \nabla^2 s - k_8 n \frac{s^2}{k_9 + s^2}, \tag{5.13}$$

where n, c and s are the cell density, the concentration of the chemoattractant and of the stimulant respectively. There are three diffusion coefficients, three initial values (n, c and s at $t = 0$) and nine parameters k in the model. We have estimates for some of these parameters while others can be estimated with reasonable confidence. There are several, however, which, with our present knowledge of the biology, we simply do not know. We discuss parameter estimates below.

Mathematical Model for Bacterial Pattern Formation in a Liquid Medium

It seems reasonable to assume that the production of chemoattractant and the chemotactic response of the cells are governed by the same functions in both the liquid and semi-solid experiments. The difference between the two groups of experiments lies more in the timescale and in the role of the stimulant. As mentioned earlier the cells do not have time to proliferate over the time course of the liquid experiments, so there is no growth term in this model. Also, the stimulant is not the main food source for the cells (it is externally supplied) so consumption of the stimulant is negligible. This shows that the liquid experiment model is simply a special case of the semi-solid experiment model. With cell growth, chemoattractant degradation and consumption of stimulant eliminated we are left with the simpler three-equation model

$$\frac{\partial n}{\partial t} = D_n \nabla^2 n - \nabla \left[\frac{k_1 n}{(k_2 + c)^2} \nabla c \right] \tag{5.14}$$

$$\frac{\partial c}{\partial t} = D_c \nabla^2 c + k_5 \, s \, \frac{n^2}{k_6 + n^2} \tag{5.15}$$

$$\frac{\partial s}{\partial t} = D_s \nabla^2 s \tag{5.16}$$

which has fewer unknown parameters than the semi-solid mode system. The last equation is uncoupled from the other equations. In the case of the simplest liquid experiment, in which the stimulant is uniformly distributed throughout the medium, the third equation can also be dropped.

Parameter Estimation

As mentioned, we have some of the parameter values and can derive estimates for others from the available literature; we also have estimates for some parameter combinations. The product $k_3 k_4$ is the maximum instantaneous growth rate, which is commonly determined from the generation time, t_{gen}, as

$$\text{instantaneous growth rate} = \frac{\ln 2}{t_{gen}}.$$

For the *E. coli* experiments, the generation time is of the order of 2 hours, giving an instantaneous growth rate of 0.35/hour. The grouping $k_3 k_4 / k_8$ is termed the yield coefficient, and is calculated by experimentalists as

$$Y = \frac{\text{weight of bacteria formed}}{\text{weight of substrate consumed}}.$$

Similarly, the grouping $k_3 k_4 / k_7$ is the yield coefficient for the bacteria as a function of chemoattractant (nitrogen source). The parameters k_1 and k_2 are calculated from measurements of cellular drift velocity and chemotaxis gradients made by Dahlquist et al. (1972).

Table 5.1. Dimensional parameter estimates obtained from the literature for use in the *E. coli* and *S. typhimurium* model equations (5.11)–(5.16). The other ks are unknown at this stage.

Parameter	Value	Source
k_1	3.9×10^{-9} M cm^{2}s^{-1}	Dahlquist et al. 1972
k_2	5×10^{-6} M	Dahlquist et al. 1972
k_3	1.62×10^{-9} hr ml^{-1}cell^{-1}	Budrene and Berg 1995
k_4	3.5×10^{8} cells ml^{-1}	Budrene and Berg 1995
k_9	4×10^{-6} M^2	Budrene and Berg 1995
D_n	$2 - 4 \times 10^{-6}$cm^{2}s^{-1}	Berg and Turner 1990; Berg 1983
D_c	8.9×10^{-6}cm^{2}s^{-1}	Berg 1983
D_s	$\approx 9 \times 10^{-6}$cm^{2}s^{-1}	Berg 1983
n_0	10^{8} cells ml^{-1}	Budrene and Berg 1991
s_0	$1 - 3 \times 10^{-3}$ M	Budrene and Berg 1995

Budrene and Berg (1995) measured growth rates for their experiments and this let us get reasonably precise determination of the parameters k_4, k_9 and k_3 by curve fitting. They also measured the ring radius as a function of time in the semi-solid experiments. The known parameter estimates are listed in Table 5.1 along with the sources used. We do not have estimates for the other parameters. However, since we shall analyse the equation sytems in their nondimensional form it will suffice to have estimates for certain groupings of the parameters. These are given below in the legends of the figures of the numerical solutions of the equations.

Intuitive Explanation of the Pattern Formation Mechanism

Before analysing any model of a biological problem it is always instructive to try and see intuitively what is going to happen in specific circumstances. Remember that the domain we are interested in is finite, the domain of the experimental petri dish. Consider the full model (5.11)–(5.13). In (5.13) the uptake term is a sink in a diffusion equation and so as time tends to infinity, the nutrient concentration, s, tends to zero. This in turn implies, from (5.11) and (5.12) that eventually cell growth and production of chemoattractant both tend to zero, while consumption of chemoattractant and death of cells continue. So, both the cell density and chemoattractant concentration also tend to zero as time tends to infinity. Thus the only steady state in this model is the one at which $(n, c, s) = (0, 0, 0)$ everywhere. But, of course, this is not the situation we are interested in. What it implies, though, is that it is not possible to carry out a typical linear analysis with perturbations about a uniform nonzero steady state. Instead we must look at the dynamic solutions of the equations.

Now consider the model system for the liquid experiments, equations (5.14)–(5.16). The last equation implies that eventually the stimulant will be spatially uniform since it is simply the classical diffusion equation which smooths out all spatial heterogeneities over time. By inspection there is a uniform steady state $(n, c, s) = (n_0, 0, 0)$ with n_0, the initial concentration of cells, being another parameter which can be varied experimentally. From (5.15) the source term is always positive so c will grow unboundedly.

In this case, eventually this concentration will be sufficiently high to significantly reduce the chemotaxis response in (5.14) and in the end simple diffusion is dominant and the solutions become time-independent and spatially homogeneous. So, again, with the liquid experiments, we have to look at the dynamic evolution of the solutions. Here a perturbation of any one of the steady states ($(n, c, s) = (n_0, 0, 0)$) results in a continually increasing concentration of chemoattractant. To get anything interesting from the model analyses therefore we have to look for patterns somewhere in that window of time between perturbation of the uniform initial conditions and saturation of the chemotactic response.

It is straightforward to see how the physical diffusion-chemotaxis system for the liquid model (5.14) to (5.16) could give rise to the appearance, and disappearance, of high density aggregates of cells. At $t = 0$ the cells begin secreting chemoattractant and since the cells are randomly distributed, some areas have a higher concentration of chemoattractant than others. Because of the chemotaxis these groups of higher cell concentration attract neighbouring cells, thereby increasing the local cell density, and decreasing it in the surrounding area. The new cells in the clump also produce chemoattractant, increasing the local concentration at a higher rate than it is being increased by the surrounding lower density cell population. In this way, peaks and troughs in cell density and chemoattractant concentration are accentuated. This is not the whole story since diffusion of the cells and the chemicals is also involved and this has a dispersive effect which tries to counter the aggregative chemotactic process and smooth out these peaks and troughs or rather prevent them happening in the first place. It is then the classical situation of local activation and lateral inhibition and which process dominates—aggregation or dispersion—depends on the intimate relation between the various parameters and initial conditions via n_0.

5.3 Liquid Phase Model: Intuitive Analysis of Pattern Formation

We saw in the last section in the discussion of the liquid experiments and their model system that patterns consisting of a random arrangement of spots will probably appear on a short timescale in the liquid medium experiments but eventually the aggregates fade and homogeneity again obtains. We suggested that this fading is probably due to saturation of the chemotactic response. Basically since cellular production of chemoattractant is not countered by any form of chemoattractant degradation (or inhibition), the amount of chemoattractant in the dish increases continuously. As a result, the chemotactic response eventually saturates, and diffusion takes over.

We also noted that the usual linear analysis about a uniform steady state is not possible so we have to develop a different analysis to study the pattern formation dynamics. The method (Tyson et al., 1999) we develop is very much intuitive rather than exact, but as we shall see it is nevertheless informative and qualitatively predictive and explains how transient patterns of randomly or circularly arranged spots can appear in a chemotaxis model and in experiment. We also give some numerical solutions to compare with the analytical predictions. For all of the analysis and simulations we assume zero flux boundary conditions which reflect the experimental situation.

We start with the simplest model for the liquid experiments which is just the semi-solid phase model with zero proliferation of cells, zero degradation of chemoattractant and uniform distribution of stimulant, s, which is neither consumed nor degraded and is thus just another parameter here. In these circumstances (5.14)–(5.16) become

$$\frac{\partial n}{\partial t} = D_n \nabla^2 n - k_1 \nabla \cdot \left[\frac{n}{(k_2 + c)^2} \nabla c \right] \tag{5.17}$$

$$\frac{\partial c}{\partial t} = D_c \nabla^2 c + k_5 s \frac{n^2}{k_6 + n^2}, \tag{5.18}$$

where n and c are respectively the density of cells and concentration of chemoattractant.

For simplicity the analysis we carry out is for the one-dimensional case where $\nabla^2 = \partial^2/\partial x^2$. Although we carry out the analysis for a one-dimensional domain the results can be extended with only minor changes to two dimensions (like what we did in Chapter 2 when investigating reaction diffusion pattern formation). We nondimensionalise the equations by setting

$$u = \frac{n}{n_0}, \quad v = \frac{c}{k_2}, \quad w = \frac{s}{s_0}, \quad t^* = \frac{k_5 s_0}{k_2} t, \quad x^* = \left(\frac{k_5 s_0}{D_c k_2} \right)^{1/2} x,$$
$$d = \frac{D_n}{D_c}, \quad \alpha = \frac{k_1}{D_c k_2}, \quad \mu = \frac{k_6}{n_0^2} \tag{5.19}$$

which gives, on dropping the asterisks for algebraic simplicity, the nondimensional equations

$$\frac{\partial u}{\partial t} = d \frac{\partial^2 u}{\partial x^2} - \alpha \frac{\partial}{\partial x} \left[\frac{u}{(1 + v)^2} \frac{\partial v}{\partial x} \right] \tag{5.20}$$

$$\frac{\partial v}{\partial t} = \frac{\partial^2 v}{\partial x^2} + w \frac{u^2}{\mu + u^2}. \tag{5.21}$$

The quantities n_0 and s_0 (essentially parameters that can be varied experimentally) are the average initial cell density and concentration of stimulant respectively. Since the stimulant s is neither consumed nor degraded $w = 1$. Also, in the liquid experiments, since there is neither growth nor death in the cell population we have the conservation equation

$$\int_0^l u(x, t)\, dx = u_0 l,$$

where u_0 is the average initial nondimensional cell density which is 1 if n_0 is the initial uniform density. The dimensionless parameter values, listed in Table 5.2, are calculated using the dimensional parameter values listed in Table 5.1. The parameter μ is unknown. The initial conditions of the experiment are uniform nonzero cell density and

Table 5.2. Known and estimated values for the variables and dimensionless parameters used in the study of the *E. coli* and *S. typhimurium* liquid medium model (5.20) and (5.21).

Variable	Initial Value
u_0	1.0
w_0	1.0
v_0	0.0

Parameter	Value
α	80–90
d	0.25–0.5
μ	unknown

zero concentration of chemoattractant. We want to find solutions of (5.20) and (5.21) which are heterogeneous in space and which initially grow and then decay with time.

The nontrivial (that is, $u_0 \neq 0$) spatially independent solution of (5.20) and (5.21) with initial conditions $u(x, 0) = 1$, $v(x, 0) = 0$ is

$$u(x, t) = 1 \tag{5.22}$$

If we suppose that the initial conditions for (5.20) and (5.21) are small, $O(\varepsilon)$, random perturbations about the initial cell density, we look for solutions in the form

$$u(x, t) = 1 + \varepsilon f(t) \sum_k e^{ikx}, \quad v(x, t) = \frac{1}{\mu + 1} t + \varepsilon g(t) \sum_k e^{ikx}, \tag{5.23}$$

where $0 < \varepsilon \ll 1$ and the k are the wavenumbers associated with the Fourier series of the random initial conditions. To approximate the actual experimental situation, where the initial concentration of chemoattractant is exactly zero, we set $g(0) = 0$. For illustration we choose $f(0) = 1$. We look for spatially varying solutions superimposed on the temporally growing solution.

Since we are looking for solutions on a finite domain with zero flux boundary conditions we have only sinusoidal (cosine) solutions involving only integer modes, m (cf. Chapter 2, Section 2.4), which are related to the wavenumbers, k, by

$$k^2 = \frac{m^2\pi^2}{l^2}, \tag{5.24}$$

where l is the dimensionless length of the domain. Substituting (5.23) into (5.20) and (5.21) and linearising (in ε) in the usual way we get, for each k, the $O(\varepsilon)$ equations

$$\frac{dF(\tau)}{d\tau} = -dk^2 F(\tau) + \alpha(\mu + 1)^2 \frac{k^2}{\tau^2} G(\tau) \tag{5.25}$$

$$\frac{dG(\tau)}{d\tau} = -k^2 G(\tau) + \frac{2\mu}{(\mu + 1)^2} F(\tau), \tag{5.26}$$

where $\tau = \mu + 1 + t$ (note that $\tau_0 = \tau_{t=0} = \mu + 1 > 0$), $F(\tau) \equiv f(t)$ and $G(\tau) \equiv g(t)$. The coefficient of the second term on the right-hand side of (5.25) is the only one which depends on the chemotaxis parameter α.

The analytical problem now is how to determine the solution behaviour of $F(\tau)$ and $G(\tau)$. It is clear from (5.25) that as $\tau \to \infty$ the coefficient of $G(\tau)$ tends to zero and the solution for $F(\tau)$ reduces to a decaying exponential. Once this happens, the solution of (5.26) also gives a decaying exponential. So, with the solution forms (5.23) the mechanism accounts for ultimate pattern disappearance with time. Let us now consider the growth of spatial pattern from the initial disturbance.

For τ near τ_0 (that is, t small) we can get more insight by combining (5.25) and (5.26) into a single second-order differential equation for the amplitude, $F(\tau)$, of the cell density pattern, to get

$$\frac{d^2F}{d\tau^2} + \left[k^2(d+1) + \frac{2}{\tau}\right]\frac{dF}{d\tau} + k^2\left(dk^2 + \frac{2d}{\tau} - \frac{2\alpha\mu}{\tau^2}\right)F = 0. \tag{5.27}$$

This has an exact solution in terms of confluent hypergeometric functions but it is essentially of zero practical use from the point of view of seeing what the solution behaviour is, which, after all, is what we want. Instead we use heuristic and qualitative reasoning. Without it, it would also be difficult to see what was actually going on if we simply solved the system numerically in the first instance. To begin we assume that the coefficients of the second-order ordinary differential equation (5.27) change much less rapidly than the function itself and its derivatives. This lets us compare (5.27) to a second-order equation with constant coefficients over small intervals of τ. Denote the coefficients of (5.27) by $D(\tau)$ and $N(\tau)$ and it becomes

$$\frac{d^2F}{d\tau^2} + D(\tau)\frac{dF}{d\tau} + N(\tau)F = 0, \tag{5.28}$$

where

$$N(\tau) = k^2\left(dk^2 + \frac{2d}{\tau} - \frac{2\alpha\mu}{\tau^2}\right), \quad D(\tau) = k^2(d+1) + \frac{2}{\tau}. \tag{5.29}$$

As noted, the last term in $N(\tau)$ is the only one in which the dimensionless parameter α (the grouping with the chemotaxis parameters) appears. The parameter μ appears explicitly only in that term as well but it is also contained in the expression for τ and so its effect is not so easily isolated. Note that $D(\tau)$ is positive for all $\tau > 0$, while $N(\tau)$ can be positive, negative or zero for τ near $\tau_0 = 1 + \mu$ (that is, where the dimensional time $t = 0$). For τ sufficiently large, $N(\tau) > 0$. Let us consider $N(\tau)$ and $D(\tau)$ to be constant for the moment, in which case the solution of (5.28), denoted by $\tilde{F}$, is formally

$$\tilde{F}(\tau) = L_1 e^{\lambda_+\tau} + L_2 e^{\lambda_-\tau}, \quad \lambda_\pm = \frac{1}{2}\left[-D(\tau) \pm \sqrt{D(\tau)^2 - 4N(\tau)}\right], \tag{5.30}$$

where the L's are constants of integration. Over a small interval of τ we can think of $N(\tau)$ and $D(\tau)$ to be approximately constant. Referring to the solution $\tilde{F}$, since $D(\tau) > 0\,\forall\,\tau$, we have $\mathrm{Re}\,(\lambda_-) < 0\,\forall\,\tau$. The sign of $\mathrm{Re}\,(\lambda_+)$, however, can vary depending on the sign of $N(\tau)$.

At this point, we are mainly interested in seeing how the chemotaxis coefficient α and the wavenumber k (or mode m) alter the solutions. Consider the effect of increasing α. If α is sufficiently large then $N(\tau)$ will be negative for small values of τ, including τ_0. As τ increases, $N(\tau)$ will increase through zero and become positive. The effect on λ_+ is to make the real part of the eigenvalue positive for small enough τ and negative for large τ. The point $\tau = \tau_{\text{crit}}$ at which λ_+ passes through zero is the same point at which $N(\tau)$ becomes zero. So, for small τ, one component of $\tilde{F}$ is a growing exponential, while for larger τ both exponentials are decaying. We predict therefore, that α has a destabilising influence; that is, the growth of pattern becomes more likely as α increases because it makes $N(\tau)$ more negative for $\tau < \tau_{\text{crit}}$. We could have predicted the destabilising effects of α, of course, from (5.20); it is the quantification of its destabilising influence that requires the analysis here.

Recall that the mode $m^2 = k^2 l^2/\pi^2$. For sufficiently large k^2, $N(\tau)$ in (5.29) becomes positive for all τ, resulting in solutions which are strictly decaying. This leads us to predict that the lowest frequency modes are the most unstable, and we would not expect to see modes of frequency larger than

$$K^2 = \frac{2}{d(1+\mu)}\left(\frac{\alpha\mu}{1+\mu} - d\right) \tag{5.31}$$

which is obtained by solving $N(\tau_0) = 0$. As time increases fewer and fewer modes remain unstable, and, as $\tau \to \infty$ the only unstable modes are those in a diminishing neighbourhood of 0. We can determine the fastest growing wavenumber, K_{grow} say, at any time by simply setting $\lambda(k^2) = 0 \equiv N(\tau) = 0$ and solving for k^2. This gives

$$K_{\text{grow}}^2 = \frac{2}{\tau}\left(\frac{\alpha\mu}{d\tau} - 1\right).$$

If the approximation of constant coefficients in (5.28) is reasonably valid over small but finite intervals of τ, then a series of solutions $\tilde{F}$ computed in sequential intervals $\Delta\tau$ could give rise to a solution which increases to a maximum and then decreases for all τ afterwards. The increasing phase would occur while λ_+ is positive. When we computed a numerical solution of the equation it confirmed the expected behaviour; Figure 5.4 is one such simulation for the parameter values given there.

The true location $\tau = \tau_{\text{crit}}$ of the maximum value of $F(\tau)$, $F_{\max}$, may be close to $\tilde{\tau}_{\text{crit}}$, given analytically by

$$N(\tilde{\tau}_{\text{crit}}) = 0 \Leftrightarrow \tilde{\tau}_{\text{crit}} = \frac{1}{k^2}\left[-1 + \sqrt{1 + \frac{2\alpha\mu k^2}{D}}\right]. \tag{5.32}$$

Tyson et al. (1999) compared it with τ_{crit} obtained numerically: the comparison is very close and gets even better as the mode k^2 (proportional to m^2) and parameter α increase.

The difference between $\tilde{\tau}_{\text{crit}}$ and τ_{crit} gives an indication of the size of $d^2F/d\tau^2$ at τ_{crit}. By definition, τ_{crit} is the time at which $dF/d\tau = 0$ and so (5.28) reduces to

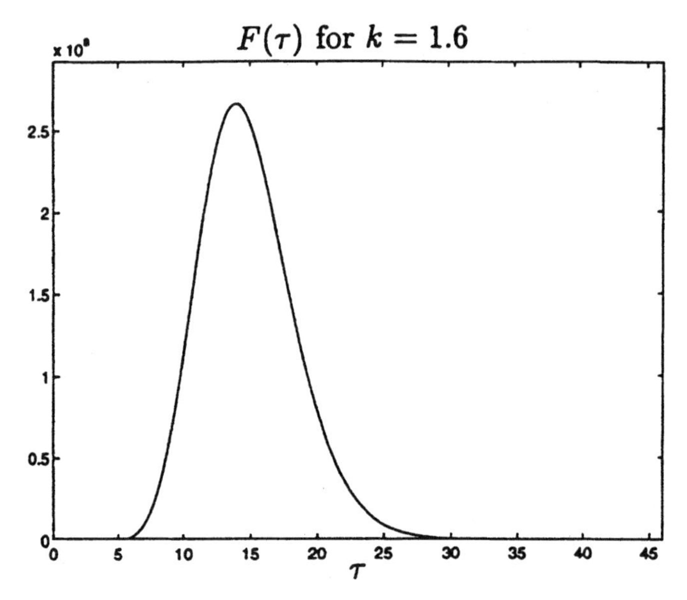

Figure 5.4. The amplitude F(τ) for the wavenumber $k = 1.6$ perturbation of the initial uniform cell density. From (5.24) and the parameters chosen this is equivalent to the mode $m = 5$. The other parameter values are: $d = 0.33$, $\alpha = 80$, $\mu = 1$, $u_0 = 1$, $w = 1$ and $l = 10$. (From Tyson 1996)

$$\left.\frac{d^2F}{d\tau^2}\right|_{\tau_{\text{crit}}} = -N(\tau_{\text{crit}})F.$$

Since the second derivative of a function is negative at a maximum we know that $N(\tau_{\text{crit}})$ is positive. Thus τ has already increased past the point where $N(\tau)$ changes sign, and $\tilde{\tau}_{\text{crit}}$ gives a minimum estimate for τ_{crit}. Since $\tilde{\tau}_{\text{crit}}$ and τ_{crit} are reasonably close, this suggests that $N(\tau_{\text{crit}})$ may be close to zero. In turn, this indicates that $d^2F/d\tau^2$ may be numerically small at the maximum, $F_{\max}$.

We are thus encouraged to solve (5.28) with the second derivative term omitted. After some straightforward algebra we get the solution of the resulting first-order ordinary differential equation as

$$F_1(\tau) = \left[\frac{(d+1)k^2\tau_0 + 2}{(d+1)k^2\tau + 2}\right]^{\alpha\mu k^2 - (2d(d-1)/(d+1)^2)} \left[\frac{\tau}{\tau_0}\right]^{\alpha\mu k^2} e^{[d/(d+1)]k^2(\tau_0-\tau)}, \tag{5.33}$$

which satisfies $F(\tau_0) = 1(= f(0))$. Plots of the first- and second-order equation solutions $F(\tau)$ and $F_1(\tau)$ are shown in Figures 5.5 and 5.6.

At first glance we notice the marked difference in the height of the two functions. Apart from this difference however, the two functions have many similarities. The peaks occur at approximately the same value of τ, and the peak interval, defined as the time

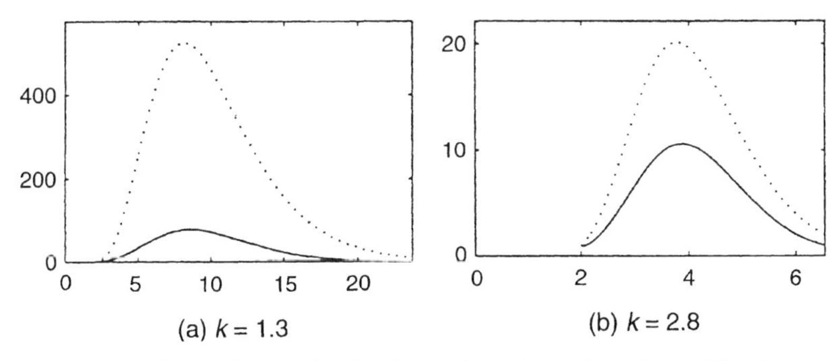

Figure 5.5. $F(\tau)$ (solid line) and $F_1(\tau)$ (dotted line) plotted together against τ for $\alpha = 30$ and wavenumbers: (**a**) $k = 1.3$, which corresponds to the mode $m = 4$, (**b**) $k = 2.8$, which corresponds to the mode $m = 9$. The parameter values are the same as in Figure 5.4. (From Tyson 1996)

during which $F(\tau) > F(\tau_0)$, is about the same, especially for the lower frequencies, and the two curves appear to be similarly skewed to the left. Increasing α results in a large increase in both $F_{\max}$ and $F_{1_{\max}}$. The two solutions are also similar with respect to the behaviour of the different modes investigated. The larger the value of k^2, the earlier τ_{crit} is reached, and the shorter the interval over which F or F_1 is larger than the initial value F_0.

If we normalise the data for $F(\tau)$ and $F_1(\tau)$ so that they lie in the interval [0, 1] we see in Figure 5.6 that the two solutions map almost directly on top of each other. So, the main difference between the approximate and numerical solutions is simply a scaling factor. This scaling factor is large, which suggests that the second-order derivative term is not small outside the neighbourhood of the maximum.

At this stage we have an intuitive understanding of the behaviour of the solutions $F(\tau)$ of (5.28). We also have an approximate analytic solution, $F_1(\tau)$ in (5.33), which we can use to predict the effect of changing various parameters.

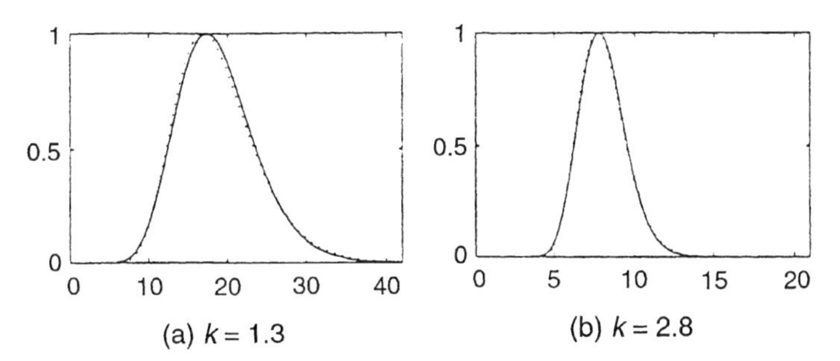

Figure 5.6. The solutions $F(\tau)$ (solid line) and $F_1(\tau)$ (dotted line) plotted against τ and normalised to lie between 0 and 1. (**a**) $k = 1.3$, which corresponds to the mode $m = 4$, (**b**) $k = 2.8$, which corresponds to the mode $m = 9$. The parameter values are the same as in Figure 5.4. (From Tyson 1996)

5.4 Interpretation of the Analytical Results and Numerical Solutions

We are particularly interested in the model's predictions as regards the number of aggregates which will form, and how long they will be visible, that is, when $F(\tau)$ is sufficiently large. If the nonlinear effects are not too strong, we should expect that the number of aggregates will be determined by the combined effect of the solutions corresponding to the various modes.

Some numerical results are shown in Figure 5.7. Note that there is one wavenumber (mode) which reaches a higher amplitude than any other. We refer to this wavenumber as k_{max}; here $k_{max} = 2.20$. Also note that every wavenumber k larger than k_{max}, initially has a slightly higher growth rate than k_{max}. These high frequencies quickly begin decaying, however, while the amplitude of the k_{max} pattern is still growing rapidly. We surmise that the k_{max} mode could be the first one to dominate the solution of the full nonlinear system of partial differential equations.

Once the k_{max} solution begins to decay, solutions corresponding to small wavenumbers become largest in decreasing order. The amplitude of each solution with $k < k_{max}$ is always in the process of decaying, once it supersedes the next highest mode. Thus we should see a continuous decrease in wavenumber of the observed pattern as time t increases, accompanied by a decrease in amplitude. This corresponds to the biologically observed coalescing of aggregates and eventual dissipation of pattern. Note that for these figures the wavenumber with the maximum growth is $K_{grow} = 5.47$ and so

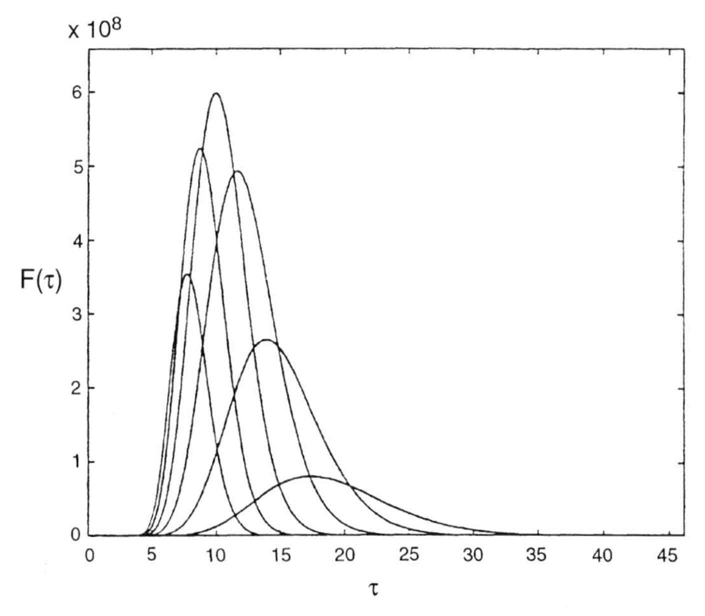

Figure 5.7. $F(\tau)$ for discrete values of $k = m\pi/l$, $m = 4$ to 9; that is, $k = 1.26$ to 2.83. The $m = 9$ curve decays the fastest, then the $m = 8$ curve, and so on. The curve corresponding to $m = 7$ has the highest peak. The parameter values are: $d = 0.33$, $\alpha = 80$, $\mu = 1$, $u_0 = 1$, $w = 1$ and $l = 10$. From (5.31) the maximum wavenumber $K = 10.9$. (From Tyson 1996)

k_{max} is less than K_{grow} by a factor of 2. For all of the numerical solutions observed by Tyson (1996) and Tyson et al. (1999) in this study k_{max} was consistently much less than K_{grow}.

One-Dimensional Numerical Simulation Results

We can now compare the predictions of the linear theory with the actual solution behaviour of the partial differential equations. Zero flux boundary conditions were used in all the simulations. The initial condition in chemoattractant concentration is zero everywhere on the domain, and in cell density it is a random perturbation about $u_0 = 1$. Among other numerical checks all of the solutions were checked against the integral form (5.35) of the conservation of bacteria since there is neither growth nor death in the liquid model.

A representative time sequence for $\alpha = 80$ is shown in Figure 5.8. The sequences were truncated at the time beyond which little change was observed in the number of peaks in cell density, and the pattern amplitude simply decreased. The plots in the left-hand column of each figure are the cell density profiles at various times τ, while the plots in the right-hand column are the corresponding power spectral densities. The density axes for the latter plots are restricted to lie above the mean value of the initial power spectral density, at $\tau = \tau_0$. This highlights the pattern modes which grow.

As predicted by (5.32), the power spectral density plots indicate that spatial patterns of modes higher than $K = 10.9$ do not grow. Also, the spread of 'nonzero' modes decreases as time increases. In the actual cell density distribution, the pattern observed initally has many peaks, and the number of these decreases over time. Our prediction that k_{max} is the spatial pattern mode which will dominate the solution is off by a factor of two in these figures. The mode which actually dominates the solution is $k_{max} \approx 1.1$ while the predicted value is 2.2.

Two-Dimensional Numerical Simulation Results

In two dimensions we obtain the same sort of behaviour we found in the one-dimensional case. Again we started with an initial condition consisting of small random perturbations about a uniform distribution of cells and patterns consisting of a random arrangement of spots were generated as shown in Figure 5.8. The surface plot in Figure 5.9 clearly shows the comparative densities between the aggregates and the regions between them. The number of spots is large at first and then decreases over time as neighbouring aggregates coalesce. Eventually, all of the spots disappear.

Recall that this is exactly what is observed in the bacterial experiments. To begin with, bacteria are added to a petri dish containing a uniform concentration of succinate. The mixture is well stirred, and then allowed to rest. At this point, the state of the solution in the petri dish is mimicked by the initial condition for our model, namely, small perturbations of a uniform distribution of cells and succinate. After a short time, of the order of 20 minutes, the live bacteria aggregate into numerous small clumps which are very distinct from one another. This behaviour corresponds to the random arrangement of spots separated by regions of near zero cell density observed in the model solutions. Experimentally, the bacterial aggregates are seen to join together, forming fewer and larger clumps. This is also the situation in the mathematical model, and is particularly

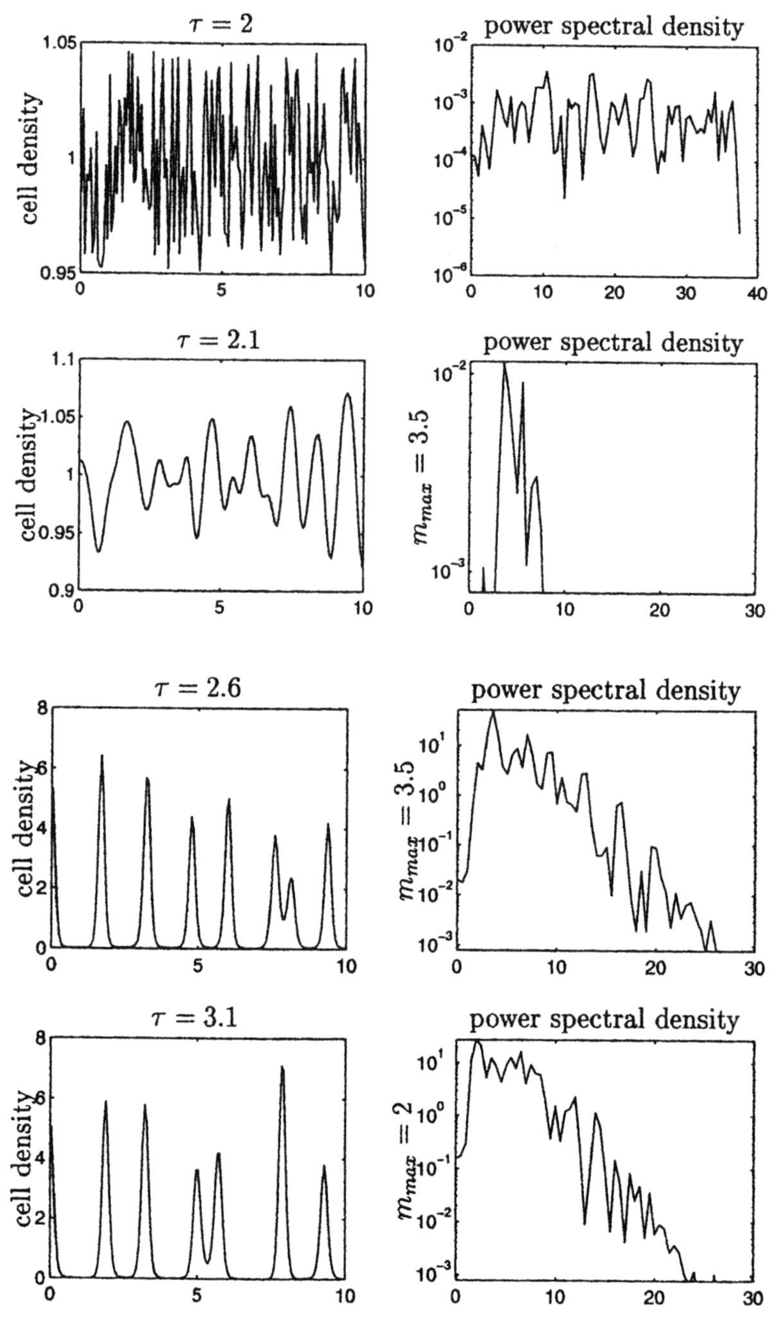

Figure 5.8. Numerical solution in one dimension of the liquid model system of equations (5.20) and (5.21). Left-hand plots show the bacterial cell density plotted against space at various times, τ. Right-hand plots show the corresponding power spectral density functions. Initially the cells are uniformly distributed over the one-dimensional domain and disturbed with a small perturbation of $O(0^{-1})$. Parameter values are the same as for Figure 5.7: $d = 0.33$, $\mu = 1$, $u_0 = 1$, $w_0 = 1$, $l = 10$ and $\alpha = 80$. (From Tyson 1996)

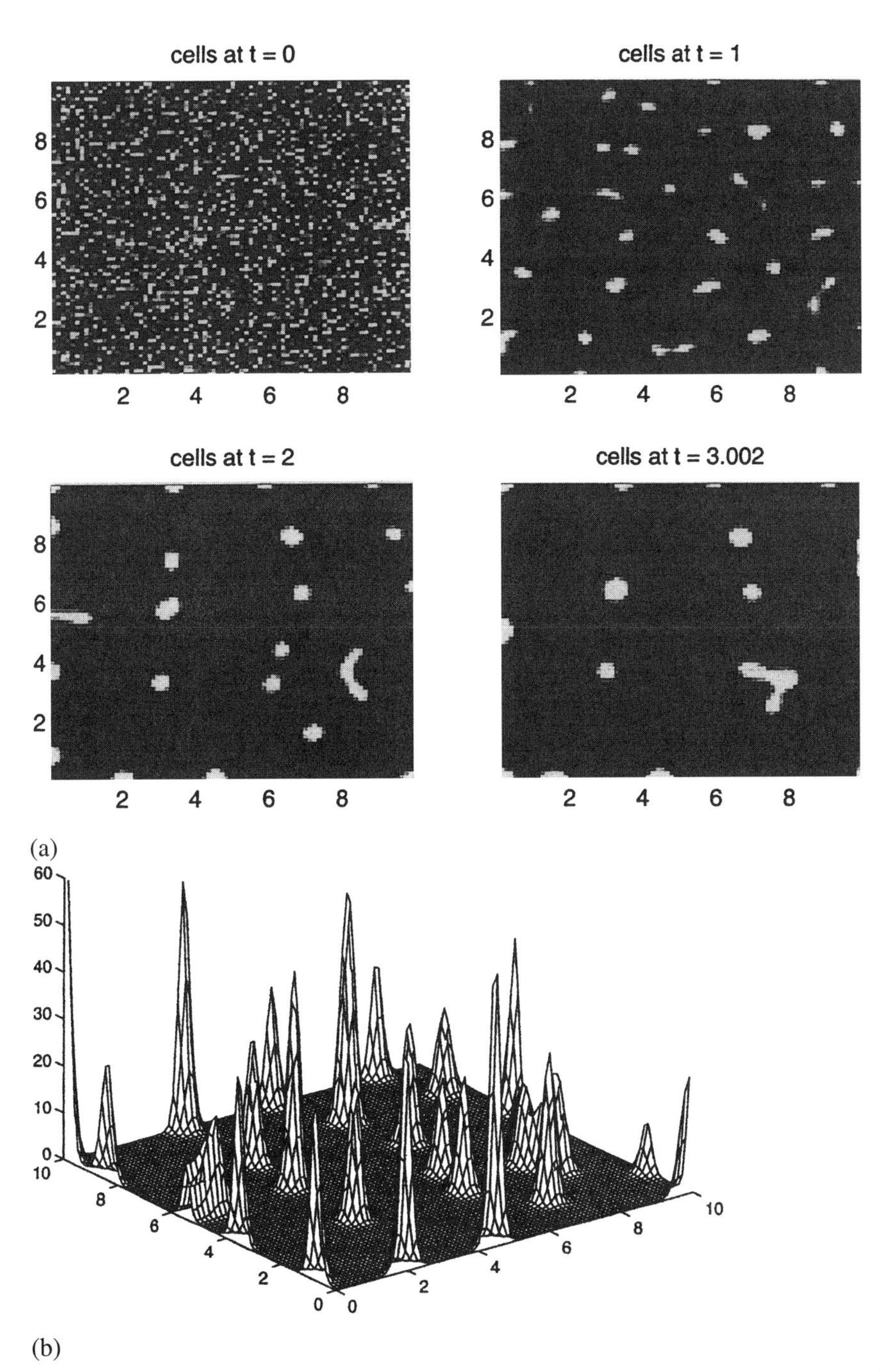

Figure 5.9. (**a**) Time evolution of two-dimensional cell density patterns arising from a uniform distribution of stimulant on a square domain. White corresponds to high cell density and black to low cell density. $\tau = 0$, the initial conditions, $\tau = 1$, $\tau = 2$, $\tau = 3.002$. (**b**) Surface plot of the solution for $\tau = 2$ showing the high density of the aggregates and the low density between them. Parameter values: $d_u = 0.33$, $\alpha = 80$, $\mu = 1$, $u_0 = 1$, $w = 1$ and $l = 10$. (From Tyson 1996)

clearly seen when the solutions are displayed as a movie with the frames separated by small time increments. In both the model and experiment, the spots eventually disappear and cannot be induced to re-form. This is again explained by the mathematical model as a saturation of the chemotactic response, which no longer has any effect as the production of chemoattractant increases continually.

Nonuniform Distribution of Stimulant

Up to now we have considered the cell density patterns which emerge in response to a spatially uniform distribution of the stimulant w; the above analysis qualitatively captures the experimentally observed behaviour in this case. Other patterns however, have been observed in experiment when the stimulant is added to the medium as a localised drop as we mentioned. If the model is essentially correct, it should also reproduce these patterns. This we briefly examine here.

With a nonuniform distribution of stimulant, we have to include diffusion of stimulant in the model. So, the model, given by equations (5.20) and (5.21), has to be expanded to include the equation for stimulant, namely, the system (5.14)–(5.16) in dimensionless form. This more general model is therefore

$$\frac{\partial u}{\partial t} = d\nabla^2 u - \alpha\nabla \cdot \left[\frac{u}{(1+v)^2}\nabla v\right] \tag{5.34}$$

$$\frac{\partial v}{\partial t} = \nabla^2 v + w\frac{u^2}{\mu + u^2} \tag{5.35}$$

$$\frac{\partial w}{\partial t} = d_s\nabla^2 w, \tag{5.36}$$

where the dimensionless $d_s = D_s/D_c$ (cf. (5.19)). With this formulation, we now require that the average value of $u(\mathbf{x}, t)$ and $w(\mathbf{x}, t)$ at all times be equal to 1.

Numerical results from a simulation of this model are shown in Figure 5.10. As observed experimentally in Figure 5.1, a ring of high cell density develops around the point where the stimulant was added. Some aggregates form inside the ring (also observed experimentally in Figure 5.1), but these are nowhere near as dense as the ring. The reason for this is that the ring recruits cells from outside its circumference, and so the number of cells available to it is much larger than that available to the aggregates in the centre.

Over time, simulations indicate that the radius of the ring decreases. Eventually the radius becomes so small that the ring is essentially one spot. It would be interesting to see if this behaviour is observed experimentally.

Before discussing the modelling of the more complex patterns in the semi-solid experiments it is perhaps helpful to recap what we have done in the above sections. We showed how a relatively simple but intuitively revealing analysis explains how evolving patterns of randomly or circularly arranged spots appear transiently in the chemotaxis model for the experimental arrangement. The central idea is to consider the rate of growth of individual modes over small time intervals, and extrapolate from this to the combined behaviour of all disturbance frequencies. Low mode number perturbations to

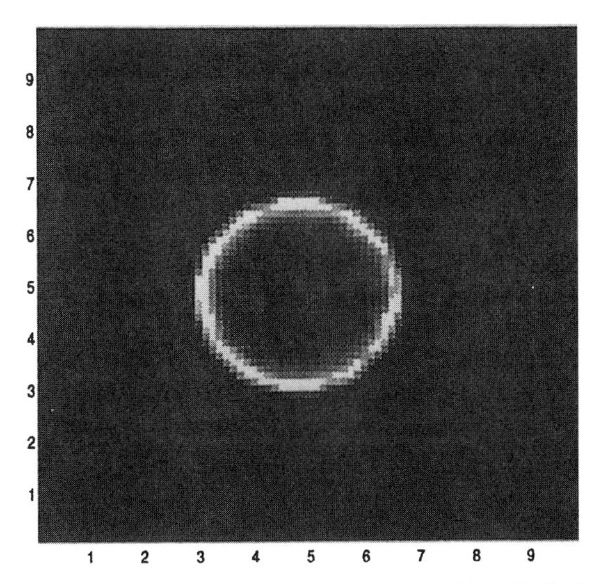

Figure 5.10. Two-dimensional cell density pattern arising from a nonuniform distribution of stimulant on a square domain. The stimulant was added as a single-humped function at the centre of the domain. Parameter values are the same as for Figure 5.9. (From Tyson 1996)

the uniform solution are unstable and grow in magnitude, but eventually these stabilize and decay with the larger mode numbers stabilizing first. This not only agrees qualitatively but also to a considerable extent quantitatively, with what is observed experimentally and numerically: clumps form, coalesce into larger aggregates and eventually disappear.

5.5 Semi-Solid Phase Model Mechanism for *S. typhimurium*

As we discussed in Section 5.1 in the semi-solid experiments, Budrene and Berg (1991) observed two very different pattern forming mechanisms. With *S. typhimurium* a thin bacterial lawn spreads out from the inoculum, and rings of more concentrated bacteria form well behind the lawn edge. Each ring may eventually break up into spots, but usually not until several more rings have formed at larger radii as shown in Figure 5.2. The second pattern forming mechanism is exhibited by *E. coli*, and involves first the formation of an expanding ring of high bacterial density, referred to as a swarm ring. As this swarm ring expands, it leaves behind smaller aggregates of bacteria, which form the striking patterns shown in Figure 5.3.

The bacterial lawn well ahead of the *S. typhimurium* pattern suggests that a spatially and temporally uniform steady state is first established, and then the pattern forms on it. As we showed, the model for the semi-solid experiments has no nonzero steady state. If consumption of nutrient is sufficiently slow, however, we can neglect it and the model reduces to two equations which admit the necessary steady state. We suppose then, that in the first pattern formation mechanism, consumption of nutrient is negligible.

This assumption is further supported by the fact that, experimentally, food is present in quantities well above the saturation level for the cells.

The second patterning mechanism appears to involve a vigorous nutrient consumption rate (Budrene and Berg 1995). There is little left behind the expanding swarm ring. In this case, all three equations of the semi-solid model are important.

We start by nondimensionalising the model system of equations (5.11)–(5.13) for the semi-solid experiments by setting

$$u = \frac{n}{n_0}, \quad v = \frac{c}{k_2}, \quad w = \frac{s}{\sqrt{k_9}}, \quad t^* = k_7 n_0 t, \quad \nabla^{*2} = \frac{D_c}{k_7 n_0}\nabla^2,$$
$$d_u = \frac{D_n}{D_c}, \quad d_w = \frac{D_s}{D_c}, \quad \alpha = \frac{k_1}{D_c k_2}, \quad \rho = \frac{k_3}{k_7}, \quad \delta = \frac{k_4}{n_0},$$
$$\beta = k_5 \frac{\sqrt{k_9}}{k_7 k_2 n_0}, \quad \kappa = \frac{k_8}{k_7\sqrt{k_9}}, \quad \mu = \frac{k_6}{n_0^2} \tag{5.37}$$

and we obtain the dimensionless model, (with parameter estimates given in Table 5.3) where for algebraic convenience we have omitted the asterisks,

$$\frac{\partial u}{\partial t} = d_u \nabla^2 u - \alpha \nabla \cdot \left[\frac{u}{(1+v)^2}\nabla v\right] + \rho u \left(\delta \frac{w^2}{1+w^2} - u\right) \tag{5.38}$$

$$\frac{\partial v}{\partial t} = \nabla^2 v + \beta w \frac{u^2}{\mu + u^2} - uv \tag{5.39}$$

$$\frac{\partial w}{\partial t} = d_w \nabla^2 w - \kappa u \frac{w^2}{1+w^2}. \tag{5.40}$$

Recall that in the *S. typhimurium* experiments there are two distinct steps in the pattern forming process in the first of which there is a thin, disk-shaped bacterial lawn, suggesting that a spatially and temporally uniform steady state is temporarily present. During the second step a high density cluster of bacteria forms in the shape of a ring which appears well behind the leading edge of the bacterial lawn, suggesting that the ring pattern forms on top of the intermediate steady state. Since pattern is still forming long after the lawn has been established, consumption of nutrient in the lawn must be

Table 5.3. Known dimensionless parameter values calculated from the dimensional parameter values listed in Table 5.1.

Parameter	Value
d_u	0.2–0.5
d_w	0.8–1.0
α	87
δ	3.5

negligible. So, we can approximate the full system by neglecting the dynamics of food consumption. Also, since the concentration of nutrient is large, but the concentration of cells is small, we can study the simplified two-equation model

$$\frac{\partial u}{\partial t} = d_u \nabla^2 u - \alpha \nabla \cdot \left[\frac{u}{(1+v)^2} \nabla v \right] + \rho u \left(\delta \frac{w}{1+w} - u \right) \tag{5.41}$$

$$\frac{\partial v}{\partial t} = \nabla^2 v + \beta w \frac{u^2}{\mu + u^2} - uv. \tag{5.42}$$

The analysis of these equations is much easier than the analysis in Section 5.3 in that there is a homogeneous steady state in which both u and v are nonzero. So, we can use the usual linear analysis to determine whether or not this steady state is unstable and whether or not spatial patterns are likely to form. We also carried out a thorough nonlinear analysis some of whose results we discuss below since they are highly relevant to the specific patterns that are formed.

5.6 Linear Analysis of the Basic Semi-Solid Model

The linear analysis is the same as we discussed at length in Chapter 2 and is now straightforward. We linearize (5.41) and (5.42) about the nonzero steady state (u^*, v^*) given by

$$(u^*, v^*) = \left(\delta \frac{w}{1+w}, \beta w \frac{u^*}{\mu + u^{*2}} \right). \tag{5.43}$$

It is algebraically simpler in what follows to use general forms for the terms in the model equations (5.41) and (5.42): w is in effect another parameter here. We thus consider

$$\frac{\partial u}{\partial t} = d_u \nabla^2 u - \alpha \nabla \cdot [u \chi(v) \nabla v] + f(u, v) \tag{5.44}$$

$$\frac{\partial v}{\partial t} = d_v \nabla^2 v + g(u, v), \tag{5.45}$$

which on comparison with (5.41) and (5.42) define

$$\chi(v) = \frac{1}{(1+v)^2}, \quad f(u, v) = \rho u \left(\delta \frac{w}{1+w} - u \right),$$
$$g(u, v) = \beta w \frac{u^2}{\mu + u^2} - uv. \tag{5.46}$$

We now linearise the system about the steady state in the usual way by setting

$$u = u^* + \varepsilon u_1, \quad v = v^* + \varepsilon v_1, \tag{5.47}$$

where $0 < \varepsilon \ll 1$. Substituting these into (5.44) and (5.45) we get the linearized equations

$$\frac{\partial u_1}{\partial t} = d_u \nabla^2 u_1 - \alpha u^* \chi^* \nabla^2 v_1 + f_u^* u_1 + f_v^* v_1 \tag{5.48}$$

$$\frac{\partial v_1}{\partial t} = d_v \nabla^2 v_1 + g_u^* u_1 + g_v^* v_1 \tag{5.49}$$

since $f^* = 0$ and $g^* = 0$. Here the superscript * denotes evaluation at the steady state. We write the linear system in the vector form

$$\frac{\partial}{\partial t}\begin{pmatrix} u_1 \\ v_1 \end{pmatrix} = (A + D\nabla^2)\begin{pmatrix} u_1 \\ v_1 \end{pmatrix}, \tag{5.50}$$

where the matrices A and D are defined by

$$A = \begin{bmatrix} f_u^* & f_v^* \\ g_u^* & g_v^* \end{bmatrix}, \quad D = \begin{bmatrix} d_u & -\alpha u^* \chi^* \\ 0 & d_v \end{bmatrix}. \tag{5.51}$$

We now look for solutions in the usual way by setting

$$\begin{pmatrix} u_1 \\ v_1 \end{pmatrix} = \begin{pmatrix} c_1 \\ c_2 \end{pmatrix} e^{\lambda t + i\mathbf{k}x}, \tag{5.52}$$

where $\mathbf{k}$ is the wavevector, the c's constants and the dispersion relation $\lambda(\mathbf{k})$, giving the growth rate, is to be determined. Substituting this into the matrix equation gives

$$\left[\lambda I + D|\,\mathbf{k}\,|^2 - A\right]\begin{bmatrix} u_1 \\ v_1 \end{bmatrix} = \begin{bmatrix} 0 \\ 0 \end{bmatrix},$$

which has nontrivial solutions if and only if the determinant of the coefficient matrix, $|\,\lambda I + D|\,\mathbf{k}|^2 - A\,| = 0$ (recall Chapter 2). So, the dispersion relation $\lambda(k^2)$, $k^2 = |\,\mathbf{k}\,|^2$ is given by the characteristic equation

$$\lambda^2 + \lambda\left[(d_u + d_v)k^2 - (f_u^* + g_v^*)\right] + \left[d_u d_v k^4 - (d_u g_v^* + d_v f_u^* + \alpha u^* \chi^* g_u^*)k^2 + f_u^* g_v^* - f_v^* g_u^*\right] = 0$$

with solutions denoted by λ^+ and λ^-.

We are interested in determining pattern modes which have at least one positive growth rate; that is, at least one solution has $Rl\lambda > 0$ so we focus on the larger of the two solutions, λ^+, given by

$$\lambda^+ = \tfrac{1}{2}\left[-b(k^2) + \sqrt{[b(k^2)]^2 - 4c(k^2)}\right], \tag{5.53}$$

where

$$b(k^2) = (d_u + d_v)k^2 - (f_u^* + g_v^*)$$
$$c(k^2) = d_u d_v k^4 - (d_u g_v^* + d_v f_u^* + \alpha u^* \chi^* g_u^*)k^2 + f_u^* g_v^* - f_v^* g_u^*. \quad (5.54)$$

Recalling the discussion in Chapter 2, if $Rl\lambda^+$ is positive (negative) then perturbations about the steady state will grow (decay). We look for solutions which are stable ($\lambda^+ < 0$) to purely temporal perturbations (that is, $k^2 = 0$), but unstable ($\lambda^+ > 0$) to at least one spatial mode, which is just like the pure diffusionally driven instability situation for some nonzero k_1 and k_2. Mathematically, we look for a range of parameters such that

$$\lambda^+(0) < 0, \quad \lambda^+(k^2) > 0 \text{ for all } k \text{ such that } 0 \le k_1^2 < k^2 < k_2^2. \quad (5.55)$$

To satisfy the first of (5.55) we must have

$$b(0) = f_u^* + g_v^* < 0, \quad c(0) = f_u^* g_v^* - f_v^* g_u^* > 0. \quad (5.56)$$

Note that these conditions imply that λ^- is always negative so we need only focus on λ^+. Graphically, these conditions yield an inverted parabolic curve for $\lambda(k^2)$, which has its maximum to the right of $k^2 = 0$ and so is the most basic dispersion relation which gives diffusion-chemotaxis-driven instability discussed in detail in Chapter 2. Typical dispersion relations are shown in Figure 5.11, where by way of example we have shown how they vary with the parameter μ: for μ small enough there is no range k_1^2, k_2^2.

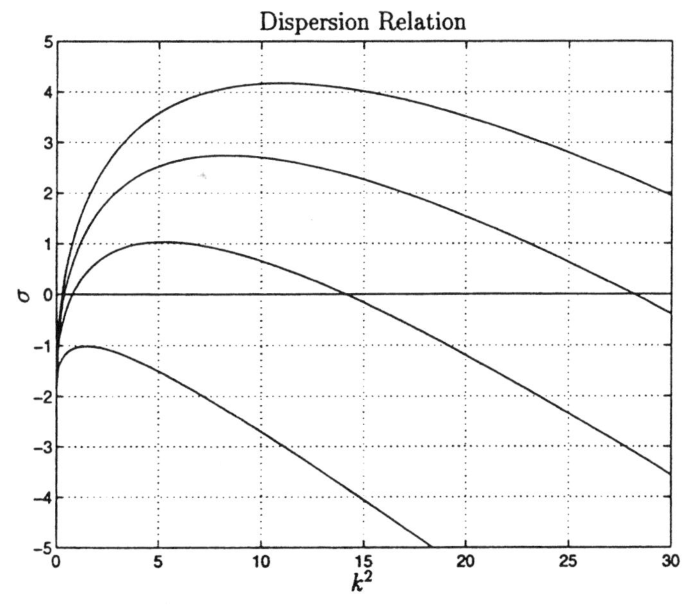

Figure 5.11. The dispersion relation, $\lambda(k^2)$ for parameter values $d_v = 1.0$, $d_u = 0.3$, $\alpha = 80$, $\beta = 2$, $\mu = 4$, 6, 8 and 10, $\delta = 2$, $\rho = 1$ and $w = 10$. The curve corresponding to $\mu = 4$ is the lowest one.

We are interested in the intersection of the curve with $\lambda = 0$ which gives k_1^2 and k_2^2. These correspond to a boundary in parameter space, which from (5.54) is given by the two solutions k^2 of

$$d_u d_v k^4 - (d_u g_v^* + d_v f_u^* + \alpha u^* \chi^* g_u^*)k^2 + f_u^* g_v^* - f_v^* g_u^* = 0. \tag{5.57}$$

In general, for each set of parameter values there are two, one or zero values of k^2 which satisfy (5.57). At bifurcation, where there is one value, we have

$$k_c^2 = \frac{d_u g_v^* + d_v f_u^* + \alpha u^* \chi^* g_u^*}{2 d_u d_v} \tag{5.58}$$

and

$$(d_u g_v^* + d_v f_u^* + \alpha u^* \chi^* g_u^*)^2 - 4 d_u d_v (f_u^* g_v^* - f_v^* g_u^*) = 0, \tag{5.59}$$

where k_c is the critical wavenumber. Solving the last equation for α and substituting it into (5.58) we get the crucial values

$$\alpha = \frac{-(d_u g_v^* + d_v f_u^*) + 2\sqrt{d_u d_v (g_v^* f_u^* - g_u^* f_v^*)}}{u^* \chi^* g_u^*}, \quad k_c^2 = \sqrt{\frac{(f_u^* g_v^* - f_v^* g_u^*)}{d_u d_v}}, \tag{5.60}$$

which give the parameter spaces for diffusion-chemotaxis-driven spatial instability in terms of the other parameters in the model system via the functions χ, f and g, here defined by (5.46), and their derivatives evaluated at the steady state.

For these results the positive solutions to the quadratic were chosen. Equation (5.60) defines a critical boundary set in parameter space which separates the two regions of positive and negative λ. The curve (5.59) is the bifurcation where $\lambda^+ = 0$ and wave patterns will neither grow nor decay. A sequence of these curves is shown in Figure 5.12 for various values of μ. On the upper side of each curve, $\lambda^+ > 0$ and all unstable pattern modes will grow, that is, for all $k_1^2 < k^2 < k_2^2$ obtained from (5.57). We do not know at this point what patterns to expect, only that spatial patterns are possible. To go further and determine which patterns will emerge is a nonlinear problem. To date the only analytical way to determine these is by what is called a weakly nonlinear analysis which means an anlysis near where the solutions bifurcate from spatial homogeneity to spatial heterogeneity. Some of the references where this has been done for reaction diffusion equations were given in Chapter 2. Exactly the same procedures are used in these types of equations but are just a little more complex. Zhu and Murray (1995) carried out a nonlinear analysis and evaluated, analytically, the parameter spaces for both reaction diffusion and diffusion-chemotaxis systems to compare their potential for generating spatial pattern. As already mentioned, Tyson (1996) used the method to analyze the specific model equations we have discussed here.

We do not carry out the (rather complicated) nonlinear analysis here (refer to Zhu and Murray 1995 and Tyson 1996 for a full discussion). Here we just sketch the proce-

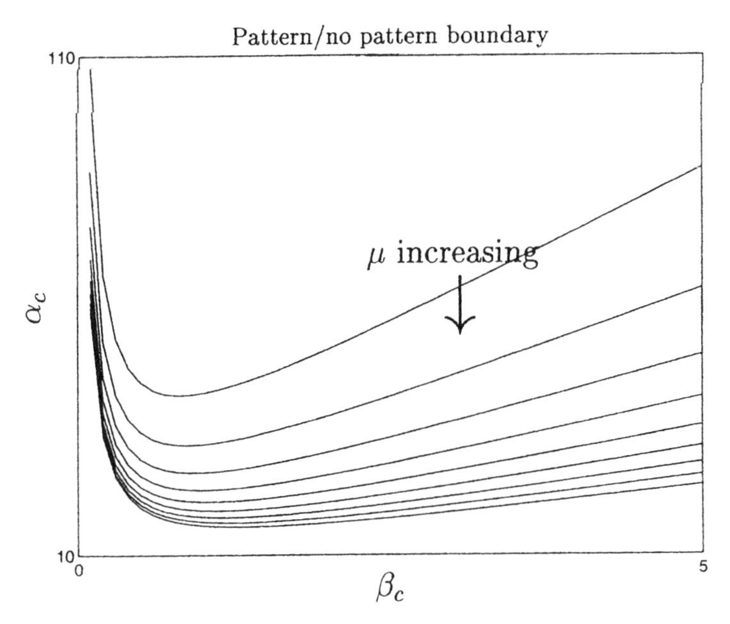

Figure 5.12. The boundary in (α_c, β_c) parameter space between regions where patterns are possible and where they are not. For (α, β) pairs above each line patterns are possible $(\lambda > 0)$ while below, they are not. The parameter values are $d_{uc} = 0.3$, $d_{vc} = 1.0$, $\mu_c = 4$ (uppermost line), to 8 (lowermost line) in steps of 0.5, $\delta_c = 2$, $\rho_c = 0.5$ and $w_c = 5$. (From Tyson 1996)

dure and give the results. From the linear analysis we first determine in what parameter regions patterns are possible. The nonlinear analysis gives, in effect, the type of patterns which will form. The analysis is based on the assumption that the parameters are such that we are close to the bifurcation curve in parameter space. We develop an asymptotic analysis based on one parameter (the one of specific interest), for example, the chemotaxis parameter α, which is close to its bifurcation value, α_c, on the bifurcation curve but whose value moves the system into the pattern formation space as it passes through its critical value. We start with the linear solution to the boundary value problem; for example, in one dimension it involves a solution including $e^{\lambda(t)} \cos kx$, where k is in the unstable range of wavenumbers. On a linear basis this solution will start to grow exponentially with time with a growth rate $e^{\lambda(k)t}$ with $\lambda(k)$ the dispersion relation. Intuitively, by examining the undifferentiated terms in the system of equations (5.38)–(5.40), the solutions can not grow unboundedly. In a weakly nonlinear analysis we first consider the linear solution to be the solution to the linear boundary value problem with $k^2 = k_c^2$; that is, the parameters are such that we are close to the bifurcation curve from no pattern to pattern. An asymptotic perturbation method is then used to study the solutions in the form where the magnitude (or amplitude) of the solution is a slowly varying function of time. The conditions under which the amplitude is bounded as $t \to \infty$ are determined. This procedure then determines which of the various solution possibilities will evolve as stable solutions. We give a few more details to further explain the general procedure when we discuss the forms of the possible linear solutions to the boundary value problem.

Linear Boundary Value Problem

A necessary prerequisite for the nonlinear analysis is solving the linear boundary value problem relevant to the nonlinear analysis. The pattern types which are possible depend on the number of different wavevectors **k** allowed by the boundary conditions. From our point of view we are interested in having the experimental domain tesselated by repeating patterns. In the numerical simulations a square domain was chosen for numerical simplicity and so we are interested in square or rectangular tiles giving stripes and spots. So, each square or rectangular unit is subject to periodic boundary conditions. In general, with regular tesselations, we can have squares, stripes, hexagons, and so on as we discussed in Chapter 2.

Consider the rectangular domain, S, defined by $0 \leq x \leq l_x, 0 \leq y \leq l_y$ with the sides denoted by S_1:$x = 0, 0 \leq y \leq l_y$, S_2:$x = l_x, 0 \leq y \leq l_y$, S_3:$y = 0, 0 \leq x \leq l_x$, S_4:$y = l_y, 0 \leq x \leq l_x$. The spatial eigenvalue problem with periodic boundary conditions is then

$$\nabla^2 \psi + k^2 \psi = 0, \qquad \begin{cases} \psi|_{S_2} = \psi|_{S_4} \\ \psi|_{S_1} = \psi|_{S_3} \end{cases}. \tag{5.61}$$

The possible eigensolutions of the partial differential equation (5.61) are

$$\psi = \mathrm{A} \cos(\mathbf{k}_n \cdot \mathbf{x}) + \mathrm{B} \sin(\mathbf{k}_n \cdot \mathbf{x}), \tag{5.62}$$

where the $\mathbf{k}_n^2$ are allowable eigenvectors, which we discuss below. Substituting the boundary conditions in (5.61) into the solutions (5.62) we obtain

$$\begin{cases} \cos(\mathbf{k}_n \cdot (0, y)) = \cos(\mathbf{k}_n \cdot (l_x, y)) \\ \sin(\mathbf{k}_n \cdot (x, 0)) = \cos(\mathbf{k}_n \cdot (x, l_x)) \\ \cos(\mathbf{k}_n \cdot (0, y)) = \cos(\mathbf{k}_n \cdot (l_y, y)) \\ \sin(\mathbf{k}_n \cdot (x, 0)) = \cos(\mathbf{k}_n \cdot (x, l_y)) \end{cases}$$

which after using some trigonometric identities can be written as

$$\begin{cases} \cos(k_n^y y)\left[1 - \cos(k_n^x l_x)\right] + \sin(k_n^y y) \sin(k_n^x l_x) = 0 \\ \sin(k_n^y y)\left[1 - \cos(k_n^x l_x)\right] - \cos(k_n^y y) \sin(k_n^x l_x) = 0 \\ \cos(k_n^x x)\left[1 - \cos(k_n^y l_y)\right] + \sin(k_n^x x) \sin(k_n^y l_y) = 0 \\ \sin(k_n^x x)\left[1 - \cos(k_n^y l_y)\right] - \cos(k_n^x x) \sin(k_n^y l_y) = 0, \end{cases} \tag{5.63}$$

where $\mathbf{k}_n^T = (k_n^x, k_n^y)$. The general solutions of (5.63) are

$$\mathbf{k}_1 = \begin{bmatrix} 0 \\ 2m\pi/l_y \end{bmatrix}, \quad \mathbf{k}_2 = \begin{bmatrix} 2n\pi/l_x \\ 0 \end{bmatrix}, \quad \mathbf{k}_3 = \begin{bmatrix} 2p\pi/l_x \\ 2q\pi/l_y \end{bmatrix},$$

where m, n, p and q are all integers.

We are concerned with parameters such that we are near the bifurcation curve so we are particularly interested in the solutions $\mathbf{k}_n$ which satisfy $|\mathbf{k}_n|^2 = k_c^2$.

The number of such solution vectors which equations (5.63) admit depends on the relationship between l_x and l_y. Suppose $k_c^2 = (2M\pi/l)^2$ and also that S is square with sides $l_x = l_y = l$. If $M = 1$ then there are two possible solution vectors:

$$\mathbf{k}_1 = \begin{bmatrix} 0 \\ 2\pi/l \end{bmatrix}, \quad \mathbf{k}_2 = \begin{bmatrix} 2\pi/l \\ 0 \end{bmatrix}. \tag{5.64}$$

If $M = 5$ then there are four possible solution vectors:

$$\begin{aligned} \mathbf{k}_1 &= \begin{bmatrix} 0 \\ 2\cdot 5\pi/l \end{bmatrix}, \quad \mathbf{k}_2 = \begin{bmatrix} 2\cdot 5\pi/l \\ 0 \end{bmatrix} \\ \mathbf{k}_3 &= \begin{bmatrix} 2\cdot 3\pi/l \\ 2\cdot 4\pi/l \end{bmatrix}, \quad \mathbf{k}_4 = \begin{bmatrix} 2\cdot 4\pi/l \\ 2\cdot 3\pi/l \end{bmatrix}. \end{aligned} \tag{5.65}$$

These solution vectors are important in the nonlinear analysis.

5.7 Brief Outline and Results of the Nonlinear Analysis

With the linear analysis we can only determine the small amplitude initial behaviour of u and v about the uniform steady state when the steady state is driven unstable to spatially heterogenous perturbations. These spatially inhomogeneous solutions initially grow exponentially and are clearly not valid for all time. For the class of problems here we can carry out a nonlinear asymptotic analysis and obtain the solutions to $O(\varepsilon)$ (and in principle to higher orders but the algebra is prohibitive) which are valid for all time. As mentioned the details of the procedure are given by Zhu and Murray (1995) for several pattern formation mechanisms including a diffusion-chemotaxis one. For the more complex chemotaxis mechanism the analysis has been carried out by Tyson (1996). Here we sketch the analytical procedure, namely, a multi-scale asymptotic analysis, for determining small perturbation solutions valid for all time of the system of equations (5.41) and (5.42) the general forms of which are (5.44) and (5.45). We start by writing

$$\begin{aligned} u &= u^* + \hat{u} = u^* + (\varepsilon u_1 + \varepsilon^2 u_2 + \varepsilon^3 u_3 + \cdots) \\ v &= v^* + \hat{v} = v^* + (\varepsilon v_1 + \varepsilon^2 v_2 + \varepsilon^3 v_3 + \cdots), \end{aligned} \tag{5.66}$$

where (u^*, v^*) is the spatially homogeneous steady state which depends on the model parameters and which, as we saw in the last section, can be driven unstable as a parameter passes through the bifurcation value which results in spatially unstable solutions. We scale time by writing

$$T = \hat{\omega} t, \quad \text{where } \hat{\omega} = \varepsilon\omega_1 + \varepsilon^2\omega_2 + \cdots, \tag{5.67}$$

where the ω_i, $i = 1, 2, \ldots$ have to be determined.

Consider equation (5.44). Substituting the expansions (5.66) and (5.67) into the individual terms, we get

$$\begin{aligned}\frac{\partial u}{\partial t} &= \hat{\omega}\frac{\partial \hat{u}}{\partial T}, \quad \nabla^2 u = \nabla^2 \hat{u}\\ \alpha\nabla\cdot[u\chi(v)\nabla v] &= \alpha\nabla\cdot\left[(u^*+\hat{u})\left(\chi^*+\chi_v^*\hat{v}+\tfrac{1}{2}\chi_{vv}^*\hat{v}^2+\cdots\right)\nabla\hat{v}\right]\\ &= u^*\chi^*\nabla^2\hat{v}+\left(u^*\chi^*\nabla\hat{v}+\chi^*\nabla\hat{u}\right)\cdot\nabla\hat{v}\\ &\quad+\left(u^*\tfrac{1}{2}\chi_{vv}^*\nabla(\hat{v}^2)+\chi_v^*\nabla(\hat{u}\hat{v})\right)\cdot\nabla\hat{v}+\cdots\\ f(u,v) &= f^*+(f_u^*+f_v^*)\hat{u}+\tfrac{1}{2}(f_{uu}^*+f_{vv}^*)\hat{u}^2+f_{uv}^*\hat{u}\hat{v}\\ &\quad+\tfrac{1}{6}(f_{uuu}^*+f_{vvv}^*)\hat{u}^3+\cdots.\end{aligned} \tag{5.68}$$

The first two expressions have linear terms in $\hat{u}$, while the last two expressions have linear, quadratic, cubic and so on with higher-order terms in $\hat{u}$ and $\hat{v}$; $f^*=0$ by definition of the steady state. So (5.44) transforms into another equation with linear, quadratic and higher-order terms in $\hat{u}$ and $\hat{v}$. Equation (5.45) transforms in an equivalent way. In general then, (5.44) and (5.45) with (5.67) and (5.68) take the form

$$\hat{\omega}\frac{\partial}{\partial T}(\vec{\hat{u}}) = \left[D^*\nabla^2+A^*\right]\vec{\hat{u}}+\mathcal{Q}^*(\vec{\hat{u}})+\mathcal{C}^*(\vec{\hat{u}})+\cdots, \quad \vec{\hat{u}}=\begin{bmatrix}\hat{u}\\ \hat{v}\end{bmatrix} \tag{5.69}$$

and where $*$ denotes evaluation at the steady state (u^*, v^*).

The quantities $A\vec{\hat{u}}$, $\mathcal{Q}(\vec{\hat{u}})$ and $\mathcal{C}(\vec{\hat{u}})$, represent the linear, quadratic and cubic terms respectively, of the expansion of the chemotaxis and reaction functions about the steady state. The matrices $A\vec{\hat{u}}$ and D were determined above in the linear analysis. We are interested in situations where the model parameters have particular values such that $\lambda=0$. This occurs when the parameters are defined by (5.60); we call this set a critical parameter set. Basically, it means the parameter set is sitting on the boundary between growing spatially heterogeneous solutions and spatially homogeneous solutions.

If we now perturb one of the model parameters a, say, which can be any one of the parameters in (5.44) and (5.45), about its value in a critical set, the eigenvalue for temporal growth becomes

$$\begin{aligned}\lambda(a) &= \lambda(a_c)+\left.\frac{\partial\lambda}{\partial a}\right|_{a_c}\Delta a_c+\cdots\\ &= \left.\frac{\partial\lambda}{\partial a}\right|_{a_c}\Delta a_c+\cdots\end{aligned}$$

since $\lambda(a_c)=0$ by definition of the critical parameter a_c. We take the perturbation to be such that

$$\left.\frac{\partial k_c^2}{\partial a}\right|_{a_c} = 0,$$

so that the perturbation effect on the solution $e^{\lambda t + i\vec{k}\cdot\vec{x}}$ is restricted to a change in the temporal growth rate λ. Then, if the change in a_c makes $\mathrm{Re}(\lambda(a))$ positive, the pattern mode corresponding to k_c^2 is predicted to grow according to linear theory. Depending on the parameters, the result can be a stable or unstable spatially heterogeneous solution. If this growth is sufficiently slow, we can predict whether or not it will develop into a temporally stable pattern and furthermore, what the characteristics of the pattern will be such as spots or stripes. We start by perturbing the steady state model (5.69) about the critical set. To keep the analysis simple we perturb only one parameter, and to keep the analysis general we call the parameter a. Tyson (1996) carried out the analysis with the actual parameters from the model equations and it is her results we give below.

Consider an expansion of the form

$$a = a_c + \hat{a} = a_c + (\varepsilon a_1 + \varepsilon^2 a_2 + \cdots). \tag{5.70}$$

Substituting this into (5.69) we get the system

$$\begin{aligned}\hat{\omega}\frac{\partial}{\partial T}(\vec{\hat{u}}) &= \left[D^{*c}\nabla^2 + A^{*c}\right]\vec{\hat{u}} + \mathcal{Q}^{*c}(\vec{\hat{u}}) + \mathcal{C}^{*c}(\vec{\hat{u}})\\ &\quad + \hat{a}\left[D_a^{*c}\nabla^2 + A_a^{*c}\right]\vec{\hat{u}} + \hat{a}\mathcal{Q}_a^{*c}(\vec{\hat{u}})\\ &\quad + \text{higher-order terms,}\end{aligned} \tag{5.71}$$

where the superscript c denotes evaluation at the critical set. The change in the critical parameter a only occurs in $\hat{a}$ and so its effect can be isolated in the analysis.

Substituting the expansions for all of the small variables $\hat{()}$, and collecting and equating terms of like order in ε, we obtain systems of equations for each order in ε. For notational simplicity, the superscript c is omitted in the result, and for the remainder of the analysis all parameter values are from a critical set. To show what these equations look like we just give the $O(\varepsilon)$ and $O(\varepsilon^2)$ systems although to carry out the nonlinear analysis it is necessary to also have the $O(\varepsilon^3)$ system which is algebraically extremely complicated. We do not need them to sketch the procedure. The $O(\varepsilon)$ equations are

$$\begin{bmatrix}0\\0\end{bmatrix} = \begin{bmatrix} d_u\nabla^2 + f_u^* & -\alpha u^*\chi^*\nabla^2 + f_v^*\\ g_u^* & d_v\nabla^2 + g_v^*\end{bmatrix}\begin{bmatrix}u_1\\v_1\end{bmatrix} = L\begin{bmatrix}u_1\\v_1\end{bmatrix} \tag{5.72}$$

which are linear and define the coefficient matrix as the linear operator L. The order $O(\varepsilon^2)$ equations are

$$\begin{aligned}L\begin{bmatrix}u_2\\v_2\end{bmatrix} &= \omega_1\frac{\partial}{\partial T}\begin{bmatrix}u_1\\v_1\end{bmatrix} + \begin{bmatrix}\alpha(\chi^*\nabla u_1 + u^*\chi_v^*\nabla v_1)\cdot\nabla v_1\\ 0\end{bmatrix}\\ &\quad - \begin{bmatrix}\frac{1}{2}f_{uu}^*u_1^2 + f_{uv}^*u_1v_1D + \frac{1}{2}f_{vv}^*v_1^2\\ \frac{1}{2}g_{uu}^*u_1^2 + g_{uv}^*u_1v_1 + \frac{1}{2}g_{vv}^*v_1^2\end{bmatrix}\\ &\quad - a_1\begin{bmatrix}(f_u^*)_a & -(\alpha u^*\chi^*)_a\nabla^2 + (f_u^*)_a\\ (g_u^*)_a & (g_v^*)_a\end{bmatrix}\begin{bmatrix}u_1\\v_1\end{bmatrix}.\end{aligned} \tag{5.73}$$

The analysis, which needs the $O(\varepsilon^3)$ equations, requires the solutions of these linear systems of equations. This was done by Tyson (1996) (and an equivalent analysis by Zhu and Murray 1995). The algebra is horrendous but necessary to get the uniformly valid (for all time) solution to $O(\varepsilon)$ and to determine which specific patterns will be stable.

So as to be able to give the results and explain what the nonlinear analysis gives, we need the solutions to the $O(\varepsilon)$ system (5.72). To get them we look for solutions in the form

$$\begin{bmatrix} u_1 \\ v_1 \end{bmatrix} = \sum_{l=1}^{N} \vec{V}1_l A_l, \tag{5.74}$$

where

$$A_l = a_l(T)e^{i\vec{k}_l\cdot\vec{x}} + \bar{a}_l(T)e^{-i\vec{k}_l\cdot\vec{x}} \tag{5.75}$$

is the sinusoidal part of the solution. Relating this form to (5.62), $\bar{a}_l(T) + a_l(T) \propto B$. Substituting (5.74) into (5.72) we obtain an expression for $\vec{V}1_l$ to within an arbitrary constant multiple. This is usually chosen such that the magnitude of the vector is unity and so

$$\vec{V}1_l = \begin{bmatrix} V1_{l1} \\ V1_{l2} \end{bmatrix} = \frac{1}{\sqrt{(d_v|\vec{k}_l|^2 - g_v^*)^2 + (g_u^*)^2}} \begin{bmatrix} d_v|\vec{k}_l|^2 - g_v^* \\ g_u^* \end{bmatrix}. \tag{5.76}$$

Since we only consider critical sets of parameters, we know that $|\vec{k}_l|^2 = k_c^2 \forall l$ and so $\vec{V}1_l = \vec{V}1 \forall l$.

At this stage we do not know $a_l(T)$ and $\bar{a}_l(T)$, the complex amplitudes (which are functions of T the slowly varying time defined by (5.67)) of the $O(\varepsilon)$ solution; the solution amplitude is $|a_l(T)|$. The key to the nonlinear analysis is the determination of the amplitude. So, we have to solve the $O(\varepsilon^2)$ equations. Since these linear equations, with the same operator L, contain undifferentiated terms on the right-hand side it is possible to have solutions with secular terms, which are terms which involve expressions like $x \sin x$ which become unbounded for large x.

It is easy to see how secular terms arise if we consider the simple equation for $u(x)$:

$$u'' + u = -\varepsilon u',$$

where primes denote differentiation with respect to x, $0 < \varepsilon \ll 1$ and, to be specific, let us require $u(0) = 1, u'(0) = 0$. If we write $u = u_0 + \varepsilon u_1 + \cdots$ we assume all the $u_i, i = 1, 2, \ldots$ are all $O(1)$. Substituting this into the equation and collecting like terms in ϵ we get $u_0(x) = \cos x$ and the equation and boundary conditions for u_1, the $O(\varepsilon)$ terms, as

$$u_1'' + u_1 = \sin x, \quad u_1(0) = u_1'(0) = 0,$$

the solution of which is $u_1(x) = (1/2)(\sin x - x\cos x)$. So, $u_0 + \varepsilon u_1 + \cdots$ is not a uniformly valid solution since $u_1(x)$ is not $O(1)$ for all x because of the $x\cos x$ term: it is the secular term. The asymptotic procedure for obtaining uniformly valid solutions of this type of equation is pedagogically described in detail in the book on asymptotic analysis by Murray (1984).

To go back to the above discussion of the $O(\varepsilon^2)$ equations, it turns out that these do not give rise to secular terms and so the amplitude functions $a_l(T)$ and $\bar{a}_l(T)$ remain undetermined at this order. However, at $O(\varepsilon^3)$ secular terms do appear. It is at this stage that the equations for the amplitude are determined: they are chosen so that these secular terms do not occur in the $O(\varepsilon^3)$ solutions even though we do not actually find the solutions at this order. It is the algebra involved in obtaining these equations, known as the Landau equations, that is so complicated and detailed. The equations crucially involve the number N in (5.74) which is the number of modes in the solution which have $|\, k_l^2 \,| = k_c^2$. We saw how the solutions and eigenvectors varied with this number in the discussion of the boundary value problem in the last section.

By way of example, let us suppose $N = 2$; then Tyson (1996) showed that the amplitude, or Landau, equations are

$$\frac{d|\, a_1 \,|^2}{dT} = |\, a_1 \,|^2(X_A|\, a_1 \,|^2 + X_B|\, a_2 \,|^2) + Y|\, a_1 \,|^2 \tag{5.77}$$

$$\frac{d|\, a_2 \,|^2}{dT} = |\, a_2 \,|^2(X_B|\, a_1 \,|^2 + X_A|\, a_2 \,|^2) + Y|\, a_2 \,|^2, \tag{5.78}$$

where the X_A, X_B and Y are complicated functions of the parameters of the original system (5.44) and (5.45) (and hence (5.41) and (5.42)). The $|\, a_1 \,|$ and $|\, a_2 \,|$ directly relate to the A and B in equation (5.62) except that here they are functions of time. Whether or not a stable spatially heterogeneous solution exists depends on the solutions of these amplitude equations as $t \to \infty$. They are just ordinary differential equations with constant coefficients. They have the following possible steady state solutions and their existence depends on the signs of the coefficients,

$$\begin{array}{lll}
(1) & |\, a_1 \,|^2 = 0, & |\, a_2 \,|^2 = 0 \\[2ex]
(2) & |\, a_1 \,|^2 = 0, & |\, a_2 \,|^2 = -\dfrac{Y}{X_A} \\[2ex]
(3) & |\, a_1 \,|^2 = -\dfrac{Y}{X_A}, & |\, a_2 \,|^2 = 0 \\[2ex]
(4) & |\, a_1 \,|^2 = -\dfrac{Y}{X_A + X_B}, & |\, a_2 \,|^2 = -\dfrac{Y}{X_A + X_B}
\end{array} \tag{5.79}$$

The first steady state corresponds to a zero amplitude pattern, or no pattern at all. The second and third correspond to a zero amplitude in one direction and a nonzero amplitude in the other, and this gives stripes. The fourth steady state has a nonzero amplitude in each direction and therefore gives spots. If none of these steady states is stable, then the analysis does not determine the type of pattern formed. This is referred to

Table 5.4. Conditions for stability of the patterns possible for the four steady states when the number of critical wavenumber vectors $\vec{k}_n$ is $N = 2$.

Steady State $(\vert a_1 \vert^2, \vert a_2 \vert^2)$	Pattern	Stability Conditions
(0,0)	none	$Y < 0$
$(0, -Y/X_A)$	horizontal stripes	$Y > 0$ $X_B/X_A > 1$
$(-Y/X_A, 0)$	vertical stripes	$Y > 0$ $X_B/X_A > 1$
$\left(-\dfrac{Y}{X_A + X_B}, -\dfrac{Y}{X_A + X_B}\right)$	spots	$Y > 0$ $X_A \pm X_B < 0$

as the undetermined region in the parameter plots which we show below. The conditions for stability of these steady states are summarized in Table 5.4. By computing X_A, X_B and Y over a given parameter space, we can use Table 5.4 to divide this space into regions where spots, stripes, no pattern or an undetermined pattern can occur. These parameter spaces were calculated by Tyson (1996) for the system under study here and by Zhu and Murray (1995) for a simpler reaction diffusion-chemotaxis system.

5.8 Simulation Results, Parameter Spaces and Basic Patterns

We present in this section some simulations of two models: equations (5.41) and (5.42) and a moderately simplified version. For these simulations we used initial conditions which strictly apply to the analysis. That is, the simulations are begun with a small (order ε) perturbation of the spatially and temporally unvarying steady state solution, and a smaller (order ε^2) perturbation of one of the parameters. Recall that the boundary conditions are periodic.

The Simplified Model

Consider first the following simplified model of (5.41) and (5.42),

$$\frac{\partial u}{\partial t} = d_u \nabla^2 u - \alpha \nabla \left[\frac{u}{(1+v)^2} \nabla v \right] + \rho u(\delta - u)$$
$$\frac{\partial v}{\partial t} = \nabla^2 v + \beta u^2 - uv. \tag{5.80}$$

This model has only five parameters, and so the parameter space is a little easier to explore than for equations (5.41) and (5.42). If we take d_u and ρ to be fixed, then we can vary δ and β, and determine α from the first of (5.61). Each point in the δ, β plane will thus correspond to a critical set of parameters which we can use to determine X_A,

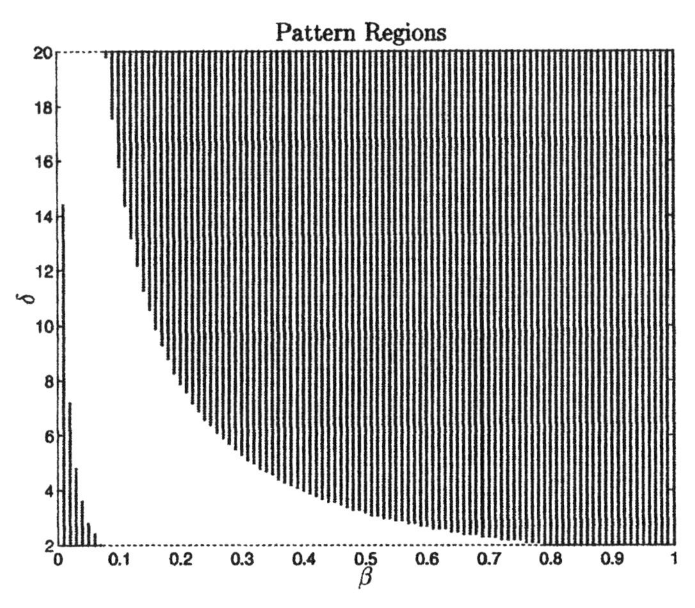

Figure 5.13. The pattern domain for the simple model (5.80) with $d_u = 0.25$, $\rho = 0.01$, $\mu = 1.0$ and $w_0 = 1.0$. As β is varied, the corresponding value of α is determined from equation (5.80). The perturbation parameter is β. The pattern regions based on the evaluated X_A, X_B and Y for Table 5.4 give stripes in the striped (dark) areas and spots in the clear (white) area. (From Tyson 1996)

X_B and Y. This approach creates the pattern regions shown in Figure 5.13. On this plot, curves of constant α are hyperbolas given by

$$\beta = \frac{(\sqrt{d_u} + \sqrt{\rho d_v})^2}{\alpha} \frac{(1+\delta)^2}{\delta}.$$

Numerical solutions were obtained for two sets of parameters: one from the upper stripes region, and one from the spots region. Simulations were performed on square domains just large enough to hold one full period of the pattern. This required the choice

$$l_x = l_y = \frac{2\pi}{\sqrt{k_c^2}}.$$

So, we expected to obtain one stripe for the stripe parameters and one spot for the spot parameters. These were found, and typical results for the cell density are shown in Figure 5.14. Each simulation has two plots which indicate the initial conditions and the steady state pattern.

Note that since the boundary conditions are periodic, the maximum of the spot or stripe pattern can occur anywhere in the domain. Also, for the stripe pattern, the orientation of the stripe is not determined from the analysis, since we are working with a square domain. So, for one set of random initial conditions the stripe will appear vertical, whereas for another set it will appear horizontal.

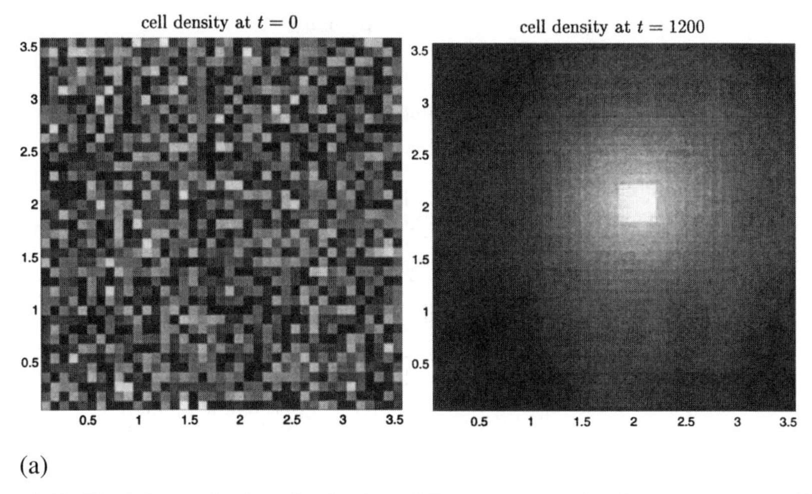

(a)

Figure 5.14. Simulation results from the simple model parameter domain where spots were predicted in Figure 5.13. Cell density is plotted as an image at $t = 0$ and at steady state. Cell density profiles are shown to demonstrate that the pattern has indeed reached a steady state, and the predicted cell density profile is also shown for comparison. The perturbation was $\varepsilon = 0.1$. The times given are in nondimensional units. White indicates high cell density, black the opposite. The parameter values are: (**a**) $d_u = 0.25$, $\alpha = 1.50$, $\beta = 0.1$, $\delta = 15.0$, $\rho = 0.01$, $w = 1.0$, $u_0 = u^* = 15.0$, $v_0 = v^* = 1.50$, $k_c^2 = 3.0$ and $l_x = l_y = l = 3.6276$ (domain which can sustain one 2π oscillation), a spot.

The Full Model

We consider now the more biologically accurate model

$$\begin{aligned}\frac{\partial u}{\partial t} &= d_u \nabla^2 u - \alpha \nabla \left[\frac{u}{(1+v)^2} \nabla^* v \right] + \rho u (\delta \frac{w^2}{1+w^2} - u) \\ \frac{\partial v}{\partial t} &= \nabla^2 v + \beta w \frac{u^2}{\mu + u^2} - uv, \end{aligned} \tag{5.81}$$

which has seven parameters, including w and for which we have experimental estimates for α, d_u and δ. One of the four remaining parameters can be determined from the bifurcation condition (5.58). In the simpler model we solved for the critical value of α given all of the other parameters. Since we know α, however, we would like to fix its value and solve for one of the unknown parameters. The simplest one to solve for is ρ, which is given by

$$\rho = \left[\frac{\sqrt{\alpha\beta\delta(\mu - \delta^2)}}{\mu + \delta(\beta + \delta)} - \sqrt{d_u} \right]^2 . \tag{5.82}$$

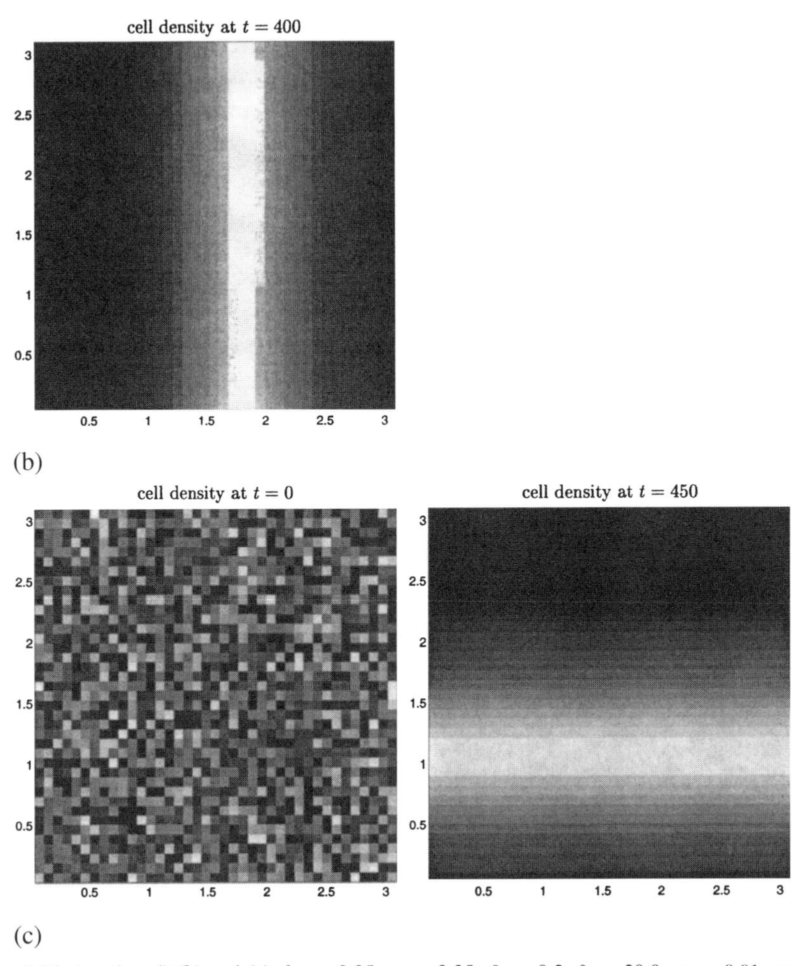

Figure 5.14. (continued) (**b**) and (**c**) $d_u = 0.25$, $\alpha = 2.25$, $\beta = 0.2$, $\delta = 20.0$, $\rho = 0.01$, $w = 1.0$, $u_0 = u^* = 20.0$, $v_0 = v^* = 4.0$, $k_c^2 = 4.0$ and $l_x = l_y = l = 3.1416$ (domain which can sustain one 2π oscillation) but a stripe. In (**a**) and (**b**) the same seed was used while in (**c**) a different seed was used for the random initial conditions. The same amplitude and size of stripe is obtained, but this time it is horizontal instead of vertical. (From Tyson 1996)

The remaining parameters are μ, β and w. So, we need to explore the (β, μ), (μ, w) and (β, w) parameter spaces. We are dealing with a four-dimensional parameter space but it is three-dimensional for given ρ. At this point we can marginally simplify the system by setting

$$\bar{\delta} = \frac{\delta w^2}{1 + w^2}, \quad \bar{\beta} = \beta w \tag{5.83}$$

and rewriting the model as

$$\frac{\partial u}{\partial t} = d_u \nabla^2 u - \alpha \nabla \left[\frac{u}{(1+v)^2} \nabla v \right] + \rho u(\bar{\delta} - u)$$

$$\frac{\partial v}{\partial t} = \nabla^2 v + \bar{\beta} \frac{u^2}{\mu + u^2} - uv. \tag{5.84}$$

This simplifies the algebra considerably, but still lets us determine the effect of increasing or decreasing w by mapping points (β, δ) to $(\bar{\beta}, \bar{\delta})$ as w increases using (5.81). This mapping is a sigmoidal curve for each (β, δ) pair in the $(\bar{\beta}, \bar{\delta})$ plane.

With this formulation the unknown parameters are β, δ and μ. If we fix μ to be large compared with u^2, we recover the simplified model (5.80) that we just considered. Surprisingly, $\bar{\beta}$ does not need to be very much larger than β for this pattern domain to be recovered. For smaller values of μ, the pattern domain is different. An example is shown in Figure 5.15.

Tyson (1996) carried out simulations for parameters from the stripes, spots and indeterminate regions and these confirmed the analytical predictions as to which pattern type will be found in small periodic domains subject to small random perturbations of the steady state. Interestingly she also found that steady state patterns exist in at least part of the indeterminate region, where the analysis can not, as yet, predict what patterns will occur. An explanation for the latter is an interesting analytical problem. Examples of these patterns are shown in Figure 5.16.

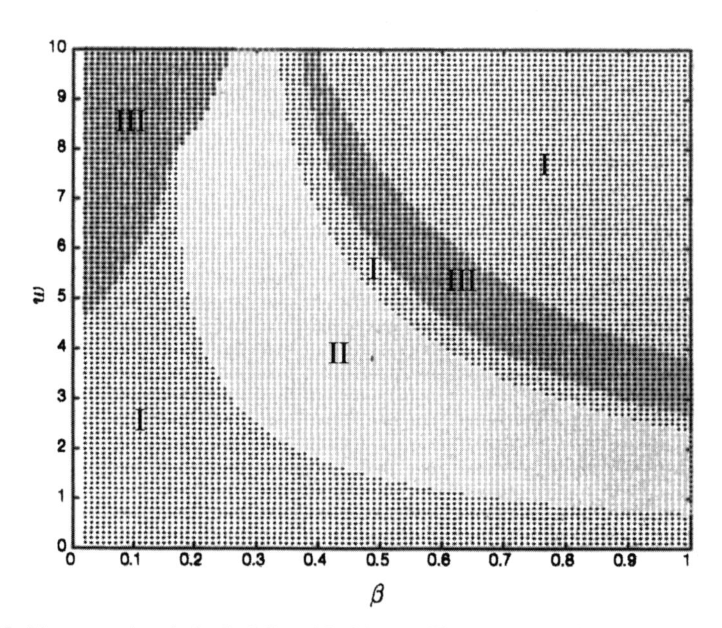

Figure 5.15. The pattern domain for the full model without w. The parameter values are $d_u = 0.25$, $\alpha = 90.0$ and $\mu = 1000.0$. As $\bar{\beta}$ and $\bar{\delta}$ (defined by (5.83))are varied, the corresponding value of ρ is determined from equation (5.82). The perturbation parameter is β. The pattern regions give stripes (I), spots (II) and indeterminate patterns (III). (From Tyson 1996)

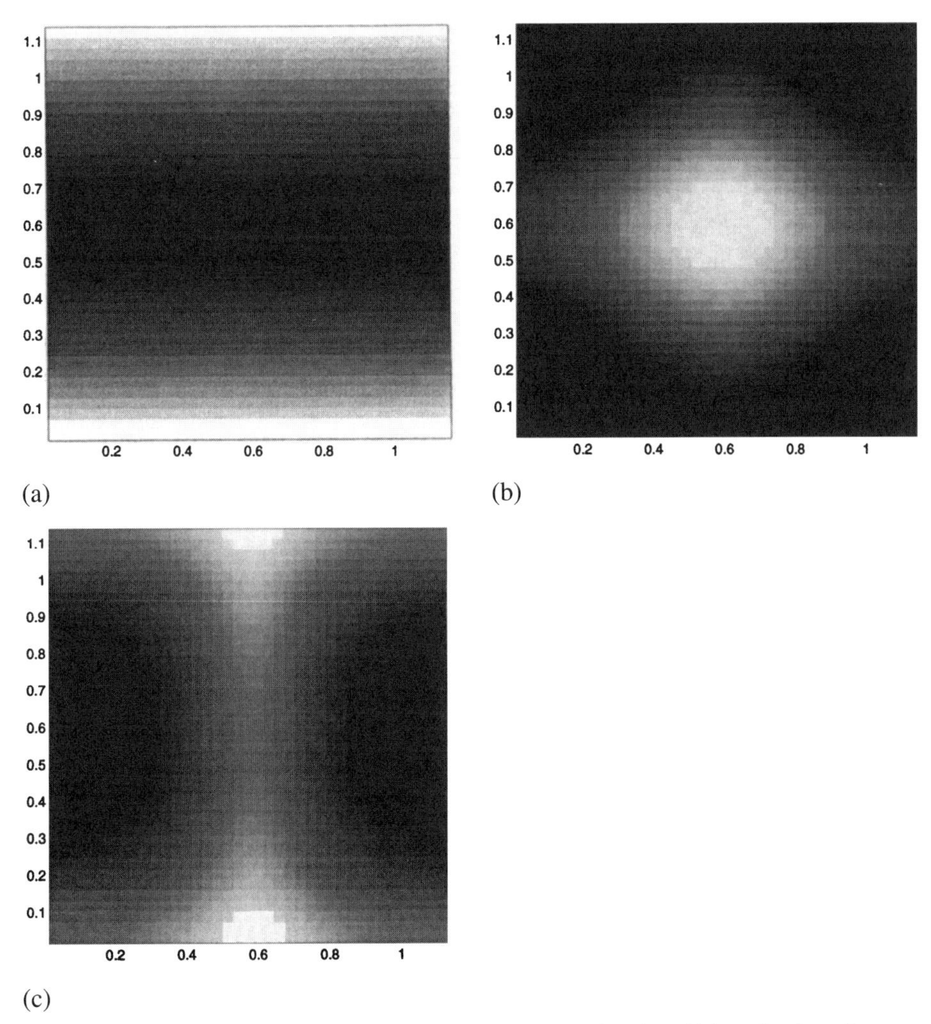

Figure 5.16. Simulations of the full equations for parameters in the domains which predict the various patterns. Light regions denote high cell density. In each case the domain chosen can sustain one 2π oscillation. The perturbation was $\varepsilon = 0.1$. The parameter values are $d_u = 0.25$, $\alpha = 90.0$, $\bar{\beta} = 10.0$, $\mu = 100.0$; then (**a**) Stripe: $\bar{\delta} = 4.6$, $\rho = 8.5133$, $u_0 = u^* = 4.6$, $v_0 = v^* = 0.3797$ and $l_x = l_y = l = 1.213$; (**b**) Spot: $\bar{\delta} = 5.1$, $\rho = 7.797$, $u_0 = u^* = 5.1$, $v_0 = v^* = 0.4747$ and $l_x = l_y = l = 1.177$; (**c**) Indeterminate pattern: $\bar{\delta} = 5.5$, $\rho = 7.139$, $u_0 = u^* = 5.5$, $v_0 = v^* = 0.4233$ and $l_x = l_y = l = 1.159$. (From Tyson 1996)

5.9 Numerical Results with Initial Conditions from the Experiments

The nonlinear analysis of the semi-solid models we discussed above only applies to small random perturbations of the uniform positive steady state. The initial conditions in the experiments are, as described in Section 5.1, completely different. Initially there is no chemoattractant present, and only a small localised inoculum of cells which is a relatively large perturbation of the uniform zero steady state. We also have zero flux boundary conditions on a large domain, rather than periodic boundary conditions on a small domain. The bacterial lawn preceding the pattern establishes conditions relevant to the nonlinear behaviour. So, with the experimental initial conditions we expect to find rings in the stripes region of the pattern domain, and broken (spotted) rings in the spots region. Tyson (1996) carried out extensive simulations in both one and two dimensions with parameters in the various regions which give spots and stripes. Here we give some of her results and some of those presented by Tyson et al. (1999).

We start with the simplified semi-solid model (5.80) in one dimension. With the same parameter values as in Figure 5.14 which gave a stripe pattern we again get a series of concentric rings. In the case of the parameter values which gave a spot pattern only a few small pulses appeared near the the initial disturbance which slowly decayed. It seems that the spotted ring patterns obtained experimentally arise either from the stripes region of the nonlinear analysis pattern domain, or from some pattern region completely outside the predictions of the nonlinear analysis. In no way did we obtain the complete patterning scenario from the nonlinear analysis.

Even a cursory investigation of parameter space showed that we need not restrict ourselves to parameters from the nonlinear analysis to find interesting patterns and how patterns vary with parameter variation. For example, if we increase the chemotaxis coefficient α, the amplitude increases as well as the wavelength of the stripes. A small increase in nutrient concentration w, increases the propagation speed and causes the pulses behind the wavefront to decay. Only a slightly larger increase in w makes the pulses disappear altogether, as the cell density rapidly approaches the uniform positive steady state. Pulses still form for decreased levels of w, but the pattern propagates more slowly. The rate of growth, ρ, affects the pattern in exactly the same way as nutrient concentration. A factor of two decrease in chemoattractant production β increases both the amplitude and the frequency of pulses. The carrying capacity δ is directly related to the amplitude of the pattern and also the length of time it takes to form. The permutations are endless.

In two dimensions, simulations were run with the parameters which gave stripes from the nonlinear analysis and the simulation gave concentric rings. The wavelength of the radial pattern is smaller in two dimensions than that in one dimension, and the amplitude is about the same. The spacing between rings does not change as the radius increases as shown in Figure 5.17.

Increasing the chemotaxis parameter from $\alpha = 2.25$ to $\alpha = 5$ in two dimensions we get a series of concentric rings made up of spots. A time sequence of the pattern is shown in Figure 5.18 together with a surface plot of the solution at $t = 70$. As in one dimension, increasing the chemotaxis coefficient increases both the wavelength and amplitude of the pattern. Importantly we found that nutrient consumption does not change a spot pattern to a stripe pattern, nor does it change the wavelength.

We now consider the full model system (5.84) which introduces one more parameter, namely, μ which measures saturation level of chemoattractant production. From the simple model we found that varying the parameters α, β, ρ, δ and κ can change a pattern of concentric rings. Basically the full model gives the same patterns found in the simple model but with certain quantitative differences as we would expect. Trivially, for the saturation level of chemoattractant production, μ, sufficiently large and little consumption of nutrient, the full model essentially reduces to the simple one and behaves in exactly the same way. More importantly, for parameters which allow more of the

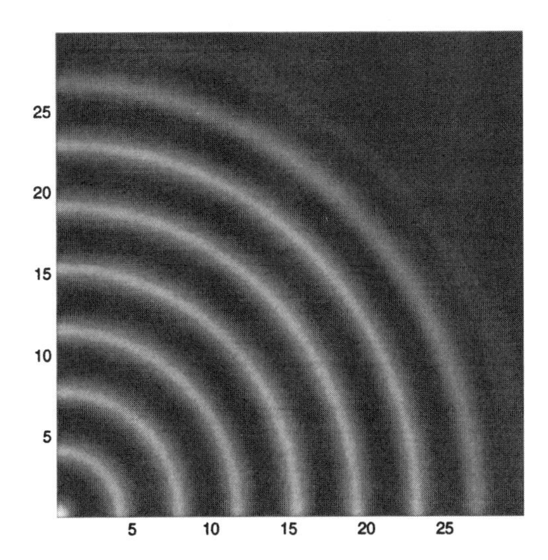

Figure 5.17. Simulation results for the model system (5.80) with the same parameters as in Figure 5.14(b) but with initial conditions like those of the semi-solid experiments. The parameter values are: $d_u = 0.25$, $\alpha = 2.25$, $\beta = 0.2$, $\mu = 1.0$, $\delta = 20.0$, $\rho = 0.01$, $w = 1.0$, $u_0 = u^* = 5.0$ at the bottom left corner and $v_0 = 0.0$. (From Tyson 1996)

middle and saturating portions of the aspartate production curve to play a part in the simulations, we also find continuous and spotted rings as shown in Figures 5.19(a) and (b). Adding consumption of food does not change the nature of the ring pattern, except to make it gradually disappear from the centre outwards.

Relation of the Simulations to the Experiments

The simulation results compare remarkably well with the experimental results obtained for *S. typhimurium* and described in Section 5.1. The pattern is preceded by a bacterial lawn of low cell density. Each ring forms at a discrete radial distance from the previous one, and then remains stationary. The spotted rings form first as continuous rings which subsequently break up into spots. All of these traits obtained from the model mechanism are characteristic of the *S. typhimurium* patterns.

If we take the parameter values from Figure 5.19(b) we get

$$k_7 = \frac{k_3}{\rho} = 1.6 \times 10^{-9} \text{ ml (cell hr)}^{-1}$$

$$x = x^* \sqrt{\frac{D_c}{k_7 n_0}}.$$

Four rings form in a distance $x^* = 10$ which corresponds to $x = 1.4$ cm, which is close to the experimentally observed value of $x \approx 1$ cm (Woodward et al. 1995).

5.10 Swarm Ring Patterns with the Semi-Solid Phase Model Mechanism

The most dramatic patterns observed by Budrene and Berg (1991) arise from an expanding high-density ring of bacteria called a swarm ring. These patterns were described and

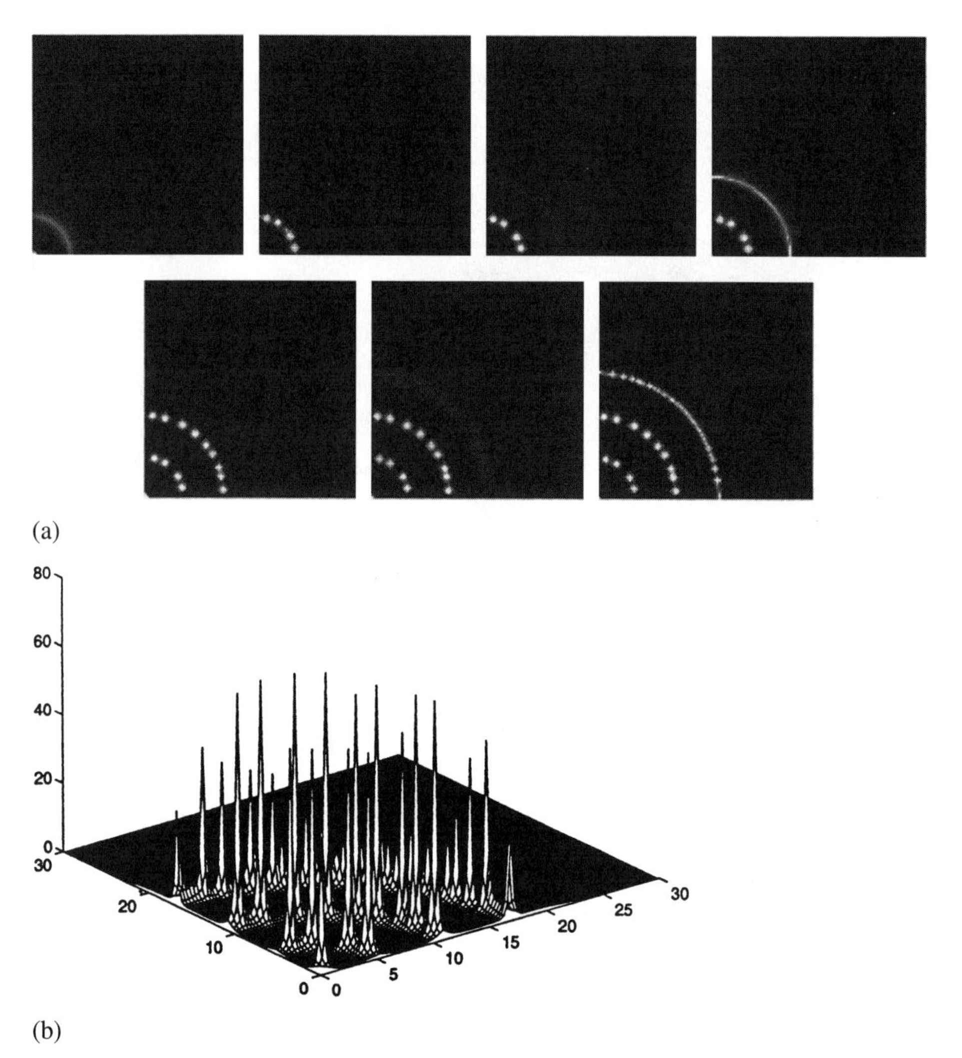

Figure 5.18. Simulation results for the model system (5.80) with the same parameters as in Figure 5.16(b) (and in Figure 5.14) but with initial conditions like those of the semi-solid experiments except that here $\alpha = 5$. Again the initial cell density was zero everywhere except for a small inoculum of maximum density 5 at the origin; the initial chemoattractant concentration was zero everywhere and the initial nutrient concentration was 1 everywhere. Solutions are shown in time increments of $t = 5$ from $t = 40$ to 70. The last figure is a surface plot of the data presented in the final image when $t = 70$. (From Tyson 1996)

illustrated in Section 5.1. The initial conditions are a localised inoculum of cells on a dish containing a uniform distribution of food and no chemoattractant, and the boundary conditions are zero flux.

Since the patterns begin with a continuous swarm ring, which only later leaves behind a pattern with angular variation, it is natural to study first a travelling wave or pulse in the one-dimensional version. The initial configuration is a uniform concentration of food and a dense inoculum of bacteria at one place. There is no chemoattractant present.

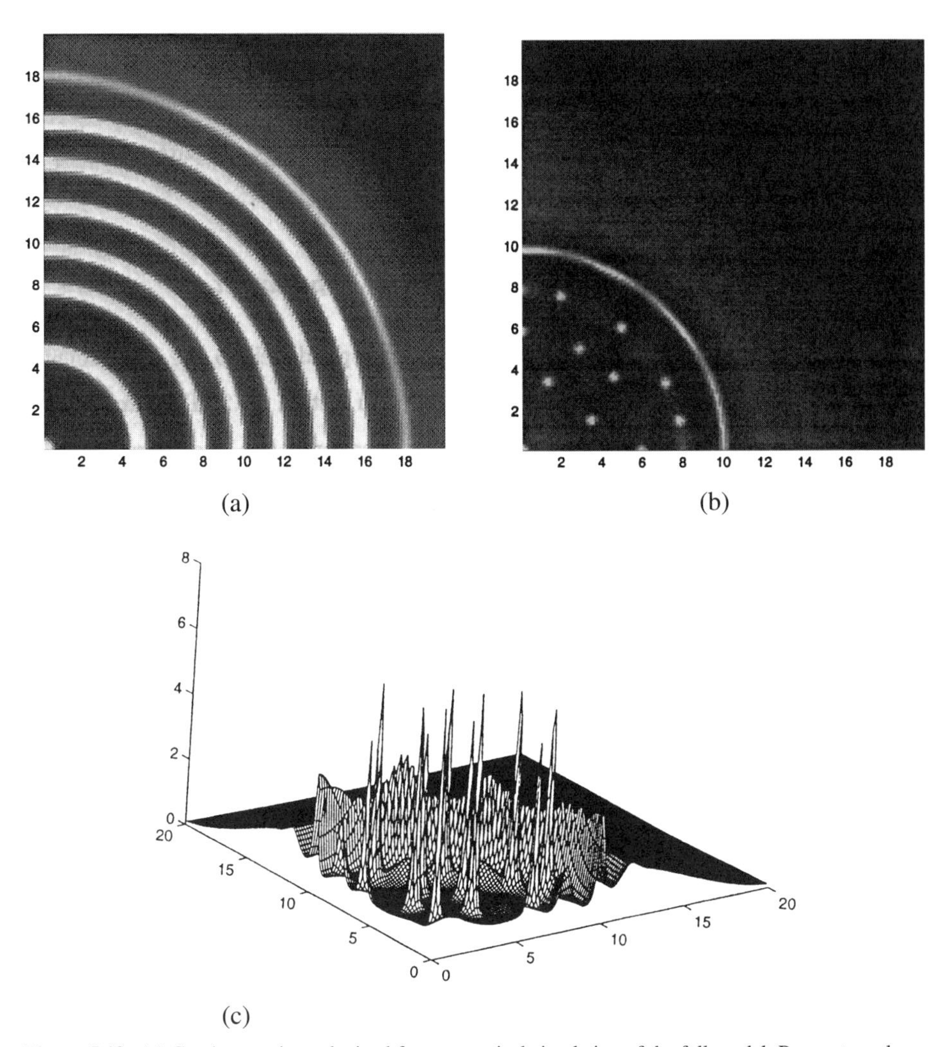

Figure 5.19. (**a**) Continuous rings obtained from numerical simulation of the full model. Parameter values: $d_u = 0.25$, $\alpha = 7.0$, $\beta = 10.0$, $\mu = 250$, $\delta = 10.0$, $\rho = 0.1$, $w = 5.0$. (**b**) Spotted rings obtained with parameter values: $d_u = 0.25$, $\alpha = 30.0$, $\beta = 10.0$, $\mu = 50$, $\delta = 10.0$, $\rho = 1.0$, $w = 0.8$, $\kappa = 0.1$. Note the interdigitation of the spots in successive rings. (**c**) Surface plot of the spotted ring pattern in (**b**). (From Tyson 1996)

The simplest swarm ring can form in the absence of chemoattractant production. As time increases, the bacteria consume the food and diffuse outwards. Those left in the middle become non-motile. At the outer edge of the diffusing mass of cells, the cell density is low and the food concentration is high and as a result the cells proliferate, increasing the local cell density. Meanwhile, at the centre of the spot where the cells were initially placed, the food has been consumed and its concentration reduced to the point where cell death dominates. The result is that cell numbers decrease at the location of the initial inoculum, and increase at the diffusing front. This situation evolves to produce a travelling pulse.

The addition of chemoattractant can make the smooth swarm ring unstable. It is reasonable to suppose that the instability can nucleate more complicated geometries in one, and even more, so in two dimensions, and in particular can give rise to the spots observed trailing the swarm ring in experiments.

We studied this phenomenon analytically in detail. We began by looking for one-dimensional travelling pulse solutions in the simplified version of the model and looked for solutions in terms of the travelling wave coordinate $z = x - ct$, where c is pulse propagation speed which has to be determined. We start with the semi-solid model equations (5.38)–(5.40) and, writing $u(x,t) = U(z)$, $v(x,t) = V(z)$ and $w(x,t) = W(z)$, the travelling waveforms of the equations become

$$d_u U'' + cU' - \alpha \left[\frac{U}{(1+V)^2} V' \right]' + \rho U \left(\delta \frac{W^2}{1+W^2} - U \right) = 0$$

$$V'' + cV' + \beta W \frac{U^2}{\mu + U^2} - UV = 0$$

$$d_w W'' + cW' - \kappa U \frac{W^2}{1+W^2} = 0, \tag{5.85}$$

where prime denotes differentiation with respect to z. Equations (5.85) can be written as a first-order system of 6 equations for U, U', V, V', W, W' with steady states

$$(U, U', V, V', W, W') = (0, 0, 0, 0, 0, 0) \tag{5.86}$$

and

$$(U, U', V, V', W, W') = (0, 0, 0, 0, W_0, 0). \tag{5.87}$$

The first is the steady state which exists behind the pulse, while the second is that which exists in front. Realistic solutions U, V and W must be non-negative and bounded, and so as we approach the two steady states the eigenvalues must be real so that there are no oscillations. Linearising equations (5.85) about (5.86) and (5.87) and solving for the eigenvalues, we find that the first steady state is always a focus. For the second to be a focus as well, we must have

$$c \geq c_{\text{min}} = 2\sqrt{d_u \rho \delta W_0} \tag{5.88}$$

For the Fisher–Kolmogoroff equation, with appropriate initial conditions (namely, compact support) that we studied in detail in Chapter 13, Volume I we showed that a stable travelling wave solution evolves with speed c_{min}. The experiments effectively imply such initial conditions so we suppose that the travelling pulse solution to equations (5.85) will also travel at or near c_{min}. Interestingly neither the rate of food consumption, κ, nor the chemotaxis coefficient, α, have any effect on c_{min}; according to the linear analysis it is purely the kinetics of the diffusing and proliferating bacteria which determine the pulse speed.

Stability of the Swarm Ring

Once the swarm ring has formed and starts to expand across the petri dish, it has been observed experimentally that the ring periodically breaks up, leaving behind a pattern of spots. This suggests that mathematically we should look for a swarm ring solution which is locally stable to perturbations in the radial direction and locally unstable to perturbations in the angular direction.

In the following we sketch how it might be possible to get some stability information on these swarm rings. So that the analytical suggestions mimic the experimental arrangement, we consider a rectangular domain (for analytical convenience), with the initial inoculum of cells placed along one edge of length l. Perpendicular to that edge we expect the solution to be a travelling pulse. We rewrite the two-dimensional model in (x, y, t) coordinates in terms of one travelling pulse coordinate, z, the coordinate parallel to the front, y, and so the equations (5.38)–(5.40) become

$$\begin{aligned}
\frac{\partial u}{\partial t} - c\frac{\partial u}{\partial z} &= d_u\left(\frac{\partial^2 u}{\partial z^2} + \frac{\partial^2 u}{\partial y^2}\right) - \alpha\left(\frac{\partial}{\partial z}, \frac{\partial}{\partial y}\right)\cdot\left[\frac{u}{(1+v)^2}\left(\frac{\partial v}{\partial z}, \frac{\partial v}{\partial y}\right)\right] \\
&\quad + \rho u\left(\delta\frac{w^2}{1+w^2} - u\right) \\
\frac{\partial v}{\partial t} - c\frac{\partial v}{\partial z} &= \left(\frac{\partial^2 v}{\partial z^2} + \frac{\partial^2 v}{\partial y^2}\right) + \beta\frac{wu^2}{\mu + u^2} - uv \\
\frac{\partial w}{\partial t} - c\frac{\partial w}{\partial z} &= d_w\left(\frac{\partial^2 w}{\partial z^2} + \frac{\partial^2 w}{\partial y^2}\right) - \kappa\frac{uw^2}{1+w^2}.
\end{aligned} \tag{5.89}$$

Now suppose that in the z-direction we have a travelling wave solution $U(z, y) = U(z)$, $V(z, y) = V(z)$, $W(z, y) = W(z)\ \forall\ y \in [0, l]$. Then by definition,

$$\begin{aligned}
-c\frac{\partial U}{\partial z} &= d_u\left(\frac{\partial^2 U}{\partial z^2}\right) - \alpha\left(\frac{\partial}{\partial z}\right)\cdot\left[\frac{U}{(1+V)^2}\left(\frac{\partial V}{\partial z}\right)\right] \\
&\quad + \rho U\left(\delta\frac{W^2}{1+W^2} - U\right) \\
-c\frac{\partial V}{\partial z} &= \left(\frac{\partial^2 V}{\partial z^2}\right) + \beta\frac{WU^2}{\mu + U^2} - UV \\
-c\frac{\partial W}{\partial z} &= d_w\left(\frac{\partial^2 W}{\partial z^2}\right) - \kappa\frac{UW^2}{1+W^2}.
\end{aligned} \tag{5.90}$$

If this solution is stable to perturbations in z, as it is likely to be for some range of parameters, then we could get some idea of the effect of perturbations perpendicular to the direction of the wave by looking at spatiotemporal perturbations involving only the spatial coordinate y. So as a first attempt at such a stability analysis we could consider the solutions u, v and w of the form

$$
\begin{aligned}
u(z, y, t) &= U(z) + \bar{u}e^{\lambda t + iky} \\
v(z, y, t) &= V(z) + \bar{v}e^{\lambda t + iky} \\
w(z, y, t) &= W(z) + \bar{w}e^{\lambda t + iky},
\end{aligned}
\tag{5.91}
$$

where $\bar{u}$, $\bar{v}$ and $\bar{w}$ are small constants. Substituting these forms into equations (5.89) and collecting terms of $O(\bar{u})$, $O(\bar{v})$ and $O(\bar{w})$ we would obtain a linear system of equations of the form

$$
A \begin{bmatrix} \bar{u} \\ \bar{v} \\ \bar{w} \end{bmatrix} e^{\lambda t + iky} = 0,
\tag{5.92}
$$

where the matrix A is given by

$$
A = \begin{bmatrix}
\lambda + d_u k^2 - H_1 & H_2 & -\rho\delta \frac{2UW}{(1+W^2)^2} \\
-\beta\mu \frac{2WU}{(\mu+U^2)^2} + V & \lambda + k^2 + U & -\beta \frac{U^2}{\mu+U^2} \\
\kappa \frac{W^2}{1+W^2} & 0 & \lambda + d_w k^2 + \frac{2\kappa UW}{(1+W^2)^2}
\end{bmatrix},
\tag{5.93}
$$

where

$$
\begin{aligned}
H_1 &= \rho \left(\frac{\delta W^2}{1 + W^2} - 2U \right) + \alpha \left[\frac{V'}{(1 + V)^2} \right]', \\
H_2 &= -\frac{\alpha U k^2}{(1 + V)^2} - 2\alpha \left[\frac{UV'}{(1 + V)^3} \right]'.
\end{aligned}
\tag{5.94}
$$

In the now usual way, nonzero solutions exist for the perturbations $\bar{u}$, $\bar{v}$ and $\bar{w}$ if and only if $| A | = 0$ which gives the characteristic equation for the dispersion relation λ. Setting $| A | = 0$ we get a cubic equation for λ of the form

$$
\lambda^3 + A\lambda^2 + B\lambda + C = 0,
\tag{5.95}
$$

where the A, B and C are functions of the parameters, the wavenumber k and functions of z via the travelling wave solutions U, V, W and their derivatives. If we suppose, for the moment that A, B and C are constants, the Routh-Hurwitz conditions (see Appendix A, Volume 1) which guarantee $Rl(\lambda) < 0$ are

$$
A > 0, \quad C > 0, \quad AB - C > 0.
\tag{5.96}
$$

So, to ensure instability, namely $Rl(\lambda) > 0$ for some $k^2 \neq 0$ at least one of these conditions must be violated. We rewrite the coefficient matrix A of (5.93) as

$$\begin{bmatrix} \lambda + c_1 & c_4 & -c_5 \\ c_6 & \lambda + c_2 & -c_7 \\ +c_8 & 0 & \lambda + c_3 \end{bmatrix}, \tag{5.97}$$

where the terms which are always negative have a minus sign in front and those which are clearly positive have a plus sign. With this notation (5.95) then has

$$\begin{aligned} A &= c_1 + c_2 + c_3, \\ B &= c_1c_2 + c_1c_3 + c_2c_3 - c_4c_6 + c_5c_8, \\ C &= c_1c_2c_3 - c_4(c_6c_3 + c_7c_8) + c_5c_2c_8. \end{aligned} \tag{5.98}$$

Now consider the three necessary conditions for stability. With A, the only term which can be negative is c_1, and this can occur only if H_1 in (5.94) is sufficiently large. But this depends on z and hence where we are on the travelling wavefront. Similar arguments apply to the other conditions of stability. It is clear then that whether or not the travelling wavefront, the swarm ring, is unstable to transverse perturbations depends on the travelling wave variable z. It would be astonishing if, for at least some z, it was not possible to violate the conditions (5.96) and hence have $Rl\lambda > 0$ for nonzero wavenumbers and transverse spatial instabilities which imply the breaking up of the swarm ring into spots. Since the chemotaxis parameter α appears in H_1 and H_2 it is clear that once again chemotaxis plays a critical role. A full analysis of the stability of these swarm rings is a challenging unsolved problem.

Tyson (1996) solved the one-dimensional equations and obtained a single travelling pulse or a train of two to four travelling pulses from the full model for a variety of parameter values, and nonnegligible consumption of nutrient. The effect of the model parameters α, β, μ, δ, ρ and w on the travelling pattern is analogous to their effect on the stationary patterns that we discussed above. The wavelength of the pulses in the pulse train is affected chiefly by the chemotaxis coefficient α.

The predicted and numerically computed wavespeeds compare very well. The computed wavespeed was always larger (around 5–10%) than the predicted value, $c_{\min}$, as it should be. Since necessarily boundary conditions had to be imposed on a finite domain, we do not expect the wavespeed to be as small as $c_{\min}$ in our simulations.

Numerical Results for Two-Dimensional Swarm Rings

In two dimensions, we found that the travelling pulse trains become a swarm ring followed by one or two rings of spots. The spots arise from inner rings which develop angular instabilities and subsequently break up into spots. Image plots of a well-developed swarm ring pattern are shown in Figure 5.20 at various times in their development.

This model development and analysis was the first mathematical model of *E. coli* and *S. typhimurium* which yielded a swarm ring spawning spots without assuming any extraneous biological activity. Woodward et al. (1995) obtained interesting patterns by assuming the presence of a chemoattracting nutrient. Experiments with *S. typhimurium* have been performed under such circumstances, but in the *E. coli* experiments no such nutrient is present. With this model, we have shown that chemotaxis toward the aspar-

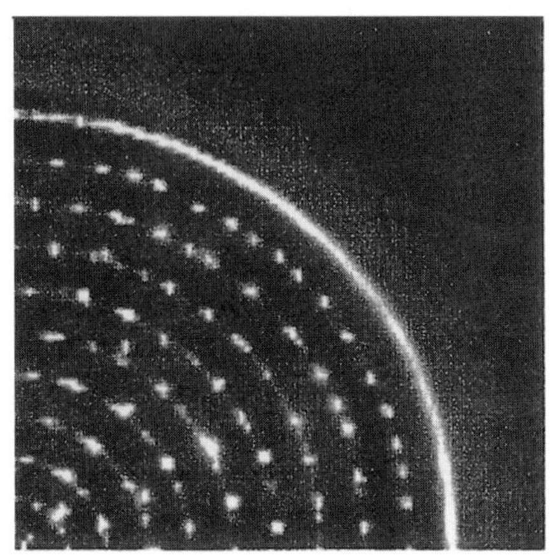

Figure 5.20. Swarm ring patterns in *S. typhimurium* obtained from a numerical simulation of the full system (5.38)–(5.40): (**a**) Concentric rings showing transition to a spotted ring pattern. Parameter values: $d_u = 0.25$, $d_w = 0.8$, $\alpha = 40$, $\beta = 10$, $\delta = 70$, $\rho = 1$, $\kappa = 4.5 \times 10^{-3}$, $\mu = 10^2$. (**b**) Concentric spotted rings. Parameter values: $d_u = 0.26$, $d_w = 0.89$, $\alpha = 88.9$, $\beta = 8$, $\delta = 7$, $\rho = 1$, $\kappa = 10^{-3}$, $\mu = 10^2$. (From Tyson 1996)

tate produced by the cells coupled with consumption of food are sufficient to generate the experimentally observed behaviour. This is in keeping with the experimentalists' intuition.

5.11 Branching Patterns in *Bacillus subtilis*

The patterns we have discussed above are complex, but fairly regular patterns formed by spots and rings. The bacterium *Bacillus subtilis*, when inoculated onto a agar medium which has little nutrient, can exhibit quite different fractal-like patterns not unlike those found in diffusion-limited aggregation (see, for example, Matsuyama and Matsushita 1993). When the agar is semi-solid, however, the bacterial colonies formed by *Bacillus subtilis* are dense-branching patterns enclosed by a smooth envelope. The stiffness of the medium affects the patterns formed. Shigesada and her colleagues (Kawasaki et al. 1997) have studied this particular bacterium and constructed a relatively simple reaction diffusion model which captures many of the pattern characteristics found experimentally: they compare the results with experiments. Here we briefly describe their model and show some of their results. Although their model is a reaction diffusion one it is original and fundamentally different to those reaction diffusion systems we have studied up to now. It highlights, once again, the richness of pattern formation by such relatively simple systems.

They propose a model consisting of a conservation equation for the bacterial cells and the nutrient given by

$$\frac{\partial n}{\partial t} = D_n \nabla^2 n - \frac{knb}{1 + \gamma n} \tag{5.99}$$

$$\frac{\partial b}{\partial t} = \nabla \cdot (D_b \nabla b) + \theta \frac{knb}{1+\gamma n}, \quad D_b = \sigma nb, \tag{5.100}$$

where n and b are the concentration of the nutrient and bacterial cell densities respectively. Here the function $knb/(1+\gamma n)$, where k and γ are constants, is the consumption rate of the nutrient by the bacteria and $\theta(knb/(1+\gamma n))$ is the growth rate of the cells with θ the conversion rate factor. D_n and D_b are the diffusion coefficients of the nutrient and cells respectively. We now motivate the form given for D_b.

The reasoning behind the form $D_b = \sigma nb$ is based on the work of Ohgiwara et al. (1992) who observed the detailed movement of the bacteria and found that the cells did not move much in the inner region of the expanding colony where the level of nutrient was low but that they moved vigorously at the periphery of the colony where the nutrient level is much higher. They also noted that at the outermost front of the colony, where the cell density is quite low, the cells were again fairly inactive. Kawasaki et al. (1997) then argued that the bacteria are immobile where either the nutrient n or the bacteria density b are small. They modelled these effects by taking the bacterial diffusion as proportional to nb with the proportionality factor σ. It was also observed that although each cell moves in a typical random way some of them exhibit stochastic fluctuations. They quantified this by setting $\sigma = 1 + \Delta$ where the parameter Δ is a measure of the stochastic fluctuation from the usual random diffusion.

Kawasaki et al. (1997) studied the pattern formation potential of these model equations in two dimensions subject to initial conditions

$$n(\mathbf{x}, 0) = n_0, \quad b(\mathbf{x}, 0) = b_0(\mathbf{x}), \tag{5.101}$$

where n_0 is the concentration of the initial uniformly distributed nutrient and $b_0(\mathbf{x})$ is the initial inoculum of bacteria. Since the nutrient concentration in the experiments is relatively low the saturation effect, accounted for by the γn term, is negligible so the consumption of nutrient can be taken as approximately knb, the functional form we use below.

We nondimensionalise the equations by setting

$$n^* = \left(\frac{\theta}{D_n}\right)^{1/2} n, \quad b^* = \left(\frac{1}{\theta D_n}\right)^{1/2} b, \quad \gamma^* = \left(\frac{D_n}{\theta}\right)^{1/2} \gamma \tag{5.102}$$

$$t^* = k(\theta D_n)^{1/2} t, \quad \mathbf{x}^* = \left(\frac{\theta k^2}{D_n}\right)^{1/4} \mathbf{x}, \tag{5.103}$$

with which the model mechanism becomes, using the above consumption approximation with $\gamma = 0$, and omitting the asterisks for algebraic simplicity,

$$\frac{\partial n}{\partial t} = \nabla^2 n - nb \tag{5.104}$$

$$\frac{\partial b}{\partial t} = \nabla \cdot (\sigma nb \nabla b) + nb, \tag{5.105}$$

which has only one parameter σ, with initial conditions

$$n(\mathbf{x}, 0) = \left(\frac{\theta}{D_n}\right)^{1/2} n_0 \equiv \nu_0, \quad b(\mathbf{x}, 0) = \left(\frac{1}{\theta D_n}\right)^{1/2} b_0(\mathbf{x}) \equiv \beta_0(\mathbf{x}). \tag{5.106}$$

Kawasaki et al. (1997) solved this system under a variety of different situations and found that the solutions exhibited a remarkable spectrum of complex patterns. Figure 5.21 shows some examples.

Kawasaki et al. (1997) also investigated the patterns formed when the stochastic parameter $\Delta = 0$. Patterns still form and still give rise to branching-like patterns but since there is no anisotropy they are much more regular and symmetric. In reality, of course, the small random perturbations in nutrient and bacterial densities would not result in such regular patterns as pointed out by Kawasaki et al. (1997).

Although the pattern evolves in two dimensions each tip essentially grows in one dimension except when they branch. This makes it possible to obtain some approximate analytical results for tip growth, and hence the colony growth, using the one-dimensional version of the model equations, namely,

$$\frac{\partial n}{\partial t} = \frac{\partial^2 n}{\partial x^2} - nb \tag{5.107}$$

$$\frac{\partial b}{\partial t} = \frac{\partial}{\partial x}\left(\sigma nb \frac{\partial b}{\partial x}\right) + nb. \tag{5.108}$$

The numerical simulation of these one-dimensional equations gave growth rates which compared well with those obtained from the two-dimensional equations. They made a further approximation to the model by substituting $\sigma_0 \nu_0 b$ and $\nu_0 b(1 - b/K)$ for σnb and nb in (5.108) to obtain the scalar equation in b, namely,

$$\frac{\partial b}{\partial t} = \frac{\partial}{\partial x}\left(\sigma_0 \nu_0 b \frac{\partial b}{\partial x}\right) + \nu_0 b(1 - b/K). \tag{5.109}$$

Here the growth of the bacteria is limited by the nutrient according to a typical logistic growth where K is the saturating level of the bacteria. If we consider (5.107) and (5.108) in the absence of diffusion we can add the equations, integrate and with $n + b = \nu_0$ we get the logistic form $db/dt = \nu_0 b(1 - b/\nu_0)$ for the bacteria and so we relate K to ν_0. The form of (5.109) is then the same as the equation we discussed in detail in Section 13.4 in Chapter 13, Volume I and which has an exact travelling wave solution with the wavespeed $v_{\text{colony growth}}$ given by

$$v_{\text{colony growth}} = \left(\frac{\sigma_0 \nu_0^3}{2}\right)^{1/2}. \tag{5.110}$$

This velocity is a very good approximation (a slight overestimate) for the velocity of colonial growth obtained from (5.107) and (5.108).

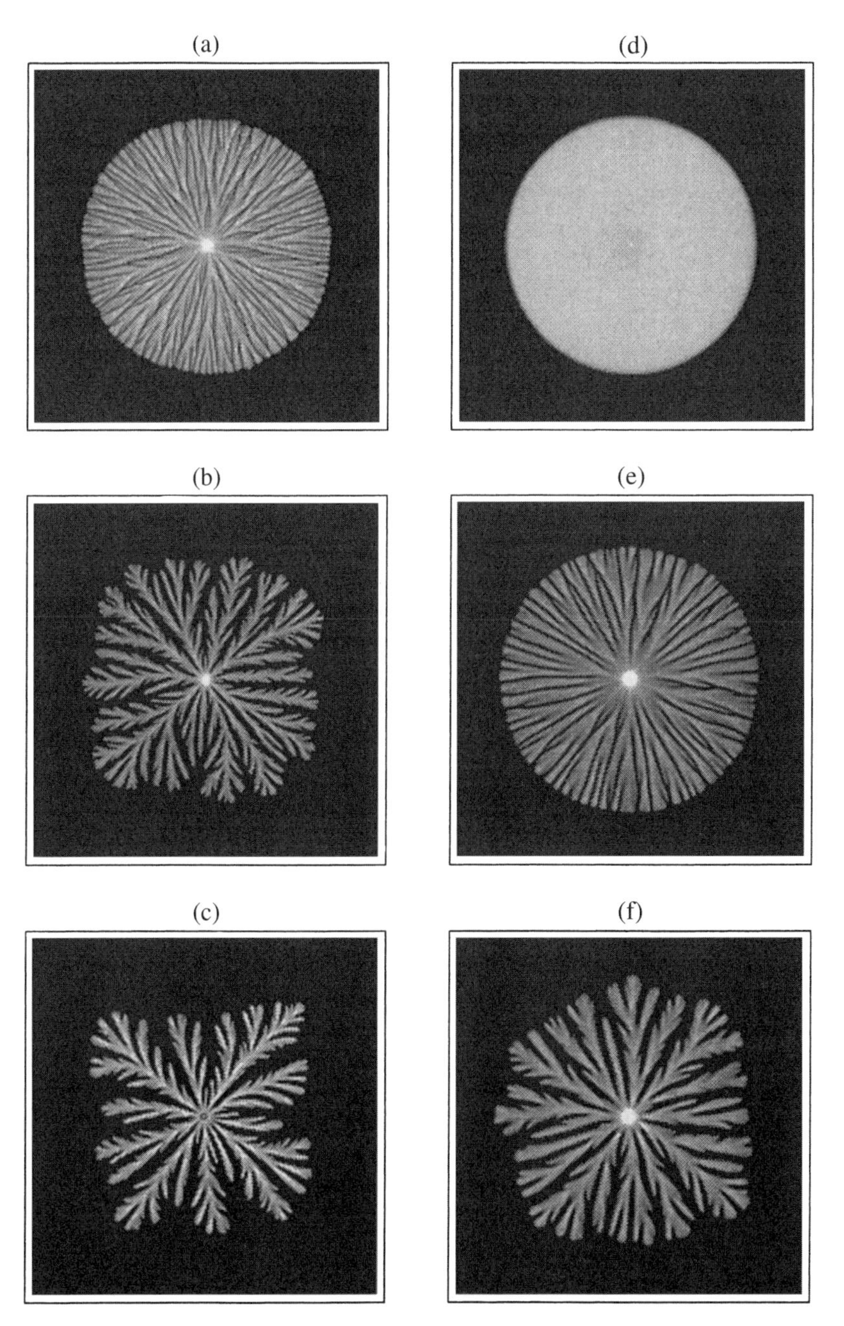

Figure 5.21. Typical dense-branching bacterial patterns for the *Bacillus subtilis* model obtained from a numerical simulation of (5.104)–(5.105). The parameter σ was perturbed about the mean σ_0 with the random variable Δ. Parameter values and times are: (**a**) $\sigma_0 = 1$, $\nu_0 = 1.07$, $t = 396$; (**b**) $\sigma_0 = 1$, $\nu_0 = 0.71$, $t = 2828$; (**c**) $\sigma_0 = 1$, $\nu_0 = 0.35$, , $t = 19233$; (**d**) $\sigma_0 = 4$, $\nu_0 = 1.07$, $t = 127$; (**e**) $\sigma_0 = 4$, $\nu_0 = 0.71$, $t = 566$, (**f**) $\sigma_0 = 4$, $\nu_0 = 0.35$, $t = 4525$. In real time the pattern is quite dense after about two days. (From Kawasaki et al. 1997 and reproduced with the permission of Dr. N. Shigesada)

When the nutrient level is not low the full model (5.99) and (5.100) has to be used with $\gamma \neq 0$. Kawasaki et al. (1997) considered such situations and found that as γ increases the branch width increases and the degree of complexity decreases. Their way of including a stochastic element is interesting and important since it allows for some stochasticity without the usual complexities involved in such studies. The concept clearly has a much wider application such as to many of the models studied in this book.

6. Mechanical Theory for Generating Pattern and Form in Development

6.1 Introduction, Motivation and Background Biology

In spite of the seemingly endless series of exciting new discoveries in other areas of biology from mapping the genome to cloning animals the major problems in developmental biology are still essentially unsolved. In an interesting survey conducted by the journal *Science* in 1994 (Barinaga 1994) more than 100 leading developmental biologists were asked what they thought were the most important unanswered problems in development and where they thought the biggest breakthroughs would come in the following five years.

Among the 66 responses, the most important unanswered question was that of how the body's specialized organs and tissues are formed. The formation of structure in embryology is known as morphogenesis. Coming second after morphogenesis was how the actual mechanisms evolved and how evolution acted on the mechanisms to effect change and generate new species. What is more, morphogenesis came in second on the list of areas where rapid progress is expected in the next five years. By a significant margin the largest number of votes on 'Development's greatest unsolved mysteries' was, 'What are the molecular mechanisms of morphogenesis?' Also high up on their lists was the question, 'How are patterns established in the early embryo?'

Now, at the start of the 21st century, just over five years after the survey, although there have indeed been many new discoveries in morphogenesis we still do not know of any actual mechanism for generating spatial pattern in a developing embryo. The mechanisms for laying down an animal's body plan is still unsolved. Of the 'Dozen hot areas for the next half decade' listed in the survey, the mechanisms of morphogenesis came second on the list. Morphogenesis encompasses pattern formation from the initial mass of cells to the final body form. There is clearly no need to justify studying possible mechanisms for generating biological pattern and form. It is, without question, still (and for the foreseeable future) a central issue in embryology.

Brief Historical Aside

It is interesting to recall a fairly widely held scientific view in the latter half of the 19th century, namely that if we understand the development of one animal then we can extend this understanding to other animals. This is no different, in a general sense, to the current view which justifies the intense research into fruit fly development, salamander development and so on. A major, highly influential and justly controversial naturalist in the second half of the 19th and early 20th centuries, Ernst Haeckel (1834–1919),[1] drew a number of embryos at what he said were parallel developmental stages to indicate their similarities and help prove, among other things, his theory that "ontogeny recapitulates phylogeny." In other words this means that developing organisms pass through their evolutionary history: for example early embryonic slits for gills in the human embryo purport to reflect the evolutionary descent from fish. The theory is wrong, of course, but it is one of Haeckel's theories for which he is still well known; even now, it is still not universally discarded. One of his figures is reproduced in Figure 6.1. Haeckel, an extremely talented technical artist, simply falsified some of the figures. In many cases he drew the embryos from real specimens but, in some instances, left out crucial elements, such as limb buds, so that he could say there were no traces of limbs at the stage of development he was purporting to show.

The fraudulent manipulation by Haeckel was known to many scientists of the time but since the mid-1990's there has been a resurgence of interest in Haeckel, not only for his dishonest manipulation of biological images, but for some of his other ideas, such as that of a superior race, a view which was warmly embraced by the eugenicists and others in the first half of the 20th century. In a general article about Haeckel, Gould (2000) discusses some of the disreputable facts and puts them in a scientific and historical context: see also the brief article by Richardson and Keuck (2001) and references there.

Pattern generation models are generally grouped together as morphogenetic models. These models provide the embryologist with possible scenarios as to how pattern is laid down and how the embryonic form might be created. Although genes of course play a crucial role in the control of pattern formation, genetics says nothing about the actual *mechanisms* involved nor how the vast range of pattern and form that we see evolves from a homogeneous mass of dividing cells.

Broadly speaking the two prevailing views of pattern generation that have dominated the thinking of embryologists in the past 20 years or so are the long-standing Turing chemical prepattern approach (that is, the reaction diffusion-chemotaxis mechanisms, which we have discussed at length in previous chapters, and the Murray–Oster mechanochemical approach developed by G.F. Oster and J.D. Murray and their colleagues (for example, Odell et al. 1981, Murray et al. 1983, Oster et al. 1983, Murray and Oster 1984a,b, Lewis and Murray 1991). General descriptions of the mechanical approach have been given, for example, by Murray and Maini (1986), Oster and Murray (1989), Murray et al. (1988), Bentil (1991), Cruywagen (1992), Cook (1995) and Maini (1999). Specific components have also been studied, for example, by Barocas and Tranquillo (1994, 1997a,b), Barocas et al. (1995), Ferrenq et al. (1997) and Tranqui and Tracqui (2000). Numerous other references involving the theory and its practical use will be given at appropriate places in this and the following four chapters.

In this chapter we develop in some detail the Murray–Oster mechanical approach to biological pattern formation, which, among other things, considers the role that mechanical forces play in the process of morphogenetic pattern formation, and apply it to several specific developmental problems of current widespread interest in embryology. A clear justification for the need for a mechanical approach to the development of pat-

[1] Haeckel, incidently, coined the word ecology for the discipline as we know it.

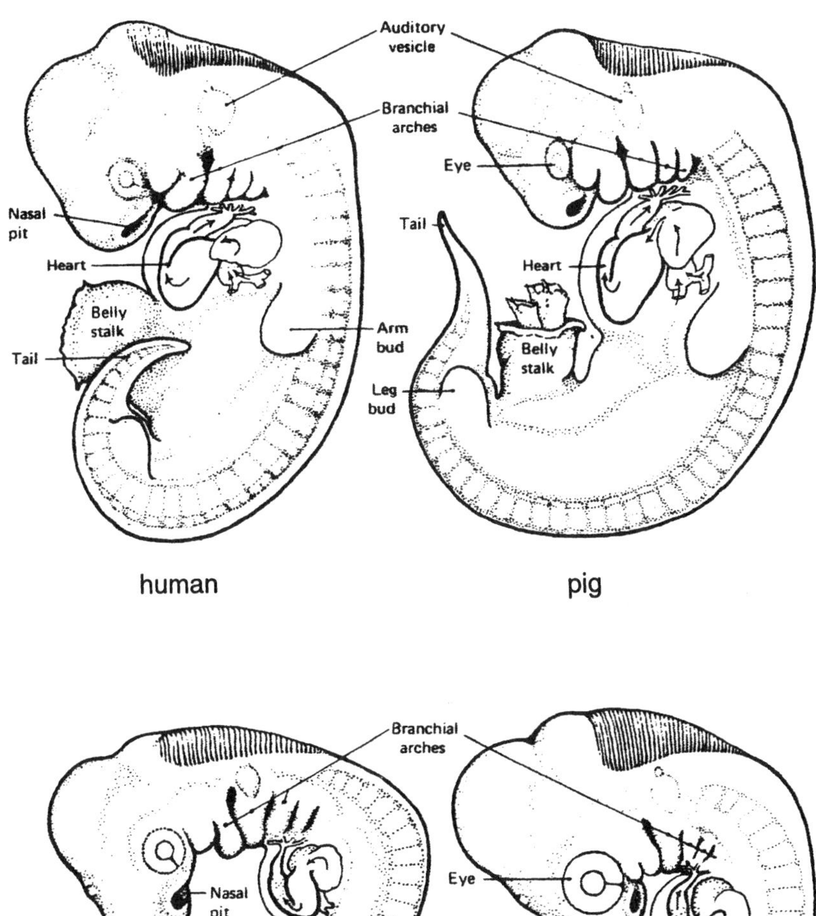

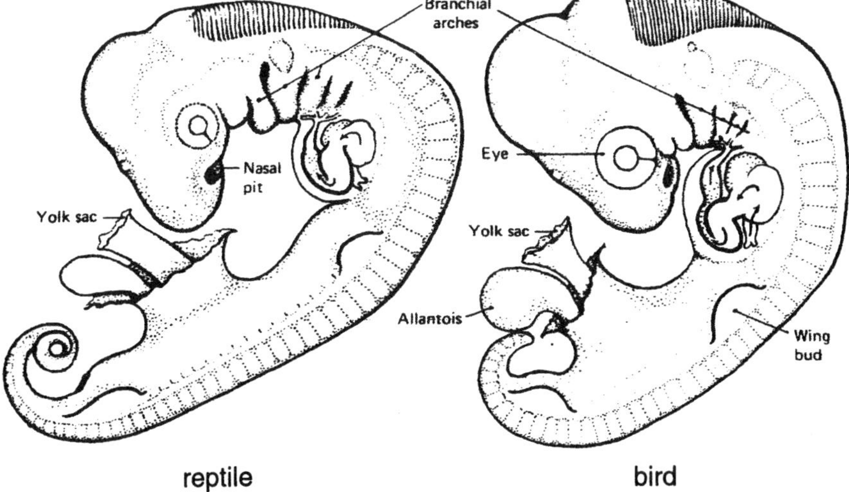

Figure 6.1. Mid-19th century drawings by Ernst Haeckel (1834–1919) of a series of embryos at similar developmental stages as he wanted to portray them. He was trying to make the point that embryos were fairly similar at equivalent stages in their development (which they are not, of course).

tern in cellular terms is inferred from the following quote from Wolpert (1977). 'It is clear that the egg contains not a description of the adult, but a program for making it, and this program may be simpler than the description. Relatively simple cellular forces can give rise to complex changes in form; it seems simpler to specify how to make complex shapes than to describe them.'

I feel it is important to make the case irrefutable for the inclusion of forces in thinking about morphogenetic processes. The following is essentially the preface my friend and colleague George Oster and I had intended for a book (in the mid-1980s) on the mechanochemical theory of biological pattern formation. The intention is certainly not to denigrate other models and mechanisms discussed in detail in the book and which also have major roles to play, but rather to add perspective to what are fundamental questions in morphogenesis.

> Looking at any time-lapse film of cells or developing embryos perhaps the one overwhelming impression is the constant motion. Cells move and embryos twitch and jerk about incessantly. Indeed, the very notion of morphogenesis (*morphos* = shape, genesis = change) implies motions—the motions that shape the embryo.
>
> All motions require forces to generate them. It is surprising that this fundamental law of nature has largely been ignored by embryologists and cell biologists. Very few books on embryology even mention forces. There may be good reasons for this, for only recently has it become possible to actually measure mechanical forces at the cell and tissue level. And what good to ponder immeasurable quantities while the sirens of chemistry and genetics beckon with tangible rewards?
>
> However one chooses to ignore mechanics, nevertheless, presiding over every embryonic twitch and jerk are Newton's laws. And whatever role chemistry and genetics play in embryogenesis, they must finally submit their programs for Newtonian execution. Therefore, we have adopted the philosophy that, since morphogenesis is—at least proximally—a mechanical event, it is reasonable to start analyses of morphogenetic processes by examining the forces that produced them, and then, working backwards, add chemistry and genetics as needed.
>
> Aside from personal prejudices and a certain aesthetic parsimony, we feel that there is a deeper rationale for viewing morphogenesis from a mechanical perspective. This reason arises from considerations of stability and evolutionary economy.
>
> It is certainly possible to construct an organism by first laying down a chemical prepattern, and then have the cells execute their internally programmed instructions for mechanical behaviour (for example, shape change) according to the chemically specified recipe. In this view, mechanics is simply a slave process to chemistry. Indeed, there is a large number of biologists who think that embryogenesis works in just this way.
>
> However, building an organism in this way would be a very unreliable and unstable enterprise. How would it correct for the inevitable chemical and

mechanical perturbations an embryo must face? Once the chemical prepattern has been laid down, the cells must execute these instructions regardless of any new contingencies that arise. In the parlance of control theory, the system is 'open loop': there is no feedback from the mechanical state of the tissue to its chemical state. Such control systems are notoriously unstable, for they are unable to correct and compensate for external disturbances. The only protection against disturbances according to this scheme is for the genetic system to anticipate and code for any possible mechanical or chemical perturbation. This would obviously place an enormous burden of complexity on the genetic control system, and it is hard to see how such a gadget could evolve.

A more reasonable alternative, in our view, is that nature (that is, evolution) has 'closed the loop' so that the mechanical state of a cell or tissue can influence its chemical state. Thus, mechanical disturbances can be compensated and genetic programs executed reliably without the burden of overhead programming that an open loop system requires.

In this view, embryogenesis is not primarily a problem in specifying a chemical prepattern; rather it is a mechanochemical process wherein it is artificial to separate 'pattern formation' from 'morphogenesis.' The mechanics and the chemistry act in concert to create spatial patterns directly, neither being the slave of the other, but both participating in a coordinated feedback scheme.

If one accepts this view, then a certain modelling philosophy follows. The laws of chemical kinetics are remarkably unconfining: with no constraints on the number of reactions it is easy to design chemical networks that will accomplish virtually any desired dynamical behaviour. The laws of Newtonian mechanics and physical chemistry, however, are not so flexible. They constrain what is possible quite severely. These constraints are just what is needed to prune down the mechanisms one might conceive for a particular embryonic system.

In the mechanochemical models we have adopted this philosophy, and commenced our analyses of embryogenic phenomena by focusing first on the mechanical forces which drive the observed behaviour. Then we can add chemistry in the simplest way consistent with the facts so that the resulting mechanochemical model can stably reproduce the physical pattern.

This is not to say that nature always acts in the simplest possible way: parsimony is a human construct, and evolution is an opportunistic process which builds on the available materials, not according to any global optimization scheme. However, when building models, it ill behooves the modeller to capriciously add complexity when simple mechanisms will do the job.

The best way to decide between competing models is through experiment. We feel that one of the modeller's jobs is to present the experimental biologist with a shopping list of possibilities which are consistent not only with the observations, but with the known laws of physics and chemistry.

These models will suggest experiments, and guide further model building. We see modelling and experiment cooperating in a feedback loop—just as chemistry and mechanics do in our models—the combination being a more efficient tool for research than either one acting alone.

How complicated should a model be? Consider the task of explaining to someone how a clock works. It would help, of course, if they understood the mechanics of gears and levers; however, to understand the clock you would have to simply describe it: this gear turns that one, and so on.

Now this is not a very satisfactory way to understand a phenomenon; it is like having a road map with a scale of one mile equals one mile. 'Understanding' usually involves some simplified conceptual representation that captures the essential features, but omits the details or secondary phenomena. This is as good a definition as any of what constitutes a model.

Just how simplified a model can be and still retain the salient aspects of the real world depends not only on the phenomenon, but how the model is to be used. In these chapters on mechanical aspects of morphogenesis we deal only with mathematical models; that is, phenomena which can be cast in the form of equations of a particular type.

Mathematical models can be used to make detailed predictions of the future behaviour of a system (as we have seen). This can be done only when the phenomenon is rather simple; for complex systems the number of parameters that must be determined is so large that one is reduced to an exercise in curve fitting. The models we deal with in this book have a different goal. We seek to explain phenomena, not simply describe them.

If one's goal is explanation rather than description then different criteria must be applied. The most important criterion, in our view, was enunciated by Einstein: 'A model should be as simple as possible. But no simpler.' That is, a model should seek to explain the underlying principles of a phenomenon, but no more. We are not trying to fit data nor make quantitative predictions. Rather we seek to understand. Thus we ask only that our models describe qualitative features in the simplest possible way.

Unfortunately, even with this modest (or ambitious) goal, the equations we deal with are probably more complicated than even most physical scientists are accustomed to. This is because the phenomena we are attempting to describe are generally more complex than most physical systems, although it may reflect our own ineptness in perceiving their underlying simplicity.

The reaction diffusion (and chemotaxis) approach is basically quite different to the mechanical approach. In the chemical prepattern approach, pattern formation and morphogenesis take place sequentially. First the chemical concentration pattern is laid down, then the cells interpret this prepattern and differentiate accordingly. So, in this approach, morphogenesis is essentially a slave process which is determined once the chemical pattern has been established. Mechanical shaping of form which occurs during embryogenesis is not addressed in the chemical theory of morphogenesis. The elusiveness of these chemical morphogens has proved a considerable drawback in the acceptance of such a theory of morphogenesis. There is, however, absolutely no doubt that chemicals play crucially important roles in development.

In the mechanochemical approach, pattern formation and morphogenesis are considered to go on simultaneously as a single process. The patterning and the form-shaping movements of the cells and the embryological tissue interact continuously to produce the observed spatial pattern. Another important aspect of this approach is that the models associated with it are formulated in terms of measurable quantities such as cell densities, forces, tissue deformation and so on. This focuses attention on the morphogenetic process itself and in principle is more amenable to experimental investigation. As we keep repeating, the principal use of any theory is in its predictions and, even though each theory might be able to create similar patterns, they are mainly distinguished by the different experiments they suggest. We discuss some of the experiments associated with the mechanical theory later in this and the next four chapters. The chapters on vasculogenesis and dermal wound healing rely heavily on the concepts developed in this chapter.

A particularly telling point in favour of simultaneous development is that such mechanisms have the potential for self-correction. Embryonic development is usually a very stable process with the embryo capable of adjusting to many outside disturbances. The process whereby a prepattern exists and then morphogenesis takes place is effectively an open loop system. These are potentially unstable processes and make it difficult for the embryo to make the necessary adjustment to such disturbances as development proceeds.

In this chapter we discuss morphogenetic processes which involve coordinated movement or patterning of populations of cells. The two types of early embryonic cells we are concerned with are fibroblast, or dermal or mesenchymal cells and epidermal, or epithelial cells. Fibroblast cells are capable of independent movement, due to long finger-like protrusions called filopodia or lamellapodia which grab onto adhesive sites, which can be other cells, and pull themselves along (you can think of such a cell as something like a minute octopus); spatial aggregation patterns in these appear as spatial variations in cell number density. Fibroblasts can also secrete fibrous material which helps to make up the extracellular matrix (ECM) tissue within which the cells move. Epidermal cells, on the other hand, in general do not move but are packed together in sheets; spatial patterns in their population are manifested by cell deformations. Figure 6.2 schematically sets out some of the key properties of the two types of cells. A good description of the types of cells, their movement properties and characteristics and their role in embryogenesis, is given, for example, in the textbook by Walbot and Holder (1987). The definitive text exclusively on the cell is by Alberts et al. (1994). These books are particularly relevant to the material in this chapter.

We first consider mesenchymal (fibroblast) cell pattern formation in early embryogenesis. In animal development the basic body plan is more or less laid down in the first few weeks, such as the first 4 weeks in man, where gestation is about 280 days, and not much more in the case of a giraffe, for example, which has a gestation period of nearly 460 days. It is during this crucial early period that we expect pattern and form generating mechanisms, such as we propose here, to be operative. We saw in Chapter 4 that the alligator embryo looked very much like a small alligator very early in gestation.

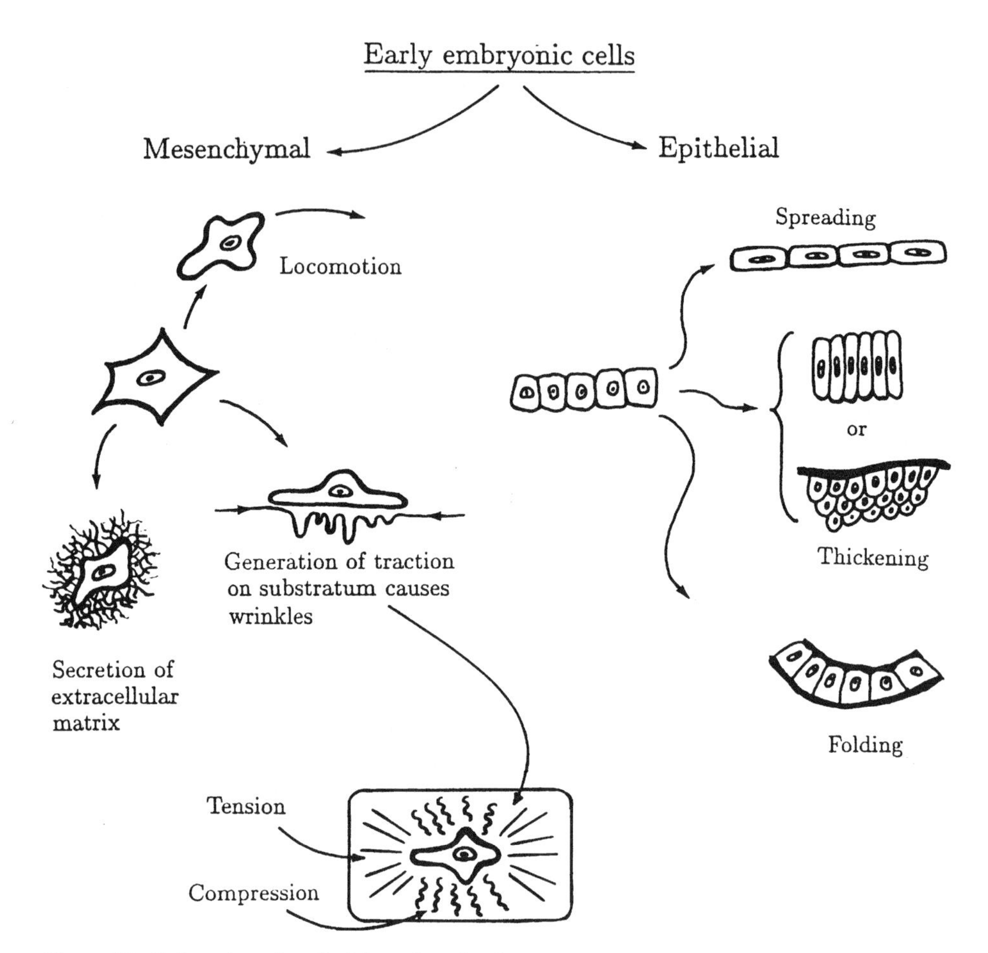

Figure 6.2. Early embryonic cells. Mesenchymal cells are motile, generate large traction forces and can secrete extracellular matrix which forms part of the tissue within which the cells move. When these cells are placed on a thin silicon rubber substratum their traction forces deform the rubber sheet; see the photograph in Figure 6.3. Epithelial cells do not move about but can spread or thicken when subjected to forces; this affects cell division (see, for example, Folkman and Moscona 1978).

The models we discuss here take into account considerably more biological facts than most of those we have considered up to now. Not surprisingly this makes the models more complicated. It is essential, however, for mathematical biologists genuinely concerned with real biology to appreciate the complexity of biology. So, it is appropriate that we should now discuss the modelling of mechanisms for some of the more complex but realistic aspects of development of pattern and form and to which experimentalists can specifically and concretely relate. All of the models we propose in this chapter are firmly based on macroscopic experimentally measurable variables and on generally accepted properties of embryonic cells. The preoccupation of many theoreticians and experimentalists to look for unrealistically simple models is often counterproductive.

In the following section we derive a fairly general model and subsequently deduce simpler versions. This is rather different to the approach we have adopted up to now and reflects, in part, the complexity of real modelling in embryology and in part on the assumption that the readers are now more sophisticated in their approach.

We should add here that these models pose numerous challenging mathematical, both analytical and numerical, and biological modelling problems which have not yet been investigated in any depth.

6.2 Mechanical Model for Mesenchymal Morphogenesis

Several factors affect the movement of embryonic mesenchymal cells. Among these factors are: (i) convection, whereby cells may be passively carried along on a deforming substratum; (ii) chemotaxis, whereby a chemical gradient can direct cell motion both up and down a concentration gradient; (iii) contact guidance, in which the substratum on which the cells crawl suggests a preferred direction; (iv) contact inhibition by the cells, whereby a high density of neighbouring cells inhibits motion; (v) haptotaxis, which we describe below, where the cells move up an adhesive gradient; (vi) diffusion, where the cells move randomly but generally down a cell density gradient; (vii) galvanotaxis, where movement from the field generated by electric potentials, which are known to exist in embryos, provides a preferred direction of motion. These effects are all well documented from experiment. Haptotaxis can be somewhat more complex than is implied here since chemical processes can be involved. Recently Perumpanani et al. (1998) showed that ECM-mediated chemotactic movement can actually impede migration.

The model field equations we propose in this section encapsulate key features which affect cell movement within its extracellular environment. We shall not include all of the effects just mentioned and others related to chemical influences, but it will be clear how they can be incorporated and their effect quantified with enough experimental knowledge. The subsequent analysis of the field equations will show how regular patterned aggregates of cells come about. Later in the chapter we describe several practical applications of the model, such as the highly organised patterns on skin like the primordia which become feathers and scales and the condensation of cells which mirror the cartilage pattern in developing limbs and fingerprints.

The basic mechanical model hinges on two key experimentally determined properties of mesenchymal cells *in vivo*: (i) cells migrate within a tissue substratum made up of fibrous extracellular matrix, which we often refer to as the ECM, and other cells (Hay 1981); (ii) cells can generate large traction forces (Harris et al. 1981, Ferrenq et al. 1997, Tranqui and Tracqui 2000). Figure 6.3 is a photograph of cells on a thin silicone substratum: the tension and compression lines they generate are clearly seen; see also Harris et al. (1980). The basic mechanism we shall develop models the mechanical interaction between the motile cells and the elastic substratum, within which they move.

Mesenchymal cells move by exerting forces on their surroundings, consisting of the elastic fibrous ECM and the surface of other cells. They use their cellular protrusions, the filopodia or lamellapodia, which stretch out from the cell in all directions, gripping whatever is available and pulling. The biology of these protrusions is discussed by Trinkaus (1980); see also the book by Trinkaus (1984) which is useful background

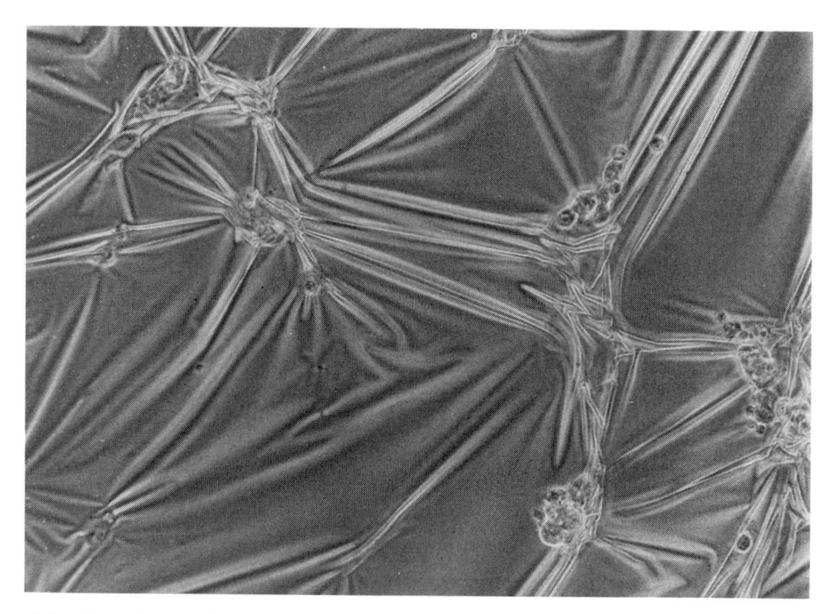

Figure 6.3. Mesenchymal cells on an elastic substratum. The strong tractions generated deform the substratum and create compression and tension wrinkles. The tension wrinkles can extend several hundreds of cell diameters. (Photograph courtesy of Albert K. Harris)

reading for morphogenetic modelling. Oster (1984) specifically discusses the mechanism of how an individual cell crawls. As the cells move through the ECM they deform it by virtue of their traction forces. These deformations in the ECM induce anisotropy effects which in turn affect the cell motion. The resulting coordination of the various effects, such as we have just mentioned, result in spatially organised cell aggregations. The basic model is essentially that proposed by Murray et al. (1983) and Murray and Oster (1984a,b), with a detailed biological description by Oster et al. (1983).

The model, a continuum one, consists of three equations governing (i) the conservation equation for the cell population density, (ii) the mechanical balance of the forces between the cells and the ECM, and (iii) the conservation law governing the ECM. Let $n(\mathbf{r}, t)$ and $\rho(\mathbf{r}, t)$ denote respectively the cell density (the number of cells per unit volume) and ECM density at position $\mathbf{r}$ and time t. Denote by $\mathbf{u}(\mathbf{r}, t)$ the displacement vector of the ECM; that is, a material point in the matrix initially at position $\mathbf{r}$ undergoes a displacement to $\mathbf{r} + \mathbf{u}$. We derive forms for each of these equations in turn.

Cell Conservation Equation

The general form of the conservation equation is (recall Chapter 11, Volume I)

$$\frac{\partial n}{\partial t} = -\nabla \cdot \mathbf{J} + M, \tag{6.1}$$

where $\mathbf{J}$ is the flux of cells, that is, the number crossing a unit area in unit time, and M is the mitotic or cell proliferation rate; the specific form is not important at this stage.

For simplicity only we take a logistic model for the cell growth, namely, $rn(N-n)$, where r is the initial proliferative rate and N is the maximum cell density in the absence of any other effects. We include in $\mathbf{J}$ some of the factors mentioned above which affect cell motion.

Convection

With $\mathbf{u}(\mathbf{r}, t)$ the displacement vector of the ECM, the convective flux contribution $\mathbf{J}_c$ is

$$\mathbf{J}_c = n\frac{\partial \mathbf{u}}{\partial t}. \tag{6.2}$$

Here the velocity of deformation of the matrix is $\partial \mathbf{u}/\partial t$ and the amount of cells transported is simply n times this velocity. It is likely that the convective flux is the most important contribution to cell transport. This form is based on the fact that we do not consider gross movement of the tissue here; see the discussion on more accurate forms in Chapter 10.

Random Dispersal

Cells tend to disperse randomly when in a homogeneous isotropic medium. Classical diffusion (see Chapter 11, Volume I) contributes a flux term $-D_1\nabla n$ which models the random motion in which the cells respond to *local* variations in the cell density and tend to move down the density gradient. This results in the usual diffusion contribution $D_1\nabla^2 n$ to the conservation equation and represents local, or *short range* random motion.

In developing embryos the cell densities are relatively high and classical diffusion, which applies to dilute systems is not, perhaps, sufficiently accurate. The long filopodia extended by the cells can sense density variations beyond their nearest neighbours and so we must include a *nonlocal* effect on diffusive dispersal since the cells sense more distant densities and so respond to neighbouring *averages* as well. Figure 6.4 schematically illustrates why this long range sensing could be relevant. This long range diffusion is probably not very important. The concept, however, is important in at least haptotaxis.

The Laplacian operator acting on a function reflects the difference between the value of the function at position $\mathbf{r}$ and its local average as can be seen on writing it in its simplest finite difference approximation. Alternatively, the Laplacian can be written in the form

$$\nabla^2 n \propto \frac{n_{av}(\mathbf{r}, t) - n(\mathbf{r}, t)}{R^2}, \quad \text{as} \quad R \to 0, \tag{6.3}$$

where n_{av} is the average cell concentration in a sphere of radius R about $\mathbf{r}$ defined by

$$n_{av}(\mathbf{r}, t) = \frac{3}{4\pi R^3}\int_V n(\mathbf{r}+\mathbf{s}, t)\, d\mathbf{s}, \tag{6.4}$$

where V is the volume of the sphere. If the integrand in (6.4) is expanded in a Taylor series and n_{av} substituted in (6.3), the proportionality factor is $10/3$. Again recall the full discussion and analysis in Chapter 11, Volume I.

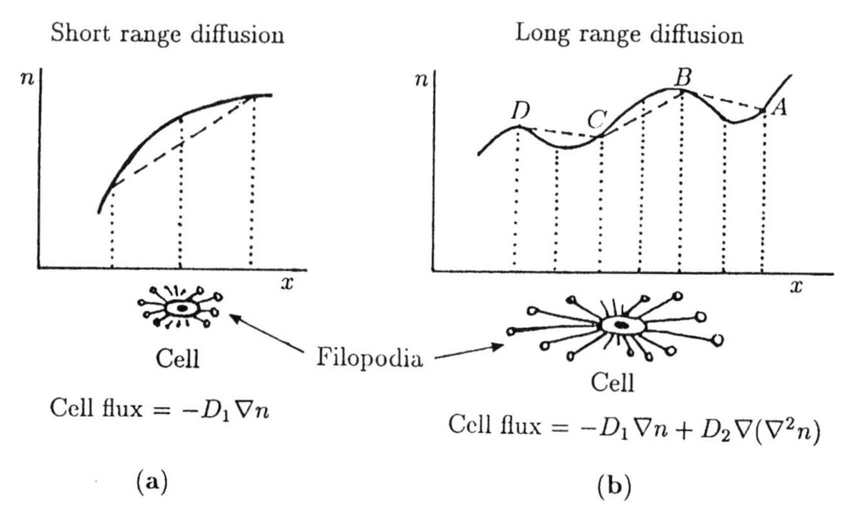

Figure 6.4. In (**a**) the filopodia only sense the immediate neighbouring densities to determine the gradient (the broken line) and hence disperse in a classical random manner giving a flux of $-D_1\nabla n$. With the situation in (**b**) the long filopodia can sense not only neighbouring densities but also neighbouring averages which contribute a long range diffusional flux term $D_2\nabla(\nabla^2 n)$. This contributes to directed dispersal which is not necessarily in the same direction as indicated by short range (again denoted by broken lines) diffusion. Long range diffusion suggests general movement of cells from A to D whereas short range diffusion implies movement from D to C, B to C and B to A.

The flux of cells is thus given by

$$\mathbf{J}_D = -D_1\nabla n + D_2\nabla(\nabla^2 n), \tag{6.5}$$

where $D_1 > 0$ is the usual Fickian diffusion coefficient and $D_2 > 0$ is the long range diffusion coefficient. The long range contribution gives rise to a biharmonic term in (6.1). In the morphogenetic situations we consider, we expect the effect of diffusion to be relatively small. Nonlocal diffusive dispersal was considered by Othmer (1969); his work is particularly apposite to the cell situation. Cohen and Murray (1981) derived and considered a related model in an ecological context.

Recall from Section 11.5 (Chapter 11, Volume I) that this long range diffusion has a stabilising effect if $D_2 > 0$. We can see this immediately if we consider the long range diffusion equation, obtained by substituting (6.5) into (6.1) and, omitting the mitotic term M, to get

$$\frac{\partial n}{\partial t} = -\nabla\cdot\mathbf{J}_D = D_1\nabla^2 n - D_2\nabla^4 n.$$

We now look for solutions of the form $n(\mathbf{r},t)\propto\exp[\lambda t + i\mathbf{k}\cdot\mathbf{r}]$, where $\mathbf{k}$ is the usual wave vector. Substituting this into the last equation gives the dispersion relation as $\lambda = -D_2k^4 - D_1k^2 < 0$ for all wavenumbers $k(=|\mathbf{k}|)$. So $n\to 0$ as $t\to\infty$, which implies $n = 0$ is stable. If the biharmonic term had $D_2 < 0$, $n = 0$ would be unstable for wavenumbers $k^2 > -D_1/D_2$. The question of what the diffusion contribution should be can, in fact, be more complicated. We derive other forms in Chapter 10 below.

Haptotaxis or Mechanotaxis

The traction exerted by the cells on the matrix generates gradients in the matrix density $\rho(\mathbf{r}, t)$. We associate the density of matrix with the density of adhesive sites for the cell lamellapodia to get a hold of. Cells free to move in an adhesive gradient tend to move up it since the cells can get a stronger grip on the denser matrix. This results in a net flux of cells *up* the gradient which, on the simplest assumption, is proportional to $n\nabla\rho$. It is very similar to chemotaxis (recall Chapter 11, Volume I, Section 11.4). As mentioned, this can be more complex (Perumpanani et al. 1998). Because of the physical properties of the matrix and the nonlocal sensing properties of the cells we should also include a long range effect, similar to that which gave the biharmonic term in (6.5). In this case the haptotactic flux is given by

$$\mathbf{J}_h = n(a_1\nabla\rho - a_2\nabla^3\rho), \tag{6.6}$$

where $a_1 > 0$ and $a_2 > 0$. There is considerably more justification for including long range effects here than with long range diffusion.

The cell conservation equation (6.1), with the flux contributions to $\mathbf{J}$ from (6.2), (6.5) and (6.6) with the illustrative logistic form for the mitosis M, becomes

$$\begin{aligned} \frac{\partial n}{\partial t} = & - \underbrace{\nabla \cdot \left[n\frac{\partial \mathbf{u}}{\partial t} \right]}_{\text{convection}} + \underbrace{\nabla \cdot [D_1\nabla n - D_2\nabla(\nabla^2 n)]}_{\text{diffusion}} \\ & - \underbrace{\nabla \cdot n[a_1\nabla\rho - a_2\nabla^3\rho]}_{\text{haptotaxis}} + \underbrace{rn(N-n)}_{\text{mitosis}}, \end{aligned} \tag{6.7}$$

where D_1, D_2, a_1, a_2, r and N are positive parameters.

We have not included galvanotaxis nor chemotaxis as such in (6.7) but we can easily deduce what such contributions in their basic forms could look like. If ϕ is the electric potential then the galvanotaxis flux can be written as

$$\mathbf{J}_G = gn\nabla\phi, \tag{6.8}$$

where the parameter $g > 0$. If c is the concentration of a chemotactic chemical, a chemotactic flux can then be of the form

$$\mathbf{J}_C = \chi n\nabla c,$$

where $\chi > 0$ is the chemotaxis parameter. Another effect which could be important but which we shall also not include here, is the guidance cues which come from the directional cues in the ECM. For example, there is experimental evidence that matrix strain results in aligned fibres which encourages movement along the directions of strain as opposed to movement across the strain lines. This effect can be incorporated in the equation for n by making the diffusion and haptotactic coefficients functions of the elastic strain tensor (see, for example, Landau and Lifshitz 1970) of the ECM defined by

$$\boldsymbol{\varepsilon} = \frac{1}{2}(\nabla \mathbf{u} + \nabla \mathbf{u}^T). \tag{6.9}$$

In principle the qualitative form of the dependence of, for example, $D_1(\boldsymbol{\varepsilon})$ and $D_2(\boldsymbol{\varepsilon})$ on $\boldsymbol{\varepsilon}$ can be deduced from experiments. We discuss this in depth in Chapter 10 and its effect on patterning in Chapter 8. In the following, however, we take D_1, D_2, a_1 and a_2 to be constants.

In (6.7) we modelled the mitotic, or cell proliferation rate, by a simple logistic growth with linear growth rate r. The detailed form of this term is not critical as long as it is qualitatively similar. It is now well known from experiment (for example, Folkman and Moscona 1978) that the mitotic rate is dependent on cell shape. So, within our continuum framework, r should depend on the displacement $\mathbf{u}$. A brief review of the ECM and its effect on cell shape, proliferation and differentiation is given by Watt (1986). At this stage, however, we shall also not include this potentially important effect.

One of the purposes of this section is to show how possible effects can be incorporated in the model. Although the conservation equation (6.7) is clearly not the most general possible, it suffices to show what can be expected in more realistic model mechanisms for biological pattern generation.

The analysis of such models lets us compare the various effects as to their pattern formation potential and hence to come up with the simplest realistic system which can generate pattern and which is experimentally testable. Simpler systems are discussed later in Section 6.4. Perhaps we should mention here that only the inclusion of convection in the cell conservation equation is essential. Intuitively this is what we might have expected at least as regards transport effects.

Cell–Matrix Mechanical Interaction Equation

The composition of the fibrous extracellular matrix, the ECM, within which the cells move is complex and moreover, its constituents change as development proceeds. Its mechanical properties have not yet been well characterised. Here, however, we are interested only in the mechanical interaction between the cells and the matrix. Also the mechanical deformations are small so, as a reasonable first approximation, we take the composite material of cells plus matrix to be modelled as a linear, isotropic viscoelastic continuum with stress tensor $\boldsymbol{\sigma}(\mathbf{r}, t)$.

The timescale of embryonic motions during development is very long (hours) and the spatial scale is very small (less than a millimetre or two). We are thus in a very low Reynolds number regime (cf. Purcell 1977) and so can ignore inertial effects in the mechanical equation for the cell–ECM interaction. Thus we assume that the traction forces generated by the cells are in mechanical equilibrium with the elastic restoring forces developed in the matrix and any external forces present. The mechanical cell–matrix equation is then (see, for example, Landau and Lifshitz 1970)

$$\nabla \cdot \boldsymbol{\sigma} + \rho \mathbf{F} = 0, \tag{6.10}$$

where $\mathbf{F}$ is the external force acting on the matrix (per unit matrix) and $\boldsymbol{\sigma}$ is the stress tensor. (This equation applied to a spring loaded with a weight simply says the applied force is balanced by the elastic force from the extended spring.) We must now model the various contributions to $\boldsymbol{\sigma}$ and $\mathbf{F}$.

Consider first the stress tensor $\boldsymbol{\sigma}$. It consists of contributions from the ECM and the cells and we write

$$\boldsymbol{\sigma} = \boldsymbol{\sigma}_{\text{ECM}} + \boldsymbol{\sigma}_{\text{cell}}. \tag{6.11}$$

The usual expression for a linear viscoelastic material (Landau and Lifshitz 1970) gives the stress–strain constitutive relation as

$$\begin{aligned} \boldsymbol{\sigma}_{\text{ECM}} &= \underbrace{[\mu_1 \boldsymbol{\varepsilon}_t + \mu_2 \theta_t \mathbf{I}]}_{\text{viscous}} + \underbrace{E'[\boldsymbol{\varepsilon} + \nu' \theta \mathbf{I}]}_{\text{elastic}}, \\ \text{where} \quad E' &= E/(1+\nu), \quad \nu' = \nu/(1-2\nu). \end{aligned} \tag{6.12}$$

The subscript t denotes partial differentiation, $\mathbf{I}$ is the unit tensor, μ_1 and μ_2 are the shear and bulk viscosities of the ECM, $\boldsymbol{\varepsilon}$ is the strain tensor defined above in (6.9), $\theta(= \nabla \cdot \mathbf{u})$ is the dilation and E and ν are the Young's modulus and Poisson ratio respectively.

The assumption of isotropy is certainly a major one. While the ECM may be isotropic in the absence of cell tractions (and even this is doubtful) it is probably no longer isotropic when subjected to cellular forces. Although we do not specifically consider in this chapter a nonisotropic model, we should be aware of the kind of anisotropy that might be included in a more sophisticated model. When a fibrous material is strained the fibres tend to align in the directions of the principal stresses and the effective elastic modulus in the direction of strain increases. With the main macroscopic effect of fibre alignment being to strengthen the material in the direction of strain we can model this by making the elastic modulus E an increasing function of the dilation θ, at least for small θ. It does not of course increase indefinitely since eventually the material would yield. Figure 6.5 is a typical stress–strain curve. It is possible that ν is also a function of θ; here, however, we take it to be constant.

Fibrous materials are also characterised by nonlocal elastic interactions since the fibres can transmit stress between points in the ECM quite far apart. By arguments analogous to those which lead to the biharmonic term in the cell conservation equation (6.7) we should include long range effects in the elastic stress for the composite material. The anisotropic effect discussed in the last paragraph and this nonlocal effect can be modelled by writing in place of the elastic contribution in (6.12)

$$\begin{aligned} \boldsymbol{\sigma}_{\text{ECM}}]_{\text{elastic}} &= E'(\theta)[\boldsymbol{\varepsilon} + \beta_1 \nabla^2 \boldsymbol{\varepsilon} + \nu'(\theta + \beta_2 \nabla^2 \theta)\mathbf{I}], \\ \text{where} \quad E' &= E(\theta)/(1+\nu), \quad \nu' = \nu/(1-2\nu) \end{aligned} \tag{6.13}$$

and the β's are parameters which measure long range effects. However at this stage of modelling it is reasonable to take $\beta_1 = \beta_2 = 0$ and $E(\theta)$ to be a constant. We consider anisotropy in detail in Chapter 8 and in even more detail in Chapter 10.

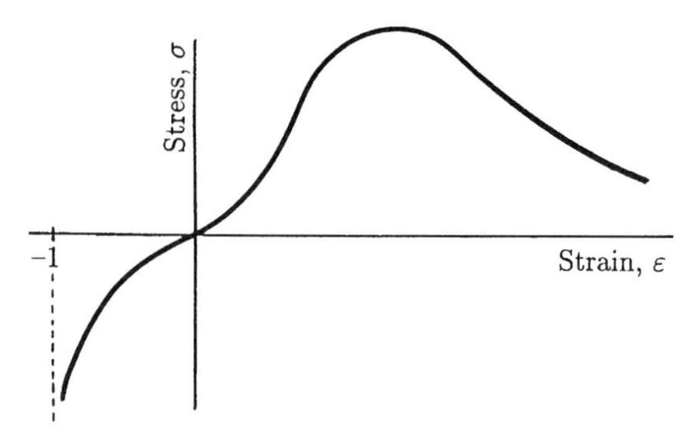

Figure 6.5. The effect of straining the extracellular matrix is to align the fibres and stiffen the material. If we think of a one-dimensional situation the strain from (6.9) is $\varepsilon = \partial u/\partial x$ and the dilation $\theta = \partial u/\partial x$. The effective elastic modulus E is the gradient of the stress–strain curve. It increases with strain until the yield point whereupon it levels off and drops for large enough strains as the material tears. The ECM is in compression when $\varepsilon < 0$ (also $\theta < 0$). Because a given amount of material (cells + matrix) cannot be squeezed to zero, there is a lower limit of $\varepsilon = -1$ (also $\theta \geq -1$) where the stress tends to $-\infty$.

Now consider the contribution to the stress tensor from the cell tractions, that is, $\boldsymbol{\sigma}_{\text{cell}}$. The more cells there are the greater the traction force. There is, however, experimental evidence indicating cell–cell contact inhibition with the traction force decreasing for large enough cell densities. This can be simply modelled by assuming that the cell traction forces, $\tau(n)$ per unit mass of matrix, initially increase with n but eventually decrease with n for large enough n. Here we simply choose

$$\tau(n) = \frac{\tau n}{1 + \lambda n^2}, \tag{6.14}$$

where τ (dyne–cm/gm) is a measure of the traction force generated by a cell and λ is a measure of how the force is reduced because of neighbouring cells; we come back to this below. Experimental values for τ are of the order of 10^{-3} dyne/μm of cell edge, which is a very substantial force (Harris et al. 1981). The actual form of the force generated per cell, that is, $\tau(n)/n$, as a function of cell density can be determined experimentally as has been done by Ferrenq et al. (1997).

Even though cell traction plays such a central role in pattern formation in development it has proved very difficult to quantify the cellular forces involved because of the complexity of the cell–matrix interactions and the difficulty of separating out the various mechanical effects in real biological tissue. Ferrenq et al. (1997) describe a new experimental technique and general approach for quantifying the forces generated by endothelial cells on an extracellular matrix. They first developed a mathematical model, based on the Murray–Oster mechanochemical theory described in detail in this chapter, and in which different forms for the cell generated stress are proposed. They then used these as the basis for a novel experimental device in which cells are seeded on a biogel of fibrin (the matrix) held between two holders one of which can move and is attached to a force sensing device. By comparing the displacement of the gel calculated from the model expressions with the experimental data recorded from the moving holder they were able to justify specific expressions for the cell traction stress; they did this for various experimental setups. They were then able to compare different plausible analytical expressions for the cell traction stress with the corresponding force quantification and to compare the results with experiment and other reported measurements for different kinds of cells by similar and different experimental devices. They show how experimentally justifiable forms for the cell–gel traction stress can be derived and give estimates for each of the parameters involved. They found that the expression

$$\boldsymbol{\sigma}_{\text{cell}} = \tau \rho n (N_2 - n) \mathbf{I},$$

where τ is the cell traction and the parameter N_2 controls the inhibition of cell traction as the cell density increases was validated by experiments and estimates given for their values. This paper is an excellent example of genuine interdisciplinary mathematical biology research with theory and experiment each playing an important role in the outcome. This interdisciplinary approach was exploited by Tranqui and Tracqui (2000) in their investigation of mechanical signalling in angiogenesis. They again used the mechanical theory with the viscous stress tensor given by (6.12) and the elastic stress tensor given by (6.13) which includes long range elastic effects.

If the filopodia, with which the cells attach to the ECM, extend beyond their immediate neighbourhood, as they probably do, it is not unreasonable to include a nonlocal effect analogous to the long range diffusion effect we included in the cell conservation equation. For our analysis we take the contribution $\boldsymbol{\sigma}_{\text{cell}}$ to the stress tensor to be

$$\boldsymbol{\sigma}_{\text{cell}} = \frac{\tau n}{1 + \lambda n^2} (\rho + \gamma \nabla^2 \rho) \mathbf{I}, \tag{6.15}$$

where $\gamma > 0$ is the measure of the nonlocal long range cell–ECM interactions. The long range effects here are probably more important than the long range diffusion and haptotaxis effects in the cell conservation equation.

If the cells are densely packed the nonlocal effect would primarily be between the cells and in this case a more appropriate form for (6.15) should perhaps be

$$\boldsymbol{\sigma}_{\text{cell}} = \frac{\tau \rho}{1 + \lambda n^2} (n + \gamma \nabla^2 n) \mathbf{I}. \tag{6.16}$$

There are various possible forms for the cell traction all of which might reasonably be used. One way of resolving the issue might be to use a molecular method developed by Sherratt (1993) in his derivation of the actin generated forces involved in embryonic wound healing; we discuss his technique in Chapter 9.

Finally let us consider the body force $\mathbf{F}$ in (6.10). With the applications we have in mind, and discussed below, the matrix material is attached to a substratum of underlying tissue (or to the epidermis) by what can perhaps best be described as being similar to guy ropes. We model these restraining forces as body forces proportional to the density of the ECM and the displacement of the matrix from its unstrained position and therefore take

$$\mathbf{F} = -s\mathbf{u}, \tag{6.17}$$

where $s > 0$ is an elastic parameter characterising the substrate attachments.

In the model we analyse we shall not include all the effects we have discussed but only those we feel are the more essential at this stage. So, the force equation we take for the mechanical equilibrium between the cells and the ECM is (to be specific) (6.10), with (6.11)–(6.17), which gives

$$\nabla \cdot \left[\underbrace{\mu_1 \boldsymbol{\varepsilon}_t + \mu_2 \theta_t \mathbf{I}}_{\text{viscous}} + \underbrace{E'(\boldsymbol{\varepsilon} + \nu' \theta \mathbf{I})}_{\text{elastic}} + \underbrace{\tau n (1 + \lambda n^2)^{-1} (\rho + \gamma \nabla^2 \rho) \mathbf{I}}_{\text{cell traction}} \right] - \underbrace{s \rho \mathbf{u}}_{\text{external forces}} = 0, \tag{6.18}$$

where

$$E' = \frac{E}{1 + \nu}, \quad \nu' = \frac{\nu}{1 - 2\nu}. \tag{6.19}$$

Matrix Conservation Equation

The conservation equation for the matrix material, $\rho(\mathbf{r}, t)$, is

$$\frac{\partial \rho}{\partial t} + \nabla \cdot (\rho \mathbf{u}_t) = S(n, \rho, \mathbf{u}), \tag{6.20}$$

where matrix flux is taken to be mainly via convection and $S(n, \rho, \mathbf{u})$ is the rate of secretion of matrix by the cells. Secretion and degradation is thought to play a role in certain situations involving mesenchymal cell organisation, and it certainly does in wound healing, an important application discussed in Chapter 10. However, on the timescale of cell motions that we consider here we can neglect this effect and shall henceforth assume $S = 0$. Experimental evidence (Hinchliffe and Johnson 1980) indicates that $S = 0$ during chondrogenesis and pattern formation of skin organ primordia.

Equations (6.7), (6.18) and (6.20) with $S = 0$ constitute the field equations for our model pattern formation mechanism for fibroblast cells that we now examine. The three dependent variables are the density fields $n(\mathbf{r}, t)$, $\rho(\mathbf{r}, t)$ and the displacement field $\mathbf{u}(\mathbf{r}, t)$. The model involves 14 parameters, namely, D_1, D_2, a_1, a_2, r, N, μ_1, μ_2, τ, λ, γ, s, E and ν, all of which are in principle measurable and some of which have been investigated experimentally with others currently under study.

As usual, to assess the relative importance of the various effects, and to simplify the analysis, we nondimensionalise the equations. We use general length and timescales L and T, a uniform initial matrix density ρ_0 and set

$$\begin{gathered}
\mathbf{r}^* = \frac{\mathbf{r}}{L}, \quad t^* = \frac{t}{T}, \quad n^* = \frac{n}{N}, \quad \mathbf{u}^* = \frac{\mathbf{u}}{L}, \quad \rho^* = \frac{\rho}{\rho_0}, \\
\nabla^* = L\nabla, \quad \theta^* = \theta, \quad \boldsymbol{\varepsilon}^* = \boldsymbol{\varepsilon}, \quad \gamma^* = \frac{\gamma}{L^2}, \quad r^* = rNT, \\
s^* = \frac{s \rho_0 L^2 (1 + \nu)}{E}, \quad \lambda^* = \lambda N^2, \quad \tau^* = \frac{\tau \rho_0 N (1 + \nu)}{E},
\end{gathered} \tag{6.21}$$

$$a_1^* = \frac{a_1\rho_0 T}{L^2}, \quad a_2^* = \frac{a_2\rho_0 T}{L^4}, \quad i = 1, 2$$
$$\mu_i^* = \frac{\mu_i(1+\nu)}{TE}, \quad i = 1, 2, \quad D_1^* = \frac{D_1 T}{L^2}, \quad D_2^* = \frac{D_2 T}{L^4}. \tag{6.21 (continued)}$$

The nondimensionalisation has reduced the 14 parameters to 12 parameter groupings. Depending on what timescale we are particularly concerned with we can reduce the set of 12 parameters further. For example, if we choose T as the mitotic time $1/rN$, then $r^* = 1$; this means we are interested in the evolution of pattern on the mitotic time scale. Alternatively we could choose T so that $\gamma^* = 1$ or $\mu_i^* = 1$ for $i = 1$ or $i = 2$. Similarly we can choose a relevant length scale and further reduce the number of groupings.

With the nondimensionalisation (6.21) the model mechanism (6.7), (6.18) and (6.20), with matrix secretion $S = 0$ (the effect of a matrix source term S on the subsequent analysis is currently under investigation), becomes, on dropping the asterisks for notational simplicity,

$$n_t = D_1\nabla^2 n - D_2\nabla^4 n - \nabla\cdot[a_1 n\nabla\rho - a_2 n\nabla(\nabla^2\rho)] - \nabla\cdot(n\mathbf{u}_t) + rn(1-n), \tag{6.22}$$

$$\nabla\cdot\left\{(\mu_1\boldsymbol{\varepsilon}_t + \mu_2\theta_t\mathbf{I}) + (\boldsymbol{\varepsilon} + \nu'\theta\mathbf{I}) + \frac{\tau n}{1+\lambda n^2}(\rho + \gamma\nabla^2\rho)\mathbf{I}\right\} = s\rho\mathbf{u}, \tag{6.23}$$

$$\rho_t + \nabla\cdot(\rho\mathbf{u}_t) = 0. \tag{6.24}$$

Note that the dimensionless parameters, all of which are positive, are divided into those associated with the cell properties, namely, a_1, a_2, D_1, D_2, r, τ, λ and those related to the matrix properties, namely, μ_1, μ_2, ν', γ and s.

Although the model system (6.22)–(6.24) is analytically formidable the model's conceptual framework is quite clear, as illustrated in Figure 6.6. As we have noted, this

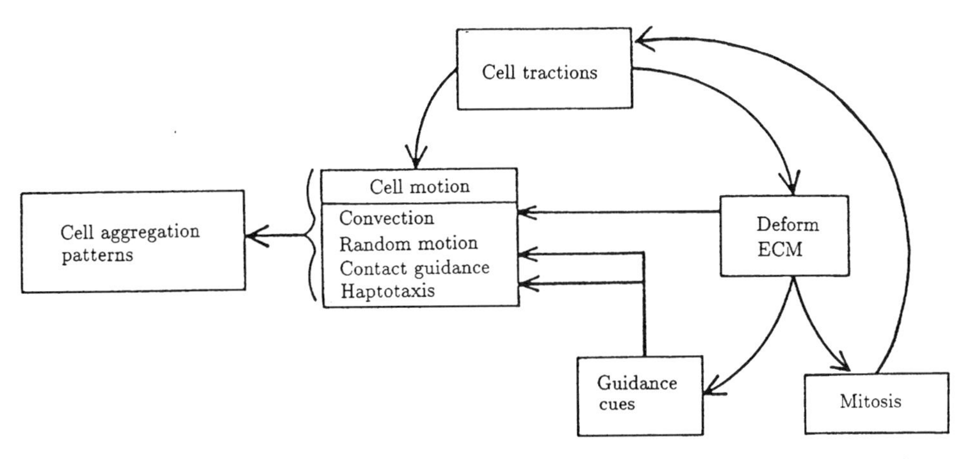

Figure 6.6. Conceptual framework for the mechanical models. Cell tractions play a central role in orchestrating pattern formation.

model does not include all the effects that might be relevant, such as the effect of matrix secretion and strain-dependent diffusion and contact guidance. Although later we shall derive considerably simpler systems we should have some idea of what a model looks like that incorporates many of the features that biologists might feel are important. As we also said above, one of the major roles of such modelling and subsequent analysis is to indicate what features are *essential* for pattern formation. So, in the initial linear analysis that follows we retain all of the terms in the model (6.22)–(6.24) and only set various parameters to zero in the general results to see what effects may be redundant or dwarfed by others.

6.3 Linear Analysis, Dispersion Relation and Pattern Formation Potential

To model spatial aspects observed in embryonic development the equation system (6.22)–(6.24) must admit spatially inhomogeneous solutions. Considering their complexity and with the experience gained from the study of the pattern forming models in earlier chapters we have little hope, at this stage, of finding useful analytical solutions to such nonlinear systems. We know, however, that much of the pattern formation potential is predicted by a linear analysis about uniform steady state solutions. We also now know that such linear predictions are not infallible and must be backed up by numerical simulations if finite amplitude structures far from homogeneity are required. (It will be helpful in the following to recall in detail the material and discussions relating to spatial pattern formation in Chapter 2, particularly Sections 2.3 to 2.6.)

Before carrying out a linear analysis let us note that one of the applications of this theory will be to the pattern formation process that accompanies the formation of skin organ primordia for feathers, scales and teeth, for example; see Section 6.5. The initial cell aggregations which appear in the dermis, that is, the layer just under the epidermis on which the scales and feathers start, only differ in cell density from the surrounding tissue by fairly small amounts. Therefore it is worthwhile from the practical biological application viewpoint to carry out a detailed linear analysis of the field equations, not only as a first analytical step to indicate spatial pattern potentialities and guide numerical work, but also because the patterns themselves may involve solutions that effectively fall within the linear regime. The latter are often effectively those from a nonlinear theory close to bifurcation from uniformity. We come back to the biological applications in more detail below.

The uniform steady state solutions of (6.22)–(6.24) are

$$n = \mathbf{u} = \rho = 0; \quad n = 1, \mathbf{u} = \rho = 0; \quad n = \rho = 1, \mathbf{u} = 0. \tag{6.25}$$

The first two solutions are not relevant as $\rho = 0$ is not relevant in the biological situation. The third solution is relevant (ρ here is normalised to 1 by the nondimensionalisation) and the linear stability of this solution is found in the usual way (recall, in particular, Section 2.3 in Chapter 2) by seeking solutions of the linearised equations from (6.22)–(6.24). We thus consider $n - 1$, $\rho - 1$ and $\mathbf{u}$ to be small and, on substituting into the nonlinear system and keeping only linear terms in $n - 1$, $\rho - 1$ and $\mathbf{u}$ and their derivatives, we get the following linear system, where for algebraic convenience we have written n and ρ for $n-1$ and $\rho-1$ respectively.

$$n_t - D_1\nabla^2 n + D_2\nabla^4 n + a_1\nabla^2\rho - a_2\nabla^4\rho + \theta_t + rn = 0, \tag{6.26}$$

$$\nabla\cdot\Big[(\mu_1\boldsymbol{\varepsilon}_t + \mu_2\theta_t\mathbf{I}) + (\boldsymbol{\varepsilon} + \nu'\theta\mathbf{I}) + (\tau_1 n + \tau_2\rho + \tau_1\gamma\nabla^2\rho)\mathbf{I}\Big] - s\mathbf{u} = 0, \tag{6.27}$$

$$\rho_t + \theta_t = 0, \tag{6.28}$$

where

$$\tau_1 = \frac{\tau}{1+\lambda}, \qquad \tau_2 = \frac{\tau(1-\lambda)}{(1+\lambda)^2}. \tag{6.29}$$

Note that if $\lambda > 1$, $\tau_2 < 0$, λ, which is nonnegative, is a measure of the cell–cell contact inhibition.

We now look for solutions to these linearised equations by setting

$$(n, \rho, \mathbf{u}) \propto \exp[\sigma t + i\mathbf{k}\cdot\mathbf{r}], \tag{6.30}$$

where $\mathbf{k}$ is the wavevector and σ is the linear growth factor (not to be confused with the stress tensor). In the usual way (cf. Chapter 2) substitution of (6.30) into (6.26)–(6.28) gives the dispersion relation $\sigma = \sigma(k^2)$ as solutions of the polynomial in σ given by the determinant

$$\begin{vmatrix} \sigma + D_1k^2 + D_2k^4 + r & -a_1k^2 - a_2k^4 & ik\sigma \\ ik\tau_2 & ik\tau_1 - ik^3\tau_1\gamma & -\sigma\mu k^2 - (1+\nu')k^2 - s \\ 0 & \sigma & ik\sigma \end{vmatrix} = 0,$$

where $k = |\mathbf{k}|$. A little algebra gives $\sigma(k^2)$ as the solutions of

$$\begin{aligned} &\sigma[\mu k^2\sigma^2 + b(k^2)\sigma + c(k^2)] = 0, \\ &b(k^2) = \mu D_2k^6 + (\mu D_1 + \gamma\tau_1)k^4 + (1 + \mu r - \tau_1 - \tau_2)k^2 + s, \\ &c(k^2) = \gamma\tau_1 D_2k^8 + (\gamma\tau_1 D_1 - \tau_2 D_2 + D_2 - a_2\tau_1)k^6 \\ &\qquad + (D_1 + sD_2 - \tau_1 D_1 + \gamma\tau_1 r - a_1\tau_2)k^4 \\ &\qquad + (r + sD_1 - r\tau_1)k^2 + rs. \end{aligned} \tag{6.31}$$

Here we have set $\mu = \mu_1 + \mu_2$ and τ_1, τ_2, μ and s replace $\tau_1/(1+\nu')$, $\tau_2/(1+\nu')$, $\mu/(1+\nu')$ and $s/(1+\nu')$ respectively. The dispersion relation is the solution of (6.31) with the largest $\operatorname{Re}\sigma \geq 0$, so

$$\begin{aligned} \sigma(k^2) &= \frac{-b(k^2) + \{b^2(k^2) - 4\mu k^2 c(k^2)\}^{1/2}}{2\mu k^2}, \\ \sigma(k^2) &= \frac{-b(k^2)}{\mu k^2}, \quad \text{if} \quad c(k^2) \equiv 0. \end{aligned} \tag{6.32}$$

Spatially heterogeneous solutions of the linear system are characterised by a dispersion $\sigma(k^2)$ which has $\text{Re}\,\sigma(0) \le 0$ but which exhibits a range of unstable modes with $\text{Re}\,\sigma(k^2) > 0$ for $k^2 \neq 0$. From (6.31), if $k^2 = 0$, the spatially homogeneous case, we have $b(0) = s > 0$ and $c(0) = rs > 0$ since all the parameters are positive. So $\sigma = -c/b < 0$ and hence stability obtains. Thus we require conditions for $\text{Re}\,\sigma(k^2) > 0$ to exist for at least some $k^2 \neq 0$. All the solutions (6.30) with these k's are then linearly unstable and grow exponentially with time. In the usual way we expect these unstable heterogeneous linear solutions will evolve into finite amplitude spatially structured solutions. Heuristically we see from the nonlinear system (6.22) that such exponentially growing solutions will not grow unboundedly—the quadratic term in the logistic growth prevents this. In models where the mitotic rate is not set to zero, the contact inhibition term (6.23) ensures that solutions are bounded. Numerical simulations of the full system, for example, by Perelson et al. (1986), Bentil (1990) and Cruywagen (1992) bear this out.

The linearly unstable solutions have a certain predictive ability as to the qualitative character of the finite amplitude solutions. The predictability again seems to be limited to only small wavenumbers in a one-dimensional situation. As we saw in Chapters 2 and 3 this was usually, but not always, the case with reaction diffusion systems.

From the dispersion relation (6.32), the only way a solution with $\text{Re}\,\sigma(k^2) > 0$ can exist is if $b(k^2) < 0$ or $c(k^2) < 0$ or both. Since the only negative terms involve the traction parameter τ, occurring in τ_1 and τ_2, a necessary condition for the mechanism to generate spatially heterogeneous solutions is that the cell traction $\tau > 0$. Note from (6.29) that it is possible that τ_2 can be negative. It is also clear heuristically from the mechanism that τ must be positive since the cell traction forces are the only contribution to the aggregative process in the force-balance equation (6.23). So, with $\tau > 0$, sufficient conditions for spatial structured solutions to exist are when the parameters ensure that $b(k^2) < 0$ and/or $c(k^2) < 0$ for some $k^2 > 0$. Because of the central role of the cell traction we shall use τ as the bifurcation parameter. There is also a biological reason for choosing τ as the bifurcation parameter. It is known that, *in vitro*, the traction generated by a cell can increase with time (for a limited period) typically as illustrated in Figure 6.7.

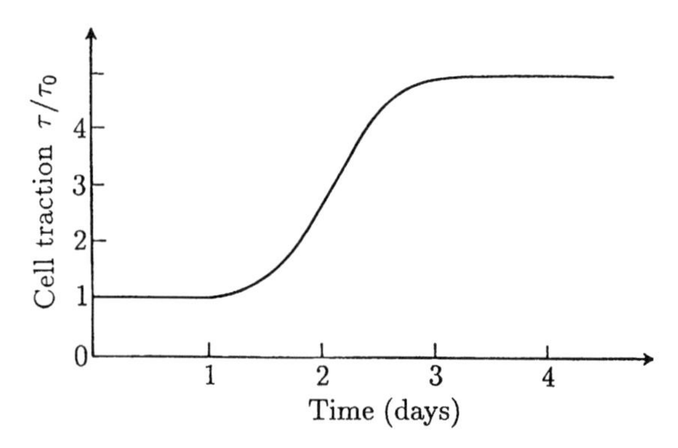

Figure 6.7. Qualitative *in vitro* behaviour of fibroblast cell traction with time after placing the cells on a dish. τ_0 is a base value, typically of the order of 10^{-2}Nm^{-1} of cell edge.

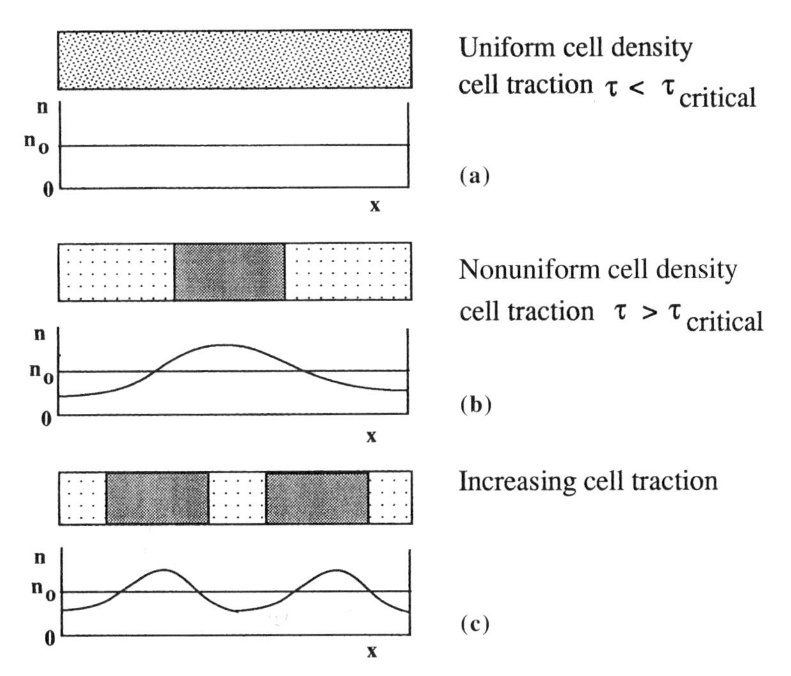

Figure 6.8. How the mechanical patterning process works. (**a**) Any perturbation in cell density is smoothed out because the cell traction is not sufficient to overcome the elastic resistance in the ECM. (**b**) Once the cell traction passes through the critical value the homogeneous steady state is linearly unstable and a pattern forms: the specific pattern that develops depends on the dispersion relation. Here we have assumed it is a basic mode. (**c**) Further increase in the cell traction gives rise to more complex patterns, again determined in general by the dispersion relation.

We can see intuitively how the patterning process works if we use the cell traction as the bifurcation parameter. Refer to Figure 6.8. In Figure 6.8(a) the cell traction is less than the critical traction and is not sufficient to overcome the elastic resistance of the ECM and any heterogeneity in cell density simply smooths out. As the traction increases it passes through the critical value where the cell generated forces are greater than the resistance of the ECM and spatial inhomogeneities start to form as in Figure 6.8(b). As the cell traction increases further we then have to use the dispersion relation to determine which pattern starts to grow as in Figure 6.8(c). Because of the form of the equations (recall the discussion in Chapter 2) we also have similar scale and geometry effects with more complex patterns possible the larger the domain.

We can deduce the qualitative effects of some of the various terms in the model, as regards their pattern formation potential, by simply looking at the expressions for $b(k^2)$ and $c(k^2)$. Since we require b or c to be negative before spatial pattern will evolve from random initial cell densities, we see, for example, that if the tethering, quantified by s, is increased it tends to stabilise the solutions since it tends to make both b and c more positive. The long range effect of the cells on the matrix, quantified by the parameter γ, is also a stabilising influence; so also is the viscosity. On the other hand long range haptotaxis, via a_2, is always destabilising. Considerable quantitative information can be obtained simply from the polynomial coefficients in the dispersion relation. Much of it, of course, is intuitively clear. Where the parameters combine, however, further analysis is necessary to draw biological implications; examples are given in the following section.

The expressions for $\sigma(k^2)$ in (6.32), and $b(k^2)$ and $c(k^2)$ in (6.31), determine the domains in parameter space where spatially inhomogeneous linearly unstable solutions exist. They also give the bifurcation surfaces in parameter space, that is, the surfaces which separate homogeneous from inhomogeneous solutions. It is algebraically very complicated to determine these surfaces in general. In any case, because of the dimensionality of the parameter space, it would be of little conceptual help in understanding the basic features of the pattern formation process. It is more instructive to consider various special cases whereby we assume one or more of the various factors affecting cell motion and matrix deformation to be negligible. One result of this is to produce several much simpler model mechanisms which are all capable of generating spatial patterns. Which mechanism is most appropriate for a given biological situation must be determined by the biology.

With the polynomial complexity of $b(k^2)$ and $c(k^2)$ in the dispersion relation $\sigma(k^2)$ in (6.32) we can expect complex linear growth behaviour. In the following section we consider some particular models, all of which are capable of generating spatial patterns. We also display the remarkable variety of dispersion relations which mechanical models can produce from relatively simple mechanisms.

6.4 Simple Mechanical Models Which Generate Spatial Patterns with Complex Dispersion Relations

In this section we consider some special cases of (6.22)–(6.24) where one or more of the factors affecting cell motion or the mechanical equilibrium are assumed to be negligible; each highlights something new. These are deduced by simply setting various parameters to zero and examining the resultant dispersion relation $\sigma(k^2)$ from (6.32). It is not, of course, a haphazard procedure: we examine the effect on $b(k^2)$ and $c(k^2)$ and determine, a priori, the likely outcome.

(i) $D_1 = D_2 = a_1 = a_2 = 0$: no cell diffusion and no haptotaxis, $r = 0$: no cell division.

From the general model (6.22)–(6.24) the mechanism becomes

$$\begin{gathered} n_t + \nabla\cdot(n\mathbf{u}_t) = 0, \\ \nabla\cdot\left\{(\mu_1\boldsymbol{\varepsilon}_t + \mu_2\theta_t\mathbf{I}) + (\boldsymbol{\varepsilon} + \nu'\theta\mathbf{I}) + \frac{\tau n}{1+\lambda n^2}(\rho + \gamma\nabla^2\rho)I\right\} = s\rho\mathbf{u}, \qquad (6.33) \\ \rho_t + \nabla\cdot(\rho\mathbf{u}_t) = 0. \end{gathered}$$

The implication of the simple conservation equations for n and ρ is that the cells and matrix are simply convected by the matrix. As mentioned in Section 6.3 this is thought to be the major transport process. This is particularly evident in network formation (see Chapter 8). The one-dimensional version of the model mechanism is

$$
\begin{aligned}
& n_t + (nu_t)_x = 0, \\
& \mu u_{xxt} + u_{xx} + \left[\frac{\tau n}{1 + \lambda n^2} (\rho + \gamma \rho_{xx}) \right]_x = s\rho u, \qquad (6.34) \\
& \rho_t + (\rho u_t) x = 0,
\end{aligned}
$$

where we have set $\mu = (\mu_1 + \mu_2)/(1 + \nu')$, $\tau = \tau(1 + \nu')$, $s = s/(1 + \nu')$. This system linearises to

$$
\begin{aligned}
& n_t + u_{tx} = 0, \\
& \mu u_{xxt} + u_{xx} + [\tau_1 \rho + \tau_2 n + \tau_1 \gamma \rho_{xx}]_x = su, \qquad (6.35) \\
& \rho_t + u_{tx} = 0,
\end{aligned}
$$

where, from (6.24) and (6.31),

$$
\tau_1 = \frac{\tau}{(1 + \lambda)}, \qquad \tau_2 = \frac{\tau(1 - \lambda)}{(1 + \lambda)^2},
$$

with, as we just said, τ replacing $\tau/(1 + \nu')$.

For the system (6.33) we have from (6.31), $c(k^2) \equiv 0$ and so the dispersion relation from (6.32) is

$$
\begin{aligned}
& \sigma(k^2) = \frac{-b(k^2)}{\mu k^2}, \\
& b(k^2) = \gamma \tau_1 k^4 + (1 - \tau_1 - \tau_2) k^2 + s.
\end{aligned} \qquad (6.36)
$$

The only way we can have $\operatorname{Re} \sigma > 0$, and here, of course, σ is real, is if $b(k^2) < 0$ for some $k^2 > 0$. This requires $\tau_1 + \tau_2 > 1$ and from the second of (6.36)

$$
b_{\min} = s - \frac{(\tau_1 + \tau_2 - 1)^2}{4 \gamma \tau_1} < 0. \qquad (6.37)
$$

In terms of τ, λ, γ and s this becomes

$$
\tau^2 - \tau(1 + \lambda)^2 [1 + \gamma s(1 + \lambda)] + \tfrac{1}{4}(1 + \lambda)^4 > 0, \qquad (6.38)
$$

which implies that spatial patterns will evolve only if

$$
\tau > \tau_c = \tfrac{1}{2}(1 + \lambda)^2 \left[1 + \gamma s(1 + \lambda) + \{[1 + \gamma s(1 + \lambda)]^2 - 1\}^{1/2} \right]. \qquad (6.39)
$$

The other root is not relevant since it implies $\tau_1 + \tau_2 < 1$ and so from (6.36) $b(k^2) > 0$ for all k. The surface $\tau = \tau_c(\lambda, \gamma, s)$ is the bifurcation surface between spatial homogeneity and heterogeneity. In view of the central role of cell traction, and the form of the traction versus time curve in Figure 6.7, it is natural to take τ as the bifurcation parameter. As soon as τ increases beyond the critical value τ_c, the value which first makes

$b(k^2)$ zero, the uniform steady state bifurcates to a spatially unstable state. The natural groupings are $\tau/(1+\lambda)^2$ and $\gamma s(1+\lambda)$ and the bifurcation curve $\tau/(1+\lambda)^2$ versus $\gamma s(1+\lambda)$ is a particularly simple monotonic one.

When (6.39) is satisfied, the dispersion relation (6.36) is a typical basic dispersion relation (cf. Figure 2.5(b)) which initiates spatial patterns as illustrated in Figure 6.10(a) which is given towards the end of this section. All wavenumbers k in the region where $\sigma(k^2) > 0$ are linearly unstable: here that is the range of k^2 where $b(k^2) < 0$, which from (6.36) is given by

$$k_1^2 < k^2 < k_2^2$$

$$k_1^2, k_2^2 = \frac{(\tau_1 + \tau_2 - 1) \pm \{(\tau_1 + \tau_2 - 1)^2 - 4s\gamma\tau_1\}^{1/2}}{2\gamma\tau_1} \tag{6.40}$$

$$\tau_1 = \frac{\tau}{1+\lambda}, \quad \tau_2 = \frac{\tau(1-\lambda)}{(1+\lambda)^2},$$

where τ, γ, λ and s must satisfy (6.39). There is a fastest growing linear mode which again predicts, in the one-dimensional model with random initial conditions, the ultimate nonlinear spatial pattern (cf. Chapter 2, Section 2.6). Other ways of initiating the instability results in different preferred modes. We discuss nonlinear aspects of the models later and present some simulations of a full nonlinear model in the context of a specific biological application.

Before considering another example, note the form of the cell conservation equation in (6.35). With no cell mitosis there is no natural cell density which we associated with the maximum logistic value N in (6.7). Here we can use N (or n_0 to highlight the different situation) for the nondimensionalisation in the usual way but now it comes in as another arbitrary parameter which can be varied. It appears, of course, in several dimensionless groupings defined by (6.21) and so offers more potential for experimental manipulation to test the models. Later we shall describe the results from experiments when the cell density is reduced. With (6.21), we can thus determine how dimensionless parameters which involve N vary, and hence predict the outcome, from a pattern formation point of view, by investigating the dispersion relation. All models without any cell proliferation have this property. Another property of such models is that the final solution will depend on the initial conditions; that is, as cell density is conserved, different initial conditions (and hence different total cell number) will give rise to different patterns. However, as the random perturbations are small, the differences in the final solution will be small. This is biologically realistic since no two patterns are *exactly* alike.

(ii) $D_2 = \gamma = 0$: no long range diffusion and no cell–ECM interactions. $a_1 = a_2 = 0$: no haptotaxis. $r = 0$: no cell proliferation.

From the general model (6.22)–(6.24) the cell equation involves diffusion and convection and the model mechanism is now

$$n_t = D_1\nabla^2 n - \nabla\cdot(nu_t)$$

$$\nabla\cdot\left\{(\mu_1\boldsymbol{\varepsilon}_t + \mu_2\theta_t\mathbf{I}) + (\boldsymbol{\varepsilon} + \nu'\theta\mathbf{I}) + \frac{\tau\rho n\mathbf{I}}{1+\lambda n^2}\right\} = s\rho\mathbf{u}, \tag{6.41}$$

$$\rho_t + \nabla\cdot(\rho\mathbf{u}_t) = 0,$$

From (6.31) and (6.32) the dispersion relation for the system is

$$\begin{aligned}\sigma(k^2) &= \frac{-b \pm [b^2 - 4\mu k^2 c]^{1/2}}{2\mu k^2},\\ b(k^2) &= \mu D_1 k^4 + (1 - \tau_1 - \tau_2)k^2 + s,\\ c(k^2) &= D_1 k^2 [k^2(1 - \tau_2) + s].\end{aligned} \tag{6.42}$$

As noted above, in this model the homogeneous steady state cell density $n = N$ where N is now simply another parameter.

The critical value of τ, with τ_1 and τ_2 defined by (6.29), which makes the minimum of $b(k^2)$ zero is

$$\begin{aligned} b_{\min} = 0 \quad &\Rightarrow \quad \tau_1 + \tau_2 = 1 + 2(\mu s D_1)^{1/2}\\ &\Rightarrow \quad \tau_{b=0} = \tau_c = (1+\lambda)^2\left[\tfrac{1}{2} + (\mu s D_1)^{1/2}\right]\end{aligned} \tag{6.43}$$

and $c(k^2)$ becomes negative for a range of k^2 if $\tau_2 > 1$; that is,

$$c(k^2) < 0 \quad \Rightarrow \quad \tau_{c=0} = \tau_c = \frac{(1+\lambda)^2}{1-\lambda}, \quad \lambda \neq 1. \tag{6.44}$$

(The special case when $\lambda = 1$ makes $\tau_2 = 0$ exactly; such specific cases are unlikely to be of biological interest.) Now, as τ increases, whether b or c becomes zero first depends on other parameter groupings. In the case of the minimum of b becoming zero first, this occurs at the critical wavenumber which is obtained from the expression for $b(k^2)$ in (6.42) with (6.43) as

$$[k]_{b=0} = \left(\frac{s}{\mu D_1}\right)^{1/4}. \tag{6.45}$$

If τ is such that $c(k^2)$ becomes zero first, then from (6.42)

$$c(k^2) < 0 \quad \text{for all} \quad k^2 > \frac{s}{\tau_2 - 1}. \tag{6.46}$$

The linear and nonlinear solution behaviour depend critically on whether $b(k^2)$ or $c(k^2)$ becomes zero first, that is, whether $\tau_{c=0}$ is greater or less than $\tau_{b=0}$. Suppose that, as τ increases from zero, $b = 0$ first. From the dispersion relation (6.42), σ is complex at $\tau_c = t_{b=0}$ and so for τ just greater than τ_c, the solutions

$$n, \rho, \mathbf{u} \sim O(\exp[\operatorname{Re}\sigma(k_{b=0}^2 t + i\operatorname{Im}\sigma(k_{b=0}^2)t + i\mathbf{k}_{b=0}\cdot\mathbf{r}]). \tag{6.47}$$

These solutions represent exponentially growing *travelling* waves and the prediction is that no steady state finite amplitude solutions will evolve.

On the other hand if $\tau_{c=0}$ is reached first then σ remains real, at least near the critical k^2, and it would seem that spatial structures would evolve in the usual way. In any simulation related to a real biological application, a very careful analysis of the dispersion relation and the possible parameter values which are to be used is an absolutely essential part of the process since there can be transitions from normal evolving spatial patterns through unstable travelling waves to spatial patterning again. With the nonstandard type of partial differential equation we are dealing with such behaviour might be thought to be an artifact of the numerical simulations.

An even simpler version of this model, which still exhibits spatial structure, has $D_1 = 0$. The system in this case is, from (6.41),

$$\begin{aligned} &n_t + \nabla \cdot (nu_t) = 0 \\ &\nabla \cdot \left\{ (\mu_1 \boldsymbol{\varepsilon}_t + \mu_2 \theta_t \mathbf{I}) + (\boldsymbol{\varepsilon} + \nu' \theta \mathbf{I}) + \frac{\tau \rho n \mathbf{I}}{1 + \lambda n^2} \right\} = s\rho u, \\ &\rho_t + \nabla \cdot (\rho \mathbf{u}_t) = 0. \end{aligned} \tag{6.48}$$

Here, from (6.31), $c(k^2) \equiv 0$ and

$$\sigma(k^2) = \frac{-b(k^2)}{\mu k^2}, \quad b(k^2) = (1 - \tau_1 - \tau_2)k^2 + s.$$

So, we require τ and λ to satisfy

$$\begin{aligned} \tau_1 + \tau_2 > 1 \quad &\Rightarrow \quad \tau > \frac{(1+\lambda)^2}{2} \\ &\Rightarrow \quad \sigma(k^2) > 0 \quad \text{for all} \quad k^2 > \frac{s}{2\tau(1+\lambda)^{-2} - 1}. \end{aligned} \tag{6.49}$$

The dispersion relation here, and illustrated in Figure 6.11(a) given at the end of this section, is fundamentally different to that in Figure 6.10(a); there is an *infinite* range of unstable wave numbers. That is, perturbations with very large wavenumbers, which correspond to very small wavelengths, are unstable. This is because the version (6.48) of the model has no long range effects included: such effects tend to smooth out small wavelength patterns. It is not clear what pattern will evolve from random initial conditions. The ultimate spatial structure depends intimately on the initial conditions. Asymptotic analyses are still lacking for systems with such dispersion relations.

(iii) $D_1 = D_2 = 0$: no cell diffusion. $a_1 = a_2 = 0$: no haptotaxis. $\mu_1 = \mu_2 = 0$: no viscoelastic effects in the ECM.

Here the system (6.22)–(6.24) reduces to

$$\begin{aligned} &n_t + \nabla \cdot (n\mathbf{u}_t) = rn(1-n), \\ &\nabla \cdot \left\{ (\boldsymbol{\varepsilon} + \nu' \theta \mathbf{I}) + \frac{\tau n}{1 + \lambda n^2} (\rho + \gamma \nabla^2 \rho) \mathbf{I} \right\} = s\rho \mathbf{u}, \\ &\rho_t + \nabla \cdot (\rho \mathbf{u}_t) = 0, \end{aligned} \tag{6.50}$$

which, in one space dimension, is

$$
\begin{aligned}
&n_t + (nu_t)_x = rn(1-n)\\
&[u_x + \tau n(1+\lambda n^2)^{-1}(\rho + \gamma\rho_{xx})]_x = s\rho u \qquad (6.51)\\
&\rho_t + (\rho u_t)_x = 0,
\end{aligned}
$$

where again we have incorporated $(1+\nu')$ in τ and s. The dispersion relation in this case is, using (6.31) and (6.32),

$$
\begin{aligned}
\sigma(k^2) &= -\frac{c(k^2)}{b(k^2)},\\
b(k^2) &= \gamma\tau_1 k^4 + (1-\tau_1-\tau_2)k^2 + s, \qquad (6.52)\\
c(k^2) &= \gamma\tau_1 r k^4 + r(1-\tau_2)k^2 + rs.
\end{aligned}
$$

Here, as τ increases $b(k^2)$ becomes zero first, so the bifurcation traction value is given by τ_c where $b(k^2) = 0$. This expression for $b(k^2)$ is the same as that in (6.36) and so the critical τ_c is given by (6.39). Here, however, $c(k^2)$ is not identically zero and so the dispersion relation is quite different as τ increases beyond τ_c. Since $\tau_2 = \tau(1-\lambda)/(1+\lambda)^2$ if $\lambda > 1$, $c(k^2) > 0$ for all $k^2 > 0$. For the purposes of our discussion however, we assume $\lambda < 1$ so that $c(k^2)$ can also become negative. Denote the critical τ_c when $b(k^2)$ and $c(k^2)$ first become zero by $\tau_c^{(b)}$ and $\tau_c^{(c)}$ respectively: here $\tau_c^{(b)} < \tau_c^{(c)}$. In this case the $\sigma(k^2)$ behaviour as τ increases is illustrated in Figure 6.9, which exhibits a fundamentally different dispersion relation to what we have found and discussed before.

First note in Figure 6.9 that the range of unstable wavenumbers is finite and that there are two bifurcation values for the traction parameter τ. The pattern formation potential of a system with such a dispersion relation is much richer than is possible with the standard dispersion form in Figure 6.10(a). Linear theory, of course, is not valid where the linear growth is infinite. So, from an analytical point of view, we must include other effects which effectively round off the discontinuities in $\sigma(k^2)$. This in turn implies the existence of a singular perturbation problem. We do not consider such problems here but we can see intuitively that such dispersion relations, with large linear growth rates, imply a 'fast focusing' of modes with preferred wavenumbers. For example, in Figure 6.9(e) we would expect the modes with wavenumbers at the lower end of the lower band and the upper end of the upper band to be the dominant modes. Which mode eventually dominates in the nonlinear theory will depend critically on the initial conditions.

With large linear growth we can see how necessary it is to have some cell–cell inhibition in the full nonlinear system. If there is fast focusing there is the possibility of unlimited growth in, for example, the cell density. This in turn implies the appearance of spikelike solutions. The effect of the inhibition terms, as measured by the parameter λ, is therefore essential. There are some potentially interesting analytical aspects with these fast focusing models.

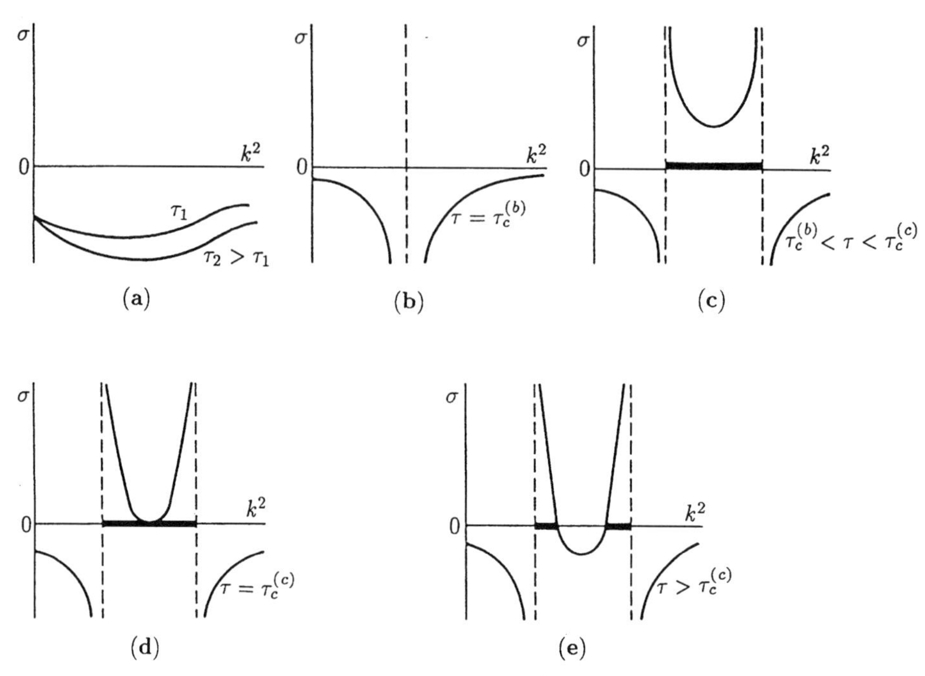

Figure 6.9. Qualitative variation in the dispersion relation $\sigma(k^2)$ in (6.52) for the model system (6.50) (and (6.51)) as the traction parameter τ increases. The bifurcation values $\tau_c^{(b)}$ and $\tau_c^{(c)}$ denote the values of τ where $b(k^2) = 0$ and $c(k^2) = 0$ respectively. The wavenumbers of the unstable modes are denoted by the heavy line on the k^2-axis.

It is now clear how to investigate various simpler models derived from the more complicated basic model (6.22)–(6.24). Other examples are left as exercises.

Figures 6.10 and 6.11 indicate the richness of dispersion relation types which exist for the class of mechanical models (6.22)–(6.24). Figure 6.10 shows only some of the dispersion relations which have finite ranges of unstable modes while Figure 6.11 exhibits only some of the possible forms with infinite ranges of unstable modes. A nonlinear analysis in the vicinity of bifurcation to spatial heterogeneity, such as has been done by Maini and Murray (1988), can be used on the mechanisms which have a dispersion relation of the form illustrated in Figure 6.10(a). A nonlinear theory for models with dispersion relations with an infinite range of unstable modes, such as those in Figure 6.11, is, as we noted, still lacking as is that for dispersion relations which exhibit infinite growth modes (Figures 6.10(b),(e)–(g)). Although we anticipate that the pattern will depend more critically on initial conditions than in the finite range of unstable mode situations, this has also not been established.

Mechanical models, as we noted above, are also capable of generating travelling waves: these are indicated by dispersion relations with complex σ. Table 6.3 gives examples of models which admit such solutions.

From a biological application point of view two- and three-dimensional patterns are naturally of great interest. With the experience gained from the study of the numerous reaction diffusion chemotaxis and neural models in the book, we expect the simulated

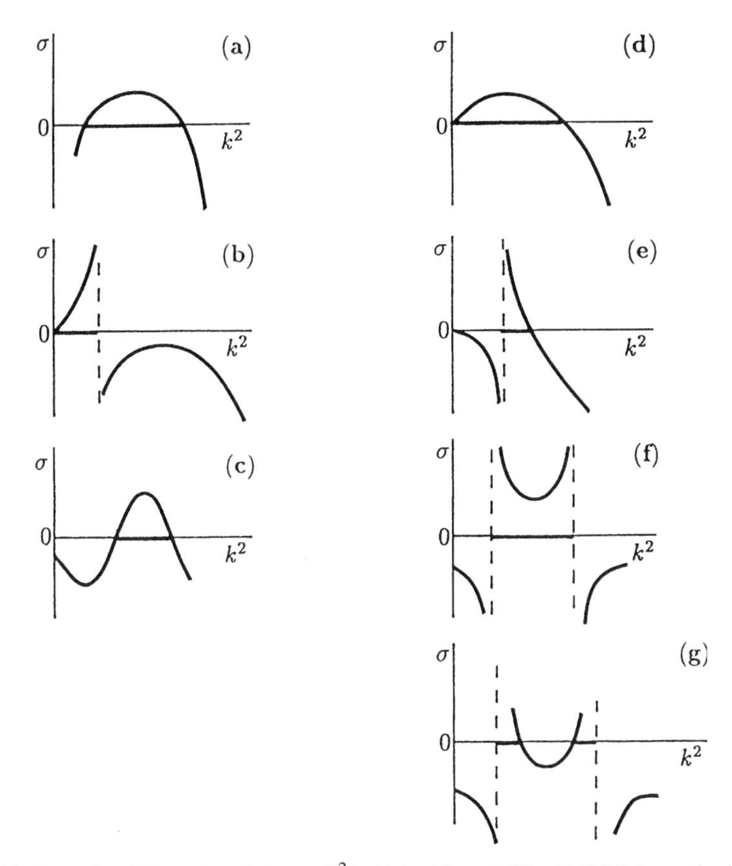

Figure 6.10. Examples of dispersion relations $\sigma(k^2)$, obtained from (6.32) with (6.31) for mechanical models based on the mechanism (6.22)–(6.24). The various forms correspond to the specific conditions listed in Table 6.1. Realistic models for those with infinite growth must be treated as singular perturbation problems, with small values for the appropriate parameters in terms which have been omitted so as to make the linear growth finite although large.

Table 6.1. Mechanical models, derived from the basic system (6.22)–(6.24) with positive nonzero parameters denoted by •, which have dispersion relations with a finite range of unstable wavenumbers. The corresponding dispersion relation forms are given in Figure 6.10, with $\lambda = 0$, $\tau_1 = \tau_2 = \tau$.

Figure 6.10	D_1	D_2	a_1	a_2	r	μ	s	γ	λ	Condition on τ
(a)	○	○	○	○	○	•	•	•	○	$\tau > \{1 + \gamma s + [(1 + \gamma s)^2 - 1]^{1/2}\}/2$
(b)	•	○	○	○	○	•	○	•	○	$1 > \tau > 1/2$
(c)	•	○	○	○	○	○	○	•	○	$1 > \tau > 1/2$
(d)	•	○	○	○	○	○	○	•	○	$1 < \tau;\ D_1 > (2\tau - 1)/(\tau - 1)$
(e)	•	•	•	•	○	○	•	○	○	$1/2 > \tau;\ [\tau(D_1 + a_1) - D_1 - sD_2]^2$
		○		•						$> 4sD_1[D_2 + \tau(a_1 a_2 - D_2)]$
(f)	○	○	○	○	•	○	•	•	○	$1/2 < \tau < 1;\ (2\tau - 1)^2 > 4s\gamma\tau$
(g)	○	○	○	○	•	○	•	•	○	$\tau > 1;\ (\tau - 1)^2 > 4s\gamma\tau$

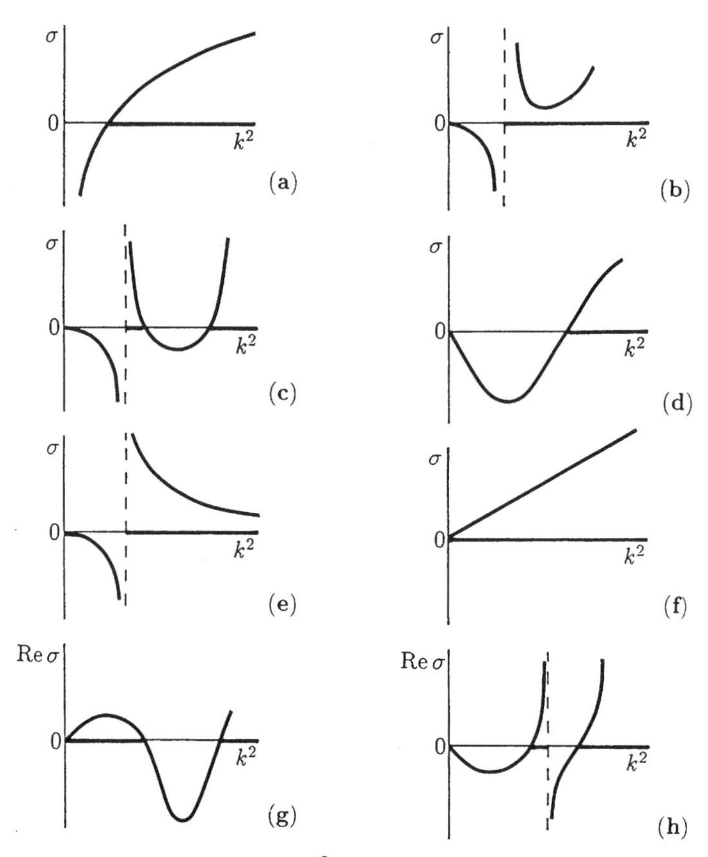

Figure 6.11. Examples of dispersion relations $\sigma(k^2)$, obtained from (6.32) with (6.31) for mechanical models based on the mechanism (6.22)–(6.24), with an infinite range of unstable modes. The conditions listed in Table 6.2 relate the models to specific forms. In (**g**) and (**h**) the imaginary part of σ is nonzero.

Table 6.2. Mechanical models, derived from the basic system (6.22)–(6.24) with positive nonzero parameters denoted by •, which have dispersion relations with an infinite range of unstable wavenumbers. The corresponding dispersion relation forms are given in Figure 6.11, with $\lambda = 0$, $\tau_1 = \tau_2 = \tau$.

Figure 6.11	D_1	D_2	a_1	a_2	r	μ	s	γ	λ	Condition on τ
(a)	○	○	○	○	○	•	•	○	○	$\tau > 1/2$
(b)	•	○	○	○	○	○	•	○	○	$^*\tau < 1/2$
(c)*	•	•	•	•	○	○	•	○	○	$\tau > 1/2$, plus quadratic condition on τ
(d)	•	○	•	○	○	○	•	○	○	$D_1/(D_1 + a_1) > \tau > 1/2$
(e)	○	○	○	○	•	○	•	○	○	$1/2 < \tau < 1$
(f)	•	○	○	○	○	○	○	○	○	$1/2 < \tau < 1$

* This is only one of the possibilities, depending on $c(k^2)$ and its zeros.

Table 6.3. Mechanical models, derived from the basic system (6.22)–(6.24) with positive nonzero parameters denoted by •, which have dispersion relations with a finite range of unstable wavenumbers and which admit temporal oscillations. The corresponding dispersion relation forms are given in Figure 6.11, with $\lambda = 0$, $\tau_1 = \tau_2 = \tau$.

Figure 6.11	D_1	D_2	a_1	a_2	r	μ	s	γ	λ	Conditions on τ Are Algebraically Complicated
(g)	•	○	○	○	○	•	•	○	○	
(h)	○	○	○	○	•	•	•	○	○	

patterns for the full nonlinear models here to reflect some of the qualitative features of the linearised analysis of the basic model equations (6.26)–(6.28). This is motivation for looking at possible symmetries in the solutions. We do this by taking the divergence of the linear force balance equation (6.27) which, with (6.26) and (6.28), gives, using the identity

$$\operatorname{div} \boldsymbol{\varepsilon} = \operatorname{grad}(\operatorname{div} \mathbf{u}) - \tfrac{1}{2}\operatorname{curl}\operatorname{curl}\mathbf{u},$$

$$n_t - D_1\nabla^2 n + D_2\nabla^4 n + a_1\nabla^2\rho - a_2\nabla^4\rho + \theta_t + rn = 0, \tag{6.53}$$

$$\nabla^2[(\mu\theta_t + (1+\nu')\theta + \tau_1 n + \tau_2\rho + \tau_1\gamma\nabla^2\rho)] - s\theta = 0, \tag{6.54}$$

$$\rho_t + \theta_t = 0, \tag{6.55}$$

where $\tau_1 = \tau/(1+\lambda)$, $\tau_2 = \tau(1-\lambda)/(1+\lambda)^2$ and $\mu = \mu_1 + \mu_2$. To determine a reasonably full spectrum of relevant solutions of this set of equations is not trivial. However, we can look for periodic solutions, which tessellate the plane, for example. Such solutions (cf. Chapters 2 and 12) satisfy

$$\Gamma(\mathbf{r} + m\boldsymbol{\omega}_1 + l\boldsymbol{\omega}_2) = \Gamma(\mathbf{r}), \tag{6.56}$$

where $\Gamma = (n, \mathbf{u}, \rho)$, m and l are integers and $\boldsymbol{\omega}_1$ and $\boldsymbol{\omega}_2$ are independent vectors. A *minimum* class of such periodic solutions of the linear system (6.53)–(6.55) includes at least the eigenfunctions of

$$\nabla^2\psi + k^2\psi = 0, \quad (\mathbf{n}\cdot\nabla)\psi = 0 \quad \text{for} \quad \mathbf{r} \text{ on } \partial B, \tag{6.57}$$

where $\mathbf{n}$ is the unit normal vector on the boundary ∂B of the domain B. With these boundary conditions the solutions are periodic. As we saw in Chapter 2, Section 2.4 and Chapter 12 regular plane periodic tessellation has the basic symmetry group of the hexagon, square (which includes rolls) and rhombus solutions given respectively by equations (2.47), (2.48) and (2.49). For convenience the solutions in polar coordinate form (r, ϕ) are reproduced again here:

$$\text{Hexagon:} \quad \psi(r,\theta) = \tfrac{1}{3}\left[\cos\left\{kr\sin\left(\phi + \frac{\pi}{6}\right)\right\} + \cos\left\{kr\sin\left(\phi - \frac{\pi}{6}\right)\right\} + \cos\left\{kr\sin\left(\phi - \frac{p}{2}\right)\right\}\right]. \tag{6.58}$$

$$\text{Square:} \qquad \psi(r,\phi) = \tfrac{1}{2}\left[\cos\{kr\cos\phi\} + \cos\{kr\sin\phi\}\right] \tag{6.59}$$

$$\text{Rhombus:} \qquad \psi(r,\phi;\delta) = \tfrac{1}{2}\left[\cos\{kr\cos\phi\} + \cos\{kr\cos(\phi-\delta)\}\right], \tag{6.60}$$

where δ is the rhombic angle. Such symmetric solutions are illustrated in Figure 6.12.

Small Strain Approximation: A Caricature Mechanical Model for Two-Dimensional Patterns

In many embryological situations the strain, cell density and ECM density changes during the pattern formation process are small. Such assumptions can lead to the linear model system (6.26)–(6.28). Linear systems, however, pose certain problems regarding long term stability. We can exploit the small strain approximation to derive a simple scalar equation model which retains certain key nonlinearities which allow us to carry out a nonlinear analysis in the two-dimensional situation and obtain stable nonlinear solutions of biological relevance.

To illustrate this let us consider the nonlinear dimensionless system (6.33), the nontrivial steady state of which we take as $n = \rho = 1$, $\mathbf{u} = 0$. Because of the small strains we linearise the cell and matrix conservation equations, the first and third of (6.33), to get

$$\begin{aligned} n_t + \nabla\cdot\mathbf{u}_t = 0 \quad &\Rightarrow \quad n_t + \theta_t = 0, \\ \rho_t + \nabla\cdot\mathbf{u}_t = 0 \quad &\Rightarrow \quad \rho_t + \theta_t = 0, \end{aligned}$$

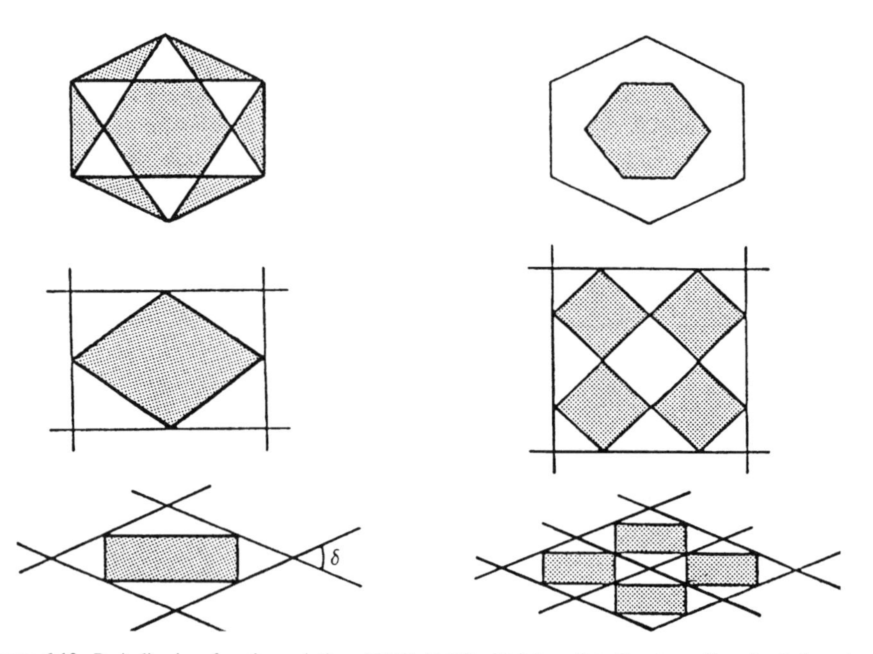

Figure 6.12. Periodic eigenfunction solutions (6.58)–(6.60) which tessellate the plane. Here the dark regions represent higher densities.

since the dilation $\theta = \nabla \cdot \mathbf{u}$. Integrating with respect to t and using the fact that when $\theta = 0$, $n = \rho = 1$ we get

$$n(\mathbf{r}, t) = 1 - \theta(\mathbf{r}, t) = \rho(\mathbf{r}, t). \tag{6.61}$$

Since we consider θ small, certainly $\theta < 1$, n and ρ remain positive, as is necessary of course.

Because of (6.61) we now replace the external force $s\rho\mathbf{u}$ in the force balance equation, the second of (6.33), by its linear approximation $s\mathbf{u}$. Now substitute the linear forms relating n and ρ to the dilation from (6.61) into the second of (6.33), take the divergence of the resulting equation and use the tensor identity

$$\nabla \cdot \boldsymbol{\varepsilon} = \text{grad div } \mathbf{u} - \tfrac{1}{2}\text{curl curl } \mathbf{u}.$$

This yields the following scalar equation for the dilation θ,

$$\mu\nabla^2\theta_t + \nabla^2\theta + \tau\nabla^2[(1-\theta)^2 - \gamma(1-\theta)\nabla^2\theta] - s\theta = 0, \tag{6.62}$$

where we have incorporated $1 + \nu'$ into redefinitions for μ, τ and s and for algebraic simplicity taken $\lambda = 0$. The effect of $\lambda \neq 0$ is simply to introduce a multiplicative term $[1 + 2\lambda\theta/(1+\lambda)]$ in the square bracket in (6.62) and have $\tau/(1+\lambda)$ in place of τ.

Maini and Murray (1988) carried out a nonlinear analysis on the caricature model (6.62) and obtained roll and hexagonal solutions. The significance of the latter is discussed in the following section on a biological application to skin organ morphogenesis.

Perhaps it should be mentioned here that the spectrum of spatial patterns possible with the mechanism (6.22)–(6.24) and its numerous simplifications is orders of magnitude greater than with a reaction diffusion system—even three-species systems. The implications of a paper by Penrose (1979) are that tensor systems have solutions with a wider class of singularities than vector systems. Since the cell–matrix equation is a tensor equation, its solutions should therefore include a wider class of singularities than reaction diffusion vector systems. Even with the linear system (6.26)–(6.28) there are many analytical and numerical problems which remain to be investigated.

In the following sections we consider biologically important and widely studied pattern formation problems using mechanical models of pattern generation. As always we reiterate that the actual mechanism is not yet known but a mechanical mechanism, we suggest, is certainly a strong candidate or at least a mechanism which involves some of the major ingredients.

6.5 Periodic Patterns of Feather Germs

Generation of regular patterns occurs in many situations in early embryogenesis. These are particularly evident in skin organ morphogenesis such as in the formation of feather and scale primordia and are widely studied (see, for example, Sengel 1976 and Davidson 1983). Feather formation has much in common with scale formation during early development of the primordia. Here we concentrate on feather germ formation with particular reference to the chick, and fowl in general. Feather primordial structures are distributed across the surface of the animal in a characteristic and regular hexagonal fashion. The application of the Murray–Oster mechanical theory to feather germ primordia was first put forward by Murray et al. (1983) and Oster et al. (1983) and it is their scenario we describe here. It was also investigated with a tissue interaction model by Cruywagen et al. (1992, 2000); we discuss their model in Section 6.10. We first present the biological background which suggests using a mechanical model.

Vertebrate skin consists essentially of two layers: an epithelial epidermis overlays a much thicker mesenchymal dermis and is separated from it by a fibrous basal lamina. The layer of epithelial cells, which in general do not move, can deform as we described in Section 6.1. Dermal cells are loosely packed and motile and can move around in the extracellular matrix, the ECM, as we described earlier. The earliest observable developmental stages of feather and scale germs begin the same way. We concentrate here on the initiation and subsequent appearance of feather rudiments in the dorsal pteryla—the feather forming region on the chick back.

In the chick the first feather rudiments become visible about 6 days after egg fertilization. Each feather germ, or primordium, consists of a thickening of the epidermis with one or more layers of columnar cells, called a placode, beneath which is an aggregation of dermal (mesenchymal) cells, called a papilla. Excellent pictures of papillae and placodes are given by Davidson (1983). The dermal condensations are largely the result of cell migration, with localised proliferation playing a secondary role. There is still no general agreement as to whether the placodes form prior to the dermal papillae or the other way round. There is considerable experimental work going on to determine the order of appearance or, indeed, whether the interaction between the epidermis and dermis produces the patterns simultaneously. The dermis seems to determine the spatial patterning—as shown by epidermal–dermal recombination experiments (Rawles 1963, Dhouailly 1975). We come back to this below when we describe tissue interaction systems. The model we discuss here is for the formation of dermal papillae. Subsequent development, however, is a coordinated process involving both the epidermal and dermal layers (Wessells 1977, Sengel 1976, Cruywagen et al. 1992).

Davidson (1983) demonstrates that chick feather primordia appear sequentially. A central column of dermal cells forms on the dorsal pteryla and subsequently breaks up into a row of papillae. As the papillae form, tension lines develop joining the cell aggregation centres. With the above mechanical models this is consistent with the cells trying to align the ECM. Now lateral rows of papillae form sequentially from the central column outwards—in the ventral direction—but these are interdigitated with the papillae in the preceding row; see Figures 6.13(a)–(d). These lateral rows spread out from the central midline almost like a wave of pattern initiation. Experiments by Davidson (1983) tend to confirm this wave theory; later we show how these results can be explained by our model and we present corroborative numerical results.

These observations suggest that it is reasonable first to model the pattern formation process for the initial row of papillae by a one-dimensional column of cells and look for the conditions for spatial instability which generate a row of papillae. This is stage 1 and is represented by the sequential process illustrated in Figures 6.13(a)–(d).

We have seen in the previous sections that the mechanical model (6.22)–(6.24), and simpler models derived from it as in Section 6.4, that as the cell traction parameter τ increases beyond some critical value τ_c the uniform steady state becomes spatially

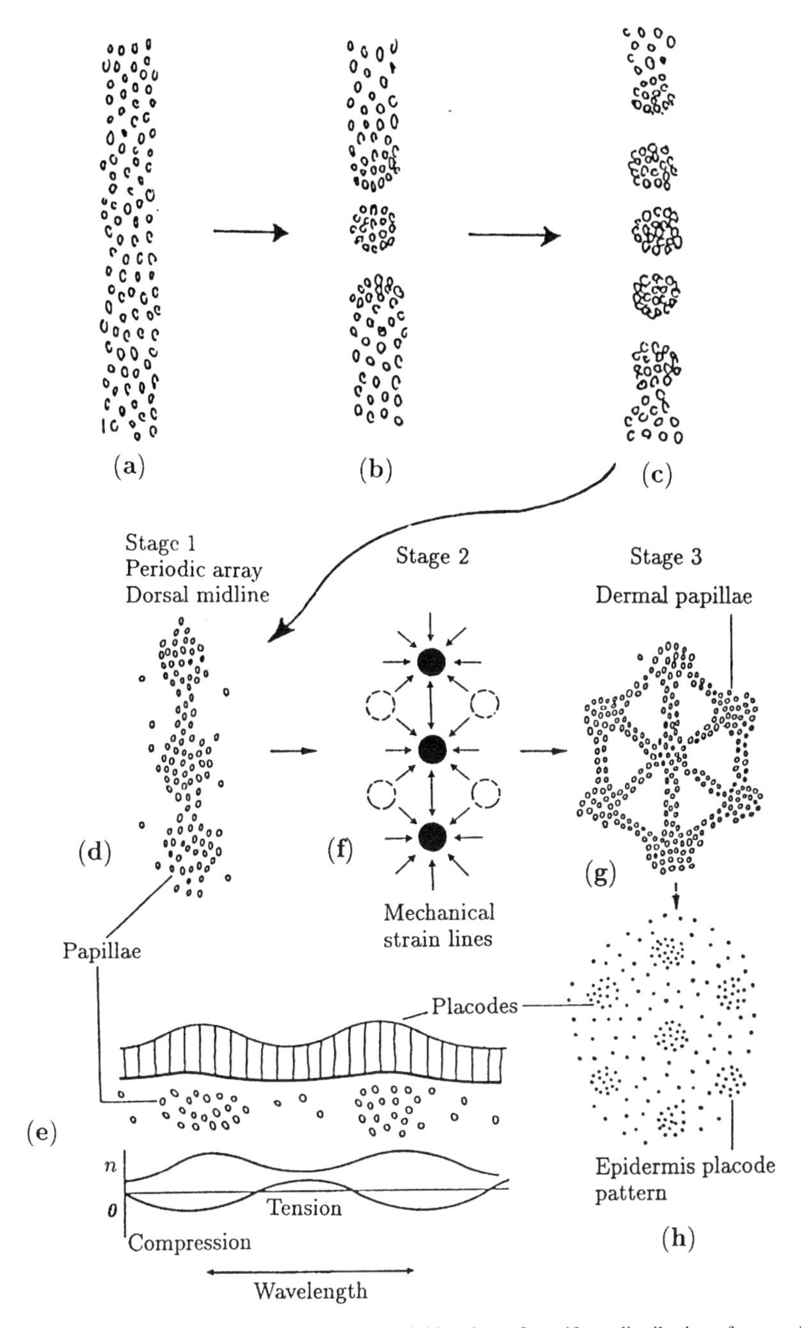

Figure 6.13. (**a**)–(**d**) These show the predicted sequential breakup of a uniform distribution of mesenchymal (dermal) motile cells into regular cell condensations with a wavelength determined by the parameters of the model mechanism (stage 1). These cell aggregations are the primordial papillae for feathers and scales. (**e**) Vertical cross-section qualitatively showing the feather germ primordia. The placodes in the epidermis are underlain by the papillae which create the stress field. (**f**) Subsequent aggregations form laterally. The prestressed strain field from the first line of condensations induces a bias so that the neighbouring line of papillae interdigitate with the first line (stage 2). The resulting periodic array is thus hexagonal, the basic unit of which is illustrated in (**g**) (stage 3). (**h**) The epidermal placode pattern that mirrors the dermal pattern.

unstable. With the standard basic dispersion relation as in Figure 6.10(a), the mode (6.30) with a specific wavenumber k_c, that is, with wavelength $2\pi/k_c$, first becomes unstable and a spatial pattern starts to evolve; this generates a regular pattern of dermal papillae.

Simulations of the one-dimensional version of the full nonlinear mechanical model (6.22)–(6.24) with negligible long range haptotaxis, $a_2 = 0$ and, reflecting the biological situation of small cell mitosis, $r = 0$, have been carried out, for example, by Perelson et al. (1986), Bentil (1990) and Cruywagen (1992). Perelson et al. (1986) particularly addressed the problem of mode selection in models with many parameters, and proposed a simple scheme for determining parameter sets to isolate and 'grow' a specific wavelength pattern. Bentil and Murray (1991) developed an even simpler and easy to use scheme for doing this. Figure 6.14(a) shows a typical steady state pattern of cell aggregations (the papillae), ECM displacement and density variations. As we would expect intuitively, the cell aggregations are in phase with the ECM density variation ρ, and both are out of phase with the ECM displacement u. The reason is that the cell aggregations pull the matrix towards the areas of higher cell density thus stretching the matrix between them; Figure 6.13(e) illustrates what is going on physically.

Patterns of the type illustrated in Figure 6.14(a) occur only if the cell traction parameter is above a certain critical value (see Section 6.4). So, a possible scenario for the formation of the pattern along the dorsal midline is that there is a wave of initiation that sweeps down the column and this could be related to tissue age; in this case, *in vitro* experiments show that the cell traction parameter increases; refer to Figure 6.7. As the cells become stronger τ passes through the critical value τ_c and pattern is initiated. Note that in the model, one-dimensional pattern develops simultaneously whereas the experiments suggest sequential development. This is reminiscent of the mode of pattern formation illustrated in Figure 2.15(d) in Chapter 2.

Let us now consider the formation of the distinctive hexagonal two-dimensional pattern of papillae. We described above how a wave of pattern initiation seems to spread out from the dorsal midline. This suggests that the pattern of matrix strains set up by the initial row of papillae biases the formation of the secondary condensations at positions displaced from the first line by half a wavelength. Figure 6.14(b) shows the appropriate numerical simulation based on such a scenario: note how the patterns are out of phase with those in Figure 6.14(a). If we now look at Figures 6.13(f) and (g) we see how this scenario generates a regular hexagonal pattern in a sequential way like a wave emanating from the central dorsal midline.

This 'wave' is, however, not a wave in the usual sense since if the dermal layer is cut along a line parallel to the dorsal midline the wave simply starts up again beyond the cut *ab initio*. This is consistent with Davidson's (1983) experimental observations; he specifically investigated the qualitative effect on spacing of stretching and cutting the epidermis.

This quasi-one-dimensional scenario, although suggested by linear theory, was to a certain extent validated by the nonlinear simulations. A better verification would be from a simulation of the two-dimensional model. However, using our scenario, it is possible to make predictions as to the change in wavelength as the experimental parameters are changed. For example, one version of the models predicts a spacing that increases as

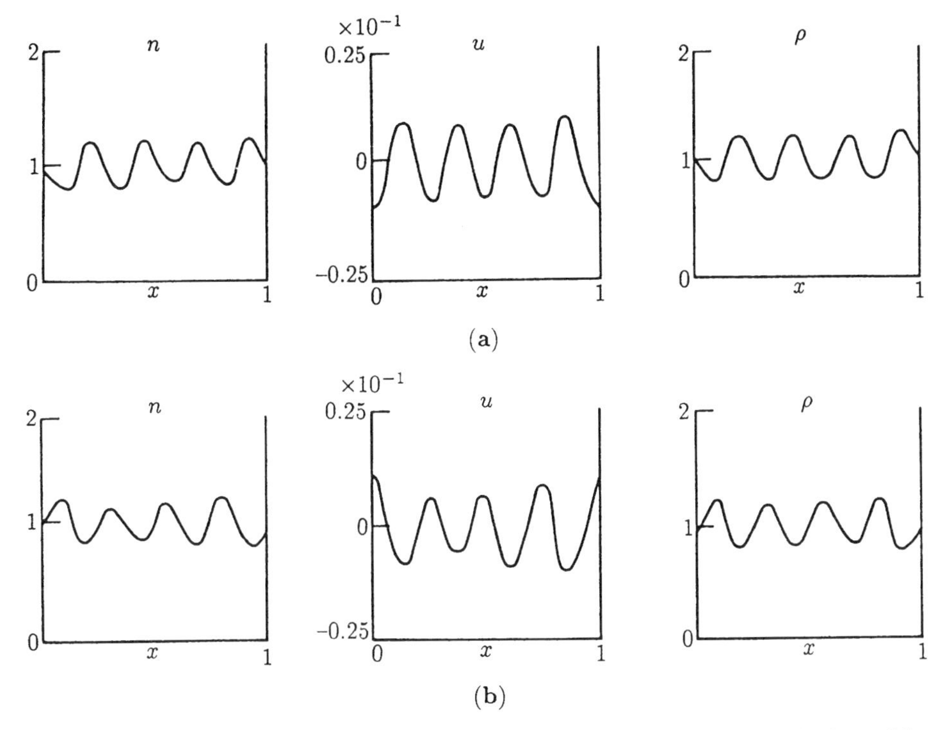

Figure 6.14. Steady state solutions, for the cell density n, ECM displacement u and ECM density ρ of the nonlinear one-dimensional version of the mechanical model (6.22)–(6.24). (**a**) Periodic boundary conditions were used and initial conditions were random perturbations about the uniform steady state $n = \rho = 1$, $u = 0$. (**b**) Heterogeneous steady state solutions with the initial displacement pattern in (**a**). Parameter values: $D_1 = a_1 = \gamma = 10^{-3}$, $\lambda = 0.12$, $\tau = 1.65$, $\sigma = 400$, $\mu = 1$. Note that n and ρ are in phase and both are out of phase with u.

the total number of cells N decreases. This agrees with experimental observation (Dr. Duncan Davidson, personal communication 1983).

One of the most useful aspects of a nondimensional analysis, with resulting nondimensional groupings of the parameters, is that it is possible to assess how different physical effects, quantified by the parameters in the nondimensionalisation, trade off against one another. For example, with the nondimensional groupings (6.21) we see from the definition of the dimensionless traction, namely, $\tau^* = \tau\rho_0 N(1+\nu)/E$, that with the model (6.22)–(6.24) the effect of a reduction in the cell traction τ is the same as a reduction in the cell density or an increase in the elastic modulus E. To see clearly the overall equivalence the bifurcation surfaces in parameter space must be considered. An important caveat in interpreting results from experimental manipulation is that quite different cell or matrix alterations can produce compensating and thus equivalent results. Although such a caveat is applicable to any model mechanism it is particularly apposite to experimentation with mechanical models since the morphogenetic variables are unquestionably real.

We should perhaps mention here that an alternative model based on a reaction diffusion theory was proposed by Nagorcka (1986), Nagorcka and Mooney (1985) for the initiation and development of scale and feather primordia, and by Nagorcka and Mooney (1982) for the formation of hair fibres. See also Nagorcka and Adelson (1999), who among other things discuss some experimental tests, and other references there.

The modelling here does not cast light on the controversy regarding the *order* of formation of placodes and papillae. However, since the traction forces generated by the dermal cells can be quite large the model lends support to the view that the dermis controls the pattern even if it does not initiate it. Current thinking tends towards the view that initiation requires tissue interaction between the dermis and epidermis. It is well known that mechanical deformations affect mitosis and so tissue interaction seems natural with mechanical models; see also the discussion in Oster et al. (1983). Nagorcka et al. (1987) investigated a tissue interaction mechanism specifically with the complex patterns of scales in mind. We describe some of these complex patterns and their model in Section 6.10. Later in the chapter we investigate tissue interacting models and some of the implications from their study. An experimental paper by Nagawa and Nakanishi (1987) confirms the importance of mechanical influences of the mesenchyme on epithelial branching morphology.

6.6 Cartilage Condensations in Limb Morphogenesis and Morphogenetic Rules

The vertebrate limb is one of the most widely and easily studied developmental systems and such studies have played a major role in embryology; see, for example, the book by Hinchliffe and Johnson (1980), the general paper by Thorogood (1983) and the review by Tickle (1999), who subscribes to the positional information chemical prepattern scenario, which describes some of the recent biochemical evidence for such an approach. We shall discuss some evolutionary aspects of limb development in the next chapter. Here we show how a mechanical model could generate the pattern of cell condensations which evolve in a developing limb bud and which eventually become cartilage; it was first put forward by Murray et al. (1983) and Oster et al. (1983). A related mechanochemical model was later proposed by Oster et al. (1985a,b).

The pattern in developing limb buds which determines the final cartilage patterns, which later ossify into bones, involves aggregations of chondrocyte cells, which are mesenchymal cells such as we have been considering. The basic evolution of chondrocyte patterns takes place sequentially as the limb bud grows, which it does from the distal end. Figure 6.15 gives an explanation of how, with geometry and scale as bifurcation parameters, chondrogenesis could proceed. The actual sequence of patterns for the developing chick limb is illustrated in Figure 6.15(c); Figure 6.15(d) is a photograph of a normal adult limb. The detailed explanation of the process based on a mechanical mechanism is the following.

As the limb bud grows, through cell proliferation in the apical ectodermal ridge, which is at the distal end, the cross-section of the tissue domain, which includes the ECM and mesenchymal cells, is approximately circular but with an elliptical bias. Let us consider this to be the two-dimensional domain for our mechanical model with zero

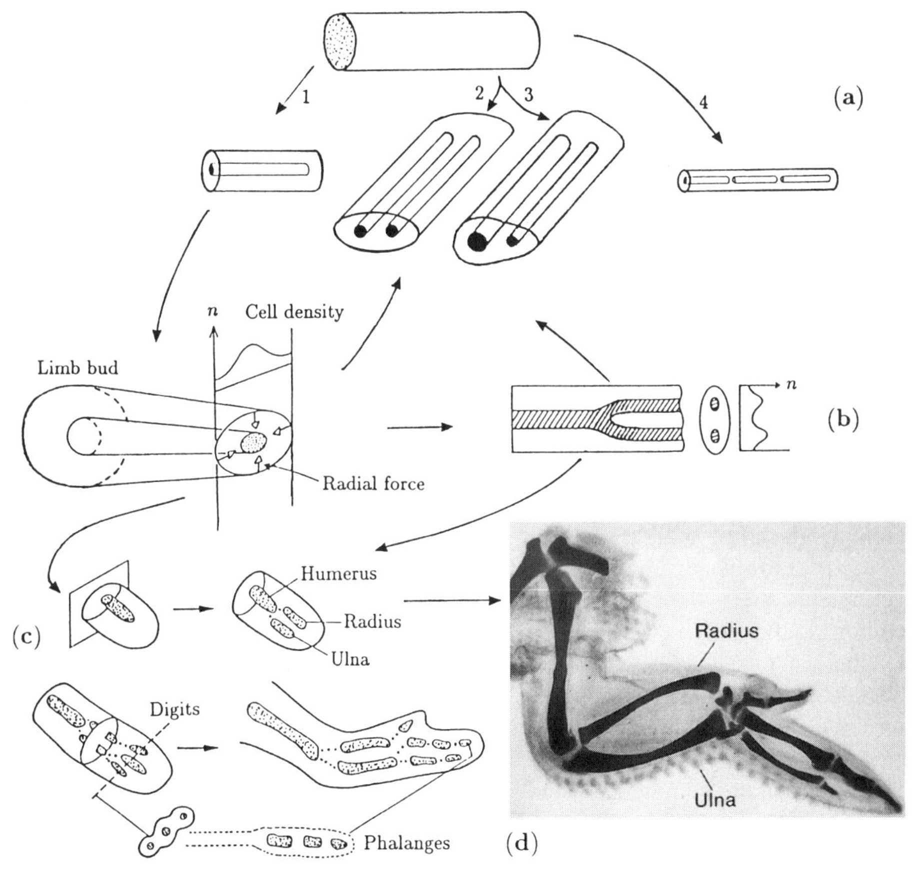

Figure 6.15. (**a**) The type of axial condensation is influenced by the cross-sectional shape of the limb. Initially a single condensation, path 1, will be produced (for example, the humerus in (**c**)). A more elliptical cross-section allows two aggregations to form with an aerofoil-shaped domain producing unequal condensations, paths 2 and 3 (for example, the radius and ulna in (**c**)). In a long thin cylinder the axial condensations form segmental units, path 4 (for example, the phalanges in (**c**)). (**b**) This shows how the mechanical mechanism influences cross-sectional form thereby inducing the required sequence of chondrogenic patterns. As the cells form the central condensation their tractions deform the limb thus making it more elliptical. At a critical ellipticity the pattern bifurcates to two condensations. How three condensations are formed is important and explained in the text; refer also to Figure 6.18(d). (**c**) The schematic bifurcation sequence of chondrocyte (mesenchymal) cell aggregations which presage cartilage formation in the developing chick limb. (**d**) Photograph of the normal cartilage pattern in the limb of a 10-day chick. (Photograph courtesy of L. Wolpert and A. Hornbruch)

flux boundary conditions for the cells n and matrix ρ. The condition for $\mathbf{u}$ is an imposed restraining force which comes from the epidermis—the sleeve of the limb bud. The dispersion relation for the mechanism with such a domain is reminiscent of that for reaction diffusion mechanisms with similar geometry, as discussed in detail in Chapter 2, particularly Section 2.5. Let us suppose that as the cells age the traction increases as in Figure 6.7 and eventually passes through the critical value τ_c. The detailed form of the dispersion relation is such that in the appropriate parameter space this first bifurcation produces a single central aggregation of cells recruited from the surrounding tissue.

The axial cell aggregations are influenced by the cross-sectional shape as shown in Figure 6.15(a). As the cells condense into a single aggregation they generate a strong centrally directed stress as in Figure 6.15(b). This radial stress deforms the already slightly elliptical cross-section to make it even more elliptical. This change in geometry in turn induces a secondary bifurcation to two condensations because of the changed flatter geometry of the cross-section. An aerofoil section gives rise to two condensations of different size as in Figure 6.15(a), path 3: these we associate with, for example, the radius and ulna in forelimbs as in the photograph in Figure 6.15(d).

We should interject, here, that this behaviour is directly equivalent to the situation with the patterns generated by reaction diffusion (and chemotaxis) mechanisms as illustrated in Figures 2.14(c) and 2.17. A similar scenario for sequential laying down of a prepattern for cartilage formation by reaction diffusion models equally applies. However, what is fundamentally different is that with a mechanical model, the condensation of cells influences the *shape* of the domain and can actually *induce* the sequence of bifurcations shown in Figure 6.15(c).

After a two-condensation state has been obtained, further growth and flattening can generate the more distal patterns. By the time the limb bud is sufficiently flat, cell recruitment effectively isolates patterning of the digits. Now subsequent growth induces longitudinal or segmental bifurcations with more condensations simply fitted in as the domain, effectively linearly now, increases and we get the simple laying down of segments, for example, the phalanges in Figure 6.15(c), as predicted by Figure 6.15(a), path 4.

It is important to reiterate here, that the sequence of cell pattern bifurcations need not be generated by a changing geometry; it can result from a variation of other parameters in the model. Also asymmetric condensations can result from a spatial variation or asymmetry in a parameter across the limb cross-section. There is well-documented experimental evidence for asymmetric properties, which, of course, are reflected in the different bone shapes and sizes in the limb such as the radius and ulna in Figure 6.15(d). Whatever triggers the bifurcations as we move from the proximal to distal part of the limb, the natural sequence is from a single condensation, to two condensations and then to several as in Figure 6.15(c); see also Figure 6.18.

Much of the extensive experimental work on chondrogenesis has been to investigate the chondrogenic patterns which result from tissue grafts. The major work on chick limbs was initiated and carried out by Lewis Wolpert and his co-workers and it has been the principal stimulus for much of the current research in this line, on other animals as well as chicks; see, for example, Wolpert and Hornbruch (1987), Smith and Wolpert (1981) and the more review-type article by Wolpert and Stein (1984). One set of experiments involves taking a piece of tissue from one part of a limb bud and grafting it onto another as in Figure 6.16(a). The region from the donor limb is referred as the zone of polarising activity—the ZPA. This results in the double limb as in Figure 6.16(b).

Consider now the double limb in Figure 6.16(b) in the light of our model and let us examine how this can arise. Let us be specific and take geometry and scale as the bifurcation parameters. The effect of the tissue graft is to increase the width of the limb cross-section by increasing the cell division in the apical ectodermal ridge, the distal

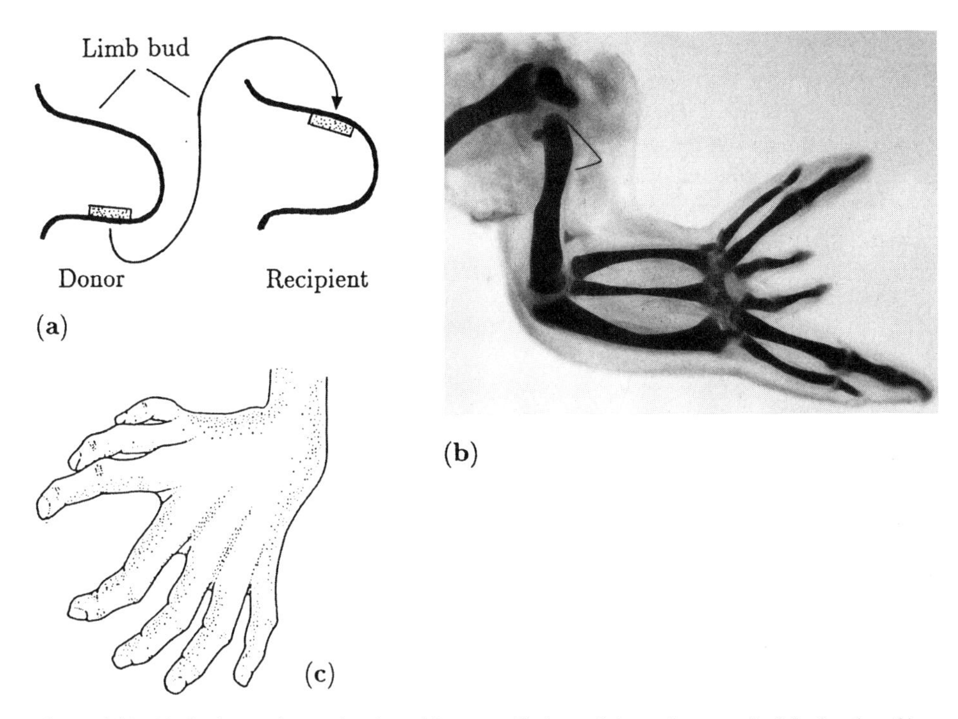

Figure 6.16. (**a**) Graft experiments involve taking a small piece of tissue from one limb bud and grafting it onto another. The effect of such a graft is to induce increased cell proliferation and hence increase the subsequent size of limb. The result is to induce growth commensurate with a domain in which multiple cell condensations can be fitted in at each stage of growth and hence result in double limbs. (**b**) Photograph of a double limb in a 10-day chick following an anterior graft of tissue from the posterior region, the zone of polarising activity (ZPA), of another limb as in (**a**). The grafted tissue creates the appropriate symmetry which results in a mirror image limb. (From Wolpert and Hornbruch 1987; photograph courtesy of L. Wolpert and A. Hornbruch) (**c**) A natural example of a double hand of a Boston man: note the lack of thumb and the mirror symmetry. (After Walbot and Holder 1987)

edge of the wing bud (Smith and Wolpert 1981). This means that at each stage, after the graft, the domain is sufficiently wide for a double set of cell condensations to form and thus generate a double limb as shown in Figure 6.16(b). Not all grafts result in a double limb. Different double patterns can be obtained, and which depends on where and when in development the graft is inserted. Figure 6.16(c) shows an example of a natural double hand. It is not uncommon for people to have six fingers on a hand, often an inherited trait. (Anne Boleyn, one of Henry VIII's wives had six fingers—unfortunately she had the wrong appendage cut off!) An experimental prediction of the model then is that if the limb bud with a graft is geometrically constrained within a scale commensurate with a single limb, it would not be able to undergo the double bifurcation sequence necessary for a double limb to form. The book by Walbot and Holder (1987) gives a good description of such graft experiments and discusses the results in terms of a positional information approach with, in effect, a chemical prepattern background.

A new model for patterning of limb cartilage development has been presented by Dillon and Othmer (1999) in which they incorporate the interactions between morphogens in the ZPA (zone of polarizing activity) and the AER (apical epidermal ridge). They demonstrate the importance of the interaction and importantly explicitly include growth. The numerical scheme they develop can be used to explore the effects of various experimental interventions.

Some interesting experiments have been carried out which show that many of the results of graft experiments can be achieved by subjecting the limb bud, during development, to doses of retinoic acid sequestered in small beads inserted into the limb. The retinoic acid is slowly released in the tissue. A quantitative analysis of the effects is given by Tickle et al. (1985); see also Tickle (1999). Disruption of the chondrogenic process by chemicals and drugs is well known—the thalidomide affair is a tragic example. This work highlights one of the important uses of any theory which might help in unravelling the mechanism involved in chondrogenesis. Until we understand the process it is unlikely we shall be able to understand how drugs, chemicals and so on will disrupt the process during development.

When we look at the cartilage patterns after the initial pattern has been laid down we see, as in Figure 6.15(c), that there is a gap between the bifurcations, for example, between the humerus and the radius and ulna. Whether or not there is a gap in cell condensation as the pattern bifurcates from a single to a double aggregation depends on the dispersion relation; these are the same possibilities we discussed regarding Figures 2.17(a) and (b) in Chapter 2. It has been shown from experiment in the case of cartilage patterning in the developing limb for a large number of animals that the bifurcation is a clear branching process as seen in Figure 6.17. In fact, with the mechanical model the bifurcation is continuous. The separation comes from subsequent recruitment of cells to form the observed gaps. This bifurcation patterning puts certain constraints on allowable dispersion relations which, in turn, imply certain constraints which any model mechanism must satisfy.

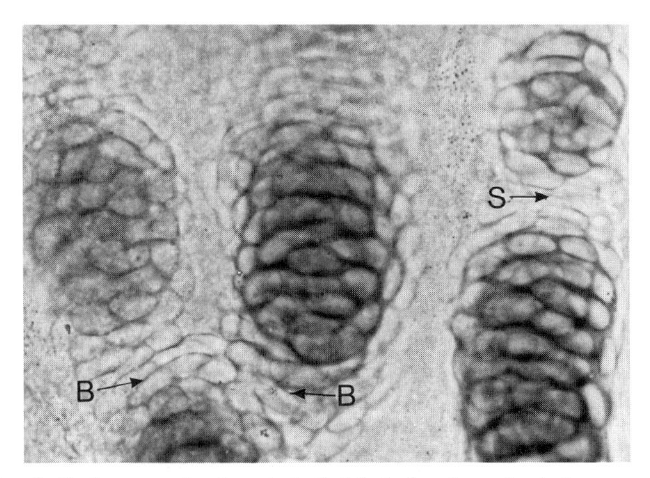

Figure 6.17. Longitudinal cross-section through the limb bud of a salamander *Ambystoma mexicanum*: note the branching bifurcation of cell condensation. At a later stage the branches of the Y separate from the main stem through cell recruitment at the ends; these are marked B. A segmental bifurcation can be seen starting at S. (Photograph courtesy of P. Alberch)

Morphogenetic Rules for Cartilage Morphogenesis in the Limb

With a completely symmetric geometry and tissue isotropy it is possible to move through the bifurcation space of parameters from one aggregation, to two, to three and so on. With reaction diffusion mechanisms, such as we considered in Chapter 2, it is possible to choose a path in some parameter space to achieve this; refer to Figure 2.14(b) for a specific example. It is also possible to do this with mechanical models. However, with the natural anisotropy in embryological tissue such isotropy does not exist. The question then arises as to how the pattern sequence from a double to a triple condensation is effected. We believe that for all practical purposes the process must be that in which one branch of the double condensation itself undergoes a branching bifurcation while near the other branch either a focal condensation appears or it undergoes a segmental bifurcation (see Figure 6.18). Let us now note another experimentally observed fact, namely, that during chondrogenesis there appears to be little cell division (Hinchliffe and Johnson 1980). This implies that condensations principally form through recruitment of cells from neighbouring tissue. Thus, as the limb bud grows the pattern bifurcation that takes place following a branching bifurcation is as illustrated in Figure 6.18(c). Figures 6.18(a) and (b) show the other two basic condensation elements in setting up a cell condensation pattern in a developing limb.

If we now take the bifurcating pattern elements in Figures 6.18(a)–(c) as the three allowable types of cell condensations we can see how to construct any limb cartilage pattern by repeated use of the basic condensation elements in Figures 6.18(a)–(c). Figure 6.18(e), which is the forelimb of a salamander, is just one example to illustrate the process. So, even without considering any specific mechanism, we hypothesise an important *set of morphogenetic rules* for the patterning sequence of cartilage in the development of the vertebrate limb. This hypothesis, encapsulated in the theory put forward by Murray et al. (1983a) and Oster et al. (1983), was exploited by Oster et al. (1988) who present extensive experimental evidence for its validity.

In the above discussion we had in mind a mechanical model for pattern formation in mesenchymal cells. The morphogenetic rules which we deduced equally apply to reaction diffusion models of pattern formation. In fact we suggest they are model-independent, or rather any model mechanism for chondrogenic pattern formation must be capable of generating such a sequence of bifurcating patterns.

Model Predictions and Biological Implications

If we look at the bifurcation in cell condensations in Figure 6.17 we see that the bifurcation is continuous, essentially like that shown in Figure 6.18(b). This imposes constraints on any proposed model mechanism for cartilage patterning. Let us suppose we have a pattern formation mechanism which gives rise to a dispersion relation which varies as shown in Figure 6.19(a) as a parameter changes. If domain size is such a parameter then the implication for the evolution of the pattern is then illustrated in the second and third diagrams of Figure 6.19(a). In other words as the limb bud grows the initial condensation bifurcates as predicted but with a homogeneous region separating the initial condensation and the Y-bifurcation. If we now consider the dispersion variation scenario in Figure 6.19(b) the biological implication is that the bifurcation from

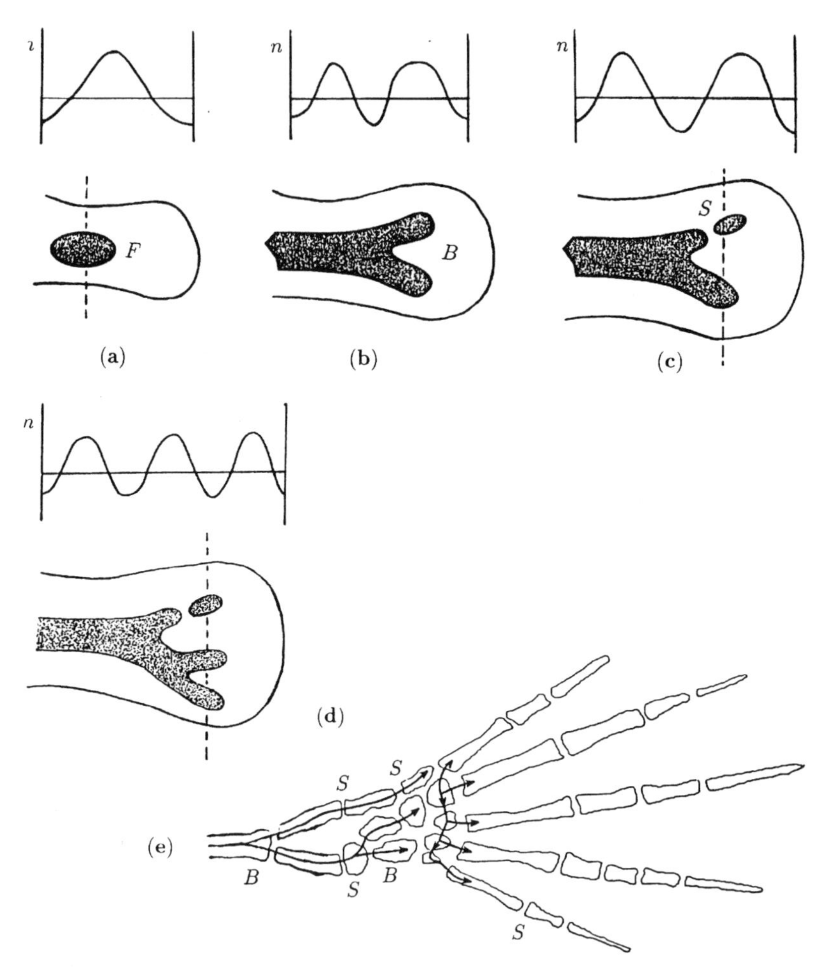

Figure 6.18. (**a**)–(**c**) The three basic types of cell condensations which generate cartilage patterns in the developing vertebrate limb. These are postulated as the morphogenetic rules for cartilage pattern generation for all vertebrate limbs. (**a**) Focal condensation, F; (**b**) Branching bifurcation, B; (**c**) Segmentation condensation, S. (**d**) Formation of more patterns is by further branching or independent foci. (**e**) An example of a branching sequence showing how the cartilage patterns in the limb of a salamander can be built up from a sequence of F, B and S bifurcations.

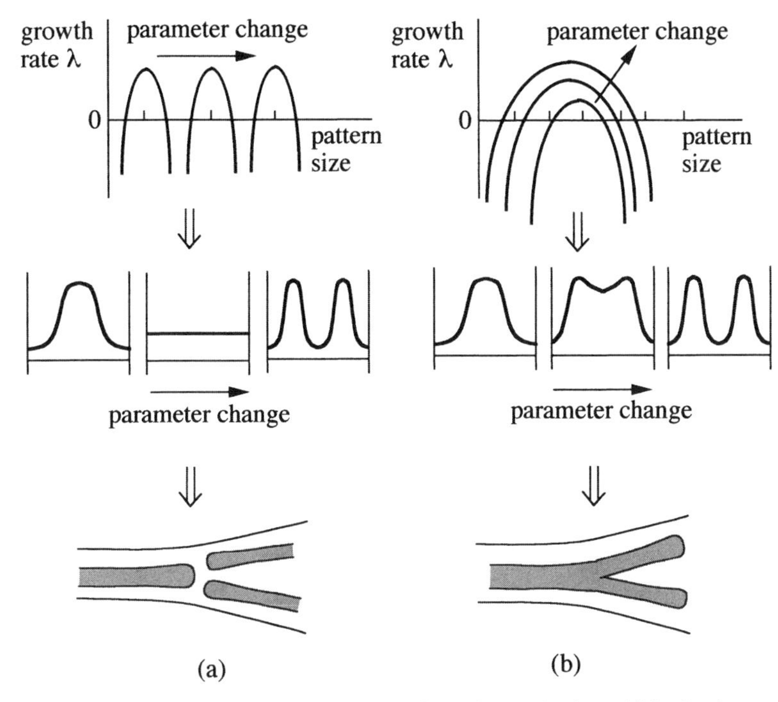

Figure 6.19. When the dispersion relation for a pattern formation mechanism exhibits the characteristics in (**a**) as a critical parameter varies, the implication for the subsequent branching is that prior to the Y-bifurcation there is a region of homogeneity. In (**b**), on the other hand, the dispersion scenario implies a continuous bifurcation. These have very specific biological implications for the developing limb.

one to two condensations is continuous. This clearly puts developmental constraints on the model mechansim.[2]

Experimental evidence from amphibians suggests that osmotic properties of the ECM may be important in morphogenesis. Hyaluronate is a principal component of the ECM and can exist in a swollen osmotic state. As the condensation of chondrocytes starts the cells secrete an enzyme, hyaluronidase, which degrades the hyaluronate. This could lead to the osmotic collapse of the matrix thus bringing the cells into close enough contact to initiate active contractions and thus generate cell aggregations. Cell motility is probably not important in this scenario. A modification of the mechanical model to incorporate these chemical aspects and the added forces caused by osmotic pressure was proposed and analysed by Oster et al. (1985b). They showed that such a mechanochemical model would generate similar chondrogenic patterning for the developing limb.

[2]Prior to the experiment which produced Figure 6.17 I separately asked two major developmental biologists with decades of experience and expertise on cartilage patterning in the developing limb what they thought happened when the humerus bifurcated in the radius and ulna. Does it happen as in Figure 6.19(a) or (b)? Each of them answered without any hesitation whatsoever and with complete conviction: the first said that it was, of course, like that shown in Figure 6.19(a) while the second, with equal conviction, said it was as in Figure 6.19(b) (the correct answer). Interestingly, if you look at the developing limb a short time after the photograph in Figure 6.17 was taken the condensations have recruited cells from the continuous bifurcating region in Figure 6.19(b) and created a homogeneous region as in Figure 6.19(a). The first had simply not looked at the embryonic limb bud during this small window of time.

A new model for patterning of limb cartilage development has been proposed by Dillon and Othmer (1999) in which they incorporate the interactions between morphogens in the ZPA (zone of polarizing activity) and the AER (apical epidermal ridge). They demonstrate the importance of the interaction and importantly explicitly include growth. The numerical scheme they develop can be used to explore the effects of various experimental interventions. A general interest article by Riddle and Tabin (1999) on how limbs develop brings in genetic aspects, which play a major role of course in controlling the mechanisms.

A major role of theory in morphogenesis is to suggest possible experiments to distinguish different models each of which can generate the appropriate sequence of patterns observed in limb chondrogenesis. Mechanical models lend themselves to experimental scrutiny more readily than reaction diffusion models because of the elusiveness of chemical morphogens. In the next chapter, where further examples are given of the application of these general rules, we shall see how they introduce developmental constraints which lets us make certain practical predictions. It also relevant to the existence and nonexistence of certain developmental teratologies. We shall describe some of the experimental results based on these predictions; they have important evolutionary implications regarding vertebrate limb development. These morphogenetic rules constitute another example of a biologically practical result which arose from specific modelling concepts but which is model-independent as in the case of alligator striping.

6.7 Embryonic Fingerprint Formation

The study and classification of fingerprint patterns—dermatoglyphics—has a long history and their widespread use in genetic, clinical, pathological, embryonic, anthropological and forensic studies (not to mention palmistry) has produced an extensive descriptive literature. Figure 6.20(a) (see also Figure 12.5(c)) is a photograph of a human fingerprint on which various common traits are marked. The descriptive methods used may loosely be described as topological, that is, those associated with the study of properties that are unchanged by continuous deformation (Penrose 1979), and statistical (Sparrow and Sparrow 1985) and depend on the area of application. Topological methods have been found to be especially efficient for genetic and diagnostic purposes because ridges appear in their definitive forms during embryogenesis and the basic patterns do not change under continuous deformation during growth (Elsdale and Wasoff 1976, Loesch 1983, Bard 1990).[3]

In a study of two-dimensional patterns created on a confluent dish of normal fibroblasts, and dermatoglyphic patterns of primate palms and soles, Elsdale and Wasoff (1976) observed that both patterns were characterized by different types of interruptions or discontinuities in fields of otherwise parallel aligned elements. They concluded that because the discontinuities were invariant under plastic deformations as well as rigid motions, topological considerations in addition to analyses of cell behaviour were necessary to understand dermatoglyphic pattern development. However, topological clas-

[3]The old literature abounds with interesting studies. In one 19th century book the author describes in great detail the effect on his fingerprints of self-mutilation. When he had used up all his fingers he moved on to his feet. The results of 40 years of this were completely inconclusive but he had a lot of interesting scars.

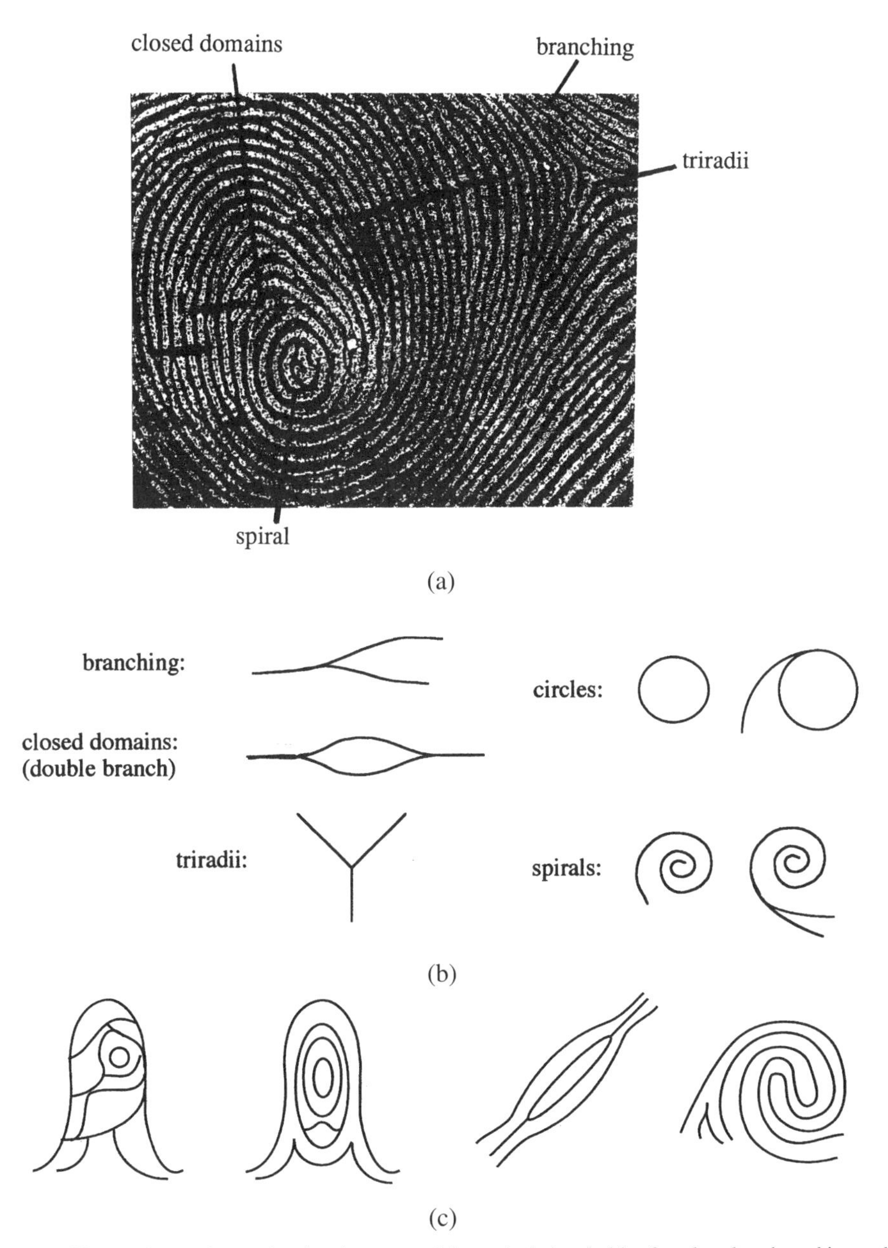

Figure 6.20. (**a**) Human fingerprint showing some of the typical singularities found such as branching and triradii; see also Figure 6.22. (**b**) Sketches of some unusual bifurcation patterns associated with dermatoglyphic patterning. A common feature is the bifurcation to two or more ridges and 'melting' of such ridges to form a single one. (**c**) Such bifurcation patterns are known to exist in patients with chromosomal abberations (Trisomy 21). (Dr. K. de Braganca, personal communication 1991).

sification adopts systems of rules to quantify features of patterns and is limited since it ignores much of the geometry and focuses only on fingerprint identification techniques. Others believe that due to the elastic nature of the human skin a combination of a topological and statistical description is required. It is thought, however, that successive rolled impressions usually suffer from a degree of relative distortion (translation, rotation and stretching) and topological systems should be free from the detrimental effects of plastic distortions caused by changes in physical circumstances under which sophisticated patterns are formed. Such an approach was successfully adopted by the National Bureau of Standards in the U.S.A. in the matching of fingerprint patterns: various comparison algorithms were developed for use on rolled impressions that were invariably inexpensive to implement (see Sparrow and Sparrow 1985). Since then several new more sophisticated pattern recognition computer-oriented techniques are used in classification.

There has been a number of studies on the development of general ridge patterns for which tissue culture and microscopy have been used (for example, Cummins and Midlo 1943, Schaumann and Alter 1976, Green and Thomas 1978, Elsdale and Wasoff 1976, Okajima 1982, Okajima and Newell-Morris 1988 and Bard 1990; see other references in these). Observations by Loesch (1983) indicate that in humans, for example, the development of epidermal ridge patterns seems to start around the third month of gestation. The ultimate patterns are dependent on the degree of asymmetry and, to a considerable extent, on pad formation: symmetrical pads produce whorls; asymmetrical pads produce loops while other forms of pad development lead to arches.

There is little general agreement as to how dermatoglyphic patterns are formed or what mechanism creates them. As with other skin appendages a key question is whether the dermis or epidermis alone has the capacity to form a particular type of derivative. Various grafting experiments (for example, Davidson 1983) support the view that the dermis and epidermis interact concurrently in the formation of related patterns such as feather germs. The results of heterospecific recombination of mouse plantar (sole of the foot) dermis with epidermis from other skin regions, and vice versa (Kollar 1970) suggest that lack of hair follicles and the appearance of dermatoglyphic patterns may be determined by the dermis rather than the epidermis.

Experiments by Okajima (1982) led to the development of a method to inspect dermatoglyphic patterns on the exposed dermis and have confirmed that ridged structures first appear on the dermis. These ridges appear as a result of opposing movement of cell masses. A ridge may form from cells ejected from the deeper cell layers. The shape of the ridge frequently suggests an accumulation of cells in specific regions and discontinuities could depend on the curvature or geometry of the region. In sequential photographs (Green and Thomas 1978) short ridges often seem to fuse into longer ones. Long ridges may undergo uniform lateral displacement, or a local displacement to produce an arch; see Figure 6.21(a). Sometimes arched ridges curve into whorls, suggesting that the 'resistance' of the two sides of the arch is unequal. The size of the whorl is determined by the size of the field available for organization, confirming that the process of dermatoglyphic pattern formation can be divided into two stages: first, the formation of ridges and second the curving of these ridges into whorls, loops and so on. Patterns may, therefore, develop by a process of cell movement in the dermis, which first produces ridges and then curves these ridges into bifurcating patterns of increasing

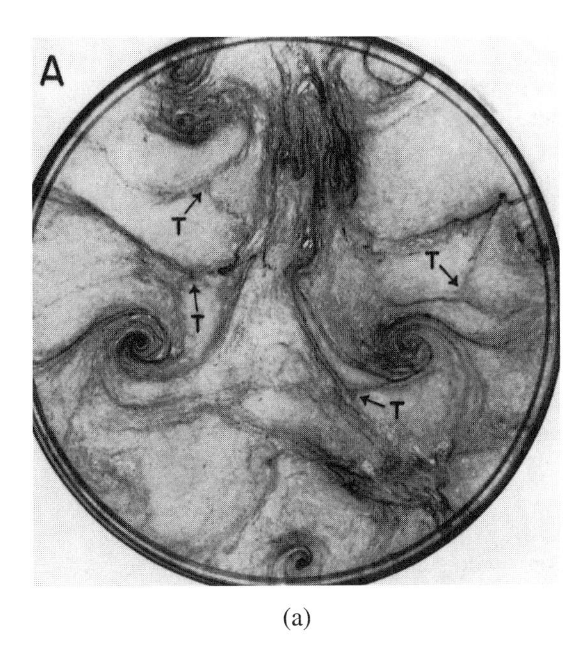

(a)

displacement field **u** (=u,v)

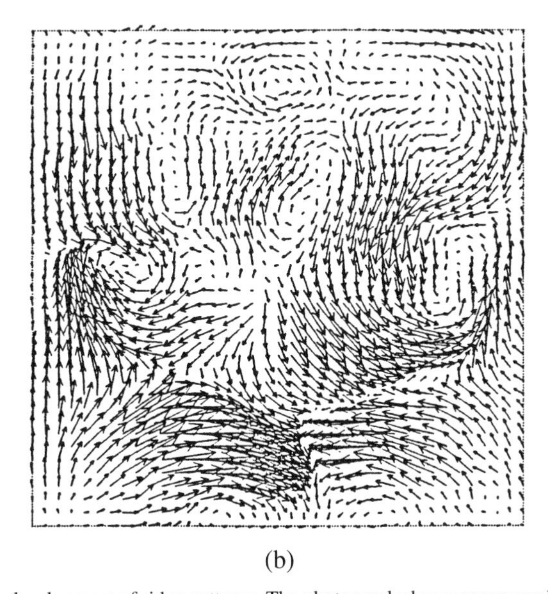

(b)

Figure 6.21. (**a**) The development of ridge patterns. The photograph shows numerous left- and right-handed whorls. The field is frequently divided into regions of different patterns in which triradial lines meet at angles of approximately 120° relative to each other. (From Green and Thomas 1978; photograph courtesy of Howard Green) (**b**) Numerical simulation of a heterogeneous steady state solution of the model system with homogeneous boundary conditions on a square domain with a random initial perturbation in cell density. Here the displacement field associated with the passive movement of the ECM reflects a typical scenario during the onset of embryonic dermal patterning. Parameter values: $\tau = 1.1$, $\mu = \mu_1 + \mu_2 = 0.7$, $D = 0.002$, $a_1 = 0.003$, $s = 140.0$, $\gamma = 0.02$, $m = r = 0$, $\eta_i = \mu_i/2(1+\nu')$, $\eta_1 = \eta_2 = 1.0$.

complexity, ultimately whorls, loops and triradii. Genetic disorders can also produce unusual patterns as illustrated in Figure 6.20(c).

A particularly challenging aspect of dermatoglyphic patterning, from the modelling point of view, is not only how normal patterns are generated but also how they form the unusual ridge bifurcation patterns such as shown in Figure 6.20(b). These bifurcation patterns are usually associated with some genetic diseases (see, for example, Karev 1986 and Okajima 1982), and are especially known to exist in patients with chromosomal abberations (Trisomy 21) (Dr. K. de Braganca, personal communication 1991). In de Braganca's studies, abrupt changes of ridge directions (bifurcations) on the dermal patterns of the fingertips were noticed. There were cases where two or more ridges could bifurcate and 'fuse' together at a later stage to form a single minutiae as seen in Figure 6.20(b) and were explained as possible shortening of the phalanges in their development. These bifurcation forms suggest that the underlying mechanism could arise from the mechanical motion of cells on the extracellular matrix, which is constantly being convected.

Given the possibility that the mechanical motion of cells on the extracellular matrix generates bifurcation patterns during the onset of embryonic dermatoglyphic patterns, Daniel Bentil and I (see also Bentil 1990) decided to use the mechanical approach to see if a mechanical mechanism could create similar dermatoglyphiclike patterns: the full two-dimensional system involves the stress tensors for the ECM and the cell traction forces.

Since experimental evidence suggests that the formation of ridges involves a mechanical deformation of the dermis and subsequently the epidermis, we investigate a model for ridge formation based on the mechanical forces associated with the dermis and the tendency for the epidermis to buckle and fold in line with the dermis. We consider two of the essential processes that occur during cell motion, namely, the spreading of the extracellular matrix (ECM) with the convection of cells and adhesive sites with it, and the growth of tissues and cells during dermatoglyphic pattern formation. The full model we propose for dermatoglyphic patterning includes an ECM source term dependent on cell and matrix densities.We suggest that cell motion and cell traction forces, together with proliferation, could produce branching patterns on the continuum ECM substratum and cell density. Migrating cells follow the branching patterns as they are carried by the tissue matrix. We include a nonlocal (long range) traction term in the cell–matrix interaction (ECM) equation. As we saw above the inclusion of this nonlocal term ensures stable spatial patterning. With these, using (6.7) with $D_2 = a_2 = 0$, (6.18) with $\tau n/(1 + \lambda n^2) = \tau n$ since cell densities are generally small, and (6.20) with $S(n, \rho, \mathbf{u}) = mn\rho(\rho_0 - \rho)$ with m and ρ_0 constants, the model mechanism is then

$$\frac{\partial n}{\partial t} = \underbrace{D_1 \nabla^2 n}_{\text{diffusion}} - \underbrace{\nabla \cdot (n\mathbf{u}_t)}_{\text{convection}} - \underbrace{a_1 \nabla \cdot n \nabla \rho}_{\text{haptotaxis}} + \underbrace{rn(N - n)}_{\text{mitosis}}, \tag{6.63}$$

$$\nabla \cdot \left[\underbrace{\mu_1 \boldsymbol{\varepsilon}_t + \mu_2 \theta_t \mathbf{I}}_{\text{viscous}} + \underbrace{E'(\boldsymbol{\varepsilon} + \nu' \theta \mathbf{I})}_{\text{elastic}} + \underbrace{\tau n(\rho + \gamma \nabla^2 \rho)\mathbf{I}}_{\text{cell traction}} \right] - \underbrace{s\rho \mathbf{u}}_{\text{external forces}} = 0, \tag{6.64}$$

$$\frac{\partial \rho}{\partial t} + \underbrace{\nabla \cdot (\rho \mathbf{u}_t)}_{\text{convection}} = \underbrace{mn\rho(\rho_0 - \rho)}_{\text{matrix secretion}}. \tag{6.65}$$

The matrix source term is probably necessary to induce branching.

From the above sections we know that this system, given by (6.63)–(6.65), can generate a spectrum of complex spatial patterns if the parameters are in the appropriate ranges. After nondimensionalising the system using (6.21), linearising about the relevant steady state $\mathbf{u} = 0, n = \rho = 1$ and looking for solutions proportional to $e^{\sigma t + i\mathbf{k}.\mathbf{x}}$ we get (refer to Section 6.3) the dispersion relation, which is similar to (6.31) but with an extra contribution because of the matrix secretion term in (6.65) which makes the characteristic polynomial a cubic, namely,

$$\begin{aligned}
&a(k^2)\sigma^3 + b(k^2)\sigma^2 + c(k^2)\sigma + d(k^2) = 0,\\
&a(k^2) = \mu k^2, \quad b(k^2) = (\mu D_1 + \gamma\tau)k^4 + [1 + \mu(r+m) - 2\tau]k^2 + s,\\
&c(k^2) = \gamma\tau D_1 k^6 + [(1 + m\mu - \tau)D_1 + (\gamma r - a_1)\tau]k^4\\
&\qquad\quad + [r + m + sD_1 + m\mu r - (r+m)\tau]k^2 + (r+m)s,\\
&d(k^2) = m[D_1 k^4 + (r + sD_1)k^2 + rs],
\end{aligned} \tag{6.66}$$

for which $\sigma = 0$ is not a solution as in Section 6.3. Here a discussion of linear stability requires using the Routh–Hurwitz conditions for a cubic (see Appendix A, Volume 1). In this case the steady state is unstable if, for any $k^2 \neq 0$, $a(k^2) \neq 0$ and the coefficients in (6.67) satisfy the inequality

$$H(k^2) = \frac{1}{[a(k^2)]^2}\left[b(k^2)c(k^2) - a(k^2)\,d(k^2)\right] < 0. \tag{6.67}$$

The Routh–Hurwitz condition (6.67) is difficult to use analytically since there are high-order polynomials in k^2 and several parameters. So, we backed up our model analysis with optimization and graphical techniques in the following way. First, we choose the cell traction, τ, as the crucial parameter in the dispersion relation that drives the pattern formation. If $\tau = 0$, then $b(k^2) > 0$ and $c(k^2) > 0$ for all wavelengths k, and the system (6.63)–(6.65) cannot generate spatial structures. The critical traction, τ_c, appears in the form of a quadratic. Second, we used the Logical Parameter Search (LPS) method developed by Bentil and Murray (1991) to generate parameter sets which gave dispersion curves we wanted. The LPS method is an online search procedure which scans given parameter ranges to generate parameter sets satisfying some given logical conditions, namely, conditions for instability and spatial patterning, and it easily extends to higher-dimensional problems. To isolate specific integer modes, we chose biologically realistic parameter ranges (as far as we could with our current knowledge of the biology) for each of the parameters in question. We set up iterative procedures for the parameter ranges and evaluated the roots of the dispersion relation by Newton's method and used a numerical algorithm to find the roots of equation (6.67) (with an equality sign). The LPS procedure then checked whether the conditions for instability and, therefore, spatial patterning were satisfied at specific modes for various sets of parameter values within the parameter space. The search was continued until all the parameter ranges had been examined. This method seemed to be particularly useful in the search for parameter sets at critical bifurcation values where the cell traction $\tau = \tau_c$.

In line with linear predictions (Bentil and Murray 1993) any of the generated sets is guaranteed to give numerical results that exhibit the desired spatially heterogeneous patterns. The initial conditions consisted of steady state values plus a random perturbation about the steady state cell population. As a first step, we considered square and rectangular domains as representative domains for palms and soles.

Comparison of Numerical Results with In Vitro Experiments

Here we describe some of the patterns that were obtained from a direct solution of the model equations in two spatial dimensions and relate them to results of specific *in vitro* experiments and actual dermatoglyphic patterns.

As we have seen these dermal mechanical mechanisms can generate spatial patterns in the dependent variables. We can obtain various vector displacement (strain) fields that correspond to the magnitude and direction of displacement patterns of cells on the cell–matrix composite. The displacement vector $\mathbf{u}$ represents the direction of preferred orientation of the cell–matrix tissue in the neighborhood of any point after a random initial perturbation in cell density. Below we relate our simulation results with specific experiments.

(i) *Development of Curved Ridges from Cultured Human Epidermal Cells. In vitro* experiments by Green and Thomas (1978) show that it is possible to enhance a serial cultivation of human epidermal keratinocytes by including in the culture lethally irradiated fibroblasts (usually 3T3 cells). The presence of 3T3 cells induces epidermal cell colonies to grow and eventually fuse to make confluent layers after about 21 days. At a later stage (between 30 to 40 days) some cells begin to form thickened ridges which resemble arches, loops and whorls. Thus, cultures made from disaggregated human epidermal cells grow to a confluent cell layer followed by the emergence of patterns resembling those of human dermatoglyphs. Although these are not dermal cells it highlights, among other things, the importance of cell division in the patterning process. Figure 6.21(a) shows an advanced ridge pattern of human keratinocytes. For comparison Figure 6.21(b) is a numerical solution of the model system (6.63)–(6.65) for dermal patterning on a square domain. Here, we suggest that cell movement on the ECM substratum, which is passively dragged along by convection and the final displacement and strain fields due to convective and other effects, presages dermal patterning.

(ii) *Development of Curved Ridges from Fibroblast Culture.* Elsdale and Wasoff (1976) investigated a way in which densely packed cells organize. A pedagogical discussion has been given by Bard (1990). Normal human diploid lung fibroblasts were cultured and cell movement and patterning were studied with time-lapse cinemicrography. The cells spread and eventually stabilized forming a dense patchwork of arrays of fibroblasts as confluences were approached. As a result, the confluent culture formed a patchwork of numerous parallel arrays. The arrays merged at a confluence where the cells in two adjacent arrays shared the same orientation to within about 20°. Merging was inhibited where the orientation differed significantly and discontinuities of characteristic forms arose as empty ditches between raised banks of cells. Experiments by Erickson (1978) complement those of Elsdale and Wasoff (1976) and indicate that when cells come into contact with each other at a small angle, only a small portion of the filopodial protrusions are inhibited and neighbouring cells glide along and adhere to each other. At large angles of contact cells may crawl over each other or move away from each other. The angle of contact that produces this feature was suggested for different tissues. For example, a fetal lung fibroblast can realign at an angle of less than 20°. The work of Edelstein-Keshet and Ermentrout (1990) on the formation of parallel arrays of cells is pertinent to these studies.

Elsdale and Wasoff (1976) examined fibroblast organization using the techniques of geometric topology. They calculated the topological index of pattern elements and characterized the geometry of patterns according to their indices.

The article by Penrose (1979) discusses ridge patterns in general and how to calculate their indices in detail; he gives numerous examples. An important point to note is that ridge patterns are not vector fields. Suppose, by way of example, we consider a vector field classical centre or star as in Figures 6.22(a) and (b) (these are Figures A.1(d) and (f) in Appendix B, Volume 1). First draw a circle round the singularity, start at a point on this circle and move round it in an anticlockwise direction. The vector changes direction as we move round, increasing by 2π when we get back to where we started. The index is given by dividing the angle turned by 2π. In these two cases the vector has turned through an angle 2π so these singularities each have an index $+1$. On the other hand if we consider the saddle point illustrated in Figure 6.22(c) the angle changes by -2π and so it has an index -1. A dipole singularity, for example, has an index $+2$: it is like the juxtaposition of two singularities like that in Figure 6.22(a). If there is a group of singularities and you draw a circle round them all the index is then the sum of the indices of each singularity inside.

Now consider ridge patterns where the lines have no direction as in the singularities illustrated in Figures 6.22(d) to (f). None of them can occur with vector fields. The inference (Penrose 1979) is that the ridges are not produced by a field of force or as a line along which some quantity is constant (like a streamline in incompressible inviscid fluid flow). Penrose (1979) concluded that these singularities are probably formed by 'something of a tensor character, such as a stress or strain, or perhaps a curvature of a surface.' In the case of the model mechanisms we have proposed for fingerprint patterns this is what we have. The fingerprint shown in Figure 12.5 in Chapter 12 is a particularly clear example of the double loop singularity shown in Figure 6.22(f).

Let us now consider the singularities in Figures 6.22(d) to (f). If we move round our circle enclosing the singularity in Figure 6.22(d), for example, it is invariant under a rotation of π and so has an index of $+(1/2)$. In the case of the triradius in Figure 6.22(e) the singularity has index $-(1/2)$ while the multiple situation in Figure 6.22(f) has index $1 = (+1/2) + (+1/2)$.

The singularities in Figures 6.22(d) and (e) are the simplest type of ridge singularities. Others can essentially be made up of a combination of these: for example, Fig-

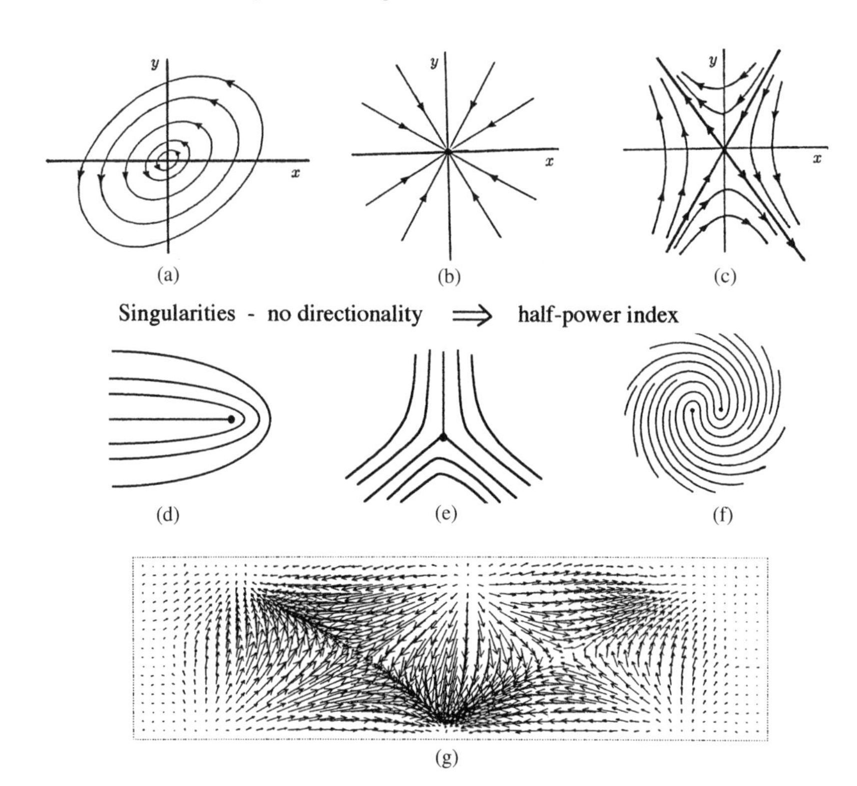

Figure 6.22. Typical vector singularities: (**a**) Centre: index $+1$. (**b**) Star: index $+1$. (**c**) Saddle point: index -1. (**d**) Ridge singularity: index $+(1/2)$. (**e**) Triradial ridge singularity: index $-(1/2)$. (**f**) Multiple ridge singularity, namely, two singularities as in (**d**), has index $+1$. (**g**) Solution of the model system of equations for the displacement field on a rectangular domain with homogeneous boundary conditions. Again the displacement field associated with the passive movement of the ECM is typical of embryonic dermal patterning. The simulations give patterns that resemble a quadriradius and have a topological index of $n = -1$. Parameter values: $\tau = 0.76$, $\mu = \mu_1 + \mu_2 = 0.04$, $D = 0.08$, $a_1 = 0.005$, $s = 100$, $m = r = 0.0$, $\gamma = 0.042$, $\eta_i = \mu_i/2(1 + \nu')$, $\eta_1 = \eta_2 = 1.0$.

ure 6.22(f) is made up of two loops. Instead of having these fractional indices and to distinguish them from vector singularities we can define them as multiples of π rather than 2π and denote them by N. So, the triradius has an index $N = -1$, the loop has an index $N = +1$ and the complex double loop arrangement in Figure 6.22(f) has $N = 2$. If we now have a ridge pattern made up of L loops and T triradii within a closed domain, the index $N = L - T$.

Penrose (1979) applied these ideas to the normal human hand which has numerous loops and triradii (as well as other singularities as in Figure 6.22). Now consider the hand to be a plane with the ridges on the boundary being simple ridge patterns normal to the boundary except at the fingertips where it is parallel to the tip (that is, there is no change in angle as we go round the tip). As we move round the boundary of a single finger it is only in the gap between the fingers that the angle changes, by $-\pi$. So, going all the way round the hand $N = -(D - 1)$, where D is the number of digits. If we consider there are only loop and triradial fingerprint patterns we then obtain, using $N = L - T$, the relation between the digits and the singularities as $D + L - T = 1$

or for the normal hand, $T - L = 4$. There are, however, more complex singularities in normal fingerprints such as illustrated in Figures 6.20 and 6.22.

In Figure 6.22(g) we give a typical solution of the model equations on a rectangular domain. From a numerical point of view, realignment is influenced by the shape of the domain. Initially, the cells were randomly distributed on the ECM but after some time elapsed, the interaction of the cells and matrix produced cell–matrix displacement fields; these types of patterns are very similar to the experimental patterns of Elsdale and Wasoff (1976); see also Bard (1990).

As a preliminary step towards understanding the pattern behaviour and formation of early dermatoglyphic patterns, in this section we focused on displacement fields produced in the dermal ECM since experimental evidence suggests, as we noted, that dermatoglyphic patterns are formed on the dermis (Okajima, 1982) and the dermis transmits morphogenetic messages across the basal lamina to the epithelium.[4] The mechanical model mechanism, (6.63)–(6.65) can generate stable displacement fields that have similar characteristics to some of the patterns obtained from cultured fibroblasts. The solutions exhibit features that reaction diffusion model mechanisms, for example, do not seem to have. It would be interesting to see what quantitative effect cell mitosis and matrix growth would have on the patterns generated by the model. With our model, the orientations in the displacement fields can change continuously and in a systematic manner from point to point within the tissue matrix except at singularities. Elsdale and Wasoff (1976) pointed out a basic feature that must be present in such continuum models of cell–matrix generated singularities, namely, that the patterns must be invariant under rotation through π to ensure that they possess the appropriate symmetry.

With regard to the boundary geometry, it would be more realistic to study the model mechanism on semi-circular and finger-like domains resembling palms and soles of primates. Other types of boundary conditions and their effect on dermatoglyphic patterning would also be interesting to investigate.

As a footnote, since the final pattern depends on the initial conditions (which *in vivo* always involve a stochastic element) it follows from the material in this chapter that no two people can have identical fingerprints—not even identical twins.

6.8 Mechanochemical Model for the Epidermis

The models we considered in earlier sections were concerned with internal tissue, the dermis and mesenchyme. The epithelium, an external tissue of epidermal cells, is another major tissue system in the early embryo which plays an important role in regulating embryogenesis. We briefly alluded to the interaction between the dermis and the epidermis in the formation of skin organ primordia; recall the discussion on the patterning of teeth primiordia in Chapter 4. Many major organs rely on tissue interactions so their importance cannot be overemphasised; see, for example, Nagawa and Nakanishi (1987) who specifically investigate mechanical effects. As a necessary prerequisite in understanding dermal–epidermal tissue interactions we must thus have a model for the

[4]In a discussion about fingerprints with one of the senior police officers in Oxford (who was a specialist on fingerprints) he told me that one method of identifying a drowned body was by the fingerprints from the dermis. Apparently if a body is in the water a considerable length of time the epidermis sloughs off leaving only the dermis which gives an accurate fingerprint of the person when alive.

epidermis. Here we describe a model for the epithelium. As noted above the cells which make it up are quite different to the fibroblast cells we considered above. They do not actively migrate but are arranged in layers or sheets, which can bend and deform during embryogenesis; see Figure 6.2. During these deformations the cells tend to maintain contact with their nearest neighbours and importantly, unlike dermal cells, they cannot generate traction forces under normal conditions. There are noted exceptions, one of which we shall discuss later in Chapter 9 when we consider a model for epidermal wound healing.

Odell et al. (1981) modelled the epithelium as a sheet of discrete cells adhering to a basal lamina. Using a model for the contractile mechanism within the cell they showed how many morphogenetic movements of epithelial sheets could result from the mechanical interactions between the constituent cells. Basic to their model was the mechanochemistry of the cytogel, the interior of the cell, which provides an explanation for the contractile properties of the cells. The continuum model we describe in this section was proposed by Murray and Oster (1984a) and is based on the discrete model of Odell et al. (1981).

A different but related model for epithelial movement was proposed by Mittenthal and Mazo (1983). They model the epithelium as a fluid elastic shell which allows cell rearrangement and such spatial heterogeneity creates tensions which can alter the shell shape.

Some Biological Facts About Cytogel

The cell cytoplasm consists largely of a viscoelastic gel which is a network of macromolecular fibres mostly composed of actin linked by myosin crossbridges, the same elements involved in muscle contraction. This network has a number of complex responses and is a dynamic structure. The cell can contract actively by regulating the assembling and disassembling of the cross-linking of the fibres and carry out a variety of shape changes. When the fibres are strongly linked the cytoplasm tends to gel whereas when they are weakly linked it solates. Here we focus on just two key mechanical properties which subsume the complex process into a mechanochemical constitutive relation.

Chemical control of the cell's contractility, related to the sol–gel transition or degree of actomyosin cross-linking, is mainly due to the local concentration of free calcium in the cytogel. Calcium regulates the activity of the solation and gelation factors and the contractile machinery. Figure 6.23 gives a cartoon summary of the principal components of the contractile apparatus although we do not deal with it at this level of detail.

At low concentration levels calcium encourages cross-linking in the gel and, on a sliding filament concept (as in muscles), more cross-bridges become operative and this implies that the fibres tend to shorten and hence become stronger. It is similar to the increased strength of muscle in a contracted state and a stretched state—you can lift a much heavier object with your arm bent than with it outstretched. So, as free calcium concentration goes up the gel first starts to contract actively. If the concentration gets too high, however, the gel becomes solated (the network begins to break apart) and cannot support any stress. There is thus a 'window' of calcium concentration which is optimal for contractile activity. Thus the concentration of free calcium and the mechanical forces associated with the cytogel must be key variables in our mechanical model.

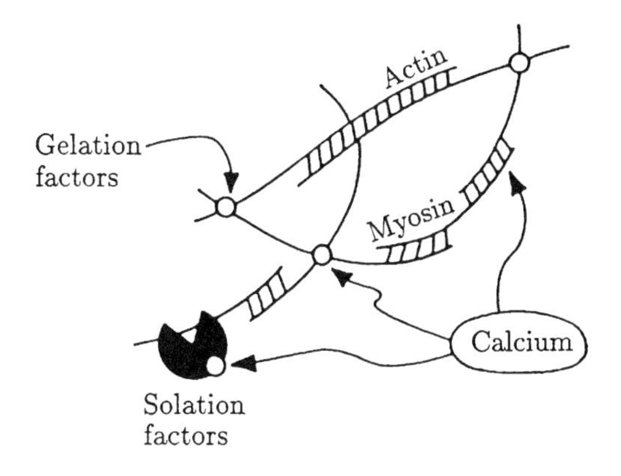

Figure 6.23. Microscopic description. The cytogel consists of actin and myosin which generate the traction forces. Solation and gelation enzymes control the connectivity of the gel and thus its viscosity and elasticity. Although there is a complex of chemicals involved, in the model, we consider free calcium to control the activity of the solation and gelation factors and hence the contractile apparatus. We model the macroscopic properties of the cytogel as a viscoelastic continuum with a viscosity, an elastic modulus and an active traction force τ.

Force Balance Equation for Cytogel Contractility

We model the epithelial sheet of cells as a viscoelastic continuum of cytogel. As with the model for mesenchymal cells, inertial forces are negligible, so the force balance mechanical equation can be taken to be

$$\nabla \cdot (\boldsymbol{\sigma}_V + \boldsymbol{\sigma}_E) + \rho \mathbf{F} = 0, \tag{6.68}$$

where $\mathbf{F}$ represents the external body forces per unit density of cytogel, ρ is the cytogel density, assumed constant, and the stress is the sum of a viscoelastic stress $\boldsymbol{\sigma}_V$ and an elastic stress $\boldsymbol{\sigma}_E$ given by (cf. (6.12))

$$\begin{aligned} \boldsymbol{\sigma}_V &= \mu_1 \boldsymbol{\varepsilon}_t + \mu_2 \theta_t \mathbf{I}, \\ \boldsymbol{\sigma}_E &= \underbrace{E(1+\nu)^{-1}(\boldsymbol{\varepsilon} + \nu' \theta \mathbf{I})}_{\text{elastic stress}} + \underbrace{\tau \mathbf{I}}_{\text{active contraction stress}}, \end{aligned} \tag{6.69}$$

where τ is the contribution of the active cytogel traction to the elastic stress and $\boldsymbol{\varepsilon}$, θ, $\mathbf{I}$, μ_1, μ_2, E, ν and ν' have the same meaning as in the cell–matrix equation (6.18) in the dermal model in Section 6.2. Here, however, the nonlinear dependence of the parameters on the dependent variables is different.

There is a relationship between the two models. In the mesenchymal model we considered the cell–matrix material as an elastic continuum in which were embedded motile contractile units, the dermal cells. In the model here the elastic continuum cytogel also has contractile units—the actomyosin cross-bridges. However, we do not need to account for any motion of these contractile units except for the deformation of the gel sheet itself. In our model for the epithelium the role of the cells is now played by the chemical trigger for contraction, namely, the free calcium concentration which we denote by $c(\mathbf{r}, t)$. Thus we model the constitutive parameters μ_1, μ_2, E and τ in the stress tensor in (6.69) as given below. In fact, as we did in Section 6.2, we incorporate

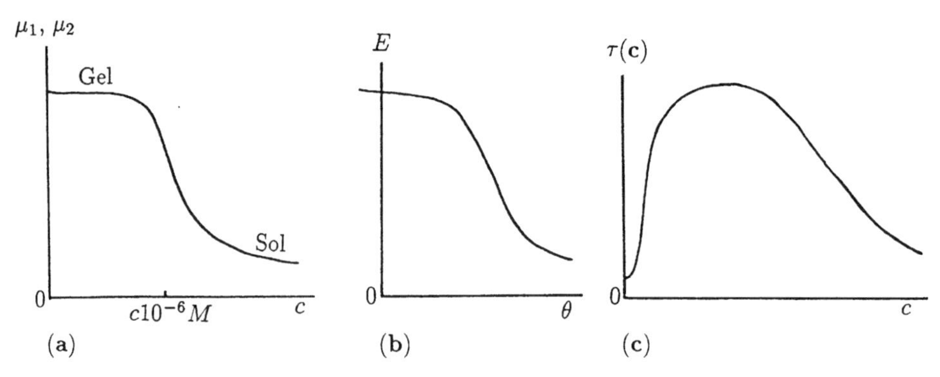

Figure 6.24. Typical nonlinear dependence on the dependent variables of the constitutive parameters in the stress tensor in (6.69). In (**a**) and (**b**) the precipitous drop in the viscosities and actomyosin network elastic modulus occurs when the gel solates. (**c**) Actomyosin traction as a function of c.

more generality than we subsequently study, not only for completeness but because the full model poses interesting and as yet unsolved mathematical problems.

(i) *Viscosity Parameters* μ_1 *and* μ_2. The severing of the gel network effected by the calcium results in a precipitous drop in the apparent viscosity and so we model $\mu_i(c)$, $i = 1, 2$ by a typical sigmoidal curve as illustrated in Figure 6.24(a).

(ii) *Elasticity.* A characteristic property of the actomyosin fibrils is that as the amount of overlap of the actin fibres increases, so does the number of cross-bridges, and so the fibre gets stronger as it contracts. Also, when a fibrous material is stretched the fibres tend to align in the direction of the stress and the effective elasticity increases. This means, as in the full dermal model, that the material is anisotropic. These are nonlinear effects which we can model by taking the elastic modulus E as a function of the dilation θ. Choosing $E(\theta)$ as a decreasing function of the dilation, as illustrated in Figure 6.24(b), is a reasonable form to start with.

(iii) *Active Traction.* The fibrous material of the cytogel starts to generate contractile forces once the actomyosin machinery is triggered to contract. The onset of contraction occurs when the free calcium is in the micromolar range. When the calcium level gets too high the fibrous material can no longer exert any contractile stress. We thus model the active stress contribution $\tau(c)$ as a function of calcium as illustrated in Figure 6.24(c).

(iv) *Body Force.* Movement of the epithelial layer is inhibited by its attachment to the basal lamina, which separates it from the mesenchyme, by restraining tethers, equivalent to the 'guy lines' in the dermal model. We assume this restraining force per unit of cytogel density, $\mathbf{F} = s\mathbf{u}$, where s is a factor reflecting the strength of the attachments and $\mathbf{u}(\mathbf{r}, t)$ is the displacement of a material point of the cytogel. The form is similar to that used in the dermal model.

We incorporate these effects into the stress tensor (6.69) and the force balance equation (6.68) for the cytogel which takes the form

$$\nabla \cdot \left\{ \mu_1 \boldsymbol{\varepsilon}_t + \mu_2 \theta_t \mathbf{I} + E(1+\nu)^{-1}[\boldsymbol{\varepsilon} + \nu' \theta \mathbf{I} + \tau(c)\mathbf{I}] \right\} - s\rho\mathbf{u} = 0. \tag{6.70}$$

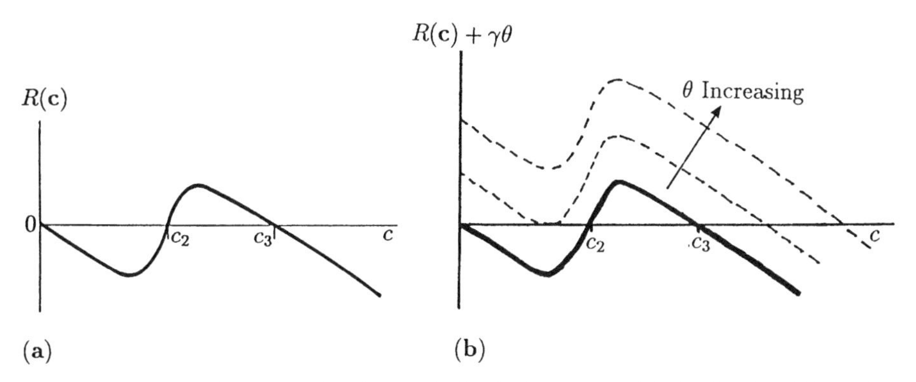

Figure 6.25. (**a**) Qualitative form of the calcium-stimulated calcium release kinetics function $R(c)$. A release of calcium can be triggered by an increase of calcium: this is a switch from the zero steady state to $c = c_3$. (**b**) A strain-induced calcium release, that is, stretch activation, based on the kinetics function $R(c) + \gamma\theta$ in (6.72), if $\gamma\theta$ exceeds a certain threshold where θ is the dilation.

For algebraic simplicity, or rather less complexity, we take the viscosities μ_i, $i = 1, 2$ and E to be constant. An analysis including variable viscosities and elastic modulus would be interesting, particularly on the wave propagation potential of the model.

Conservation Equation for Calcium

We must first describe some of the chemical aspects of the cytogel. Calcium is sequestered in membranous vesicles dispersed throughout the cytogel. It is released from the vesicles by an autocatalytic process known as calcium-stimulated calcium release (CSCR). This means that if the free calcium outside the vesicles exceeds a certain threshold value it causes the vesicles to release their store of calcium (it is like the toilet flush principle). We can model this aspect by a threshold kinetics. If we assume the resequestration of calcium is governed by first-order kinetics, we can combine the processes in a kinetics function $R(c)$ where

$$R(c) = \frac{\alpha c^2}{1 + \beta c^2} - \delta c, \tag{6.71}$$

where α, β and δ are positive constants. The form of $R(c)$ is typically S-shaped as shown in Figure 6.25(a): if $4\beta\delta^2 < \alpha^2$ there are two linearly stable steady states at $c = 0$ and $c = c_3$ and an unstable steady state at $c = c_2$.

The release of calcium can also be triggered by straining the cytogel, a phenomenon known as 'stretch activation.' We can model this by including in the kinetics $R(c)$ a term $\gamma\theta$, where γ is the release per unit strain and θ is the dilation. Figure 6.25(b) shows the effect of such a term and how it can trigger calcium release if it exceeds a certain threshold strain. (Certain insect flight muscles exhibit this phenomenon in that stretching induces a contraction by triggering a local calcium release.)

Calcium, of course, also diffuses so we arrive at a model conservation equation for calcium given by

$$\frac{\partial c}{\partial t} = D\nabla^2 c + R(c) + \gamma\theta$$
$$= D\nabla^2 c + \frac{\alpha c^2}{1+\beta c^2} - \delta c + \gamma\theta, \tag{6.72}$$

where D is the diffusion coefficient of the calcium. We have already discussed this equation in detail in Chapter 3, Section 3.3 (cf. also Chapter 6, Exercise 3) and have shown it gives rise to excitable kinetics. We should emphasise here that the kinetics in (6.72) is simply a model which captures the qualitative features of the calcium kinetics. The biochemical details of the process are not yet completely understood.

The mechanochemical model for the cytogel consists of the mechanical equilibrium equation (6.70), and the calcium conservation equation (6.72). They are coupled through the calcium-induced traction term $\tau(c)$ in (6.70) and the strain-activation term $\gamma\theta$ in (6.72). In the subsequent analysis we shall take $E(\theta)$, the viscosities μ_i, $i = 1, 2$ and the density ρ to be constants.

We nondimensionalise the equations by setting

$$r^* = \frac{r}{L}, \quad t^* = \delta t, \quad c^* = \frac{c}{c_3}, \quad \mathbf{u}^* = \frac{\mathbf{u}}{L},$$
$$\theta^* = \theta, \quad s^* = \frac{\rho s L^2(1+\nu)}{E}, \quad \mu_i^* = \frac{\mu_i \delta(1+\nu)}{E}, \quad (i = 1, 2),$$
$$\boldsymbol{\varepsilon}^* = \boldsymbol{\varepsilon}, \quad \tau^*(c^*) = \frac{(1+\nu)\tau(c)}{E}, \quad R^*(c^*) = \frac{R(c)}{\delta c_3}, \tag{6.73}$$
$$\alpha^* = \frac{\alpha c_3}{\delta}, \quad \beta^* = \beta c_3^2, \quad \gamma^* = \frac{\gamma}{\delta c_3}, \quad D^* = \frac{D}{L^2\delta},$$

where L is some appropriate characteristic length scale and c_3 is the largest zero of $R(c)$ as in Figure 6.25(b). Substituting these into (6.70) and (6.72) and omitting the asterisks for notational simplicity, we have the dimensionless equations for the cytogel continuum as

$$\nabla \cdot \{\mu_1 \boldsymbol{\varepsilon}_t + \mu_2 \theta_t \mathbf{I} + \boldsymbol{\varepsilon} + \nu'\theta\mathbf{I} + \tau(c)\mathbf{I}\} = s\mathbf{u},$$
$$\frac{\partial c}{\partial t} = D\nabla^2 c + \frac{\alpha c^2}{1+\beta c^2} - c + \gamma\theta = D\nabla^2 c + R(c) + \gamma\theta. \tag{6.74}$$

The boundary conditions depend on the biological problem we are considering. These are typically zero flux conditions for the calcium and periodic or stress-free conditions for the mechanical equation.

The linear stability of the homogeneous steady state solutions of (6.74), namely,

$$\mathbf{u} = \theta = 0, \quad c = c_i, \quad i = 1, 2, 3, \tag{6.75}$$

where c_i are the zeros of $R(c)$, can be carried out in the usual way and is left as an exercise.

With tissue interactions in mind, we can see how this model can be modified to incorporate a mechanical influence from the mechanical dermal model. This would appear as a dermal input in the calcium equation proportional to θ_D where θ_D is a dilation contribution from the dermis. If the strain activation parameter γ is not sufficiently large for the steady state $c = c_3$, $\mathbf{u} = 0$ to be unstable for $k^2 > 0$, the effect of the dermal input θ_D could initiate an epidermal instability since it enhances the $\gamma\theta$ term in the calcium equation, the second of (6.74). Inclusion of such a term also changes the possible steady states since these are now given by

$$R(c) + \theta_D = 0, \quad \mathbf{u} = 0.$$

The qualitative effect of θ_D, when it is constant, for example, can be easily seen from Figure 6.25(b); a threshold effect for large enough θ_D is evident.

It is clear how tissue interaction is a natural consequence in these mechanical models. A nonuniform dermal cell distribution can trigger the epithelial sheet to form placodes. This scenario would indicate that the papillae precede the placodes. On the other hand the epidermal model can also generate spatial patterns on its own and in turn affect the dermal mechanism by a transferred strain and hence effect dermal patterns. Also the epidermal model could be triggered to disrupt its uniform state by an influx of calcium (possibly from a reaction diffusion system). So, at this stage we can draw no conclusions as to the order of appearance of placodes and papillae solely from the study of these models without further experimental input. However, it is an attractive feature of the models that tissue interaction between dermis and epidermis can be so naturally incorporated.

Travelling Wave Solutions of the Cytogel Model

One of the interesting features of the model of Odell et al. (1981) was its ability to propagate contraction waves in the epithelium. Intuitively we expect the continuum model here to exhibit similar behaviour. The appearance of contraction waves is a common phenomenon during embryogenesis; recall the discussion in Section 13.6 on post-fertilization waves on eggs.

The one-dimensional travelling wave problem can be solved numerically but to determine the solution behaviour as a function of the parameters it is possible to consider a simplified form of the equations, which retains the key qualitative behaviour but for which we can obtain analytical solutions.

The one-dimensional version of (6.74) is

$$\begin{aligned} &\mu u_{xxt} + u_{xx} + \tau'(c)c_x - su = 0, \\ &c_t - Dc_{xx} - R(c) - \gamma u_x = 0, \end{aligned} \tag{6.76}$$

where $\mu = \mu_1 + \mu_2$ and where we have incorporated $(1 + \nu')$ into the redefined μ and s. We can then look for travelling wave solutions (cf. Chapter 13, Volume I, Section 13.1) of the form

$$u(x, t) = U(z), \quad c(x, t) = C(z); \quad z = x + Vt, \tag{6.77}$$

where V is the wavespeed; with $V > 0$ these represent waves travelling to the left. Substitution into (6.76) gives

$$
\begin{aligned}
&\mu V U''' + U'' + \tau'(C)C' - sU = 0,\\
&VC' - DC'' - R(C) - \gamma U' = 0,
\end{aligned}
\tag{6.78}
$$

where $R(C)$ and $\tau(C)$ are qualitatively as shown in Figures 6.25(a) and 6.24(c) respectively. This system gives a fourth-order phase space the solutions of which would have to be found numerically for given $R(C)$ and $\tau(C)$.

To proceed analytically we can use a technique proposed by Rinzel and Keller (1973) for waves in the FitzHugh–Nagumo system (cf. Chapter 1, Section 1.6) which involves a piecewise linearization of the nonlinear functions $R(C)$ and $\tau(C)$. This method retains the key qualitative features of the full nonlinear problem but reduces it to a linear one with a different linear system for different ranges in z. The procedure involves patching the solutions together at the region boundaries. Appropriate piecewise linear forms could be

$$
\begin{aligned}
\tau(C) &= H(C) - H(C - c_2),\\
R(C) &= -C + H(C - c_2),
\end{aligned}
\tag{6.79}
$$

where $H(C - c_2)$ is the Heaviside function, $H = 0$ if $C < c_2$ and $H = 1$ if $C > c_2$. These functional forms preserve the major qualitative feature of the nonlinear forms of $R(c)$ and the traction force $\tau(c)$. The analytical and numerical procedures used by Lane et al. (1987) in a related problem can then be used here; it has not yet been done.

Lane et al. (1987) only investigated travelling wave*fronts*. Their analytical approach can be used to consider wave*back* solutions. It is, in principle, possible to construct wavepulse solutions; it would be an algebraically formidable problem.

One of the applications of this model was to the post-fertilization waves on vertebrate eggs which we discussed in some detail in Chapter 13, Volume I, Section 13.6. There, we were specifically concerned with the calcium waves that swept over the egg. We noted that these were accompanied by deformation waves and alluded to the work in this chapter. Lane et al. (1987) took the cytogel model described here in a spherical geometry, representing the egg's surface, and investigated surface waves with post-fertilization waves in mind; see Figure 13.12, Volume I. It is informative to reread Section 13.6, Volume I in the light of the full mechanochemical cytogel model and the analysis we have presented in this section.

6.9 Formation of Microvilli

Micrographs of the cellular surface frequently show populations of microvilli, which are foldings on the cell membrane, arrayed in a regular hexagonal packing as shown in the photograph in Figure 6.26. Oster et al. (1985a) proposed a modification of the model in Section 6.8 to explain these patterns and it is their model we now discuss.

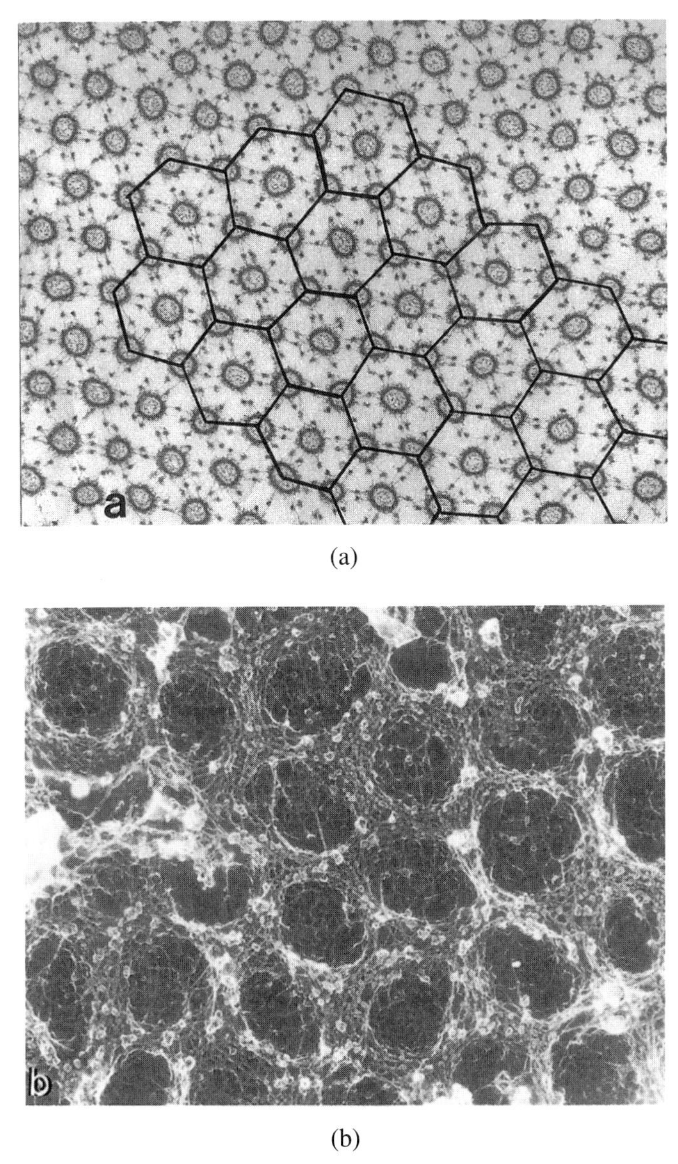

Figure 6.26. (**a**) Micrograph of the hexagonal array on a cellular surface after the microvilli (foldings) have been sheared off. The photograph has been marked to highlight the hexagonal array. (Photograph courtesy of A.J. Hudspeth) (**b**) View of a field of microvilli from the cytoplasmic side, the inside of the cell. Note the bands of aligned actin fibres enclosing the regions of sparse actin density. (Micrograph courtesy of D. Begg)

The model is based on the following sequence of events. First, the cytogel is triggered to contract, probably by an increase in the level of calcium, and as it does so spatial patterns are formed as a result of the instability of the uniform steady state. They proposed that the hexagonal patterns observed are essentially the hexagonal periodic solutions for the tension patterns which are generated. The pattern established in this way creates arrays of lacunae, or spaces, which are less dense in actomyosin. At this stage, osmotic pressure expands these regions outward to initiate the microvilli. One of the new elements in the model is the inclusion of an osmotic pressure.

The biological assumptions underlying the model are: (i) The subcortical region beneath the apical membrane of a cell consists largely of an actin-dense gel. (ii) The gel can contract by a sliding filament mechanism involving myosin cross-bridges linking the actin fibers as in the above model for the epithelium; see Figure 6.23. We shall show that these assumptions are sufficient to ensure that an actin sheet will not necessarily remain spatially homogeneous, but can form a periodic array of actin fibres. This arrangement of actin fibres could be the framework for the extrusion of the microvilli by osmotic forces.

We should interject here, that Nagawa and Nakanishi (1987) comment that dermal cells have the highest gel-contraction activity.

Here we briefly describe the model and give only the one-dimensional analysis. The extension to more dimensions is the same as in the models in earlier sections. We consider a mechanochemical model for the cytogel which now involves the sol–gel and calcium kinetics, all of which satisfy conservation equations, and an equation for the mechanical equilibrium of the various forces which are acting on the gel. The most important difference with this model, however, is the extra force from the osmotic pressure; this is fairly ubiquitous during development.

We consider the cytogel to consist of a viscoelastic continuum involving two components, the sol, $S(x,t)$, and the gel, $G(x,t)$, whose state is regulated by calcium, $c(x,t)$. There is a reversible transition from gel to sol. The actomyosin gel is made up of cross-linked fibrous components while the sol is made up of the non-crosslinked fibres. The state of cytogel is specified by the sol, gel and calcium concentration distribution plus the mechanical state of strain, $\varepsilon(= u_x(x,t))$, of the gel.

We consider the sol, gel and calcium to diffuse, although the diffusion coefficient of the gel is very much smaller, because of its cross-links, than that of the sol and calcium. We also reasonably assume there is a convective flux contribution (recall Section 6.2) to the conservation equations for the gel and sol.

With the experience gained from the study of the previous mechanochemical models the conservation equations for S, G and c are taken to be

$$\text{Sol:} \qquad S_t + (Su_t)_x = D_S S_{xx} - F(S, G, \varepsilon), \tag{6.80}$$

$$\text{Gel:} \qquad G_t + (Gu_t)_x = D_G G_{xx} + F(S, G, \varepsilon), \tag{6.81}$$

$$\text{Calcium:} \qquad c_t = D_c c_{xx} + R(c, \varepsilon), \tag{6.82}$$

where the D's are diffusion coefficients and $R(c,\varepsilon)$ is qualitatively similar to the kinetics $R(c)$ in (6.71) plus the strain activation term (cf. Figure 6.25). The specific form of the kinetics terms in (6.80) and (6.81) reflects the conservation of the sol–gel system: for example, a loss in sol is directly compensated by a gain in gel. These two equations can be collapsed into one differential equation and one algebraic: namely, $G+S =$ constant. The function $F(S,G,\varepsilon)$ incorporates the details and strain-dependence of the sol–gel reaction kinetics which we take to be

$$F(S, G, \varepsilon) = k_+(\varepsilon)S - k_-(\varepsilon)G. \tag{6.83}$$

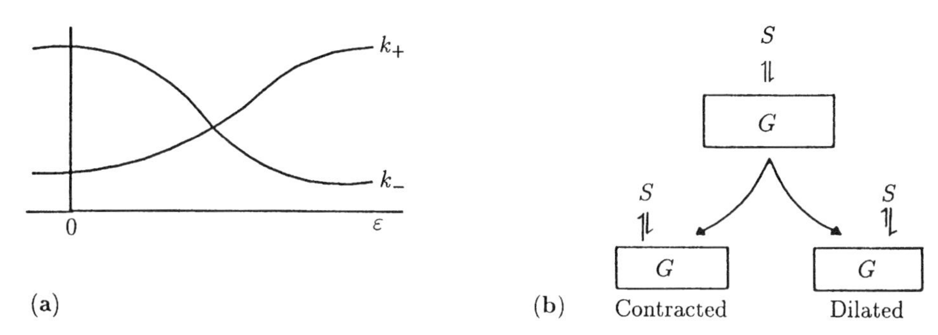

Figure 6.27. (**a**) Qualitative form of the gelation rate $k_+(\varepsilon)$ and solation rate $k_-(\varepsilon)$. (**b**) In the unstressed state the gel (G) is in chemical equilibrium with the sol (S). When the gel contracts its density goes up and the chemical equilibrium shifts towards the sol phase, whereas when the gel is dilated the gel density decreases so that the equilibrium is shifted towards the gel phase. Thus the equilibrium gel fraction is an increasing function of strain. The size of the arrows indicates the relative rate changes under contraction and dilation.

The schematic forms of the rates, $k_+(\varepsilon)$ and $k_-(\varepsilon)$, are shown in Figure 6.27(a). These forms are consistent with the sol–gel behaviour in which if the gel is dilated, that is, ε increases, its density decreases and the mass action rate of gelation increases. Conversely when the gel contracts the gel density goes up and the solation rate increases. Thus the gel concentration may increase with increasing strain; Figure 6.27(b) illustrates the sol–gel equilibrium states.

In the mechanical force balance the osmotic pressure is a major contributory force in gel and is a decreasing function of strain. The definition of strain implies $\varepsilon \geq -1$. (For example, if we consider a strip of gel of unstretched length L_0 and disturbed length L, the strain is $(L - L_0)/L_0$: the absolute minimum of L is zero so the minimum strain is -1.) As $\varepsilon \to -1$ the osmotic pressure becomes infinitely large (we cannot squeeze a finite amount of gel into no space), while for large strains the osmotic effects are small. We thus model the osmotic forces qualitatively by a stress tensor contribution

$$\sigma_0 = \frac{\pi}{1 + \varepsilon},$$

where π is a positive parameter.

The elastic forces not only involve the classical linear stress strain law but also, because of the long strand-like character of the gel, they involve long range effects (cf. Section 6.2). We model this by a stress tensor

$$\sigma_E = GE(\varepsilon - \beta\varepsilon_{xx}),$$

where E is the elastic modulus and $\beta > 0$ is a measure of the long range elastic effect; recall the discussion of long range effects in Section 6.2. The inclusion of G is because the elastic force acts through the fibres of the gel and so the more gel the stronger the force. The elastic forces oppose the osmotic pressure.

The contribution from the active contraction of the gel depends on the calcium concentration and the strain. We model this by

$$\sigma_A = \frac{G\tau(c)}{1+\varepsilon^2},$$

where $\tau(c)$ is a measure of the active traction strength which depends on the calcium concentration c, increasing with increasing c for at least low values of c. The ε-dependence is suggested by the fact that the contractile force is smaller if the gel is dilated, that is, larger ε. Again we require the multiplicative G since traction also acts through the gel fibres.

Finally the gel has an effective viscosity, which results in a viscous force

$$\sigma_V = G\mu\varepsilon_t,$$

where μ is the viscosity. This force again acts through the gel fibres.

All of the forces are in equilibrium and so we have the continuum mechanical force balance equation

$$\sigma_x = [\sigma_0 + \sigma_E + \sigma_A + \sigma_V]_x = 0,$$

$$\sigma = \underbrace{\frac{\pi}{1+\varepsilon}}_{\text{osmotic}} - \underbrace{GE(\varepsilon - \beta\varepsilon_{xx})}_{\text{elastic}} - \underbrace{\frac{G\tau(c)}{1+\varepsilon^2}}_{\text{active stress}} - \underbrace{G\mu\varepsilon_t}_{\text{viscous}}. \tag{6.84}$$

Here there are no external body forces restraining the gel. Note that only the osmotic pressure tends to dilate the gel.

Equations (6.80)–(6.82) and (6.84) together with constitutive relations for $F(S, G, \varepsilon)$ and $R(c, \varepsilon)$ along with appropriate boundary and initial conditions, constitute this mechanochemical model for the cytogel sheet.

Simplified Model System

This system of nonlinear partial differential equations can be analyzed on a linear basis, similar to what we have now done many times, to demonstrate the pattern formation potential. This is algebraically quite messy and unduly complicated if we simply wish to demonstrate the powerful pattern formation capabilities of the mechanism. The full system also poses a considerable numerical simulation challenge. So, to highlight the model's potential we consider a simplified model which retains the major physical features.

Suppose that the diffusion timescale of calcium is very much faster than that of the gel and of the gel's viscous response (that is, $D_c \gg DG, \mu$). Then, from (6.82) we can take c to be constant (because D_c is relatively large). Thus c now appears only as a parameter and we replace $\tau(c)$ by τ. If we assume the diffusion coefficients of the sol and gel are the same, (6.80) and (6.81) imply $S+G = S_0$, a constant and so $S = S_0 - G$.

If we integrate the force balance equation (6.84) we get,

$$G\mu\varepsilon_t = GE\beta\varepsilon_{xx} + H(G, \varepsilon), \tag{6.85}$$

where

$$H(G, \varepsilon) = \frac{\pi}{1+\varepsilon} - \frac{G\tau}{1+\varepsilon^2} - GE\varepsilon - \sigma_0, \tag{6.86}$$

where the constant stress σ_0 is negative, in keeping with the convention we used in (6.84).

Equation (6.85) is in the form of a 'reaction diffusion' equation where the strain, ε, plays the role of 'reactant' with the 'kinetics' given by $H(G, \varepsilon)$. The 'diffusion' coefficient depends on the gel concentration and the elastic constants E and β.

With $S = S_0 - G$ the gel equation (6.81) becomes

$$G_t + (Gu_t)_x = k_+(\varepsilon)S_0 - [k_+(\varepsilon) + k_-(\varepsilon)]G + D_G G_{xx}. \tag{6.87}$$

Now introduce nondimensional quantities

$$\begin{aligned} G^* &= \frac{G}{S_0}, \quad \varepsilon^* = \varepsilon, \quad x^* = \frac{x}{\sqrt{b}}, \quad t^* = \frac{tE}{\mu}, \\ k_+^* &= \frac{k_+\mu}{E}, \quad k_-^* = \frac{k_-\mu}{E}, \quad \sigma_0^* = \frac{\sigma_0}{S_0 E}, \\ D^* &= \frac{D_G\mu}{\beta E}, \quad \tau^* = \frac{\tau}{E}, \quad \pi^* = \frac{\pi}{S_0 E}, \end{aligned} \tag{6.88}$$

with which (6.85), (6.86) and (6.87) become, on dropping the asterisks for notational simplicity,

$$G\varepsilon_t = G\varepsilon_{xx} + f(G, \varepsilon), \qquad G_t + (Gu_t)_x = DG_{xx} + g(G, \varepsilon), \tag{6.89}$$

where

$$\begin{aligned} f(G, \varepsilon) &= -\sigma_0 + \frac{\pi}{1+\varepsilon} - \frac{G\tau}{1+\varepsilon^2} - G\varepsilon, \\ g(G, \varepsilon) &= k_+(\varepsilon) - [k_+(\varepsilon) + k_-(\varepsilon)]G, \end{aligned} \tag{6.90}$$

where the qualitative forms of $k_+(\varepsilon)$ and $k_-(\varepsilon)$ are shown in Figure 6.27(a).

The 'reaction diffusion' system (6.89) with (6.90) is similar to those studied in depth in Chapters 1 to 3 and so we already know the wide range of pattern formation potential. Although in most of these analyses there was no convection, its presence simply enhances the steady state and wave pattern formation capabilities of the system. Typical null clines $f = 0$ and $g = 0$ are illustrated in Figure 6.28. Note that there is a nontrivial steady state (ε_s, G_s). Note also that the strain 'reactant' ε can be negative; it is bounded below by $\varepsilon = -1$.

If we linearise the system about the steady state as usual by writing

$$(w, v) = (G - G_s, \varepsilon - \varepsilon_s) \propto \exp[\lambda t + ikx], \tag{6.91}$$

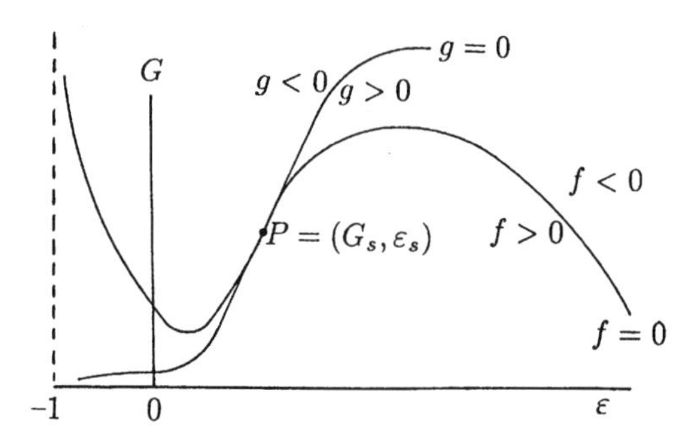

Figure 6.28. Null clines $f(G, \varepsilon) = 0$, $g(G, \varepsilon) = 0$ for the strain-gel 'reaction diffusion' system (6.89) with (6.90) with parameter values such that the system can generate steady state spatial heterogeneous patterns in G and ε.

and substitute into the linearised system from (6.89), namely,

$$
\begin{aligned}
G_s v_t &= G_s v_{xx} + f_G w + f_\varepsilon v, \\
w_t + G_s v_t &= D w_{xx} + g_G w + g_\varepsilon v,
\end{aligned}
\tag{6.92}
$$

where the partial derivatives of f and g are evaluated at the steady state (G_s, ε_s), we get the dispersion relation $\lambda(k^2)$ as a function of the wavenumber k. It is given by the roots of

$$
G_s \lambda^2 + b(k)\lambda + d(k) = 0, \tag{6.93}
$$

where

$$
\begin{aligned}
b(k) &= G_s(1 + D)k^2 - [f_\varepsilon + G_s g_G - G_s f_G], \\
d(k) &= G_s D k^4 - [D f_\varepsilon + G_s g_G]k^2 + [f_\varepsilon g_G - f_G g_\varepsilon].
\end{aligned}
\tag{6.94}
$$

To get spatially heterogeneous structures we require

$$
\begin{aligned}
&\mathrm{Re}\,\lambda(0) < 0 \quad \Rightarrow \quad b(0) > 0, \quad d(0) > 0 \\
&\mathrm{Re}\,\lambda(k) > 0 \quad \Rightarrow \quad b(k) < 0 \quad > \text{and/or} \quad d(k) < 0 \quad \text{for some} \quad k \neq 0.
\end{aligned}
\tag{6.95}
$$

The first of these in terms of the f and g derivatives at the steady state requires

$$
-[f_\varepsilon + G_s g_G - G_s f_G] > 0, \quad f_\varepsilon g_G - f_G g_\varepsilon > 0. \tag{6.96}
$$

From Figure 6.28 we see that

$$
f_\varepsilon > 0, \quad f_G < 0, \quad g_\varepsilon > 0, \quad g_G < 0, \tag{6.97}
$$

and so (6.96) gives specific conditions on the parameters in f and g in (6.90). For spatial instability we now require the second set of conditions in (6.95) to hold. Because of the first of (6.96) it is not possible for $b(k)$ in (6.94) to be negative, so the only possibility for pattern is if $d(k)$ can become negative. This can happen only if the coefficient of

k^2 is negative and the minimum of $d(k)$ is negative. This gives the conditions on the parameters for spatially unstable modes as

$$\begin{aligned} &Df_\varepsilon + G_s g_G > 0, \\ &(Df_\varepsilon + G_s g_G)^2 - 4G_s D(f_\varepsilon g_G - f_G g_\varepsilon) > 0, \end{aligned} \tag{6.98}$$

together with (6.96). The forms of f and g as functions of G and ε are such that these conditions can be satisfied.

From Chapter 2 we know that such reaction diffusion systems can generate a variety of one-dimensional patterns and in two dimensions, hexagonal structures. Although here we have only considered the one-dimensional model, the two-dimensional space system can indeed generate hexagonal patterns. Another scenario for generating hexagonal patterns, and one which is, from the viewpoint of generating a regular two-dimensional pattern, a more stable process, is if the patterns are formed sequentially as they were in the formation of feather germs in Section 6.5, with each row being displaced half a wavelength.

Let us now return to the formation of the microvilli and Figure 6.26. The tension generated by the actomyosin fibres aligns the gel along the directions of stress. Thus the contracting gel forms a tension structure consisting of aligned fibres in a hexagonal array. Now between the dense regions the gel is depleted and less able to cope with the osmotic swelling pressure, which is always present in the cell interior. The suggestion, as mentioned at the beginning of this section, then is that this pressure pushes the sheet, at these places, into incipient microvilli, patterns of which are illustrated in Figure 6.26. These are the steady state patterned solutions of the model mechanism.

6.10 Complex Pattern Formation and Tissue Interaction Models

In many reptiles and animals there are complex spatial patterns of epidermal scales and underlying osteoderms, which are bony ossified-like dermal plates, in which there is no simple one-to-one size correspondence although the patterns are still highly correlated; Figure 6.29 shows some specific examples. There are also scale patterns whereby a regular pattern appears to be made up from a superposition of two patterns with different basic wavelengths as the illustrative examples in Figure 6.30 show.

Numerous epidermal–dermal tissue recombinant studies (for example, Rawles 1963, Dhouailly 1975, Sengel 1976) clearly demonstrate the importance of instructive interaction between the epithelial and mesenchymal layers during skin pattern formation. Dhouailly (1975) studied interaction by combining interspecific epidermal and dermal tissues from three different zoological classes, namely, mammals (mice), birds (chicks) and reptiles (lizards). The results of her recombination experiments suggest that messages originating in the dermis influence the patterns formed in the epidermis. For example, chick dermis explanted with any type of epidermis forms the appendage specific to the epidermis, but the typical shape, size and distribution are similar to those seen in feather bud formation.

In Section 6.5 we proposed a mechanical mechanism for generating dermal papillae and it is also possible to generate the spatial patterns we associate with placodes with

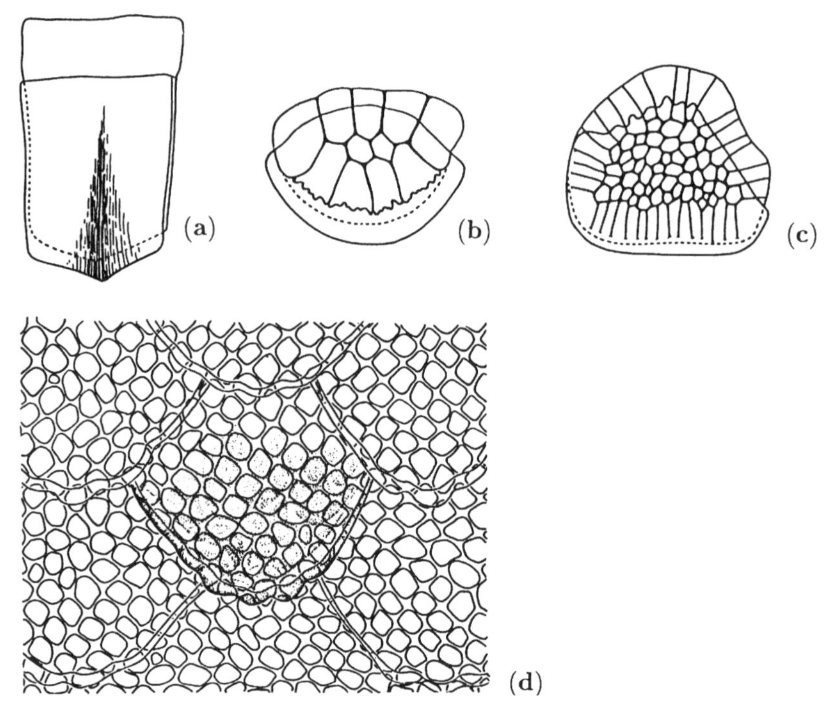

Figure 6.29. Examples of the different relationship between the osteoderms (bony dermal plates) and the overlying horny epidermal scales. (After Otto 1908). (**a**) The dorsal region of the girdle-tailed lizard *Zonurus cordylus*. (**b**) The dorsal caudal (tail) region in the skink (a small lizard) *Chalcides (Gongylus) ocellatus* of the family *Scincidae*. (**c**) The region near the cloaca (anus) of the apotheker or 'pharmacist' skink *Scincus officinalis*. (**d**) The ventral region of the common gecko, *Tarentola mauritanica*: here we have shaded one of the large epidermal scales. The small structures are osteoderms. (From Nagorcka et al., 1987)

the model in Section 6.8. Nagorcka (1986) proposed a reaction diffusion mechanism for the initiation and development of primordia (see also Nagorcka 1988 and Nagorcka and Mooney 1985, 1992) as well as for the formation of the appendages themselves such as hair fibres (Nagorcka and Mooney 1982, 1992, Nagorcka and Adelson 1999).

Until tissue interaction models were developed none of the traditional reaction diffusion mechanisms seemed to have the capacity to produce complex spatial patterns such as illustrated in Figures 6.29 and 6.30. (Making the parameters space-dependent is not an acceptable way—this is simply putting the pattern in first.) The patterns in Figure 6.30 may be viewed as a superposition of two patterns whose wavelengths differ by a factor of at least two as can be seen by comparing the distance between neighbouring small scales denoted λ_s, and neighbouring large scales denoted by λ_l.

These complex patterns suggested the need to explore the patterns which can be formed by an interactive mechanism which combines the mechanisms for the epidermis and dermis. The work of Nagawa and Nakashini (1987) substantiates this.

Since the early 1990's there have been several studies in which interaction has been taken into account. Nagorcka (1986) was the first to propose a tissue interaction mechanism to account for the initiation of skin organ primordia. His model consists of a

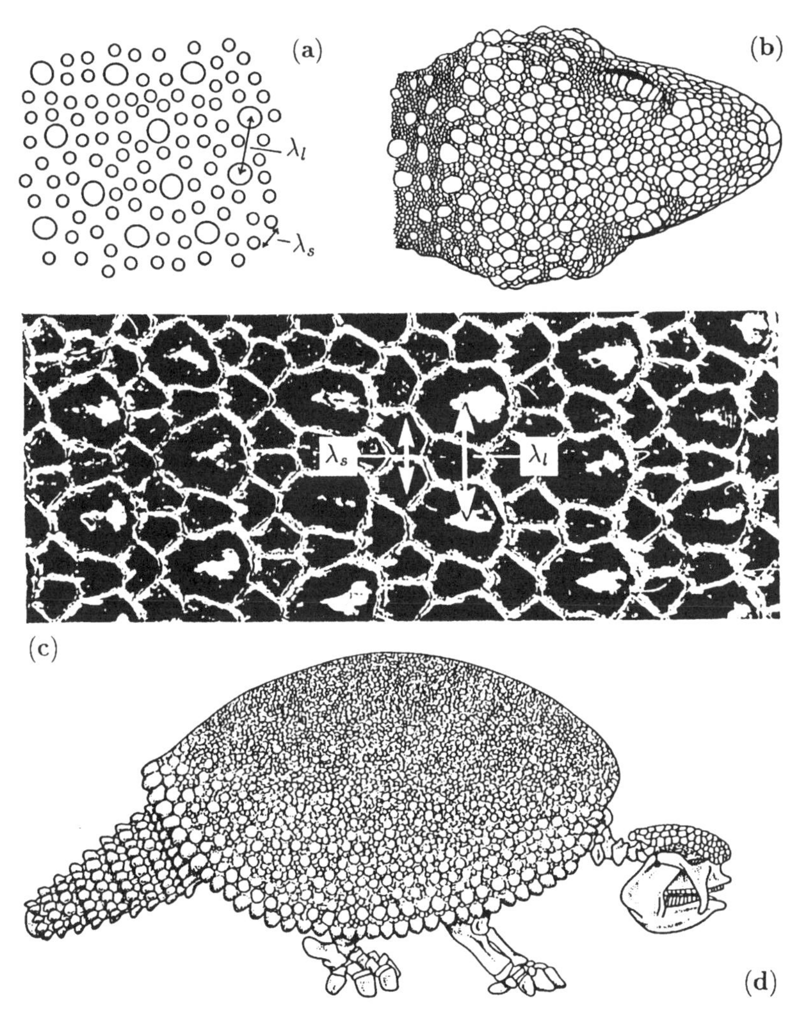

Figure 6.30. (**a**) An example of a feather pattern composed of two basic units, one of small diameter and one of large diameter, seen in the skin area under the beak of a species of common coot, *Fulic atra*, after 12 days of incubation. Associated with the pattern are two wavelengths λ_s and λ_l, namely, the distances between neighbouring small and large feather follicles, respectively. (After Gerber 1939) (**b**) Typical small and large scales in the dorsal head region of lizards, here *Cyrtodactylus fedschenkoi* of the *Gekkonidae* family. (After Leviton and Anderson 1984) The regional variation in the arrangement could be quantified by the ratio λ_s/λ_l. (**c**) In this example the small and large epidermal scales are in one-to-one correspondence with the underlying bony scutes (osteoderms) forming the secondary dermal armour in, at least, some species of armadillo, such as in *Dasypus novemcinctus* shown here. (**d**) The bony scutes seen in the carapace of the *Glyptodon* (after Romer 1962), an ancestor of the armadillo. (From Nagorcka et al. 1987)

reaction diffusion system in the epidermis controlled by a chemical switch mechanism in the dermis. The spatial prepattern in the morphogen concentration set up in the epidermis then serves to provide positional information for epidermal cell patterning and induces dermal cell condensations. Variations of the model were used, and related models developed, by Nagorcka and Mooney (for example, 1982, 1985, 1992). Nagorcka et al. (1987) considered the pattern forming properties of an integrated mechanism consisting of a dermal mechanical cell traction model and a reaction diffusion mechanism of epidermal origin, which interact with each other. In their model the morphogen concentration in the epidermis controls certain mechanical properties in the dermis. In turn, dermal cells produce a factor which causes morphogen production in the epidermis. They demonstrated numerically that their tissue interaction model can generate regular complex spatial patterns similar to those in Figure 6.30 as a single pattern by the integrated mechanism. This was confirmed by a detailed analytical study of a similar system by Shaw and Murray (1990). In these composite models, each submodel is capable of generating spatial pattern of any desired wavelength, so if coupled appropriately the full model can exhibit superposition of patterns with two distinct wavelengths thereby giving rise to some of the scale patterns on reptiles shown in Figure 6.30.

The dispersion relation of integrated mechanisms can be thought of as involving two independent dispersion relations, each with a set of unstable wavenumbers but of different ranges. If the coupling is weak it is reasonable to think of the composite mechanism as being unstable to perturbations with wavelengths approximately equal to λ_s and λ_l, which characterise the two mechanisms independently. The patterns obtained are similar to those in Figure 6.30 provided a large difference exists between the mechanisms' intrinsic wavelengths. The observed patterns may also be produced by any two mechanisms whose dispersion relations are characterised by two separated ranges of unstable wavenumbers. The mechanical model gives such dispersion relations; see, for example, Figure 6.9(e).

Let us consider a general system denoted by $\mathbf{F}(\mathbf{m}) = 0$ with $\mathbf{m}$ the vector of dependent variables. Suppose there is a uniform steady state $\mathbf{m} = \mathbf{m}_0$. In the usual way we look for perturbations about this uniform solution of the linearised system in the form

$$\mathbf{w} = \mathbf{m} - \mathbf{m}_0 \propto \exp[\lambda t + i\mathbf{k}.\mathbf{x}],$$

where $\mathbf{k}$ is the wave vector of the linear disturbances. With this we obtain a characteristic polynomial for λ, say $P(\lambda) = 0$ the solution of which gives the dispersion relation $\lambda = \lambda(\mathbf{k}, \mathbf{p})$ where $\mathbf{p}$ represents the parameter set for the system. If it is possible for the mechanism to generate spatial patterns we know that for a range of wavenumbers $\mathbf{k}$ and appropriate parameters, $\text{Re}\lambda > 0$, and that these wavenumbers grow exponentially with time. A typical situation is illustrated in Figure 6.31. These linearly unstable modes evolve into steady state spatially heterogeneous solutions when the nonlinear contributions in the mechanism are taken into account. We have seen many such characteristic polynomials, dispersion relations and parameter spaces where spatially heterogeneous patterns are generated.

In the basic scenario with a single pattern formation mechanism the pattern can be a regular tessellation of the plane or, if the domain is large enough, random initial conditions can evolve to a highly irregular pattern of spots. In one dimension the wave

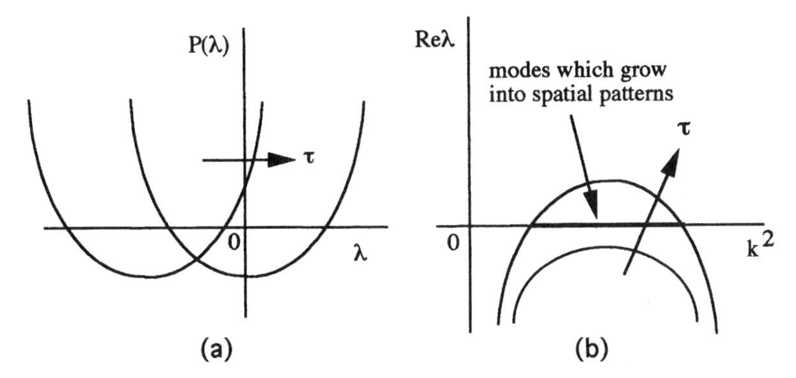

Figure 6.31. In the case of mechanical mechanisms one of the bifurcation parameters we used is the cell traction τ. As the cell traction increases a typical characteristic polynomial, $P(\lambda)$, as a function of λ eventually crosses the axis as in (**a**) and indicates that a spatially inhomogeneous solution exists for λ with a positive real part as in (**b**). The dispersion relation, $\lambda(k^2)$, gives the range of wavenumbers which are linearly unstable and in a one-dimensional situation frequently predicts the final steady state solution; it is often the mode with the fastest linear growth.

length, or rather spectrum of wavelengths of the final pattern, is generally within the band of unstable wavelengths predicted by the dispersion relation. Let us consider now the effect of coupling two pattern generators on the understanding that the formation of skin organ primordia is an interactive process between the dermis and epidermis. Both tissues are assumed to be capable of generating spatial patterns on their own.

Let us denote the patterning mechanism in the epidermis and dermis respectively by $\mathbf{F}_E(\mathbf{m}_E) = 0$ and $\mathbf{F}_D(\mathbf{m}_D) = 0$. Each mechanism can give rise to a dispersion relation which indicates pattern formation. Now suppose that a product of each tissue influences the other. For example, suppose the epidermal mechanism is a reaction diffusion one and the dermal mechanism a mechanical one. The cell traction in the dermal model can be affected by a morphogen which diffuses from the epidermis. The influence of the dermis can be tissue contraction which in turn affects the morphogen production. We denote the strength of the interactions by the parameters δ and Γ and represent the coupled epidermal–dermal system by

$$\mathbf{F}_E(\mathbf{m}_E, \mathbf{m}_D; \delta) = 0, \quad \mathbf{F}_D(\mathbf{m}_D, \mathbf{m}_E; \Gamma) = 0.$$

This system has a uniform steady state about which we can carry out a linear analysis which highlights the effect of the tissue interaction. For typical pattern generating systems, and in particular the one studied by Nagorcka et al. (1987), the characteristic polynomial for the coupled system is of the form

$$P(\lambda) = P_D(\lambda)P_E(\lambda) - \delta\Gamma I(\lambda) = 0, \tag{6.99}$$

where $I(\lambda)$ is a polynomial in λ and $P_D(\lambda)$ and $P_E(\lambda)$ are the respective characteristic polynomials of the two tissue mechanisms in isolation. Solutions of this polynomial, usually of order at least 4, give the dispersion relation for the interaction model. The strength of the interaction is quantified by the product $\delta\Gamma$. Typical characteristic polynomials and corresponding dispersion relations are sketched in Figure 6.32.

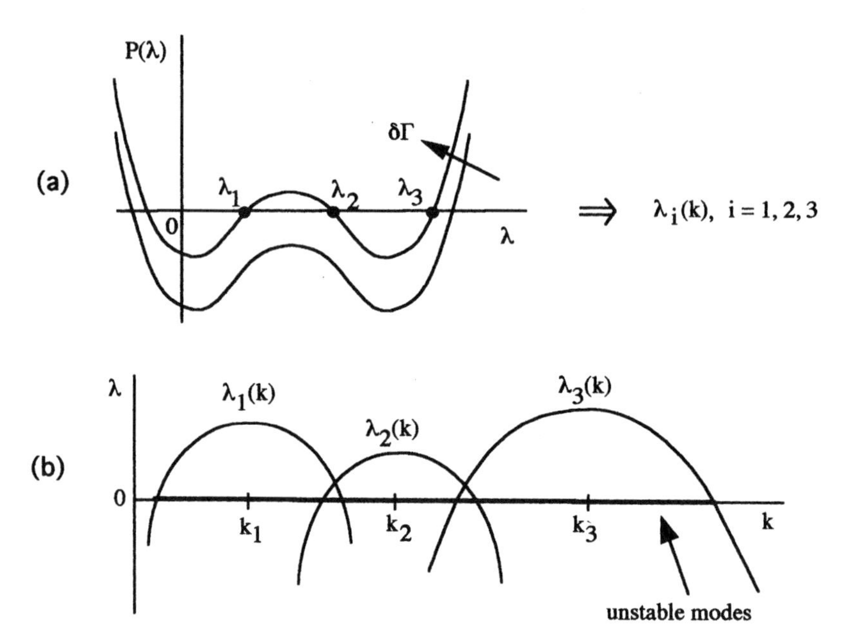

Figure 6.32. (**a**) Typical composite characteristic polynomial for a tissue interaction model mechanism with corresponding dispersion relations in (**b**). Note how an increase in the interaction parameters, δ and Γ, introduces more complex spatial patterns by exciting other $\lambda(k)$ which have linearly unstable modes.

The dominant modes associated with each $\lambda_i(k)$ are denoted by k_i and the solution of the linearized system is dominated by

$$\begin{pmatrix} \mathbf{m}_E - \mathbf{m}_E^{(0)} \\ \mathbf{m}_D - \mathbf{m}_D^{(0)} \end{pmatrix} \sim \mathbf{a}_1 e^{\lambda(k_1)t + ik_1x} + \mathbf{a}_2 e^{\lambda(k_2)t + ik_2x} + \cdots, \tag{6.100}$$

where $\mathbf{m}_E(0)$, $\mathbf{m}_D(0)$ are the uniform steady states of the coupled system. From this we see that the solution is a superposition of the several dominant unstable modes in Figure 6.32 and since growth of those with wavenumbers k_1 and k_3 is larger than that for k_2 these suggest that the final patern is a superposition of two patterns with wavelengths $2\pi/k_1$ and $2\pi/k_3$. Numerical simulation of the full nonlinear system studied by Nagorcka et al. (1987), namely, a reaction diffusion system coupled to a mechanical system, showed that this is the case when the pattern formation system is near the bifurcation from homogeneity to structure. The specific mechanical system they used in the dermis is a version of the model (6.22)–(6.24), namely,

$$\begin{aligned} &n_t = D_1 \nabla^2 n - \alpha \nabla \cdot (n \nabla \rho) - \nabla \cdot (n u_t), \\ &\nabla \cdot \left[(\mu_1 \boldsymbol{\varepsilon}_t + \mu_2 \theta_t \mathbf{I}) + (\boldsymbol{\varepsilon} + \nu' \theta \mathbf{I}) + \tau(n, V, \rho) \mathbf{I} \right] = s\rho \mathbf{u}, \\ &\rho_t + \nabla \cdot (\rho \mathbf{u}_t) = 0, \end{aligned} \tag{6.101}$$

where the traction depends on the morphogen V from the basal layer of the epidermis and is given by

$$\tau(n, V, \rho) = \tau n(\rho + \gamma \nabla^2 \rho) \frac{1 + \Gamma(V - V_0)}{1 + \lambda n^2}. \tag{6.102}$$

The parameters have the same meaning as in the sections above.

The reaction diffusion system they used in the basal layer of the epidermis is

$$\begin{aligned} V_t &= D_V \nabla^2 V + V^2 W - V + A[1 + \delta(n - n_0)], \\ W_t &= D_W \nabla^2 W - V^2 W + B, \end{aligned} \tag{6.103}$$

where A and B are constants and n_0 is a steady state solution of (6.101). The δ and Γ are measures of the interaction between the two systems. It is this interaction model which gives a characteristic polynomial of the form (6.99).

Nagorcka et al. (1987) and Shaw and Murray (1990) considered specific situations such as when $\delta \neq 0, \Gamma = 0$, which implies that the epidermis influences the dermis but not the other way round, and $\delta = 0, \Gamma \neq 0$ where the dermis influences the epidermis but not the other way round. In either of these cases only one model system of equations has to be solved. They also investigated the case where $\delta \neq 0, \Gamma \neq 0$ which is the situation whereby the epidermis and the dermis influence each other. Here the full interactive system has to be solved.

The result of solving the full integrated system with the parameters used in Figure 6.33(d) is not very different to that in Figure 6.33(b) where the low frequency variation in n causes a large change in the spatial pattern of V. The high frequency variation in V has little effect on the dermal cell concentration n. In these simulations the wavenumber $k_V \gg k_n$. On the other hand if the parameters were chosen so that $k_V \ll k_n$ we would expect the change in the variation in V to be small and that in n to be large.

What these results suggest is that, depending on the relative sizes of the wavenumbers of the two mechanisms on their own, we can often neglect the effect of either the dermis on the epidermis (the situation if k_n is significantly larger than k_V) or the other way round. Nagorcka et al. (1987) exploited this in their two-dimensional simulations related to the patterns shown in Figures 6.29 and 6.30.

Tissue Interaction Mechanism: Cell Adhesion Molecules (CAMs)

The interaction models we have just discussed are based on various hypotheses of the detailed biology. We still do not know what the actual process is, not even all the key elements. Here we describe a different model primarily based on experimental work by Gallin et al. (1986). They found that disrupting the balance of neural cell adhesion molecules (N-CAMs) in the chick skin leads to dramatic changes in the patterning of feather germs. This shows that the epidermis, in turn, can influence patterns in the dermis. Not only that, their results appear to implicate cell adhesion molecules (CAMs) in the signalling process.

Basically there are two possible ways in which instructions can be transmitted between the mesenchyme and the epithelium. One is via chemical signalling, for example, paracrine signalling, and the other is via mechanical interaction between the epithelia and dermal cells which are in direct contact with each other. One or the other or both could be involved. So, we describe here a very different mechanism which involves

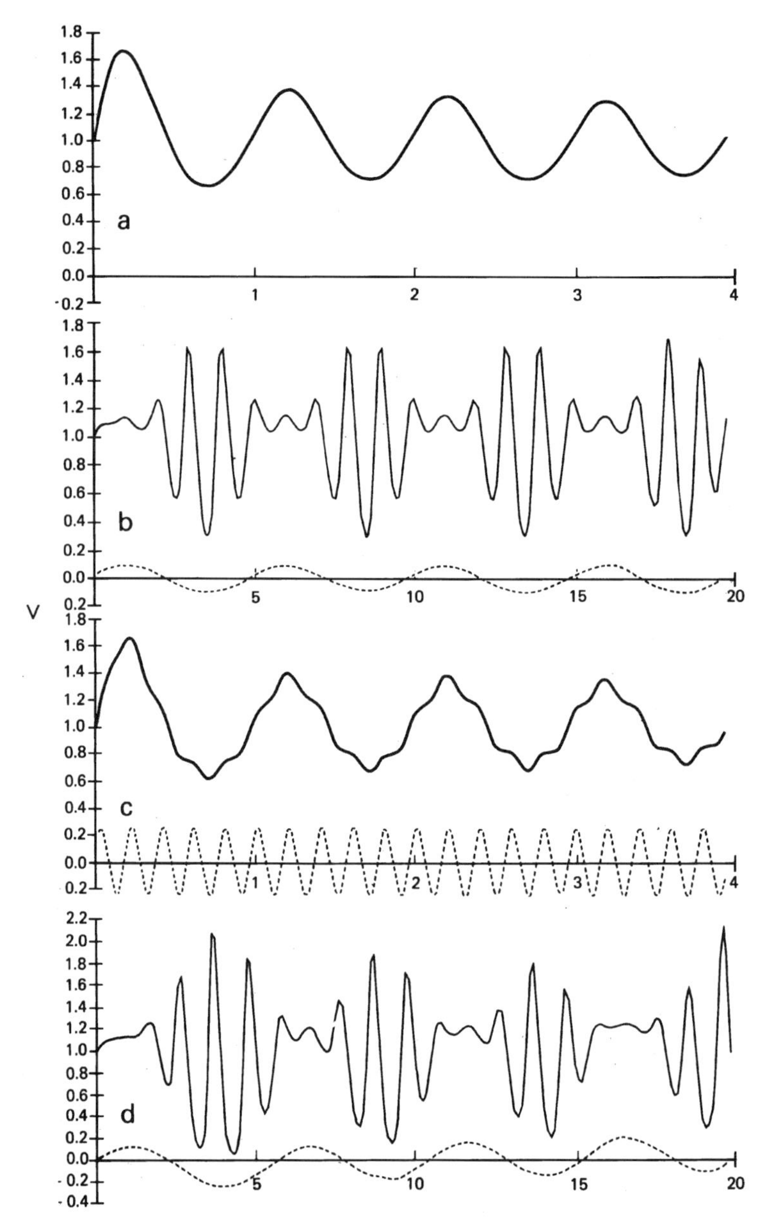

Figure 6.33. (**a**) The one-dimensional solution for V (solid line) of the reaction diffusion system (6.103) is given in (**a**), (**b**) and (**c**) and shows the influence of the forcing by the dermis. The parameter values are $D_V = 0.1$, $D_W = 4.5$, $A = 0.25$, $B = 0.75$ with fixed boundary values. In (**a**) there is no dermal influence, that is, $\delta = 0$ while in (**b**), $\delta = 0.4$ and in (**c**), $\delta = 1.0$. The dashed curves show the forcing function $A\delta(n - n_0)$. In (**d**) we show a solution of the integrated system of (6.101)–(6.103) but with a simplified version of (6.101) and (6.102) with parameter values $D_1 = 1.0$, $\mu_1 + \mu_2 = 5614$, $s = 0.22$, $\tau = 500$, $\lambda = 0.3$. The interaction parameters are $\delta = 8.0$, $\Gamma = 0.1$. In the simplified version the effect of the matrix density ρ is neglected, that is, $\alpha = \gamma = 0$, $\rho = \rho_0 = 1.0$ and the ρ equation is not needed. Again the morphogen V (solid line) is shown and the quantity $A\delta(n - n_0)$ (dashed line) but now it is part of the solution. The horizontal axis is distance in units of the wavelength of V. (From Nagorcka et al. 1987)

CAMs and is based on the experimental work of Chuong and Edelman (1985). They suggested that a specific factor produced by the L-CAM positive dermal cells, perhaps a hormone or peptide, triggers the formation of dermal condensations. This factor could act as a chemotactic agent and stimulate N-CAM expression to induce N-CAM linked papillae; this agrees with the results of Gallin et al. (1986). The recombination experiments of Dhouailly (1975) also suggest that a dermally produced signal is involved in epidermal patterning. Chuong and Edelman (1985) therefore proposed that epidermal placode formation is induced by a factor produced by the developing dermal condensations. When feather germ formation is completed, the inductive factors are modified so as to stop dermal aggregation. Since these factors can still be active in neighbouring tissue, periodic feather germ patterns could thus be formed in a self-propagating way.

Cruywagen and Murray (1992) developed an interacting tissue model described in detail below which is based on the work of Gallin et al. (1986) and schematically illustrated in Figure 6.34.

The mechanism involves seven field variables and incorporates elements of reaction diffusion chemotaxis systems and a mechanical mechanism. The epidermal variables at position $\mathbf{x}$ and time t are:

$N(\mathbf{x}, t) =$ epidermal cell density,

$\mathbf{u}(\mathbf{x}, t) =$ displacement of a material point in the epidermis which was initially at $\mathbf{x}$,

$\hat{e}(\mathbf{x}, t) =$ concentration of a signal morphogen produced in the epidermis,

$\hat{s}(\mathbf{x}, t) =$ epidermal concentration of a signal morphogen, received from the dermis.

Similarly, the variables in the dermis are:

$n(\mathbf{x}, t) =$ dermal cell density,

$e(\mathbf{x}, t) =$ dermal concentration of a signal morphogen received from the epidermis,

$s(\mathbf{x}, t) =$ dermal concentration of a signal morphogen produced in the dermis.

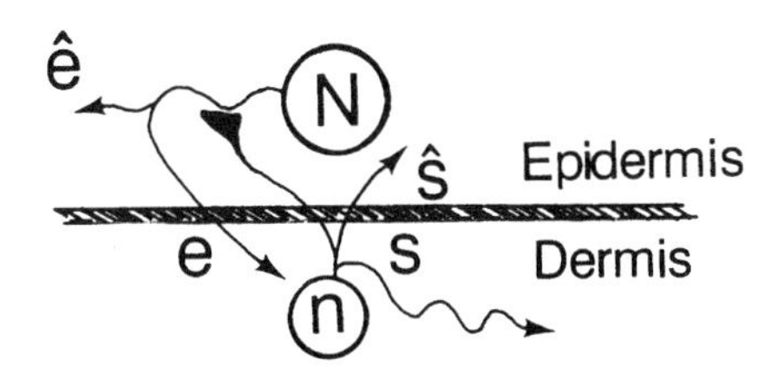

Figure 6.34. Schematic tissue interaction model of Cruywagen and Murray (1992). The dermal cells, n, produce a morphogen, s, which diffuses to the epidermis where it is denoted by $\hat{s}$. In the epidermis $\hat{s}$ increases cell traction which, in turn, causes cell aggregation. Similarly $\hat{e}$, produced by the epidermal cells, N, diffuses to the dermis where it is denoted by e. In the dermis e acts as a chemoattractant for dermal cells, causing dermal aggregation.

Morphogen variables and parameters specific to the epidermal layer are distinguished from the dermal layer by the hat symbol.

It is helpful to refer to Figure 6.34 which schematically encapsulates the following scenario. We consider the epithelial sheet to be a two-dimensional, viscoelastic continuum in equilibrium (recall Section 6.8) and consider the epidermal cells only move by convection. We assume the epidermal morphogen, $\hat{e}(\mathbf{x}, t)$, secreted by the epidermal cells diffuses from a high concentration in the epidermis, across the basal lamina, to the lower concentration in the dermis. There the morphogen, $e(\mathbf{x}, t)$, acts as a chemoattractant for dermal cells inducing dermal cell aggregations which form the papillae. Similarly the morphogen $s(\mathbf{x}, t)$ is the signal produced by the dermal cells which then diffuses through the basal lamina into the epidermal layer. There, the morphogen, represented by $\hat{s}(\mathbf{x}, t)$, increases cell traction thus causing cell aggregation which eventually leads to placode formation.

In spite of the seeming simplicity of the schematic interaction mechanism in Figure 6.34 the model is actually very complicated. We give the full system of equations, which reflects this qualitative description in part to show that oversimplification can sometimes ignore crucial aspects. The various parts of the model are based on those derived above in Sections 6.2 and 6.8 together with reaction diffusion-chemotaxis equations.

In the epidermis we have

$$\frac{\partial N}{\partial t} = \underbrace{-\nabla \cdot \left[N \frac{\partial \mathbf{u}}{\partial t} \right]}_{\text{convection}},$$

$$\nabla \cdot \left[\underbrace{\mu_1 \boldsymbol{\varepsilon}_t + \mu_2 \theta_t \mathbf{I}}_{\text{viscous}} + \underbrace{E'[(\boldsymbol{\varepsilon} - \beta_1 \nabla^2 \boldsymbol{\varepsilon}) + \nu'(\theta - \beta_2 \nabla^2 \theta)\mathbf{I}]}_{\text{elastic}} \right.$$
$$\left. + \underbrace{\tau \hat{s}^2 (1 + c\hat{s}^2)^{-1} \mathbf{I}}_{\text{cell traction}} \right] - \underbrace{\rho \mathbf{u}}_{\text{body forces}} = 0, \tag{6.104}$$

$$\frac{\partial \hat{e}}{\partial t} = \underbrace{\hat{D}_{\hat{e}} \nabla^2 \hat{e}}_{\text{diffusion}} + \underbrace{f(N, \hat{s})}_{\text{production}} - \underbrace{P_e(\hat{e} - e)}_{\text{dermal signal}} - \underbrace{\hat{\gamma} \hat{e}}_{\text{degradation}}$$

$$\frac{\partial \hat{s}}{\partial t} = \underbrace{\hat{D}_{\hat{s}} \nabla^2 \hat{s}}_{\text{diffusion}} + \underbrace{P_s(s - \hat{s})}_{\text{dermal contribution}} - \underbrace{\hat{\nu} N \hat{s}}_{\text{degradation}},$$

where $\hat{D}_{\hat{e}}$, $\hat{D}_{\hat{s}}$, β_1, β_2, P_e, P_s, ρ, τ, $\hat{\gamma}$, $\hat{\nu}$, c, are constants and the reproduction function $f(N, \hat{s})$ is a function of N and $\hat{s}$ which increases with N and decreases with $\hat{s}$; the specific form is not crucial. The first and second of (6.104) are versions of (6.7) and (6.72) with $\hat{s}$ replacing c in the traction-dependence on the morphogen; we have included long range effects in the elastic variables since these could be important. The third and fourth are typical reaction diffusion equations with terms which are motivated by the above biological scenario.

The dermal equations are taken to be

$$
\begin{aligned}
\frac{\partial n}{\partial t} &= \underbrace{D\nabla^2 n}_{\text{diffusion}} - \underbrace{\alpha\nabla\cdot n\nabla e}_{\text{chemotaxis}} + \underbrace{rn(n_0 - n)}_{\text{mitosis}} \\
\frac{\partial s}{\partial t} &= \underbrace{D_s\nabla^2 s}_{\text{diffusion}} + \underbrace{g(n, e)}_{\text{production}} - \underbrace{P_s(s - \hat{s})}_{\text{dermal loss}} - \underbrace{\hat{\nu}N\hat{s}}_{\text{degradation}} \\
\frac{\partial e}{\partial t} &= \underbrace{D_e\nabla^2 e}_{\text{diffusion}} + \underbrace{P_e(\hat{e} - e)}_{\text{epidermal signal}} - \underbrace{\gamma ne}_{\text{metabolism}},
\end{aligned}
\tag{6.105}
$$

where D_e, D_s, γ, ν are constants and $g(n, e)$, the detailed form of which is again not crucial, reflects the production of s; it increases with n. These equations are typical reaction diffusion equations but with the first including chemotaxis. Again the general forms are based on the above biological description.

These model equations were solved numerically by Cruywagen and Murray (1992) (see also Cruywagen 1992) who also carried out a nonlinear analysis on a small strain version using the concepts described in Section 6.4 which reduce the number of equations. They also investigated some of the steady state solutions. They used the LPS method (Bentil and Murray 1991) for isolating appropriate parameter values. We do not carry out any of the analysis here, but with the usual linear analysis much useful information can be obtained; it is left as an exercise (an algebraically complicated one). What comes out of the linear analysis is that the two-way interaction is *essential* for spatial patterning. In fact, for the spatial pattern to evolve, the epidermal traction and the dermal chemotaxis, induced by the signal chemicals, must be large, the respective cell metabolisms must be low, and the chemical diffusion across the basal lamina must be fast. The importance of long range elastic restoring forces also proved important in generating a coherent pattern.

This model was also studied by Cruywagen et al. (1992, 1993, 1994, 2000) who were interested in seeing whether or not the model could generate sequential patterning. As we commented on earlier, feather germs appear to spread out from the dorsal midline. Such patterning was indeed possible. They again used a caricature model. It is a challenging problem to determine the speed of spread of such a pattern and its wavelength. The problem is considerably harder than the one we discussed with chemotaxis waves in Chapter 3. In the other interaction models we mentioned above by Nagorcka et al. (1987) effectively two separate pattern generators were coupled.

Pattern Robustness and Morphogenesis

A large number of processes are involved in development many of which occur at the same time and at different levels such as the genetic, the cellular, the tissue and so on. The amazing fact is that they are so well orchestrated to produce more or less similar copies of the final animal. The whole process is remarkably robust. We have touched on this before when discussing parameter spaces for specific patterns, as in Chapter 2, and the size of the space is one of several indications of the robustness of a mechanism as we saw in Chapter 4.

A major problem for single pattern formation models with a finite range of excitable spatial modes is to generate complex patterns which are effectively made up from a combination of two (or more) different spatial patterns with different wavelengths. In the case of many scale patterns, on the integument of the armadillo and many lizards, for instance, examples of which are shown in Figure 6.30, there is a regular array of large scales surrounded by small scales. If only one size scale were present a single pattern generator with the required mode would suffice to form such a pattern for the primordia. If a single pattern formation mechanism has two more or less equally preferred modes (as we know can be found in the mechanical models discussed above) then the final pattern will be dependent on the initial conditions. Given the randomness present in developing organisms the result would be a random distribution of both scale patterns.

Since the evidence is clear that tissue interaction is crucial in the development of many organs, such as scale and feather primordia, it seemed appropriate to investigate the pattern potential of model mechanisms of tissue interaction systems which is exactly what Nagorcka et al. (1987) did. We saw some of the effects of coupling a reaction diffusion pattern generator with a spatially varying forcing term representing the input from another pattern generator. This was a caricature model for the coupling of two mechanisms each with a preferred pattern wavelength. The Cruywagen–Murray (1992) model is a full tissue interaction model system for generating complex but regular patterns of skin organ primordia. They showed that both mechanisms (one in the epidermis and one in the dermis) are *simultaneously* involved in the pattern formation process. Their model produced regular complex patterns such as those shown above on the integument of lizards.

The conference proceedings article by Murray (1990) (see also Cruywagen and Murray 1992) briefly reviewed tissue interaction models from both the biological and mathematical aspects. They showed how spatial mode selection varies with the strength of interaction between two mechanisms. When mechanisms are coupled, it would be reasonable to suppose that the pattern spectrum is larger than that which can be created by either mechanism on its own. This might also be expected from the simple linear theory above. A nonlinear analysis (see Shaw and Murray 1990) close to bifurcation from homogeneity, however, picks out a specific pattern, namely, that with the dominant linear growth, even though others may be quantitatively close to it from the point of view of linear theory's exponential growth. Extensive numerical simulation of the full nonlinear systems shows unequivocably that the spectrum of possible solutions is very much restricted: there is a greatly reduced number of excitable modes. This is one of the most interesting aspects of tissue interaction, namely, that the spectrum of patterns for the composite mechanism is usually very much *less* than the sum of the two classes of individual patterns which can be formed by the individual generators. In fact, as Murray (1990) pointed out, there seems to be a strong basin of attraction for a specific and highly restricted subset of the theoretically possible patterns. It is clear that the nonlinearity and the coupling enhance the strength of the basin of attraction of specific patterns.

There are important exceptions to the reduction in the number of possible patterns when typical tissue pattern formation mechanisms are coupled. An obvious one is the case when neither mechanism can produce a pattern on its own but coupling them together results in a pattern (Shaw and Murray 1990). The Cruywagen–Murray model is another but here the coupling is an intrinsic part of the mechanism and although it is a tissue interaction model it is essentially one mechanism.

It is well known that in the case of space-independent oscillators coupling results in phase locking into specific periods. What we have here is a kind of *space phase locking*. What we see in this section, and in the papers cited, is that coupling of multiple modes can break mode symmetry. The coupling of pattern generators, each with its own basin of attraction, instead of introducing more complexity actually has the effect of reducing the number of possible patterns. Not only that, the simulations suggest that these patterns are highly robust, more so than the patterned solutions generated by either mechanism on its own.

We suggest, therefore, that the dynamic coupling between different mechanisms significantly reduces the degrees of freedom available to the whole system, thereby contributing to the robustness of morphogenesis. If this is true, the consequences for our understanding of developmental and evolutionary processes could be profound.

We believe our conjecture may apply more generally to the coupling of multiple developmental mechanisms, each of whose eigenfunction modes breaks the symmetries of the other mechanism. As we saw, coupling distinct mechanisms was useful to introduce tissue interaction. We also saw that if two linear modes of different systems are simultaneously present they span its state space because each amplifies a different eigenvector of the underlying variables and if simultaneously amplified, should mutually break each other's symmetries, just as with nonlinear coupling between two mechanisms. This was found by Nagorcka et al. (1987), Shaw and Murray (1990), Cruywagen (1992) and Cruywagen and Murray (1992). When more than one mode is present in the domain, nonlinear couplings among these modes can lead to a final pattern which is fully independent of the orientations of the first few modes. In other words, coupling modes of one system, like the coupling of modes among more than one system, can lead to a robust morphology from a large basin in state space, and almost certainly, in parameter space. That is, the dynamic coupling of different mechanisms or tissues involved in development reduces the choices available to the system, because of bias in successive symmetry-breaking, thereby having a simplifying and stabilising effect. If this is true it has an extremely important implication on our thinking of how pattern evolves during embryologenesis and we could hypothesize that this is perhaps another example of a morphogenetic law such as we discuss in the following chapter.

In conclusion, there can be no doubt that mechanochemical processes are involved in development. The models we have described here and those in subsequent chapters represent a very different approach to reaction diffusion processes and the concepts suggest that mechanical forces could be major elements in producing the correct sequence of tissue patterning and shape changes which are found in the developing embryo. The models simply reflect the laws of mechanics as applied to tissue cells and their environment, and are based on known biological and biochemical facts: importantly the parameters involved are in principle measurable several of which have already been estimated.

We should add that these models are still basic systems (even the complex tissue intereaction model in Figure 6.34), and considerable mathematical analysis is required to investigate their potential to the full. In turn this will suggest model modifications in the usual way of realistic biological modelling. The results from the analysis here, however, are sufficient to indicate a wealth of wide-ranging patterns and mathematically challenging problems. The models have been applied realistically to a variety of morphogenetic problems of current major interest some of which we describe in subsequent chapters. The results and basic ideas have initiated considerable experimental investigation and new ways of looking at a wide spectrum of embryological problems.

Exercises

1. A mechanical model for pattern formation consists of the following equations for $n(x,t)$, the cell density, $u(x,t)$, the matrix displacement and $\rho(x,t)$ the matrix density,

$$\begin{aligned} &n_t + (nu_t)_x = 0, \\ &\mu u_{xxt} + u_{xx} + \tau[n\rho + \gamma\rho_{xx}]_x - s\rho u = 0, \\ &\rho_t + (\rho u_t)_x = 0, \end{aligned}$$

where μ, τ, γ and s are positive parameters. Briefly describe what mechanical effects are included in this model mechanism. Show, from first principles, that the dispersion relation $\sigma(k^2)$ about the uniform nontrivial steady state is given by

$$\sigma(k^2) = -\frac{\gamma\tau k^4 + (1-2\tau)k^2 + s}{\mu k^2},$$

where k is the wave number.

Sketch the dispersion relation as a function of k^2 and determine the critical traction or tractions τ_c when the uniform steady state becomes linearly unstable. Determine the wavelength of the unstable mode at bifurcation to heterogeneity.

Calculate and sketch the space in the $(\gamma s, \tau)$ plane within which the system is linearly unstable. Show that the fastest growing linearly unstable mode has wavenumber $(s/\gamma\tau)^{1/4}$. Thus deduce that for a fixed value of γs the wavelength of the fastest growing unstable mode is inversely proportional to the root of the strength of the external tethering.

2. The mechanical model mechanism governing the patterning of dermal cells of density n in an extracellular matrix of density ρ, whose displacement is measured by u, is represented by

$$\begin{aligned} &n_t = D_1 n_{xx} - D_2 n_{xxxx} - a(n\rho_x)_x - (nu_t)_x + rn(1-n), \\ &\mu u_{xxt} + u_{xx} + \tau[n(\rho + \gamma\rho_{xx})]_x = su\rho, \\ &\rho_t + (\rho u_t)_x = 0, \end{aligned}$$

where all the parameters are nonnegative. Explain what each term represents physically.

Show that the trivial steady state is unstable and determine the dispersion relation $\sigma(k^2)$ for the nontrivial steady state $n = \rho = 1$, $u = 0$.

Sketch the dispersion relation σ as a function of k^2 for various values of τ in the situation where viscous effects are negligible. Briefly discuss the implications from a spatial pattern generating point of view. Now sketch the dispersion relation when the viscosity parameter $0 < \mu \ll 1$ and point out any crucial differences with the $\mu = 0$ case.

3. Consider the dimensionless equations for the cytogel continuum given by

$$\nabla \cdot \{\mu_1 \boldsymbol{\varepsilon}_t + \mu_2 \theta_t \mathbf{I} + \boldsymbol{\varepsilon} + \nu' \theta \mathbf{I} + \tau(c)\mathbf{I}\} = s\mathbf{u},$$

$$\frac{\partial c}{\partial t} = D\nabla^2 c + \frac{\alpha c^2}{1 + \beta c^2} - c + \gamma\theta = D\nabla^2 c + R(c) + \gamma\theta,$$

where $\boldsymbol{\varepsilon}$ and θ are the strain tensor and dilation respectively, $\mathbf{u}$ is the cytogel displacement of a material point and c is the concentration of free calcium. The active traction function $\tau(c)$ is as illustrated in Figure 6.24 and μ_1, μ_2, ν', α, D, γ and β are constants.

Investigate the linear stability of the homogeneous steady state solutions c_i, $i = 1, 2, 3$ and show that the dispersion relation $\sigma = \sigma(k^2)$, where here σ is the exponential temporal growth factor and k is the wavenumber of the linear perturbation, is given by

$$\begin{aligned}
&\mu k^2\sigma^2 + b(k^2)\sigma + d(k^2) = 0,\\
&b(k^2) = \mu D k^4 + (1 + \nu' - \mu R_i')k^2 + s,\\
&d(k^2) = D(1 + \nu')k^4 + [sD + \gamma\tau_i' - (1 + \nu')R_i']k^2 - sR_i',
\end{aligned}$$

where

$$\mu = \mu_1 + \mu_2, \quad R_i' = R'(c_i), \quad \tau_i' = \tau'(c_i), \quad i = 1, 2, 3.$$

Show that it is possible for the model to generate spatial structures if γ is sufficiently large. Determine the critical wavenumber at bifurcation.

7. Evolution, Morphogenetic Laws, Developmental Constraints and Teratologies

7.1 Evolution and Morphogenesis

We shall never fully understand the process of evolution until we know how the environment affects the mechanisms that produce pattern and form in embryogenesis. Natural selection must act on the developmental programmes to effect change. We require, therefore, a morphological view of evolution, which goes beyond the traditional level of observation to a morphological explanation of the observed diversity. Later in this chapter we shall discuss some specific examples whereby morphogenesis has been experimentally influenced to produce early embryonic forms, early, that is, from an evolutionary point of view. This chapter has no mathematics *per se* and is more or less a stand-alone biological chapter. However, the concepts developed and their practical applications are firmly based on the models, and their analysis, presented and elaborated on in earlier chapters, particularly Chapters 2, 3, 4 and 6.

Natural selection is the process of evolution in which there is preferential survival of those who are best adapted to the environment. There is enormous diversity and within species such diversity arises from random genetic mutations and recombination. We must therefore ask why there is not a continuous spectrum of forms, shapes and so on, even within a single species. The implication is that the development programmes must be sufficiently robust to withstand a reasonable amount of random input. From the extensive genetic research on the fruit fly *Drosophila*, for example, it seems that only a finite range of mutations is possible, relatively few in fact.

The general belief is that evolution never moves backward, although it might be difficult to provide a definition of what we mean by direction. If evolution takes place in which a vertebrate limb moves from being three-toed to four-toed, from a morphological view of evolution there is no reason, if conditions are appropriate, that there cannot be a transition 'back' from the four-toed to the three-toed variety. From our study of pattern formation mechanisms this simply means that the sequential bifurcation programme is different. In Section 7.4 we show an example where an experimentally induced change in the parameters of the mechanism of morphogenesis results in embryonic forms which, with the accepted direction of evolutionary change, means that evolution has moved backwards.

If we take the development of the vertebrate limb we saw in Section 6.6 in the last chapter that development was sequential in that the humerus preceded the formation of the radius and ulna and these preceded the formation of the subsequent cartilage patterns such as the phalanges. As a specific example, we argued that the formation of the humerus could cue the next bifurcation by influencing the geometry of the limb bud. We also saw how graft experiments could alter the pattern sequence and we showed how the result was a natural consequence when viewed from a mechanistic viewpoint.

So, intimately associated with the concept of bifurcation programmes, are discrete events whereby there is a discrete change from one pattern to another as some parameter passes through a bifurcation value. The possibility of discrete changes in a species as opposed to gradual changes is at the root of a current controversy in evolution, between what is called *punctuated equilibrium* and *phyletic gradualism*, which has raged for about the past few decades. (Neo-Darwinism is the term which has been used for punctuated equilibrium.) Put simply, punctuated equilibrium is the view that evolutionary change, or speciation and morphological diversification, takes place effectively instantaneously on geological time, whereas gradualism implies a more gradual evolution to a new species or a new morphology. The arguments for both come from the fossil records and different sets of data are used to justify each view—sometimes even the same set of data is used. Figure 7.1 schematically shows the two extremes.

From a strictly observational approach to the question we would require a much more extensive fossil record than currently exists, or is ever likely to be. From time to time newly discovered sites are described which provide fine-scaled palaeontological resolution of speciation events. For example, Williamson (1981a,b) describes one of these in northern Kenya for molluscs and uses it to argue for his view of evolution. Sheldon (1987) gathered fossil data, from sites in mid-Wales, on trilobites (crab-like marine creatures that vary in size from a few millimetres to tens of centimetres) and on the basis of his study argues for a gradualist approach. From an historical point of view, the notion of punctuated equilibrium was very clearly put by Darwin (1873) himself in the 6th and later editions of his book, *On the Origin of Species*, in which he said (see the summary at the end of Chapter XI, p. 139), 'although each species must have passed

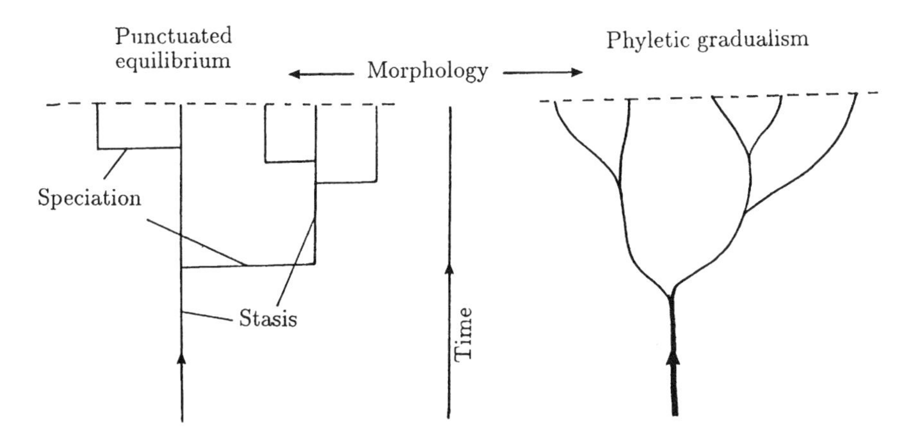

Figure 7.1. Punctuated equilibrium implies that as we move through geological time changes in speciation occur very quickly (on geological time) as compared with stasis, the period between speciation events. Phyletic gradualism says that speciation and diversification are a gradual evolutionary process.

through numerous transitional stages, it is probable that the periods, during which each underwent modification, though many and long as measured by years, have been short in comparison with the periods during which each remained in an unchanged condition.' (The corresponding passage in the first edition is in the summary of Chapter X, p. 139.)

From our study of pattern formation mechanisms in earlier chapters the controversy seems artificial. We have seen, particularly from Chapters 2 to 6, 4 and 5 that a slow variation in a parameter can affect the final pattern in a continuous and discrete way. For example, consider the mechanism for generating butterfly wing patterns in Section 3.3. A continuous variation in one of the parameters, when applied, say, to forming a wing eyespot, results in a continuous variation in the eyespot size. The expression (equation (3.24), for example) for the radius of the eyespot shows a continuous dependence on the parameters of the model mechanism. In the laboratory the varying parameter could be temperature, for example. Such a continuous variation falls clearly within the gradualist view of evolutionary change.

On the other hand suppose we consider Figures 2.14(b) and (c) which we reproduce here as Figures 7.2(a) and (b) for convenience. It encapsulates the correspondence be-

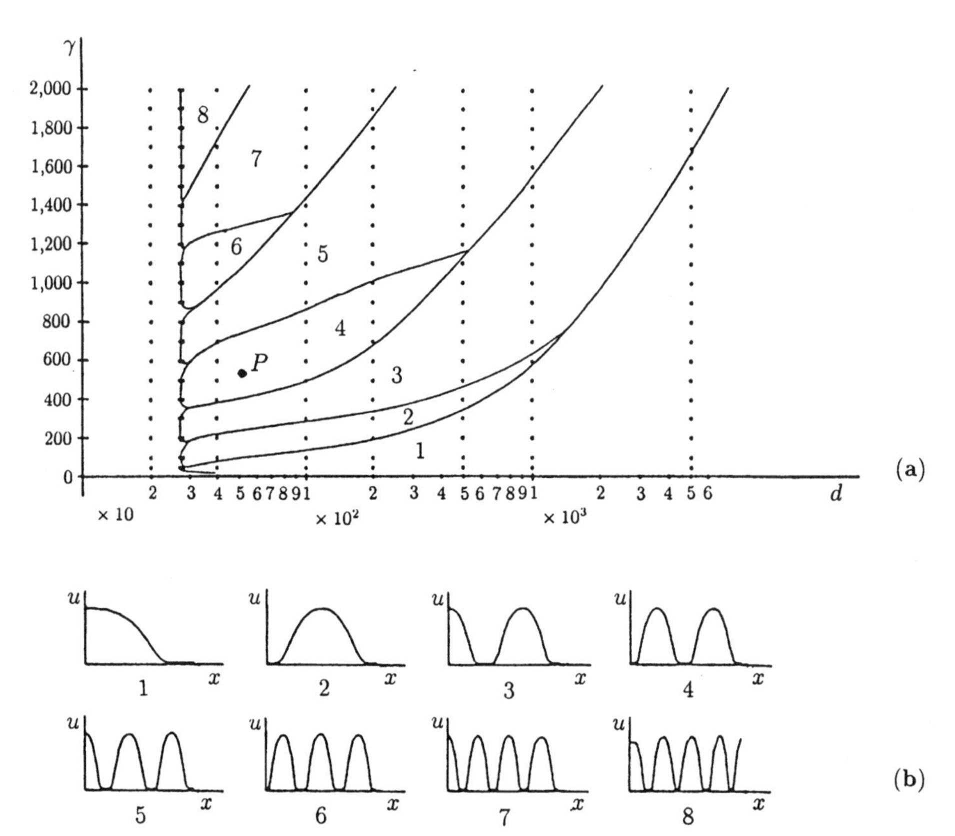

Figure 7.2. (**a**) Solution space for a reaction diffusion mechanism (system (2.8) in Chapter 2) with domain size, γ, and morphogen diffusion coefficient ratio, d. (**b**) The spatial patterns in morphogen concentration with d, γ parameter values in the regions indicated in (**a**). (From Arcuri and Murray 1986)

tween discrete patterns and two of the mechanism's dimensionless parameters. Another example is given by Figure 4.21 in Chapter 4 on the formation of teeth primordia where again the parameter space indicates abrupt changes in the pattern formed. Although Figures 7.2 and 4.21 are the bifurcation space for a specific reaction diffusion mechanism we obtain comparable bifurcation spaces for the mechanochemical models in the last chapter and other pattern formation mechanisms. In Section 6.6 in the last chapter we noted that the effect of a tissue graft on the cartilage patterns in the developing limb was to increase cell proliferation and hence the size of the actual limb bud. Let us, for illustrative purposes, focus on the development of the vertebrate limb. In Figure 7.2(a), if we associate cell number with domain size γ we see that as γ continuously increases for a fixed d, say, $d = 100$, we have bifurcation values in γ when the pattern changes abruptly from one pattern in Figure 7.2(b) to another. So, a continuous variation in a parameter here effects discontinuous changes in the final spatial pattern. This pattern variation clearly falls within a punctuated equilibrium approach to evolution.

Thus, depending on the mechanism and the specific patterning feature we focus on, we can have a gradual or discontinuous change in form. So to reiterate our comment above it is clear that to understand how evolution takes place we must understand the morphogenetic processes involved.

Although the idea that morphogenesis is important in understanding species diversity goes back to the mid-19th century, it is only relatively recently that it has been raised again in a more systematic way by, for example, Alberch (1980) and Oster and Alberch (1982); we briefly describe some of their ideas below. Oster et al. (1988) presented a detailed study of vertebrate limb morphology, which is based on the notion of the morphological rules described in the last chapter. The latter paper presents experimental evidence to justify their morphogenetic view of evolutionary change; later in the chapter we describe their ideas and some of the supporting evidence.

Morphogenesis is a complex dynamic process in which development takes place in a sequential way with each step following, or bifurcating, from a previous one. Alberch (1980, 1982), and Oster and Alberch (1982) suggest that development can be viewed as involving only a small set of rules of cellular and mechanochemical interactions which, as we have seen from previous chapters, can generate complex morphologies. Irrespective of the actual mechanisms, they see developmental programmes as increasingly complex interactions between cell populations and their gene activity. Each level of the patterning process has its own dynamics (mechanism) and it in turn imposes certain constraints on what is possible. This is clear from our studies on pattern formation models wherein the parameters must lie in specific regions of parameter space to produce specific patterns; see, for example, Figure 7.2. Alberch (1982) and Oster and Alberch (1982) encapsulate their ideas of a developmental programme and developmental bifurcations in the diagram shown in Figure 7.3.

If the number or size of the mutations is sufficiently large, or sufficiently close to a bifurcation boundary, there can be a qualitative change in morphology. From our knowledge of pattern formation mechanisms, together with Figure 7.3, we can see how different stability domains correspond to different phenotypes and how certain genetic mutations can result in a major morphological change and others do not. Not only that, we can see how transitions between different morphologies are constrained by the topology of the parameter domains for a given morphology. For example, a transition be-

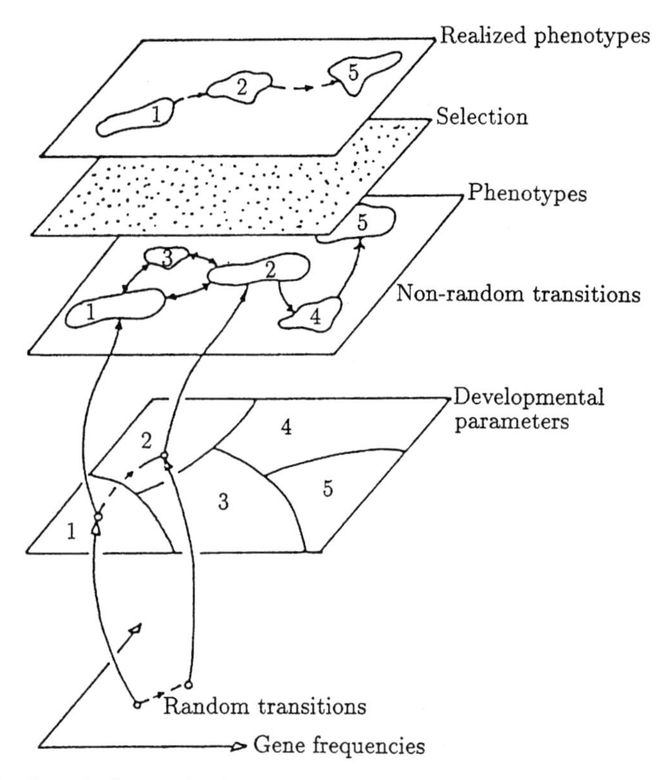

Figure 7.3. A schematic diagram showing how random genetic mutations can be filtered out to produce a stable phenotype. For example, here random genetic mutations affect the size of the various developmental parameters. With the parameters in a certain domain, 1 say, the mechanisms create the specific pattern 1 at the next level up; this is a possible phenotype. Depending on the size of the random mutations we can move from one parameter domain to another and end up with a different phenotype. There is thus a finite number of realisable forms. At the next stage selection takes place and the final result is a number (reduced) of realised phenotypes. (From Oster and Alberch 1982)

tween states 1 and 2 is more likely than between 1 and 5 and furthermore, to move from 1 to 5 intervening states have to be traversed. An important point to note is that existing morphological forms depend crucially on the history of their past forms. The conclusion therefore is that the appearance of novel phenotypic forms is *not* random, but can be discontinuous. As Alberch (1980) notes, 'We need to view the organism as an integrated whole, the product of a developmental program and constrained by developmental and functional interactions. In evolution, selection may decide the winner of a given game but development non-randomly defines the players.'

Developmental Constraints

In previous chapters we have shown that, for given morphogenetic mechanisms, geometry and scale impose certain developmental constraints. For example, in Chapter 3, Section 3.1 we noted that a spotted animal could have a striped tail but not the other way round. In the case of pattern formation associated with skin organ primordia, as discussed in the last chapter, we have mechanical examples which exhibit similar developmental constraints. Holder (1983) carried out an extensive observational study of 145 hands and feet of four classes of tetrapod vertebrates. He concluded that developmental constraints were important in the evolution of digit patterns.

Figure 7.4 (refer also to Chapter 4, Figure 4.10) shows some of the key mechanical steps in the early development of certain skin organs such as feathers, scales and teeth. In Section 6.6 we addressed the problem of generating cell condensation patterns which we associated with the papillae. In the model for epithelial sheets, discussed in Section 6.8, we saw how spatially heterogeneous patterns could be formed and even initiated by the

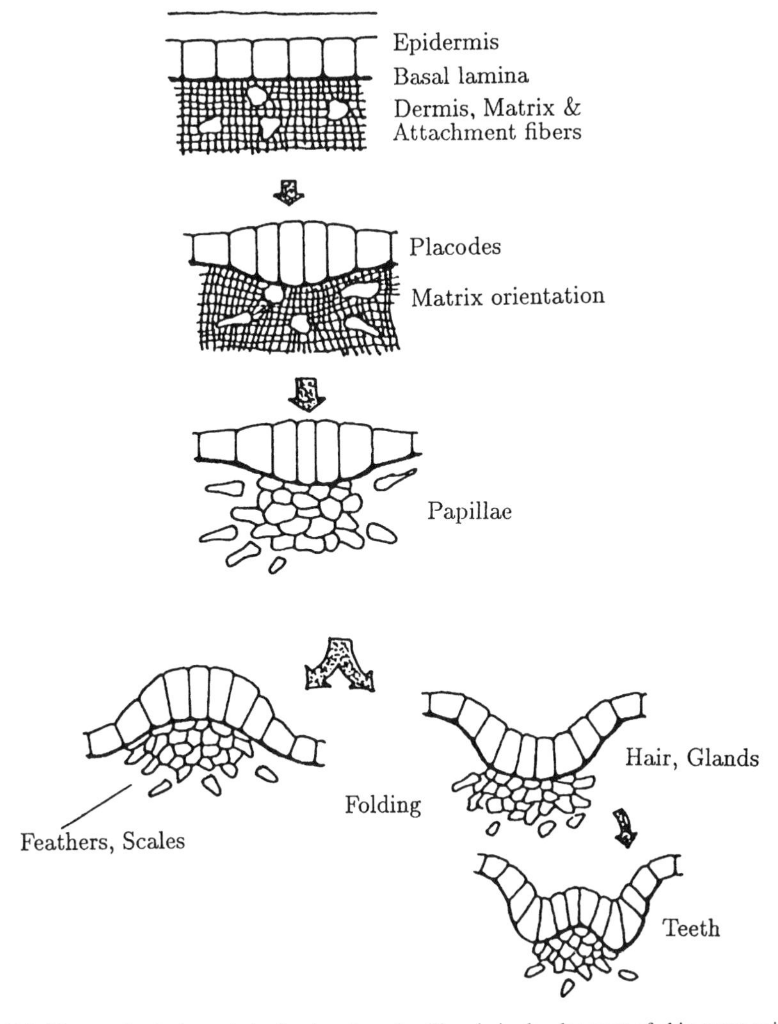

Figure 7.4. Key mechanical events in the dermis and epidermis in development of skin organ primordia. (After Oster and Alberch 1982)

dermal patterns. Odell et al. (1981) showed how buckling of sheets of discrete cells such as in Figure 7.4, could arise. From the sequential view of development we might ask whether it is possible to move onto a different developmental pathway by disrupting a mechanical event. There is experimental evidence that a transition can be effected from the scale pathway to the feather pathway, for example, Dhouailly et al. (1980), by treating the skin organ primordia with retinoic acid. In their experiments feathers were formed on chick foot scales.

7.2 Evolution and Morphogenetic Rules in Cartilage Formation in the Vertebrate Limb

In Section 6.6 in the last chapter we showed how a mechanical model could generate the cartilage patterns in the vertebrate limb. There we proposed a simple set of general morphogenetic construction rules for how the major features of limb cartilage patterns are established. Here we use these results and draw on comparative studies of limb morphology and experimental embryological studies of the developing limb to support our general theory (which is essentially mechanism-independent) of limb morphogenesis. We then put the results in an evolutionary context. The following is mainly based on the work of Oster et al. (1988) which arose from discussions between George Oster, the late Pere Alberch and myself in 1985.

Since the limb is one of the most morphologically diversified of the vertebrate organs and one of the more easily studied developmental systems it is not surprising it is so important in both embryology and evolutionary biology. Coupled with this is a rich fossil record documenting the evolution of limb diversification (see, for example, Hinchliffe and Johnson 1980 for a comprehensive biology).

Although morphogenesis appears deterministic on a macroscopic scale, on a microscopic scale cellular activities during the formation of the limb involve considerable randomness. Order emerges as an average outcome with some high probability. We argued in Section 6.6 that some morphogenetic events are extremely unlikely, such as trifurcations from a single chondrogenic condensation. Mathematically, of course, they are not strictly forbidden by the pattern formation process, be it mechanochemical or reaction diffusion, but are highly unlikely since they correspond to a delicate choice of conditions and parameter tuning. This is an example of a 'developmental constraint' although the term 'developmental bias' would be more appropriate.

Let us recall the key results in Section 6.6 regarding the 'morphogenetic rules' for limb cartilage patterning. These are summarised in Figures 6.18(a)–(c) the key parts of which we reproduce for convenience in Figure 7.5.

The morphogenetic process starts with a uniform field of mesenchymal cells from which a precartilagenous focal condensation of mesenchymal cells forms in the proximal region of the limb bud. With the mechanical model discussed in the last chapter, this is the outcome of a model involving the cells, the extracellular matrix (ECM) and its displacement. With the model of Oster et al. (1985a,b), various mechanochemical processes are also involved. Subsequent differentiation of the mesenchymal cells is intimately tied to the process of condensation. It seems that differentiation and cartilage

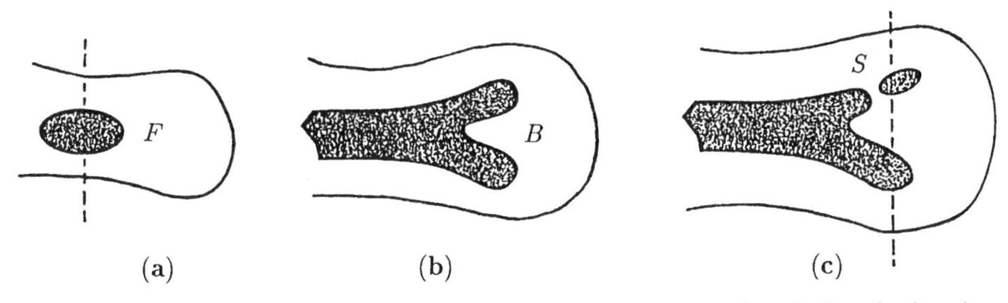

Figure 7.5. Morphogenetic rules: the three basic cell condensation types, namely, a single or focal condensation, F, as in (**a**), a branching bifurcation, B, as in (**b**) and a segmental condensation, S, as in (**c**). More complicated patterns can be built up from a combination of these basic bifurcations; see Figures 7.8, 7.11 (cf. Figure 6.18(e)) and 7.12.

morphogenesis are frequently interrelated phenomena. An alternative cell-chemotaxis model with cell differentiation whereby condensation and morphogenesis take place simultaneously was proposed by Oster and Murray (1989).

There is a zone of recruitment created around the chondrogenic focus. That is, an aggregation of cells autocatalytically enhances itself while depleting cells in the surrounding tissue. This is effectively setting up a lateral inhibitory field against further aggregation. Because nearby foci compete for cells this leads to almost cell-free regions between foci. In other words, a condensation focus establishes a 'zone of influence' within which other foci are inhibited from forming.

As the actual cartilagenous element develops, the cells seem to separate into two regions: the outer region consists of flattened cells concentrically arranged, while the cells in the inner region are rounded. The outer cells differentiate to form the perichondrium which sheaths the developing bone. As suggested by Archer et al. (1983) and Oster et al. (1985a,b), the perichondrium constrains the lateral growth of cartilage and forces its elongation. It also restricts the lateral recruitment of additional cells, so that cells are added to this initial condensation primarily by adding more mesenchymal cells at the distal end thus affecting linear growth as illustrated in Figure 7.6, which also shows the general features of the condensation process.

As we noted in Section 6.6, limb morphogenetic patterns are usually laid down sequentially, and not simultaneously over an entire tissue (Hinchliffe and Johnson 1980). The latter method would be rather unstable. Theoretical models show that sequential pattern generation is much more stable and reproducible. Recall the simulations associated with the formation of animal coat patterns in Chapter 3, Section 3.1, where the final pattern was dependent on the initial conditions, as compared with the robust formation of hexagonal feather germ and scale arrays in birds, discussed in Section 6.5 in the last chapter, and the supporting evidence from the model simulations by Perelson et al. (1986).

Although most of the pattern formation sequence proceeds in a proximo-distal direction, the differentiation of the digital arch (see Figure 7.11) occurs sequentially from anterior to posterior. The onset of the differentiation of the digital arch is correlated with the sudden broadening and flattening of the distal region of the limb bud into a paddlelike shape. From the typical dispersion relations for pattern generation mechanisms

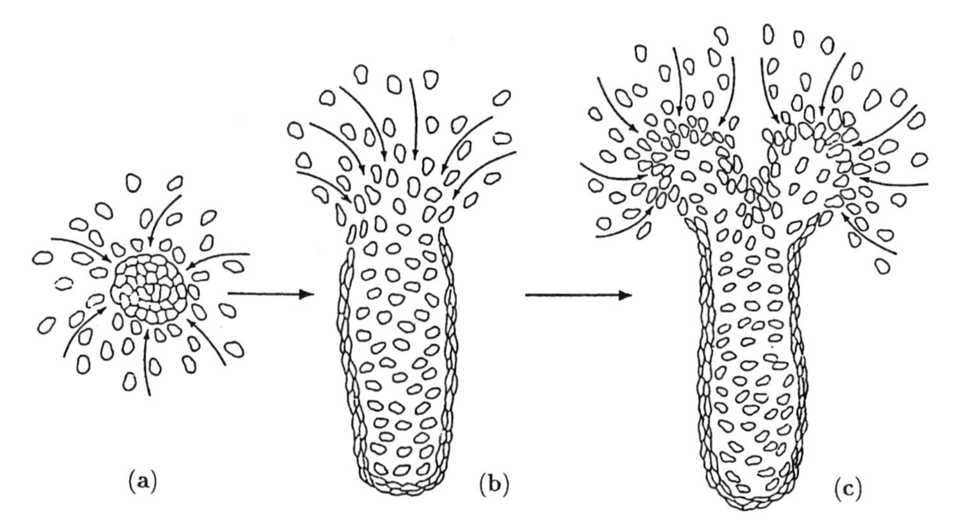

Figure 7.6. Schematic illustration of the cell condensation process. (**a**) Cells aggregate initially into a central focus. (**b**) Development of the cartilaginous element restricts cell recruitment to the distal end of the condensation. (**c**) When conditions are appropriate the aggregation undergoes a Y-bifurcation. (After Shubin and Alberch 1986)

(recall, for example, the detailed discussion in Chapter 2, Section 2.5) such a change in geometry can initiate independent patterns and is the key to understanding this apparent exception to the sequential development rule. Physically, this means that where the domain is large enough, an independent aggregation arises and is far enough away from the other aggregations that it can recruit cells to itself without being dominated by the attractant powers of its larger neighbours. Of course other model parameters are also important elements in the ultimate pattern and its sequential generation and initiation. The key point is that, irrespective of whether reaction diffusion or mechanochemical models create the chondrogenic condensations, the model parameters, which include the size and shape of the growing limb bud, are crucially important in controlling pattern. Experimental manipulations clearly confirm this importance.

Alberch and Gale (1983) treated a variety of limb buds with the mitotic inhibitor colchicine. This chemical reduces the dimensions of the limb by reducing cell proliferation. As we predicted, from our knowledge of pattern generation models and their dispersion relations, such a reduction in tissue size reduces the number of bifurcation events, as illustrated in Figure 7.7.

Note that a possibility that cannot be ruled out is that colchicine affects the timing and number of bifurcations by altering some other developmental parameter, such as cell traction or motility, in addition to the size of the recruitment domains. This alteration, of course, is still consistent with the theory. At this stage further experiments are required to differentiate between the various possibilities. The main point is that these experiments confirm the principle that alterations in developmental parameters (here tissue size) can change the normal sequence of bifurcation events, with concomitant changes in limb morphology that are significant.

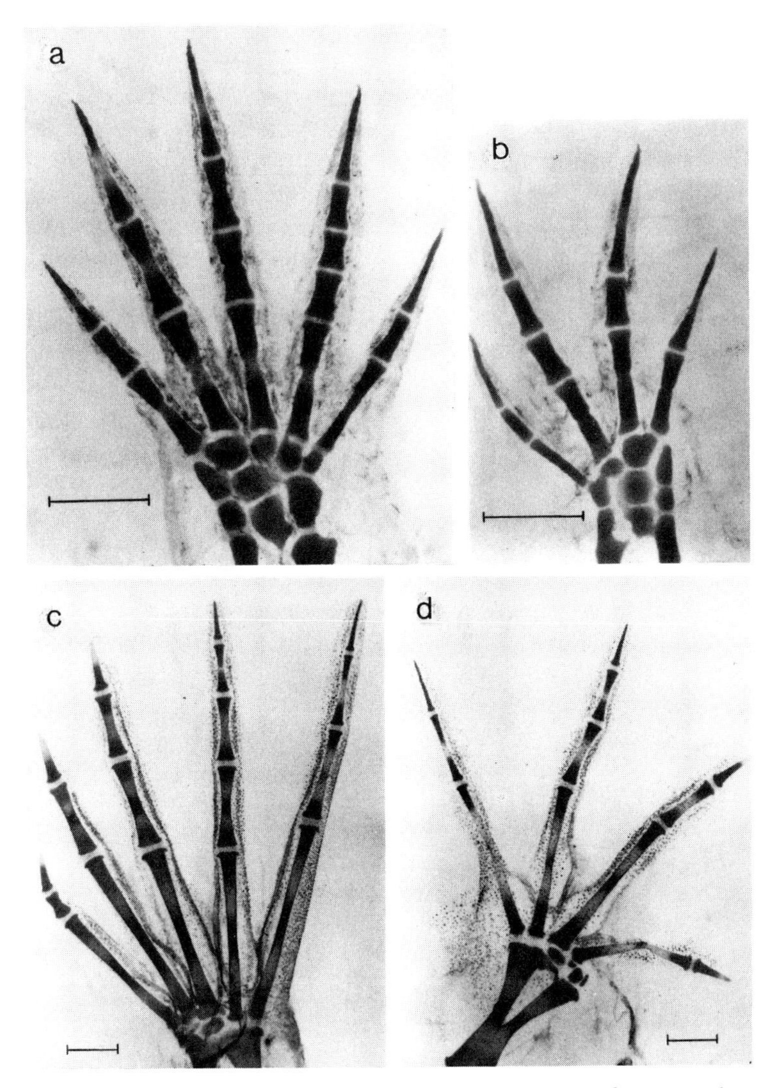

Figure 7.7. Experimentally induced alterations in the foot of the salamander *Ambystoma mexicanum* and the frog *Xenopus laevis* through treatment of the limb bud with colchicine. (**a**) Normal right foot of the salamander and (**b**) the treated left foot. (**c**) Normal right foot of the frog with (**d**) the treated left foot. (From Alberch and Gale 1983; photographs courtesy of Dr. Pere Alberch)

Using the basic ideas of cartilage pattern formation in Oster et al. (1983), Shubin and Alberch (1986) carried out a series of comparative studies with amphibians, reptiles, birds and mammals, and confirmed the hypothesis that tetrapod limb development consists of iterations of the processes of focal condensation, segmentation and branching. Furthermore, they showed that the patterns of precartilage cell condensation display several striking regularities in the formation of the limb pattern. Figure 7.8 presents just some of these results; other examples are also given in Oster et al. (1988).

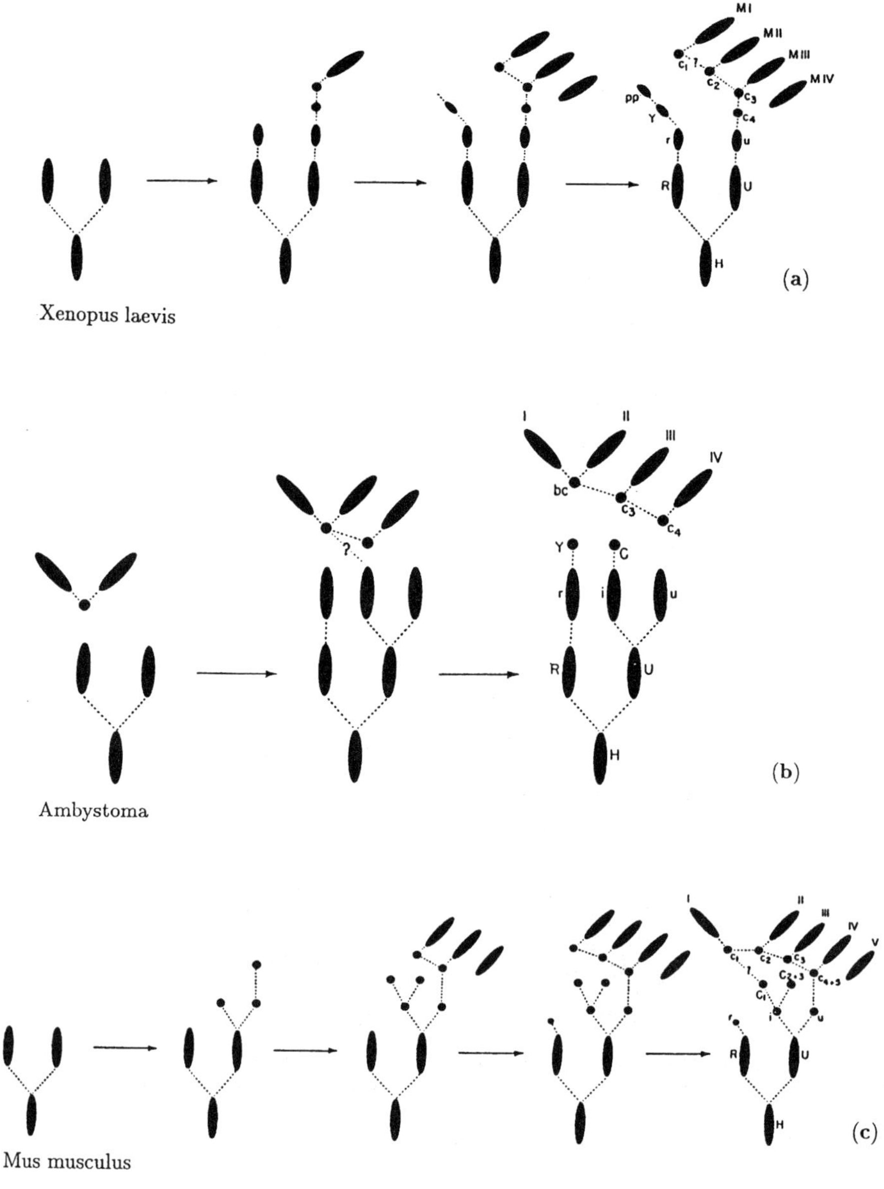

Figure 7.8. Comparative examples of the branching and segmentation in cartilage patterning. (**a**) Amphibian. The foreleg of the frog *Xenopus laevis*. (**b**) Reptile. Forelimb of the salamander *Ambystoma mexicanum*. (**c**) Mammal. Limb of the house mouse *Mus musculus*. These are all constructed from repetitive use of the three basic morphogenetic rules displayed in Figure 7.5.

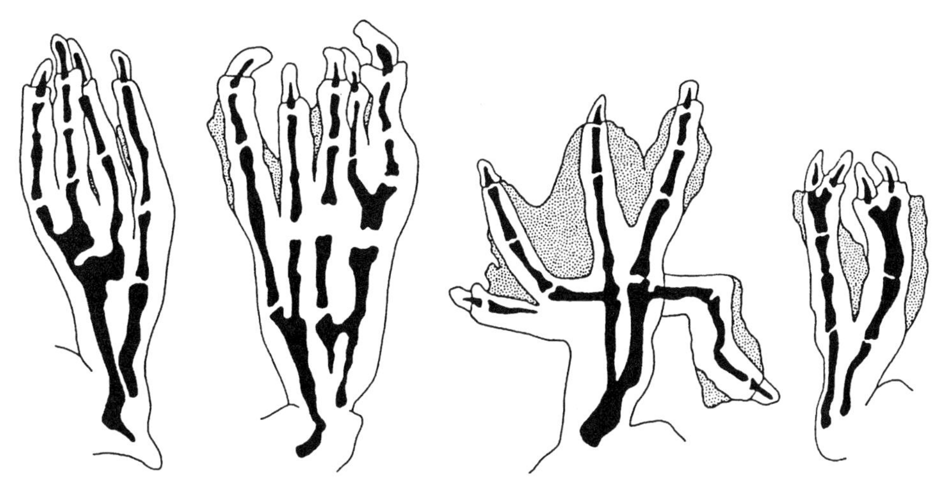

Figure 7.9. Cartilage patterns obtained after the tissue and cells from the limb buds of a duck and chick embryo were extracted, mixed together and then repacked into the limb bud sleeves. (From Patou 1973) The patterns are highly irregular but are still generated by iterations of the basic morphogenetic rules (Figure 7.5).

Condensation, branching and segmentation are intrinsic properties of cartilage forming tissue, although where and when condensation occurs depends on several factors. The stability and reproducibility of a condensation pattern depends crucially on its sequential formation. Patou (1973), in some interesting experiments, removed and disaggregated the tissue and cells involved in cartilage formation from the leg buds of duck and chick embryos. The two populations were then mixed and repacked into the empty limb bud sleeves. The resulting cartilage patterns were highly abnormal and did not display the characteristics of either species, as seen from Figure 7.9. In all cases, however, the condensation patterns were generated by iterations of the three basic processes of condensation, branching and segmentation, as shown in Figure 7.5. The results support the theoretical conclusion that branching, segmentation and *de novo* condensation events are reflections of the basic cellular properties of cartilage forming tissue.

We now see how the study of pattern formation mechanisms can define more precisely the notion of a 'developmental constraint.' The above discussion together with that in Section 6.6 on limb morphogenesis is only one example. It is based on a pattern formation sequence for laying down the cell aggregation pattern reflected in the final limb architecture.

7.3 Teratologies (Monsters)

Our study of theoretical models for pattern formation has shown that there are considerable restrictions as to the possible patterns of chondrogenesis (as well as other developmental aspects). From the morphogenetic laws described in the last chapter, for example, it is highly unlikely that a trifurcation is possible, that is, a branching of *one* element into *three* elements, or even one into many. Although subsequent growth may present the appearance of a 1-to-3 splitting, the theory suggests that all branchings are initially binary. This is because a trifurcation is possible only under a very narrow set of parameter values and conditions. If we include asymmetries it makes it even more unlikely. Such a delicate combination of requirements almost always leads to an unstable pattern even with numerical simulations of the model mechanisms.

Alberch (1989) applied this notion of the unlikelihood of trifurcations to other examples of internal constraints in development. He argued that this is the reason we do not see any three-headed monsters. There are numerous examples of two-headed snakes and other reptiles, Siamese twins and so on. Figures 7.10(a)–(c) show the three basic

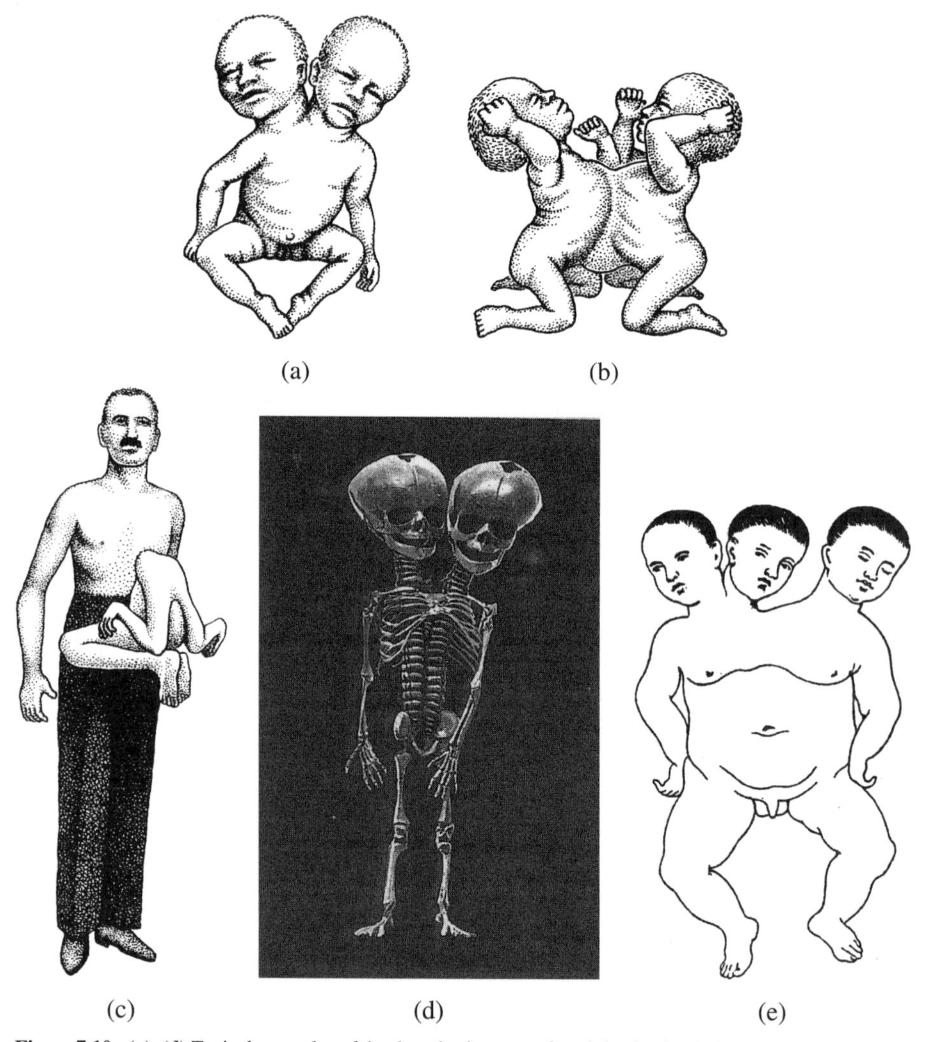

Figure 7.10. (**a**)–(**d**) Typical examples of the three basic types of conjoined twins in humans. The equivalent forms are fairly common in fish. In (**a**) the duplication arises from a bifurcation of the body axis while in (**b**)—a more common form—it is a consequence of fusing. This can occur through any part of the body. (**c**) An example of a fused fragment from the chest region. (From Stockard 1921) (**d**) The skeleton (19th century) of a *Dicephalus*, a young boy, with (**e**) a 19th century example of a *Tricephalus*.

types of conjoined twins. Very few three-headed monsters have been reported, and of these the veracity is often highly questionable. If we come back to the limited bifurcations suggested by the morphogenetic laws above, and specifically that a trifurcation is highly unlikely, we can see how a three-headed monster can arise, namely, via a bifurcation of the body axis such as we see in the skeleton in Figure 7.10(d) followed by a further bifurcation of one of the branches as seems clear in the example in Figure 7.10(e).[1]

The study of monsters—teratology—has a long history. Art (that of Hieronymus Bosch is a particularly good source) and mythology are replete with splendid monsters and new morphologies. One mediaeval description of a three-headed human is that it was born with one head human, the other a wolf's and the third a bloody mass without skin. It finally died after it appeared before the city Senate and made a series of dire predictions! There is a Dürer drawing of a calf with two bodies fused at the single head while the head of Medusa with its numerous serpents surrounding the head was a popular theme; the Rubens picture is, of course, well known.

Monsters have fascinated people for a very long time. An early interesting attempt at monster classification was given by Ambroise Paré in 1573 in his bestiary *Monstres et Marveillles*. He thought, for example, that a failure of the right mix of male and female seed gave rise to monsters. Interspecific seed he thought gave rise to mixed human and animal forms, such as the centaur. Paré suggested that constrictions of the womb could give rise to hermaphrodites. He also thought, as was (and still is) common, the wrath of God played a part. The book by Pallister (1982) discusses Paré's work and his classification of monsters.

The mediaeval and later literature abounds with descriptions of an incredible menagerie of reputed monster births which with our current knowledge are totally impossible. Some, however, are reminiscent of the thalidomide deformities. Again there are very few references to three-headed monsters (unless one counts those whose head instead consisted of several horns) but lots with multiple hands and fingers. Some of the more fanciful births are supposedly by concubines of the various Popes (they couldn't miss a chance at a dig at the papacy in England at that time). The literature is fascinating as a window onto the beliefs of people at the time. One marvelous example is the book by William Turner, MA (an Oxford graduate), who was the Vicar of Walberton in England in 1697 when his book came out. This was more than 30 years after the founding of the Royal Society in London with scientific giants like Newton (personally a rather spiteful man who incidentally believed his major contribution to knowledge was his writing on religion). It was the era of long titles; the title of Turner's book is

> 'A Compleat History Of the Most *Remarkable Providences*, both of Judgement and Mercy, Which have hapned in this Present Age. Extracted From

[1]This example was found in the mid-1980's by a colleague (a historian) following a dinner conversation in my Oxford college. I had commented on the absence of three-headed monsters other than mythical ones typically found in mediaeval bestiaries. He said that he had seen an example—the one reproduced here. I was sceptical but added that if it really had existed as a single body with three heads I could predict the general shape. We agreed that if I got it right (a rough sketch was put in a sealed envelope) I would get the agreed bottle of wine we bet on it. Having just finished working on these morphogenetic laws it seemed that the only way it could arise was by an initial bifurcation of the body axis followed by another bifurcation of one of the bifurcations as indeed it is from this figure. I won the bottle of wine.

> the Best Writers, the Author's own Observations, and the Numerous Relations sent to him from divers Parts of the Three Kingdoms. To which is Added, Whatever is Curious in the *Works of Nature and Art*' ((London: Printed for John Dunton at the *Raven*, in *Jewen Street*). The author recommends it 'as useful to ministers in Furnishing Topicks of Reproof and Exhortation and to Private Christians for their closets and Families.'

One example he gives of *Monstrous Births and Conceptions of Mankind* is

> '. . . a Child, terrible to behold, with flaming and shining Eyes; the Mouth and Nostrils were like those of an Ox; it had long horns, and a Back Hairy like a Dog's It had the Faces of Apes in the Breast where Teats should stand; it had Cats Eyes under the Navel, fasten'd to the *Hypogastrium*, and they looked hideously. It had the Heads of Dogs upon both Elbows, and at the White-Bones of each Knee, looking forwards. It was Splay-footed and Splay-handed; the Feet like Swans Feet, and a Tail turn'd upwards, that crook'd up backwards about half an Ell long. It lived four Hours from its Birth; and near its Death, it spake thus, *Watch for the Lord your God comes.*'

Geoffroy Saint-Hilaire, one of the major scientists of his time, wrote extensively on the subject of monsters and specifically commented on the absence of three-headed monsters and produced the definitive work work on teratology in 1836. He also put forward various propositions as to their cause. There is much interesting 19th century literature on monsters particularly towards the latter half of the century and the beginning of the 20th. A CIBA symposium in 1947 was specifically concerned with monsters in nature (Hamburger 1947) and art (Born 1947). Hamburger (1947) reproduces illustrations and photographs of numerous examples including one similar to that in Figure 7.10(c) known as the *Genovese Colloredo* in which there is a fragmentary duplication fused in the chest region: the fragment has a head, one leg and only three fingers on each hand.

Stockard (1921, see other references there) studied teratologies extensively and carried out some of the earliest experiments on fish by treating the embryos with chemicals such as magnesium chloride. The horrifying teratological effects of thalidomide are well documented and we have already commented in earlier chapters on the more recent use of retinoic acid to create limb cartilage teratologies. Stockard (1921) discussed in detail the incidence of hyperdactyly, that is, where the hands or feet have extra digits particularly in identical twins since, like most double teratologies, they come from a single egg and not by fusion of two eggs. Hamburger (1947) also describes the incidence of Cyclopia (one-eyed monsters) in humans. Stockard (1909) artificially produced one-eyed monsters using chemicals: magnesium chloride is just one of them.

The history of teratologies from Aristotle to the present day (manifested, for example, by the numerous horror films involving mythical monsters) is an ever fascinating subject. In the 19th century the preservation of skeletons of naturally occurring teratologies was a particular fascination: Figure 7.10(d) of a young boy is one example. The skeletons of giants and dwarfs were often coveted and anticipated when the subjects were still alive. Hamburger (1947) describes the case of Charles Byrne, an Irish giant of 7 feet 8-3/4 inches (2.36 metres). Byrne, who very much did not want his body to end up in a museum took various precautions to ensure it would not happen by arranging to have his corpse sunk in the Irish Channel by fishermen. After he died, however, his wish was thwarted by a well-known surgeon, John Hunter, who managed to get the body (by a higher bribe no doubt) and Byrne's skeleton is now exhibited in the museum of the Royal College of Surgeons in England.

Teratology highlights some of the most fundamental questions in evolution, namely, why do we not get certain forms in nature. The developmental process, as we have seen, embodies various systems of constraint which bias the evolution of the system. We thus come back to what we have mentioned before, namely, that we must understand the role of internal, as opposed to external, factors if we are to understand evolution. Alberch (1989) gives a thorough discussion of this approach and, among other things, puts it in a historical context. Teratologies, among other things, provide an excellent source of information on the potential of developmental processes. They also suggest which monstrosities are possible and which are not. It is interesting that specific morphologies are found in quite different species suggesting a certain common developmental process for part of their development.

7.4 Developmental Constraints, Morphogenetic Rules and the Consequences for Evolution

Variation and selection are the two basic components of an evolutionary process. Genetic mutations generate novelties in the population, while natural selection is limited by the amount of variability present although it is usually quite high. There is generally no direct correspondence between genetic and morphological divergence. This lack of correspondence suggested looking for constraints on the final phenotype in the mapping from genes to phenotypes, such as occurs in developmental processes. We should remember that genes do not specify patterns or structures; they change the construction recipe by altering the molecular structures or by regulating other genes that specify cellular behaviour. So, it is well to reiterate what we said at the beginning of the chapter, namely, that only with an understanding of developmental mechanisms can we address the central question of how genes can produce ordered anatomical structures.

There is considerable interest in the role of constraints in evolution which has led to the widespread usage of the term 'developmental constraint' (see, for example, the review by Levinton 1986). Unfortunately the concept of developmental constraint, as mentioned before, is often used loosely to describe a phenomenological pattern (Williamson 1981b). From the above discussion and application of the morphogenetic rules for generating limb cartilage architecture, we can see how it might be varied during evolution and thus give a more precise operational definition of 'developmental constraints' on morphological evolution. From this perspective we can resolve certain puzzles concerning the evolutionary homology (that is, the phenomenon of having the same phylogenetic origin but not the same final structure) of bone structures which appear to be geometrically related in the adult skeleton.

Figure 7.7 shows that the application of mitotic inhibitors to developing limbs produces a smaller limb with a regular reduction in the number of digits and the number of tarsal elements. Perturbing development in this way produced morphologies char-

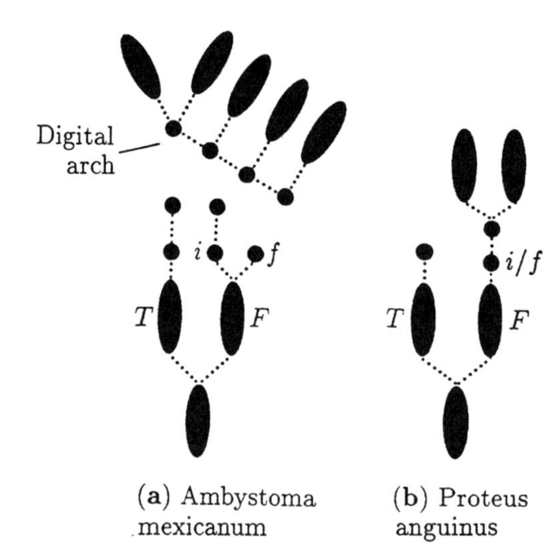

Figure 7.11. Comparison of the embryonic precartilage connections in the salamanders (**a**) *Ambystoma mexicanum* and (**b**) *Proteus anquinus*. In the former the fibula (F) branches into the fibulare (f) and intermedium (i). In *Proteus*, however, this branching event has been replaced by a segmentation of F into the single element, either the fibulare or the intermedium, although into which is neither answerable nor a proper question. (After Shubin and Alberch 1986)

acterised by the absence of certain elements. These lost elements resulted from the inability of a growing cartilage focus to undergo segmentation or branching events. The patterns produced in these experiments paralleled much of the variation of the species' limbs found in Nature.

An extreme experimental variant in the salamander (*Ambystoma*) limb (when treated with mitotic inhibitors) showed a striking resemblance to the pattern of limb evolution in the paedomorphic, or early embryological, form *Proteus* as shown in Figure 7.11. This suggests that *Proteus* and *Ambystoma* share common developmental mechanisms and hence a common set of developmental constraints. The similarity can be explained in terms of the bifurcation properties, or rather failure of aggregations to bifurcate, of the pattern formation process which restricts the morphogenetic events to the three types of spatial condensation we enumerated in our catalogue of morphogenetic rules.

It is dangerous to relate geometrically similar elements without knowledge of the underlying developmental programme, for the processes that created the elements may not correspond. For example, the loss of a digit as in Figure 7.7 may result from the failure of a branching bifurcation. It is then not sensible to ask *which* digit was lost, since the basic sequence has been altered; see also Figures 7.11 and 7.12. In the latter figure Alberch (1989) used the morphogenetic 'rules' outlined in Figure 7.5 to show how the differences can be explained.

With these examples we see that it is not easy to compare the cartilage elements themselves but rather the morphogenetic processes that created them. Thus the development of limb elements can be compared using the bifurcation patterns of pre-cartilage condensations and evolutionary changes can be resolved into iterations of condensation, branching and segmentation events.

At the beginning of this chapter we mentioned the possibility of evolution moving backward. It is clearly possible when we consider evolution of form as simply variations in mechanical (or rather mechanistic) parameters. Figure 7.12 is an unequivocal example where this has happened solely through changing the morphogenetic processes—

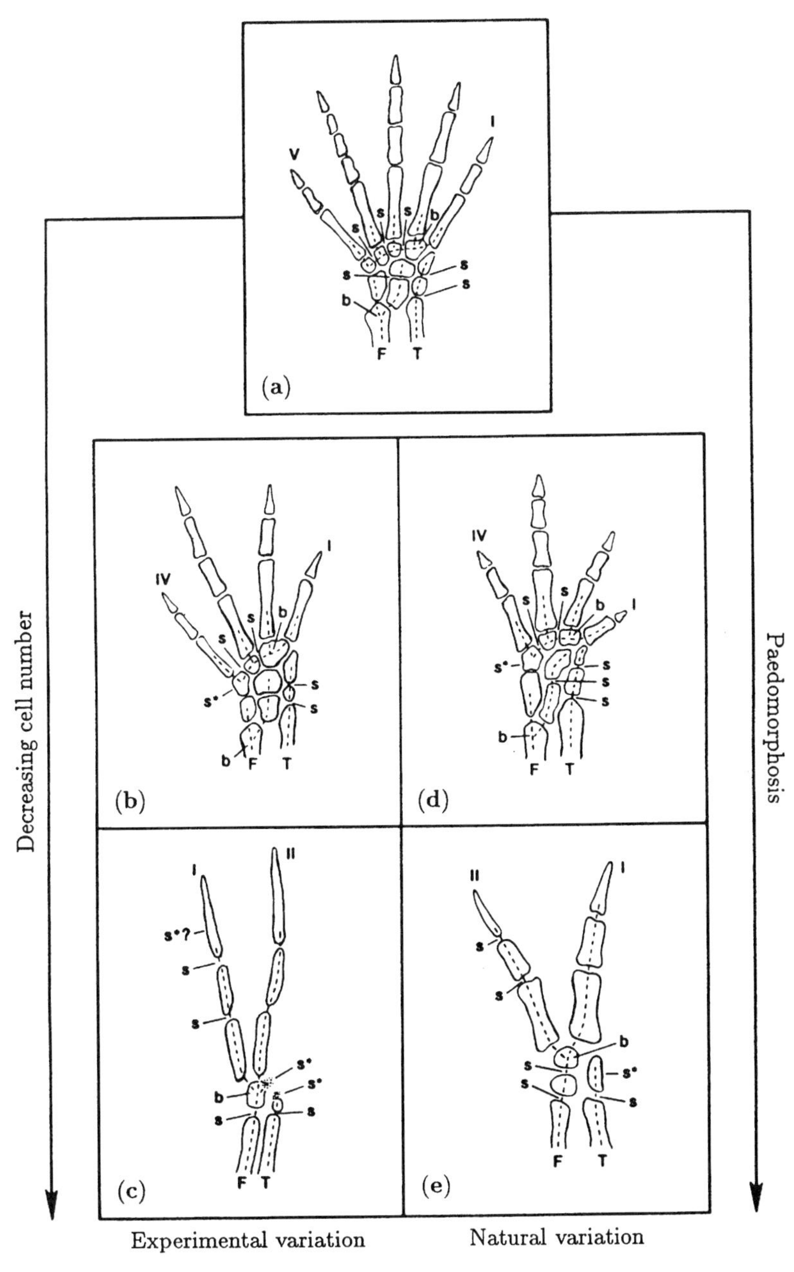

Figure 7.12. The effect of treating the foot of the salamander *Ambystoma mexicanum* with the mitotic inhibitor colchicine is to reduce the number of skeletal elements, for example, from the four-toed (**b**) to the two-toed (**c**). The effect of the colchicine is to reduce the cell number in the limb (and hence the size). The dotted lines show the type of bifurcations during development; here b and s denoting respectively Y-bifurcations and segmental bifurcations as in Figures 7.5(b) and (c) respectively. The loss of elements in the developmentally disturbed, or evolutionary derived, species is usually caused by a failure of the foci to undergo a branching bifurcation. An asterisk marks the failure of a segmentation event. (From Alberch 1988)

evolutionary change certainly need not always have the same direction. A change in environmental conditions can obviously affect mechanism parameters and hence pattern. Another example is given in an interesting paper on melanin pigmentation in birds. Price and Pavelka (1996) investigated the evolution of colour pattern from an historical, developmental and evolutionary viewpoint using the philosophy of domain restrictions and their effect on patterns. Although they base their arguments on a reaction diffusion theory they could equally well base them on the mechanical theory of pattern formation or a chemotaxis system. They carry out a comparative analysis across many bird species and show that various elements of pattern have been lost and regained during evolution. They hypothesize that a simple shift in threshold sensitivity could easily account for these shifts. They suggest that the partitioning of the different roles of selection and development could reflect how the developmental mechanism had affected immediate and distant ancestors.

Finally, we should note that the construction rules based on bifurcation sequences suggest that descriptions of limb diversity using D'Arcy Thompson's 'grid deformations' can be misleading (Thompson 1917). He showed, for example, that by superimposing a grid on a fish shape, different species could be 'derived' by simply deforming the grid structure. This gives only a phenomenological description. This kind of comparison between geometries is good only for 'topological' deformations, and precludes branching and *de novo* condensations, because these bifurcations destroy the underlying assumption of continuity upon which the method rests.

Concluding Remarks

The specific set of construction rules encapsuled in Figure 7.5 provides a scenario for the construction of the vertebrate limb. Such a scheme allows us to be more precise about what we mean by a developmental constraint. It is not suggested for a moment that this list is complete or definitive; nevertheless, such rules point to a new approach to the study of limb morphology, and suggest how the shape of the tetrapod limb has been constrained during evolution. They also show how extremely useful the study of pattern formation mechanisms can be in suggesting practical 'laws' of development.

Limbs develop initially within thin tubular boundaries. This initial domain shape defines two specific developmental constraints: (i) limb development must be largely sequential, and (ii) proximal development is initiated by a single focal condensation. Subsequent distal development must proceed from this focus by branching and segmentation using the basic elements in Figure 7.5.

The growth pattern of the limb bud frequently produces a wide distal paddle. Within the paddle region there is room for focal condensation and extensive branching and segmentation into carpal and tarsal elements. This branching and segmentation is staggered since branching inhibits colateral branching or segmentation because of the competition for cells. If the paddle region is sufficiently large, independent foci—the digital arch—can arise. Subsequent development is from anterior to posterior, that is, from the thickest part of the paddle towards the posterior 'open field.'

Given that the limb has a characteristic growing period, which limits its final size, the number of branching bifurcations is limited. This probably accounts for the fact that limbs generally have at most 5 or 6 terminal digits. As we saw in Section 6.6 in the last chapter, grafting experiments which result in duplicate limbs (cf. Figures 6.16(a) and (b)) require a much larger tissue mass to sustain the supernumerary digits.

As we learn more about the process of morphogenesis, and not only for the limb, we shall be able to add to our (still) small list of construction rules. Each of these will add further constraints on the evolution of limb morphology. We feel that the mechanistic viewpoint we have espoused here provides a more concrete definition of 'developmental constraints' in evolution.

8. A Mechanical Theory of Vascular Network Formation

8.1 Biological Background and Motivation

Vasculogenesis, *in vivo*, is the formation of (major) blood vessels by cells (endothelial cells and angioblasts) while angiogenesis refers to the development of vascular structures which sprout out from existing vessels. The word angiogenesis was first used in 1935 to describe the formation of new blood vessels in the placenta. The motivation for the modelling in this chapter is to try and determine the key elements in the underlying mechanism involved in creating such patterned structures. Since *in vivo* studies are prone to a variety of sensitivity problems much of the experimental work in this area has been on *in vitro* (biological) model systems which avoid many of the experimental difficulties with *in vivo* systems. The development of *in vitro* angiogenesis (biological) models provides a controlled means for studying blood vessel formation (Folkman and Haudenschild 1980). The reasonable assumption is that, if the *in vitro* studies replicate the type of patterns observed *in vivo*, then these models provide information on the pattern formation mechanism which operates *in vivo*. The essentially planar network patterns we study in this chapter are more akin to vasculogenesis than angiogenesis which are distinctly three-dimensional.

Angiogenesis, and vessel formation in general, is of fundamental importance in wound healing, sustaining tumour growth, morphogenesis and so on. Vascularisation is necessary for the growth of solid cancer tumours. Folkman (1972), in what is now considered as possibly one of the most important papers in the solid tumour cancer field, hypothesized that if it were possible to inhibit neovascularization it might stop the growth of the tumour or at least contain its growth to a dormant mass of around 2 to 3 mm in diameter; see also Folkman's (1976) general article on vascularisation of tumours. He speculated that such antiangiogenesis could be the basis for a new form of cancer therapy. He presented some convincing evidence that the tumour produces a chemical, a tumour-angiogenesis factor, which induces neovascularisation. His work was almost totally ignored (even ridiculed) until the late 1990's when anti-angiogenesis factors, such as endostatin, an endogenous inhibitor of angiogenesis and tumor growth, and angiostatin were found (O'Reilly et al. 1997).[1] Even thalidomide is having a come-

[1]When the news broke in the world press in 1997 it was heralded as the holy grail of cancer therapy with numerous people making ever wilder predictions. After most of the rushed interviews and media hyperbole had subsided Folkman made the pertinent comment that 'If you are a mouse with cancer we can take good care of you.'

back: it has been shown to inhibit angiogenesis (D'Amato et al. 1994). Folkman (1972) concluded with a series of highly pertinent medical questions which suggest a variety of modelling challenges still of relevance, or rather of even more relevance in the light of recent findings.[2] Folkman and Klagsbrun (1987) described some angiogenic factors that had been found in tumours. A particularly important aspect, from a cancer therapy point of view, is that antiangiogenic therapy does not induce acquired drug resistance in experimental cancer (Boehm et al. 1997) unlike chemotherapy which does induce drug resistance (see Chapter 11). Folkman's (1995) review article discusses some clinical applications of research on angiogenesis. The general article by Kerbel (1997) gives an overview of the newer developments while Sage (1997a) discusses several protein regulators of angiogenesis in particular, and in Sage (1997b) describes results on the role of the protein SPARC which suppresses tumourigenicity of human melanoma cancer cells. The field of anti-angiogenesis is now fast growing with an increasing number of areas where modelling could be of some considerable value.

After reading some of Folkman's early work in the early 1990's we initiated research into trying to model the biological processes involved in the network pattern observed in angiogenesis. Even without the cancer connection, it is, as we have noted, an important and challenging question. In this chapter we describe the application of the mechanochemical (although it is strictly just mechanical) theory to the patterning process. Although motivated by Folkman's ideas from the 1970's this application is directly related to experiments carried out by Sage and Vernon and their coworkers (Vernon et al. 1995). These experiments were also used for comparison between the theory and experiment (see below) and were the basis for most of the parameter estimates which are essential in any practical application of a model to a specific biological problem.

8.2 Cell–Extracellular Matrix Interactions for Vasculogenesis

The study of *in vitro* angiogenesis systems has shown the important mechanical role that the extracellular matrix (ECM) plays in vasculogenesis. It provides, among other things, a scaffold necessary for cell migration, cell spreading, a process which is important in controlling the cell cycle (Bray 1992), and morphogenesis (for example, Vernon 1992). Cells can not only produce and degrade their ECM but also alter its structure by applying mechanical forces. Through matrix production and degradation cells can influence the mechanical properties of their ECM and through their mechanical forces they can reorganise its fibrous components into matrix lines that the cells use as migratory pathways and movement cues (see also the full discussion in Chapter 10). Such a mechanical scenario in which cell–ECM interactions can be orchestrated to form complex spatial patterns in development makes it a serious candidate for the application of the Murray–Oster mechanical theory (see Chapter 6). The ability of the cell–matrix interaction to effect an alignment of the matrix thus influencing cell movement is quite common in development; some discussion of this is given in Manoussaki (1996), Manoussaki et al. (1996) and Murray et al. (1998). A key reference is the book of experimental and the-

[2]The papers I have read by Judah Folkman have been paragons of clarity with a plethora of original ideas. I strongly recommend that anyone thinking of working in this general area should read his articles, not only for the many modelling avenues they open up.

oretical articles edited by Little et al. (1998) on vascular morphogenesis (appropriately with an introduction by Folkman).

Mechanical and fluid mechanical forces play important roles in the overall development of vasculature. As early as 1893 Thoma suggested the importance of (fluid) mechanical factors in the growth of blood vessels during development. In a review of mechanical forces on angiogenesis, Hudlická and Brown (1993) describe Thoma's observations of how vascular sprouting in the growing embryo might occur as a result of a combination of velocity of blood and pressure. Since then many studies have highlighted the role mechanical forces play in the development of vasculature: they may also be involved in the early branching in the developing lung (Lubkin and Murray 1995).

The study of how mechanical forces influence the development of vasculature has been mainly organized along two main lines. One is focused on describing the macroscopic vessel remodelling where the term 'vessel remodelling' has been used to describe changes in vessel wall structure as well as changes in the vascular tree generated by capillary generation and regression. For example, changes in blood pressure can induce changes in the thickness of various vascular structures and elevated pressure in pulmonary arteries during hypertension results in the thickening of, for example, the intima (the innermost lining of membrane of an organ) and the external tissue layer of blood vessels (Fung and Liu 1991). Changes in the vessel wall stress could be important in remodelling of the vascular plexus (Price and Skalak 1994) and vascular shear stresses or blood flow patterns have been suggested as one of the causes of vascular network remodelling via growth in surface volume by intercalation of material already present (intussusception) (Patan et al. 1996).

The second approach concentrated on elucidating the mechanism by which mechanical forces applied to vascular cells alter gene expression (Ando and Kamiya 1996) and the molecular effects on the signalling pathway. Shear stress resulting from blood flow has also been shown to change cell shape and affect the cell's cytoskeletal organization, which, in turn, can influence the cell cycle (Ingber et al. 1995). Shear stress will stimulate endothelial cell DNA synthesis (Ando et al. 1990) and the migration (as we shall see) and proliferation of endothelial cells (Ando et al. 1987).

Reviews on some of the roles of mechanical forces in vascular development and remodelling are given, for example, by Hudlická and Brown (1993) and Skalak and Price (1996).

Although the approach is quite different, the work of Chaplain and his coworkers on capillary networks is related to the material discussed in this chapter; see Chaplain and Anderson (1999) for a review, details of earlier work and other references to theoretical work on angiogenesis. They are primarily concerned with tumour-induced angiogenesis. Their continuum models involve cells responding to chemotaxis and haptotaxis (see Chapter 6) with the haptotactic structure coming from fibronectin which is secreted by the cells. Their models are closely linked to experimental work in this specific area, a comprehensive review of which is given by Schor et al. (1999) with particular attention as to how the experimental data relate to theoretical models.

Here we describe a mechanical model mechanism based on the Murray–Oster mechanochemical modelling framework of Chapter 6 which was developed by Manoussaki (1996), Manoussaki et al. (1996) and Murray et al. (1998) which attempts to capture the key interactions between the mechanical forces generated by the cells and their extracellular matrix. We show that a purely mechanical version of the theory could be responsible for the observed pattern and how they are actually formed in development. Such a mechanical model is in effect based on very simple mechanical concepts and does not specify the type of cells and matrix involved but rather only considers possible mechanical interactions between the various components. Basically cells adhere to the matrix substratum and as a consequence of their traction forces they deform the matrix and the cells adhering to it. Aggregates of cells and matrix form together. Due to the heterogeneity in matrix and cell densities, cell traction tension lines form between the clusters. These tension lines correspond to aligned matrix fibers along which cells actively move thereby defining cellular highways between the clusters.

The recent paper by Tranqui and Tracqui (2000) is particularly relevant to the work in this chapter, but it is more especially oriented towards angiogenesis. They also use a mechanical model, based on their *in vitro* experiments, for generating capillary-like structures. They include cell isotropic diffusion and haptotaxis in the equation for cell conservation and and long range elasticity effects (see Chapter 6). They were able to obtain estimates for all of the model parameters involved. What is particularly interesting is that they show that the analytically-derived critical parameters for pattern evolution correspond to threshold values of the experimental variables. They obtain, by numerically simulating the one-dimensional version of their model, spatial patterns in the fibrin gel and cell density which are compared with those obtained experimentally.

Experimental Model of In Vitro Vascular Network Formation

Experiments, *in vitro*, by Vernon, Sage and their coworkers (Vernon et al. 1992, 1995) show that various endothelial cells (specifically bovine aortic endothelial cells, BAEC), adult human dermal fibroblasts and human smooth muscle cells when cultured on gelled basement membrane (Matrigel) reorganize the matrix and form networks. Moreover, cells cultured on different substrates, for example, type I collagen gels, also form networks provided the matrix is malleable enough. Their studies, and those of Tranqui and Tracqui (2000) suggest that the patterning mechanism lacks cell and matrix specificity and that a mechanical mechanism could provide a possible explanation of the generality of the network forming process.

The cells were plated on $60\,\mu$m–$600\,\mu$m thick layers of Matrigel. Matrigel is a matrix formed with collagen fibres in a sheet-like mesh rich in laminin which helps the cells adhere to it. The process for all cell lines studied is similar: cells adhere to the matrix and start pulling on it via their finger-like filopodia. The pulling results in movement of the matrix and the cells that have adhered to it, eventually forming cell aggregates with significant amounts of matrix accumulated underneath the aggregates. As a result cell traction tension lines appear around the clusters. The traction forces that fibroblasts exert on thin silicone rubber, and shown in Figure 6.3 in Chapter 6, generate very clear tension lines (Harris et al. 1980). In time, the matrix components appear to form fibrous cords of aligned matrix between neighboring clusters and along the tension lines. Once the lines form, the cells in contact with them become elongated and bipolar, oriented parallel to the lines, become actively motile and migrate along these matrix pathways, thus forming cellular cords. Eventually, about 24 hours after plating, the culture plane is tessellated with polygons with sides defined by the cellular cords (Vernon et al. 1992). With time some of these polygons enlarge and some close in a sphincter-like manner thereby forming larger polygons. This continuous remodelling is accompanied by corresponding distortions in the Matrigel thickness. Figure 8.1 shows a typical experimentally obtained spatiotemporal evolution of a network of cords with Figure 8.1(i) a larger scale photograph of such a network. Elongation and migration only occurred on the fibre tracks and did not start until the tracks appeared. Also cells which were not in contact with the pronounced fibre tracks never elongated but remained more or less fixed in position. The pattern evolved continuously with some fibre tracks pinching off and others elongating and rotating with the result that some polygons grew while others decreased or disappeared. After a full 24 hours of culture most of the Matrigel had been pulled into aggregates with the remainder comprised of cords connecting the aggregates.

Collagen (an ingredient of tissue matrix) content influences the formation of networks, presumably by altering matrix stiffness and thereby the effect of the cell-exerted traction onto the matrix. Thickness of the matrix layer also influences the size of the networks formed: thin layers of matrix result in small or no network formation. On a ramp of increasing matrix thickness, larger cellular networks form on the thicker areas of matrix. We come back to these experimental results below when we compare experiment with the model predictions.

These results suggest that mechanical interactions are of primary importance for the development of pattern. We quantify the interactions in the mathematical model below. As we shall see, the solutions of the equations for the proposed mechanism confirm the crucial role that mechanical forces play in the network pattern formation. The mathematical model thus provides a tool for assessing which mechanical properties of the ECM control the patterning process as well as how the parameters may modulate pattern size. It also provides various experimental scenarios to highlight the effect of experimentally variable parameters, such as matrix gel thickness, density of cells and so on. Because the thickness of the experimental gel on which the cells are embedded is very thin, and in particular relative to the cell–matrix spatial domain, a two-dimensional model is sufficient for comparison with the *in vitro* experiments.

Model Mechanical Mechanism for Vascular Network Formation

The two-dimensional mathematical model quantifies the basic experimental mechanical scenario described above. We denote the average local cell density by n (cells mm^{-2}). We do not include cell proliferation; on the Matrigel cultures the first cell aggregates and lines of tension appear after 4 hours and the networks are complete within about 24 hours. The time between two subsequent cell mitoses, however, is about 17 hours for endothelial cells so we assume that no significant changes in the cell populations occur which could influence the pattern forming process. It is easy to include cell and matrix proliferation and such studies are currently under investigation. With the experience from other chapters on the application of the mechanical mechanism for pattern formation, particularly the chapter on full depth wound healing, Chapter 10, we can anticipate the general form the model will take. In the following we follow the work of Manoussaki (1996) and Murray et al. (1998).

We consider here only the small strain approximation which although not really the situation which obtains in the experiments nevertheless should give a clear indication as to whether or not these are the processes involved in the pattern generation.

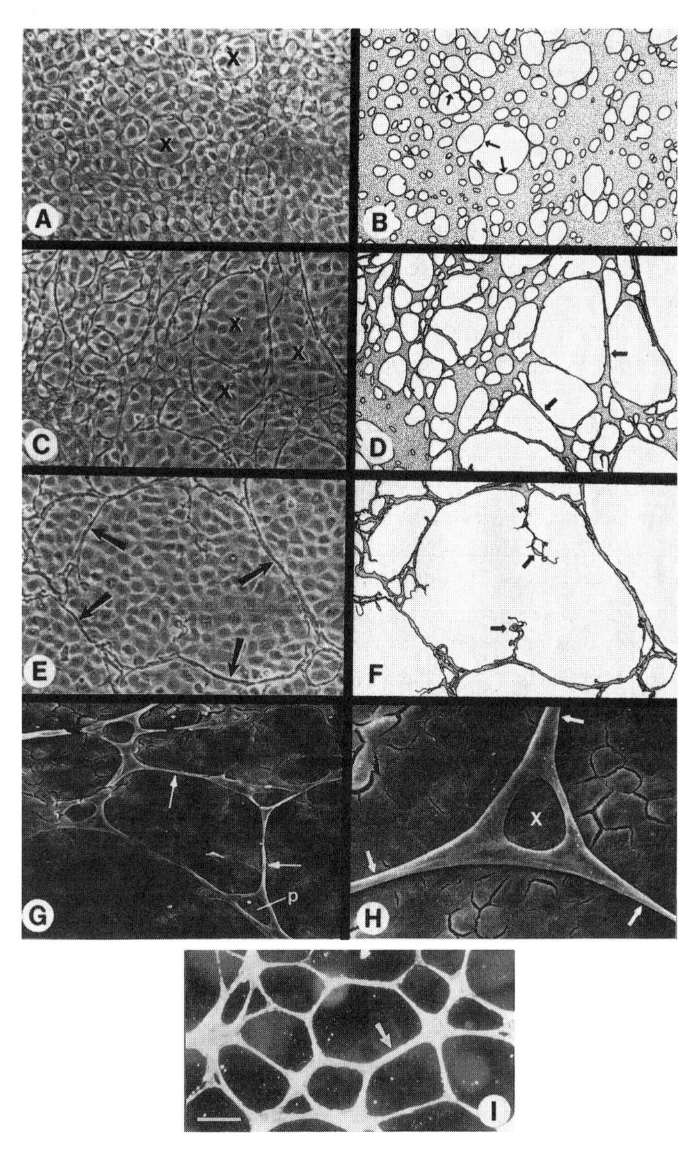

Figure 8.1. (**a**)–(**h**) Evolution of a planar network by bovine aortic endothelial cells (BAEC) cultured on a basement membrane of the artifical matrix (BMM, Matrigel). The arrows indicate cords while the $\times$ in (**a**), (**c**) and (**h**) correspond. (Photograph courtesy of Dr. R. B. Vernon and reproduced with permission) (**i**) Cells cultured for 48 hours in a subconfluent monolayer on top of a layer of Matrigel clearly displaying a typical network. Each cord (an arrow marks an example) comprises many cells. Bar $= 200\,\mu$m. Image is viewed by darkfield illumination. (Photograph courtesy of Dr. Charles Little and reproduced with permission)

We assume local changes in the cell density are a combination of mainly two movements, the convective flux and an anisotropic strain-dependent random motion (diffusion) tensor. We take the convective flux $\mathbf{J}_{\text{convection}} = n\mathbf{v}$ where $\mathbf{v}(x, y, t)$ denotes the matrix velocity at the point (x, y) at time t. Here, under the small strain approximation, $\mathbf{v} = \mathbf{u}_t$, where $\mathbf{u}$ is the vector displacement of the matrix. (See the discussion in the previous chapter on the form of the convective velocity in the situation where the strains are finite.)

We model the flux due to cellular locomotion by an anisotropic strain-dependent random motion biased along the areas of matrix alignment. We take $D(\epsilon)$ to be the random motion diffusion tensor, which is dependent on the matrix strain $\boldsymbol{\epsilon} = (1/2)(\nabla\mathbf{u} + \nabla\mathbf{u}^T)$. The flux due to the active movement is therefore taken as

$$\mathbf{J}_{\text{diffusion}} = -\nabla(\mathbf{D}(\boldsymbol{\epsilon})n) = -n\nabla\mathbf{D}(\boldsymbol{\epsilon}) - \mathbf{D}(\boldsymbol{\epsilon})\nabla n.$$

The form of $\mathbf{D}(\epsilon)$ depends on the specific assumption about how cells perform their random movements in a strained field. Its specific form for both finite and small strains was originally derived by Cook (1995) under the assumption that the movement bias increases in expansion in one direction and/or compression in the perpendicular direction. We derive the various forms in the chapter on dermal wound healing, specifically Section 10.8. For small strains $\mathbf{D}(\boldsymbol{\epsilon})$ is given by (8.2) there.

The conservation equation for the cell density is now given by

$$\underbrace{\frac{\partial n}{\partial t}}_{\substack{\text{rate of change}\\ \text{in cell density}}} = \underbrace{-\nabla\cdot\left(n\frac{\partial \mathbf{u}}{\partial t}\right)}_{\text{convection}} \underbrace{+\nabla\cdot\nabla\cdot(\mathbf{D}(\boldsymbol{\epsilon})n)}_{\substack{\text{strain-biased}\\ \text{random motion}}}, \tag{8.1}$$

where, from Chapter 10, (10.54), in the (x, y) plane,

$$\mathbf{D}(\boldsymbol{\epsilon}) = D_0 \begin{pmatrix} 1 + \dfrac{\epsilon_{11} - \epsilon_{22}}{2} & \dfrac{\epsilon_{12}}{2} \\ \dfrac{\epsilon_{21}}{2} & 1 + \dfrac{\epsilon_{22} - \epsilon_{11}}{2} \end{pmatrix}. \tag{8.2}$$

Here $\epsilon_{11}, \epsilon_{22}, \epsilon_{12}, \epsilon_{21}$ are the components of the strain tensor, $\boldsymbol{\epsilon}$, and D_0 is the motility coefficient when no strain is present. The basic idea behind the derivation of this diffusion matrix is that in the random walk of the cell it tends to move in fibre space in a direction tangential to the fibre it starts from. It is somewhat more complicated than this; refer to the discussion in Section 10.8.

Force Balance Equation

Since we consider only small strains, we model the matrix as a linear viscoelastic material whose properties remain constant during the initial stages of the patterning process. We follow the continuum formulation used elsewhere in this book with a few extra reasonable assumptions. As said above, the matrix domain is very much larger than the thickness of the matrix; we approximate the matrix as a two-dimensional material. We also assume that the traction exerted by the cells remains confined in the plane parallel to the dish (or, approximately, the matrix) surface; in other words we consider the material to be under a *plane stress* assumption. The movement of the matrix is considerably resisted by the attachment of the matrix to the dish. *In vivo* angiogenesis is, by comparison, influenced by the potential attachment and continuation of the fibrous and cellular components of the adjacent tissue.

The forces present in the tissue are: (i) the cell-exerted traction, (ii) the resistance due to the matrix–dish contact, and (iii) the viscoelastic forces of the matrix material which are resisting the deformation caused by the cells. During the pattern formation process inertial effects are negligible compared to the other forces so the forces at any point are in equilibrium and we have

$$\underset{\substack{\text{viscoelastic}\\ \text{restoring forces}}}{\mathbf{F}_{\text{matrix}}} + \underset{\substack{\text{cell-generated}\\ \text{traction}}}{\mathbf{F}_{\text{cells}}} + \underset{\substack{\text{attachment}\\ \text{on dish}}}{\mathbf{F}_{\text{anchoring}}} = 0. \tag{8.3}$$

The cell forces and the matrix response are described via the divergence of the corresponding stress tensors within the matrix medium with $F_{\text{cells}} = \nabla \cdot \sigma_{\text{cells}}$ with σ_{cells} the stress tensor in the matrix due to the cells and $F_{\text{matrix}} = \nabla \cdot \sigma_{\text{matrix}}$ with σ_{matrix} the viscoelastic stress tensor in the matrix material. $F_{\text{anchoring}}$ is the external (body) force resisting the matrix displacement.

The cell-exerted traction F_{cells} depends on the local cell density and various other factors which we have discussed at various places in the book, particularly in Chapter 6 and Chapter 10 on wound healing. For the purposes of this chapter it suffices to take a reasonable qualitative form for the cell-generated stress on the matrix, namely,

$$\boldsymbol{\sigma}_{\text{cells}} = \tau \frac{n}{1 + \alpha n^2} \mathbf{I}, \tag{8.4}$$

where τ (dynes/cell) represents the stress that one cell exerts on the matrix at low cell densities and α is the parameter which quantifies the effect of neighboring cells on the traction force; see also Chapter 6 and the discussion there on the various functional forms and their relation to experimental data. For densities with which we are concerned the stress tensor is approximately equal to $\tau n\mathbf{I}$. The parameter α quantifies the decrease in cell traction for large cell densities and reflects an upper limit on the total traction forces possible; it also prevents unrealistically large forces.

Although we use (8.4) it should be reiterated that it is only a plausible qualitative description. Cell traction, as mentioned before, has proved difficult to measure because of the complexity of the cell–matrix interactions and the difficulty of separating out the various mechanical effects in real biological tissue. The seminal paper by Ferrenq et al. (1997) describes a powerful experimental technique and general approach for quantifying the forces generated by endothelial cells on extracellular matrix; see the description of their technique in Chapter 6.

Since we consider small strains we take the viscoelastic stresses, σ_{matrix}, arising in the matrix in response to the cell traction to be the linear Voigt form used before in previous chapters (see also Fung 1993) in which the viscous and elastic stresses are added linearly. We thus have

$$\boldsymbol{\sigma}_{\text{matrix}} + \boldsymbol{\sigma}_{\text{viscous}} + \boldsymbol{\sigma}_{\text{elastic}} = \mu_1 \boldsymbol{\epsilon}_t + \mu_2 \theta_t \mathbf{I} + E' \left(\boldsymbol{\epsilon} + \nu' \theta \mathbf{I} \right), \tag{8.5}$$

where parameters μ_1, μ_2 (dynes sec^{-1}) are the bulk and shear viscosities of the matrix, $\theta = \nabla \cdot \mathbf{u}$ is the dilation of the matrix along the plane of the dish, $E' = E/(1+\nu)$ and $\nu' = \nu/(1-2\nu)$. E is the Young's modulus, which quantifies the stiffness of the material and ν (mm/mm) is the Poisson ratio which measures how much a strip of gel will contract in one direction when it is stretched by a unit length in the transverse direction.

Now consider the attachment of the matrix to the dish. Experiments indicate that some matrix fibrils remain attached to the dish, while the rest are dragged across the lower parts of the matrix. The net effect on the surface of the matrix of the attachment of the fibrils on the dish, is a resistance, which, in a two-dimensional approximation of no-slip in a three-dimensional gel of local thickness, ρ, we model as a viscous drag represented by

$$F_{\text{anchoring}} = -\frac{s}{\rho} \mathbf{u}_t. \tag{8.6}$$

Here $\rho(x, y, t)$ (mm) represents the thickness of the matrix and s is a measure of the strength of the resistance of the attachment. We assume that the more matrix layers there are (larger ρ) between the cells and the dish the less the adhesion of the matrix onto the dish will resist matrix reorganization above it. If the matrix thickness were large, as in a nonplanar *in vitro* or in an *in vivo* situation the model would be three-dimensional and this external force would be zero in most situations.

Matrix Thickness

Since the vertical stress $\sigma_{zz} = 0$ in this two-dimensional situation, we get a relation between the strain components in the three directions, $\epsilon_{xx}, \epsilon_{yy}, \epsilon_{zz}$. From Hooke's law in three dimensions we have

$$0 = \sigma_{zz} = \frac{E}{1+\nu} \left[\epsilon_{zz} + \frac{\nu}{1-2\nu} \theta \right] \Rightarrow \epsilon_{zz} = -\frac{\nu}{1-2\nu} \theta.$$

When there is a strain ϵ_{zz} in the z-direction, the thickness is then calculated using $\rho(x, y, t) = \rho_0(x, y)(1 + \epsilon_{zz})$ from which we derive the relation for the thickness, ρ, as a function of space and time, namely,

$$\rho(x, y, t) = \rho_0(x, y)\,(1 + \epsilon_{zz}) = \rho_0 \left(1 - \frac{\nu}{1-2\nu} \theta \right), \tag{8.7}$$

where θ is the dilation.

Boundary Conditions for the Model Equations

Experiments have shown that the matrix barely moves along the edge of the dish. This we incorporate into the model simulation by assuming zero displacement as the boundary condition. If the domain is denoted by B this gives $\mathbf{u}(x, y, t) = 0$ for (x, y) on ∂B. So, for a square domain (which we used rather than a circular one simply for ease of numerical programming and reduction of computer time) with sides a and b the condition becomes

$$\mathbf{u}(x, y = a, t) = \mathbf{u}(x, y = b, t) = \mathbf{u}(x = a, y, t) = \mathbf{u}(x = b, y, t) = 0, \tag{8.8}$$

where $\mathbf{u}(x, y, t)$ denotes displacement at position $\mathbf{x} = (x, y)$ at time t. The cells, of course, remain within the dish which gives a zero flux boundary condition for the cells, which here is

$$\mathbf{J}_{\text{cells}} = n\mathbf{v} - \nabla \cdot (\mathbf{D}(\boldsymbol{\epsilon})n) = 0, \quad \mathbf{x} \text{ on } \partial B, \tag{8.9}$$

where, as above, the velocity $\mathbf{v} = \mathbf{u}_t$ in the approximation we are using here and $\mathbf{D}(\boldsymbol{\epsilon})$ is given by (8.2).

The cell conservation equation (8.1) and force balance equation (8.3) with the various expressions given by (8.2) and (8.4)–(8.7) constitute the model mechanism. The equations have to be solved with boundary conditions (8.8) and (8.9) with initial conditions in cell density and matrix distribution in keeping with the experimental arrangement.

We first nondimensionalise the equations by introducing dimensionless variables

$$\begin{gathered}
n^* = \frac{n}{n_0}, \quad \rho^* = \frac{\rho}{\rho_0}, \quad \mathbf{u}^* = \frac{\mathbf{u}}{L}, \quad \mathbf{r}^* = \frac{\mathbf{r}}{L}, \quad t^* = \frac{t}{T}, \quad s^* = \frac{s(1+\nu)L^2}{ET\rho_0}, \\
\alpha^* = \alpha {n_0}^2, \quad \tau^* = \frac{\tau n_0(1+\nu)}{E}, \quad \nu^* = \frac{\nu}{1-2\nu}, \quad \boldsymbol{\epsilon}^* = \boldsymbol{\epsilon}, \quad \theta^* = \theta, \\
D^* = \frac{DT}{L^2}, \quad D_0^* = \frac{D_0 T}{L^2}, \quad {\mu_i}^* = \frac{\mu_i(1+\nu)}{ET}, \quad i = 1, 2,
\end{gathered} \tag{8.10}$$

where L and T are typical length and timescales (L, for example, could be a dimension of the dish), n_0 is the seeding initial density of cells and ρ_0 is the initial thickness of the uniform matrix layer. With these the model system becomes

$$\begin{gathered}
\frac{\partial n}{\partial t} + \nabla \cdot (n\mathbf{u}_t) = \nabla \cdot \nabla \cdot (\mathbf{D}(\boldsymbol{\epsilon})n), \\
\nabla \cdot \left[\mu_1 \boldsymbol{\epsilon}_t + \mu_2 \theta_t \mathbf{I} + \boldsymbol{\epsilon} + \nu\theta\mathbf{I} + \frac{\tau n \mathbf{I}}{1+\alpha n^2}\right] = \frac{s}{\rho}\mathbf{u}_t, \\
\rho(x, y, t) = 1 - \nu\theta,
\end{gathered} \tag{8.11}$$

where for convenience we have omitted the asterisk superscript. The above boundary conditions in nondimensional form are the same with the dimensions now scaled to L.

8.3 Parameter Values

The various parameters describing cell behaviour and the matrix properties depend in varying degree on the stress and strain (and therefore fibre orientation) of the matrix.

This is a property of biomaterials which makes parameter estimation particularly difficult. We shall see in the chapter on full depth wound healing how complex these dependences can be. However, on the assumption that anisotropy does not greatly affect the parameter values, we are able to derive some estimates from the literature for most of the parameters describing cell behavior and matrix properties such as matrix stiffness (Young's modulus), the Poisson ratio and the cellular random motility coefficient; these are given in Table 8.1. Some characterization of the material properties can be achieved by using appropriate creep and relaxation experiments for the matrix (Fung 1993). Rheological properties of collagen I gels (Barocas et al. 1995) and tissue-derived gels such as the vitreous body (Lee et al. 1993, 1994a,b) have been described by applying slow-strain techniques described in these papers. We use such results where they refer to tissues of approximately similar structure to the matrices used in the experiments of Vernon et al. (1992); there are no studies on the mechanical properties of Matrigel as such.

The *stiffness* of the matrix is represented by the elastic modulus of the material (E dynes cm^{-2}) and is determined primarily by the type, amount and organization of its fibrous components. For example, the anterior porcine vitreous body and the posterior bovine vitreous body were found to have similar collagen contents and gave elastic moduli of 26.93 dyne cm^{-2} and 8.01 dyne cm^{-2} respectively (Lee et al. 1994a,b). Since the collagen gels used in the vascular network-forming experiments have collagen content comparable to that of the vitreous bodies mentioned, we assume that these gels' elastic moduli are within a similar range.

The *shear viscosity*, μ_1, of 2.1 mg ml^{-1} collagen I gel has been measured using creep tests to be 7.4×10^6 Poise (Barocas et al. 1995) while that of the posterior bovine vitreous body was estimated to be 2.5×10^2 Poise (Lee et al. 1994). It is possible the disparity in the values is due to the different composition of the two gels with the former having a collagen content about 14 times larger. With the model it is possible, of course, to predict the effect of varying viscosities on the network pattern formed and this has been done (Manoussaki 1996, Murray et al. 1998). With regard to the bulk viscosity, we expect it to be considerably larger than the shear viscosity.

Traction per cell, τ dynes cell^{-1}, we consider to be comparable to that of human umbilical vein endothelial cells which is approximately 6.1×10^4 dynes cell^{-2} for a

Table 8.1. Parameter estimates obtained from the literature and experiments.

Parameter	Range of Values
Traction per cell, τ	0.03–0.7 dyne cell^{-1}
Young's modulus, E	8–27 dyne cm^{-2}
Poisson ratio, ν	0.2 (cm cm^{-1})
Shear viscosity, μ_1	2.5×10^2–7×10^7 dyne sec cm^{-2}
Shear viscosity, μ_2	10^5–10^9 dyne sec cm^{-2}
Random motility, D_0	4–9×10^{-9} cm^2 sec^{-1}
Average cell density, n_0	4×10^4–2×10^5 cells cm^{-2}
Anchoring parameter, s	10^6–10^{11} dyne sec cm^{-3}

confluent monolayer of cells (Kolodney and Wysolmerski 1992). From the figures given at cell densities of $2.25 - 4 \times 10^5$ cell cm^{-2} we estimate the traction force per cell to be in the region of $\tau \approx 0.15 - 0.27$ dynes cell^{-1}. Harris et al. (1980) estimated cell traction to be at least 0.03 dynes cell^{-1}.

The *Poisson ratio* (ν cm cm^{-1}), a dimensionless parameter, can, in theory, be measured by extending a material by a certain amount and by measuring its compression in the perpendicular direction. The ratio of relative extension to compression gives the magnitude of the parameter. In practice, such measurements for soft materials, such as collagen gels, can be very difficult, because the value of ν changes in time as the fibrous component separates from the water component of the gel: the two phases, liquid and solid, will have different Poisson ratios. The Poisson ratio for the fibre network of such gels, which is draining of liquid, is estimated at $\nu \approx 0.2$ (Scherer et al. 1991).

Cell motility for endothelial cells has been calculated using migration assays (Hoying and Williams 1996). The values reported for endothelial (human microvessel endothelial cells) cells on gelatin are $9.5 \pm 1.2 \times 10^{-9}$ cm^2 sec^{-1} in the presence of endothelial cell growth factor and $2.6 \pm 0.6 \times 10^{-9}$ cm^2 sec^{-1} in its absence and $19.3 \pm 4.22 \times 10^{-9}$ cm^2 sec^{-1} on fibronectin.

The *anchoring parameter*, s, was introduced in the modelling and no data are available on it. We estimate it by taking $(s/\rho)\mathbf{u}_t$ of the same order as the shear viscosity term when the matrix thickness is large, of the order of 1 cm. Again, simulations can quantify the effect of varying this parameter.

In our study, we are concerned with the initial stages of network pattern emergence and can assume that the parameters remain approximately constant.

How the Pattern Formation Mechanism Works

Let us suppose the gel and the cells are initially uniformly distributed and we introduce a small random variation in the cell density. It is a classical local activation and lateral inhibition scenario. The traction generated by the cells increases with cell density and so the traction exerted on the matrix is greater in regions of higher cell density. So the cells in these regions pull more matrix around them from neighbouring areas which have a lower cell density. As the matrix accumulates in these areas it carries the cells that have adhered to it. So, if the traction forces are large enough to overcome the elastic restoring force of the matrix, regions of higher cell density form thereby generating a spatial heterogeneity in cell and matrix densities. Inter-aggregate tension causes the fibres to reorient in directions stretching between the cell aggregates. Cells that are in contact with these regions of aligned fibres exhibit enhanced movement along the direction of alignment and the areas of highly aligned fibres fill with cells.

8.4 Analysis of the Model Equations

With the above patterning scenario and experience from several other chapters we can clearly carry out a linear, and probably a nonlinear, stability analysis on the model equations about the relevant steady state, which here is

$$n_s = 1, \; \rho_s = 1, \; \mathbf{u}_s = 0. \tag{8.12}$$

We only carry out a linear stability analysis here to determine parameter domains for pattern formation, to highlight the key elements necessary for the system to generate spatial patterns and to get some idea of the patterning potential of the mechanism.

We linearize (8.11) about the uniform steady state (8.12) by setting $n = 1 + \bar{n}$, $\rho = 1 + \bar{\rho}$, $\mathbf{u} = \bar{\mathbf{u}}$ and, omitting the bars for convenience, obtain the linearized system

$$
\begin{aligned}
n_t + \nabla \cdot \mathbf{u}_t &= D_0 \nabla^2 \left(n + \tfrac{1}{2} \nabla \cdot \mathbf{u} \right), \\
\nabla \cdot [\mu_1 \boldsymbol{\epsilon}_t + \mu_2 \theta_t \mathbf{I} + \boldsymbol{\epsilon} + \nu \theta \mathbf{I} + \tau_1 n] &= s \mathbf{u}_t, \\
\rho &= -\nu \theta,
\end{aligned} \tag{8.13}
$$

where n, ρ, $\mathbf{u}$ now denote the small perturbations from the steady state $(1, 1, 0)$ and

$$
\tau_1 = \tau \frac{(1 - \alpha)}{(1 + \alpha)^2}.
$$

In the usual way, we now look for solutions in the form

$$
(n, \rho, \mathbf{u}) \propto e^{\sigma t + i \mathbf{k} \cdot \mathbf{x}}, \tag{8.14}
$$

where σ and $\mathbf{k}$ are respectively the growth rate and the wave vector. After some messy algebra we obtain the Jacobian matrix (here 4×4 because of the two equations for the components of $\mathbf{u}$) from which we get the characteristic equation for σ given by

$$
\sigma \left[(\mu_1 k^2 + 2s)\sigma + k^2 \right] \left[(\mu k^2 + s)\sigma^2 + b(k^2)\sigma + c(k^2) \right] = 0, \tag{8.15}
$$

where $\mu = \mu_1 + \mu_2$, $k^2 = |\, \mathbf{k} \,|^2$ and

$$
b(k^2) = \mu D k^4 + (sD + 1 + \nu - \tau_1)k^2, \quad c(k^2) = Dk^4(1 + \nu - \tfrac{1}{2}\tau_1). \tag{8.16}
$$

So that we do not lose sight of what the parameters represent, in dimensional terms the parameters in (8.15) and (8.16) are given by

$$
\begin{aligned}
s &= \frac{s(1 + \nu)L^2}{ET\rho_0}, \quad \tau_1 = \frac{\tau n_0 (1 + \nu)(1 - \alpha n_0^2)}{E(1 + \alpha n_0^2)^2}, \\
\mu &= \frac{(\mu_1 + \mu_2)(1 + \nu)}{ET}, \quad \nu = \frac{\nu}{1 - 2\nu}, \quad D = \frac{D_0 T}{L^2}.
\end{aligned} \tag{8.17}
$$

There are 4 roots, σ, of (8.15), namely,

$$
\begin{aligned}
\sigma_1, \sigma_2 &= \frac{-b(k^2) \pm \sqrt{b^2(k^2) - 4(\mu k^2 + s)c(k^2)}}{2(\mu_1 k^2 + s)}, \\
\sigma_3 &= -\frac{k^2}{\mu_1 k^2 + 2s} \leq 0, \quad \sigma_4 = 0.
\end{aligned} \tag{8.18}
$$

Depending on the sign of

$$\Delta = \sqrt{b^2(k^2) - 4(\mu k^2 + s)c(k^2)} \tag{8.19}$$

the solutions given by (8.14) can be real or complex.

Pattern Formation and Parameter Domains

A fairly comprehensive picture of the pattern formation potential of the mechanism can be obtained from a detailed study of the roots σ given by (8.18). To be able to generate coherent spatial patterns we require that $\sigma(k^2 = 0) \leq 0$ and that there exist wavenumbers $k^2 > 0$ such that $Rl\sigma(k^2) > 0$; refer to Chapter 6 for a full description of the justification. Basically they ensure that the homogeneous steady state is stable but it is unstable to spatial perturbations some of which initially grow exponentially. Certainly when $k^2 = 0$ the first condition is clearly satisfied from (8.15). Let us now briefly consider the second condition.

The largest root is σ_1 in (8.18) so we can therefore concentrate on it. From (8.16) it is clear that both $b(k^2)$ and $c(k^2)$ can be positive or negative depending on the relative size of τ_1, the other parameters and the wavenumber $\mathbf{k}$. Here we only discuss a few of the several cases, leaving a full analysis and quantification of the parameter spaces for pattern formation as a pedagogical, biologically useful and practical, exercise.

Let us first consider the special case in which the cells do not diffuse, that is, $D = 0$. Using (8.16) in (8.18) we see that the largest σ is then

$$\sigma_1 = -\frac{b(k^2)}{\mu k^2 + s} = \frac{(\tau_1 - 1 - \nu)k^2}{\mu k^2 + s} \tag{8.20}$$

and so, if the cell traction parameter $\tau_1 > 1 + \nu$, $\sigma_1 > 0$ for all $k^2 > 0$ and hence all spatial modes will grow exponentially (in this linear theory) according to (8.14). Since all modes are unstable the final pattern will depend intimately on the initial conditions. As we show in the numerical simulations presented below the patterns formed are generally fairly random but with a gross coherent structure. The interesting implication here is that patterns are generated solely by the interaction of the various mechanical forces present, namely, the cell traction, the matrix and dish resistance to movement and how they move the cells around via convection. It is an often held belief that diffusion is crucial, if not essential, to create spatial patterns of relevance.

*Parameter Domain $\tau_1 > 2(1 + \nu)$: Region **I***

Let us consider σ_1 from (8.18) together with the expressions in (8.16) with $k^2 > 0$; now, of course, $D \neq 0$. If $c(k^2) < 0$ then $\sigma_1 > 0$ and the uniform steady state is linearly unstable. So, if $\tau_1 > 2(1 + \nu)$ all wavenumbers $k^2 > 0$ are unstable and again the final pattern depends on the initial conditions.

In dimensional terms, using (8.17), we have the result that small perturbations in the cell density from the uniform steady state can initiate matrix instabilities and start to form spatial patterns in the cell and matrix densities if

$$\begin{aligned} \frac{\tau n_0(1+\nu)(1-\alpha n_0^2)}{E(1+\alpha n_0^2)^2} &> 2\left(1+\frac{\nu}{1-2\nu}\right) \\ \Rightarrow \quad \tau &> \frac{2E(1-\nu)(1+\alpha n_0^2)^2}{n_0(1+\nu)(1-2\nu)(1-\alpha n_0^2)}. \end{aligned} \tag{8.21}$$

In this parameter domain all wavenumbers are unstable irrespective of the values of D and s. A typical dispersion relation in this parameter range is illustrated in Figure 8.2(a).

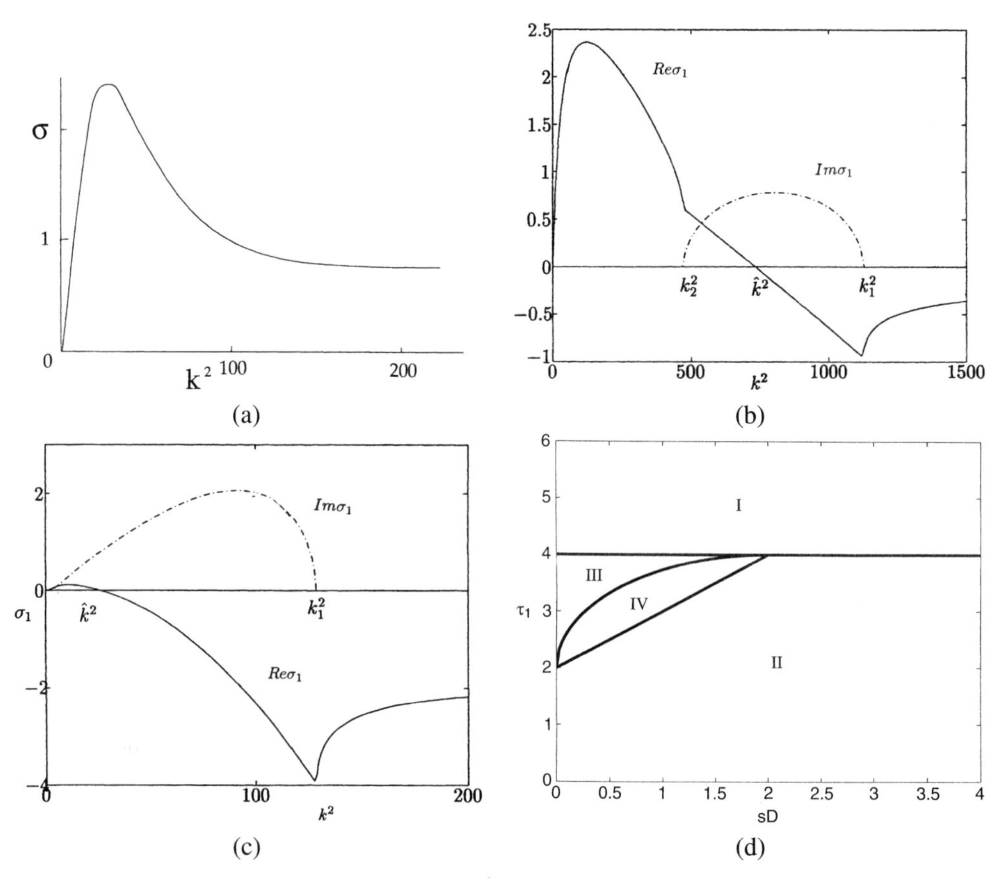

Figure 8.2. (**a**)–(**c**) Typical dispersion relations $\sigma(k^2)$, giving the linear growth rate of spatial instabilities. (**a**) An infinite range of unstable wavenumbers with $\sigma(k^2)$ real. Parameter values: $s = 10$, $\alpha = 0.01$, $\mu = 1$, $\nu = 0.2$, $\tau = 5$, $D = 0.001$. (**b**) The unstable growth is more complicated (and more interesting) with a finite part of the domain exhibiting growing oscillatory solutions. Parameter values: $s = 10, \alpha = 0.01$, $\mu = 0.3$, $\nu = 0.2$, $\tau = 1.4$, $D = 0.001$. (**c**) A finite range of unstable growing oscillations. Parameter values: $s = 10, \alpha = 0.01, \mu = 0.3, \nu = 0.2, \tau = 2.4, D = 0.001$. (From Manoussaki 1996) (**d**) Example of the parameter domains with different solution behaviour for $\nu = 1$. The regions **I, II, III** and **IV** correspond to the solution domain behaviour summarized in Table 8.2.

Since the ratio of the cell density used in experiments to the confluent density (where crowding effects come into play) is of the order of 0.1–0.4, $(1+\alpha n_0^2)^2/(1-\alpha n_0^2)\approx 1$ and the last condition becomes simply

$$\tau > \frac{2E(1-\nu)}{n_0(1+\nu)(1-2\nu)}.$$

We immediately get some qualitative results on the role of key parameters in the experiments, such as the larger the seeding density (n_0) the greater the likelihood of pattern formation. Other qualitative results on how the patterns form are obtained from the further detailed analysis below.

Parameter Domain $\tau_1 < 2(1+\nu), sD > 1+\nu$: *Region* ***II***

In this range, $c(k^2) > 0$ for all $k^2 > 0$ so instability, that is, $Rl\sigma_1 > 0$, can only obtain if $b(k^2) < 0$. If $sD > 1+\nu$ the coefficient of the $O(k^2)$ term is always positive and so is $b(k^2)$ and hence the uniform steady state is always stable and no pattern formation is possible.

Parameter Domain $\tau_1 < 2(1+\nu), sD < 1+\nu$: *Region* ***III***

If $sD < 1+\nu$ then, from (8.19) with (8.16) the roots of Δ are

$$k_0^2 = 0, k_1^2, k_2^2 = \frac{1}{\mu D}\left(1+\nu - sD \pm \sqrt{\tau_1[2(1+\nu)-\tau_1]}\right). \tag{8.22}$$

With these we can then deduce (after a little algebra) that if the parameters satisfy

$$\begin{aligned}&\tau_1 < 2(1+\nu), sD < 1+\nu,\\ &1+\nu+\sqrt{sD[2(1+\nu)-sD]} < \tau_1 < 2(1+\nu)\end{aligned} \tag{8.23}$$

the dispersion relation, σ_1, from (8.18) is complex for wavenumbers $k_1^2 < k^2 < k_2^2$. In detail we have

$$\begin{aligned}&Rl\sigma_1(k^2) > 0, 0 < k^2 < \bar{k}^2, \quad Rl\sigma_1(k^2) < 0, \bar{k}^2 < k^2 < \infty,\\ &Im\,\sigma_1(k^2) > 0, k_2^2 < k^2 < k_1^2, \quad Im\,\sigma_1(k^2) = 0, 0 < k^2 < k_2^2, k^2 > k_1^2,\end{aligned} \tag{8.24}$$

where $\bar{k}^2$ is the value of k^2 where the curve for $Rl\sigma(k^2) = 0$. A typical dispersion form in this parameter range is shown in Figure 8.2(b). Note that in this case there is a finite range of unstable modes with a fastest growing mode which can be derived from the expression for $\sigma(k^2)$. There is also the possiblity of oscillatory growing perturbations in this case since for $k_2^2 < k^2 < \bar{k}^2$, $Rl\sigma_1 > 0$, $Im\,\sigma_1 \neq 0$.

Parameter Domain $\tau_1 < 2(1+\nu), sD < 2(1+\nu)$: *Region* ***IV***

By a similar analysis we get the quantitative behaviour of the dispersion relation for the parameters in the ranges

$$\begin{aligned} &\tau_1 < 2(1+\nu), \quad sD < 2(1+\nu), \\ &1+\nu+sD < \tau_1 < 1+\nu+\sqrt{sD[2(1+\nu)-sD]}, \end{aligned} \tag{8.25}$$

and in this case we get the dispersion relation behaviour

$$\begin{aligned} &Rl\sigma_1(k^2) > 0, 0 < k^2 < \bar{k}^2, \quad Rl\sigma_1(k^2) < 0, \bar{k}^2 < k^2 < \infty, \\ &Im\,\sigma_1(k^2) > 0, 0 < k^2 < k_1^2, \quad Im\,\sigma_1(k^2) = 0, k_1^2 < k^2, \\ &\bar{k}^2 = (\tau_1 - 1 - \nu - sD)/\mu D, \end{aligned} \tag{8.26}$$

with k_1^2 given by (8.22).

In this case there is a finite range of unstable modes and all of them grow in an oscillatory manner. An example of this dispersion relation is given in Figure 8.2(c).

We now have a fairly complete picture of how the parameters determine whether or not the uniform steady state is unstable and whether or not the mechanism will produce a spatial pattern. Figure 8.2(d) shows the $\tau_1 - sD$ parameter space divided into the various regions covered by the parameter domains **I, II, III** and **IV** given above. The parameter domain is in fact 6-dimensional in the dimensionless parameters D, μ_1, μ_2, ν, τ_1 and s.

Effect of Parameter Variation on the Pattern Formation

The linear analysis provides a means of classifying the type of instabilities we can expect—exponential growth, growing exponential oscillations or a mixture of both—when the parameters are in the various parameter domains and the relative growth of the unstable wavenumbers.

Table 8.2 gives the type of growing instability for various parameter regions. The effect in terms of the dimensional parameters is then given by using the definitions

Table 8.2. Conditions on the parameters for the formation of spatial patterns with regions **I, II, III** and **IV** denoted in Figure 8.2. The parameter $\tau_1 = \tau(1-\alpha n_0^2)/(1+\alpha n_0^2)^2$, $\tau_a = 2(1+\nu)$, $\tau_b = 1+\nu+\sqrt{sD[2(1+\nu)-sD]}$, $\tau_c = 1+\nu+sD$.

Region	Parameter Range	Linear Solution Behaviour
I	$\tau_a < \tau_1$	Steady exponential growth for all nonzero wavenumbers
II	$\tau_a > \tau_1$ $sD \geq 1+\nu$	No pattern forms
III	$\tau_b < \tau_1 \leq \tau_a$ $sD < 1+\nu$	Oscillatory growth with finite range of wavenumbers
IV	$\tau_c < \tau_1 \leq \tau_b$ $sD < 1+\nu$	Growing oscillations with finite range of wavenumbers

of the dimensionless parameters in (8.17) with the τ_a, τ_b, τ_c defined in the legend of Table 8.2. A crucial parameter is clearly the cell traction which in dimensional terms involves the ratio of the cell traction, τ, to the Young's modulus, E, of the matrix. Manoussaki (1996) has analyzed the effect of varying the parameters on the growth rate, σ. All of the information to do this is contained in the above expressions for σ and simply involves plotting the dispersion relation for all but one of the parameters fixed (or two if we use a three-dimensional space). As an example, suppose we have a parameter set which gives a point in region **IV** in Figure 8.2(d); that is, the steady state is unstable by growing oscillations. Now suppose we keep all parameters fixed except the dish attachment parameter, s, and let it increase. From the figure we see that for sufficiently large s we move into the stable regime **II**. So, for this parameter set we see that increasing the attachment of the matrix to the dish can inhibit pattern formation: this is as we would expect intuitively.

Intuitively we would also expect that increasing diffusion would inhibit pattern formation. What is less clear, however, is that if D increases it enhances the growth of smaller wavenumbers: with increasing D the curve in Figure 8.2(a) moves downwards but at the same time it makes the range of fastest growing modes much narrower and so gives advantage to the small wavenumbers, or large wavelengths. Many of the effects can be expected intuitively, such as an increase in viscosity will inhibit the formation of patterns. The quantification of the effects, however, has to be done analytically. For example, it is to be expected that an increase in the cell traction makes it more likely that patterns will form. What is less clear is that as the traction decreases the growth of the highest wavenumbers is inhibited and eventually suppressed with only a finite range of unstable wavenumbers with growth a combination of steady and oscillatory growth. It should be emphasised, however, that these effects of parameter variation are based on a linear stability theory and may not be borne out quantitatively when nonlinear effects are taken into account. To date a nonlinear analysis has not been carried out. The specific technique used by Zhu and Murray (1995) might be able to be used on these equations although there are some possible difficulties because of the (biologically realistic) parameter ranges that have to be used.

8.5 Network Patterns: Numerical Simulations and Conclusions

We now have a fairly good intuitive idea as to whether a pattern will form and how the parameter values affect this. To determine the actual patterns and their dependence on the parameters we have to solve the model equations numerically. Such numerical solutions give further useful insight, help to assess the relative importance of the various elements and predict how changes in these factors affect pattern growth and shape. Extensive numerical simulations were carried out by Manoussaki (1996) and some of her results are described and presented in this section; see also Manoussaki et al. (1996) and Murray et al. (1998).

The initial conditions that were used consisted of either a small random perturbation in the cell density about the uniform steady state $n = 1$, or by introducing small patches of cells distributed in a random way. The matrix was initially at the uniform steady state $\rho = 1$. In some of the simulations the matrix thickness was varied (it affects the attachment parameter s) as were the parameters to determine their quantitative effect on both the time to create the pattern and its characteristics. We also simulated the situation where there is a density ramp of Matrigel density and obtain graded network patterns which were subsequently investigated experimentally and confirmed our predicitions. We only present here some of the results in which the initial conditions consisted of small random perturbations in cell density about its uniform steady state.

In the numerical simulations we were guided by the parameter restrictions and parameter domains obtained from the linear analysis described above. As predicted the small initial perturbation in the cells initiated instabilities in the cell and matrix densities which evolved into spatial patterns such as illustrated in Figure 8.3. These initial heterogeneities grow and continuously rearrange themselves with time as the aggregates recruit more of the neighboring cells. Eventually the cell and matrix densities formed a network of the small irregular polygons defined by the cellular cords. With time these become smaller and eventually disappear, or rather coalesce to form larger polygons. After a long time the cells start to migrate towards the boundary and accumulate there. The matrix is also pulled towards the boundary. A video made of the spatiotemporal solution behaviour shows this very clearly and, in fact, looks remarkably like a time-lapse video of the *in vitro* situation.

The results here should now be compared with the typical experimental patterns shown in Figure 8.1: the comparison is remarkably close and together with other comparative aspects described below, suggests that the mechanism we propose for the formation of these networks *in vitro* is a serious candidate for this pattern formation mechanism. At the very least we suggest that it captures some of the major elements such a mechanism must include. Further evidence for a mechanical mechanism is given by the work of Tranqui and Tracqui (2000) which we briefly described in Section 8.2.

When the cell densities are still small perturbations about the uniform steady state and the matrix is initially at its uniform steady state, the matrix thickness exactly follows

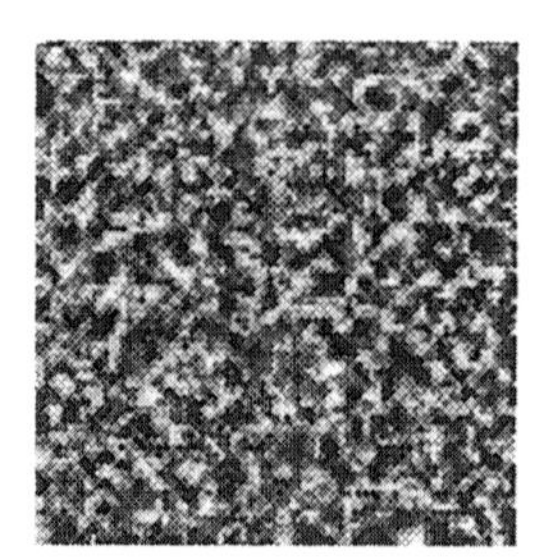
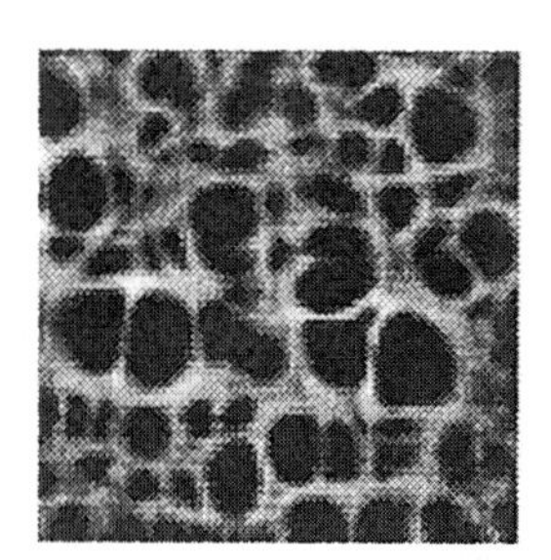
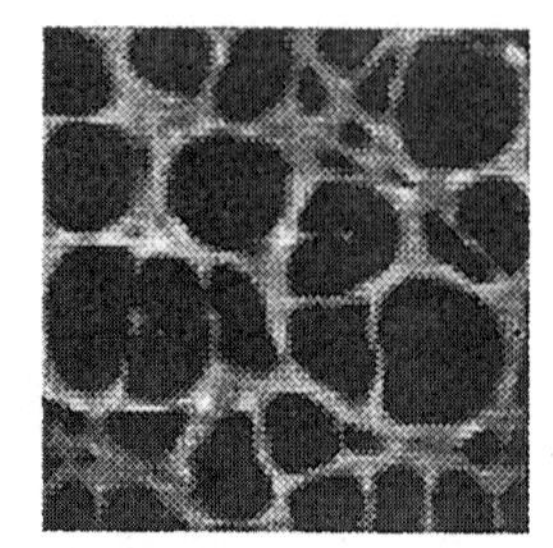

Figure 8.3. Representative numerical simulation showing the evolving network of cellular patterns. The matrix was initially uniform. From left to right the figures show the small random perturbation in cell density about the uniform steady state $n = 1$ at $t = 0$, the evolution to a well-defined network shown at dimensional times $t = 12$ hours and $t = 24$ hours. White denotes high cell densities while the black denotes areas of very low cell density. Regular patterns never formed but there is a fairly well-defined range for the sizes of the network loops. The side of the square corresponds to 400 μm. Dimensional parameter values (the dimensionless parameters can be derived using (8.10)) are $\tau = 0.06$ dyne cell^{-1}, $D = 0.4 \times 10^{-11}$ cm^2 sec^{-1}, $\nu = 0.2$, $\mu_1 = 1.5 \times 10^7$ dyne sec cm^{-2}, $\mu_2 = 0.9 \times 10^7$ dyne sec cm^{-2}, $s = 10^{10}$ dyne sec cm^{-3}, $\alpha = 4 \times 10^{-13}$ (cell cm^{-2})$^{-2}$, $E = 20$ dyne cm^{-2} with $n_0 = 2 \times 10^4$ cells cm^{-2}, Matrigel density 10 μm–14 mm. The matrix density evolution is essentially the same. (From Manoussaki et al. 1996)

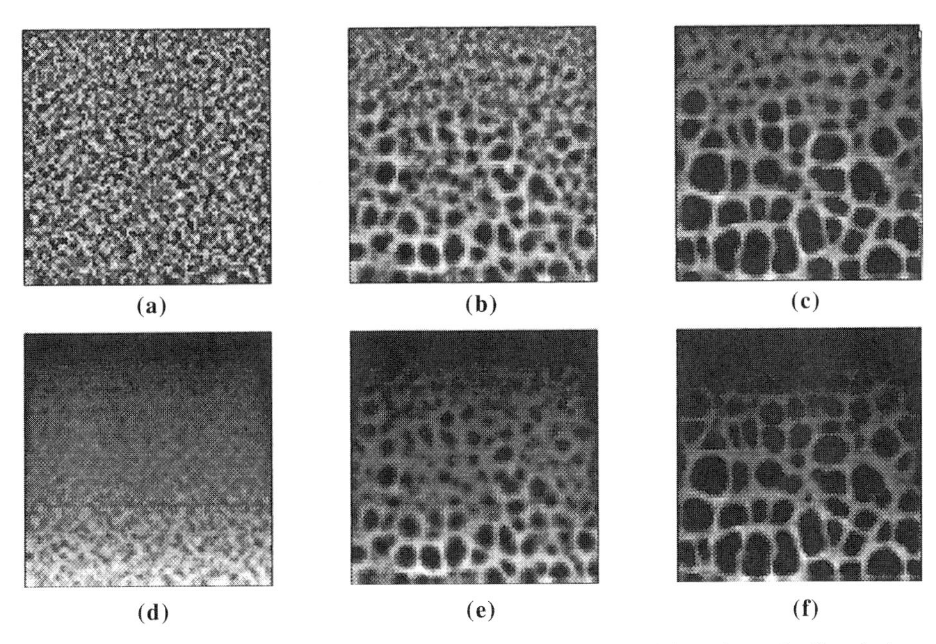

Figure 8.4. Numerical solutions for cells initially laid on a matrix with thickness increasing linearly from $\rho = 0.01$ at the top to $\rho = 0.1$ at the bottom. (**a**)–(**c**) give the cell densities with (**d**)–(**f**) the matrix densities. Dark denotes low densities (black represents very low cell density and matrix thickness) with lighter regions denoting higher densities. Parameter values are the same as those given in Figure 8.3. (From Manoussaki 1996)

the cell densities; see also Figure 8.4 where both the cell and matrix patterns are given. This is as we would expect in the linear growth regime from the linear equations (8.13) from which it can be deduced that the cells and matrix are in phase and both out of phase with the deformation. For example, from the third of (8.13) and the solution form (8.14) we have

$$\rho = -\nu\theta \propto ie^{\sigma t + i\mathbf{k}\cdot\mathbf{x}} = e^{\sigma t + i\mathbf{k}\cdot\mathbf{x} + i\pi/2}.$$

Using (8.13) it can be similarly shown that n is also out of phase with the deformation and that n and ρ are in phase.

Intuitively, even in the nonlinear regime of the numerical simulations, we would expect a fairly close connection between the cell and matrix patterns: this is exactly what we found and the matrix pattern evolution looks almost exactly the same as that for the cells shown in Figure 8.3. From the linearized model system, in the case when there is no cell diffusion the conservation equation for the cell density (on integrating with respect to t) and the matrix thickness from (8.13) are exactly the same and differ only in the parameter ν in the ρ equation. With the parameter range used for the diffusion coefficient the effect of diffusion is very small so we should expect them to show the same evolving patterning.

The similarity in pattern for the cells and matrix also makes biological sense. As the matrix is deformed by the cells, it tends to organize into cord-like networks and since the cells, initially fairly uniformly spread on the matrix, ride passively on the reorganizing matrix they will form similar networks. Of course, when the cells are not uniformly spread throughout the matrix, the cell–matrix densities are no longer in agreement. Here the cells deform the matrix around them and the resulting deformations depend quantitatively on the values of the parameters.

A further confirmation of the close connection between the cells and the matrix and what is seen experimentally is a plot of the dilation $\theta = \nabla \cdot \mathbf{u}$. Here we would expect the high regions of dilation to be where there are few cells and little matrix. In this case we would expect the pattern for the dilation to be essentially the mirror image of that shown in Figure 8.3 with the light regions where the dark regions are there. Again this is exactly what we found.

Matrix Thickness Influences the Size of Pattern

Experiments show that the matrix thickness influences the size of the networks as well as whether or not they will form at all. Thinner matrix results in smaller polygons while if the thickness falls below a critical threshold no networks are observed. In our model, matrix thickness is reflected in the dimensionless attachment parameter s; see (8.10). If we relate this result to the above analysis and specifically Figure 8.2(d) we see that if we have a parameter set such that we are in a pattern formation domain, either regions **III** or **IV** say, then increasing s, that is, decreasing ρ_0, we move into region **II** in which no pattern can be formed. The quantitative effect on the polygonal patterns of varying a parameter requires a nonlinear analysis.

Since experimentally the polygonal pattern depends on matrix thickness we solved the model equations when the initial matrix distribution had a gradual decrease in thickness. The matrix thickness varied linearly from $\rho = 0.01$ at one end of the domain to $\rho = 0.1$ at the other. Initially almost all random perturbations seemed to grow at the same rate, but quite soon, by time $t = 6$, it appeared that only the cells at the thicker end formed networks whereas the cells at the thin end remained unorganized. At the thin end, no networks formed until after a relatively long time, $t = 24$, after which they remained very small up to time $t = 40$. Figure 8.4 shows the results of these simulations for both the cell and matrix densities.

If the initial matrix density was higher, for example, with a variation from $\rho = 0.5$ to $\rho = 1.5$ larger networks formed on the thicker end of the matrix and smaller ones at the thin end but the differences were less obvious than in the lower thickness simulations.

If the cell traction is sufficiently large we saw that patterns will form: this is region **I** in Table 8.2 in which the dimensionless parameter $\tau_1 > \tau_a$. In dimensional terms this condition is given by (8.19) in which the matrix thickness, namely, ρ_0, does not appear. This suggests that, at least for early times, matrix thickness does not affect the growth rate of the pattern. As the nonlinear effects become important, matrix thickness influences the size of the patterns that are formed. The numerical simulations showed that pattern was possible at all matrix thicknesses examined, but the time required to form patterns increased with matrix thickness. This is as we should expect since more matrix had to be moved around before heterogeneity showed up.

We mentioned above, that in the experiments after a long time of cell and matrix remodeling, the cells and matrix tend to accumulate at the dish boundary. Numerical simulations (Manoussaki 1996) found similar behaviour after a long time; the cells and matrix eventually formed high density regions along the domain boundary. This phenomenon of cell accumulation at rigid boundaries is possibly an example of *desmotaxis*, a term which describes the tendency of cells moving in a fibrous environment to migrate towards rigid objects, when such are present. In our study, the boundary plays the role of that fixed object. Cells pull on the matrix that is fixed at the boundary, so all deformations sense the presence of the boundary. In fact, it was clear in the numerical simulations that the boundary influences the pattern in a major way.

Manoussaki (1996) also simulated the situation in which patches of cells were intially introduced. Intuitively we now know what happens qualitatively in a gross way. Initially the patch contracts, with a slightly higher cell density at the edge of the patch, due to recruitment, and the matrix underneath the cells accumulates, while around the patch it becomes thinner. Eventually the patch breaks up. A more interesting scenario is when two or more patches are introduced. Here, if the patches are sufficiently close to have significant influence on each other, we could expect the formation of a kind of 'super-highway' formed between them which accumulates cells.

Effect of Parameter Variation

Many of the effects are as we should expect. In the case of cell traction, simulations confirmed the prediction of the linear stability analysis, that increasing cell traction, τ, causes the patterns to form more quickly. Higher traction results in an increase in the maximum distance possible before two cell patches influence each other and thereby give rise to the formation of larger networks, as has been observed in culture experiments.

It was at first somewhat surprising that diffusion seemed to play little role in whether or not networks formed. But, since it is mainly the interaction of the various forces which generates the network, it is on reflection to be expected. Of course, if the diffusion parameter, D, is sufficiently large it does have an effect. Its effect becomes significant ($D \geq 10^{-2}$) when we observed a smearing of the networks. Even in the absence of diffusion ($D = 0$) patterns form if the traction is sufficiently large. Cells still move towards regions of matrix compression. Biased cellular migration may be a component of pattern formation *in vivo* or *in vitro*, but we have shown it is not a *necessary* feature.

One of the unexpected results of our analysis was that anisotropy was not a necessary requirement for the formation of pattern; in the model the cell traction is applied isotropically and the only component in the model that showed anisotropic behavior, namely, the cell movements associate with preferred diffusion directions, played little role in the formation of networks. The analysis and numerical simulations show that isotropic cell traction and convected motion are sufficient to generate the observed pattern formation at least *in vitro*. In Chapter 6 in the discussion of the Murray–Oster mechanochemical theory one of the strong points made for its applicability was that the cells, by virtue of their filapodial traction could change their spatial environment and thereby influence the pattern formed. The situation we discuss here is a prime example of where this happens.

Changing the values of viscosity parameters, μ_1 and μ_2, influences the rate at which pattern forms. Higher viscosity values cause the pattern to take longer to form but do not change the appearance of pattern, all again as we would expect. Increasing the dish attachment parameter, s, also has this effect: if it is large enough and we are in the appropriate parameter domain it can actually inhibit pattern formation as we mentioned before.

Since diffusion (and anisotropy) is not necessary for generating these networks we can make a significant simplification to the basic model by setting $D = 0$ in which case equations (8.11) become

$$\begin{gathered} n_t + \nabla \cdot (n\mathbf{u}_t) = 0, \\ \nabla \cdot \left[\mu_1 \boldsymbol{\epsilon}_t + \mu_2 \theta_t \mathbf{I} + \boldsymbol{\epsilon} + \nu\theta\mathbf{I} + \frac{\tau n \mathbf{I}}{1 + \alpha n^2} \right] = \frac{s}{\rho}\mathbf{u}_t, \\ \rho(x, y, t) = 1 - \nu\theta. \end{gathered} \tag{8.27}$$

If we use the small strain approximation in the equation for n, it becomes, on linearising about $n = 1$,

$$n_t + \nabla \cdot (\mathbf{u}_t) = 0 \quad \Rightarrow \quad n = 1 - \theta,$$

on integration. If we now substitute this and the expression for ρ into the force balance equation we get the single (vector) equation as the model mechanism for the network formation, namely,

$$\nabla \cdot \left[\mu_1 \boldsymbol{\epsilon}_t + \mu_2 \theta_t \mathbf{I} + \boldsymbol{\epsilon} + \nu\theta\mathbf{I} + \frac{\tau(1-\theta)\mathbf{I}}{1 + \alpha(1-\theta)^2} \right] = \frac{s}{1 - \nu\theta}\mathbf{u}_t. \tag{8.28}$$

If we linearise this equation about $\theta = \mathbf{u} = 0$ to examine its pattern formation it becomes

$$\nabla \cdot \left[\mu_1 \boldsymbol{\epsilon}_t + \mu_2 \theta_t \mathbf{I} + \boldsymbol{\epsilon} + \nu\theta\mathbf{I} - \frac{\tau(1-\alpha)}{(1+\alpha)^2}\theta\mathbf{I} \right] = s\mathbf{u}_t. \tag{8.29}$$

Using the vector relation $\nabla \cdot \boldsymbol{\epsilon} = \text{grad div}\, \mathbf{u} - (1/2)\text{curl curl}\mathbf{u}$ and taking the divergence of (8.29) we get the following scalar equation for θ,

$$\mu\nabla^2\theta_t + \left[1 + \nu - \frac{\tau(1-\alpha)}{(1+\alpha)^2} \right] \nabla^2\theta - s\theta_t = 0, \tag{8.30}$$

where $\mu = \mu_1 + \mu_2$.

Now look for solutions in the form

$$\theta \propto e^{\sigma t + i\mathbf{k}\cdot\mathbf{x}}, \tag{8.31}$$

which on substituting into (8.30) gives the dispersion relation

$$\sigma(k^2) = \frac{\left[\frac{\tau(1-\alpha)}{(1+\alpha)^2} - 1 - \nu\right]k^2}{\mu k^2 + s}. \tag{8.32}$$

So, all wavenumber perturbations are unstable if

$$\tau > (1+\nu)\frac{(1+\alpha)^2}{1-\alpha} \approx 1+\nu \quad \text{for } \alpha \ll 1, \tag{8.33}$$

which is the result we obtained before in the case when $D = 0$.

The initial conditions play a major role in the specific network formed. How important depends in part on the form of the dispersion relation, being more important if there is an infinite range of unstable wavenumbers as in the basic model we have just discussed. Although the exact form of the pattern depends on the initial perturbation its qualitative features remain the same. It is the qualitative nature and the rate of growth of the pattern which depend primarily on the model parameters.

The modelling we have discussed here is a very practical biological application of the Murray–Oster mechanical theory of pattern formation, and importantly one in which there is a directly related experimental study. It is the first mathematical description of cell–matrix interactions for the formation of vascular networks in which all of the component variables are in principle measurable. The fact that we have been able to estimate the major parameters and thereby obtain patterned networks which so closely resemble the experimental networks *in vitro* suggest that we have perhaps indeed isolated the key ingredients in this patterning process.

There are still many open problems associated with this model for cell–matrix network formation, the related experimental studies and more importantly, in *in vivo* network formation. All of the analysis here has been on a linear basis. Although nonlinear analysis based on a multi-scale asymptotic analysis, such as used by Zhu and Murray (1995), can be applied to the pattern formation near bifurcation to heterogeneity, it is limited to just that, pattern formation near bifurcation. In this specific model it can be used for bifurcation states with only some of the parameters. Ultimately, as we have described, the final pattern is one in which the cells and matrix cling to small regions on the domain boundary. However, a nonlinear theory would give an indication of which of the various two-dimensional patterns are stable or rather quasi-stable—hexagons one would guess. The basic method was developed by Ermentrout (1991) for reaction diffusion mechanisms.

Essentially the techniques presented by Zhu and Murray (1995) for several different pattern formation mechanisms assume a small growth rate of order ε^2 where $0 < \varepsilon \ll 1$. A small $O(\varepsilon)$ perturbation is then introduced to the dependent variables about the uniform steady state. By examining the stability of amplitude functions that occur in the assumed solution, the pattern that forms for long time is determined. It is the same technique used by Tyson (1996) in her study of bacterial patterns and which we discussed in more detail in Chapter 5. There are, however, some problems with applying these techniques directly to the model here, as we have mentioned, and so modifications in the technique might have to be developed. The analysis may still be possible, but is probably not very easy.

A major amendment would be to use a more realistic description of the matrix such as that developed by Cook (1995) and discussed in some detail in Chapter 10 on full depth wound healing. The model will necessarily be considerably more complex but more biologically realistic for *in vivo* studies. Such a formulation, among other things, would bring in tissue plasticity and make it possible to have strained configurations in zero stress states as we shall show in that discussion.

Clearly a major extension of the modelling would be to *in vivo* studies and also to angiogenesis in connection with the control of tumour growth mentioned in Section 8.1. Such applications will require several new features. Even in relation to further *in vitro* studies the theory can be extended to account for the effect of cell-secreted molecules on the matrix material properties, cell proliferation and cell secretion, thus providing an analytical tool for assessment of the relevant importance of these factors during vasculogenesis and angiogenesis. It can also be used to study the effect on the pattern if there is spatial variability in the parameters: the results with ramp in matrix density is one such example. We believe that the success of the current model in mimicking experimental observations gives some reason for optimism that it can be used as a guide to predict the outcome of other experiments and, optimally, the consequences of experimentally disturbing the mechanism which obtains *in vivo*. We should perhaps reiterate that the model here is a model of vasculogenesis and not the angiogenetic process of vessel formation.

In spite of the many agreements between the experimental observations and the solutions of the model equations, the model is certainly not a complete description of the phenomenon. One of the most crucial assumptions is that of small strains and the absence of matrix plasticity. Nevertheless, this basic model does capture important aspects of the cell–matrix interaction and many of its detailed predictions on the formation of irregular networks have been confirmed experimentally, such as those associated with a ramp of variable matrix thickness.

9. Epidermal Wound Healing

9.1 Brief History of Wound Healing

Every civilization with a written history has some reference to wound healing and treatment. The modern historical literature on medicine and surgery is large and medical history is a firmly established historical discipline. An overall review of western medicine is given in the book of articles by specialists edited by Loudon (1997) and the earlier classic and graphically illustrated book by Lyons and Petrucelli (1978). An interesting book by Majno (1975) is concerned with the history of surgery (and especially wounds) in the ancient world. One of the earliest descriptions of how to treat wounds is in the Edwin Smith Surgical Papyrus of ancient Egypt around the 13th Dynasty (about 2500 BC). Figure 9.1 is an example from the papyrus which describes a wound to the soft tissue of the temple but with the bone intact while Figure 9.2 gives an example (with an English translation) of a case study which starts with the title, followed by the examination, the diagnosis and finishing with the recommended treatment.

There is an enormous number of references to the fascination with the human body and what we now call surgery to remedy defects and heal wounds. After Nero murdered his mother Agrippina he reputedly rushed off to examine her corpse and between drinks he discussed the good and bad points of her limbs. This story was used in the middle ages to discredit such a well-known persecutor of early Christians since the mediaeval view was that cutting open the human body to acquire knowledge was very much taboo (and still is in many communities today). To reveal the internal organs to understand their function was considered blasphemous since it was also considered arrogant that one could learn something more than the great Galen (c129–c216) had taught about the human body. Galen was very practical in his medical advice; for example, on skull operations he commented on the *'disadvantage of using a gouge is that the head is shaken vigorously by the hammer strokes.'* In his day eight different skull fractures were classified. It was not until around the 15th century that anatomy was introduced by the surgeons, barbers and executioners who had the somewhat doubtful privilege of handling human blood and tissue. In the 14th century surgery was still considered a manual trade at the service of medicine and most of the surgeons were viewed as '*ignorami, barbers, adventurers, vagabonds, rakes, ruffians, brothel keepers, pimps and quacks*'; very few of them were literate. The state of surgery was as miserable as its practice.

Surgical operations—some pretty gruesome—have been carried out for thousands of years. There were numerous, wild, scary but sometimes effective recipes for anaes-

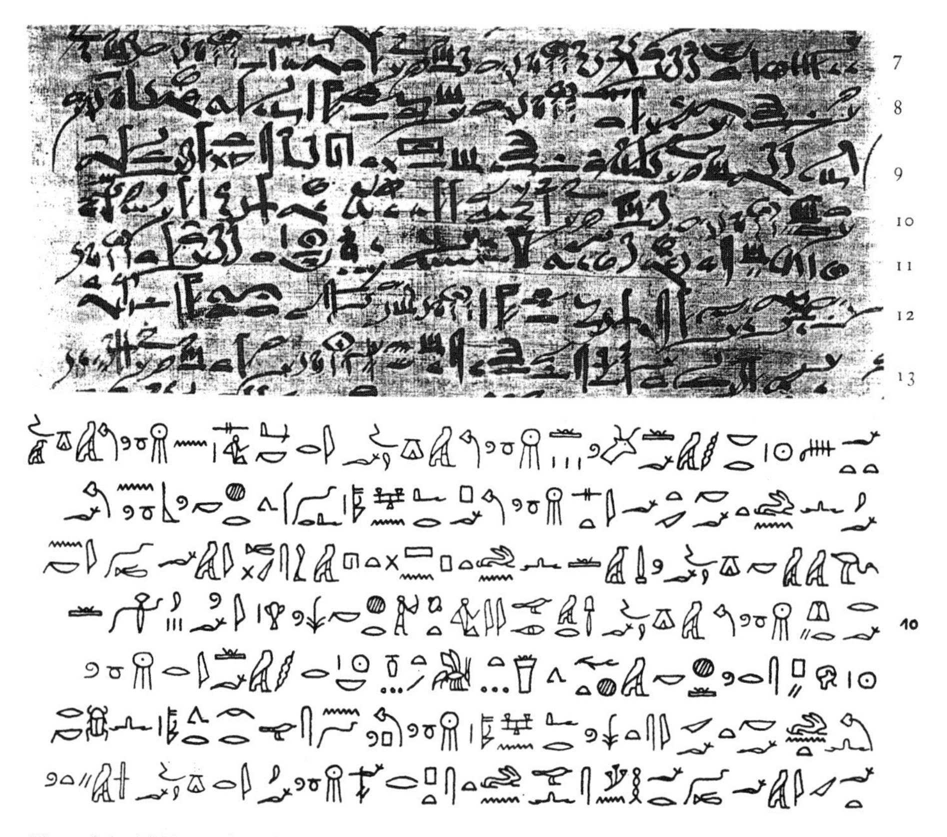

Figure 9.1. **(a)** Extract from the Edwin Smith Surgical Papyrus (c13th Dynasty). It describes a wound to the soft tissue of the temple, with the bone being uninjured. **(b)** Hieroglyphic transliteration from Breasted (1930)

thesia in the Middle Ages. One example is: sponge soaked in a mixture of opium, hyoscyamus, mulberry juice, lettuce, hemlock, mandragora and ivy, dried and, when moistened, 'inhaled' (probably swallowed by the patient), who was subsequently awakened by applying fennel juice to the nostrils. Throughout the Middle Ages mandragora was the soporific *par excellence*, preferable to opium and hemlock because, unlike these which are 'cold in the fourth degree' they are of the 'third degree.' John Donne, the rake[1] turned poet and cleric, said *'its operation is between sleep and poison'* while Marlowe in his *The Jew of Malta* (Act V, Scene i) wrote *'I drank of poppy and cold mandrake juice and being asleep, belike they thought me dead.'*

The great luminary of surgery of the mediaeval period was Henri de Mondeville (1260(?)–c1320) a surgeon and medical scholar, born in Mondeville near Caen, France. He was educated and officially a cleric. He died 'asthmaticus, tussiculosus, phthisicus'—asthmatic, coughing and tubercular. He did much to change the mediaeval view of surgery and its practice. He taught and wrote about surgery and anatomy, believing, for example, that wounds should be cleaned without probing, treated without

[1]He is reputed to have had an early prayer, 'Oh Lord, grant me chastity—but not yet awhile.'

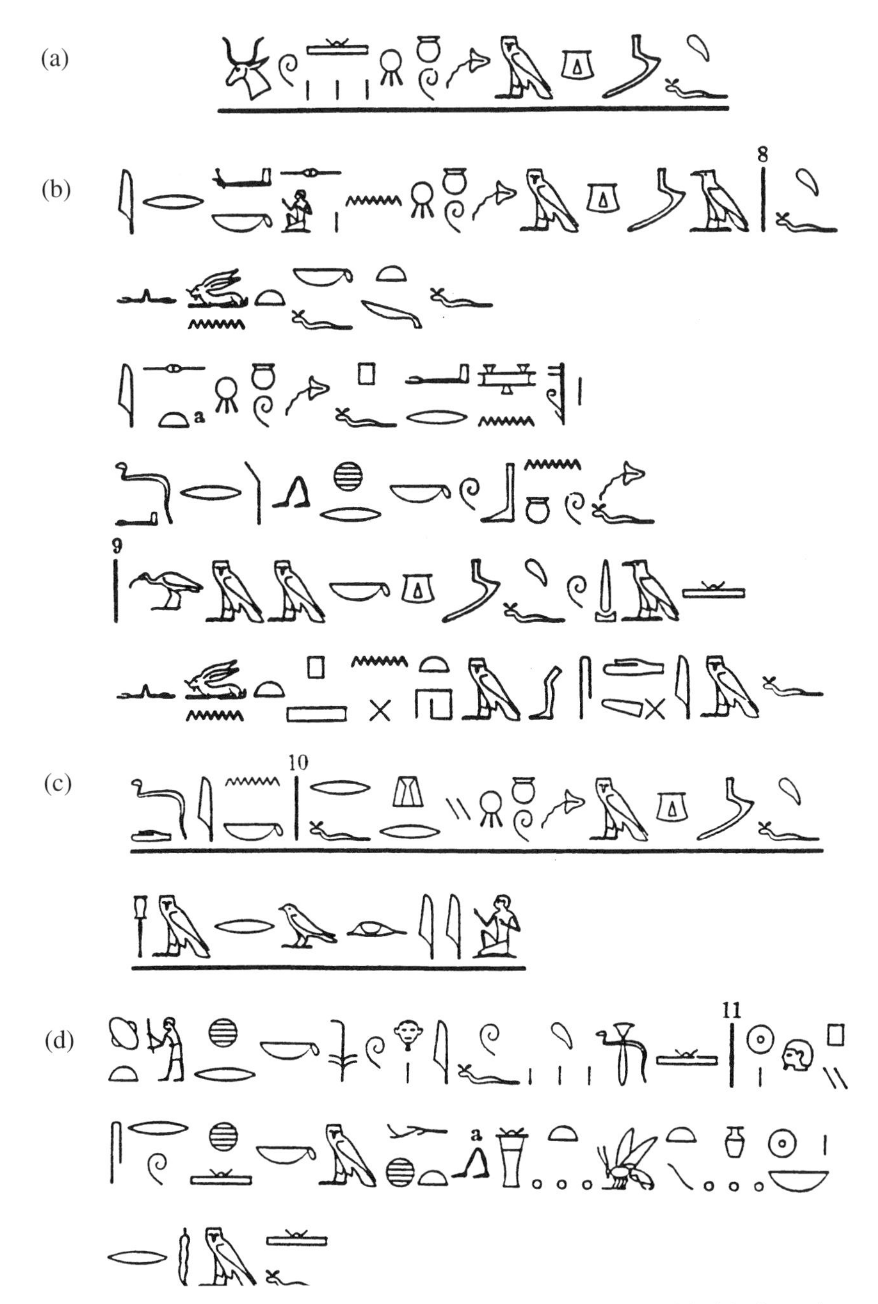

Figure 9.2. Extract from the Edwin Smith Surgical Papyrus (c13th Dynasty). **(a)** Title: Instructions concerning a wound in his temple. **(b)** Examination: If thou examinest a man having a wound in his temple, if not a gash, while that wound penetrates to the bone, thou shouldst palpate his wound. Shouldst thou find his temporal bone uninjured there being no split, (or) perforation, (or) smash on it, (conclusion of diagnosis). **(c)** Diagnosis: Thou shouldst say concerning him: 'One having a wound in his temple. An ailment which I will treat.' **(d)** Treatment: Thou shouldst bind it with fresh meat the first day, (and) thou shouldst treat afterword (with) grease, (and) honey every day, until he recovers. (From Breasted 1930).

irritant dressings and closed so that they might heal promptly. He removed pieces of iron from the flesh with magnets. Instead of cutting down on diet (the custom then) he believed in wine and other 'wound drinks.' Figure 9.3a is a copy of a mediaeval painting said to be of Mondeville (but probably not).

Mondeville's monumental and immensely influential book *Chirurgie* (most current translations from the Latin of the original are in French) was begun in 1306 and left unfinished in 1316. It is forthright, sharp-tongued and outspoken—just like its author. He had no patience with ignorant colleagues noting that *'Many more surgeons know how to cause supperation than to heal a wound.'* He argued against the already outmoded Galen who was still viewed as a god. He said, *'God surely did not exhaust all His creative powers in making Galen.'*

His writings include much practical philosophy and advice, noting that *'It would be vain for a surgeon of our day to know his art, science and operations, if he does not also have the adroitness and knowledge of how to make it pay for him.'* He also blatantly wanted to soak the rich and heal the poor without payment. Some of his practical advice is another mirror of the times, such as, *'Keep up your patient's spirits with violin music and a ten-stringed psalter, with false letters about the deaths of his enemies, or—if he is a spiritual man—by telling him he has been made a bishop.'* He also taught how to 'replace' lost virginity, a profitable business at the time. A brief summary of his savoir-faire is *'... not to shrink from stench; to know how to cut or destroy like a butcher; to be able to lie courteously; and to know how delicately to wheedle gifts and money from patients.'* He also strongly believed in surgery and was critical of the quack physicians (he thought physicians, as opposed to surgeons, were very inferior and uneducated, as, of course, many were). He wrote (*Ars Medica*): *'Surgery is both a science and an art. Surgery is surer, more preferable, noble, perfect, necessary and lucrative ....'*

9.2 Biological Background: Epidermal Wounds

At the basic level, wound healing is crucial for preserving organ integrity after tissue loss. As a subject it is vast and encompasses some aspect from practically every field in biology. Because of the diversity and complexity of the cellular, biochemical and biophysical phenomena involved, the control of this important biological process is still poorly understood in spite of its intense study. In this chapter we can give only an introduction to some of the models that have been proposed for wound repair in the epidermis and, in the following chapter, of full-depth wounds which include the dermis. Much of the current modelling on wound healing, particularly the material we discuss here, is due to Cook, Maini, Sherratt, Tracqui and their coworkers; references will be given at the appropriate places here and in the following chapter. We shall be concerned only with mammalian wounds.

The process of wound healing is traditionally divided into three stages: inflammation (blood-clot formation, influx of leucocytes and so on), wound closure and extracellular matrix remodelling in scar tissue. In a full-depth wound both the dermis and overlying epidermis are removed at injury. The second stage is accomplished by epidermal migration, in which epidermal cells spread across the wound, and wound contraction, in which the main body of the wound contracts causing the wound edges to move

(a)

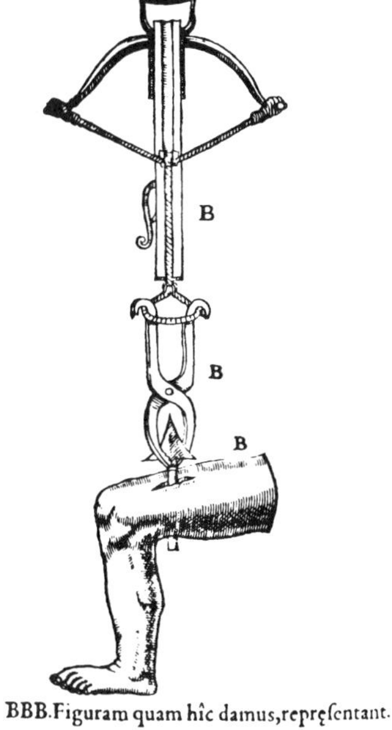

(c)

(b)

Figure 9.3. **(a)** Reputed painting of the great mediaeval surgeon Henri de Mondeville (1260(?)–c1320) whose book on surgery was arguably the most influential book in the later mediaeval period. (Reproduced with permission of the Wellcome Trust Library) **(b)** Illustrations from a mediaeval medical book. It seems that no problem was viewed as untreatable in one way or another! **(c)** Suggested method for removing an arrowhead. It was certainly quick.

inwards. In an epidermal wound, wound closure—re-epithelialisation—is entirely due to epidermal migration which thus provides a way to study this process independently of wound contraction due to the dermis. In epidermal wound healing there are fundamental differences between adult and embryonic wound healing. In the next chapter we discuss full-depth wound healing; it is considerably more complex. For the rest of this chapter we shall only be concerned with epidermal wound healing, both adult and fetal.

The mechanism of epidermal migration is still only partially understood which is, of course, one of the motivations for investigating possible models. An extensive review of epidermal wound healing is given by Sherratt (1991). Normal epidermal cells are non-motile. However, in the neighbourhood of the wound, they undergo marked phenotype alteration that gives the cells the ability to move via finger-like lamellipodia (Clark 1988). Although the main factor controlling cell movement seems to be contact inhibition (for example, Bereiter-Hahn 1986) there is probably regulation by the growth factor profile via chemotaxis, contact guidance and mitogenetic effects. Autoinhibitors are also involved and well documented; see the brief review by Sherratt and Murray (1992) who also discuss some of the clinical implications. Below we include in the models some aspects of activators and inhibitors in regulating the re-epithelialisation since they have been demonstrated in experiments.

As a whole, the spreading epithelial sheet has one or two flattened cells at the advancing margin, whereas the epidermis behind this front is between two and four cells deep. Various mechanisms have been proposed for the movement of the sheet. In one, the 'rolling mechanism,' the leading cells are successively implanted as new basal cells and take on an oval or cuboidal shape and become embedded in the wound surface. Other cells roll over these. In the 'sliding mechanism,' on the other hand, the cells in the interior of the sheet respond passively to the pull of the marginal cells. However, all of the migrating cells have the potential to be motile: for example, if a gap opens in the migrating sheet, cells at the boundary of this develop lamellipodia and move inwards to close the gap (Trinkaus 1984, a good source book on forces in embryogenesis). Although the morphological data of mammalian epidermal wound healing is plausibly explained by the rolling mechanism unequivocal evidence is lacking, whereas the sliding mechanism is well documented in simpler systems such as amphibian epidermal wound closure (Radice 1980).

After injury there is no immediate increase in the rate of cell generation above the normal mitotic rate found in epidermis; epidermal migration is essentially a spreading out of the existing cells. However, soon after the onset of epidermal migration, mitotic activity increases in a band (about 1 mm thick) of the new epidermis near the wound edge which provides an additional population of cells at this source (Bereiter-Hahn 1986). The greatest mitotic activity is actually at the wound edge, where it can be as much as 15 times the rate in normal epidermis (Danjo et al. 1987); moving outwards, activity decreases rapidly across the band. The stimulus for this increase in mitotic activity is still uncertain; various factors have been proposed. Two factors that are certainly involved are the absence of contact inhibition, which applies to mitosis as well as to cell motion (Clark 1988), and change in cell shape: as the cells spread out they become flatter, which tends to increase their rate of division (Folkman and Moscona 1978). There is also experimental evidence, which we now briefly review, for production by epidermal cells both of chemicals that inhibit mitosis and of chemicals that stimulate it.

The evidence for inhibitory growth regulators is considerable. There are two established epidermal inhibitors, which act at different points in the cell cycle. Their chemical properties are summarized by Fremuth (1984). Experimental work to investigate dose–response relationships has shown a general increase in inhibitory effect with dosage although beyond this it is inconclusive. Yamaguchi et al. (1974) investigated the variation of proliferation rate with time near the edge of wound in mice, and concluded that inhibition occurs at three distinct points in the cell cycle. Sherratt and Murray (1992) briefly summarize the arguments for an inhibitor role.

With regard to epidermal growth activators produced by epidermal cells themselves, evidence for these is given by Eisinger et al. (1988a,b), respectively *in vivo* and *in vitro* studies. In the *in vivo* study, an extract derived from epidermal cell cultures was found to increase the rate of epidermal migration when applied, on a dressing, to wounds in pigs. In the *in vitro* study, the same extract was found to increase the growth rate of cultures of epidermal cells.

More recently Werner et al. (1994) have shown that the mitotic activator keratinocyte growth factor is an important regulator in epidermal wound healing. This is produced in the dermis, but in response to epidermal signals, as an early response to injury.

9.3 Model for Epidermal Wound Healing

Two simple models were proposed by Sherratt and Murray (1990) who, based on comparison with experiment, concluded that the following was the more biologically realistic. The better of the models, and the one that compares more favorably with experiment, consists of two conservation equations, one for the epithelial cell density per unit area, (n), and one for the concentration, (c), of the mitosis-regulating chemical. In view of the above comments, we consider two cases, one in which the chemical activates mitosis and the other in which it inhibits it. The epidermis is sufficiently thin that we can reasonably consider the wound to be two-dimensional. This is a reasonable assumption since we consider wounds whose linear dimensions are of the order of centimetres while the thickness of the epidermis is $O(10^{-2})$ cm. The general word form of the model equations is:

$$\boxed{\begin{array}{c}\text{rate of change}\\ \text{of cell density, } n\end{array}} = \boxed{\begin{array}{c}\text{cell}\\ \text{migration}\end{array}} + \boxed{\begin{array}{c}\text{mitotic}\\ \text{generation}\end{array}} - \boxed{\begin{array}{c}\text{natural}\\ \text{loss}\end{array}}, \tag{9.1}$$

$$\boxed{\begin{array}{c}\text{rate of increase}\\ \text{of chemical}\\ \text{concentration, } c\end{array}} = \boxed{\begin{array}{c}\text{diffusion}\\ \text{of } c\end{array}} + \boxed{\begin{array}{c}\text{production}\\ \text{of } c \text{ by cells}\end{array}} - \boxed{\begin{array}{c}\text{decay of}\\ \text{active chemical}\end{array}}. \tag{9.2}$$

We use a diffusion term to model contact inhibition controlled cell migration. Following the representations of short range cellular diffusion in models discussed in Chapter 6 we choose a constant diffusion coefficient, independent of n. Sherratt and Murray (1990) showed that a density-dependent diffusion term, in the absence of biochemical control, is unable to capture crucial aspects of the healing process as demonstrated by the experiments of Van den Brenk (1956), Crosson et al. (1986), Zieske et al. (1987),

Lindquist (1946) and Frantz et al. (1989); these experimental results are summarized in Figure 9.6 below. We now consider the mathematical representation of each reaction term in turn.

The time decay of an active chemical is typically governed by first-order kinetics, so we model this term in the now usual way by $-\lambda c$, where λ is a positive rate constant.

The term for the production of the chemical by the cells, $f(n)$ say, is less simple. This is a function of n, such that $f(0) = 0$, since with no cells nothing can be produced, and $f(n_0) = \lambda c_0$ so that the unwounded state is a steady state. Here n_0 and c_0 denote the unwounded cell density and chemical concentration; we assume a nonzero concentration of chemical in the unwounded state. Further, $f(n)$ must reflect an appropriate cellular response to injury depending on whether the chemical activates or inhibits mitosis. Typical qualitative forms of $f(n)$ in these two cases are shown in Figure 9.4. To be specific we take simple functional forms that conform to these requirements: namely,

$$f(n) = \lambda c_0 \cdot \frac{n}{n_0} \cdot \left(\frac{n_0^2 + \alpha^2}{n^2 + \alpha^2} \right) \qquad \text{for the activator} \tag{9.3}$$

$$f(n) = \frac{\lambda c_0}{n_0} \cdot n \qquad \text{for the inhibitor,} \tag{9.4}$$

where α is a positive parameter which relates to the maximum rate of chemical production.

The rate of natural cell loss is due to the sloughing of the outermost layer of epidermal cells, and we take it as proportional to n, kn say, where k is another positive parameter.

We now have to consider the functional form for the chemically controlled cell division. We choose this term so that when $c = c_0$, the unwounded concentration, the sum of this term plus the previous one for cell decay (that is, the right-hand side kinetics term for n) is of logistic growth form, $kn(1 - n/n_0)$. This, as we have seen, is a commonly used metaphor for simple growth in population biology models; k is the linear mitotic rate. So, we model this term with $s(c)n(2-n/n_0)$, where $s(c)$ reflects the

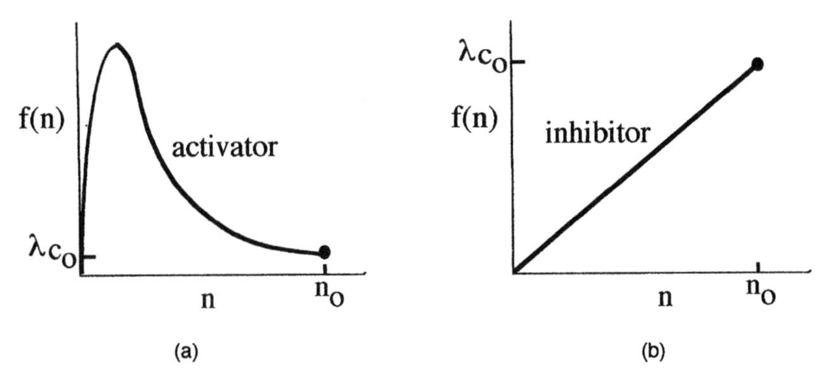

Figure 9.4. Qualitative forms of the function $f(n)$, which reflect the rate of chemical production by the epidermal cells, n: (**a**) activator; (**b**) inhibitor.

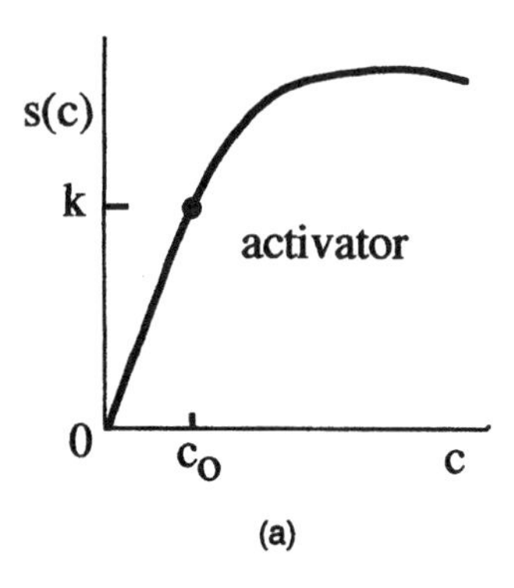

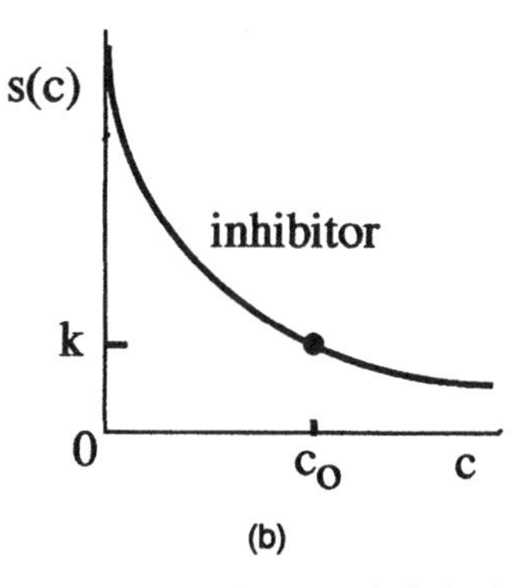

Figure 9.5. The qualitative form of the function $s(c)$, which reflects the chemical control of mitosis: (**a**) activator; (**b**) inhibitor. Here c_0 represents the steady state chemical concentration in the unwounded state, and k is a parameter equal to the reciprocal of the cell cycle time. The • marks the steady state uninjured conditions.

chemical control of mitosis, and $s(c_0) = k$; $s(c)$ is qualitatively as shown in Figure 9.5 for the two possibilities—activation and inhibition. In the case of a chemical activator, a decrease of $s(c)$ to $s(0)$ for large c is included because it is found experimentally (Eisinger 1988a); we shall see that it has little effect on the solutions of the model equations. In both cases we require $0 < s_\infty < s_{\max} = hk$, say, where h is a constant, and we take $s_\infty = k/2$. Again we take simple functional forms satisfying these criteria: namely,

$$s(c) = k \cdot \left\{ \frac{2c_m(h-\beta)c}{c_m^2 + c^2} + \beta \right\} \text{ where } \beta = \frac{c_0^2 + c_m^2 - 2hc_0c_m}{(c_0 - c_m)^2} \tag{9.5}$$

for the activator, where $c_m (> c_0)$ is another constant parameter which relates to the maximum level of chemical activation of mitosis, and

$$s(c) = \frac{(h-1)c + hc_0}{2(h-1)c + c_0} \cdot k \tag{9.6}$$

for the inhibitor.

With these specific forms for the terms in the above word equation system (9.1) and (9.2) the model equations are

$$\frac{\partial n}{\partial t} = D\nabla^2 n + s(c)n\left(2 - \frac{n}{n_0}\right) - kn \tag{9.7}$$

$$\frac{\partial c}{\partial t} = D_c\nabla^2 c + f(n) - \lambda c. \tag{9.8}$$

We now envisage a scenario which mimics many of the experiments; namely, we form a wound domain by removing the epidermis which gives initial conditions as

$$n = c = 0 \text{ at } t = 0 \text{ inside the wound domain,} \tag{9.9}$$

and boundary conditions

$$n = n_0, \, c = c_0 \text{ on the wound boundary, for all } t. \tag{9.10}$$

There is no agreement in the biological literature as to whether mitosis drives cell migration or vice versa (see, for example, Potten et al. 1984, Wright and Alison 1984). The biologically reasonable results given by the model here and discussed below are based on the assumption that, in fact, both processes are dependent on the local cell density.

9.4 Nondimensional Form, Linear Stability and Parameter Values

As usual we now nondimensionalise the model system of equations. We use a length scale L (a typical linear dimension of the wound) and choose the timescale $1/k$ (the cell cycle time seems the most relevant timescale). We then introduce the dimensionless quantities (denoted by *):

$$\begin{aligned} n^* &= \frac{n}{n_0}, \quad c^* = \frac{c}{c_0}, \quad \vec{r}^* = \frac{\vec{r}}{L}, \quad t^* = kt, \quad D^* = \frac{D}{kL^2}, \\ D_c^* &= \frac{D_c}{kL^2}, \quad \lambda^* = \frac{\lambda}{k}, \quad c_m^* = \frac{c_m}{c_0}, \quad \alpha^* = \frac{\alpha}{n_0}. \end{aligned} \tag{9.11}$$

With these definitions, the dimensionless model equations are, dropping the asterisks for notational simplicity,

$$\frac{\partial n}{\partial t} = D\nabla^2 n + s(c)n(2-n) - n \tag{9.12}$$

$$\frac{\partial c}{\partial t} = D_c \nabla^2 c + \lambda g(n) - \lambda c \tag{9.13}$$

with initial conditions

$$n = c = 0 \text{ at } t = 0 \quad \text{inside the wound domain,} \tag{9.14}$$

and boundary conditions

$$n = 1 \text{ and } c = 1 \quad \text{on the wound boundary, for all } t. \tag{9.15}$$

Here, for the activator

$$g(n) = \frac{n(1+\alpha^2)}{n^2+\alpha^2}, \quad s(c) = \frac{2c_m(h-\beta)c}{c_m^2+c^2} + \beta \text{ where } \beta = \frac{1+c_m^2-2hc_m}{(1-c_m)^2}, \tag{9.16}$$

and for the inhibitor

$$g(n) = n, \quad s(c) = \frac{(h-1)c + h}{2(h-1)c + 1}, \tag{9.17}$$

where we assume $h > 1$ and $c_m > 1$.

We require the unwounded state to be stable to small perturbations, while the wounded state is unstable. A straightforward linear analysis (which is left as an exercise) shows that these conditions are satisfied provided

$$s(0) > 1/2 \Rightarrow \begin{cases} c_m > (2h-1) + [(2h-1)^2 - 1]^{1/2} \\ h > 1/2 \end{cases} \text{for the} \begin{cases} \text{activator} \\ \text{inhibitor} \end{cases}. \tag{9.18}$$

Parameter Values

We can estimate some of the parameters, specifically λ and k, from experimental data. We estimate λ in the case of a chemical inhibitor using data of Brugal and Pelmont (1975). They found a decrease in the proliferation rate in intestinal epithelium during the 12 h after injection with epithelium extract. Also Hennings et al. (1969) were able to maintain suppression of epidermal DNA synthesis by repeated injection of epidermal extract at 12 h intervals. Based on these studies, we take the half-life of chemical decay as 12 h. If we consider only the decay term in (9.13) this gives exponential decay with a half-life of $\lambda^{-1} \ln 2$. We therefore take $\lambda = 0.05$ ($\approx (1/12) \ln 2h^{-1}$).

In the case of a chemical activator, there are few quantitative experimental data. However, comparison of the work of Eisinger et al. (1988a,b) on chemical activators in wound healing and the clinical studies of Rytömaa and Kiviniemi (1969, 1970) suggests a longer timescale for the activator activity, by a factor of about 6, so we take $\lambda = 0.3\ \text{h}^{-1}$ for the activator.

The parameter k is simply the reciprocal of the epidermal cell cycle time. This varies from species to species, but is typically about 100 h (Wright 1983), so we take $k = 0.01\ \text{h}^{-1}$. The diffusion coefficients D and D_c were estimated by Sherratt and Murray (1990) based on a best fit analysis with data on wound healing, since there are at present no direct experimental data from which they can be determined. This gave values $D = 3.5 \times 10^{-10}\ \text{cm}^2\ \text{s}^{-1}$, $D_c = 3.1 \times 10^{-7}\ \text{cm}^2\ \text{s}^{-1}$ for the activator, and $D = 6.9 \times 10^{-11}\ \text{cm}^2\ \text{s}^{-1}$, $D_c = 5.9 \times 10^{-6}\ \text{cm}^2\ \text{s}^{-1}$ for the inhibitor. These are not biologically unreasonable for cells and biochemicals of relatively low molecular weight, respectively.

9.5 Numerical Solution for the Epidermal Wound Repair Model

Sherratt and Murray (1990, 1991) numerically solved the model system of equations (9.12) and (9.13) together with (9.14)–(9.18) in a radially symmetric geometry and the results were compared with data from a variety of experimental sources both *in vitro* and *in vivo*. The results are given in Figure 9.6 together with the quantitative results from experimental studies on re-epithelialisation. For example, in Van den Brenk's (1956)

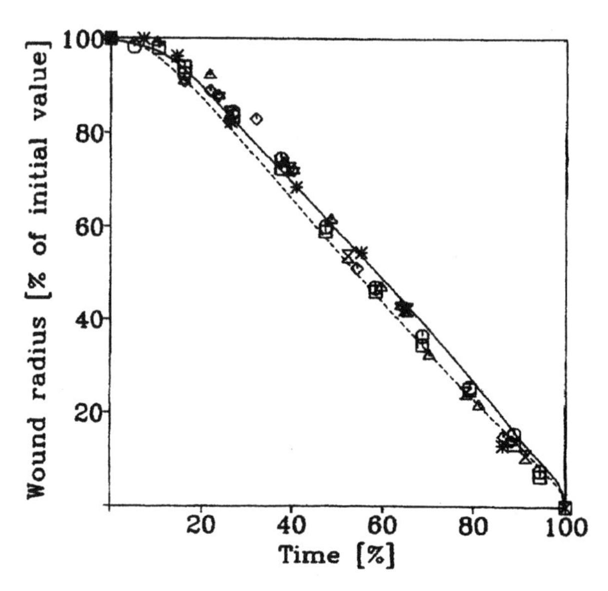

Figure 9.6. The decrease in wound radius with time for the normal healing of a circular wound, with time expressed as a percentage of total healing time (the way much of the experimental data are presented). The results are: the solid line denotes the activator mechanism and the dashed line the inhibitor mechanism. For comparison we also give the data from experiments involving a range of species and wounding location, but all involving wounds of diameter about 1 cm, which is the value used in the model solutions. The sources of the data are: □, ○ Van den Brenk (1956); △ Crosson et al. (1986); ◇ Zieske et al. (1987); ▽/△ Lindquist (1946); ∗ Frantz et al. (1989). In Lindquist's (1946) experiments there is some dermal contraction, so we extrapolated to the case of no contraction. The dimensionless parameter values used in the model simulations are: for the activator mechanism, $D = 5 \times 10^{-4}$, $D_c = 0.45$, $\lambda = 30$, $h = 10$, $\alpha = 0.1$, $c_m = 40$, and for the inhibitor mechanism, $D = 10^{-4}$, $D_c = 0.85$, $\lambda = 5$, $h = 10$. Interestingly for both types of mitotic regulation, the model solutions compare well with experimental data. (From Sherratt and Murray 1992)

experiments the full thickness of epidermis was removed from a circular region, 1 cm in diameter, in the ears of rabbits. Particular care was taken not to leave behind any hair follicles so that our model with no 'internal' sources of epidermal cells is appropriate. The change in wound radius with time was recorded. To capture the concept of 'wound radius' from our model, Sherratt and Murray (1990, 1991) considered the wound as 'healed' when the cell density reached 80% of its unwounded value, that is, when $n = 0.8$ for the nondimensional equations. The choice of this level as 80% is somewhat arbitrary, but does not significantly affect the results since the solutions for n and c have travelling waveform, which we discuss below. Figure 9.7 gives plots of n and c against r at a selection of equally spaced times.

As well as giving good overall agreement with the data, the numerical solutions exhibit the two phases (a lag phase and then a linear phase) that characterise epidermal wound healing (see, for example, Snowden 1984). The (constant) speed of the linear phase can be approximately calculated visually from the graph of n against r. For a wound radius of 0.5 cm, this gives dimensional wavespeeds of 2.6×10^{-3} mm h^{-1} for the activator and 1.2×10^{-3} mm h^{-1} for the inhibitor. These compare, for example, with the speed 8.6×10^{-3} mm h^{-1} found in the Van den Brenk (1956) study.

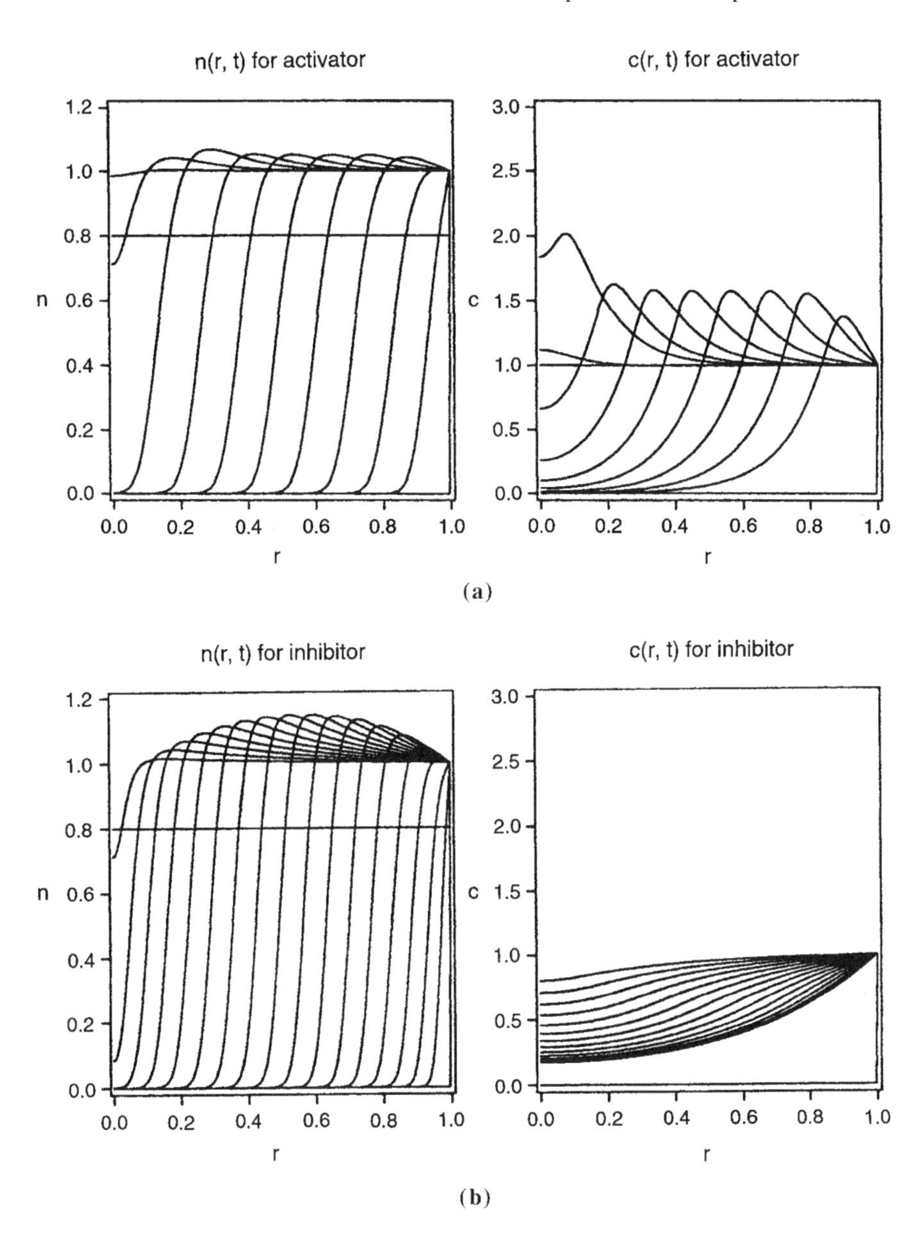

Figure 9.7. Cell density (n) and chemical concentration (c) as a function of radius (r) at a selection of equally spaced times, obtained from (9.12) and (9.13). (**a**) Biochemical activation of mitosis with parameter values $D = 5 \times 10^{-4}$, $D_c = 0.45$, $\lambda = 30$, $h = 10$, $\alpha = 0.1$, $c_m = 40$; (**b**) biochemical inhibition of mitosis with parameter values $D = 10^{-4}$, $D_c = 0.85$, $\lambda = 5$, $h = 10$. (From Sherratt and Murray 1990, 1991)

An interesting result of the comparison of the numerical solutions with the data is that with these parameter values (as reasonable as we can determine) there is little difference between the activator and inhibitor mechanisms which in part explains the lack of agreement as to which obtains in re-epithelialisation. As we show below there is a difference between the two possible mechanisms in the wound healing time and also when we discuss the effect of geometry on the healing time.

9.6 Travelling Wave Solutions for the Epidermal Model

For both types of chemical, the qualitative form of the solution in the linear phase is of a wave moving with constant shape and constant speed. Such a solution suggests that there could be a travelling wave of repair and perhaps amenable to analysis if we consider a one-dimensional geometry, rather than the two-dimensional radially symmetric geometry considered above. This situation is biologically relevant for large wounds of any fairly regular shape, since to a good approximation these are one-dimensional during much of the healing process. Numerical solutions of the model equations for this new one-dimensional geometry are not significantly different from those in Figures 9.6 and 9.22; the dimensionless wavespeeds are approximately 0.05 for the activator and 0.03 for the inhibitor.

We look for travelling wave solutions of the form $n(x,t) = N(z)$, $c(x,t) = C(z)$, $z = x + at$, where a is the wavespeed, positive here since we consider waves moving to the left. Substituting these forms into the model equations (9.12) and (9.13) we get the following system of ordinary differential equations,

$$aN' = DN'' + s(C)N(2-N) - N \tag{9.19}$$

$$aC' = D_c C'' + \lambda g(N) - \lambda C, \tag{9.20}$$

where primes denote differentiation with respect to z. Biologically appropriate boundary conditions are

$$N(-\infty) = C(-\infty) = 0,\ N(+\infty) = C(+\infty) = 1,\ N'(\pm\infty) = C'(\pm\infty) = 0. \tag{9.21}$$

Since the system of ordinary differential equations is fourth-order, and therefore not easy to analyse globally, we consider two reasonable approximations which reduce the order of the system: in the first we consider λ as infinite and in the second we take D to be zero. For the parameter values we are using, the values of these dimensionless parameters are $D = 5 \times 10^{-4}$, $\lambda = 30$ for the activator, and $D = 10^{-4}$, $\lambda = 5$ for the inhibitor. Biologically, the first approximation corresponds to the chemical kinetics always being in equilibrium, while the second corresponds to an absence of any cell diffusion, so that increase in cell density is due only to mitosis.

In the numerical solution $c \ll c_m$ for all x and t, and so it is a good approximation in the activator case to take $s(c)$ as a simple linear function. Specifically, we take

$$s(c) \approx \gamma c + 1 - \gamma \quad \text{where } \gamma = \frac{2(h-1)}{c_m - 2}.$$

The numerical solution of (9.12) and (9.13) using this linear form differs negligibly from that with the original form. We use this linear approximation in the subsequent analysis, since it makes the analysis much easier algebraically. The approximation is valid provided $c_m \gg 1$.

Travelling Wave Solutions with $\lambda = \infty$

Here we consider that the derivative terms in the second equation are negligible compared with the reaction terms. Intuitively this seems a reasonable approximation in the case of an inhibitor ($\lambda = 5$), and a good approximation in the case of an activator ($\lambda = 30$). The system (9.21) then reduces to a second-order ordinary differential equation for N, namely,

$$N'' = \frac{a}{D}N' - \frac{1}{D}\psi(N) \quad \text{where } \psi(N) = sNg(N)(2-N) - N, \tag{9.22}$$

and we look for a solution with boundary conditions, from (9.21), $N(+\infty) = 1$, $N(-\infty) = 0$ and $N'(\pm\infty) = 0$, and, of course, with $N \geq 0$ everywhere.

A plot of $\psi(N)$ on the interval (0, 1) shows that it essentially has a parabolic shape for both forms of $g(N)$ in (9.16), the activator form, and (9.17), the inhibitor form. So, this equation can be analysed in an analogous way to the standard analysis for travelling wave solutions of the Fisher–Kolmogoroff equation, $u_t = u_{xx} + u(1-u)$ discussed in detail in Chapter 13, Volume I. There is a unique solution of the required form for each wavespeed $a \geq a_{\min} = 2[D(2s(0) - 1)]^{1/2}$. In the usual way we expect the solution to evolve to a travelling wave with $a = a_{\min}$ for initial conditions such that $N = 1$ for sufficiently large z and $N = 0$ for sufficiently small z; biologically relevant initial conditions for our problem certainly satisfy these conditions. The parameter values we are using give dimensionless values for the wavespeed as $a_{\min} = 0.01$ for the activator and $a_{\min} = 0.09$ for the inhibitor. These compare to the wavespeeds 0.05 and 0.03 respectively found in the numerical solution of (9.12) and (9.13). The discrepancy indicates inadequacies in the approximation of chemical equilibrium in which the chemical decays as it is formed; intuitively we would expect this approximation to give a lower wavespeed for the activator and a higher wavespeed for the inhibitor.

In the inhibitor case an approximate analytic solution for the travelling wave can also be obtained. If we rescale the independent variable, $\zeta = z/a$, (9.22) becomes

$$\epsilon N'' - N' + \psi(N) = 0 \quad \text{where } \epsilon = D/a^2 \tag{9.23}$$

and primes denote $d/d\zeta$, $\zeta = z/a$. This looks like a singular perturbation problem in the small parameter $\epsilon \approx 0.01$, but it can in fact be solved by regular perturbation techniques. It is analogous to solving the travelling wave problem for the Fisher–Kolmogoroff equation in Chapter 13, Volume I. There are some significant differences, however, so we carry out the analysis to introduce a technique not covered elsewhere in the book. In the activator case, however, the method fails because $\epsilon \approx 5$ is large rather than small. So, following the procedure in Chapter 13, Volume I we look for a regular perturbation solution of (9.23) in the form

$$N(\zeta;\epsilon) = N_0(\zeta) + \epsilon N_1(\zeta) + \epsilon^2 N_2(\zeta) + \cdots,$$

which on substituting into (9.23) and equating coefficients of powers of ϵ we get

$$N_0' = \psi(N_0), \quad N_1' = N_1 \frac{d\psi(N_0)}{dN_0} + N_0'' \dots .$$

The boundary conditions are

$$\begin{aligned} &N_0(-\infty) = 0, \quad N_0(0) = 1/2, \quad N_0(+\infty) = 1 \\ &N_i(-\infty) = N_i(0) = N_i(+\infty) = 0 \quad \text{for } i \geq 1. \end{aligned}$$

The value of $N(0)$ is arbitrary; it must be specified to give a unique solution (this simply fixes the origin of z and ζ as well). We choose $N(0) = 1/2$, with which we then get

$$\begin{aligned} \zeta &= \int_{1/2}^{N_0} \frac{d\xi}{\psi(\xi)} \\ &= \left(\frac{1}{2h-1}\right) \ln(2N_0) - \left(\frac{2h-1}{3h-2}\right) \ln[2(1-N_0)] \\ &\quad + \frac{(4h-3)(h-1)}{(2n-1)(3h-2)} \ln\left[\frac{2(h-1)N_0 + 2(2h-1)}{5h-3}\right] \end{aligned}$$

for the inhibitor, assuming $h > 1$ as above. This cannot be inverted explicitly as was possible with the simple Fisher–Kolmogoroff equation. However, observing that ζ is a monotonically increasing function of N_0, we consider N_0, rather than ζ, as the independent variable. Now

$$\frac{dN_1}{d\zeta} = N_1 \frac{d\psi(N_0)}{dN_0} + \frac{d^2 N_0}{d\zeta^2},$$

which, on dividing by $dN_0/d\zeta$, gives

$$\begin{aligned} \frac{dN_1}{dN_0} &= \frac{N_1}{dN_0/d\zeta} \cdot \frac{d\psi(N_0)}{dN_0} + \frac{d}{dN_0}\left(\frac{dN_0}{d\zeta}\right) \\ &= \frac{N_1}{\psi(N_0)} \frac{d\psi(N_0)}{dN_0} + \frac{d\psi(N_0)}{dN_0}, \quad \text{on using } \frac{dN_0}{d\zeta} = \psi(N_0). \end{aligned}$$

Dividing through by $\psi(N_0)$ gives

$$\frac{d}{dN_0}\left[\frac{N_1}{\psi(N_0)} - \ln[\psi(N_0)]\right] = 0.$$

So, since $N_1 = 0$ when $N_0 = 1/2$ (at $\zeta = 0$),

$$N_1 = \psi(N_0) \ln\left[\frac{\psi(N_0)}{\psi(1/2)}\right].$$

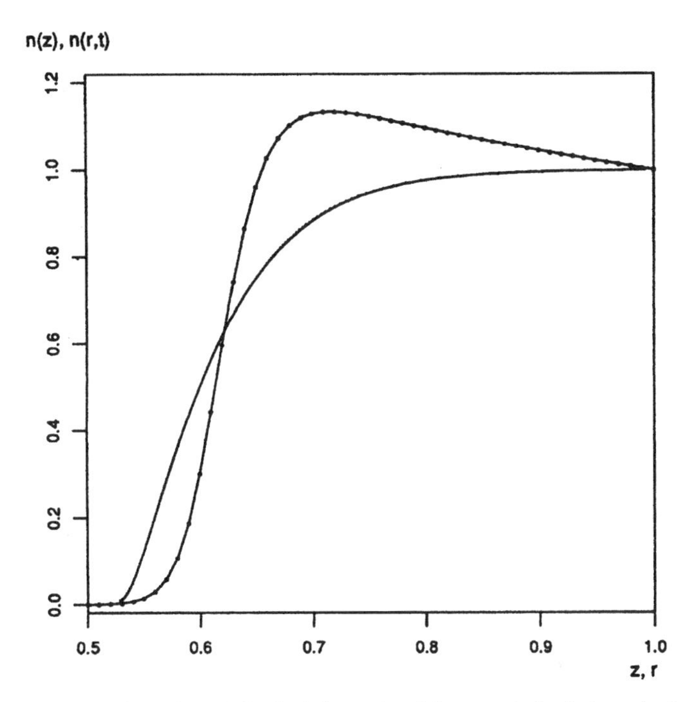

Figure 9.8. Comparison of $N_0(z)(\approx n(z))$ (the full curve) and the numerical solution, $n(r,t)$, of (9.12) and (9.13) (the curve with dots) for the inhibitor model. The $O(\epsilon)$ correction ϵN_1 to the leading order term N_0 in the asymptotic solution of (9.23) is so small that it is not visible here. The parameter values are $D = 10^{-4}$, $D_c = 0.85$, $\lambda = 5$, $h = 10$.

Plots of N_0 and $N_0 + \epsilon N_1$ against z are shown in Figure 9.8 and compared to the numerical solution of (9.12) and (9.13) at a time in the middle of the linear phase of the wound repair. These show that the first-order correction is already quite small and the analytic solution agrees reasonably with the partial differential equation solution in regard to the slope of the linear portion of the wavefront. However, it does not capture the important feature that $n > 1$ in part of the wavefront. This is an inadequacy in the approximation $\lambda = \infty$.

Travelling Wave Solutions with $D = 0$

In view of the shortcomings of the previous approximation we now consider the approximation $D = 0$. Recall that $D = 5 \times 10^{-4}$ for the activator model and $D = 10^{-4}$ for the inhibitor model. The fourth-order system (9.19) and (9.20) now reduces to a third-order system:

$$N' = -\frac{N}{a} + \frac{1}{a}s(C)N(2-N) \tag{9.24}$$

$$C'' = \frac{a}{D_c}C' + \frac{\lambda}{D_c}C - \frac{\lambda}{D_c}g(N), \tag{9.25}$$

and we look for a solution subject to boundary conditions $N(-\infty) = C(-\infty) = 0$, $N(+\infty) = C(+\infty) = 1$, $C'(\pm\infty) = 0$, with $N, C \geq 0$ for all z. It is not an easy analytical problem.

Sherratt and Murray (1991) solved this system numerically, also not a particularly easy problem (see their paper for the details). The solutions they obtained are compared with the numerical solutions of the full partial differential equation system (9.12) and (9.13) in Figure 9.9: for both the activator and inhibitor models, the reasonably close agreement indicates that the approximation $D = 0$ is a good one.

Using a phase space analysis for the activator we can obtain an upper bound on the wavespeed. The numerical solution of (9.24) and (9.25) approaches the steady state monotonically. However, for parameter values close to those we are using, using linear analysis of these equations near the steady state, we find there are two eigenvalues with negative real parts at this steady state, with the third eigenvalue real and positive. Further, the two eigenvalues with negative real parts are complex (implying oscillatory behaviour) unless $a \leq a_{\max}$, where $a_{\max}$ is the value of a at which the eigenvalue equation has two equal negative roots. After more algebra this condition gives the following cubic in $a_{\max}^2$,

$$\begin{aligned}
&\{[(1+\lambda)^2 - 4\Gamma]/(4D_c)\}a_{\max}^6 \\
&\quad + \{4\Gamma(3 + D_c^2) - 18(1+\lambda)\Gamma + 2(1+\lambda)^2(1+2\lambda)\}a_{\max}^4 \\
&\quad + D_c\{(1+\lambda)^2 + 3\Gamma(6\lambda + 2 - 9\Gamma)\}a_{\max}^2 + 4D_c^2\Gamma = 0,
\end{aligned}$$

where $\Gamma = \lambda[1 - s'(1)g'(1)]$. For the parameter values we are using, this equation has a unique solution in $(0.0, 0.1)$ and the numerical solution gives the upper bound as $a_{\max} \approx 0.0546$. This compares with the dimensionless wavespeed 0.05 found in the numerical solution of (9.12) and (9.13).

Because of the parameter values, for the activator, we can again obtain an analytic solution of (9.23) and (9.24) using a regular perturbation theory. It is, however, algebraically more complicated but in principle is the same. In this case we write the equations as

$$aN' = \epsilon(C-1)N(2-N) + N(1-N) \tag{9.26}$$

$$D_cC'' - aC' - \lambda C = -\lambda\frac{(1+\alpha^2)N}{N^2+\alpha^2} \tag{9.27}$$

but here $\epsilon (= 2(h-1)/(c_m - 2) \approx 0.47)$ is the small parameter. With this value for ϵ, we strictly require the $O(\epsilon)$ correction to the $O(1)$ solution. As before we look for solutions of the form

$$\begin{aligned}
N(z;\epsilon) &= N_0(z) + \epsilon N_1(z) + \epsilon^2 N_2(z) + \cdots \\
C(z;\epsilon) &= C_0(z) + \epsilon C_1(z) + \epsilon^2 C_2(z) + \cdots .
\end{aligned} \tag{9.28}$$

Substituting these into the ordinary differential equation system (9.26) and (9.27), changing the independent variable to $\xi = e^{z/a}$ and equating coefficients in ϵ, the $O(1)$ terms

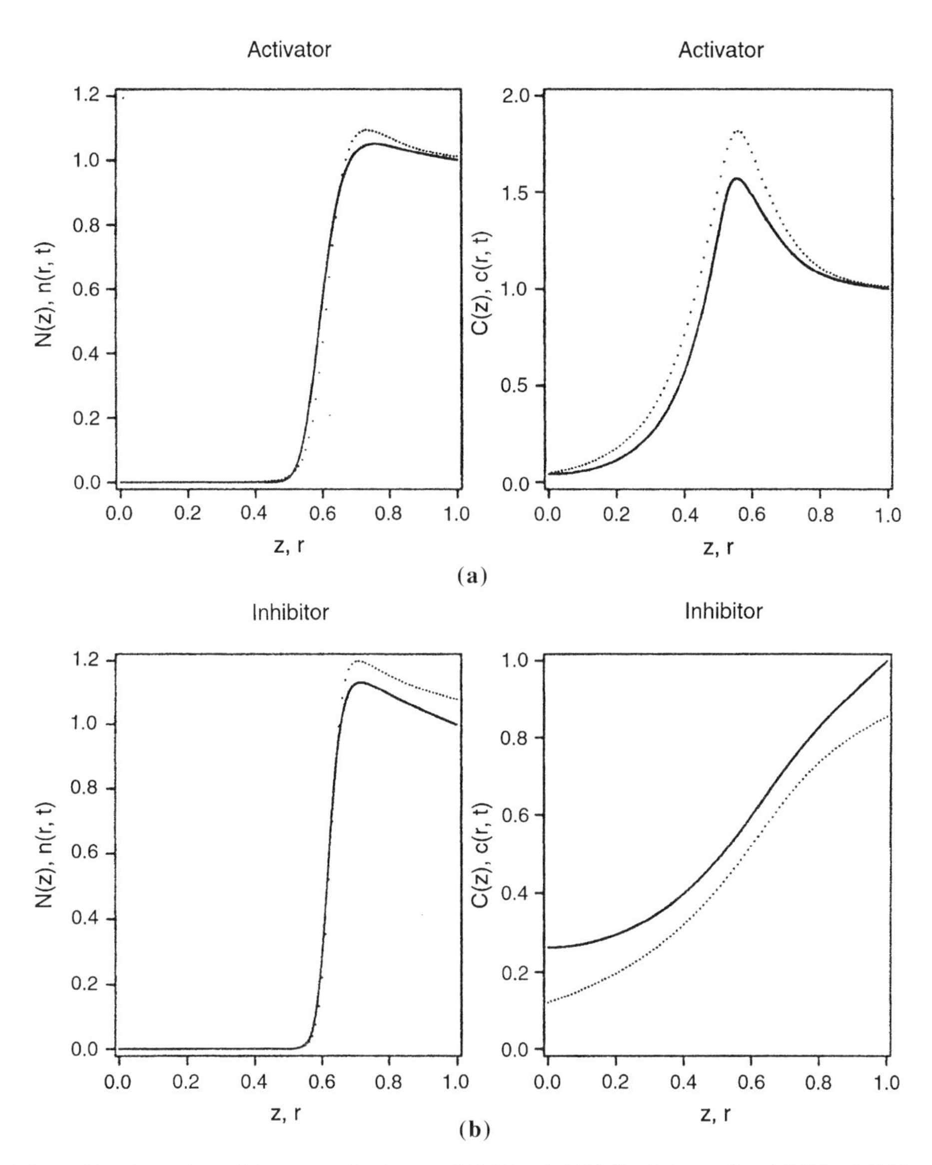

Figure 9.9. Comparison of the numerical solutions of (9.12) and (9.13) (the full curves) and (9.24) and (9.25) (the dotted curves). (**a**) Activator model with parameter values $D = 5 \times 10^{-4}$, $D_c = 0.45$, $\lambda = 30$, $h = 10$, $\alpha = 0.1$, $c_m = 40$, $a = 0.05$; (**b**) inhibitor model with parameter values $D = 10^{-4}$, $D_c = 0.85$, $\lambda = 5$, $h = 10$, $a = 0.03$. The values for the wavespeed a are those calculated from the numerical solutions of (9.12) and (9.13).

give

$$\xi N_0' = N_0 - N_0^2$$

$$\kappa\xi(\xi C_0')' - \xi C_0' - \lambda C_0 = -\lambda \frac{(1+\alpha^2)N_0}{N_0^2 + \alpha^2},$$

where $\kappa = D_c/a^2$ and prime now denotes $d/d\xi$. The relevant boundary conditions are $N_0(+\infty) = C_0(+\infty) = 1$, $N_0(0) = C_0(0) = 0$, with $N_0(1) = 1/2$ for uniqueness. The solution for N_0 is simply $N_0 = \xi/(1+\xi)$. For C_0, the standard method of variation of parameters gives the general solution of

$$\kappa\xi[\xi y'(\xi)]' - \xi y'(\xi) - \lambda y(\xi) = F(\xi) \text{ as } y(\xi) = \gamma_+(\xi)\xi^{q^+} + \gamma_-(\xi)\xi^{q^-}, \quad (9.29)$$

where

$$\gamma_\pm' = \frac{\pm 1}{\sqrt{1+4\lambda\kappa}} \cdot \frac{F(\xi)}{\xi^{(q^\pm+1)}} \quad \text{and} \quad q^\pm = \frac{1 \pm \sqrt{1+4\lambda\kappa}}{2\kappa}.$$

Using this,

$$C_0 = \frac{\lambda(1+\alpha^2)}{\sqrt{1+4\lambda\kappa}} \left[\xi^{q^+} \int_\xi^\infty \frac{N_0(x)}{x^{(q^+ +1)}[N_0^2(x)+\alpha^2]}\,dx \right.$$
$$\left. + \xi^{q^-} \int_0^\xi \frac{N_0(x)}{x^{(q^- +1)}[N_0^2(x)+\alpha^2]}\,dx \right].$$

Here the values of the two constant limits of integration are necessary (but not sufficient) for convergence at $\xi = 0$ and $+\infty$.

We now consider the boundary conditions, which imply convergence, by investigating the behaviour of the integrals in the last expression for C_0 as $\xi \to 0$ and $\xi \to +\infty$. We have

$$\lim_{\xi\to+\infty} \xi^{q^+} \int_\xi^{+\infty} \frac{N_0(x)}{x^{(q^+ +1)}[N_0^2(x)+\alpha^2]}\,dx$$
$$= \lim_{\theta\to 0} \frac{1}{\theta^{q^+}} \int_{1/\theta}^{+\infty} \frac{N_0(x)}{x^{(q^+ +1)}[N_0^2(x)+\alpha^2]}\,dx$$
$$= \lim_{\theta\to 0} \frac{-\theta^{q^+}(1+1/\theta)(-1/\theta^2)}{q^+\theta^{(q^+ -1)}[\theta^{-2} + \alpha^2(1+1/\theta)^2]}$$
$$= \frac{1}{q^+(1+\alpha^2)},$$

where we used L'Hôpital's rule and the expression for N_0 in the third step. Similarly

$$\lim_{\xi \to +\infty} \xi^{q^-} \int_0^{\xi} \frac{N_0(x)}{x^{(q^- + 1)}[N_0^2(x) + \alpha^2]} dx = \frac{-1}{q^-(1 + \alpha^2)},$$

and so

$$C_0(+\infty) = \frac{\lambda(1 + \alpha^2)}{\sqrt{1 + 4\lambda\kappa}} \cdot \frac{1}{1 + \alpha^2} \cdot \left(\frac{1}{q^+} - \frac{1}{q^-} \right) = 1$$

on using the expressions for $q^{\pm}$. Similarly the condition at $\xi = 0$ is satisfied.

Now consider the first-order perturbations which give, on equating coefficients of ϵ,

$$\xi N_1' = N_1(1 - 2N_0) + (C_0 - 1)N_0(2 - N_0)$$

$$\kappa\xi(\xi C_1')' - \xi C_1' - \lambda C_1 = -\lambda(1 + \alpha^2) \left[\frac{\alpha^2 - N_0^2}{(\alpha^2 + N_0^2)^2} \right] N_1.$$

The relevant boundary conditions are now

$$N_1(0) = N_1(+\infty) = C_1(0) = C_1(+\infty) = N_1(1) = 0.$$

In the first equation, substituting for $N_0 = \xi/(1 + \xi)$, multiplying through by the integrating factor $(1 + 1/\xi)^2$ and integrating gives

$$N_1 = \frac{1}{\xi + 2 + 1/\xi} \int_1^{\xi} [C_0(x) - 1] \left(1 + \frac{2}{x} \right) dx.$$

Using L'Hôpital's rule again, as above, shows that the boundary conditions are satisfied. Then (9.29) gives C_1, and again using L'Hôpital's rule confirms that the boundary conditions are satisfied. By repeating this process we can derive (albeit with a lot of messy algebra) all the terms in the expansion, showing in particular that λ, D_c and a occur in each term of the series (9.28), and so in the solution as a whole, only within the groupings $q^{\pm}$ and $\lambda/\sqrt{1 + 4\lambda\kappa}$.

It is encouraging that such a relatively simple model for epidermal wound healing in which the parameter values are based as far as possible on experimental fact give such good comparison with either chemical activation or inhibition of mitosis with experimental data on the normal healing of circular wounds. These results tend to support the view that biochemical regulation of mitosis is fundamental to the process of epidermal migration in wound healing. The analytical investigation of the solutions was possible because these numerical solutions approximate travelling waves during most of the healing process. Analysis of the two biologically relevant approximations gives information about the accuracy of these approximations and, usefully, the roles of the various model parameters in the speed of healing of the wound.

9.7 Clinical Implications of the Epidermal Wound Model

The available experimental evidence strongly suggests that mitogenic autoregulation is the dominant control mechanism in epidermal wound repair and the existence of autoinhibitors and autoactivators is well documented. In unwounded epidermis, homeostasis seems to be due to an interplay between growth activators and inhibitors (Watt 1988a,b), and the role of such regulators in re-epithelialisation has been demonstrated both *in vitro* and *in vivo* (Eisinger et al. 1988a,b, Madden et al. 1989, Yamaguchi et al. 1974).

The epidermal models we discussed above let us select key aspects of the underlying biology to be investigated individually, and because of the analytical results, the effect of parameter variation. Since the model results compare well with existing experimental data (recall Figure 9.6) we can now go on to make predictions which may suggest new ideas and directions for experimental studies. We do this in this section, following the work of Sherratt and Murray (1992a,b) who discussed such predictions for re-epithelialisation.

Topical Application of Mitotic Regulators

Since our model of epidermal migration focuses on chemical autoregulation of cell division we begin by investigating the effect of applying additional quantities of the mitosis-regulating chemical onto the wound surface. Sherratt and Murray (1992a,b) first considered the effects of a single, 'one off' addition of the activator and inhibitor chemical at various points during healing, but this produced no significant effect, even when the added chemical concentration was so high as to be experimentally unrealistic. This was caused, as expected intuitively, by the combination of diffusive spread of the added chemical away from the wound, and the exponential decay of active chemical. This suggested a different approach, namely, the gradual release of the regulatory chemical into the wound. Experimentally this can be done by using a dressing soaked either in a solution of isolated chemical or in an epidermal cell extract or exudate, a technique used *in vivo* by Eisinger et al. (1988a,b) and Madden et al. (1989). When such a chemical release is incorporated into the model, the effects on the healing profile are significant, as illustrated in Figure 9.10. The amended model incorporating a constant dressing of activator or inhibitor is simply (9.12) and (9.13) with an extra constant term c_{dress} on the right-hand side of the c-equation. We can then assign various levels of the applied chemical regulator. A given rate of chemical release has a greater effect in the inhibitor case than in the activator case, because experimental data suggest that the rate of chemical decay is significantly higher in the activator case (Sherratt and Murray 1990).

Unfortunately the experiments of Eisinger et al. (1988a,b) and Madden et al. (1989) are only qualitative. However, the predictions illustrated in Figure 9.10 could be tested against data from similar quantitative studies, if such data were available. Not only that, the mathematical formulation depends only on the ratio of the amount of chemical release per hour and the concentration present in unwounded skin. So, experimental measurements of the rate of chemical release required to produce a given change in the total healing time would enable the model solutions to predict, quantitatively, the concentrations of mitotic regulators that are present in unwounded epidermis *in vivo*.

Varying Wound Geometry

Although in the above we considered only circular wounds for simplicity, the model can easily be solved (numerically of course) for any initial wound shape. One of the original aims of the modelling was to investigate and hopefully quantify the effects of wound

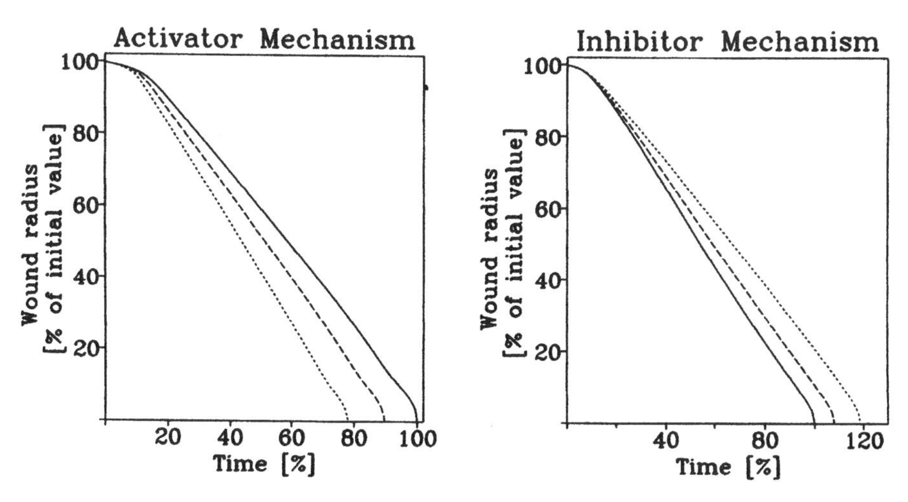

Figure 9.10. The model predictions of the effects of a constant gradual release of mitosis-regulating chemical c_{dress} onto the healing epidermal wound. The solid curves denote the healing profile for a control wound with either autoactivation or autoinhibition of mitosis, and time is expressed as a percentage of the total healing time in these cases. The dashed curves denote the healing profile when chemical is added to the wounded area throughout healing. In each case, the results are for two different rates of chemical release: $c_{\text{dress}} = 0.2$ (dashed line) and $c_{\text{dress}} = 0.5$ (dotted line) per hour in the activator case, and $c_{\text{dress}} = 0.02$ (dashed line) and $c_{\text{dress}} = 0.05$ (dotted line) per hour in the inhibitor case. In dimensional terms these are the fractions of the concentration of regulatory chemical in the unwounded epidermis, namely, c_0 as used in (9.11).

shape on the healing process. To do this we need to identify quantifiable aspects of the initial wound geometry. With rectangular wounds we can choose, for example, the inital midline ratio and thereby compare wound healing times for rectangular shapes. We can take, again by way of example, the boundary function in x, y coordinates to be $y = x^p$, where $0 < p < \infty$ which gives a diamond shape for $p = 1$ to an ovate shape and finally to a rectangular shape as $p \to \infty$. Sherratt and Murray (1992a,b) give some results for such a geometry variation and compare the effect of these shape changes on the healing time. Another approach is to take a cusp-ovate family of boundary functions quantified by the function

$$y = f_{\text{shape}}(x;\alpha) = \frac{1}{2}\left(1+\frac{1}{\alpha}\right) - \text{sign}\,(\alpha)\left[\frac{1}{2}\left(1+\frac{1}{\alpha^2}\right) - \left(x+\frac{1}{2\alpha}-\frac{1}{2}\right)^2\right]^{1/2} \tag{9.30}$$

for $-1 < \alpha < 1$. It is the functional form of an arc of circle with centre $((1/2) - (1/2\alpha), (1/2) + (1/2\alpha))$ and radius $[(1/2)(1 + (1\alpha^2))]^{1/2}$. As $\alpha \to 1$ we get a cusp shape while for $\alpha = 0$ we get a diamond shape. For positive α we get ovate shapes. Figure 9.11 shows how the shapes change with α.

The solutions for four different shapes are illustrated in Figure 9.12. Such solutions let us predict, from our model, the variation in healing time with various aspects of initial wound geometry. In one study Sherratt and Murray (1992a,b) considered the effect on

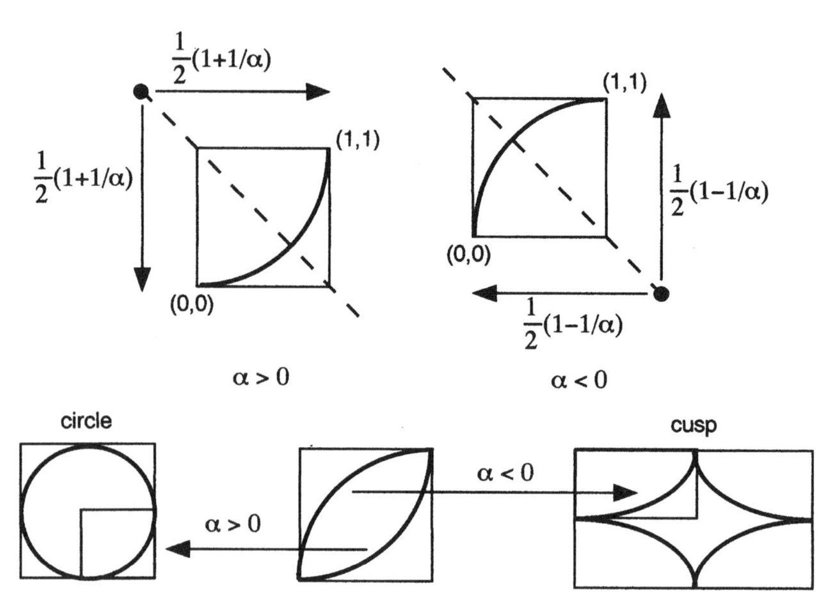

Figure 9.11. Wound shapes, parametrized by the single parameter α lying between -1 and $+1$. The case $\alpha = 0$ corresponds to a diamond shaped wound. The mathematical definition of each of the wound boundaries is given by (9.30). In our model solutions, we take the ratio of the wound midline lengths to be 3:2.

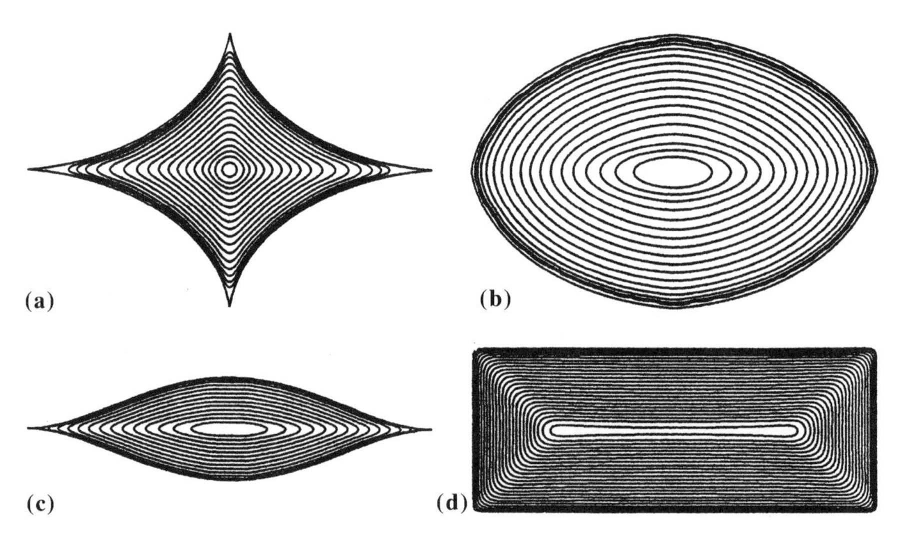

Figure 9.12. The wound edge at a selection of equally spaced times for different initial wound shapes. In the simulations all initial shapes had area 1. (**a**) The healing of a cusped wound from one of the family illustrated in Figure 9.11 with $\alpha = -0.8$ for the activator mechanism. (**b**) The healing of an ovate wound from the family illustrated in Figure 9.11 with $\alpha = +0.8$ for the activator mechanism. (**c**) The healing of an 'eye-shaped' wound, as predicted by the inhibitor mechanism. (**d**) The healing of a rectangular wound, as predicted by the inhibitor mechanism. Parameter values are the same as for Figure 9.6: for the activator mechanism, $D = 5 \times 10^{-4}$, $D_c = 0.45$, $\lambda = 30$, $h = 10$, $\alpha = 0.1$, $c_m = 40$, and for the inhibitor mechanism, $D = 10^{-4}$, $D_c = 0.85$, $\lambda = 5$, $h = 10$. (From Sherratt and Murray 1992a,b)

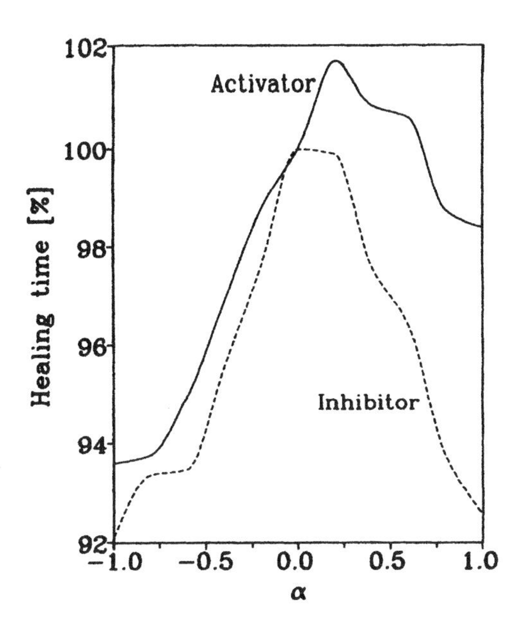

Figure 9.13. The variation in healing time with the length-to-width ratio of rectangular wounds, as predicted by our model with both autoactivation (—) and autoinhibition (- - -) of cell proliferation. Healing time is expressed as percentage of that for a diamond shaped wound given by $\alpha = 0$. The parameter values are the same as in Figure 9.12 and the initial dimensionless wound area is always taken to be 1. (From Sherratt and Murray 1992a,b)

the healing time of the length to width ratio of rectangular wounds of given initial area and they made some predictions for both chemical activation and inhibition of mitosis. As expected, the healing time decreases as the side-length ratio increases; moreover, they found no significant difference between the results for the two types of chemical control. These results are amenable to testing against quantifiable experimental data.

A more interesting investigation, however, is to vary the actual shape of the initial wound. To do this in a quantifiable way, we considered a single parameter family of wound shapes defined by (9.30) (see Figure 9.11) and illustrated in Figure 9.13. For simplicity, all the wound shapes in this figure are shown with the same midline lengths, but in our solutions, the lengths of the midlines are chosen, in a given ratio, so that the initial wound area is the same in each case, namely, normalised to 1. As the parameter α increases from -1 to $+1$, the initial wound geometry changes from a cusped shape, through a diamond at $\alpha = 0$, to an ovate shape, and finally to an ellipse at $\alpha = +1$. By solving the model equations for a range of values of α, we can then predict the dependence of healing time on the initial wound geometry. The results are illustrated in Figure 9.13 and suggest that autoregulation of mitosis via a chemical activator and inhibitor imply similar variations in healing time with wound shape when the shapes are cusped ($\alpha < 0$), but quite different variations for ovate wound shapes ($\alpha > 0$). This difference is borne out by other families of wound shapes investigated; this suggests a possible experimental approach for distinguishing between healing controlled by autoactivation and autoinhibition of mitosis.

An explanation for this difference between the two mechanisms is suggested by changes in the wound shape during healing. As shown schematically in Figure 9.14, when the wound initially has a cusped shape, the wound edge rounds up during healing, in both the activator and inhibitor cases. In contrast, for ovate wound shapes, the wound tends to flatten during healing, and this occurs to a much greater extent for the inhibitor

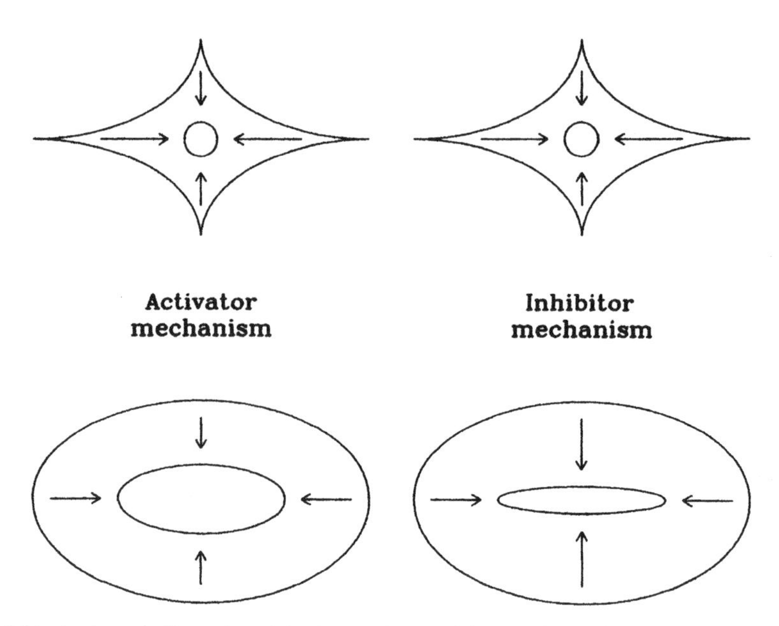

Figure 9.14. A schematic illustration of the changes in wound shape during healing. The model predicts that a cusped wound rounds up during healing, in both the activator and inhibitor cases, but that an ovate wound flattens during healing. Moreover, this flattening occurs to a much greater extent for the inhibitor mechanism than for the activator mechanism. (From Sherratt and Murray 1992a,b)

mechanism than for the activator mechanism. More pronounced flattening results in a larger wound perimeter and, thus, faster healing.

This explanation raises an important question as to what the key factors are that control the extent to which ovate wounds flatten during healing. To answer this, Sherratt and Murray (1992a,b) considered a particularly simple equation which mathematically caricatures the full model, namely,

$$\frac{\partial n}{\partial t} = \nabla^2 n + \Gamma n(1 - n), \tag{9.31}$$

where Γ is the single dimensionless parameter which reflects the relative contributions of cell mitosis and cell migration to the healing process; the larger Γ is the larger the effect of mitosis.

This equation ignores all biochemical effects and so cannot be expected to represent the healing mechanism quantitatively. However, we can use this simple caricature to investigate the dependence of change in wound shape on these two basic aspects of the healing process. The results are illustrated in Figure 9.15: an increase in Γ corresponds to an increase in the role played by mitosis relative to migration. The initial wound shape was ovate with a midline length ratio of 2. The wound edge became flatter during healing as Γ increases. This suggests that shape change during healing is a competitive process, with cell migration tending to cause the wound to round up, while cell division tends to cause the wound to flatten. This simple prediction could be tested

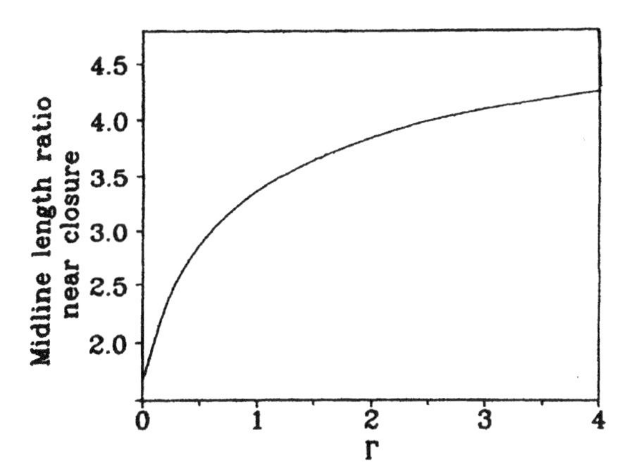

Figure 9.15. An illustration of one measure of the dependence of change in wound shape during healing on the relative importance of cell division and cell migration. The plot shows the midline length ratio of a wound near closure, when its area reaches 10% of its initial value, as predicted by the simple caricature (9.31) of the full model (9.12) and (9.13), for a range of values of Γ. As Γ increases, cell division becomes more important in the healing process relative to the role of cell migration. The wound becomes more flattened as Γ increases. In each case, the wound is initially elliptical with midline length ratio 2. The inital wound area is always 1. (From Sherratt and Murray 1992a,b)

experimentally by biochemically altering the relative importance of one of these two processes.

Although the model (9.12) and (9.13) is undoubtedly a gross simplification of *in vivo* re-epithelialisation, we feel it is nevertheless a useful first approximation which led us to make quantitative predictions on the effects of adding mitotic regulators to healing wounds, and on the variation of healing time with wound shape. This latter study has led to an understanding of a possible mechanism for the control of changes in wound shape during healing. An important point is that all the predictions are amenable to experimental verification, and all suggest new experiments that would improve our current understanding of the mechanisms responsible for epidermal wound healing.

A clinical aspect of re-epithelialisation we have not discussed here but of great importance is that of skin grafts and skin substitutes. The modelling aspects of this are interesting, challenging, timely and potentially very useful in the general area of plastic surgery. Although much of the latter is associated with full-depth wounds which implicate the underlying dermis it is appropriate to mention it here since there is an ever-increasing number of studies on skin substitutes (see, for example, Green 1991 and Bertolami et al. 1991). In the following chapter we discuss modelling of the very much more complex process of full-depth wound healing.

Although the results obtained with these models support the view that biochemical control of mitosis plays a major role in re-epithelialisation, they do not rule out a contribution from mechanical mechanisms of wound closure such as we now discuss.

9.8 Mechanisms of Epidermal Repair in Embryos

Epidermal wound healing in embryonic mammalian skin is very different from that in adult epidermal skin. Although not explicitly stated above, our modelling there is for adult re-epithelialisation. When adult skin is wounded, the epidermal cells surrounding the lesion move inwards to close the defect and the fairly well-established mechanism for closure is by lamellipodial crawling of the cells near the wound edge. Also almost all aspects of adult epidermal wound healing are regulated by biochemical growth factors. Although there is still controversy about some of the details the key elements seem fairly well established and it is these we incorporated in our models. Embryonic epidermal wound healing is very much less documented and very different. A brief review of the two processes is given by Sherratt et al. (1992).

Experimental work by Martin and Lewis (1991, 1992) suggests that embryonic epidermal wound healing may be caused by a completely different mechanism. Using a tungsten needle they made lesions on the dorsal surface of four-day embryonic chick wing buds, by dissecting away a patch of skin approximately 0.5 mm square and 0.1 mm thick. The epidermis, its basal lamina, and a thin layer of underlying mesenchyme were removed. The wounds healed perfectly and rapidly, typically in about 20 hours. Although mesenchymal contraction played some role in healing, the epidermis also moved actively across the mesenchyme. However, unlike adult epidermal cells, there was no evidence of lamellipodia at the wound front of the epidermis as shown in Figure 9.16(a). This absence of lamellipodia was consistent with a second important observation in a related study, namely, when a small island of embryonic skin was grafted onto a denuded region of the limb bud surface, the grafted epidermis, instead of expanding over the adjacent exposed mesenchyme, actually contracted, leaving its own mesenchyme exposed. This phenomenon suggested that the mechanism underlying epidermal movement might be a circumferential tension at the free edge, acting like a purse string that pulls the edges inwards. Such a mechanism would cause shrinkage of both an epidermal island and a hole in the epidermis.

Searching for the cause of such a purse string effect, Martin and Lewis (1991, 1992) examined the distribution of filamentous actin in the healing wounds, by staining specimens with fluorescently tagged phalloidin, which binds selectively to filamentous actin. This revealed a thick cable of actin around the epidermal wound margin as shown in Figures 9.16(b) and (c) localized within the leading row of basal cells: it is a very narrow band. The cable appeared to be continuous from cell to cell, presumably via adherens junctions, except at a very few points; it was present within an hour of wounding, and persisted until the wound was closed. Preliminary evidence suggests that an actin cable may also form at the periphery of the contracting epidermis in the skin grafts mentioned above. Simple incisional wounds made by a slash, Figure 9.17, along the proximodistal axis of the limb take only seconds to make, as opposed to the 5–10 minutes required to dissect a square of skin from the limb bud. As described in the review article by Sherratt et al. (1992) the slash lesion data indicate that the cells at the wound edge begin to organize their actin into a cable within minutes of wounding, although the cable takes an hour or more to attain its full thickness. Below we discuss models which incorporate the concept of an actin cable and its role in embryonic wounds.

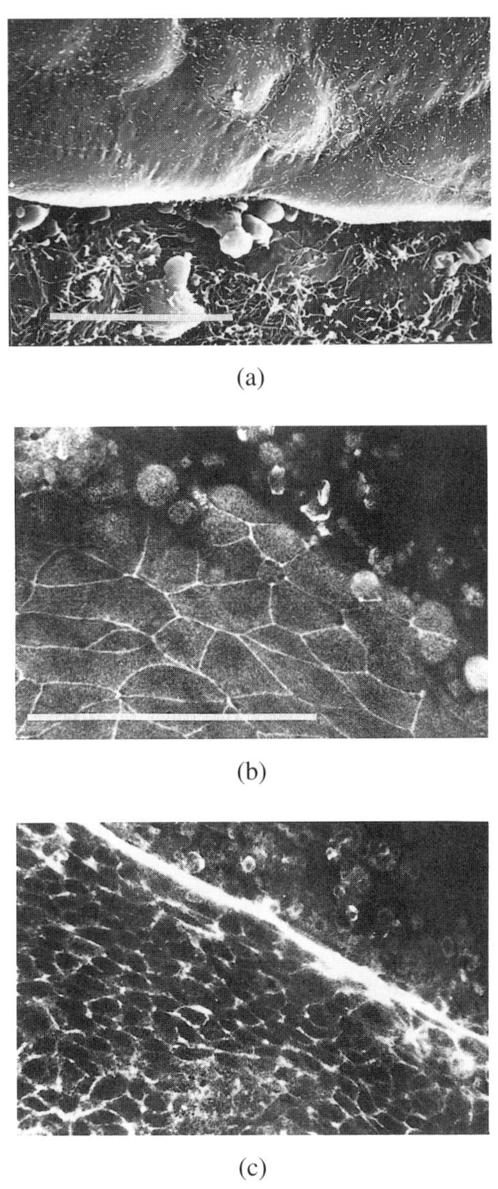

Figure 9.16. (a) Scanning electron micrograph of a wound edge in chick embryonic epidermis 12 hours after the operation. The wound was made by removing a square patch of embryonic skin from the dorsal surface of a chick wing bud at 4 days of incubation. The epidermis is above, with the flattened surface of the exposed mesenchyme below. Note the smooth edge of the epidermis and the absence of finger-like lamellipodia. Scale bar = 10 μm. (**b**) and (**c**) The epidermal wound front on the dorsum of a chick wing bud after 12 hours of healing, as seen in optical section by confocal scanning laser microscopy. The tissue has been stained with rhodamine-labelled phalloidin, which binds to filamentous actin, and the sections are parallel to the plane of the epidermis. The epidermis is at the lower left, with the exposed mesenchyme (largely below the plane of section) at the upper right. (**b**) Superficial section, in the plane of the periderm, whose broad flat cells are outlined by their cortical actin. (**c**) Section about 4 μm deeper, in the plane of the basal epidermal cells, showing the acting cable at the wound front. Scale bar for (**b**) and (**c**), = 50 μm. (From Sherratt et al. 1992, originally presented in Martin and Lewis 1991)

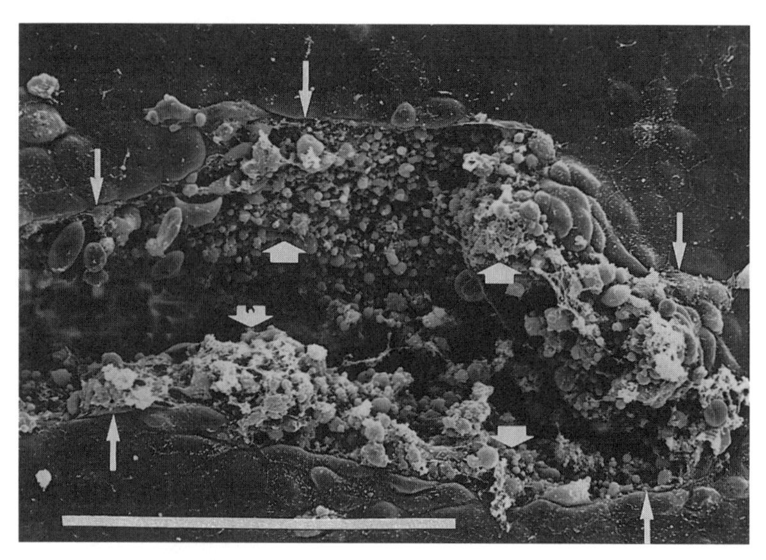

Figure 9.17. Scanning electron micrograph of a linear slash wound in chick embryonic epidermis, immediately after the operation. The wound was made on the dorsal surface of the wing bud at 4 days of incubation (stage 22/23). The thin arrows indicate the epidermal wound edge, while the thicker arrows denote the edge of the underlying mesenchyme. The epidermis has retracted over the underlying mesenchyme by about 30–40 μm. Scale bar = 100 μm. (From Sherratt et al. 1992)

The different mechanisms of healing in adult and embryonic epidermis raise many problems that call for mathematical modelling. Perhaps the most important aspect of embryonic wound healing is that they heal without scarring, and not only epithelial wounds. An understanding of how the processes differ could have far-reaching implications for clinical wound management (see, for example, Martin 1997). Fetal surgery is a high-stakes field, has many attractions and dangers and is highly controversial with many complex ethical issues. It has already been used to treat spina bifida (since 1998) and this is only the beginning. There are many advocates for treating nonlethal conditions such as facial deformities and other defects. We have already mentioned the potential for cleft palate repair in Chapter 4. The review article by Longaker and Adzick (1991) and their edited book of contributed articles specifically on fetal wound healing (Adzick and Longaker 1991) is a good place to start. The article in this collection by Ferguson and Howarth (1991) specifically discusses scarless wound healing with marsupials and lists the many features which distinguish fetal from adult wound healing. A reason for studying marsupials is that unlike most mammals their young are very immature at birth. The article by Martin (1997) is also particularly relevant; he discusses wound healing from the point of view of perfect (scarless) skin regeneration. There is a large literature in the area with an abundance of data.

Another aspect of wound healing associated with the skin are the techniques now available for making cell layers from suspensions of disassociated cells which proliferate rapidly and form confluent monolayers for treatment of burns, plastic surgery and so on. For a view of the general picture, see the *Scientific American* article by Green

(1991). Again there is an increasing body of knowledge in this area with many modelling problems associated with the formation of these monolayers of cells.

9.9 Actin Alignment in Embryonic Wounds: A Mechanical Model

The experimental results of Martin and Lewis (1991, 1992) raise two major questions: first, how does the actin cable form, and, second, how does it cause the wound to close, if in fact it has this function? Here we consider the first of these questions. The actin cable forms in response to the creation of a free boundary at the wound edge. In this section we consider possible explanations for the aggregation and pronounced alignment of filamentous actin at the wound edge, which give rise to the actin cable, in terms of a mechanical response to this free boundary.

At the developmental stages we consider, the embryonic epidermis is two cell layers thick, consisting of a superficial, pavement-like peridermal layer and a cuboidal basal layer. The actin cable develops in the basal layer, and it is to this layer that our modelling considerations apply; refer also to Figure 9.16. The basal cells form a confluent sheet, attached to the underlying basal lamina. Following wounding, the cytoskeleton of these cells undergoes rapid changes, on the timescale of a few minutes, and reaches a new quasi-equilibrium state with a cable of actin at the wound margin. It is this state that we try to analyse. Of course, as the healing proceeds, this new 'equilibrium' state will change but over a timescale of hours; we do not consider here the processes taking place over this longer timescale.

Since forces clearly play a crucial role in the whole process we model the initial response to wounding by amending the mechanochemical model for the deformation of epithelial sheets initally proposed by Murray and Oster (1984). They considered the epithelium as linear, isotropic, viscoelastic continuum and derived a force balance equation for the various forces present; this basic model was discussed in Chapter 6. In their model the cell traction is mediated by calcium and so they included an equation for the calcium. Here we modify this model to investigate the new equilibrium reached by the basal cell sheet after wounding. Viscous effects can be neglected since we are only interested here in the short term 'equilibrium' state, and this simplifies the model considerably. However, an important addition to the Murray–Oster model, introduced by Sherratt (1991), are the effects of a microfilament anisotropy, which, as we shall see, plays a crucial role in our system and the wound healing process.

The mechanical properties of confluent cell sheets are largely determined by the intracellular actin filaments (Pollard 1990). In epithelial sheets, cell–cell adherens junctions serve as connection sites for actin filaments, and so the intracellular actin filaments are linked, via transmembrane proteins at these junctions, in an effectively two-dimensional transcellular network. Our model addresses the equilibrium state of this network. The forces exerted on an epidermal cell by the epidermal cells around it, via this actin filament network, can be divided into two types: elastic forces and active contraction forces; we include any effects of osmotic pressure within 'contraction forces.' The elasticity of the actin filament network arises from the extensive interpenetration of the long actin filaments, which tends to immobilize them (Janmey et al. 1988). As we discussed at length in Chapter 6 cell traction forces have been observed in a range of cell types and they play a major role in pattern formation processes. There we discussed in detail the form of elastic forces in the pattern generating mechanisms.

At equilibrium, these elastic and traction forces balance the elastic restoring forces that arise from attachment to the underlying mesenchyme. Let us recap the force balance scenario in the basal cell sheet. Following wounding, the epidermis retracts relative to the underlying mesenchyme, until a mechanical equilibrium is reached. Elastic and traction stresses are exerted by the surrounding cells at each point in the sheet, and in the postwounding equilibrium; these elastic and traction forces balance the restoring forces due to substratum attachments. The movement of the 'springs' represents local deformation of the superficial mesenchyme due to tension in the epidermis. In reality, the mesenchyme and epidermis are separated by a basal lamina, and our modelling assumes that the attachments are fixed in the basal lamina, and that this becomes wrinkled (so that the attachments are compressed) in response to wounding, to an extent reflecting the compaction of the cell sheet. Thus, the model equation, which predicts the new equilibrium configuration attained by the actin filament network of the epidermis after wounding, has the following general form,

$$\begin{aligned}&\text{Elastic forces within the actin filament network}\\&\quad + \text{Traction forces exerted by actin filaments}\\&\quad = \text{Elastic restoring forces due to substratum attachment.}\end{aligned}$$

We now have to quantify the various terms in this equation; it is helpful to review the relevant discussion in Section 6.2 in Chapter 6.

At equilibrium we model the stress tensor $\boldsymbol{\sigma}$ by

$$\boldsymbol{\sigma} = \underbrace{G[E\boldsymbol{\epsilon} + \Gamma\nabla\cdot\mathbf{u}\mathbf{I}]}_{\text{elastic stresses}} + \underbrace{\tau G\mathbf{I}}_{\substack{\text{active}\\\text{contraction}\\\text{stress}}}, \tag{9.32}$$

where $\mathbf{u}$ is the displacement of the material point which was initially at position $\mathbf{r}$, the strain tensor $\boldsymbol{\epsilon} = (1/2)(\nabla\mathbf{u} + \nabla\mathbf{u}^T)$, $G(\mathbf{r})$ is the density of the intracellular actin filaments at the material point initially at $\mathbf{r}$, E and Γ are positive (elastic) parameters and $\mathbf{I}$ is the unit tensor. In general, as we discuss below, the traction τ is a function of the local compaction of the tissue. Unlike the general form used in Chapter 6 we have no viscous terms here since we are only dealing with the equilibrium situation.

We now consider the restoring forces due to the underlying substratum, via the cellular attachments, the importance of which has been demonstrated experimentally by Hergott et al. (1989). Following the concepts in Chapter 6 (and also Murray and Oster 1984a,b), we model these restoring forces by $\lambda G\mathbf{u}$, where λ is a measure of the strength of the attachments. The proportionality factor, G, reflects the actin filament density and, as we shall see, in effect their filament orientation. So, the new equilibrium equation to be solved is

$$\nabla\cdot\boldsymbol{\sigma} - \lambda G\mathbf{u} = \mathbf{0}, \tag{9.33}$$

where $\boldsymbol{\sigma}$ is given by (9.32). Since, in comparison with the experimental wound size, the cell sheet is infinite, the boundary conditions are

$$\mathbf{u}(\infty) = \mathbf{0}, \quad \boldsymbol{\sigma} \cdot \mathbf{n} = \mathbf{0} \quad \text{on a free edge,} \tag{9.34}$$

where $\mathbf{n}$ is the unit vector normal to the wound edge. Initially we assume the boundary condition in the unwounded state is $\mathbf{u} = \mathbf{0}$ everywhere. In a developing embryo this last assumption is not a totally trivial one. It is based on the exeprimental fact that the cell density is uniform and the substratum attachments are forming sufficiently fast so that the restoring forces are essentially negligible.

A crucial feature of the modelling is how to relate the density of microfilaments, G, at a point and the amount of expansion or compaction of the tissue resulting from the displacements, $\mathbf{u}$, of that and neighbouring points, from their initial positions; we denote this compaction by Δ. There could, for example, be some actin polymerization at the edge of the wound. There does not seem to be any evidence of this, so we assume that in reponse to wounding there is no such polymerization and so we assume that the amount of filamentous actin is constant in a given region as it is deformed as a consequence of the wound injury. This implies that the actin density function G satisfies $G(\mathbf{r})(1 - \Delta) = \kappa$, where κ is a constant. As a first approximation we take $\Delta \approx -\nabla \cdot \mathbf{u}$ and so conservation of actin becomes

$$G(\mathbf{r})(1 + \nabla \cdot \mathbf{u}) = \kappa, \tag{9.35}$$

where κ is a constant. The term $\nabla \cdot \mathbf{u}$ is the dilation, denoted by θ.

There is abundant experimental evidence, such as the seminal work by Kolega (1986) and the earlier study by Chen (1981), that actin filaments tend to align themselves parallel to the maximum applied stress. The former showed that fish epidermal cells when subjected to a tension aligned themselves in a matter of seconds while the latter showed that chick fibroblasts reoriented their actin filaments on the order of 15 seconds. If we think of a network of unstrained actin fibres (intuitively something like steel wool) and then subject it to an applied tension the effect of alignment is to concentrate the actin density by alignment along the tension lines (think again of a bundle of steel wool stretched: the strands tend to align themselves in the direction of the applied pull).

When actin filaments are compacted the sum of the traction forces exerted individually is less than the total force when acting together because of the formation of myosin cross-bridges (see, for example, the book by Alberts et al. 1994). Following the models of Oster (1984), Oster and Odell (1984) and Oster et al. (1985a) we take the specific form $\tau = \tau_0/(1 - \beta\Delta)$ to quantify the effect of compacting the actin filaments where β is a parameter. With this form a decrease in dilation (that is, an increase in compaction) results in an increase in the cell traction stress because of the increase in actin fibre density and, because of the synergy phenomenon, an increase in the traction stress per filament. Since we intuitively expect the former effect to be the larger effect we therefore require the parameter $\beta < 1$. With this restriction, τ is bounded since $\Delta < 1$ because a region cannot be compressed to point. In our analysis below we consider $0 < \beta < 1$. Murray and Oster (1984a,b), who first proposed a continuum model for epithelial morphogenesis, took $\beta = 0$. Here we again approximate the compaction by $\Delta \approx -\nabla \cdot \mathbf{u}$ and hence take the active traction term τ to be a function of the dilation and modelled by

$$\tau = \frac{\tau_0}{1 + \beta \nabla \cdot \mathbf{u}}, \quad \beta < 1, \tag{9.36}$$

where $0 < \beta < 1$ is a parameter, a crucial one as we shall see. With this form the restriction on the parameter β ensures τ is finite as long as the approximation $\Delta \approx -\nabla \cdot \mathbf{u}$ is not unreasonable as would be the case, for example, if $\nabla \cdot \mathbf{u} < -1$, a situation which arises in the following section and with which we have to deal.

With these forms for G and τ, together with (9.32), the model equation (9.33) becomes

$$\nabla \cdot \left[\frac{\kappa}{1 + \nabla \cdot \mathbf{u}} \left(E\boldsymbol{\epsilon} + \Gamma \nabla \cdot \mathbf{u}\mathbf{I} + \frac{\tau_0 \mathbf{I}}{1 + \beta \nabla \cdot \mathbf{u}} \right) \right] - \frac{\lambda \kappa \mathbf{u}}{1 + \nabla \cdot \mathbf{u}} = 0. \tag{9.37}$$

We now nondimensionalise (9.37) by introducing the nondimensional parameters,

$$\mathbf{r}^* = \mathbf{r}/L \quad \mathbf{u}^* = \mathbf{u}/L, \quad E^* = E/\tau_0, \quad \Gamma^* = \Gamma/\tau_0, \quad \lambda^* = \lambda L^2/\tau_0, \tag{9.38}$$

where dimensionless quantities are denoted by $*$ and L is a typical length associated with the wound, such as the initial radius in a circular wound. Two of these dimensionless parameters, E^* and Γ^*, reflect the mechanical properties of the microfilament network in the epithelium, while another, λ^*, represents the extent to which displacements of the epithelium are resisted by its tethering to the underlying mesenchyme. The fourth (dimensionless) parameter, β, reflects the extent to which compaction of the actin filament network increases the traction stress exerted per filament. As mentioned a synergy phenomenon is intuitively expected to some extent in that the actin filaments together exert a greater traction force than the sum of the traction forces they exert separately, since as the degree of filament overlap increases, additional myosin cross-bridges form (Oster and Odell 1984). Unlike some other applications of mechanochemical models, the experimental data are simply not available to provide estimates for parameter values. Substitution of these quantities into (9.37) gives the nondimensional model equation as

$$\nabla \cdot \left[\frac{1}{1 + \nabla \cdot \mathbf{u}} \left(E\boldsymbol{\epsilon} + \Gamma \nabla \cdot \mathbf{u}\mathbf{I} + \frac{\mathbf{I}}{1 + \beta \nabla \cdot \mathbf{u}} \right) \right] - \frac{\lambda \mathbf{u}}{1 + \nabla \cdot \mathbf{u}} = 0, \tag{9.39}$$

where for convenience we have omitted the *. We now want the solution of this equation subject to the dimensionless boundary conditions given by (9.34); it is the same in dimensionless form. The force balance equation proposed by Murray and Oster (1984a,b) for their epidermal morphogenesis model is (9.37) with $\beta = 0$ but in which the traction τ is a function of calcium (refer to Chapter 6).

We should reiterate the importance of stress-induced alignment of the actin filaments in our model. The actin cable consists of an aggregation of microfilaments at the wound edge, all aligned parallel to this edge. The experimental evidence suggests that this alignment could be a direct result of the anisotropic stress field near the wound edge, which results from this edge being a free boundary. An in depth mathematical analysis of this is given by Sherratt and Lewis (1993) and Sherratt (1993). We now analyze (9.39) mainly following the exposition of Sherratt (1993).

One-Dimensional Model Solution and Parameter Restrictions

The most detailed quantitative data on the initial response of embryonic epidermis to wounding is from slash wound experiments which can be modelled by the one-dimensional form of the equation, which from (9.39) becomes

$$\frac{d}{dx}\left[\frac{1}{1+u_x}\left((E+\Gamma)u_x+\frac{1}{1+\beta u_x}\right)\right]-\frac{\lambda u}{1+u_x}=0 \tag{9.40}$$

with boundary conditions from (9.34),

$$(E+\Gamma)u_x+\frac{1}{1+\beta u_x}=0 \quad \text{at } x=0, \qquad u=0 \quad \text{at } x=\infty. \tag{9.41}$$

Here we take the wound edge to be at $x = 0$. We now analyze this system and show that there is a unique solution, u, if the parameters satisfy certain constraints; the technique involved is basic but the algebra is messy but worth it for the resulting biological implications.

Integrating (9.40) and using the second boundary condition at $x = \infty$ we get

$$\frac{1}{2}\lambda u^2 = P(u_x), \tag{9.42}$$

where $P(u_x)$ is given by

$$\begin{aligned} P(u_x) &= (E+\Gamma)[u_x-\log(1+u_x)]+\frac{1}{1-\beta}\log(1+u_x) \\ &\quad -\frac{2-\beta}{\beta(1-\beta)}\log(1+\beta u_x)-\frac{1}{\beta(1+\beta u_x)}+\frac{1}{\beta}, \end{aligned} \tag{9.43}$$

where $\beta < 1$, which is the biologically realistic range.

The boundary condition at the wound edge $x = 0$ from (9.41) can be written as

$$Q(u_x) \equiv (E+\Gamma)\beta {u_x}^2+(E+\Gamma)u_x+1=0 \tag{9.44}$$

which only has real roots when $u_x < 0$. But, from (9.42) and (9.43), $u_x = 0$ only if $u = 0$ and so the boundary condition (9.41) at $x = \infty$ implies that the solution u must be monotonically decreasing with x. Remember, the physical constraint that a region cannot be compressed to a point means that we are only interested in a solution in which $-1 < u_x \leq 0$.

Let us now consider the role of β. For $(E+\Gamma) > (1/(1-\beta))$, $P(u_x) \to +\infty$ as $u_x \to -1^+$ while for $(E+\Gamma) < (1/(1-\beta))$, $P(u_x) \to -\infty$ as $u_x \to -1^+$. When

$(E+\Gamma) = (1/(1-\beta))$ we have, after a little algebra,

$$P(-1) = \frac{2-\beta}{\beta(1-\beta)}\left[\frac{-2\beta}{2-\beta} - \log(1-\beta)\right],$$

which is strictly positive when $0 < \beta < 1$ since then $((2-\beta)/(\beta(1-\beta))) > 0$ and

$$\frac{d}{d\beta}\left[\frac{-2\beta}{2-\beta} - \log(1-\beta)\right] = \frac{\beta^2}{(2-\beta)^2(1-\beta)} > 0,$$

with $[(-2\beta)/(2-\beta) - \log(1-\beta)]_{\beta=0} = 0$ so that

$$\left[\frac{-2\beta}{2-\beta} - \log(1-\beta)\right] > 0$$

for $0 < \beta < 1$. Differentiating $P(u_x)$ with respect to u_x gives

$$P'(u_x) = \frac{(E+\Gamma)u_x}{1+u_x} - \frac{\beta u_x}{(1+\beta u_x)^2} - \frac{u_x}{(1+u_x)(1+\beta u_x)}$$

which gives $P'(0) = 0$ and, differentiating again, $P''(0) = E + \Gamma - (1+\beta)$. So, for small u_x, $P(u_x)$ has the same sign as $E + \Gamma - (1+\beta)$.

So, except for the root at $u_x = 0$ we have

$$\begin{aligned} P'(u_x) = 0 &\Longleftrightarrow (E+\Gamma)(1+\beta u_x)^2 = \beta(1+u_x) + (1+\beta u_x) \\ &\Longleftrightarrow u_x = \frac{1}{\beta(E+\Gamma)}\left[1 - (E+\Gamma) \pm \sqrt{(E+\Gamma)(\beta-1)+1}\right] \end{aligned}$$

and so these roots are real if and only if $E + \Gamma < (1/(1-\beta))$ in which case the larger root lies in $[-1, 0)$ if $E + \Gamma > 1 + \beta$ and is positive if $E + \Gamma < 1 + \beta$; the smaller more negative root is less than -1. For

$$\begin{aligned} &\frac{1}{\beta(E+\Gamma)}\left[1 - (E+\Gamma) + \sqrt{(E+\Gamma)(\beta-1)+1}\right] < 0 \\ &\qquad \Longleftrightarrow \sqrt{(E+\Gamma)(\beta-1)+1} < E + \Gamma - 1 \\ &\qquad \Longleftrightarrow E + \Gamma > 1 \quad \text{and} \quad (E+\Gamma)(\beta-1) + 1 < (E+\Gamma-1)^2 \\ &\qquad \Longleftrightarrow E + \Gamma > 1 + \beta, \\ &\frac{1}{\beta(E+\Gamma)}\left[1 - (E+\Gamma) + \sqrt{(E+\Gamma)(\beta-1)+1}\right] > -1 \\ &\qquad \Longleftrightarrow \sqrt{(E+\Gamma)(\beta-1)+1} > -[(E+\Gamma)(\beta-1)+1] \end{aligned}$$

which is trivially satisfied if $E + \Gamma \neq (1/(1-\beta))$, in which case the root equals -1 and

$$\frac{1}{\beta(E+\Gamma)}\left[1-(E+\Gamma)-\sqrt{(E+\Gamma)(\beta-1)+1}\right]<-1$$
$$\Longleftrightarrow -\sqrt{(E+\Gamma)(\beta-1)+1}<-(E+\Gamma)(\beta-1)-1$$
$$\Longleftrightarrow (E+\Gamma)(\beta-1)+1<1 \quad \text{since} \quad E+\Gamma\leq\frac{1}{1-\beta}$$
$$\Longleftrightarrow \beta<1.$$

Now consider the roots $q_\pm$ of the boundary condition (9.44), namely,

$$q_\pm=\frac{1}{2\beta(E+\Gamma)}\left[-(E+\Gamma)\pm\sqrt{(E+\Gamma)^2-4\beta(E+\Gamma)}\right],$$

which are real if and only if $E+\Gamma>4\beta$ in which case they are both negative. But $Q(0)=1$ and $Q(-1)=(E+\Gamma)(\beta-1)+1$ so if $E+\Gamma>(1/(1-\beta))$ exactly one root of the quadratic Q lies in $(-1,0)$, that is,

$$q_+\in(-1,0) \quad \text{and} \quad q_-<-1 \quad \text{if} \quad E+\Gamma>\frac{1}{1-\beta}. \tag{9.45}$$

On the other hand, if $4\beta<E+\Gamma<(1/(1-\beta))$, a little algebra shows

$$-1<q_+,q_-<0 \quad \text{if} \quad \beta>\frac{1}{2}, \quad q_+,q_-<-1 \quad \text{if} \quad \beta<\frac{1}{2}. \tag{9.46}$$

We now have the preliminary results we need to consider the properties of solutions, $u(x)$, of (9.42)–(9.44) when $\beta<1$. From (9.46) and the form of $P(u_x)$, if $E+\Gamma>(1/(1-\beta))$ there is a unique monotonically decreasing solution of (9.42) subject to (9.44) since then $P(u_x)$ decreases monotonically from $+\infty$ to zero in the interval $(-1,0]$. This means that there is a $1-1$ correspondence between $u\geq0$ and $u_x\in(-1,0]$ given by the governing equation (9.42) and exactly one of the roots for u_x of the boundary conditions (9.44) lies in range $(-1,0)$.

If $1+\beta<E+\Gamma<(1/(1-\beta))$, there is a unique solution if $\beta>1/2$, $E+\Gamma>4\beta$; the latter ensures $q_\pm$ are real and $q_+<p_{\max}<q_+$. Here $p_{\max}$ is the value of u_x which gives the maximum $P(u_x)$ on $(-1,0)$, that is,

$$p_{\max}=\frac{1}{\beta(E+\Gamma)}\left[1-(E+\Gamma)+\sqrt{(E+\Gamma)(\beta-1)+1}\right].$$

$P(u_x)$ decreases to zero on $[p_{\max},0]$. So, there is a $1-1$ correspondence between $u\in[0,P(p_{\max})]$ and $u_x\in[p_{\max},0]$ given by (9.42). Also, one of the roots for u_x of the boundary condition (9.44) lies in $(p_{\max},0)$. Remember that $\beta>1/2$ and $E+\Gamma>4\beta$ imply that the condition $E+\Gamma>1+\beta$. These three conditions are necessary when $1+\beta<E+\Gamma<(1/(1-\beta))$ except for the special case of $E+\Gamma=4\beta$. In this situation, if $\beta>1/2$, $p_{\max}=q_+=q_-=-(1/2\beta)$, and so there is a unique solution, while if $\beta<1/2$, $q_+=q_-=-(1/2\beta)<p_{\max}$ and so there is no solution.

We now show that when $4\beta < E + \Gamma < (1/(1-\beta))$, $\beta > 1/2$ we have $q_- < p_{\max} < q_+$. We have

$$
\begin{aligned}
q_- < p_{\max} &\iff -\sqrt{(E+\Gamma)^2 - 4\beta(E+\Gamma)} \\
&< \ 2-(E+\Gamma)+2\sqrt{(E+\Gamma)(\beta-1)+1} \\
&\iff (E+\Gamma)^2 - 4(E+\Gamma)\beta \\
&> \ \left[2-(E+\Gamma)+2\sqrt{(E+\Gamma)(\beta-1)+1}\right]^2 \quad \text{or} \\
&\quad\ 2-(E+\Gamma)+2\sqrt{(E+\Gamma)(\beta-1)+1} > 0 \\
&\iff 2[(E+\Gamma)(\beta-1)+1] \\
&< \ (E+\Gamma-2)\sqrt{(E+\Gamma)(\beta-1)+1} \quad \text{or} \\
&\quad\ 2\sqrt{(E+\Gamma)(\beta-1)+1} > E+\Gamma-2 \\
&\iff 2\sqrt{(E+\Gamma)(\beta-1)+1} \neq E+\Gamma-2 \\
&\iff E+\Gamma \neq 4\beta
\end{aligned}
$$

and

$$
\begin{aligned}
p_{\max} < q_+ &\iff 2-(E+\Gamma)+2\sqrt{(E+\Gamma)(\beta-1)+1} \\
&< \ \sqrt{(E+\Gamma)^2 - 4\beta(E+\Gamma)} \\
&\iff \left[2-(E+\Gamma)+2\sqrt{(E+\Gamma)(\beta-1)+1}\right]^2 \\
&< \ (E+\Gamma)^2 - 4\beta(E+\Gamma) \quad \text{or} \\
&\quad\ 2-(E+\Gamma)+2\sqrt{(E+\Gamma)(\beta-1)+1} < 0 \\
&\iff 2[(E+\Gamma)(\beta-1)+1] \\
&< \ (E+\Gamma-2)\sqrt{(E+\Gamma)(\beta-1)+1} \quad \text{or} \\
&\quad\ 2\sqrt{(E+\Gamma)(\beta-1)+1} < E+\Gamma-2 \\
&\iff E+\Gamma > 4\beta,
\end{aligned}
$$

where we have used the fact that $E + \Gamma > 4\beta$ and $\beta > 1/2$ which together imply $E + \Gamma > 2$.

The only case we have not yet considered for $\beta < 1$ is when $E + \Gamma = (1/(1-\beta))$. Here $P(-1)$ is finite and positive and $P'(-1) = 0$. However, when $E+\Gamma = (1/(1-\beta))$, the roots of Q given by (9.44) are

$$
q_+ = -1 \quad \text{and} \quad q_- = 1 - \frac{1}{\beta} < -1 \quad \text{if} \quad \beta < \frac{1}{2}
$$

$$q_- = -1 \quad \text{and} \quad q_+ = 1 - \frac{1}{\beta} \in (-1, 0) \quad \text{if} \quad \beta > \frac{1}{2}.$$

In this case $P'(-1) = 0$ and so from (9.42), $u(0) > 0$ and $u_x(0) = -1$ which imply that $u_{xx}(0)$ is undefined and hence inadmissable since the original (9.40) is of second order. Finally we have that when $E + \Gamma = (1/(1-\beta))$, there is a unique solution which satisfies $u_x > -1$ if and only if $\beta > 1/2$.

To summarize, we have shown that in the biologically realistic situation where $\beta < 1$ there is a unique monotonically decreasing solution of (9.40) with boundary conditions (9.44) and which satisfies the physical constraint $u_x > -1$ if and only if the paramaters E, Γ and β satisfy

$$\begin{aligned} &\beta < 1 - \frac{1}{E + \Gamma}, \qquad 1 \leq E + \Gamma \leq 2 \\ &\beta \leq \tfrac{1}{4}(E + \Gamma), \qquad E + \Gamma > 2. \end{aligned} \tag{9.47}$$

Sherratt (1993) also carried out the analysis, for completeness, for $\beta \geq 1$.

The numerical solutions were compared with the experimental data of Martin and Lewis (1991) for a range of parameters satisfying conditions (9.47). The solutions had to capture not only the intense actin aggregation at the wound edge but also the extent of retraction of the cell layers. The retraction is of the order of 60 μm. For a wide range of parameters Sherratt (1991) obtained solutions which captured this retraction phenomenon (in this one-dimensional situation). However, to capture the intense actin aggregation near the wound boundary, which means that $G(x)$ has a sharp peak close to $x = 0$, it was necessary to have not only $\beta < 1/2$ but also $E + \Gamma - 1/(1-\beta)$ small and positive: from the first of (9.47), $E + \Gamma > 1/(1-\beta)$. The solutions gave a retraction in the case of slash wounds of approximately 14% over a few cell lengths and the actin cable formed over approximately two cell lengths. The solutions become progressively more sensitive the closer $E + \Gamma$ is to the limit $1/(1-\beta)$. So, for a given $E + \Gamma$ there is a critical β given, from (9.47), by $\beta_{\text{crit}} = 1 - 1/(E + \Gamma)$.

For the situation in which $\beta > 1/2$ it was not possible to obtain solutions which captured both of the key observations, the approximate amount of retraction and the high actin aggregation at the boundary. What this implies is that the parameters must be close to the critical bifurcation values. Before discussing a modification to the model to remedy these problems when compared with experiment we must first consider the radially symmetric solution. It is also relevant when general two-dimensional wounds are considered.

Two-Dimensional Radially Symmetric Solution

Before discussing the biological relevance of the parameters and the conditions (9.47) we consider the radially symmetric case of a circular wound. Circular wounds are more difficult to perform experimentally since removing a piece of the epidermis involves a time commensurate with the formation of the actin cable around the wound edge. In this situation $\mathbf{u} = u(r)\vec{\mathbf{r}}$, where $\vec{\mathbf{r}}$ is the unit vector in the radial direction and the nondimensional stress tensor in plane polar coordinates has the form

$$\boldsymbol{\sigma} = \begin{bmatrix} p(r) & 0 \\ 0 & q(r) \end{bmatrix}, \tag{9.48}$$

where $p(r)$ and $q(r)$ are the radial and tangential principal values of the stress. From (9.32) in nondimensional form using (9.38) we get (from the square bracketed terms in (9.39))

$$\begin{aligned} p(r) &= \frac{Eu' + \Gamma(u' + u/r) + [1 + \beta(u' + u/r)]^{-1}}{1 + u' + u/r} \\ q(r) &= \frac{Eu/r + \Gamma(u' + u/r) + [1 + \beta(u' + u/r)]^{-1}}{1 + u' + u/r}, \end{aligned} \tag{9.49}$$

where prime denotes differentiation with respect to r and we used the fact that $\nabla \cdot \mathbf{u} = u' + u/r$ in radially symmetric plane polar coordinates. So

$$\begin{aligned} \nabla \cdot \boldsymbol{\sigma} &= \left(\vec{\mathbf{r}} \frac{\partial}{\partial r} + \frac{1}{r} \vec{\boldsymbol{\theta}} \frac{\partial}{\partial \theta} \right) \cdot \left(p(r) \vec{\mathbf{r}} \vec{\mathbf{r}} + q(r) \vec{\boldsymbol{\theta}} \vec{\boldsymbol{\theta}} \right) \\ &= p' \vec{\mathbf{r}} + \frac{p}{r} \left(\vec{\boldsymbol{\theta}} \cdot \frac{\partial \vec{\mathbf{r}}}{\partial \theta} \right) \vec{\mathbf{r}} + \frac{q}{r} \left(\vec{\boldsymbol{\theta}} \cdot \frac{\partial \vec{\boldsymbol{\theta}}}{\partial \theta} \right) \vec{\boldsymbol{\theta}} + \frac{q}{r} \frac{\partial \vec{\boldsymbol{\theta}}}{\partial \theta} \\ &= \left(p' + \frac{p}{r} - \frac{q}{r} \right) \vec{\mathbf{r}}, \end{aligned} \tag{9.50}$$

where we have used the relations

$$\frac{\partial \vec{\mathbf{r}}}{\partial r} = \mathbf{0}, \quad \frac{\partial \vec{\boldsymbol{\theta}}}{\partial r} = \mathbf{0}, \quad \frac{\partial \vec{\mathbf{r}}}{\partial \theta} = \vec{\boldsymbol{\theta}}, \quad \frac{\partial \vec{\boldsymbol{\theta}}}{\partial \theta} = -\vec{\mathbf{r}}.$$

We now substitute these forms for $p(r)$ and $q(r)$ into the model equation (9.39) and we obtain

$$\begin{aligned} r \frac{d}{dr} &\left[\frac{Eu' + \Gamma(u' + u/r) + [1 + \beta(u' + u/r)]^{-1}}{1 + u' + u/r} \right] \\ &+ E \frac{u' - u/r}{1 + u' + u/r} = \frac{r \lambda u}{1 + u' + u/r}. \end{aligned} \tag{9.51}$$

We now take the dimensional L in the nondimensionalisation to be the initial wound radius and so the boundary conditions (9.34) become

$$Eu' + \Gamma(u' + u/r) + [1 + \beta(u' + u/r)]^{-1} = 0 \quad \text{at } r = 1 \tag{9.52}$$

$$u = 0 \quad \text{at } r = \infty. \tag{9.53}$$

To give some idea of the dimensions, we consider wounds with a radius of about 500 μm and at the developmental stage of the chick embryos in the experiments a typical cell is about 10 μm in diameter which gives a dimensionless cell diameter of around 0.02.

Sherratt (1991) studied (9.51) in detail and showed that the solutions of the equation with constant traction, that is, $\beta = 0$, cannot capture the experimentally observed high aggregation of actin close to the wound edge. It was for this reason that he investigated the more realistic model in which $\beta \neq 0$ (see also Sherratt 1993).

There are four parameters, E, Γ, λ and β and a detailed numerical study of the solution dependence on these parameters showed that there is, as in the one-dimensional case, a unique monotonically decreasing solution provided the parameters satisfy certain constraints. As we would expect from the one-dimensional results there is again a critical $\beta = \beta_{crit}$ which is dependent on the other parameters above which solutions could not be found. The two-dimensional β_{crit} is smaller than in the one-dimensional situation. By considering λ large, Sherratt (1993) derived an analytical asymptotic estimate for β_{crit} in terms of the other parameters which compares well with the value obtained numerically.

The actin density at the wound edge, $G(1)$, in this radially symmetric case is given by $G(1) = 1/[1 + u'(1) + u(1)]$. With $u(1) > 0$ this means that for $G(1)$ to be large we require $u'(1) < -1$. In the solutions, however, this still results in a dilation $\nabla \cdot \mathbf{u} > -1$ as it must. A typical solution of (9.51)–(9.53) is shown in Figure 9.18.

Biological Interpretation of the Parameters

The parameters E and Γ are directly related to the elastic properties of the epidermal sheet while λ relates to the strength of the attachment of the epidermis to the basal

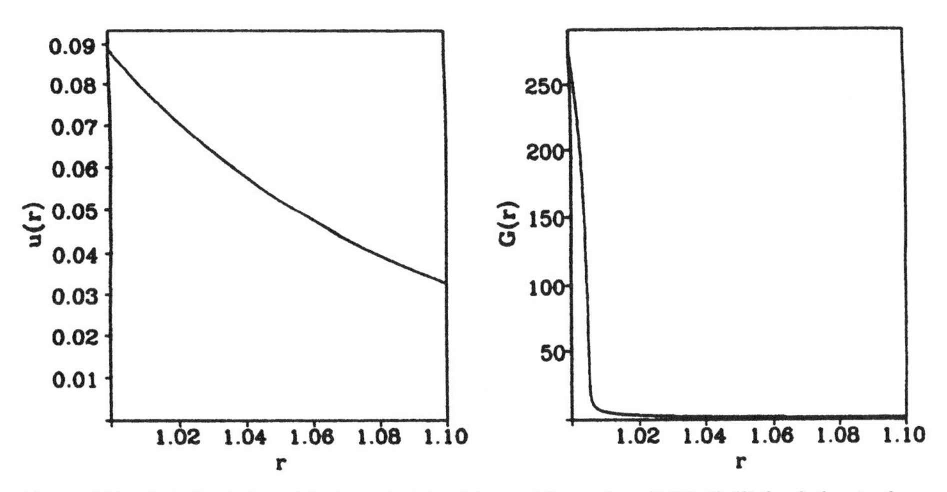

Figure 9.18. Typical solution $u(r)$, shown in (**a**), of the model equations (9.51)–(9.53) for β close to β_{crit}: the dimensionless actin density $G(r) = 1/(1 + \nabla \cdot \mathbf{u})$ is shown in (**b**). The parameter values are $\lambda = 3.0$, $E = 0.5$, $\Gamma = 0.8$, $\beta = 0.255$. The dimensional wound radius is $L = 500\,\mu$m. Note the intense concentration of actin at the wound edge. (From Sherratt 1991)

layer; their interpretation is intuitively clear. The parameter β plays a crucial role in the formation of the actin cable and one possible interpretation is the following. Recall that this parameter is a measure of how the actin fibre density affects the traction force. As mentioned this is probably achieved by the formation of myosin cross-bridges between the actin filaments as proposed by Oster and Odell (1984) in their mechanical model for cytogels. Since the cable formation process takes of the order of several minutes we can anticipate that β actually increases with time. So, the model equation solutions as β increases can be thought of as the time evolution of the epidermal displacement taking place in the first few minutes after wounding. The parameter β is prevented from reaching its critical value β_{crit} by the physical constraints imposed on the actin filament packing. This interpretation of β would imply that the densities of actin filaments and myosin at the wound edge increase in parallel following wounding, a suggestion that has been confirmed experimentally by Bement et al. (1993).

9.10 Mechanical Model with Stress Alignment of the Actin Filaments in Two Dimensions

Although the model for embryonic wound healing in the above section captures the cable formation to the extent that it reflects the experimentally observed intense actin density near the wound edge, it does *not* capture the retraction of the wound edges (Figure 9.18(a)): the retraction is over many cell lengths (in nondimensional terms a cell size is about 0.02). This means that we have to reconsider the model for the radially symmetric situation. Also there is a conceptual difficulty with the model in the last section since $u'(1) < -1$ implies that the cells near the wound edge change places after wounding; in other words the leading cells before wounding do not remain the leading cells after wounding, a phenomenon not observed experimentally. This problem arises from how we modelled the synergy effect of actin filament compaction Δ. Recall that we modelled this by taking the traction τ in the model equation (9.36) to be an increasing function of the local compaction and took the specific form $\tau = \tau_0/(1 - \beta\Delta)$ and used the approximation $\Delta \approx \nabla \cdot \mathbf{u}$. Here Δ is the fraction of the pre-wounding volume by which the cytogel contracts in the initial response to wounding. If we consider a small infinitesimal rectangle whose sides are oriented along the principal axes of the local two-dimensional stress tensor, $\boldsymbol{\epsilon}$, a deformation changes the rectangle into another rectangle. So, $\Delta = 1-(1+p_1)(1+p_2)$ where p_1 and p_2 are the components of the displacement vector $\mathbf{u}$ along the principal strain; the book by Segel (1977) is a good source for this material. In the above model we chose $\Delta \approx -(p_1+p_2) = -\nabla\cdot\mathbf{u}$ all according to linear elastic theory. As soon as either of the p's approach -1, as in the last section on the radially symmetric wound, the approximation is no longer sufficiently accurate. So, in this section we consider the model (9.33) in the radially symmetric two-dimensional situation with

$$\Delta = -(p_1 + p_2 + p_1 p_2) = -(u' + u/r + uu'/r). \tag{9.54}$$

This more accurate form for Δ does not alter the model equation for the one-dimensional case because only one of the p's is nonzero. The linear form of the stress tensor given by (9.32) is still valid (see Segel 1977). In two dimensions it affects G and τ in (9.35) and (9.36). The stress tensor given by (9.48) has the principal values of the stress, denoted by Σ_r and Σ_θ to distinguish them from $p(r)$ and $q(r)$. Now, instead of the expressions in (9.49), we have

$$\begin{aligned} \Sigma_r &= \frac{Eu' + \Gamma(u' + u/r) + [1 + \beta(u' + u/r + uu'/r)]^{-1}}{1 + u' + u/r + uu'/r} \\ \Sigma_\theta &= \frac{Eu/r + \Gamma(u' + u/r) + [1 + \beta(u' + u/r + uu'/r)]^{-1}}{1 + u' + u/r + uu'/r} \end{aligned} \tag{9.55}$$

and the model equation becomes, instead of (9.51) (and using (9.50)),

$$\begin{aligned} r\frac{d}{dr}&\left[\frac{Eu' + \Gamma(u' + u/r) + [1 + \beta(u' + u/r + uu'/r)]^{-1}}{1 + u' + u/r + uu'/r}\right] \\ &+ E\frac{u' - u/r}{1 + u' + u/r + uu'/r} = \frac{r\lambda u}{1 + u' + u/r + uu'/r}. \end{aligned} \tag{9.56}$$

The boundary conditions, in place of (9.52) and (9.53), are now

$$Eu' + \Gamma(u' + u/r) + [1 + \beta(u' + u/r + uu'/r)]^{-1} = 0 \quad \text{at } r = 1 \tag{9.57}$$

$$u = 0 \quad \text{at } r = \infty. \tag{9.58}$$

When Sherratt (1991) solved this equation with these boundary conditions he again found, as we would now expect, a critical β_{crit} above which solutions could not be found. He also found that the solutions for a range of parameter values exhibited the intense aggregation of actin at the wound edge. However, it was not possible to obtain the actin cable formation and the experimentally observed retraction of the wound edges of around 10%–15% with only the leading 4–5 cell layers contracted. So, the model has to be amended even further and the question is how. Although we have briefly discussed the effect of compaction on the actin fibre network its effect was modelled as a scalar and so does not include any preferential alignment of the filaments, which intuitively could have a significant effect. We now modify the model further to include this aspect again following Sherratt (1991); the approach is briefly reviewed by Sherratt et al. (1992).

Figure 9.19 is an idealized representation of alignment of an actin filament network as a result of an applied stress. If the stress is not isotropic the filaments are oriented in all directions but generally tend to orient themselves in the direction of the principle stresses.

The experimental evidence of Chen (1981) and Kolega (1986) shows that actin fibre alignment or reorganisation takes place on the order of minutes so we again consider the process of actin alignment in response to stress to be instantaneous, as we did above. The inclusion of fibre alignment in a two-dimensional geometry has a major effect.

As we described, when a microfilament network is subject to a nonisotropic stress field, the filaments tend to align along the directions of maximum stress. In the ab-

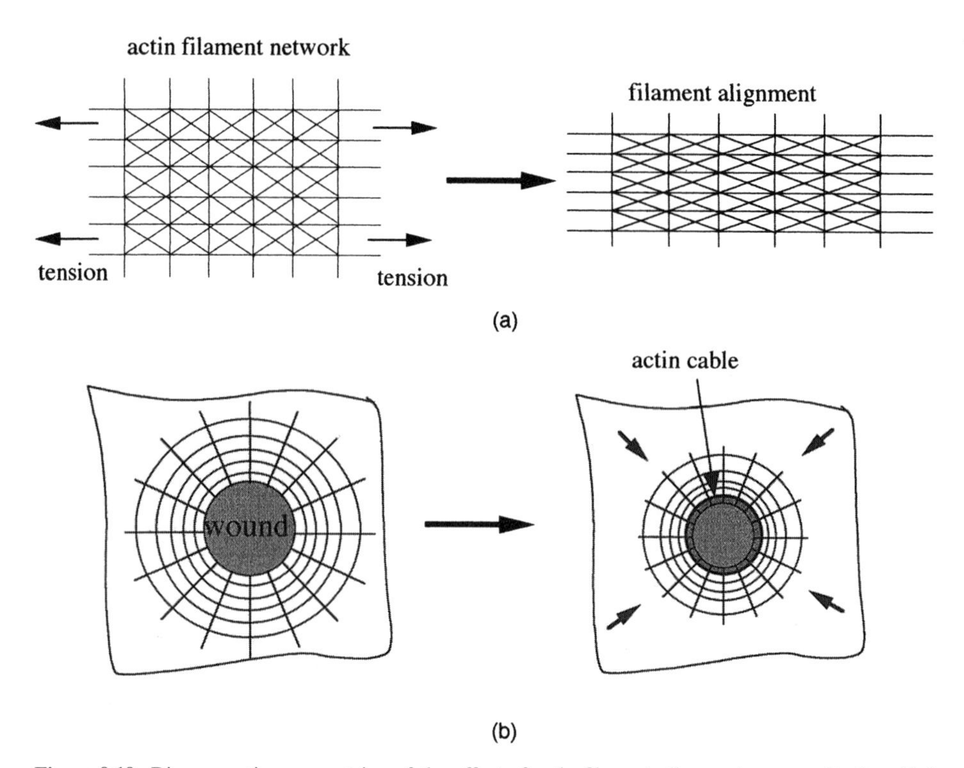

Figure 9.19. Diagrammatic representaion of the effect of actin filament alignment as a result of applied stresses. In (**a**) we idealize the two-dimensional situation while in (**b**) we show a similar scenario for a circular wound and the actin cable.

sence of detailed information on the biological mechanisms responsible for this stress alignment, we develop a model for the phenomenon by making a number of intuitively plausible assumptions (Sherratt 1991, Sherratt et al. 1992, Sherratt and Lewis 1993). (A related but very different approach is described in detail in the following chapter on full-depth dermal wound healing.) The most important assumption here is that the alignment occurs as a direct response to the ratio of the principal components of stress, σ_1 and σ_2 say. These principal components of the stress tensor are (cf. (9.32))

$$\begin{aligned}\sigma_1 &= G_1[E\epsilon_1 + \Gamma\nabla\cdot\mathbf{u}] + G_1\tau,\\ \sigma_2 &= \underbrace{G_2[E\epsilon_2 + \Gamma\nabla\cdot\mathbf{u}]}_{\text{elastic stress}} \quad \underbrace{+\, G_2\tau}_{\substack{\text{active}\\ \text{contraction}\\ \text{stress}}}\ ,\end{aligned} \tag{9.59}$$

where G_1 and G_2 are the effective densities of the actin filaments along the principal axes of stress. Here, as before, $\mathbf{u}(\mathbf{r})$ is the displacement of the material point initially at $\mathbf{r}$, ϵ_1 and ϵ_2 are the principal components of the strain tensor $\boldsymbol{\epsilon} = (1/2)(\nabla\mathbf{u}+\nabla\mathbf{u}^T)$, and E and Γ are positive parameters again as before. The stress and strain tensors $\boldsymbol{\sigma}$ and $\boldsymbol{\epsilon}$ have the same principal axes. The form (9.59) is the standard representation of the stress at equilibrium in the Murray–Oster mechanical theory of morphogenesis we discussed in detail in Chapter 6 and is the form used by Murray and Oster (1984a,b) but with the important difference that here we take τ to be a function of the local compaction and specific forms for the Gs are derived, as described below.

We model the actin filament density functions (Sherratt et al. 1992) by

$$G_1 = G_0 \int_0^{\pi/2} F(\phi; \sigma_1/\sigma_2) \cos\phi \, d\phi, \quad G_2 = G_0 \int_0^{\pi/2} F(\phi; \sigma_1/\sigma_2) \sin\phi \, d\phi, \tag{9.60}$$

where $G_0(\mathbf{r})$ is the local 'scalar' actin filament density, and $F(\phi; \sigma_1/\sigma_2)\delta\phi$ is the fraction of actin filaments at the point concerned that are oriented at an angle between ϕ and $\phi + \delta\phi$ to the $\vec{\mathbf{r}}$-axis, for a given ratio of the principal components of stress, σ_1/σ_2, which in the radially symmetric case is $\sigma_{rr}/\sigma_{\theta\theta}$. We require that the function F satisfies the following conditions.

(i) $F(\phi; 0) = \delta(\phi)$: when the stress is unidirectional, all the actin filaments are oriented in that direction. (Here δ denotes the Dirac delta-function.)

(ii) $F(\phi; 1)$ is constant: when the stress is isotropic, the microfilament network is randomly oriented.

(iii) $\int_0^{\pi/2} F(\phi; \sigma_1/\sigma_2)\, d\phi = 1$ for all σ_1 and σ_2: filament alignment does not affect the total amount of filamentous actin. Remember that F is a probability density function.

(iv) $F(\phi; \sigma_1/\sigma_2) = F((1/2)\pi - \phi; \sigma_1/\sigma_2)$: a symmetry condition.

This last condition implies that

$$\int_0^{\pi/2} F(\phi; \sigma_1/\sigma_2) \sin\phi \, d\phi = \int_0^{\pi/2} F(\phi; \sigma_2/\sigma_1) \cos\phi \, d\phi.$$

Although these conditions are very restrictive they do not determine a specific form for F. Sherratt and Lewis (1993) developed a form for F which satisfies these conditions, and have shown that, to a good approximation,

$$\int_0^{\pi/2} F(\phi; \sigma_1/\sigma_2) \cos\phi \, d\phi = \frac{\pi\sigma_1^p}{2\sigma_1^p + (\pi - 2)\sigma_2^p}. \tag{9.61}$$

The positive parameter p reflects the sensitivity of the microfilament network to changes in the stress field. In what follows we take $p = 1$ for algebraic simplicity. The model without fibre alignment simply has $p = 0$. So, in the postwounding equilibrium we are considering, the principal components of the stress tensor in the cell sheets are given by (9.59) with the actin filament densities G_1 and G_2 given by (9.60) with (9.61) as

$$G_1 = G_0 \frac{\pi\sigma_1}{2\sigma_1 + (\pi - 2)\sigma_2}, \quad G_2 = G_0 \frac{\pi\sigma_2}{2\sigma_2 + (\pi - 2)\sigma_1}. \tag{9.62}$$

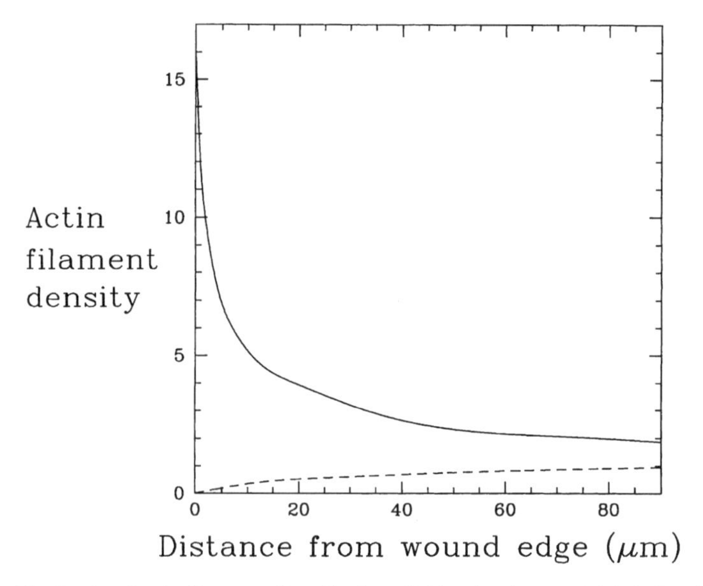

Figure 9.20. The density of actin filaments aligned in the radial (- - -) and tangential (—) directions near the edge of a circular wound, as predicted by the model, for parameter values close to the edge of the parameter domain in which a solution exists. Distance from the wound edge is plotted horizontally, and actin filament density is expressed relative to the (uniform) pre-wounding density. The dimensionless parameters are: $\lambda = 3.0$, $E = 0.5$, $\Gamma = 0.8$, $\beta = 0.48$, where the length scale L (the wound radius) is $L = 500\,\mu$m. (These radial symmetric results were kindly provided by Professor J.A. Sherratt)

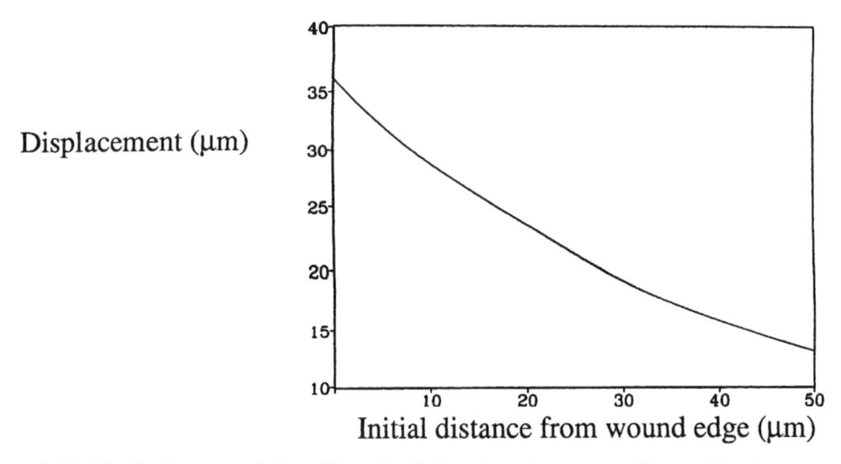

Figure 9.21. The displacement of the epidermal cell sheet from its pre-wounding position in response to a linear slash wound, as predicted by the model. The predicted displacement of the wound edge (36 μm) is roughly the same as that observed experimentally (see Figure 9.16). Analytical investigation of the model equations, which is discussed in detail by Sherratt (1993), shows that, as expected intuitively, the displacement of the wound edge increases as the elasticity parameters E and Γ increase, and decreases as the adhesion and traction parameters λ and τ_0 are increased. The dimensionless elasticity and adhesion parameter ratios are: $\lambda = 3.0$, $E = 0.2$, $\Gamma = 0.75$, $\beta = 0.177$, where the length scale $L = 500\,\mu$m. These parameter values are close to the edge of the parameter domain in which a solution exists. (From Sherratt et al. 1992)

The new model is now (9.33) but with $\boldsymbol{\sigma}$ given by (9.59) and the actin densities G_1 and G_2 given by (9.62). In the case of the radially symmetric wound the principal axes of the stress tensor are aligned with the radial and tangential coordinate axes, namely, the r and the θ directions. In this case $\sigma_1 = \sigma_{rr}, \sigma_2 = \sigma_{\theta\theta}, \epsilon_1 = \epsilon_{rr}, \epsilon_2 = \epsilon_{\theta\theta}$ and the radial and tangential component actin filament densities $G_1 = G_r$ and $G_2 = G_\theta$. Here, from (9.59) and (9.62), we solve for the stress components and, after some elementary algebra, get

$$\begin{aligned}\sigma_{rr} &= \frac{E}{4-\pi}[2u' - (\pi-2)u/r] + \Gamma(u' + u/r) + \left[1 + \beta\left(u' + \frac{u}{r} + \frac{uu'}{r}\right)\right]^{-1}\\ \sigma_{\theta\theta} &= \frac{E}{4-\pi}[2u/r - (\pi-2)u'] + \Gamma(u' + u/r) + \left[1 + \beta\left(u' + \frac{u}{r} + \frac{uu'}{r}\right)\right]^{-1}.\end{aligned} \tag{9.63}$$

Substituting these and the resulting expressions for G_r and G_θ, using (9.62), into (9.33) we get, in place of (9.56), after some more algebra, the following highly nonlinear ordinary equation for u,

$$\begin{aligned}u'' = {} & \frac{1}{[\Psi + 2E/(4-\pi)](r+u) + \Gamma(r - u^2/r)} \cdot \left[\lambda r u\left(1 + u' + \frac{u}{r} + \frac{uu'}{r}\right)\right.\\ & \left. - \left(u' - \frac{u}{r}\right)\left\{\frac{2E}{4-\pi}\left(1 + \frac{u}{r} - u'^2 + \frac{uu'}{r}\right) + \Gamma(1 - u'^2) + \Psi(1 + u')\right\}\right],\end{aligned} \tag{9.64}$$

where

$$\Psi = \frac{(\pi-2)Eu}{(4-\pi)r} - \frac{1 + \beta + 2\beta(u' + \frac{u}{r} + \frac{uu'}{r})}{[1 + \beta(u' + \frac{u}{r} + \frac{uu'}{r})]^2}. \tag{9.65}$$

The boundary conditions here are $\sigma_{rr} = 0$ at $r = 1$, the wound edge, and $u(\infty) = 0$.

The actin density distributions $G_r(r)$ and $G_\theta(r)$ in the radial and azimuthal directions are now given by (9.62) with $\sigma_1 = \sigma_{rr}$ and $\sigma_2 = \sigma_{\theta\theta}$ obtained from (9.63) and the retraction at the wound edge by $u(1)$. Of course (9.64) has to be solved numerically as was done by Sherratt (1991) (see also Sherratt et al. 1992, Sherratt 1993).

The model equation has a solution provided the parameter values are again restricted to a well-defined parameter domain and in particular for β near one edge of this domain. The model solutions exhibit both an intense aggregation of filamentous actin at the wound edge, and pronounced alignment of these aggregated filaments with the wound edge, as shown in Figure 9.22 and Figure 9.23. Moreover, for appropriate parameter values, the model solutions also predict a similar degree of retraction of the epidermis over the underlying mesenchyme to that found experimentally, namely, about 30 to 40 μm (Figures 9.16 and 9.21). To show more visually the correspondence between the predicted actin filament density and the experimental results of Martin and

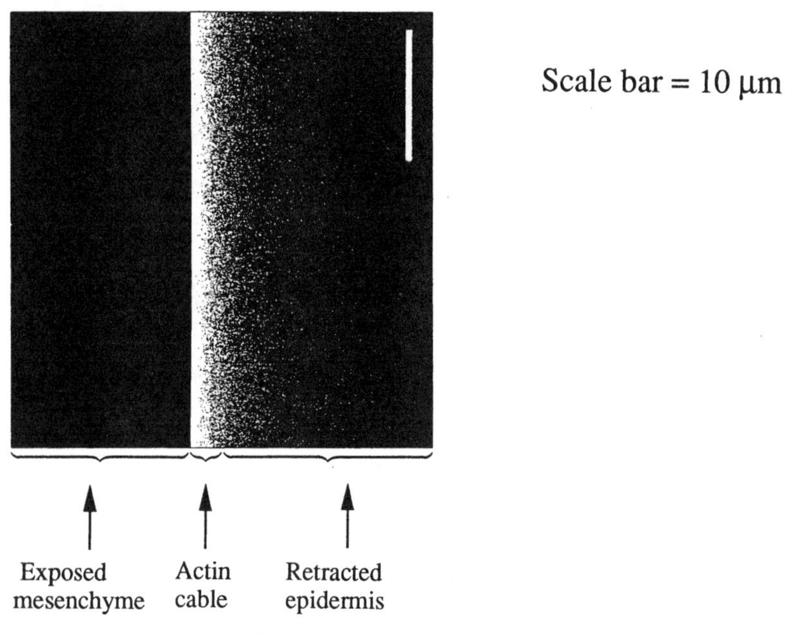

Figure 9.22. A simulation of an optical section through the basal layer of the epidermis in response to a circular wound of radius 500 μm, after staining with fluorescently labelled phalloidin, as predicted by the model solutions. We treat the filamentous actin distribution predicted by the model, appropriately normalised, as a probability distribution for phalloidin staining. The figure shows the post-wounding positions $[u(r)+r]\vec{\mathbf{r}}$ of 9 million points, with radial coordinate chosen at random from this distribution, and angular coordinate chosen at random from a uniform distribution. The dimensionless parameter ratios are: $\lambda = 1.5$, $E = 0.29$, $\Gamma = 0.77$, $\beta = 0.2876$ which imply that $\beta_{\text{crit}} \approx 0.2878$. The scale bar $= 10\,\mu$m. (From Sherratt et al. 1992)

Lewis (1991), Sherratt et al. (1992) used the model solutions to simulate an optical section through the basal layer of the epidermis of a specimen stained with fluorescently labelled phalloidin. To do this, we treated the filamentous actin distribution predicted by the model, appropriately normalized, as an intensity distribution for phalloidin staining. Our simulation is shown in Figure 9.22; it compares well with the corresponding experimental result (see Figure 9.16(b)).

As pointed out by Sherratt et al. (1992), our modelling suggests that the initial formation of the actin cable in embryonic epidermal wounds may occur simply as a byproduct of the post-wounding mechanical equilibrium in the epidermal cell sheet. Central to the formation of the cable is the phenomenon of stress-induced actin filament alignment, as illustrated in Figure 9.22 and Figure 9.23. This alignment phenomenon may play an important part in a number of processes in normal morphogenesis but the response to wounding represents a rather different case, in that alignment is caused by a change in external conditions, namely, the generation of a free edge by wounding.

Some Concluding Remarks on Epidermal Wound Healing

Epidermal wound healing provides a rich source of practical and useful modelling opportunities. Here we have focused on two key issues: the role of biochemical control of

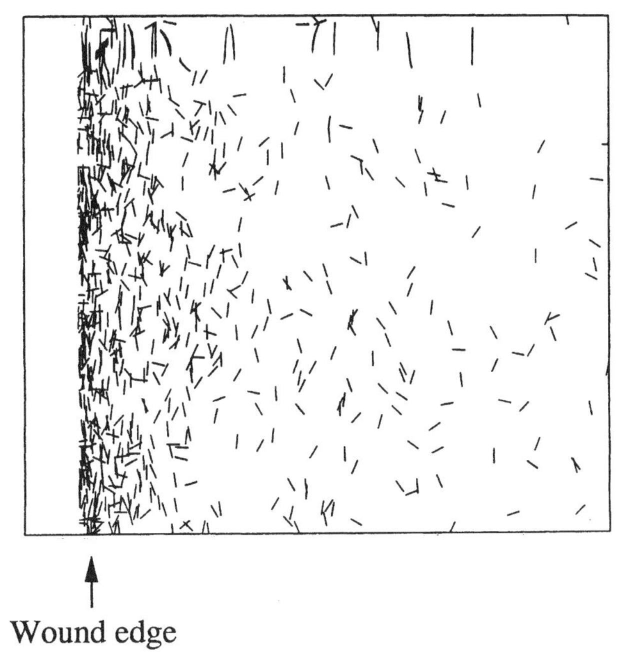

Figure 9.23. An illustration of the stress-induced preferential alignment of the microfilament network in response to a linear slash wound, as predicted by the model solutions. We treat the filamentous actin distribution predicted by the model, appropriately normalised, as an intensity distribution for the location of small line segments. The figure shows 800 such line segments, centred at a point whose orthogonal distance from the wound edges is chosen at random from this distribution, with position parallel to the wound edge chosen at random from a uniform distribution. The orientation of the line segments is chosen at random from the distribution F of actin filament orientations, which is discussed above. The dimensionless parameter ratios are: $\lambda = 3.0$, $E = 0.26$, $\Gamma = 0.7$, $\beta = 0.22$, where the length scale $L = 500\,\mu$m. (From Sherratt et al. 1992)

mitosis in adult epidermis and the mechanism of formation of the actin cable in the embryonic system. They present very different scenarios and require very different types of model: in the former a reaction diffusion model for the conservation of cells and growth factors as the wound front advances while in the latter case, a mechanical model representing the force balance in the cytoskeleton. As noted by Sherratt et al. (1992) these two models do not represent mutually exclusive concepts of the way wounds heal; rather, they deal with different aspects of this complex process. Nevertheless, the two models draw attention to some marked differences between embryonic and adult epidermal healing. Certainly the motile force in adult epidermal wound healing is not provided by an actin cable, and the process of lamellipodial crawling prohibits the existence of a permanent cable around the wound margin. Conversely, embryonic cells are competent to crawl via lamellipodia, since wounds in embryonic sheets *in vitro* heal by crawling rather than by a purse string mechanism. As for the role of mitotic control in embryonic versus adult wound healing, important differences may arise from differences in the sizes of the lesions we consider. Another significant difference is that growth factor levels appear to be much lower in fetal wounds (Whitby and Ferguson 1991). But that phenomenon, hinging on the behaviour of the connective tissue in the healing wound, would call for yet another sort of mathematical model.

The existence of these strong mechanical forces is amply shown during several embryogenetic processes. For example, if a small part of the embryo bursts through the epidermal layer it is eventually amputated by the constricting actin cable (Dr. Susan Bryant, personal communication 2000) indicating a 'boa-constrictor' aspect and the strength of the mechanical forces involved. Even a cursory study of epidermal wound healing reveals many extant modelling problems. For example, with regard to the development of an actin filament cable, geometry clearly plays a major role. Here we have considered only one-dimensional and radially symmetric geometries. Nonsymmetric two-dimensional geometries could give rise to different healing scenarios. In the discussions on embryonic wound healing we only considered the equilibrium state. It would be interesting to include cell division and the closure of the wound subsequent to the cable formation. In the next chapter we deal with actual wound closure but in the case of full-depth wounds which include the dermis.

10. Dermal Wound Healing

10.1 Background and Motivation—General and Biological

Unlike epidermal wounds discussed in the last chapter, dermal wounds are deep wounds to the skin and their healing is crucial for preserving organ integrity after tissue loss. Such wounds frequently lead to a severe scar, graphic evidence of the fact that wound healing in postnatal mammals is a process of repair rather than regeneration. The tissue within the wound is continually remodelled during the initial wound healing response and for months, even years afterwards the mature scar remains visibly and mechanically distinct from the surrounding skin. Furthermore, wound healing frequently involves significant contraction, or reduction in the area of the wound, which can accentuate the unsightliness of the scar and can cause loss of functionality. It is mainly these two aspects of wound healing, contraction and scar formation, that have been the focus of most mathematical modelling to date. Even compared with epidermal wound healing the diversity of biological processes is even greater and our understanding even less. Deep wounds involve not only dermal wound healing but also re-epithelialisation. Many of the following general comments apply equally to epidermal wound healing.

The importance of wound healing is obvious. For example, serious burns are dermal wounds; more than 2 million people sustain thermal injuries in the U.S.A. alone each year. However, wound healing has far wider implications. For example, there is evidence that scar formation is related to fibroplastic diseases such as Dupuytren's contracture (Gabbiani and Majno 1972) and the formation of stretch marks during pregnancy (Shuster 1979) is another example of residual tissue deformation after stretching; we briefly discuss a model for this below. Wounds in the dermis have much in common with those in the cornea, so important in radial keratotomy (Jester et al. 1992, Petroll et al. 1993). Restenosis (blocking of the arteries) after treatment of atherosclerosis by balloon angioplasty involves a fibrotic wound healing response (Forrester et al. 1993). Other diseases involving cell-mediated tissue contraction are, for example, hepatic cirrhosis, pulmonary fibrosis, scleroderma and diseases that result in tractional detachment of the retina; the list goes on. There is, however, an even broader context in which we can judge the value of wound healing studies. Wound healing is unique in its position at the meeting point of medicine and developmental biology. Dermal wound healing involves processes of cell differentiation, migration and proliferation, the creation of new tissue and the mechanical deformation of that tissue by the cells within it. Each of these processes has analogues in development and, even if the results of tissue repair were not important to us, the study of wound healing would certainly be worth pursuing for the insights it offers into related developmental processes. These include bone development, the formation of somites, skin primordia and nephrons in the kidney, vascularisation and others. The challenging biological and mathematical modelling problems are endless and the justification and need for interdisciplinary research is very clear. The Holy Grail of dermal wounds is scarless healing.

As mentioned in the brief history at the beginning of the previous chapter research into wound healing has a very very long history and is one of the oldest scientific disciplines (Majno 1975). As a result, a vast literature documents the healing of deep (dermal) wounds through the inflammatory response, granulation tissue formation and remodelling of the scar. Along with the strictly scientific study of wound healing there is also an extremely well-developed art of wound management particularly with regard to controlling infection and in making intentional wounds, grafting and so on. However, many of the original questions that concerned early researchers remain unanswered.[1] For example, why do circular wounds seem to heal more slowly[2] and with less satisfactory results? We still do not know what causes pathological scars—hypertrophic scars and keloids—the former are very likely to result after severe burn injuries.

The progress of wound healing has been quantified in a number of ways, including measuring wound area and contraction rate and measuring wound strength. Scar quality is primarily a reflection of the organization of the collagen in the extracellular matrix. Experiments on animals and observation of wounds in humans have led to rules of thumb such as wounds aligned with the lines of skin tension tend to heal with better results. There is no doubt that such knowledge is extremely useful. However, as the number of ways in which we can intervene in the natural process continues to increase, the need to understand the mechanisms involved in dermal wound healing and the need to be able to predict alternative outcomes become more and more urgent. We have to conclude that, although our knowledge of the details is increasing rapidly, most of our current knowledge is on a scale which is difficult to relate to the clinically important macroscopic properties of the scar. It is precisely here that we believe mathematical modelling has a lot to offer.

Wound healing research is even more active now than ever. This is in part a result of advances in diverse areas and is also certainly aided by wound healing's position on the boundary between developmental biology and medicine. Certain particular events have prompted the current excitement in the field. Firstly, it was discovered in the early 1980's that mammalian fetal wounds heal without scars. In the last chapter we discussed in detail some of the work by Sherratt (see Sherratt 1991, 1993 and Sherratt et al. 1992)

---

[1] I attended the 4th Annual European Tissue Repair Society Meeting in 1994 at which I was invited to give a talk; there were several hundred 10-minute contributed talks with the order of a thousand people at the meeting. What was clear was the incredible number of unsolved problems in the field and so little fundamental science done on them. So many of the short contributed talks were essentially marking their territory (many with patents in mind) and consisted of a description of what chemical dressing was put on what kind of wound and what was observed.

[2] This was remarked on by Hippocrates himself. Ambroise Paré, a 17th century French surgeon commented 'L'ulcère rond ne reçoit cure, s'il ne prend une autre figure'; in other words a round wound will not heal unless it takes on another shape.

and his coworkers on embryonic wound healing. This gives many researchers hope that it may be possible to 'turn back the clock' and induce adult wounds to heal without scars. Secondly, a vast amount of research effort is now focused on growth factors. These chemicals are involved in the control of many of the most important processes of wound healing. Clinical trials involving either promotion or inhibition of growth factors are currently in progress. Finally, as mentioned in the last chapter, the development of artificial skin for use in grafting holds great promise.

In the following sections we discuss some of mathematical models for dermal wound healing and in part follow the expositions of Murray et al. (1998), Tanquillo and Murray (1992, 1993), Tracqui et al. (1993) and Cook (1995) (this thesis, unfortunately most of it unpublished, is a particularly full, original and in depth study of dermal wound healing and arguably the most comprehensive and definitive treatment to date). After providing the biomedical context we give a brief critical review of some of the mathematical models published to date; it is not possible to give here a comprehensive review of the current literature. We pay particular attention to those aspects which relate to the biomechanical modelling of a tissue whose structure evolves over time which puts wound healing in a broader context. We also discuss other directions for future research both experimental and theoretical and conclude with some general remarks on the specific value of mathematical models for dermal wound healing.

Observations and Questions from Biomedicine

The gross time course of dermal wound healing is as follows (Clark 1991). Immediately after wounding the wound edges retract due to natural skin tension and there follows an inflammatory response. This induces specialized cells called fibroblasts to proliferate, invade the wound and lay down a tissue matrix on which they exert a contractile force, thereby filling and closing the wound. Angiogenesis (the formation of blood vessels) occurs at the same time and the bright red tissue filling the wound is referred to as granulation tissue. The wound is gradually covered by an invading epithelium but the dermal and epidermal processes are essentially independent (Peacock 1984). The extracellular matrix (ECM) is continually remodelled, rapidly at first and at a lower rate for many months thereafter. In most cases the result is a scar which differs in appearance from the surrounding skin.

Wound Area

Many authors have measured the area and perimeter of the wound over time or have noted the maximum extent of wound retraction and contraction, particularly McGrath and Simon (1983) whose data we use for quantitative comparison with our theoretical results; see Figure 10.5. Typically, directly after wounding the wound edges retract due to skin tension. Following a latent period of 5 to 10 days the wound area reduces rapidly (5 to 10% per day) as a result of contractile activity within the wound. The contraction curve eventually bottoms out and the area then changes very little over a long period. The final contracted area may be around $1/5$ of the original area (this varies with species and with wound). Sometimes there is late retraction of the wound edges (the scar expands slightly).

Wound Shape

There is some controversy about the effect of wound shape on contraction rates (McGrath and Simon 1983, Cook 1995). The change of shape that accompanies wound contraction is highly dependent on the initial shape, on the orientation of the wound to the skin tension lines, and to a lesser extent on the laxity of the surrounding tissue (Cook 1995 gives an extensive discussion). Circles heal by decreasing their perimeter (taking on an elliptical or oval shape) while polygonal wounds tend to have little change in perimeter length; the edges collapse inwards forming star-shaped scars. Bertolami et al. (1991), among other things, investigated tissue contraction following skin grafts by noting the movement of tatoo marks on the skin.

Tension

There is general agreement that wounds parallel to the skin tension lines heal with more satisfactory results and surgeons use various techniques to minimize the tension across a wound (Hinderer 1977 discusses surgical measures to try and prevent gross scarring). Skin tension affects the extent and course of contraction as well as retraction and it has also been implicated in the formation of pathological scars. Excessive tension results in wide scars but application of modest tension can improve scar strength (Timmenga et al. 1991).

Mechanical Properties

A wound becomes stiffer, stronger and more extensible throughout the wound healing process and the shape of the stress–strain curve (the stress–strain relationship) changes over time. What determines these changes is not known although various correlations with matrix organization (collagen, in particular) have been observed.

Pathological Scars

Pathological scars, termed either hypertrophic scars or keloids, mature very slowly or not at all (Kischer et al. 1990). The former are almost guaranteed to occur in severe burn wounds (Kisher et al. 1982). Both types are characterized by abnormal collagen organization, often involving 'nodules,' unusual fibre orientation and abnormal mechanical properties. Tension on the wound appears to play a part in determining the structure of the extracellular matrix (ECM).

Theories

Cook (1995) gave an extensive critical review of the various theories for wound contraction and scar formation at the time. The majority of recent authors believe that the contractile force resides in myofibroblasts, although a significant number maintain that regular fibroblasts are equally, if not more, important (Ehrlich 1989, Grierson et al. 1988). There are clear differences in organization (including fibre orientation) between scar tissue and normal skin but the causes of pathological scar formation remain unclear.

10.2 Logic of Wound Healing and Initial Models

As far as we know the first mathematical models for the possible mechanism involved in dermal wound contraction were those of Murray and Tranquillo (Murray et al. 1988, Tranquillo and Murray 1992, 1993). These were based on the Murray–Oster mechanical theory discussed above in detail in Chapter 6 and applied elsewhere in the book. Although this theory was originally proposed in the context of developmental pattern formation it extends quite naturally to wound healing since in both instances a key role is played by the mechanical interaction between cells and the viscoelastic extracellular matrix (ECM) which they inhabit and through which they move. Variants of the basic models allow for cell movement (diffusion, chemotaxis, haptotaxis, haptokinesis) ECM turnover, cell proliferation and death, osmotic forces, and mechanical interaction between tissue layers.

One of the achievements of mathematical modelling to date has been the isolation of the critical component processes of wound contraction and scar formation; Sherratt's (1991, 1993) seminal contribution to the latter, discussed in the last chapter, is one example. An outline of the logic of wound contraction and scar formation is schematically shown in Figure 10.1 (Murray et al. 1998). Specialized cells called fibroblasts invade the wound and are capable of deforming (contracting) the extracellular matrix through which they move. The invasion may be enhanced by chemotaxis, haptotaxis or contact guidance (a crucial aspect of vasculogenesis and angiogenesis which is discussed in Chapter 8). The fibroblasts also play an important role in remodelling (creating and degrading) the ECM. Subdermal attachments and the ECM itself resist the contractile stresses set up by the fibroblasts (or the specialized phenotypic variants, myofibroblasts).

We translate the various key processes into mathematical form below. Since the resulting equations are complicated we again first give word equations which demon-

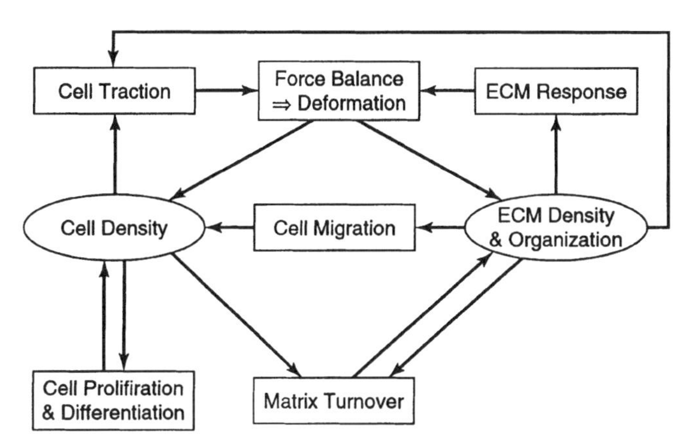

Figure 10.1. Key variables and processes in dermal wound healing. This structure underlies the mathematical model: each variable requires an evolution equation with terms corresponding to the various processes. The force balance box involves both a variable (momentum) and a conservation equation. This constraint determines another variable, deformation. ECM refers to extracellular matrix.

strate how the various pieces fit together. In the models which deal with ECM structure, further equations are required (see Section 10.6 in particular, but also the following four sections).

$$\begin{array}{l}\text{rate of change}\\\text{of cell density}\end{array} = \begin{array}{l}\text{(passive) movement due}\\\text{to ECM deformation}\end{array} + \begin{array}{l}\text{(active) movement}\\\text{relative to ECM}\end{array}$$

$$+ \text{proliferation/differentiation}$$

$$\begin{array}{l}\text{rate of change}\\\text{of ECM density}\end{array} = \begin{array}{l}\text{movement due}\\\text{to deformation}\end{array} + \text{synthesis} - \text{degradation}$$

$$\text{body forces} = \text{traction forces} + \text{ECM resistive forces}.$$

In the last equation the forces on the right-hand side are usually in the form of the divergence of a stress, here the traction stress plus the ECM stress.

We now give the core equations for the model of Murray et al. (1988) as an example of its type and to highlight why more complex models had to be developed. Most of the later models involving cell forces have many of these core elements. The equations of the base model are derived respectively from conservation of cells, ECM and momentum of the combined continuum with respect to fixed (Eulerian) coordinates. These correspond to the word equations above. Cells are convected, diffuse and multiply and ECM is convected. The continuum is subject to active traction forces and also body forces due to elastic tethering to the subdermal layer. The ECM response is taken to be linear viscoelastic. In view of the timescales involved, inertial effects are ignored.

The cell conservation equation is

$$\frac{\partial n}{\partial t} + \nabla \cdot (n\mathbf{u}_t) = \nabla \cdot (D\nabla n) + rn(n_0 - n), \tag{10.1}$$

the ECM conservation equation is

$$\frac{\partial \rho}{\partial t} + \nabla \cdot (\rho \mathbf{u}_t) = 0, \tag{10.2}$$

and the force equilibrium is given by

$$\nabla \cdot \left[\mu_1 \boldsymbol{\epsilon}_t + \mu_2 \theta_t \mathbf{I} + E'(\boldsymbol{\epsilon} + \nu' \theta \mathbf{I}) + \frac{\tau \rho n}{1 + \lambda n^2} \mathbf{I} \right] = s\rho \mathbf{u}. \tag{10.3}$$

Here n is the cell density with n_0 the uninjured steady state value, ρ the matrix density, $\mathbf{u}$ the ECM displacement vector, $\boldsymbol{\epsilon} = (1/2)(\nabla\mathbf{u} + \nabla\mathbf{u}^T)$ the strain tensor, $\theta = \nabla \cdot \mathbf{u}$ the dilation, $\mathbf{I}$ the unit tensor, $E' = E/(1 + \nu)$ where E is the Young's modulus, ν the Poisson ratio, $\nu' = (\nu/(1 - 2\nu))$, μ_1, μ_2 the shear and bulk viscosities, D the cell motility coefficient, r the maximum mitotic rate, s the tethering elasticity coefficient and λ is a parameter quantifying how the cell traction depends on n. The last term on the left of the last equation is the active cell traction term while the term on the right of the equation is the body force due to subdermal attachments. Note that there is no turnover of ECM in these equations.

The question of how to quantify the cell traction force is an important one currently being investigated experimentally. The one we choose is qualitatively plausible and we use it solely by way of example. A major contribution to this aspect is that by Ferrenq et al. (1997) who give specific forms for the force in terms of biological parameters. The seminal work by Barocas and Tranquillo (1997a,b) and Barocas et al. (1995) is concerned with several aspects of the interplay between cell traction and fibrillar network deformation. The latter is also particularly relevant to the cell–ECM interaction modelling approach by Cook (1995) a part of which we discuss below.

We now introduce nondimensional quantities

$$
\begin{gathered}
n^* = \frac{n}{n_0}, \quad \rho^* = \frac{\rho}{\rho_0}, \quad \mathbf{u}^* = \frac{\mathbf{u}}{L}, \quad \mathbf{r}^* = \frac{\mathbf{r}}{L}, \quad t^* = \frac{t}{T}, \quad s^* = \frac{s\rho_0 L^2(1+\nu)}{E}, \\
\lambda^* = \lambda {n_0}^2, \quad \tau^* = \frac{\tau\rho_0 n_0(1+\nu)}{E}, \quad r^* = r n_0 T, \quad \boldsymbol{\epsilon}^* = \boldsymbol{\epsilon}, \quad \theta^* = \theta, \\
D^* = \frac{DT}{L^2}, \quad {\mu_i}^* = \frac{\mu_i(1+\nu)}{ET}, \quad i = 1, 2,
\end{gathered} \tag{10.4}
$$

where L and T are whatever length and timescales we choose as the most relevant. The equations then become

$$
\begin{aligned}
\frac{\partial n}{\partial t} + \nabla \cdot (n\mathbf{u}_t) &= \nabla \cdot (D\nabla n) + rn(1-n), \\
\frac{\partial \rho}{\partial t} + \nabla \cdot (\rho \mathbf{u}_t) &= 0, \\
\nabla \cdot \left[\mu_1 \boldsymbol{\epsilon}_t + \mu_2 \theta_t \mathbf{I} + (\boldsymbol{\epsilon} + \nu' \theta \mathbf{I}) + \frac{\tau\rho n}{1+\lambda n^2}\mathbf{I}\right] &= s\rho\mathbf{u},
\end{aligned} \tag{10.5}
$$

where we have omitted the * for algebraic convenience.

Relevant initial conditions for this system are $\mathbf{u} = 0$, $n = 1$, $\rho = 1$ for $\mathbf{r}$ outside the initial wound boundary and $n = 0$ inside the wound. Since there is no ECM production we must assume that $\rho(\mathbf{r}, 0) = 1$ inside the wound as well since we are in effect assuming the (fibrin) blood clot forms instantaneously relative to the contraction phase: this is not unreasonable (Clark 1985). Further, Murray et al. (1988) assumed that the mechanical properties of the wound matrix are the same as the surrounding matrix and remain so as it becomes modified by the cell biosynthesis.

As with all models the parameter values play an important role. In the case of this base model parameter sets must satisfy certain constraints. In particular the uniform steady state must be stable to perturbations in the dependent variables. With the number of stability analyses we have carried out in the book it is a now routine procedure to do a linear analysis on the model set of equations (10.5) about the uniform (unwounded) steady state $n = 1$, $\rho = 1$, $\mathbf{u} = 0$. It is left as an exercise to show that the necessary and sufficient conditions for stability are:

$$s + r\left(1 - \frac{\tau}{1+\lambda}\right) < 0 \quad \text{or} \quad 1 + \mu r - \frac{2\tau}{(1+\lambda)^2} < 0.$$

Most quantitative studies on wound healing have been done on animals. There is a basic difference between animal and human wound healing. Animals possess a skin organ (the panniculus carnosus) under the dermis which is absent in humans. It confers greater mobility to the skin in response to stress. In animals up to 100% of animal wound area may be closed by contraction; values of 20 to 40% are more the norm in humans. Nevertheless, the belief is that the mechanisms of wound healing are similar. Animal wounds are studied also to try and determine the wound geometry which closes the fastest and with least scarring. This is often done by measuring the wound area following an excisional wound, typically using tatoo marks to delineate the wound boundary and distinguish contraction from epithelialisation (see, for example, the article by Bertolami et al. 1991). Among the best quantitative studies for comparing our results with experiment are those by McGrath and Simon (1983) who studied dermal wounds on rats. They showed there is a rapid retracting phase, then a plateau phase followed by a contracting phase, which can be described by a simple exponential dependence on time given by

$$A(t) = A_f + (A_0 - A_f)\exp(-k_c t), \tag{10.6}$$

where A_0 is the wound area when contraction begins, A_f is the area remaining after contraction is complete, both areas being scaled to the excised area and k_c is the contraction rate constant. Figure 10.2 shows the experimental results from McGrath and Simon (1983).

Murray et al. (1988) showed that this base model can produce a stable nonuniform steady state solution which evolves from the initial wounded state. However, it was not qualitatively consistent with the extant data on a contracting wound for the parameter ranges investigated and particularly the data shown in Figure 10.2. Typically they found a transient retraction of the wound boundary outwards (that is, in one dimension $u > 0$) and a relaxation inwards back to the undisturbed state $u = 0$ rather than any contraction of the wound boundary inward ($u < 0$); this is certainly a crucial deficiency. It appears therefore that what we considered to be a minimal set of properties applicable to dermal wound healing are inadequate for simulating wound contraction.

We certainly made some gross assumptions in deriving the model system (10.5), such as a small strain approximation, no production of ECM and so on. Tranquillo and Murray (1992) discuss in more detail the deficiencies of this model and investigate amended versions. Tranquillo and Murray (1993) discuss the clinical implications of the modelling. Here we briefly describe these amendments for pedagogical reasons. In the following sections we discuss more realistic versions but with the some of the same basic concepts. The question Murray et al. (1988) addressed was how to amend the above model to make it more biologically realistic. All of the amendments added to the cell functions and the ECM properties. With the exception of a term for the biosynthesis of the ECM they were reasonably motivated by the recognised influence of inflammatory-derived biochemical mediators on fibroblast functions. These can be extremely complex and much is still unknown about them. One possibility to incorporate some aspects of

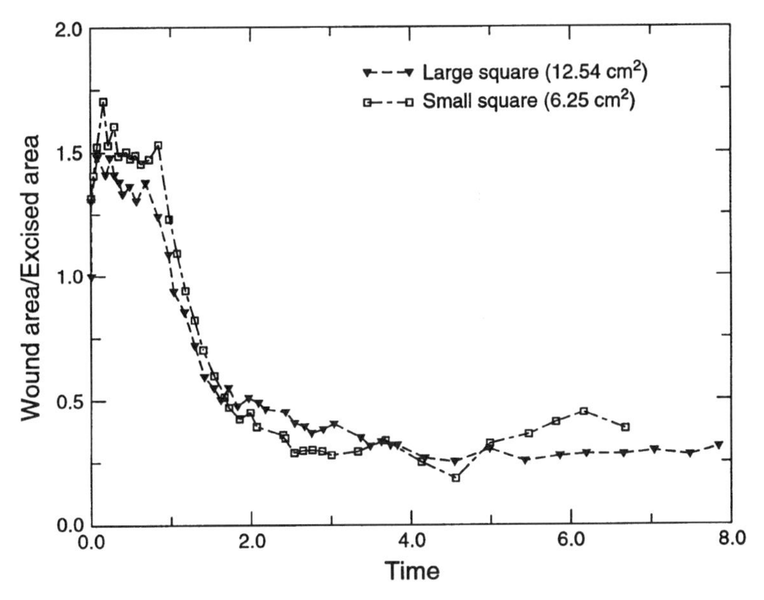

Figure 10.2. A graph of the wound area relative to the excised area for a full-depth rat wound. (Data from McGrath and Simon 1983) A_0 is the wound area when contraction begins, A_f is the area remaining after the contraction is complete and $A_c = A_0 - A_f$ is the contracted area.

this is to include a generic equation for such biochemical mediators. Instead, based on experimental evidence, Murray et al. (1988) and Tranquillo and Murray (1992, 1993) incorporated a mediator dependence in the cell traction term by setting

$$\tau = \tau_0 \left(1 + \frac{\tau_f c}{1 + c} \right),$$

where τ_f is a traction enhancing parameter and τ_0 is the base cell traction. They then took the biochemical mediator to be a given function, namely,

$$c = c_0 \exp\left(-x^2/\sigma\right),$$

where c_0 is the concentration at the wound centre, x measures distance from the wound centre and σ is a parameter specifying the spatial influence of the mediator from the wound centre. Tranquillo and Murray (1992) studied this modified system in detail and found that for appropriate parameter ranges the solutions for the tissue displacement agreed reasonably well with experiment. They went on to discuss other possible variations such as including chemotaxis, ECM biosynthesis and cell growth variation.

A major criticism of these models is the prescription of the mediator concentration gradient in lieu of it being autonomously determined from a model of inflammation. It makes the comparison of the results with experimental data somewhat unconvincing. It also circumvents the whole question of the causal factors and the ultimate regulatory control of wound contraction. What it does show, however, is how the modulation of fibroblast function by an inflammatory mediator could result in the wound healing scenario we described above. Tranquillo and Murray (1992) investigated the effect of cell traction forces within this modelling scenario.

There are several obvious deficiencies in the above formulation other than that associated with the inflammatory response, such as the omission of biosynthesis of the cells and the ECM which are not known. Also the theory is based on a small strain viscoelastic formulation when *in vivo* wound contraction involves finite strains. So, more realistic stress–strain constitutive relations are required, in particular one that accounts for aniosotropic fibre orientation, a topic we discussed above in the chapter on embryonic epidermal wound healing. It is much more complex in full-depth wounds. In spite of these difficulties the simple models discussed within this framework provided a means of considering the effects of known and speculated cell proliferation, migration, and traction responses and ECM rheological properties both individually and collectively. In the following sections we describe some of the more realistic models which have been proposed and studied based on the above modelling concepts.

10.3 Brief Review of Subsequent Developments

Here we briefly recap the review given by Murray et al. (1998). Since the first wound healing models were published mathematical modelling has greatly increased. More complicated versions of the model described above include multiple cell types (with differentiation between types) and multiple types or phases of ECM (which interact and possess different mechanical properties and have different effects on cells). Other models include additional equations for chemicals (such as growth factors) which modulate cell proliferative, motile and contractile behaviour. When the mechanical properties of the evolving ECM are modelled with greater realism further equations are required.

Maini, Sherratt and their coworkers have presented some very interesting results which include the following. Olsen et al. (1995) extended the Murray–Tranquillo model to include both explicit interconversion between fibroblasts and traction-enhancing myofibroblasts and more complex inflammatory mediator dynamics. The model captures the experimentally observed temporal changes in the densities of myofibroblasts and growth factors. A wound remains transiently contracted while ECM remodelling (slow) is in progress. Olsen et al. (1996) related pathological scarring to the existence of alternative steady states using a mechanical model and a simpler caricature. A high rate of growth factor production, for example, could cause a spread of the pathological state across the wound and into surrounding tissue. Complementary to this study, Dale et al. (1996) used a non-mechanical model to investigate the ratio of collagen I to collagen III in scar tissue (this ratio is known to be related to fibre thickness and scar quality) focusing on the regulation by TGF isoforms (the latter growth can be topically applied).

Recent work by the Maini–Sherratt group (Dallon et al. 1999) on extracellular matrix dynamics is an important contribution to the study of tissue regeneration and reorganization and scar formation in general. They consider the cells to be discrete and the matrix to be a continuum: cell and fibre orientation dominate the process. The work of Edelstein-Keshet and her colleagues (see Edelstein-Keshet and Ermentrout 1990,

Mogilner and Edelstein-Keshet 1996 and other references there) on cell orientation is particularly apposite to this approach. Dallon et al. (1999; see other references there) show how cell movement is directed by the matrix substrate and in turn how they in turn reorient the extracellular matrix. They are able to quantify, by extensive numerical simulation of their models under different biological assumptions, a variety of effects such as the rate of movement of the cells, the influence of cell contact guidance, the original orientation of the matrix fibres, fibre production and degradation and so on. They present visually graphic results which show the effect of these different factors on the final fibre alignment pattern. They conclude that in wound healing cell flux is a particularly important factor in tissue alignment.

Tracqui et al. (1995) extended the Murray–Oster mechanical formulation to include two distinct types of extracellular matrix, an early provisional matrix being replaced by the collagenous matrix characteristic of mature scars. A key feature of this model is that ECM turnover results in plastic behaviour whereby the wound remains permanently contracted after the inflammatory response has subsided. Barocas and Tranquillo (1994, 1997a) accounted separately for the interstitial fluid and the fibrous network making up the ECM (this is potentially important in collagen gel assays which can contract much more rapidly than wounds). In their wound healing models they use the limit of zero drag which effectively reduces the system back to a single phase model. An important extension remains however: they were able to account for matrix anisotropy via an orientation tensor which is based on the strain tensor. The fibre orientation tensor governs cell movement (contact guidance) and traction (greatest in the direction of greatest fibre alignment). In circular wounds, contact guidance can reduce the extent of wound contraction as cells align (and pull) circumferentially (Barocas and Tranquillo 1997).

Cook (1995) extended the Murray et al. (1988) and Tranquillo and Murray (1992) model primarily by introducing more realistic mechanics and structure of an evolving, anisotropic ECM, effective strain and by introducing a fibre orientation tensor which is related to it. In addition to providing a measure of scar quality, fibre orientation feeds back to affect (i) cell movement (the flux of cells due to contact guidance depends upon effective strain in a similar way to the Barocas and Tranquillo 1997a model, although Cook's derivation was based on a biased random walk) and (ii) cell traction (greatest contractile force in the direction of fibre orientation). Cook (1995) also allowed for orthotropic skin tension (and wound orientation relative to skin tension lines). In contrast to previous models contractile forces were assumed to be zero outside the wound and late in scar formation. Finally, this was the first study to attempt numerical solution of the equations in two dimensions (radially symmetric solutions are quite different).

Cook's results (1995) included the following: (i) a study of changes in wound shape, contraction rates, and fibre alignment (wound orientation proves important); (ii) true plasticity: wounds remain contracted (and fibre alignment is anisotropic) in the absence of inflammatory mediators or any other externally imposed differences between skin and mature scar; (iii) strong contact guidance: this can give rise to instabilities (pattern formation) suggestive of nodules of aligned collagen fibres in pathological scars. Finally, a highly simplified model based on linear elasticity suggested that circular wounds should contract most slowly (agreeing with experiments). The role of wound shape is yet to be addressed completely in the full model.

Biomechanics of Evolving Tissues

The early wound healing models were derived from the Murray–Oster mechanical models of Chapter 6 for developmental processes such as the positioning of skin primordia (hairs, feathers, scales) and pre-cartilage patterning in the developing limb. Wound healing can be seen as just one of many morphogenetic processes which involve tissue growth and remodelling both in development and in adults. In a comprehensive review Taber (1995) describes how related processes occur in bone, skeletal muscle, the heart and arteries. He also describes a number of theoretical advances which have allowed the biomechanics of growth and remodelling to be brought into a continuum-mechanical framework.

In classical elasticity theory a strain field characterizes the mapping between material points in a reference state and locations in a deformed state. The mechanical properties of the material are incorporated in a constitutive equation which relates stresses to strains. Finally, applied body and boundary forces determine the global deformation via the equations of equilibrium which constrain the stresses in the material. Although finite strains and nonlinear constitutive equations provide complications, applying classical viscoelasticity theory to biological materials is no different from applying it to nonliving materials. Once experiments have been devised to determine the appropriate constitutive equations for each mechanically distinct component (a highly nontrivial task) a whole range of biomechanical problems can be solved (Fung 1993 gives many examples).

However, there are two qualities of living tissues which require that classical continuum mechanics be extended. Firstly, biomaterials can support residual stresses: even when all external forces are removed stresses may exist within the tissue. Residual stresses can be indirectly observed by cutting out sections of tissue and recording deformations (for example, Fung and Liu 1989). The existence of residual stresses violates the assumption of classical elasticity theory that the reference state is globally stress-free. The second problem is that living tissues can change their form, their structure and even their material properties, either as part of the natural developmental process (growth is an example) or in response to other signals (such as injury or applied stress). Thus, classical mechanics must be extended to accommodate tissue growth and tissue remodelling.

A number of scientists have addressed these problems including Cook (1995) in the particular case of dermal wound healing. Rodriguez et al. (1994) described, using the language of tensors, a general finite strain theory for elastic biomaterials which incorporates both residual stresses and volumetric tissue growth and remodelling (surface growth, such as occurs in bone, has been described by Skalak et al. 1982; see Taber 1995 for other references). The Rodriguez theory has application to a wide range of important biomedical processes, including dermal wound healing.

It is not possible to discuss all of these interesting advances on wound healing modelling. For anyone seriously interested in modelling wound healing the above articles are required reading. We restrict our discussion primarily to some of the work of Tracqui et al. (1995) and particularly Cook (1995) since most of it remains unpublished.

10.4 Model for Fibroblast-Driven Wound Healing: Residual Strain and Tissue Remodelling

In this section we discuss the mechanical model of Tracqui et al. (1993) for the contraction and relaxation phases of healing dermal wounds based on the extracellular control of the traction forces exerted by the migrating fibroblastic cells. We assume that these cell–extracellular medium interactions are controlled by the viscoelastic properties of both the provisional matrix formed initially in the lesion and the newly synthesised collagenous matrix secreted at the same time. In addition, and importantly, we include the plastic response of the collagenous matrix to the cell traction. This mechanical model accounts for the different experimental phases of the wound boundaries' movement with time. Furthermore, it provides a quantification of the residual strain and stress resulting from the plastic response and remodelling of the extracellular matrix which could help to characterise the scar tissue formation in wound healing.

In the model we describe here we concentrate on wound contraction and the related displacement of the wound margins with time. The model tries to characterise some of the major biological factors which, through their interactions, are sufficient at least to qualitatively take into account the dynamics of the wound margins. In this section we consider only the one-dimensional version.

Let us briefly summarize the wound healing process we consider: further biological details are given below. The movement of the original wound boundaries toward the centre of the lesion (wound contraction) results from the mechanical interaction between the fibroblastic cells (fibroblasts and myofibroblasts) and the surrounding extracellular matrix: a good overview is given by Jennings et al. (1992). In cutaneous wounds, this reduction in the wound area takes place after an initial retraction of the wound boundaries, due to the intrinsic tension that is usually present in skin. Next a latent phase occurs, during which no gross movement takes place. The contraction phase is then characterised by an exponential decay of the wound area as shown in Figure 10.2. Later, an incomplete relaxation phase takes place, leading to a sustained contraction of the wound. Figure 10.2, taken from the data of McGrath and Simon (1983) illustrates each of these phases. The latent phase lasts about a week with the exponential phase lasting until around the beginning of the 6th week. The epithelialisation starts around the beginning of the 3rd week eventually merging with the dermal curve about week 8.

We have already discussed various hypotheses to account for the inward wound boundary movement; see also Welch et al. (1990). Experimental results seem to favour the pull theory, in which inward movement of the wound margins is due to forces lying in the regenerated tissue in the wound bed. In this section, we describe a model for wound contraction based on the modulation of the traction force exerted by the fibroblastic cells through the difference in the properties of the different wound extracellular matrices.

Various experimental data support this conceptual framework. First, the distortion by fibroblast traction of various substrata, including collagen, is well documented in the classic paper by Harris et al. (1981) which has been frequently referred to in this book (see also Guidry 1992). Also several steps in the reorganisation of the extracellular medium in the early stages of the wound healing have been characterised by Guidry and Grinnell (1986). Coagulation activates fibrin, which cross-links to form an initial matrix, the fibrin clot. In addition, various plasma proteins are trapped within this porous gel-like network. Fibronectins, initially deposited in a soluble form by the plasma and later supplied locally by fibroblasts, can bind to fibrin, collagen, hyaluronic acid (HA) and fibroblast surface receptors and thus provide some anchoring for fibroblast movement and traction (for example, Jennings et al. 1992). Alternatively, the binding of HA to fibrin stabilises and increases the volume of the fibrin gel, creating a more porous medium as suggested by Stern et al. (1992) and Tranquillo and Murray (1992). The constructed HA–fibrin matrix provides within the wound a viscoelastic deformable medium in which fibroblasts and myofibroblasts can migrate and proliferate. At the same time, these cells degrade this provisional matrix and secrete a new collagenous matrix with different properties (Welch et al. 1990).

It is this biological scenario just described that was considered in the modelling framework developed in Chapter 6 for dealing with cell condensations in a deformable extracellular medium and used above in Section 10.1. We pointed out some of the drawbacks in that formulation, prime among which is the externally imposed biochemical gradient. Here we investigate how the wound contraction dynamics could result from the nonlinear intrinsic mechanical properties of the wound extracellular matrices.

Plastic Response of ECM to Cell Traction Forces

This model again takes into account cell migration, mitosis and cell traction forces on the ECM. There are, however, some fundamental differences to those models we have discussed up to now. There are four dependent variables, namely, the fibroblastic cell concentration, $n(x,t)$, the collagenous extracellular matrix concentration, $\rho(x,t)$, the provisional matrix concentration inside the wound, $m(x,t)$, and the displacement $u(x,t)$ of the ECM at position x at time t. We consider here a one space-dimension wound along the x-axis where $x = 0$ is the wound centre and $x = 1$ defines the original wound margin inside a domain of half-size L.

The model assumes the customary conservation and force equilibrium equations. For the fibroblasts we assume there is random migration, passive convection and logistic mitotic growth. The inclusion of an equation for the provisional matrix is based on the above discussion and the experimental literature. For it we assume that it degrades, is a viscoelastic isotropic medium with distributed cell traction stress and without subdermal attachments. For the collagenous matrix, the ECM of earlier models, we consider biosynthesis and assume it is an isotropic, nonlinear elastoplastic viscous medium with distributed cell traction stress and elastic subdermal attachments.

The crucial incorporation of elastoplastic behaviour in the collagenous matrix in response to forces generated by cell traction is mainly supported by experiments in which fibroblasts contract hydrated collagen gels. Guidry and Grinnell (1986) have shown that the removal of cells from contracted gels holds in place the framework of the bundles of collagen fibrils that has been organised around and in between the cells. The same behaviour can be observed in the absence of cells, when the gel is centrifuged. It is the modelling of this behaviour that is quite different to the above force equilibrium equations.

Cell–ECM Mechanical Interactions

The general force balance equation of the matrix is again given by the divergence of the stresses set equal to the external forces; that is,

$$\nabla \cdot [\boldsymbol{\sigma}_{\mathrm{ECM}}(x,t) + \boldsymbol{\sigma}_{\mathrm{cell}}(x,t)] = s\rho u(x,t), \tag{10.7}$$

where $\boldsymbol{\sigma}_{\mathrm{ECM}}(x,t)$ denotes the stress tensor of the ECM and $\boldsymbol{\sigma}_{\mathrm{cell}}(x,t)$ is the traction stress exerted by the cells on the ECM. The term $s\rho u(x,t)$ models the restriction of the collagenous matrix by its connections to the external substratum via subdermal attachments (essentially like a simple spring): the form is analogous to (10.5) with the positive parameter s reflecting the strength of the attachments.

The stress $\boldsymbol{\sigma}_{\mathrm{ECM}}$ is made up of an elastoplastic and a viscous part:

$$\boldsymbol{\sigma}_{\mathrm{ECM}} = \boldsymbol{\sigma}_{\mathrm{elastoplastic}} + \boldsymbol{\sigma}_{\mathrm{viscous}}. \tag{10.8}$$

It is the modelling of the tensor $\boldsymbol{\sigma}_{\mathrm{elastoplastic}}$ that is different and crucial. This stress is attributed to the passive elastic properties of the provisional matrix and to the nonlinear elastic and plastic characteristics of the collagenous matrix. To describe the nonlinear elastic behaviour of the matrix, we use the hypoelastic formulation based on the incremental form of the stress–strain relationship. In this formulation, the state of stress depends on the current state of strain as well as the stress path followed up to that state. A general form of the constitutive equation in tensor notation is given by Chen and Mizuno (1990), as

$$d\sigma_{ij} = C_{ijkl}(\epsilon_{pq})\, d\epsilon_{kl}, \tag{10.9}$$

where C_{ijkl} is usually called the tangential stiffness tensor of the material.

We use the simplest class of hypoelastic models where the incremental stress–strain relationships are formulated directly as an extension of the isotropic linear elastic model by replacing the elastic constants by variable tangential moduli which are taken to be functions of the strain invariants. We consider the incremental (one-dimensional) relationship between the stress σ and the strain ϵ to be given by

$$d\sigma(x,t) = E_T(\epsilon, \rho, m)\, d\epsilon(x,t), \tag{10.10}$$

where the tangential stiffness modulus $E_T(\epsilon, \rho, m)$ is a function of the strain ϵ and of the time-varying density of ECM. In addition, experiments on collagen gels (Guidry and Grinnell 1986) suggest that the plasticity effect of the collagenous matrix depends on the matrix density ρ and the provisional matrix, m. We thus describe the spatiotemporal variation of the tangential stiffness modulus $E_T(\epsilon, \rho, m)$ by the relation

$$\frac{\partial E_T}{\partial t} = (S - P)E_{0\rho} - QE_{0m} + \alpha\left(\rho\frac{\partial|\epsilon|}{\partial t} + S|\epsilon|\right). \tag{10.11}$$

The constants $E_{0\rho}$ and E_{0m} are the characteristic elasticity moduli of the collagenous and provisional matrix respectively. They can be related to the Lamé coefficients λ and

μ of a normalised density of each material (see, for example, Chen and Mizuno 1990 or Landau and Lifshitz 1970). The terms $S(n, \rho, m)$ and $P(n, \rho, m)$ correspond to the variation in the amount of collagenous matrix secreted and degraded respectively, while $Q(n, \rho, m)$ is the amount of provisional matrix degraded. The coefficient α is a positive constant defining the amplitude of the plastic response of the collagenous matrix under increasing stretching or increasing compression (loading). When stretching or compression decreases (unloading), α is zero.

Under these hypotheses, the expression of the stress $\sigma_{\text{elastoplastic}}$ is

$$\sigma_{\text{elastoplastic}}(x, t) = \int_0^t E_T(\epsilon, \rho, m) \frac{\partial \epsilon(x, t')}{\partial t'} dt' - \sigma_R(x, t). \tag{10.12}$$

The increase in the stiffness modulus $E_T(\epsilon, \rho, m)$ generates within the ECM a residual stress $\sigma_R(x, t)$ which can be evaluated from (10.10) at the beginning of the unloading stage.

The stress tensor σ_{viscous} is given by the general form used before (as in (10.3), for example):

$$\sigma = \mu_1 \frac{\partial \epsilon}{\partial t} + \mu_2 \frac{\partial \theta}{\partial t}, \tag{10.13}$$

where μ_1 and μ_2 are the shear and bulk viscosity respectively, and θ is the dilation. The viscosity coefficients μ_1 and μ_2 depend on the ECM density and on the strain ϵ. For simplicity, we assume that these coefficients are proportional to the elastic coefficients through a single parameter. This approximation, which is often used, takes advantage of the symmetric structure in ϵ and $\partial \epsilon / \partial t$ of the elastic and viscous stress respectively (Landau and Lifshitz 1970).

We model the traction force exerted by the cells on the matrix by taking it to be proportional to the wound matrix densities $\rho(x, t)$ and $m(x, t)$, according to the relationship:

$$\sigma_{\text{cell}} = \tau_{\text{cell}} = \frac{n}{1 + \gamma n^2} (\tau_0 \rho + \tau_1 m), \tag{10.14}$$

where the constants τ_0 and τ_1 monitor the strength of the traction, while γ measures the saturation of the traction stress as the cell density increases.

Matrix and Cell Conservation Equations

We model the collagenous matrix density conservation by

$$\frac{\partial \rho}{\partial t} + \frac{\partial}{\partial x} \left(\rho \frac{\partial u}{\partial t} \right) = bn(\rho_0 - \rho), \tag{10.15}$$

in which we consider the rate of the collagenous matrix biosynthesis by the cells to be self-regulated around a saturation threshold ρ_0 through a positive proliferation rate b.

The equation for the provisional matrix initially present in the wound is convected and degraded by the cells according to a first-order removal given by

$$\frac{\partial m}{\partial t} + \frac{\partial}{\partial x}\left(m\frac{\partial u}{\partial t}\right) = -\omega n m, \tag{10.16}$$

where ω is the positive decay constant. We assume that there is no continuous creation of the provisional matrix; it is only formed in the very early stages after the wound injury.

The cell conservation consists of the usual terms, namely, diffusion with coefficient D, convection along with the ECM at velocity $\partial u/\partial t$ together with inhibition of cell mitosis for high cell density which is qualitatively again modelled by a logistic type growth curve $rn(n_0 - n)$, where r is the mitotic rate. The velocity, $\partial u/\partial t$, is an approximation to the convection velocity; see the discussion below. The equation is then

$$\frac{\partial n}{\partial t} + \frac{\partial}{\partial x}\left(\frac{\partial u}{\partial t}\right) = D\frac{\partial^2 n}{\partial x^2} + rn(n_0 - n). \tag{10.17}$$

The model mechanism is then given by the four equations (10.7), (10.15)–(10.17) for the four dependent variables n, ρ, m and u with the various terms in (10.7) given by (10.8)–(10.14).

10.5 Solutions of the Model Equations and Comparison with Experiment

As usual it is first necessary to nondimensionalise the equations and decide on the appropriate boundary and intitial conditions. The nondimensionalisation is left as an exercise; it is quite standard. We take the initial state of the wound ($0 \le x \le 1$) to be filled with the provisional matrix which is devoid of fibroblasts: this is why we do not have a production term in the provisional matrix equation (10.16). The surrounding dermis is approximated as a medium with size $L > 1$. We assume that there are initially cells and collagenous matrix outside the wound only. As boundary conditions we take

$$u(0,t) = 0, \quad \frac{\partial n(0,t)}{\partial x} = 0, \quad \frac{\partial \rho(0,t)}{\partial x} = 0, \quad \frac{\partial m(0,t)}{\partial x} = 0,$$
$$u(L,t) = 0, \quad \frac{\partial n(L,t)}{\partial x} = 0, \quad \frac{\partial \rho(L,t)}{\partial x} = 0, \quad \frac{\partial m(L,t)}{\partial x} = 0.$$

There are few data currently available from which to assign values for the model parameters. Following Tranquillo and Murray (1992), we choose parameter values which are qualitatively consistent with the experimental curve of McGrath and Simon (1983) given in Figure 10.2. As described above the contracting phase of excised full-thickness wounds on rats could be described with a simple exponential time-dependence for the wound area, $A(t)$, given by (10.6). In the absence of adequate information for the rheological properties of the provisional matrix, in the numerical solutions of Tracqui et al. (1995) the same value was chosen for the elasticity moduli $E_{0\rho}$ and E_{0m} in (10.11). They took different values of the parameter α to investigate the influence of the collagenous matrix plasticity on the wound contraction dynamics. It is their results we present below.

We first consider the case when there is no plasticity so $\alpha = 0$ in (10.11). As fibroblastic cells repopulate the wound (Figure 10.3(a)), they degrade the transitional wound matrix (Figure 10.3(d)) and secrete a new collagenous matrix (Figure 10.3(c)). Figure 10.3(b) shows the plot of the wound margin movement over time. The preliminary expansion is followed by a lag phase and then an exponential contraction phase which can be described by the relation (10.6). A crucial deficiency of this model, as we knew to expect, is that it does not exhibit the sustained contraction, $u(1, \infty) < 0$, ob-

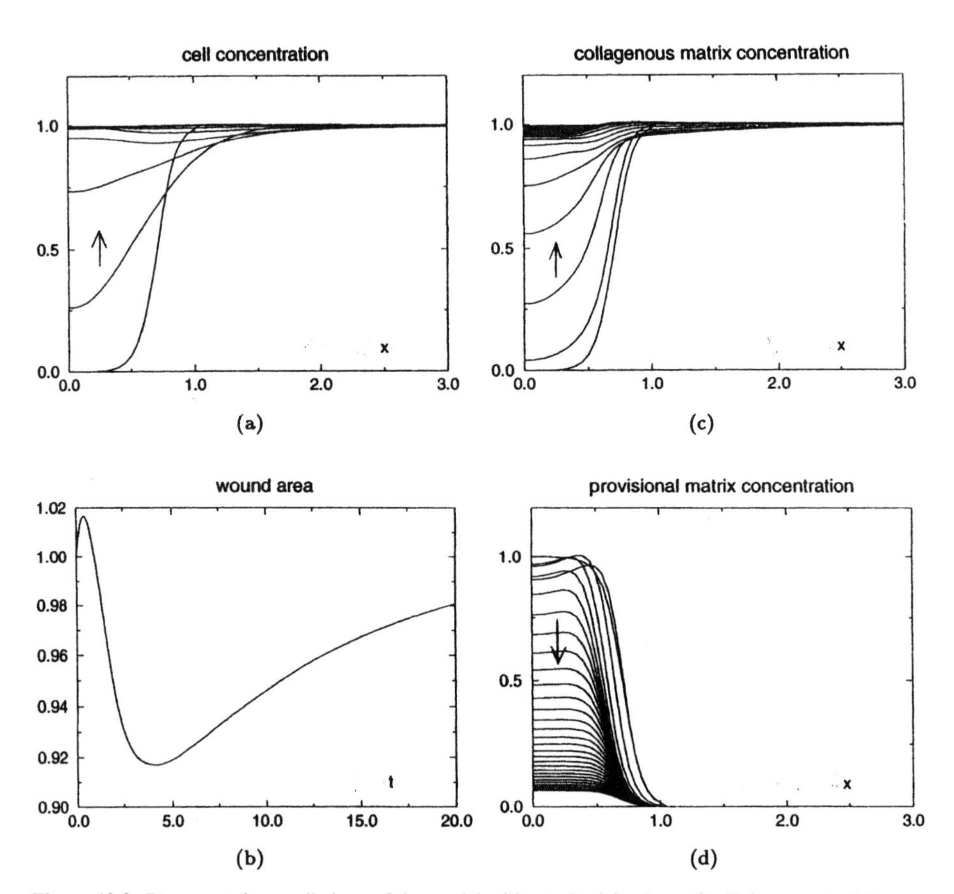

Figure 10.3. Representative predictions of the model without plasticity ($\alpha = 0$). Enlargement of the spatiotemporal profiles of (**a**) fibroblastic cell concentration; (**c**) collagenous matrix concentration; and (**d**) provisional matrix concentration all plotted against distance from the centre of the wound for a series of times after wounding. The curves correspond to times $t = 1, 2, 3, \ldots$. The size of the half wound is 1 inside a domain of size $L = 5$. (**b**) The associated relative wound area as a function of time. Parameter values are $b = 0.5$, $\omega = 0.1$, $D = 0.1$, $\tau_0 = 0.5$, $\tau_1 = 1.0$, $r = 1.0$, $s = 1.0$, $\gamma = 1.0$, $E_{0m} = E_{0\rho} = 1.0$. (From Tracqui et al. 1995)

served experimentally and the wound area relaxes toward its initial value. In this case, the stress–strain curve at the wound margin is, as expected, almost a straight line.

In the more interesting and biologically relevant nonlinear elastoplastic case $\alpha \neq 0$, the newly secreted matrix does not relax completely to a stress-free state when the active contraction relaxes. The first stages of the change in wound area over time in Figure 10.4(a) are still qualitatively consistent with the experimental curve in Figure 10.2. More interesting is the fact that the overall simulated time evolution of the wound area, including the final expansion, corresponds to the behaviour which has been observed in several experimental situations (for example, McGrath and Simon 1983 and Madison and Gronwall 1992). The contraction of the ECM composite is sustained after the wound has been populated by the migrating cells and the provisional matrix replaced by the collagenous matrix. This is due to the change in the tangential modulus E_T (Figure 10.4(c)) and to the corresponding build-up of the residual stress σ_R (Figure 10.4(d)).

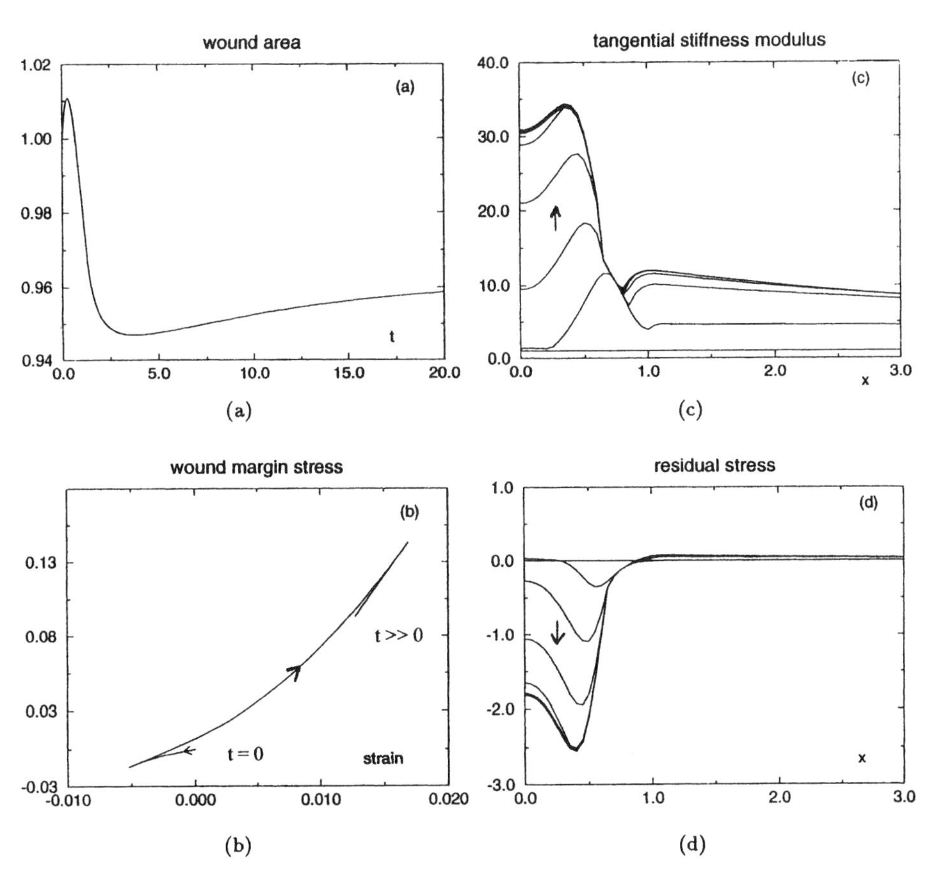

Figure 10.4. Representative predictions of the model with plasticity; that is, $\alpha \neq 0$. (**a**) The relative wound area as function of time; (**b**) The stress–strain profile at the wound margin. Spatiotemporal profiles of the tangential stiffness modulus (**c**) and the residual stress (**d**). The arrows denote increasing time. Notation and parameter values are the same as those defined in Figure 10.3 except $\alpha = 500$ and $\tau_1 = 2.0$. (From Tracqui et al. 1995)

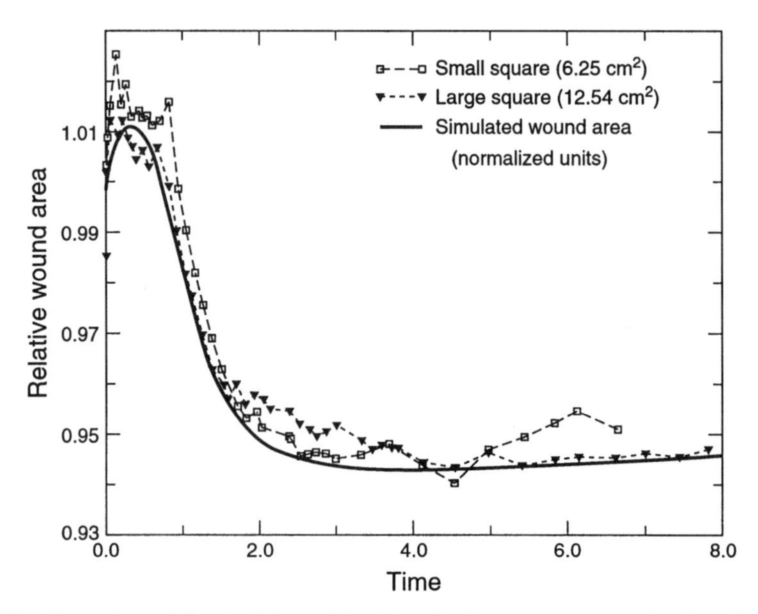

Figure 10.5. Comparison of the plasticity model solution for the wound area as a function of time from Figure 10.4(b) with the data of McGrath and Simon (1983) from Figure 10.2. Notation and parameter values are the same as those defined in Figure 10.3 with the exceptions given in Figure 10.4.

Figure 10.4(b) shows the associated nonlinear stress–strain curve at the wound boundary; note the direction of time.

The solution in Figure 10.4(a) is compared with the experimental data given by McGrath and Simon (1983) in Figure 10.5.

The models we have discussed in this section crucially included some viscoplasticity effects which we saw are essential if we are to obtain a final steady state deformation of the skin when the wound has healed. The model also simulates the expansion, time lag, contraction and relaxation phases of cutaneous wound healing as observed experimentally.

Determination of the residual stress and strain, associated with a given temporal variation of the wound area, provides a measure of the balance between traction forces leading to wound contraction and the tension forces which could prevent the complete restoration of the tissue functional properties (Dunn et al. 1985). We suggest that the strain remaining in the collagenous matrix after the fibroblastic phase has been completed gives a measure of the severity of scarring. Figure 10.6 shows the difference between the linear elastic response, following a wound, and the nonlinear plastic response which results in a residual displacement of the tissue. It would be extremely useful, of course, if this type of model constituted a general basis for investigating various hypotheses, with the corresponding estimation of some of the model parameters from *in vivo* and *in vitro* data. It is certainly possible that refinements to the model could lead to the design of new therapeutic strategies for wound contraction management. In the following Sections 10.6–10.10 we discuss another approach, but with some similar concepts, to determine the residual strain in the tissue matrix. It also addresses some of the deficiencies in the model discussed here.

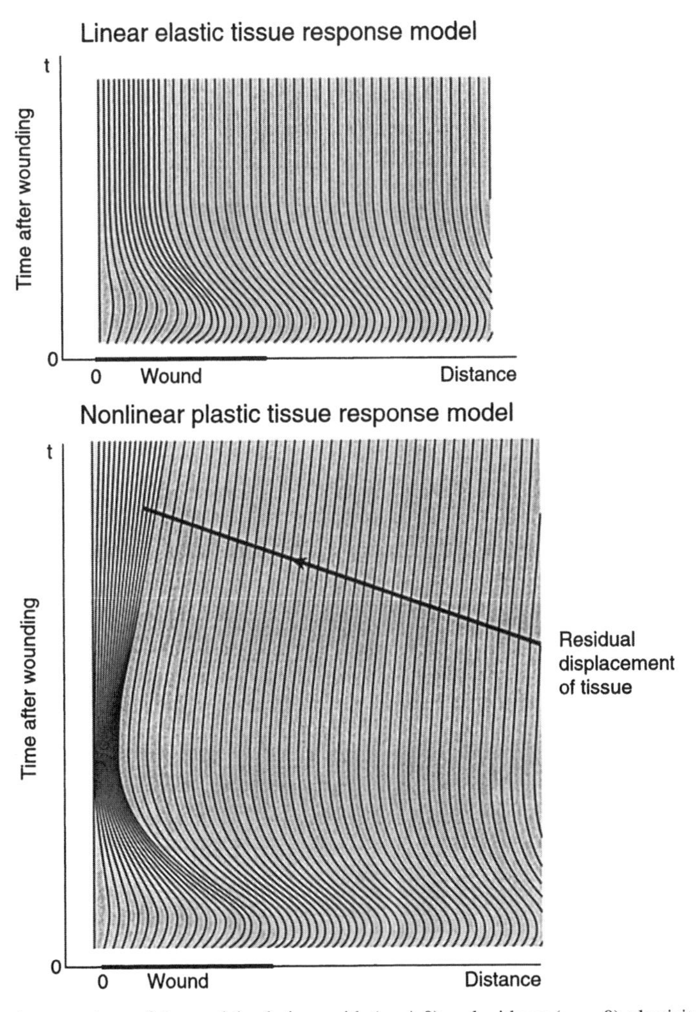

Figure 10.6. A comparison of the model solutions with ($\alpha \neq 0$) and without ($\alpha = 0$) plasticity. Parameter values are the same as those defined in Figure 10.3 with the exceptions given in Figure 10.4.

10.6 Wound Healing Model of Cook (1995)

Most of the models for dermal wounds, beginning with that of Murray et al. (1988), involved the following simplifying assumptions, some of which we have already mentioned: (1) small strain, linear viscoelasticity (based on the Voigt model, which can not be plastically deformed), (2) isotropic material properties for the ECM, (3) isotropic contractile stresses, (4) no attempt to relate ECM turnover and remodelling to changes in mechanical properties (including zero stress states and residual stresses and strains). The model discussed in the last section is a particular exception with regard to number (4) and in the more complex stress–strain constitutive relation developed. By accounting for residual strains and fibre orientation and by considering finite deformations Cook

(1995) was the first to address all of these limitations. It is some of his modelling that we now discuss. The work we describe here is two-dimensional which makes it considerably more complicated but, importantly, can be directly used to quantify the effect of wound healing geometry with scarring. The material in this section gives a much fuller discussion of the wound healing work reported in Murray et al. (1998).

In outline, with each point in space is associated a matrix density, a residual strain and a fibre orientation distribution. In Cook's (1995) model these characterize the scar, determine the mechanical properties of the ECM (via a microstructurally based constitutive law), determine the anisotropic contractile stress tensor and, finally, determine cell motion. In principle, evolution equations are required for ECM density, orientation and residual strain to account for remodelling and growth. A simplification is achieved by approximating the orientation distribution as being uniform in the zero stress state: the single extra equation is that for residual strain. We shall derive the equations, in an Eulerian frame, for the small strain approximation for simplicity of exposition but touch on the finite strain situation. Before doing so we need to derive some preliminary results.

A key ingredient here is the quantification of the residual strain and how it relates to the matrix. It is not possible to go into all the details of Cook's (1995) analysis here, suffice it to say that he develops a comprehensive nonlinear theory of the evolving structure of the extracellular matrix and the effect of orientation and motion of its constituent fibres. Here we give only a relatively brief sketch of his exposition and derive one of his model systems and give another for reference and possible future study. To do this we require a few preliminary results involving some matrix algebra.

Preliminary Matrix Deformation Results

Under a general deformation which maps points x_i to $y_i = x_i + u_i$, where u_i is the displacement, a small neighbourhood of a point is transformed according to

$$\begin{bmatrix} dx_1 \\ dx_2 \end{bmatrix} \rightarrow \begin{bmatrix} 1+\dfrac{\partial u_1}{\partial x_1} & \dfrac{\partial u_1}{\partial x_2} \\ \dfrac{\partial u_2}{\partial x_1} & 1+\dfrac{\partial u_2}{\partial x_2} \end{bmatrix} \begin{bmatrix} dx_1 \\ dx_2 \end{bmatrix} = (I + \nabla \mathbf{u})\, d\mathbf{x} = d\mathbf{y}, \tag{10.18}$$

where I is the unit matrix. We can express this in terms of the deformed state, namely,

$$d\mathbf{x} = (I - \nabla_{\mathbf{y}} \mathbf{u})\, d\mathbf{y} = M\, d\mathbf{y},$$

where the matrix M is given by

$$M = (I - \nabla_{\mathbf{y}} \mathbf{u}) = \begin{bmatrix} 1-\dfrac{\partial u_1}{\partial y_1} & -\dfrac{\partial u_1}{\partial y_2} \\ -\dfrac{\partial u_2}{\partial y_1} & 1-\dfrac{\partial u_2}{\partial y_2} \end{bmatrix} = \frac{\partial x_i}{\partial y_j}. \tag{10.19}$$

For example, under this transformation a circle $d\mathbf{x}^T I \, d\mathbf{x} = 1$ is mapped (deformed) into an ellipse $d\mathbf{y}^T M^T M \, d\mathbf{y} = 1$. We can further find a rotation matrix P satisfying $P^T P = I$ that diagonalises the symmetric $M^T M$, which is the 'strain' matrix. If we now let $d\mathbf{z} = P^T \, d\mathbf{y}$ we then have

$$(d\mathbf{z})^T P^T M^T M P \, d\mathbf{z} = d\mathbf{z}^T \Lambda \, d\mathbf{z} = 1,$$

where the matrix $\Lambda = \text{diag}\,(\Lambda_1, \Lambda_2)$ is diagonal where the Λ_i are the eigenvalues of the strain matrix $M^T M$ and P is made up of normalised eigenvectors.

Cook (1995) derived the orientation and stretch distributions, obtained under a general deformation, by an initially uniform distribution of relaxed fibres. He went on to derive evolution equations for the fibre density, fibre orientation distribution and stretch distribution based on the (Lagrangian) velocity $\nu = D\mathbf{u}/Dt$ (the total derivative of $\mathbf{u}$), the material velocity, with a linear approximation $\nu = \partial u/\partial t$.

Plasticity, Zero Stress State and Effective Strain

The problem with realistic modelling of dermal wound healing, as we have mentioned before, is that we must include plasticity of the ECM. Mechanical plasticity involves the formation and breaking of chemical bonds during deformation in such a way that the tissue does not return to its original state before deformation. More relevant to wound healing, however, is remodelling plasticity which results from the cells remodelling the tissue not only while it is being deformed but also afterwards. In the model discussed in the last section we had new matrix formed and provisional matrix (which originally filled the wound area) continuously being degraded. We have already mentioned that there are well-recognised skin tension lines[3] in the human body (and exploited by plastic surgeons) so the skin does not completely relax after wound healing is finished. Wound healing models therefore, have to allow for stresses and a nonuniform distribution of fibres to exist at equilibrium. At this stage of our knowledge of tissue plasticity *in vivo* and *in vitro* it is necessary to derive evolution equations for the ECM structure variables which rely on many simplifying approximations. The fine details of stretch and orientation distributions are not included directly but rather implied by the effective strain and plastic effects via the evolution of the zero stress state.

It is possible locally to deform the matrix such that there is no stress which means that the *zero stress state* is the deformation state in which the local matrix is stress free. This is at each point; it does not imply that we can construct a deformation that makes the whole tissue return to a stress free state.

In the reconstruction of the wounded area there is matrix production and matrix degradation and so, after a while, there is little connection with the original reference state. We introduce the concept of an *effective strain* which is the strain relative to the zero stress state. We further introduce the *residual strain* which is the zero stress

[3]The anatomy of these skin tension lines was studied in considerable detail by Langer (the German papers date from 1862; see the translations in Langer 1978a,b,c,d). He talked about cleavability and tension lines and mapped the directions on the human body with some precision; they are known as Langer lines. These cleavage lines reflect collagen fibre orientation and are conserved throughout adult life except under extended stress, such as by pregnancy.

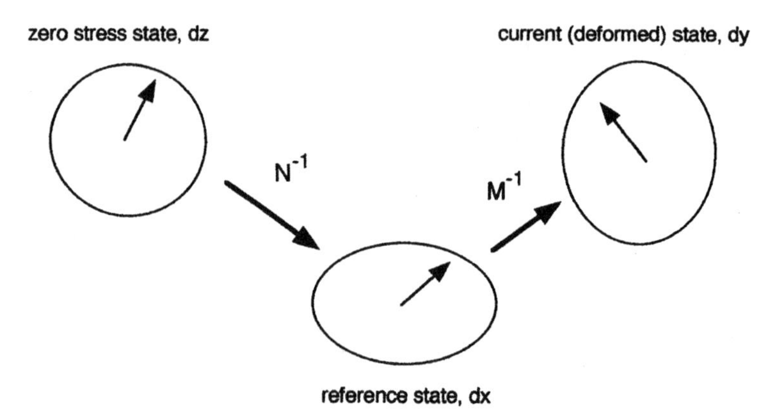

Figure 10.7. Schematic deformations of a tissue element in the reference state to the current deformed state and from the zero stress state. The deformation matrices are denoted by M and N as explained in the text.

state relative to the original configuration. The effective strain is computed from the residual strain and the actual deformation with respect to the original reference state as schematically shown in Figure 10.7.

Rather than having to deal with fibre densities we make a further approximation and assume, not unreasonably, that the fibres are uniformly distributed and unstretched when the tissue is in the zero stress state. Cook (1995) discussed the more sophisticated approach in which this assumption was relaxed.

Refer now to Figure 10.7 and let points in the reference configuration have coordinates denoted by x_i, those in the zero stress state by z_i and those in the current deformed state by y_i. Consider a small element, dx, about a particular point in the reference state with the corresponding elements dz and dy in the zero stress and deformed states respectively as schematically shown in Figure 10.7. We write the linear transformations in the form

$$dz = N\,dx, \quad dx = M\,dy, \tag{10.20}$$

where, if $y_i = x_i + u_i$ and $x_i = z_i + v_i$ the matrices M and N are given by (cf. (10.19))

$$\begin{aligned} M = (I - \nabla_{\mathbf{y}}\mathbf{u}) &= \begin{bmatrix} 1 - \dfrac{\partial u_1}{\partial y_1} & -\dfrac{\partial u_1}{\partial y_2} \\ -\dfrac{\partial u_2}{\partial y_1} & 1 - \dfrac{\partial u_2}{\partial y_2} \end{bmatrix}, \\ N = (I - \nabla_{\mathbf{x}}\boldsymbol{\nu}) &= \begin{bmatrix} 1 - \dfrac{\partial v_1}{\partial x_1} & -\dfrac{\partial v_1}{\partial x_2} \\ -\dfrac{\partial v_2}{\partial x_1} & 1 - \dfrac{\partial v_2}{\partial x_2} \end{bmatrix}. \end{aligned} \tag{10.21}$$

Now the effective deformation from the zero stress state to the current state has to satisfy

$$dz = NM\,dy \tag{10.22}$$

which then gives the *effective strain* matrix as

$$M^T N^T N M = M^T Z M, \tag{10.23}$$

where $Z = N^T N$ defines the *residual strain* matrix Z and relates the zero stress state to the reference state; the residual strain Z is what we want to calculate.

Small Strain Approximation

Let us now compute the expressions for small strains. Let

$$Z = \delta_{ij} + 2z_{ij}, \quad M = \delta_{ij} - u_{i,j}, \quad |z_{ij}|, |u_{i,j}| \ll 1, \tag{10.24}$$

where $\delta_{ij} = 1$ if $i = j$, $\delta_{ij} = 0$ if $i \neq j$ and we use the comma notation to denote differentiation. In this case

$$M^T Z M = \delta_{ij} + 2z_{ij} - u_{i,j} - u_{j,i}. \tag{10.25}$$

But the infinitesimal strain, ϵ_{ij} is related to the effective strain matrix, $M^T Z M$ by

$$M^T Z M = \delta_{ij} - 2\epsilon_{ij} \Rightarrow \epsilon_{ij} = \frac{1}{2}(u_{i,j} + u_{j,i}) - z_{ij} = e_{ij} - z_{ij}. \tag{10.26}$$

So, in this small strain approximation situation when both residual and actual strains are small, the effective strain is equal to the actual strain, here denoted by e_{ij}, minus the effective strain, z_{ij}. We use this in our wound healing model equations below.

10.7 Matrix Secretion and Degradation

Before we can write the model equations we have to consider the addition and removal of new matrix fibres; again we use averages instead of using the various fibre densities to derive the relevant equations.

Loss of fibres affects the stress but not the residual strain if we assume that it affects all fibres in the same way. If all fibres were added in a relaxed state and secretion and degradation occurred continually, eventually the tissue matrix would become totally relaxed, a situation not observed in skin and scar tissue. So, we hypothesise that as the matrix density increases a larger and larger proportion of new collagen is added to the preexisting fibres. We assume that collagen not added to existing fibres forms new unstressed fibres.

Recalling the discussion above we can think of the effective strain as being represented by the ellipse $dy^T M^T Z M\, dy = 1$ with its principal axes in the directions of the eigenvectors of $M^T Z M$ and of length $1/\sqrt{\Lambda_i}$. The effect of adding new fibres is to relax the eigenvalues of the effective strain towards 1: the eigenvectors remain the same. We now derive the change in residual strain as we add new fibres, a crucial step in the model development.

Let the effective strain at time t be $M^T Z(t) M$ with eigenvalues $\Lambda_i(t)$ and assume that

$$\frac{D\Lambda}{Dt} = f(\Lambda), \tag{10.27}$$

where $f(\Lambda_i)$ is a nonincreasing function satisfying $f(1) = 0$ that depends, in general, on the matrix secretion rate and density; we derive a form for $f(\Lambda)$ below. Here $D\Lambda/Dt$ is the total derivative. The rotation matrix, P satisfies

$$P^T M^T Z(t) M P = \Lambda(t)^T, \tag{10.28}$$

where $\Lambda(t)$ is the diagonal matrix of the corresponding eigenvalues $\Lambda_i(t)$. From the assumed equation (10.27) we have

$$\Lambda_i(t + dt) = \Lambda_i(t) + dt[f(\Lambda_i(t))] \tag{10.29}$$

so that

$$P^T M^T Z(t + dt) M P = \Lambda(t + dt) = \Lambda(t) + dt\{\text{diag}\,[f(\Lambda_1), f(\Lambda_2)]\}. \tag{10.30}$$

So

$$Z(t + dt) = (M^{-1})^T P[\Lambda(t) + dt\{\text{diag}\,[f(\Lambda_1), f(\Lambda_2)]\}]P^T M^{-1} \tag{10.31}$$

and the evolution equation for Z due to the relaxation in the eigenvalues becomes

$$\frac{DZ}{Dt} = (M^{-1})^T P\{\text{diag}\,[f(\Lambda_1), f(\Lambda_2)]\}P^T M^{-1}. \tag{10.32}$$

This expression, after simplifying the inner three matrices (done below) is the basis for the evolution equation for the residual strain. In Eulerian form we have to add a convection term and, of course, specify the function $f(\Lambda_i)$ which we do below.

Now consider $f(\Lambda_i)$. We can approximate the matrix secretion by considering it to average the eigenvalues, where the latter are defined as

$$l_i = \Lambda_i^{-1/2}. \tag{10.33}$$

At time t the length of the principal axes is $l_i(t)$. Over a small increment in time, dt, the density of newly added unstressed fibres is $q(S)(dS/dt)\,dt$ and the density of collagen added to existing fibres is $[1 - q(S)](dS/dt)\,dt$ where $\partial S/\partial t$ is the local secretion rate. The function $q(S)$ specifies the fraction of newly added fibres which are added to the unstressed state and is a function of the matrix density, S, which is a measure of tissue maturity; we come back to this later. The principal eigenvalue at time $t + dt$ is then

$$l_i(t + dt) = \frac{\left(S + [1 - q(S)]\frac{dS}{dt}\,dt\right) l_i(t) + \left(q(S)\frac{dS}{dt}\,dt\right)(1)}{S + \frac{dS}{dt}\,dt} \tag{10.34}$$

which leads to

$$\frac{Dl_i}{Dt} = \frac{q(S)}{S}\frac{\partial S}{\partial t}(1 - l_i) \tag{10.35}$$

which shows how the length of the principal axis relaxes towards 1. This last equation and (10.33) lead to an expression for the rate of change of the eigenvalues, namely,

$$\frac{D\Lambda_i}{Dt} = 2\frac{q(S)}{S}\frac{dS}{dt}(1 - \sqrt{\Lambda_i})\Lambda_i = f(\Lambda_i) \tag{10.36}$$

which determines the function we need in (10.32). An alternative approach is to assume the averaging takes place in Λ instead of in l and a similar argument produces

$$\frac{D\Lambda_i}{Dt} = \frac{q(S)}{S}\frac{dS}{dt}(1 - \Lambda_i) = f(\Lambda_i) \tag{10.37}$$

which has a similar form but is a little simpler. In this case the evolution equation for the residual strain matrix Z is given by

$$\frac{DZ}{Dt} = \frac{q(S)}{S}\frac{dS}{dt}[(M^{-1})^T M^{-1} - Z] \tag{10.38}$$

since $\mathrm{diag}\,[f(\Lambda_1), f(\Lambda_2)] = (q(S)/S)(dS/dt)[1 - \mathrm{diag}\,(\Lambda_1, \Lambda_2)]$. As we shall see, (10.36) and (10.37) agree in the limit of small strain and (10.38) is equal to the small strain approximation for the evolution of Z. A simple form for the function $q(S)$ is simply a pair of straight lines such as illustrated in Figure 10.8.

Small Strain Approximation for the Residual Strain

Let us focus on (10.38) and use the same small strain definitions as in (10.24) with M given by (10.21). Using the fact that $(M^{-1})^T M^{-1} = (MM^T)^{-1}$ and that $Z = \delta_{ij} + 2z_{ij}$

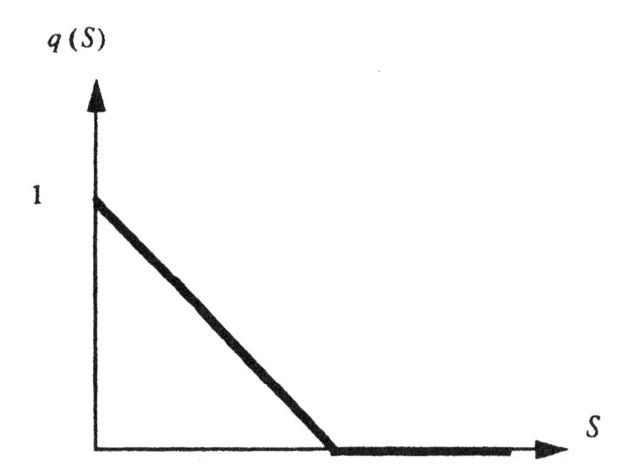

Figure 10.8. Typical function for the fraction, $q(S)$, of new collagen that contributes to the new unstressed fibres as a function of the matrix density S. At high densities all the new collagen is added to the existing fibres so that there is no further relaxation due to the process of secretion and degradation of collagen.

a little algebra shows that the linear form of (10.38) is given by

$$\frac{Dz_{ij}}{Dt} = \frac{q(S)}{S}\frac{dS}{dt}[e_{ij} - z_{ij}], \tag{10.39}$$

which clearly shows that the residual strain z_{ij} relaxes towards the actual strain $e_{ij} = (1/2)(u_{i,j} + u_{j,i})$. Although we have considered only (10.38) as the equation to linearise, in fact, a linearisation of (10.37) gives exactly the same result.

Cook (1995) discusses all this in considerably more detail and he goes on to derive structural constitutive equations for soft tissues. For the purposes of the material in this chapter (10.39) is the evolution equation we require for our model.

10.8 Cell Movement in an Oriented Environment

It is clear that for the model we need to know how cells actually invade the wound after injury. The process is not at all well understood since a variety of processes could be involved such as chemotaxis, haptotaxis (discussed in Chapter 6) and contact guidance. Contact guidance is the process by which cells migrate along pathways in the extracellular matrix. Its role has been discussed by numerous authors; see, for example, the book by Bard (1990) and references given there. Manoussaki et al. (1996) and Murray et al. (1998) include contact guidance in their models for the formation of vascular networks which we discussed in Chapter 8. As we know, fibroblast cells exert traction forces (see Chapter 6) which are implicated in a wide variety of morphogenetic situations and directed movement in many of these is by haptotaxis, essentially movement up an adhesion gradient; some recent modelling along these lines has been shown to mimic scale arrangement on butterfly wings (Sekimura et al. 1999).

Although orientational cues are bidirectional, with cell movement expected in either direction, it turns out that unidirectional movement has been observed in many developmental systems. Here we briefly discuss the phenomenon of contact guidance and, again following Cook (1995), show that directional movement should be expected in a matrix in which orientational cues vary even if the overall matrix density were constant (thus ruling out haptotaxis) and even in the absence of any chemoattractants. In other words orientational cues can give rise to unidirectional movement. It has been well established that matrix deformation can result from cell traction (Harris et al. 1984; see also Chapter 6) and this in turn gives rise to passive convection as we have seen (refer to Chapter 6). So, both active and passive cell movements, coupled to matrix deformation, are likely in wound healing.

A key step is to understand how information on the matrix deformation and its remodelling relate to cell movement. It is fundamental, complicated and necessary. Here we consider only a fairly straightforward situation which suffices for the exposition in this chapter. Again, Cook (1995) has gone into the whole question in depth. In the following we continue with the same notation.

In a homogeneous environment a simple random walk gives an evolution equation for cell density, n, as

$$\frac{\partial n}{\partial t} = n_{,ii}, \tag{10.40}$$

where, for simplicity, we have taken the diffusion matrix to be the identity matrix. Now suppose that the diffusion domain is deformed and we measure densities with respect to the deformed state. In other words cells undergo a random walk in one space while we observe them in another, the deformed state. We want to know what diffusion looks like in this space. Let us suppose that the original space in which diffusion takes place is deformed according to a transformation $x_i \to y_i$ with deformation matrix $m_{ij} = \partial x_i/\partial y_j$. The area scale factor is $1/|\, m_{ij}\,|$ so the cell densities are related by

$$n(\mathbf{x}, t) = \frac{N(\mathbf{y}, t)}{|\, m_{ij}\,|}. \tag{10.41}$$

We now obtain an equation for the evolution of the cell density, N, by substituting for $n(\mathbf{x}, t)$ in (10.40) to get

$$\frac{\partial N}{\partial t} = m^{sj}|\, m_{ij}\,| \left[\left(\frac{N}{|\, m_{ij}\,|} \right)_{,r} m^{rj} \right]_{,s}, \tag{10.42}$$

where the notation m^{ij} denotes the inverse of m_{ij} and repeated suffixes denote summation. We now need to put this into the standard conservation form

$$\frac{\partial N}{\partial t} = [A_{ij}N_{,j} + B_i N]_{,i}. \tag{10.43}$$

Let $m^{ij} = C_{ij}/\Delta$ where the determinant $\Delta = |\, m_{ij}\,|$ and

$$C_{ij} = \begin{bmatrix} x_{2,2} & -x_{1,2} \\ -x_{2,1} & x_{1,1} \end{bmatrix}. \tag{10.44}$$

From this definition we have

$$C_{ij,i} = 0. \tag{10.45}$$

On using this last equation it can be verified that

$$m^{sj}\Delta \left[\left(\frac{N}{\Delta} \right)_{,r} m^{rj} \right]_{,s} = \left[m^{ik}m^{jk}N_{,j} - m^{ik}m^{jk}\frac{\Delta_{,j}}{\Delta}N \right]_{,j}. \tag{10.46}$$

If we now define

$$D_{ij} = m^{ik}m^{jk} \tag{10.47}$$

the equation for N, namely, (10.42), can be written in the form

$$\frac{\partial N}{\partial t} = \left[D_{ij} N_{,j} + \left(\frac{D_{ij} | D |_{,j}}{2 | D |} \right) N \right]_{,i} = \left[\frac{D_{ij}}{| D |^{1/2}} \left(| D |^{1/2} N \right)_{,j} \right]_{,i} \tag{10.48}$$

which is in conservation form. Here, $M = m_{ij}$ and

$$D_{ij} = m^{ik} m^{jk} = (M^T M)^{-1} \quad \text{and} \quad | D | = | D_{ij} | = \Delta^{-2}. \tag{10.49}$$

As opposed to (10.43) the usual form of a diffusion equation is

$$\frac{\partial N}{\partial t} = \left[D_{ij} N_{,j} \right]_{,i} . \tag{10.50}$$

The fundamental difference is the presence of the convection term in (10.48), specifically the second term in the first of the right-hand sides. It implies that there is a drift of cells towards regions of high compression: $N =$ constant is no longer a solution. The belief that bidirectional guidance cues cannot give unidirectional movement at the population level is simply wrong. Note that the steady states in $\mathbf{x}$ space are $n \equiv$ constant while those in $\mathbf{y}$ space are $N =$ constant $| D |^{1/2}$.

Diffusion Coefficient Matrix in a Strained Field

We can use the nonlinear form of the strain matrix $M^T M$ to obtain an expression for D_{ij} in (10.49). Here, however, we again only evaluate the small strain approximation. For small strains, since $e_{ij} = \epsilon_{ij}$,

$$M = m_{ij} = \delta_{ij} \approx \delta_{ij} - \epsilon_{ij} = \delta_{ij} - u_{i,j}, \tag{10.51}$$

we have the following approximation

$$M^T M = \delta_{ij} - 2\epsilon_{ij}, \quad \text{where} \quad \epsilon_{ij} = \frac{1}{2} \left(\frac{\partial u_i}{\partial y_j} + \frac{\partial u_j}{\partial y_i} \right) \tag{10.52}$$

is the linear strain tensor. Then, from (10.49) with (10.52)

$$m^{ik} m^{jk} = (M^T M)^{-1} = \frac{1}{2} \begin{bmatrix} 2 + \epsilon_{11} - \epsilon_{22} & 2\epsilon_{12} \\ 2\epsilon_{12} & 2 + \epsilon_{22} - \epsilon_{11} \end{bmatrix}, \tag{10.53}$$

and so we obtain the expression for the diffusion coefficient matrix $\mathbf{D}$ as

$$\mathbf{D} = \frac{D}{4} \begin{bmatrix} 2 + \epsilon_{11} - \epsilon_{22} & 2\epsilon_{12} \\ 2\epsilon_{12} & 2 + \epsilon_{22} - \epsilon_{11} \end{bmatrix}. \tag{10.54}$$

Cook (1995) derives various forms for the cell flux, $\mathbf{J}$, again from the microscopic properties of the fibres, in situations where the cells are in an environment with a distribution of fibre orientation and also when they do not follow the fibres. He uses probability distributions in which he defines a probability of cells moving in a given direction, then derives forms for the flux and hence a relevant diffusion equation for N. In the case of cells which do not follow the fibre directions he obtains the following diffusion form,

$$\frac{\partial N}{\partial t} = -\nabla \cdot \mathbf{J} = [DD_{ij}N]_{,ij}. \tag{10.55}$$

Again, a uniform cell density is not a steady state solution of the equation. Another point about this form is that the variable diffusion coefficient appears inside both derivatives. This form of the diffusion terms is used in the first model mechanism below and in numerical simulations discussed later.

10.9 Model System for Dermal Wound Healing with Tissue Structure

We now give the governing equations for Cook's (1995) finite strain model mechanism. We then give the small strain approximation. A key element in the modelling, as it was in the Tracqui et al. (1993) model in Sections 10.4 and 10.5, is the structure of the ECM and in particular its influence on cell movement. Although it is the underlying basis of this discussion on deformations and stress fields that it is cell traction which generates them, we must also keep in mind that there is passive convection with velocity $\boldsymbol{v}$; we discuss this aspect further below.

The equations (the specific details of which we discuss immediately below) for cell density (N), ECM density (S), displacement (u_i), residual strain (Z_{ij}) and force equilibrium are quite different in detail to those used before and are given by

$$\begin{aligned}
\frac{\partial N}{\partial t} &= \left[D_{ij}N\right]_{,ij} - [N\nu_i]_{,i} + f(N) \\
\frac{\partial S}{\partial t} &= -\,[S\nu_i]_{,i} + g_1(N,S) - g_2(N,S) \\
\frac{\partial u_i}{\partial t} &= \nu_i - \nu_j u_{i,j} \\
\frac{\partial Z_{ij}}{\partial t} &= \frac{q(S)}{S} g_1(N,S)\left[(MM^T)^{-1} - Z_{ij}\right] - \nu_k Z_{ij,k} \\
\left[\sigma_{ij} + \tau_{ij}\right]_{,j} &= b_i(S,u_i).
\end{aligned} \tag{10.56}$$

The expressions for the actual deformation gradient (with respect to the deformed state), effective strain, stress, diffusion matrix and traction are here:

$$\begin{aligned}
&M_{ij} = \delta_{ij} - u_{i,j}, \quad \epsilon_{ij} = \tfrac{1}{2}(\mathbf{I} - M^T Z M), \quad \sigma_{ij} = S\chi\left[2\epsilon_{ij} + \delta_{ij}\epsilon_{\alpha\alpha}\right] \\
&D_{ij} = \frac{D}{Tr + 2\sqrt{\Delta}}\left[(Tr + \sqrt{\Delta})\delta_{ij} - M^T Z M\right], \quad \tau_{ij} = T(N,S)D_{ij}/D \\
&Tr = Trace[M^T Z M], \quad \Delta = Det[M^T Z M],
\end{aligned} \tag{10.57}$$

where the repeated α implies summation; here $\epsilon_{\alpha\alpha} = \epsilon_{11} + \epsilon_{22}$ with M the deformation matrix defined by (10.19) and the strain matrix, Z, defined by (10.23). Before discussing the various functions in (10.56), and $T(N, S)$ in (10.57), let us recall what the various terms in the equations represent.

Equation for the Cell Density N

The first term, involving the diffusion matrix, incorporates the effect of contact guidance by the fibres and is a function of the strain and is of the form used in (10.55). If we consider the contact guidance to be along fibre directions this term would be different: the diffusion matrix would then be given by (10.48) with (10.49). The second term on the right is the contribution from the passive convective flux of cells as the cell–matrix continuum deforms with material velocity $\mathbf{v}$. The last term, $f(N)$, is the cell proliferation contribution and includes cell death, differentiation and dedifferentiation. It is also probably a function of the strain since it is well known that cell shape can influence mitosis (for example, if a cell is too flat it tends not to divide).

The evolution equation for N is based on a conservation law and as written is in an Eulerian frame of reference (as opposed to a Lagrangian frame of reference); that is, the coordinate system is fixed and the material flows past it. This is the usual way of writing an equation for a conservation law. There is a difference between $\boldsymbol{v}$, which is the material velocity, and the term $\partial u/\partial t$ which is the derivative of the displacement for a point in the fixed frame of reference. When we consider finite strains these are not the same and we need a further equation, namely, the third equation in (10.56). In the case of a small strain approximation, however, they are equal.

Equation for the Matrix Density, S

The tissue is also convected since it is part of the cell–matrix continuum. We have separated the matrix secretion, g_1, and degradation, g_2 terms since in the model only the secretion term affects the mechanical plasticity via the strain matrix Z.

Equation for the Residual Strain Matrix, Z

This equation, which shows how the residual strain tensor, Z, changes, uses the specific form we derived above for the plastic remodelling. The function $q(S)$ describes the fraction of newly secreted matrix that forms new fibres in the tissue. The last term, the convection term, is like that in the cell equation and is the contribution from moving from a Lagrangian frame of reference to the Eulerian frame. The second definition in (10.57) determines the effective strain once we have obtained the residual strain Z and the deformation matrix M.

Force Balance Equation

This is the now customary equation which says that the forces are in quasi-equilibrium at all times. The form of the stress–strain relation we use is given by the third equation in (10.57) which defines the stress tensor σ_{ij}. The specific form we have chosen for the cell traction, τ_{ij} is given in (10.57). We discuss the body force, b_i and $T(N, S)$ below.

Let us now consider the specific functions and some reasons for the forms chosen in (10.57). A crucial ingredient is the cell traction. We first assume that it is directly related to cell movement in the presence of fibre orientation in a similar way that the diffusion matrix is associated with fibre orientation. There is considerable experimental justification on cell–matrix interaction for this in that cells tend to align themselves with their fibrous environment (see, for example, the experimental work by Vernon et al. 1992, 1995 and the discussion in Chapter 8). Other factors no doubt play a role but we do not yet have enough experimental evidence to incorporate these in the formulation; the effect we do include is certainly a major one. We thus model the (nonisotropic) cell traction tensor by

$$\tau_{ij} = T(N, S)\frac{D_{ij}}{D},$$

where the magnitude of the traction, $T(N, S)$, has to be specified.

Cell traction is certainly a function of cell density but its dependence is such that outside of the wound the traction is essentially zero. We do not have any biological data on how the traction depends on matrix density, S but we assume a major effect on a cell's traction is the proximity of its neighbours. We thus consider the traction to be given by

$$\tau_{ij} = T(N, S)\frac{D_{ij}}{D}, \quad T(N, S) = \tau_0 SN\left(1 - \frac{N}{\bar{N}}\right)^{\theta}, \tag{10.58}$$

where θ is a positive (even number) constant and τ_0 is the traction force per cell. With this form, when $N = \bar{N}$ the traction force is zero.

As with so much about wound healing the synthesis and degradation of the ECM is also not well understood so we take very simple linear forms in which production is proportional to the cell density and the degradation is proportional to the matrix density S. We thus take

$$g_1(N, S) = \kappa_1 N, \quad g_2(N, S) = \kappa_2 NS. \tag{10.59}$$

The first of these means that we assume the cells in the intact dermis (away from the wound) continue to make ECM at the same rate as within the wound. However, as the ECM matures we assume it has a weaker effect and so we use the matrix density S as a measure of tissue maturity. The effect of maturity is reflected in the production function $q(S)$ which is a decreasing function of the newly secreted collagen that makes up the new fibres. As $q(S) \to 0$ the only effect of ECM turnover is then to maintain the density at a constant level. A simple function that reflects this is

$$q(S) = \begin{cases} 1 - \frac{S}{\beta} & \text{when } S < \beta \\ 0 & \text{otherwise} \end{cases}, \tag{10.60}$$

where β is another parameter. Note that from the second of (10.56) the forms in (10.59) imply the uniform steady state matrix density $\bar{S} = \kappa_1/\kappa_2$. This implies that we do not consider any difference between the matrix density within the wound and that outside. The orientation of the fibres that make up the ECM via the effective strain is the sole measure of scar quality.

Finally we have to model the subdermal attachment forces reflected in b_i: these resist the tissue deformation. We assume they are proportional to the displacement and so we take

$$b_i(S, \mathbf{u}) = \alpha S u_i, \tag{10.61}$$

where α is a constant. With the form of the force balance equation where ECM response and traction are positive, this means that $\alpha > 0$. Although we have taken α to be constant it no doubt depends on the wound depth in three-dimensional versions of the models.

Finally, gathering all these functions together, the finite deformation Cook (1995) model for dermal wound healing is given by (10.56) and (10.57) where the functions for cell proliferation, matrix secretion, matrix degradation, traction magnitude, body force and new fibre fraction are defined by

$$\begin{gathered} f = rN\left(1 - \frac{N}{\bar{N}}\right), \quad g_1 = \kappa_1 N, \quad g_2 = \kappa_2 NS \\ T = \tau_0 SN\left(1 - \frac{N}{\bar{N}}\right)^{\theta}, \quad b_i = \alpha S u_i, \quad q = \begin{cases} 1 - \dfrac{S}{\beta} & \text{when } S < \beta \\ 0 & \text{otherwise} \end{cases} \end{gathered} \tag{10.62}$$

which involve the parameters listed, for convenience, in Table 10.1.

To complete the problem the initial conditions consist of a piece of tissue which is orthotropically stretched, which means there is an initial strain with respect to the prestretched state, and which corresponds to normal skin which is under some tension and accounts for the retraction on wounding. In Cook's (1995) simulations the tissue boundary was kept fixed during retraction and contraction. This is quite different to the scenario we discussed in relation to epidermal wound healing.

In the first phase of the process the wound was created by cutting away tissue, which means lowering the ECM density, and then solving the force balance equations as the wound retracts. Since retraction is taken to be instantaneous the only processes taking

Table 10.1. Definitions of the symbols and parameters for the finite strain model of dermal wound healing given by (10.56), (10.57) with (10.62).

Symbol	Definition
$\bar{N}$	steady state cell density
D	diffusion coefficient
r	low density growth rate
κ_1	ECM secretion rate
κ_2	ECM degradation rate
β	matrix maturity threshold
τ_0, θ	traction parameters
χ	matrix stiffness
α	strength of body force

place are the convection of cells and the ECM. The actual process of wound healing only starts with the expanded area of low ECM density, which defines the wound. Outside the wound the cells are ready to proliferate and then move into the wound and start to build up new ECM. The model implies that cells are produced in the wound via $f(N)$ in the cell density equation. Throughout there is zero flux of cells and matrix across the tissue margins.

Small Strain Model for Wound Healing

In the case of small strains the model equations are somewhat simplified. From (10.56) and (10.57) they are given by

$$\begin{aligned}
\frac{\partial N}{\partial t} &= \left[D_{ij}N\right]_{,ij} - [N v_i]_{,i} + f \\
\frac{\partial S}{\partial t} &= -[S v_i]_{,i} + g_1 - g_2 \\
\frac{\partial u_i}{\partial t} &= v_i \\
\frac{\partial z_{ij}}{\partial t} &= \frac{q}{S} g_1 \left(e_{ij} - z_{ij}\right) - v_k z_{ij,k} \\
\left[\sigma_{ij} + \tau_{ij}\right]_{,j} &= b_i,
\end{aligned} \tag{10.63}$$

where the various functions are the same and given by (10.62), but with the stress in the ECM, diffusion matrix, effective strain, actual strain and active contractile stress tensors now given by

$$\begin{aligned}
\sigma_{ij} &= S\chi\left[2\varepsilon_{ij} + \delta_{ij}\epsilon_{\alpha\alpha}\right] \\
D_{ij} &= D\left[2\varepsilon_{ij} + (2 - \epsilon_{\alpha\alpha})\,\delta_{ij}\right] \\
\epsilon_{ij} &= e_{ij} - z_{ij} \\
e_{ij} &= \frac{1}{2}\left(\frac{\partial u_i}{\partial x_j} + \frac{\partial u_j}{\partial x_i}\right) \\
\tau_{ij} &= T\frac{D_{ij}}{D}.
\end{aligned} \tag{10.64}$$

We should perhaps give a brief recap at this stage. The important extensions in particular to the Murray et al. (1988) framework (and to a lesser extent the model of Tracqui et al. 1993) are that the ECM stress, σ_{ij}, the diffusion matrix, D_{ij}, and the active contractile stress, τ_{ij}, all depend on the *effective* strain, ϵ_{ij}, which in small strain theory is simply the actual Cauchy strain, e_{ij}, minus the residual strain, z_{ij}. The residual strain represents the 'correction' to the effective strain due to changes in the zero stress state. The evolution equation for z_{ij} involves a term which depends on ECM turnover and a convection term (since we are in an Eulerian frame). The justification for the form of the ECM turnover term is that only some fraction, q, of new ECM is assumed to be added to the existing fibrous network in the unstrained state (so that skin tension can be maintained even with continual ECM turnover).

The diffusionlike term for active cell movement represents a form of contact guidance. Since the 'diffusion matrix,' D_{ij}, appears within both derivatives it has a convective as well as a diffusive component. This expression is derived from a random walk argument in which the direction of movement depends on the orientation of fibres (which depends upon the effective strain). Note that the same tensor appears in the contractile stress term, τ_{ij}. Just as cells move in the direction of fibre alignment, they exert a greater force in that direction.

Cook (1995) studied this small strain model numerically with the initial and boundary conditions discussed above. It is the two-dimensional situation which is of particular interest, in view of the extension to the more complex geometries that we have in mind. Simulations show how the fibre orientation within the scar would be expected to evolve over time with different wound orientation to the direction of skin tension. Figure 10.9 is an example of how fibre orientation varies across an elliptical wound. The elliptical wound is oriented perpendicular to the skin tension (which is vertical). In Figure 10.9(a) the wound is shown in its retracted state before contraction has started while Figure 10.9(c) shows the contracted state 25 days after wounding. Each figure shows the alignment field. The gray-scale indicates the extent of anisotropy with the lighter areas denoting more anisotropy with the scale on the right quantifying the difference.

Figure 10.9(a) shows the retracted state and the circumferential alignment at the margins; the wound has opened rather than lengthening when the wound is oriented with its long axis parallel to the direction of skin tension. By day 5, Figure 10.9(b), the wound has started to contract with the fibre alignment tending to be along the length of the wound. Note the alignment outside the wound and how far away the effect of contraction shows up. By day 25 the wound has retracted slightly. The main alignment inside the wound is in line with the skin tension. The temporal behaviour of a wound initially oriented with its long axis parallel to the direction of skin tension has a different healing scenario.

10.10 One-Dimensional Model for the Structure of Pathological Scars

It is interesting to study whether or not these models have the potential for pattern formation by contact guidance and cell traction to see if we can get some insight into the clumping of collagen and cells which are observed in pathological scars such as hypertrophic scarring (Kischer et al. 1982). Contact guidance directs cells towards compressed regions and since compressed regions result from high cell densities there is the now usual possibility of positive feedback (or local activation and lateral inhibition) to enhance the heterogeneity and eventually generate stable spatial patterns. We saw this was possible with the network models discussed in Chapter 8. Pregnancy stretch marks (striae gravidarum) give rise to scarring as a consequence of stretching during the third trimester of pregnancy: they often form fairly regular almost parallel lines. We feel the models we have discussed here could offer a partial explanation. These stretch marks

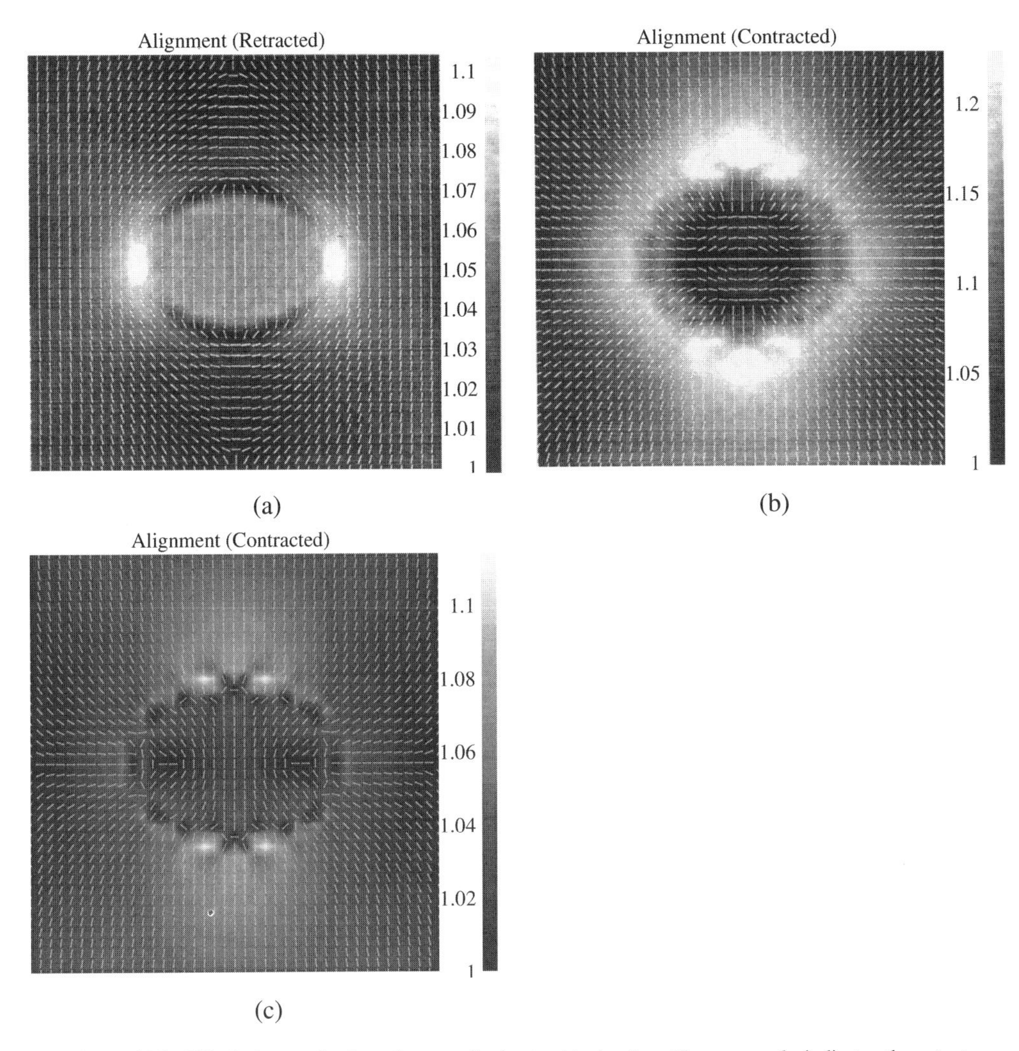

Figure 10.9. Elliptical wound oriented perpendicular to skin tension. The gray-scale indicates the extent of anisotropy with the lighter areas more strongly aligned (note the scale on the right). (**a**) The retracted wound before healing has started. (**b**) The wound 5 days after wounding. (**c**) Day 25 postwounding. Parameter values are: domain $= 10$ cm $\times$ 10 cm, wound $= 4.8$ cm $\times$ 3.6 cm, initial strains $= 2.5\%$ and 1.25%, $D = 4 \times 10^{-6}$ m^2/day, $\theta = 2$, $\tau_0 = 3$, $\chi = 1$, $r = 0.05$/day (corresponding to a doubling time of 14 days), and $\kappa_1 = \kappa_2 = 0.04$/day. Initial conditions: no cells and 80% loss of ECM in wound.

could be an example of tissue reorganisation directly related to the process of tissue reorganisation during full-depth wound repair. There are many physiological changes during pregnancy and stretch marks are not restricted only to the abdomen (see, for example, the article by Elling and Powell 1997).

Using these models Cook (1995) investigated one-dimensional wounds and also deformation of a plastic matrix by cells. Here we again examine in detail the one-dimensional situation since it involves only cell motion relative to the ECM, deformation of the cell–matrix continuum due to the balance between the cell traction and matrix resistance and cell and matrix convection as a consequence of that deformation. We also do not include any cell proliferation and matrix synthesis and degradation. With f, g_1, and g_2 zero we do not need the equation for residual strain z_{ij} and so it can be ignored in which case the effective and actual strain are one and the same. In these circumstances, with the modelling philosophy embedded in the model system (10.56) with (10.57), a reasonable first model to investigate pattern formation potential is

$$\begin{aligned} N_t &= (D(u_x)N)_{xx} - (Nv)_x \\ S_t &= -(Sv)_x \\ u_t &= v(1 - u_x) \\ [\sigma(u_x, N, S) + \tau(u_x)]_x &= \alpha S u, \end{aligned} \tag{10.65}$$

where now

$$D(u_x) = D_0(1 + \gamma u_x), \quad \sigma(u_x) = S\chi u_x, \quad \tau_{ij} = \tau_0 SN(1 + \delta u_x) \tag{10.66}$$

with γ and δ measures, respectively, of the effect of fibre alignment on movement and on traction, while $D(u_x)$ again occurs inside both derivatives so as to reflect source-dependent movement (in the random walk derivation) reflecting contact guidance; here D_0 is a constant.

We now carry out a standard linear stability analysis of (10.65) with (10.66) as we have done many times earlier in the book (see, for example, Chapters 3 or 6). We first nondimensionalise the system and then look for perturbations about a uniform steady state solution. Introduce a representative length, L and time T and write

$$\begin{aligned} n^* &= N/\bar{N}, \quad s^* = S/S_0, \quad t^* = t/T, \quad x^* = x/L, \\ D_0^* &= D_0 T/L^2, \quad \chi^* = \chi/\alpha L^2, \quad \tau_0^* = \tau_0 \bar{N}/\alpha L^2, \end{aligned} \tag{10.67}$$

where $\bar{N}$ and S_0 are the uniform undisturbed cell and matrix densities. We could choose the length and the timescales to be related to the dimensional parameters and reduce the number of parameters even further. For example, we could choose the length scale, L, as that associated with the matrix stiffness, χ, and the strength of the subdermal attachment, α, by choosing $L = \sqrt{\chi/\alpha} \Rightarrow \chi^* = 1$ and the time, T, that associated with diffusion time by choosing $T = L^2/D_0 = \chi/(\alpha D_0) \Rightarrow D_0^* = 1$. We would then be left with only one dimensionless parameter $\tau_0^* = \tau_0 \bar{N}/\chi$. We retain the flexibility of choosing the length and timescales as in (10.67) and the model system (10.65) becomes, on dropping the * for convenience,

$$\begin{aligned} n_t &= [D_0(1 + \gamma u_x)n]_{xx} - (nu_t)_x \\ s_t &= -(su_t)_x \\ [s\chi u_x + \tau_0(1 + \delta u_x)sn]_x &= su, \end{aligned} \tag{10.68}$$

where we have already used the linearisation of the displacement equation, the third of (10.65), by taking $v = u_t$.

The relevant uniform steady state solution of (10.68) representing the undisturbed skin is $(n, s, u) = (1, 1, 0)$ and we look for solutions of the form $[n, s, u]^T \propto \exp(\lambda t + ikx)$ of the linearised sytem of (10.68); that is,

$$
\begin{aligned}
n_t &= D_0 u_{xxx} + D_0 \gamma n_{xx} - u_{xt}, \\
s_t &= -u_{xt}, \qquad\qquad (10.69) \\
\chi u_{xx} + \tau_0(n_x + s_x + \delta u_{xx}) &= u.
\end{aligned}
$$

The dispersion relation for the growth rate λ for each mode k is then given by

$$
\begin{vmatrix} \lambda + D_0 k^2 & 0 & ik\lambda + iD_0\gamma k^3 \\ 0 & \lambda & ik\lambda \\ ik\tau_0 & ik\tau_0 & -k^2(\chi + \delta\tau_0) - 1 \end{vmatrix} = 0, \qquad (10.70)
$$

which simplifies to

$$
\lambda = \frac{-D_0 k^2 [k^2(\chi + \delta\tau_0 - \gamma\tau_0 - \tau_0) + 1]}{k^2(\chi + \delta\tau_0 - 2\tau_0) + 1}. \qquad (10.71)
$$

Parameter ranges and wavenumbers k which result in $\lambda(k^2) > 0$ give rise to growing spatial patterns.

From (10.71) we see that if k^2 is small but nonzero, $\lambda < 0$ and so small wavenumbers, which correspond to long wavelengths, are stable. If $k^2 \gg 1$ then

$$
\lambda \approx \frac{-D_0 k^2 (\chi + \delta\tau_0 - \gamma\tau_0 - \tau_0)}{(\chi + \delta\tau_0 - 2\tau_0)}. \qquad (10.72)
$$

So, if the parameters χ, τ_0, γ, and δ are in appropriate domains it is possible for λ to be positive and hence the steady state to be unstable. For example, if $\gamma + 1 > \delta > 2$ the condition for $\lambda > 0$ from the last equation requires $\tau_0 > \chi/(\gamma - \delta + 1)$.

It is clear that contact guidance together with cell traction if sufficiently large can destabilise the tissue and give rise to spatially growing instabilities manifested by cell and tissue clumping. From the dispersion relation we can make some qualitative observations. The larger γ, that is, the greater the response of movement to alignment the lower the traction necessary for instability. Intuitively this is as we would expect since we believe contact guidance is destabilising. This was also the situation in Chapter 8 where networks were formed from the instabilities. On the other hand traction response to alignment via the parameter δ has a stabilising effect since with compression traction tends to decrease in the direction of the compression.

Because of the form of the dispersion relation (10.72) if the parameters do give $\lambda = 0$ for some critical k_c^2 then all modes $k^2 > k_c^2$ are also unstable and the fastest growing modes are the largest possible modes, that is, perturbations with the smallest wavelengths. Suppose, for example, $\delta > 2$; then if $\tau_0 > \chi/[1 + \gamma - \delta]$, (10.71) shows that all wavenumbers greater than $k_c^2 = 1/[\tau_0(1+\gamma-\delta)-\chi]$ are unstable. Linear theory, as we have seen before, cannot predict the final spatial pattern: the initial conditions play a crucial role. Even in the finite domain problem initial conditions play a major role in the final pattern. The problem of unbounded growth of high wavenumbers appeared in some of the mechanical models discussed in Chapter 6 and there the traction was in effect averaged over some neighbourhood at the point. Here this could be achieved by introducing a long range effect with a factor $s + \zeta s_{xx}$ to the traction term (see Chapter 6) where ζ is another parameter.

Let us now come back to the question of stretch marks in pregnancy. The above one-dimensional analysis shows that spatial instabilities can form and from our experience with pattern formation mechanisms these will form a series of waves alternating between high and low cell densities. In Chapter 8 we saw that contact guidance gave rise to random networks of tissue along which cells moved. There are some similarities with that situation here with the crucial difference that *in vivo* the skin is under tension and so the effect of the various strains we have discussed here must come into play in the equivalent two-dimensional problem. We suggest that the interaction of the skin tissue and tension caused by the deformation makes for directed scar stretch marks; the irregularity of the lines *in vivo* is probably caused by the usual random initial conditions and of course the general two-dimensional heterogeneity of the distended (and initially undistended) skin.

10.11 Open Problems in Wound Healing

The above finite strain model deals more completely with the problem of material anisotropy so crucial to scar structure and formation. In the more general theory, orientation and effective strain are decoupled. An enormous amount of work remains to be done in this area. Complicated as this model is, it would achieve considerable generality and flexibility if Cook's (1995) work were to be combined with the finite strain (but isotropic growth) theory of Rodriguez et al. (1994) and the fabric tensor theory of Cowin (1986) and Cowin et al. (1992); see also Tozeren and Skalak 1988). The Rodriguez et al. (1994) description of growth and deformation under finite strains has far-ranging implications for wound healing and many other mechanically constrained morphogenetic processes. Murray et al. (1998) very briefly (or rather cryptically) describe this general theory of growth and remodelling. The functional forms of some of the terms are key ingredients in these models. Some of them come from experiments while with others, as in the case of the force function dependence on filament orientation in the work of Sherratt and Lewis (1993) on embryonic wounds discussed in the last chapter, from a basic theoretical approach. Forms for others can sometimes be derived from a more microscopic approach using the ideas of Edelstein-Keshet and Ermentrout (1998) and Ermentrout and Edelstein-Keshet (1998) who model actin filament dynamics with both polymerization and fragmentation (see other references in these papers). Spiros and Edelstein-Keshet (1998) importantly relate their model for the dynamics of actin structures with extant experimental data and estimate parameters. The work by Mogilner and Oster (1996) on how cell motility can be driven by actin polymerisation is also pertinent to the modelling we have described here.

As we have said above, wound healing is immensely complicated and immensely important. The problems are legion. Cook's (1995) work is a seminal contribution to the field and has opened up numerous new avenues to study full-depth wounds in a manageable but highly relevant way. With his approach there are many avenues that have not been explored: Murray et al. (1998) suggested the following partial list of some open problems and other areas for future research, both theoretical and experimental.

Cell Origin, Proliferation, Differentiation and Migration

In all the models discussed cells have migrated into the wound from the surrounding tissue (in the plane of the skin). However, the origin of fibroblasts and myofibroblasts is not fully understood and they may originate in small islands of less-damaged tissue or even migrate into the wound from below. Also, although multiple phenotypes seem to be involved, the control of differentiation and dedifferentiation and the differences in properties between groups are poorly understood. We also need to know more about the signals directing cell movement. Random motion, chemotaxis, haptotaxis, haptokinesis and contact guidance have all been suggested. We need to know how cells respond to the entire range of environments that are possible during the healing of a full thickness wound. Controlled collagen gel experiments (such as by Barocas and Tranquillo 1994) should provide insights here.

Contractile Stress

We have discussed the differences between static and dynamic generation of contractile stress. We need to know much more about its timing and control (such as what causes saturation and what are the feedbacks from ECM structure) as well as distinguishing between cell contraction and the building of stress into the ECM via semi-permanently local reorganization. The work of Tracqui and his coworkers (for example, Ferrenq et al. 1997 and Tranqui and Tracqui 2000) on cell traction quantification and their integration of cell–ECM interactions is a major new approach and particularly relevant to wound healing.

ECM Structure, Turnover and Mechanics

We have dealt with theoretical aspects of the relationships between ECM turnover, structure and mechanics. Ideally we want the material properties of the evolving tissue to be subject to biomechanical testing at all stages of the healing process. On what timescale is scar tissue (passively) viscoelastic, for example? And what are the residual stresses? Experiments with a series of small cuts would seem to be the only option here. Do we need to consider multiple phases or fibre types? We also need to know what determines the alignment and stretch of new fibres and to separate production from degradation. There is very little experimental information available here. Corneal wounds and experiments with collagen gels should provide insights. The work of Maini, Sherratt and their colleagues is particularly relevant here (for example, Dale et al. 1995, 1996, Olsen et al. 1995, 1996 and Dallon et al. 1999). Theoretical extensions are required to allow for changes in the binding of scar tissue to the hypodermis. A similar approach to that for remodelling of the ECM within the plane of the scar tissue would seem appropriate. That this is an important aspect of the plasticity of wound contraction is demonstrated by experiments in which the edge of a contracting wound is undermined and the wound edges retract.

Biochemical Interactions

There are certainly a large number of reacting and diffusing chemicals (for example, growth factors) involved in controlling the various subprocesses of wound healing. Isolating the effects of each of these will be a long and laborious process but there are a large number of groups involved in this work and mathematical models will be required to interpret the results.

Geometry

We need to carry out more complete two-dimensional theoretical studies, examining different shapes, sizes and orientations, and the effects of grafts and skin substitutes. With respect to the latter the work of Green is directly related; see, for example, the general article by Green (1991). He developed a new method for culturing human cells for use as grafts in burns and other wounds. His earlier work (Green and Thomas 1978) on fibroblast patterns is also of direct interest. With three-dimensional simulations we will be able to look more realistically at variations in depth and binding to the subdermal layers. We will also be able to investigate pathological scars which rise above the surface of the skin (and pressure therapy would be examined by changing the boundary conditions).

Computation and Analysis

It would be useful theoretical work to build general models for growing tissues within the finite element methodology. All models to date have been simulated using finite differences but it could be argued that finite elements would be more natural. On the analytical front we need to investigate the possibility of instabilities within wound healing models. Linear stability analyses of simplified models suggest that under conditions of high contractile stress, for example, homogeneous solutions would not be expected even in a near-homogeneous scar environment. If two-dimensional studies demonstrate that dense bands of aligned fibres are predicted then we may have an explanation for the nodules observed in pathological scars; we touched on this briefly in the last section. If the work on vasculogenesis, discussed in Chapter 8 is typical of mechanochemical systems then we should be optimistic in this regard. There contact guidance plays a major role in network structuring of cells embedded in a matrix. In principle, instabilities could give rise to numerical problems in that the scale of the pattern must be resolvable on the grid. Also, the models themselves may not be robust if the instabilities do not give rise to stable patterns which are dominated by a finite length scale.

The type of problems mathematical models can address include a considerable array of highly relevant biological and clinical problems. Some we have already mentioned. In the case of the standard wound we can use them to investigate: (i) retraction (extent and shape of the retracted wound), (ii) the contraction curve (the lag, whether the rapid retraction is linear or exponential or some other form and the late retraction), (iii) the shape of the contracted wound, (iv) scarring, both inside and outside of the wound (measured, for example, by the maximum fibre alignment and strain). Basic studies consider the: (i) effect of wound size, (ii) effect of domain size (that is, the size of surrounding tissue), (iii) effect of skin tension, (iv) effect of wound orientation (relative to skin tension lines), (iv) effect of wound shape (circles, triangles, rectangles and so on), (v) changing cell proliferation (growth enhancing factors), (vi) changing cell migration (including chemotaxis effects and the role of filament-driven motility), (vii) changing matrix turnover, (viii) changing traction relative to matrix stiffness and (ix) changing the qualitative forms of the various functions in the models. Other aspects which could be addressed are: (i) grafting and leaving an island within the wound, (ii) splinting a wound and then releasing it, (iii) pathological scar formation associated with mechanical instabilities and (iv) inclusion of subdermal attachments with plasticity to reflect matrix evolution. Of course several of these require significant amendments to the models but with the experience of the modelling in this and the previous chapter we can at least make biologically plausible preliminary suggestions as to what these additions could be.

10.12 Concluding Remarks on Wound Healing

Although we have said it before (Murray et al. 1998) it bears repeating to ask why we use mathematics to study something so complicated and badly understood as wound healing. We argue that mathematics must be used if we hope to truly convert an understanding of the underlying mechanisms into a predictive science. Mathematics is required to bridge the gap between the level on which most of our knowledge is accumulating (cellular and below) and the macroscopic level of the scar itself which is of primary concern to the surgeon and the patient. A mathematical approach allows one to explore the logic of wound healing. Even if the mechanisms were well understood—and they certainly are not at this stage—mathematics would be required to explore the consequences of manipulating the various parameters associated with any particular wound management scenario. The number of options that are fast becoming available to wound managers will become overwhelming unless we can find a way to simulate particular treatment protocols before applying them in practice.

We should also remember that an important point arising from the models is that a scar contains its own history. Consider an engineering analogy (Murray et al. 1998) of our role in managing the wound healing process. It is one thing to suggest that a bridge requires a thousand tons of steel, that any less will result in too weak a structure, and any more will result in excessive rigidity. It is quite another matter to instruct the workers on how best to put the pieces together. It is conceivable that the cells involved in tissue repair have enough ‘expertise’ that given the right set of ingredients and initial instructions they could be persuaded to repair a deep skin wound with a result that looked more like skin than scar tissue. This is perhaps the hope of those who are searching for the optimal cocktail of growth factors to apply topically to a wound. However, it seems to us very likely that the global effect of all this sophisticated cellular activity would be critically sensitive to the sequence of events occurring during tissue repair. As managers we should concern ourselves with how to take advantage of the limited opportunities we have for communicating with the workforce so as to direct the wound healing process towards an acceptable end-product. This may sound rather philosophical, but even a cursory look at the theories of scar formation in the literature reveal a fixation on simplistic explanations. For example, it has variously been suggested that pathological scars result from defects in collagen synthesis or in collagen degradation.

We suggest that there need be no misfunctioning of any particular subprocess to explain scar formation. Tissue repair is, after all, fundamentally different from development. The body has only a limited set of 'rules' with which it must confront an unlimited number of possible repair 'problems.'

We are without question a long way from being able to reliably simulate actual wounds. Not only is the active cellular control of the key processes poorly understood, but also wounds are difficult to reproduce, with similar wounds on different parts of the same body, healing at different rates and with different results. Despite these limitations, we argue that exploring the logic of wound healing is worthwhile even in our present state of knowledge. It allows us to take a hypothetical mechanism and examine its consequences in the form of a mathematical model, making predictions and suggesting experiments that would verify or invalidate the model; even the latter casts light on the biology. Indeed, the very process of constructing a mathematical model can be useful in its own right. Not only must we commit to a particular mechanism, but we are also forced to consider what is truly essential to the process of wound healing, the key players (variables) and the key processes by which they evolve. We are thus involved in constructing frameworks on which we can hang our understanding of wound healing. The equations, the mathematical analysis and the numerical simulations that follow serve to reveal quantitatively as well as qualitatively the consequences of that logical structure.

The use of mathematical model mechanisms for the study of wound healing has already shed light on several hitherto poorly understood topics. The rapid increase in biological technology has led to more sophisticated experiments and such research builds on advances in our understanding of the cell–matrix interactions, the use of growth factors and biomaterials, more powerful computers and advanced cell culturing techniques. The experimental work such as that of Green (1991) mentioned above and the techniques by Bertolami et al. (1991) on skin substitutes are just two examples. The available information now lets us try and realistically relate microscopic and macroscopic phenomena with mathematical models. The work of Cook (1995) along these lines provides a basis for studying the major problem of granulation tissue and scar formation during wound contraction and matrix remodelling. Pathologic scars are related to abnormal structure of the ECM (Dunn et al. 1985). The work of Dale et al. (1995) and Olsen et al. (1995) is particularly relevant here. These models also let us begin to seriously address the surgical issues of wound orientation with respect to tension lines, wound shape and geometry so as to minimize scarring. We believe that the approach of building models from the microscopic (cell level) to determine the macroscopic models such as we have done here is a fruitful one. Realistic models must eventually incorporate finite strains which makes the development of appropriate constitutive equations for biomaterials particularly pressing. Crucial to this are microscopic and macroscopic measurements of cell traction within biological tissue, some of which has been done by, for example, Delvoye et al. (1991), Ferrenq et al. (1997) and Tranqui and Tracqui (2000).

None of the individual models we have discussed and not even all of them put together could be considered a complete model of dermal wound healing, even if we think only in terms of wound contraction and scar formation. However, each model has shed light on different aspects of the process and we can now say what must be the most important elements of a complete model. These studies have served to highlight where our knowledge is deficient and to suggest directions in which fruitful experimentation might lead us. Indeed, a critical test of these theoretical constructs is in their impact on the experimental community. The field has now achieved some level of maturity and we believe that future dialogue between experimentalists and applied mathematicians will lead us most rapidly towards the goal of scarless wound healing.

11. Growth and Control of Brain Tumours

Brief Historical Perspective on Brain Surgery

Surgery on the brain or skull has an incredibly long history and is arguably the most ancient of surgical interventions. The books of articles edited by Greenblatt (1997) and Walker (1951) make fascinating reading. There is an early description of how to treat head wounds in the Edwin Smith Surgical Papyrus of ancient Egypt (about 2500 BC) in which trepanning, or trephining, is described as a procedure. Trephining is the surgical removal of a piece of bone from the skull. Numerous skulls have been found showing healed craniotomies; the procedure was certainly far from always being fatal and the large number of such healed craniotomies suggests some success.

Verano and his colleagues (see Verano et al. 1999, 2000 and Arriaza et al. 1998 and other references there) discuss interesting examples, from archaeological finds, of trephination, and other major surgeries such as foot amputation, by the coastal cultures of Peru and the Incas. There is a remarkable photograph in Arriaza et al. (1998) of an Inca cranium (from the archaeological museum in Cuzco) showing four trephinations which were clearly well healed while the patient was alive. The trephinations on this skull are all of the order of 5 cm in diameter.

There are numerous pictures and woodcuts from the mediaeval period (see also the historical discussion on surgery and wound healing in Chapter 9) showing surgical operations on the head. The procedure for carrying out such trephining varies widely. Figure 11.1(a) is from a mediaeval manuscript showing how it might have been carried out while Figure 11.1(b) is from another mediaeval manuscript showing the end of a succesful completion of the surgery (albeit with a somewhat dubious and unhappy looking patient). Figure 11.1(c) is a caricature example of a late mediaeval operation, this time being carried out by a group of animal surgeons which indicates what people thought of surgeons at the time. As mentioned in the chapter on epidermal wound healing, Galen, who clearly carried out trephining, was very practical in his advice such as noting that the *'disadvantage of using a gouge is that the head is shaken vigorously by the hammer strokes.'*

Richard of Parma in his *Practica Chirugiae* (Practice of Surgery) of 1170 (see Valenstein 1997) writes: *'For mania or melancholy a cruciate incision is made in the top of the head and the cranium is penetrated to permit the noxious material to exhale to the outside. The patient is held in chains and the wound is treated, as above.'* The classic picture by the late 15th and early 16th century enigmatic Hieronymous Bosch called *The Cure of Folly* (or Madness) or *The Stone Operation* in the Prado Museum in Madrid

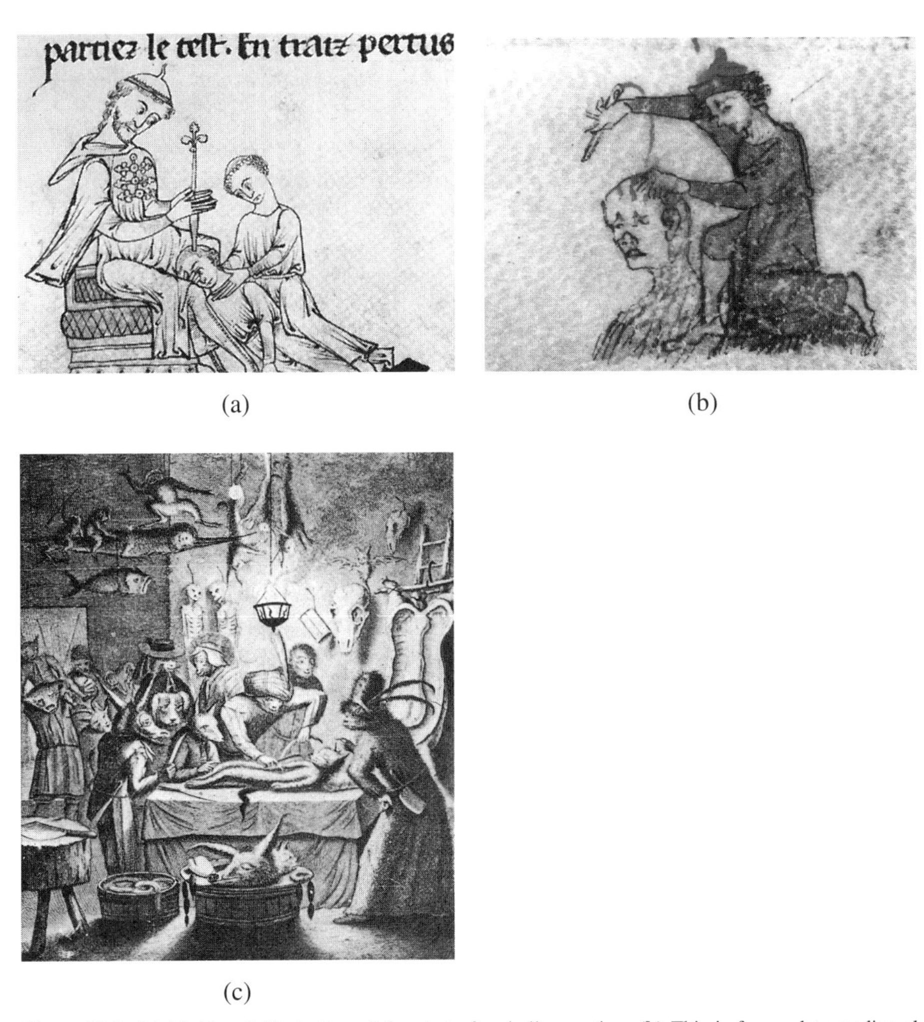

Figure 11.1. (**a**) Mediaeval illustration of the start of a skull operation. (**b**) This is from a late mediaeval manuscript showing post-surgery suturing. (**c**) Brain surgery being carried out by a number of animal 'surgeons.' It highlights the low opinion people had of surgeons in mediaeval times. (Courtesy of the Wellcome Trust Medical Library, London and reproduced with permission)

shows a man making an incision on the scalp of a seated man. The inscription (translated by Cinotto 1969) has been translated as *'Master, dig out the stones of folly, my name is "castrated dachshund."'* At the time a 'castrated dachshund' was frequently the name used for a simpleton. Madness was often believed to be caused by stones in the head. Art historians and others have discussed and disagreed about the interpretation of this painting for a very long time; some, for example, say it ridicules the itinerant mediaeval charlatans who went around 'curing' madness and other ailments by trephining while others say it is an allegory of human gullibility and stupidity which is so common in Bosch's paintings. The classic book *Anatomy of Melancholy* by Robert Burton (1652) includes *'tis not amiss to bore the skull with an instrument to let out the fuliginous vapours ... Guierius curted a nobleman in Savoy by boring alone, leaving the hole open a month together by means of which, after two years melancholy and madness he was delivered.'*

The practice of trephining declined from the beginning of the 19th century when the surgical operation was moved into the hospitals because the mortality went up dramatically since such hospitals were infection paradises at the time. (Even now infection in hospitals is still a major killer, even if the prognosis of survival is much greater.)

Trephination by indigenous peoples has been used (and still is today) for a variety of problems, such as insanity, depression, behavioural disorders (lobotomies are still carried out by professional surgeons in many hospitals) to relieve the results of head injuries from blows to the head (such as depressed fractures) or falls from trees or simply from banging their head on a door lintel. Margetts (1967) and in particular the interesting article by Furnas et al. (1985) (see earlier references there) describe (with many photographs of actual operations) in fascinating medical detail craniotomies regularly performed by traditional craniotomists of the Kisii tribe in Kenya. The craniotomist (called an omobari) is highly respected and very skillful. They use a variety of herbs and drugs to aid healing and prevent infection. The operations are often carried out in the open. They use a series of instruments to scrape through the skull, roughly a 5×5 cm^2 patch, until a very thin layer is left. They then use a fine pick to puncture this last membrane thereby exposing the dura. Some, however, also use a drill or hacksaw. The aim of the treatment can be, for example, to remove bone splinters (resulting from a blow to the head) or to relieve depression or severe headaches. It appears that the mortality rate is low and patients seem satisfied with the results.

11.1 Medical Background

Cancerous tumours, or neoplasms, originate from the mutation of one or more cells which usually undergo rapid uncontrolled growth thereby impairing the functioning of normal tissue. There are many different cancers each with their own characteristics. In this chapter we shall only be concerned with brain tumours, and in particular gliomas or glioblastomas, which make up about half of all primary brain tumours diagnosed; they are particularly nasty tumours with a depressingly dismal prognosis for recovery. Gliomas are highly invasive and infiltrate the surrounding tissue. The impressive increased detection capabilities (but as we shall see, still woefully inadequate) in computerized tomography (CT) and magnetic resonance imaging (MRI) over the past 20 years have resulted in earlier detection of glioma tumours. Despite this progress, the benefits of early treatment have been minimal (see, for example, Silbergeld et al. 1991, Alvord and Shaw 1991, Kelley and Hunt 1994, Cook et al. 1995 and Burgess et al. 1997). For example, even with extensive surgical excision well beyond the grossly visible tumour boundary, regeneration near the edge of resection ultimately results in eventually leading to death (see, for example, Matsukado et al. 1961, Kreth et al. 1993, Woodward et al. 1996 and other references there). This failure of resection is analogous to trying to put out a forest fire from behind the advancing front. The action of the fire (tumour growth) is primarily at the periphery.

The brain basically consists of two types of tissue: grey matter and white matter. Grey matter is composed of neuronal and glial cell bodies that control brain activity while the cortex (like the 'bark') is a coat of grey matter that covers the brain. White matter fibre tracts are myelinated neuron axon bundles located throughout the inner regions of the brain. These fibres establish pathways between grey matter regions. The corpus callosum is a thick band of white matter fibres connecting the left and right cerebral hemispheres of the brain. Within each hemisphere, there are several white matter pathways connecting the cortex to the nuclei deep within the brain; see Figure 11.2. Figure 11.3 shows two photographs of a human brain showing grey and white matter distribution and the corpus collosum.

Gliomas are neoplasms of glial cells (neural cells capable of division) that usually occur in the upper cerebral hemisphere but which can be found throughout the brain (Alvord and Shaw 1991). Astrocytomas, originating from an abnormally multiplying astrocyte glial cell, are the most common gliomas. Depending on their aggressiveness (grade), astrocytomas are further divided into several subcategories. Astrocytomas are the least aggressive or lowest grade, anaplastic astrocytomas are the more aggressive or mid-grade and glioblastomas (multiforme) are the most aggressive or highest grade. Tumour grade indicates the level of malignancy and is based on the degree of anaplasia (or deformity in behaviour and form) seen in the cancerous cells under a microscope. Gliomas often contain several different grade cells with the highest or most malignant grade of cells determining the grade, even if most of the tumour is lower grade. There is still no general clinical agreement on the grading.

Generally, the higher-grade cancer cells are more capable of invading normal tissue and so are more malignant. However, even with their invasive abilities, gliomas very rarely metastasize outside the brain.

The prognosis for patients with neoplasms affecting the nervous system depends on many factors. A major element in the prognosis is the quantitative evaluation of the spatiotemporal infiltration of the tumour, taking into account the anatomic site of the tumour as well as the effectiveness of the various treatments.

Since we believe that the modelling developed in this chapter has a practical bearing on patient treatment it is necessary to give more detailed medical information which we believe is an important part of realistic medical modelling.

Difficulties in Treating Brain Tumours

An enormous amount of experimental and some theoretical work has been devoted to trying to understand why gliomas are so difficult to treat. Unlike many other tumours, gliomas can be highly diffuse. Experiments indicate that within 7 days of tumour implantation in a rat brain, glioma cells can be identified throughout the central nervous system (Silbergeld and Chicoine 1997). A locally dense tumour growth remains where the cancerous tissue was initially implanted but there are solitary tumour cells throughout the central nervous system (Silbergeld and Chicoine 1997 and Silbergeld, personal communication, 1998). Most glioma treatments are directed locally to the bulk mass when, in fact, the action of tumour growth and invasion is elsewhere.

There are various, regularly used, treatments for gliomas, mainly chemotherapy, radiation therapy and surgical intervention. Resection, the surgical removal of an accessible tumour, has a wretched history of success. Recurrence of tumour growth at the

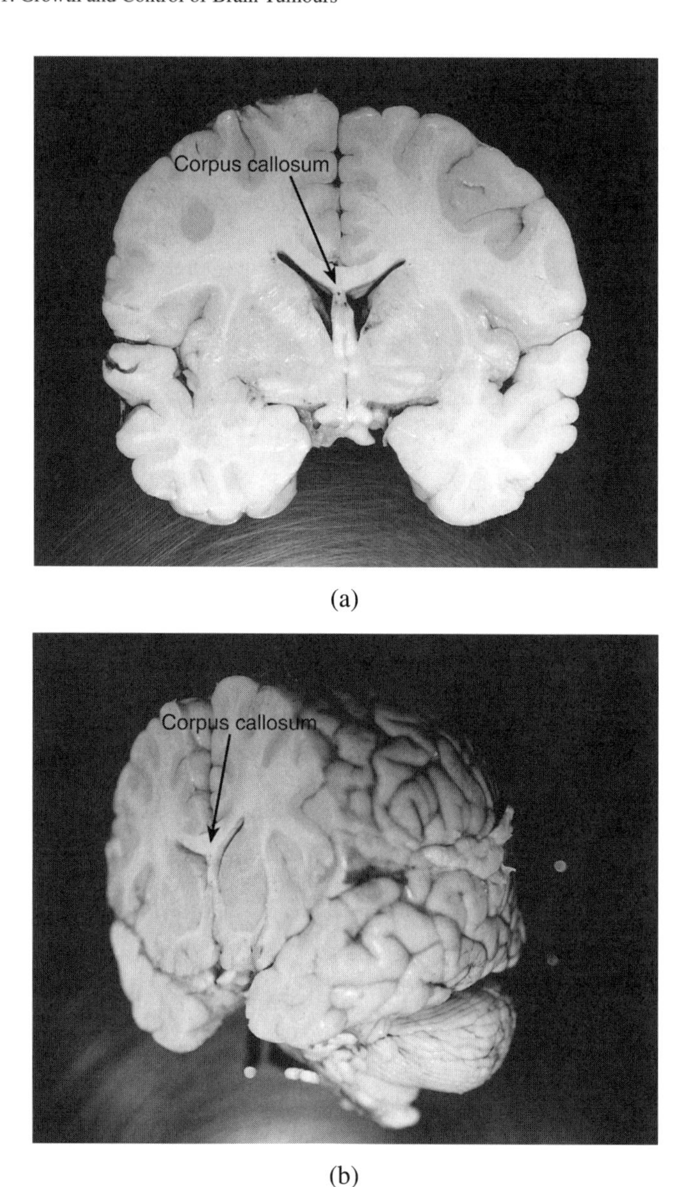

Figure 11.2. Two cross-sections of a human brain showing fibrous white matter and the corpus callosum which connects the left and right cerebral hemispheres. (Figure courtesy of Dr. E.C. Alvord, Jr.)

resection boundary is a well-documented phenomenon in glioma research (see, for example, Silbergeld and Chicoine 1997, Woodward et al. 1996, Kreth et al. 1993, Kelley and Hunt 1994 and other references there). Both experimentalists and theoreticians believe that the distantly invaded cells are responsible for tumour regeneration following surgery (Chicoine and Silbergeld 1995, Silbergeld and Chicoine 1997). Since the density of cancerous cells (remaining after resection) is highest at the resection boundary,

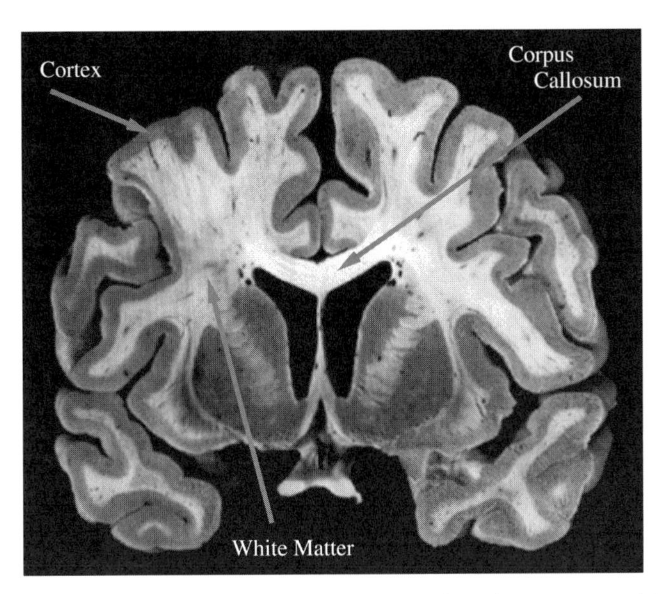

Figure 11.3. Cross-section of the human brain. The cortex consists of grey matter and is connected to other grey matter regions by white matter fibre bundles. The corpus callosum is a white matter tract connecting the left and right cerebral hemispheres. (Figure reproduced with the permission of Professor Paul Pietsch; derived from T.H. Williams, N. Gluhbegovic and J.Y. Jew Virtual Hospital Figure 5-30. http://www.vh.org/Providers/Textbooks/BrainAnatomy/BrainAnatomy.html)

regrowth seems most probable at this location. Alternatively, Silbergeld and Chicoine (1997) suggested, and are presently testing, the hypothesis that damaged brain tissue at the resection site releases cytokines that recruit the diffusely invaded tumour cells. Nevertheless, both explanations are consistent with the argument that the diffuse nature of gliomas is fundamentally responsible for tumour recurrence near the resection boundary. The difference is that the former is a physical model and the later is more biochemical. In this chapter we shall study a model (basically an incredibly simple one—even linear) for resection therapy and show why it generally fails. As we shall show, it increases life expectancy minimally; we compare the results and predictions with patient data.

Chemotherapy essentially uses specialized chemicals to poison the tumour cells. The brain is naturally defended from these and other types of chemicals by the intricate capillary structure of the blood–brain barrier. Water-soluble drugs, ions and proteins cannot permeate the blood–brain barrier but lipid-soluble agents can. Recently, agents have been devised to temporarily disrupt the blood–brain barrier. Many chemotherapeutic treatments are cell-cycle-dependent: the drugs are triggered by certain phases of the cell cycle. Silbergeld and Chicoine (1997) have observed that the motile cells distant from the bulk tumour do not appear to enter mitosis so cell-cycle specific drugs and standard radiation therapy have limited effectiveness. Not only that, gliomas are often heterogeneous tumours. Those drugs that do reach the cancerous cells are hindered by drug resistance commonly associated with cancer cell heterogeneity. While one cell type is responsive to treatment and dies off, other types are waiting to dominate. This phenomenon requires a model which includes cell mutation to drug resistance cells, in other words a polyclonal model. Below we describe and analyze such a model for chemotherapy and again compare the results with patient data.

The biological complexity of gliomas makes treatment a difficult undertaking. For planning effective (or seemingly so) treatment strategies, information regarding the growth rates and invasion characteristics of tumours is crucial. The use of mathematical modelling can help to quantify the effects of resection, chemotherapy and radiation (it tries to kill the tumour cells with radiation) on the growth and diffusion of malignant gliomas. In this chapter we shed some light on certain aspects of brain tumour treatment with the aim of helping to determine better, or even optimal, therapeutic regimes for patients. A major goal is the development of interactive computer models with which the effects of various treatment strategies for specific tumours could be examined. We believe that the work described in this chapter goes some way in achieving this goal. Having said this, however, all of the treatments mentioned above have a very poor record of success. There is a pressing need for a totally different approach to the treatment of gliomas, several of which are currently being investigated.

11.2 Basic Mathematical Model of Glioma Growth and Invasion

Like all tumours, the biological and clinical aspects of gliomas are complex and the details of their spatiotemporal growth are still not well understood. In constructing models therefore we have to make some major assumptions. In doing so we have relied heavily on the medical input of Dr. E.C. Alvord (Professor of Neuropathology, University of Washington) with whom J.D. Murray has an ongoing collaboration which started in the early 1990s. The group has included many students and postdoctorals, all of whom are on one or another of the papers referred to below and in which more of the medical details are given. It is mainly their work and relevant medical studies on brain tumours that we discuss in detail here.

With the philosophy espoused in this book we start with as simple a model as is reasonable and build up from it. The simplest theoretical models involve only the total number of cells in the tumour, with growth of the tumour usually assumed to be exponential, Gompertzian, or logistic (Swan 1987, Marusic et al. 1994). Such models do not take into account the spatial arrangement of the cells at a specific anatomical location or the spatial spread of the cancerous cells. These spatial aspects are crucial in estimating tumour growth since they determine the invasiveness and the apparent border of the tumour. It is necessary to try and determine the extent of infiltration of the tumour in most treatment situations, such as estimating the likely benefit of surgical resection (Alvord 1991). Until the work described in this chapter, the absence of even a simple model explaining the growth and recurrence of human gliomas made it difficult to explain why results of treatment by surgical excision are so disappointing (Nazzarro and Neuwelt 1990, Kreth et al. 1993). One of the surprising aspects of the work we describe here is how a very simple (linear) deterministic model can provide meaningful and helpful clinical information with a direct bearing on patient care.

In this section we develop a mathematical model for the spatiotemporal dynamics of tumour growth. Importantly we can estimate the model parameters, including the proliferation, or growth rate, and the diffusion coefficient of the cells from clinical data obtained from successive computerized tomography (CT) scans of patients and independent experimental work. This is described below in Section 11.10 on chemotherapy where we also describe what these CT scans show and how we use them.

Once we have established the feasibility of reconstructing some of the kinetic events in invasion from histological sections, it will be possible to investigate other gliomas with different characteristic growth patterns, geometries and the effects of various forms of therapy (surgery, chemo- and/or radiotherapy) using the same types of data from other patients. The growth patterns essentially define the gross and microscopic characteristics not only of the classical tumours of different degrees of malignancy but also of 'mixed-gliomas' and 'multi-centric gliomas' (Alvord 1992).

Previous mathematical modelling (Tracqui et al. 1995, Cruywagen et al. 1995, Cook et al. 1995, Woodward et al. 1996, Burgess et al. 1997) used a theoretical framework to describe the invasive nature of gliomas, both with and without treatment regimes, by isolating two characteristics: proliferation and diffusion. Here diffusion represents the active motility of glioma cells. These models showed that diffusion is more important in determining survival than the proliferation rate of the tumour. *In vivo* studies show that malignant glioma cells implanted in rats quickly invade the contralateral hemisphere of the brain dispersing via white matter tracts (Kelley and Hunt 1994, Silbergeld and Chicoine 1997). The diffusion of glioma cells in white matter is different from that in the grey matter and is included in a more realistic model.

The basic model considers the evolution of the glioma tumour cell population to be mainly governed by proliferation and diffusion. Tumour cells are assumed to grow exponentially. This is a reasonable reflection of the biology for the timescale with which we are concerned, namely, the time to the patient's death. Although necrotic core (a region of dead cells) formation is evident in some gliomas, only the highly proliferative, slowly diffusive tumours are significantly affected by necrosis; we do not include this but the model could be modified to include it. Also, a typical logisticlike growth would strictly be more accurate but we are most interested in understanding the diffusive nature of the tumour behaviour and the medical timescale considered. Logistic growth can be trivially incorporated. (In the case of breast cancer Hart et al. 1998 and Shochat et al. 1999 have shown that exponential growth is not a valid approximation.) Silbergeld and Chicoine (1997) suggested that diffusion is a good approximation for the tumour cell motility. A very good review of glioma invasion is given by Giese and Westphal (1996). We show later that diffusion reasonably models the cell spreading dynamics observed *in vitro*.

Let $\bar{c}(\bar{\mathbf{x}}, \bar{t})$ be the number of cells at a position $\bar{\mathbf{x}}$ and time $\bar{t}$. We take the basic model, in dimensional form, as a conservation equation

$$\frac{\partial \bar{c}}{\partial \bar{t}} = \bar{\nabla} \cdot \mathbf{J} + \rho \bar{c}, \tag{11.1}$$

where ρ (time^{-1}) represents the net rate of growth of cells including proliferation and death (or loss). The diffusional flux of cells, $\mathbf{J}$, we take as proportional to the gradient of the cell density:

$$\mathbf{J} = \bar{D} \bar{\nabla} \bar{c}, \tag{11.2}$$

where $\bar{D}$ (distance^2/time) is the diffusion coefficient of cells in brain tissue. The theoretical models, referred to above, considered the brain tissue to be homogeneous so the diffusion and growth rates of the tumour cells are taken to be constant throughout the brain. This is not the case, of course, when considering tumour invasion into white matter from grey. With constant diffusion the governing equation (11.1) with (11.2) is then

$$\frac{\partial \bar{c}}{\partial \bar{t}} = \bar{D}\bar{\nabla}^2\bar{c} + \rho\bar{c}. \tag{11.3}$$

As we shall see this model gives reasonable agreement with the CT scans on which the model is based and has given surprisingly good results in predicting survival times under various treatment scenarios (Tracqui et al. 1995, Cruywagen et al. 1995, Cook et al. 1995, Woodward et al. 1996, Burgess et al. 1997). We give a full discussion of some of these models and results below. Although the models gave surprisingly (in view of the gross simplifying assumptions) good results, they contain several basic simplifications which can now be reconsidered, as was done by Swanson (1999) and Swanson et al. (2000). For example, given a source of glioma cells at a given location, most of the previous models, for numerical simplicity (and ignoring anatomic boundaries), considered the 'front' of detectable cells propagates symmetrically out from the source. They all knew, of course, that clinical and experimental observation indicate that, in fact, symmetry in growth of the tumour is not the case. The first model we discuss here deals with this aspect as well as tissue heterogeneity.

White matter serves as a route of invasion between grey matter areas for glioma cells. The diffusion coefficient (motility) for glioma cells is larger in the white matter than in the grey matter. *In vivo* studies show that malignant glioma cells implanted in rats quickly invade the contralateral hemisphere of the brain dispersing via white matter tracts (Chicoine and Silbergeld 1995, Silbergeld and Chicoine 1997, Kelley and Hunt 1994). The model we study now incorporates the effects of the heterogeneous tissue on the cell diffusion and tumour growth rates to emulate more accurately the clinically and experimentally observed asymmetries of the gross visible tumour boundaries.

Model with Spatial Heterogeneity

We can account for spatial heterogeneity in our model by taking the diffusion $\bar{D}$ to be a function of the spatial variable, $\bar{\mathbf{x}}$, thereby differentiating regions of grey and white matter. This gives, in place of (11.3),

$$\frac{\partial \bar{c}}{\partial \bar{t}} = \bar{\nabla} \cdot (\bar{D}(\bar{\mathbf{x}})\bar{\nabla}\bar{c}) + \rho\bar{c}. \tag{11.4}$$

We take zero flux boundary conditions on the anatomic boundaries of the brain and the ventricles. So, if B is the brain domain on which the equation (11.4) is to be solved, the boundary conditions are

$$\mathbf{n} \cdot \bar{D}(\bar{\mathbf{x}})\bar{\nabla}\bar{c} = 0 \quad \text{for } \mathbf{x} \text{ on } \partial B, \tag{11.5}$$

where $\mathbf{n}$ is the unit normal to the boundary ∂B of B. With the geometric complexity of an anatomically accurate brain (which we shall in fact use) it is clearly a very difficult analytical problem and a nontrivial numerical problem, even in two dimensions. We studied this problem in both two and three dimensions with anatomically accurate white and grey matter distribution obtained from EMMA (Extensible MATLAB Medical Analysis). This is a programme package developed by Collins et al. (1998) at the McConnel Brain Imaging Centre (Montreal Neurological Institute) to aid the analysis of medical imaging. Figure 11.4 shows an example of the white and grey matter distribution in a horizontal slice of the brain produced by EMMA.

In two dimensions, Swanson (1999) mapped out the regions of white and grey matter to simulate the clinical data from the CT scans used as the basis for the original two-dimensional model (11.3). Her initial goal was to simulate and produce a more accurate qualitative reproduction of these CT scan images than the original models which considered only homogeneous brain tissue. As mentioned, even with this simplification the results compared well with various gross medical quantities and patient data. We are particularly interested in the effects of grey and white matter distributions on glioma growth and invasion.

We first use the simplified anatomical domain of the rat brain. Numerical simulation and analytic results of this simpler case will be compared with experimental data procured and analyzed in the Neuropathology Laboratory at the University of Washington Medical School by (our collaborators) Dr. Ellsworth Alvord, Jr. (Head of Neuropathology) and Dr. Daniel Silbergeld (Neurological Surgery). We shall then consider, in detail, the model on an anatomically accurate human domain in both two and three dimensions.

We first nondimensionalise the spatially heterogeneous model, which as usual, also decreases the number of effective parameters in the system, and get some idea of the relative importance of various terms (without regard to units). To give some concept of the numbers involved there can be 10^{11} cancerous cells in a small tumour while the diffusion coefficient can be of the order 10^{-4} cm^2/day. Parameter estimates are given later in various tables.

We consider the diffusion coefficients to be constant, but different, in each of the two tissues, the white matter and the grey matter. So, we have to solve

$$\frac{\partial \bar{c}}{\partial \bar{t}} = \bar{\nabla} \cdot (\bar{D}(\bar{\mathbf{x}})\bar{\nabla}\bar{c}) + \rho\bar{c}, \tag{11.6}$$

where

$$\bar{D}(\bar{\mathbf{x}}) = \begin{cases} D_w & \text{for } \bar{\mathbf{x}} \text{ in white matter} \\ D_g & \text{for } \bar{\mathbf{x}} \text{ in grey matter} \end{cases} \tag{11.7}$$

subject to the zero flux boundary conditions

$$\mathbf{n} \cdot \bar{D}(\bar{\mathbf{x}})\bar{\nabla}\bar{c} = 0 \quad \text{for } \mathbf{x} \text{ on } \partial B \tag{11.8}$$

and initial condition $\bar{c}(\bar{\mathbf{x}}, 0) = \bar{f}(\bar{\mathbf{x}})$: we come back to the appropriate form of $\bar{f}(\bar{\mathbf{x}})$ below.

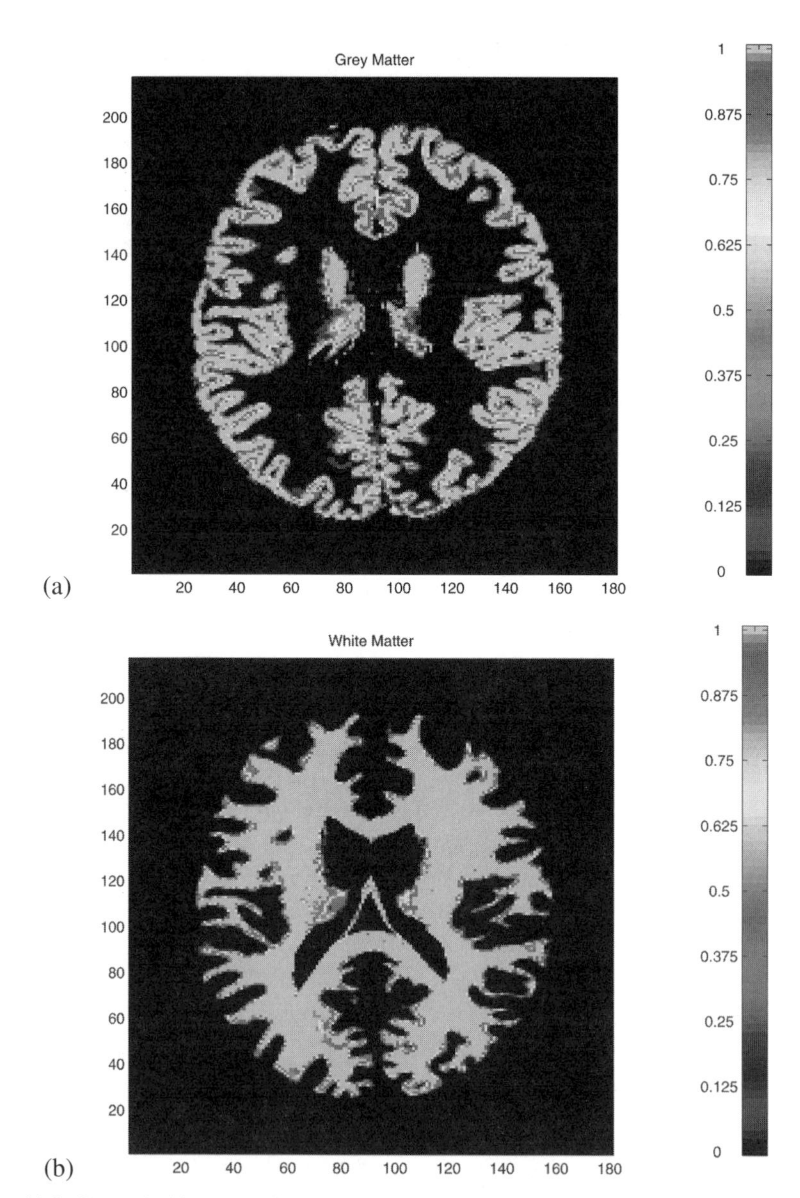

Figure 11.4. Grey and white matter distributions in a horizontal slice of the brain. The figure was generated using EMMA (Extensible MATLAB Medical Analysis) developed by Collins et al. (1998). (See the comment in parentheses in Figure 11.10 below.)

Introduce the nondimensional variables

$$\mathbf{x} = \sqrt{\frac{\rho}{D_w}}\bar{\mathbf{x}}, \quad t = \rho\bar{t}, \quad c(\mathbf{x}, t) = \frac{D_w}{\rho N_0}\bar{c}\left(\sqrt{\frac{\rho}{D_w}}\bar{\mathbf{x}}, \rho\bar{t}\right), \tag{11.9}$$

where $N_0 = \int \bar{f}(\bar{x})d\bar{x}$ represents the initial number of tumour cells in the brain at model time $\bar{t} = 0$. With these (11.6) becomes

$$\frac{\partial c}{\partial t} = \nabla \cdot (D(\mathbf{x})\nabla c) + c, \tag{11.10}$$

where

$$D(\mathbf{x}) = \begin{cases} 1 & \text{for } \mathbf{x} \text{ in white matter} \\ \gamma = \frac{D_g}{D_w} & \text{for } \mathbf{x} \text{ in grey matter} \end{cases} \tag{11.11}$$

with $c(\mathbf{x}, 0) = f(\mathbf{x}) = (D_w/\rho N_0)\bar{f}(\sqrt{(\rho/D_w)}\,\bar{\mathbf{x}})$ and $\mathbf{n} \cdot D(\mathbf{x})\nabla c = 0$ for $\mathbf{x}$ on ∂B. With this nondimensionalisation, diffusion is on the spatial scale of diffusion in white matter with time measured on the timescale of the tumour growth.

Let us now consider the form of the initial condition and appropriate forms for $f(\mathbf{x})$. Theoretically, the tumour started out as one cancerous cell, but the time of appearance of this original cell and the type of growth and the spread of the early cancer cells are unknown. We assume that, at the time of the first scan, the diffusion process has already broken any previous, possibly uniform, distribution of the cells. So, the cells are normally distributed with a maximum cell density, a, at the centre, $\mathbf{x}_0$, of the tumour; that is,

$$c(\mathbf{x}, 0) = a\exp\left(\frac{-|\,\mathbf{x} - \mathbf{x}_0\,|^2}{b}\right), \tag{11.12}$$

where b is a measure of the spread of tumour cells.

Initial Analytical Estimate of Survival Time

Clinically, tumour cells can not be detected at very low densities. On a CT scan, the profile of the tumour is defined by some nonzero level of resolution corresponding to a cellular density c^* (roughly $40,000$ cells/cm^2). Below this threshold, cancer cells are not detected by the imaging technique. A typical serial CT scan is in essence a series of pictures at several different levels of the brain where the tumour is located. From these an approximate three-dimensional shape is constructed. Figure 11.5 is an idealized scan of a tumour with three levels. With these we can approximate the size of the tumour and an equivalent spherical radius; see Section 11.10 where patient scans are used directly to estimate parameters to determine tumour size and growth.

In the case of a constant growth rate ρ and homogeneous diffusion $\bar{D}(\bar{\mathbf{x}}) = D$, equation (11.3) in two spatial dimensions with an initial delta function source of N_0

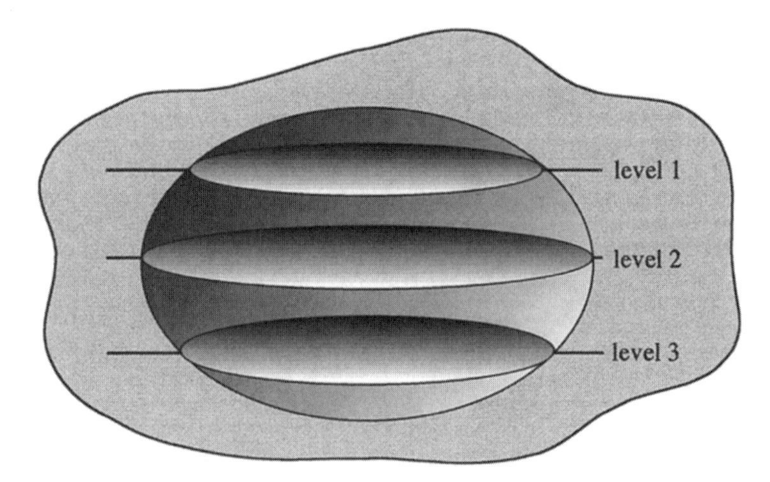

Figure 11.5. Idealized 3-level computerized tomography (CT) scan. With the three areas of detected cancer cells we can reconsruct an approximate tumour shape and volume. An example of an actual scan is shown in Figure 11.12.

tumour cells at $\bar{\mathbf{x}} = 0$ has the solution

$$\bar{c}(\bar{\mathbf{x}}, \bar{t}) = \frac{N_0}{4\pi D\bar{t}} \exp\left(\rho\bar{t} - \frac{\bar{r}^2}{4D\bar{t}}\right), \tag{11.13}$$

where $\bar{r}$ is the axially symmetric radial coordinate. If the detectable threshold density is $\bar{c}^*$ then the radius $\bar{r}^*$ of the tumour profile, given by (11.13) with $\bar{c}(\bar{\mathbf{x}}, \bar{t}) = \bar{c}^*$ is

$$\bar{r}^* = 2\sqrt{D\rho}\,\bar{t}\sqrt{1 - \frac{1}{\rho\bar{t}}\ln\left(4\pi D\bar{t}\frac{\bar{c}^*}{N_0}\right)} \sim 2\sqrt{D\rho}\,\bar{t} \quad \text{for } t \text{ large.} \tag{11.14}$$

The last expression is, of course, just the asymptotic form of the radial travelling wave of the axisymmetric Fisher–Kolmogoroff equation; it has velocity $2\sqrt{D\rho}$ (refer to Chapter 13, Volume I). So, if the tumour is identified when it has a radius $\bar{r}_{\text{detect}}$ and the tumour is fatal when it has a radius $\bar{r}_{\text{lethal}}$ we can approximate the untreated survival time by

$$\begin{aligned} \text{survival time} =& \bar{t}_{\text{lethal}} - \bar{t}_{\text{detect}} \\ \approx& \frac{1}{\sqrt{D\rho}}\left(\bar{r}_{\text{lethal}} - \bar{r}_{\text{detect}}\right). \end{aligned}$$

This shows that D and ρ are both important parameters in determining survival time: increasing either ρ or D will decrease survival time. The average radius (or rather equivalent radius) at which a tumour is identified is 1.5 cm and lethal is 3.0 cm (Burgess et al. 1997). The estimates for the detectable and lethal radii are only average values observed clinically (Blankenberg et al. 1995); the range is quite large. With these radii, the untreated survival time is given approximately by $3/(2\sqrt{D\rho})$. For a high grade tumour (parameter values are given in Section 11.5, Table 11.5 below), the untreated survival time suggested by the model is approximately 200 days. This is consistent with the observed 6 to 12 month median survival time for treated glioblastoma (Alvord 1991).

In terms of the nondimensional variables defined in (11.9)

$$r^* = 2t\sqrt{1 - \frac{1}{t}\ln \psi t}, \quad \psi = 4\pi \frac{D\bar{c}^*}{\rho N_0}. \tag{11.15}$$

Depending on the parameter value ψ, the observed radius of the tumour either increases monotonically with time or there is a delay between tumour initiation and visible progression; this time lapse is sometimes referred to as the establishment phase. Basically it is the time it takes the tumour to get established before it starts to spread.

The length of the establishment phase t_e can be calculated by simply setting $r^* = 0$ and $t = t_e$ in (11.15) which gives

$$t_e = \ln \psi t_e. \tag{11.16}$$

An establishment phase only exists for $\psi > 1$. We can see this immediately if we plot the curve $\psi = e^{t_e}/t_e$, note that the minimum is at $t_e = e$ and so a solution exists only for for $\psi > e$. This defines a relationship between all the parameters and can be useful in relating experimental observations and model parameters. From (11.16) the length of the establishment phase increases with ψ and, since $\psi \propto D/\rho$, it increases as D/ρ increases.

For a fixed ψ, the nondimensional detectable tumour radius r^* is given by a single curve. So, we can imagine a high grade tumour (that is, high ρ, high D) and a low grade tumour (low ρ, low D) such that the ratio D/ρ (and thus ψ) is fixed. The detectable tumour radii r^* increase in the same manner for both tumours but it takes the low grade tumour longer to reach a lethal size. Although the spatial invasion characteristics are the same for a fixed ψ, the timescale on which they occur can be very different.

Spatial dispersal in a heterogeneous environment has mainly been studied in an ecological context. For example, Cantrell and Cosner (1991) modelled the dynamics of a population inhabiting a heterogeneous environment with a diffusive logistic equation with spatially varying growth rates. They were interested in the effect of the spatial arrangement of favorable and unfavorable patches on the overall suitability of the environment. Cruywagen et al. (1996) modelled the risk of spread of a genetically engineered organism using a reaction diffusion model on a spatially heterogeneous environment. Conditions for invasion of the organism were determined for the case of spatially periodic diffusion rates and carrying capacity. Their work is discussed in Chapter 1. Shigesada and Kawasaki (1997) in their book on biological invasions discuss dispersal with a piecewise-constant spatially varying diffusion coefficient. The environment consists of an arrangement of favorable and unfavorable patches. With diffusion and logistic growth spatially periodic, they show the existence of traveling periodic wave solutions.

With respect to the spatial spread of brain tumours in a heterogeneous domain, Swanson (1999) carried out a one-dimensional analysis for the model equation (11.10) with (11.11) in which the spatial domain consisted of a small $O(\epsilon)$ domain of white matter embedded in an infinite domain of grey matter. The boundary conditions consist of continuity of the cell density and the cell flux at the boundaries between the grey and white regions. From an asymptotic point of view, with $\epsilon \ll 1$, since the diffusion coefficient in the white region is larger than that in the grey regions intuitively we would expect that the cell density is approximately constant in the white region as was confirmed analytically.

11.3 Tumour Spread *In Vitro*: Parameter Estimation

In vitro experiments are frequently used to help characterize *in vivo* behaviour. Necessarily, several parameter estimates come from such *in vitro* experiments. Underlying the parameter estimation from experimental data is the general assumption that our modelling approach and the model apply to these experiments. In this section, we consider two such experiments to determine parameter estimates. We discuss this in some detail since parameter estimation is particularly important when making predictions as to possible patient treatment.

Normal glial cells have a very low motility rate (Silbergeld and Chicoine 1997) but glioma cells can exhibit abnormally high motility rates (Chicoine and Silbergeld 1995, Giese et al. 1996a,b,c, Giese and Westphal 1996, Pilkington 1997a,b, Silbergeld and Chicoine 1997, Amberger et al. 1998, Giese et al. 1998). Using time-lapse video microscopy and other techniques, Chicoine and Silbergeld (1995) quantified brain tumour cell motility and invasion capabilities *in vivo* and *in vitro* and their results suggest an average linear velocity of 12.5 μm/hr for human glioma cells *in vitro* and a minimum linear velocity of 4.8 μm/hr *in vivo*.

In Vitro Experiments of Chicoine and Silbergeld (1995): Cell Motility

Chicoine and Silbergeld (1995) developed a tumour cell motility assay known as the 'radial dish assay.' Basically 2×10^4 cells are plated in a central 2-cm diameter disk of an 8-cm diameter petri dish. Cell mitosis is inhibited and daily microscopy of the dish shows the spatial spread of the cell population. We can use the above model as it applies to this experiment to estimate the diffusion coefficient of the glioma tumour cells *in vitro*.

In this situation there is no growth so $\rho = 0$ and no heterogeneity in diffusion so $\bar{D}(\bar{x}) = D$. The dimensional model (11.3) (with radial symmetry) is then

$$\frac{\partial \bar{c}}{\partial \bar{t}} = D\bar{\nabla}^2\bar{c} \tag{11.17}$$

with zero flux boundary conditions

$$\bar{\mathbf{n}} \cdot \bar{\nabla}\bar{c}(\bar{r}, \bar{t}) = 0 \quad \text{for } \bar{r} = R_0, \tag{11.18}$$

where R_0 is the radius of the petri dish; in the assay $R_0 = 4$ cm. Initially cells are uniformly distributed in a central circular area of radius R:

$$\bar{c}(\bar{r}, 0) = \bar{c}_0 H(R - \bar{r}), \tag{11.19}$$

where H denotes the Heaviside function and $R = 1$ cm in the Chicoine et al. (1995) experiments. Since mitosis is blocked, there is no growth and a total of $N = \bar{c}_0 \pi R^2$ cells are in the petri dish throughout the experiment.

If we now nondimensionalise the model by setting

$$\mathbf{x} = \frac{\bar{\mathbf{x}}}{R_0}, \quad t = \frac{D}{R_0^2}\bar{t}, \quad c(\mathbf{x}, t) = \frac{\bar{c}\left(\frac{\bar{\mathbf{x}}}{R_0}, \frac{D}{R_0^2}\bar{t}\right)}{\bar{c}_0}, \tag{11.20}$$

where $N = \bar{c}_0 \pi R^2$ is the initial number of tumour cells at time $t = 0$ the nondimensional model, which, with radial symmetry involves only the radial coordinate r, becomes

$$\begin{aligned} \frac{\partial c}{\partial t} &= \frac{\partial^2 c}{\partial r^2} + \frac{1}{r}\frac{\partial c}{\partial r} \quad \text{for } 0 < r < 1 \\ c(r, 0) = H(\lambda - r), & \quad \frac{\partial c}{\partial r} = 0 \quad \text{at } r = 1, \end{aligned} \tag{11.21}$$

where $\lambda = R/R_0$; in the case of the radial dish assay experiments, $\lambda = 1/4$.

Asymptotic Approximation for the Cell Density $c(r, \theta, t)$

If we can assume that the petri dish is sufficiently large, $R_0 \gg R$, we can approximate the solution by (11.13) with $\rho = 0$, the solution of the simple diffusion equation on an infinite domain. However, $\lambda = 1/4$ is not small and so we have to use a more accurate approximation.

The solution of (11.21) is a classical one involving Bessel functions and is obtained by standard methods; see, for example, Kevorkian (1999) or Carslaw and Jaeger (1959). It can be derived by superposition of the basic solution for a ring at radius r_0 integrated from 0 to λ.

Since we are primarily interested in large t all we need is the asymptotic approximation to the solution. The solution for a ring of cells at r_0 is given by

$$c_{\text{ring}}(r, t; r_0) = \frac{1}{4\pi t} \exp\left(-\frac{r^2 + r_0^2}{4t}\right) I_0\left(\frac{r r_0}{2t}\right), \tag{11.22}$$

where I_0 is the modified Bessel function. Asymptotically, for $r r_0/2t$ small, this solution is given approximately by

$$c_{\text{ring}}(r, t; r_0) \sim \frac{1}{4\pi t} \exp\left(-\frac{r^2 + r_0^2}{4t}\right) \left[1 + \frac{1}{4}\left(\frac{r r_0}{2t}\right)^2 + O\left(\left(\frac{r r_0}{2t}\right)^4\right)\right]. \tag{11.23}$$

When $r_0 = 0$ this gives, of course, the exact solution for a point source of cells at the origin.

For a disk of cells of radius R we have

$$c(r,0) = H(\lambda - r) = 2\pi \int_0^\lambda c_{\text{ring}}(r,0;r_0) r_0 \, dr_0.$$

Combining this with the approximation (11.23), the solution of the full problem (11.21) with disk initial conditions is then

$$\begin{aligned} c(r,t) &= 2\pi \int_0^\lambda c_{\text{ring}}(r,t;r_0) r_0 \, dr_0 \\ &\sim e^{-\frac{r^2}{4t}}\left(1 - e^{-\frac{\lambda^2}{4t}}\right) + \frac{r^2}{4t} e^{-\frac{r^2}{4t}}\left[1 - \left(1 + \frac{\lambda^2}{4t}\right) e^{-\frac{\lambda^2}{4t}}\right] + \cdots \end{aligned} \tag{11.24}$$

for $\nu = r\lambda/2t$ small. In the case of the Chicoine and Sibergeld (1995) experiments $\nu = r/8t$.

We now use the approximate solution (11.24) to obtain an *in vitro* estimate for the diffusion coefficient for the glioma cells. We expect the estimate to be most accurate for small r and $t \sim O(1)$ or larger. Figure 11.6(a) is a plot of the experimental measurements of cell densities in the radial dish assay from Chicoine and Silbergeld (1995) with the (dimensional) form of the asymptotic solution (11.24) for $D = 0.002$ cm^2/hr at the different experimental times. Swanson (1999) solved the full problem (11.22) numerically for $D = 0.002$ cm^2/hr with the same experimental data from Figure 11.6(a) and this is shown in Figure 11.6(b). Comparing the two figures, the asymptotic approximation (11.24) is quite accurate for determining parameter estimates.

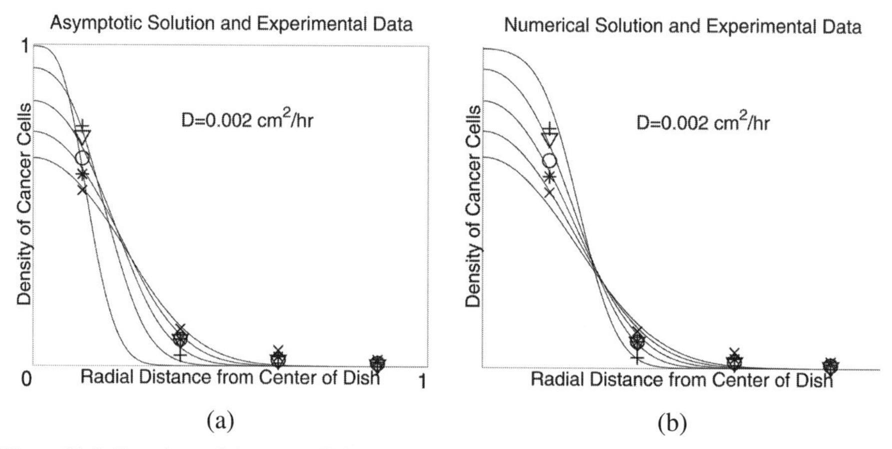

Figure 11.6. Experimental *in vitro* cell density results from Chicoine and Silbergeld (1995) as compared with the (**a**) asymptotic approximation from (11.24) and (**b**) the numerical solution of (11.22) (from Swanson 1999). The cell densities were observed for mixed glioma cells in the radial dish assay experiment. The asymptotic approximation is strictly only valid for moderate t and small r (that is, $r\lambda/2t$ small). The different curves correspond to experimental observation times $t = 24, 48, 72, 96, 120$ hours.

Asymptotic Approximation for the Average Distance, $\langle r \rangle$, of Cells from the Initial Inoculum

Chicoine and Silbergeld (1995) also calculated the average distance from the origin, $\langle r \rangle$, for the glioma cells in the radial dish assay method. To determine $\langle r \rangle$ we have to integrate the solution for all r. The asymptotic solution (11.24) is only strictly valid for $r\lambda/2t$ small which is not small for arbitrarily large r. So, we really have to calculate $\langle r \rangle$ in full.

If $c(r, t)$ is the distribution of the cells at time t, then the mean radial distance of cells from the origin $\langle r \rangle$ is given by:

$$\langle r \rangle = \frac{\int_0^\infty r^2 c(r, t)\, dr}{\int_0^\infty r c(r, t)\, dr} = \int_0^\infty r^2 c(r, t)\, dr, \tag{11.25}$$

where $c(r, t)$ is given by the integral in (11.24). The answer is given in terms of an integral which we approximate for large and small t. It involves some algebra and a straightforward application of Laplace's method for evaluating integrals involving small and large parameters (see, for example, Murray 1984 for a pedagogical discussion). For small t with $A = 1/4t$ we obtain

$$\begin{aligned}
\langle r \rangle &\sim \frac{3 + 2A\lambda^2}{6A\lambda}\left(1 + \operatorname{erf}(\sqrt{A}\,\lambda)\right) - \frac{5}{6\lambda^2\sqrt{A^3\pi}} + \frac{5 + 2A\lambda^2}{6\sqrt{\pi}\,A\lambda^2} e^{-A\lambda^2} \\
&= \frac{6t + \lambda^2}{3\lambda}\left[1 + \operatorname{erf}\left(\frac{\lambda}{2\sqrt{t}}\right)\right] - \frac{20}{3\lambda^2}\sqrt{\frac{t^3}{\pi}} + \frac{10t + \lambda^2}{3\sqrt{\pi}\,\lambda^2} e^{-\frac{\lambda^2}{4t}} \\
&\sim \frac{2\lambda}{3} + \frac{4t}{\lambda} + \cdots \quad \text{as } A = \frac{1}{4t} \to \infty.
\end{aligned} \tag{11.26}$$

As $t \to 0$ ($A \to \infty$), $\langle r \rangle$ approaches $2\lambda/3$ which corresponds to the exact mean radial distance associated with the (nondimensional) initial distribution of cells.

For large t (small A), the mean displacement is given by

$$\langle r \rangle = \sqrt{\pi t}\left(1 + \frac{\lambda^2}{16t} - \frac{\lambda^4}{768t^2} + \cdots\right). \tag{11.27}$$

For t large, the mean radius converges to the case of a point source of cells at the origin: $\langle r \rangle = \sqrt{\pi t}$ (in dimensional form, $\langle \bar{r} \rangle = \sqrt{\pi D \bar{t}}$). What it means is that after a long time diffusion will have spread the population out sufficiently so that the original distribution can not be identified. However, for large time, the effects of the boundaries of the petri dish become important. Since the cells in the petri dish are not allowed to grow during the Chicoine and Silbergeld (1995) experiments, as $t \to \infty$, the cell density approaches a uniform steady state $c \to \lambda^2$ (in dimensional form, $\bar{c} \to \bar{c}_0\lambda^2$). So a more accutate asymptotic estimate for the large time behaviour should use this steady state to compute the mean displacement:

$$\langle r \rangle = 2\pi \int_0^\infty r^2 c(r, t)\, dr \to 2\pi \int_0^1 \lambda^2 r^2\, dr = \frac{2\pi}{3}\lambda^2, \tag{11.28}$$

which is the long time limit for the petri dish experiments. When compared with the experimental results neither of these approximations is very good although the small t approximation is better.

Chicoine and Silbergeld (1995) actually calculated a 'mean radius' in a slightly different way to what we have done here. Their calculation neglected the contribution of the cells within the initial ring of radius R. This corresponds to simply replacing the lower limit 0 in the integral in (11.25) with λ. If we denote this new 'mean radius' by $\langle r^* \rangle$ it satisfies

$$\langle r^* \rangle = \frac{\int_\lambda^\infty r(r-\lambda)c(r, t)\, dr}{\int_\lambda^\infty rc(r, t)\, dr} = \frac{2\pi}{1-\lambda^2}\int_\lambda^\infty r(r-\lambda)c(r, t)\, dr. \tag{11.29}$$

Swanson (1999) has shown, using similar approximation methods to those above for $\langle r \rangle$,

$$\langle r^* \rangle \sim \frac{2\pi t^2}{1-\lambda^2} + \frac{40}{3(1-\lambda^2)}\sqrt{\pi t^3} + \cdots \qquad \text{for small } t \text{ (large } A\text{)}. \tag{11.30}$$

Remember that $R = 1$ cm which in the experiments gives $\lambda = 1/4$.

The glioblastoma cells are observed to reach the edge of the petri dish by 96 hours thus defining the overestimation of the asymptotic result for large t values in Figure 11.7(c).

In Figure 11.7 we plot the mean radius $\langle r^* \rangle$ computed from the experimental observations with the asymptotic expression (11.30) which is strictly only valid for small t. This figure suggests estimates of 1.6×10^{-4} cm^2/hr, 2×10^{-3} cm^2/hr and 3×10^{-3} cm^2/hr for the diffusion coefficients for anaplastic astrocytoma, mixed glioma, and glioblastoma multiforme cells, respectively. Therefore, with increasing malignancy, the cell motility increases. In fact, these results suggest that there is approximately a twofold difference between the motility of mid-grade anaplastic astrocytoma cells and high grade glioblastoma cells *in vitro*.

In Figure 11.7 we see that for glioblastoma cells the asymptotic expansion (11.26) overestimates the mean radius for $\bar{t} > 96$ hours. This is as we should expect because Chicoine and Silbergeld (1995) observed that the glioblastoma cells move so quickly they are capable of reaching the edge of the dish in 96 hours and so boundary effects become important. The anaplastic astrocytoma and mixed glioma cells are not capable of reaching the petri dish edge until later times.

From the calculations of $\langle r \rangle$, there is a steady state limit for the tumour cell population in the petri dish, namely, $c \to \lambda^2$ for large time. We can calculate the effect this steady state has on the limiting value of $\langle r^* \rangle$:

$$\langle r^* \rangle \to \frac{\int_\lambda^1 r(r-\lambda)\lambda^2\, dr}{\int_\lambda^1 \lambda^2 r\, dr} = \frac{2 - 3\lambda + \lambda^3}{3(1-\lambda^2)} \tag{11.31}$$

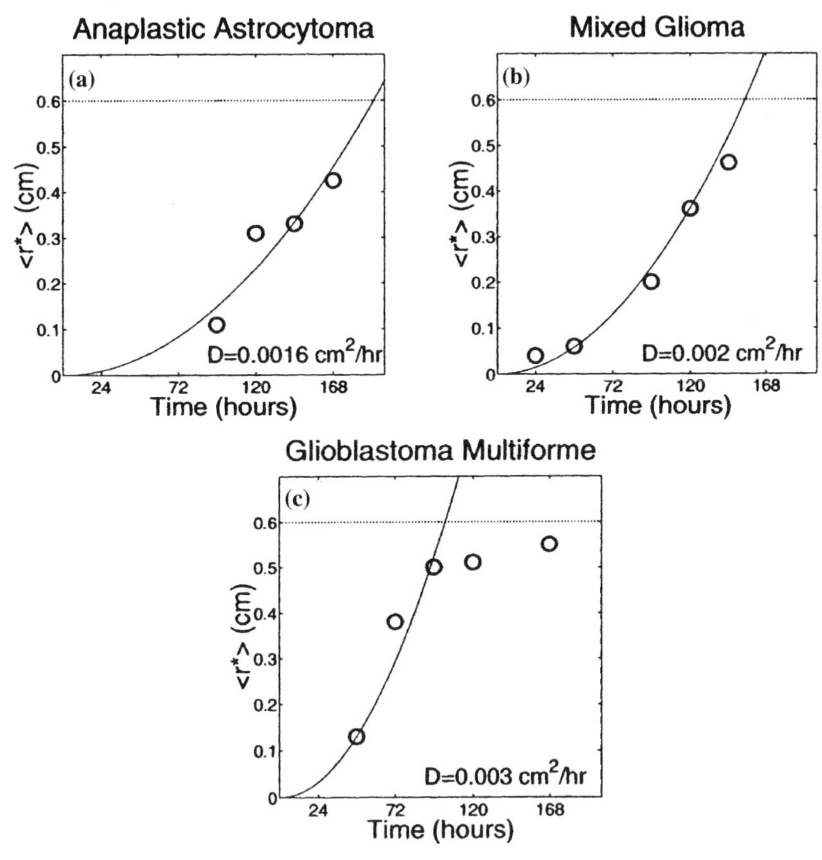

Figure 11.7. Asymptotic approximation for the mean radius $\langle r^* \rangle$ from (11.30) versus time for (**a**) anaplastic astrocytoma ($D = 1.6 \times 10^{-3}$ cm^2/hr), (**b**) mixed glioma ($D = 2 \times 10^{-3}$ cm^2/hr) and (**c**) glioblastoma multiforme ($D = 3 \times 10^{-3}$ cm^2/hr) cells. The uniform steady state solution defines the long time limit (dotted line) of $\langle r^* \rangle$ given by (11.31). (From Swanson 1999)

as $t \to \infty$. This limiting value is included in Figure 11.7 and defines the overestimation of $\langle r^* \rangle$ by the asymptotic approximation (11.30) for glioblastoma multiforme cells in Figure 11.7(c).

In Vitro Experiments of Giese et al. (1996a,b,c): Cell Motility and Proliferation

As we have mentioned, glioma cell motility in white matter is higher than that in grey matter. Giese et al. (1996b) carried out *in vitro* experiments to characterize the increased motility of glioma cells on myelin, a key component of white matter. These experiments were similar to those of Chicoine and Silbergeld (1995) but on a shorter timescale, namely, 40 to 100 hours, with the main difference being that Giese et al. (1996b) allowed the cells to proliferate during the experiment. They also tabulated the increase in radius of the visible front of tumour cells. In addition, the initial inoculum of cells covered a smaller region than in the Chicoine and Silbergeld (1995) experiments. So, we expect the experiment to be modelled more appropriately by

$$\frac{\partial \bar{c}}{\partial \bar{t}} = \bar{D}\bar{\nabla}^2\bar{c} + \rho\bar{c} \quad \text{for } \bar{r} \leq R_0, \tag{11.32}$$

where R_0 is the radius of the dish in which the cells are allowed to migrate, the Laplacian is axisymmetric and there are zero flux boundary conditions

$$\mathbf{n} \cdot \bar{\nabla}\bar{c}(\bar{r}, \bar{t}) = 0 \quad \text{for } \bar{r} = R_0, \tag{11.33}$$

with $\mathbf{n}$ the outward pointing normal at the edge of dish. Initially, there is a point source of N cells at the origin:

$$\bar{c}(\bar{r}, 0) = N\delta(\bar{r}). \tag{11.34}$$

For sufficiently large R_0, the solution is approximately (11.13); that is,

$$\bar{c}(\bar{r}, \bar{t}) \sim \frac{N}{4\pi D\bar{t}} \exp\left(\rho\bar{t} - \frac{\bar{r}^2}{4D\bar{t}}\right). \tag{11.35}$$

To estimate diffusion coefficients from the experimental data we use the Fisher–Kolmogoroff approximation. In the case of the Fisher–Kolmogoroff equation in one space dimension studied in some detail in Chapter 13, Volume I we saw that the travelling wavespeed is given by $v = 2\sqrt{\rho D}$ where D and ρ are the diffusion coefficient and the linear growth rate respectively. So, a population governed by growth and diffusion alone expands at a rate of $2\sqrt{\rho D}$ after a large time. In the two-dimensional case defined by (11.32)–(11.34) the tumour cell concentration is given by (11.35) if the boundary R_0 is far enough away to approximate an infinite domain. Since experimentally we can track the profile of tumour cells above some detection level c^*, substituting this into the last equation, and solving for D, gives

$$\begin{aligned} D &= \frac{\bar{r}^2}{4\bar{t}}\left[\rho\bar{t} - \ln\left(4c^*\pi D\bar{t}/N\right)\right]^{-1} \\ &\approx \frac{\bar{r}^2}{4\rho\bar{t}^2} \quad \text{for large } \bar{t} \text{ since } \rho\bar{t} \gg \ln\bar{t}. \end{aligned}$$

Let $v = \bar{r}/\bar{t}$ be the velocity of the profile front; then

$$D \approx \frac{v^2}{4\rho}. \tag{11.36}$$

We therefore associate the diffusion coefficients in white and grey matter by $D_w = v_w^2/4\rho$ and $D_g = v_g^2/4\rho$ with the experimentally observed linear velocities v_w and v_g, respectively. To deduce estimates for the diffusion coefficients we need to determine the growth rate of the tumour cells. We now use this approximation (11.36) with the experimental results.

We should introduce here a note of caution. With this derivation, the Fisher–Kolmogoroff estimate implies that a very low growth rate with a large linear velocity gives very high estimates for the diffusion rate for the tumour cells in both grey and white matter. Since the proliferation and motility rates are generally (positively) correlated with malignancy as we show below, the tumour cell populations with high linear velocities could have a high (or at least not small) growth rate. Of course, certain midgrade tumours with small ρ and large D could spawn subpopulations with even larger D's with ρ only increasing marginally, if at all.

Giese et al. (1996b) determined the range expansion of the population over time on myelin. That is, given some threshold of detection c^*, the radius of the detectable tumour cell region was recorded as a function of time. From (11.14) the detectable radius r^* of the tumour cell population is given by

$$r^* \sim 2\sqrt{D\rho}\, t \tag{11.37}$$

for large t. Figure 11.8 is a linear least squares fit of the Giese et al. (1996b) experimental observations of the detectable radius versus time for three glioblastoma cell lines (G-112, G-140, G-168). The slope of the line is taken to be equal to $2\sqrt{\rho D}$. So, with an estimate for the growth rate ρ we can deduce a value for the diffusion coefficient, D, from these linear least squares fits to the data.

In Table 11.1 we give the slopes of the linear fits shown in Figure 11.8. By assuming ρ does not vary when the cells are migrating on the control extracellular matrix material (ECM) or on the myelin, we can use the slopes in Table 11.1 to deduce a relationship between the diffusion coefficient on ECM and myelin. Denote by D_{ECM} and v_{ECM} the diffusion coefficient and linear velocity of the cells on ECM, respectively. Similarly, denote by D_m and v_m the diffusion coefficient and linear velocity of the cells on myelin, respectively. The linear velocities then satisfy

$$v_{\text{ECM}} = 2\sqrt{\rho D_{\text{ECM}}}, \quad v_m = 2\sqrt{\rho D_m} \Rightarrow \frac{D_m}{D_{\text{ECM}}} = \left(\frac{v_m}{v_{\text{ECM}}}\right)^2. \tag{11.38}$$

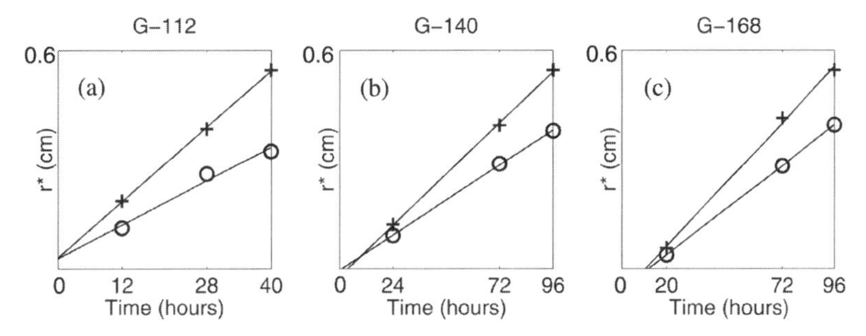

Figure 11.8. Model prediction of the detectable radius r^* for the experimental observations of Giese et al. (1996b) for three different glioblastoma cell lines: (**a**) G-112, (**b**) G-140, (**c**) G-168 on the control substrate; o denotes extracellular matrix (ECM) and + denotes white matter myelin. (From Swanson 1999)

Table 11.1. Velocity v of detectable tumour radius as defined by the slope of linear least squares fit to the Giese et al. (1996a) data shown in Figure 11.8. This suggests there is a 2 to 3 fold difference between the diffusion coefficient on ECM and myelin for these experimental conditions.

Cell Line	Velocity on Myelin (cm/hr)	Velocity on ECM (cm/hr)	Ratio of Diffusion Coefficients in Myelin and ECM
	v_m	v_{ECM}	D_m/D_{ECM}
G-112	2.6×10^{-3}	1.5×10^{-3}	3.00
G-140	1.9×10^{-3}	1.3×10^{-3}	2.14
G-168	3.5×10^{-3}	2.5×10^{-3}	1.96

From the velocity values given in Table 11.1 we deduce that the ratio of the diffusion coefficient on myelin to that on ECM diffusion coefficient is approximately 2. If we associate D_{ECM} with the diffusion coefficient in grey matter then we can suggest that the diffusion coefficient in white matter is twice that in grey matter. Clearly, *in vitro* experimental conditions are different to those observed *in vivo* but from this result we expect at least a twofold difference between the diffusion coefficient in grey and white matters.

As with all experimental results, there is the question as to whether the behaviour observed in the petri dish is analogous to that *in vivo*. To deal with this problem a range of experimental conditions is often considered. In Table 11.2 we give the results of another series of experiments by Giese et al. (1996b). Each row of Table 11.2 corresponds to slightly different experimental conditions defined by the amount of fetal calf serum supplied to the cells. With increasing concentrations of serum (Experiments $1 \rightarrow 4$), the growth rate of the tumour cells increases with serum concentration but the motility reaches a maximum and then decreases. From the various experimental conditions we now use the approximation for the rate of expansion given by the Fisher–Kolmogoroff approximation for the wavespeed for large t, namely, $2\sqrt{\rho D}$ (essentially the derivative with respect to t of (11.37)) to deduce the diffusion coefficient for the glioblastoma cells from $D = v^2/4\rho$.

Table 11.2. Giese et al. (1996b) experimental observations of the detectable tumour radius velocity v_{ECM} in ECM and the growth rate ρ. From v_{ECM} and ρ, the diffusion coefficient is deduced using the Fisher–Kolmogoroff approximation $D = v^2/4\rho$. Each row corresponds to slightly different experimental conditions (defined by the amount of fetal calf serum supplied to the cells).

Experiment Number	Velocity on ECM v_{ECM} (cm/hr)	Growth Rate ρ (1/day)	Diffusion Coefficient D_{ECM} (cm^2/hr)
1	6×10^{-4}	.075	3×10^{-5}
2	1.2×10^{-3}	.1	9×10^{-5}
3	1.55×10^{-3}	.2	7×10^{-5}
4	1.15×10^{-3}	.575	10^{-5}

Table 11.3. *In vitro* parameter estimates for mid-grade (anaplastic astrocytoma) to high-grade (glioblastoma multiforme) gliomas. (From Swanson 1999)

Parameter	Symbol	Range of Values	Units
Linear velocity on ECM	v_{ECM}	$0.6 - 2.1 \times 10^{-3}$	cm/hr
Linear velocity on myelin	v_m	$1.8 - 3.0 \times 10^{-3}$	cm/hr
Diffusion coefficient on ECM	D_{ECM}	$1.0 - 9.0 \times 10^{-5}$	cm^2/hr
Diffusion coefficient on myelin	D_m	$> 0.2 - 2.0 \times 10^{-4}$	cm^2/hr
Net growth rate	ρ	$0.075 - 0.575$	day^{-1}

We now incorporate all of the above results to determine parameter estimates for glioma cells *in vitro*. Table 11.3 summarizes the various estimates we have obtained for the *in vitro* experimental conditions discussed above.

11.4 Tumour Invasion in the Rat Brain

Rats are commonly used to study the *in vivo* dynamics of tumour spread. With gliomas, typically a tumour is implanted in the cortex of the rat brain and allowed to grow and spread. At some later time, the rat is sacrificed and the tumour dynamics are approximated. Experimental studies of gliomas in rats provide estimates for the minimum linear velocity of tumour cells. Although glioma cells do not travel a linear path, if cells are identified at some distance from the original implantation location at sacrifice, the cells must have traveled at some minimum linear velocity to reach that distance. We can take this linear velocity to correpond to the velocity of the detectable radius of the tumour where the threshold concentration for detection is very low. The model equation describing tumour cell dynamics *in vivo* in rats is the spatially heterogeneous model derived in Section 11.2 and given by equation (11.4) in dimensional form.

In Vivo Parameter Estimation for the Rat Brain

Using time-lapse video microscopy and other techniques, Chicoine and Silbergeld (1995) and Silbergeld and Chicoine (1997) quantified brain tumour cell motility and invasion capabilities *in vivo* in rats. Their results suggest a minimum linear velocity of 4.8 μm/hr *in vivo* in rats.

We are particularly interested in the variation in parameter values in white and grey matter regions. A mean linear velocity of 36 μm/hr and 70 μm/hr in grey (v_g) and white (v_w) matter, respectively, has been suggested (D.L. Silbergeld, personal communication 1998). These values are higher than the estimates given in the last section but represent the evolution of new experimental techniques for assessing motility. This highlights an ongoing problem in determining definitive parameter estimates. Even with the best possible theoretical techniques we have to rely on the ever-changing best experimental techniques and so parameter estimates inevitably change.

Glioma cells do not travel a linear path which is why we model glioma motility by a random walk. To relate linear velocity v, proliferation ρ and a random walk diffusion

Table 11.4. Parameter estimates for the rat model using the Fisher–Kolmogoroff approximation (11.36) with the linear velocities from D.L. Silbergeld (personal communication 1998) and the doubling time from Alvord and Shaw (1991).

Parameter	Symbol	Range of Values	Units
Linear velocity in grey matter	v_g	36 ± 12	μm/hr
Linear velocity in white matter	v_w	70 ± 15	μm/hr
Diffusion coefficient in grey matter	D_g	$.008 - 3.3$	cm^2/day
Diffusion coefficient in white matter	D_w	$.032 - 12.0$	cm^2/day
Tumour doubling time	t_d	$\frac{1}{4} - 12$	months
Net growth rate	ρ	$0.001 - 0.1$	day^{-1}

D we again use the Fisher–Kolmogoroff approximation $D \approx v^2/4\rho$ from (11.36) as was also used by Burgess et al. (1997). We again associate the diffusion coefficients in white and grey matter by $D_w = v_w^2/4\rho$ and $D_g = v_g^2/4\rho$ with the experimentally observed linear velocities v_w and v_g, respectively. To deduce estimates for the diffusion coefficients we need to determine the growth rate of the tumour cells. Alvord and Shaw (1991) cite doubling times of one week to one month for gliomas *in vivo*. The resulting parameter estimates for rats *in vivo* are presented in Table 11.4. With these, using the Fisher–Kolmogoroff estimate (as described in the last section) with a very low growth rate and with a large linear velocity gives very high estimates for the diffusion rate for the tumour cells in both grey and white matter. Since the proliferation and motility rates are generally (positively) correlated with malignancy as we see from the above, the tumour cell populations with high linear velocities could have a high (or at least not small) growth rate. So, the very large diffusion coefficients given in Table 11.4 may not be that accurate.

Numerical Simulations of Tumour Invasion in the Rat Brain

In earlier studies of untreated tumour invasion (Burgess et al. 1997) the brain was considered homogeneous. Swanson (1999) investigated the increased invasion capabilities associated with the introduction of heterogeneity as a consequence of the white and grey matter. We first focused on the simple rat brain topology illustrated in Figure 11.9. For illustrative purposes she assumed a diffusion coefficient in white matter of approximately 10 times that in grey matter. From the parameter estimates in Table 11.4, experimental results indicate a diffusion coefficient in white matter from 2 to 100 times greater than the diffusion coefficient in grey matter.

It is generally accepted that the corpus callosum is a common pathway for invasion of tumour cells to the contralateral hemisphere. In the coronal plane of the simpler rat brain topology in Figure 11.9, the corpus callosum is a compact arch of white matter fibres connecting the left and right cerebral hemispheres. In the simulations we approximate the corpus callosum of the rat brain by the regular area enclosed by the curve in Figure 11.10 that looks like a pair of horns.

In the preliminary simulations the corpus callosum is connected to the grey matter cortex by white matter fibres radiating from the corpus callosum, represented by the

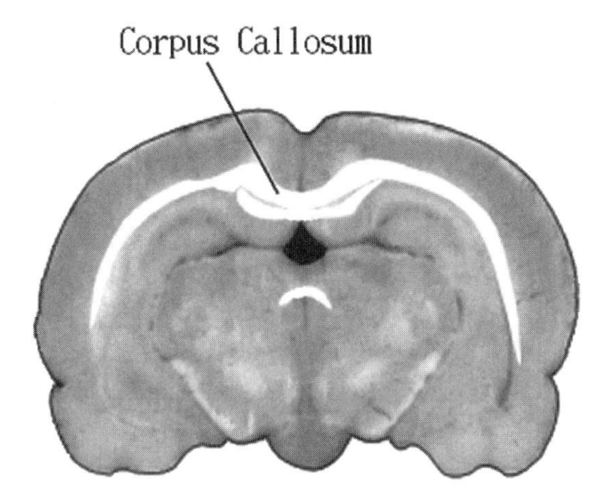

Figure 11.9. Geometry of a coronal slice of a rat brain. White matter appears white and grey matter appears dark grey. The corpus callosum is the arch of white matter connecting the left and right cerebral hemispheres. The width of the brain is about 1.5 mm. (Redrawn from Toga et al. 1995)

boomerang shape, as shown in Figure 11.10(a)–(d). Remember that the corpus callosum is an extensive bundle of white matter fibre. These fibre tracts running from the corpus callosum to the cortex are much narrower than the corpus callosum. They are represented as straight lines extending from the corpus callosum. In the human brain some of these (curvilinear) tracts connecting the cortex to the corpus callosum are identified by a thin trail of cancer cells. Swanson (1999) simulated the model system defined by (11.10) and (11.11) with zero flux boundary conditions and continuity and conservation conditions discussed in Section 11.2 to demonstrate how the tumour invasion is facili-

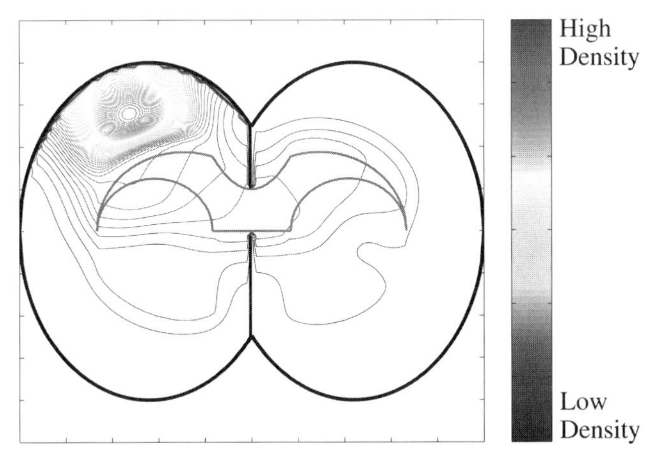

Figure 11.10. A small tumour was introduced and allowed to diffuse according to the model. There is clear enhanced invasion via white matter tracts; the corpus calossum fairly quickly fills with cancer cells. The model parameters are: $D_g = 1.3 \times 10^{-3}$ cm^2 day^{-1}, $D_w = 10D_g$, $\rho = 1.2 \times 10^{-2}$ day^{-1}. (From Swanson 1999) (The scale is a black and white copy of the original colour scale (blue at low density via yellow to red at high density). The interpretation is an increase in density from the outer lines towards the tumour origin.)

tated by these thin fibrous tracts and the corpus callosum. These fibres help to facilitate the invasion by the tumour cells to more distant regions more quickly. The review article by Giese and Westphal (1996) shows some examples where the tumour spread has been facilitated by the corpus collosum in the human brain.

The simulations in Figure 11.10, in which a small tumour was introduced and allowed to diffuse and grow according to the model, show a dramatic increase in invasion capability by the presence of white matter to invade more distant locations. In a relatively short time tumour cells essentially fill the corpus callosum as is observed experimentally and also clinically in the human brain (see Giese and Westphal 1996).

It is commonly noted in experiments that tumour cells migrate across the corpus callosum laterally. It is because of this lateral invasion, that it is thought that gliomas cells move preferentially along the fibres of the white matter corpus callosum (see, for example, Giese and Westphal 1996). The simulations in Figure 11.10 show that anisotropy in the diffusion is *not* necessary to result in a bulk lateral movement of the tumour margin across the corpus calossum. It can be argued that the different diffusivi-

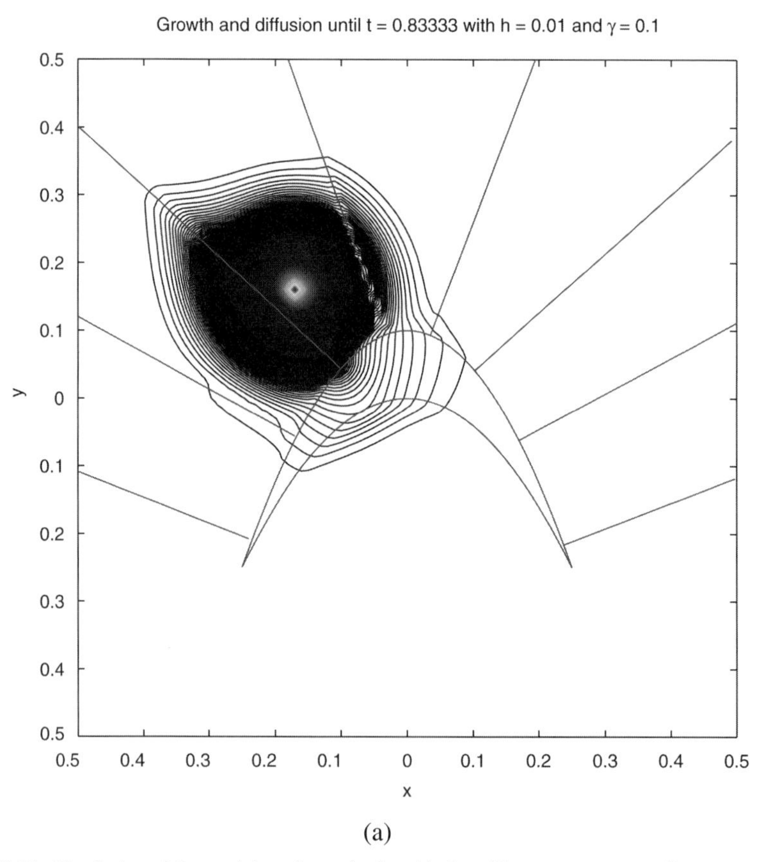

(a)

Figure 11.11. Simulation of the model on the rat brain with the white matter corpus callosum present. Parameter values are: $D_g = 1.3 \times 10^{-3}$ cm^2 day^{-1}, $D_w = 10D_g$, $\rho = 1.2 \times 10^{-2}$ day^{-1} at times (**a**) $t \approx 0.83$, (**b**) $t \approx 1.67$, (**c**) $t \approx 2.5$, (**d**) $t \approx 3.3$. (From Swanson 1999)

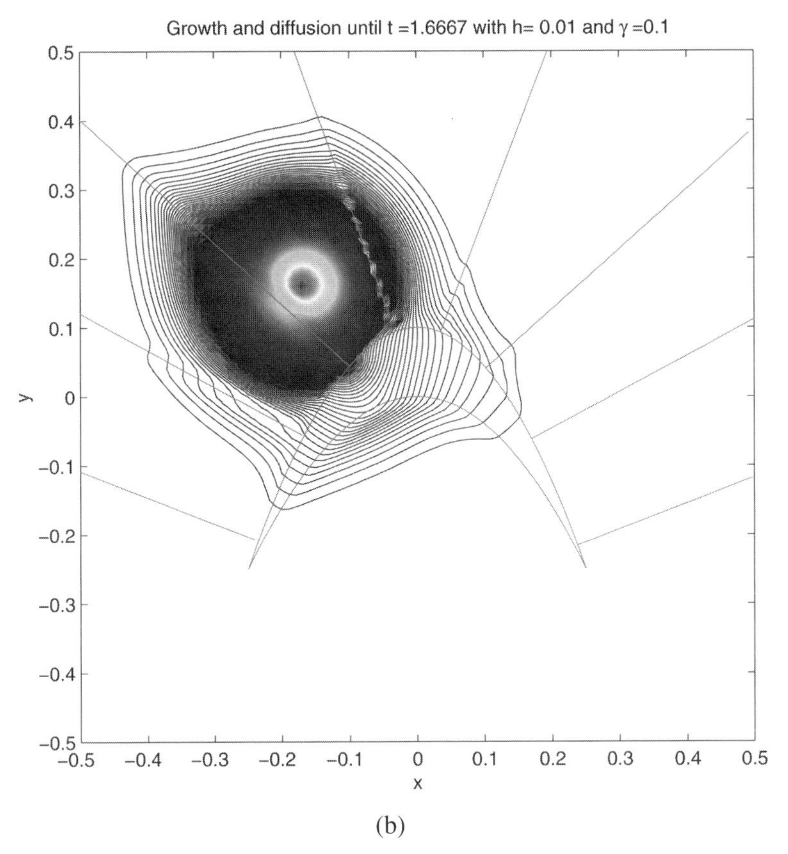

(b)

Figure 11.11. (continued) (See the comment in parentheses in Figure 11.10.)

ties in the grey and white matter are a kind of anisotropy. This lateral invasion is accentuated by the physical structure of the brain in the region of the corpus callosum—the ventricles below and the fissure of the cortex above the corpus callosum help to facilitate the lateral invasion.

Figure 11.11 is a simulation of the rat brain with anatomical boundaries, along with spatially heterogeneous grey and white matter distributions as given in Figure 11.9. Because of the restriction of the rat brain cortex, there is an accumulation of tumour cells near the boundaries. In addition, since the corpus callosum is situated between the ventricles and the cortex we expect to see a marked increase in the invasion of cells across the corpus callosum to the contralateral hemisphere.

11.5 Tumour Invasion in the Human Brain

The simple geometry of the rat brain is a good model for fetal human brain topology where the cortex is smooth. The complex geometries of the adult human brain only accentuate the effects we have seen in the rat brain geometry. We expect increased bound-

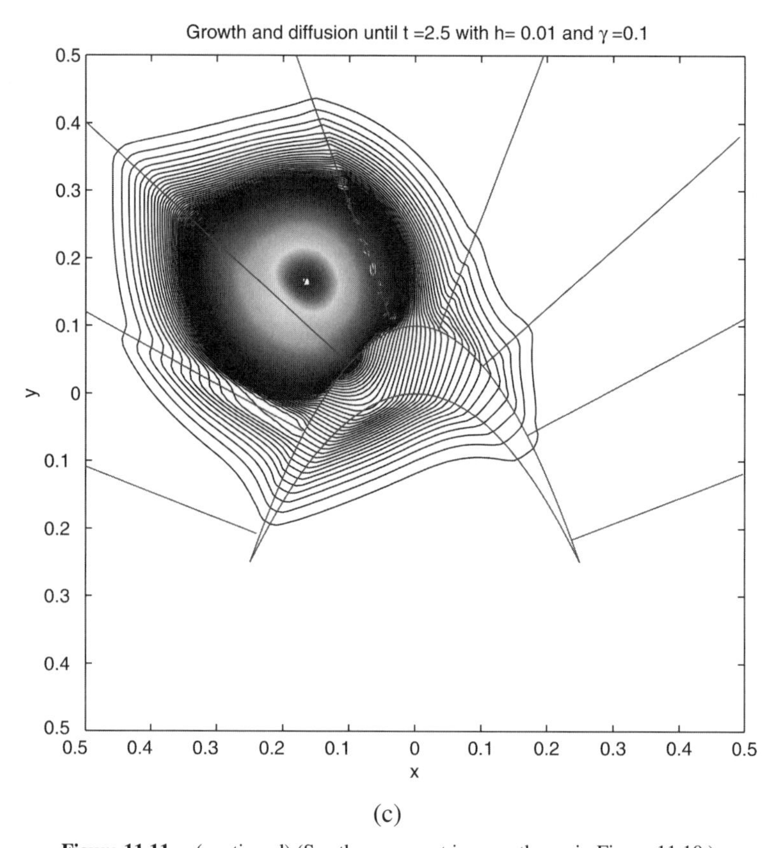

(c)

Figure 11.11. (continued) (See the comment in parentheses in Figure 11.10.)

ary effects and more complex patterns of invasion due to the heterogeneous distribution of grey and white matter. We now discuss the model application to the human brain geometry and investigate the implications of heterogeneous diffusion on the spread of an introduced virtual tumour.

A major development which let us apply our model of tumour invasion to the human brain was provided by the BrainWeb database[1] (Collins et al. 1998) and EMMA, mentioned above. Briefly, EMMA (Extensible MATLAB Medical Analysis) is a tool for manipulating medical images. The Brain Web database was created using an MRI simulator and defines the locations and distribution of grey and white matter in the brain in three spatial dimensions on a $181 \times 217 \times 181$ grid. It was created to visualize frozen images in MATLAB so that they could be manipulated and studied. In this section, we describe some of the numerical simulations of the model system (11.10) and (11.11) presented by Swanson et al. (2000). The results are for two-dimensional slices of the brain for ease of representation. Three-dimensional studies have been carried out and some of these will be discussed below.

[1] http://www.bic.mcgill.ca/brainweb.

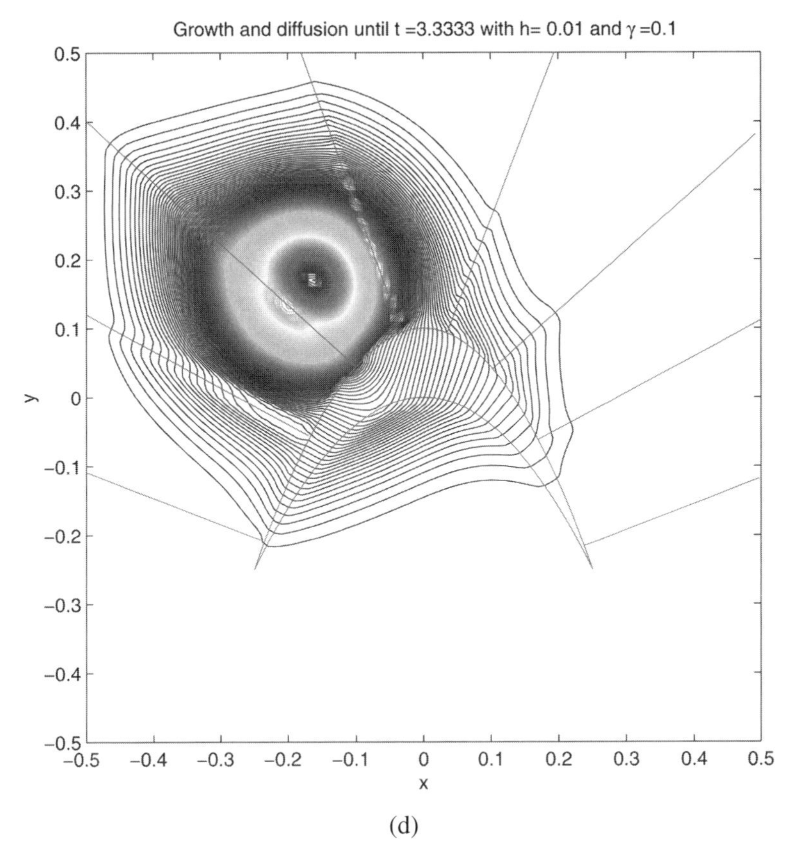

(d)

Figure 11.11. (continued) (See the comment in parentheses in Figure 11.10.)

Parameter Estimates

Parameter estimates are obtained from data on mid to high grade astrocytomas (glioblastoma multiforme). High grade astrocytomas account for $\sim$ 50% of all astrocytomas, which in turn make up $\sim$ 50% of all primary brain tumours (Alvord and Shaw 1991).

We use the Alvord and Shaw (1991) estimate of one week to 12 months as the doubling times for gliomas. So, for cell proliferation of a high grade astrocytoma, we take a growth rate of $\rho = 0.012$/day which corresponds to a 60-day doubling time.

In the last section, we discussed the marked motility of glioma cells *in vitro* and *in vivo* in rats and used experimental data of Chicoine and Silbergeld (1995) and Silbergeld and Chicoine (1997) to estimate the parameters in the model.

For the human brain model, we are interested in the variation in parameter values in white and grey matter regions. Practically, the data describing the tumour growth and invasion for a patient are obtained from medical images such as CT scans and MRI (magnetic resonance imaging). As mentioned these scans approximately define the profile of the *detectable* portion of the tumour. While Figure 11.5 is the idealized form of a CT scan an actual scan is very much less clear. This study originally began with analysis

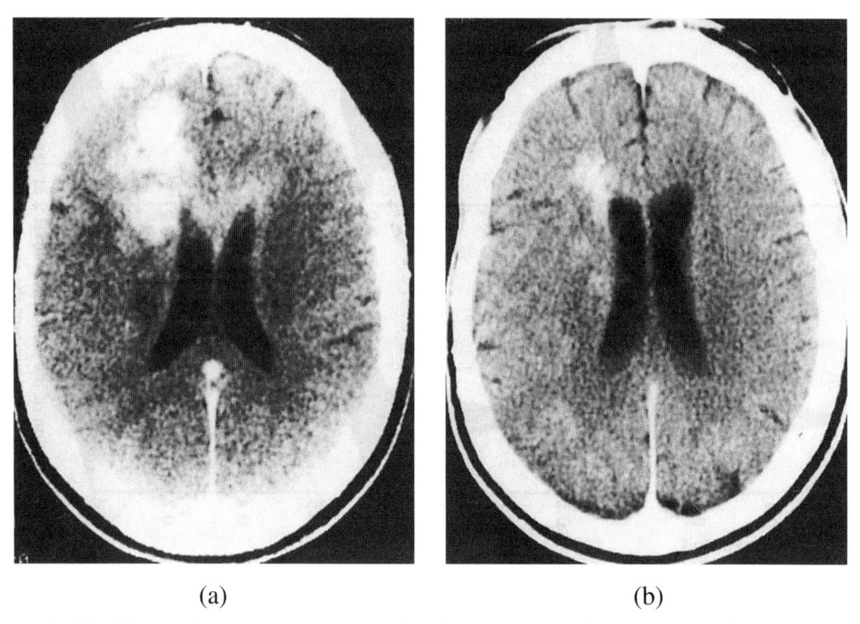

(a) (b)

Figure 11.12. CT scans during the terminal year of a patient with anaplastic astrocytoma who was undergoing chemotherapy and radiation treatment. The image on left was taken approximately 180 days after the image on the right. (Figures from Tracqui et al. 1995)

of serial CT scans during the terminal year of a patient with anaplastic astrocytoma. The patient underwent various chemotherapy and radiation treatments which we discuss in detail in Section 11.10; an example of one of these scans is shown in Figure 11.12.

As in the previous section, we use the Fisher–Kolmogoroff approximation to approximate D in terms of the linear velocity v and proliferation ρ. We then have $D_w = v_w^2/4\rho$ and $D_g = v_g^2/4\rho$ with the experimentally observed linear velocities v_w and v_g, respectively. We now have to determine v_w and v_g for a given patient. The most accessible information regarding tumour infiltration for a human patient is given by CT, MRI or other imaging scans. We use the CT scans used in the development of the original model (see Figure 11.12) to determine the velocities of the tumour front in grey and white matter. Within the right hemisphere the margin of detectable tumour moved about 1.5 cm in 180 days (Tracqui et al. 1995, Woodward et al. 1996), that is, an average speed of $v = 8.0 \times 10^{-3}$ cm/day. For the growth rate $\rho = 0.012$/day, the Fisher–Kolmogoroff approximation then gives the diffusion coefficient $D = v^2/4\rho = 1.3 \times 10^{-3}$ cm^2/day. Due to the proximity of the invasion to the deep cerebral nuclei (grey matter in the right hemisphere), we associate this value with grey matter diffusion so $v_g \approx 8.0 \times 10^{-3}$ cm/day (Tracqui et al. 1995) and $D_g \approx 1.3 \times 10^{-3}$ cm^2/day. From the CT scans, the speed of advance of the tumour margin across the corpus callosum (white matter) was two to three times as fast as that in (predominately) grey matter so we estimate $v_w > 2v_g \approx 1.6 \times 10^{-2}$ cm/day and $D_w > 4D_g \approx 4.2 \times 10^{-3}$ cm^2/day. In our simulations, we generally assumed a five-fold difference in the diffusion coefficient in grey and white matter $D_w = 5D_g$.

Table 11.5. Parameter estimates from patient data for a high-grade human glioma. (From Swanson 1999)

Parameter	Symbol	Range of Values
Linear velocity in grey matter	v_g	8.0×10^{-3} cm/day
Linear velocity in white matter	v_w	$> 1.6 \times 10^{-2}$ cm/day
Diffusion coefficient in grey matter	D_g	1.3×10^{-3} cm^2/day
Diffusion coefficient in white matter	D_w	$> 4.2 \times 10^{-3}$ cm^2/day
Tumour doubling time	t_d	2 months
Net growth rate	ρ	1.2×10^{-2} day^{-1}

We can also estimate the diffusion coefficient from the linear velocity of individual glioma cells by using the root-mean-square distance (in two dimensions) from the origin,

$$\langle \bar{r}^2 \rangle = 4D\bar{t}.$$

Since growth is not included in this definition, the value of D, calculated from this formula using our experimental data regarding the progression of the tumour profile, would be an overestimate of the actual diffusion coefficient. Although only valid for large $\bar{t}$, the Fisher–Kolmogoroff estimate is probably a better approximation for the glioma diffusion coefficients. Parameter estimates (patient generated) for a high grade glioma are listed in Table 11.5.

Numerical Simulations of the Model System: Invasion of Virtual Tumours

Solving the model equations on an anatomically accurate domain is a fairly complicated numerical undertaking. For our simulations (Swanson 1999, Swanson et al. 2000) we allow a 10-fold variation in ρ and D to simulate different tumour grades: high grade (high ρ and high D), intermediate grade (high ρ and low D or low ρ and high D) and low grade (low ρ and low D). Woodward et al. (1996) found that this range accords very well with the various types of glioma tumour growth and invasion. The numerical simulation procedure lets us track the invasion of virtual tumours of any initial size and distribution from any site.

Figure 11.13 shows a coronal section of the human brain obtained with EMMA as the domain for our model simulations. In the figure, white and grey matter appear white and grey. The three tumour sites we specifically consider are labeled in Figure 11.13: Position 1 represents an inferior fronto-parietal tumour, Position 2 corresponds to a superior fronto-parietal tumour and Position 3 a temporal lobe tumour. At each of these locations we consider the 4 tumour grades representing 10-fold variations in the growth rate ρ and the diffusion coefficient D: High Grade (high ρ, high D), Intermediate Grade (high ρ, low D or low ρ, high D) and Low Grade (low ρ, low D).

Tumour Invasion as a Function of Initial Tumour Location

Due to the heterogeneity of the brain tissue composition and geometry of the cortex, which is very clear in Figure 11.13, the dynamics of invasion can be very different

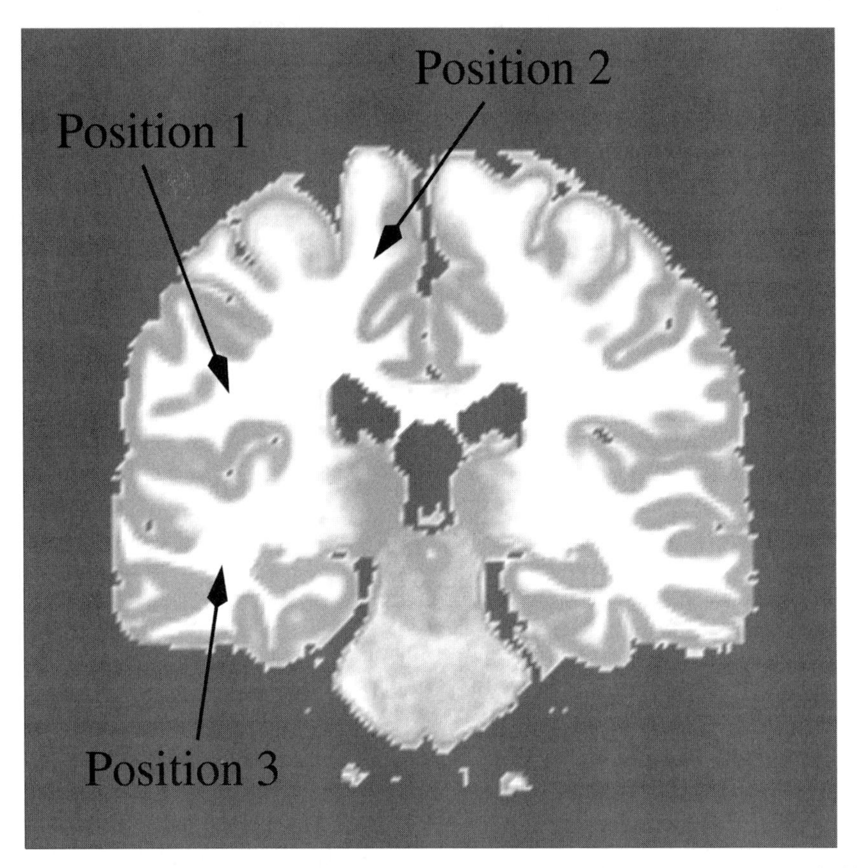

Figure 11.13. Computational domain with tumour locations. Grey and white matter appear grey and white. Position 1 is the area location of an inferior fronto-parietal tumour, Position 2 is the location for a superior fronto-parietal tumour, while Position 3 is for a temporal lobe tumour.

depending on the location of the bulk tumour. For a given tumour grade, that is, fixed ρ and D, the extent of tumour invasion is different at each tumour position as seen in Figures 11.14–11.16, and depends on the local distribution of grey and white matter and the proximity to anatomical boundaries.

Figure 11.15 represents simulations of a high grade (high ρ, high D) tumour at Position 2 (defined in Figure 11.13) within the brain. Figure 11.15(a) represents the portion of the detectable tumour on enhanced CT at diagnosis. Here we assume that diagnosis occurs when the detectable tumour, that is, the portion of tumour with density above the threshold of detection, covers an area equivalent to a circle of diameter 3 cm. Although the tumour looks fairly localized, by increasing our detection abilities by a factor of 20 (mathematically we can set the threshold at any positive value), which corresponds to 500 cells/cm^2, we see in Figure 11.15(c) that the tumour has dramatically invaded throughout the right cerebral lobe and across the corpus callosum to the contralateral hemisphere. After 140 days, Figure 11.15(b) represents the portion of the tumour detectable on enhanced CT at the time of death. Here death is assumed to occur when the

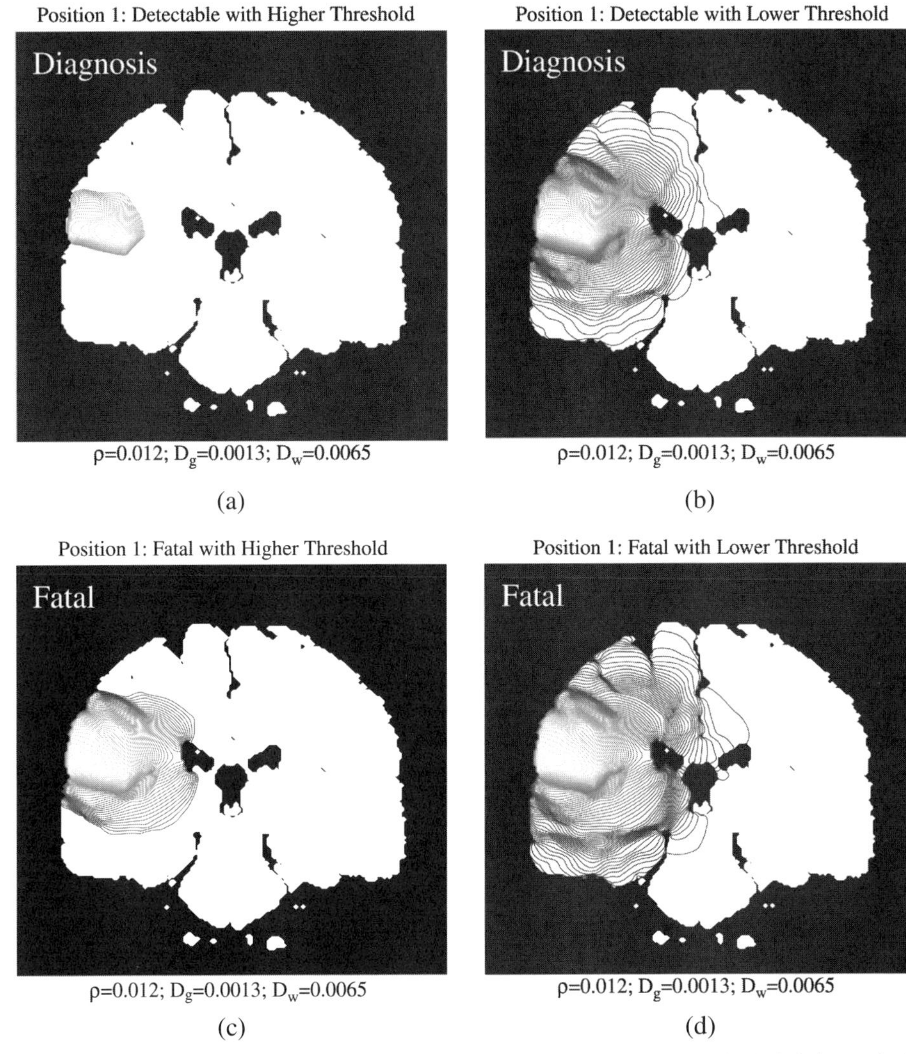

Figure 11.14. Simulation of tumour invasion of a High Grade Glioma in Position 1 in the inferior cerebral hemisphere: (**a**) and (**b**) at diagnosis; (**c**) and (**d**) at death; (**a**) and (**c**) as seen on the CT scan; (**b**) and (**d**) as calculated out to 5% of the threshold (boundary) cell concentration defined by CT. (From Swanson et al. 2000) (See the comment in parentheses in Figure 11.10.)

detectable tumour (again the portion of tumour with density above the threshold of detection) covers an area equivalent to a circle of diameter 6 cm. By increasing detection capabilities to correspond to a resolution of 500 tumour cells/cm^2, Figures 11.15(b) and (d) clearly show the augmented invasion not apparent with the present technology of enhanced CT. Note that Figures 11.15(b) and (d) do not represent the full extent of tumour invasion but rather an increase in detection ability associated with a theoretical imaging technique.

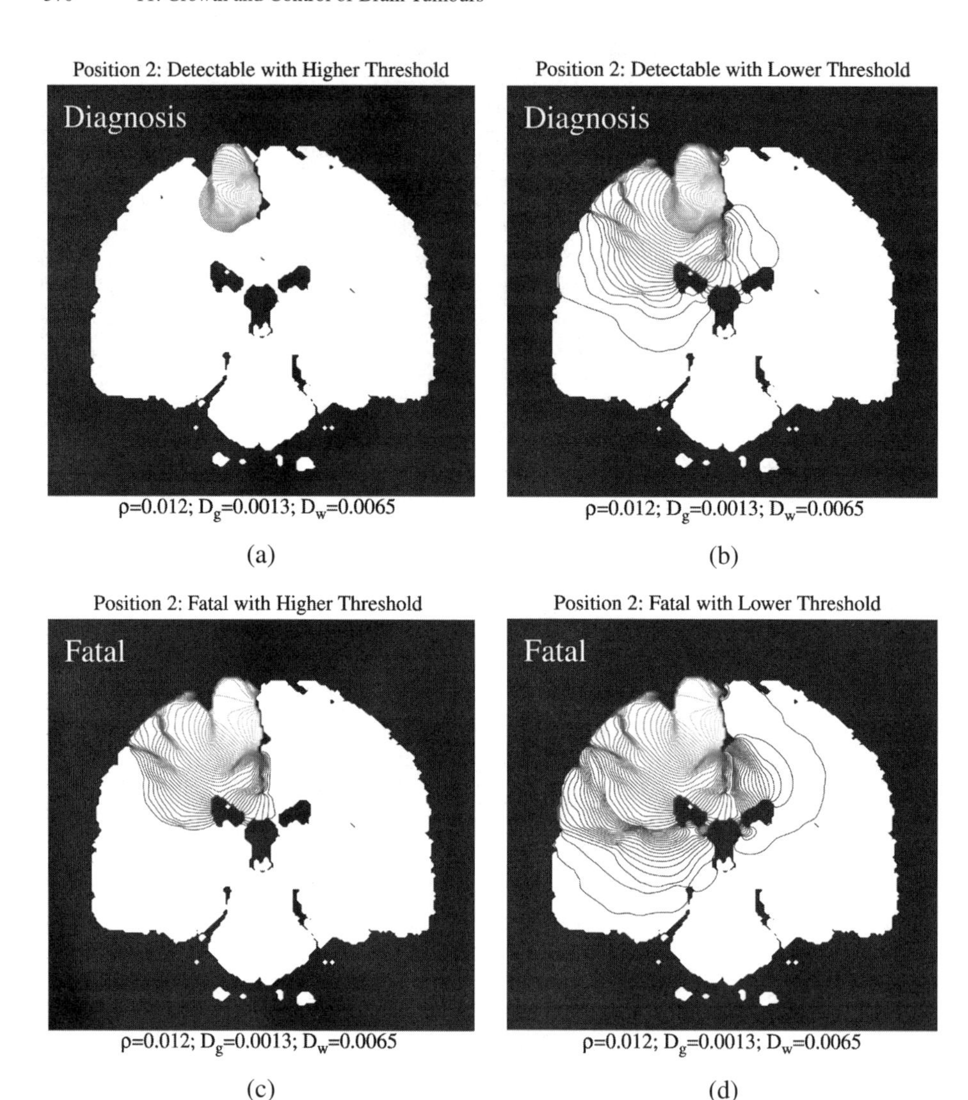

Figure 11.15. Simulation of tumour invasion of a High Grade Glioma in Position 2 in the superior cerebral hemisphere: (**a**) and (**b**) at diagnosis; (**c**) and (**d**) at death; (**a**) and (**c**) as seen on the CT scan; (**b**) and (**d**) as calculated out to 5% of the threshold (boundary) cell concentration defined by the CT scan. (From Swanson et al. 2000) (See the comment in parentheses in Figure 11.10.)

Swanson (1999) also compared the simulated tumour invasion in the case of a high grade tumour initially at Position 3 in Figure 11.13 with clinically observed invasion from microscopic postmortem analysis of the entire brain by Burger et al. (1988) and found shapes similar to the subthreshold-CT invasion depicted by the high grade tumour simulated invasion shown in Figures 11.16(b) and (d).

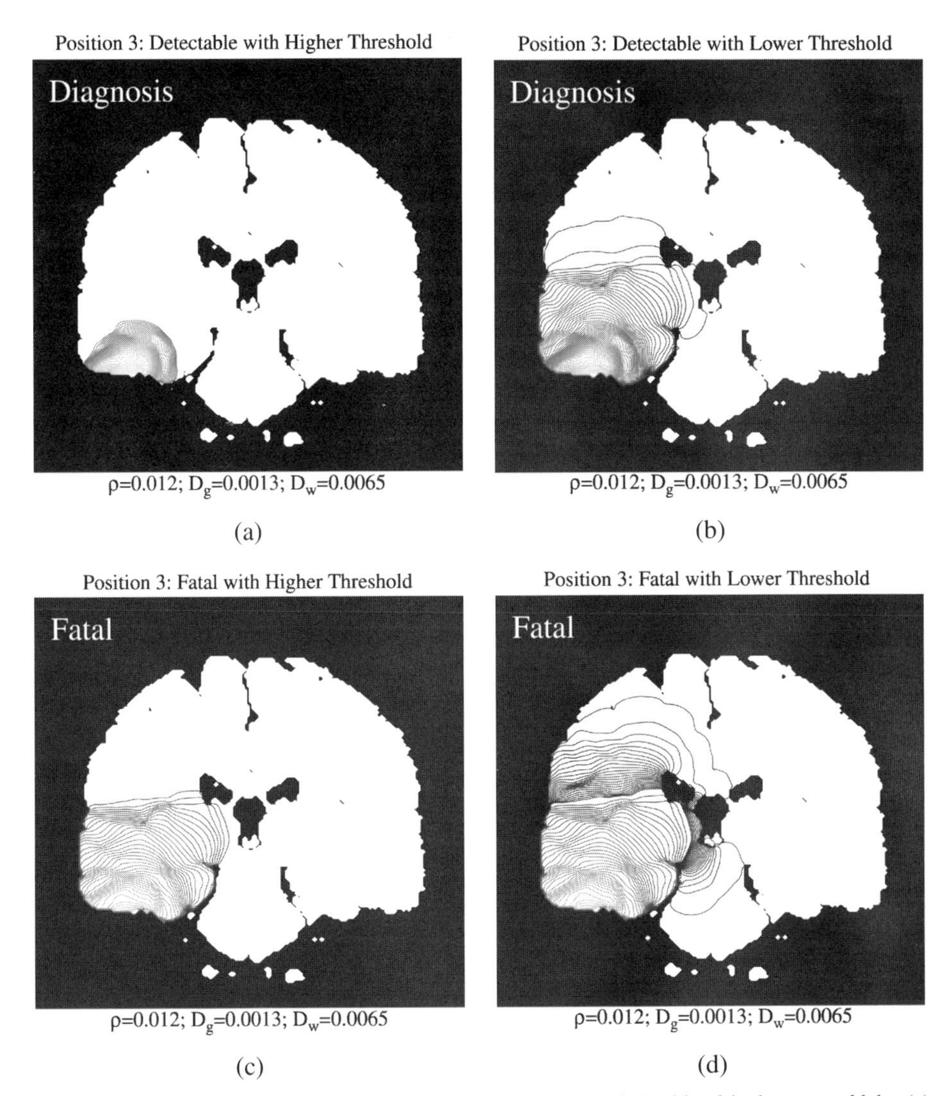

Figure 11.16. Simulation of tumour invasion of a High Grade Glioma in Position 3 in the temporal lobe: (**a**) and (**b**) at diagnosis; (**c**) and (**d**) at death; (**a**) and (**c**) as seen on the CT scan; (**b**) and (**d**) as calculated out to 5% of the threshold (boundary) cell concentration defined by the CT scan. (From Swanson et al. 2000) (See the comment in parentheses in Figure 11.10.)

Untreated Survival Time as a Function of Tumour Location

Table 11.6 quantifies the effect of tumour position on survival time for the 4 hypothetical tumour grades representing 10-fold variations in the growth rate, ρ, and diffusion coefficient, D_g and D_w, parameters.

In Table 11.6, the time to diagnosis is defined to be when the detectable tumour covers an area of $\pi(1.5)^2$ cm^2, that is, equivalent to a circular tumour of radius 1.5 cm. The time of death is assumed to be when the detectable tumour covers an area of $\pi(3)^2$ cm^2, which is equivalent to a circular tumour of radius 3.0 cm. The clinically definable sur-

Table 11.6. Survival periods for various grades of gliomas at different sites in the human brain given in Figure 11.13. The diffusion coefficient in white matter, D_w, is 5 times the diffusion coefficient in grey matter, D_g, for each tumour grade. (From Swanson 1999)

	Growth Rate ρ (1/day)	Diffusion Coefficient Grey Matter D_g (cm^2/day)	Survival Period t_S (day) Position		
Grade			1	2	3
High (HH)	1.2×10^{-2}	1.3×10^{-3}	109.7	137.5	172.7
Intermediate (HL)	1.2×10^{-2}	1.3×10^{-4}	398.2	494.9	581.9
Intermediate (LH)	1.2×10^{-3}	1.3×10^{-3}	55.5	259.3	347.2
Low (LL)	1.2×10^{-3}	1.3×10^{-4}	1097.2	1375.0	1726.9

vival time t_S is the time between death and diagnosis. Note that the diffusion coefficient in white matter is simply five times that in grey matter for each tumour grade listed in Table 11.6. Note also that tumours located in the temporal lobe are associated with increased untreated life expectancy following diagnosis.

On the assumption of a uniform diffusion coefficient in homogeneous brain tissue we obtained an approximate survival in Section 11.2; namely,

$$\text{survival time} \approx \frac{1}{\sqrt{D\rho}}\left(\bar{r}_{\text{lethal}} - \bar{r}_{\text{detect}}\right). \tag{11.39}$$

We can make a rough comparison between the results in Table 11.6 if we take $\bar{r}_{\text{lethal}} - \bar{r}_{\text{detect}} = 1.5$ cm and, in the case of a high grade tumour, $\rho = 1.2 \times 10^{-2}$/day and a diffusion coefficient $1.3 \times 10^{-3} \leq D \leq 6.5 \times 10^{-3}$ cm^2/day, that is, a motility between the grey and white matter. We get an estimate of between 170 days and 380 days. For a low grade tumour with $\rho = 1.2 \times 10^{-3}$/day and $1.3 \times 10^{-4} \leq D \leq 6.5 \times 10^{-4}$ cm^2/day we get between 1698 days and 3798 days. The lower of these survival times is reasonably close to the values given in Table 11.6 for a high and low grade tumour in Position 3. In general this very rough estimate is not too unrealistic if we use the lower of the two (grey and white matter) diffusion coefficients. Burgess et al. (1997) solved the full spherically symmetric three-dimensional model and obtained, for the high grade tumour with $\rho = 1.2 \times 10^{-2}$/day and $D = 1.3 \times 10^{-3}$ cm^2/day, a survival duration of 179 days and for a low grade tumour with $\rho = 1.2 \times 10^{-3}$/day and $D = 1.3 \times 10^{-4}$ cm^2/day of 1796 days. When we compare the approximation with the homogeneous situation the comparison is very good. Position 3 has the best prognosis as to survival duration so, for tumours in other positions the approximate survival duration is very much an overestimate. In the case of a tumour located at Position 1 in the high grade case it overestimates survival duration by approximately 50% and for a low grade tumour by approximately 85%.

From Figure 11.13, we see that the geometry of the temporal lobe white matter tends to lead tumour cells into the deep cerebral nuclei (grey matter), thereby constraining the invasion of the tumour from Position 3 and providing the best prognosis; see Table 11.6. Similarly comparing Figures 11.13 and 11.15, we see that the tumour in Position 2 is constricted superiorly and medially and has direct access to the white matter corona radiata corpus callosum. The tumour in Position 1 has the worst prognosis because it is adjacent to corona radiata white matter and has little interaction with anatomic boundaries, being restricted only laterally.

The results in Table 11.6, as might be expected, indicate that high grade gliomas generally have the worst prognosis while low grade tumours have the best with intermediate grade tumours having intermediate survival times. However. the highly proliferative intermediate grade glioma has a better prognosis than the highly diffuse intermediate grade tumours and that the highly diffuse intermediate grade tumour in Position 1 has a prognosis even worse than that of the high grade gliomas. This suggests that the diffusion abilities of a tumour are a stronger indicator of prognosis than the growth rate. None of the prognoses are encouraging.

We now extrapolate theoretical survival curves based on tumour grade and location. For each tumour grade, we vary growth, ρ, and motility, D, by $\pm 50\%$ from their values in Table 11.6. This defines 9 hypothetical patients for each of the 4 growth and diffusion coefficient combinations. Survival times for tumours at various locations are given in Figure 11.17. Each of the 3 curves for a given tumour grade defines the survival times of the 9 hypothetical patients with untreated gliomas. Figure 11.17(a) represents high grade tumours (HH) defined by a high growth rate ρ and a high diffusion coefficient D. The opposite extreme (LL) is shown in Figure 11.17(d). The intermediate grade populations with high–low (HL) and low–high (LH) combinations of ρ and D are presented in Figures 11.17(b) and (c). Note that temporal lobe tumours (Position 3) consistently have the best prognoses. Although the curves look very similar, the timescales are very different in each plot. Clinical data are, to our knowledge, not available for comparison due to the fact that our simulated tumours are untreated. (We discuss the model amendments to consider various treatments below.) However, as noted above, even following surgical resection we can find now clinical results relating to different sites of gliomas.

The proportion of tumour detectable on enhanced CT decreases with increasing diffusion coefficient and decreasing growth rate, as might be expected. For a fixed growth rate, as diffusion increases, tumour cells migrate larger distances and, so, diffusion tends to smooth out the spatial distribution of glioma cells. For a fixed diffusion coefficient, as the growth rate decreases, tumour cells have less chance to build up to detectable levels. Interestingly, if the ratio of the growth rate to the diffusion coefficient is fixed while ρ and D change, the proportion of tumour detectable is fixed. This type of relationship can be very important in the study of individual patient cases where independent estimates for the diffusion and growth rate parameters are unrealistic.

We use a threshold of detection corresponding to a tumour cell density of 40,000 cells/cm^2. Above this threshold, the tumour is observable on enhanced CT while below it, it is not. Because of the diffuse nature of gliomas, we can only ‘detect’ a small portion of the actual tumour with currently available imaging techniques. This fact has been substantiated by Chicoine and Silbergeld (1995) and Silbergeld and Chicoine (1997) with their cultured tumour cells from histologically normal brain far from the bulk tumour location.

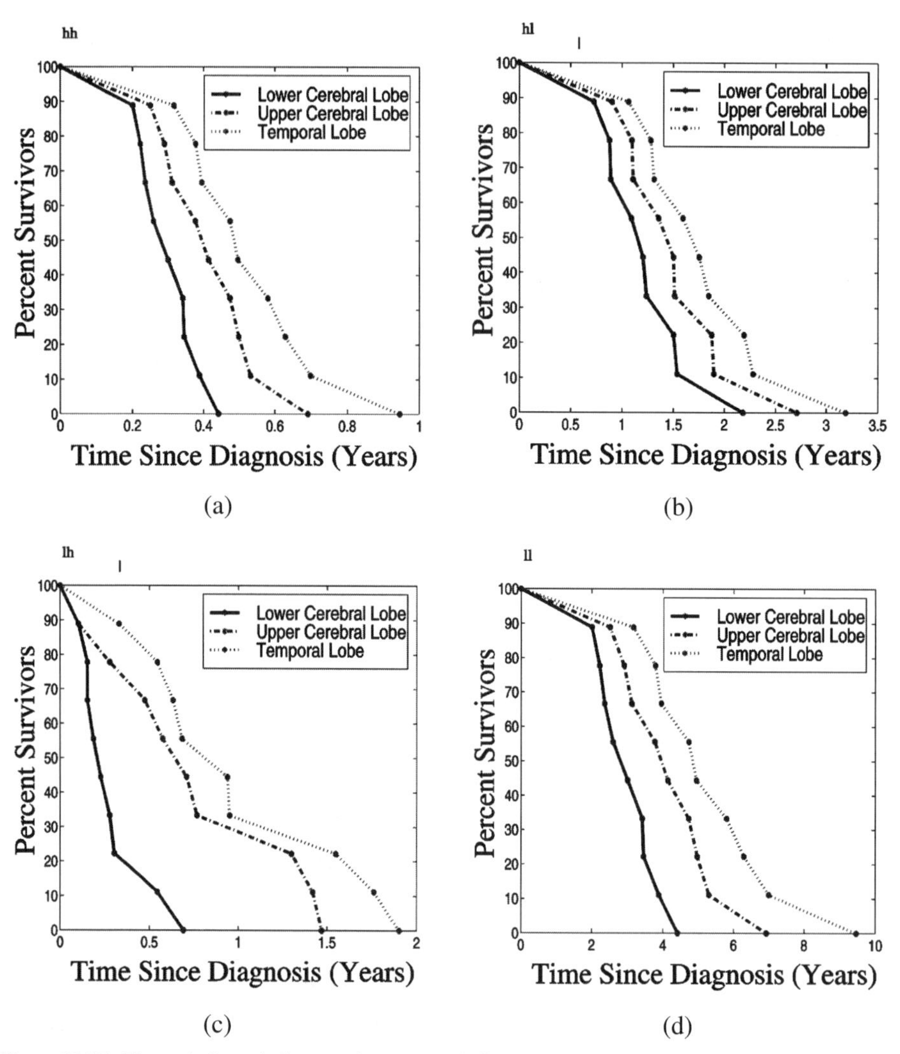

Figure 11.17. Theoretical survival curves for untreated gliomas of all grades: (**a**) represents high grade tumours with high growth (ρ) and high diffusion (D_g, D_w); (**b**) represents intermediate grade tumours with high growth (ρ) and low diffusion (D_g, D_w); (**c**) represents intermediate grade tumours with low growth (ρ) and high diffusion (D_g, D_w); (**d**) represents low grade tumours with low growth (ρ) and low diffusion (D_g, D_w).

An advantage of mathematical modelling is the ability to alter theoretically any parameter and analyze its effect thereby simulating hypothetical experiments. By mathematically decreasing the threshold of detection in our model, we can visualize the usefulness of increasing detection capabilities in medical imaging. Figure 11.18 illustrates the effect of decreasing the threshold of detection on the proportion of the tumour we actually detect. Augmented detection abilities invariably increase the amount of tumour identifiable but the rate of increase varies among tumour positions and grade. Notice that detection of high and low grade tumours coincides. This suggests that the ratio be-

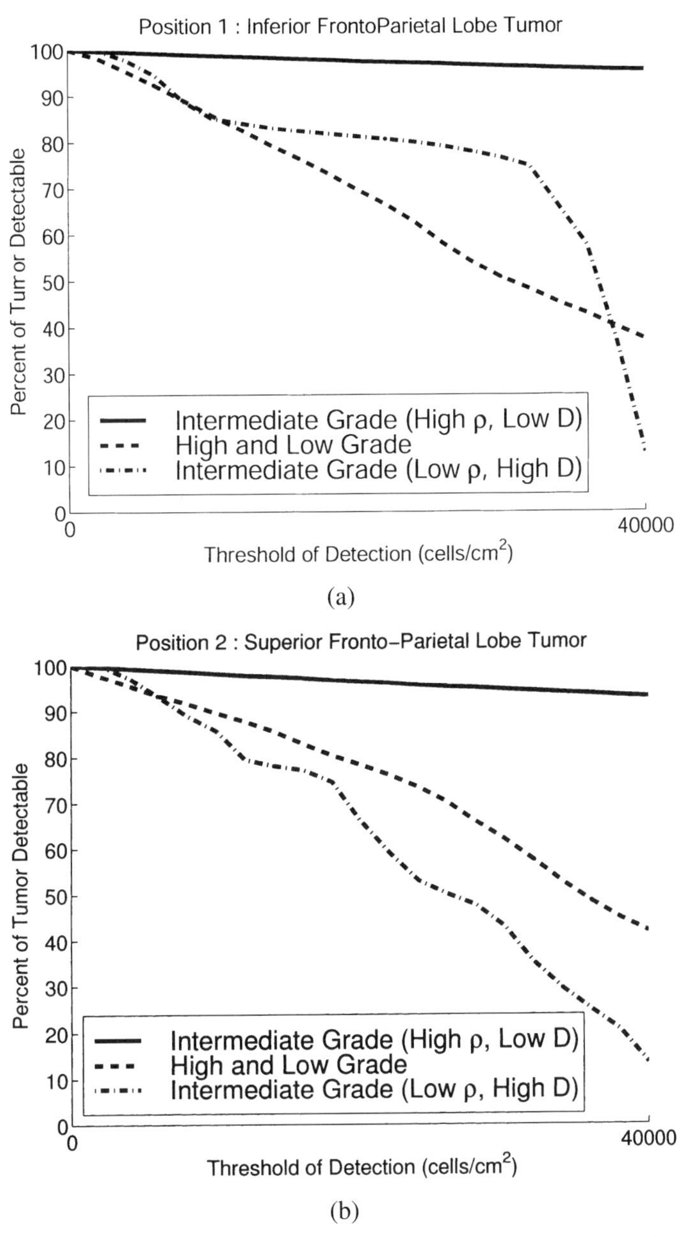

Figure 11.18. Percent of detectable tumour volume as a function of the detection threshold at various positions. The 40,000 cells/cm^2 represents the threshold of detection for a CT scan. (From Swanson et al. 2000)

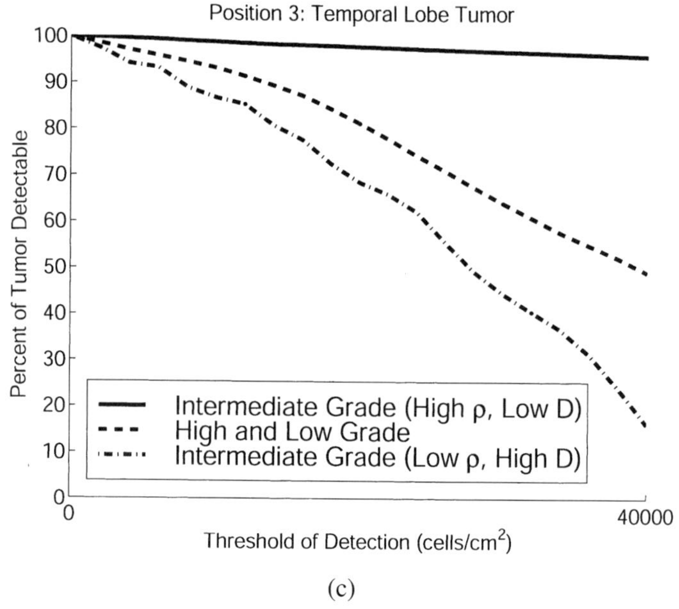

(c)

Figure 11.18. (continued)

tween the growth rate ρ and the diffusion coefficient D is critical to the determination of the proportion of tumour detectable. Again, it suggests that the invasive nature of the tumour is defined by the *ratio* D/ρ. So, another potential practical application of this model for patients is simplified since separate estimates for D and ρ are not necessary. The problem here, of course, is how to obtain an estimate for D/ρ.

Note in Figure 11.18 that the slowly growing and quickly diffusing intermediate grade tumours are associated with the least proportion of tumour that is detectable at diagnosis. This is clearly due to the extensive invasive abilities of those tumours. High and low grade tumours are associated with the same proportion of tumour detectable as a function of the threshold of detection. The difference between the high and low grade tumours is the timescale on which the growth and invasion occurs. It takes high grade tumours a very short time to invade. Low grade tumours may follow the invasion path of high grade tumours but much more slowly.

All of the results presented in this section can be obtained for any other sagittal, coronal or horizontal slice of the brain.

The key question we must now ask is how a three-dimensional simulation of the model might affect these results. The numerical simulation is clearly much more complex. Burgess et al. (1997) investigated the three-dimensional constant diffusion spherically symmetric situation and studied the interaction of growth rates and diffusion coefficients on the spatial spread both with and without resection. They were particularly interested in quantifying the effect of the size of the resected volume on the increased life expectancy. With the more accurate model here, three-dimensional simulations have also been carried out. Figure 11.19 is an example of such a simulation

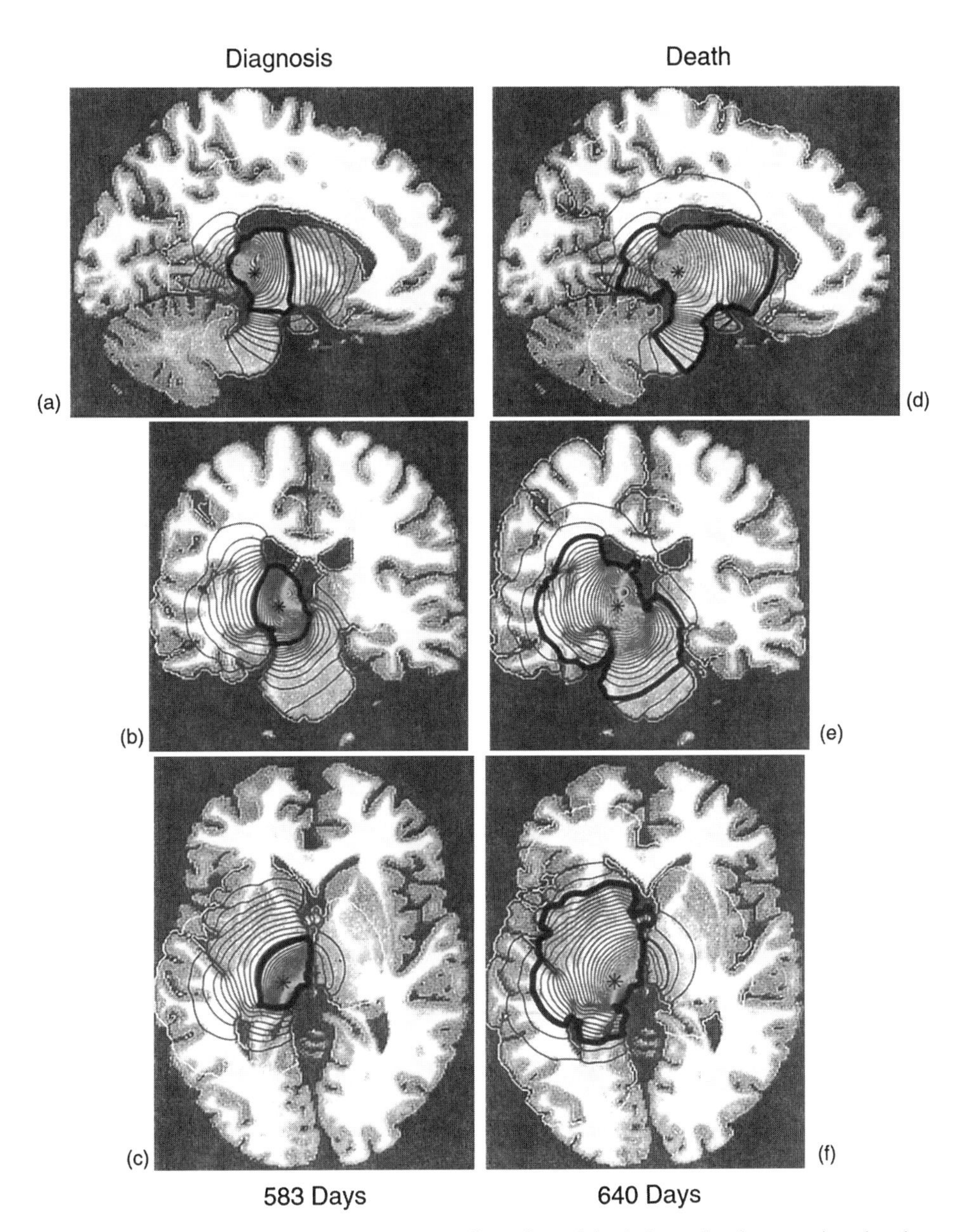

Figure 11.19. The spatial spread of the cancer cells at time of death for a virtual tumour introduced at Position 3 (Figure 11.13), denoted by the asterisk, for three cross-sections through the three dimensional tumour, namely sagittal, coronal and horizontal which intersect at the tumour site. (**a**)–(**c**) are at diagnosis while (**d**)–(**f**) are at death. The dark line defines the edge of the tumour detected by a CT scan in which the threshold is 40,000 cells/cm^2. (From Swanson et al. 2000)

for an untreated tumour initiated at Position 3 (see Figure 11.13) and shows the cell distribution at time of death.

Recap of the Model Assumptions and Limitations

Before discussing various treatment scenarios we should perhaps recall the various assumptions in the modelling. The model assumes that glioma tumour cells grow with a constant exponential rate, ρ. We have not included any necrotic core formation (which can be incorporated by using a logistic growth dynamics for the cells): it is evident in some gliomas, but only the highly proliferative, slowly diffusive tumours are significantly affected by necrosis. In fact, our experimental data were taken from a tumour without necrosis (anaplastic astrocytoma). Tracqui et al. (1995), Woodward et al. (1996) and Burgess et al. (1997) have shown that results from our model with logistic growth for the cells differ little from the model with exponential growth. Tracqui et al. (1995) showed that logistic growth decreased the amount of time before identification of the tumour and decreased the life expectancy as the necrotic cell density decreased.

Diffusion begins after about 23 generations when the tumour contains approximately 4×10^3 cells; at this point we consider the model equation (11.6) to start governing the tumour growth and invasion. Without this limitation no local tumour mass appears to form, corresponding to certain cases of gliomatosis cerebri.

We have taken the diffusion coefficient as approximately 5 times larger in white matter than in grey matter (by way of example). There is probably some anisotropy to the diffusion along individual white matter fibres but the network of fibres is very complicated on the whole brain scale. There are several pathways which help facilitate lateral diffusion such as the corpus callosum and, to a limited degree, fronto-parietally. A change in the overall diffusion coefficient could perhaps be sufficient to simulate the increased motility of gliomas in white matter. There is some justification for this possibility in view of the good comparison the results of Burgess et al. (1997) had with patient data.

The tumour is detectable above a critical density of 40,000 cells/cm^2 corresponding to the detectable threshold on enhanced CT. This value was determined by the analysis of patient data from histological sections obtained at autopsy and terminal CT scans (Tracqui et al. 1995). The tumour is diagnosed, on average, when the detectable area covers $\pi(1.5)^2$ cm^2 (the area equivalent to a 1.5-cm radius circular tumour). The tumour is fatal when it is detected at twice the diameter at diagnosis (that is, covering an area of 9π cm^2). These choices were based on average size of gliomas in a large number of scanned or autopsied patients, but the range is very wide.

Although brain boundaries and tissue heterogeneity are taken into account, we do not define any destruction to the normal tissue by the tumour cell invasion. It is presently a point of contention experimentally whether the tumour cells have the ability to break down the normal brain parenchyma distant from the tumour site and affect the time to death.

Our simulations can be done in both two and three dimensions. Three dimensions are, of course, better and are much easier to visualize with respect to two-dimensional CT or MRI scans. Practically, we expect to fit the model to data obtained from two-dimensional CT scans (schematically shown in Figure 11.5) for patients. The use of two-dimensional computations is certainly not inappropriate.

In spite of the many assumptions involved the model has allowed us to obtain many clinically potentially useful results. In particular, we have identified that temporal lobe tumours may have a better prognosis than fronto-parietal lobe tumours. Although the difference in survival times between temporal and fronto-parietal lobe tumours may be statistically insignificant in clinical studies, it may account for at least some of the clinical variability. It has also let us include a significant measure of heterogeneity in brain tissue to account for asymmetric tumour geometries. We have also been able to simulate clinically observed tumour geometries and suggest survival times of patients. This, of course, requires further clinical analysis to see whether or not they are consistent with clinical data. With the availability of a detailed description of the local composition of grey and white matter throughout the brain, the model has still a lot of unexplored potential some of which we discuss in the following sections.

11.6 Modelling Treatment Scenarios: General Comments

As we have already commented, the prognosis for a patient with a glioma is not encouraging and depends on many factors, among them the type of neoplasm and the grade of malignancy. The relatively simple mathematical model we propose takes into account the growth and diffusion rates of the glioma and tissue heterogeneity. As we have seen, the results in the previous sections are reasonably consistent with clinical data (Cruywagen et al. 1995, Tracqui et al. 1995, Cook et al. 1995, Woodward et al. 1996, Burgess et al. 1997, Swanson et al. 2000), so we believe that the model could be used to make predictions regarding the survival time of the patient following various types of treatment including surgical resection, radiation and chemotherapy.

A major problem with present treatment strategies is the local focus of therapy when the action of the tumour growth and invasion is elsewhere (see, for example, Gaspar et al. 1992 and Liang and Weil 1998). The failure of glioma treatment regimes represents a large portion of glioma clinical and experimental literature (for example, Yount et al. 1998 and other references there). Local therapies are desirable, of course, to reduce the bulk tumour which is mostly responsible for pressure-induced symptoms. However, they can not control the motile invading cells responsible for recurrence (Silbergeld and Chicoine 1997).

The most appropriate treatment for a given glioma tumour is often not at all clear, irrespective of whether or not the subsequent quality of remaining life is taken into account. Because of their invasive characteristics, malignant gliomas can rarely be cured by surgical or radiological resection alone. There is much new research on quite different treatments, such as tumour-attacking viruses. Varying degrees of resection have been shown to increase survival time only marginally for glioblastoma multiforme although the increase is generally more significant for the lower grade (anaplastic) astrocytomas. With the general failure of surgical resection alone, to increase the survival time of patients, multi-modality treatments have been developed combining resection, radiation and chemotherapies and other therapies. Giese and Westphal (1996) in their review article discuss the perspectives for antiinvasive therapy. Numerous clinical studies have attempted to demonstrate the effectiveness of various treatments and combinations (see, for example, Ramina et al. 1999 and numerous other references there). Counterintuitively, some combinations of therapies have been shown, for various reasons, to be less effective than each of the therapies separately. Fairly recently, it was found that ionizing radiation can inhibit chemotherapy-induced cell death (apoptosis) in certain glioblastoma cells (Yount et al. 1998). (There have also been suggestions that antioxidants may in fact be exploited by the cancer cells to help prevent their destruction.) This type of multimodality treatment failure can be attributed to the induced mutation of cells exposed to harsh chemicals or radiation. Cancer cells are, by definition, mutated 'normal' cells, so with the accumulation of mutations, the cancerous cells progressively become more malignant and treatment-resistant. We discuss polyclonal models later when we discuss a modification of the basic model to consider chemotherapy treatment.

Clearly, there is a threshold of treatment sustainable by a given patient and determining the optimal minimal strategies is of major importance. Not only that, even with similar tumour sizes, histologic malignancy and anatomic location, certain treatments work better on some patients than on others. There is no one universal treatment, but in deciding the best course of treatment for patients it is necessary to take into account all of the available information regarding the cancer before proceeding. We believe that realistic modelling can help to address this complicated issue by quantifying certain virtual glioma behaviours with and without treatment.

As we shall see, with relatively slight alterations, our model can take into account the effects of chemotherapy, radiation and resection on the spatiotemporal behaviour of the tumour. This capability lets us compare projected growths and invasion of the tumour under various treatment scenarios, thereby giving some insight into the optimal therapeutic course. Given a sense of the location, size, shape, diffusion coefficient and growth rate of a specific tumour, our model can help to suggest the best type of therapy to maximize survival time, that is, the difference between the time of diagnosis and death.

11.7 Modelling Tumour Resection in Homogeneous Tissue

Earlier uses of our model by Cook (Dr. J. Cook, personal communication 1994), Woodward et al. (1996) and Burgess et al. (1997) considered the effect of resection on patient survival time for the spatially homogeneous model (11.3) in two spatial dimensions. Their analyses suggest that surgery adds little more than two extra months of life. The recurrence (on a CT scan) of the tumour is due to the low density infiltrated tumour cells far afield from the gross tumour site.

Mathematically, surgical resection is simulated by the removal of all tumour cells within a defined location. Death was assumed to occur either when the visible tumour reached a defined radius or when the total cancer cell count reached a critical value. Given an initial tumour size, grade and location these models can then simulate the effects of various resection sizes and geometries on survival time. Woodward et al. (1996) took as initial conditions (11.12) and then, when the tumour had a visibly (CT) detectable area equivalent to a circle of radius 1.5 cm a region was resected. Woodward et al. (1996) estimated the diffusion coefficient differently to the methods discussed above; the method is described in detail in Cruywagen et al. (1995) and Tracqui et al. (1995). They took patient CT scans from three levels through the tumour (idealized in

Figure 11.5) to calculate the tumour area at each level. With the proliferation parameter, ρ, given they then numerically solved the model equation and determined the diffusion coefficient which gave the best fit when the cell density was at the observed threshold on the scans at each level. This gave three estimates for the diffusion coefficient, one at each level and they took the average: interestingly (and fairly crucial in fact) the variation in values for D did not vary markedly for each scan level. We discuss the procedure in more detail in Section 11.10.

With the estimated values for the model parameters, Woodward et al. (1996) then carried out hypothetical resections of different diameters on the tumour when it had reached the CT detectable size of 3 cm in diameter. With this resected state as the initial condition they numerically solved the model equation until the recurring tumour had the equivalent size to the lethal size, that is, of a circle of diameter 6 cm. The shape of the tumour of course was no longer a disk but rather an annulus with low cell density inside and far afield outside. The simulations were carried out with anatomical constraints including the ventricles (as was also done by Tracqui et al. 1995 and Cruywagen et al. 1995 in their modelling of chemotherapy treatment); these constraints had a major effect on the results as we found and illustrated in the previous sections. The accumulation of cells within a restricted area effectively slows down the spatial spread of the tumour. Cook (Dr. J. Cook, personal communication 1994) developed an analytical procedure to study this problem and obtained limits on the parameter values when the prediagnostic distribution of cells was defined by the same growth characteristics as occur later. We discuss his results in the next section.

Woodward et al. (1996) considered three excision diameters, denoted by S, namely, $S = 3, 4,$ and 5 cm. In each case the cell density outside the resected area is, of course, less than the detectable level. Medically $S = 3$ cm simulates the 'gross total resection' of the tumour when diagnosed and treated. So, the larger resections are much more extensive. Death, assumed to occur when the visually detectable area is equivalent to a circle of radius 3 cm, represents a volume of about 113 cm^3.

Figure 11.20(a) shows cumulative survival curves by assuming a uniform population by nine idealized patients with all possible combinations of three values of ρ and three values of D. Each population is subjected to four extents of surgery: $S = 0$ represents 'biopsy only' (no resection), $S = 3$ represents 'gross total resection' while $S = 4$ and $S = 5$ represent increasing resections of the margin of the tumour; we cannot consider $S = 6$ since that is considered fatal. Figure 11.20(b) compares the computed survival curves with those of 115 actual patients reported by Kreth et al. (1993), two groups of patients with glioblastoma, one subjected to biopsy plus radiation (58 patients) and the other subjected to surgical resection plus radiation (57 patients). All of them died within 100 weeks. The median survival times for the biopsy group was 32 weeks and for the resection group, 39.5 weeks. The median increase in survival time of 7.7 weeks between $S = 0$ cm and $S = 4$ cm resection compares well with the median value of 7.5 weeks found by Kreth et al. (1993). In our simulations, however, we required a 1-cm more extensive resection which could be accounted for in our estimate of the threshold number of cells or, of course, how extensive the resections of Kreth et al. (1993) really were. It is surprising that even with such a relatively simple model, simulated with homogeneous tissue but including brain boundaries, the results are so well correlated with patient data.

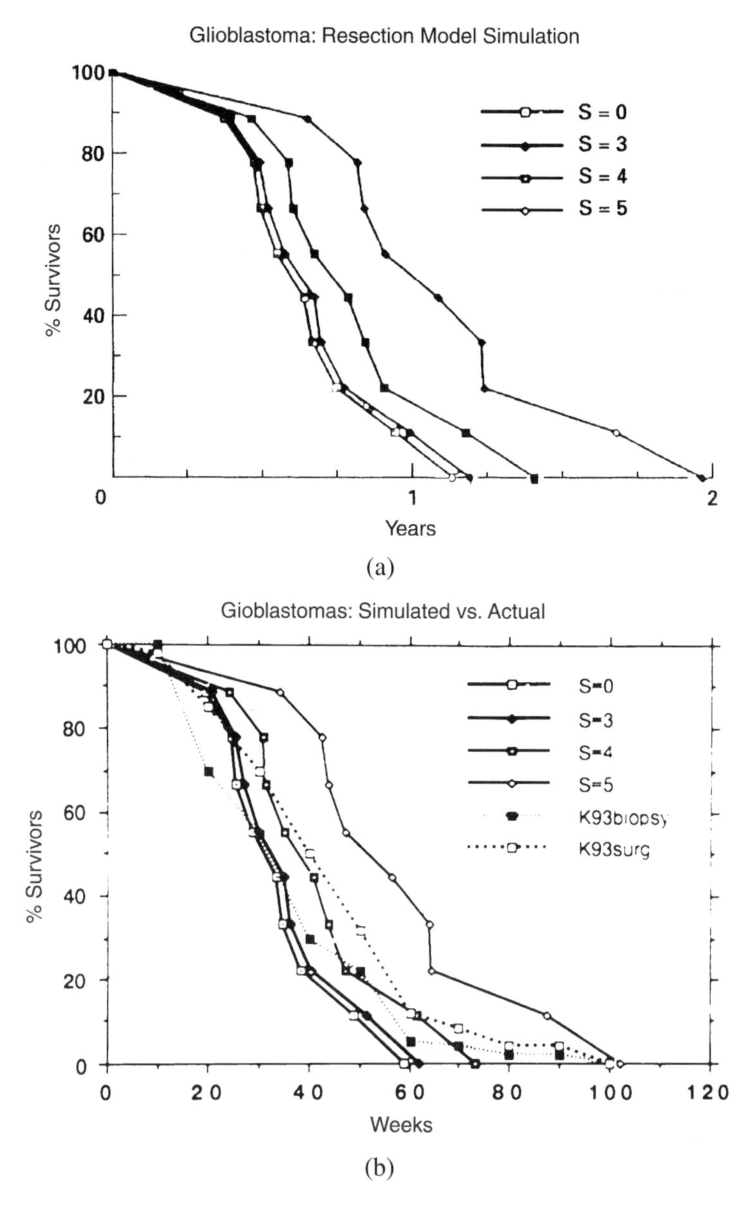

Figure 11.20. (**a**) Simulated cumulative survival times of 9 patients with high rates of growth and high diffusion. The 4 curves represent resected areas of $S = 0, 3, 4, 5$-cm diameter of the tumour and surrounding infiltrated tissue. Each curve is composed of 9 patients with all possible combinations of growth rate $\rho = 0.8 \times 10^{-2}$, 1.2×10^{-2} or 2.4×10^{-2}/day and $D = 4.32 \times 10^{-4}$, 1.0×10^{-4} or 1.3×10^{-3} cm^2/day. The median survival times are 33.4 weeks without resection ($S = 0$) and 35.0, 41.1 and 52.6 weeks with resection. (**b**) Superimposed actual survival curves from Kreth et al. (1993) following biopsy only (K93biopsy) or extensive surgical resection (K93surg), both groups of patients also receiving postoperative X-irradiation. (From Woodward et al. 1996)

Woodward et al. (1996) went on to explore the effect of varying ρ and D tenfold from the low and high values they used in Figure 11.20 to mimic the four tumour grades discussed above, starting with high ρ and high D (HH) down to the low grade tumour (LL). Populations of nine hypothetical patients for each of the four combinations were formed by allowing 50% variation in each of these values and survival times calculated. Figure 11.21 shows the results. Figure 11.21(a), the highest grade tumour, is the same as Figure 11.20(a), included for comparison, while Figure 11.21(d) gives the results for the lowest grade.

Although the curves in Figure 11.21 look quite similar the timescales are very different. The curves have progressively decreasing slopes from the high ρ – high D grade tumour to the low ρ – low D one. When all curves are put on the same timescale they can then be compared directly with the actual patient results summarised by Alvord

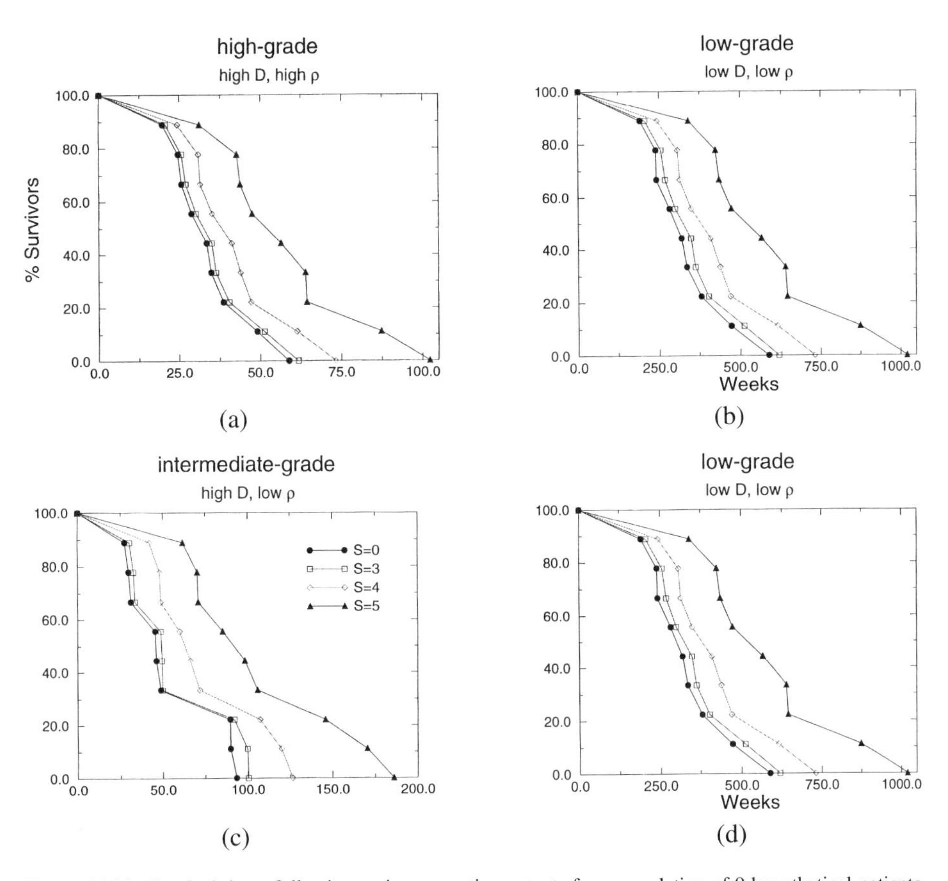

Figure 11.21. Survival times following various resection extents for a population of 9 hypothetical patients formed by allowing 50% variation in the growth and diffusion rates. Each family of four curves includes the various extents of resection $S = 0, 3, 4, 5$-cm diameter of the tumour and surrounding infiltrated tissue. (**a**) High grade, that is, a high–high combination (HH) of ρ and D, (**b**) intermediate grade, high–low (HL), (**c**) intermediate grade, low–high (LH) and (**d**) low grade tumour, low–low (LL), with a low ρ and low D. (From Woodward et al. 1996)

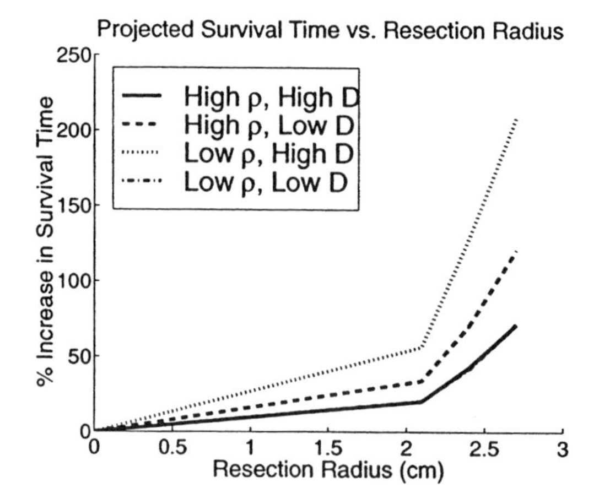

Figure 11.22. Spatially homogeneous model projected increase in survival time following various extents of resection. (From Woodward et al. 1995)

(1992). His results show a very wide range of glioma behaviour which is also found, not surprisingly, by the model results here. We should add that there is no general agreement as to how to grade gliomas medically. Perhaps another possible use of a theory is to provide a basis for such a classification.

Assuming a radius of 3 cm of detectable tumour size as fatal, Figure 11.22 shows the effects of various resection radii on subsequent survival time. The figure suggests that only slowly diffusing gliomas are significantly affected by resection as we would expect from the above. Again, we find that diffusion is fundamental to glioma treatment difficulties and recurrence.

One of the major conclusions that come out of the above detailed analysis and comparison with patient data is that there is no clear evidence that resection surgery increases survival, at least in patients with intermediate and high grade astrocytomas. This was also the conclusion of Nazzaro and Neuwalt (1990) who reviewed 33 major reports over the previous 50 years. Kreth et al. (1993) noted a statistically insignificant increase of 7.5 weeks. Kreth et al. (1993) said they found it equally difficult to define the benefit of resection even in low grade tumours.

Although we have varied the growth rate of the tumour cells we have assumed that for a given tumour it is constant. As commented by Woodward et al. (1996) it is generally believed that the tumour margin contains the less malignant-looking cells, but they are still cancerous and grow no matter how much of the tissue outside the (CT) visible tumour is excised.

11.8 Analytical Solution for Tumour Recurrence After Resection

Cook (Dr. J. Cook, personal communication 1994) considered the effect of resection in a homogeneous tissue analytically. As we noted above, mathematically, surgical resection is simulated by the removal of all tumour cells within a defined location centred on the tumour. It is his analysis that we present here. As explained, the cancer is assumed to be fatal either when the visible size reaches an equivalently defined circle of radius 3 cm or when the total cancer cell count reaches a critical value. We expect the bulk effect of resection is reasonably quantified by the homogeneous case but subtleties associated with tumour location and local tissue heterogeneity could adjust these results; below we discuss this situation.

The dimensionless model is that derived in Section 11.2 and defined by (11.10) and (11.11) with boundary conditions (11.8). The nondimensionalisation is given by (11.9). Without spatial heterogeneity in the diffusion coefficient, that is, $D(\mathbf{x}) \equiv 1$, the model equation is

$$\frac{\partial c}{\partial t} = \nabla^2 c + c. \tag{11.40}$$

Before and after resection the tumour cell population satisfies this equation but with different initial conditions.

For simplicity, and in view of the surprisingly good results (when compared with patient data) from the two-dimensional studies above, we consider the post-resection problem in two spatial dimensions in polar coordinates. We consider the tumour was initiated by a point source of N cells at the origin and so the pre-resection problem satisfies (11.38) for $0 < t < t_r$, where t_r denotes the time of the resection, with initial conditions $c_{\text{preresect}}(r, \theta, 0) = N\delta(r)$. On the infinite domain the solution is given by (11.13), which, in dimensionless form at time t_r, is

$$c_{\text{preresect}}(r, \theta, t_r) = \frac{N}{4\pi t_r} \exp\left(t_r - \frac{r^2}{4t_r}\right). \tag{11.41}$$

At resection, a central core of radius R_r is removed, so the post-resection problem satisfies (11.38) for $t > t_r$ with initial conditions ($t = t_r$),

$$\begin{aligned} c_{\text{postresect}}(r, \theta, t_r) &= F(r, \theta) \\ &= NH(r - R_r)c_{\text{preresect}}(r, \theta, t_r) \\ &= H(r - R_r)\frac{N}{4\pi t_r} \exp\left(t_r - \frac{r^2}{4t_r}\right), \end{aligned} \tag{11.42}$$

where H is the Heaviside function.

By superposition, the solution of the post-resection problem with these initial conditions can be represented as the integral

$$c_{\text{postresect}}(r, \theta, t) = \int_0^{2\pi} \int_0^{\infty} K(r, \theta, t; \xi, \alpha, t_r)F(\xi, \alpha)\xi \, d\xi \, d\alpha \quad \text{for} \quad t > t_r,$$

where K is the fundamental solution of (11.38) for a point source at $(r, \theta) = (\xi, \alpha)$ introduced at $t = t_r$,

$$K(r, \theta, t; \xi, \alpha, t_r) = \frac{1}{4\pi(t-t_r)} \exp\left(t - t_r - \frac{r^2 + \xi^2 - 2r\xi\cos(\theta-\alpha)}{4(t-t_r)}\right).$$

Now

$$c_{\text{postresect}}(r, \theta, t) = N \frac{\exp\left(t - \frac{r^2}{4(t-t_r)}\right)}{(4\pi)^2 t_r (t-t_r)} \int_{R_r}^{\infty} \xi \exp\left(-\frac{\xi^2}{4t_r} - \frac{\xi^2}{4(t-t_r)}\right)$$
$$\left[\int_0^{2\pi} \exp\left(\frac{2r\xi}{4(t-t_r)}\cos(\theta-\alpha)\right) d\alpha\right] d\xi.$$

Consider the second integral, in square brackets, which we denote by $\mathcal{I}$ and introduce the change of variables $\nu = \theta - \alpha$, and so

$$\mathcal{I} = \int_0^{2\pi} e^{A\cos(\theta-\alpha)}\, d\alpha = \int_{\theta-2\pi}^{\theta} e^{A\cos v}\, dv, \tag{11.43}$$

where $A = 2r\xi/(4(t-t_r))$. Since we are interested in eventually integrating $\mathcal{I}$ from $\xi = R_r$ to ∞ and A is proportional to ξ, A large is the limit we are interested in. Additionally, from numerical simulations, we find that the nondimensional time $t = \rho\bar{t}$ is at most O(1) for the range of growth rates ρ we consider. These two observations support exploiting the limit for A large. For A large, we can use Laplace's method (see, for example, Murray 1984) to approximate $\mathcal{I}$; this method exploits the fact that the integral is dominated by the contribution over a region near the maximum of the exponent: the maximum of $\cos v = 1$ so $v = 0$ or 2π. Rescaling near $v = 0$ we have, for large A,

$$\mathcal{I} = 2\int_0^{\epsilon} e^{A(1-(v^2/2)\ldots)}\, dv \sim e^A \int_{-\infty}^{\infty} e^{-A(v^2/2)}\, dv \sim e^A\sqrt{\frac{2\pi}{A}}.$$

With this

$$c_{\text{postresect}}(r, \theta, t) \sim N \frac{\sqrt{2\pi}\, e^{\left(t - \frac{r^2}{4(t-t_r)}\right)}}{16\pi^2 t_r (t-t_r)} \int_{R_r}^{\infty} \frac{e^{-\frac{\xi^2}{4t_r} - \frac{\xi^2}{4(t-t_r)} + \frac{r\xi}{2(t-t_r)}}}{\sqrt{\frac{r\xi}{2(t-t_r)}}} \xi\, d\xi$$
$$= N \frac{e^t}{8t_r\sqrt{r\pi^3(t-t_r)}} \int_{R_r}^{\infty} \sqrt{\xi} \exp\left(-\frac{\xi^2}{4t_r} - \frac{(r-\xi)^2}{4(t-t_r)}\right) d\xi$$

as $A \to \infty$. Let

$$\mathcal{J} = \int_{R_r}^{\infty} \sqrt{\xi} \exp\left(-\frac{\xi^2}{4t_r} - \frac{(r-\xi)^2}{4(t-t_r)}\right) d\xi = \int_{R_r}^{\infty} g(\xi) e^{-xh(\xi)}\, d\xi,$$

where $g(\xi) = \sqrt{\xi}$, $h(\xi) = (\xi^2/r) + (rt_r/t) - 2(\xi t_r/t)$ and $x = rt/(4t_r(t - t_r))$ which we can approximate for x large again using Laplace's method. This is equivalent to assuming $t \approx t_r$ and $r \sim O(1)$ or $t \gg t_r$ and $r \gg 4t_r$. The asymptotic contribution comes from around the minimum of $h(\xi)$ which is at $\xi = rt_r/t$. There are two possibilities.

(i) The minimum lies in the range of integration: $(rt_r/t) > R_r$. Introducing a new variable $w = \xi - (rt_r/r)$ and expanding for large x gives

$$\mathcal{J} \sim \sqrt{4\pi r(t - t_r)}\,\frac{t_r}{t} e^{-r^2/4t_r} + O\left(\left[\frac{4t_r(t - t_r)}{rt}\right]^{3/2} e^{-r^2/4t_r}\right) \quad \text{as } x \to \infty. \tag{11.44}$$

(ii) The minimum does not lie in the range of integration: $rt_r/t < R_r$. Introducing a new variable $w = \xi - R_r$ and expanding for large x gives

$$\mathcal{J} \sim \frac{2t_r(t - t_r)\sqrt{R_r}}{R_r t - rt_r} \exp\left(-\frac{R_r^2}{4t_r} - \frac{(r - R_r)^2}{4(t - t_r)}\right) + O\left(\exp\left(-\frac{R_r(R_r t - t_r r)}{4t_r(t - t_r)}\right)\left[\frac{4t_r(t - t_r)}{rt}\right]^2\right) \quad \text{as} \quad x \to \infty. \tag{11.45}$$

We now gather these results together to obtain the solution, following resection, as

$$c_{\text{postresect}}(r, \theta, t) \sim \begin{cases} \dfrac{N}{4(R_r t - rt_r)}\sqrt{\dfrac{(t - t_r)R_r}{r\pi^3}}\, e^{t-(R_r^2/4t_r)-(r-R_r)^2/(4(t-t_r))} & \text{for } r < \dfrac{t}{t_r} R_r \\ \dfrac{N}{4\pi t} e^{(t-(r^2/4t_r))} & \text{for } r > \dfrac{t}{t_r} R_r \end{cases} \tag{11.46}$$

which is strictly only valid for $(t/2t_r)(rR_r/(2(t - t_r)))$ and $(rR_r/(2(t - t_r)))$ large. It is quite clear from this solution that irrespective of the resected area the solution will eventually grow (exponentially) to a critical size.

We are mainly interested in how different the tumour cell population growth and invasion is, with and without resection. So, we look at the front of tumour cells invading the tissue beyond the bulk tumour margin. From the asymptotic solution (11.43) we can deduce how the invading front of tumour cells has been slowed down as a consequence of resection. For $r > (t/t_r)R_r$, we can write the asymptotic solution following resection as

$$c_{\text{postresect}}(r, \theta, t) \sim \frac{N}{4\pi t} \exp\left(t - \frac{r^2}{4t_r}\right) = c_{\text{noresect}}(r, \theta, t) \exp\left(-\frac{r^2}{4t_r} + \frac{r^2}{4t}\right), \tag{11.47}$$

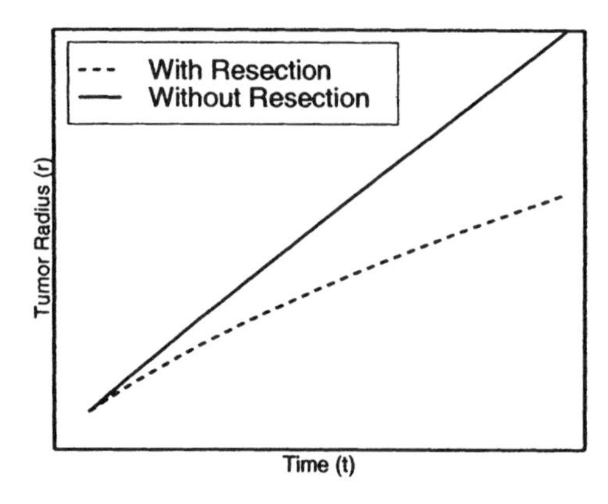

Figure 11.23. Tumour radius, with and without resection, from (11.49).

where (cf. (11.41)),

$$c_{\text{noresect}}(r, \theta, t) = \frac{N}{4\pi t} \exp\left(t - \frac{r^2}{4t}\right) \tag{11.48}$$

is the analytic solution if resection had not been performed. So, following resection, the tumour cell concentration ahead of the resected area is suppressed by a factor of $\exp(-(r^2/4t_r) + r^2/4t)$; this is only for the asymptotic parameters (defined above) sufficiently large. If we assume cell density is detectable (by CT) at the threshold c^*, the outer radius of the detectable tumour then satisfies

$$r^2_{\text{postresect}} \sim 4tt_r\left(1 - \frac{\ln(4\pi tc^*/N)}{t}\right) < 4t^2\left(1 - \frac{\ln(4\pi tc^*/N)}{t}\right) = r^2_{\text{noresect}} \tag{11.49}$$

for $t_r < t < (r/R_r)t_r$. Figure 11.23 gives the detectable tumour radius with and without resection using the asymptotic form in the last equation.

11.9 Modelling Surgical Resection with Brain Tissue Heterogeneity

Swanson (1999) reconsidered resection in the case of an anatomically accurate brain such as we described above. The principle is the same as in the last section but the mathematical problem is considerably more complex. The difference is in the spatial heterogeneity of the diffusion coefficient due to the white and grey matter distribution in the brain. As before, following resection, the tumour cell population continues to be governed by the model system developed in Section 11.2. The mathematical problem with resection is given by the dimensionless system

$$\frac{\partial c}{\partial t} = \nabla \cdot (D(\mathbf{x})\nabla c) + c \tag{11.50}$$

for $t > t_r$, where

$$D(\mathbf{x}) = \begin{cases} 1 & \text{for } \mathbf{x} \in \text{ White Matter} \\ \gamma = \dfrac{D_g}{D_w} & \text{for } \mathbf{x} \in \text{ Grey Matter} \end{cases} \tag{11.51}$$

with $c(\mathbf{x}, t_r) = F(\mathbf{x})$, the initial distribution of cells after resection and $\mathbf{n} \cdot D(\mathbf{x})\nabla = 0$ for $\mathbf{x}$ on ∂B (the boundary of the brain).

We are interested here in the effects of resection on tumour growth and invasion. With the procedures described above for the untreated tumour growth we can look at the effects of resection on tumours at different locations in the brain. The solutions are necessarily numerical. As a detailed example we focus specifically on Position 1 (inferior fronto-parietal) tumours and consider only a high grade tumour, that is, high growth rate and high diffusion; Swanson (1999) gives further results for all grades of tumour. The growth rate ρ does not appear in the dimensionless model but, of course, in the nondimensionalisation (refer to (11.9)).

Figure 11.24 is the simulated response of a high grade glioma at Position 1 (see Figure 11.13) following gross total resection, that is, removal of the detectable (by CT) portion of the tumour. Figure 11.25 shows the evolution of the same tumour following extensive resection, that is, removal of an area twice the size of the detectable portion of the tumour. Recall that a gross total resection corresponds to removal of an area of tumour equivalent to a circle with radius 1.5 cm; in other words the total resected area (as seen on a CT scan) is $\pi(1.5)^2$ cm^2. As mentioned above, the term 'gross total resection' is commonly used by neurosurgeons to describe a successful removal of the tumour such that just after surgery there is no detectable tumour observable on enhancement of a CT scan. The maximum extensive resection corresponds to removal of an area of tumour equivalent to a circle with radius 3 cm; that is, the total resected area (as seen on a CT scan) is $\pi 3^2$ cm^2. In the work of Woodward et al. (1996) the maximum resection they considered was equivalent to a circle with radius 2.5 cm. This scenario defines the recent trend to remove not only the detectable tumour but some radius of normal-appearing tissue surrounding the grossly visible tumour. It makes little difference to the final outcome.

Figure 11.24(a) represents the detectable portion (by a CT scan) of the tumour at diagnosis; that is, $t = t_r$. This is immediately followed by gross total resection, Figure 11.24(b), which shows the actual extent of the tumour using the threshold we used in Section 11.7. We see that the resection succeeded in removing only a fraction of tumour, 36.9% in fact, of the total tumour volume (see Table 11.7). After 60 days, the detectable portion of the tumour has increased from an area of 0 to $\pi(1.5)^2$ cm^2; in other words the tumour has recurred. The recurrent tumour is represented by a ring of detectable tumour surrounding the bed of resection, which is a common characteristic of human gliomas. The CT detectable portion of the recurrent tumour and the actual tumour are shown in Figures 11.24(c) and (d), respectively. Assuming no other treatment is considered, the tumour is fatal 130 days following resection.

If we carry out an extensive resection (see Figure 11.25(b)), 86.7% of the total tumour volume is removed (see Table 11.7). The tumour recurs after 225 days. The

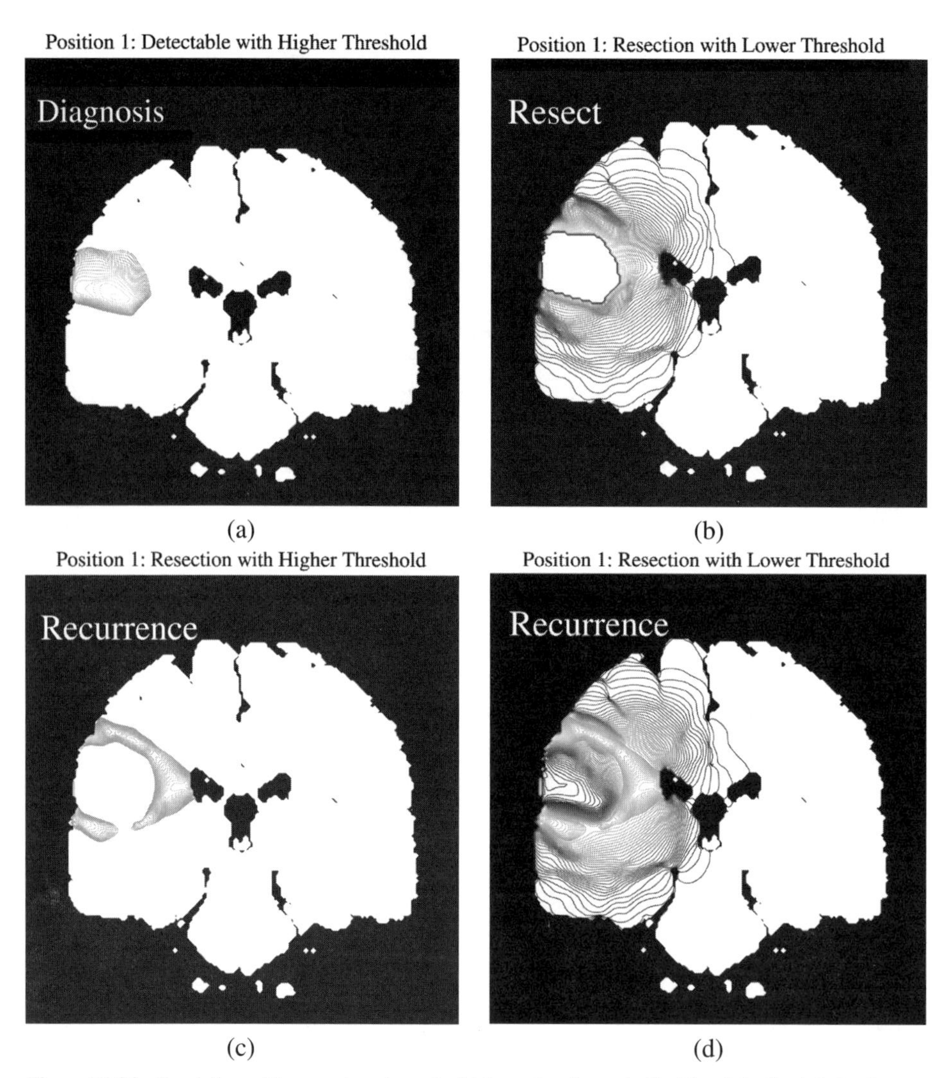

Figure 11.24. Simulation of tumour invasion of a high grade glioma in Position 1 in the inferior fronto-parietal lobe following gross total resection: (**a**) at diagnosis (preresection); (**b**) just after resection of the identifiable tumour, that is, gross total resection; (**c**) and (**d**) at recurrence (60 days later); (**a**) and (**c**) are as seen on the CT scans; (**b**) and (**d**) are as calculated out to 5% of the threshold (boundary) cell concentration defined by CT. (From Swanson 1999) (See the comment in parentheses in Figure 11.10.)

recurrent tumour does not form a ring around the tumour bed but forms isolated tumour islands near the edge of the resection bed (see Figure 11.25(c)). Clearly recurrence following extensive resection is significantly affected by the boundaries of the brain. The three islands of recurrent tumour are located at or near the boundary. Since the concentrations of tumour cells are fairly low after the resection, a single bulk tumour does not form. The tumour cells that remain after resection continue to diffuse and, on encountering boundaries, there is a build-up of tumour cells. An example of clinically

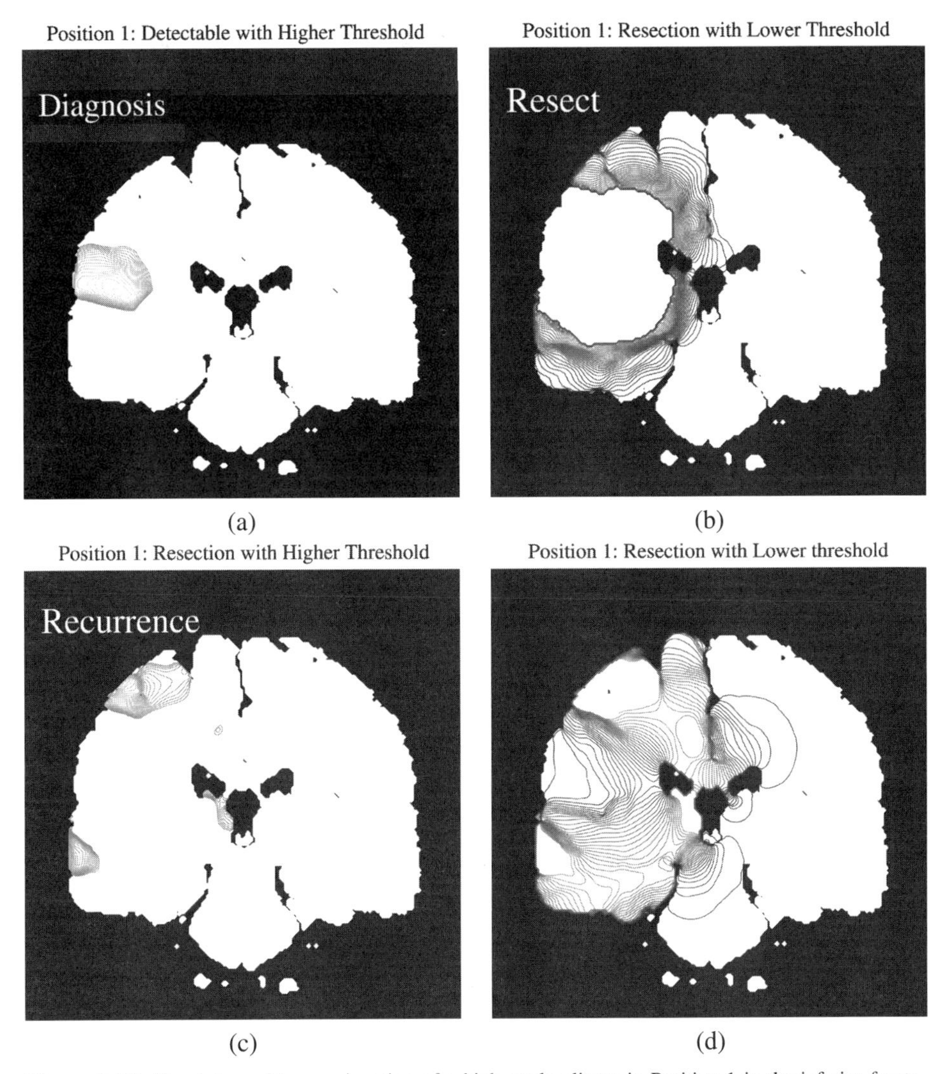

Figure 11.25. Simulation of tumour invasion of a high grade glioma in Position 1 in the inferior fronto-parietal lobe following extensive resection beyond the (CT) detectable margin of the tumour: (**a**) at diagnosis (preresection); (**b**) after resection of the extended region outside the detectable tumour; (**c**) and (**d**) at recurrence (253 days later); (**a**) and (**c**) are as seen on the CT scan; (**b**) and (**d**) are as calculated out to 5% of the threshold (boundary) cell concentration defined by CT. (From Swanson 1999) (See the comment in parentheses in Figure 11.10.)

observed multifocal recurrence is shown in Figure 11.26. The simulated tumour is fatal 253 days following resection, assuming no other treatment regime is attempted, that is, an increase in survival time of 123 days or about 100%. This type of recurrence after *extensive* resection is often considered a 'new' tumour separate from that which was resected. The model suggests that, in fact, this regrowth is part of the same tumour, namely, the diffusively invaded portion that has now become detectable.

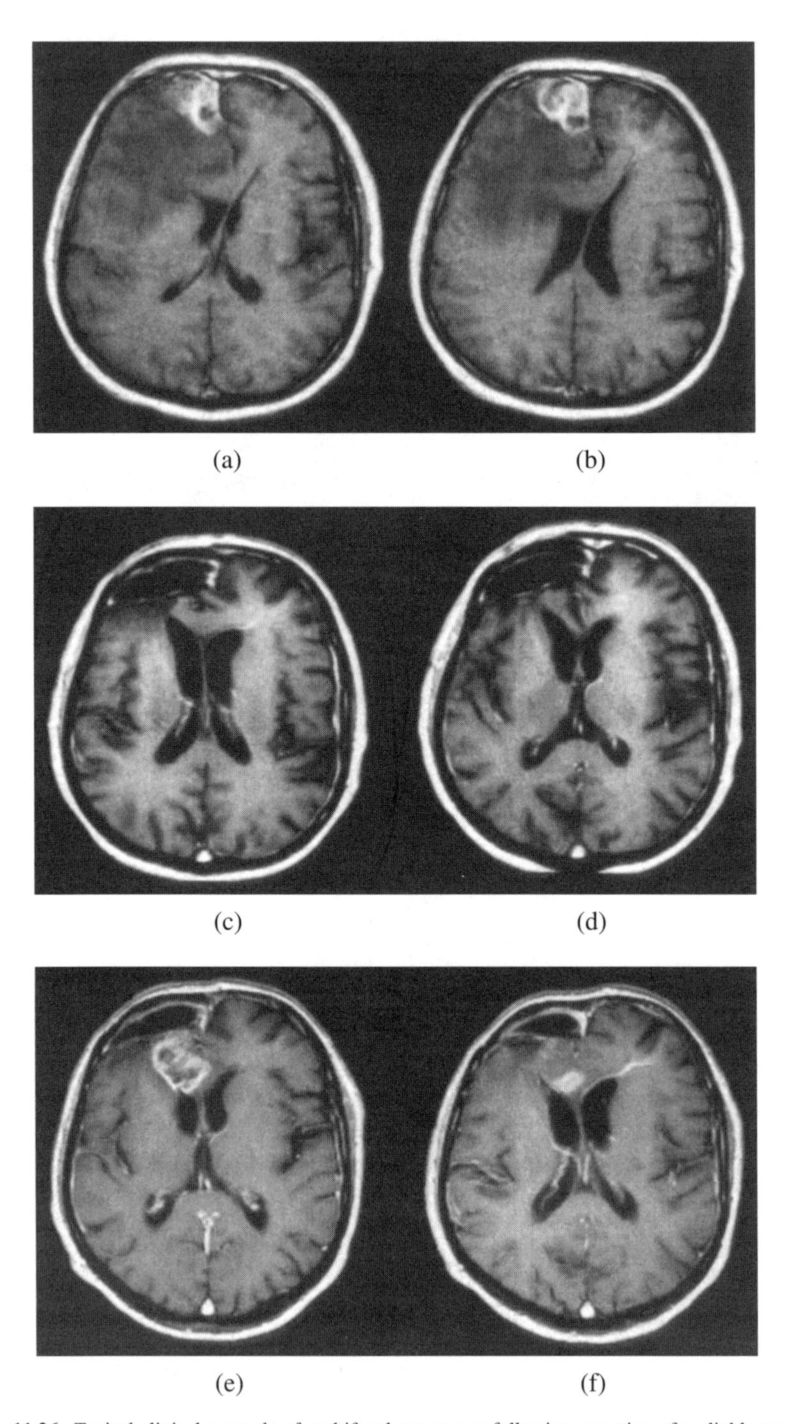

Figure 11.26. Typical clinical example of multifocal recurrence following resection of a glioblastoma. The images are of a 58-year-old man diagnosed with a glioblastoma. Images (**a**) and (**b**) show the histological invasion. Images (**c**) and (**d**) show the gross total resection which was followed by radiation and chemotherapy. Images (**e**) and (**f**) show local recurrence after 6 months. (Data and images were kindly provided by Dr. Alf Giese)

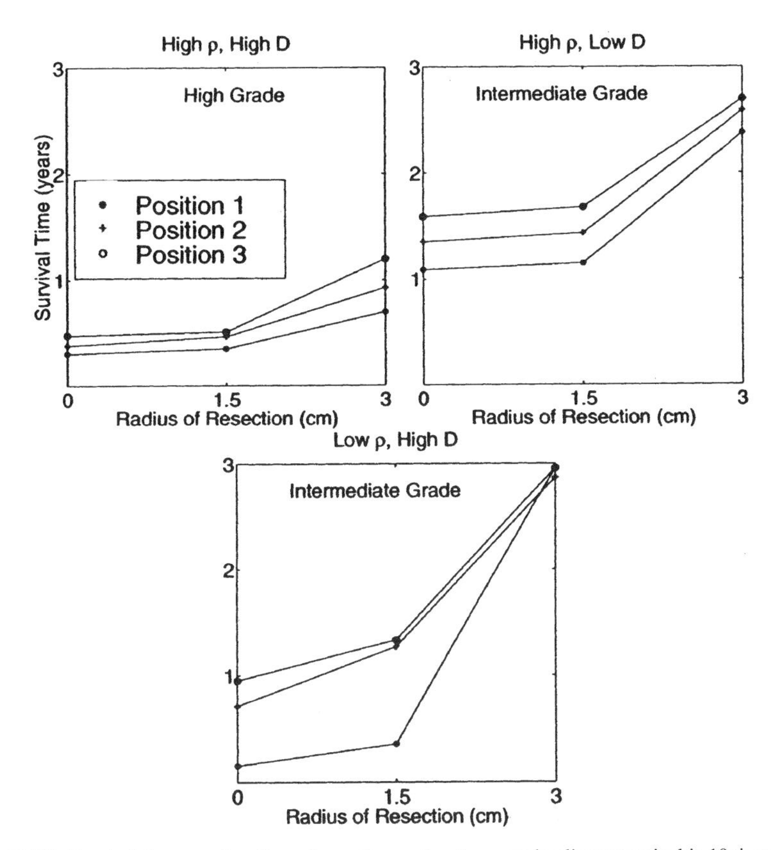

Figure 11.27. Survival time as a function of resection radius. Low grade glioma survival is 10 times that for high grade gliomas. (From Swanson 1999)

Survival Time Following Tumour Resection in Heterogeneous Medium

Although it is clear that diffusely invaded tumour cells contribute directly to the survival time of patients it is instructive to consider the model implications of the idealized fatal tumour size we have used throughout this work. We performed simulations similar to those described above for tumours in the other positions marked in Figure 11.13, namely Positions 2 and 3; these simulation results are given in Swanson (1999). Figure 11.27 shows the survival time associated with tumours at all three positions for high grade and two intermediate grade tumours. Low grade gliomas are not given since their behaviour is essentially the same as that of high grade gliomas but on a much slower timescale.

Tables 11.7 and 11.8 summarize the results on survival time and resection. We tabulate the portion of the total tumour volume removed following gross total or extensive resection. Under our model assumptions, a gross total resection can remove as little as 12% of the total tumour since it is already so diffuse at diagnosis. An extensive resection can remove greater than one-half of the tumour. The effectiveness of the resection depends, of course, on the ratio of the diffusion coefficient D and the growth rate ρ. Also, as we have commented on above, the location of the tumour with respect to the boundaries significantly affects the survival times predicted by our model.

Table 11.7. Percent of tumour volume resected with survival time for all positions with gross total resections for high grade and two intermediate grade tumours. (From Swanson 1999)

Position	ρ (/day)	D (cm^2/day)	Gross Total % Resected	Survival Time (days)
1	1.2×10^{-2}	1.3×10^{-3}	36.9	127.3
	1.2×10^{-2}	1.3×10^{-3}	95.5	420.8
	1.2×10^{-4}	1.3×10^{-4}	12.5	129.6
2	1.2×10^{-2}	1.3×10^{-3}	41.3	169.0
	1.2×10^{-2}	1.3×10^{-3}	92.9	525.5
	1.2×10^{-4}	1.3×10^{-4}	13.1	462.9
3	1.2×10^{-2}	1.3×10^{-3}	48.3	185.6
	1.2×10^{-2}	1.3×10^{-3}	95.8	613.9
	1.2×10^{-4}	1.3×10^{-4}	15.9	486.1

Table 11.8. Percent of tumour volume resected with survival time for all positions with extensive resections. (From Swanson 1999)

Position	ρ (/day)	D (cm^2/day)	Extensive % Resected	Survival Time (days)
1	1.2×10^{-2}	1.3×10^{-3}	86.7	253.7
	1.2×10^{-2}	1.3×10^{-3}	99.9	868.5
	1.2×10^{-4}	1.3×10^{-4}	55.7	1078.7
2	1.2×10^{-2}	1.3×10^{-3}	92.4	337.0
	1.2×10^{-2}	1.3×10^{-3}	99.9	945.0
	1.2×10^{-4}	1.3×10^{-4}	44.2	1046.3
3	1.2×10^{-2}	1.3×10^{-3}	96.2	438.0
	1.2×10^{-2}	1.3×10^{-3}	99.8	985.6
	1.2×10^{-4}	1.3×10^{-4}	52.8	1078.7

The relationship between the amount of tumour resected and survival time clearly depends on many factors. In particular, it is very important to have a sense of physical environment of the tumour location such as proximity to boundaries and the white and grey matter distributions. With such information we can use this approach to deduce more accurately the effectiveness of resection treatment. There are others and we discuss a major one of these, chemotherapy below.

11.10 Modelling the Effect of Chemotherapy on Tumour Growth

In this section we describe the model developed by Cruywagen et al. (1995) and Tracqui et al. (1995) to quantify the effect of chemotherapy on tumour growth. It is based in part on quantitative image analysis of histological sections of a human brain glioma and especially on cross-sectional area/volume measurements of serial CT images while the patient was undergoing chemotherapy. We estimated the model parameters using optimization techniques to give the best fit of the simulated tumour area to the CT scan data. We carried out numerical simulations on a two-dimensional domain, which took into account the geometry of the brain (only the ventricles and the skull) and its natural barriers to diffusion. The results were used to determine the effect of chemotherapy on the spatiotemporal growth of the tumour. (Shochat et al. (1999) have used computer-based simulations of various basic ordinary differential equation models to evaluate the efficacy of chemotherapy protocols on breast cancer.)

One of the reasons for discussing this work is that it is a somewhat different approach and it shows how the model parameters are estimated from clinical data obtained from successive computerized tomograms (CT) of a patient with an anaplastic astrocytoma who received radiotherapy and chemotherapy. This approach is a feasible one when other methods or independent data are not available. As before the time course of the treatments was incorporated in the model, and, by using optimization techniques to fit the model response to the experimental data, an evaluation of the effectiveness of the chemotherapy was obtained by identifying parameters characterizing the rate of death of tumour cells.

Just as with the above discussions, once the feasibility of reconstructing some of the kinetic events in invasion from histological sections has been established, the hope is to be able to use such modelling to investigate other gliomas with different characteristic growth patterns and geometries and the effects of various forms of therapy using the same types of data from other patients.

Experimental Data

Since many patients with brain tumours die too soon after detection of their tumour without a sufficient number of follow-up scans, we were fortunate to be able to study one patient with an anaplastic astrocytoma who had been diagnosed and treated with X-irradiation three years previously. The tumour recurred and was re-treated chemotherapeutically with moderate success, while CT scans were taken repeatedly during the 12 months before the patient's death. From these scans one or more of the pieces of tumour were observed. The tumour area of the largest piece (piece a) was measured at three different levels (referred to as levels 1, 2 and 3; recall Figure 11.3) of the brain using a digitizing tablet and computerized area measurement technique. Since the levels at which the areas had been recorded were not always exactly the same for all the scans, some data points were obtained by linear interpolation between the neighboring upper and lower levels.

During the same terminal year, the patient received two different chemotherapies (Figure 11.28). The first one was a course of six drugs (6-thioguanine, procarbazine, dibromodulcitol, CCNN, 5-florouracil, and hydroxyurea given over 15 days and repeated every six to eight weeks to allow for recovery of bone marrow) applied five times. The second consisted of two courses of cisplatinum given at month intervals. In addition, the patient received neutron beam irradiation during the last three weeks.

The tumour cell density was determined by image analysis of a biopsy obtained at the time of the first scan. The number of nuclei in the tumour tissue detected per mm^2

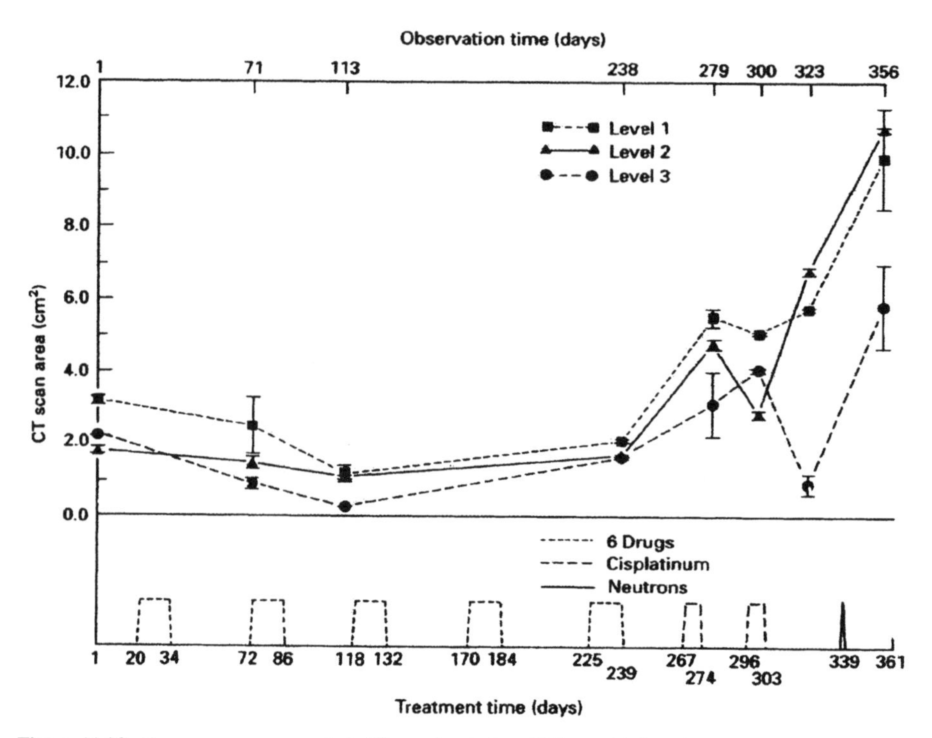

Figure 11.28. Tumour area measured at different levels (recall Figure 11.5), referred to as levels 1, 2 and 3 of the brain, from 8 CT scans taken during the 12 months before the patient's death. During the same terminal year, the patient received 5 cycles of 6 drugs (UW protocol), two cycles of cisplatinum, and neutron beam irradiation. (From Tracqui et al. 1995)

was averaged over five images. This cell density was used as a starting condition for the model. Another evaluation of tumour cell density was performed on autopsy material. From the autopsy tissue, the number of nuclei detected per mm^2 in normal tissue was also averaged over five images. This cell density was used as the threshold in the model over which the area was determined to be tumorous.

Mathematical Model

Here we consider the tissue to be homogeneous, that is, we take the diffusion coefficient and growth rate parameter to be constant. The basic mathematical model, without including the effect of chemotherapy, is given by (11.3) with zero flux boundary conditions on the brain boundaries.

We now have to quantify the effect of the chemotherapy. We do this by assuming that the cell death due to the chemotherapy can be modelled by a linear removal rate, $K(t)c(\mathbf{x}, t)$, where the function $K(t)$ describes the temporal profile of the treatment. We further assume that, for the timescale considered here, the action of the chemotherapy can be modelled, as a first approximation, by a step function with a constant amplitude, such that

$$K(t) = \begin{cases} k_1 & \text{if } t_{1,i} \le t \le t_{1,i+1} \quad i = 1, 3, \ldots, 9 \\ k_2 & \text{if } t_{2,j} \le t \le t_{2,j+1} \quad j = 1, 3 \\ 0 & \text{otherwise} \end{cases}, \tag{11.52}$$

where $t_{1,i}$ and $t_{1,i+1}$ are the starting and finishing times, respectively, of each of the five treatments of the first six-drug course of chemotherapy, while $t_{2,j}$ and $t_{2,j+1}$ correspond to the starting and finishing times of the second cisplatinum chemotherapy course as illustrated in Figure 11.29. With this and equation (11.3) the basic (single cell type) equation is

$$\frac{\partial c}{\partial t} = D\nabla^2 c + \rho c - K(t)c. \tag{11.53}$$

We now need initial and boundary conditions. Theoretically, the tumour started out as one cancerous cell, but the time of appearance of this original cell and the type of

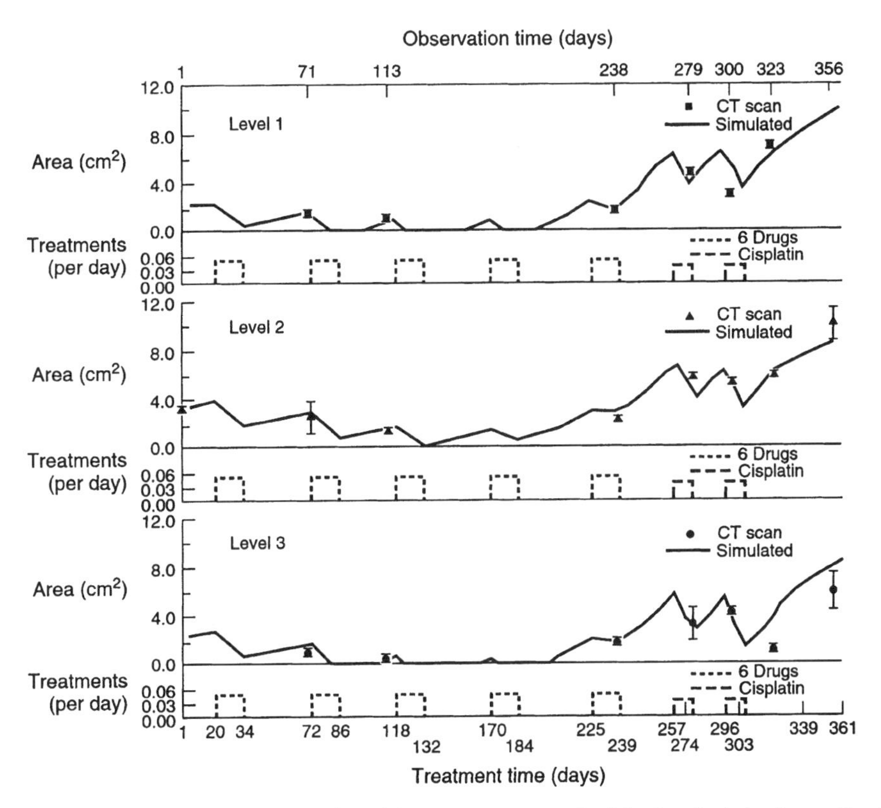

Figure 11.29. The simulated time evolution of the tumour area at each of the three brain levels considered. The model system (11.56)–(11.59) was solved numerically on a two-dimensional grid which takes into account the different brain and ventricle boundaries at each level. The optimisation routine determined the parameters which give the best least squares fit of the simulated tumour area to the CT scan data. The effect of the chemotherapies on each level is also shown. (From Tracqui et al. 1995)

growth and the spread of the early cancer cells are unknown. Since approximate initial conditions must be provided, we assumed, as above, that at the time of the first scan the diffusion process has already broken any previous possibly uniform distribution of the cells. So, we assumed that the cells were normally distributed with a maximum cell density, a, at the centre, $\mathbf{x}_0$, of the tumour for the considered level; that is,

$$c(\mathbf{x}, 0) = a \exp\left(\frac{-|\mathbf{x} - \mathbf{x}_0|^2}{b}\right), \tag{11.54}$$

where b is a measure of the spread of tumour cells.

As boundary conditions, we require zero flux of cells to the outside of the brain or into the ventricles, so, at the boundaries of the domain we again have

$$\mathbf{n} \cdot \nabla c(\mathbf{x}, t) = 0, \tag{11.55}$$

where $\mathbf{n}$ is the unit vector normal to the boundary.

The initial attempt to represent the spatiotemporal variation of the tumour under the hypothesis of a single cell population with the above model was unsuccessful because it could not account for the qualitative changes in behaviour seen in the clinical data. In particular, it was not possible to capture the sharp increase observed in the tumour area at the end of the first chemotherapy as seen in Figure 11.28, which indicates the failure of this treatment. This suggested that a polyclonal model, which incorporated cell mutations, was necessary and a reasonable first modification.

It was necessary, therefore, to modify the model to include two cell populations. The first cell type, denoted by $c_1(\mathbf{x}, t)$ and which is sensitive to both the first course of chemotherapy (UW protocol of six drugs) and the second course (cisplatinum), comprises most (actually calculated to be over 90%) of the tumour at the time of the first scan. The second cell type, denoted by $c_2(\mathbf{x}, t)$ and which is assumed resistant to the first course of chemotherapy, but possibly sensitive to the second course, comprises the remainder of the cancerous cell population. To model the effect of chemotherapy we considered that cell death is proportional to the number of cells present during the chemotherapy that are sensitive to the particular treatment. The second type of tumour cells could have originated from an earlier genetic mutation, either spontaneous or induced by the radiation therapy three years earlier, or by a similar mutation induced at the beginning of the chemotherapy itself. Under this hypothesis, the proliferation of the second cell population is determined by the initial density of this cell type. A different model could consider a different hypothesis to include the probability of mutation per cell generation as suggested by Goldie and Coldman (1979). We further assumed that both populations of cells have the same properties with regard to diffusion but they may have different growth rates.

Two-Cell Population Model

In the model here we do not view the second cell population as a mutation but rather a cell line in its own right. In the following section on polyclonality we describe such a model. Here, we consider two cell populations, $c_1(\mathbf{x}, t)$ and $c_2(\mathbf{x}, t)$, with the same diffusion coefficient D, but possibly different growth rates, r_1 and r_2. Then, with the same model (11.53) as the basis, the model we take for the two cell populations is now

$$\begin{aligned}\frac{\partial c_1}{\partial t} &= D\nabla^2 c_1 + r_1 c_1 - K_1(t)c_1 - K_2(t)c_1\\ \frac{\partial c_2}{\partial t} &= D\nabla^2 c_2 + r_2 c_2 - K_2(t)c_2.\end{aligned} \tag{11.56}$$

Here we assume the first cell type, c_1, is sensitive to both treatments, and the second cell type, c_2, is resistant to the first chemotherapy but sensitive to the second course of drugs; that is,

$$\begin{aligned}K_1(t) &= \begin{cases} k_1 & \text{if } t_{1,i} \le t \le t_{1,i+1} \quad i = 1,3,5,7,9\\ 0 & \text{otherwise}\end{cases}\\ K_2(t) &= \begin{cases} k_2 & \text{if } t_{2,1}+4 \le t \le t_{2,2}+6\\ k_2 & \text{if } t_{2,3} \le t \le t_{2,4}+2\\ 0 & \text{otherwise}\end{cases}\quad .\end{aligned} \tag{11.57}$$

The times, $t_{1,1}$, $i = 1, 10$ and $t_{2,j}$, $j = 1, 4$, are directly related to those given in Figure 11.29. Here (and in the numerical simulations) we consider the first chemotherapy to act from the time treatment begins to the time treatment ends, but the second chemotherapy does not start acting until 0 to 4 days after the date of treatment and is still effective 2 days after the last administration. The exact timing is unknown but these assumptions give a better fit to the clinical data.

The associated boundary conditions are (cf. (11.55))

$$\mathbf{n}\cdot\nabla c_1 = 0, \qquad \mathbf{n}\cdot\nabla c_2 = 0 \tag{11.58}$$

and the initial spatial conditions (cf. (11.54))

$$c_1(\mathbf{x},0) = a_1 \exp\left(\frac{-|\,\mathbf{x}-\mathbf{x}_0\,|^2}{b}\right), \quad c_2(\mathbf{x},0) = a_2 \exp\left(\frac{-|\,\mathbf{x}-\mathbf{x}_0\,|^2}{b}\right). \tag{11.59}$$

The parameter $a = a_1 + a_2$ is the maximum initial density of cells (located at the centre of the tumour, $\mathbf{x}_0$), and the parameter, b, measures the spread of the tumour cells.

Initial values of a and b are estimated from the cell densities determined by the biopsy at the time of the first scan and the first scan area, so that at the time of the first scan the simulated value of the area of the tumour is close to the one determined experimentally. The parameter a_1 (the maximum initial density of cells in the first population) still has to be estimated. Assuming an initial homogeneous three-dimensional diffusion of the tumour away from the boundaries, different initial conditions were chosen for the initial cell concentration at the different levels corresponding to the different initial areas as measured at the first scan. The initial cell density a at the centre of the tumour at level 2 is thus larger than those at levels 1 and 3. Accordingly, the coefficient b, characterizing the initial dispersion of the cells is slightly larger for the level 2 compared to

Table 11.9. The model parameters for each level determined by the optimisation routine to give the best least squares fit to the simulated area to the CT scan areas. Here D is the cell diffusion coefficient, r_1 and r_2 are the growth rates and Td_1 and Td_2 are the cell doubling times. The parameters k_1 and k_2 are the cell death rates caused by the two chemotherapies. We assume that at the time of the first scan ('initial' conditions) the cells were normally distributed with the parameter, a, measuring the density of cells at the centre of the tumour and parameter, b, characterising the initial dispersion of the cells. The initial percentage of type 2 cells is determined by the optimisation routine. The weighted least-square error sum is used as a measure of the fit of the simulated tumour area to the CT scan data. (From Tracqui et al. 1995)

	Level 1	Level 2	Level 3
Model Parameters			
D (cm^2/day)	1.11×10^{-2}	1.08×10^{-2}	1.05×10^{-2}
D (cm^2/sec)	1.29×10^{-7}	1.24×10^{-7}	1.21×10^{-7}
r_1 (1/day)	1.05×10^{-2}	1.03×10^{-2}	1.07×10^{-2}
Td_1 (days)	66.0	67.4	64.7
r_2 (1/day)	1.24×10^{-2}	1.10×10^{-2}	1.14×10^{-2}
Td_2 (days)	55.7	63.2	60.6
k_1 (1/day)	3.91×10^{-2}	5.5×10^{-2}	5.10×10^{-2}
k_2 (1/day)	2.85×10^{-2}	2.99×10^{-2}	3.23×10^{-2}
Initial Conditions			
a	706	1120	814
b (cm^2)	5.68	5.68	5.68
initial % of type 2	6.66%	7.95%	9.09%
least square error sum	1480	158	155

the other two levels. Table 11.9 shows how these parameter estimates compare at each level.

Numerical Methods and Results

The two cell populations, $c_1(\mathbf{x}, t)$ and $c_2(\mathbf{x}, t)$ were obtained by the numerical integration of the model system (11.56)–(11.59). An immediate problem is the values to assign to the parameters. As in the evaluation of the tumour area from the CT scan, the numerical determination of the simulated area involves a threshold, c_{th}, in cell density. Mathematically we can choose that to be whatever we want. However, since we are using the model here in conjunction with the data we have to use a different approach. According to the experimental data obtained from the comparisons of histological sections obtained at autopsy with terminal CT scans, the retained value of the normalized threshold is taken as 40% of the biopsy density. This is a reasonable estimate, but obviously it would be much better to have an independent estimation of the threshold from brain slice scans. We can, however, illustrate quantitatively the effect of a variation of the threshold value on the model parameters. For example, a rather large decrease of 25% in the detection threshold has less than a 10% effect on the other parameters, characterized by slight increases in the values of D and k_1 and slight decreases in the values of the growth rates, r_1 and r_2; we come back to this below. The numerical simulation details are given in Tracqui et al. (1995) and Cruywagen et al. (1995).

Initial values of the model parameters, which gave results in good agreement with the observed data, were determined heuristically; we have some idea of these values from the literature such as described in earlier sections but here they were originally estimated from experimental data on cell doubling times from Alvord and Shaw (1991) and *in vitro* motility from Chicoine and Silbergeld (1995). The model parameters, given in Table 11.9, are those obtained by the optimization procedure. Subsequently, a global procedure was used to improve the fit of the model solution to the data through an optimization[2] of the six unknown model parameters.

With this model the unknown parameters are the diffusion coefficient, D, the two growth rates r_1 and r_2 of the first and second cell populations respectively, and the values k_1 and k_2 characterising the death rates of the two cell populations during the two different chemotherapies. The percentage of type 1 cells and type 2 cells at the starting time (that is, at the time of the first scan) is also an unknown parameter.

The total density of cancerous cells, $c(\mathbf{x}, t) = c_1(\mathbf{x}, t) + c_2(\mathbf{x}, t)$ was obtained by the numerical solution. The anatomic boundaries of the skull and of the ventricles associated with each level were taken into account.

Figure 11.29 shows the simulated time evolution of the tumour area at each of the levels 1, 2 and 3. In the simulations, the tumour area increases up to the beginning of the first chemotherapy. As expected, the decrease in the tumour area is larger for a higher value of the effect of the first chemotherapy, k_1, whereas increasing the value of the diffusion coefficient, D, or the growth rate, r_1, has the opposite effect. To generate an increase of the tumour area before the end of the first chemotherapy, the effect of k_1 must be counterbalanced by the growth rate of the second cell population, r_2. The effect of the second chemotherapy is measured by k_2. The value of these parameters, as well as the initial composition of the tumour, are thus critical for determining the spatial-temporal variation of the tumour. The inclusion of the physical brain boundaries and the position of the ventricles in the model has a considerable effect on the results: not only do the boundaries contribute towards a higher tumour cell density due to the accumulation of cells within a restricted area, but also there is clearly a maximum value for the tumour area, which is the brain area without the ventricles.

The evolution of the tumour area simulated by the model is in good agreement with the experimental data. A good fit is obtained for all three levels, which is encouraging for the model validity considering the wide variations in the data points observed in the latter part of the experimental curves, that is, during the second chemotherapy (see Figure 11.28). It is also encouraging to see from Table 11.9 that similar values are obtained for the model parameters for each of the three levels considered, as expected if no specific differences in cell behaviour and properties between the different levels are assumed. Larger inter-level variations are observed for the parameters k_1 and k_2

[2]The minimized criterion was a least square function F, where each term is weighted by the variance of the data point (Ottaway 1973), with

$$F = \sum_{i=1}^{m} \frac{\sum_{j=1}^{n} (S_{ij} - S_i^*)^2}{\sum_{j=1}^{n} (S_{ij} - \bar{S}_i)^2},$$

where m is the number of data points, n is the number of measurements at each data point, $\bar{S}_i$ is the mean value of the area and S_i^* is the simulated area at each data point. The optimization algorithm was coupled with the numerical integration procedure which provided the simulated values at each of the observation times.

measuring the efficiency of the chemotherapies. However, the sawtoothlike shape of the curve in Figure 11.30, resulting from the successive treatments, is sensitive to any small change in the starting and finishing time of each step function modelling each stage of the chemotherapy. So, such variations are to be expected in the absence of more detailed information on the kinetics of the treatment, which would allow for a more precise modelling of the chemotherapy action with time.

Tracqui et al. (1995) reran the optimisation with a 25% decrease in the detection threshold and essentially the same optimisation criterion value to see what effect this had on the parameter estimates. They found that it causes a much smaller change, less than 10%, in the other parameters estimated, with slightly increased values of D and k_1 and slightly decreased values of the growth rates, r_1 and r_2. They also carried out simulations to give the spatiotemporal variation in the tumour in terms of the total cancerous cells and also the two subpopulations. They found, as expected from the model basis that although type 1 cells decreased, type 2 cells increased and it is clear that the patient's death was due to the emergence of a subpopulation of type 2 cells.

Figure 11.30 shows the two-dimensional images of the simulated tumour, and, for comparison, the pictures of the tumour obtained from scans at days 113 and 300. The assumption of a homogeneous medium leads to a simulated tumour shape more regular than the observed one. The variation in the scan levels, with the concomitant modification of the location of the ventricles, prevents a precise tracking of the evolution in tumour shape, a problem not encountered with the approach of Swanson (1999) discussed above.

As mentioned above, the model parameters were estimated from the clinical data. The basic parameters, the cell proliferation and diffusion parameters, are robust in that small changes to their values give quantitatively similar results. On the other hand, the parameters measuring the strength and duration of the chemotherapy treatments are more sensitive to small changes, as described. Even more important than the agreement between the parameters for each level is the agreement between simulated and observed values determined in other experiments and also from various other biological data as given in Tables 11.5 and 11.6. This lends further credence to the method used to estimate the parameters.

The initial cell density (Day 1) corresponds to an average value of 1026 cells/mm^2 measured by the biopsy carried out after the first scan. The cell density given by the model at the end of the simulation time (Day 356) gave an average of 710 cells/mm^2, which is very close to the one measured at autopsy, specifically an average of 750 cells/mm^2.

The growth rate of the second mutant cell population estimated by the model is not significantly different from that of the first cell population. The associated doubling times are 66 days for the first cell population and 60 days for the second cell type (Table 11.9), which is in agreement with the values reported for this type of neoplasm (Alvord and Shaw 1991).

The value of the diffusion coefficient in Table 11.9 (here given in cm^2/day) was determined by using a scale factor of $(5/2.2)^2$ between the scan and the real brain size. The mean value identified from the model is 1.25×10^{-7} cm^2/sec. This value can be compared with estimation of the rate of migration of cancerous cells *in vitro* (Chicoine and Silbergeld 1995). Using the experimental speed value of $v = 15$ microns/hour and

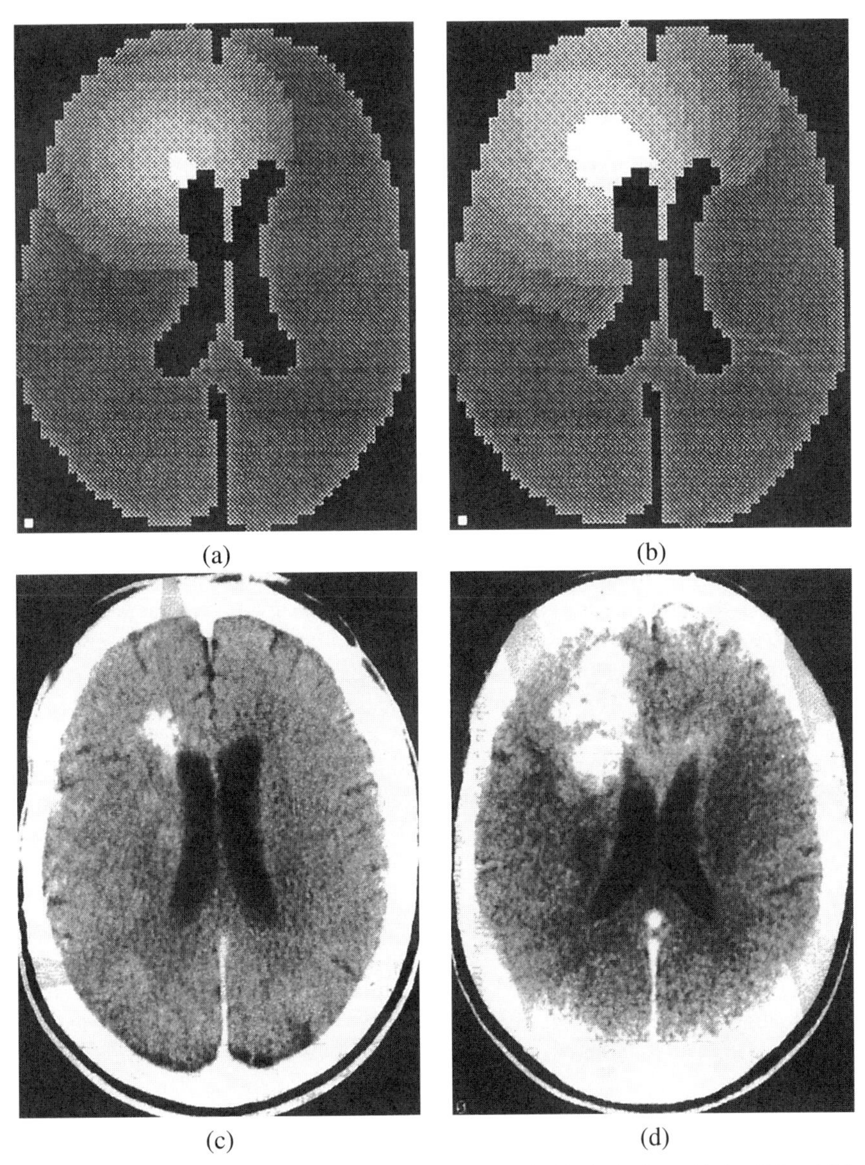

Figure 11.30. The image of the spatial spreading of the tumour inside the brain at (**a**) day 113 and (**b**) day 300 and, for comparison, the CT scans themselves at the same times are given in (**c**) and (**d**). (From Tracqui et al. 1995)

the mean value of the growth rates identified from the model, $\rho = 1.2 \times 10^{-2}$/day, we get an estimate of $D = v^2/4\rho = 3.13 \times 10^{-7}$ cm^2/s; see also Table 11.5.

Obviously other cases must be analysed to determine the ranges of the various parameters defined in the present case. To analyse other cases, at least a few MRI or CT scans over a time period (preferably at least a year) must be available, as well as histologic samples of the tumour and its margins from biopsy or autopsy. Since astrocytomas have estimated doubling times varying from one week to one year or more, in general correlating with their histological degree of malignancy, we anticipate that corresponding differences in r_1 and r_2 will be definable by the model in other cases. Of course, astrocytomas also respond to radio- and/or chemo-therapies in unpredictable degrees, so we anticipate corresponding differences in the parameters, k_1 and k_2.

Perhaps one of the most critical questions is how much the diffusion coefficient D varies from one glioma to another. Another question is whether the cell concentration or the cell type contributes independently to D.

Current therapies of gliomas are limited by the tolerance of normal brain for X-irradiation and of hematopoetic tissues for chemotherapies, but it is a reasonable expectation that, as additional therapies are developed, definitions of the parameters in the model will be helpful in planning the dose and timing of these therapies.

The prediction of tumour response to therapy is a goal which mathematical models can help to reach. This benefit should appear especially true for chemotherapy, which is a treatment suffering from a great lack of knowledge concerning drug pharmacokinetics and the existence of drug-resistant subpopulations within large populations of cells (both factors varying from patient to patient). A knowledge of the generic values of the model parameters for a given type of brain tumour will provide a first estimation of the tumour growth response to the routine treatment. Online adaptation of the parameters in space with the specific data obtained from each new CT or MRI scan of the treated patient will provide necessary corrections needed to approach optimal patterns of drug delivery.

We should voice, again, a word of caution on the use of models. A large series of cautious explorations and validations of the model predictions have to be made before an optimal clinical strategy can be reached. But this approach allows an improvement of cancer treatment which does not imply or depend on the discovery of completely new drugs, but rather suggests an optimal use and *in situ* evaluation of the already existing ones.

There are, of course, some obvious limitations of the model. Tumour heterogeneity can be taken into account by considering various populations with specific kinetics and possibly competitive interactions which can affect tumour growth dynamics (Michelson et al. 1987). The effect of different therapy scheduling can easily be investigated by shifting the time boundary values of the step functions modelling chemotherapy, and additional knowledge on the pharmacokinetics of the drugs can furthermore be introduced to modulate the shape of these latter functions. The effect of different extents of surgical excision was explored with this type of model by Woodward et al. (1996) and more recently by Swanson (1999) and Swanson et al. (2000), which crucially included facilitated diffusion in the direction of white fibre tracts. Although we model motility by diffusion it is not likely that tumour cells diffuse passively through normal brain tissue, which is too viscous to allow much passive diffusion. Instead, as shown experimentally

(Chicoine and Silbergeld 1995), tumour cells probably migrate by active amoeboid processes. The mathematical framework in which the present model has been developed is general enough to formalize these specificities, as exemplified by different models in population dynamics such as described in this book. Tracqui (1995) discusses this aspect by comparing the approach discussed in this section to the minimal hypotheses relating cell motility and cell traction forces. Other limitations and suggestions for improving the model are given by Tracqui et al. (1995).

With the approach here to parameter estimation, additional work is needed to determine to what extent modelling the tumour as a three-dimensional object significantly modifies the estimation of the model parameters, especially the diffusion coefficient. A major contribution to this is given by the work of Swanson (1999) and Swanson et al. (2000). She has also applied her approach (Swanson 1999) to the problem of chemotherapy treatment. However, the benefit of a more accurate description of the phenomena will necessitate a three-dimensional analysis of the brain and ventricular boundaries together with the heterogeneous distribution of grey and white matter.

11.11 Modelling Tumour Polyclonality and Cell Mutation

With the exception of the last section we have assumed that the tumour cell population is homogeneous involving only one cell type. In spite of the good quantitative comparison with data, such as regards life expectancy after resection, in the last section we saw that variable sensitivity to chemotherapeutic drugs exhibited by gliomas necessitated a model with tumour cell heterogeneity. Gliomas are known to be heterogeneous (polyclonal) with heterogeneity generally increasing with grade. The more malignant cells are believed to have a stronger propensity to mutate thus increasing heterogeneity. Therefore, we expect to see different types of cells within the tumour (for example, Pilkington 1992). The basic model can be extended to consider the case of a polyclonal tumour as we saw in the last section by simply creating two (or more) cell populations within the tumour that may have different diffusivity and growth rates. In the model equation (11.3) (the homogeneous tissue situation) the cell density $\bar{c}$ becomes the vector of cell densities $\bar{\mathbf{c}}$ and the diffusion coefficient D and the growth rate ρ now become diagonal matrices of diffusion coefficients and growth rates, respectively. The total tumour population, at a given point in space and time $(\bar{\mathbf{x}}, \bar{t})$, is the sum of the components of the $\bar{\mathbf{c}}$ vector. In the chemotherapy model in the last section we took the two cell populations to be independent. Often, however, the subpopulations are not independent but are connected by mutational events transferring cancer cells from cell population i to cell population j. A fairly general but basic model to account for population polyclonality can be written in the dimensional form

$$\frac{\partial \bar{\mathbf{c}}}{\partial \bar{t}} = \bar{\nabla} \cdot (D\bar{\nabla}\bar{\mathbf{c}}) + \rho\bar{\mathbf{c}} + T\bar{\mathbf{c}}, \tag{11.60}$$

where T is a matrix representing transfer between subpopulations. To complete the mathematical problem we use initial and boundary conditions

$$\bar{\mathbf{n}} \cdot D\bar{\nabla}\bar{\mathbf{c}} = 0 \qquad \text{for } \bar{\mathbf{x}} \text{ on } \partial B \text{ (the boundary of the brain)}$$

$$\bar{\mathbf{c}}(\bar{\mathbf{x}}, 0) = \bar{\mathbf{f}}(\bar{\mathbf{x}}) \qquad \text{for } \bar{\mathbf{x}} \text{ in } B \text{ (the brain domain).}$$

We expect that the introduction of multiple cell populations in the tumour introduces more heterogeneity in the growth patterns of the simulated tumour. Clinical and experimental results have shown fingering and branching of the visible tumour.

In this section we consider, by way of example, the existence of two clonal subpopulations within a tumour with mutational transfer. To be specific we assume one population has a high growth and low diffusion coefficient while the second population has a moderate growth rate and high diffusion coefficient; there are various other scenarios we could take. Let

$$\bar{\mathbf{c}} = \begin{pmatrix} \bar{u} \\ \bar{v} \end{pmatrix}, \quad D = \begin{pmatrix} D_1 & 0 \\ 0 & D_2 \end{pmatrix}, \quad \rho = \begin{pmatrix} \rho_1 & 0 \\ 0 & \rho_2 \end{pmatrix}, \quad T = \begin{pmatrix} -k & 0 \\ k & 0 \end{pmatrix},$$

where the D's, ρ's and k's are constant parameters, and (11.60) becomes

$$\begin{aligned} \frac{\partial \bar{u}}{\partial \bar{t}} &= \bar{\nabla} \cdot (D_1 \bar{\nabla} \bar{u}) + \rho_1 \bar{u} - k\bar{u}, \\ \frac{\partial \bar{v}}{\partial \bar{t}} &= \bar{\nabla} \cdot (D_2 \bar{\nabla} \bar{v}) + \rho_2 \bar{v} + k\bar{u} \end{aligned} \tag{11.61}$$

with given initial conditions $\bar{u}(\bar{\mathbf{x}}, 0) = \bar{f}(\bar{\mathbf{x}})$ and $\bar{v}(\bar{\mathbf{x}}, 0) = \bar{g}(\bar{\mathbf{x}})$. With $\bar{u}$ the more rapidly proliferating population and $\bar{v}$ the more rapidly diffusing population, $D_2 > D_1$ and $\rho_1 > \rho_2$. We further assume $\bar{u}$ cells are the only tumour cells initially present ($\bar{f}(\bar{\mathbf{x}}) > 0$ and $\bar{g}(\bar{\mathbf{x}}) = 0$). With some small probability, $k \ll \rho_1$, $\bar{u}$ cells mutate to form $\bar{v}$ cells. Although we do not include it here, each of the cell populations could retain the ability to diffuse faster in white matter regions.

Initially, let us suppose there is a source of $\bar{u}$ tumour cells that have mutated from healthy cells and can proliferate faster then the neighboring normal cells thereby starting to form a tumour. We can think of k as a measure of the probability of $\bar{u}$ tumour cells mutating to become the rapidly diffusing tumour cell population $\bar{v}$.

Introduce the nondimensional variables

$$\mathbf{x} = \sqrt{\frac{\rho_1}{D_1}}\bar{\mathbf{x}}, \quad t = \rho_1 \bar{t}, \quad \beta = \frac{\rho_2}{\rho_1} < 1, \quad \alpha = \frac{k}{\rho_1}, \quad \nu = \frac{D_2}{D_1}, \tag{11.62}$$

$$u(\mathbf{x}, t) = \frac{D_1}{\rho_1 u_0}\bar{u}\left(\sqrt{\frac{\rho_1}{D_1}}\bar{\mathbf{x}}, \rho_1 \bar{t}\right), \quad v(\mathbf{x}, t) = \frac{D_1}{\rho_1 u_0}\bar{v}\left(\sqrt{\frac{\rho_1}{D_1}}\bar{\mathbf{x}}, \rho_1 \bar{t}\right), \tag{11.63}$$

where $u_0 = \int \bar{f}(\bar{\mathbf{x}})d\bar{\mathbf{x}}$, the total original cancer cell population. Growth is measured on the timescale of the $\bar{u}$ population proliferation and diffusion is on the spatial scale of the $\bar{u}$ cell diffusion.

Equations (11.61) now become

$$\frac{\partial u}{\partial t} = \nabla^2 u + u - \alpha u, \tag{11.64}$$

$$\frac{\partial v}{\partial t} = \nu \nabla^2 v + \beta v + \alpha u. \tag{11.65}$$

The parameter $\alpha = k/\rho_1 \ll 1$ is the proportion of the first subpopulation's growth lost to mutation.

In one space dimension on an infinite domain we can write down the analytical solutions from which some interesting and highly relevant conclusions can be deduced as we shall see. In one space dimension

$$\frac{\partial u}{\partial t} = \frac{\partial^2 u}{\partial x^2} + u - \alpha u, \tag{11.66}$$

$$\frac{\partial v}{\partial t} = \nu \frac{\partial^2 v}{\partial x^2} + \beta v + \alpha u. \tag{11.67}$$

Let us take the initial source of u tumour cells to be $u(x, 0) = \delta(x)$ and take $v(x, 0) = 0$. The v population diffuses faster than u so $\nu > 1$. The growth rate of the u population is larger than v so $\beta < 1$. The growth rate of u is much higher than the probability of mutation so $\alpha \ll 1$.

The solution of (11.66) can be solved separately from the v equation and has the solution

$$u(x, t) = \frac{1}{\sqrt{4\pi t}} \exp\left((1 - \alpha)t - \frac{x^2}{4t} \right), \tag{11.68}$$

which on substituting this result into the v equation (11.67) gives

$$\frac{\partial v}{\partial t} = \nu \frac{\partial^2 v}{\partial x^2} + \beta v + \alpha \frac{1}{\sqrt{4\pi t}} \exp\left((1 - \alpha)t - \frac{x^2}{4t} \right). \tag{11.69}$$

We use a Fourier transform in the spatial variable x to solve this equation. With the transform and its inverse defined by

$$\mathcal{F}[f(x, t)](t; \omega) = \int_{-\infty}^{\infty} f(x, t) e^{-i\omega x}\, dx,$$

$$\mathcal{F}^{-1}[F(t; \omega)](x, t) = \frac{1}{2\pi} \int_{-\infty}^{\infty} F(t; \omega) e^{i\omega x}\, d\omega$$

the transformed equation for v is then

$$\frac{\partial V}{\partial t} = \nu (i\omega)^2 V + \beta V + \alpha e^{(1-\alpha-\omega^2)t} \quad \text{with} \quad V(t = 0; \omega) = 0,$$

where $V(t; \omega) = \mathcal{F}[v(x, t)]$. The solution for V is then given by

$$V(t; \omega) = \frac{\alpha \left[e^{(1-\alpha-\omega^2)t} - e^{(\beta-\nu\omega^2)t} \right]}{1 - \alpha - \beta + \omega^2(\nu - 1)}. \tag{11.70}$$

Taking the inverse transform gives, after some algebra, $v(x,t)$ as

$$\begin{aligned} v(x,t) &= \mathcal{F}^{-1}[V(t;\omega)] \\ &= \alpha\mathcal{F}^{-1}\left[e^{(1-\alpha-\omega^2)t} - e^{(\beta-\nu\omega^2)t}\right] * \mathcal{F}^{-1}\left[\frac{1}{1-\alpha-\beta+\omega^2(\nu-1)}\right] \\ &= \alpha e^{(1-\alpha)t}\int_{\xi=-\infty}^{\infty} \frac{\exp\left(-\frac{(x-\xi)^2}{4t}\right)}{\sqrt{4\pi t}} \frac{\exp(-A|\xi|)}{A(\nu-1)}\,d\xi \\ &\quad - \alpha e^{\beta t}\int_{\xi=-\infty}^{\infty} \frac{\exp\left(-\frac{(x-\xi)^2}{4\nu t}\right)}{\sqrt{4\pi\nu t}} \frac{\exp(-A|\xi|)}{A(\nu-1)}\,d\xi. \end{aligned} \tag{11.71}$$

If we now assume $1-\alpha-\beta>0$, integration of the convolution integrals gives

$$\begin{aligned} v(x,t) = \frac{\alpha\exp\left(\frac{\nu(1-\alpha-\beta)}{\nu-1}t\right)}{2A(\nu-1)}&\left[e^{-Ax}\left(\operatorname{erf}\left[\frac{x-2At}{2\sqrt{t}}\right] - \operatorname{erf}\left[\frac{x-2A\nu t}{2\sqrt{\nu t}}\right]\right)\right. \\ &\left. - e^{Ax}\left(\operatorname{erf}\left[\frac{x+2At}{2\sqrt{t}}\right] - \operatorname{erf}\left[\frac{x+2A\nu t}{2\sqrt{\nu t}}\right]\right)\right], \end{aligned} \tag{11.72}$$

where $A=\sqrt{(1-\alpha-\beta)/(\nu-1)}$. An important implication of this solution is that for certain values of α, we see that the initially nonexistent subpopulation v can eventually dominate the growth of the tumour. Swanson (1999) computed the two populations using this solution and Figure 11.31 shows the transition of dominance from the rapidly growing u population to the rapidly diffusing v population. Although the initial tumour consisted of only u cells, for large time the v subpopulation can dominate depending on the value of the mutation probability parameter α. Yet again, we find that the diffusion is more important to glioma growth and invasion then proliferation.

There are two relevant ways to view transition of dominance: (i) the total tumour cell population volume is dominated at later times by the second, more aggressive, tumour subpopulation v; or, (ii) v dominates at a single point or neighborhood of points, say, near the centre of the tumour, but does not necessarily fill the volume.

When performing a biopsy of a tumour, the sample of tissue is extracted from a fairly random location within the tumour. Analysis of this piece of tissue is expected to give an accurate portrayal of the tumour composition. Since it is often the case that different grades of tumour cells are histologically (physically) distinct, a pathologist could quantify the proportion of the tumour that is of a certain grade from the tissue biopsy. In particular, we expect that a biopsy would reveal information about the clonal distribution of the total tumour volume and the results would not be significantly different from that if the entire tumour were analyzed (as is possible postmortem). From Swanson's (1999) analysis below and our discussion of transmission of dominance at a given location and in volume, it is not necessarily clear that a biopsy will, in fact, define an accurate representation of the total tumour cell composition (tumour volume).

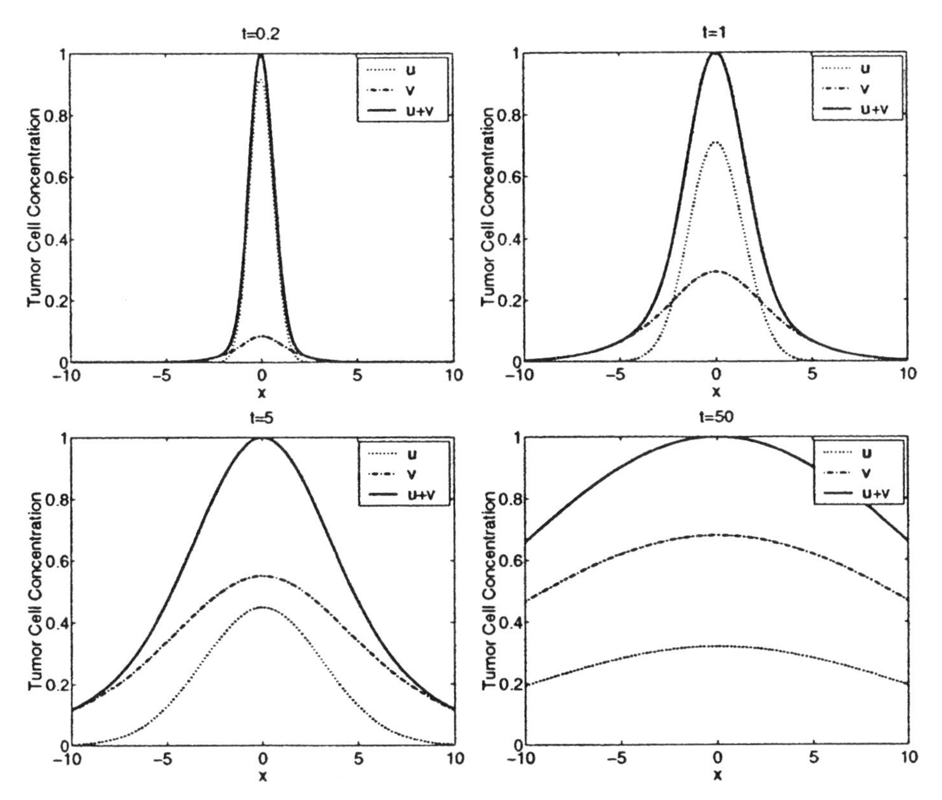

Figure 11.31. Transition of dominance from u to v. Between $t = 0.2$ and $t = 50$, there is a transition in dominance from the highly proliferative u subpopulation to rapidly diffusing v subpopulation. Parameters: $\alpha = 0.5$, $\nu = 10$, and $\beta = 0.1$. (From Swanson 1999)

Transition of Dominance at a Given Location

To determine if the subpopulation v eventually dominates the tumour composition at a given location, say, the centre of the tumour $x = 0$ where a biopsy would be performed, we consider the ratio of the two subpopulations $v(x,t)/u(x,t)$. Certainly, for small t, this ratio is less than 1 since u is initially the only population present. The v subpopulation becomes dominant if this ratio $v/u > 1$.

For large t, the ratio of the two subpopulations is approximated by

$$\begin{aligned}\frac{v(x,t)}{u(x,t)} &\sim \frac{2\alpha}{1-\alpha-\beta} - \frac{\alpha(\nu-1)}{(1-\alpha-\beta)^2}\left(\frac{1}{t}\right)\\ &\quad - \frac{\alpha x^2[x^2(1-\alpha-\beta)-4(\nu-1)]}{16(1-\alpha-\beta)}\left(\frac{1}{t^2}\right) + O\left(\frac{1}{t^3}\right). \qquad (11.73)\end{aligned}$$

So, for large time, the ratio tends to a constant value $2\alpha/(1-\alpha-\beta)$. If $2\alpha/(1-\alpha-\beta) > 1$ then the second subpopulation v dominates for large time. The condition for the diffuse population v dominance at a point ($x = 0$) is then given by

$$\frac{2\alpha}{1-\alpha-\beta} > 1 \implies \beta > 1 - 3\alpha. \tag{11.74}$$

Note that this parameter condition does not depend on ν, the ratio of the diffusion coefficients of u and v. We can also use the asymptotic expression (11.73) to estimate the time at which the transition of dominance occurs. Setting (11.73) equal to 1 and solving for t, we find that the time at which transition of dominance at $x = 0$ occurs $t_{\text{dominance}}$, say, is given approximately by

$$t_{\text{dominance}} \sim -\frac{\alpha(\nu - 1)}{(2\alpha + \beta - 1)^2 - \alpha}, \tag{11.75}$$

which does depend on ν and is positive if the condition (11.74) for transition of dominance at a point is satisfied. For large ν, the diffusion coefficient of the v cells is much higher than the u cells and the time to transition of dominance is large. However, if the parameters α and β satisfy the parametric condition (11.74), the time to this transition can be quite short if the diffusion coefficients are nearly equivalent ($\nu \approx 1$).

Transition of Dominance in Volume

Let us now consider the situation in which we can analyze the entire tumour and determine the proportion of the tumour volume occupied by u and v cells separately. This is biologically analogous to a detailed postmortem analysis of the entire brain. In this case, transition of dominance occurs if the volume occupied by v cells exceeds that occupied by u cells.

The model equations (11.66) and (11.67) define the spatiotemporal dynamics of the u and v subpopulations. By integrating these equations over all space, we can determine the temporal behavior of the tumour cell subpopulation volumes. Define $V_u(t)$ and $V_v(t)$ to be the volume of the tumour occupied by u and v cells, respectively, then

$$V_u(t) = \int_{-\infty}^{\infty} u(x,t)\,dx \quad \text{and} \quad V_v(t) = \int_{-\infty}^{\infty} v(x,t)\,dx. \tag{11.76}$$

Integrating equations (11.66) and (11.67) over all space (the infinite domain) we get

$$\frac{dV_u}{dt} = (1-\alpha)V_u \qquad \text{with} \quad V_u(0) = 1,$$

$$\frac{dV_v}{dt} = \beta V_v + \alpha V_u \qquad \text{with} \quad V_v(0) = 0.$$

Solving the first equation gives

$$V_u(t) = e^{(1-\alpha)t} \tag{11.77}$$

and V_v then satisfies

$$\frac{dV_v}{dt} = \beta V_v + \alpha e^{(1-\alpha)t} \quad \text{with} \quad V_v(0) = 0 \tag{11.78}$$

which has solution

$$V_v(t) = \frac{\alpha\left(e^{(1-\alpha)t} - e^{\beta t}\right)}{1-\alpha-\beta}. \tag{11.79}$$

Transition of dominance in the volume occurs if the ratio $V_v/V_u = 1$ at some finite time. So,

$$\begin{aligned}\frac{V_v}{V_u} = 1 \implies e^{(1-\alpha-\beta)t} &= \frac{\alpha}{2\alpha+\beta-1}\\ \implies t_{\text{dominance}} &= \frac{1}{1-\alpha-\beta}\ln\left(\frac{\alpha}{2\alpha+\beta-1}\right),\end{aligned} \tag{11.80}$$

where the time of transition of dominance exists (that is, positive) if

$$1-\alpha-\beta > 0 \quad \text{and} \quad 2\alpha+\beta-1 > 0.$$

Equivalently, the condition on the parameters for transition of dominance in volume is

$$1-2\alpha < \beta < 1-\alpha. \tag{11.81}$$

Figure 11.32 shows the parameter domains (in the α–β plane) for transition of dominance at a point, in volume and both. The shaded region of Figure 11.32(a) represents the parameter domain for which transition of dominance at a point (say, the origin) occurs while Figure 11.32(b) represents the parameter values for which transition of dominance by volume is possible. Clearly, there are values of α and β for which transition of dominance at a point occurs and that of volume does not. This indicates a failure for accurate biopsies of tumours. Analysis of a portion of tissue extracted from the centre of a tumour will not necessarily reveal the actual total tumour composition.

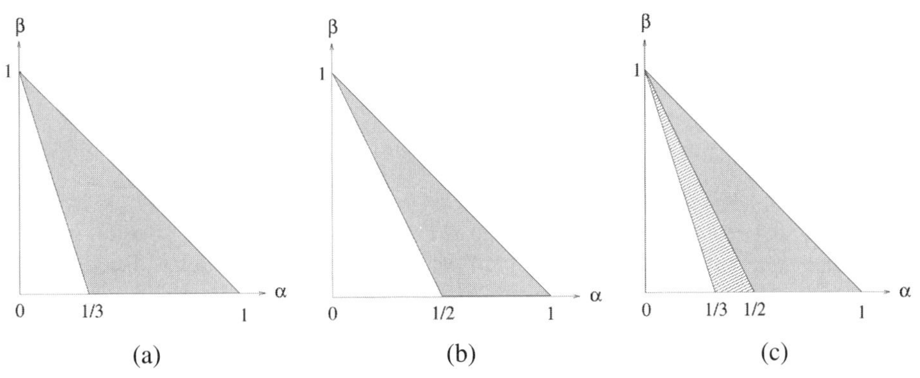

Figure 11.32. Parameter domains for which transition of dominance occurs: (**a**) at a point (taken to be $x = 0$ here); (**b**) by volume; (**c**) both. (From Swanson 1999)

Multi-Cell Model with Tumour Polyclonality

We now briefly consider a simple model with homogeneous diffusion in which there is mutation from one cell line to another but in which there are n lines: the formulation mimics what we did above but there is the possibility of very much more complex spatiotemporal dynamics. The general problem is then, in appropriate dimensionless form,

$$\frac{\partial \mathbf{c}}{\partial t} = \nabla \cdot (D\nabla \mathbf{c}) + P\mathbf{c} + T\mathbf{c} \tag{11.82}$$

with $\mathbf{c}$ satisfying

$$\mathbf{n} \cdot D\nabla \mathbf{c} = 0 \quad \text{for } \mathbf{x} \text{ on } \partial B, \qquad \mathbf{c}(\mathbf{x}, 0) = \mathbf{f}(\mathbf{x}), \tag{11.83}$$

where D, P and T are respectively the diffusion coefficient matrix, growth rate matrix and transfer matrix between the populations.

Define the eigenvalue problem by

$$\nabla \cdot (D\nabla \mathbf{W}) + K^2\mathbf{W}(\mathbf{x}) = 0, \quad \mathbf{n} \cdot D\nabla \mathbf{W} = 0, \tag{11.84}$$

where K denotes the eigenvalues. We now look for solutions (cf. Chapter 2) in the form

$$\mathbf{c}(\mathbf{x}, t; K_i) \propto e^{(P+T-K^2I)t}, \tag{11.85}$$

where the K_i are the discrete eigenvalues and I is the identity matrix. Keep in mind that the exponent of the exponential is a matrix. The spatiotemporal behaviour on a linear basis is that determined by the eigenvalues $\sigma(\rho_i, K_i, D_i)$, where the ρ_i and D_i are components in the growth matrix and diffusion matrix respectively. The eigenvalues are given by

$$|\, P + T - (K_i^2 + \sigma)I \,| = 0. \tag{11.86}$$

The analysis of the spatiotemporal behaviour of the solutions here is a generalisation of the two-species model systems we have discussed extensively in many chapters of the book. The implications for tumour growth and invasion governed by this type of model are still unexplored.

Some Concluding Comments

Cancer is no single disease. We have only considered some of the basic models for the growth and control of brain tumours. We have seen that, even with such simple linear models as discussed in this chapter, there is a surprising richness of potentially applicable results. There is clearly much still to be done with this modelling approach.

There is a large literature on models for other tumours and it is impossible to mention them all. One place to start is with several of the articles on cancer and related topics in the book edited by Chaplain et al. (1999). Other tumours can involve a wide range of biological phenomena such as angiogenesis and capillary networks (Chaplain and Anderson 1999), pattern formation in cancer macrophage dynamics (Owen and Sherratt 1997), role of inhibitors and cell-adhesion (Byrne and Chaplain 1995, 1996) and cell traction (Holmes and Sleeman 2000). What seems very clear is that the use of realistic models can help to elucidate many of the complex processes (or at the very least pose informative questions) that take place in the growth of malignant tumours; the closer the connection to the medical situation the more useful are the results.

A very good example of how such a close connection between experiment and theory can be so productive is given by the new approach to cancer treatment, notably melanoma, which has been developed by Jackson and her colleagues (Jackson 1998, Jackson et al. 1999a,b,c). Her work is directed to improving the efficacy of chemotherapeutic agents for treating cancer. It involves a two-step process which is designed to minimize the toxicity to the body while at the same time maximizing the toxicity in the cancer. The treatment involves enzyme-conjugated antibodies (ECA) in combination with prodrugs. Basically an antibody that binds to a tumour-associated antigen is conjugated to an enzyme not present in the host and then injected into the bloodstream. The ECA distributes itself in the tissue throughout the body and after some time is localized within the tumour because of high binding affinity. The prodrug is then injected into the body and the enzyme then converts the prodrug back into the toxic form. In this way the drug penetrates the tumour with minimal penetration into normal tissues and the bloodstream. The modelling involves taking this scenario and the basic biochemistry and constructing a realistic model which can be tested experimentally *in vitro*. The model mechanism involves space and time and gives rise to a different type of system of nonlinear coupled partial differential equations. Because of the experimental interdisciplinary collaboration, Jackson (1998) was able to assign realistic parameter values to her model, which makes the comparison with experiment so relevant and her approach important. The close quantitative agreement she was able to get with the experiments (several of which were motivated by the theory) is particularly encouraging for this approach to chemotherapy regimens for various cancers, but not, unfortunately for brain tumours.

12. Neural Models of Pattern Formation

Perhaps the most obvious complex spatial patterning processes are those associated with the nervous system such as pattern recognition and the transmission of visual information to the brain. This is a vast field of study. In this chapter we give only an introduction to some of the models, involving nerve cells, which have been proposed as pattern generators. Basic to the concept of neural activity is the nerve cell, or neuron. The neuron consists of a cell body with its dendrites, axon and synapses. It is a bit like a tree with the roots the dendrites, the base the cell body, the trunk the axon and the numerous branches the synapses. The cell receives information, from other cells, through its dendrites, passes messages along the axon to the synapses which in turn pass signals on to the dendrites of other cells. This neuronal process is central to brain functioning. The axons are the connectors and make up the white matter in the brain with the dendrites and synapses making up the grey matter; Figure 11.4 in the last chapter shows the white and grey matter distribution in the human brain.

We start pedagogically in Section 12.1 with a simple scalar model to introduce the basic concepts and show how spatial pattern evolves from a neural type of model. In Section 12.2 we derive a model for stripe formation, the ocular dominance stripes, in the visual cortex based on neural activation while in Section 12.3 we discuss a theory of hallucination patterns. In Section 12.4 we describe and analyse an application of a neural activity theory to the formation of mollusc shell patterns. Finally in Section 12.5 we discuss the possible connection between drug-induced hallucinogenic patterns and palaeolithic rock paintings and shamanism.

12.1 Spatial Patterning in Neural Firing with a Simple Activation–Inhibition Model

Nerve cells can fire spontaneously; that is, they show a sudden burst of activity. They can also fire repeatedly at a constant rate. Whether or not a cell fires depends on its autonomous firing rate and the excitatory and inhibitory input it gets from neighbouring cells: this input can be from other than nearest neighbours, that is, long range interaction. Such input can be positive, which induces activity, or negative which inhibits it. In Section 11.5 in Chapter 11, Volume I we briefly discussed an integral equation, namely,

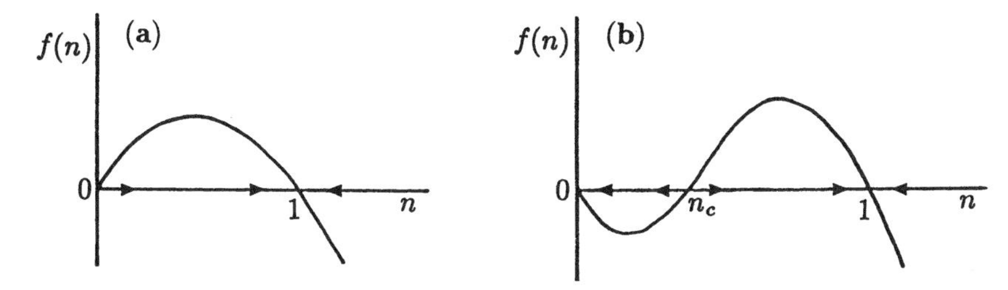

Figure 12.1. Functional form for the rate of change of the firing rate $f(n)$: (**a**) Situation with a single stable positive steady rate of firing, a representative $f(n) = rn(1-n)$; (**b**) Example with bistable threshold kinetics with linearly stable steady state rates $n = 1$ and the quiescent state $n = 0$. A perturbation from $n = 0$ to $n > n_c$ implies $n \to 1$: a representative $f(n) = rn(n - n_c)(1 - n)$.

(11.41), incorporating similar concepts. We showed how it related to a diffusion-type model and used it to introduce the idea of long range diffusion. Here we discuss it again and in more detail since it introduces the key ingredient for the models we discuss later in this chapter. Although this model is primarily for pedagogical purposes and is not a model for any specific phenomenon it turns out to be closely related, mathematically, to a special case of the neural model we discuss in the following section.

Consider the one-dimensional situation in which the cells are functions only of x and t. Denote by $n(x,t)$ the firing rate of the cells. In the absence of any neigbourhood influences we assume the cells can be in a quiescent state or fire autonomously at a uniform rate which we normalise to 1. If the cells' firing rate is perturbed we assume it evolves according to

$$\frac{dn}{dt} = f(n), \tag{12.1}$$

where $f(n)$ has zeros at $n = 0$ and $n = 1$, the steady state rates, with a functional form as in Figure 12.1(a). Here the only steady state is $n = 1$. $f(n)$ might exhibit bistable threshold kinetics whereby there is an unstable threshold steady state, n_c say, such that if $n > n_c$, $n \to 1$ and if $n < n_c$, $n \to 0$. A typical bistable form for $f(n)$ is illustrated in Figure 12.1(b). The kinetics of the firing dynamics (12.1) determines the subsequent firing rate given any initial rate n. In the case of bistable kinetics if $n(x,0) = 0$, input from neighbouring cells could temporarily raise the firing rate to $n > n_c$ in which case the cells could eventually fire at a constant rate $n = 1$.

Let us now include spatial variation and incorporate the effect on the firing rate of cells at position x, of neighbouring cells at position x'. We assume the effect of close neighbours is greater than that from more distant ones; the spatial variation is incorporated in a weighting function w which is a function of $|x - x'|$. We must integrate the effect of all neighbouring cells on the firing rate and we model this by a convolution integral involving an influence kernel.

To be specific let us take $f(n)$ as in Figure 12.1(a) where the positive steady state is $n = 1$. Now modify (12.1) so that if $w > 0$ there is a positive contribution to the firing rate from neighbours if $n > 1$ and a negative one if $n < 1$. Thus a possible first attempt at modelling the mechanism is the integrodifferential equation

$$\begin{aligned}\frac{\partial n}{\partial t} &= f(n) + \int_D w(x-x')[n(x',t)-1]\,dx' \\ &= f(n) + w*(n-1),\end{aligned} \tag{12.2}$$

where D is the spatial domain over which the influence kernel $w(x)$ is defined and $*$ denotes the convolution defined by this equation. The form (12.2) ensures that $n = 1$ is a solution. To complete the formulation of the model we must specify the influence kernel w, which we assume to be symmetric and so

$$w(|x-x'|) = w(x-x') = w(x'-x).$$

A nonsymmetric kernel could arise, for example, as the result of some superimposed gradient. To ensure that n is always nonnegative we must ensure that $n_t > 0$ for small n. This requires

$$\int_D w(|x-x'|)\,dx' < 0 \tag{12.3}$$

since $f(n) \to 0$ with n and (12.2) reduces to

$$\frac{\partial n}{\partial t} \sim -\int_D w(|x-x'|)\,dx' > 0.$$

This condition is satisfied with the typical kernels we envisage in the practical applications; see, for example, the analysis of the illustrative example in (12.13) below.

In this model we envisage a cell to have a short range activation effect and a long range inhibitory effect. This is typical of the cell behaviour in the models discussed in the following sections. Pattern formation concepts based on local activation and lateral, or long range, inhibition have been discussed fully in earlier chapters. Such a cell–cell influence is incorporated in a kernel of the form illustrated in Figure 12.2. On the infinite domain, $w(z)$ is a continuous symmetric function of the variable $z = x - x'$ such that

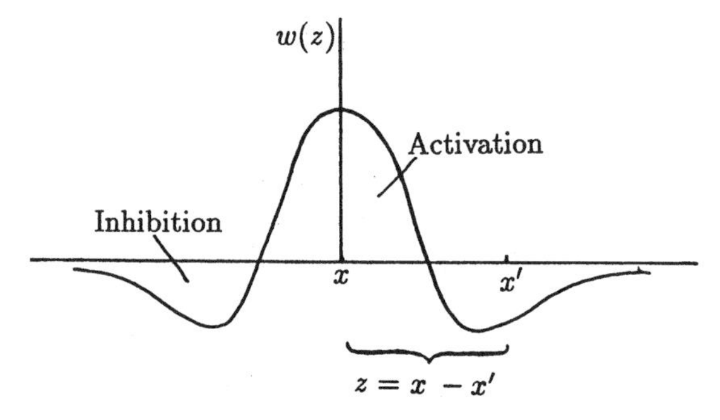

Figure 12.2. Typical kernel $w(z)$ which exhibits local activation—long range inhibition. $w(z)$ measures the effect of cells at position x' on cells at position x; here $z = x - x'$.

$$w(z) \to 0 \quad \text{as} \quad |z| \to \infty; \quad z = x - x'. \tag{12.4}$$

We can see intuitively how the mechanism (12.2) with $w(z)$ as in Figure 12.2 can start to create spatial patterns. To be specific consider $f(n)$ as in Figure 12.1(a) so $n = 1$ is the spatially uniform steady state solution of (12.2). Now impose small spatially heterogeneous perturbations about $n = 1$. If, in a small region about x where $w > 0$, we have $n > 1$, the effect of the integral term is to increase n autocatalytically in this region while at the same time inhibiting the nearest neighbourhood where $w < 0$. Thus small perturbations above the steady state will start to grow while those less than the steady state will tend to decrease further thus enhancing the heterogeneity. It is the classic activation–inhibition situation for generating spatial pattern. What we want, of course, is to quantify analytically the growth and pattern wavelength in terms of the model parameters.

Consider the infinite domain and linearise (12.2) about the positive steady state $n = 1$ by setting

$$\begin{aligned} &u = n - 1, \quad |u| \ll 1 \\ \Rightarrow \quad &u_t = -au + \int_{-\infty}^{\infty} w(|x - x'|)u(x', t)\,dx', \quad a = |f'(1)|. \end{aligned} \tag{12.5}$$

Now look for solutions, as usual, in the form

$$u(x, t) \propto \exp[\lambda t + ikx], \tag{12.6}$$

where k is the wavenumber and λ the growth factor. Substituting into (12.5), setting $z = x' - x$ in the integral and cancelling $\exp[\lambda t + ikx]$ gives λ as a function of k, that is, the dispersion relation, from (12.5) as

$$\lambda = -a + \int_{-\infty}^{\infty} w(z)\exp[ikz]\,dz = -a + W(k), \tag{12.7}$$

where $W(k)$, defined by the last equation, is simply the Fourier transform of the kernel $w(z)$. Remember that $w(z) = w(|z|)$ with the kernels we consider. We could of course solve (12.5) in general by taking the Fourier transform and using the convolution theorem on the integral term. This is equivalent to summing u in (12.6) over all k.

A simple symmetric kernel of the form illustrated in Figure 12.2 can, for example, be constructed from a combination of exponentials of the form

$$\exp[-bx^2], \quad b > 0 \tag{12.8}$$

which has Fourier transform

$$\int_{-\infty}^{\infty} \exp[-bx^2 + ikx]\,dx = \left(\frac{\pi}{b}\right)^{1/2} \exp\left[\frac{-k^2}{4b}\right]. \tag{12.9}$$

This exponential and its transform are sketched in Figure 12.3(a).

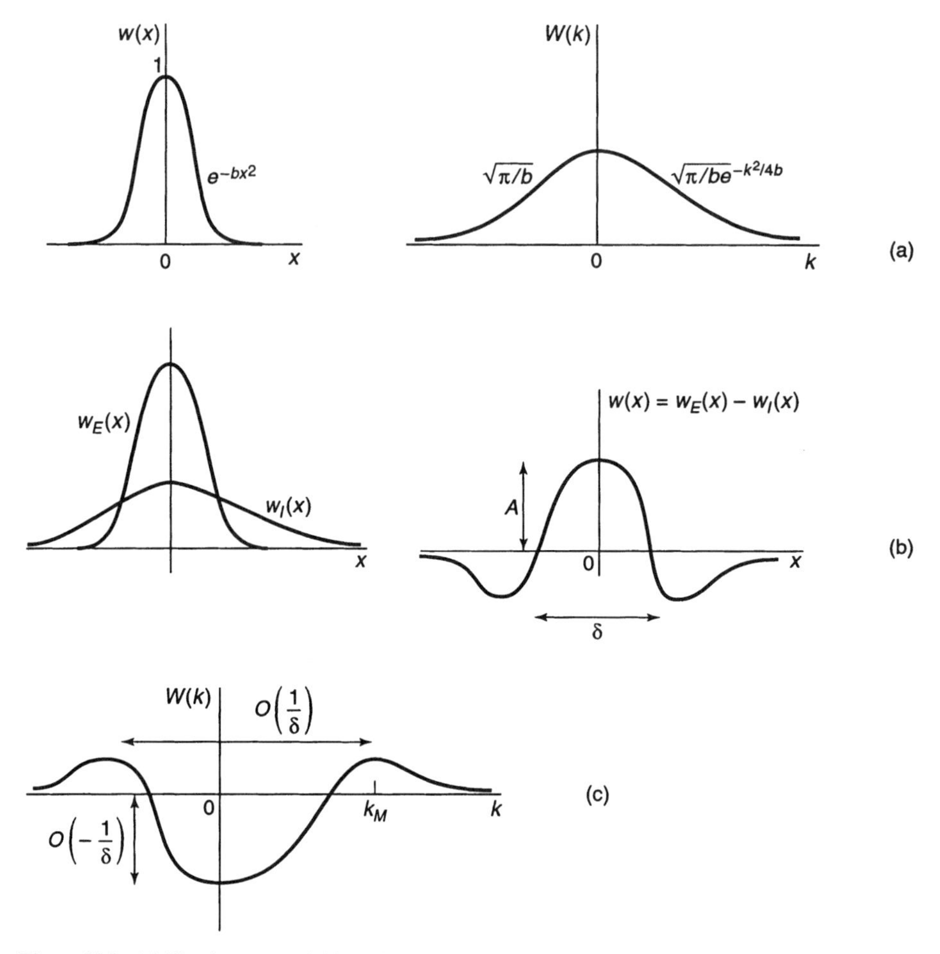

Figure 12.3. (**a**) Simple exponential kernel and its Fourier transform. Note how the widths and height in x- and k-space relate to each other. (**b**) Typical activation (w_E) and inhibition (w_I) kernels which give a composite kernel $w = w_E - w_I$ with a local activation ($w > 0$) and long range inhibition ($w < 0$). (**c**) Sketch of $W(k) = W_E(k) - W_I(k)$, the Fourier transform of $w_E - w_I$. Note again how the height and width of the kernel in (**b**) and its transform are related.

Two particularly relevant properties of Fourier transforms for our purposes are that the taller and narrower the function is in x-space the shorter and broader is its transform in k-space. Figure 12.3(b) shows how to construct a short range activation and long range inhibition kernel from two separate kernels while Figure 12.3(c) is a sketch of its Fourier transform. The transform kernel $W(k)$ is similar in shape to the original kernel $w(x)$ but upside down.

Now consider the dispersion relation (12.7) giving the growth factor $\lambda = \lambda(k)$. From the full discussion of dispersion relations in Chapter 2, specifically Section 2.5, we can determine much of the pattern generation potential of the model. From the form of the transform of the activation–inhibition kernel in Figure 12.3(c) we can plot λ as a function of k as shown in Figure 12.4. We see, from (12.7) that if the parameter $a(> 0)$

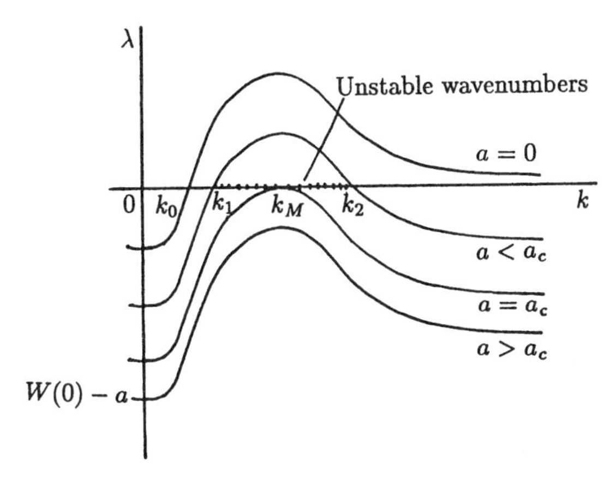

Figure 12.4. Dispersion relation $\lambda(k)$ as a function of the wavenumber k, from (12.7), for various values of the bifurcation parameter a. $W(k)$, sketched in Figure 12.3(c), defines $W(k_M)$ and k_M.

is large enough (and it is clearly a stabilising factor in equation (12.5)), $\lambda < 0$ for all wavenumbers k and the uniform solution $n = 1$ is linearly stable to all spatial disturbances. As the parameter a decreases, a critical bifurcation value a_c is reached where $\lambda = 0$ with $\lambda > 0$ for a finite range of k when $a < a_c$. The critical bifurcation a_c depends on the structure of the kernel: it is obtained from Figure 12.3(c), which defines k_M from (12.7) as $a_c = W(k_M)$, with k_M from (12.10). When $a < a_c$, for each $k_1 < k < k_2$ in Figure 12.4 $\lambda(k) > 0$ and the solution $u(x, t)$ in (12.6) is exponentially growing with time, like $\exp[\lambda(k)t]$. There is a fastest growing mode, with wavenumber k_M from (12.10), which dominates the linear solution. Eventually linear theory no longer holds and the nonlinear terms in (12.2) bound the solution and it evolves to a spatially heterogeneous steady state. In the case of the one-dimensional situation the final steady state finite amplitude structure is generally closely related to the wavelength of the fastest growing linear mode as was the case with other pattern formation mechanisms discussed in earlier chapters.

Suppose $0 < a < a_c$. Then, as t increases the dominant part of the solution $u(x, t)$ of (12.5) is given by the sum of all exponentially growing modes of the form (12.6), that is, those with wavenumbers bounded by k_1 and k_2 in Figure 12.4, namely,

$$u(x, t) \sim \int_{k_1}^{k_2} A(k) \exp[\lambda(k)t + ikx]\, dk,$$

where $A(k)$ is determined from the initial conditions. All other modes tend to zero exponentially since $\lambda(k) < 0$. Using Laplace's method the asymptotic expansion of the integral (see Murray 1984) gives

$$u(x, t) \sim A(k_M) \left\{ \frac{-2\pi}{t\lambda''(k_M)} \right\}^{1/2} \exp[\lambda(k_M)t + ik_M x].$$

The maximum λ is given by $\lambda(k_M)$ where $\lambda'(k_M) = 0$ and k_M is the maximum of $W(k)$; that is,

$$\lambda_M = \lambda(k_M), \quad W'(k_M) = 0, \quad W''(k_M) < 0. \tag{12.10}$$

The appearance of spatial pattern according to the dispersion relation in Figure 12.4 has almost, but crucially not quite, the no-frills basic dispersion relation form discussed in Chapter 2. It differs in a major way as $a \to 0$. From (12.7), $\lambda = W(k)$ when $a = 0$, which, as is clear from Figure 12.3(c), gives an *infinite* range of unstable wavenumbers to the right of k_0 where $W(k) = 0$. That is, disturbances with very large wavenumbers—very small wavelengths—are all unstable. The ultimate steady state spatial structures in this situation usually depend critically on the initial conditions. From the numerical simulations of the models in the following sections, even in two dimensions, irregular stripe patterns can dominate. The ultimate pattern here is strictly a nonlinear phenomenon.

In Chapter 11, Volume I we related the integral formulation (12.2) to the differential equation reaction diffusion formulation in which the spatial interaction is via diffusion. To relate our analysis to the discussion in Chapter 11, Volume I we start with (12.2) on an infinite domain, with w symmetric, and set $z = x' - x$ to get

$$n_t = f(n) + \int_{-\infty}^{\infty} w(z)[n(x+z,t) - 1]\,dz. \tag{12.11}$$

If the kernel's influence is restricted to a narrow neighbourhood around $z = 0$ we can expand $n(x - z)$ in a Taylor series to get

$$\begin{aligned} n_t = f(n) &+ [n(x,t) - 1]\int_{-\infty}^{\infty} w(z)\,dz + n_x \int_{-\infty}^{\infty} zw(z)\,dz \\ &+ \frac{n_{xx}}{2}\int_{-\infty}^{\infty} z^2 w(z)\,dz + \frac{n_{xxx}}{3!}\int_{-\infty}^{\infty} z^3 w(z)\,dz + \cdots . \end{aligned}$$

The integrals are the moments of $w(z)$. Exploiting the symmetry property of $w(z)$, the odd moments of the kernel, that is, those with odd powers of z in the integrand, are zero, so

$$\begin{aligned} n_t &= f(n) + w_0(n-1) + w_2 n_{xx} + w_4 n_{xxxx} + \cdots \\ w_{2m} &= \frac{1}{(2m)!}\int_{-\infty}^{\infty} z^{2m} w(z)\,dz, \quad m = 0, 1, 2, \ldots, \end{aligned} \tag{12.12}$$

where w_{2m} is the $2m$th moment of the kernel w. The signs of the moments depend critically on the form of w.

As an example, suppose we choose the kernel to be

$$w(z) = b_1 \exp\left[-\left(\frac{z}{d_1}\right)^2\right] - b_2 \exp\left[-\left(\frac{z}{d_2}\right)^2\right], \quad b_1 > b_2, \quad d_1 < d_2 \tag{12.13}$$

with b_1, b_2, d_1 and d_2 positive parameters; this form is of the type in Figure 12.3(b). Its transform (compare with (12.9)) is

$$\begin{aligned} W(k) &= \int_{-\infty}^{\infty} w(z)\exp[ikz]\,dz \\ &= \sqrt{\pi}\left\{b_1 d_1 \exp\left[-\frac{(d_1 k)^2}{4}\right] - b_2 d_2 \exp\left[-\frac{(d_2 k)^2}{4}\right]\right\}, \end{aligned} \tag{12.14}$$

which is of the type in Figure 12.3(c) if

$$W(0) < 0 \Rightarrow b_1 d_1 - b_2 d_2 < 0. \tag{12.15}$$

In this case k_M in Figure 12.3(c) and (12.10) exists and is given by the nonzero solutions of $W'(k) = 0$; namely,

$$k_M^2 = \frac{4}{(d_2^2 - d_1^2)} \ln\left[\frac{b_2}{b_1}\left(\frac{d_2}{d_1}\right)^3\right] > 0 \tag{12.16}$$

since $d_2/d_1 > 1$ and $b_2 d_2/b_1 d_1 > 1$.

The moments w_{2m}, defined by (12.12), can be evaluated exactly for the kernel (12.13) by noting that if

$$I(b) = \int_{-\infty}^{\infty} \exp[-bz^2]\,dz = \left(\frac{\pi}{b}\right)^{1/2}, \tag{12.17}$$

differentiating successively with respect to b immediately gives the even moments. For example,

$$I'(b) = -\tfrac{1}{2}\left(\frac{\pi}{b^3}\right)^{1/2} = -\int_{-\infty}^{\infty} z^2 \exp[-bz^2]\,dz.$$

With $w(z)$ in (12.13) we thus determine the moments as

$$w_0 = \sqrt{\pi}(b_1 d_1 - b_2 d_2), \quad w_2 = \frac{\sqrt{\pi}}{4}(b_1 d_1^3 - b_2 d_2^3), \ldots \tag{12.18}$$

and so on; they can be either positive or negative depending on the parameters. The decision as to where the expansion should be terminated in (12.12) depends on the relative magnitude of the moments. The wider the kernel's spatial influence the more terms that are required in the expansion.

If (12.15) holds for (12.13), $w_0 < 0$ and $w_2 < 0$. With $w_0 < 0$ this kernel satisfies the criterion (12.3) which ensures that the solution for n in (12.2) is always nonnegative. It is also clear from (12.12) in the space-independent situation for n small since

$$f(n) + w_0(n-1) \sim -w_0 > 0.$$

If $w_2 < 0$, that is, negative diffusion, this term is *destabilising* while if $w_4 < 0$, that is, negative long range diffusion, the 4th-order term is *stabilising*. This situation arises in an offshoot of the models discussed in Section 12.4 below. The evolution of spatially structured solutions of such higher-order equations like (12.12) with $w_2 < 0$, $w_4 < 0$ was analysed in detail by Cohen and Murray (1981) whose method of analysis can also be applied to the model mechanism (12.2).

Now that we have introduced the basic ideas let us consider some specific practical neural models.

12.2 A Mechanism for Stripe Formation in the Visual Cortex

Visual information is transmitted via the optic nerve from the retinal cells in each eye through a kind of relay station, called the lateral geniculate nucleus, to the visual cortex. The inputs from each eye are relayed separately and are distributed within a specific layer—layer IVc—of the cortex. Experiments involving electrophysiological recordings from the visual cortex of cats and monkeys have shown that the spatial pattern of nerve cells, neurons, in the visual cortex that are stimulated by the right eye form spatial bands, which are interlaced with bands of neurons which can be stimulated by the left eye. As mentioned, these neurons branch and form synapses (regions where nervous impulses move from one neuron to another). By moving the electrode within the cortex, and relating the response to the input from the eyes, the pattern of stripes is found to be about 350 μm wide and about the same apart. These are the *ocular dominance stripes* the explanation for which is associated with the names of D.H. Hubel and T.N. Wiesel (Hubel and Wiesel 1977) and for which they were awarded a Nobel Prize in 1981. Figure 12.5(a) shows a reconstruction of the spatial pattern of these ocular dominance stripes obtained from a macaque monkey.

The appearance of ocular dominance stripes is not universal in mammals; mice, rats and American monkeys, for example, do not seem to have them. It is possible that they are not necessary for the visual process but are associated with some other process. Even where they exist they are not always so clearly delineated as is the case with the cat and macaque monkey.

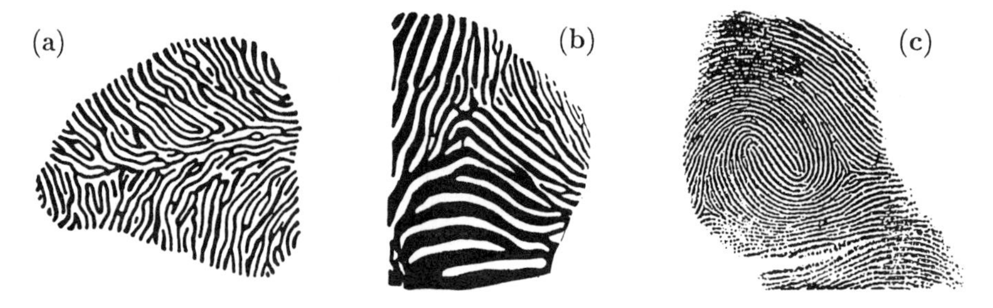

Figure 12.5. (**a**) Spatial pattern of ocular dominance stripes in the visual cortex of a macaque monkey. The dark bands are areas which receive input from one eye while the unshaded regions receive input from the other eye. (After Hubel and Wiesel 1977) (**b**) Stripe pattern on the rear flank of a Grevy's zebra (see also Figure 3.3(a)). (**c**) Typical human fingerprint pattern.

It is difficult not to be struck by the resemblance of ocular dominance stripes to many others in the natural world, such as the stripes on zebras as in Figure 12.5(b) and fingerprints as in Figure 12.5(c). We must, of course, not read too much into this similarity. Although it is possible that a neural-type model could be responsible for animal coat markings (although I very much doubt it) it is certainly not the case for fingerprint formation. The mechanochemical mechanism discussed earlier in Chapter 6 is a good candidate where it was used in Section 6.7.

In the case of the monkey it seems that the initial patterning process starts before birth, and is apparent just before birth. The pattern formation process is complete by six weeks of age. The actual stripe pattern can be altered as long as it has not been completely formed (Hubel et al. 1977). For example, removal or closure of one eye for about 7 weeks after birth causes the width of the bands, which reflect input from that eye, to become narrower while those from the other eye get broader. If both eyes are blindfolded there seems to be little effect on the stripes. There seems to be a critical period for producing such monocular vision since blindfolding seems to have no effect after the stripes have been formed. If these experiments are carried out after about 2 months of age, when the pattern generation is complete, there seems to be no effect on the pattern.

Hubel et al. (1977) suggested that the stripes are formed during development as a consequence of competition between the eye terminals in the cortex. That is, if a region is dominated by input synapses associated with the right eye it inhibits the establishment of left eye synapses, at the same time enhancing the establishment of right eye synapses. This is like an activation–inhibition mechanism involving two species. Intuitively we can see how this could produce spatial patterns in an initial uniform distribution of unspecified synapses on the cortex. It is also an attractive idea since it is in keeping with the experimental results associated with eye closure during development. For example, if one eye is closed there is no inhibition of the other eye's terminals and so there are no stripes. Swindale (1980), in a seminal paper, proposed a model mechanism based on this idea of Hubel et al. (1977) and it is his model we discuss in detail in this section.

Model Mechanism for Generating Ocular Dominance Stripes

The model is effectively an extension of that discussed in the last section except that here we consider two different classes of cells. Consider the layer (IVc) in the visual cortex—the layer on which the ocular dominance stripes are formed—to be a two-dimensional domain D and denote by $n_L(\mathbf{r}, t)$ and $n_R(\mathbf{r}, t)$ the left and right eye synapse densities on this cortical surface at position $\mathbf{r}$ at time t. The stimulus function, which measures the effect of neighbouring cells (neurons), is made up of positive and negative parts which respectively enhance and inhibit the cells' growth. As before this is represented by the sum of the product of a weighted kernel (w) and the cell density (n). The stimulus, both positive and negative, from neighbouring synapses on the rate of growth of right eye synapses n_R, for example, is taken to be the sum, s_R, of the appropriate convolutions; that is,

$$w_{RR} * n_R = \int_D n_R(|\mathbf{r} - \mathbf{r}^*|) w_{RR}(\mathbf{r}^*)\, d\mathbf{r}^*,$$

$$w_{LR} * n_L = \int_D n_L(|\mathbf{r} - \mathbf{r}^*|) w_{LR}(\mathbf{r}^*)\, d\mathbf{r}^*, \quad s_R = w_{RR} * n_R + w_{LR} * n_L. \tag{12.19}$$

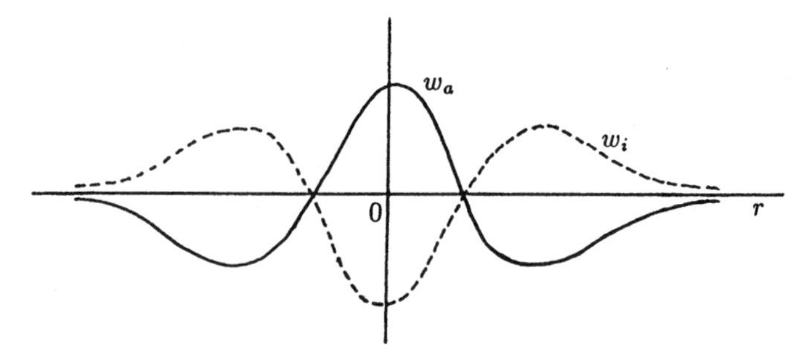

Figure 12.6. Typical activation, w_a, and inhibition, w_i, kernels, as functions of r, for ocular dominance stripe formation. The activation region around the origin is of the order of 200 μm with the inhibition region of order 400 μm.

The stimulus function s_L for the left synapses n_L is defined similarly. It is reasonable to assume that under normal developmental conditions the activation and inhibition effects from each type of synapse is symmetric; that is,

$$w_{RR} = w_{LL} = w_a, \quad w_{RL} = w_{LR} = w_i.$$

We envisage these weighting kernels to exhibit short range activation with long range, or lateral, inhibition such as illustrated in Figure 12.2, where the activation region is of the order of half the stripe width, say, 200 μm, with the inhibition region from about 200 μm to 600 μm. With the above idea of right synapses locally enhancing growth of right synapses and laterally inhibiting left synapses, and vice versa, the qualitative forms of w_a and w_i are illustrated in Figure 12.6.

The cell densities n_R and n_L must, of course, be nonnegative and bounded above by some cell density, N say. We must include such restraints in the model formulation and do so with density-dependent multiplicative factors on the stimulation functions. We thus propose as the model mechanism

$$\begin{aligned} \frac{\partial n_R}{\partial t} &= f(n_R)[w_a * n_R + w_i * n_L] \\ \frac{\partial n_L}{\partial t} &= f(n_L)[w_a * n_L + w_i * n_R], \end{aligned} \tag{12.20}$$

where $f(n)$ has zeros at $n = 0$ and $n = N$ with $f'(0) > 0$ and $f'(N) < 0$; a logistic form for $f(n)$ is a reasonable one to take at this stage.

Analysis of the Model

The mechanism (12.20), with $f(n)$ qualitatively as described, clearly limits the growth of synapses and reflects, for example, the existence of some factor which reduces the rate of synapse growth with increasing density of that synapse density. The model also implies that if the synapses from a given eye type ever disappear from a region of the cortex they can never reappear.

The last property is the same as saying that the steady states $n_R = 0$, $n_L = N$ and $n_L = 0$, $n_R = N$ are stable. For example, suppose we consider n_R and look at the stability of $n_R = 0$, $n_L = N$ from the first of (12.20). To be specific let us take $f(n) = n(N - n)$. For small n_R in a given region, any growth is inhibited by the neighbouring presence of n_L; that is, $w_i * n_L < 0$ in the first of (12.20). So, in this region, retaining only first-order terms in n_R, (12.20) gives

$$\frac{\partial n_R}{\partial t} \approx N n_R w_i * N < 0$$

and so $n_R = 0$, $n_L = N$ is stable. Symmetry arguments show that $n_L = 0$, $n_R = N$ is also linearly stable.

Many of the numerical simulations were carried out with the initial distribution such that the total density at any point, namely, $n_R + n_L$, is a constant and equal to N. This would be the case if there were a fixed number of postsynaptic sites on the cortex which are always occupied by one or other eye synapse. This assumption implies that $\partial n_R/\partial t = -\partial n_L/\partial t$. The two equations in (12.20) when written in terms of one of the densities must be the same. This requires $w_a = -w_i$. This simplifies the model considerably, which then reduces to

$$\begin{aligned} \frac{\partial n_R}{\partial t} &= f(n_R)[w_a * (2n_R - N)] = f(n_R)[2w_a * n_R - K], \\ n_L &= N - n_R, \quad K = N \int_D w_a(|\mathbf{r} - \mathbf{r}^*|)\, d\mathbf{r}^*, \end{aligned} \tag{12.21}$$

namely, a scalar equation for n_R with certain common features to that considered in detail in the last section; $w_a(r)$, the solid line kernel in Figure 12.6, is similar to that in Figure 12.2.

Equation (12.21) has 3 steady states:

$$n_R = 0, n_L = N; \quad n_R = N, n_L = 0; \quad n_R = n_L = \frac{N}{2}. \tag{12.22}$$

We have already shown that the first two are linearly stable.

Consider now the stability of the third steady state in which the synapses from each eye are equally distributed over the cortex. The model intuitively implies this state is unstable. Consider a small perturbation $u(\mathbf{r}, t)$ about the steady state, substitute into the first equation in (12.21) and keep only first-order terms in u to obtain

$$u_t = \tfrac{1}{2} N^2 [w_a * u]. \tag{12.23}$$

This equation is similar to that in (12.5) if $a = 0$. Now look for solutions in a similar way to that in (12.6), but now in two space dimensions; that is, we take

$$u(\mathbf{r}, t) \propto \exp[\lambda t + i\mathbf{k} \cdot \mathbf{r}], \tag{12.24}$$

where $\mathbf{k}$ is the eigenvector with wavelength $2\pi/k$ where $k = |\mathbf{k}|$. Substitution into (12.23) gives the dispersion relation as

$$\lambda = \frac{1}{2}N^2 W_a(k), \quad W_a(k) = \int_D w_a(\mathbf{r} - \mathbf{r}^* |) \exp[i\mathbf{k} \cdot \mathbf{r}^*]\, d\mathbf{r}^*. \tag{12.25}$$

$W_a(k)$ is simply the Fourier transform of the kernel w_a over the domain of the cortex.

If we consider the one-dimensional situation, $W_a(k)$ is similar in form to $W(k)$ in Figure 12.3(c) since w_a is similar to $W(k)$ there. Referring to Figure 12.4 we see that there is an infinite range of unstable wavenumbers. Although there is a wavenumber k_M giving a maximum growth rate $\lambda = \lambda_M$ it is not clear that it will dominate when nonlinear effects are included. In fact numerical simulations indicate a strong dependence on initial conditions, as appears to be the case whenever there is an infinite range of unstable modes. The two-dimensional situation is similar in that there is an infinite range of wavevectors $\mathbf{k}$ which give linearly unstable modes in (12.24). The dominant solution is given by an integral of the modes in (12.24), with $\lambda(k)$ from (12.25), over the $\mathbf{k}$-space of all unstable wavenumbers, namely, where $W_a(k) > 0$: it is infinite in extent. The instability evolves to a spatially heterogeneous steady state.

Numerical simulation of the model mechanism confirms this and generates spatial patterns such as in Figure 12.7(a), which shows the result of one such computation.

The development of ocular stripes takes place over a period of several weeks during which considerable growth of the visual cortex takes place. This can have some effect on the stripe pattern formed by the model mechanism. Numerical simulations indicate that the stripes tend to run in the direction of growth as shown in Figure 12.7(b).

With the model developed here it is possible to carry out a variety of 'experiments' such as monocular deprivation. This can be done by restricting the input from one eye for a period of time during development. Swindale (1980) carried out simulations in this situation and found certain critical periods where deprivation played a major role; these results are in keeping with many of the real experimental results.

Figure 12.7. (**a**) Numerical solution of the model with constant total synapse density N with antisymmetric kernels ($w_a = -w_i$); that is, (12.21). The kernel chosen, similar to that in (12.13), was $w_a = A\exp[-r^2/d_1^2] - B\exp[-(r-h)^2/d_2^2]$ with parameter values $A = 0.3$, $B = 1.0$, $d_1 = 5$, $d_2 = 1.4$, $h = 3.7$. The dark and light regions correspond to the final steady state synapse densities n_R and n_L. (**b**) Simulation giving the pattern formed when the domain is subjected to unidirectional growth of 20% during pattern formation: here n_R and n_L could vary independently. The same kernel as in (**a**) was used for w_a and w_i, with parameter values $A = 1$, $B = 0.9$, $d_1 = 3$, $d_2 = 10$, $h = 0$ for w_a and $A = -1$, $B = -1$, $d_1 = 4$, $d_2 = 10$, $h = 0$ for w_i. (Redrawn from Swindale 1980)

It is interesting to note that this, and other neural models seem to have a preference for stripe pattern formation whereas reaction diffusion models in large domains seem to favour a more spotted pattern. It is interesting mathematically why this is the case. Conditions for spots and stripes formed by several different pattern formation mechanisms (although not this model) are derived by Zhu and Murray (1995). From a linear analysis we see that there is an *infinite* range of unstable modes with the eigenvalues bounded away from $k = 0$. From a linear instability point of view this is one of the main differences between the dispersion relation here and those in Chapter 2. It would be interesting to determine when a model with a dispersion relation with an unbounded range of unstable wavenumbers, $0 < k < \infty$, has a preference for irregular stripes as opposed to isolated individual patches.

12.3 A Model for the Brain Mechanism Underlying Visual Hallucination Patterns

Hallucinations occur in a wide variety of situations such as with migraine headaches, epilepsy, advanced syphilis and, particularly since the 1960's, as a result of external stimulus by drugs such as the extremely dangerous LSD and mescaline (derived from the peyote cactus). General descriptions are given by Oster (1970) and, with much more detail, by Klüver (1967).

Hallucinogenic drugs acquired a certain mystique, since users felt the drugs could alter their perceptions of reality. From extensive studies of drug-induced hallucinations by Klüver (1967), it appears that in the early stages the subject sees a series of simple geometric patterns which can be grouped into four pattern types. These four categories (Klüver 1967) are: (i) lattice, network, grating honeycomb; (ii) cobweb; (iii) spiral; (iv) tunnel, funnel, cone. Figure 12.8 shows typical examples of these pattern types. In Figure 12.8(a) the fretwork type is characterised by regular tesselation of the plane by a repeating unit; that in Figure 12.8(b), the spiderweb, is a kind of distorted Figure 12.8(a).

The hallucinations are independent of peripheral input: for example, experiments showed that LSD could produce visual hallucinations in blind subjects. These experiments, and others such as those in which electrodes in the subcortical regions generated visual experiences, suggest that the hallucinogenic patterns are generated in the visual cortex. Ermentrout and Cowan's (1979) seminal paper is based on the assumption that the hallucinations are cortical in origin and proposed and analysed a neural net model for generating the basic patterns; see also the discussion on large scale nervous activity by Cowan (1982) and the less technical, more physiological, exposition by Cowan (1987). Ermentrout and Cowan (1979) suggest the patterns arise from instabilities in neural activity in the visual cortex; we discuss their model in detail in this section.

Geometry of the Basic Patterns in the Visual Cortex

A visual image in the retina is projected conformally onto the cortical domain. The retinal image, which is described in polar coordinates (r, θ), is distorted in the process of transcription to the cortical image where it is described in (x, y) Cartesian coordinates. It is a mechanism for the creation of these cortical projection patterns that we need to

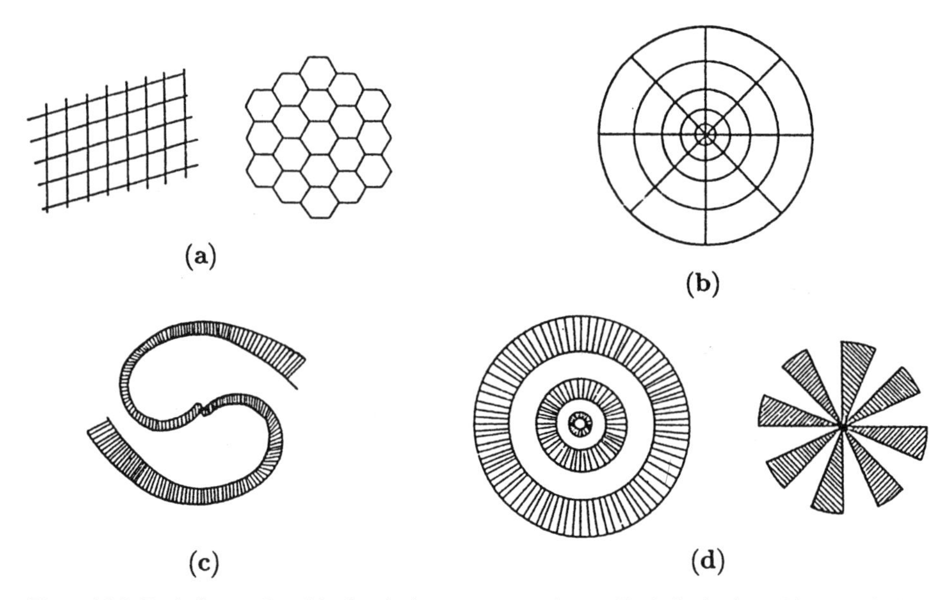

Figure 12.8. Typical examples of the four basic pattern types observed by hallucinating subjects: (**a**) lattice; (**b**) cobweb; (**c**) spiral; (**d**) tunnel and funnel. (After Ermentrout and Cowan 1979)

model. The packing of retinal ganglion cells (the ones that transmit the image via the lateral geniculate nucleus which relays it to the visual cortex) decreases with distance from the centre and so small objects in the centre of the visual field are much bigger when mapped onto the cortical plane. So, a small area $dx\,dy$ in the cortical plane corresponds to $Mr\,dr\,d\theta$ in the retinal disk, where M is the cortical magnification parameter which is a function of r and θ. Cowan (1977) deduced the specific form of the visuo-cortical transformation from physiological measurements; it is defined by

$$x = \alpha \ln[\beta r + (1 + \beta^2 r^2)^{1/2}], \quad y = \alpha\beta r\theta(1 + \beta^2 r^2)^{-1/2}, \tag{12.26}$$

where α and β are constants. Close to the centre (the fovea) of the visual field, that is, r small, the transformation is approximately given by

$$x \sim \alpha\beta r, \quad y \sim \alpha\beta r\theta, \quad r \ll 1, \tag{12.27}$$

whereas for r far enough away from the centre (roughly greater than a solid angle of $1°$)

$$x \sim \alpha \ln[2\beta r], \quad y \sim \alpha\theta. \tag{12.28}$$

Thus, except very close to the fovea, a point on the retina denoted by the complex coordinate z is mapped onto the point with complex coordinate w in the visual cortex according to

$$w = x + iy = \alpha \ln[2\beta r] + i\alpha\theta = \alpha \ln[z], \quad z = 2\beta r \exp[i\theta]. \tag{12.29}$$

This is the ordinary complex logarithmic mapping. It has been specifically discussed in connection with the retino-cortical magnification factor M. Figure 12.9 shows typical patterns in the retinal plane and their corresponding shapes in the cortical plane as a result of the transformation (12.29); see any complex variable book which discusses conformal mappings in the complex plane or simply apply (12.29) to the various shapes such as circles, rectangles and so on. We can thus summarise, from Figure 12.9, the cortical patterns which a mechanism must be able to produce as: (i) cellular patterns of squares and hexagons, and (ii) roll patterns along some constant direction. All of these patterns tessellate the plane and belong to the class of doubly periodic patterns in the plane. In Section 2.4 in Chapter 2 we saw that reaction diffusion mechanisms can generate similar patterns at least near the bifurcation from homogeneity to heterogeneity.

Model Neural Mechanism

The basic assumption in the model is that the effect of drugs, or any of the other causes of hallucinations, is to cause instabilities in the neural activity in the visual cortex and these instabilities result in the visual patterns experienced by the subject. Ermentrout and Cowan's (1979) model considers the cortical neurons, or nerve cells, to be of two types, excitatory and inhibitory, and assume that they influence each other's activity or firing rate (recall the discussion in Section 12.1). We denote the continuum spatially distributed neural firing rates of the two cell types by $e(\mathbf{r}, t)$ and $i(\mathbf{r}, t)$ and assume that cells at position $\mathbf{r}$ and time t influence themselves and their neighbours in an excitatory and inhibitory way much as we described in the last section with activation and inhibition kernels.

Here the activity at time t strictly depends on the time history of previous activity and so in place of the dependent variables e and i we introduce the time coarse grained activities

$$\begin{pmatrix} E(\mathbf{r}, t) \\ I(\mathbf{r}, t) \end{pmatrix} = \int_{-\infty}^{t} h(t-\tau) \begin{pmatrix} e(\mathbf{r}, t) \\ i(\mathbf{r}, t) \end{pmatrix} d\tau, \tag{12.30}$$

where $h(t)$ is a temporal response function which incorporates decay and delay times; $h(t)$ is a decreasing function with time which is typically approximated by a decaying exponential $\exp[-at]$ with $a > 0$.

There is physiological evidence (see, for example, Ermentrout and Cowan 1979) that suggests the activity depends on the self-activation through E and inhibition through I. The activity of E and of I also decay exponentially with time, so the model mechanism can be written as

$$\begin{aligned} \frac{\partial E}{\partial t} &= -E + S_E(\alpha_{EE} w_{EE} * E - \alpha_{IE} w_{IE} * I), \\ \frac{\partial I}{\partial t} &= -I + S_I(\alpha_{EI} w_{EI} * E - \alpha_{II} w_{II} * I), \end{aligned} \tag{12.31}$$

where, from physiological evidence, the functions S_E and S_I are typical threshold functions of their argument, such as the S shown in Figure 12.10(a) and the α's are constants related to the physiology and, for example, drug dosage; note that S is bounded

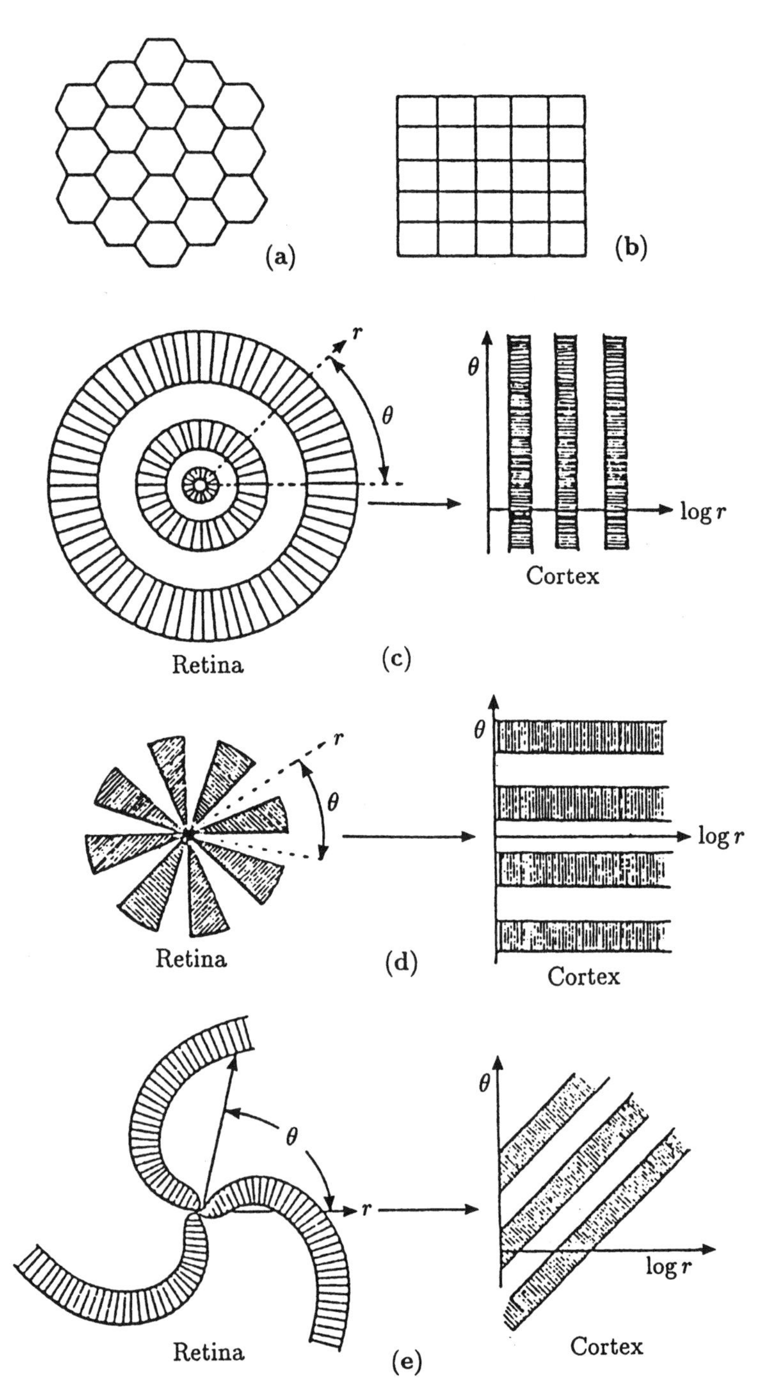

Figure 12.9. Corresponding patterns under the visuo-cortical transformation. (**a**) The lattice patterns in Figure 12.8(a), except for distortions, are effectively unchanged. The other visual field hallucination patterns are on the left with their corresponding cortical images on the right: (**c**) tunnel; (**d**) funnel; (**e**) spiral. (After Ermentrout and Cowan 1979)

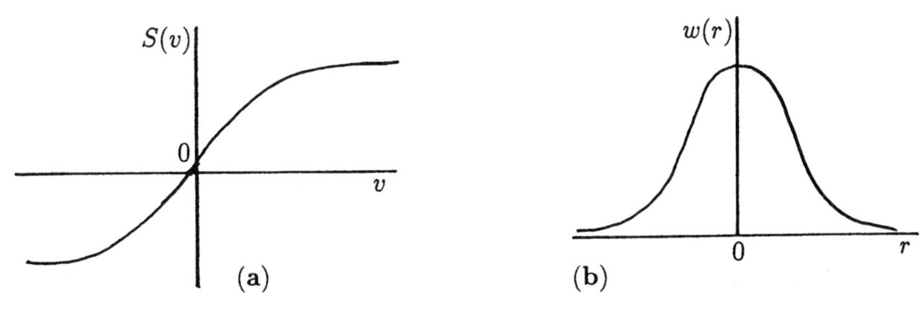

Figure 12.10. (**a**) Typical threshold response function $S(v)$; (**b**) Typical kernel function $w(\mathbf{r})$ in the model mechanism (12.31).

for all values of its argument with $S(0) = 0$. The convolutions are taken over the two-dimensional cortical domain and the kernels here are nonnegative, symmetric and decaying with distance, as illustrated in Figure 12.10(b): a symmetric decaying exponential such as $\exp[-(x^2 + y^2)]$ is an example. The argument in the interaction functions, S_E, for example, represents the difference between the weighted activation of the local excitation and the local inhibition due to the presence of inhibitors. The inhibitors are enhanced through the argument of the S_I function, in the I equation, via the w_{EI} convolution. The inhibitors also inhibit their own production via the w_{II} convolution. There are similarities with the model discussed in the last section except there the activation and inhibition were included in each kernel.

Stability Analysis

Let us now examine the linear stability of the spatially uniform steady state of (12.31), namely, $E = I = 0$, that is, the rest state. The nonlinearity in the system is in the functions S so the linearised form of (12.31), where now E and I are small, is

$$\begin{aligned} \frac{\partial E}{\partial t} &= -E + S_E'(0)(\alpha_{EE} w_{EE} * E - \alpha_{IE} w_{IE} * I) \\ \frac{\partial I}{\partial t} &= -I + S_I'(0)(\alpha_{EI} w_{EI} * E - \alpha_{II} w_{II} * I), \end{aligned} \tag{12.32}$$

where, because of the forms in Figure 12.10(a), the derivatives $S_E'(0)$ and $S_I'(0)$ are positive constants. We now look for spatially structured solutions in a similar way to that used in the last two sections except that here we are dealing with a system rather than single equations ((12.5) and (12.23)), by setting

$$\begin{pmatrix} E(\mathbf{r}, t) \\ I(\mathbf{r}, t) \end{pmatrix} = \mathbf{V} \exp[\lambda t + i\mathbf{k} \cdot \mathbf{r}] = \mathbf{V} \exp[\lambda t + ik_1 x + ik_2 y], \tag{12.33}$$

where $\mathbf{k}$ is the wave vector with wavenumbers k_1 and k_2 in the (x, y) coordinate directions; λ is the growth factor and $\mathbf{V}$ the eigenvector. If $\lambda > 0$ for certain $\mathbf{k}$, these eigenfunctions are linearly unstable in the usual way.

Substituting (12.33) into the linear system (12.32) gives a quadratic equation for $\lambda = \lambda(k)$, the dispersion relation, where $k = |\mathbf{k}| = (k_1^2 + k_2^2)^{1/2}$. For example, with (12.33)

$$\begin{aligned} w_{EE} * E &= \int_D w_{EE}(|\mathbf{r} - \mathbf{r}^*|) \exp[\lambda t + i\mathbf{k} \cdot \mathbf{r}^*] \, d\mathbf{r}^* \\ &= \exp[\lambda t] \int_D w_{EE}(|\mathbf{u}|) \exp[i\mathbf{k} \cdot \mathbf{u} + i\mathbf{k} \cdot \mathbf{r}] \, d\mathbf{u} \\ &= \exp[\lambda t + i\mathbf{k} \cdot \mathbf{r}] \int_D w_{EE}(|\mathbf{u}|) \exp[i\mathbf{k} \cdot \mathbf{u}] \, d\mathbf{u} \\ &= W_{EE}(\mathbf{k}) \exp[\lambda t + i\mathbf{k} \cdot \mathbf{r}], \end{aligned} \tag{12.34}$$

where $W_{EE}(\mathbf{k})$ is the two-dimensional Fourier transform of $w_{EE}(\mathbf{r})$ over the cortical domain D. A typical qualitative form for the w-kernels and its transform is

$$\begin{aligned} w(\mathbf{r}) &= \exp[-b(x^2 + y^2)] \\ &\Rightarrow W(\mathbf{k}) = \frac{\pi}{b} \exp[-k^2/4b], \quad k^2 = k_1^2 + k_2^2. \end{aligned} \tag{12.35}$$

Setting (12.33) into (12.32) and cancelling $\exp[\lambda t + i\mathbf{k} \cdot \mathbf{r}]$, the quadratic for λ is given by the characteristic polynomial

$$\begin{vmatrix} -\lambda - 1 + S'_E \alpha_{EE} W_{EE} & -S'_E \alpha_{IE} W_{IE} \\ S'_I \alpha_{EI} W_{EI} & -\lambda - 1 - S'_I \alpha_{II} W_{II} \end{vmatrix} = 0. \tag{12.36}$$

For algebraic convenience let us incorporate the derivatives $S'_E(0)$ and $S'_I(0)$ into the α-parameters. In anticipation of a bifurcation to spatially structured solutions as some parameter, p say, is increased, again for simplicity let us assume that the mechanism is modulated by p multiplying the α-parameters and so we write pa for α. In the case of drug-induced hallucinations p could be associated with drug dosage. With this notation

$$\begin{aligned} S'_E \alpha_{EE} &= p a_{EE}, \quad S'_E \alpha_{IE} = p a_{IE}, \\ S'_I \alpha_{II} &= p a_{II}, \quad S'_I \alpha_{EI} = p a_{EI}, \end{aligned} \tag{12.37}$$

and (12.36) for λ is

$$\begin{aligned} &\lambda^2 + L(\mathbf{k})\lambda + M(\mathbf{k}) = 0, \\ L(\mathbf{k}) &= 2 - p a_{EE} W_{EE}(\mathbf{k}) + p a_{II} W_{II}(\mathbf{k}), \\ M(\mathbf{k}) &= 1 + p^2 a_{IE} a_{EI} W_{IE}(\mathbf{k}) W_{EI}(\mathbf{k}) - p a_{EE} W_{EE}(\mathbf{k}) \\ &\quad - p^2 a_{EE} a_{II} W_{EE}(\mathbf{k}) W_{II}(\mathbf{k}) + p a_{II} W_{II}(\mathbf{k}). \end{aligned} \tag{12.38}$$

Solutions of the type (12.33) are linearly stable in the usual way if $\mathrm{Re}\,\lambda < 0$ and unstable if $\mathrm{Re}\,\lambda > 0$. Since here we are interested in the spatial patterns which arise from spatially structured instabilities we require the space-independent problem to be stable; that is, $\mathrm{Re}\,\lambda(0) < 0$: recall the related discussion in Section 2.3 in Chapter 2. Here this is the case if

$$\mathrm{Re}\,\lambda(k=0) < 0 \Rightarrow L(0) > 0, \quad M(0) > 0, \tag{12.39}$$

which impose conditions on the parameters in (12.38).

Let us now look at the dispersion relation $\lambda = \lambda(k)$, the solution of (12.38). From the typical kernel forms, such as in (12.35), we see that the $W(\mathbf{k})$ tends to zero as $\mathbf{k} \to \infty$ in which case (12.38) becomes

$$\lambda^2 + 2\lambda + 1 \approx 0 \Rightarrow \lambda < 0 \quad \text{for} \quad k \to \infty$$

and so large wavenumber solutions, that is, those with $k \gg 1$ and hence small wavelengths, are linearly stable. This, plus condition (12.39), points to the basic spatial pattern formation type of dispersion relation like that illustrated in Figure 12.11. Here we have used the parameter p as the bifurcation parameter which we specifically consider.

The mechanism for spatial pattern creation is then very like that with the other pattern formation mechanisms we have so far discussed. That is, the pattern is generated when a parameter passes through a bifurcation value, p_c say, and that for larger p there is a finite range of unstable wavenumbers which grow exponentially with time, $O(\exp[\lambda(k)t])$, where $\lambda(k) > 0$ for a finite range of k. For $p = p_c + \varepsilon$, where $0 < \varepsilon \ll 1$, the spatially patterned solutions are approximately given by the solutions of the linear system (12.32) just like the linear eigenvalue problems we had to solve in Chapter 2, such as that posed by (2.46). The asymptotic procedure to show this is now standard; for example, Lara-Ochoa and Murray (1983) used it in the equivalent reaction diffusion situation and Zhu and Murray (1995) applied it to several other pattern generation mechanisms including chemotaxis-diffusion and mechanical systems.

Let us now consider the type of patterns we can generate with the linear system (12.32) and relate them to the hallucinogenic patterns illustrated in Figure 12.9. Near

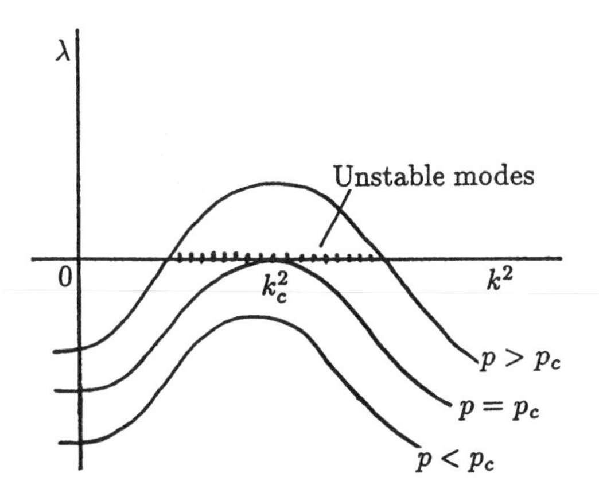

Figure 12.11. Basic dispersion relation giving the growth rate $\lambda(k)$ as a function of the wavenumber $k = |\mathbf{k}|$. The bifurcation parameter is p (for example, a measure of drug dosage); at bifurcation, where $p = p_c$, from spatially uniform to spatially heterogeneous solutions, the critical wavenumber is $k_c = (k_{1c}^2 + k_{2c}^2)^{1/2}$.

bifurcation, spatially heterogeneous solutions are constructed from the exponential form in (12.33), namely,

$$\mathbf{V}\exp i[k_1x+k_2y], \quad k_1^2+k_2^2=k_c^2, \tag{12.40}$$

where $\mathbf{V}$ is the eigenvector corresponding to the eigenvalue $\mathbf{k}_c$ and k_1 and k_2 are the wavenumbers in the coordinate directions. We are specifically interested in solution units which can tesselate the plane, such as rolls, hexagons and so on. It will be helpful to recall the latter part of Section 2.4 where we examined such solutions and their relation to the basic symmetry groups of hexagon, square and rhombus. All such solutions can be constructed from combinations of the specific basic units

$$\exp i[k_cx], \quad \exp[k_cy], \quad \exp[k_c(y\cos\phi+x\sin\phi)] \tag{12.41}$$

which are respectively periodic with period $2\pi/k_c$ in the x-direction, y-direction and perpendicular to the lines $y\cos\phi+x\sin\phi$ which make angles of $\pm\phi$ with the x-axis.

Let us first consider the simplest periodic structures in the visual cortex, namely, the right-hand forms in Figures 12.9(c),(d) and (e). The small amplitude steady state expressions for E and I are of the form

$$\begin{pmatrix} E(x,y) \\ I(x,y) \end{pmatrix} = \mathbf{V}(p_c,k_c^2)\cos(a+k_cx) \tag{12.42}$$

in the case of vertical stripes, where a is a constant which simply fixes the origin. The patterns have period $2\pi/k_c$ and are illustrated in Figure 12.12(a); they correspond to those on the right in Figure 12.9(c). E and I are constant along lines $x=$ constant. In a similar way

$$\begin{pmatrix} E(x,y) \\ I(x,y) \end{pmatrix} = \mathbf{V}(p_c,k_c^2)\cos(a+k_cy) \tag{12.43}$$

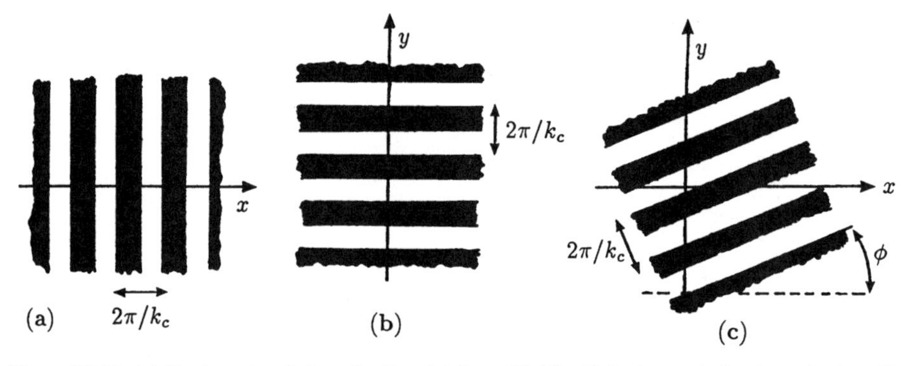

Figure 12.12. (**a**) Steady state solutions for E and I from (12.42) which give vertical stripes, that is, rolls. (**b**) and (**c**) respectively illustrate horizontal rolls and those at an angle ϕ obtained from the solutions (12.43) and (12.44). Note the comparison between these and the visual cortex patterns in Figure 12.9 and their related visual images.

represents the horizontal striping shown in Figure 12.12(b), which in turn corresponds to those on the right in Figure 12.9(d) while those corresponding to Figure 12.9(e) are reproduced in Figure 12.12(c) and given by

$$\begin{pmatrix} E(x, y) \\ I(x, y) \end{pmatrix} = \mathbf{V}(p_c, k_c^2) \cos(a + k_c x \cos\phi + k_c y \sin\phi), \tag{12.44}$$

where a is again a constant.

Let us now consider unit solutions with hexagonal symmetry, that is, with the patterns in Figure 12.9(a) in mind. We wish to construct solutions for E and I which are invariant under the hexagonal rotation operator H. In polar coordinates (r, θ) this means that

$$H[E(r, \theta)] = E(r, \theta + \frac{\pi}{3}) = E(r, \theta). \tag{12.45}$$

Such hexagonal solutions involve the specific exponential forms

$$\begin{aligned} &\exp\left[ik_c\left(\frac{\sqrt{3}y}{2} \pm \frac{x}{2}\right)\right]; \quad k_{1c} = \pm\frac{k_c}{2}, \quad k_{2c} = \frac{k_c\sqrt{3}}{2} \\ &\exp[ik_c x]; \quad k_{1c} = k_c, \quad k_{2c} = 0 \end{aligned}$$

and the relevant E and I are given by (see also (2.47) in Section 2.4)

$$\begin{aligned} \begin{pmatrix} E(x, y) \\ I(x, y) \end{pmatrix} = \mathbf{V}(p_c, k_c^2) &\left\{\cos\left[a + k_c\left(\frac{\sqrt{3}y}{2} + \frac{x}{2}\right)\right]\right. \\ &\left. + \cos\left[b + k_c\left(\frac{\sqrt{3}y}{2} - \frac{x}{2}\right)\right] + \cos[c + k_c x]\right\}, \end{aligned} \tag{12.46}$$

where a, b and c are constants. In polar coordinates, a form which makes it clear that the solution is invariant under the hexagonal rotation (12.45) is

$$\begin{aligned} \begin{pmatrix} E(x, y) \\ I(x, y) \end{pmatrix} = \mathbf{V}(p_c, k_c^2) &\{\cos[a + k_c r \sin(\theta + \pi/6)] \\ &+ \cos[b + k_c r \sin(\theta - \pi/6)] + \cos[c + k_c r \cos(\theta - \pi/6)]\}, \end{aligned} \tag{12.47}$$

The procedure for generating the other patterns, the square and rhombus, is clear. In the case of the square lattice associated with Figure 12.9(b) the relevant solution is (compare also (2.48) in Section 2.4)

$$\begin{pmatrix} E(x, y) \\ I(x, y) \end{pmatrix} = \mathbf{V}(p_c, k_c^2)\{\cos[a + k_c x] + \cos[b + k_c y]\} \tag{12.48}$$

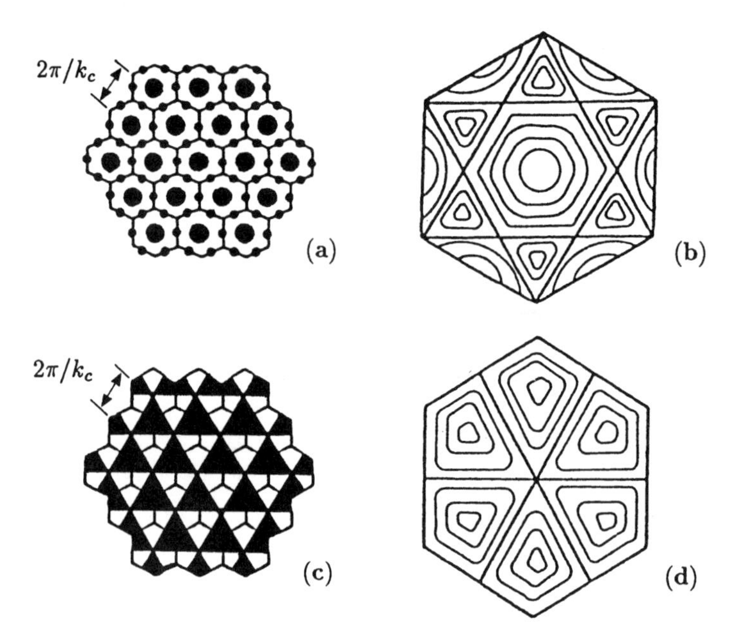

Figure 12.13. Hexagonal patterns from the near-bifurcation solution (12.46) with two different parameter sets for a, b and c. The shaded areas have $E > 0$ and the unshaded $E < 0$; on the contour lines both $I(x, y)$ and $E(x, y)$ are constant. (**a**) $a = b = c = 0$ with the contours of a single hexagonal cell shown in (**b**). (**c**) $a = \pi/2$, $b = c = 0$ with (**d**) showing the corresponding contour lines for a single cell. Compare these patterns with the hallucinogenic pattern in Figure 12.9(a). (After Ermentrout and Cowan 1979)

while the rhombus tesselation solution (compare with (2.49)) is

$$\begin{pmatrix} E(x, y) \\ I(x, y) \end{pmatrix} = \mathbf{V}(p_c, k_c^2)\{\cos[a + k_c x] + \cos[b + k_c(x\cos\phi + y\sin\phi)]\}, \qquad (12.49)$$

where again a and b are constants. Patterns from these solutions are illustrated in Figure 12.14.

The above linear analysis of the neural net model (12.31) for generating spatial patterns in the density of excitatory and inhibitory nerve cell synapses in the visual cortex shows that the mechanism can generate the required patterns which correspond to the basic hallucinogenic patterns in Figure 12.9. These patterns are initiated when a physiological parameter p passes through a bifurcation value in the usual way of creating spatial patterns with a dispersion relation like that in Figure 12.11. The modelling here is similar in many respects to that in the last section. This suggests that physiologically we would expect actual pattern dimensions associated with hallucinogenic patterns to be comparable, roughly 2 mm. Further discussion on the applications and physiological implications is given by Cowan (1987).

If the model mechanism is to be an explanation for hallucinogenic patterns, the solutions we have described must be stable. The question of stability is not an easy one and in general, at this stage anyway, the best indication seems to be from the numerical simulation of the full nonlinear equations. Bifurcation and asymptotic analyses, how-

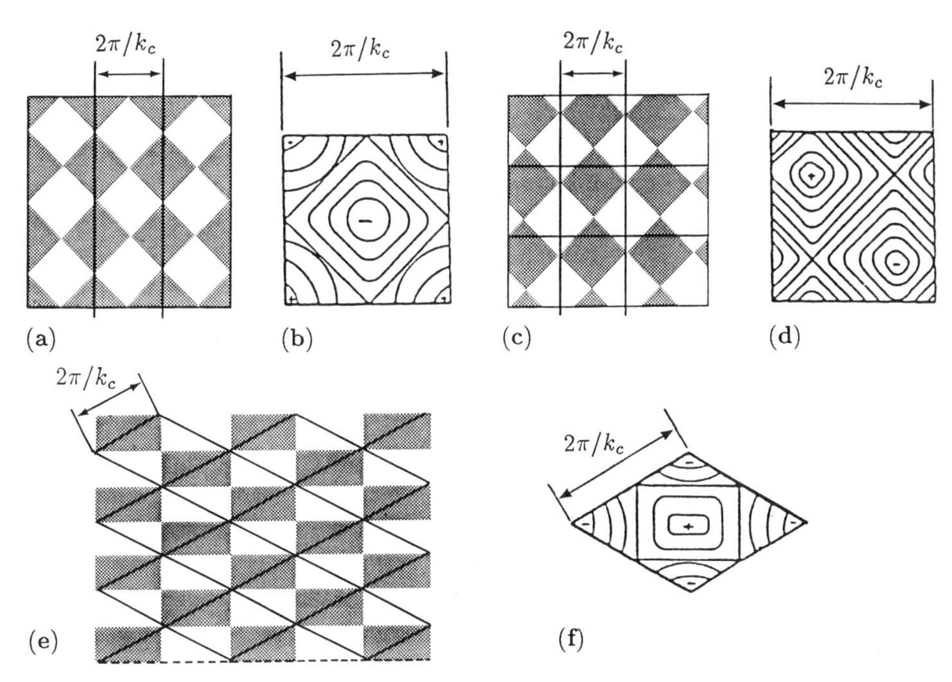

Figure 12.14. Lattice patterns which tessellate the domain with square and rhombic cell units. The shaded areas have $E > 0$ and the unshaded $E < 0$; on the contour lines, $I(x, y)$ and $E(x, y)$ are constant. (**a**) Solution (12.48) with $a = b = 0$ with (**b**) showing the contours for a single cell. (**c**) $b = -a = \pi/2$ with the contours of a single cell shown in (**d**). Compare these patterns with the hallucinogenic pattern in Figure 12.9(b). (**e**) Rhombus solution (12.49) with $a = b = \pi$ with (**f**) showing the corresponding contour lines for a single rhombic cell. (After Ermentrout and Cowan 1979)

ever, provide strong indications (see Zhu and Murray 1995). This has been done by Ermentrout and Cowan (1979).

Tass (1995) used a related model to study the spontaneous pattern formation in epileptic seizures. Whereas in the Cowan and Ermentrout (1979) model, in which the excitability of the activator and inhibitor was increased, Tass (1995) reduced the influence of the inhibitor on the activator. His reason for this was the experimental findings (Klee et al. 1991) that a reduction in the influence of the inhibitory neurons on the excitatory neurons was the cause of epileptic seizures. Tass (1995), in his seminal paper, carried out extensive numerical simulations of his model and obtained the patterns exhibited by the activator and the corrresponding hallucinogenic images. He also investigated how noise affects the cortical images and found, for example, that stars and spirals rotate stochastically. On the other hand, rings pulsate stochastically. His work provides an interesting and important dynamic element of these hallucinogenic patterns and greatly extends the effect range of patterns.

12.4 Neural Activity Model for Shell Patterns

The intricate and colourful patterns on mollusc shells are almost as dramatic as those on butterfly wings; see, for example, Figure 12.15. The reasons for these patterns, unlike the case with butterflies, is somewhat of a mystery since many of these species with spectacular patterns spend their life buried in mud.

Ermentrout et al. (1986) suggested that, since these markings probably do not appear to serve any adaptive purpose, this is why there are so many extreme polymorphic patterns observed in certain species. The novel model proposed by Ermentrout et al. (1986), which is the one we discuss in mathematical detail in this section, combines elements of discrete time models (refer to Chapters 2–4, Volume I) and continuous spatial variation as in the previous sections. Several other modelling attempts have been made to reproduce some of the observed mollusc shell patterns: Waddington and Cowe (1969) and Wolfram (1984), for example, used a cellular automata phenomenological approach while Meinhardt and Klingler (1987) employed an activator–inhibitor reaction diffusion model which Meinhardt (1995) also used in his visually beautiful book on shell patterns. All of these models can mimic many of the more common shell patterns and so, as we keep reiterating, the only way to determine which mechanism has a better claim must be through the different experiments that each suggests and it is only through these that an informed decision can be made. Having said that, however, shell patterns are formed over several years and it would be surprising if a reaction diffusion system could sustain the necessary coherence over such a long period. The nervous system, on the other hand, is an integral part of the mollusc's physiology throughout

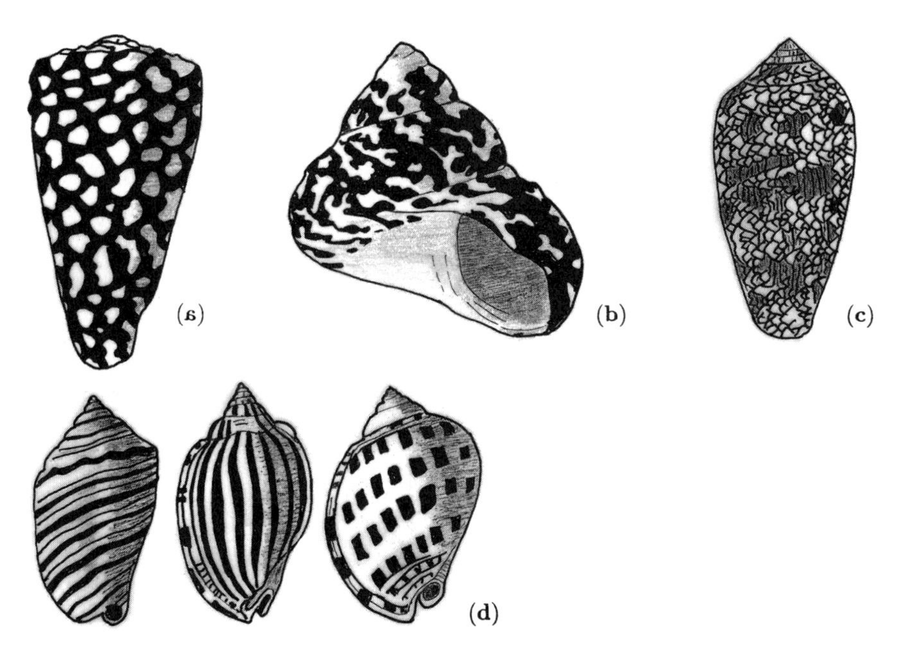

Figure 12.15. Examples of common shell patterns: (**a**) *Conus marmoreus*; (**b**) *Citarium pica*; (**c**) *Conus textus*; (**d**) common patterns frequently seen, for example, on snail shells.

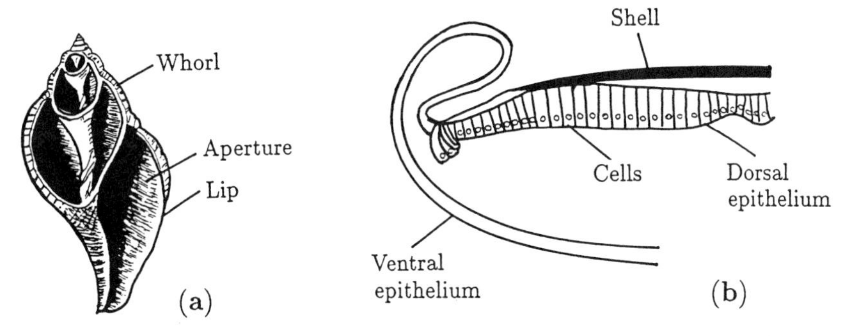

Figure 12.16. (**a**) Typical spiral shell structure. (**b**) Basic anatomical elements of the mantle region on a shell. (Drawn after Ermentrout et al. 1986)

its life. As regards cellular automata models, they make no connection with any of the underlying biological processes involved in the mollusc's growth and development.

A typical shell is a conical spire made up of tubular whorls which house the visceral mass of the animal; see Figure 12.16(a). These whorls are laid down sequentially and wind round a central core and eventually terminate at the aperture; see Figure 12.16(a). A readable introductory survey of the biology of molluscs is given in the textbook by Barnes (1980). Just under the shell is the mantle with epithelial cells which secrete the material for shell growth. Figure 12.16(b) is a simplified diagram of the anatomy of the mantle region. The basic assumption in the Ermentrout et al. (1986) model is that the secretory activity of the epithelial cells is controlled by nervous activity and that the cells are enervated from the central ganglion (concentrated masses of nerve cells—a kind of brain). The cells are activated and inhibited by the neural network which joins the secretory cells to the ganglion.

Neural Model for Shell Pattern Formation

The specific assumptions of the model, which we now enlarge upon below, are:

(i) Cells at the mantle edge secrete material intermittently.

(ii) The secretion depends on (a) the neural stimulation, S, from surrounding regions of the mantle and (b) the accumulation of an inhibitory substance, R, present in the secretory cell.

(iii) The neural net stimulation of the secretory cells consists of the difference between the excitatory and inhibitory inputs from the surrounding tissue (recall the discussions in the previous two sections).

It is known that the shell is laid down intermittently. At the start of each secretory period the assumption is that the mantle aligns with the previous pattern and extends it. The alignment is effected through a *sensing* (a kind of tasting) of the pigmented (or unpigmented) areas of the previous secretion session. Alternatively it is possible that the pigmented shell laid down in the previous session stimulates the mantle neurons locally to continue with that pattern.

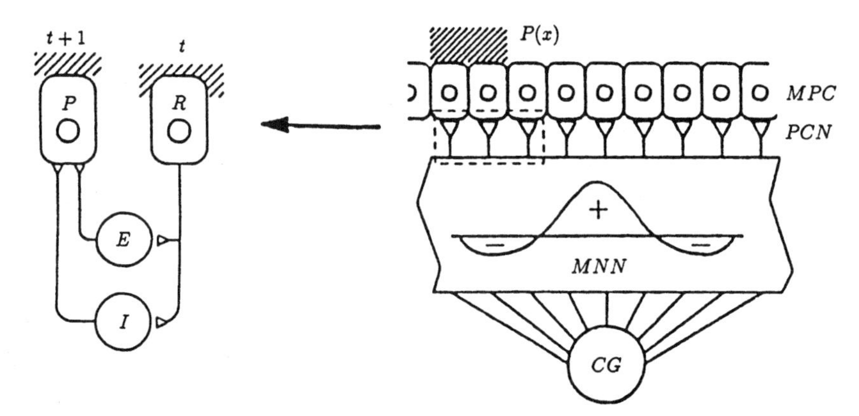

Figure 12.17. Schematic model for neural activation and control of pigment secretion; see text for an explanation. Here MPC denotes the mantle pigment cells, PCN the pigment cell neurons, MNN the mantle neural net and CG the central ganglion. R denotes the receptor cells sensing pigment laid down in the time period t, P denotes the pigment cells secreting pigment in time period $t+1$, E denotes the excitatory neurons and I the inhibitory neurons. (After Ermentrout et al. 1986)

We now incorporate these assumptions into a model mechanism with discrete time and continuous space. It is helpful to refer to Figure 12.17. Consider the edge of the mantle to consist of a line of secretory cells, or mantle pigment cells, with coordinate x measured along the line. Let $P_t(x)$ be the amount of pigment secreted by a cell at position x at time t (the unit of period is taken as 1). Let $A_t(x)$ be the average activity of the mantle neural net, $R_t(x)$ the amount of inhibitory substance produced by the cell and the functional $S[P]$, the net neural stimulation, which depends on P_{t-1}, the pigment secreted during the previous session. Although the final patterns are two-dimensional, the way they are laid down lets us consider, in effect, a one-dimensional model.

The model equation for the neural activity is taken to be

$$A_{t+1}(x) = S[P_t(x)] - R_t(x) \tag{12.50}$$

which simply says that the neural activity is stimulated by the net neural stimulation in the earlier session ($S[P_t]$) and inhibited by the previous session's inhibitory substance (R_t). We assume the inhibitory material, R_t, depends linearly on the amount of pigment secreted during the previous period while at the same time degrading at a constant linear rate δ. The governing conservation equation for R_t is then

$$R_{t+1}(x) = \gamma P_t(x) + \delta R_t(x), \tag{12.51}$$

where the rate of increase $\gamma < 1$ and the rate of decay $\delta < 1$ are positive parameters. Because this is a discrete equation (recall Chapter 2, Volume I) note that the decay term is positive. For example, if $\gamma = 0$, R_t decreases with each time-step if $0 < \delta < 1$; it increases if $\delta > 1$.

We now assume that pigment will be secreted only if the mantle activity is stimulated above some threshold, A^* say, and so we write

$$P_t(x) = H(A - A^*), \tag{12.52}$$

where H is the Heaviside function; $H = 0$ when $A < A^*$ and $H = 1$ when $A > A^*$.

It is reasonable, at this stage of the modelling, to assume that the pigment secretion P_t is simply proportional to the activity A_t and subsume the threshold behaviour into the stimulation function $S[P_t]$. With this, the simpler model is then

$$P_{t+1}(x) = S[P_t(x)] - R_t(x) \tag{12.53}$$

$$R_{t+1}(x) = \gamma P_t(x) + \delta R_t(x) \tag{12.54}$$

which we now study in detail.

Consider now the neural stimulation functional $S[P_t]$. This consists of excitatory and inhibitory effects. Although secretion at a given time $t + 1$ depends only on the excitation during the time period t to $t + 1$, each period's excitation depends on the stimulation from sensing the previous period's pigment pattern. It is reasonable to assume that the time constants for neural interactions are much faster than those for shell growth and so we use an average neural firing rate in the mantle. We thus define the excitation $E_{t+1}(x)$ and inhibition $I_{t+1}(x)$ by the convolution integrals

$$\begin{aligned} E_{t+1}(x) &= \int_\Omega w_E(|\, x' - x\,|) P_t(x')\, dx' = w_E * P_t \\ I_{t+1}(x) &= \int_\Omega w_I(|\, x' - x\,|) P_t(x')\, dx' = w_I * P_t, \end{aligned} \tag{12.55}$$

where Ω is the mantle domain and may be circular, or a finite length, depending on the shell we are considering. The excitatory and inhibitory kernels w_E and w_I are measures of the effect of neural contacts between cells at x' and those at x. They represent nonlocal spatial effects which we are now familiar with from the previous sections of this chapter. Figure 12.18(a) illustrates the general form of these kernels while Figure 12.17 caricatures their combined effect, namely, a positive excitation (+) and a negative inhibition (−), in the mantle neural net (MNN).

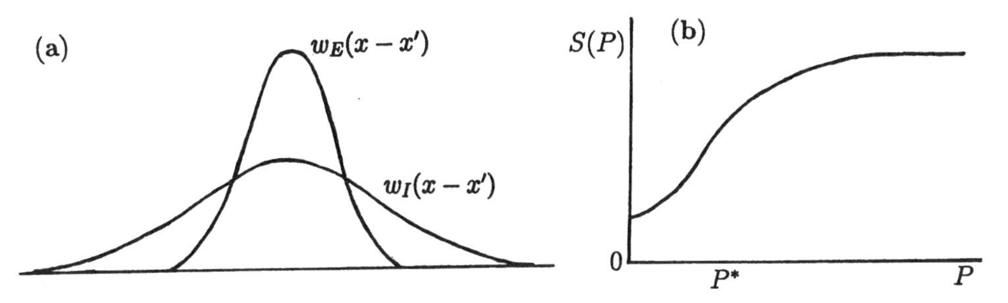

Figure 12.18. (**a**) Schematic form of the activation, w_E, and inhibition, w_I, kernels in (12.55). The width of the inhibition kernel is greater than that of the activation kernel; compare with Figure 12.3(b). (**b**) Typical threshold form for the stimulation function $S[P]$ in the model equation (12.53): P^* is the approximate threshold value equivalent to the activation threshold A^* in (12.52).

From the analysis in Section 12.1–12.3 (and intuitively with the experience we now have) we require the width of the inhibition kernel w_I to be greater than that of the activation kernel w_E; see Figure 12.18(a). For illustrative numerical simulation purposes only, Ermentrout et al. (1986) chose the following kernels,

$$\begin{aligned} w_j &= 0 \quad \text{for} \quad |x| > \sigma_j, \quad j = E, I \\ w_j &= q_j \left\{2^p - [1 - \cos(\pi x/\sigma_j)]^p\right\} \quad \text{for} \quad |x| \leq \sigma_j, \quad j = E, I, \end{aligned} \tag{12.56}$$

where q_j are chosen such that

$$\int_\Omega w_j(x)\,dx = \alpha_j, \quad j = E, I. \tag{12.57}$$

The parameters σ_j measure the range of the kernels; $\sigma_I > \sigma_E$ in our model. The parameter p, which of course could also be a p_j, controls the sharpness of the cut-off. For p small the kernels are sharply peaked while for p large they are almost rectangular. The amplitude of the excitatory and inhibition kernel functions is controlled by the α_j where, in our system, we expect $\alpha_E > \alpha_I$.

The stimulation functional S is composed of the difference between the excitatory and inhibitory elements. For analysis a fairly general form as in Figure 12.18(b) suffices but for numerical simulation a specific form for S which exhibits a threshold-type behaviour is required and Ermentrout et al. (1986) chose

$$\begin{aligned} S[P_t(x)] &= S_E[E_t(x)] - S_I[I_t(x)] \\ S_j(u) &= \{1 + \exp[-\nu_j(u - \theta_j)]\}^{-1}, \quad j = E, I, \end{aligned} \tag{12.58}$$

where the ν_j controls the sharpness of the threshold switch which is located at θ_j. The sharpness ($S'_j(\theta_j)$) increases with ν_j; θ_j corresponds to P^* in Figure 12.18(b).

The complete model system used in the numerical simulations is given by (12.53)–(12.58): it has 11 parameters, or 12 if we have different p in (12.56). Rescaling reduces the number while other parameters appear only as products. The influence-function parameters are the α, σ and p, those associated with the firing threshold are the ν and θ, while the dynamic or refractory parameters, in (12.53) and (12.54), are δ and γ. We now examine analytically the way spatial patterns are generated by the model.

Linear Stability Analysis

Equations (12.53) and (12.54) can be combined into the single scalar equation by iterating (12.53) once in time and using (12.54) and (12.53) again to get

$$\begin{aligned} P_{t+2} &= S[P_{t+1}] - \gamma P_t - \delta(S[P_t] - P_{t+1}) \\ &= S[P_{t+1}] + \delta P_{t+1} - \delta S[P_t] - \gamma P_t. \end{aligned} \tag{12.59}$$

For notational simplicity we have written P_t for $P_t(x)$ and so on.

In the usual linear stability way, we now look for the homogeneous steady state, P_0 say, perturb it linearly and analyse the resulting linear equation. From (12.59)

$$\begin{aligned} P_0 &= S[P_0] + \delta P_0 - \delta S[P_0] - \gamma P_0 \\ \Rightarrow P_0 &= \frac{(1-\delta)S[P_0]}{1+\gamma-\delta} \end{aligned} \tag{12.60}$$

with the form for $S[P]$ in Figure 12.18(b) at least one positive steady state solution exists: simply draw each side as a function of P_0 to see this. Now linearise about P_0 by writing

$$P_t(x) = P_0 + u_t(x), \quad | u_t | \text{ small.} \tag{12.61}$$

On substitution into (12.59) and retaining only linear terms in u, we get

$$\begin{aligned} &u_{t+2} - L_0[u_{t+1}] - \delta u_{t+1} + \delta L_0[u_t] + \gamma u_t = 0, \\ &L_0[u] = S'_E(P_0) w_E * u - S'_I(P_0) w_I * u, \end{aligned} \tag{12.62}$$

where L_0 is a linear integral (convolution) operator. The eigenfunctions of L_0 on a periodic domain of length L are $\exp[2\pi inx/L]$, $n = 1, 2, \ldots$. Since we consider the finite linear size L of the domain, that is, the length of the shell lip, to be very much larger than the range of the kernels, these eigenfunctions are approximately those for L.

Since (12.62) is a linear equation, discrete in time and continuous in x, we look for solutions in the form

$$u_t \propto \lambda^t \exp[ikx], \tag{12.63}$$

which on substituting into (12.62) gives

$$\begin{aligned} \lambda^{t+2} \exp[ikx] - \lambda^{t+1} L_0[\exp[ikx]] - \delta\lambda^{t+1} \exp[ikx] \\ + \delta\lambda^t L_0[\exp[ikx]] + \gamma\lambda^t \exp[ikx] = 0. \end{aligned} \tag{12.64}$$

Here

$$\begin{aligned} L_0[\exp[ikx]] &= S'_E(P_0) \int_\Omega w_E(| x' - x |) \exp[ikx'] \, dx' \\ &\quad - S'_I(P_0) \int_\Omega w_I(| x' - x |) \exp[ikx'] \, dx', \\ &= \exp[ikx]\{S'_E W_E(k) - S'_I W_I(k)\} \\ &= \exp[ikx] L^*(k), \end{aligned} \tag{12.65}$$

which defines $L^*(k)$, where $W_E(k)$ and $W_I(k)$, the Fourier transforms of $w_E(x)$ and $w_I(x)$ over the domain Ω, are defined by

$$W_j(k) = \int_\Omega w_j(z) \exp[ikz]\, dz, \quad j = E, I. \tag{12.66}$$

Using (12.65) in (12.64) and cancelling $\lambda^t \exp[ikx]$ we get the characteristic equation for λ as

$$\lambda^2 - \lambda(L^*(k) + \delta) + (\delta L^*(k) + \gamma) = \lambda^2 + a(k)\lambda + b(k) = 0, \tag{12.67}$$

which define $a(k)$ and $b(k)$. The solutions $\lambda = \lambda(k)$ give the dispersion relations for the linear problem (12.62).

Unlike previous models, the time variation in the solutions is discrete and so linear stability here requires $|\lambda(k)| < 1$ (recall the analysis in Chapter 2). That is, for the solutions (12.62) to be stable for all k, $\lambda(k)$ from (12.67) must lie within the unit circle in the complex λ-plane. From (12.67),

$$\lambda = \tfrac{1}{2}[-a \pm (a^2 - 4b)^{1/2}] \tag{12.68}$$

and so we deduce, using a little elementary algebra, that the condition $|\lambda| < 1$ is satisfied if $a(k)$ and $b(k)$ lie within the triangle in the (a, b) plane shown in Figure 12.19(a).

Let us now consider the kind of spatial instabilities which appear as $|\lambda|$ passes through the bifurcation value 1. This means that the solutions λ must move across the unit circle in the complex λ-plane. The pattern evolution from such dispersion relations is a little more subtle than those we have considered up to now, with the novel aspects directly related to the discrete time element in the model. Typical dispersion relations are illustrated in Figure 12.19(b), one for $\lambda > 1$ and one for $\lambda < -1$. With the latter, the instability (12.63) increases with time in an oscillatory way with fastest growth for modes with wavenumbers $k = 0$; in other words there is no spatial homogeneity. We come back to this case below.

The model parameters, which of course include the shape of the activation and inhibition kernels w_E and w_I, define a point, (a_0, b_0) say, in (a, b) parameter space using the definitions of a and b in (12.67). From Figure 12.19(a), if (a_0, b_0) lies within the triangular domain, $|\lambda| < 1$ and the steady state P_0 is stable since from (12.63) $u_t(x) \to 0$ as $t \to \infty$. Instability is initiated as a parameter (or parameters) varies so that the related point in the (a, b) plane moves out of the stability triangle. Recalling the models and analyses in the previous sections, the total effect of the excitation and inhibition kernels introduced here is like a single kernel, $w(x)$ say, with a local, or short range, activation and a long range, or lateral, inhibition. $w(x)$ and its Fourier transform $W(k)$ have qualitatively similar properties, respectively, to a composite kernel in the linear convolution $L_0[u(x)]$ defined by (12.62) and its Fourier transform $L^*(k)$ defined by (12.65). Figure 12.20(a) is a sketch of a typical composite kernel, $L(x)$ say, equivalent to the two contributions in $L_0(x)$ together with its Fourier transform $L^*(k)$. Recall the similar forms in Figures 12.3(b) and (c).

Now consider what type of pattern is created as a parameter passes through a critical bifurcation value, thus moving λ out of the unit disc in Figure 12.19(a). There are basically three different ways λ crosses the unit circle, namely, bifurcation through: (i)

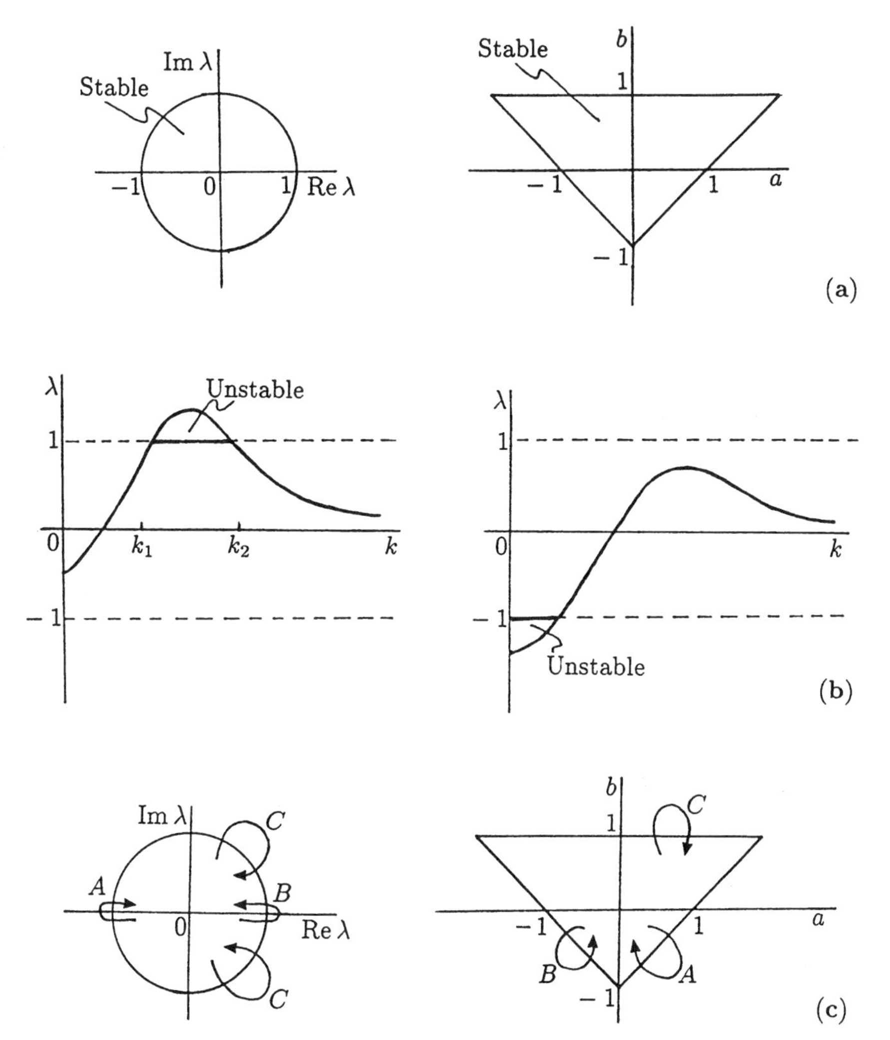

Figure 12.19. (**a**) The unit disk in the λ-plane and the stability domain in the (a, b) parameter plane. (**b**) Typical dispersion relations $\lambda(k)$ for spatial instability: since the solution (12.63) is discrete in time, instability occurs if $|\lambda| > 1$. For illustrative purposes these forms have λ real; see Figure 12.20(b) for complex λ. (**c**) Instability path in the λ-plane for each type of bifurcation from the stability triangle. These are obtained using the solutions for λ in (12.68) and considering the signs of a and b.

$\lambda = 1$, (ii) $\lambda = -1$ and (iii) $\lambda = \exp[i\phi]$, $\phi \neq 0$, $\phi \neq \pi$. We consider the pattern evolution implications of each of these in turn.

(i) *Bifurcation through* $\lambda = 1$

From Figure 12.19(c), which was obtained from an analysis of (12.68), we see that this occurs if both $a(k) < 0$ and $b(k) < 0$ and the point (a, b) crosses the bifurcation line in the (a, b) plane in the 3rd quadrant. From the definitions in (12.67) this means

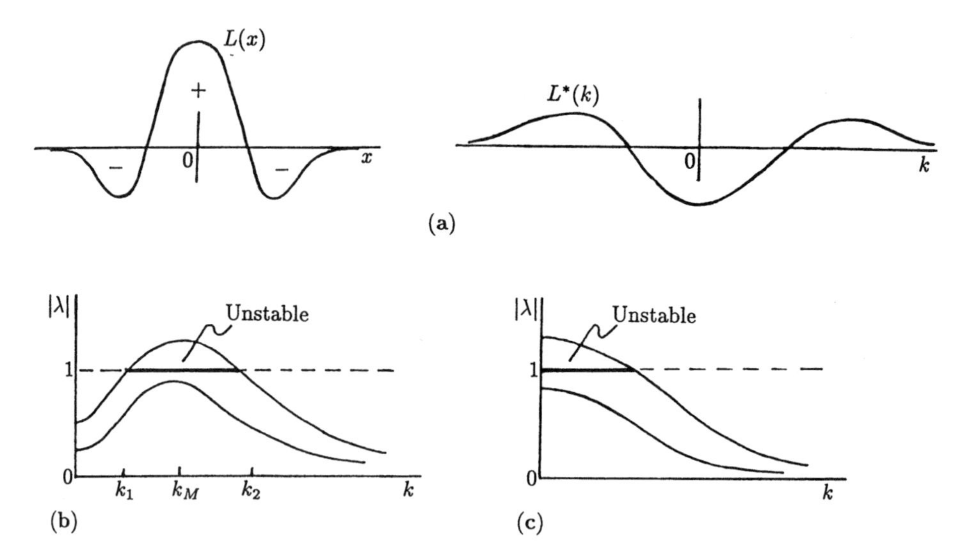

Figure 12.20. (**a**) Qualitative shape of the composite kernel $L(x)$ in the convolution $L_0(x)$ in (12.62), that is, the difference between excitatory and inhibitory convolutions, and its Fourier transform $L^*(k)$; compare with Figure 12.3(c). (**b**) and (**c**) Typical dispersion relations which result in spatial patterns. In (**b**) eigenfunctions (12.63) for a finite range of wavenumbers $0 < k_1 < k < k_2$ are unstable while in (**c**) the unstable modes include $k = 0$ as the fastest growing eigenfunction.

that for a range of wavenumbers k,

$$\delta > -L^*(k) > \frac{\gamma}{\delta} \Rightarrow \delta^2 > \gamma. \tag{12.69}$$

Referring to the right-hand figure in Figure 12.20(a) this means that the possible range of wavenumbers k which can satisfy (12.69) is in the vicinity of $k = 0$.

From (12.61) and (12.62),

$$P_{t+1}(x) - P_0 \propto \lambda^t \exp[ikx] \tag{12.70}$$

which grows with time for all eigenfunctions with wavenumbers k in the range in which (12.69) is satisfied. The fastest growing solution is dominated by the minimum k. If $k = 0$ we have

$$P_{t+1}(x) - P_0 \propto \lambda^t(0). \tag{12.71}$$

This in fact does form a spatial pattern on the shell consisting of regularly spaced horizontal stripes—homogeneous stripes parallel to the shell edge. These are incremental lines and are simply homogeneous lines of pigment laid down in each time-step; see Figure 12.21(a) and also the middle shell pattern in Figure 12.15(d). If $k \neq 0$ then spatial heterogeneity arises as in the next case.

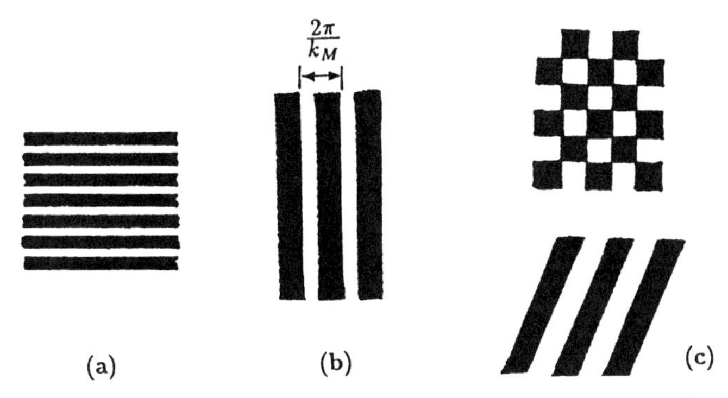

Figure 12.21. (**a**) Horizontal stripe pigment pattern arising from a bifurcation at $\lambda = +1$ and generated by solutions (12.71). (**b**) Spatial pattern of vertical stripes from (12.74) which arises when the bifurcation is at $\lambda = -1$. (**c**) Patterns which arise from complex bifurcations with solutions from (12.76).

(ii) *Bifurcation through* $\lambda = -1$

From Figure 12.19(c) we see that this situation arises if $a(k) > 0$ and $b(k) < 0$ and the bifurcation line in the (a, b) plane is crossed in the 4th quadrant. From the definitions in (12.67) this requires

$$-L^*(k) > \max\left[\delta, \frac{\gamma}{d}\right] \quad \text{for} \quad 0 < k_1 < k < k_2. \tag{12.72}$$

Now from (12.61) and (12.62),

$$P_{t+1}(x) - P_0 \propto \lambda^t \exp[ikx] \tag{12.73}$$

which grows with time for eigenfunctions with wavenumbers k in the range bounded by k_1 and k_2 with a maximum λ, $\lambda_M = \lambda(k_M)$; refer also to Figure 12.20(b). So, after some time the dominant solution is

$$P_{t+1}(x) - P_0 \propto \lambda_M^t \exp[ik_M x]. \tag{12.74}$$

Since λ_M is real and greater than 1, λ_M^t is always positive so the pigment laid down at each step is in line with the pigment laid down at the previous step. Thus (12.74) forms a vertical, regularly spaced, stripe pattern with wavelength $2\pi/k_M$ as shown in Figure 12.21(b): these are basic longitudinal bands.

(iii) *Bifurcation through* $\lambda = \exp[i\phi]$, $\phi \neq 0$, $\phi \neq p$

We see from Figure 12.19(c) that in this case (a, b) has to leave the stability triangle in the 1st or 2nd quadrant, so $b(k) > 0$ and $a(k)$ can be positive or negative. To be specific take the case $a(k) > 0$. From (12.67) this requires

$$-L^*(k) > \delta, \quad \delta L^*(k) + \gamma > 0 \quad \Rightarrow \quad \delta^2 < \gamma. \tag{12.75}$$

In this type of bifurcation

$$P_{t+1}(x) - P_0 \propto \lambda^t \exp[ikx]. \tag{12.76}$$

Since here λ is complex, each time-step moves the pattern along the x-axis by an amount equal to $\arg \lambda$. Thus this solution generates stripe patterns, which are at a fixed angle, or checkered patterns, such as illustrated in Figure 12.21(c). An oblique stripe pattern is one of the basic shell patterns.

The values of the parameters along with the scale determine which patterns are created. One example of the role of parameters, for example, is immediately obtained from the restrictions on δ and γ in (12.69) and (12.75). The former generates incremental lines, as in Figure 12.21(a), while the latter creates oblique stripes as in Figure 12.21(c). δ and γ are respectively the degradation and production rates of the inhibitory material R_t in the model (12.53) and (12.54). There is thus a bifurcation line in (δ, γ) space, namely, $\delta^2 = \gamma$, across which the pattern generated changes from incremental lines to oblique and checkered patterns. Suppose the shell pattern which is evolving is a checkered one; that is, $\delta^2 < \gamma$. If there is a sudden reduction in the production of the inhibitory substance (R_t), that is, γ is reduced, the pattern formed afterwards is a horizontal striped one. Many shells exhibit such abrupt pattern changes; see the example in Figure 12.22(d).

Spatial Patterns Generated by the Neural Model

Of course the pattern predictions in Figure 12.21 are based on a linear theory and the final stable patterns come from the full nonlinear model and are obtained numerically (Ermentrout et al. 1986). The linear analysis, however, is a good predictor for the finite amplitude patterns formed by the full nonlinear system. In the simulations presented below the kernels w in (12.56) and stimulation functions S in (12.58), with the same ν, were used. The mechanism was solved in the form of the single second-order difference equation (12.59) with random initial conditions.

The patterns depend on the parameter values. In parameter space there are bifurcation lines and surfaces across which pattern bifurcation takes place. Figures 12.22(a)–(c) show examples of longitudinal bands, checkered patterns and oblique stripes in one species while the middle shell in Figure 12.15(d) is a good example of incremental lines. If, during development, a bifurcation surface is crossed then there is an abrupt change of pattern; Figure 12.22(d) shows an example.

The direction of the oblique stripes also depends on the parameter values. In the model we assumed that the parameters were constant as the mantle grew. It is not difficult to imagine a situation where there is a gradient in parameter values in the mantle and this can give a bias to the stripe direction. This gradient could change its direction as growth proceeds thus causing a change in the stripe direction. These give rise to divaricate patterns, examples of which are shown in Figure 12.23 together with simulations which mimic them.

The model can generate a wide spectrum of patterns which are commonly observed (see Ermentrout et al. 1986 for examples of other patterns not discussed here). They include, for example, wavy stripes, checks, irregular stripe patterns, tents and so on. Figure 12.24(a) shows an example of irregular wandering stripes while Figure 12.24(b) shows typical tent patterns common in the courtly cones. The latter simulation has

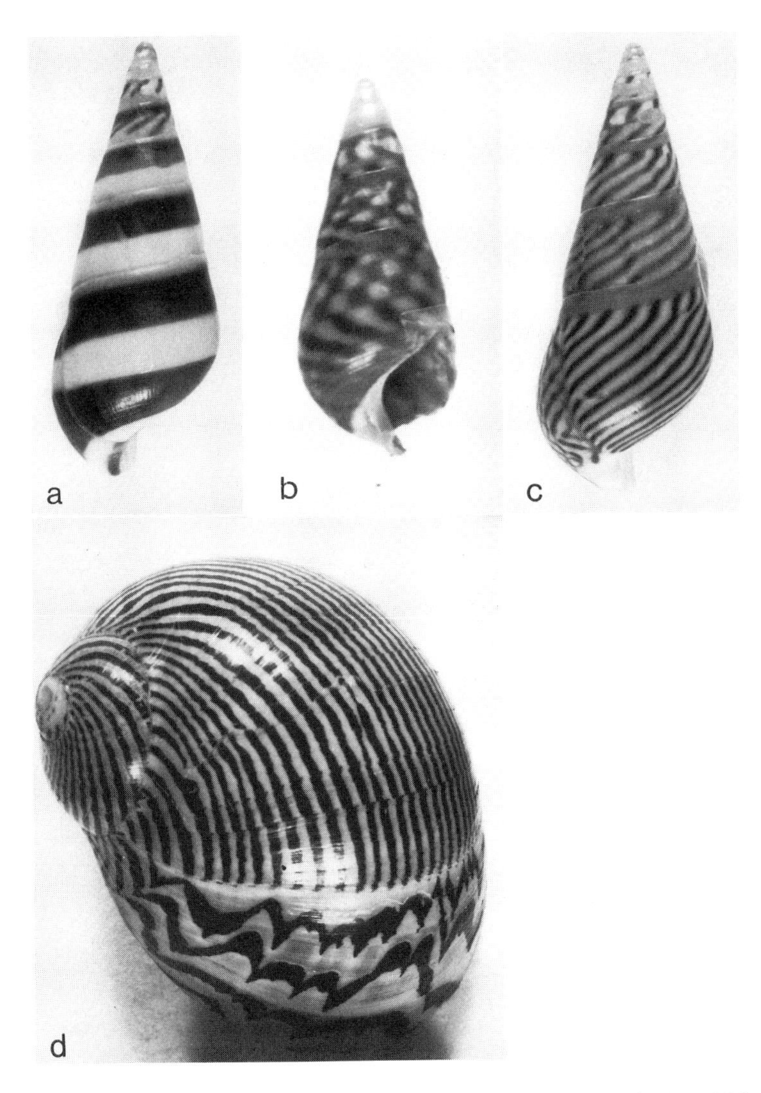

Figure 12.22. (**a**)–(**c**) are examples of basic shell pigment patterns in *Bankivia fasciata*: (**a**) longitudinal bands; (**b**) chequered pattern; (**c**) oblique or diagonal stripes. (**d**) Abrupt reorganisation of pattern on a shell of *Neritta turrita* after a break in the shell. (From Ermentrout et al. 1986; photographs courtesy of J. Campbell)

somewhat longer range kernels. From simulations it appears that the excitation threshold and kinetics parameters in the equation for the refractory substance R_t play particularly important roles in the patterns created.

The model (12.53) and (12.54) under certain limiting circumstances can be shown (Ermentrout et al. 1986) to reduce to a cellular automata mechanism which has been studied by Wolfram (1984) and which exhibits chaotic and tent patterns. It is to be expected that similar chaotic pattern behaviour can be generated by the system here: the patterns in Figure 12.24 could be examples. The basic pattern formation potential of

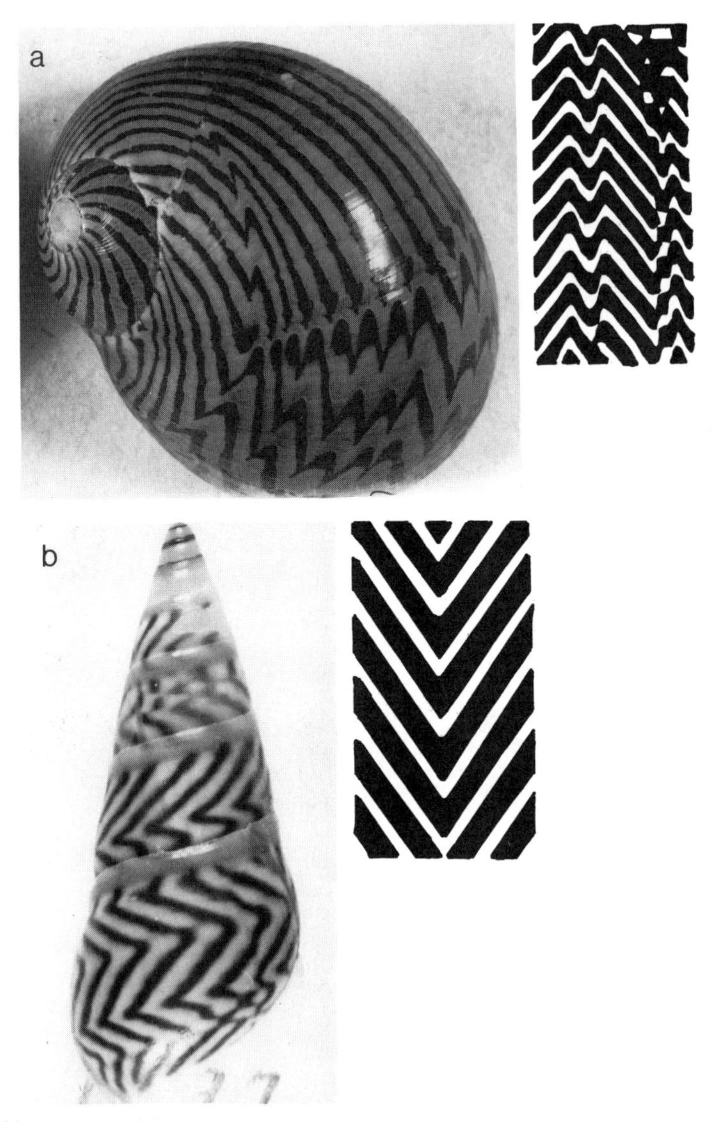

Figure 12.23. Examples of divaricate patterns: (**a**) Wavy bands on the shell *Nerita turrita* and a model simulation. Parameter values: $\theta_E = 1$, $\theta_I = 100$, $\alpha_E = 5$, $\alpha_I = 4$, $\sigma_E = 0.05$, $\sigma_I = 0.2$, $\gamma = 0.8$, $\delta = 0.4$, $\nu = 2$. (**b**) Patterns on the shells of *Bankivia fasciata* and a corresponding simulation. (From Ermentrout et al. 1986; photographs courtesy of J. Campbell)

such discrete time neural models has only been explored here for basic patterns. If such models are those which govern shell pattern formation then the shell gives a hardcopy printout of the neural activity of the mollusc's mantle and how it interacts with the shell geometry.

As mentioned in the introduction similar patterns can be generated with reaction diffusion models. This is not surprising since, as we showed before, this type of mech-

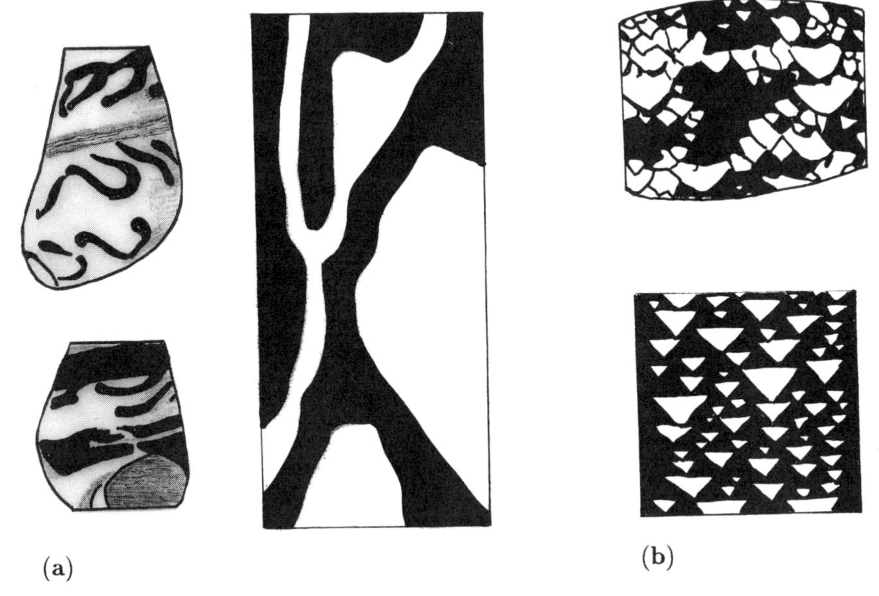

Figure 12.24. (**a**) Two examples of irregular wandering stripe patterns in *Bankivia fasciata* and a model simulation. Parameter values: $\theta_E = 4.5$, $\theta_I = 0.32$, $\alpha_E = 15$, $\alpha_I = 0.5$, $\sigma_E = 0.1$, $\sigma_I = 0.15$, $\gamma = 0.1$, $\delta = 0.8$, $\nu = 8$. (**b**) Typical tent characteristics of the textile and courtly cones (the section shown is from a *Conus episcopus*; see also Figure 12.15(c)) and a related simulation. Parameter values: $\theta_E = 5.5$, $\theta_I = 5.5$, $\alpha_E = 10$, $\alpha_I = 4$, $\sigma_E = 0.1$, $\sigma_I = 0.2$, $\gamma = 0.3$, $\delta = 0.2$, $\nu = 8$. (Drawn from Ermentrout et al. 1986)

anism can be couched in terms of short range activation and long range inhibition. The mechanochemical models for pattern formation discussed in Chapter 6 are also capable of generating similar patterns.

To see what connection the discrete time model has with a continuous time model we now briefly consider the analogue we get from a continuous approximation to the discrete model we have just studied.

Continuous Time Model Analogue

If we subtract P_t and R_t from each side of (12.53) and (12.54) respectively we have

$$P_{t+1}(x) - P_t(x) = S[P_t(x)] - R_t(x) - P_t(x)$$

$$R_{t+1}(x) - R_t(x) = \gamma P_t(x) + \delta R_t(x) - R_t(x)$$

which suggests an analogous continuous time model of the form

$$\frac{\partial P}{\partial t} = S[P] - R - P, \tag{12.77}$$

$$\frac{\partial R}{\partial t} = \gamma P - (1 - \delta)R, \tag{12.78}$$

where $R(x, t)$ and $P(x, t)$ are now functions of continuous space and time, and $S[P]$ has a sigmoid form as a function of its argument as in Figure 12.18(b). Remember from the original formulation that $\gamma < 1$ and $\delta < 1$. If we take the excitatory and inhibitory functional forms S_E and S_I in (12.58) to be the same, then with (12.55), we can take

$$S[P] = S(w_E * P - w_I * P) = S(w * P), \tag{12.79}$$

where $w(x)$ is a typical local activation lateral inhibition kernel such as illustrated in Figure 12.20(a). The continuous time model mechanism, given by (12.77) to (12.79), is an integrodifferential equation system.

If we now consider the kernel's influence to be restricted to a small neighbourhood around x we can use the same procedure as in Section 12.1 to expand the integral in (12.79). This gives

$$\begin{aligned} \int w(|\, x' - x \,|) P(x')\, dx' &= \int w(z) P(x+z)\, dz \\ &\approx M_0 P + M_2 \frac{\partial^2 P}{\partial x^2} + M_4 \frac{\partial^4 P}{\partial x^4} + \cdots , \end{aligned} \tag{12.80}$$

where we have used the symmetry properties of the kernel and where the moments

$$M_{2m} = \frac{1}{(2m)!} \int z^{2m} w(z)\, dz, \quad m = 1, 2, \ldots . \tag{12.81}$$

If the kernel is very narrow then higher-order moments $|\, M_{2m} \,|$, $m \geq 2$ are small compared with $|\, M_0 \,|$ and $|\, M_2 \,|$ and expanding S in (12.79) in a Taylor series; (12.77) and (12.78) reduce to a familiar reaction diffusion form, namely,

$$\frac{\partial P}{\partial t} = S(M_0 P) - R - P + D \frac{\partial^2 P}{\partial x^2} = f(P, R) + D \frac{\partial^2 P}{\partial x^2}, \tag{12.82}$$

$$\frac{\partial R}{\partial t} = \gamma P - (1 - \delta) R = g(P, R), \tag{12.83}$$

which define the kinetics functions f and g and where the diffusion coefficient $D = M_2$. With the form of S from Figure 12.18(b) the null clines are schematically shown in Figure 12.25: at least one positive steady state exists.

We analysed reaction diffusion systems in depth in Chapter 2 and know that under appropriate circumstances spatial patterns can be generated (see Section 2.3 in Chapter 2). We briefly investigate here whether or not the system (12.82) and (12.83) can be diffusion-driven unstable and hence produce spatial patterns. We can say at the outset that if the single steady state is at B the system cannot generate spatial patterns under any circumstances (see Chapter 2). In the following we consider the steady state to be at a point A, where $P_A < P_M$ in Figure 12.25.

It is left as a revision exercise to show that the dispersion relation $\lambda = \lambda(k)$ about the steady state A in Figure 12.25 is given by the characteristic equation

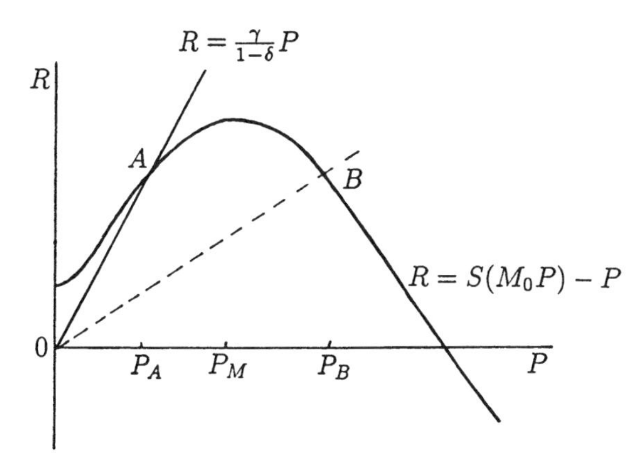

Figure 12.25. Schematic null clines $f(P, R) = 0$, $g(P, R) = 0$ in the phase plane for the reaction diffusion system (12.82) and (12.83). As γ decreases, the steady state moves to the right from A towards B.

$$\lambda^2 + a(k)\lambda + b(k) = 0, \tag{12.84}$$

where

$$\begin{aligned} a(k) &= Dk^2 - (f_P + g_R), \\ b(k) &= -g_R Dk^2 + (f_P g_R - f_R g_P), \end{aligned} \tag{12.85}$$

where the derivatives of $f(P, R)$ and $g(P, R)$ are evaluated at the uniform steady state. From (12.83) $g_R = -(1 - \delta) < 0$ and f_P can be positive or negative. Since we want the system to generate spatial patterns in a Turing sense we require (see Section 2.3)

$$f_P + g_R < 0, \quad f_P g_R - f_R g_P > 0 \tag{12.86}$$

which makes $a(k) > 0$ and, since $g_R < 0$, $b(k) > 0$ for all k. In this case the solutions of (12.84) have $\mathrm{Re}\,\lambda < 0$ and so this specific system cannot form spatial patterns. This is not, however, the end of the matter.

The usual reaction diffusion form was obtained by terminating the integral expansion in (12.80) at the second moment M_2. If the range of the excitation and inhibition is less local then we must include higher-order moments as we did in Chapter 6. Not only that, in (12.83) we tacitly assumed that $D = M_2$ is positive. This need not be the case; it depends on the form of the activation–inhibition kernel. If the lateral inhibition is longer range and more severe then it is possible for M_2 in (12.81) to be negative. If the kernel is such that this is the case then we must include at least one higher-moment term, namely, $M_4\partial^4 P/\partial x^4$, and more if M_4 is positive. The reason for this is that if $D = M_2 < 0$ then diffusion in (12.82) is destabilising (recall the discussion in Section 11.5 in Chapter 11, Volume I and in Chapter 6) whereas if we include long range diffusion which is associated with M_4 it is stabilising if $M_4 < 0$. If we include some long range diffusion then, in place of (12.82) and (12.83) we obtain

$$\frac{\partial P}{\partial t} = S(M_0 P) - R - P - D_1 \frac{\partial^2 P}{\partial x^2} - D_2 \frac{\partial^4 P}{\partial x^4} \tag{12.87}$$

$$\frac{\partial P}{\partial t} = \gamma P - (1-\delta) R = g(P, R), \tag{12.88}$$

where now $D_1 = -M_2 > 0$, $D_2 = -M_4 > 0$ represent diffusion coefficients, with the higher 4th-order operator representing the nonlocal, or long range, diffusion element.

Carrying out a linear analysis for (12.87) and (12.88) the dispersion relation is again given by (12.84) but with

$$\begin{aligned} a(k) &= D_2 k^4 - D_1 k^2 - (f_P + g_R), \\ &= D_2 k^4 - D_1 k^2 - (f_P - 1 + \delta), \\ b(k) &= -D_2 g_R k^4 + g_R D_1 k^2 + (f_P g_R - f_R g_P), \\ &= D_2(1-\delta) k^4 - (1-\delta) D_1 k^2 - [(1-\delta) f_P + \gamma f_R]. \end{aligned} \tag{12.89}$$

If either of $a(k)$ or $b(k)$ becomes negative for a range of nonzero wavenumbers k then $\operatorname{Re} \lambda > 0$ and the steady state is spatially unstable. Both of $a(k)$ and $b(k)$ are quadratic in k^2 with the coefficient of k^2 negative. Figure 12.26 sketches $a(k)$ by way of illustration: $b(k)$ behaves qualitatively the same. It is clear that if D_1 is large enough, that is, the destabilising term, the system becomes linearly unstable.

From (12.89), since $g_R < 0$ the minimum of $a(k)$ and $b(k)$ is at $k_m^2 = D_1/2D_2$ and

$$\begin{aligned} a(k_m) &= -\frac{D_1^2}{4D_2} - (f_P - 1 + \delta), \\ b(k_m) &= -(1-\delta)\frac{D_1^2}{4D_2} - [f_P(1-\delta) + \gamma f_R]. \end{aligned} \tag{12.90}$$

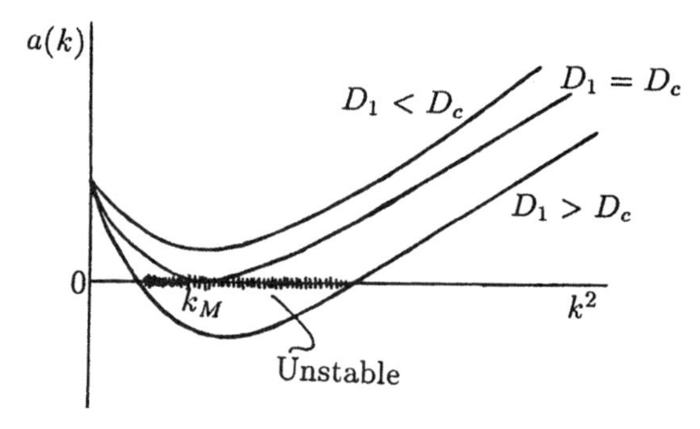

Figure 12.26. Sketch of $a(k)$ from (12.89). When $a(k) < 0$ the dispersion relation $\lambda(k)$ for the system (12.87) and (12.88) has $\operatorname{Re} \lambda > 0$ for a range of wavenumbers and the modes with these wavenumbers are linearly unstable. When $a(k) < 0$, or $b(k) < 0$, $\operatorname{Re} \lambda > 0$ and spatial patterns evolve.

So, bifurcation of the steady state occurs when $D_1 = D_c$ where

$$D_c^2 = \min\left[-4D_2(f_P - 1 + \delta), 4D_2\left(f_P + \frac{\gamma f_R}{1-\delta}\right)\right]. \tag{12.91}$$

Because of (12.86) both terms on the right in (12.91) are positive. For $D_1 > D_c$ there is thus a range of unstable modes with wavenumbers k such that $\mathrm{Re}\,\lambda(k) > 0$ for $k \neq 0$ since $a(k) < 0$; see Figure 12.26. The system (12.87) and (12.88) is thus capable of generating spatial patterns. With this continuous time analogue we could envisage the shell patterns to evolve as the shell grows because of a slow change in the parameters, such as D_1, the bifurcation parameter we used for illustration here. However, since the growth and pattern formation are on a similar time scale, it would perhaps be more appropriate to use the type of models discussed in Chapter 4.

All of the models discussed in this chapter rely on a local activation and longer range inhibition element. In many ways an integral formulation is preferable to a traditional differential equation formulation since the former is intuitively more easily related to the biological concepts. We can also see intuitively how spatial pattern can arise. The mechanisms proposed here, in Sections 12.2–12.4, all have some physiological justification but at this stage need considerably more experimental studies to justify possible acceptance.

12.5 Shamanism and Rock Art

In Section 12.3 we described a model mechanism, involving excitatory and inhibitory neurons—an activator and inhibitor model—for generating visual hallucination patterns; some of these images and their corresponding shapes in the visual cortex are illustrated in Figure 12.9. Visual perception does not depend on light, of course, as is easily demonstrated by shutting your eyes and relaxing. More complex patterns are obtained, with your eyes closed, if you press a finger into the corner of each eye: depending on the pressure you get different patterns. These self-illuminating patterns are called *phosphenes* (see the general articles by Oster 1970 and Klüver 1967 and other references below) and are images which are generated in the eye and brain by some light-independent mechanism. As mentioned, hallucinogenic drugs, epilepsy, migraine, certain mental illnesses and various diseases also give rise to hallucinogenic patterns. There seems to be a firm connection between phosphenes and these drug-induced hallucinations. A review from a physiological, biochemical and psychological point of view is given by Asaad and Shapiro (1986).

Aboriginal people in many parts of the world have known for a very long time that certain plants, such as mescaline, induce hallucinations and trances. The narcotic and psychedelic properties of these plants and their derivatives have been used by shamans for both medicinal and cult purposes in widely different and unrelated cultures from prehistoric times. Wellmann (1978) suggested that North American Indian rock paintings may have been made by shamans while they were under the influence of these hallucinogenic drugs. He specifically focused on two areas. One is the Chumash and Yokuts Indian region of California, where polychrome paintings show motifs similar to those visualized during trances induced by jimsonweed (*Datura* species). The other area is the lower Pecos River region of Texas. Here shamanistic figures exhibit aspects which are believed to be conceptual analogues of the mescal bean (*Sophora secondiflora*) cult which was practiced by Indians of the Great Plains.

Kellogg et al. (1965) studied the scribblings of a large number of young children from two to four years old from a spectrum of ethnic origins and found that their drawings have distinct phosphene characteristics. They were able to classify the recurring patterns into 15 distinct groups. Kellogg and her colleagues suggested that a child's scribbles are derived from phosphenes and are similar to those induced electrically in adults. Kellogg et al. (1965) concluded that the young children only develop the ability to draw, or rather scribble, geometric patterns over the age of three but after this time they quickly develop the ability to draw a wide number of different patterns but using only a few basic ones. They further suggested that the activation of such drawings comes from the activation of preformed neuronal networks in the visual system (recall the description in Section 12.2 on the formation of cortical stripes). These basic patterns progress from basic scribbles through shapes, their combination and aggregation leading eventually to actual pictorial figures; some examples are shown in Figure 12.27.

Figure 12.27. Examples of children's scribbles. The progression from basic scribbles such as in (**a**), to diagrams as in (**b**), to combinations then aggregations, such as in (**c**) finally leading to human and animal pictorial representations. The first 'human' example in (**d**) is a piece of cloth cut out and drawn (to represent me) by our daughter, Sarah Robertson (Murray), when she was six, while (also to represent me) the second is by her daughter, Isabella Mazowe Robertson, aged 4 and a half, who also drew the 'cat.' Kellogg et al. (1965) gives a complete classification with examples.

The ubiquity of some of these recurring patterns or phosphenes induced by hallucinogenic plants or drugs has led to a reexamination of some of the European and North American cave paintings. These signs of Upper Palaeolithic art were, for a long time, interpreted ethnographically in the archaeological literature. During the latter part of the 20th century many authors (see, for example, Wellmann 1978, Lewis-Williams and Dowson 1988, Hedges 1992, 1993 and the numerous references given in these) have suggested that this art may have been produced by shamans while under the influence of hallucinogenic drugs. These authors present convincing evidence of the correspondence between basic phosphenes, rock art and designs similar to those induced by hallucinogenic compounds.

Lewis-Williams and Dowson (1988), as well as reviewing the now large literature, proposed a model for classifying Upper Palaeolithic signs without any ethnographic analogy. They also studied shamanistic practices in Southern Africa to substantiate their hypotheses. Since the limited range of possible visual patterns (whether drug-induced or not) arise from the human nervous system it is not unreasonable to suppose that, irrespective of cultural backgrounds, people who are in an hallucinogenic state will see similar patterns. The shamans in different cultures, however, interpret these patterns within the context of their experience. The patterns are essentially variations and local

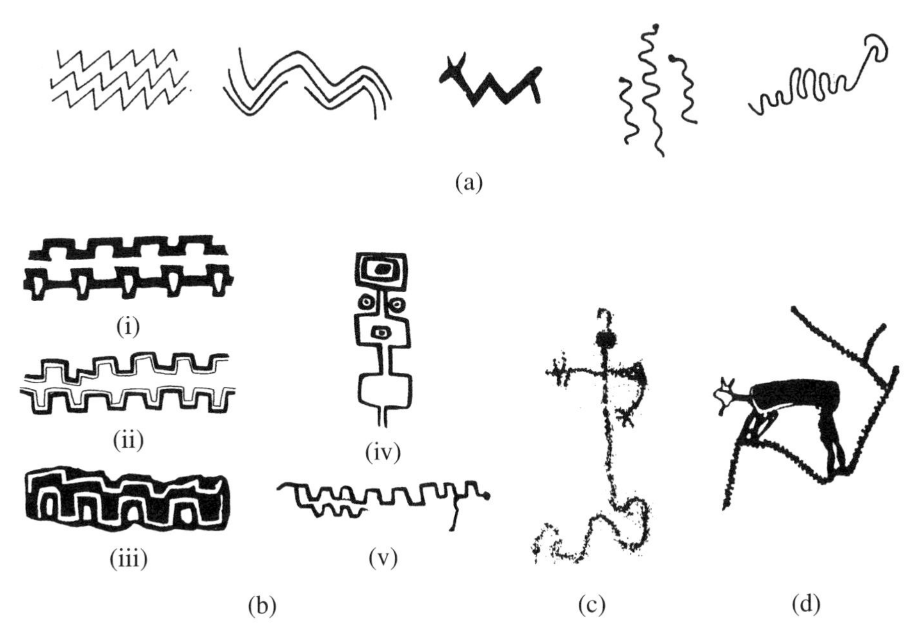

Figure 12.28. (**a**) Example of a basic pattern evolving into cult drawings found in San rock paintings (second, third and fifth) in Southern Africa and the Shonean Coso rock painting (fourth) in California. (Redrawn from Lewis-Williams and Dowson 1988). (**b**) Examples of basic crenalated phosphene designs and related cave motifs. Here (i) is a migraine pattern, (ii) is an hallucinogenic pattern from Tukano shamanistic art in Colombia, (iii) is a rock painting from Montevideo, Baja, California, (iv) is Hohokam rock art from central Arizona with (v) a petroglyph from Grapevine Canyon, Nevada. (**c**) A figure from McCoy Spring, California. (From Hedges 1992) (**d**) A fun San rock (South Africa) painting of a man looking over a cliff (from Lewis-Williams and Dowson 1988). (Figures (**b**) and (**c**) courtesy of K. Hedges)

cultural interpretations of the basic patterns (and their combination) based on the model we discussed in Section 12.3 earlier in the chapter and on the important extension by Tass (1995). His investigation as to how noise affects the cortical images greatly increases the practical range of possible hallucinogenic patterns and it is some of these that are reflected in many rock art images.

It is interesting, although with our current knowledge less surprising, that these motifs frequently appear in rock art in as widely dispersed geographic regions as the west of North America, European Palaeolithic caves and in Southern Africa. There seems a clear underlying universality of basic forms which have a plausible scientific explanation. In Figure 12.28(a) we see how a basic pattern can easily evolve into an example of a common theme in San rock art in Southern Africa and Shoshonean Coso rock art of the California Great Basin and possibly even that of the figure shown in Figure 12.28(c). Figures 12.28(b) to (d) give other examples found on ancient rock paintings and drawings in diverse areas. A more complex example relating a phosphene pattern and an anthropomorph is shown in Figure 12.29. Many of the basic (rock art) patterns and related variations are common in present-day American Indian art.

Although deformations and slight or topologically plausible changes in simple patterns and their dynamics variations can give rise to patterns found in petroglyphs and rock art they do not in themselves justify the hypotheses that they arise from a few basic motifs, the evidence produced by an extensive and carefully researched body of work makes for a very convincing justification. It clearly changes the whole interpretation

Figure 12.29. An example of an anthropomorph (**b**) from Renegade Canyon, Coso Range, California which has a clear resemblance to a phosphene pattern (**a**) (Walker 1981) which is made up of patterns which can be derived from those shown in Section 12.3. (From Hedges 1992; figures courtesy of K. Hedges)

of many firmly held views in the archaeological literature. As Hedges (Dr. K. Hedges, personal communication 2000) says,

> 'Much of what has been poorly understood in rock art can be better interpreted in the context of shamanism and visionary images. Most rock art is representational—we simply have to determine what it is that is being represented. . . .The shaman enters a separate reality when he undergoes transformation in the trance state. In rock art we have images from that separate reality which we are only beginning to understand.'

As a last comment, the correspondence between many of the rock art forms and phosphene patterns, by whatever manner they are induced, does not confirm the connection between these rock art forms and the basic patterns which are generated by a relatively simple mechanistic approach described in Section 12.3. It does, however, suggest potentially important implications for furthering our understanding of certain patterning phenomena of the human mind from children to shamans.

Exercises

1. Consider the integral equation model

$$\frac{\partial n}{\partial t} = f(n) + \int_{-\infty}^{\infty} w(x - x')[n(x', t) - 1]\,dx',$$

where $n = 1$ is a zero of $f(n) = 0$ with $f'(1) < 0$. Construct activation–inhibition kernels w from (i) simple square waves and (ii) exponentials of the form $\exp[-|x|/a]$. Sketch the resulting dispersion relation $\lambda(k)$ and determine the critical parameter values for spatially structured solutions. Find the wavelength of the fastest growing mode in a linearly unstable solution.

2. Consider a model for ocular dominance stripes with left and right eye synapse densities n_R and n_L given by the system

$$\frac{\partial n_R}{\partial t} = f(n_R)[w_{RR} * n_R + w_{LR} * n_L],$$

$$\frac{\partial n_L}{\partial t} = f(n_L)[w_{LL} * n_L + w_{RL} * n_R],$$

where $*$ denotes the convolution and $f(n) = n(N - n)$ where N is the constant total synapse density; that is, $n_R + n_L = N$ and the interaction kernels w are as illustrated in Figure 12.6.

Show first that conservation of total synapse density implies

$$w_{RR} = -w_{RL}, \quad w_{LL} = -w_{LR}.$$

Setting

$$u = n_R - n_L, \quad w = w_{RR} + w_{LL}, \quad K = N * (w_{RR} - w_{LL})$$

show that the model mechanism reduces to the single convolution equation

$$\frac{\partial u}{\partial t} = (N^2 - u^2)(w * u + K).$$

[When there is eye symmetry $w_{RR} = w_{LL}$ and so $K = 0$.]

If initially $n_R = n_L$ then for small t, u is small and hence

$$\frac{\partial u}{\partial t} = N^2(w * u + K).$$

If the domain is one-dimensional, $x \in (-\infty, \infty)$, show that spatial patterned solutions for $u(x, t)$ will evolve with time by determining and analysing the dispersion relation.

In the case of symmetric eye inputs determine the wavelength of the fastest growing mode in terms of the parameters when the kernel w is given by (12.13).

3. A model for shell patterns is described by the system (12.53) and (12.54). Choose Heaviside functions for the stimulation functions S in (12.56) and rectangular forms for the excitation (w_E) and inhibition (w_I) kernels in (12.55); keep in mind the requirement that the inhibition is of longer range than the excitation. Hence derive the piecewise linear system governing the pigmentation pattern $P_t(x)$ and the refractory substance $R_t(x)$.

Starting with this piecewise linear model derive the equation governing $P_t(x)$ for large t when the build-up of R_t by P_t is zero; that is, the parameter $\gamma = 0$.

13. Geographic Spread and Control of Epidemics

The geographic spread of epidemics is less well understood and much less well studied than the temporal development and control of diseases and epidemics. The usefulness of realistic models for the geotemporal development of epidemics be they infectious disease, drug abuse fads or rumours or misinformation, is clear. The key question is how to include and quantify spatial effects. In this chapter we describe a diffusion model for the geographic spread of a general epidemic which we then apply to a well-known historical epidemic, namely, the ever fascinating mediaeval Black Death of 1347–50. We then discuss practical models for the current rabies epidemic which has been sweeping through continental Europe and is now approaching the north coast of France. These types of models, of course, are not restricted to one disease.

13.1 Simple Model for the Spatial Spread of an Epidemic

We consider here a simpler version of the epidemic model discussed in detail in Chapter 10, Volume I, Section 10.2. We assume the population consists of only two populations, infectives $I(\mathbf{x}, t)$ and susceptibles $S(\mathbf{x}, t)$ which interact. Now, however, I and S are functions of the space variable $\mathbf{x}$ as well as time. We model the spatial dispersal of I and S by simple diffusion and initially consider the infectives and susceptibles to have the same diffusion coefficient D. As before we consider the transition from susceptibles to infectives to be proportional to rSI, where r is a constant parameter. This form means that rS is the number of susceptibles who catch the disease from each infective. The parameter r is a measure of the transmission efficiency of the disease from infectives to susceptibles. We assume that the infectives have a disease-induced mortality rate aI; $1/a$ is the life expectancy of an infective. With these assumptions the basic model mechanism for the development and spatial spread of the disease is then

$$\begin{aligned}
\frac{\partial S}{\partial t} &= -rIS + D\nabla^2 S, \\
\frac{\partial I}{\partial t} &= rIS - aI + D\nabla^2 I,
\end{aligned} \tag{13.1}$$

where a, r and D are positive constants. These equations are simply (10.1) and (10.2) in Chapter 10, Section 10.2, Volume I with the addition of diffusion terms. The problem we are now interested in consists of introducing a number of infectives into a uniform population with initial homogeneous susceptible density S_0 and determining the *geo*temporal spread of the disease.

Here we consider only the one-dimensional problem; later we present results of a two-dimensional study. We nondimensionalise the system by writing

$$I^* = \frac{I}{S_0}, \quad S^* = \frac{S}{S_0}, \quad x^* = \left(\frac{rS_0}{D}\right)^{1/2} x,$$
$$t^* = rS_0 t, \quad \lambda = \frac{a}{rS_0}, \tag{13.2}$$

where S_0 is a representative population and the model (13.1) becomes, on dropping the asterisks for notational simplicity,

$$\frac{\partial S}{\partial t} = -IS + \frac{\partial^2 S}{\partial x^2},$$
$$\frac{\partial I}{\partial t} = IS - \lambda I + \frac{\partial^2 I}{\partial x^2}. \tag{13.3}$$

The three parameters r, a and D in the dimensional model (13.1) have been reduced to only one dimensionless grouping, λ. The basic *reproduction rate* (cf. Chapter 10, Volume I, Section 10.2) of the infection is $1/\lambda$; it has several equivalent meanings. For example, $1/\lambda$ is the number of secondary infections produced by one primary infective in a susceptible population. It is also a measure of the two relevant timescales, namely, that associated with the contagious time of the disease, $1/(rS_0)$, and the life expectancy, $1/a$, of an infective.

The specific problem we investigate here is the spatial spread of an epidemic wave of infectiousness into a uniform population of susceptibles. We want to determine the conditions for the existence of such a travelling wave and, when it exists, its speed of propagation.

We look for travelling wave solutions in the usual way (cf. Chapter 1) by setting

$$I(x,t) = I(z), \quad S(x,t) = S(z), \quad z = x - ct, \tag{13.4}$$

where c is the wavespeed, which we have to determine. This represents a wave of constant shape travelling in the positive x-direction. Substituting these into (13.3) gives the ordinary differential equation system

$$I'' + cI' + I(S - \lambda) = 0, \quad S'' + cS' - IS = 0, \tag{13.5}$$

where the prime denotes differentiation with respect to z. The eigenvalue problem consists of finding the range of values of λ such that a solution exists with positive wavespeed c and nonnegative I and S such that

$$I(-\infty) = I(\infty) = 0, \quad 0 \leq S(-\infty) < S(\infty) = 1. \tag{13.6}$$

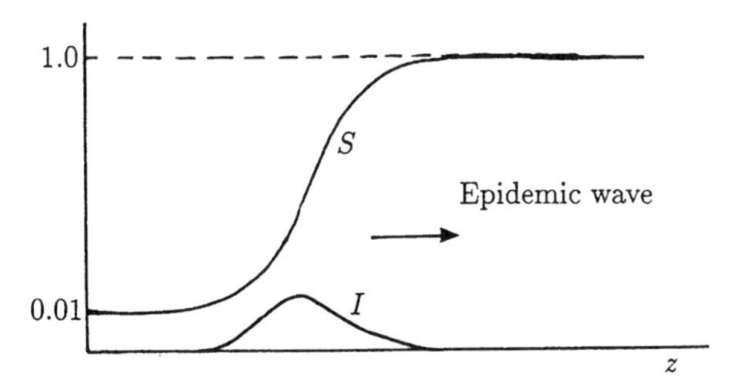

Figure 13.1. Travelling epidemic wave of constant shape, calculated from the partial differential equation system (13.3) with $\lambda = 0.75$ and initial conditions (that is, with compact support) compatible with (13.6). Here a pulse of infectives (I) moves into a population of susceptibles (S) with speed $c = 1$ which in dimensional terms from (13.2) is $(rS_0D)^{1/2}$ which agrees with the analytical wavespeed (13.11) with $\lambda = a/rS_0 = 0.75$.

The conditions on I imply a pulse wave of infectives which propagates into the uninfected population. Figure 13.1 shows such a wave; Figure 13.5 below, which is associated with the spread of a rabies epidemic wave, is another example, although there, only the infectious population I diffuses.

The system (13.5) is a fourth-order phase space system. We can determine the lower bound on allowable wavespeeds c by using the same technique we employed in Chapter 13, Volume I, Section 13.2 in connection with wave solutions of the Fisher–Kolmogoroff equation. Here we linearise the first of (13.5) near the leading edge of the wave where $S \to 1$ and $I \to 0$ to get

$$I'' + cI' + (1 - \lambda)I \approx 0, \tag{13.7}$$

solutions of which are

$$I(z) \propto \exp\left[(-c \pm \{c^2 - 4(1-\lambda)\}^{1/2})z/2\right]. \tag{13.8}$$

Since we require $I(z) \to 0$ with $I(z) > 0$ this solution cannot oscillate about $I = 0$; otherwise $I(z) < 0$ for some z. So, if a travelling wave solution exists, the wavespeed c and λ must satisfy

$$c \geq 2(1-\lambda)^{1/2}, \quad \lambda < 1. \tag{13.9}$$

If $\lambda > 1$ no wave solution exists so this is the necessary threshold condition for the propagation of an epidemic wave. From (13.2), in dimensional terms the threshold condition is

$$\lambda = \frac{a}{rS_0} < 1. \tag{13.10}$$

This is the same threshold condition found in Chapter 10, Volume I, Section 10.2 for an epidemic to exist in the spatially homogeneous situation.

With our experience with the Fisher–Kolmogoroff equation we expect such travelling waves computed from the full nonlinear system will, except in exceptional conditions, evolve into a travelling waveform with the minimum wavespeed $c = 2(1-\lambda)^{1/2}$. In dimensional terms, using (13.2), the wave velocity, V say, is then given by

$$V = (rS_0D)^{1/2}c = 2(rS_0D)^{1/2}\left[1 - \frac{a}{rS_0}\right]^{1/2}, \quad \frac{a}{rS_0} < 1. \tag{13.11}$$

The travelling wave solution $S(z)$ cannot have a local maximum, since there $S' = 0$ and the second of (13.5) shows that $S'' = IS > 0$, which implies a local minimum. So $S(z)$ is a monotonic increasing function of z. By linearising the second equation of (13.5) as $z \to \infty$, where $S = 1 - s$, with s small, we have

$$s'' + cs' - I = 0,$$

which, with $I(z)$ from (13.8), shows that

$$S(z) \sim 1 - O\left(\exp[\{-c \pm [c^2 - 4(1-\lambda)]^{1/2}\}z/2]\right)$$

and so, as $z \to \infty$, $S(z) \to 1$ exponentially.

The threshold result (13.10) has some important implications. For example, we see that there is a minimum critical population density $S_c = a/r$ for an epidemic wave to occur. On the other hand for a given population S_0 and mortality rate a, there is a critical transmission coefficient $r_c = a/S_0$ which, if not exceeded, prevents the spread of the infection. With a given transmission coefficient and susceptible population we also get a threshold mortality rate, $a_c = rS_0$, which, if exceeded, prevents an epidemic. So, the more rapidly fatal the disease is, the less chance there is of an epidemic wave moving through a population. All of these have implications for control strategies. The susceptible population can be reduced through vaccination or culling; we discuss this and immunity effects below. For a given mortality and population density S_0, if we can, by isolation, medical intervention and so on, reduce the transmission factor r of the disease, it may be possible to violate condition (13.10) and hence again prevent the spread of the epidemic. Finally with $a/(rS_0) < 1$ as the threshold criterion we note that a sudden influx of susceptible population can raise S_0 above S_c and hence initiate an epidemic.

Here we have considered only a simple two-species epidemic model. We can extend the analysis to a three-species *SIR* system. It becomes, of course, more complicated. In Sections 13.5–13.9 we discuss in some detail such a model for the current European epidemic of rabies.

13.2 Spread of the Black Death in Europe 1347–1350

Historical Aside on the Black Death and Plague in the 20th Century

The fascination with the Black Death, the catastrophic plague pandemic that swept through Europe in the mid-14th century, has not abated with the passage of time. Albert

Camus' *The Plague*, published in 1947, is one example, in a modern context. In the many accounts of the Black Death over the centuries, whether factual or romanticised, a vision has been conjured up of horrific carnage, wild debauchery, unbelievable acts of courage and altruism, and astonishing religious excesses.

The Black Death, principally bubonic plague, was caused by an organism (*Bacillus pestis*) and was transmitted by fleas, mainly from black rats, to man. It was generally fatal. The article by Langer (1964) gives a graphic description and some of the relevant statistics. The historical article by McEvedy (1988) discusses the pandemic's progress and surveys some of the current thinking on the periodic occurrences of bubonic plague. The plague was introduced to Italy in about December 1347, brought there by ship from the East where it had been raging for years. During the next few years it spread up through Europe at approximately 200–400 miles a year. About a quarter to a third of the population died and approximately 80% of those who contracted the disease died within 2–3 days. Figure 13.2 shows the geotemporal spread of the wavefront of the disease.

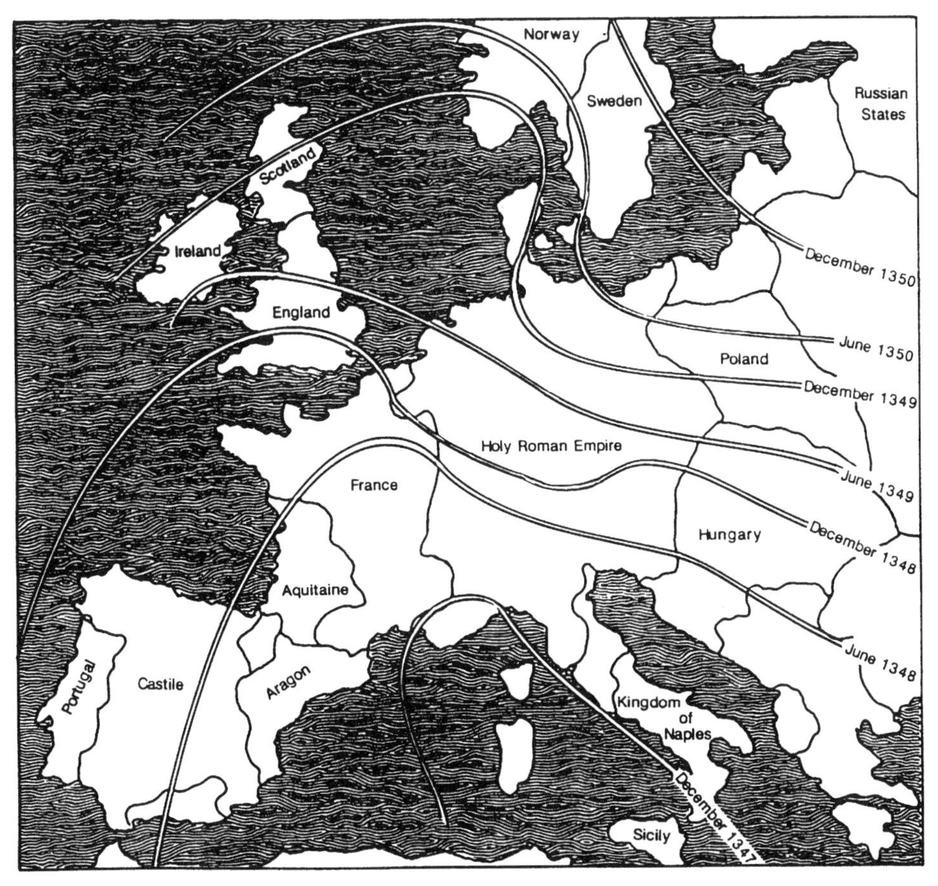

Figure 13.2. Approximate chronological spread of the Black Death in Europe from 1347–1350. (Redrawn from Langer 1964).

After the Black Death had passed, around 1350, a second major outbreak of plague appeared in Germany in 1356. From then on periodic outbreaks seemed to occur every few years although none of them were in the same class as regards severity as the Black Death epidemic of 1347. In Section 13.4 we shall describe an obvious extension to the simple model in this section which takes into account the partial recovery of the population after the passage of an epidemic wave. Including this results in periodic outbreaks, smaller ones, appearing behind the main front: see Figure 13.6 and Figure 13.7. Figures 13.9 and 13.10 in Section 13.5, which consider a three-species model for the spatial spread of a rabies epidemic, exhibit even more dramatic periodic epidemic waves which follow the initial outbreak. There we can estimate the period of the recurring outbreaks analytically.

There was a great variety of reactions to plague in mediaeval Europe and later (just as there is to plague today and the current AIDS epidemic).[1] Groups of penitents, vigorously flagellating their half-naked bodies and preaching the coming of the end of the world, wandered about the countryside; some of the elegant and beautifully carved ivory handles of the flails of the richer flagellants still survive. Cures for the plague abounded during this period. One late 15th century cure, recently discovered in Westfalen-Lippe in northwest Germany, involved the following preparation. The tip of an almost hatched egg was cut off and the brood allowed to run out. The remaining egg yolk was mixed with raw saffron and the egg refilled and resealed with the shell pieces originally removed. The egg was afterwards fried until it turned brown. The recipe then called for the same amount of white mustard, some dill, a crane's beak and theriak (a popular quack medicine of the time). The mixture had to be swallowed by the victim who had to eat nothing more for 7 hours. There is no record of how effective this cure was!

The disease, of which there are three kinds, bubonic, pneumonic and septicemic, is caused by a bacillus carried primarily by fleas which are in turn carried by rats, mice and a host of other animals. Septicemic plague involves the bacilli multiplying extremely rapidly in the victim's blood and is almost invariably fatal (even now), whether treated or not; the victim usually dies very quickly and often suddenly. Septicemic plague often develops from the pneumonic form which is extremely contagious. There are descriptions of plague victims who suddenly sat down and simply keeled over dead. These could well have been septicemic cases who contracted it from the coughs of pneumonic victims. Children at the time of the Great Plague of London from 1664–1666, which peaked in 1665, used to sing the well-known English nursery rhyme

> 'Ring-a ring o'roses
> A pocket full of posies
> A-tishoo, A-tishoo
> We all fall down.'

---

[1] John Calvin, the scholar, theologian, unsurpassed killjoy and a man with a monumental ego and self-righteousness was convinced witches acting as agents of the devil brought it to Geneva where he was. He fled, terror stricken, from the epidemic and managed to survive (unfortunately). His baleful influence is still abundantly evident in the Scottish psyche and society even today and has, I am in no doubt, contributed to the massive emigration of Scots over the centuries to escape his depressing, deterministic and unforgiving view of the world.

which is believed to date from that period. Onions and garlic were held to the nose, in 'posies' or the bird-like masks of the physicians, to keep out the bad odours that were thought to be the cause of the disease.

There are considerably more data and information about the Great Plague of London in 1665 than are available about the Black Death. The people's reaction, however, seems not to have been dissimilar—fewer overt extreme penitents perhaps. The diarist Samuel Pepys describes the scene in my own university town of Oxford as one of 'lewd and dissolute behaviour'. Plus ça change. . . ! Daniel Defoe's journal (1722) of the epidemic vividly conjures up a contemporary image and makes fascinating reading: 'It was then indeed, that man withered like the grass and that his brief earthly existence became a fleeting shadow. Contagion was rife in all our streets and so baleful were its effects, that the church-yards were not sufficiently capacious to receive the dead. It seemed for a while as though the brand of an avenging angel had been unloosed in judgment.'

There is a widely held belief that plague more or less ceased to be a problem after the Great Plague of London. This is far from the case, however, as clearly documented in the book by Gregg (1985). The last plague pandemic started in Yunnan in China about 1850 and only finished officially, according to the World Health Organisation, in 1959: more than 13 million deaths have been attributed to it, and it affected most parts of the world. The reported cases (and through ignorance or political expediency the figures must clearly be considered lower bounds) since 1959 makes it clear that plague epidemics are still with us. The thousands who died of it during the Vietnam war, particularly between 1965 and 1975, is a dramatic case to point.

Plague was brought by ship to the Northwest of America around 1900. About 200 deaths were recorded in the three-year San Francisco epidemic which started just after the earthquake in 1906. The article by Risse (1992) is specifically on this San Francisco epidemic. As a result of this epidemic, the western part of the U.S.A., particularly New Mexico, is now one of the two largest residual foci of plague (in mice and voles particularly) in the world—the other is in Russia. The plague bacillus has spread steadily eastwards from the west coast and in 1984 was found among animals in the midwest. The wavefront has moved on average about 35 miles a year. The disease is carried by a large number of native wild animals. Rats are by no means the sole carrier: it has been found in nearly 30 different mammals including, for example, squirrels, chipmunks, coyotes, prairie dogs, mice, voles, domestic pets and bats. The present complacency about the relatively small annual number of plague deaths is hardly justified. If, or rather *when*, plague reaches the east coast of the U.S.A. with its large urban areas, the potential for a serious epidemic will be considerable. New York, for example, has an estimated rat population of one rat per human; and mice—also effective disease carriers — probably number more. The prevailing lack of both concern and knowledge about the plague is dangerous. Plague symptoms are often not recognised or, at best, only belatedly diagnosed. Therefore the victim is free to expose a substantial number of people to the disease, particularly if it is pneumonic plague which is one of the most infectious diseases known.

To return now to our modelling, let us apply our simple epidemic model to the spread of the Black Death. We first have to estimate the relevant parameters, not a simple task with the paucity of hard facts about the social conditions of the time. Noble (1974) used such a model to investigate the spread of the plague and, after a study of the known facts, suggested approximate values for the parameters, some of which we use.

There were about 85,000,000 people in Europe in 1347 which gives a population density $S_0 \approx 50/\text{mile}^2$. It is particularly difficult to estimate the transmission coefficient r and the diffusion coefficient D. Let us suppose that the spread of news is governed by diffusion with a diffusion coefficient D. The time to cover a distance L miles purely by diffusion is then $O(L^2/D)$ years. Suppose, with the limited communications that existed at the time, that news and minor gossip, say, travelled at approximately 100 miles/year; this gives a value of $D \approx 10^4$ miles^2/year. To transmit the disease the fleas have to jump from rats to humans and humans have to be close enough to infect other humans; this is reflected in the value for r. Noble (1974) estimated r to be 0.4 mile^2/year. He took an average infectious period of two weeks (too long probably), which gives a mortality rate $a \approx 15/\text{year}$. These give $\lambda = a/(rS_0) \approx 0.75$. With the wavespeed given by (13.11) in terms of the model parameters, we then get the speed of propagation, V, of the plague as

$$V = 2(rS_0D)^{1/2}\left[1 - \frac{a}{rS_0}\right]^{1/2} \approx 140 \text{ miles/year}.$$

Although this is somewhat lower than the speed of 200 to 400 miles/year, quoted by Langer (1964), it is not an unreasonable comparison in view of the gross estimates used for the unknown, and what are undeterminable, parameters.

Of course, such a model is extremely simple and does not take into account a number of factors, such as the nonuniformity in population density, the stochastic element and so on. Nevertheless it does indicate certain global features of the geographic spread of an epidemic. As we noted in Chapter 10, Section 10.2, Volume I the stochastic model studied by Raggett (1982) for the plague epidemic of 1665 to 1666 in the village of Eyam did not give as good comparison with the data as did the deterministic model. Stochastic elements, however, are more important in spatial models, particularly when the numbers involved are small.

Keeling and Gilligan (2000) have recently proposed an interesting new model for the spatiotemporal spread of the bubonic plague incidence (there are several thousand deaths each year).[2] Plague is a zoonosis (a disease which spreads from animals to humans) and in many areas where it is prevalent rats are clearly implicated. Their model crucially incorporates the rat, as well as human, populations and includes stochasticity. They show that the disease can reside in rat subpopulations thereby letting the disease persist for many years. They discuss both deterministic and stochastic versions of their model and use a cellular automaton model to incorporate spatial stochasticity. From an analysis of their models they obtain, among other things, criteria for the spread in the human population in terms of the rat population. They use data on rodent populations from North America and use current estimates for the parameters.

[2]When I was visiting the Los Alamos National Laboratory in New Mexico in 1985 a 14-year-old boy in the neighbouring village died from getting infected while moving logs from a wood pile in which infected chipmunks had recently died of the disease. He contracted the septicemic form and died within three days.

13.3 Brief History of Rabies: Facts and Myths

Rabies—A Mediaeval View

Rabies is arguably the most horrifying disease; the patient undergoes the most frightening nightmarish experiences before dying in prolonged and terrifying agony. St. Augustine of Hippo included it in a list of disasters he compiled (which included such things as insanity, imprisonment and bankruptcy). In spite of the fact that an effective rabies vaccination is now available, even totally reliable if given soon after being in contact with a rabid animal, the horror of rabies is almost as rampant today as it ever was. If a person reaches the actual rabid stage, that is, displays the clinical symptoms, there is no cure, nor has there ever been a reliably recorded case of a cure. Dr. Patricia Morrison (personal communication, 1992), who has studied historical aspects of the disease, related the following story about the 8th century bishop of Liège, St. Hubert.[3]

A hundred years after Hubert died his corpse was dug up (and reputedly found not to be in a decayed state) and carted off to a poor monastery in the Ardennes. The abbot and monks of the monastery were in dire need of some relic to attract pilgrims, give it some kudos and generally boost morale. Hubert's corpse was just the relic and not surprisingly its arrival was the start of many miraculous cures for the visiting pilgrims. The town, St. Hubert des Ardennes, sprang up around the monastery. In the tourist office, now in the abbot's house of the monastery, you can obtain leaflets about the saint and his conversion. Hubert, a young nobleman, when hunting saw a white stag with a crucifix between its antlers. Christ approached him and he decided to go into the church. Apparently it is complete plagiarism; it was lifted directly from *The Life of St. Eustace.* St. Hubert is the patron saint of hunters. In all the ceremonies that currently take place in the town little is ever mentioned of his role in the rabies story.

In the 11th century a monk wrote that it was customary to take people bitten by a rabid (or supposedly rabid) animal, a dog or wolf generally, to St. Hubert's shrine. The procedure was for the priest to cut the pilgrim's forehead and insert a thread taken from St. Hubert's episcopal stole. The monks later said the stole had been woven by the Virgin Mary and hand delivered by an angel. The connection between the thread ritual (called *la taille*) and rabies is possibly due to the fact that rabies was thought to be caused by worms in a dog's anus or under its tongue. So, the thread from the saint's stole was thought to be a kind of inoculation against rabies. The mediaeval cloak, which never apparently got any smaller, can still be seen in a reliquary in the church, which is enormous and testament to its popularity among pilgrims. Early in the 18th century, during the height of the developing vampire legend (see below), records show that 1956 people were given *la taille*. It still had its followers in the 1920's and I suspect even

[3]The conversation took place in the unlikely venue over dinner in Corpus Christi College, my Oxford College. Dr. Morrison, the art critic of the *Financial Times* (London) subsequently wrote an article for the newspaper ('A saintly 'cure' for rabies,' *Financial Times*, 29–30 August, 1992) which discussed the modelling below, the then current view and the St. Hubert story. I feel it is yet another justification for easy intermingling of the academic disciplines. It greatly helps, of course, if *haute cuisine* is the common fare as it was at 'High Table' where only faculty and their guests can dine.

now.[4] In the church, close to St. Hubert's altar, there is a large iron ring on the wall to which were tied the poor wretches who were writhing, shouting, groaning, barking and convulsing. Occasionally they were 'cured,' in which case St. Hubert had interceded on their behalf. Rabies hysteria, where people thought they had rabies and displayed many of the symptoms of furious rabies, has been well documented. If the poor wretch, who was tied to the ring, after the required nine days wait died then St. Hubert had decided not to intercede. It was very much a win–win situation for St. Hubert and the monastery.

Rabies and the Vampire Legend

Vampires were first mentioned, and widely believed in, during the last quarter of the 17th century. They were thought to be reanimated corpses which rose from their graves, seeking nourishment by sucking the blood of sleeping persons. In the 18th century they were greatly feared, particularly in the Balkans, and were, according to Voltaire 'the sole topic of conversation between 1730 and 1735.' Gómez-Alonso (1998), in an interesting paper on the legend, puts forward a possible explanation for the original belief by suggesting that rabies may have played a role and it is from his work that the following main elements have been taken; his article has a comprehensive list of references both new and old. (Those of a sqeamish bent should perhaps not read the following even though I have omitted many of the even more horrifying descriptions.)

The legend began with a late 17th century report describing the existence of cadavers full of liquid blood, supposedly taken from people and animals. Some villagers alleged to have seen a ghost in the form of a dog, with others a hideous man, that attacked people, seizing them by the throat. Other gory and graphic details are provided in the report. The stories grew. In the village of Medvedja in Serbia in 1731 to 1732 some peasants' deaths were attributed to a vampire. Signs of vampirism were found in 17 exhumed corpses which were pierced with stakes, decapitated and cremated: this was the motivating event for the 'sole topic of conversation' mentioned by Voltaire and picked up by many well-known members of the Enlightenment. Dracula, of literary and film fame, only came into being at the very end of the 19th century.

Vampires were attributed with a wide array of habits, such as leaving their graves to have sexual intercourse as well as to kill innocent victims for their blood. A person could become a vampire if they had been attacked by a vampire or eaten animals that had been killed by one, had died of rabies or plague or even having been a great lover; there were numerous other ways you could become a vampire. Not surprisingly there were also numerous remedies for avoiding becoming a vampire. Two of the visible signs that a corpse was a vampire were prominent genitalia and a body swollen with blood that flowed out of its mouth.

Rabies is a zoononis, that is, a disease that can be transmitted from vertebrates to humans, and in the rabid stage can cause unpredictable violent and aggressive behaviour. Certain diseases of the limbic system can also affect sexual behaviour. Cadavers buried in cold and humid places can often delay decomposition of the corpse by causing the subcutaneous tissue to become waxlike. As to the liquid blood, certain dis-

[4]Dr. Morrison said that when she visited the church and asked the church warden about *la taille* he said that no one believed that nonsense now but did add that he had known of a case and that some people who had been vaccinated after being bitten by what seemed to be a rabid dog had come asking for *la taille*.

eases prolong the liquid stage but when decomposition does set in with dissolution of internal organs, the resulting gases distend various parts of the body such as the genitalia and face and cause the tongue to protrude giving rise to blood frothing from the mouth.

Let us now consider some of the symptoms of human rabies. Most humans develop the 'furious' form of the disease, rather than the paralytic form, and insomnia, uncontrolled agitation, hydrophobia (the former name for the disease), muscular spasms, fear of seeing themselves in a mirror and other extremely ghastly and bizarre manifestations. The spasms can cause the facial muscles to retract the lips showing teeth in a grimace and the emission of unintelligible sounds. There are stories of rabid humans, like rabid dogs, rushing at people and aggressively assaulting (including biting) them. During intermediate quiet times patients drool blood from their mouths. Hypersexuality can also occur with days of permanent erections, frequent intercourse and violent rape attempts (see, for example, Warrell, 1977).

Rabies can be transmitted person-to-person in a variety of ways such as animal (or human) bites, genital mucosae and so on (Warrell, 1977). Numerous theories have been put forward for the legend from simple superstition to schizophrenia. During times of epidemics bodies were sometimes buried in shallow graves and dug up by dogs and wolves, thus giving rise to the idea that vampires rose from their graves. Wandering rabid people pre-19th century could also have given support to the vampire legend with their aggressiveness and hypersexuality. Gómez-Alonso (1998), in his well-documented article, makes a fascinating a highly plausible case for rabies giving rise to the vampire legend. The history of rabies, and attitudes towards it, is fascinating, riveting, often horrifying and sometimes funny.

Rabies: 18th and 19th Century England

Ritvo (1987), in her excellent historical book on animal–human sociology in Victorian England, gives some amazing facts and presents some fascinating insights into the attitudes to rabies in 19th century England. Some of the views and beliefs are hilarious.

In England in the 19th century there were several outbreaks (the numbers were in fact very small) which wreaked havoc and spawned some hilarious laws and views. George Fleming, a distinguished veterinarian, said in his treatise *Rabies and Hydrophobia* that 'there may be some foundation for the supposition that intense sexual excitement may produce rabies ....' The interconnection of sex and sin and rabies could be discreetly suggested: the Victorians loved it. The poaching dogs—the 'curs and lurchers'—of the lower classes were considered particlarly prone to rabies.

The geographical range of rabies and its frequency of incidence increased during the 19th century, although the death toll was never very high. For example, 79 people died in 1877, 35 in 1879, 47 in 1875, (which worked out to two rabies deaths per million of population). The average English citizen of the later 19th century was more than 10 times as likely to be murdered as to die of rabies. Ritvo (1987) quotes some hilarious comments at the time such as in the *Kennel Review* which defined hydrophobia (rabies) as 'a peculiar madness that seizes men and impels them to destroy dogs.' A similar view was expressed by the Surgeon General, Charles Alexander Gordon, in his testimony to the House of Lords Committee on rabies.

Remnants of some of the attitudes from a much earlier age, even from mediaeval times, were still evident in the second half of the 20th century and no doubt still exist.

Rabies: Current Situation

Rabies is still a very serious disease that exists with varying degrees of severity in practically all countries of the world except for Britain, Ireland, Sweden, Australia, New Zealand and a few others. The World Health Organisation (WHO 1998) is an excellent source of disease data and it is the source for the latest (1994) global statistics which give some idea of the extant problem with rabies. The estimated number of human cases worldwide is approximately 35,000–45,000 with about 10–20 in Europe, 4–8 in North America, 200–400 in Latin America, 500–5000 in Africa, 35,000–45,000 in Asia and 30,000–40,000 in India. In Bangladesh, for example, in 1994 there were 3000 cases while in the U.S.A. there were 6 with 4 there in each of 1995, 1996 and 1997. France had 1 in 1994 and 3 in each of 1995 and 1996. The number of rabies cases in animals, of course, is very much higher: in the U.S.A. in 1994 there were 8224 confirmed rabies cases while in Bangladesh there were 960 laboratory confirmed cases with 3500 not laboratory confirmed.

Vaccination has been a major control strategy for rabies in parts of Europe. Aubert (1997) describes the status of it in France as a consequence of vaccination (administered via bait in the spring and summer for two years in a row) carried out since 1986. From 1989 to 1996 animal rabies was almost totally eradicated in the regions treated. France created an immunological barrier from the English channel to Switzerland thus stopping the southern progression of the disease. He notes that all cases of canine rabies in the previous 20 years were observed in imported animals, with the last in 1995 and could have been prevented with stricter border controls.

Pastoret (1998) discusses the rabies scene in Belgium and notes that after a period of elimination it appeared again in 1994. Barrat and Aubert (1993) comment on the decline of rabies in France from its peak in 1989 which they partly attribute to the oscillations such as those exhibited by the model.

With such widespread global movement of people and animals it is inevitable that rabies will continue to be introduced into countries hitherto free of the disease. Britain's paranoia about rabies has not been helped with the Channel Tunnel and the fact that bats can carry the disease. Infected bats have been found in parts of Belgium. Teulières and Saliou (1995) noted that although there were no domestic case of rabies from 1970 to 1993, 14 patients contaminated in enzootic areas died from rabies in France. Human vaccination protocol, since 1988, is now based on intramuscular injections, two on day 0 at two different sites with boosters on day 7 and day 21; no failures have been reported. The Center for Disease Control (CDC) in Atlanta recommend booster injections on days 0, 7, 28 and 365 for people in exposed areas; the protection lasts for three years.

The vampire bat is an important reservoir for rabies in, for example, Mexico and Latin America where it has been the origin of rabies outbreaks in cattle. In Asia, Latin America and Africa it is mainly enzootic dog rabies that is the serious problem. Most humans contract the disease through direct bite or scratch from a rabid animal although aerosol transmission in caves with infected bats is also possible. Although rabies is rare in the U.S. when it occurs it is almost always from a bite from an infected bat. Of the 25 cases between 1980 and 1999 all but three contracted it from bats. People bitten by a bat when they are awake will feel the pinch. When asleep, however, the needle-like teeth make practically no wound and may not even be felt; this is probably the likely cause of most of these cases. The Center for Disease Control (CDC) recommends having vaccination (now five shots in the arm over a four-week period) if you wake up and find a bat in your room.

There have been some bizarre and tragic transmissions of the disease including a 14-year-old girl who contracted it from an infected dog that licked her genitals. Human-to-human transmission can also occur: the case of a woman who contracted it from a corneal transplant from a man who was infected (Houff et al. 1979) is particularly ghastly. It was only after both had died from the paralytic form of the disease that the rabies virus was found in their eyes. Corneal transplantation was also implicated in human-to-human transmission in the case of Creutzfeldt–Jakob disease (Duffy et al. 1974) which is considered the human form of BSE—mad-cow disease— contracted from eating beef from infected cows.

13.4 The Spatial Spread of Rabies Among Foxes I: Background and Simple Model

Rabies, as mentioned in the last section, is widespread throughout the world and epidemics are quite common. During the past few hundred years, Europe has been repeatedly subjected to rabies epidemics. It is not known why rabies died out in Europe some 50 or so years before the current epidemic started. The analysis of the models here, however, provides one possible scenario.

The present European epizootic (an epidemic in animals) seems to have started about 1939 in Poland and it has moved steadily westward at a rate of 30–60 km per year. It has been slowed down, only temporarily, by such barriers as rivers, high mountains and autobahns. The red fox is the main carrier and victim of rabies in the current European epidemic. The spread of rabies is like a travelling wave as shown in Figure 13.3.

Rabies, a viral infection of the central nervous system, is transmitted by direct contact, and the dog is the principal transmitter of the disease to man. As mentioned, the incidence of rabies in man, at least in Europe and America, is now rare, with only very few deaths a year, but with considerably more in underdeveloped countries. The effect of rabies on other mammals, domestic and wild, however, is serious. In France, in 1980 alone, 314 cases of rabies in domestic animals were reported and 1280 cases in wild animals. Rabies justifiably gives cause for concern and warrants extensive study and development of control strategies, a subject we discuss later in Section 13.6.

Figure 13.3 shows the advance of the rabies epidemic in France obtained from data from the French *Centre National d'Études sur la Rage* every two years between 1969 and 1977 on the northeastern part of the country. Macdonald (1980) discusses the situation at this time in France in more detail and describes the effects of a vaccination control and what happened when it was stopped. Since this time, however, there has been a concerted effort to control the spread by vaccination through bait and it has been quite successful in several countries in Europe.

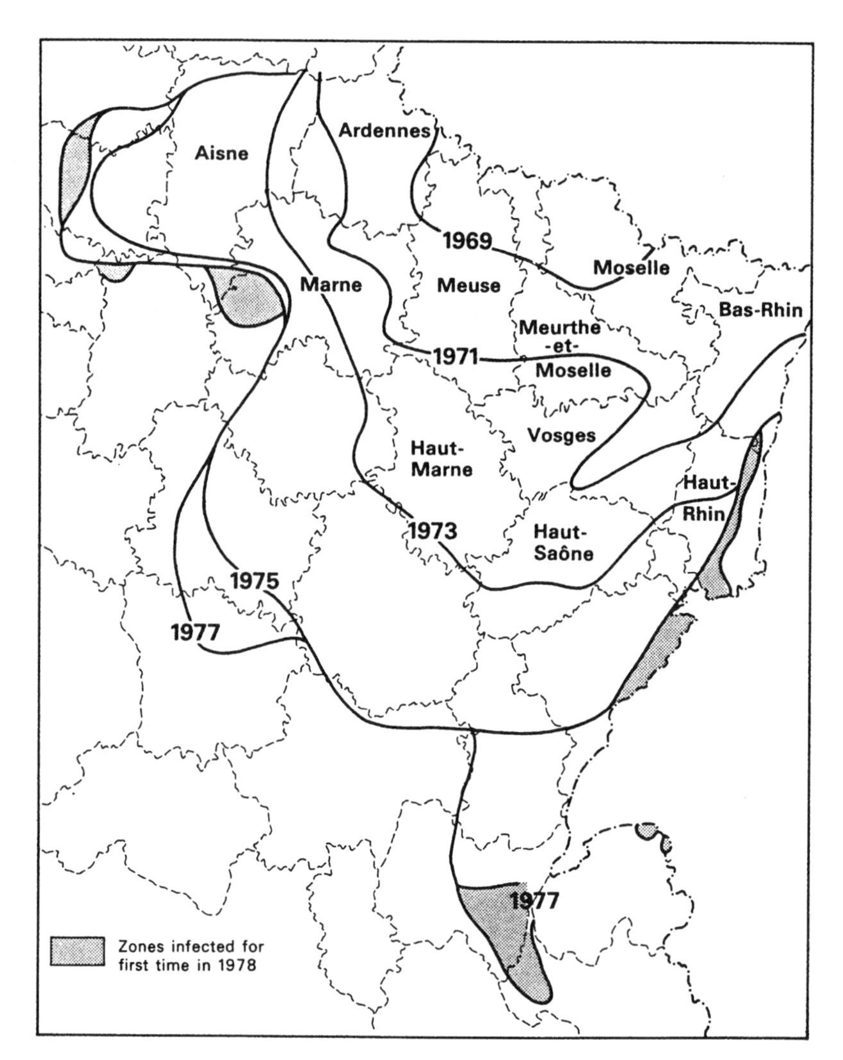

Figure 13.3. Spatial advance of the rabies epizootic in France from 1969 to 1977: note the (heterogeneous) wavelike characteristic of the spatial spread. (Data from *Centre National d'Études sur la Rage*)

A rabies epidemic is also moving rapidly up the east coast of America: the main vector here is the racoon. In this epidemic, the progress was considerably enhanced by the importation into Virginia (by hunting clubs) of infected racoons from Georgia and Florida.

If we refer to Figure 13.1 again, we see that, just as in the spatially uniform epidemic system situation discussed in Chapter 10, Volume I, after the epidemic has passed a proportion of the susceptibles have survived. It would be useful to be able to estimate this survival fraction analytically in a spatial context. This we can do in the following very simple but still illuminating model for the spatial spread of rabies.

Red foxes account for about 70% of the recorded cases in Western Europe. Although Britain has effectively been free from rabies since about 1900, the disease could be reintroduced in the near future through the illegal importation of pets or even by infected bats from the continent. The problem would be particularly serious in Britain because of the high rural and urban density of foxes, dogs and cats. In Bristol, for example, the fox density is of the order of 12 foxes/km^2 as compared with a rural population of 2–4 foxes/km^2. The book on the fox and rabies by Macdonald (1980) provides many of the facts and data for Britain. General data on rabies in Europe is available from the *Centre National d'Etudes sur la Rage* in France. The books edited by Kaplan (1977) and Bacon (1985) are specifically concerned with the population dynamics of rabies and provide biological and ecological background together with some data on the disease.

It is important to understand how the rabies epizootic wavefront progresses into uninfected regions, what control methods might halt it and how the various parameters affect them. The remaining sections of this chapter will be concerned with these specific spatial problems. The material primarily comes from the model of Murray et al. (1986) and, in this section, from the much simpler, but less realistic, model of Källén et al. (1985). The models and control strategies we propose in Sections 13.6 and 13.9 are specifically related to the current European fox epizootic but the type of model is applicable to many other spatially propagating epidemics.

The spatial spread of epidemics is usually a very complex process, and rabies is no exception. In modelling such a complex process we can try to incorporate as many of the facts as possible, which necessarily involves many parameters, estimations of which are difficult to obtain with extant data. An alternative approach is to start with as simple a model as possible but which captures the key elements and for which it is possible to determine estimates for the fewer parameters. There is a trade-off between comprehensiveness and thus complexity, and the difficulty of estimating many parameters and a simpler approach in which parameter values can be reasonably assessed. For the models in this chapter, we have opted for the latter strategy. In spite of their simplicity, they nevertheless pose highly relevant practical questions and give estimates for various characteristics of importance in the spatial spread of diseases. Although in this section we describe and analyse a particularly simple model, it is one for which we can obtain useful analytical results.

Although many animals are involved, a basic, and reasonable, assumption is that the ecology of foxes, the principal vectors, determines the dynamics of the spread of rabies. We further assume that the spatial spread of the epizootic is due primarily to the random erratic migration of rabid foxes. Uninfected foxes do not seem to wander far from their territory (Macdonald 1980). We divide the fox population into two groups—susceptible and rabid. Although the resulting model captures certain aspects of the spatial spread of the epizootic front, it leaves out a basic feature of rabies, namely, the long incubation period of between 12 and 150 days from the time of an infected bite to the onset of the clinical infectious stage. We include this in the more realistic model presented in Section 13.5.

To control, and ideally prevent, the spread of the disease, it is important to have some understanding of how rabies spreads so as to assess the effects of possible control strategies. It is with this in mind that we first study a particularly simple modified version of the epidemic model system (13.1), which captures some of the key elements in the spread of rabies in the fox population. We shall then use it to derive some estimates of essential facts about the epizootic wave.

We consider the foxes to be divided into two groups, infectives I, and susceptibles S; the infectives consist of rabid foxes and those in the incubation stage. The principal assumptions are: (i) The rabies virus, contained in the saliva of the rabid fox, is transmitted from the infected fox to the susceptible fox. Foxes become infected at an average rate per head, rI, where r is the transmission coefficient which measures the rate of contact between the two groups. (ii) Rabies is invariably fatal and foxes die at a per capita rate a; that is, the life expectancy of an infected fox is $1/a$. (iii) Foxes are territorial and divide the countryside into non-overlapping ranges. (iv) The rabies virus enters the central nervous system and induces behavioral changes in the fox. If the virus enters the spinal cord it induces paralysis whereas if it enters the limbic system it induces transient aggression during which it loses its sense of territory and the fox wanders about in a more or less random way. So, we assume that it is only the infectives which disperse with diffusion coefficient $D\text{km}^2/\text{year}$. With these assumptions our model is then (13.1) except that the susceptible foxes do not disperse. We exclude here the migration of cubs seeking their own territory. When they do move they try to stay as close to their original territory as possible. The model system in one dimension is then

$$\begin{aligned} \frac{\partial S}{\partial t} &= -rIS, \\ \frac{\partial I}{\partial t} &= rIS - aI + D\frac{\partial^2 I}{\partial x^2}. \end{aligned} \tag{13.12}$$

From the analysis in the last section we expect this system to possess travelling wave solutions, whose speed of propagation depends intimately on the parameter values. The realistic estimation of these few parameters is important but still not easy.

Using the nondimensionalisation (13.2), the system (13.12) becomes (cf. (13.3))

$$\begin{aligned} \frac{\partial S}{\partial t} &= -IS, \\ \frac{\partial I}{\partial t} &= IS - \lambda I + \frac{\partial^2 I}{\partial x^2}, \end{aligned} \tag{13.13}$$

where now S, I, x and t are dimensionless, and, as in the last section, $\lambda = a/rS_0$ is a measure of the mortality rate as compared with the contact rate. As before the contact rate is crucial and is not known with any confidence. We expect the threshold value to be again $\lambda = 1$ but we now verify this (see also Exercise 2).

Travelling wavefront solutions of (13.13) are of the form

$$S(x,t) = S(z), \quad I(x,t) = I(z), \quad z = x - ct, \tag{13.14}$$

where c is the wavespeed and we look for solutions satisfying the boundary conditions

$$S(\infty) = 1, \quad S'(-\infty) = 0, \quad I(\infty) = I(-\infty) = 0. \tag{13.15}$$

Refer back to Figure 13.1 for the type of wave anticipated. Note that it is the *derivative* of $S(z)$ which tends to zero as $z \to -\infty$ since we anticipate a residual number, as yet undetermined, of susceptible foxes to survive the epidemic. With (13.14) the system (13.13) becomes

$$\begin{aligned} cS' &= IS, \\ I'' + cI' + I(S - \lambda) &= 0. \end{aligned} \tag{13.16}$$

Linearising about $I = 0$ and $S = 1$ exactly as we did in the last section and requiring I to be always nonnegative, we find that this requires $\lambda < 1$, in which case the wavespeed

$$c \geq 2(1 - \lambda)^{1/2}, \quad \lambda < 1. \tag{13.17}$$

With this specific model we are able to take the analysis further and find the actual fraction of susceptibles which survives the epidemic. From the first of (13.16), $I = cS'/S$, which on substituting into the second equation gives

$$I'' + cI' + \frac{cS'(S - \lambda)}{S} = 0.$$

Integration gives

$$I' + cI + cS - c\lambda \ln S = \text{constant}.$$

Using the boundary conditions as $z \to \infty$ from (13.15), where $S = 1$, $I = 0$ and with $I' = 0$, we determine the constant to be c. If we now let $z \to -\infty$, again using (13.15) with $I = I' = 0$, we get the following transcendental equation for the surviving susceptible population, σ say, after the passage of the epizootic wavefront,

$$\sigma - \lambda \ln \sigma = 1, \quad \lambda < 1, \quad \sigma = S(-\infty), \tag{13.18}$$

which is independent of c. Writing this in the form

$$\frac{\sigma - 1}{\ln \sigma} = \lambda < 1 \Rightarrow 0 < \sigma < \lambda < 1. \tag{13.19}$$

From (13.19), with $\lambda = 0.4$, $\sigma = 0.1$ for example, whereas with $\lambda = 0.7$, $\sigma = 0.5$. λ is a measure of the severity of the epidemic. The smaller λ the fewer susceptibles survive; in other words, the worse the epidemic. Figure 13.4 illustrates the surviving susceptible fraction σ as a function of λ obtained from (13.18); the curve was obtained by plotting λ as a function of σ.

The critical bifurcation value for λ is $\lambda = 1$, which in dimensional terms, from (13.2), means $a/(rS_0) = 1$. If $\lambda > 1$ no epidemic wave can propagate. This is to be expected since if $a > rS_0$ it means the mortality rate is greater than the rate of recruitment of new infectives. As before this bifurcation result says that given r and a,

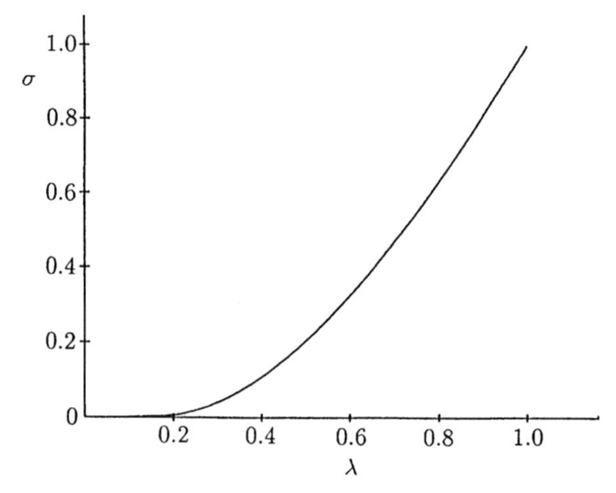

Figure 13.4. The fraction, σ, of the original susceptible fox density which survive, after the passage of the epidemic wave, as a function of the epidemic severity: here, in terms of the original dimensional variables, $\sigma = S(-\infty)/S_0$ and $\lambda = a/(rS_0)$.

there is a critical minimum fox density $S_c = a/r$ below which rabies cannot persist in the population and any infectives introduced will not cause an epidemic.

When rabies does persist, that is, $\lambda < 1$, the computed speed of propagation of the epidemic wave is the minimum of the allowable speeds, namely, $c = 2(1-\lambda)^{1/2}$, which in dimensional terms from (13.17) and (13.2) is

$$c = 2[D(rS_0 - a)]^{1/2}. \tag{13.20}$$

Figure 13.5 shows an example of the computed travelling front solutions for S and I, from (13.13), for $\lambda = 0.5$. From Figure 13.4 with $\lambda = 0.5$, the surviving fraction of susceptibles $\sigma \approx 0.2$.

Let us now compare the qualitative form of the susceptible fox population in the epidemic in Figure 13.5 with that obtained from data from continental Europe as illus-

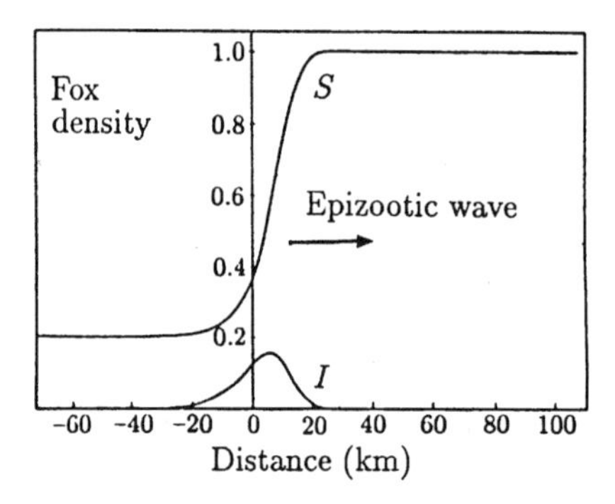

Figure 13.5. Dimensionless epidemic wavefront solution for the susceptible (S) and infected (I) fox populations computed from (13.13): here $\lambda = 0.5$. The wavespeed is $c = \sqrt{2}$. Note the qualitative similarity with Figure 13.1.

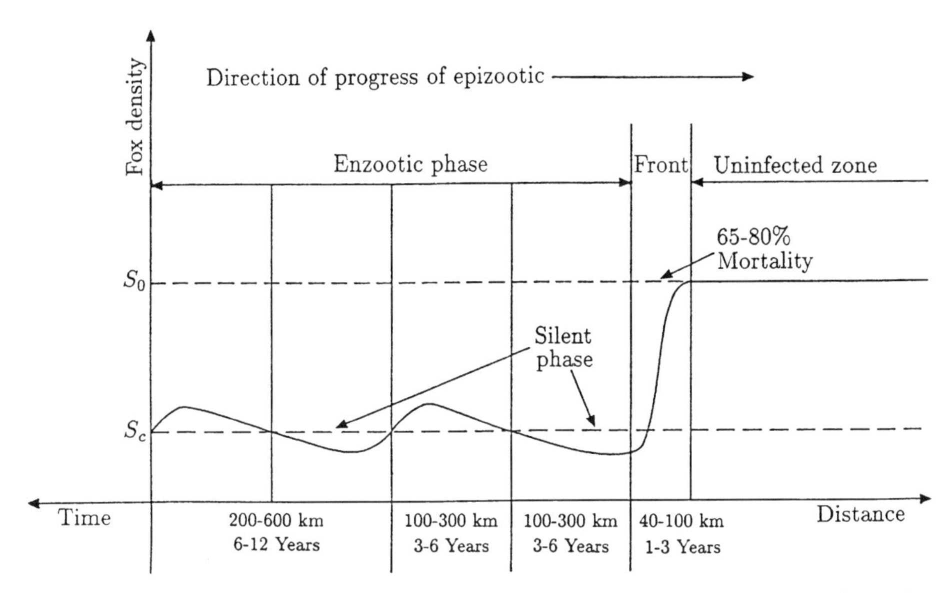

Figure 13.6. Fluctuations in the susceptible fox population density as a function of the passage of the rabies epizootic obtained from data from *Centre National d'Etudes sur la Rage* 1977. In dimensional terms, S_0 is the uninfected susceptible population ahead of the epidemic wave. Note the periodic, but decreasing, fluctuations in S, which follow the main wavefront, as S tends to its steady state. (Redrawn from Macdonald 1980).

trated in Figure 13.6. There is a clear schematic difference in the behaviour behind the front in the two figures. The model (13.13) is only intended to cover the passage of an epidemic front. Clearly after the passage of the wavefront the susceptible population will start to increase again since the foxes find themselves in an environment which admits a larger carrying capacity. In other words, the timescale of the model (13.13) is considerably shorter than that associated with the oscillations in Figure 13.6. To include in our model the situation which obtains after the front has passed we must include a term for the fox reproduction. If we model this by a simple logistic growth, the equation for the susceptibles in place of the first of (13.13) becomes

$$\frac{\partial S}{\partial t} = -rIS + BS\left(1 - \frac{S}{S_0}\right), \tag{13.21}$$

where B is the linear growth rate. With the same nondimensionalisation (13.2) as before, the model now becomes

$$\begin{aligned} \frac{\partial S}{\partial t} &= -IS + bS(1 - S), \\ \frac{\partial I}{\partial t} &= I(S - \lambda) + \frac{\partial^2 I}{\partial x^2}, \end{aligned} \tag{13.22}$$

where $b = B/rS_0$, that is, the ratio of linear birth rate to the basic rate of infection per infective. Figure 13.7 shows an example of the resulting epidemic wave of susceptibles

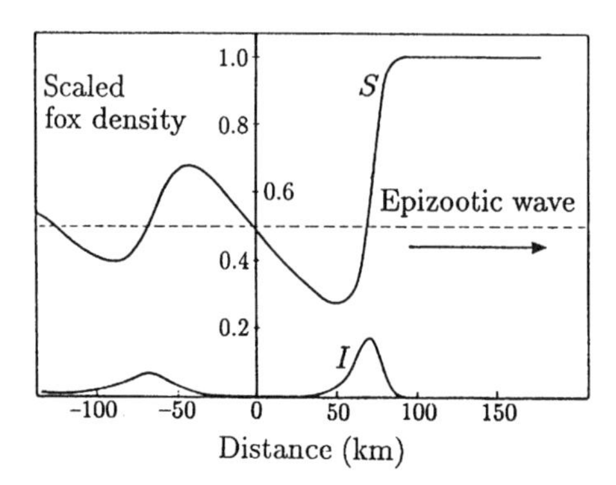

Figure 13.7. Travelling epidemic wave solution for the susceptible (S) and infective (I) foxes from (13.22) when logistic growth is taken into account in the susceptible fox population: parameter values $b = 0.05$, $\lambda = 0.5$. Note the qualitative similarity with the data illustrated in Figure 13.6. The initial front is succeeded by recurring, but smaller, outbreaks of the disease. (After Källén et al. 1985).

and infectives, obtained by numerically solving (13.22); there is now good qualitative comparison between the results from this model and the data recorded in Figure 13.6. The oscillations are decaying and eventually $S \to \lambda$ and $I \to b(1 - \lambda)$, the steady state solutions of (13.22), far behind the front.

Although the wavelength of the quasi-periodic outbreaks in both time and space are given by the numerical solutions, we can obtain some useful analytical results even for this more complex model (13.22). Let us start with the dimensional version of (13.22), namely,

$$\begin{aligned} \frac{\partial S}{\partial t} &= -rSI + BS\left(1 - \frac{S}{S_0}\right), \\ \frac{\partial I}{\partial t} &= rSI - aI + D\frac{\partial^2 I}{\partial x^2}. \end{aligned} \tag{13.23}$$

If we now introduce the nondimensional quantities

$$\begin{aligned} &U = \frac{S}{S_0}, \quad V = \frac{rI}{BS_0}, \quad t^* = BT, \quad x^* = \left(\frac{B}{D}\right)^{1/2} x, \\ &\lambda = \frac{a}{rS_0}, \quad \alpha = \frac{rS_0}{B}, \end{aligned} \tag{13.24}$$

the dimensionless model becomes, on omitting the asterisks for notational simplicity,

$$\begin{aligned} U_t &= U(1 - U - V), \\ V_t &= \alpha V(U - \lambda) + V_{xx}. \end{aligned} \tag{13.25}$$

These equations are the same as equations (1.3) in Chapter 1, but without prey diffusion and with α and λ here in place of a and b there, and is the system we studied in detail in Section 1.2. The steady state solutions of (13.25) are $(0, 0)$, $(1, 0)$ and $(\lambda, 1 - \lambda)$ with the latter existing in the positive quadrant only if $\lambda < 1$. If we look for travelling wave solutions in the usual way (cf. equations (1.5)), the analysis of the three-dimensional phase space (U, V, W), where $W = V'$, is given in Section 1.2. There we showed that with $\lambda < 1$ a travelling wave solution exists which joins the steady states $(1, 0)$ and $(\lambda, 1 - \lambda)$. We also showed that a threshold $\alpha = \alpha^*$ exists such that if $\alpha > \alpha^*$ the approach to the steady state $(\lambda, 1 - \lambda)$ is oscillatory, whereas if $\alpha < \alpha^*$ it is monotonic (cf. Figure 1.3). The computed solution in Figure 13.7 is an example with $\alpha > \alpha^*$.

Let us now return to the observation in Section 13.2 about the subsequent outbreaks of plague which followed the initial Black Death epidemic. If we modify the susceptible equation in the model (13.1) to take into account the recovery of the population we again get subsequent periodic outbreaks of the disease following the initial epidemic similar to those shown in Figures 13.6 and 13.7. This is just a bit too facile an explanation since it was the interaction of populations which governed the Black Death, people, fleas, rats and so on. In spite of the simplicity of the model discussed here the results qualitatively capture some of the major phenomena observed. As with so many of the models we have discussed, even such a simple approach can elicit relevant questions.

13.5 The Spatial Spread of Rabies Among Foxes II: Three-Species (*SIR*) Model

To be of practical use in developing control strategies to contain the spatial spread of an epidemic, we have to consider more realistic and hence more complex models, which allow for quantitative comparison with known data and let us make practical predictions with more confidence. The model in the last section, although capturing certain aspects of the spread of an epizootic front, is rather too primitive for quantitative purposes. One of the major exclusions from the previous model is the long incubation period, which, as mentioned above, can be from 12 to 150 days, before the fox becomes rabid. In this section we consider a more realistic model which takes this, among other things, into account. With it we can obtain quantitative estimates for various times and distances of epidemiological and public health significance.

In this section we consider a three-species model where again the rabid foxes are considered the main cause of the spatial spread. The data on the movement of rabid foxes in the wild, although rather scant, is not zero; some of it will be used later when we estimate the crucial diffusion coefficient for rabid foxes.

The model we develop is still comparatively simple, but, even so, some of the parameters are difficult to estimate from the available data. Such parameter estimates will be required in any realistic models, so it is important to learn more about fox ecology and the impact of rabies on fox behaviour in order to improve on the estimation of the more critical parameters.

The model, analysis and results we give here are based on the work of Murray et al. (1986) who give further details and results. It extends the work of Anderson et al. (1981) (who considered only the time-dependent situation) by including spatial effects, specifically the crucial spatial dispersal of rabid foxes and later in Section 13.9 the dispersal of all foxes.

We consider a three-species *SIR* model in which we divide the fox population into susceptible foxes, S, infected, but noninfectious, foxes, I, and infectious, rabid foxes, R.

The need for at least three species is primarily based on the long incubation period of from 12 to 150 days (and in some cases longer) that the rabies virus undergoes in the infected animal, during which time the animal appears to behave normally and does not seem to transmit the disease, and on the relatively short period (1 to 10 days) of clinical disease which follows.

The basic model assumptions are closely linked to those in the last section (we use a slightly different notation) but we reiterate them here for convenience. The assumptions are:

(i) The dynamics of the fox population in the absence of rabies can be approximated by the simple logistic form

$$\frac{dS}{dt} = (a - b)S\left(1 - \frac{S}{K}\right),$$

where a is the linear birth rate, b is the intrinsic death rate and K is the environmental carrying capacity. The parameters a, b and K may vary according to the habitat but at this stage we take them to be constants. Later, when we present the numerical results for the English 'experiment,' we shall consider K to vary, as it does in a major way in England.

(ii) Rabies is transmitted from rabid to susceptible fox by direct contact between foxes, usually by biting. Susceptible foxes become infected at an average per capita rate βR, which is proportional to the number of rabid foxes present, where the transmission coefficient β, taken to be constant, measures the rate of contact between the two species.

(iii) Infected foxes become infectious (rabid) at an average per capita rate, σ, where $1/\sigma$ is the average incubation time.

(iv) Rabies is invariably fatal, with rabid foxes dying at an average per capita rate α: $1/\alpha$ is the average duration of clinical disease.

(v) Rabid and infected foxes continue to put pressure on the environment, and die of causes other than rabies, but they have a negligible number of healthy offspring. These effects are small but are included for completeness.

To take into account the spatial effects we make the following further assumptions.

(vi) Foxes are territorial, and divide the countryside up into nonoverlapping ranges.

(vii) Rabies acts on the central nervous system with about half of infected foxes having the so-called 'furious rabies', and exhibit the ferocious symptoms typically associated with the disease, while with the rest the virus affects the spinal cord and causes paralysis. Foxes with furious rabies may become aggressive and confused, losing their sense of direction and territorial behaviour, and wandering randomly. It is these we consider the main cause of the spatial spread of the disease.

These assumptions suggest the following model for the spatial and temporal evolution of the rabies epizootic.

$$\begin{aligned}
\frac{\partial S}{\partial T} &= aS - bS - \frac{(a-b)NS}{K} - \beta RS, \\
\frac{\partial I}{\partial T} &= -bI - \frac{(a-b)NI}{K} + \beta RS - \sigma I, \\
\frac{\partial R}{\partial T} &= -bR - \frac{(a-b)NR}{K} + \sigma I - \alpha R + D\frac{\partial^2 R}{\partial X^2},
\end{aligned} \tag{13.26}$$

where the total population

$$N = S + I + R. \tag{13.27}$$

We have written the equations in this form to highlight what each term means. The only source term comes from the birth of susceptible foxes. All die naturally; the life expectancy is $1/b$ years. The term $(a-b)N/K$ in each equation represents the depletion of the food supply by all foxes. The transition from susceptible to infectious foxes is accounted for by the βRS term and from the infected to the infectious group by σI. Rabid foxes also die from rabies and thus are represented by the αR term; the life expectancy of a rabid fox is $1/\alpha$. Rabid foxes also diffuse with diffusion coefficient D. Typical parameter values, except for the crucially important D, are given in Table 13.1. If, in the absence of any spatial effects, we add equations (13.26) we get

$$\frac{dN}{dt} = aS - bN - \frac{(a-b)N^2}{K} - \alpha R, \tag{13.28}$$

which is the equivalent logistic form for the total population.

We have written the equations in one-dimensional form but we shall use the full two-dimensional form when we apply the model to the spread of the disease from a hypothetical outbreak in England, which we discuss later.

This model neglects the spatial dispersal of rabies by young itinerant foxes, who may get bitten while in search of a territory and carry rabies with them before they become rabid. There is some justification for this since rabies is much less common in the young than in adults (Artois and Aubert 1982, Macdonald 1980).

Table 13.1. Parameter values for rabies among foxes (from Anderson et al. 1981).

Parameter	Symbol	Value
Average birth rate	a	1 fox year^{-1}
Average intrinsic death rate	b	0.5 fox year^{-1}
Average duration of clinical disease	$1/\alpha$	5 days
Average incubation time	$1/\sigma$	28 days
Critical carrying capacity	K_T	1 fox km^{-2}
Disease transmission coefficient	β	80 km^2 year^{-1}
Carrying capacity	K	0.25–4.0 foxes km^{-2}

The spatially homogeneous steady state solutions of (13.26), other than the zero steady state, are given, after some algebra, by

$$\begin{aligned}
S_0 &= \beta^{-1}[\sigma\beta K - a(a-b)]^{-2}\{[(\alpha+b)\beta K \\
&\quad + (a-b)(\alpha+a)][\sigma\beta K(\sigma+b) + \alpha(a-b)(\sigma+a)]\}, \\
I_0 &= [\sigma\beta K - a(a-b)]^{-1}[(\alpha+b)\beta K + (a-b)(\alpha+a)]R_0, \\
R_0 &= \{\beta[\sigma\beta K - a(a-b)]\}^{-1}(a-b)[\sigma\beta K - (\sigma+a)(\alpha+a)].
\end{aligned} \tag{13.29}$$

In the spatially uniform situation ($D = 0$), when rabies is introduced into a stable population of healthy foxes three possible behaviours are possible. Which behaviour occurs depends on the size of K relative to the critical carrying capacity K_T which is given by the condition for a nonzero value for the steady state R_0 in the last equation, namely,

$$K_T = \frac{(\sigma+a)(\alpha+a)}{\sigma\beta}. \tag{13.30}$$

If $K < K_T$, the epidemic threshold value of the carrying capacity, rabies eventually disappears ($R \to 0, I \to 0$), and the population returns to its initial value $S = K$. On the other hand, if K is larger than K_T, then the population oscillates about the steady state. From a standard linear stability analysis of the steady state (S_0, I_0, R_0), the equivalent of which we do below for $K > K_T$, it can be shown (after some algebra) that if K is not too much bigger than K_T the steady state is stable and perturbations die out in an oscillatory way. On the other hand, if K is sufficiently larger than K_T limit cycle solutions exist. There are thus 2 bifurcation values for K, namely, K_T and the critical K between a limit cycle oscillation and a stable steady state.

From the epidemiological evidence, rabies seems to die out if the carrying capacity is somewhere between 0.2 and 1.0 foxes/km^2 (WHO Report 1973, Macdonald 1980, Steck and Wandeler 1980, Anderson et al. 1981, Boegel et al. 1981). The parameter β, which is a measure of the contact rate between rabid and healthy foxes, cannot be estimated directly given the difficulties involved in observing these contacts. Anderson et al. (1981) used the expression (13.30) as an indirect way to estimate β since we have estimates for K_T and all the other parameters except β. Parameter estimation is always an important aspect of any realistic modelling. Murray et al. (1986) discuss in some detail how they affect the spatial spread of rabies: the model is quite robust to variations in many of the parameters within a band around the estimates used. Another method for getting parameter information with observational limitations is given by Bentil and Murray (1991).

With $K > K_T$ the parameter choices listed in Table 13.1, give 3–5 year periods for the oscillations and 0 to 4% *equilibrium persistence*, p, of rabies, where p is defined by

$$p = \frac{R_0 + I_0}{S_0 + I_0 + R_0}. \tag{13.31}$$

These figures are in agreement with the available epidemiological evidence (Toma and Andral 1977, Macdonald 1980, Steck and Wandeler 1980, Jackson and Schneider 1984).

Travelling Epizootic Wavefronts and Their Speed of Propagation

We introduce nondimensional quantities by setting

$$\begin{aligned} s &= \frac{S}{K}, \quad q = \frac{I}{K}, \quad r = \frac{R}{K}, \quad n = \frac{N}{K}, \\ \varepsilon &= \frac{a-b}{\beta K}, \quad \delta = \frac{b}{\beta K}, \quad \mu = \frac{\sigma}{\beta K}, \quad d = \frac{\alpha + b}{\beta K}, \\ x &= \left(\frac{\beta K}{D}\right)^{1/2} X, \quad t = \beta K T, \end{aligned} \tag{13.32}$$

with which the model equations (13.26) with (13.27) become

$$\begin{aligned} \frac{\partial s}{\partial t} &= \varepsilon(1-n)s - rs, \\ \frac{\partial q}{\partial t} &= rs - (\mu + \delta + \varepsilon n)q, \\ \frac{\partial r}{\partial t} &= \mu q - (d + \varepsilon n)r + \frac{\partial^2 r}{\partial x^2}, \\ n &= s + q + r, \end{aligned} \tag{13.33}$$

which have a positive uniform steady state solution (s_0, q_0, r_0) given by equations (13.29) on dividing (S_0, I_0, R_0) by K. The condition (13.30) for an epidemic to occur, namely, $K > K_T$, is then

$$0 < d < \left[1 + \frac{\delta + \varepsilon}{\mu}\right]^{-1} - \varepsilon. \tag{13.34}$$

The system (13.33) now depends on only 4 dimensionless parameters ε, δ, μ and d as compared with the original dimensional system's 7 parameters. Values for these dimensionless parameters are obtained from the parameter estimates of a, b, α, σ, K and β in Table 13.1. If we choose a representative carrying capacity of $K = 2$ foxes/km^2 we get $\varepsilon = \delta = 0.003$, $\mu = 0.08$ and $d = 0.46$. The fact that ε and δ are relatively small numbers compared to any of 1, μ, δ and $1-d$ can be used to simplify the analysis of the model system (13.33) and lets us derive useful *analytical* results; see below and Murray et al. (1986).

It is perhaps appropriate here to reiterate yet again the major benefit of nondimensionalisation, namely, that the parameter groupings show equivalent effects of variations in actual field parameters. For example, with ε and δ small this means that, during the epidemic, the infectious rate is relatively very much larger than the birth and death rates from causes other than rabies.

Let us now look for epizootic wave solutions to the system (13.33), which travel at a constant velocity v into an undisturbed, rabies-free region. (For algebraic simplicity we use a different notation from what we have used earlier in this chapter.) So, we look for solutions s, q and r as functions of the single variable $\xi = x + vt$, which thus satisfy

$$\begin{aligned} vs' &= \varepsilon(1-n)s - rs, \\ vq' &= rs - (\mu + \delta + \varepsilon n)q, \\ vr' &= \mu q - (d + \varepsilon n)r + r'', \\ n &= s + q + r, \end{aligned} \tag{13.35}$$

where prime denotes differentiation with respect to ξ and where $s \to 1, q \to 0, r \to 0$ as $\xi \to -\infty$, that is, far ahead of the wavefront. As usual, of course, we are interested only in nonnegative solutions. In the following we use the fact that $\varepsilon \ll 1$ and $\delta \ll 1$.

The system (13.35) has three possible steady state solutions for (s, q, r) in the positive quadrant, namely, $(1, 0, 0)$, $(0, 0, 0)$ and (s_0, q_0, r_0), where s_0, q_0 and r_0 are given by (13.29) on dividing by K. Since ε and δ are small, to first-order in ε and δ,

$$s_0 = d + \left[\varepsilon + \frac{\varepsilon d + \delta}{\mu}\right] d, \quad q_0 = \frac{\varepsilon d(1-d)}{\mu}, \quad r_0 = \varepsilon(1-d). \tag{13.36}$$

From the full expressions for (s_0, q_0, r_0) all of s_0, q_0 and r_0 are nonnegative only if the threshold condition (13.34) is satisfied.

A travelling wave solution to (13.35), with the required properties, is a trajectory in the 4-dimensional phase space of (13.35), which goes from the equilibrium at $s = 1$, $q = r = 0$ to one of the other two equilibrium points, $(0, 0, 0)$ or (s_0, q_0, r_0). We do not carry out all the algebra since the procedure is the customary one used throughout both Volume I and II; we just briefly describe the various steps and leave the details to be worked out as an exercise.

We write (13.35) as a 4-dimensional first-order system in (s, q, r, r') and first linearise about the critical point $(s, q, r, r') = (1, 0, 0, 0)$. In the usual way, this gives a linear system whose solutions are linear combinations of the eigensolutions $\mathbf{x}_i \exp(\lambda_i \xi)$, where $\mathbf{x}_i$ and λ_i are the four eigenvectors and eigenvalues of the stability matrix. We can thus determine the solution behaviour near the critical point by looking at all possible linear combinations of the eigensolutions. If $\operatorname{Re}\lambda_i < 0$, then $\mathbf{x}_i \exp(\lambda_i \xi) \to 0$ as $\xi \to \infty$ and the trajectory approaches the critical point, while if $\operatorname{Re}\lambda_i > 0$ the trajectory comes out of the critical point. Trajectories leaving the critical point thus correspond to linear combinations of those eigensolutions with $\operatorname{Re}\lambda_i > 0$. If an eigenvalue is complex, then its eigensolution is oscillatory. After some algebra we find that the four eigenvalues for the linear system near $(1, 0, 0, 0)$ are $\lambda = -\varepsilon/v < 0$ and the roots of the cubic

$$\begin{aligned} f(\lambda) = \lambda^3 &+ \left(\frac{\mu + \delta + \varepsilon}{v} - v\right)\lambda^2 - (d + \mu + \delta + 2\varepsilon)\lambda \\ &+ \frac{\mu(1 - d - \varepsilon) - (\delta + \varepsilon)(d + \varepsilon)}{v}. \end{aligned} \tag{13.37}$$

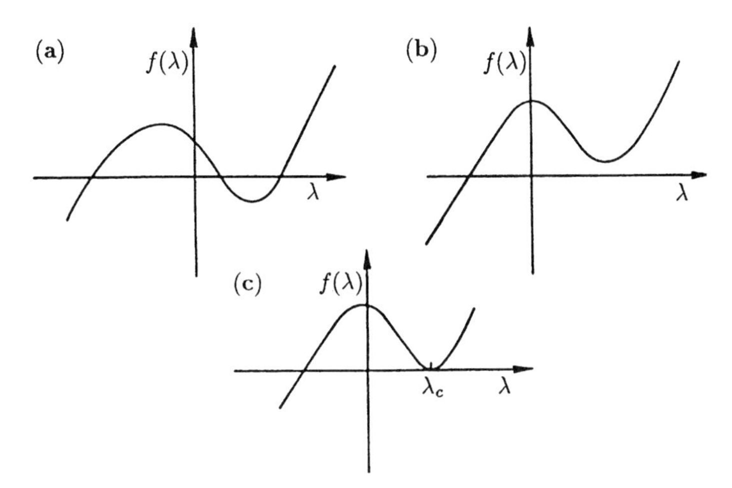

Figure 13.8. The eigenfunction cubic $f(\lambda)$ in (13.37), the zeros of which are eigenvalues of the linearised system about $(1, 0, 0)$. The solutions have either two positive roots as in (**a**), two complex roots with $\mathrm{Re}\,\lambda > 0$ as in (**b**), or a double real root at λ_c as in (**c**).

Note that $f(\lambda) \to \infty$ as $\lambda \to \infty$ and $f(\lambda) \to -\infty$ as $\lambda \to -\infty$. Further, if (13.36) holds, then $f(0) > 0$ and f has a negative slope at $\lambda = 0$. Depending on the values of the various parameters, $f(\lambda)$ can look like any of the forms illustrated in Figure 13.8.

With all the parameters fixed, as the velocity v is varied $f(\lambda)$ sequentially looks like each of these shapes. Thus, as long as the threshold condition (13.34) holds, f has one negative real root and, depending on the value of the velocity of the wave, it has either two positive real roots, or two complex roots. When the velocity v is such that (13.37) has complex roots, these represent oscillatory solutions which imply negative populations and such waves are physically unrealistic. The bifurcation value for v, v_c say, between realistic and unrealistic solutions is the value when (13.37) has a double root as in Figure 13.8(c). Thus the range of allowable wavespeeds of travelling waves is determined by v_c. This is given by setting $f = 0$ and $df/d\lambda = 0$ and eliminating λ to get an equation for v_c in terms of the parameters. After considerably more algebra we find that, to first-order in ε and δ, v_c is given by the positive real roots of $g(v_c^2)$, where $g(z)$ is given by

$$\begin{aligned} g(z) = {} & \left[4\mu + (d-\mu)^2\right] z^3 + 2\left[3\mu(1-d)(3d+\mu) + (d+\mu)^2(2d+\mu)\right] z^2 \\ & + \mu^2\left[(d+\mu)^2 - 6(1-d)(3d+\mu) - 27(1-d)^2\right] z \\ & - 4\mu^4(1-d). \end{aligned} \tag{13.38}$$

When the threshold criterion (13.34) holds, $g(z)$ is negative and d^2g/dz^2 is positive at $z = 0$. A rough sketch of $g(z)$ shows it has a unique positive root which corresponds to the minimum possible velocity for an epizootic wave.

We now show that it is not possible for a trajectory to go from the critical point at $s = 1, q = 0, r = 0$ to that at the origin where $s = q = r = 0$. On linearising (13.35) about the origin we find (after more algebra) the eigensolutions

$$\begin{pmatrix} s \\ q \\ r \\ r' \end{pmatrix} = \mathbf{a} \exp\left[-\frac{(\mu+\delta)\xi}{v}\right], \quad \mathbf{b} \exp\left[\frac{v}{2} \pm \left(d + \frac{v^2}{4}\right)^{1/2}\right]\xi, \quad \mathbf{c} \exp\left[\frac{\varepsilon\xi}{v}\right],$$

where

$$\mathbf{a}^T = \left[0, \frac{d-\mu-\delta}{\mu} - (\mu+\delta)^2\mu v^2, 1, -\frac{\mu+\delta}{v}\right],$$

$$\mathbf{b}^T = \left[0, 0, 1, \frac{v}{2} \pm \left(d + \frac{v^2}{4}\right)^{1/2}\right], \quad \mathbf{c}^T = [1, 0, 0, 0],$$

and the superscript T denotes the transpose. Sufficiently close to the origin, trajectories which approach the origin are linear combinations of the two eigensolutions with negative exponents, and so they approach the origin in the plane $s = 0$. For the system (13.35) 'time' is reversible, in the sense that we can replace ξ by $-\xi$ and trace backwards along any trajectory. Setting $\tau = -\xi$ in (13.35) and taking $s = 0$ initially, we see that $s = 0$ for all positive τ irrespective of the initial values of r and q. This implies that a trajectory which has $s = 0$ for any ξ had $s = 0$ for all previous ξ, and has $s = 0$ for all subsequent ξ. So, a trajectory cannot come from $s = 1$, enter the $s = 0$ plane, and approach the origin.

This implies that a travelling wave can only occur if there is a trajectory from $s = 1$ to the critical point (s_0, q_0, r_0) and this requires, as we expected, that condition (13.34) must hold. To determine the behaviour of the wave as it approaches this critical point, we now linearise (13.35) about (s_0, q_0, r_0) to get (after more algebra) the eigenvalues

$$\lambda_1, \lambda_2 = \frac{1}{2}\left\{v - \frac{\mu}{v} \pm \left[\left(v - \frac{\mu}{v}\right)^2 + 4(\mu+d)\right]^{1/2}\right\} \tag{13.39}$$

to first-order in ε and δ, and

$$\begin{aligned} \lambda_3, \lambda_4 = &\pm \frac{i}{v}\left[\frac{\varepsilon\mu d(1-d)}{\mu+d}\right]^{1/2} \\ &- \varepsilon d\left[2v(\mu+d)^2\right]^{-1}\left[\mu(1-d)\left(\frac{\mu}{v^2} - 1\right) + (\mu+d)^2\right] \end{aligned} \tag{13.40}$$

to second-order in ε and δ. λ_1 is positive and so, near the critical point, any solution which approaches (s_0, q_0, r_0) as $\xi \to \infty$ is a linear combination of the eigensolutions corresponding to λ_2, λ_3 and λ_4. Since $|\lambda_2| \gg |\mathrm{Re}(\lambda_3, \lambda_4)|$, the amplitude of its eigensolution decays much more rapidly than that of the eigensolutions of the complex eigenvalues. Thus, sufficiently far back in the tail of the wave (that is, for sufficiently large ξ), the solutions corresponding to the complex eigenvalues govern the behavior of the travelling wave. The eigenvectors corresponding to these eigenvalues are given by

$$\begin{pmatrix} s - s_0 \\ q - q_0 \\ r - r_0 \\ r' \end{pmatrix} = \begin{pmatrix} 1 \\ \pm i \left[\dfrac{\varepsilon d(1-d)}{\mu(\mu+d)} \right]^{1/2} \\ \pm i \left[\dfrac{\varepsilon \mu(1-d)}{d(\mu+d)} \right]^{1/2} \\ \dfrac{\varepsilon \mu(1-d)}{v(\mu+d)} \end{pmatrix}$$

which, on taking an arbitrary real linear combination of the eigensolutions, gives, for sufficiently large ξ,

$$\begin{aligned} s - s_0 &\sim [A \cos \omega\xi/v + B \sin \omega\xi/v] \exp(-\lambda\xi/v), \\ q - q_0 &\sim \frac{\omega}{\mu}[A \sin \omega\xi/v - B \cos \omega\xi/v] \exp(-\lambda\xi/v), \\ r - r_0 &\sim \frac{\omega}{d}[A \sin \omega\xi/v - B \cos \omega\xi/v] \exp(-\lambda\xi/v). \end{aligned} \tag{13.41}$$

Here ω is the period of the waves, given by the imaginary part of the complex eigenvalues divided by v, and λ is the decay rate of the amplitude, given by the real part of these eigenvalues divided by v. A and B are constants, which depend on the way the trajectory approaches (s_0, q_0, r_0), which of course cannot be determined from a linear analysis.

Let us now exploit the smallness of ε and δ to obtain certain useful asymptotic analytical approximations (Murray et al. 1986). From the approximate steady state forms (13.36) we note that the rabid fox density $r_0 = \mu q_0/d$. From (13.41) we see that far back in the tail of the wave, that is, ξ large, we also have $r - r_0 \sim \mu(q - q_0)/d$. That is, the profiles for the infected and rabid fox densities are similar, differing only in scale. In the simulations, such as those in Figures 13.9 and 13.10, of the full nonlinear system, the striking profile similarity holds for the *entire* wave. This surprising fact suggests that, in view of the complexity of the three-species model, it would be of considerable benefit if we could obtain analytically the conditions under which the travelling wave problem for the three-species model could be modelled to a high degree of approximation with a two-species model. That is, we could replace, for example, the three-species *SIR* system of susceptible, infectious, rabid populations by a two-species system of only susceptible and rabid populations. The infected, but not yet rabid, fox population is then given by a simple scaling of the rabid population, namely,

$$q(\xi) \sim \frac{dr(\xi)}{\mu}. \tag{13.42}$$

Under certain conditions it is possible to give an analytical explanation as to why this phenomenon occurs. It is not obvious from the model system (13.35). The mathemat-

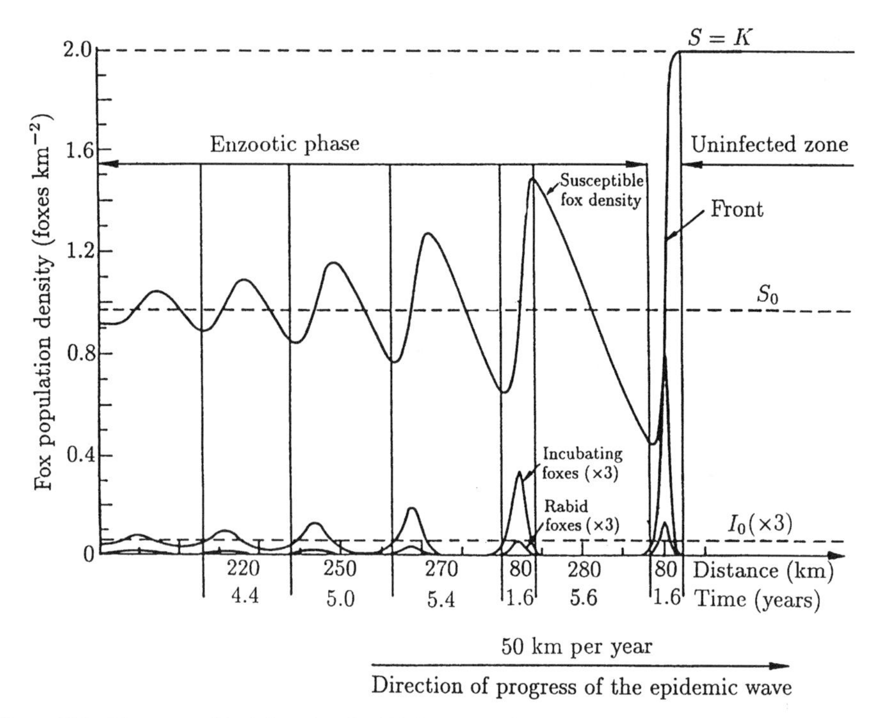

Figure 13.9. The susceptible, infected and rabid fox populations due to the passage of a rabies epidemic wave from a numerical simulation of the model mechanism (13.37)–(13.40). The fox density in the uninfected region ahead of the front of the epidemic is taken to be at a carrying capacity of 2 foxes/km^2, a typical value (averaged over the yearly cycle) for much of continental Europe. The time and distance between the recurring outbreaks and the wavespeed were obtained from the model using estimates for the field parameters given in Table 13.1 and a diffusion coefficient $D = 200$ km^2/yr. (From Murray et al. 1986)

ical analysis is based on μ being small compared with both d and the nondimensional wavespeed v, but large compared with ε and δ. The singular perturbation analysis is quite complicated and is given in detail by Murray et al. (1986).

To get the actual travelling wavefront solutions for the epizootic we must solve the partial differential equation system (13.33) numerically, starting with $s = 1$ (that is, dimensionally the susceptible population is $S = K$, the undisturbed carrying capacity) everywhere and with a small concentration of rabid foxes at the origin. When the threshold criterion (13.34) is satisfied an epidemic wave forms and travels outward from the initial concentration of rabid foxes with near constant velocity. If, of course, the threshold inequality (13.34) is violated, then rabies dies out, and the fox population returns to the carrying capacity of the environment. Figure 13.9 is an example of the travelling wavefront which evolves for parameter values appropriate to the current epizootic on continental Europe. The wave consists of the rabies front, in which the largest number of foxes die from the disease, followed by an oscillatory tail in which each successive outbreak of rabies is smaller than the preceding one. The oscillations gradually die out and the populations approach constant nonzero values with the rabid and infected fox population zero. Figure 13.10 illustrates the fluctuations in fox density for a travelling

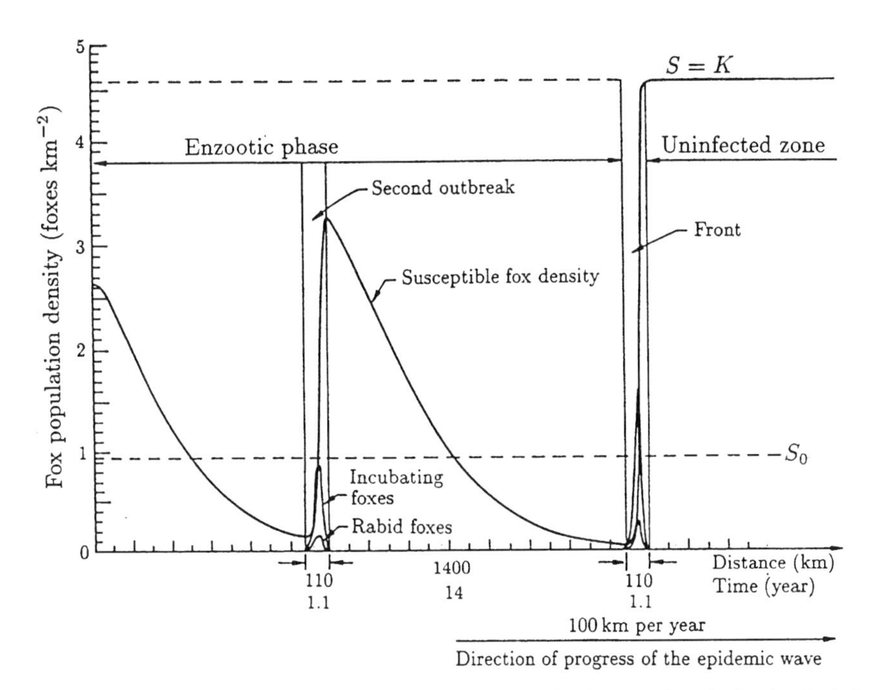

Figure 13.10. Fox populations during the passage of the rabies epidemic wave when the fox density in front of the epidemic is at the carrying capacity of 4.6 foxes/km^2, which is common in parts of England. The diffusion coefficient $D = 200$ km^2/yr, and the other parameters were taken from Table 13.1. Compare the different wavelengths and periods of the recurring epidemics following the front with those in Figure 13.9. The epidemics for England are more severe. (From Murray et al. 1986)

epizootic wave with parameters appropriate for England; note the wilder fluctuations in the susceptible population compared with those in Figure 13.9.

With the parameter values in Table 13.1, the threshold condition (13.34) is satisfied and

$$\varepsilon \quad \text{and} \quad \delta \ll 1, \quad d, \quad \mu \quad \text{and} \quad 1 - d. \tag{13.43}$$

Under these circumstances the wavefront is followed by an oscillatory tail. Analytically the minimum speed is given by $v = z^{1/2}$, where z is the unique positive root of the cubic (13.38). A contour plot for this root v is shown in Figure 13.11 for $0 \le d \le 1$. All of the waves found numerically appear to travel at this minimum speed which, from (13.32), is given in dimensional form as

$$V = (D\beta K)^{1/2} v. \tag{13.44}$$

For example, with the parameter values in Table 13.1, a diffusion coefficient of 200 km^2/yr and a carrying capacity of 2 foxes/km^2, we evaluate d and μ from (13.32) and then read off the appropriate v from Figure 13.11; this gives the dimensional speed of propagation as $V = 51$ km/yr.

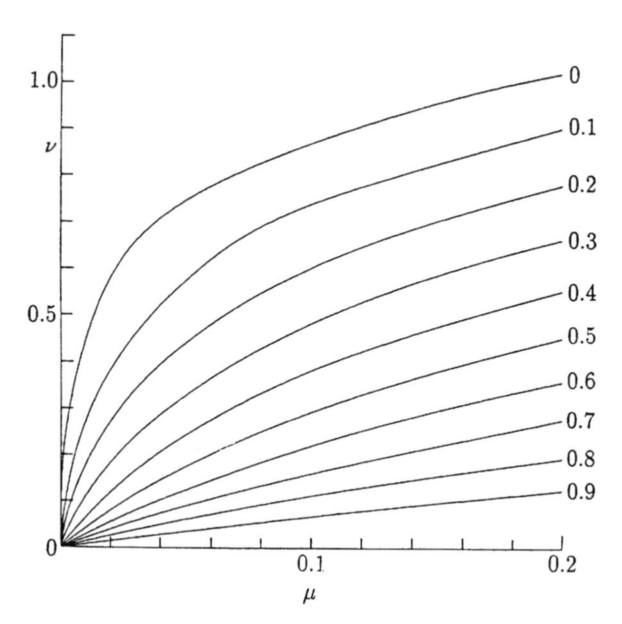

Figure 13.11. The dimensionless velocity of propagation, v, of the epidemic front as a function of the dimensionless parameter μ for various values of d. Recall that μ is related to the incubation time for the rabies virus, and d is related to the duration time of the symptomatic, infectious, stage; see (13.32). The dimensional wavespeed is given by $V = (\beta K D)^{1/2} v$. Note that $v = 0$ for $d \geq 1$, which corresponds to a carrying capacity less than the critical value. (From Murray et al. 1986).

The linear analysis near the steady state (s_0, q_0, r_0), described above, shows that for sufficiently large times the wave tends to decaying oscillations given by (13.41). In terms of the original (x, t) variables these solutions can be written in the form

$$\begin{aligned} s(x,t) &= s_0 + A\cos[\omega(t + x/v) + \psi]\exp[-\lambda(t + x/v)], \\ q(x,t) &= q_0 + \frac{1}{\mu}(s - s_0)', \\ r(x,t) &= r_0 + \frac{\mu}{d}(q - q_0), \end{aligned} \tag{13.45}$$

to first order in ε and δ, where the prime denotes differentiation with respect to $(t + x/v)$ and the nondimensional wavenumber ω is given by

$$\omega = \varepsilon^{1/2}\left[\frac{\mu d(1-d)}{\mu + d}\right]^{1/2} + O(\varepsilon^{3/2}), \tag{13.46}$$

with the decay rate λ given by

$$\lambda = \frac{\varepsilon d}{2(\mu + d)^2}\left[\mu\left(\frac{\mu}{v^2} - 1\right)(1-d) + (\mu + d)^2\right]. \tag{13.47}$$

A and ψ are constants. Note that the oscillations in the susceptible population are 90° out of phase with both the infected and, as noted above, the rabid populations. (This symmetry is broken if the oscillations are calculated to the next order in ε and δ.) As we also noted above, the $r-q$ proportionality relationship (13.42) seems to hold universally as indicated by the numerical simulations when physically reasonable parameters are used.

The singular perturbation analysis of Murray et al. (1986) yields several useful approximations regarding the epidemic. For example, the maximum densities of infected and rabid foxes in the first outbreak, are given by

$$\begin{aligned} r_{max} &\approx \mu\left(\ln d + \frac{1-d}{d}\right), \\ q_{max} &\approx d\left(\ln d + \frac{1-d}{d}\right), \end{aligned} \tag{13.48}$$

which in dimensional terms are

$$\begin{aligned} R_{max} &\approx \frac{\sigma K_T}{\alpha}\left[\ln\left(\frac{K_T}{K}\right) + \frac{K}{K_T} - 1\right], \\ Q_{max} &\approx K_T\left[\ln\left(\frac{K_T}{K}\right) + \frac{K}{K_T} - 1\right]. \end{aligned} \tag{13.49}$$

Since no epidemic ensues if $K \leq K_T$, the threshold carrying capacity, both R_{max} and Q_{max} are zero for $K = K_T$. Note that both R_{max} and Q_{max} increase as K increases above K_T.

Once we have the dimensionless wavespeed $v(= z^{1/2})$ from (13.38) we can then determine the decay rate λ from (13.47). It turns out that λ is always positive. This means that the limit cycle behaviour which the diffusionless version of (13.26), that is, with $D = 0$, can exhibit for sufficiently large $K > K_T$ disappears when diffusion is taken into account: the oscillations always decay to the constant state (s_0, q_0, r_0). The dimensional decay rate is $\beta K\lambda$. The dimensional period of the recurring epidemics is

$$\tau = \frac{2\pi}{\beta K\omega},$$

where ω is given by (13.46); in terms of the original dimensional parameters, using (13.32), the period T is

$$T = 2\pi\left\{(\alpha+\sigma+b)\left[(a-b)(\alpha+b)\sigma\left\{1-\frac{\alpha+b}{\beta K}\right\}\right]^{-1}\right\}^{1/2}. \tag{13.50}$$

Note that T decreases with K. So, in general, the greater the fox density before the appearance of rabies, the less frequently rabies outbreaks will appear far behind the front; this agrees with some observations (Macdonald 1980). However, numerically it was found that close to the front, where nonlinearities are important, the time between outbreaks may increase with K; see Figures 13.9 and 13.10. Once we have the dimensional velocity V and period T we get the dimensional wavelength $L = VT$.

Estimate for the Diffusion Coefficient D and Sensitivity of Wavespeed and Epidemic Wavelength to Variations in D

To calculate the real dimensional speed V of the epizootic, and hence the period and wavelength of the recurring epidemics which follow the main front, we need an estimate for the diffusion coefficient, which is a measure of the rate at which a rabid fox covers ground in its wanderings. Little is known about the behaviour of rabid foxes in the wild, making it very difficult to estimate D.

Andral et al. (1982) tracked three rabid adult foxes in the wild. They accomplished this by inoculating captured foxes with rabies virus, equipping them with signal-emitting collars, and releasing them at the point of capture. They traced the fox movements first during the incubation period, to determine their home ranges and normal behaviour, and then during the rabid period, to observe the changes induced by the disease. Once the foxes became rabid their pattern of daily activity changed. Drawings showing, for each fox, the incubation period range and the principal displacements during the rabid period indicate that all three left their home range at some point during the rabid phase, but none travelled very far away.

Murray et al. (1986) used the results of Andral et al. (1982) to estimate the diffusion coefficient, in a rather primitive way, from the formula

$$D \approx \frac{1}{N} \sum_{j=1}^{N} \frac{(\text{straight line distance from the start})^2}{4 \times (\text{time from the start})},$$

where the sum is over the number of all foxes involved. Using the distance between the start of the rabid period and the point of death, along with the approximate length of the rabid period, gives an estimate of 50 km^2/yr for D. Since two of the three foxes happened to die much closer to their starting position than their mean distance away from it, this is most likely a lower bound on D. An extremely rough idea of an upper bound can be gained from the maximum distance that any one fox travelled away from its starting point. About halfway through the rabid phase, one fox got as far away from its starting point as 2.7 km, giving an estimate of 330 km^2/yr as an upper bound on D.

There are other ways of estimating diffusion coefficients. For example, D can be estimated as the product of the average territory size A and the average rate k at which a rabid fox leaves home. For their two-species model, Källén et al. (1985) supposed that infected foxes leave home at the end of the incubation period of one month, that is, when they were assumed to become rabid. Taking an average territory size to be about 5 km^2, they obtained $D = 60\ \text{km}^2$. To determine D for our 3-species model, we need an estimate for the average rate at which foxes leave their territories *after* the onset of clinical disease. If N infected foxes are observed, and the jth one leaves its territory a time interval t_j after becoming rabid, then k can be estimated by

$$\frac{1}{N} \sum_{j=1}^{N} \frac{1}{t_j}.$$

Since roughly half of all infected foxes develop paralytic rabies and presumably never leave their home range, t_j is infinite for about $N/2$ foxes. For the furiously rabid foxes, if we suppose that half also never leave, and that the rest leave evenly spread out over the 6 days that the disease may take to run its course, then we can estimate

$$k \approx \frac{1}{N}\sum_{j=1}^{N/4}\frac{1}{t_j} = \frac{1}{24}\sum_{j=1}^{6}\frac{1}{j \text{ days}} = 40 \text{ yr}^{-1}.$$

Keeping the estimate of 5 km^2 for an average territory size (Toma and Andral 1977; Macdonald 1980), this gives $D = 190 \text{ km}^2/\text{yr}$.

An alternative method is to estimate the mean free path and velocity of rabid foxes. The average total distance covered daily by the foxes observed by Andral et al. (1982) was 9 km during the rabid period. Suppose that this is not atypical and that, for example, a rabid fox goes 100 m at a stretch before becoming distracted and setting off in another direction. Then $D =$ (velocity) $\times$ (pathlength) gives a diffusion coefficient of 330 km^2/yr, the same as the upper bound that we estimated previously. All of these methods for estimating D should, in principle, be consistent if enough observations of rabid fox behaviour could be made. At this stage there is simply not enough known about rabid fox behaviour to get much better estimates.

Since the speed of the wave is proportional to $D^{1/2}$, changing D from 50 to 330 km^2/yr increases V by a factor of 2.6. Table 13.2 shows the sensitivity of the wavespeed and wavelength as a function of the carrying capacity for a given $D = 200 \text{ km}^2/\text{yr}$.

Another difficult parameter to estimate is the disease transmission coefficient β. As we said above, this can be estimated by inverting the threshold expression (13.30). But, absolute values of fox population densities are in practice difficult to obtain; they are usually estimated from the numbers of foxes reported dead, shot or gassed, and some assumption on the percentage of the total population that this sample represents, or else by comparison of terrain with areas of known fox densities. This in turn means K_T is particularly difficult to estimate, and values of anywhere from 0.2 to 1.2 foxes/km^2 can be estimated from the values given in the literature (WHO Report 1973, Steck and

Table 13.2. Dependence of the wavespeed and asymptotic wavelength (that is, the distance between recurring outbreaks) on the carrying capacity, calculated with $D = 200 \text{ km}^2/\text{yr}$ and other parameter values from Table 13.1.

K (foxes/km^2) or K/K_T* Carrying Capacity	V (km/yr) Velocity of the Epidemic Front	L (km) Distance Between Successive Outbreaks/Peaks
1.5	35	150
2.0	50	210
2.5	70	220
3.0	80	250

* The parameters β and K only appear as the product βK in the calculations for the values in Tables 13.1 and 13.2. From equation (13.30), $\beta K = (K/K_T)(\sigma + a)(\alpha + a)/\sigma$, so that only a knowledge of the ratio of the actual carrying capacity to the critical value is necessary to obtain the results.

Wandeler 1980, Macdonald et al. 1981, Gurtler and Zimen 1982). Since finding K/K_T only involves comparison of population sizes, this ratio might be easier to obtain than K and K_T separately.

A relevant question at this point is how sensitive the quantitative results are to the uncertainties in the parameters. This aspect and difficulties in estimating other parameters are discussed by Murray et al. (1986).

13.6 Control Strategy Based on Wave Propagation into a Nonepidemic Region: Estimate of Width of a Rabies Barrier

We discuss here one possible control strategy as developed by Murray et al. (1986), namely, that of a possible protective barrier against the rabies epizootic which can be achieved by reducing the susceptible fox population below the critical density K_T in areas ahead of the advancing wave. This, for example, has been successful in Denmark, specifically Jutland. It has also been carried out in some regions of Italy and Switzerland, where it has been pursued with diligence, but it has had mixed results (Macdonald 1980, Westergaard 1982). Such a barrier can be created either by killing or vaccination. Since killing releases territories, there could be a more rapid colonization by young foxes which could in fact enhance the spread of the disease. Vaccination causes less disruption in the ecology, is almost certainly more effective and is also probably more economic.

For a rabies 'break' to be effective we must have reasonable estimates of both the width and the allowable susceptible fox density within it. Here we derive estimates analytically for how wide the protective break region needs to be to keep rabies from reaching the areas beyond. We also present some of the results from numerical simulations of the full equation system (13.33). In what follows, we use the term 'infected fox' to refer to all foxes with rabies, whether infectious or not.

If we observe the passage of the rabies epizootic wave at a fixed place we note that each outbreak of the disease is followed by a long quiescent period, during which very few cases of rabies occur; refer to Figures 13.9 and 13.10. The spatial and temporal dimensions are such that the secondary epidemic wave is sufficiently far behind so that the first wave will either have moved past the break, or have effectively died out by the time the second one arrives. Each successive outbreak is weaker than the previous one. So, it seems reasonable to assume that the same population reduction schemes which eradicate the first outbreak will also be effective in stopping all subsequent outbreaks from passing through. We thus only need to consider how wide the break needs to be to stop the first outbreak. The width of the break is dependent on the size of the susceptible fox population density within it.

Since we model spatial dispersal by a deterministic diffusion mechanism it is, from a strict mathematical viewpoint, not possible for the density of infected foxes to vanish anywhere. This arises from treating the fox densities as continuous in space and time, rather than dealing with individual foxes, and from using classical diffusion to model the rabid fox dispersal. Thus we cannot simply have the epizootic wave move into a break of finite width and determine whether or not the density of infected foxes remains zero on the other side; it will always be positive, although exponentially small. Thus no matter how wide the break is, eventually enough infected foxes will in time leak through for the epizootic to start off again on the other side. Thus we must think instead of determining when the probability is acceptably small that an infected fox will reach the far side of the break.

Since the aim of any control scheme is to keep the density of foxes small, we treat the break region as one with a carrying capacity below K_T, the critical threshold value (13.30) for an epidemic, and we assume that the fox density has been reduced to this value well before the epizootic front arrives. To obtain estimates for the width of the break we investigate the behaviour of the model when the region of lowered susceptible fox density starts at $x = 0$ and extends to infinity. We first give here the results of the numerical simulations of the full system (13.33) and later in the section obtain approximate analytic results.

Figures 13.12 and 13.13 show what happens when the epizootic wave, coming in from the left, impinges on the break region. Remember that the epizootic wave cannot propagate when the carrying capacity is below the critical value K_T. Also, the point of maximum infected fox density will be at $x = 0$. As the infection wave moves into the region $x > 0$ it spreads out, decays in amplitude and the total number of infected foxes decreases. Eventually there will be less than p infected foxes/km^2 remaining, where p is some small number. Let $t_c(p)$ be the time at which this occurs. We now choose p sufficiently small that the probability of a rabid fox encountering a susceptible one after this critical time is negligible. Since the wave cannot propagate in the break region it simply decays, so, for all time the density of infected foxes is greatest at the edge of the break and decays with x, exponentially as x^2 in fact, as we show later. We choose the width of the break to be the point x_c where the infected fox density is a given (small) fraction m of the value at the origin, that is,[5]

$$I(x_c, t_c) + R(x_c, t_c) = m[I(0, t_c) + R(0, t_c)]. \tag{13.51}$$

Available evidence suggests that it has never been possible to eliminate all foxes from a region—a 70% reduction in population is about the best that can be achieved (Macdonald 1980). Figure 13.14 shows the dependence of the break width in terms of the percentage population reduction in the break, for different choices of the average duration time of clinical disease, $1/\alpha$.

In the numerical simulations for the curves in Figure 13.14, the value of βK outside the break was held at 160 yr^{-1}, the number of infected foxes at the critical time was taken to be $p = 0.5$ foxes/km^2, the ratio m in (13.51) was arbitrarily chosen to be 10^{-4} and all other parameters except α are from Table 13.1. With these assumptions, for any given choice of α, the nondimensional forms in (13.32) give $d = (\alpha + 0.5\text{ yr}^{-1})/(160\text{ yr}^{-1})$ and (13.30) gives a carrying capacity outside the break region of $K = 149/(\alpha + 0.5\text{ yr}^{-1})$ foxes km^{-2} yr^{-1}. For example, if we assume that the rabid period lasts an average of 3.8 days, then $d = 0.6$ and $K = 1.5$ foxes/km^2 outside of the break. If a reduction scheme can reduce the carrying capacity to 0.4 foxes/km^2 inside the break region well before the epidemic arrives, then $s_b = 0.26$ and Figure 13.14 gives $x_b = 15$. Assuming a diffusion coefficient of 200 km^2/ yr, (13.32) gives the predicted

[5]Strictly the x_c and t_c are dimensionless here.

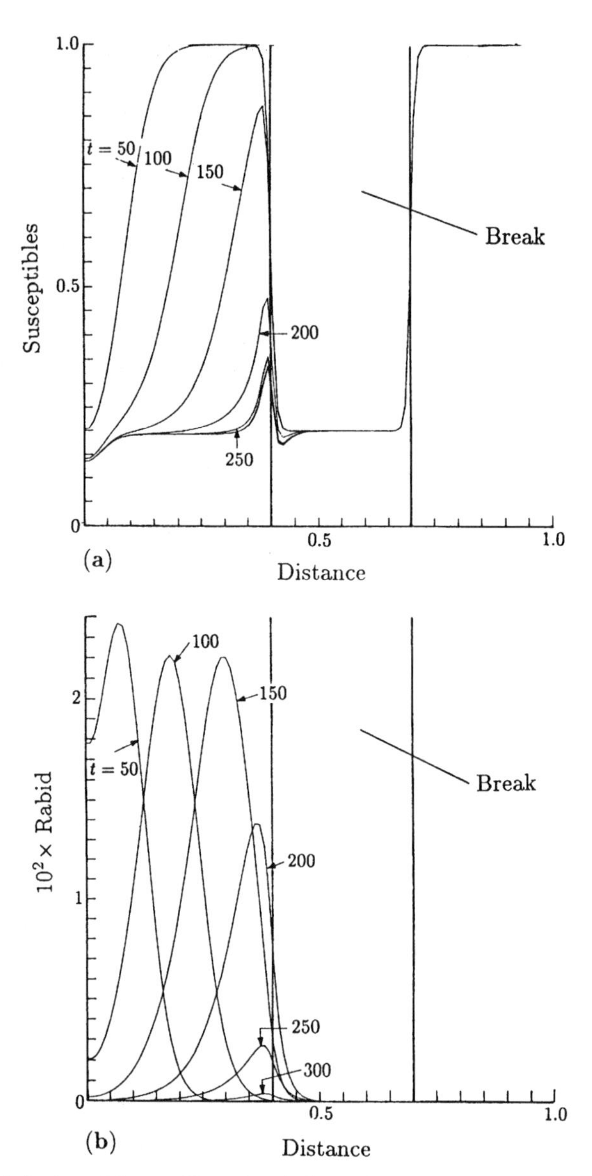

Figure 13.12. The behaviour of the travelling epizootic front when it encounters a break in the susceptible fox population. These plots show (**a**) the susceptible and (**b**) the rabid fox population densities for a sequence of times as the wave approaches the break region, stops and dissipates. They were obtained by solving equations (13.37)–(13.40) numerically with a carrying capacity of 2 foxes/km^2 in the region outside the vertical lines and of 0.4 foxes/km^2 in the region between them. Other parameter values were taken from Table 13.1. Note that the susceptible population just outside the break remains slightly higher than elsewhere, since few rabid foxes wander into this region from the right. The density of incubating foxes is proportional to the rabid population as we noted in Section 13.5: with the parameter values used, the incubating fox density is 5.6 times the rabid fox density. The times and distances are normalised values within the computer model. (From Murray et al. 1986)

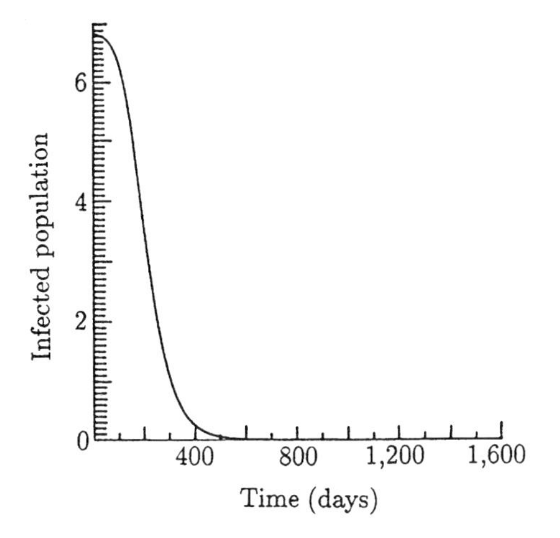

Figure 13.13. This plot shows the total infected fox density per km (the integral over x of the infected foxes $I + R$) as a function of time for the case shown in Figure 13.12 starting when the epidemic front first reaches the break. (From Murray et al. 1986)

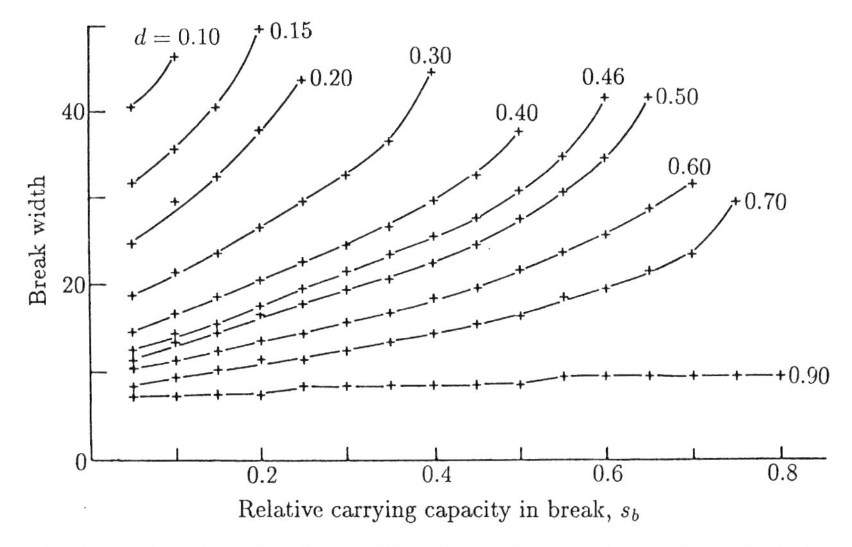

Figure 13.14. The dependence of the break width on the initial susceptible population inside the break, as predicted by the model. The break width, in nondimensional terms, is plotted against the ratio of the carrying capacity in the break to the carrying capacity outside the break for various values of the duration time of clinical disease, $1/\alpha (d \approx \alpha/\beta K)$. The curves were obtained by solving (13.33) numerically until the total infected fox population in the first outbreak is 0.5 fox/km. As described in the text, we use these curves to calculate the break width, which can be put into dimensional form using relations (13.32). The dimensional break width X_c is given by $(D/\beta K)^{1/2} x_c$, where x_c is the nondimensional break width with $m = 10^{-4}$. βK was set at 160 yr^{-1}, and all other parameters, except α, were taken from Table 13.1. For example, if we assume $1/\alpha = 5$ days then $d = 0.46$ and the carrying capacity outside the break is 2 foxes/km^2. If the carrying capacity inside the break is assumed to be 0.4 foxes/km^2 then $s_b = 0.2$ and this figure predicts $x_c = 18$. Assuming $D = 200\ \text{km}^2/\text{yr}$, the predicted break width X_c is then 20 km. (From Murray et al. 1986)

break width as 17 km. Of course, the choice of p and m depends on how cautious we want to be: Murray et al. (1986) discuss the sensitivity of the model to variations in these. The maximum value of $I + R$ at t_c for all of the calculations was less than 0.15 foxes/km^2. Even with m as large as $m = 10^{-2}$ there are fewer than 0.0015 infected foxes per square kilometre on the protected side of the break.

13.7 Analytic Approximation for the Width of the Rabies Control Break

We can determine analytically an approximate functional dependence of the break width on the parameters. The behaviour of the various fox population densities in the break region after the epizootic wave has reached it should be similar to the situation in which a concentrated localised density of infected and rabid foxes at time $t = 0$ (with the same total number of I and R as for the epizootic wave) is introduced at $x = 0$ in a domain where the carrying capacity is everywhere equal to the initial fox density in the break. We can then obtain an estimate of the break width by looking at the following idealised problem. Suppose that the carrying capacity is zero for all x, which implies that the susceptible fox density $s = 0$. At time $t = 0$, take $r = r_0\delta(x)$ and $q = q_0\delta(x)$, where $\delta(x)$ is a Dirac delta function (that is, we consider all of the r_0 rabid foxes are initially concentrated at $x = 0$).

We start by assuming that for $x \geq 0$, all of the susceptible foxes have been eliminated, for example, by immunization or killing. In our analysis here we make the added approximation that the nonlinear terms in the equations for the incubating and rabid foxes can be neglected. Since ε and δ are small parameters, this should be a reasonable approximation. A further justification for these approximations comes from the numerical computations of the break width, where it was found that the computed break width did not change if these terms were neglected. With these assumptions, equations (13.33) reduce to the linear form

$$\begin{aligned}\frac{\partial q(x,t)}{\partial t} &= -\mu q(x,t),\\ \frac{\partial r(x,t)}{\partial t} &= \mu q(x,t) - dr(x,t) + \frac{\partial^2 r(x,t)}{\partial x^2}.\end{aligned} \tag{13.52}$$

By symmetry, instead of considering the problem of a δ-function source of infected foxes at $x = 0$ and $t = 0$ which then move into the region $x \geq 0$, the initial conditions can be replaced by

$$q(x,0) = 2q_0\delta(x), \quad r(x,0) = 2r_0\delta(x) \tag{13.53}$$

and we then consider instead the region $-\infty < x < \infty$. The propagation of infected foxes into the break is described by equations (13.52) with initial conditions (13.53). The specific quantities of interest are the time t_c at which the population in the break has decayed to a given level, p, defined implicitly by the formula

$$\left(\frac{KD}{\beta}\right)^{1/2}\int_0^\infty [q(x,t_c)+r(x,t_c)]\,dx = p \tag{13.54}$$

and the break width, x_c, which, as discussed above, is given implicitly by

$$q(x_c,t_c)+r(x_c,t_c) = m[q(0,t_c)+r(0,t_c)]. \tag{13.55}$$

We first estimate t_c. Integrating equations (13.52) with respect to x from 0 to ∞, we get the two ordinary differential equations

$$\begin{aligned}\frac{dQ^*(t)}{dt} &= -\mu Q^*(t),\\ \frac{dF^*(t)}{dt} &= -dF^*(t)+dQ^*(t),\end{aligned} \tag{13.56}$$

where

$$Q^*(t)=\int_0^\infty q(x,t)\,dx, \quad F^*(t)=\int_0^\infty [q(x,t)+r(x,t)]\,dx.$$

The initial conditions for (13.56) are $F^*(0)=q_0+r_0$, $Q^*(0)=q_0$. The first of equations (13.56) is trivially solved for $Q^*(t)$ and Q^* which is then used in the second equation to obtain the following equation for F^*, namely, the (scaled) total number of foxes present in the region $x>0$,

$$\frac{dF^*}{dt} = -dF^* + dq_0e^{-\mu t}. \tag{13.57}$$

With the given initial conditions, the solution to this equation is

$$F^*(t)=\left[q_0+r_0-\frac{dq_0}{d-\mu}\right]e^{-dt}+\frac{dq_0}{d-\mu}e^{-\mu t}. \tag{13.58}$$

The critical time t_c can then be determined from (13.54) by solving the equation

$$F^*(t_c)=p\left(\frac{\beta}{KD}\right)^{1/2}.$$

Note that each of the two terms on the right-hand side of (13.58) involves an exponential factor. Since, for reasonable values of the field parameters, $d>\mu$ and $d-\mu=o(1/t_c)$, the first of those terms can be neglected in comparison with the second if t_c is sufficiently large. Let us assume this is the case, and verify it *a posteriori*. So, neglecting the first term, the resulting algebraic equation can be solved to give

$$t_c \approx \frac{1}{\mu}\ln\left[\frac{d\left(\frac{KD}{\beta}\right)^{1/2}q_0}{p(d-\mu)}\right]. \tag{13.59}$$

Typical values for δ and μ are 0.46 and 0.08, respectively. $(KD/\beta)^{1/2}q_0$ can be approximated from Figure 13.13 and the fact that $q \approx dr/\mu$, so that the total number of infected foxes satisfies

$$\int_{-\infty}^{\infty} (I + R)\, dX = \left(\frac{KD}{\beta}\right)^{1/2} \left(1 + \frac{\mu}{d}\right) q_0.$$

From Figure 13.13,

$$\int_{-\infty}^{\infty} (I + R)\, dX \approx 6.9 \text{ foxes/km},$$

giving $(KD/\beta)^{1/2}q_0 \approx 5.9$ foxes/km. For $p = 0.5$ fox/km, (13.59) gives an estimate of $t_c \approx 33$ for these values of the parameters, and so the ratio of the two exponentials $\exp[-dt_c]$ and $\exp[-\mu t_c]$ is approximately 3×10^{-6}, which justifies neglecting the smaller exponential in (13.58) in the above analysis.

We now derive an estimate for the break width x_c. This involves solving the problem posed by (13.52) with (13.53). The first of (13.52) gives

$$q(x, t) = 2q_0\delta(x)e^{-\mu t}. \tag{13.60}$$

Substituting this into the second equation gives

$$\frac{\partial r}{\partial t} = -dr + \frac{\partial^2 r}{\partial x^2} + 2q_0\mu\delta(x)e^{-\mu t} \tag{13.61}$$

the solution of which, with initial conditions (13.53), is of the form

$$r(x, t) = \frac{2r_0}{\sqrt{\pi t}} \exp\left[-\frac{x^2}{4t} - dt\right] + e^{-\mu t} r^*(x, t),$$

where $r^*(x, t)$ is the solution of

$$\frac{\partial r^*}{\partial t} = (\mu - d)r^* + \frac{\partial^2 r^*}{\partial x^2} + 2q_0\mu\delta(x) \tag{13.62}$$

with homogeneous initial data. This equation can be solved using Laplace transforms.

Denote the Laplace transform of r^* by ρ, that is,

$$\rho(x, s) = \int_0^{\infty} r^*(x, t)e^{-st}\, dt, \quad \operatorname{Re} s > 0.$$

Then ρ satisfies the inhomogeneous ordinary differential equation

$$\frac{d^2\rho}{dx^2} + (\mu - d - s)\rho = -\frac{2q_0\mu\delta(x)}{s}, \quad -\infty < x < \infty, \quad \operatorname{Re} s > 0. \tag{13.63}$$

We are only interested in the solution for $x > 0$; it is given by

$$\rho(x, s) = \mu q_0 \frac{\exp[-(s + d - \mu)^{1/2} x]}{s(s + d - \mu)^{1/2}}.$$

So, inverting the transform, we get

$$r^*(x, t) = \frac{\mu q_0}{2\pi i} \int_C \frac{\exp[-(s + d - \mu)^{1/2} x] e^{st}}{s(s + d - \mu)^{1/2}} \, ds, \tag{13.64}$$

where C is the Bromwich contour. The singularities of the integrand are a pole at $s = 0$ and a branch point at $s = -(d - \mu)$. The branch cut can be taken along the negative real axis to the left of the branch point, and so the contour of integration can be deformed to lie above and below the negative real axis. Since it is only necessary to evaluate $r^*(x, t)$ for $t = t_c$, it can be assumed that $t \gg 1$ in the integral (13.64). If we now use the method of steepest descents (see, for example, Chapter 6 in the book by Murray 1984) the main contribution to the integral is given by the residue at the pole $s = 0$; the contribution from the branch cut is exponentially small in comparison, provided that

$$\left(\frac{x}{2t}\right)^2 \ll d - \mu. \tag{13.65}$$

This inequality is shown to hold below. We thus arrive at the asymptotic solution for $r(x, t)$ given by

$$r(x, t) \sim \frac{r_0}{\sqrt{\pi t}} \exp\left[-\frac{x^2}{4t} - dt\right] + \frac{\mu q_0}{\sqrt{d - \mu}} \exp\left[-\mu t - (d - \mu)^{1/2} x\right]. \tag{13.66}$$

To estimate the break width, note that the formula (13.55) cannot be directly used since, with (13.60), $q(x, t)$ always involves a δ-function. Instead, we replace (13.55) by

$$r(x_c, t_c) = m r(0, t_c). \tag{13.67}$$

The assumptions (13.65) and $t \gg 1$ can again be used to justify neglecting the first term in (13.66) as compared with the second. So, from (13.67) and (13.66), an estimate for the break width is given by

$$x_c \sim (d - \mu)^{1/2} \ln\left(\frac{1}{m}\right). \tag{13.68}$$

If we take $m = 10^{-4}$ together with the parameters used previously to estimate t_c, then assumption (13.65) is easily verified to be valid for $t = t_c$ and $x = x_c$ since $(x_c/2t_c)^2 \approx 0.05$ and $d - \mu \approx 0.38$.

Note that, at least to leading order, the formula for x_c is independent of the critical time t_c. The calculation of t_c was only necessary for the purpose of verifying the 't large' assumption that was made throughout the analysis.

In dimensional terms, (13.68) gives, using (13.32),

$$X_c \sim -\frac{1}{\beta K} - \left(\frac{D}{\alpha + b - \sigma}\right)^{1/2} \ln m, \tag{13.69}$$

with typical values for these parameters given in Table 13.1.

In the expression (13.68), the dependence of x_c on δ and m roughly agrees with Figure 13.14. It also suggests that the break width should not be very sensitive to p, which, as shown by Murray et al. (1986) is the case when the carrying capacity in the break is not too close to the critical value.

13.8 Two-Dimensional Epizootic Fronts and Effects of Variable Fox Densities: Quantitative Predictions for a Rabies Outbreak in England

In general fox populations are not uniform, but instead vary according to the hospitality and carrying capacity of the local environment. This is very much the case in England, where interestingly some of the highest densities (by a factor of two to three) are in cities such as Bristol.

We first present the results of what happens when the epizootic wave encounters a localised region of different carrying capacity from the surrounding environment. The model system is still (13.26) except that in this two-dimensional situation, the diffusion term in the equation for the rabid population in (13.26) is replaced by $D\nabla^2 R$. Suppose that the carrying capacity, K, and the initial susceptible fox density are equal to a uniform value everywhere on a square region, except for a small patch in the centre of the square, where they have different values. We now introduce a uniform distribution of rabid foxes along one edge of the square, so that a one-dimensional epidemic front starts off across the square, and solve the model equations numerically. Figure 13.15 shows the resulting rabid and susceptible fox population densities for the case of a higher initial susceptible density in the patch.

From Figure 13.15(b) we see that the front moves faster through the region of higher carrying capacity as we would expect heuristically. The residual fox population, once the first outbreak has moved past, is slightly lower in the pocket of higher K than in the surrounding region. The converse of these effects is obtained if the wave encounters a pocket of lower susceptible fox density. One interesting feature is that the pocket of lowered density provides a sort of protection to the region just adjoining it. There are never as many cases of rabies in a ring around the outside of this region, and the final susceptible population density is higher there than farther away. The break region of the previous section also exhibits this feature, which arises because the region of lower density does not provide as many rabid foxes to diffuse into this area—there is, in effect, a preferential direction for the diffusion. The pocket of higher density has the opposite effect. Here the epidemic moves ahead of the epidemic front into the pocket of higher density; see the central figure in Figure 13.15(c). This focusing effect could account for some of the cases when outbreaks of rabies appear in advance of the front. These effects are the likely cause of the tortuous form of the epizootic front shown in Figure 13.3.

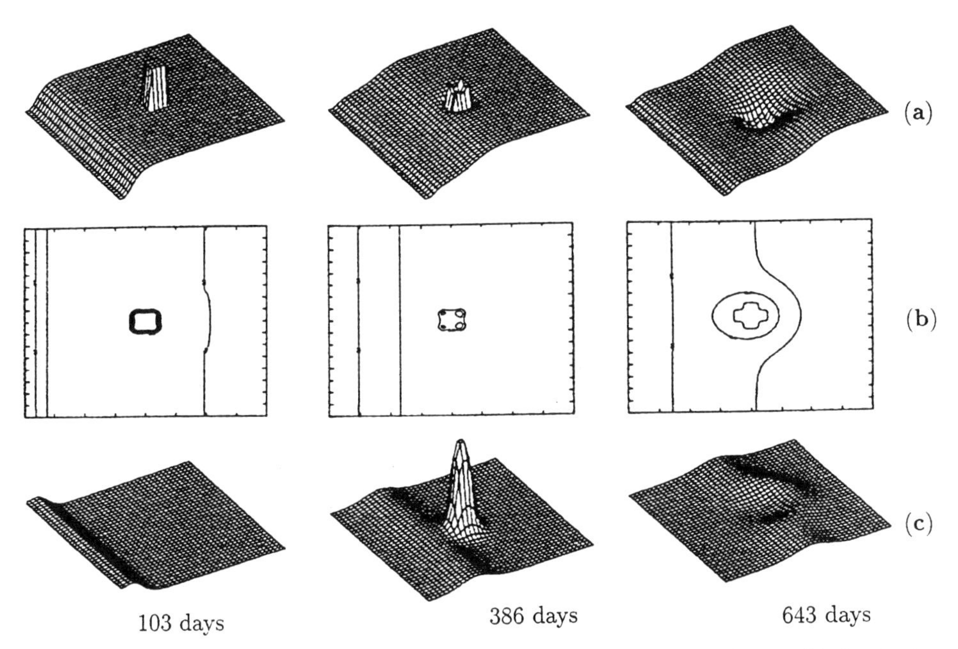

Figure 13.15. Effect on the epidemic front on encountering a pocket of higher initial susceptible fox density (and hence carrying capacity). Equations (13.37)–(13.40) were solved on a square, with the initial rabies-free fox density and the carrying capacity uniform everywhere except in a rectangular region in the centre, where they were raised by a factor of 1.7. The results are shown for a sequence of three times, namely, as the wave comes in from one side, as it passes the higher density pocket and after it passes. (**a**) Three-dimensional plot of the susceptible fox population density. (**b**) Contour plot of the susceptible fox density, with contour intervals of 0.1, where the density is normalised to have a maximum of 1. (**c**) Three-dimensional plot of the rabid fox density at each point in the square. (After Murray et al. 1986)

As mentioned before, England has remained rabies-free (except for a minor epidemic after World War I) due mainly to the strict quarantine laws[6] and high public awareness of the potential dangers. With the proximity of the disease in the north of France and the increased private boat traffic between continental Europe and Britain it seems inevitable that the disease will be brought into Britain in the near future. The appearance of rabies in Britain would be particularly serious, as we mentioned above, because of the high density of foxes, both urban and rural, in England. An additional cause for concern is the apparent compatibility of these urban foxes with cats (Macdonald 1980). If no control measures are applied, which admittedly would certainly not be the case, the epidemic would move quickly through England. We can use the model here to obtain a rough estimate for the position of the epidemic front if rabies is introduced into the fox population.

Macdonald (1980) gives a map of estimated fox densities in England (but excluding high urban pockets). Murray et al. (1986) covered the lower half of England with a grid, and assigned a density to each square based on the values given on his map. Contour

[6]In 2000 these were relaxed for selected countries if the animals had been vaccinated and underwent a series of other measures, such as blood tests and tagging, to ensure the animals are unquestionably rabies free.

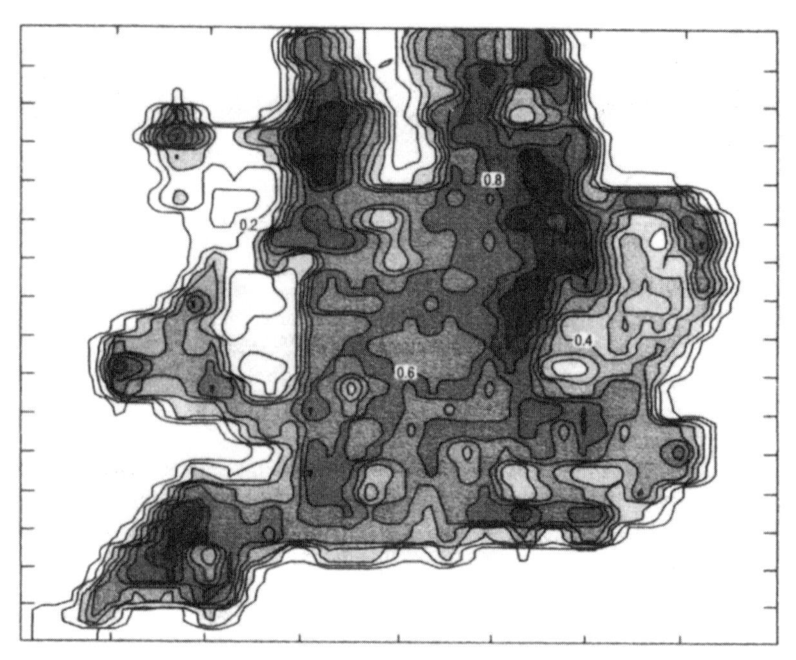

Figure 13.16. Fox densities in the southern half of England which were used in our numerical simulations. In the actual contour plots values are scaled to lie between 0 and 1, with 1 corresponding to 2.4 adult foxes/km^2 in Springtime, or to an average of 4.6 foxes/km^2 throughout the year. These values are based on Macdonald's (1980) estimates, who emphasises that the density map is probably not very accurate but is based on educated estimates. (From Murray et al. 1986)

lines of these densities, normalised from 0 to 1 were used; a shaded density map is shown in Figure 13.16. A value of 1 corresponds to 2.4 adult foxes per square kilometre in Springtime. The model studied is, in fact, in terms of fox densities averaged over the yearly cycle. Prior to the introduction of rabies, the population increases to its yearly high just after whelping, then gradually returns to the adult Springtime population. The average density is roughly the mean between the populations just before and just after whelping. The ratio of males to females is about 1.2:1, and females have an average of 3.7 to 4.2 cubs each year (Lloyd et al. 1976). Thus the average population is about 1.9 times the springtime adult population, and 1 corresponds to a carrying capacity of 4.6 foxes/km^2, the darkest shading, in Figure 13.16.

Using the carrying capacities (and initial fox densities) shown in Figure 13.16, and supposing, by way of illustration, that the rabies epidemic starts near Southampton, the two-dimensional form of (13.33) was solved numerically. The parameter values given in Table 13.1 were used, and the diffusion coefficient was taken to be 200 km^2/yr. The numerical simulations took about 120 minutes on a CRAY XMP-48 at the Los Alamos National Laboratory. The results are shown in Figures 13.17 and 13.18. The position of the front every 120 days is shown in Figure 13.17. We see that with such high fox densities the epidemic very quickly reaches most of the region studied. Within 4 years the front has effectively reached Manchester. The sequence in Figure 13.18 shows that, just as in the uniform density case, most of the cases of rabies are concentrated in a

Figure 13.17. The position of the wavefront every 120 days predicted by the model (13.33) and the spatially heterogeneous fox densities in Figure 13.16; that is, some of the parameters are space-dependent. Here a diffusion coefficient of 200 km^2/yr was taken with the other parameter values from Table 13.1. (From Murray et al. 1986)

narrow band at the front; the susceptible population is effectively decimated by the epidemic and partially regenerates before another wave starts again. Figure 13.18 shows the second outbreak starting off from Southampton, about 7 years after the first one.

These quantitative predictions can, of course, only be rough estimates. Macdonald (1980) emphasises that the fox densities in his map are only educated guesses, based on his knowledge of fox ecology. As we said above, not enough is known about the behaviour of rabid foxes to obtain a sharp estimate for the diffusion coefficient, which means that the speed of the wave may be anywhere from a half to four-thirds of our calculated result. We have also neglected such geographical factors as rivers, which tend to provide a channel for the epidemic, speeding its movement parallel to the banks and temporarily halting its direct passage. However, this relatively simple *SIR*-model provides a plausible quantitative first estimate for the progression of rabies in England if an epidemic were allowed to move unchecked. The model also provides a means of estimating realistic break widths which, at the very least, would seriously impede the spread of the disease.

The model we have investigated incorporates many of the salient features of the disease and the ecology of foxes. The model is sufficiently simple to enable us to obtain fairly reliable estimates for all of the parameters except the diffusion coefficient, for which we obtained a range of possible values. Analysis of the model produces certain predictions for the behaviour of the epidemic wave, in different environments, which provides some quantitative insight into the spatial spread of the epidemic and the transmission mechanisms responsible for its spread. For example, it is not known whether the primary reason for the spatial spread of the epidemic is the encroachment of confused

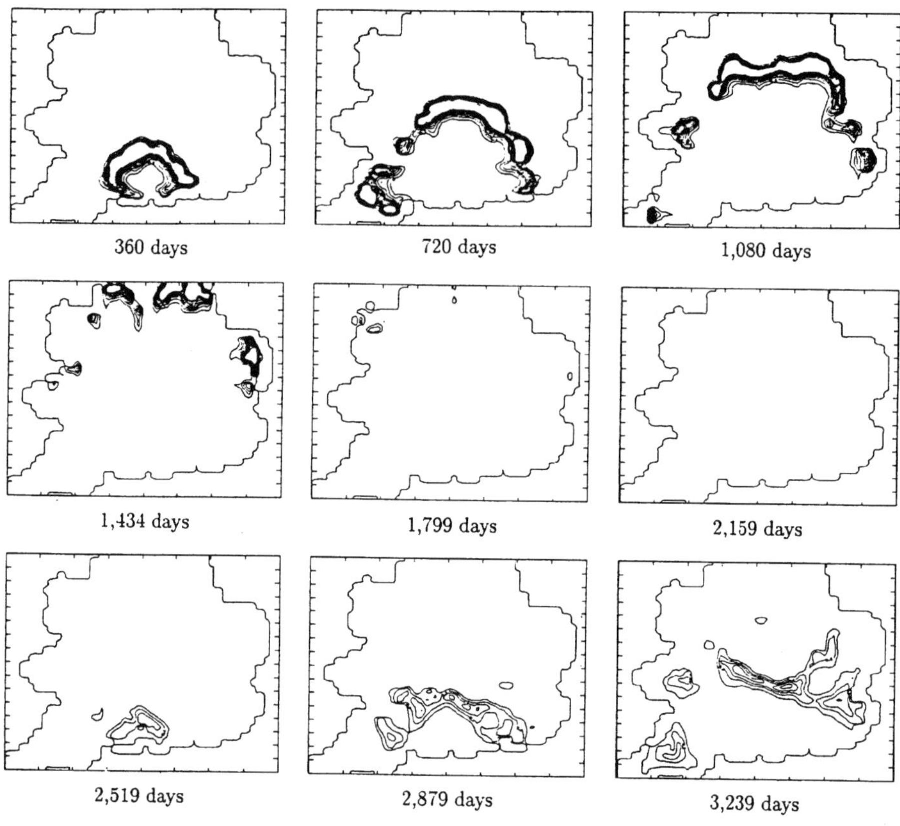

Figure 13.18. The epidemic front as it moves through the southern part of England. This was obtained by numerically solving (13.33) with the local carrying capacities and initial susceptible fox densities shown in Figure 13.16. A localised density of rabid foxes was initially introduced at Southampton on the south coast and allowed to spread. Contour plots of the rabid fox densities are given at a sequence of times, as the wave moves outward from its source. Note that, just as in the one-dimensional case, there are few rabid foxes in the region behind the front. Note also the reappearance of the second epidemic wave of lower intensity, which starts about 7 years after the initial outbreak and moves outward at the same speed. Here $D = 200 \text{ km}^2/\text{yr}$. and other parameter values are taken from Table 13.1. (From Murray et al. 1986)

rabid foxes onto their neighbour's territories, as we have assumed, or the migration of young foxes who carry the disease with them while healthy, or if both mechanisms are equally important. By isolating one of these mechanisms, we can determine how the epidemic wave behaves if that is the primary factor in its spatial spread, and compare the results with observation in continental Europe to see if it is possible for it to be the dominating factor. Our results indicate that the confused wandering of rabid foxes is sufficient to account for much of the behaviour of the current epidemic. It would be interesting to investigate a model in which migrating young foxes are the primary cause for spatial spread of rabies. It is also known that a certain percentage of foxes are immune to rabies. Such effects as these can be incorporated into the model framework here and this we investigate in the following section.

The agreement of our model with the available epidemiological evidence is quite good, despite the uncertainty in the size of the diffusion coefficient. For an initial fox density of 2 foxes/km^2, which is similar to densities reported for much of the continent, and for any reasonable choice of diffusion coefficient, the speed of the epidemic front, 25–65 km/yr, obtained from the model, encompasses the range of 30–60 km/yr usually observed. The speed of the wave increases with fox density, and drops to zero as the fox density decreases to the critical value. The model also predicts that rabies will essentially disappear for a period of about 5 years after the first outbreak, and then reappear, with the second outbreak weaker than the first. This correlates well with what has happened in many parts of Europe. Another interesting feature which emerges from the model is the enhanced movement of the rabies epidemic into regions of higher density *in advance* of the rest of the front. As we suggested, this may help to explain why outbreaks seemingly far in advance of the main epidemic occasionally occur.

It is possible for a strip of lowered susceptible fox population to check the progression of the epidemic, and protect an uninfected region ahead of the front. For this method of control to be efficiently applied, it is essential to have an indication of how wide an effective break region must be. For our model control scheme, Figure 13.14 gives nondimensional estimates for this width. If there are 2 foxes/km^2 initially, and the reduction scheme is 80% effective, then Figure 13.14 gives a break width of 10–25 km, depending on the diffusion coefficient. This is of the right order of magnitude when compared with the protective break which has proved effective in Denmark and parts of Switzerland. In Denmark, intensive control measures were applied to a strip 20-km wide with less intensive measures used in an adjoining 20-km strip.

The question of what method should be used to contain an outbreak is interesting. The model here suggests that vaccination would be more effective than gassing or poisoning since the former would help to restrict the spread of infective foxes whereas the latter would enhance the spread. It seems that chicken heads impregnated with vaccine have proved reasonably effective in Ontario: it relies on efficient scavenging by the foxes. This is not necessarily the case with urban populations[7] (personal communication from Dr. Stephen Harris, 1988). Another problem with vaccination in general is that sometimes the level of vaccination in one species may induce the disease in another as seems to be the case with the red and grey fox.

The probability that rabies will eventually reach England and other uninfected regions is not small. It is clearly of considerable importance to understand as much as possible about the disease, its transmission and how it spreads, well before it arrives. The density of foxes in England is much greater in many areas than on the continent, and the epidemic may proceed differently there. Figures 13.17 and 13.18 summarise some of the model's predictions for a particular choice of diffusion coefficient, and some estimates for the current fox populations in the southern half of England. Perhaps the most disturbing aspect of these results is the rapidity with which the epidemic would move through the central region, namely, at speeds of around 100 km/yr. No less disturbing is the reappearance of the disease several years after the passage of the epidemic front: a relatively free rabies period would certainly give rise to complacency.

---

[7] Interestingly the life expectancy of urban foxes is significantly less than for rural foxes—all the fast food in their 'menu du jour' no doubt.

13.9 Effect of Fox Immunity on the Spatial Spread of Rabies

It is known that a certain proportion of foxes develops a natural immunity to rabies. It is of interest to try and quantify its effect on the spatial spread of the disease. This was done by Murray and Seward (1992) and it is their modification of the models discussed in the above sections that we briefly consider here; see their paper for full details and more complete comparative results with those we obtained above. They showed that with realistic estimates for the size of the immune class, immunity has little effect on the propagation speed of the initial wave of the rabies epidemic but it does affect the behavior of the periodic outbreaks associated with the oscillating tail of the wave. They also studied the effect on the width of a rabies break which would be required to contain the epidemic and included the effect of spatial dispersal of susceptible and infected foxes. They finally investigated the hypothesis that the required break width might depend on whether the break was created by killing foxes or by vaccinating them against rabies. They found that the break width does not change significantly unless the rate of spatial dispersal is large and, of course unless the immune class increases significantly. We discuss their model since the methodology is not restricted to the spatial spread of rabies. With this in mind we also include diffusion of all species.

In the *SIR* model in the above sections we assumed that all rabid foxes die. In fact some foxes have been found to recover from rabies and a certain proportion of the survivors develop immunity to the disease. Steck and Wandeler (1980) found that about 2% of all infected red foxes in a number of experimental studies developed immunity. It is more difficult to evaluate the immune status of the fox population in the wild. Steck and Wandeler (1980) also presented data which suggest that not more than 8% of foxes living through the passage of an epidemic front are actually immunized against the disease. However, some estimates put the proportion of immune foxes as high as 20% in the U.S.A., where rabies is spread by both the grey (*Urocyon cinereoargenteus)* and the red fox. (Due to the mixture of species, our model is not directly valid in this case, although it could be extended by the appropriate choice of parameters.) Wandeler (1987) noted that the rarity of documented survival of clinical disease in experimentally and naturally infected animals is in contrast to frequent reports of rabies-neutralizing antibodies in sera collected from wild animals, and also that it is not clear whether these reports actually demonstrate survival of the clinical disease or not. Overall, it appears that the development of immunity is possible but that it has little effect on the spread of the disease. It is of interest, therefore, to investigate the effect of introducing an immune class into the model in Section 13.5. We expect that, for small values of the proportion of foxes developing immunity, the results produced by the modified model will differ little from those of the original. It is interesting to develop such a model not only for pedagogical reasons but also to confirm this belief, quantify the effect on the spatial spread as the immune class increases and the effect of immunity transfer to cubs.

We now divide the fox population into four groups: susceptible foxes, with a population density S; infected, but noninfectious, foxes, with a density I; infectious, rabid foxes, R; and immune foxes, Z. This division is again based on the long incubation period of 12–135 days that the rabies virus undergoes in the infected animal. As stated above, during this time the animal seems to behave normally and does not transmit the disease. Recall that the clinical period is a relatively short period of 1–10 days.

The key assumptions are those listed in Section 13.5, numbered (i)–(vi), together with the following.

(i) Rabies is not always fatal. Rabid foxes die at an average per capita rate α (where $1/\alpha$ is the average duration of clinical disease) and recover to develop immunity at an average per capita rate γ. Values for γ are obtained from the percentage p of rabid foxes that become immune, where

$$p = \frac{\gamma}{\alpha + \gamma}.$$

Foxes that recover without developing immunity would form only a small proportion of the susceptible class so we do not include them here.

(ii) Rabid and infected foxes continue to put pressure on the environment and die of causes other than rabies but they have a negligible number of healthy offspring.

(iii) Immune foxes may have susceptible or immune offspring. We examine the two extreme cases: either all the offspring are assumed to be susceptible or all the offspring are assumed to be immune.

These new assumptions (i)–(iii) together with those in Section 13.5 suggest the following amended model in place of (13.26).

$$\begin{aligned}
\frac{\partial S}{\partial T} &= (a-b)\left[1-\frac{N}{K}\right] + a^*Z - \beta RS, \\
\frac{\partial I}{\partial T} &= \beta RS - \sigma I - \left[b + (a-b)\frac{N}{K}\right] I, \\
\frac{\partial R}{\partial T} &= \sigma I - \alpha R - \gamma R - \left[b + (a-b)\frac{N}{K}\right] R + D_R \frac{\partial^2 R}{\partial X^2}, \\
\frac{\partial Z}{\partial T} &= \gamma R + (a - a^*)Z - \left[b + (a-b)\frac{N}{K}\right] Z,
\end{aligned} \tag{13.70}$$

where now the total population is

$$N = S + I + R + Z.$$

If immune foxes are assumed to have only susceptible offspring then $a^* = a$; if they have only immune offspring, $a^* = 0$. We consider here only the one-dimensional problem since our primary concern is to investigate the changes in the spatial spread of rabies due to the inclusion of an immune class. In particular, we can consider the effect of including an immune class on (i) the speed of the epizootic wave, (ii) the behaviour of the recurring epidemics after the passage of the main epidemic front as shown in the above sections and (iii) control measures.

Speed of Propagation of the Rabies Epizootic

We now nondimensionalise the model system using the same dimensionless variables and parameters as in (13.29) but with the addition of one which measures immunity. For ease of reference we give these again here, namely,

$$
\begin{aligned}
s &= \frac{S}{K}, \quad q = \frac{I}{K}, \quad r = \frac{R}{K}, \quad z = \frac{Z}{K}, \quad n = \frac{N}{K}, \\
\varepsilon &= \frac{a-b}{\beta K}, \quad \delta = \frac{b}{\beta K}, \quad \mu = \frac{\sigma}{\beta K}, \quad d = \frac{\alpha + b}{\beta K}, \quad v = \frac{\gamma}{\beta K}, \\
x &= \left(\frac{\beta K}{D}\right)^{1/2} X, \quad t = \beta K T,
\end{aligned}
\tag{13.71}
$$

which give

$$
\begin{aligned}
\frac{\partial s}{\partial t} &= \varepsilon(1-n)s + (\varepsilon + \delta)^* z - rs, \\
\frac{\partial q}{\partial t} &= rs - (\mu + \delta + \varepsilon n)q, \\
\frac{\partial r}{\partial t} &= \mu q - (v + d + \varepsilon n)r + \frac{\partial^2 r}{\partial x^2}, \\
\frac{\partial z}{\partial t} &= vr + [(\varepsilon + \delta) - (\varepsilon + \delta)^*]z - (\delta + \varepsilon n)z,
\end{aligned}
\tag{13.72}
$$

where $n = s + q + r + z$ and $(\varepsilon + \delta)^* = (\varepsilon + \delta)$ if the immune class produce only susceptible offspring, or $(\varepsilon + \delta)^* = 0$ in the case of all immune offspring.

We wish to investigate a range of values for the percentage, p, of rabid foxes that recover and become immune. The parameter γ in the v grouping in (13.71) is given by

$$
p = \frac{\gamma}{\alpha + \gamma}.
$$

The other dimensional parameter values in (13.70) are given in Table 13.1 together with the diffusion coefficient $D_R = 200$ km^2 year^{-1}; recall the difficulty in estimating the diffusion coefficient.

We first consider the case in which immune foxes have susceptible offspring. The four-class model was solved numerically by Murray and Seward (1992) for five values of the percentage of immune foxes: $p = 2\%, 5\%, 10\%, 15\%, 20\%$. Although we believe the larger values are not appropriate in this model for the European situation, we include them to see the effect on the model predictions. The shapes of both the initial wave and the recurrent outbreaks were found to vary only slightly between the three-class and the four-class models: they look like the shapes in Figures 13.9 and 13.10. However, they found that the effects of introducing an immune population are:

(i) the speed of the initial wave decreases;

(ii) the levels of the infected and rabid populations are not as high in the initial outbreak;

(iii) the susceptible population is not reduced as severely when the rabies outbreak occurs;

(iv) the time between recurrent outbreaks is reduced.

The first three of these are as we would expect intuitively while the fourth point follows from them. These effects become more marked as the immune percentage increases. From the asymptotic analysis and numerical results for the three-class system in Section 13.5 the speed of the initial wave is about 51 year^{-1} when $K = 2$ fox km^{-2} and 103 km year^{-1} when $K = 4.6$ fox km^{-2}. The computed wavespeeds in the four-class cases are given in Table 13.3.

Table 13.3. Speed of the rabies epizootic front for various immunity levels and fox densities. (From Murray and Seward 1992)

$K = 2$ fox km^{-2}		$K = 4.6$ fox km^{-2}	
Immune	Wave Speed (year^{-1})	% Immune	Wave Speed (km year^{-1})
0	51	0	103
2	49	2	102
5	47	5	100
10	43	10	96
15	40	15	92
20	36	20	89

The main effect of the inclusion of the immune population (but with susceptible offspring) is in the tail of the initial wave. With the three-class model, it takes about 5 years for the susceptible fox population to recover sufficiently for a secondary outbreak of rabies to occur when $K = 2$ fox km^{-2} and 11 years when $K = 4.6$ fox km^{-2}. In the four-class case, the first recurrent outbreak occurs after a much shorter period, as shown in Table 13.4. Murray and Seward (1992) found that the susceptible population is not

Table 13.4. Time to secondary outbreak and dependence on susceptible offspring. (From Murray and Seward 1992)

$K = 2$ fox km^{-2}		$K = 4.6$ fox km^{-2}	
Immune	Time to Recover (years)	% Immune	Time to Recover (years)
0	5.0	0	11.0
2	4.8	2	7.8
5	4.3	5	5.8
10	3.8	10	4.1
15	3.2	15	3.2
20	3.0	20	2.7

reduced as much by the initial outbreak in the four-class model as in the three-class case and also that the population level builds up again much more rapidly, which accounts for the reduced time between recurrent outbreaks.

Another noticeable change found in the four-class case is the increased damping of the oscillations in the tail of the wave. In the spatially uniform situation, that is, (13.70) with $D_R = 0$, there is a nontrivial steady state defined by the (13.70) with $a^* = a$. In the three-class model, that is, (13.70) with $Z = 0$ and $\gamma = 0$, there is a critical carrying capacity K_T given in (13.30), for this nontrivial steady state (13.29) to exist; both the steady-state values and K_T were easily obtained after some elementary algebra. In the four-class model, the analytical solution is difficult to find but straightforward to estimate the steady states and critical carrying capacity, K_T, numerically. Murray and Seward (1992) found that as the immune percentage increases, the computed solution tends to its steady state value in a shorter time period.

In the case in which immune foxes have immune offspring only, the model was solved for the same values of carrying capacity K and immune percentage p as in the previous case. The assumption of immune offspring has no effect on the propagation of the initial outbreak—the wavespeeds are the same as those given in Table 13.2 as was confirmed numerically. This is to be expected since the immune population does not exist until that outbreak has passed.

Again, the main effect is seen in the tail of the initial wave and here the differences from the four-class model with susceptible offspring and the original three-class model are significant. After the passage of the initial epizootic wave, the immune foxes form a substantial proportion of the population and this has a considerable damping effect on successive outbreaks of the disease. With a carrying capacity of $K = 2$ fox km^{-2}, the time until a second outbreak occurs is larger than in the three-class model and increases as the immune percentage increases, as shown in Table 13.5. With $K = 4.6$ fox km^{-2} and 2% immune foxes, there is a second outbreak of rabies after 18 years; with 5% immune foxes, there is a second outbreak of rabies after 21 years. The graphs obtained are again very similar for all immune percentages greater than 5% when $K = 4.6$ fox km^{-2}.

These differences can be explained by considering the steady-state solution of the system (13.70) with $a^* = 0$. In this case, there is no physically realistic steady state

Table 13.5. Time to secondary outbreak with immune offspring. (From Murray and Seward 1992)

$K = 2$ fox km^{-2}	
% Immune	Time to Recover (years)
0	5.0
2	5.4
5	5.6
10	6.2
15	6.8
20	8.1

with positive values of I and R. The model requires that the rabies epidemic die out for a steady state to be attained, in which case (13.70) reduces to a logistic growth law for the remaining total fox population, $S + Z$. The steady state solution gives $S_0 + Z_0 = K$ and the relative values of S_0 and Z_0 are determined by their initial values. The time required for the system to approach the steady state depends on the carrying capacity and the percentage of immune foxes. As either K or p increases, the system tends to the steady state faster.

With either susceptible or immune offspring, we see that the use of the four-class model has little effect on the propagation of the initial rabies outbreak. The wavespeed only changes significantly if a high proportion of immune foxes is assumed. The main effect then of the four-class model is seen in the tail of the initial wave. The effect is very different depending on whether the immune foxes are assumed to have susceptible or immune offspring.

As we said, the assumption of only immune or only susceptible offspring is a simplification. Among other effects, immune and susceptible foxes will interbreed and the proportion of the offspring with immunity will depend on the relative sizes of the immune and susceptible populations. Equations for heritability could be included in the model but this would lead to a much more complicated system than either the three- or four-class models. The prediction from the model that a low rate of natural immunity (2%–5%) has little effect on the wavespeed is in agreement with the observation that a lack of immunity is one of the contributing factors to the spread of rabies among foxes (see, for example, Blancou 1988). The most significant effect at low immunity rates is the reduction of the time period before the secondary outbreak when $K = 4.6$ fox km^{-2}.

Immunity Effects of Control Measures Associated with a Rabies 'Break'

A major use of a model for the spread of an infectious disease is to assess various control strategies to contain the disease. As discussed in Section 13.6 one possible method is to introduce a rabies 'break' ahead of the initial wave. Recall that we consider a break to be a region where the susceptible fox population is reduced below the critical carrying capacity and hence will not sustain a propagating epizootic wave. The model here can again be used to estimate the required break width for a range of parameter values.

In practice, a break can be created by killing foxes—the method used in Denmark—by intensified hunting and gassing fox dens during whelping season (Wandeler 1987), or by vaccination, which has been used successfully in Switzerland (Wandeler et al. 1987). We can compare these two approaches using the model. As mentioned a potential difficulty with killing the foxes is that a reduced population density may encourage dispersal of foxes into the area, thus reducing the effectiveness of the break. This dispersal can be modelled by introducing diffusion terms for both the susceptible and infected fox populations, that is, including terms $D_S(\partial^2 S/\partial X^2)$ and $D_I(\partial^2 I/\partial X^2)$ in the first two equations of the three-class model. The effect of the diffusion terms will be to average the populations inside and outside the break. Yachi et al. (1989) investigated the three-class model with diffusion of susceptible and infected foxes, assuming the same diffusion rates for all three populations, and found that this could result in a significant increase in the speed of propagation of the epizootic. Murray and Seward (1992) considered smaller diffusion rates for the susceptible and infected foxes than for the rabid population. The vaccination approach can be modelled by setting the initial immune population Z in the four-class model to a nonzero value to represent vaccinated foxes.

We can again estimate the required break width by solving the system of equations numerically with data such that the region of lowered fox population density starts at $x = 0$ and extends to infinity. As the epizootic wave moves into the region, eventually the total number of infected foxes remaining will be less than F infected fox km^{-1}, where 'infected fox' now refers to any fox with rabies, whether infectious or not. The number F is chosen to be sufficiently small, that is, well below the critical carrying capacity for the disease. Let $t_c(F)$ be the time at which this occurs. Again the break width is chosen to be the point x_c where the infected fox density is a given (small) fraction m of the value at $x = 0$; that is (recall (13.51)),

$$q(x_c, t_c) + r(x_c, t_c) = m[q(0, t_c) + r(0, t_c)].$$

As noted there are two ways to model a region of lowered population density: by reducing the carrying capacity K or by reducing the initial susceptible fox population S. Reducing K corresponds to an ongoing programme of control, where the susceptible population is held at a lower level over some period of time, and is the approach used in Section 13.6. The dependence of the break width on the carrying capacity in the break, the parameter d and on F and m is shown in Figure 13.14 in Section 13.6.

When only the initial value of S is reduced, the situation is that of a 'one-off' attempt to create a break, rather than an ongoing control. Intuitively, this second approach seems less likely to be an effective means of stopping the spread of the epidemic wave but the computed results show similar break widths in both cases. We discuss this effect after explaining our method for computing the break widths in a little more detail: it is essentially the same as in Section 13.6.

Murray and Seward (1992) first considered the effect of dispersal of all foxes using the three-class model, that is, without an immune population. The model in this case is simply (13.70) with $Z = \gamma = 0$ but with diffusion terms for S and I, namely,

$$\begin{aligned}
\frac{\partial S}{\partial T} &= (a - b)\left(1 - \frac{N}{K}\right) S - \beta R S + D_S \frac{\partial^2 S}{\partial X^2}, \\
\frac{\partial I}{\partial T} &= \beta R S - \sigma I - \left[b + (a - b)\frac{N}{K}\right] I + D_I \frac{\partial^2 I}{\partial X^2}, \\
\frac{\partial R}{\partial T} &= \sigma I - \alpha R - \left[b + (a - b)\frac{N}{K}\right] R + D_R \frac{\partial^2 R}{\partial X^2}.
\end{aligned} \tag{13.73}$$

This is also the system considered by Yachi et al. (1989). As discussed above it is very difficult to estimate diffusion coefficients. A rough estimate can be made, following the procedure in Section 13.5, by taking the product of the average territory size and the average rate at which a fox leaves home. Using 5 km^2 as the average territory size and the birth rate of 1 fox year^{-1} as the rate of leaving home, gives $D_S = D_I = 5$ km^2 year^{-1}. This is most likely an underestimate; for example, Garnerin et al. (1986) estimated a limit of 8 km on the dispersion distance of young foxes, based on a discrete model for the spread of rabies. Numerical results have been computed for values 5 km^2 year^{-1}, 20 km^2 year^{-1}, 50 km^2 year^{-1} and, for interest, 200 km^2 year^{-1}.

We need to extend the scheme in Section 13.6 for calculating break widths to model dispersal from a region of normal population density into a break. Murray and Seward (1992) recalculated the break widths without diffusion using the method of Section 13.6 taking $F = 0.5$ fox km^{-1} and $m = 10^{-4}$. Given these values for break width, they then set up their problem data so that a break of the calculated width occurs between two regions of normal population and let the model run to observe the epizootic wave flowing into the break. Of course, as discussed above, due to the diffusion, rabies will eventually appear across the break but this will take much longer than when the break is not present. They then evaluated two control scenarios with $K = 2$ fox km^{-2}, namely, a 'one-off' approach in which the susceptible population in the break was reduced and the other in which the carrying capacity in the break was kept low, that is, an ongoing control. The break widths were similar in both cases; in fact, with diffusion coefficients $D_S = D_I = 200$ km^2 year^{-1}, it is apparently more effective simply to reduce the initial susceptible population. The two strategies are equally effective in that the epizootic wave takes approximately the same time to cross the break regardless of the approach used to create it. The integration time over which the results were computed corresponds to a physical time of about one year and the break is set up approximately three months ahead of the time the wave reaches it. If the 'one-off' approach were used farther in advance of the impact of the epizootic wave, it would be less effective. Also, under the 'one-off' approach, it was not possible to reduce the total infected fox population to $F = 0.5$ fox km^{-1} when the susceptible population in the break was 1.2 fox km^{-2}. Numerous graphs of the various break scenarios for various diffusion coefficients and carrying capacities are given in Murray and Seward (1992).

By way of illustration of the effects of vaccination we consider the effect of vaccination by using the four-class model with the immune population Z representing vaccinated foxes. In this case, the break width is estimated using the basic scheme in Section 13.6. We solved the model equations for propagation of the wave into a region of reduced susceptible fox density $S(X, 0)$ but with an initial population of vaccinated foxes, $Z(X, 0) = K - S(X, 0)$. If we assume that these vaccinated foxes have susceptible offspring, then the break has essentially been created by the 'one-off' approach. Vaccination is applied only once to the initial population. By letting the vaccinated foxes have 'immune' offspring, we can roughly model the effect of an ongoing vaccination programme.

Results are shown in Figure 13.19 which compares the basic break widths (that is, no diffusion, no natural immunity) from the three-class model and the four-class model when $K = 2$ fox km^{-2}. We see that the break widths from the four-class model are generally larger. The case in which the vaccinated foxes have susceptible offspring is similar to having susceptible foxes diffusing into the break. As shown in Figure 13.20, with diffusion rates in the three-class model between 20 and 50 km^2 year^{-1}, the break widths are similar to those from the four-class model. It is less clear why the ongoing vaccination programme is not as effective as killing the foxes. As we commented above, in the four-class model, the infected and rabid populations are not as high in the initial outbreak. However, it seems that these populations then drop off more slowly in this case than in the three-class model. As a result, it takes longer for the total infected

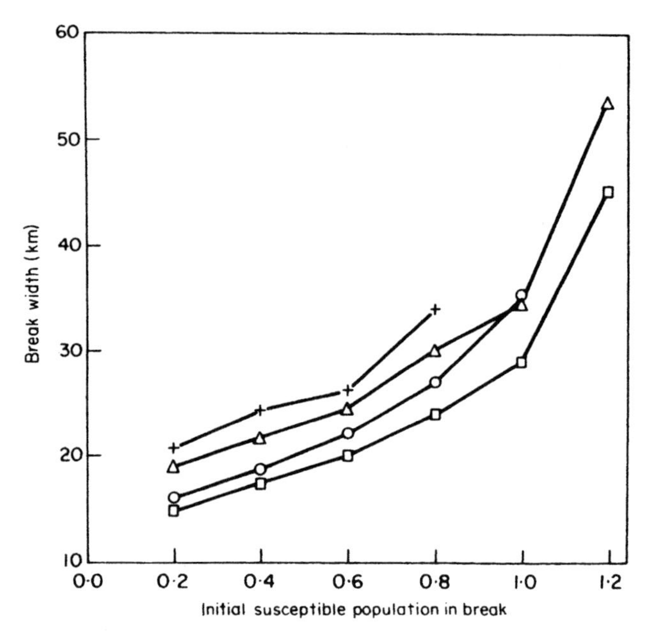

Figure 13.19. A comparison of break widths predicted by the three-class model (killing the foxes to create a break) and the four-class model (vaccination used to create the break) for a carrying capacity of $K = 2$ fox km^{-2} outside the break. Here the break width (in kilometres) is plotted against the susceptible fox population in the break, for both the ongoing control programme and the 'one-off' approach. These are for the most basic case, that is, only rabid fox diffusion in the three-class case and no natural immunity in the four-class model. The nomenclature is: $\Box$: three-class ongoing control, $\circ$: three-class, one-off control, $\triangle$: four-class, ongoing control, $+$: four-class, one-off control. (From Murray and Seward 1992)

fox population to drop below F, the disease spreads a little farther even in the reduced susceptible population and the calculated break width is larger. It is difficult to judge how much of this apparent spread is due to the nature of the differential equations and how much is actually due to the effect of the immune foxes on the environment. We note that there are two advantages of an ongoing vaccination programme over a 'one-off' approach—the break widths are slightly smaller and, probably more importantly, it is possible to create a break at higher population densities.

When $K = 4.6$ fox km^{-2}, it is more difficult to create a break. In the case in which vaccinated foxes have susceptible offspring, a break could not be created above a susceptible population density of 0.4 fox km^{-2}. Assuming 'immune' offspring, the break width when the reduced population density is 1.38 fox km^{-2} and $F = 1.5$ fox km^{-1} is 36 km. It was not possible to create a break using $F = 0.5$ km^{-1} at this population density.

If naturally immune foxes are included in the four-class model, the break widths decrease but the change is small if a small percentage (2–5%) is used. At 20% natural immunity, assuming either susceptible or 'immune' offspring, the breaks are 5–10 km narrower. The results are shown in Figure 13.20 for the case of 'immune' offspring (ongoing control programme).

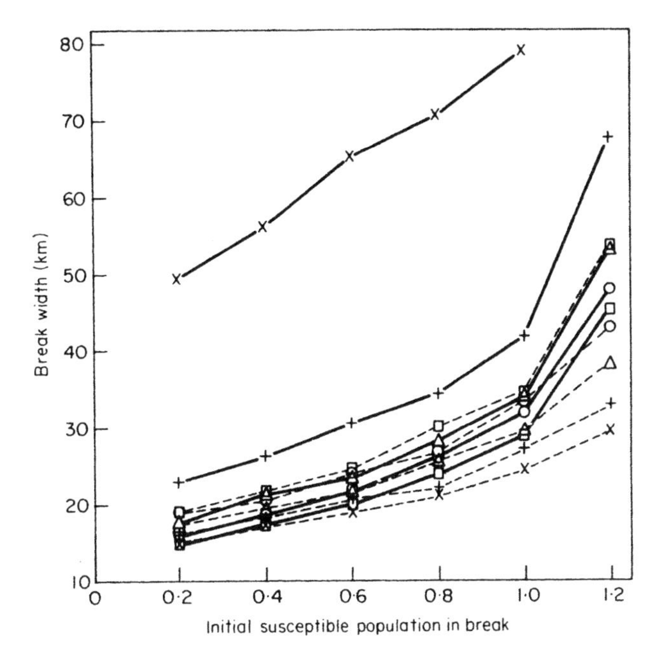

Figure 13.20. A comparison of break widths from the three-class model with varying diffusion rates (solid lines) to those from the four-class model subject to varying rates of natural immunity (dashed lines). The break width, in kilometres, is plotted against the initial susceptible fox population in the break. The carrying capacity outside the break is $K = 2$ fox km^{-2}. In both cases the break was formed by an ongoing control programme, that is, by reducing the carrying capacity in the three-class model and by assuming immune offspring in the four-class model. We see that the breaks become wider with increasing diffusion in the three-class model. In the four-class model, the break widths decrease with increasing natural immunity, more or less as we would expect. Nomenclature: □: no diffusion or immunity, o: 5 km^2 year^{-1} or 5%, △: 20 km^2 year^{-1} or 10%, +: 50 km^2 year^{-1} or 15%, ×: 200 km^2 year^{-1} or 20%. (From Murray and Seward 1992)

Keeping in mind that these mathematical models yield only rough estimates for required break widths, we make the following observations. The most effective strategy for creating a break is to carry out a programme of culling foxes. Only if this leads to a significant dispersal of neighboring foxes into the break will it be more effective to use a program of vaccination. It is difficult to create a break in a region with a large carrying capacity. An ongoing program of control generally yields slightly smaller break widths and the possibility of creating a break at a higher fox population density than a 'one-off' attempt to create a break.

We have been primarily concerned here with the spatial propagation of an epidemic. There are important and interesting problems associated with control strategies when rabies, for example, is already in a community. An interesting and very practical model to deal with this situation was proposed by Frerichs and Prawda (1975) to deal with an urban area in Colombia: the model they proposed was for canine rabies. We discussed a modified application of their approach in Chapter 10, Volume I, when discussing control strategies for bovine tuberculosis and the interaction of cattle with badgers, which provide a reservoir of the disease.

The type of models we have discussed in this chapter have wider applicability, such as to the spatial spread of pests, killer bees (see Taylor 1977 for data on the South American spread), animals, plants and so on.

Some Caveats

What is clear from the study of these kinds of models for the spatial spread of an epidemic is that although they are quite complicated we still have had to make some major assumptions. The whole question of a break brings up the discussion in Section 13.6 on what we mean by extinction of a population in a break. That discussion is just as relevant with the modelling in this section.

Although these continuous models have helped our understanding of the transmission and spatial dynamics of disease and have resulted in some useful qualitative and often quantitative predictions, many justified criticisms can be levelled at them. For example, Mollison (1991) makes some pertinent points regarding deterministic continuous versus stochastic models. Even the (relatively) simple models we have discussed here would become orders of magnitude more difficult if we included stochasticity. In many cases the distinction between the two approaches is being able to do something or nothing with the situation under investigation; a study of the effect of stochasticity would clearly be illuminating.

It can reasonably be argued that discrete models can be made more realistic (although parameter estimation is more difficult). Also foxes live in family groups with distinct territories and so infection is more likely to spread within the whole family. Reproduction is discrete rather than continuous and so on with other aspects of fox behaviour. Such effects could be included in continuous space and discrete time models.

In all modelling there is always a trade-off between simplicity and the ability to estimate parameters and inclusion of more aspects with the diminishing ability to estimate parameters and evaluate and interpret what is predicted in the solutions. One obvious defect is the problem of a reservoir of the disease behind an outbreak wave when the calculated population is essentially at the level of extinction of the infected population; this is a problem with discrete models as well. Another related problem is the assumption that there is a mean incubation time ($1/\sigma$) whereas in practice it is a distributed incubation time which can vary quite widely. There have been cases with dogs in quarantine in England that developed rabies even after 6 months, albeit very rarely. These last two points are clearly interrelated. Recently, in an interesting paper, Fowler (2000) has revisited the model discussed here in Section 13.6 and investigated these two aspects, namely incubation and extinction, and showed that inclusion of a distributed incubation time can explain why extinction does not occur. He further obtained asymptotic estimates for the minimum infected fox density. So, although there is obvious stochasticity in the field even extinction can be incorporated in a continuous model.

Exercises

1. Consider the dimensionless form of the epidemic model

$$S_t = -IS + S_{xx}, \quad I_t = IS - \lambda I + I_{xx},$$

where $\lambda > 0$ and look for travelling wave solutions $S(z)$ and $I(z)$, with $z = x - ct$, such that

$$S'(-\infty) = 0, \quad S(\infty) = 1, \quad I(-\infty) = I(\infty) = 0,$$

where prime denotes differentiation with respect to z.

Prove that, for all finite z, $0 < S < 1$ by showing that $S'(z) > 0$ is monotonic for all $-\infty < z < \infty$. Show also that $(S + I)' > 0$ and hence that for all $-\infty < z < \infty$, $S(z) + I(z) < 1$.

Prove that

$$\int_{-\infty}^{\infty} I(z')\,dz' > \int_{-\infty}^{\infty} I(z')S(z')\,dz' = \lambda \int_{-\infty}^{\infty} I(z')\,dz'$$

and hence deduce that the threshold criterion for a travelling epidemic wave solution to exist is $\lambda < 1$.

2. A rabies model which includes a logistic growth for the susceptibles S and diffusive dispersal for the infectives is

$$\frac{\partial S}{\partial t} = -rIS + bS\left(1 - \frac{S}{S_0}\right), \quad \frac{\partial I}{\partial t} = rIS - aI + D\frac{\partial^2 I}{\partial x^2},$$

where r, b, a, D and S_0 are positive constant parameters. Nondimensionalise the system to give

$$u_t = u_{xx} + uv - \lambda u, \quad v_t = -uv + bv(1 - v),$$

where u relates to I and v to S. Look for travelling wave solutions with $u > 0$ and $v > 0$ and hence show, by linearising far ahead of a wavefront where $v \to 1$ and $u \to 0$, that a wave may exist if $\lambda < 1$ and if so the minimum wavespeed is $2(1 - \lambda)^{1/2}$. What is the steady state far behind the wave?

14. Wolf Territoriality, Wolf–Deer Interaction and Survival

14.1 Introduction and Wolf Ecology

Territoriality is a fundamental aspect in the ecology of many mammals, particularly predatory animals such as wolves, lions, hyenas, African wild dogs and badgers, and it has been widely studied. In the case of wolves, whose prey are mainly moose and deer, an immediate question arises as to how the predator and prey coexist if the land is divided up into predator territories.[1] This in turn leads to the question of how territories are determined and maintained. It is clearly important in the ecology of such predatory animals. In this chapter we consider the question of mammalian territory formation, specifically as it applies to wolves, and its role in wolf–deer survival for which there is a considerable amount of data. In spite of the numerous studies on how pack territories are formed and maintained it was not addressed mechanistically until the mid-1990's with the mathematical modelling work of Lewis and Murray (1993), White (1995), White et al. (1996a,b), Lewis et al. (1997, 1998), Moorcroft et al. (1999) and Lewis and Moorcroft (2001) who studied the spatiotemporal effects on territory formation, territory maintenance and wolf–deer survival. Most of the material we describe in detail in this chapter is based on their work. First we give some background ecology on wolves.

The book (which has many beautiful photographs) by Mech (1991), who has studied wolves for nearly 40 years, is the best general introduction to the biology and ecology of wolves. It gives an excellent overview of the major aspects of wolf behaviour and social organisation; he also discusses some practical aspects of wolf conservation. Through his work, Mech has done much to change the often held traditional (erroneous) view of these splendid animals. He also points out that the stories of wolves attacking humans are mainly myths. He notes 'I have no doubt that if a single wolf—let alone a pack—wanted to kill someone, it could do so without trouble. When I have watched wolves close-up killing prey, they were swift and silent. A few good bites, and a human would be dead. The fact remains, however, that there is no record of an unprovoked, non-rabid wolf in North America seriously injuring a person.'

[1] I first became intrigued by this question during a visit to the University of British Columbia in the late 1970's when, over dinner, in a discussion on animal intelligence it was mentioned how particularly clever and intelligent wolves are, as has been noted regularly since at least Roman times. In Canada their main food source is often the moose. I started to wonder how, if wolves are so clever, did the moose manage to survive. It was not until the early 1990's that Mark Lewis and I started to look at the question from a mathematical modelling point of view being joined soon afterwards by Jane White.

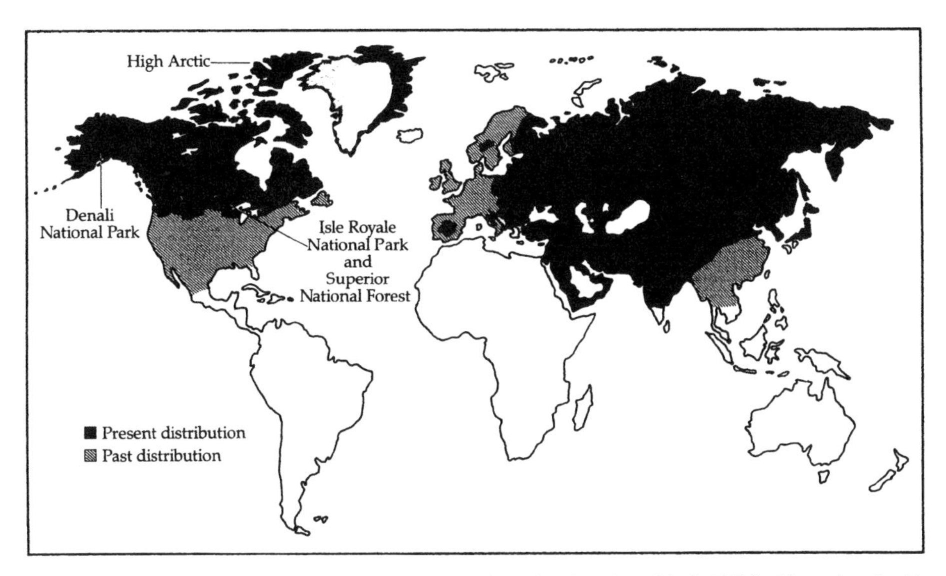

Figure 14.1. The past and present geographic distribution of wolves from Mech (1991). (Reproduced with permission of Voyageur Press Inc. (copyright holder)) With the recent reintroduction of wolf packs into the western U.S., in particular into Yellowstone National Park, the range has been extended below the 49th parallel, which separates Canada and the U.S.

Wolves used to be one of the most widely distributed animals in the northern hemisphere. Figure 14.1 shows the present and past distribution of wolves.

Wolves are social carnivores typically belonging to a pack which is a family unit consisting of 3–15 wolves. The extensive field studies which have been carried out on different packs in northeastern Minnesota, often using radiotracking techniques, have provided information on the land use by wolves from different packs. Pack territories are maintained over several years and are spatially segregated, rarely overlapping in the boundary regions; see Figure 14.2.

Territorial boundaries are usually avoided to lessen the chance of interpack conflict which often leads to death of one or more of the pack leaders (the alpha pair) which, in turn, can result in pack disintegration. In northeastern Minnesota, territories range in size from 100–310 km^2. These boundary regions between neighboring packs, known as 'buffer zones' are rarely visited by pack members. The buffer zones, which are a kind of 'no-mans-land,' are about 2 km wide and can account for as much as 25–40% of the available area.

Some of the most striking spatial patterns evident in wolf territories have been described by Mech (1973) and van Ballenberghe et al. (1975) and these provided the modelling basis for a mechanism as to how wolves and their prey may coexist in relatively close proximity (Lewis and Murray 1993, White 1995, White et al. 1996a,b, 1998). Our goal was to develop a mechanistic, spatially explicit model incorporating wolf movement, scent marking and wolf interactions that produces the spatial patterns evident in a wolf ecosystem specifically in northeastern Minnesota. Understanding pack territory formation and home range patterns is crucial if we are to understand the ecol-

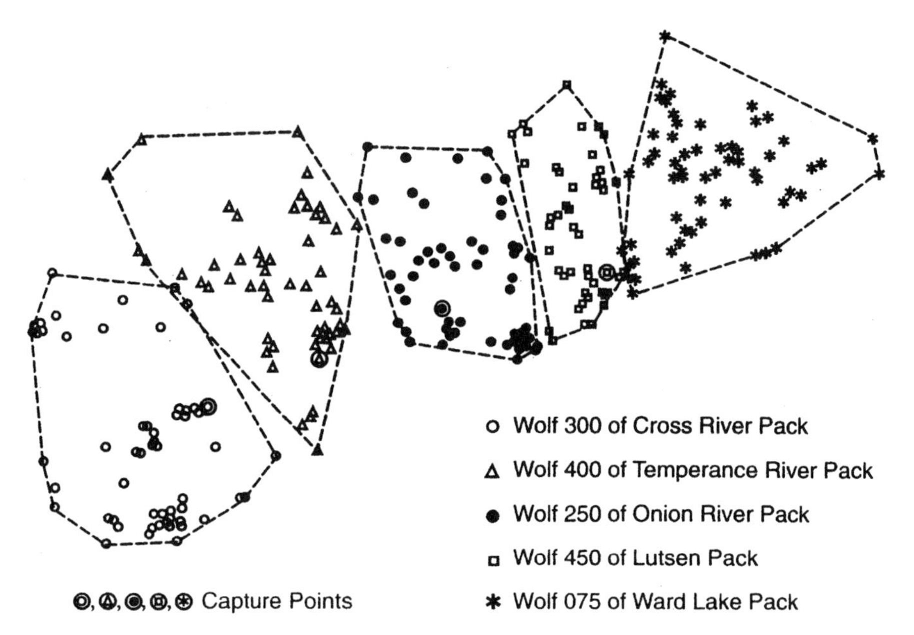

Figure 14.2. Radiolocations and home range boundaries of five adult and yearling timber wolves radiotracked in northeastern Minnesota during the summer of 1971. (Redrawn from van Ballenberghe et al. (1975) and reproduced with permission from the Wildlife Society (copyright holder))

ogy of many mammalian societies. Of course, other aspects are also important such as the social organization, mating and demography (see, for example, Clutton-Brock 1989). The literature is large and diverse; we give only a few references (see the numerous references cited in these). White (1995) gives an extensive review of the literature and modelling studies and the articles by Lewis et al. (1998) and Moorcroft and Lewis (2001) review some of the more recent theoretical studies with the latter presenting some interesting field data on coyotes using the mechanistic models discussed in this chapter.

The biological background and data used in constructing the mechanistic models is largely based on the wide-ranging radiomarking studies of wolves (*Canis lupus*) in northeastern Minnesota over the past 25 years. These have greatly facilitated the observation of wolf territories. The well-known and the most detailed quantitative studies of wolf numbers, however, is from the Isle Royale Project.

The Isle Royale National Park, an island of just over 200 square miles, is in Lake Superior close to the Minnesota–Ontario border. During a particularly cold winter in 1949, when the island was joined to the mainland by ice, some wolves crossed over to the island. They established themselves on the island where their main prey is the moose. Since 1959 the actual wolf and moose numbers, among many other things (such as beaver colonies and otter numbers), have been recorded thereby providing a remarkable data set on the wolf and moose interaction and their survival. Dr. Rolf Peterson (School of Forestry and Wood Products, Michigan Technical University, Houghton, Michigan 49931-1295, U.S.A.) is the director of the project. The report (Peterson 1999) for the 1998 to 1999 year gives an overview and some quantitative details of the ecology of the island. Some of the data on wolf–moose numbers and wolf pack territories are given below. This long term study of more than 40 years is immensely important and has allowed a wide variety of studies to be carried out. For example, other than wolf population and territorial distribution, the effect of inbreeding, disease pathology and so on are of major current interest as are the dynamics of other animal populations and interactions on the island. The data are an excellent source for modelling investigations of population interactions. Figure 14.3 shows the wolf–moose populations, the wolf pack territories and the moose distribution on Isle Royale in 1999.

Wolves can cover around 50 km in a 24-hour period and so could cover a significant portion of their territory in a day. Even so, relative to pack size, the size of territories in northeastern Minnesota really means that physical presence can not provide a sufficient defensive mechanism to protect the territory. Based on many years of field observations, Mech (1991) suggests that wolf territories are formed and maintained by interpack aggression in conjunction with two warning systems: scent marking and howling, and that the result is a mosaic of territories covering the wolves' range. While howling may provide temporary information on a pack's location, elaborate spatial patterns of scent marks serve to advertise precise information about territorial claims even in the absence of any pack members. We shall include scent marking in our models.

As with other carnivores, olfaction (smell) is the primary sense.[2] Wolves use a variety of olfactory signs but behavioral studies indicate that raised leg urination (RLU) is the most important one in territory marking and maintenance. RLU markings occur throughout the territory along wolf trails but, more importantly, they increase significantly around the buffer region giving rise to high concentrations of RLU markings from all packs in this region: Figure 14.4 sketches typical RLU markings around a pack territory. Unlike the other olfactory signs used by wolves, RLU shows little correlation with pack size because they are made by only a few mature dominant wolves in each pack. These are primarily the alpha pair who reproduce and who dominate the other pack members: wolf packs are highly structured socially. Observations also indicate aversion to the scent from RLUs made by neighbouring packs.

White-tailed deer are the main prey for wolves in northeastern Minnesota and their distribution varies seasonally. During the summer months, deer are dispersed on large home ranges but in the winter months they tend to congregate in yards as shown in Figure 14.5. In spite of the relative homogeneity in food and habitat across the study region, the deer in both summer and winter tend to remain in the buffer zones between pack territories. It has been suggested that this deer heterogeneity could be due to differential predation rates caused by the territorial nature of the wolf.

As mentioned, the main motivation for the work described here is from the wolf–deer data from northeastern Minnesota. By following the movements of radiomarked individual members from a pack, or cooperative extended family group, it has been possible to deduce distinct spatial patterns in wolf distribution. Wolves typically remain within well-defined territories (Mech 1973, Van Ballenberghe et al. 1975) that, to some

[2] As noted earlier, the domesticated male silk moth (*Bombyx mori*) uses an optimally designed antenna filter to detect molecules of the sex-attractant chemical, *Bombykol*, emitted by the female. The male silk moth cannot fly so it has to walk up the concentration gradient to find the female: in some experiments it walked upwind as much as a kilometre!

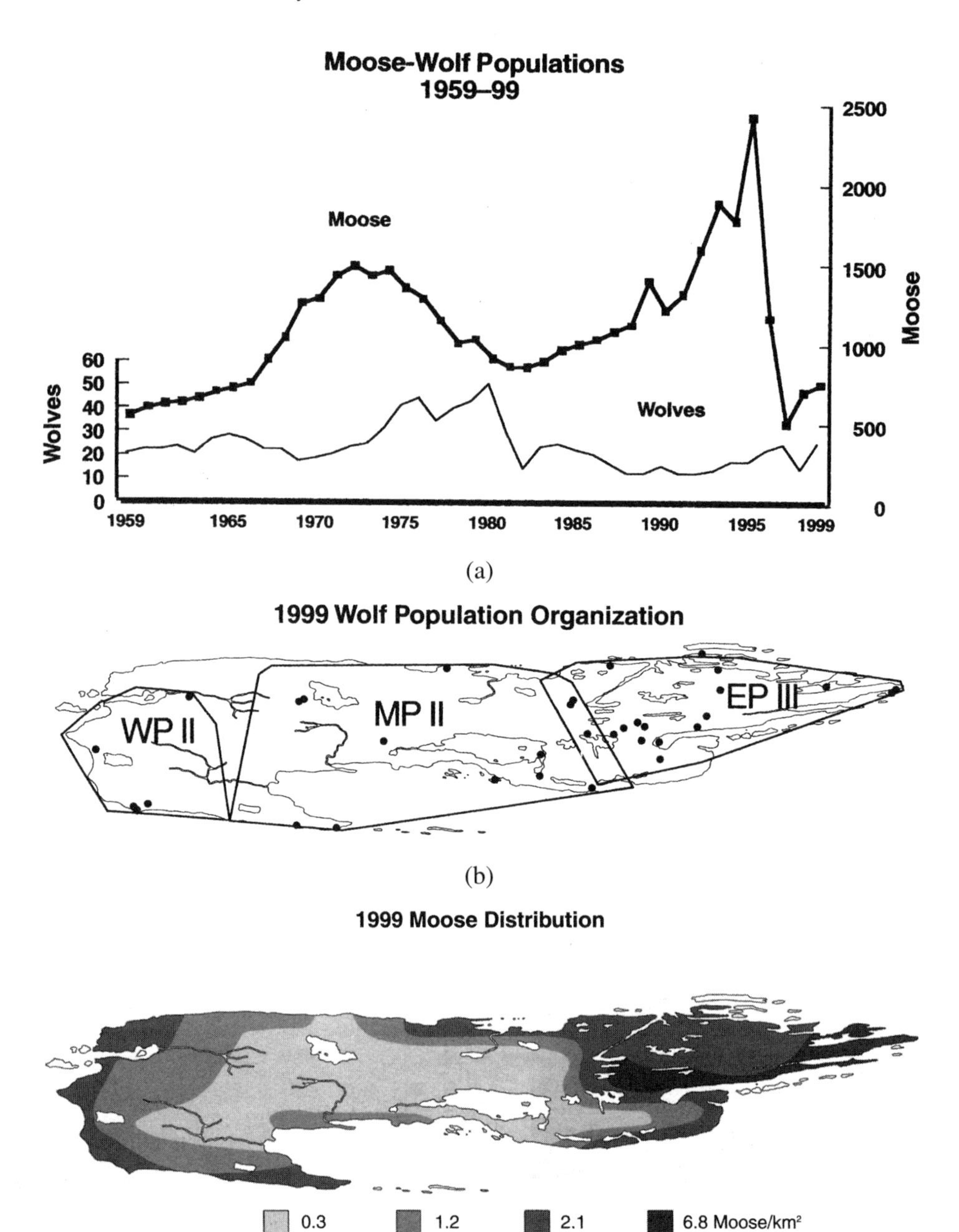

Figure 14.3. Isle Royale National Park: (**a**) Wolf and moose populations since 1959. (**b**) Wolf pack territories in the winter of 1999. In that year the sizes of the packs were: 11 in East Pack III (EP III), 10 in Middle Pack II (MP II) and 2 in West Pack II (WP II). (**c**) Moose distribution in the February 1999 census. (All figures are from Peterson 1999 and reproduced with permission of Dr. Rolf Peterson)

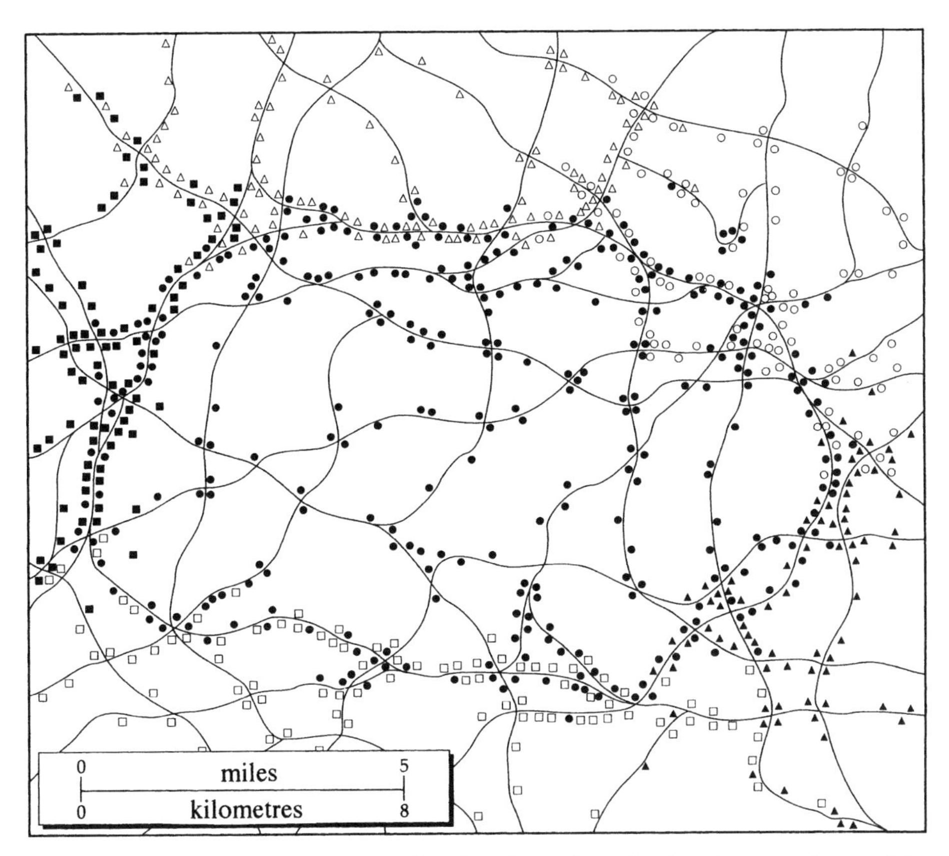

Figure 14.4. Typical distribution of RLU markings in and around a wolf pack territory in northeastern Minnesota. The levels of scent marking both due to the resident pack and its neighbours are greatest around the territory edges. The different shapes (filled squares and circles, open squares, circles and triangles) denote markings from different packs. (From Peters and Mech (1975) and reproduced with permission of Dr. David Mech)

extent, overlap only along their edges as shown in Figures 14.2 and 14.3. These territories effectively partition jurisdiction over spatially distributed prey resources.

The precise details of wolf behavior and ecology depend on local habitat conditions: there are, for example, basic differences between the habitat on Isle Royale and on the mainland. Although we concentrate on northeastern Minnesota, we believe the main results have applicability to other areas and other territorial mammals. This has been shown to be the case in the interesting study by Moorcroft et al. (1999) on coyotes (*Canis latrans*) which we discuss later. Wolf activities occur over various timescales—yearly, seasonally and daily. Thus a key element in modelling these wolf activities is the determination of an appropriate timescale.

Seasonality plays an important role in both wolf and deer ecology as is particularly evident in the reproductive behaviours of both species. Wolves produce young in the spring; pups arrive in April or May and activity centres around the den throughout the summer. Deer produce fawns in the early summer. Throughout the rest of the year,

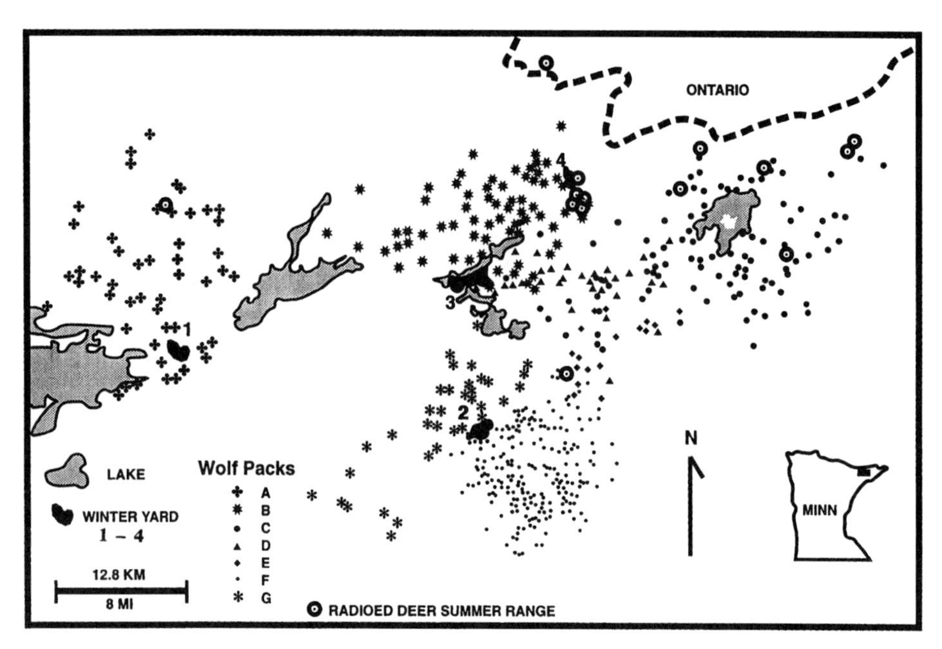

Figure 14.5. Winter yards and summer ranges of radio-collated deer in relation to wolf pack territories. (From Hoskinson and Mech 1976 and reproduced with permission of the Wildlife Society (copyright holder))

any changes in population levels are due to mortality, emigration or immigration. The entire wolf pack helps with feeding the pups (Mech 1970); adults make daily excursions and return with food. In late summer, as the pups become stronger, the den may be abandoned in favor of above-ground rendezvous sites. In the fall and winter, pups are able to move and travel widely with the pack, rarely returning to the den or rendezvous sites. In our modelling of pack territory dynamics we shall not include the yearly birth and death processes (see White 1995), but concentrate on the short term behavioral and movement dynamics.

In formulating the model we make no underlying assumptions about the size and extent of the wolf territories themselves; we show that the territorial patterns actually arise naturally as stable steady state solutions to the model equations. These mathematically generated territorial patterns share key features with field observations including buffer zones between adjacent packs, where wolves are scarce and increased levels of scent marking near territorial boundaries. The material in this chapter develops the model and analyzes it in detail. Among other aspects we show how behavioral responses to foreign scent marks determines the qualitative form of resulting spatial territories.

Very few quantitative models have been derived to explain the spatial dynamics of territories when competition for space is a key factor. As far as we are aware, the model and variations we describe here (Lewis and Murray 1993, White 1995, White et al. 1996a,b) comprise the only spatially explicit formulation designed to show how pack territories form over time based on behavioral interactions. On the other hand, field studies of pack territoriality have been extensive, and include observations of a variety of predatory mammals other than wolves, such as lions, badgers, hyenas and African wild dogs (references to all of these are given in White et al. 1998).

14.2 Models for Wolf Pack Territory Formation: Single Pack—Home Range Model

Despite the complexity of wolf and deer behaviours and ecology, the stability of the pack territories and wolf–deer distribution observed in northeastern Minnesota suggests that there may be basic mechanisms underlying the spatial structure and dynamics of the ecosystem. The background details provided in the last section form the basis for the modelling we now discuss. The principal modelling motivation is whether or not simple behavioral rules can help elucidate the following questions (not all of which we address here).

- **(i)** Can we show how pack territories form, determine their size and explain why they are stable for many years?
- **(ii)** When deer, as prey, are included can we show why they are found mainly in the buffer zones between pack territories?
- **(iii)** With seasonal changes can we explain winter increase in buffer zone trespass, wolf–wolf altercation, wolf starvation and territory change?
- **(iv)** Can we predict wolf dynamics with low winter deer populations?
- **(v)** Can we quantify our predictions of population dynamics, territory and buffer zone sizes and seasonal changes, based on behavioral parameters? How sensitive are these predictions to behavioral changes?
- **(vi)** Do buffer zones stabilize wolf–deer interactions by providing a refuge for the deer, and if so, does a refuge act to dampen population oscillations or prevent extinction or both?
- **(viii)** Do biannual migrations act as a stabilizing factor in wolf–deer interactions?

As mentioned there are seasonal changes in wolf ecology. Since we are primarily interested here in territory formation we consider the formation and maintenance of territories during the summer months and so we do not include yearly birth processes. Consequently, the models focus on wolf movement patterns, which in later sections we couple to deer mortality caused by wolf predation, and aspects of the deer movement.

Due to the small numbers of both species there are potentially significant periods of time during which areas of territory are not occupied by a wolf (or deer). In view of this, it makes sense to use a probabilistic approach in which state variables are taken to be expected densities of wolves at a point $\mathbf{x}$ and time t; direct field observations typically will not yield the exact densities.

In view of the probabilistic approach and the choice of RLU marking as the method of territory delineation, a two wolf pack model could include the following state variables.

$u(\mathbf{x}, t)$ = expected density of wolves from pack number 1

$v(\mathbf{x}, t)$ = expected density of wolves from pack number 2

$p(\mathbf{x}, t)$ = expected density of RLUs from wolf pack number 1

$q(\mathbf{x}, t)$ = expected density of RLUs from wolf pack number 2.

During the summer months, pack members focus their movements around the den but they must necessarily spend time away from the den foraging for food. At the simplest level, we anticipate that wolf movement, independent of responses to other wolf packs, is dominated by (i) dispersal as the wolves search for food and other activites (like RLU marking) and (ii) movement back towards the den as the wolves return to the social organizing centre, the den, to care for the pups. So, a typical word equation for a single pack without RLU and deer input with this scenario is

Rate of change in expected wolf density

= Rate of change due to movement of wolves towards the den

+ Rate of change due to dispersal of wolves away from high density regions in search of food.

The key question is how to model the spatial movements.

Field studies indicate that wolves use cognitive maps and are aware of their relative locations within the territory. Consequently, movement back towards the den site tends to be more or less in a straight line. Mathematically, such movement can be represented by directed motion, or convection, with a flux, $\mathbf{J}_u$ which takes the form, for the u-pack,

$$\mathbf{J}_u|_{\text{convection}} = -c_u(\mathbf{x} - \mathbf{x}_u)u, \tag{14.1}$$

where $\mathbf{x}_u$ denotes the location of the den and $c_u(\mathbf{x} - \mathbf{x}_u)$ is the space-dependent velocity of movement; Okubo (1986) used a similar form in his model for insect dispersal that we discussed in Chapter 11, Volume I in which we used the discontinuous function $c_u(x - x_u) = c_u \operatorname{sgn}(x - x_u)$ where c_u, the speed of movement, is constant. A continuous version of (14.1) which describes slowing and eventual stopping as wolves approach the den site is

$$c_u(\mathbf{x} - \mathbf{x}_u) = c_u \tanh(\beta r)\frac{\mathbf{x} - \mathbf{x}_u}{r}, \tag{14.2}$$

where $r = \|\mathbf{x} - \mathbf{x}_u\|$. The parameter c_u now measures the maximum speed of the wolf when moving towards the den and β measures the change in the rate of convective movement as the den is approached. In the limit as $\beta \to \infty$ (14.2) approaches the discontinuous form. In the presence of foreign RLUs the coefficient describing the speed of movement may be modified to include a response to the foreign RLU marking as described later.

Let us now consider movement due to foraging activity. In the first case we assume a plentiful and homogeneous food supply and in the second, discussed below, the deer density is explicitly incorporated into the model.

In the first case, the simplest assumption is that there is no preferred direction of motion for foraging and so is a random walk process as could occur if the food supply were uniformly distributed throughout the region. An extension to this assumes that movement may be density-dependent. As we now know, mathematically such movement can be represented by a diffusion flux, $\mathbf{J}_u$, which for the wolf pack u is

$$\mathbf{J}_u|_{\text{diffusion}} = -D(u)\nabla u, \tag{14.3}$$

where $D(u) = d_u u^n$, with constant d_u and $n > 0$, is the density-dependent diffusion coefficient. For n positive, density-dependence can be interpreted as an increased rate of movement in regions which are more familiar to the wolf pack.

Consider the simplest scenario of a single isolated wolf pack. Combining movement back to the den to care for the young, (14.1), with movement away from the den to forage, (14.3), the model conservation equation

$$\frac{\partial u}{\partial t} + \nabla \cdot \mathbf{J}_u = 0 \Rightarrow \frac{\partial u}{\partial t} = \nabla \cdot [c_u(\mathbf{x} - \mathbf{x}_u)u + D_u(u)\nabla u]. \tag{14.4}$$

We now have to consider appropriate initial and boundary conditions. Biologically realistic boundary conditions may involve local migration dynamics. However, the simplest possible boundary conditions are when we assume that wolves neither immigrate nor emigrate from the domain of interest denoted by Ω and which has to be determined. That is, we impose zero-flux boundary conditions for u, namely,

$$\mathbf{J}_u.\mathbf{n} = 0 \quad \text{on the boundary} \quad \partial\Omega, \tag{14.5}$$

where $\mathbf{n}$ is the outward unit normal to the boundary, $\partial\Omega$, of the domain. Initial conditions, describing the expected spatial distributions of wolves at the beginning of a study period, is given by

$$u(\mathbf{x}, t) = u_0(\mathbf{x}). \tag{14.6}$$

At any given time, the total number of wolves, Q, in the domain Ω is

$$Q = \int_\Omega u(\mathbf{x}, t)\, d\mathbf{x}. \tag{14.7}$$

Using (14.4) we see that

$$\frac{\partial}{\partial t}\int_\Omega u(\mathbf{x}, t)\, d\mathbf{x} = \int_\Omega \frac{\partial}{\partial t} u(\mathbf{x}, t)\, d\mathbf{x} = -\int_\Omega \nabla \cdot \mathbf{J}_u\, d\mathbf{x} = -\int_{\partial\Omega} \mathbf{J}_u \cdot \mathbf{n}\, ds = 0. \tag{14.8}$$

So, the zero-flux boundary condition (14.5) guarantees a constant number of wolves in the pack within the domain Ω.

We obtain the average density, U_0, of wolves in the pack throughout the region Ω as

$$U_0 = \frac{1}{A}\int_{\Omega} u_0(\mathbf{x})\, d\mathbf{x}, \tag{14.9}$$

where A is the area of the territory Ω. The mathematical problem is now completely defined.

Suppose we consider the time-independent problem. Equation (14.4) becomes

$$0 = \nabla \cdot \left[c_u(\mathbf{x} - \mathbf{x}_u)u + D_u(u)\nabla u\right]. \tag{14.10}$$

By way of illustration let us consider the one-dimensional situation with the zero-flux boundary condition (14.5) and with the continuous convection form (14.2) and obtain the steady state density distribution and territory size as a function of pack size. Let us further take the density-dependent diffusion coefficient to be given by $D_u(u) = d_u u^n$. The last equation then becomes, on integrating with respect to x,

$$c_u u \tanh \beta(x - x_u) + d_u u^n \frac{du}{dx} = \text{constant}. \tag{14.11}$$

Linear Diffusion, $n = 0$

Here $D(u) = d_u$ a constant. Using zero-flux boundary conditions integration immediately gives the steady state solution, $u_s(x)$, in one space dimension as

$$u_s(x) = \frac{B}{\left[\cosh \beta\,(x - x_u)\right]^{c_u/(d_u\beta)}}, \tag{14.12}$$

where B, a constant of integration, is determined by the conservation condition (14.7) with a given number of wolves, Q, in the pack, namely,

$$B\int_{\Omega} \frac{dx}{\left[\cosh \beta\,(x - x_u)\right]^{c_u/(d_u\beta)}} = Q. \tag{14.13}$$

Nonlinear Diffusion, $n > 0$

Here integration of (14.11) gives

$$u_s(x) = \begin{cases} \left[\dfrac{c_u n}{d_u \beta} \ln\left(\dfrac{\cosh \beta x_b}{\cosh \beta(x - x_u)}\right)\right]^{1/n} & |x - x_u| \le x_b \\ 0 & \text{otherwise,} \end{cases} \tag{14.14}$$

where the range radius of the pack, x_b, is given implicitly by

$$\int_{x_u - x_b}^{x_u + x_b} \left[\frac{c_u n}{d_u \beta} \ln\left(\frac{\cosh \beta x_b}{\cosh \beta(x - x_u)}\right)\right]^{1/n} dx = Q, \tag{14.15}$$

where Q is the number of wolves in the pack as defined above by (14.7). (Equation (14.14) is a weak solution of (14.11), in the sense that it satisfies (14.11) at all points except $x = \pm x_b$.) Similar results were obtained in Chapter 11, Volume I for positive choices of the power of the diffusion coefficient n.

The crucial difference between the two solutions ($n = 0$ and $n > 0$) is that when foraging activity is described using regular Fickian diffusion ($n = 0$) no definite territory boundary is formed whereas with the density-dependent diffusion ($n > 0$), territories with finite boundaries are formed (recall the detailed discussion in Chapter 11, Volume I). Figure 14.6(a) shows an example of the time evolution to the steady state wolf distribution for $n > 0$ obtained by numerically solving (14.4).

Figure 14.6(d) is an example of the relationship between pack and territory size given by equations (14.14) and (14.15) for representative values of the parameters. White (1995) estimated parameter values from field data and showed that values of $0.25 \leq n \leq 0.5, 0.006 \leq \beta \leq 0.02, 0.5 \leq d_u \leq 2.08$ were reasonable estimates with the higher diffusion coefficients associated with larger packs and hence larger territories.

The relationship shown in Figure 14.6(d) is reminiscent of ideas of McNab (1963) and Okubo (1980) who suggested that home range size, R, for mammals is related to

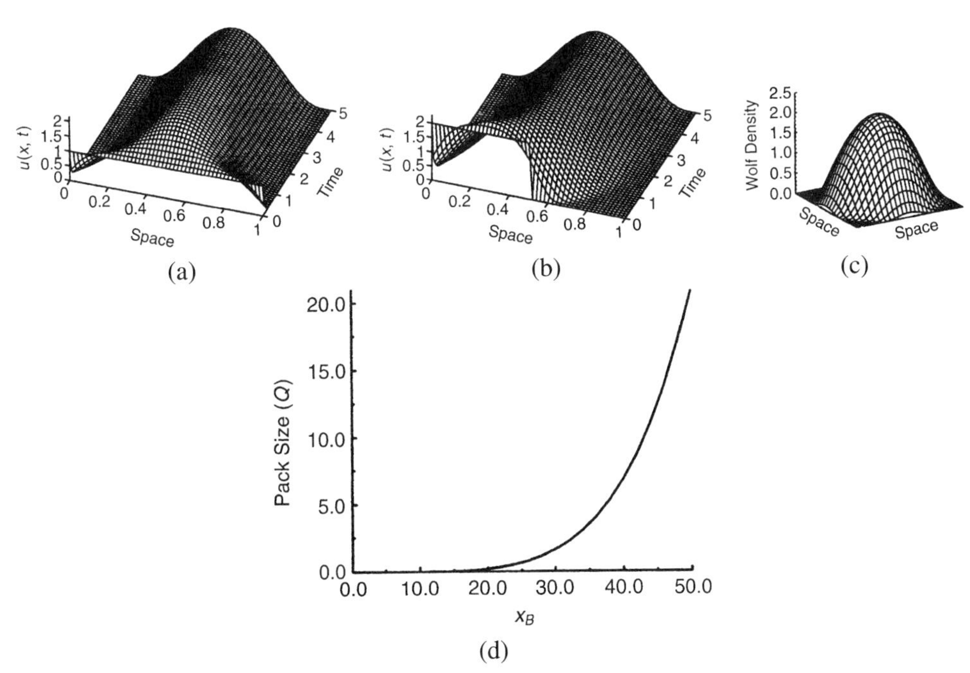

Figure 14.6. Time evolution to the steady state solution for a single pack in one space dimension with a density-dependent diffusion ($n = 0.5$): (**a**) and (**b**) have different initial conditions but the steady state solution is the same. Parameter values: $\beta = 1, c_u = 1, d_u = 0.03$. (**c**) Radially symmetric steady state solution for a single pack. Parameter values: $n = 1, \beta = 1, c_u = 1, d_u = 0.05$. (**d**) The relation between the pack territory ($2x_b$) in terms of wolf density obtained from the implicit relation (14.15) with representative parameter values $n = 0.5, \beta = 0.001, c_u = 1, d_u = 2$. (From White 1995)

the energy intake required per animal and hence the body weight, W, through the power law, $R = aW^b$, where a is a constant and $b \approx 0.75$.

14.3 Multi-Wolf Pack Territorial Model

Response to neighboring packs occurs primarily through RLU marking. The particular nature of the response (in terms of wolf movement) is not well understood and has been investigated in two different ways. In the first, the presence of foreign RLU marks increases the speed of movement back towards the den (central territory) area while increasing the production of familiar RLUs. In the second, wolves respond to gradients in foreign RLU markings by moving away from regions of high density at the same time increasing production of their own scent marks. Although similar behaviours are observed with both these scenarios there are some differences which we come back to below.

Because RLUs are made by a few mature dominant wolves in each pack, the location of these wolves is key in determining the RLU marking patterns. For the purposes of this model we can describe the location of such a dominant wolf by a probability density function denoting the chance of finding the wolf at point $\mathbf{x}$ and time t. For any given pack, we sum these probability density functions over the number of RLU-marking wolves. This provides a measure of the *expected* density of RLU-marking wolves in the pack at a point $\mathbf{x}$ and time t. From now on we refer to this quantity as the expected local density of wolves in a pack.

For a model involving two adjacent, interacting (in effect competing) wolf packs, the relevant state variables are the expected local densities of wolves in pack number 1, $u(\mathbf{x}, t)$; wolves in pack number 2, $v(\mathbf{x}, t)$; RLUs from pack number 1, $p(\mathbf{x}, t)$; and RLUs from pack number 2, $q(\mathbf{x}, t)$.

We must now include equations for the RLU densities which reflect the wolf responses to foreign RLUs from other packs. Based on the above, we assume that when members of a pack encounter RLUs from an adjacent pack, they move away from these foreign RLUs and back towards the den while also increasing their rate of RLU marking. Although mortal strife may occur when adjacent packs interact, for the purpose of modelling the populations we assume that such fatal interactions are very rare and that the number of wolves remains constant over the time period of the model. Remember that we are only considering the summer months.

The word equation for the wolf dynamics (of pack 1) is now

Rate of change in expected density of wolves (pack 1)

= Rate of change due to movement of pack 1 wolves towards their den

+ Rate of change due to dispersal of pack 1 wolves

+ Rate of change due to movement of pack 1 wolves away from the RLUs made by pack 2

the terms of which we must now quantify. Let us first consider movement in response to foreign RLU markings. We consider two ways to model movement induced by RLU levels. In the first, the response is assumed to increase the rate of movement back towards the den site. At the most extreme, this movement is assumed only to occur in the presence of competing RLUs but can be modified to allow movement independent of neighboring packs. In either of these, the convection flux, $\mathbf{J}_{c_u}$, described in the last section, (14.1) is modified to

$$\mathbf{J}_{c_u} = -c_u(\mathbf{x} - \mathbf{x}_u, q)u, \tag{14.16}$$

where, to show the dependence on foreign RLUs, q, we write as $c_u(\mathbf{x} - \mathbf{x}_u, q)$ which is a function of q such that $dc_u/dq \geq 0$ since in the presence of foreign RLUs, the wolves retreat towards the den site. The function $c_u(\mathbf{x} - \mathbf{x}_u, q)$ is typically a bounded monotonically increasing function of q; a function qualitatively like $Aq/(B + q)$ with A and B constants is reasonable.

In the second case, the response to RLUs is to make the wolves move down gradients of foreign RLU density. In this case, the movement is modelled mathematically by a flux, $\mathbf{J}_{a_u}$ given by

$$\mathbf{J}_{a_u} = a_u(q)u\nabla q, \tag{14.17}$$

where $a_u(q)$ is another monotonically non-decreasing function. Gathering these together we now have the conservation equation for the wolves in pack 1 as

$$\frac{\partial u}{\partial t} + \nabla \cdot \left[\mathbf{J}_{c_u} + \mathbf{J}_{d_u} + \mathbf{J}_{a_u}\right] = 0, \tag{14.18}$$

where the fluxes are given by

$$\begin{aligned}
\mathbf{J}_{c_u} &= -uc_u(\mathbf{x} - \mathbf{x}_u, q), & c_u(0) \geq 0, \quad & \frac{dc_u}{dq} \geq 0 \\
\mathbf{J}_{d_u} &= -d_u(u)\nabla u, & d_u(0) \geq 0, \quad & \frac{dd_u}{du} \geq 0 \\
\mathbf{J}_{a_u} &= a_u(q)u\nabla q, & a_u(0) \geq 0, \quad & \frac{da_u}{dq} \geq 0.
\end{aligned}$$

The equation for movement of the second wolf pack mirrors that for wolf pack 1 and is given by

Rate of change in expected density of wolves (pack 2)

= Rate of change due to movement of pack 2 wolves towards their den

+ Rate of change due to dispersal of pack 2 wolves

+ Rate of change due to movement of pack 2 wolves away from the RLUs made by pack 1

and is represented mathematically as

$$\frac{\partial v}{\partial t} + \nabla \cdot \left[\mathbf{J}_{c_v} + \mathbf{J}_{d_v} + \mathbf{J}_{a_v} \right] = 0, \tag{14.19}$$

where

$$\begin{aligned}
\mathbf{J}_{c_v} &= -v c_v(\mathbf{x} - \mathbf{x}_v, p), \quad & c_v(0) \geq 0, \quad & \frac{dc_v}{dp} \geq 0 \\
\mathbf{J}_{d_v} &= -d_v(v) \nabla v, & d_v(0) \geq 0, \quad & \frac{dd_v}{dv} \geq 0 \\
\mathbf{J}_{a_v} &= a_v(p) v \nabla p, & a_v(0) \geq 0, \quad & \frac{da_v}{dp} \geq 0.
\end{aligned}$$

We must now model the changes in the RLU densities p and q. Spatial distribution of RLU marks is a direct consequence of the spatial location of RLU-marking wolves. Field studies indicate that there is some low level of continuous RLU marking throughout the territory (along wolf trails) and that foreign RLU marking induces an increased rate of marking in the vicinity of the alien mark. (There is also an increase in the vicinity of a kill.) In addition, the strength of the RLU decays over time and although a first-order kinetics decay rate is assumed here, this rate will also depend fundamentally on the environmental conditions (such as rainfall, heat, snow cover and so on). Combining these three components gives the governing equation for RLU density distribution for pack 1 as

$$\frac{\partial p}{\partial t} = u[l_p + m_p(q)] - f_p p, \tag{14.20}$$

where l_p and f_p are constants describing low-level RLU marking and first-order decay kinetics respectively. The function $m_p(q)$ is plausibly assumed to be a bounded and monotonically nondecreasing function, again typically like $Aq/(B+q)$ with A and B constants. This means that there cannot be an infinite rate of urine production and a greater level of foreign RLU elicits a stronger response, at least at low levels. A similar equation holds for pack 2, namely,

$$\frac{\partial q}{\partial t} = v[l_q + m_q(p)] - f_q q. \tag{14.21}$$

To complete the mathematical formulation of our two-wolf pack model we require boundary and initial conditions. As in the single pack model biologically realistic boundary conditions may involve local migration dynamics. The simplest possible boundary conditions result, however, when we again assume that wolves neither immigrate to, nor emigrate from, the region Ω. Again we have zero-flux boundary conditions for u and v on the boundary, $\partial\Omega$, that is,

$$\left[\mathbf{J}_{c_u} + \mathbf{J}_{d_u} + \mathbf{J}_{a_u} \right] \cdot \mathbf{n} = 0 \quad \text{on } \partial\Omega \tag{14.22}$$

and

$$\left[\mathbf{J}_{c_v} + \mathbf{J}_{d_v} + \mathbf{J}_{a_v} \right] \cdot \mathbf{n} = 0 \quad \text{on } \partial\Omega, \tag{14.23}$$

where $\mathbf{n}$ is the outward unit normal to the boundary, $\partial\Omega$, of the solution domain. Initial conditions describe the expected spatial distributions of wolves and RLU markings at the beginning of a study period and are given by

$$u(\mathbf{x},0)=u_0(\mathbf{x}),\quad v(\mathbf{x},0)=v_0(\mathbf{x}),\quad p(\mathbf{x},0)=p_0(\mathbf{x}),\quad q(\mathbf{x},0)=q_0(\mathbf{x}). \tag{14.24}$$

We can again show that the zero-flux boundary conditions (14.22) and (14.23) guarantee a constant number of wolves for each pack within the domain Ω. At any given time, the total number of wolves from wolf pack 1 in the domain Ω is

$$\int_\Omega u(\mathbf{x},t)\,d\mathbf{x}.$$

Then, from (14.18) and an application of the divergence theorem, for wolf pack 1 we have

$$\begin{aligned}\frac{\partial}{\partial t}\int_\Omega u(\mathbf{x},t)\,d\mathbf{x}&=\int_\Omega \frac{\partial}{\partial t}u(\mathbf{x},t)\,d\mathbf{x}\\&=-\int_\Omega \nabla\cdot\left[\mathbf{J}_{c_u}+\mathbf{J}_{d_u}+\mathbf{J}_{a_u}\right]d\mathbf{x}=-\int_{\partial\Omega}\left[\mathbf{J}_{c_u}+\mathbf{J}_{d_u}+\mathbf{J}_{a_u}\right]\cdot\mathbf{n}\,ds=0.\end{aligned}$$

An analogous argument holds for pack 2.

The area of Ω is given by

$$A=\int_\Omega d\mathbf{x}.$$

The average density of wolves from pack 1 and pack 2 throughout the region Ω is then given by

$$U_0=\frac{1}{A}\int_\Omega u_0(\mathbf{x})\,d\mathbf{x},\quad V_0=\frac{1}{A}\int_\Omega v_0(\mathbf{x})\,d\mathbf{x}. \tag{14.25}$$

We now nondimensionalise the model system (14.18)–(14.21) and their boundary and initial conditions (14.22)–(14.24). This lets us normalise the wolf density and domain size as well as reduce the number of parameters in the usual way. Defining a length $L=A^{1/m}$, where m is the dimension of the solution domain ($m=1$ or $m=2$), we introduce dimensionless quantities, denoted by an asterisk, by

$$u^*=\frac{u}{U_0},\quad v^*=\frac{v}{V_0},\quad p^*=\frac{pf_p}{U_0l_p},\quad q^*=\frac{qf_p}{V_0l_q},\quad t^*=tf_p,\quad \mathbf{x}^*=\frac{\mathbf{x}}{L}, \tag{14.26}$$

$$c_u^*=\frac{c_u}{Lf_p},\quad c_v^*=\frac{c_v}{Lf_p},\quad d_u^*=\frac{d_u}{L^2f_p},\quad d_v^*=\frac{d_v}{L^2f_p} \tag{14.27}$$

$$a_u^*=\frac{a_uV_0l_q}{L^2f_p^2},\quad a_v^*=\frac{a_vU_0l_p}{L^2f_q^2},\quad m_p^*=\frac{m_p}{l_p},\quad m_q^*=\frac{m_q}{l_q},\quad \phi=\frac{f_q}{f_p}. \tag{14.28}$$

For the nondimensionalised quantities to be well defined, we implicitly assume that wolves from both packs are present originally ($U_0 > 0$, $V_0 > 0$), that the domain Ω has a size greater than zero ($L > 0$), that both wolf packs have a nonzero low level of RLU marking ($l_p > 0, l_q > 0$) and that the RLU intensity decays with time ($f_p > 0$, $f_q > 0$). Dropping the asterisks for notational simplicity, we then have the nondimensionalised system as

$$\frac{\partial u}{\partial t} + \nabla \cdot \left[\mathbf{J}_{c_u} + \mathbf{J}_{d_u} + \mathbf{J}_{a_u}\right] = 0, \tag{14.29}$$

$$\frac{\partial v}{\partial t} + \nabla \cdot \left[\mathbf{J}_{c_v} + \mathbf{J}_{d_v} + \mathbf{J}_{a_v}\right] = 0, \tag{14.30}$$

$$\frac{\partial p}{\partial t} = u[1 + m_p(q)] - p, \tag{14.31}$$

$$\frac{\partial q}{\partial t} = v[1 + m_q(p)] - \phi q, \tag{14.32}$$

where the fluxes are given by

$$\mathbf{J}_{c_u} = -uc_u(\mathbf{x} - \mathbf{x}_u, q), \quad \mathbf{J}_{d_u} = -d_u(u)\nabla u, \quad \mathbf{J}_{a_u} = a_u(q)u\nabla q \tag{14.33}$$

$$\mathbf{J}_{c_v} = -vc_v(\mathbf{x} - \mathbf{x}_v, p), \quad \mathbf{J}_{d_v} = -d_v(v)\nabla v, \quad \mathbf{J}_{a_v} = a_v(p)v\nabla p \tag{14.34}$$

and where the functions $c_u, c_v, d_u, d_v, a_u, a_v$ are all nonnegative functions (or constants) as described above.

The boundary conditions (14.22) and (14.23) are unchanged and an appropriate nondimensionalisation of the initial data is

$$u_0^* = \frac{u_0}{U_0}, \quad v_0^* = \frac{v_0}{V_0}, \quad p_0^* = \frac{p_0 f_p}{U_0 l_p}, \quad q_0^* = \frac{q_0 f_p}{V_0 l_q},$$

which leaves the initial conditions (14.24) also unchanged after omitting the asterisks. Note too that the nondimensionalisation of space has made the dimensionless domain Ω equal to unity. Also, with this nondimensionalisation

$$\int_\Omega u(\mathbf{x}, t)\, d\mathbf{x} = \int_\Omega v(\mathbf{x}, t)\, d\mathbf{x} = 1 \tag{14.35}$$

and so, at any given time, $u(\mathbf{x}, t)$ and $v(\mathbf{x}, t)$ are probability density functions for the location of wolves.

We now have to specify appropriate forms for the interaction functions in the model equations. Lewis et al. (1997) showed that if the increased marking function m is typically as we described above (specifically a concave down function for the scent-marking density) then the time-independent solutions of (14.29)–(14.32) satisfy a system of ordinary differential equations with space as the independent variable. The integral conditions (14.35) are transformed into initial conditions for the ordinary differential equations. The resulting expected wolf density functions decrease monotonically with distance away from the den site. A sufficient condition for the buffer zone (that is, a minimum in the value of $u + v$ between the den sites) is that the movement function, c_u, is also a concave down function of foreign scent-mark density.

For analytical simplicity and demonstration they considered a one-dimensional system with dens at opposite ends of the domain ($x_u = 0$, $x_v = 1$) and the movement response to foreign RLUs omitted. So, steady state solutions of (14.29)–(14.32) satisfy

$$0 = [J_u]_x, \quad J_u = -d_u u_x - c_u(r_u, q)u, \tag{14.36}$$

$$0 = [J_v]_x, \quad J_v = -d_v v_x + c_v(r_v, p)v, \tag{14.37}$$

$$0 = u[1 + m_p(q)] - p, \tag{14.38}$$

$$0 = v[1 + m_q(p)] - \phi q, \tag{14.39}$$

where r_u, r_v are distance measured from the respective dens. Boundary conditions (14.22) are now

$$J_v, J_u = 0 \text{ at } x = 0, 1 \tag{14.40}$$

and conservation conditions (14.35)

$$\int_0^1 u(x)\,dx = \int_0^1 v(x)\,dx = 1. \tag{14.41}$$

Generally, for any fixed values of u and v the assumption on the functional dependence of the m means that p and q can be uniquely determined as functions of u and v.

In summary what they proved is that if $m_p(q)$ and $m_q(p)$ are concave down functions then territories are determined by a system of two ordinary differential equations with the initial values at $x = 0$ specified. They showed that the expected wolf density for each pack is bounded above and below which, in turn, means that the expected scent mark density for each pack is positive and bounded above. One has to be careful with the choice of the form of these functions. The case described previously by Lewis and Murray (1993), where m_p and m_q are linear functions, can result in 'blow-up' for p and q, for certain parameter ranges. This is not surprising biologically since linear m_p and m_q imply that arbitrarily high scent marking rates are possible.

Existence of a Buffer Zone Between the Packs

We now show that a buffer zone, that is, an interior minimum for $u + v$, will arise under fairly general assumptions on the movement response function. For the sake of algebraic simplicity and illustration we consider one space dimension and two identical interacting packs (that is, $d_u = d_v = d$, $\phi = 1$ and so on), again with dens at opposite ends of the domain and we assume no explicit spatial dependence in the movement response function c. We also reasonably assume the same form for the marking functions m. Equations (14.36)–(14.39) are then, on integrating (14.36) and (14.37) and applying the boundary conditions (14.40),

$$\frac{du}{dx} = -\frac{1}{d}c(q)u, \qquad \frac{dv}{dx} = \frac{1}{d}c(p)v,$$
$$p = u[1 + m(q)], \qquad q = v[1 + m(p)],$$

subject to the integral constraints (14.41).

The solution to this system is invariant when $x \to 1 - x$, $u \leftrightarrow v$, $p \leftrightarrow q$ and so is symmetric about the midpoint $x = 1/2$. Thus, at $x = 1/2$ we have

$$u = v,\ p = q, \quad 0 > \frac{du}{dx} = -\frac{dv}{dx}, \quad \frac{dp}{dx} = -\frac{dq}{dx},$$
$$\frac{d(u+v)}{dx} = 0, \quad \frac{dq}{dx} = \frac{1 + m(p)}{1 + vm'(p)}\frac{dv}{dx} > 0,$$

and

$$\begin{aligned}
&(u+v)_{xx} \\
&= \frac{1}{d}\{c(p)v - c(q)u\}_x \\
&= \frac{2}{d}\left\{c'(p)up_x - c(p)u_x\right\} \\
&= \frac{2u^2}{d}\frac{d}{dx}\left\{\frac{c(p)}{p}\frac{p}{u}\right\} \\
&= \frac{2u^2}{d}\frac{d}{dx}\left\{\frac{c(p)}{p}(1 + m(q))\right\} \\
&= \frac{2u^2}{d}\left\{\frac{d}{dp}\left(\frac{c(p)}{p}\right)[1 + m(q)]p_x + \frac{c(p)}{p}m'(q)q_x\right\}.
\end{aligned}$$

A sufficient condition for the right-hand side of the last line to be positive is that $c(p)$ is convex. In this case $x = 1/2$ is a minimum for $u + v$ and this corresponds to a buffer zone for the interacting packs.

Lewis et al. (1997) discuss other analytical aspects of these models, for example, the dependence of territories on behavioral responses such as: (i) no marking response to foreign RLUs, (ii) marking response to foreign RLUs, (iii) switching in movement response to foreign RLUs and (iv) switching in marking response to foreign RLUs. By switching we mean, in the case of movement, for example, that there is essentially no movement back to the den (that is, $c(\mathbf{x} - \mathbf{x_u}, q) = 0$) until the foreign scent mark has reached a critical value, q_c say, after which c jumps up to a final value. In this situation we can take $c(q) = c_\infty H(q - q_c)$. A similar switching response to foreign RLUs can be incoroporated by a comparable marking response function $m(q)$. These forms are alternatives to the constant slope type functions for $c(q)$ and $m(q)$ in Figure 14.7. White (1995), White et al. (1996a,b) and Lewis et al. (1997) also investigate in some

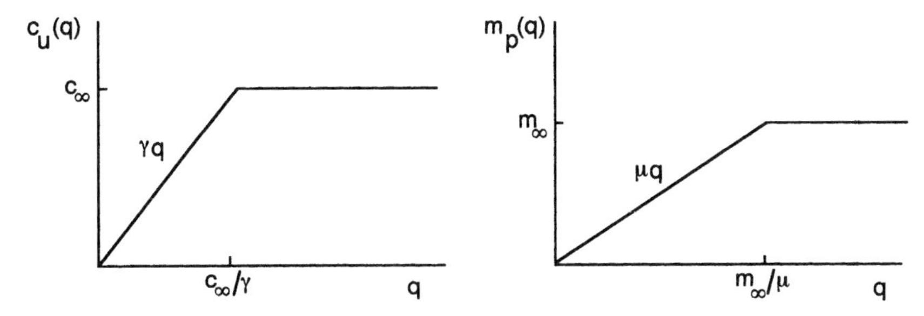

Figure 14.7. Piecewise linear forms for the movement functions $c_u(q)$, with a similar form for $c_v(p)$, and the RLU-marking functions $m_p(q)$, with a similar form for $m_q(p)$.

detail the numerical solutions of these model systems in both one and two dimensions. Some of their numerical results we give in Figure 14.9 below and the following section.

Let us now consider two specific examples, based on these general forms, discussed by White et al. (1998).

Foreign RLUs Influence Movement Back to the Den

Here we consider an encounter with foreign RLU markings causes increased movement back towards the den location in addition to increased RLU production. We also assume simple diffusion as the dispersal effect for foraging for food. With his scenario equations (14.29) to (14.32) become

$$\frac{\partial u}{\partial t} = \nabla \cdot [c_u(\mathbf{x} - \mathbf{x}_u, q)u + d_u \nabla u], \tag{14.42}$$

$$\frac{\partial v}{\partial t} = \nabla \cdot [c_v(\mathbf{x} - \mathbf{x}_v, p)v + d_v \nabla v], \tag{14.43}$$

$$\frac{\partial p}{\partial t} = u[1 + m_p(q)] - p, \tag{14.44}$$

$$\frac{\partial q}{\partial t} = v[1 + m_q(p)] - \phi q. \tag{14.45}$$

We consider a simplified case, where c and m are given by piecewise linear functions as shown in Figure 14.7 and again dens are at opposing ends of a one-dimensional domain ($x_u = 0, x_v = 1$). Steady state solutions to (14.42)–(14.45) with zero-flux boundary conditions and the functional forms in Figure 14.7 are given by

$$p = \frac{u(1 + \mu v)}{1 - \mu^2 uv}, \quad q = \frac{v(1 + \mu u)}{1 - \mu^2 uv} \tag{14.46}$$

and, after integrating (14.42) and (14.43),

$$0 = \gamma \frac{uv(1 + \mu u)}{1 - \mu^2 uv} + du_x, \quad 0 = -\gamma \frac{uv(1 + \mu v)}{1 - \mu^2 uv} + dv_x, \tag{14.47}$$

where μ and γ are the slopes of the functions defined in Figure 14.7. If we write

$$\Gamma(w) = \int_0^w \frac{dw}{1+\mu w} = \frac{1}{\mu}\log(1+\mu w) \tag{14.48}$$

we see from (14.47) that

$$\Gamma(u) + \Gamma(v) = \Gamma(u(0)) + \Gamma(v(0)) = k(u(0), v(0)), \quad \text{a constant.} \tag{14.49}$$

So

$$(1+\mu u)(1+\mu v) = \exp(\mu k) \tag{14.50}$$

gives u in terms of v and vice versa. In the special case $\mu = 0$, (14.47) implies

$$0 = \gamma uv + du_x, \quad 0 = -\gamma uv + dv_x \Rightarrow du_x + \gamma u(K-u) = 0,$$

where $K = [u(0) + v(0)]/d$ is a positive constant. This equation has solution

$$u(x) = \frac{L}{1 + Me^{\gamma Kx/d}},$$

where M and $L = (1+M)u(0)$ are constants; this solution is a monotonically decreasing function of x.

If we now return to the $\mu \neq 0$ case, substituting (14.50) into (14.47) gives a pair of decoupled differential equations for u and v which can then be solved. Differentiating (14.50) gives

$$0 = \frac{v_x}{1+\mu v} + \frac{u_x}{1+\mu u}$$

which can be used to simplify the expressions for p_x and q_x, derived from (14.46), which become

$$p_x = \frac{(1+\mu u)(\mu u - 1)}{(1-\mu uv)^2} v_x, \quad q_x = \frac{(1+\mu v)(\mu v - 1)}{(1-\mu uv)^2} u_x. \tag{14.51}$$

Since $u(x)$ and $v(x)$ are monotonically decreasing functions of distance away from their den sites, interior maxima for $p(x)$ and $q(x)$ are given only when $u(x) = 1/\mu$ and $v(x) = 1/\mu$ respectively. So, there is an interior maximum for p if and only if $u(0) \geq 1/\mu \geq u(1)$ and there is an interior maximum for q if and only if $v(0) \leq 1/\mu \leq v(1)$. In other words, if the behavioral response function m is sufficiently steep then $1/\mu$ is sufficiently large and bowl-shaped scent marking densities arise as illustrated in Figure 14.8.

In the second illustrative model discussed by White et al. (1998) they incorporate a movement response to foreign RLU markings using chemotaxis with respect to foreign RLUs, movement back to the den and foraging movement based on diffusion. The

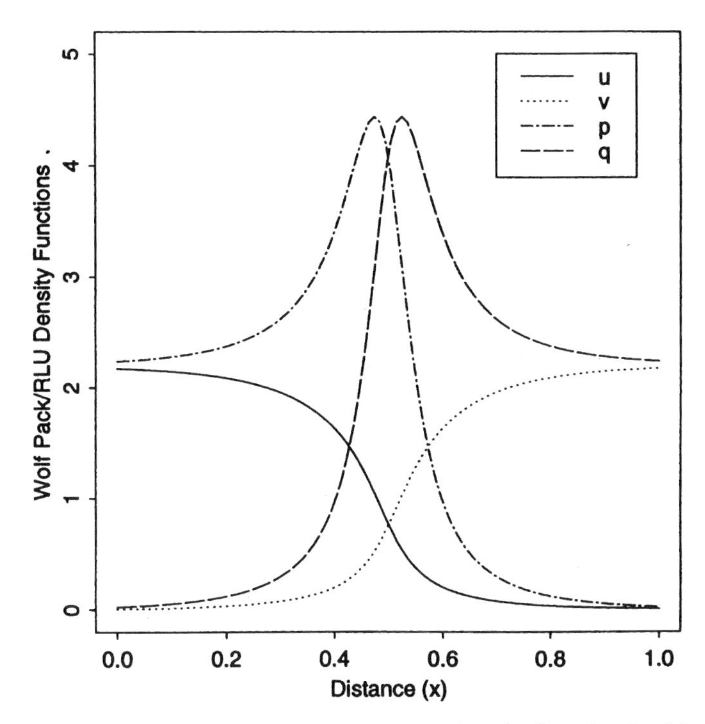

Figure 14.8. Steady state solutions of (14.42)–(14.45) with the piecewise linear functional forms shown in Figure 14.7. The solutions correspond to the analytical solutions of (14.46) and (14.47) for the wolf and the RLU densities. Note the increased scent marking near the pack boundaries. From the last equation the intersection of the line $u = 1/\mu$ with $u(x)$ gives the location of the maximum value for p, the RLU of the u-pack. Similarly the intersection of $v = 1/\mu$ with $v(x)$ gives the maximum for q, the RLU density of the v-pack. Model parameters used were $d_u = d_v = 0.333$, $\mu = 1.1$ and $\gamma = 1$ where γ is the slope of the $c(\cdot)$-functions in Figure 14.7. (From Lewis et al. 1997)

model they examined is given from (14.29)–(14.34) by

$$\begin{aligned}
\frac{\partial u}{\partial t} &= \nabla \cdot [c_u(\mathbf{x} - \mathbf{x}_u)u + D_u(u)\nabla u - a_u(q)u\nabla q],\\
\frac{\partial v}{\partial t} &= \nabla \cdot [c_v(\mathbf{x} - \mathbf{x}_v)v + D_v(v)\nabla v - a_v(p)v\nabla p],\\
\frac{\partial p}{\partial t} &= u[l_p + m_p(q)] - f_p p,\\
\frac{\partial q}{\partial t} &= v[l_q + m_q(p)] - f_q q.
\end{aligned} \tag{14.52}$$

We use the same nondimensionalisation given in (14.26)–(14.28) and get

$$\frac{\partial u}{\partial t} = \nabla \cdot [c_u(\mathbf{x} - \mathbf{x}_u, q)u + d_u\nabla u - a_u(q)u\nabla q], \tag{14.53}$$

$$\frac{\partial v}{\partial t} = \nabla \cdot [c_v(\mathbf{x} - \mathbf{x}_v, p)v + d_v \nabla v - a_v(p) v \nabla p] \tag{14.54}$$

$$\frac{\partial p}{\partial t} = u[1 + m_p(q)] - p, \tag{14.55}$$

$$\frac{\partial q}{\partial t} = v[1 + m_q(p)] - \phi q. \tag{14.56}$$

Once again analysis of the time-independent system produces a set of ordinary differential equations and generates certain criteria for the evolution of a buffer region between the two packs. Further investigation indicates that the above form of movement response to RLU markings shifts the location of the maximum expected wolf density from the den location to a position farther from the neighbouring pack.

The model also illustrates how pack splitting may occur as a response to foreign RLU markings so as to avoid overlap with neighbouring territories. This can be illustrated in a simplified case where only one pack, pack 2 say, responds to foreign RLU marking through movement avoidance, and neither pack increases its RLU marking in the presence of foreign RLU marks. If we take

$$a_u(q) = 0,\ a_v(p) = \chi_v,\ m_p(q) = 0,\ m_q(p) = 0,\ d_u(u) = d_u,\ d_v(v) = d_v \tag{14.57}$$

we get, for the one-dimensional time-independent system (from (14.55) and (14.56))

$$p(x) = u(x), \quad q(x) = \frac{v(x)}{\phi},$$

and, using (14.2), from (14.53) and (14.54),

$$\begin{aligned} 0 &= c_u u \tanh \beta(x - x_u) + d_u u_x \\ 0 &= c_v v \tanh \beta(x - x_v) + d_v v_x + \chi_v v u_x \end{aligned} \tag{14.58}$$

the solutions of which are

$$u(x) = \frac{A}{[\cosh \beta(x - x_u)]^{c_u/(\beta d_u)}}, \tag{14.59}$$

$$v(x) = e^{-\psi[\cosh \beta(x - x_u)]^{-\gamma_u}} \frac{C}{[\cosh \beta(x - x_v)]^{\gamma_v}}, \tag{14.60}$$

with

$$\psi = \frac{A\chi_v}{d_v}, \quad \gamma_u = \frac{c_u}{d_u \beta}, \quad \gamma_v = \frac{c_v}{d_v \beta}. \tag{14.61}$$

Conservation of wolf pack size (14.35) then gives

$$\int_{\Omega} \frac{A}{[\cosh \beta(x - x_u)]^{c_u/(\beta d_u)}}\, dx = 1,$$

$$\int_{\Omega} e^{-\psi \cosh^{-\gamma_u} \beta(x-x_u)} \frac{C}{\cosh^{\gamma_v} \beta(x - x_v)}\, dx = 1. \tag{14.62}$$

The function $v(x)$ can take one of two forms as shown in Figures 14.9(a) and (b) with either one or two maxima; in both cases there is a maximum value for some $x > x_u$ (assuming that $x_u < x_v$). Note that the distribution for pack 1 remains symmetric about the den location. A single maximum for pack 2 is ensured if

$$A = u(x_u) < \frac{c_v d_u}{c_u \chi_v}$$

which suggests that there is a critical relative strength of adhesion between packs beyond which packs which have the greater response to foreign RLU marking could be forced to split their territories.

Figures 14.9(c) and (d) show the steady state solution of (14.53)–(14.56) for three equal packs together with the cumulative RLU density (from all three packs) which clearly highlights the area of high RLU markings along the pack boundaries.

14.4 Wolf–Deer Predator–Prey Model

We must now include the deer as a dynamic variable. With the explicit inclusion of a deer population, we can be more specific about wolf foraging which we modelled earlier by random diffusion. Here we represent movement associated with foraging by a response of the wolves directly to the deer density. In the simplest form, the prey-taxis describes a local response of the wolves to a 'deer gradient.' In other words, wolves move towards regions of higher deer density (which assumes that there is a higher probability of a successful hunt when the deer population is more dense). This is clearly a gross simplification but it provides an initial framework from which more realistic responses to the deer can be formulated. Mathematically, this form of taxis is expressed as a flux; for example, for wolves in pack 1 (u)

$$J_{\text{deer}} = \sigma_u u \nabla h, \tag{14.63}$$

where h is the expected density of the deer and σ_u is a parameter quantifying the strength of the taxis.

The model equation governing expected deer density is somewhat simpler. Given that there is no evidence of active avoidance of wolf populations (except on the scale of escaping attack), we assume that deer do not have any largescale movements once within their summer ranges. Their density distribution is therefore dominated by wolf predation levels and so the deer population can be modelled by

$$\frac{\partial h}{\partial t} = -(\alpha_u u + \alpha_v v) g(h), \tag{14.64}$$

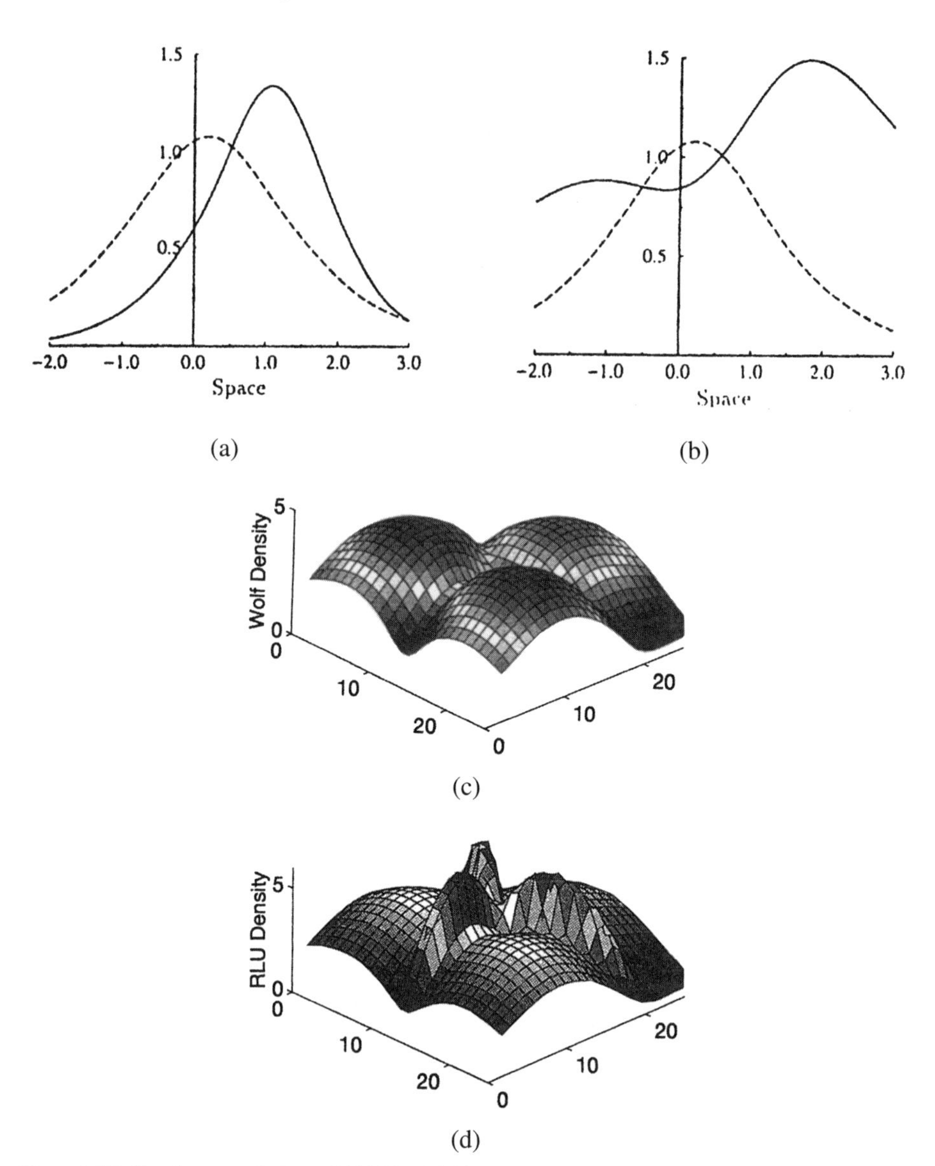

Figure 14.9. Steady state solutions of (14.53)–(14.56) where pack 2 (with den at location $x_v = 0.8$) only responds to foreign RLU marking through movement. The distribution of pack 1 is therefore unaffected by the presence of pack 2 and is symmetric about the den location at $x_u = 0.2$. In (**a**) the relative strength of adhesion of pack 1 is not strong enough to split pack 2 and hence there is a single maximum density for pack 2 located near the den for pack 2. In (**b**), however, the density distribution for pack 2 admits two maxima corresponding to the idea of pack splitting. Model parameters: $c_u = d_u = d_v = \chi_v = \beta = 1$ and in (**a**) $c_v = 2$, in (**b**) $c_v = 0.5$. The dashed line shows $u(x)$ and the solid line, $v(x)$. (**c**) and (**d**) Three-dimensional graphs of the steady state wolf and RLU densities for three identical packs obtained from a numerical simulation of the three-pack version of (14.53)–(14.56) with $c_u = 0.5, \beta = 0.5, n = 0, \alpha_u(q) = 0.25q/(1+q), \phi = 1, m(q) = 2q/(5+q)$. Note the high levels of RLUs in the pack boundary regions. (From White et al. 1996a, 1998)

where α_u and α_v are constants and $g(h)$ is some typical nonlinear saturating function such as $g = ah/(1 + bh)$ or $g = ah^m/(1 + bh^m), m > 1$ with a and b positive constants. We could add a natural mortality term (like $-kh$, for example) but natural mortality during the summer months is dwarfed by the wolf predation (mainly of the fawns).

We can now write down a basic model for wolf–deer interactions and their role in territoriality. We combine elements of the above wolf–wolf interaction models with the last two equations involving the deer. In the case of two wolf packs we obtain (in dimensional form)

$$\begin{aligned}
\frac{\partial u}{\partial t} &= \nabla \cdot [c_u(x - x_u)u - \sigma_u u \nabla h], \\
\frac{\partial v}{\partial t} &= \nabla \cdot [c_v(x - x_v)v - \sigma_v v \nabla h], \\
\frac{\partial p}{\partial t} &= u[l_p + m_p(q, h)] - f_p p, \\
\frac{\partial q}{\partial t} &= v[l_q + m_q(p, h)] - f_q q, \\
\frac{\partial h}{\partial t} &= -(\alpha_u u + \alpha_v v)g(h).
\end{aligned} \tag{14.65}$$

This model differs further from the general forms given above in that we include a response to deer density in the production of RLU markings. This comes from the field observations which suggest that there is an increase in RLU marking at kill sites (Peters and Mech 1975, Schmidt, personal communication 1994).

As with the deer, we could reasonably add other terms to the equations, for example, terms representing wolf death due to starvation and interpack conflict. In the u-equation, for example, these could be of the form $-\alpha_u u f_u(h)$ (f_u is a positive decreasing function of h) and $-k_u uv$ respectively added to the right-hand side. The models we have been considering in this chapter, however, have been for the summer period when starvation and interpack conflict are rare so it is reasonable to set these terms to zero in the analysis. The simulations presented below show the time evolution of the solutions for only a limited period. From the last equation in (14.65) it is clear that the deer population decreases with time and eventually would simply die out. For example, the simulation in Figure 14.10 is equivalent to a 24-week period. A fuller model which includes seasonal deer reproduction is discussed by White et al. (1996a); in this context they also discuss the question of deer extinction.

Parameter Estimation

The parameters play a crucial role in the wolf–deer interaction with respect to both territoriality and survival. Estimation—even a rough estimation—of some of the parameters is difficult since it involves a knowledge of behavioral response and the social organisation of the animals. White (1995) obtained some estimates from the extant literature.

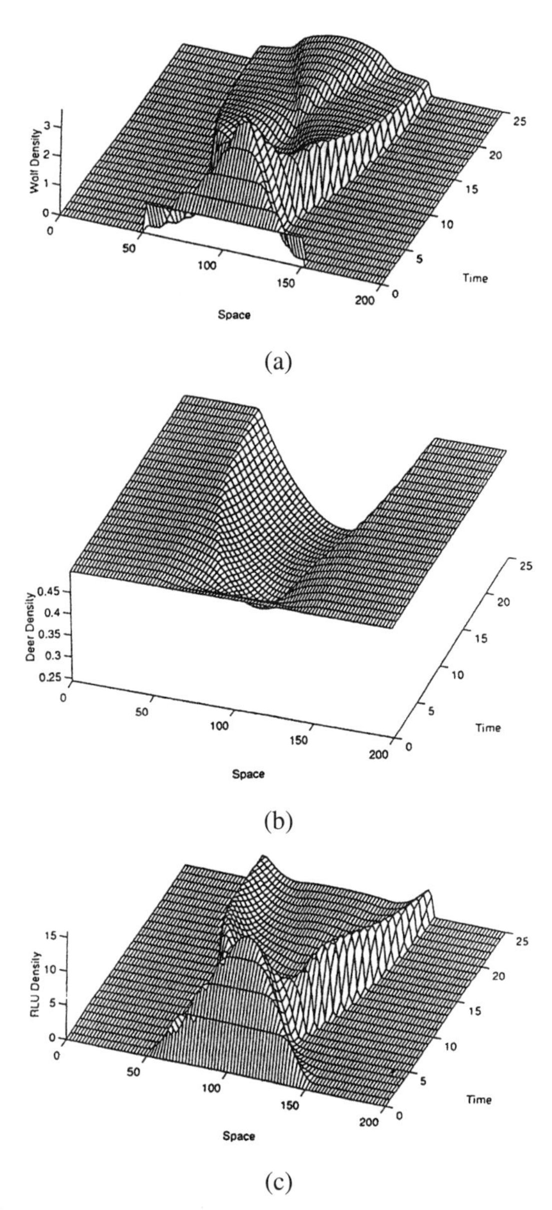

Figure 14.10. Solutions to (14.65) (with $v = q \equiv 0$) for a single wolf pack in one space dimension with $c_u = 3.5, \beta = 0.001, \sigma_u = 0.2, \alpha_u = 0.02$ and $m_p(h) = 10h^{10}/(0.45^{10} + h^{10})$. In (**a**) we show the time-dependent wolf density distribution, in (**b**) the time-dependent deer distribution and in (**c**) the time-dependent RLU density distribution. The simulation ran for the equivalent of 24 weeks and gave a mortality rate of approximately 18.5% over this period. (From White et al. 1996a)

We have reasonable estimates of pack size, approximately 3–15 wolves, which give a range of values for Q in (14.7) and territories, from 100–300 km^2, which give the area $A = (100 - 300)n$ where n is the number of wolf packs.

From field data (Peters and Mech 1975) the fresher an RLU the more likely it is to elicit further RLUs. This suggests that RLUs are typically detectable for about a week and so an estimate for the decay rate parameter is $f_p \approx (1/7)/\text{day}$. Wolves often travel at speeds of 5–8 km/hour. Since the timescale we consider is in days we get an estimate for $c_u \approx$ 5–30 km/day: it is unlikely that a wolf will travel farther than the territory diameter in a day. We can also reasonably suppose that, except for the actual kill, the speed of the movement back to the den is not very different to the speed of movement in search of deer. This implies that we can take the prey-taxis parameter, σ_u, to be small since prey-taxis is probably not large on the scale of km^2/day.

To estimate the deer mortality rate, α_u, suppose we take $g(h) = h$ and $v = 0$ in (14.64) in which case

$$h(\mathbf{x}, t) = h(\mathbf{x}, 0) \exp[-\alpha_u \int_0^t u(\mathbf{x}, s)\, ds].$$

Summer deer survival is relatively high but fawn survival is low. Using the data of Nelson and Mech (1991) an overall mortality rate in the summer months of 30% is a reasonable estimate. Taken on a daily basis this gives a mortality rate of about 0.002%. If we assume a constant pack size this gives $\alpha_u = O(10^{-2})$. This, of course, is an estimate for the case of a single pack predating the deer. It has to be scaled to account for the number of packs considered.

We should reiterate that these estimates are only rough guides to the size of the parameters. They are used in some of the numerical simulations below. Parameter estimates, from detailed field studies of territoriality among coyotes, have been obtained by Moorcroft et al. (1999); we briefly discuss their work below.

White (1995), White et al. (1996a,b, 1998) and Lewis et al. (1997) carried out extensive numerical simulations of the various model equation systems discussed above. An example of such a simulation for a single wolf pack with deer prey and RLU marking, that is, the model obtained from the first, third and fifth equations of (14.65), is shown in Figure 14.10.

To get a clear picture of how territories evolve and are delineated as well as the essential features of wolf–deer interaction and their respective survival it is necessary to consider at least three wolf packs and the deer population. This means that we require 7 coupled partial differential equations in time with two space dimensions: there are three for the expected wolf densities, three for their associated RLU densities and one for the deer population. Figure 14.11 shows one such simulation with three (identical) wolf packs and clearly shows the spatial distribution of territories together with where the deer are primarily found. They mainly stay in the buffer zones between the mutually antagonistic wolf packs.

What is clearly suggested from Figure 14.10 is that food resources play an important role in forming and maintaining territorial structure and add strong support to the explanation for survival of both the wolves and the deer. Extensive numerical simulations of this wolf–deer system were carried out by White (1995), White et al. (1996a,b)

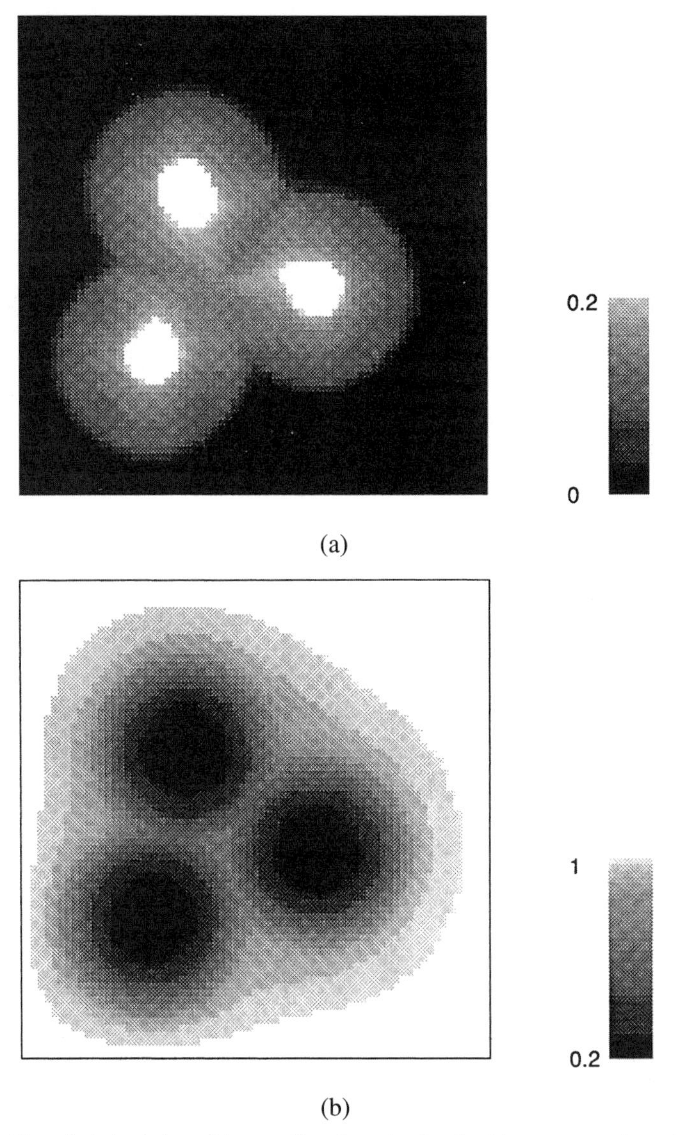

Figure 14.11. Numerical simulation of three wolf packs and a deer herd in two space dimensions showing the cumulative densities of wolf, in (**a**), and deer, in (**b**). The equations are the three-pack equivalent of (14.65) together with the form (14.2) for the movement back to the den. Note how the deer primarily congregate in the buffer zones which are regions of high cumulative RLUs from all packs. Parameter values are identical for each pack: $c_u = 0.2$, $\beta = 0.5$, $\alpha_u = 0.25$, $\sigma_u = 0.1$, $m(p, h) = 10h^{10}/(0.45^{10} + h^{10})$. Dens are located at (19.5, 19.5), (24.5, 44.5) and (44.5, 29.5) on a 70×70 grid. (From White et al. 1996a)

and Lewis et al. (1997) and indicate the following interesting features of wolf–deer systems.

(i) Initially the wolves tend to congregate in the den region before spreading farther as illustrated in Figure 14.10. This occurs because the initial density distribution of the deer is uniform and in this case the equation governing the wolf movement is dominated by the convective term back towards the den. If this did not set up a gradient of deer, the wolves would aggregate at the den location. In this sense the interaction between the wolves and deer provides a mechanism to produce the wolf pack territory.

(ii) The RLU density is greatest around the edge of the territory. This occurs because of the greater density of deer in this region and consequently a greater chance of deer kill, locations of which are often scent marked. Although it is unlikely that kill sites are the reason that greater levels of RLU marking are found around the territory edge, it may play some role in this spatial distribution.

(iii) The symmetry in density distributions which is observed for the single pack model is destroyed when several packs interact as shown in Figure 14.11 for three neighboring packs. The interesting outcome in this case is the occurrence of a buffer region between the pack territories where deer density is greatest. In fact, the interaction between the packs and deer is sufficient to produce this pattern without the presence of RLU marking. This occurs in such a model because wolves move up local deer gradients and when two neighbouring packs approach a deer population from different sides this sets up such a gradient with its peak between the packs. Ecologically, there is no reason why a wolf from one pack would then move across this prey gradient because deer density would be lower on the other side and moreover there would be greater risk of interpack conflict.

14.5 Concluding Remarks on Wolf Territoriality and Deer Survival

The mechanistic models we have discussed up to now have primarily been motivated by the well-documented wolf–deer interactions and wolf territoriality in northeastern Minnesota. All of the various models were based on simple behavioral rules for the animals and the solutions were compared with field observations regarding territory formation, shape, size and maintenance. Further analysis on these models and other variants are given in the references listed throughout the chapter. There are still many different aspects that warrant further study.

One of the major aspects of the explicit spatiotemporal nature of the partial differential equation models is that territories form naturally without prescribed boundaries. When an isolated pack moves both towards and away from a den site the simplest model (14.4) discussed in Section 14.2 predicts the size of the home range as a function of pack size (Figure 14.6). This result is of potential interest in the process of wolf reintroduction currently being considered or underway in many parts of North America such as in Yellowstone National Park. There is field evidence which suggests that strong pack adhesion still occurs for all isolated packs. The single pack home range model mimics this observation and the adhesion which occurs may be explained by optimal pack sizes which are both large enough to hunt large prey and still provide the necessary social interactions. With this simple model, of course, the territories formed are symmetric about the den site.

We also discussed several multi-pack models which differ in the nature of wolf responses to RLU marking. In all cases, interpack interactions break the territorial symmetry observed in the single pack system. Perhaps the most important aspect in these models is their capacity to produce the buffer regions between pack territories. Detailed analysis (some of it in the papers cited by White et al. 1996a,b and Lewis et al. 1997) shows that the presence or absence of this zone depends upon the shape and steepness of the scent mark response function. Also, numerical simulations in two space dimensions indicate that switching in both the movement and scent marking response functions are necessary to produce realistic territories with buffer zones in which higher densities of scent mark are present. These results suggest that field experiments might usefully be carried out to investigate responses of wolves to different RLU levels both in RLU production and aversion to foreign marks. If these responses are indeed important in territory formation, switching behavior should be observed in the field in both cases with wolves already familiar with foreign marks.

The analysis of the prey-taxis model discussed in Section 14.3 shows how the models can be used to investigate differences between packs in their responses to RLU marking. In the example there, it appears that a pack can be divided if it responds to foreign RLU marking at a significantly higher level than a strongly adhesive neighboring pack (one where there is a high probability of being found at the den site). Further theoretical study on this is presented by White (1995).

Although scent marking plays an important role in territory maintenance, howling, as we have mentioned before, is also important as a mechanism for territory defense. Future investigation concerning the effects of this short-lived, long-distance signal would clearly add to our understanding of territoriality and could be incorporated in more sophisticated models.

Our analysis of wolf–deer interactions suggests that much of the territorial structure observed in northeastern Minnesota can be explained by them. The movement of wolves towards regions of higher deer density results in spatial segregation of competing predators (neighboring packs) and their prey (deer) by setting up prey gradients between the packs. Moreover, the increase in RLU marking around the buffer region may be due, in part, to the increased deer density and hence wolf kill (which induces some level of RLU marking). In keeping with the philosophy in this book we have tried to keep the initial models to well-documented behavioral features with a view to gaining some understanding of the possible processes involved. Although these models involve some fairly basic assumptions they still lack much sophistication. Nevertheless they do pose highly relevant questions to the field ecologist regarding the interaction of predators and prey when the predators are territorial. More sophisticated models can be constructed once we have some idea of what is required and what needs futher ecological study.

Of course, as we have seen in this book, nonlinear partial differential equations have been used in a variety of ecological contexts. However, we feel that the modelling described in this chapter presents a new approach to describing and understanding the behavioral aspects of territoriality. The choice of model components was influenced by other ecological studies rather than from a derivation based on individual movements.

Some of the current work along these lines with a more quantitative bent is described in the following section. The models presented here, although involving some fairly general assumptions, suggest that the apparently complex nature of wolf territory formation and maintenance and wolf–deer survival can be explained by the application of a few relatively simple behavioral rules. The work described in the next section on coyotes, based on the above modelling, lends strong support to this view. A new ingredient has been introduced in an interesting article by Lewis and Moorcroft (2001) who introduce aspects of game theory into the above mechanistic theory for home range models in wolves. They estimate relevant parameters and show that appropriate choices of the parameters result in territories that guard against invading groups with alternative behaviours.

14.6 Coyote Home Range Patterns

Although the qualitative features of land distribution between wolves and deer as predicted by the model analyses is in broad agreement with field observations, a more quantitative practical application of the above modelling has been carried out in the seminal paper by Moorcroft et al. (1999; see also Moorcroft 1997) who studied the home range patterns of the coyote (*Canis latrans*). Their work also provides further evidence that the distribution of the land plays a significant role in the spatial distribution of both the predator animals and their prey. It is the first application of the theory to empirical home range models in which the parameters can be estimated from the field studies and tends to confirm the general mechanistic approach we have described above.

Moorcroft et al. (1999) point out several advantages of such combination (theory and empirical) studies. One advantage is that model fits can be used to evaluate various hypotheses made with regard to the spatial distribution of the land resources and dynamics of the species studied. Another is to be able to predict the effects of external perturbations on the animal societies and resource use.

Moorcroft et al. (1999) used the above models, specifically the one proposed by Lewis and Murray (1993), to characterize the home ranges of coyote in the Hanford Arid Lands Ecological Reserve in Washington State, U.S.A. Basically what Moorcroft et al. (1999) did was to show what the key model ingredients which influenced the coyote movement were, namely, that encountering foreign RLUs had the effect of making them move back towards the den and that the effect of these foreign RLUs was to make them increase their own RLU production. A detailed analytical and ecological study specifically associated with the coyotes is given by Moorcroft (1997).

Moorcroft et al. (1999) carried out two separate analyses. In the first, using radio-tracking, they followed individual coyote movements of a single pack and fitted the data to the Lewis and Murray (1993) model for its home range. In the second study they again used the Lewis and Murray (1993) model to study the spatial patterns of six contiguous packs. They used the fit of the single pack solutions to predict the expected distribution of scent markings throughout the territory, the spatial patterns of individual movement and the effect of removing the pack from its home range. They showed that the model captures the observed spatial pattern of home ranges including the location and boundaries between adjacent packs. As we saw in Figure 14.9(d) in the case of three packs we expect the highest concentration of scent markings to lie in the central region. This was also observed by Moorcroft et al. (1999).

The modelling framework discussed in this chapter and the analytical results derived for home ranges, RLU marking and spatial distribution of the land between the carnivore predators and their prey as they apply to wolf–deer systems gives reasonable qualitative results when compared with observations. The importance of the work of Moorcroft et al. (1999) is that they show that 'a mechanistic framework for home range analysis provides a method for directly integrating theoretical and empirical studies of animal home range patterns. Formulating and applying models, in which predicted patterns of space use are formally scaled from an individual-level description of movement and interaction behavior, in contrast to earlier descriptive approaches, provides a methodology for directly testing hypotheses regarding the factors governing home range patterns. This in conjunction with an ability to make predictions for individual behaviour and changes in home range patterns following perturbation, allows for the development of a quantitative, reductionist understanding of animal home range patterns.' A discussion of this work and related mechanistic models for territories is given in Moorcroft and Lewis (2001).

14.7 Chippewa and Sioux Intertribal Conflict c1750–1850

There is a well-documented human application of the general mechanistic theory we have proposed in this chapter which tends to justify intertribal warfare as a traditional means of survival. Morgan (1887) suggested that buffer zones or disputed areas between tribes was a universal feature of tribal societies.[3] These buffer zones between accepted tribal territories were not generally occupied by members of either tribe and tended to be entered only by hunting groups of considerable strength (15 to 20 men) since the risk of intertribal conflict was high. The interesting article by Hickerson (1965) discussed the situation specifically as it applied to intertribal buffer zones in the upper Mississippi valley in the second half of the 18th century and the first half of the 19th century with the Virginia deer as the game. It is from his work that the following has been extracted.

In the case of the traditional enemies the Chippewa and Sioux in Wisconsin and Minnesota there was an extensive wooded buffer zone which gave refuge to the animal prey, in particular the Virginia deer. Figure 14.12 shows the approximate intertribal boundaries and buffer zone between the Chippewa and Sioux villages: the buffer zone was generally wider than 20 miles. It was only during the rare times of truce that hunters could enter the buffer zone to hunt and trap. As pointed out by Hickerson (1965) even during times of economic and ecological stasis before the reservation period these zones were probably not occupied by either group for more than a few days. The buffer zone between the Chippewa and Sioux was an area of abundant game.

As noted by Hickerson (1965) this buffer zone was very stable and was in existence from about 1750 to 1850 which is roughly from the time of the Chippewa settlement to the time of the reservations. Hickerson (1965) suggests that it was the deer that deter-

[3] As a small boy growing up in rural Scotland a remnant of mediaeval territory marking and boundary maintenance occurred every year with the formal 'Riding of the Marches' in which a group of local horsemen rode around the official boundaries of the small town.

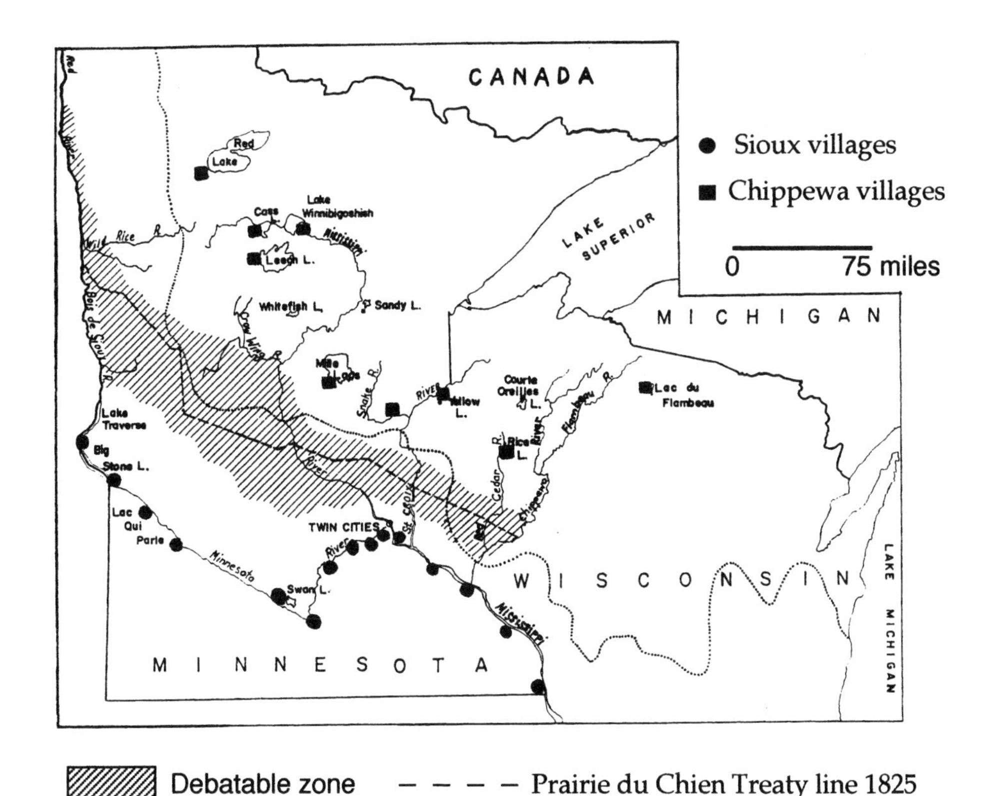

Figure 14.12. The intertribal buffer zone between the Chippewa and Sioux settlements in Wisconsin and Minnesota from about 1750 to the mid-1880's. (Redrawn from Hickerson 1965 and reproduced with permission of the American Association for the Advancement of Science (copyright holder))

mined the disputed area. Warfare between the two tribes—over the game in the buffer zone—prevented the depletion of the deer which was their most important food source.

In 1825 the boundary treaty at Prairie du Chien was established between the two tribes. However, there were numerous reports of violation of the treaty in which the tribes accused each other of encroaching on the agreed territorial boundary. War broke out again in 1831, started by the Sioux (by far the more aggressive) and only prevented from a full scale outbreak by the Chippewa's patience.

Famine was reported in both the Sioux and Chippewa villages from as early as 1828, only three years after the treaty which established a sort of truce. The famine reached a climax in 1831 and again in 1835 to 1838. During these famines there were frequent reports of boundary encroachments by both tribes in search of game. From descriptions of the agents responsible for the treaty observance the conditions of both tribes in the area were appalling. The attacks and counterattacks were often severe. In the summer of 1839 after a Chippewa incident in 1838, the Sioux killed over a hundred Chippewa who were returning from a visit to their agent. After this attack the warfare continued and the buffer zone resorted to what it was in 1826 and remained so until the reservation period in the 1850's.

What is clear is that during periods of truce of any length of time the game—mainly the Virginia deer because they were easier to catch—in the buffer zone quickly became depleted and famine ensued. Hostilities were then resumed and the buffer zone restored. As long as the buffer zone existed, which was maintained essentially by continuous intertribal warfare, deer were able to survive in large enough numbers in this disputed area to provide enough food for both tribes on each side of the disputed area. As Hickerson (1965) concludes 'The maintenance of the buffer, that is, the warfare which kept a large portion of the best deer habitat a buffer, was a function of the maintenance requirements of the Chippewa and Sioux. During times of extended truce, even in very limited regions like the St. Croix River valley, when hunting was carried out in the buffer, the supply of deer meat became depleted and the war was revived as a response to famine.'

Appendix A. General Results for the Laplacian Operator in Bounded Domains

In Chapter 2, Section 2.9, we used the result that a function $u(x)$ with $u_x = 0$ on $x = 0, 1$ satisfies

$$\int_0^1 u_{xx}^2 \, dx \geq \pi^2 \int_0^1 u_x^2 \, dx \tag{A4.1}$$

and the more general result

$$\int_B | \nabla^2 \mathbf{u} |^2 \, d\mathbf{r} \geq \mu \int_{\partial B} \| \nabla \mathbf{u} \|^2 \, d\mathbf{r}, \tag{A4.2}$$

where B is a finite domain enclosed by the simply connected surface ∂B on which zero-flux (Neumann) conditions hold; namely, $\mathbf{n} \cdot \nabla \mathbf{u} = 0$ where $\mathbf{n}$ is the unit outward normal to ∂B. In (A4.2), μ is the least positive eigenvalue of $\nabla^2 + \mu$ for B with Neumann conditions on ∂B and where $\| \cdot \|$ denotes a Euclidean norm. By the Euclidean norm here we mean, for example,

$$\| \nabla \mathbf{u} \| = \max_{r \in B} \left[\sum_{i,j} \left(\frac{\partial u_i}{\partial x_j} \right)^2 \right]^{1/2} \tag{A4.3}$$

$$\mathbf{r} = (x_j), \quad j = 1, 2, 3; \quad \mathbf{u} = (u_i), \quad i = 1, 2, \ldots, n.$$

We prove these standard results in this section: (A4.1) is a special case of (A4.2) in which $\mathbf{u}$ is a single scalar and $\mathbf{r}$ a single space variable.

By way of illustration we first derive the one-dimensional result (A4.1) in detail and then prove the general result (A4.2).

Consider the equation for the scalar function $w(x)$, a function of the single space variable x, given by

$$w_{xx} + \mu w = 0, \tag{A4.4}$$

where μ represents the general eigenvalue for solutions of this equation satisfying Neumann conditions on the boundaries; namely,

$$w_x(x) = 0 \quad \text{on} \quad x = 0, 1. \tag{A4.5}$$

The orthonormal eigenfunctions $\{\phi_k(x)\}$ and eigenvalues $\{\mu_k\}$, where $k = 0, 1, 2, \ldots$, for (A4.4) and (A4.5) are

$$\phi_k(x) = \cos \mu_k^{1/2} x, \quad \mu_k = k^2\pi^2, \quad k = 0, 1, \ldots. \tag{A4.6}$$

Any function $w(x)$, such as we are interested in, satisfying the zero-flux conditions (A4.5) can be written in terms of a series (Fourier) expansion of eigenfunctions $\phi_k(x)$ and so also can derivatives of $w(x)$, which we assume exist. Let

$$w_{xx}(x) = \sum_{k=0}^{\infty} a_k \phi_k(x) = \sum_{k=0}^{\infty} a_k \cos(k\pi x), \tag{A4.7}$$

where, in the usual way,

$$a_k = 2\int_0^1 w_{xx}(x)\cos(k\pi x)\,dx, \quad k > 0$$

$$a_0 = \int_0^1 w_{xx}(x)\,dx = [w_x(x)]_0^1 = 0.$$

Then, integrating (A4.7) twice and using conditions (A4.5) gives

$$w(x) = \sum_{k=1}^{\infty} -\frac{a_k}{\mu_k}\phi_k(x) + b_0\phi_0,$$

where b_0 and ϕ_0 are constants. Thus, since $a_0 = 0$,

$$\begin{aligned}
\int_0^1 w_x^2(x)\,dx &= [ww_x]_0^1 - \int_0^1 ww_{xx}\,dx \\
&= -\int_0^1 ww_{xx}\,dx \\
&= \int_0^1 \left[\sum_{k=1}^{\infty} \frac{a_k}{\mu_k}\cos(k\pi x)\right]\left[\sum_{k=1}^{\infty} a_k\cos(k\pi x)\right] dx \\
&\quad + b_0\phi_0 \int_0^1 \left[\sum_{k=1}^{\infty} a_k\cos(k\pi x)\right] dx \\
&= \frac{1}{2}\sum_{k=1}^{\infty}\frac{a_k^2}{\mu_k} \\
&\le \frac{1}{2\mu_1}\sum_{k=1}^{\infty} a_k^2 \\
&= \frac{1}{\mu_1}\int_0^1 w_{xx}^2\,dx = \frac{1}{\pi^2}\int_0^1 w_{xx}^2\,dx,
\end{aligned}$$

which is (A4.1); μ_1 is the smallest positive eigenvalue μ_k for all k.

The proof of the general result (A4.2) simply mirrors the one-dimensional scalar version.

Again let the sequence $\{\boldsymbol{\phi}_k(\mathbf{r})\},\ k = 0, 1, 2, \ldots$ be the orthonormal eigenvector functions of

$$\nabla^2 \mathbf{w} + \mu \mathbf{w} = 0,$$

where $\mathbf{w}(\mathbf{r})$ is a vector function of the space variable $\mathbf{r}$ and μ is the general eigenvalue. Let the corresponding eigenvalues for the $\{\boldsymbol{\phi}_k\}$ be the sequence $\{\mu_k\},\ k = 0, 1, \ldots,$ where they are so ordered that $\mu_0 = 0,\ 0 < \mu_1 < \mu_2 \cdots$. Note in this case also that $\boldsymbol{\phi}_0 = \text{constant}$.

Let $\mathbf{w}(\mathbf{r})$ be a function defined for $\mathbf{r}$ in the domain B and satisfying the zero-flux conditions $\mathbf{n} \cdot \nabla \mathbf{w} = 0$ for $\mathbf{r}$ on ∂B. Then we can write

$$\begin{aligned}
\nabla^2 \mathbf{w} &= \sum_{k=0}^{\infty} a_k \boldsymbol{\phi}_k(\mathbf{r}), \\
a_k &= \int_B \langle \nabla^2 \mathbf{w}, \boldsymbol{\phi}_k \rangle \, d\mathbf{r}, \qquad\qquad \text{(A4.8)} \\
a_0 &= \langle \boldsymbol{\phi}_0, \int_B \nabla^2 \mathbf{w} \, d\mathbf{r} \rangle = \langle \boldsymbol{\phi}_0, \int_{\partial B} \nabla \mathbf{w} \, d\mathbf{r} \rangle = 0.
\end{aligned}$$

Here $\langle \cdot \rangle$ denotes the inner (scalar) product. Integrating $\nabla^2 \mathbf{w}$ twice we get

$$\mathbf{w}(\mathbf{r}) = \sum_{k=1}^{\infty} -\frac{a_k}{\mu_k} \boldsymbol{\phi}_k(\mathbf{r}) + b_0 \boldsymbol{\phi}_0,$$

where b_0 and $\boldsymbol{\phi}_0$ are constants. With this expression together with that for $\nabla^2 \mathbf{w}$ we have, on integrating by parts,

$$\begin{aligned}
\int_B \| \nabla \mathbf{w} \|^2 \, d\mathbf{r} &= \int_{\partial B} \langle \mathbf{w}, \mathbf{n} \cdot \nabla \mathbf{w} \rangle d\mathbf{r} - \int_B \langle \mathbf{w}, \nabla^2 \mathbf{w} \rangle d\mathbf{r} \\
&= \sum_{k=1}^{\infty} \frac{a_k^2}{\mu_k} \\
&\le \frac{1}{\mu_1} \sum_{k=1}^{\infty} a_k^2 \\
&= \frac{1}{\mu_1} \int_B | \nabla^2 \mathbf{w} |^2 d\mathbf{r},
\end{aligned}$$

which gives the result (A4.2) since μ_1 is the least positive eigenvalue.

Bibliography

[1] N.S. Adzick and M.T. Longaker, editors. *Fetal Wound Healing*. Elsevier, New York, 1991.

[2] K.I. Agladze and V.I. Krinskii. Multi-armed vortices in an active chemical medium. *Nature*, 286:424–426, 1982.

[3] K.I. Agladze, E.O. Budrene, G. Ivanitsky, and V.I. Krinskii. Wave mechanisms of pattern formation in microbial populations. *Proc. R. Soc. Lond. B*, 253:131–135, 1993.

[4] P. Alberch. Ontogenesis and morphological diversification. *Amer. Zool.*, 20:653–667, 1980.

[5] P. Alberch. Developmental constraints in evolutionary processes. In J.T. Bonner, editor, *Evolution and Development, Dahlem Conference Report*, volume 20, pages 313–332. Springer-Verlag, Berlin-Heidelberg-New York, 1982.

[6] P. Alberch. The logic of monsters: evidence for internal constraint in development and evolution. *Geo. Bios, Mémoire Spéciale*, 12:21–57, 1989.

[7] P. Alberch and E. Gale. Size dependency during the development of the amphibian foot. Colchicine induced digital loss and reduction. *J. Embryol. Exp. Morphol.*, 76:177–197, 1983.

[8] B. Alberts, D. Bray, J. Lewis, M. Raff, K. Roberts, and J.D. Watson. *Molecular Biology of The Cell (3rd edition)*. Garland, New York and London, 1994.

[9] M.A. Allessie, F.I.M. Bonke, and F.G.J. Schopman. Circus movement in rabbit atrial muscle as a mechanism of tachycardia. *Circ. Res.*, 33:54–62, 1973.

[10] M.A. Allessie, F.I.M. Bonke, and F.G.J. Schopman. Circus movement in rabbit atrial muscle as a mechanism of tachycardia. II. The role of nonuniform recovery of excitability in the occurrence of unidirectional block, as studied with multiple microelectrodes. *Circ. Res.*, 39:168–177, 1976.

[11] M.A. Allessie, F.I.M. Bonke, and F.G.J. Schopman. Circus movement in rabbit atrial muscle as a mechanism of tachycardia. III. The 'leading circle' concept: a new model of circus movement in cardiac tissue without the involvement of an anatomical obstacle. *Circ. Res.*, 41:9–18, 1977.

[12] W. Alt and D.A. Lauffenburger. Transient behaviour of a chemotaxis system modelling certain types of tissue inflammation. *J. Math. Biol.*, 24:691–722, 1987.

[13] E.C. Alvord, Jr. Simple model of recurrent gliomas. *J. Neurosurg.*, 75:337–338, 1991.

[14] E.C. Alvord, Jr. Is necrosis helpful in grading of gliomas? *J. Neuropathol. and Exp. Neurology*, 51:127–132, 1992.

[15] E.C. Alvord, Jr. and C.M. Shaw. Neoplasms affecting the nervous system in the elderly. In S. Duckett, editor, *The Pathology of the Aging Human Nervous System*, pages 210–281. Lea and Febiger, Philadelphia, 1991.

[16] V. R. Amberger, T. Hensel, N. Ogata, and M. E. Schwab. Spreading and migration of human glioma and rat C6 cells on central nervous system myelin *in vitro* is correlated with tumor malignancy and involves a metalloproteolytic activity. *Cancer Res.*, 58:149–158, 1998.

[17] T. Amemiya, S. Káár, P. Kettunen, and K. Showalter. Spiral wave formation in three-dimensional excitable media. *Phys. Rev. Lett.*, 77:3244–3247, 1996.

[18] R.M. Anderson, H.C. Jackson, R.M. May, and A.M. Smith. Population dynamics of fox rabies in Europe. *Nature*, 289:765–771, 1981.

[19] J. Andersson, A.-K. Borg-Karlson, and C. Wiklund. Sexual cooperation and conflict in butterflies: A male-transferred anti-aphrodisiac reduces harassment of recently mated females. *Proc. R. Soc. Lond. B*, 267:1271–1275, 2000.

[20] J. Ando and A. Kamiya. Flow-dependent regulation of gene expression in vascular endothelial cells. *Japanese Heart J.*, 37:19–32, 1996.

[21] J. Ando, T. Komatsuda, C. Ishikawa, and A. Kamiya. Fluid shear stress enhanced DNA synthesis in cultured endothelial cells during repair of mechanical denudation. *Biorheology*, 27:675–684, 1990.

[22] J. Ando, H. Nomura, and A. Kamiya. The effect of fluid shear stress on the migration and proliferation of cultured endothelial cells. *Microvascular Res.*, 33:62–70, 1987.

[23] L. Andral, M. Artois, M.F.A. Aubert, and J. Blancou. Radio-tracking of rabid foxes. *Comp. Immun. Microbiol. Infect. Dis.*, 5:285–291, 1982.

[24] J.L. Aragón, C. Varea, R.A. Barrio, and P.K. Maini. Spatial patterning in modified Turing systems: application to pigmentation patterns on marine fish. *Forma*, 13:213–221, 1998.

[25] C. Archer, P. Rooney, and L. Wolpert. The early growth and morphogenesis of limb cartilage. In J. Fallon and A. Kaplan, editors, *Limb Development and Regeneration*, pages 267–276, part A. A.R. Liss, New York, 1983.

[26] P. Arcuri and J.D. Murray. Pattern sensitivity to boundary and initial conditions in reaction-diffusion models. *J. Math. Biol.*, 24:141–165, 1986.

[27] B.T. Arriaza, P. Cárdenas-Arroyo, E. Kleiss, and J.W. Verano. South American mummies: culture and disease. In A. Cockburn, E. Cockburn, and T.A. Reyman, editors, *Mummies, Diseases and Ancient Cultures*, pages 190–236. Cambridge University Press, Cambridge, UK, 1998.

[28] M. Artois and N.F.A. Aubert. Structure des populations (age et sexe) de renard en zones indemnés ou atteintés de rage. *Comp. Immun. Microbiol. Infect. Dis.*, 5:237–245, 1982.

[29] G. Asaad and B. Shapiro. Hallucinations: theoretical and clinical overview. *Amer. J. Psychiatry*, 143:1088–1097, 1986.

[30] M. Aubert. Current status of animal rabies in France. *Medicine Tropicale*, 57:45–51, 1997.

[31] P.J. Bacon, editor. *Population Dynamics of Rabies in Wildlife*. Academic, New York, 1985.

[32] J.T. Bagnara and M.E. Hadley. *Chromatophores and Colour Change: The Comparative Physiology of Animal Regenration*. Prentice-Hall, New Jersey, 1973.

[33] J.B.L. Bard. A unity underlying the different zebra striping patterns. *J. Zool. (Lond.)*, 183:527–539, 1977.

[34] J.B.L. Bard. A model for generating aspects of zebra and other mammalian coat patterns. *J. Theor. Biol.*, 93:363–385, 1981.

[35] J.B.L. Bard. *Morphogenesis: The Cellular and Molecular Processes of Developmental Anatomy*. Cambridge University Press, Cambridge, UK, 1990.

[36] M. Barinaga. Looking to development's future. *Science*, 266:561–564, 1994.

[37] F.R. Barkalow, R.B. Hamilton, and R.F. Soots. The vital statistics of an unexploited gray squirrel population. *J. Wildl. Mgmnt.*, 34:489–500, 1970.

[38] F.S. Barkalow. A record Grey squirrell litter. *J. Mammal.*, 48:141, 1967.

[39] R.D. Barnes. *Invertebrate Zoology*. Saunders, Philadelphia, 1980.

[40] V.H. Barocas and R.T. Tranquillo. Biphasic theory and in vitro assays of cell-fibril mechanical interactions in tissue-equivalent gels. In V.C. Mow, F. Guilak, R. Tran-Son-Tay, and R.M. Hochmuth, editors, *Cell Mechanics and Cellular Engineering*, volume 119. Springer-Verlag, New York, 1994.

[41] V.H. Barocas and R.T. Tranquillo. An isotropic biphasic theory of tissue-equivalent mechanics: the interplay among cell traction, fibrillar network deformation, fibril alignment and cell contact guidance. *J. Biomech. Eng.*, 119:137–145, 1997a.

[42] V.H. Barocas and R.T. Tranquillo. A finite element solution for the anisotropic biphasic theory of tissue-equivalent mechanics: the effect of contact guidance on isometric cell traction measurement. *J. Biomech. Eng.*, 119:261–268, 1997b.

[43] V.H. Barocas, A.G. Moon, and R.T. Tranquillo. The fibroblast-populated collagen microsphere assay of cell traction force—Part 2: Measurement of the cell traction parameter. *J. Biomech. Eng.*, 117:161–170, 1995.

[44] J. Barrat and M.F. Aubert. Current status of fox rabies in Europe. *Onderstepoort J. Veterinary Res.*, 60:357–363, 1993.

[45] R.A. Barrio, C. Varea, and J.L. Aragón. A two-dimensional numerical study of spatial pattern formation in interacting systems. *Bull. Math. Biol.*, 61:483–505, 1999.

[46] W.M. Bement, P. Forscher, and M.S. Mooseker. A novel cytoskeletal structure involved in purse string wound closure and cell polarity maintenance. *J. Cell Biol.*, 121:565–578, 1993.

[47] E. Ben-Jacob. From snowflake formation to growth of bacterial colonies II: Cooperative formation of complex colonial patterns. *Contemporary Physics*, 38:205–241, 1997.

[48] E. Ben-Jacob, I. Cohen, I. Golding, and Y. Kozlovsky. Modeling branching and chiral colonial patterning of lubricating bacteria. In P.K. Maini and H.G. Othmer, editors, *Mathematical Models for Biological Pattern Formation*, pages 211–253. Springer-Verlag, New York, 2000.

[49] E. Ben-Jacob, O. Shochet, I. Cohen, A. Tenenbaum, A. Czirók, and T. Vicsek. Cooperative strategies in formation of complex bacterial patterns. *Fractals*, 3:849–868, 1995.

[50] G. Ben-Yu, A.R. Mitchell, and B.D. Sleeman. Spatial effects in a two-dimensional model of the budworm-balsam fir ecosystem. *Comp. and Maths. with Appls. (B)*, 12:1117–1132, 1986.

[51] D.E. Bentil. *Aspects of Dynamic Pattern Formation in Embryology and Epidemiology*. PhD thesis, University of Oxford, 1990.

[52] D.E. Bentil and J.D. Murray. Pattern selection in biological pattern formation mechanisms. *Appl. Maths. Letters*, 4:1–5, 1991.

[53] D.E. Bentil and J.D. Murray. On the mechanical theory of biological pattern formation. *Physica D*, 63:161–190, 1993.

[54] C. Berding. On the heterogeneity of reaction-diffusion generated patterns. *Bull. Math. Biol.*, 49:233–252, 1987.

[55] J. Bereiter-Hahn. Epidermal cell migration and wound repair. In J. Bereiter-Hahn, A.G. Matoltsy, and K.S. Richards, editors, *Biology of the Integument*, volume 2 (*Vertebrates*), pages 443–471. Springer-Verlag, Berlin-Heidelberg-New York, 1986.

[56] H.C. Berg. *Random Walks in Biology*. Princeton University Press, Princeton, NJ, 1983.

[57] H.C. Berg and L. Turner. Chemotaxis of bacteria in glass capillary arrays. *Biophys. J.*, 58:919–930, 1990.

[58] C.N. Bertolami, V. Shetty, J.E. Milavec, D.G. Ellis, and H.M. Cherrick. Preparation and evaluation of a nonpropietary bilayer skin subsitute. *Plastic and Reconstructive Surg.*, 87:1089–1098, 1991.

[59] J. Blancou. Ecology and epidemiology of fox rabies. *Rev. Infect. Dis.*, 10(Suppl. 4):S606–S609, 1988.

[60] F.G. Blankenberg, R.L. Teplitz, W. Ellis, M.S. Salamat, B.H. Min, L. Hall, D.B. Boothroyd, I.M. Johnstone, and D.R. Enzmann. The influence of volumetric tumor doubling time, DNA ploidy, and histologic grade on survival of patients with intracranial astrocytomas. *Amer. J. Neuroradiology*, 16:1001–1012, 1995.

[61] K. Boegel, H. Moegle, F. Steck, W. Krocza, and L. Andral. Assessment of fox control in areas of wildlife rabies. *Bull. WHO*, 59:269–279, 1981.

[62] T. Boehm, J. Folkman, T. Browder, and M.S. O'Reilly. Antiangiogenic therapy of experimental cancer does not induce acquired drug resistance. *Nature*, 390:404–407, 1997.

[63] S. Bonotto. *Acetabularia* as a link in the marine food chain. In S. Bonotto, F. Cinelli, and R. Billiau, editors, *Proc. 6th Intern. Symp. on Acetabulria. Pisa, 1984*, pages 67–80, Mol, Belgium, 1985. Belgian Nuclear Center, C.E.N.-S.C.K.

[64] W. Born. Monsters in Art. *CIBA Symp.*, 9:684–696, 1947.

[65] R.H. Brady. The causal dimension of Goethe's morphology. *J. Social Biol. Struct.*, 7:325–344, 1984.

[66] P.M. Brakefield and V. French. Eyespot development on butterfly wings: the epidermal response to damage. *Dev. Biol.*, 168:98–111, 1995.

[67] D. Bray, editor. *Cell Movements*. Garland Publishing, New York, 1992.

[68] J.H. Breasted. *Edwin Smith Surgical Papyrus*. University of Chicago Press, Chicago, 1930.

[69] M.P. Brenner, L.S. Levitov, and E.O. Budrene. Physical mechanisms for chemotactic pattern formation by bacteria. *Biophys. J.*, 74:1677–1693, 1998.

[70] J.F. Bridge and S.E. Angrist. An extended table of roots of $J_n'(x)Y_n(bx) - J_n(bx)Y_n'(x) = 0$. *Math. Comp.*, 16:198–204, 1962.

[71] N.F. Britton. *Reaction-Diffusion Equations and Their Applications to Biology*. Academic, New York, 1986.

[72] N.F. Britton and J.D. Murray. Threshold wave and cell-cell avalanche behaviour in a class of substrate inhibition oscillators. *J. Theor. Biol.*, 77:317–332, 1979.

[73] G. Brugal and J. Pelmont. Existence of two chalone-like substances in intestinal extract from the adult newt, inhibiting embryonic intestinal cell proliferation. *Cell Tiss. Kinet.*, 8:171–187, 1975.

[74] E.O. Budrene and H.C. Berg. Complex patterns formed by motile cells of *Escherichia coli*. *Nature*, 349:630–633, 1991.

[75] E.O. Budrene and H.C. Berg. Dynamics of formation of symmetrical patterns of chemotactic bacteria. *Nature*, 376:49–53, 1995.

[76] B. Bunow, J.-P. Kernevez, G. Joly, and D. Thomas. Pattern formation by reaction-diffusion instabilities: application to morphogenesis in *Drosophila*. *J. Theor. Biol.*, 84:629–649, 1980.

[77] P.C. Burger, E.R. Heinz, T. Shibata, and P. Kleihues. Topographic anatomy and CT corrrelations in the untreated glioblastoma multiforme. *J. Neurosurg.*, 68:698–704, 1988.

[78] P. K. Burgess, P. M. Kulesa, J. D. Murray, and E. C. Alvord, Jr. The interaction of growth rates and diffusion coefficients in a three-dimensional mathematical model of gliomas. *J. Neuropath. and Exp. Neurology*, 56(6):704–713, June 1997.

[79] R. Burton. *The Anatomy of Melancholy*. J.M. Dent, London, 1652.

[80] H.M. Byrne and M.A.J. Chaplain. Mathematical models for tumour angiogenesis: numerical simulations and nonlinear wave solutions. *Bull. Math. Biol.*, 57:461–486, 1995.

[81] H.M. Byrne and M.A.J. Chaplain. On the role of cell-cell adhesion in models for solid tumour growth. *Math. Comp. Modelling*, 24:1–17, 1996.

[82] R. S. Cantrell and C. Cosner. The effects of spatial heterogeneity in population dynamics. *J. Math. Biol.*, 29:315–338, 1991.

[83] G.A. Carpenter. Bursting phenomena in excitable membranes. *SIAM J. Appl. Math.*, 36:334–372, 1979.

[84] S.B. Carroll, J. Gates, D.N. Keyes, S.W. Paddock, G.R.F. Panganiban, J.E. Selegue, and J.A. Williams. Pattern formation and eyespot determination in butterfly wings. *Science*, 265:109–114, 1994.

[85] H.S. Carslaw and J.C. Jaeger. *Conduction of Heat in Solids*. Clarendon Press, Oxford, second edition, 1959.

[86] V. Castets, E. Dulos, J. Boissonade, and P. De Kepper. Experimental evidence of a sustained standing Turing-type nonequilibrium chemical pattern. *Phys. Rev. Lett.*, 64:2953–2956, 1990.

[87] M.A.J. Chaplain and A.R.A. Anderson. Modelling the growth and form of capillary networks. In M.A.J. Chaplain, G.D. Singh, and J.C McLachlan, editors, *On Growth and Form. Spatio-Temporal Pattern Formation in Biology*, pages 225–249. John Wiley, New York, 1999.

[88] M.A.J. Chaplain, G.D. Singh, and J.C. McLachlan, editors. *On Growth and Form. Spatio-Temporal Pattern Formation in Biology*. John Wiley, New York, 1999.

[89] G. Chauvet. Hierarchical functional organisation of formal biological systems: a dynamical approach. I, II and III. *Phil Trans. Roy. Soc. Lond. B*, 339:425–481, 1993.

[90] W.F. Chen. Mechanism of retraction of the trailing edge during fibroblast movement. *J. Cell Biol.*, 90:198–200, 1981.

[91] W.F. Chen and E. Mizuno. *Nonlinear Analysis in Soil Mechanics*. Elsevier, New York, 1990.

[92] M. R. Chicoine and D. L. Silbergeld. Assessment of brain tumor cell motility *in vivo* and *in vitro*. *J. Neurosurg.*, 82:615–622, 1995.

[93] D.G. Christopherson. Note on the vibration of membranes. *Quart. J. Math. Oxford Ser.*, 11:63–65, 1940.

[94] C.M. Chuong and G.M. Edelman. Expression of cell-adhesion molecules in embryonic induction. I, Morphogenesis of nestling feathers. *J. Cell Biol.*, 101:1009–1026, 1985.

[95] M. Cinotti. *The Complete Works of Bosch*. Rizzoli, New York, 1969.

[96] R.A.F. Clark. Cutaneous tissue repair: basic biological considerations. *J. Amer. Acad. Dermatol.*, 13:701–725, 1985.

[97] R.A.F. Clark. Overview and general considerations of wound repair. In R.A.F. Clark and P.M. Henson, editors, *The Molecular and Cellular Biology of Wound Repair*, pages 3–33. Plenum, New York, 1988.

[98] R.A.F. Clark. Wound repair. *Curr. Op. Cell Biol.*, 1:1000–1008, 1989.

[99] R.A.F. Clark. Cutaneous wound repair. In L.A. Goldsmith, editor, *Physiology, Biochemistry, and Molecular Biology of the Skin*, pages 576–601. Oxford University Press, New York, 1991.

[100] R.A.F. Clark and P.M. Henson, editors. *The Molecular and Cellular Biology of Wound Repair*. Plenum, New York, 1988.

[101] T. Clutton-Brock. Mammalian mating systems. *Proc. R. Soc. Lond. B*, 236:339–372, 1989.

[102] G. Cocho, R. Pérez-Pascual, J.L. Rius, and F. Soto. Discrete systems, cell-cell interactions and color pattern of animals. I. Conflicting dynamics and pattern formation. II. Clonal theory and cellular automata. *J. Theor. Biol.*, 125:419–447, 1987.

[103] E.A. Coddington and N. Levinson. *Theory of Ordinary Differential Equations*. McGraw-Hill, New York, 1972.

[104] D.S. Cohen and J.D. Murray. A generalized diffusion model for growth and dispersal in a population. *J. Math. Biol.*, 12:237–249, 1981.

[105] D.S. Cohen, J.C. Neu, and R.R. Rosales. Rotating spiral wave solutions of reaction-diffusion equations. *SIAM J. Appl. Math.*, 35:536–547, 1978.

[106] D. L. Collins, A. P. Zijdenbos, V. Kollokian, J. G. Sled, N. J. Kabani, C. J. Holmes, and A. C. Evans. Design and construction of a realistic digital brain phantom. *IEEE Trans. Medical Imaging*, 17(3):463–468, June 1998.

[107] J. Cook. *Mathematical Models for Dermal Wound Healing: Wound Contraction and Scar Formation*. PhD thesis, Department of Applied Mathematics, University of Washington, Seattle, WA, 1995.

[108] J. Cook, D. E. Woodward, P. Tracqui, and J. D. Murray. Resection of gliomas and life expectancy. *J. Neuro-Oncol.*, 24:131, 1995.

[109] J.D. Cowan. Some remarks on channel bandwidths for visual contrast detection. *Bull. Neurosci. Res.*, 15:492–515, 1977.

[110] J.D. Cowan. Spontaneous symmetry breaking in large-scale nervous activity. *Intl. J. Quantum Chem.*, 22:1059–1082, 1982.

[111] J.D. Cowan. Brain mechanisms underlying visual hallucinations. In D. Paines, editor, *Emerging Syntheses in Science*. Addison-Wesley, New York, 1987.

[112] S.C. Cowin. Wolff's law of trabecular architecture at remodelling equilibrium. *J. Biomech. Engr.*, 108:83–88, 1986.

[113] S.C. Cowin, A.M. Sadegh, and G.M. Luo. An evolutionary Wolff law for trabecular architecture. *J. Biomech. Engr.*, 114:129–136, 1992.

[114] E.J. Crampin, E.A. Gaffney, and P.K. Maini. Reaction diffusion on growing domains: scenarios for robust pattern formation. *Bull. Math. Biol.*, 61:1093–1120, 1999.

[115] J. Crank. *The Mathematics of Diffusion*. Clarendon Press, Oxford, 1975.

[116] C.E. Crosson, S.D. Klyce, and R.W. Beuerman. Epithelial wound closure in the rabbbit cornea wounds. *Invest. Ophthalmol. Vis. Sci.*, 27:464–73, 1986.

[117] G.C. Cruywagen. *Tissue Interaction and Spatial Pattern formation*. PhD thesis, University of Oxford, 1992.

[118] G.C. Cruywagen and J.D. Murray. On a tissue interaction model for skin pattern formation. *J. Nonlinear Sci.*, 2:217–240, 1992.

[119] G.C. Cruywagen, P. Kareiva, M.A. Lewis, and J.D. Murray. Competition in a spatially heterogeneous environment: modelling the risk of spread of a genetically engineered population. *Theor. Popul. Biol.*, 49:1–38, 1996.

[120] G.C. Cruywagen, P.K. Maini, and J.D. Murray. Sequential pattern formation in a model for skin morphogenesis. *IMA J. Maths. Appl. in Medic. and Biol.*, 9:227–248, 1992.

[121] G.C. Cruywagen, P.K. Maini, and J.D Murray. Sequential and synchronous skin pattern formation. In H.G. Othmer, P.K. Maini, and J.D. Murray, editors, *Experimental and Theoretical Advances in Biological Pattern Formation*, volume 259 of *NATO ASI Series A: Life Sciences*, pages 61–64. Plenum, New York, 1993.

[122] G.C. Cruywagen, P.K. Maini, and J.D. Murray. Travelling waves in a tissue interaction model for skin pattern formation. *J. Math. Biol.*, 33:193–210, 1994.

[123] G.C. Cruywagen, P.K. Maini, and J.D. Murray. An envelope method for analyzing sequential pattern formation. *SIAM J. Appl. Math.*, 61:213–231, 2000.

[124] G.C. Cruywagen, D.E. Woodward, P. Tracqui, G.T. Bartoo, J.D. Murray, and E.C. Alvord, Jr. The modeling of diffusive tumours. *J. Biol. Systems*, 3(4):937–945, 1995.

[125] H. Cummins and C. Midlo. *Fingerprints, Palms and Soles. An Introduction to Dermatoglyphics*. Blakiston, Philadelphia, 1943.

[126] A.I. Dagg. External features of giraffe. *Extrait de Mammalia*, 32:657–669, 1968.

[127] T.F. Dagi. The management of head trauma. In S.H. Greenblatt, editor, *A History of Neurosurgery*, pages 289–344. American Association of Neurological Surgeons, Park Ridge, IL, 1997.

[128] F.W. Dahlquist, P. Lovely, and D.E. Koshland. Qualitative analysis of bacterial migration in chemotaxis. *Nature, New Biol.*, 236:120–123, 1972.

[129] P.D. Dale, J.A. Sherratt, and P.K. Maini. Corneal epithelial wound healing. *J. Biol. Sys.*, 3:957–965, 1995.

[130] P.D. Dale, J.A. Sherratt, and P.K. Maini. A mathematical model for collagen fibre formation during foetal and adult dermal wound healing. *Proc. R. Soc. Lond. B*, 263:653–660, 1996.

[131] J.C. Dallon and H.G. Othmer. A discrete cell model with adaptive signalling for aggregation of *Dictyostelium discoideum. Phil. Trans. R. Soc. Lond. B*, 352:391–417, 1997.

[132] J.C. Dallon and H.G. Othmer. A continuum analysis of the chemotactic signal seen by *Dictyostelium discoideum. J. Theor. Biol*, 194:461–484, 1998.

[133] J.C. Dallon, J.A. Sherratt, and P.K. Maini. Mathematical modelling of extracellular matrix dynamics using discrete cells: fiber orientation and tissue regeneration. *J. Theor. Biol*, 199:449–471, 1999.

[134] R.J. D'Amato, M.S. Loughmman, E. Flynn, and J. Folkman. Thalidomide is an inhibitor of angiogenesis. *Proc. Nat. Acad. Sci. (U.S.)*, 91:4082–4085, 1994.

[135] S. Danjo, J. Friend, and R.A. Throft. Conjunctival epithelium in healing of corneal epithelial wounds. *Invest. Opthalmol. Vis. Sci.*, 28:1445–1449, 1987.

[136] C. Darwin. *The Origin of Species*. John Murray, London, sixth edition, 1873.

[137] D. Davidson. The mechanism of feather pattern development in the chick. I. The time of determination of feather position. II. Control of the sequence of pattern formation. *J. Embryol. Exp. Morph.*, 74:245–273, 1983.

[138] P. De Kepper, Q. Ouyang, J. Boissonade, and J.C. Roux. Sustained coherent spatial structures in a quasi-1D reaction-diffusion system. *React. Kinet. Cat. Lett.*, 42:275–288, 1990.

[139] P. De Kepper, J.-J. Perraud, B. Rudovics, and E. Dulos. Experimental study of stationary Turing patterns and their interaction with traveling waves in a chemical system. *Intern. J. Bifurcation & Chaos*, 4:1215–1231, 1994.

[140] G. Dee and J.S. Langer. Propagating pattern selection. *Phys. Rev. Letters*, 50:383–386, 1983.

[141] D.C. Deeming and M.W.J. Ferguson. Environmental regulation of sex determination in reptiles. *Phil. Trans. R. Soc. Lond. B*, 322:19–39, 1988.

[142] D.C. Deeming and M.W.J. Ferguson. The mechanism of temperature dependent sex determination in crocodilians: a hypothesis. *Am. Zool.*, 29:973–985, 1989a.

[143] D.C. Deeming and M.W.J. Ferguson. In the heat of the nest. *New Scientist*, 25:33–38, 1989b.

[144] D.C. Deeming and M.W.J. Ferguson. Morphometric analysis of embryonic development in *Alligator mississippiensis, Crocodylus johnstoni* and *Crocodylus porosus*. *J. Zool. Lond.*, 221:419–439, 1990.

[145] Daniel Defoe. In E.W. Brayley, editor, *A Journal of the Plague Year; or Memorials of the Great Pestilence in London in 1665*. Thomas Tegg, London, 1722.

[146] P. Delvoye, P. Wiliquet, J.L. Leveque, B. Nusgens, and C. Lapiere. Measurement of mechanical forces generated by skin fibroblasts embedded in a three-dimensional collagen gel. *J. Invest. Dermatol.*, 97:898–902, 1991.

[147] E.J. Denton and D.M. Rowe. Bands against stripes on the backs of mackerel *Scomber scombrus L. Proc. R. Soc. Lond. B*, 265:1051–1058, 1998.

[148] D. Dhouailly. Formation of cutaneous appendages in dermoepidermal recombination between reptiles, birds and mammals. *Wilhelm Roux Arch. EntwMech. Org.*, 177:323–340, 1975.

[149] D. Dhouailly, M. Hardy, and P. Sengel. Formation of feathers on chick foot scales: a stage-dependent morphogenetic response to retinoic acid. *J. Embryol. Exp. Morphol.*, 58:63–78, 1980.

[150] R. Dillon and H.G. Othmer. A mathematical model for outgrowth and spatial patterning of the vertebrate limb bud. *J. Theor. Biol.*, 197:295–330, 1999.

[151] M.R. Duffy, N.F. Britton, and J.D. Murray. Spiral wave solutions of practical reaction-diffusion systems. *SIAM J. Appl. Math.*, 39:8–13, 1980.

[152] P. Duffy, J. Wolf, et al. Possible person-to-person transmission of Creitzfeldt–Jakob disease. *N. Engl. J. Med.*, 290:692–693, 1974.

[153] S.R. Dunbar. Travelling wave solutions of diffusive Lotka–Volterra equations. *J. Math. Biol.*, 17:11–32, 1983.

[154] S.R. Dunbar. Travelling wave solutions of diffusive Lotka–Volterra equations: a heteroclinic connection in R^4. *Trans. Amer. Math. Soc.*, 268:557–594, 1984.

[155] M.G. Dunn, F.H. Silver, and D.A. Swann. Mechanical analysis of hypertropic scar tissue: structural basis for apparent increased rigidity. *J. Investig. Derm.*, 84:9–13, 1985.

[156] G.M. Edelman. Cell adhesion molecules in the regulation of animal form and tissue pattern. *Annu. Rev. Cell Biol.*, 2:81–116, 1986.

[157] B.B. Edelstein. The dynamics of cellular differentiation and associated pattern formation. *J. Theor. Biol.*, 37:221–243, 1972.

[158] L. Edelstein-Keshet and G.B. Ermentrout. Models for contact-mediated pattern formation: cells that form parallel arrays. *J. Math. Biol.*, 29:33–58, 1990.

[159] L. Edelstein-Keshet and G.B. Ermentrout. Models for the length distributions of actin filaments: I. Simple polymerization and fragmentation. *Bull. Math. Biol.*, 60:449–475, 1998.

[160] A.G. Edmund. Dentition. In C. Gans, A.d'A. Bellairs, and T.S. Parson, editors, *Biology of the Reptilia I*, volume *Morphology A*, pages 115–200. Academic, London, 1960a.

[161] A.G. Edmund. Evolution of dental patterns in the lower vertebrates. In *Evolution: Its Science and Doctrine. R. Soc. Can. Studia Varia. Ser. 4*, pages 45–52, 1960b.

[162] H.P. Ehrlich. Wound closure: evidence of cooperation between fibroblasts and collagen matrix. *Eye*, 2:149–157, 1989.

[163] M. Eisinger, S. Sadan, I.A. Silver, and R.B. Flick. Growth regulation of skin cells by epidermal cell-derived factors: implications for wound healing. *Proc. Nat. Acad. Sci. U.S.A.*, 85:1937–1941, 1988a.

[164] M. Eisinger, S. Sadan, R. Soehnchen, and I.A. Silver. Wound healing by epidermal-derived factors: Experimental and preliminary chemical studies. In A. Barbul, E. Pines, M. Caldwell, and T.K. Hunt, editors, *Growth Factors and Other Aspects of Wound Healing*, pages 291–302. Alan R. Liss, New York, 1988b.

[165] S.V. Elling and F.C. Powell. Physiological changes in the skin during pregnancy. *Clin. Dermatol.*, 15:35–43, 1997.

[166] T. Elsdale and F. Wasoff. Fibroblast cultures and dermatoglyphics: the topology of two planar patterns. *Wilhelm Roux Arch.*, 180:121–147, 1976.

[167] I.R. Epstein and K. Showalter. Nonlinear chemical dynamics: oscillations, patterns, and chaos. *J. Phys. Chem.*, 100:13132–13147, 1996.

[168] C.A. Erickson. Analysis of the formation of parallel arrays in BHK cells in vitro. *Exp. Cell Res.*, 115:303–315, 1978.

[169] B. Ermentrout, J. Campbell, and G. Oster. A model for shell patterns based on neural activity. *The Veliger*, 28:369–388, 1986.

[170] G.B. Ermentrout. Stable small amplitude solutions in reaction-diffusion systems. *Q. Appl. Math.*, 39:61–86, 1981.

[171] G.B. Ermentrout. Stripes or spots? Nonlinear effects in bifurcation of reaction diffusion equations on the square. *Proc. R. Soc. Lond. A*, 434:413–417, 1991.

[172] G.B. Ermentrout and J. Cowan. A mathematical theory of visual hallucination patterns. *Biol. Cybern.*, 34:137–150, 1979.

[173] G.B. Ermentrout and L. Edelstein-Keshet. Models for the length distributions of actin filaments: II. Polymerization and fragmentation by gelsolin acting together. *Bull. Math. Biol.*, 60:477–503, 1998.

[174] C.R. Etchberger, M.A. Ewert, J.B. Phillips, and C.E. Nelson. Environmental and maternal influences on embryonic pigmentation in a turtle (*Trachemys scripta elegans*). *J. Zool. Lond.*, 230:529–539, 1993.

[175] R.F. Ewer. *The Carnivores.* Cornell University Press, Ithaca, NY, 1973.

[176] M.W.J. Ferguson. The structure and composition of the eggshell and embryonic membranes of *Alligator mississippiensis. Trans. Zool. Soc. Lond.*, 36:99–152, 1981a.

[177] M.W.J. Ferguson. The structure and development of the palate in *Alligator mississippiensis. Arch. Oral Biol.*, 26:427–443, 1981b.

[178] M.W.J. Ferguson. Developmental mechanisms in normal and abnormal palate formation with particular reference to aetiology, pathogenesis and prevention of cleft palate. *Brit. J. Orthodont.*, 8(3):115–137, 1981c.

[179] M.W.J. Ferguson. Review: The value of the American alligator (*Alligator mississippiensis*) as a model for research in craniofacial development. *J. Craniofacial Genetics*, 1:123–144, 1981d.

[180] M.W.J. Ferguson. Reproductive biology and embryology of the crocodilians. In C. Gans, F. Billet, and P. Maderson, editors, *Biology of the Reptilia*, volume 14A, pages 329–491. John Wiley, New York, 1985.

[181] M.W.J. Ferguson. Palate development. *Development Suppl.*, 103:41–61, 1988.

[182] M.W.J. Ferguson. Craniofacial malformations: towards a molecular understanding. *Nature Genetics*, 6:329–330, 1994.

[183] M.W.J. Ferguson and G.F. Howarth. Marsupial models of scarless fetal wound healing. In N.S. Adzick and M.T. Longaker, editors, *Fetal Wound Healing*, pages 95–124. Elsevier, New York, 1991.

[184] J.A. Feroe. Existence and stability of multiple impulse solutions of a nerve equation. *SIAM J. Appl. Math.*, 42:235–246, 1982.

[185] I. Ferrenq, L. Tranqui, B. Vailhé, P.Y. Gumery, and P. Tracqui. Modelling biological gel contraction by cells: Mechanocellular formulation and cell traction force quantification. *Acta Biotheoretica*, 45:267–293, 1997.

[186] R.J. Field and M. Burger, editors. *Oscillations and Travelling Waves in Chemical Systems*. John Wiley, New York, 1985.

[187] R.A. Fisher. The wave of advance of advantageous genes. *Ann. Eugenics*, 7:353–369, 1937.

[188] J. Folkman. Anti-angiogenesis: new concept for therapy of solid tumors. *Ann. Surg.*, 175:409–416, 1972.

[189] J. Folkman. The vascularization of tumors. *Sci. Amer.*, 234:58–73, 1976.

[190] J. Folkman. Clinical applications of research on angiogenesis. *New Eng. J. Med.*, 333:1757–1763, 1995.

[191] J. Folkman and C. Haudenschild. Angiogenesis *in vitro*. *Nature*, 288:551–556, 1980.

[192] J. Folkman and M. Klagsbrun. Angiogeneic factors. *Science*, 235:442–447, 1987.

[193] J. Folkman and A. Moscona. Role of cell shape in growth control. *Nature*, 273:345–349, 1978.

[194] R.M. Ford and D.A. Lauffenburger. Analysis of chemotactic bacterial distributions in population migration assays using a mathematical model applicable to steep or shallow attractant gradients. *Bull. Math. Biol.*, 53:721–749, 1991.

[195] J.S. Forrester, M. Fishbein, T.R. Helfan, and J. Fagin. A paradigm for restenosis based on cell biology: clues for the development of new preventive therapies. *J. Amer. Coll. Cardiology*, 17:758–769, 1993.

[196] A.C. Fowler. The effect of incubation time distribution on the extinction characteristics of a rabies epizootic. *Bull. Math. Biol.*, 62:633–655, 2000.

[197] J.M. Frantz, B.M. Dupuy, H.E. Kaufman, and R.W. Beuerman. The effect of collagen shields on epithelial wound healing in rabbits. *Am. J. Ophthalmol.*, 108:524–8, 1989.

[198] F. Fremuth. Chalones and specific growth factors in normal and tumor growth. *Acta Univ. Carol. Mongr.*, 110, 1984.

[199] V. French. Pattern formation on butterfly wings. In M.A.J. Chaplain, G.D. Singh, and J.C. McLachlan, editors, *On Growth and Form. Spatio-Temporal Pattern Formation in Biology*, pages 31–46. John Wiley, New York, 1999.

[200] V. French and P.M. Brakefield. Eyespot development on butterfly wings: The focal signal. *Dev. Biol.*, 168:112–123, 1995.

[201] R.R. Frerichs and J. Prawda. A computer simulation model for the control of rabies in an urban area of Colombia. *Management Science*, 22:411–421, 1975.

[202] Y.C. Fung. *Biomechanics. Mechanical Properties of Living Tissue*. Springer-Verlag, Berlin, 1993.

[203] Y.C. Fung and S.Q. Liu. Change of residual strains in arteries due to hypertrophy caused by aortic constriction. *Circ. Research*, 65:1340–1349, 1989.

[204] Y.C. Fung and S.Q. Liu. Changes of zero-stress state of rat pulmonary arteries in hypotoxic hypertension. *J. Appl. Physiol.*, 70:2455–2470, 1991.

[205] D.W. Furnas, M. Ashraf Sheikh, P. van den Hombergh, and I.M. Nunda. Traditional craniotomies of the Kisi tribe of Kenya. *Annals Plastic Surg.*, 15:538–556, 1985.

[206] G. Gabbiani and G. Majno. Dupuytren's contracture: fibroblast contraction? An ultrastructural study. *Amer. J. Pathol.*, 66:131–146, 1972.

[207] W.J. Gallin, C.-M. Chuong, L.H. Finkel, and G.M. Edelman. Antibodies to liver cell adhesion molecules perturb inductive interactions and alter feather pattern and structure. *Proc. Natl. Acad. Sci. USA*, 83:8235–8239, 1986.

[208] P. Garnerin and A.-J. Hazout, S. Valleron. Estimation of two epidemiological parameters of fox rabies: the length of incubation period and the dispersion distance of cubs. *Ecol. Mod.*, 33:123–135, 1986.

[209] L. E. Gaspar, B. J. Fisher, D. R. Macdonald, D. V. LeBer, E. C. Halperin, S. C. Schold, and J. G. Cairncross. Supratentorial malignant glioma: patterns of recurrence and implication for external beam local treatment. *Intern. J. Radiation Oncol. Biol. Phys.*, 24:55–57, 1992.

[210] V. Gáspár, J. Maselko, and K. Showalter. Transverse coupling of chemical waves. *Chaos*, 1:435–444, 1991.

[211] I. Geoffroy Saint-Hilaire. *Traité de Tératologie*, volume 1–3. Bailliére, Paris, 1836.

[212] A. Gerber. Die embryonale und postembryonale Pterylose der Alectromorphae. *Rev. Suisse Zool.*, 46:161–324, 1939.

[213] R.G. Gibbs. Travelling waves in the Belousov–Zhabotinskii reaction. *SIAM J. Appl. Math.*, 38:422–444, 1980.

[214] A. Gierer and H. Meinhardt. A theory of biological pattern formation. *Kybernetik*, 12:30–39, 1972.

[215] A. Giese and M. Westphal. Glioma invasion in the central nervous system. *Neurosurgery*, 39(2):235–252, 1996.

[216] A. Giese, L. Kluwe, B. Laube, H. Meissner, M. Berens, and M. Westphal. Migration of human glioma cells on myelin. *Neurosurgery*, 38(4):755–764, 1996a.

[217] A. Giese, B. Laube, S. Zapf, U. Mangold, and M. Westphal. Glioma cell adhesion and migration on human brain sections. *Anticancer Res.*, 18:2435–2448, 1998.

[218] A. Giese, M. A. Loo, N. Tran, D. Haskett, S. W. Coons, and M. E. Berens. Dichotomy of astrocytoma migration and proliferation. *Intern. J. Cancer*, 67:275–282, 1996b.

[219] A. Giese, R. Schroder, A. Steiner, and M. Westphal. Migration of human glioma cells in response to tumour cyst fluids. *Acta Neurochirurgica*, 138:1331–1340, 1996c.

[220] J.H. Goldie and A.J. Coldman. A mathematical model for realting the drug sensitivity of tumors to their spontaneous mutation rate. *Cancer Treatment Rep.*, 66:439, 1979.

[221] R. Goldschmidt. *Die quantitativen Grundlagen von Vererbung und Arbildung*. Springer-Verlag, Berlin, 1920.

[222] S. Goldstein and J.D. Murray. On the mathematics of exchange processes in fixed columns. III. The solution for general entry conditions, and a method of obtaining asymptotic expressions. IV. Limiting values, and correction terms, for the kinetic-theory solution with general entry conditions. V. The equilibrium-theory and perturbation solutions, and their connection with kinetic-theory solutions, for general entry conditions. *Proc. R. Soc. Lond. A*, 257:334–375, 1959.

[223] J. Gómez-Alonso. Rabies. a possible explanation for the vampire legend. *Amer. Acad. Neurology*, 51:856–859, 1998.

[224] B.C. Goodwin, J.D. Murray, and D. Baldwin. Calcium: the elusive morphogen in *Acetabularia*. In S. Bonotto, F. Cinelli, and R. Billiau, editors, *Proc. 6th Intern. Symp. on Acetabularia. Belgian Nuclear Center, C.E.N.-S.C.K. Mol, Belgium, 1984*, pages 101–108, Pisa, 1985.

[225] Gould, S.J. Anscheulich (Atrocious). *Natural History*, March:42–49, 2000.

[226] H. Green. Cultured cells for the treatment of disease. *Sci. Amer.*, 265(5):96–102, 1991.

[227] H. Green and J. Thomas. Pattern formation by cultured human epidermal cells: development of curved ridges resembling dermatoglyphs. *Science*, 200:1385–1388, 1978.

[228] J.M. Greenberg. Spiral waves for $\lambda - \omega$ systems. *Adv. Appl. Math.*, 2:450–455, 1981.

[229] S.H. Greenblatt, editor. *A History of Neurosurgery*. American Association of Neurological Surgeons, Park Ridge, IL, 1997.

[230] H.W. Greene. *Snakes: The Evolution of Mystery in Nature*. University of California Press, Berkeley, CA, 2000.

[231] C.T. Gregg. *Plague: An Ancient Disease in the Twentieth Century.* University of New Mexico Press, Albuquerque, 1985.

[232] I. Grierson, J. Joseph, M. Miller, and J.E. Day. Wound repair: the fibroblasts and the inhibition of scar formation. *Eye*, 2:135–148, 1988.

[233] P. Grindrod. *The Theory and Applications of Reaction-Diffusion Equations—Patterns and Waves.* Oxford University Press, New York, 1996.

[234] P. Grindrod, M.A. Lewis, and J.D. Murray. A geometrical approach to wave-type solutions of excitable reaction-diffusion systems. *Proc. R. Soc. Lond. A*, 433:151–164, 1991.

[235] P. Grindrod, J.D. Murray, and S. Sinha. Steady state spatial patterns in a cell-chemotaxis model. *J. Maths. Appl. Medic. and Biol.*, 6:69–79, 1989.

[236] C. Guidry. Extracellular matrix contraction by fibroblasts: peptide promoters and second messengers. *Cancer Metast. Rev.*, 11:45–54, 1992.

[237] C. Guidry and F. Grinnell. Contraction of hydrated collagen gels by fibroblasts: evidence for two mechanisms by which collagen fibrils are stabilized. *Collagen Rel. Res.*, 6:515–529, 1986.

[238] G.H. Gunaratne, Q. Ouyang, and H.L. Swinney. Pattern formation in the presence of symmetries. *Phys. Rev. E*, 50:2802–2820, 1994.

[239] W. Gurtler and E. Zimen. The use of baits to estimate fox numbers. *Comp. Immun. Microbiol. Infec. Dis.*, 5:277–283, 1982.

[240] E. Haeckel. *Die Radiolaren.* Georg von Reimer, Berlin, 1862. (Vol. I) 1862, (Vol. II) 1887.

[241] E. Haeckel. *Art Forms in Nature.* Dover, New York, 1974.

[242] P.S. Hagan. Spiral waves in reaction diffusion equations. *SIAM J. Appl. Math.*, 42:762–786, 1982.

[243] P.A. Hall and D.A. Levinson. Assessment of cell proliferation in histological material. *J. Clin. Pathology*, 43:184–192, 1990.

[244] V. Hamburger. Monsters in Nature. *CIBA Symposium*, 9:666–683, 1947.

[245] A.K. Harris. Traction, and its relations to contraction in tissue cell locomotion. In R. Bellairs, A. Curtis, and G. Dunn, editors, *Cell Behaviour*, pages 109–134. Cambridge University Press, Cambridge, UK, 1982.

[246] A.K. Harris, D. Stopak, and P. Warner. Generation of spatially periodic patterns by a mechanical instability: A mechanical alternative to the Turing model. *J. Embryol. Exp. Morph.*, 80:1–20, 1984.

[247] A.K. Harris, D. Stopak, and D. Wild. Fibroblast traction as a mechanism for collagen morphogenesis. *Nature*, 290:249–251, 1981.

[248] A.K. Harris, P. Ward, and D. Stopak. Silicon rubber substrata: a new wrinkle in the study of cell locomotion. *Science*, 208:177–179, 1980.

[249] L.G. Harrison, J. Snell, R. Verdi, D.E. Vogt, G.D. Zeiss, and B.D. Green. Hair morphogenesis in *Acetabularia mediterranea*: temperature-dependent spacing and models of morphogen waves. *Protoplasma*, 106:211–221, 1981.

[250] D. Hart, E. Shochat, and Z. Agur. The growth law of primary breast cancer as inferred from mammography screening trials data. *British J. Cancer*, 78:382–387, 1998.

[251] H. Hasimoto. Exact solution of a certain semi-linear system of partial differential equations related to a migrating predation problem. *Proc. Japan Acad.*, 50:623–627, 1974.

[252] A. Hastings, S. Harrison, and K. McCann. Unexpected spatial patterns in an insect outbreak match a predator diffusion model. *Proc. R. Soc. Lond. B*, 264:1837–1840, 1997.

[253] E. Hay. *Cell Biology of the Extracellular Matrix.* Plenum, New York, 1981.

[254] K. Hedges. Shamanistic aspects of California rock art. In L. Bean, editor, *California Indian Shamanism*, pages 67–88. Ballena, Menlo Park, CA, 1992.

[255] K. Hedges. Origines chamaniques de l'art rupestre dans l'Ouest Américain. *L'Anthropologie (Paris)*, 97:675–691, 1993.

[256] K. Henke. Vergleichende und experimentelle Untersuchungen an Lymatria zur Musterbildung auf dem Schmetterlingsfluegel. *Nachr. Akad. Wiss. Goettingen, Math.-Physik. (KI)*, pages 1–48, 1943.

[257] H. Hennings, K. Elgio, and O.H. Iversen. Delayed inhibition of epidermal DNA synthesis after injection of an aqueous skin extract (chalone). *Virchows Arch. Abt. B Zellpath.*, 4:45–53, 1969.

[258] I. Herán. *Animal Colouration: The Nature and Purpose of Colours in Invertebrates.* Hamlyn, London, 1976.

[259] G.J. Hergott, M. Sandig, and V.I. Kalnins. Cytoskeletal organisation of migrating retinal pigment epithelial cells during wound healing in organ culture. *Cell Motil*, 13:83–93, 1989.

[260] H. Hickerson. The Virginia deer and intertribal buffer zones in the upper Mississippi valley. In A. Leeds and A. Vayta, editors, *Man, Culture and Animals. The Role of Animals on Human Ecological Adjustments*, pages 43–65. Amer. Assoc. Adv. Sci., Washington, DC, 1965.

[261] J. Hinchliffe and D. Johnson. *The Development of the Vertebrate Limb*. Clarendon, Oxford, 1980.

[262] U.T. Hinderer. Prevention of unsatisfactory scarring. *Clinics in Plast. Surg.*, 4(2):199–205, 1977.

[263] N. Holder. Developmental constraints and the evolution of the vertebrate digit patterns. *J. Theor. Biol.*, 104:451–471, 1983.

[264] J. Holm. *Squirrels*. Whittet, London, 1987.

[265] M.J. Holmes and B.D. Sleeman. A mathematical model of tumour angiogenesis incorporating cellular traction and viscoelastic effects. *J. Theor. Biol.*, 202:95–112, 2000.

[266] R.L. Hoskinson and L.D. Mech. White-tailed deer migration and its role in wolf predation. *J. Wildl. Manag.*, 40:429–441, 1976.

[267] Y. Hosono. Singular perturbation analysis of travelling waves for diffusive Lotka–Volterra competitive models. In *Numerical and Applied Mathematics Part II (Paris, 1988) (IMACS Ann. Comput. Appl. Math., 1,2)*, pages 687–692, Baltzer, Basel, 1989.

[268] S.A. Houff and R.C. Burton. Human-to-human transmission of rabies virus by corneal transplantation. *N. Engl. J. Med.*, 300:603–604, 1979.

[269] L.N. Howard and N. Kopell. Slowly varying waves and shock structures in reaction-diffusion equations. *Studies in Appl. Math.*, 56:95–145, 1977.

[270] A.H. Howe. *A Theoretical Inquiry into the Physical Cause of Epidemic Diseases*. J. Churchill, London, 1865.

[271] J.B. Hoying and S.K. Williams. Measurement of endothelial cell migration using an improved linear migration assay. *Microcirculation*, 3(2):167–174, 1996.

[272] D.H. Hubel and T.N. Wiesel. Functional architecture of macaque monkey visual cortex. *Proc. R. Soc. Lond. B*, 198:1–59, 1977.

[273] D.H. Hubel, T.N. Wiesel, and S. LeVay. Plasticity of ocular dominance columns in monkey striate cortex. *Phil. Trans. R. Soc. Lond. B*, 278:131–163, 1977.

[274] J. Hubert. Embryology of the Squamata. In C. Gans and F. Billet, editors, *Biology of the Reptilia*, volume 14, pages 1–34. John Wiley, New York, 1985.

[275] J. Hubert and J.P. Dufaure. Table de developpment de la vipère aspic: *Vipera aspis*. *Bull. Soc. Zool. France*, 93:135–148, 1968.

[276] O. Hudlická and M.D. Brown. Physical forces and angiogenesis. In G.M. Rubanyi, editor, *Mechanoreception by the Vascular Wall*, pages 197–241. Futura, Mount Kisco, NY, 1993.

[277] A. Hunding. Turing structures of the second kind. In M.A.J. Chaplain, G.D. Singh, and J.C. McLachlan, editors, *On Growth and Form. Spatio-temporal Pattern Formation in Biology*, pages 75–88. John Wiley, New York, 1999.

[278] A. Hunding and R. Engelhardt. Early biological morphogenesis and nonlinear dynamics. *J. Theor. Biol.*, 173:401–413, 1995.

[279] A. Hunding and P.G. Sørensen. Size adaption of Turing prepatterns. *J. Math. Biol.*, 26:27–39, 1988.

[280] N. Ikeda, H. Yamamoto, and T. Sato. Pathology of the pacemaker network. *Math. Modelling*, 7:889–904, 1986.

[281] D.E. Ingber, D. Prusty, Z. Sun, H. Betensky, and N. Wang. Cell shape, cytoskeletal mechanics, and cell cycle control in angiogenesis. *J. Biomechanics*, 28:1471–1484, 1995.

[282] H.C. Jackson and L.G. Schneider. Rabies in the Federal Republic of Germany, 1950–81: the influence of landscape. *Bull. WHO*, 62:99–106, 1984.

[283] T.L. Jackson. *Mathematical Models in Two-Step Cancer Chemotherapy*. PhD thesis, Department of Applied Mathematics, University of Washington, Seattle, WA, 1998.

[284] T.L. Jackson, S.R. Lubkin, and J.D. Murray. Theoretical analysis of conjugate localization in two-step cancer treatment. *J. Math. Biol.*, 39:335–376, 1999a.

[285] T.L. Jackson, S.R. Lubkin, S.R. Siemens, N.O. Kerr, P.D. Senter, and J.D. Murray. Mathematical and experimental analysis of localization of anti-tumor anti-body-enzyme conjugates. *Br. J. Cancer*, 80:1747–1753, 1999b.

[286] T.L. Jackson, P.D. Senter, and J.D. Murray. Development and validation of a mathematical model to describe antibody enzyme conjugates. *J. Theor. Medic.*, 2:93–111, 1999c.

[287] P.A. Janmey, S. Hvidt, J. Peetermans, J. Lamb, J.D. Ferry, and T.P. Stoggel. Viscoelasticity of F-actin and F-actin/gelsolin complexes. *Biochemistry*, 27:8218–27, 1988.

[288] R.W. Jennings and T.K. Hunt. Overview of postnatal wound healing. In N.S. Adzick and M.T. Longaker, editors, *Fetal Wound Healing*, pages 25–52. Elsevier, New York, 1991.

[289] J.V. Jester, W.M. Petroll, W. Feng, J. Essepian, and H.D. Cavanagh. Radial keratotomy. 1. The wound healing process and measurement of incisional gape in two animal models using in vivo confocal microscopy. *Investig. Opthal. and Visual Sci.*, 33:3255–3270, 1992.

[290] D.W. Jordan and P. Smith. *Nonlinear Ordinary Differential Equations*. Oxford University Press, Oxford, third edition, 1999.

[291] A.K. Jowett, S. Vainio, M.W.J. Ferguson, P.T. Sharpe, and I. Thesleff. Epithelial-mesenchymal interactions are required for msx1 and msx2 gene expression in the developing murine molar tooth. *Development*, 117:461–470, 1993.

[292] A. Källén, P. Arcuri, and J.D. Murray. A simple model for the spatial spread and control of rabies. *J. Theor. Biol.*, 116:377–393, 1985.

[293] C. Kaplan, editor. *Rabies: The Facts*. Oxford University Press, Oxford, 1977.

[294] P. Kareiva. Population dynamics in spatially complex environments—Theory and data. *Phil. Trans. R. Soc. Lond. B*, 330:175–190, 1990.

[295] G.B. Karev. Digital dermatoglyphics of Bulgarians from northeastern Bulgaria. *Amer. J. Phys. Anthrop.*, 69:37–50, 1986.

[296] W.L. Kath and J.D. Murray. Analysis of a model biological switch. *SIAM J. Appl. Math.*, 45:943–955, 1986.

[297] S.A. Kauffman, R. Shymko, and K. Trabert. Control of sequential compartment in *Drosophila*. *Science*, 199:259–270, 1978.

[298] K. Kawasaki, A. Mochizuki, M. Matsushita, T. Umeda, and N. Shigesada. Modeling spatio-temporal patterns generated by *Bacillus subtilis*. *J. Theor. Biol.*, 188:177–185, 1997.

[299] M.J. Keeling and C.A. Gilligan. Bubonic plague: a metapopulation model of a zoonosis. *Proc. R. Soc. Lond. B*, 267:2219–2230, 2000.

[300] J. Keener and J. Sneyd. *Mathematical Physiology*. Springer, New York, 1998.

[301] J.P. Keener. Waves in excitable media. *SIAM J. Appl. Math.*, 39:528–548, 1980.

[302] J.P. Keener. A geometrical theory for spiral waves in excitable media. *SIAM J. Appl. Math.*, 46:1039–1056, 1986.

[303] J.P. Keener and J.J. Tyson. Spiral waves in the Belousov–Zhabotinskii reaction. *Physica D*, 21:307–324, 1986.

[304] E.F. Keller and G.M. Odell. Necessary and sufficient conditions for chemotactic bands. *Math. Biosci.*, 27:309–317, 1975.

[305] E.F. Keller and L.A. Segel. Initiation of slime mold aggregation viewed as an instability. *J. Theor. Biol.*, 26:399–415, 1970.

[306] E.F. Keller and L.A. Segel. Travelling bands of chemotactic bacteria: a theoretical analysis. *J. Theor. Biol.*, 30:235–248, 1971.

[307] P. J. Kelley and C. Hunt. The limited value of cytoreductive surgery in elderly patients with malignant gliomas. *Neurosurgery*, 34:62–67, 1994.

[308] R. Kellogg, M. Knoll, and J. Kugler. Form-similarity between phosphenes of adults and pre-school children's scribblings. *Nature*, 208:1129–1130, 1965.

[309] C.R. Kennedy and R. Aris. Traveling waves in a simple population model involving growth and death. *Bull. Math. Biol.*, 42:397–429, 1980.

[310] R.S. Kerbel. A cancer therapy resistant to resistance. *Nature*, 390:335–336, 1997.

[311] L.D. Ketchum, I.K. Cohen, and F.W. Masters. Hypertrophic scars and keloids: A collective review. *Plast. Recon. Surg.*, 53:140–154, 1974.

[312] J. Kevorkian. *Partial Differential Equations: Analytical Solution Techniques (2nd edition).* Springer-Verlag, New York, 1999.

[313] J. Kevorkian and J.D. Cole. *Multiple Scale and Singular Perturbation Methods.* Springer-Verlag, New York, 1996.

[314] J. Kingdon. *East African Mammals. An Atlas of Evolution in Africa. IIIA. Carnivores.* Academic, London, 1978.

[315] J. Kingdon. *East African Mammals. An Atlas of Evolution in Africa. IIIB. Large mammals.* Academic, London, 1979.

[316] C.W. Kischer, J. Pindur, P. Krasovitch, and E. Kischer. Characteristics of granulation tissue which promote hypertrophic scarring. *Scan. Microscopy*, 4:877–888, 1990.

[317] C.W. Kischer, M.R. Shetlar, and M. Chvapil. Hypertrophic scars and keloids: a review and new concept concerning their origin. *Scan. Electron Microscopy*, 4:1699–1713, 1982.

[318] L.M. Klauber. *Rattlesnakes: Their Habits, Life Histories, and Influence on Mankind.* University of California Press, Berkeley, CA, 1998.

[319] M.R. Klee, H.D. Lux, and E.-J. Speckmann, editors. *Physiology, Pharmacology and Development of Epileptogeneic Phenomena.* Springer-Verlag, Berlin, Heidelberg, New York, 1991.

[320] H. Klüver. *Mescal and Mechanisms of Hallucinations.* University of Chicago Press, Chicago, 1967.

[321] A.J. Koch and H. Meinhardt. Biological pattern formation: from basic mechanisms to complex structures. *Rev. of Modern Phys.*, 66:1481–1507, 1994.

[322] S. Koga. Rotating spiral waves in reaction-diffusion systems-phase singularities of multi-armed spirals. *Prog. Theor. Phys.*, 67:164–178, 1982.

[323] J. Kolega. Effects of mechanical tension on protrusive activity and microfilament and intermediate filament organization in an epidermal epithelium moving in culture. *J. Cell Biol.*, 102:1400–11, 1986.

[324] E.J. Kollar. The induction of hair follicles in embryonic dermal papillae. *J. Invest. Dermatol.*, 55:374–378, 1970.

[325] M.S. Kolodney and R.B. Wysolmerski. Isometric contraction by fibroblasts and endothelial cells in tissue culture: a quantitative study. *J. Cell Biol.*, 117:73–82, 1992.

[326] S. Kondo and R. Asai. A reaction-diffusion wave on the skin of the marine angelfish *Pomacanthus*. *Nature*, 376:765–768, 1995.

[327] N. Kopell and L.N. Howard. Plane wave solutions to reaction-diffusion equations. *Studies in Appl. Math.*, 42:291–328, 1973.

[328] N. Kopell and L.N. Howard. Target patterns and spiral solutions to reaction-diffusion equations with more than one space dimension. *Adv. Appl. Math.*, 2:417–449, 1981.

[329] F. W. Kreth, P. C. Warnke, R. Scheremet, and C. B. Ostertag. Surgical resection and radiation therapy versus biopsy and radiation therapy in the treatment of glioblastoma multiforme. *J. Neurosurg.*, 78:762–766, May 1993.

[330] V.I. Krinsky. Mathematical models of cardiac arrhythmias (spiral waves). *Pharmac. Ther. (B)*, 3:539–555, 1978.

[331] V.I. Krinsky, A.B. Medvinskii, and A.V. Parfilov. Evolutionary autonomous spiral waves (in the heart). *Mathematical Cybernetics. Popular Ser. (Life Sciences)*, 8:1–48, 1986. In Russian.

[332] J.E. Kronmiller, W.B. Upholt, and E.J. Kollar. EGF antisense oligodeoxynucleotides block murine odontogenesis *in vitro*. *Dev. Biol.*, 147:485–488, 1991.

[333] A. Kühn and A. von Engelhardt. Über die Determination des Symmetriesystems auf dem Vorderflügel von Ephestia kuehniella. *Z. Wilhelm Roux Arch. Entw. Mech. Org.*, 130:660–703, 1933.

[334] P.M. Kulesa. *A Model Mechanism for the Initiation and Spatial Patterning of Teeth Primordia in the Alligator.* PhD thesis, Department of Applied Mathematics, University of Washington, Seattle, WA, 1995.

[335] P.M. Kulesa and J.D. Murray. Modelling the wave-like initiation of teeth primordia in the alligator. *Forma*, 10:259–280, 1995.

[336] P.M. Kulesa, G.C. Cruywagen, S.R. Lubkin, M.W.J. Ferguson, and J.D. Murray. Modelling the spatial patterning of teeth primordia in the alligator. *Acta Biotheoretica*, 44:349–358, 1996a.

[337] P.M. Kulesa, G.C. Cruywagen, S.R. Lubkin, P.K. Maini, J. Sneyd, M.W.J. Ferguson, and J.D. Murray. On a model mechanism for the spatial patterning of teeth primordia in the alligator. *J. Theor. Biol.*, 180:287–296, 1996b.

[338] P.M. Kulesa, G.C. Cruywagen, S.R. Lubkin, P.K. Maini, J. Sneyd, and J.D. Murray. Modelling the spatial patterning of the teeth primordia in the lower jaw of *Alligator mississippiensis*. *J. Biol. Systems*, 3:975–985, 1993.

[339] K. Kuramoto. Instability and turbulence of wavefronts in reaction-diffusion systems. *Prog. Theor. Phys.*, 63:1885–1903, 1980.

[340] K. Kuramoto and S. Koga. Turbulized rotating chemical waves. *Prog. Theor. Phys.*, 66:1081–1085, 1981.

[341] L. Landau and E. Lifshitz. *Theory of Elasticity*. Pergamon, New York, second edition, 1970.

[342] D.C. Lane, J.D. Murray, and V.S. Manoranjan. Analysis of wave phenomena in a morphogenetic mechanochemical model and an application to post-fertilisation waves on eggs. *IMA J. Math. Applied Medic. and. Biol.*, 4:309–331, 1987.

[343] K. Langer. On the anatomy and physiology of the skin. I. the cleavability of cutis (original 1862, in German). *Br. J. Plast. Surg.*, 31:3–8, 1978a.

[344] K. Langer. On the anatomy and physiology of the skin. II. Skin tension (original 1862, in German). *Br. J. Plast. Surg.*, 31:93–106, 1978b.

[345] K. Langer. On the anatomy and physiology of the skin. III. The elasticity of the cutis (original 1862, in German). *Br. J. Plast. Surg.*, 31:185–199, 1978c.

[346] K. Langer. On the anatomy and physiology of the skin. IV. The swelling capabilities of the skin (original 1862, in German). *Br. J. Plast. Surg.*, 31:185–199, 1978d.

[347] W.L. Langer. The Black Death. *Sci. Amer.*, pages 114–121, February 1964.

[348] I.R. Lapidis and R. Schiller. Model for the chemotactic response of a bacterial population. *Biophys. J.*, 16:779–789, 1976.

[349] F. Lara-Ochoa. A generalized reaction diffusion model for spatial structureformed by mobile cells. *Biosystems*, 17:35–50, 1984.

[350] F. Lara-Ochoa and J.D. Murray. A nonlinear analysis for spatial structure in a reaction-diffusion model. *Bull. Math. Biol.*, 45:917–930, 1983.

[351] D.A. Lauffenburger and C.R. Kennedy. Localised bacterial infection in a distributed model for tissue inflammation. *J. Math. Biol.*, 16:141–163, 1983.

[352] N.M. Le Douarin. *The Neural Crest*. Cambridge University Press, Cambridge, UK, 1982.

[353] B. Lee, L. Mitchell, and G. Buchsbaum. Rheology of the vitreous body. Part 1: Viscoelasticity of human vitreous. *Biorheology*, 29:521–533, 1993.

[354] B. Lee, L. Mitchell, and G. Buchsbaum. Rheology of the vitreous body: Part 2. Viscoelasticity of bovine and porcine vitreous. *Biorheology*, 31:327–338, 1994a.

[355] B. Lee, L. Mitchell, and G. Buchsbaum. Rheology of the vitreous body: Part 3. Concentration of electrolytes, collagen and hyaluronic acid. *Biorheology*, 31:339–351, 1994b.

[356] R. Lefever and O. Lejeune. On the origin of tiger bush. *Bull. Math. Biol.*, 59:263–294, 1997.

[357] O. Lejeune and M. Tlidi. A model for the explanation of vegetation stripes (tiger bush). *J. Veg. Sci.*, 10:201–208, 1999.

[358] T. Lenoir. The eternal laws of form: morphotypes and the conditions of existence in Goethe's biological thought. *J. Social Biol. Struct.*, 7:317–324, 1984.

[359] S.A. Levin. Models of population dispersal. In S. Busenberg and K. Cooke, editors, *Differential Equations and Applications to Ecology, Epidemics and Population Problems*, pages 1–18. Academic, New York, 1981a.

[360] S.A. Levin. The role of theoretical ecology in the description and understanding of populations in heterogeneous environments. *Amer. Zool.*, 21:865–875, 1981b.

[361] J. Levinton. Developmental constraints and evolutionary saltations: a discussion and critique. In G. Stebbins and F. Ayala, editors, *Genetics and Evolution*. Plenum, New York, 1986.

[362] A.E. Leviton and S.C. Anderson. Description of a new species of *Cyrtodactylus* from Afghanistan with remarks on the status of *Gymnodactylus longpipes* and *Cyrtodactylus fedtschenkoi*. *J. Herpes*, 18:270–276, 1984.

[363] M.A. Lewis. Variability, patchiness, and jump dispersal in the spread of an invading population. In D. Tilman and P. Kareiva, editors, *Spatial Ecology. The Role of Space in Population Dynamics and Interspecific Interactions*, pages 46–69. Princeton University Press, Princeton, NJ, 1997.

[364] M.A. Lewis and P. Moorcroft. ESS analysis of mechanistic models for territoriality: The value of signals in spatial resource partitioning. *J. Theor. Biol.*, 210:449–461, 2001.

[365] M.A. Lewis and J.D. Murray. Analysis of stable two-dimensional patterns in contractile cytogel. *J. Nonlin. Sci.*, 1:289–311, 1991.

[366] M.A. Lewis and J.D. Murray. Modelling territoriality and wolf-deer interactions. *Nature*, 366:738–740, 1993.

[367] M.A. Lewis and G. Schmitz. Biological invasion of an organism with separate mobile and stationary states: modeling and anlaysis. *Forma*, 11:1–25, 1996.

[368] M.A. Lewis, G. Schmitz, P. Kareiva, and J.T. Trevors. Models to examine containment and spread of genetically engineered organisms. *J. Mol. Ecol.*, 5:165–175, 1996.

[369] M.A. Lewis, K.A.J. White, and J.D. Murray. Analysis of a model for wolf territories. *J. Math. Biol.*, 35:749–774, 1997.

[370] J.D. Lewis-Williams and T.A. Dowson. The signs of all times. Entopic phenomena in upper palaeolithic art. *Current Anthropol.*, 29:201–245, 1988.

[371] B. C. Liang and M. Weil. Locoregional approaches to therapy with gliomas as the paradigm. *Current Opinion in Oncology*, 10:201–206, 1998.

[372] S. Lindow. Competitive exclusion of epiphytic bacteria. *Appl. Envt. Microbiol.*, 53:2520–2527, 1987.

[373] G. Lindquist. The healing of skin defects. an experimental study of the white rat. *Acta Chirurgica Scandinavica*, 94:1–163, 1946. Supplement 107.

[374] P.L. Lions. On the existence of positive solutions of semilinear elliptic equations. *SIAM. Rev.*, 24:441–467, 1982.

[375] C.D. Little, V. Mironov, and E.H. Sage, editors. *Vascular Morphogenesis: In Vivo, In Vitro and In Mente*. Birkhaüser, Boston, 1998.

[376] H.G. Lloyd. Past and present distribution of red and grey squirrels. *Mamm. Res.*, 13:69–80, 1983.

[377] H.G. Lloyd, B. Jensen, J.L. Van Haaften, F.J.J. Niewold, A. Wandeler, K. Boegel, and A.A. Arata. Annual turnover of fox populations in Europe. *Zbl. Vet. Med.*, 23:580–589, 1976.

[378] D.Z. Loesch. *Quantitative Dermatoglyphics: Classification, Genetics, and Pathology*. Oxford University Press, Oxford, 1983.

[379] M.T. Longaker and N.S. Adzick. The biology of fetal wound healing: A review. *Plastic and Reconstructive Surg.*, 87:788–798, 1991.

[380] I. Loudon, editor. *Biomathematics and Cell Kinetics*. Oxford University Press, Oxford, 1997.

[381] S.R. Lubkin and J.D. Murray. A mechanism for early branching in lung morphogenesis. *J. Math. Biol.*, 34:77–94, 1995.

[382] D. Ludwig, D.G. Aronson, and H.F. Weinberger. Spatial patterning of the spruce budworm. *J. Math. Biol.*, 8:217–258, 1979.

[383] A.S. Lyons and R.J. Petrucelli. *Medicine: An Illustrated History*. Harry N. Abrams, New York, 1978.

[384] D.W. MacDonald. *Rabies and Wildlife: A Biologist's Perspective*. Oxford University Press, Oxford, 1980.

[385] A. MacKenzie, M.W.J. Ferguson, and P.T. Sharpe. Expression patterns of the homeobox gene, Hox-8, in the mouse embryo suggest a role in specifying tooth initiation and shape. *Development*, 115:403–420, 1992.

[386] A. MacKenzie, J.L. Leeming, A.K. Jowett, M.W.J. Ferguson, and P.T. Sharpe. The homeobox gene, Hox-7.1 has specific regional and temporal expression patterns during early murine craniofacial embryogenesis, especially tooth development *in vivo* and *in vitro*. *Development*, 111:269–285, 1991.

[387] K. MacKinnon. Competition between red and grey squirrels. *Mamm. Res.*, 8:185–190, 1978.

[388] M.R. Madden, E. Nolan, J.L. Finkelstein, R.W. Yurt, J. Smeland, C.W. Goodwin, J. Hefton, and L. Staiano-Coico. Comparison of an occlusive and a semi-occlusive dressing and the effect of the wound eudate upon keratinocyte proliferation. *J. Trauma*, 29:924–31, 1989.

[389] P.F.A. Maderson. Some developmental problems of the reptilian integument. In C. Gans, F. Billet, and P.F.A. Maderson, editors, *Biology of the Reptilia*, volume 14, pages 523–598. John Wiley, New York, 1985.

[390] J.B. Madison and R.R. Gronwall. Influence of wound shape on wound contraction in horses. *Am. J. Vet. Res.*, 53:1575–1577, 1992.

[391] P.K. Maini. Travelling waves in biology, chemistry, ecology and medicine. Part 1. *FORMA (Special Issue: P.K. Maini, editor)*, 10:145–280, 1995.

[392] P.K. Maini. Travelling waves in biology, chemistry, ecology and medicine. Part 2. *FORMA (Special Issue: P.K. Maini, editor)*, 11:1–80, 1996.

[393] P.K. Maini. Bones, feathers, teeth and coatmarkings: a unified model. *Science Progress*, 80:217–229, 1997.

[394] P.K. Maini. Some mathematical models for biological pattern formation. In M.A.J. Chaplain, G.D. Singh, and J.C. McLachlan, editors, *On Growth and Form. Spatio-Temporal Pattern Formation in Biology*, pages 111–128. John Wiley, New York, 1999.

[395] P.K. Maini and J.D. Murray. A nonlinear analysis of a mechanical model for biological pattern formation. *SIAM J. Appl. Math.*, 48:1064–1072, 1988.

[396] P.K. Maini and H.G. Othmer, editors. *Mathematical Models for Biological Pattern Formation.* Springer-Verlag, New York, 2000.

[397] P.K. Maini, M.R. Myerscough, K.H. Winters, and J.D. Murray. Bifurcating spatially heterogeneous solutions in a chemotaxis model for biological pattern formation. *Bull. Math. Biol.*, 53:701–719, 1991.

[398] G. Majno. *The Healing Hand. Man and Wound in the Ancient World.* Harvard University Press, Cambridge, 1975.

[399] V.S. Manoranjan and A.R. Mitchell. A numerical study of the Belousov–Zhabotinskii reaction using Galerkin finite element methods. *J. Math. Biol.*, 16:251–260, 1983.

[400] D. Manoussaki. *Modelling Formation of Planar Vascular Networks in vitro.* PhD thesis. Department of Applied Mathematics, University of Washington, Seattle, WA, 1996.

[401] D. Manoussaki, S.R. Lubkin, R.B. Vernon, and J.D. Murray. A mechanical model for the formation of vascular networks in vitro. *Acta Biotheoretica*, 44:271–282, 1996.

[402] E.L. Margetts. Trepanation of the skull by the medicine-men of primitive cultures, with particular reference to present-day native East African practice. In D. Brothwell and A.T. Sandison, editors, *Diseases in Antiquity*, pages 673–701. Charles C. Thomas, Springfield, IL, 1967.

[403] P. Martin. Wound healing—Aiming for perfect skin regeneration. *Science*, 276:75–81, 1997.

[404] P. Martin and J. Lewis. The mechanics of embryonic skin wound healing-limb bud lesions in mouse and chick embryos. In N.S. Adzick and M.T. Longaker, editors, *Fetal Wound Healing*, pages 265–279. Elsevier, New York, 1991.

[405] P. Martin and J. Lewis. Actin cables and epidermal movement in embryonic wound healing. *Nature*, 360:179–183, 1992.

[406] M. Marusic, Z. Bajzer, J.P. Freyer, and S. Vuk-Pavlovic. Analysis of growth of multicellular tumour spheroids by mathematical models. *Cell Prolif.*, 27:73–94, 1994.

[407] H. Matano. Asymptotic behaviour and stability of solutions of semilinear diffusion equations. *Publ. Res. Inst. Math. Sci. Kyoto*, 15:401–454, 1979.

[408] Y. Matsukado, C.S. McCarthy, and J.W. Kernohan. The growth of glioblastoma multiforme (asytrocytomas, grades 3 and 4) in neurosurgical practice. *J. Neurosurg.*, 18:636–644, 1961.

[409] M.M. Matsushita, J. Wakita, H. Itoh, T. Arai, T. Matsuyama, H. Sakaguchi, and M. Mimura. Formation of colony patterns by a bacterial population. *Physica A*, 274:190–199, 1999.

[410] M.M. Matsushita, J. Wakita, H. Itoh, I. Rifols, T. Matsuyama, H. Sakaguchi, and M. Mimura. Interface growth and pattern formation in bacterial colonies. *Physica A*, 249:517–524, 1998.

[411] T. Matsuyama and M. Matsushita. Fractal morphogenesis by a bacterial cell population. *Crit. Rev. Microbiol.*, 19:117–135, 1993.

[412] C. McEvedy. The bubonic plague. *Sci. Amer.*, pages 74–79, February 1988.

[413] M.H. McGrath and R.H. Simon. Wound geometry and the kinetics of wound contraction. *Plast. and Reconst. Surg.*, 72:66–73, 1983.

[414] H.P. McKean. Nagumo's equation. *Adv. in Math.*, 4:209–223, 1970.

[415] B.K. McNab. Bio-energetics and the determination of home range size. *Am. Nat.*, 97:133–140, 1963.

[416] L.D. Mech. *The Ecology and Behavior of an Endangered Species*. Natural History Press, Garden City, NY, 1970.

[417] L.D. Mech. Wolf numbers in the Superior National Forest of Minnesota. Technical Report, US Forest Service, 1973. Research Paper NC-97.

[418] L.D. Mech. *The Way of the Wolf*. Voyageur, Stillwater, MN, 1991.

[419] H. Meinhardt. *Models of Biological Pattern Formation*. Academic, London, 1982.

[420] H. Meinhardt. Hierarchical inductions of cell states: a model for segmentation of *Drosophila*. *J. Cell Sci.*, 4:357–381, 1986. (Suppl.).

[421] H. Meinhardt. *The Algorithmic Beauty of Sea Shells*. Springer-Verlag, Berlin, 1995.

[422] H. Meinhardt. On pattern and growth. In M.A.J. Chaplain, G.D. Singh, and J.C. McLachlan, editors, *On Growth and Form. Spatio-Temporal Pattern Formation in Biology*, pages 129–148. John Wiley, New York, 1999.

[423] H. Meinhardt. Beyond spots and stripes: generation of more complex patterns and modifications and additions of the basic reaction. In P.K. Maini and H.G. Othmer, editors, *Mathematical Models for Biological Pattern Formation*, pages 143–164. Springer-Verlag, New York, 2000.

[424] H. Meinhardt and M. Klingler. A model for pattern generation on the shells of molluscs. *J. Theor. Biol.*, 126:63–89, 1987.

[425] J.H. Merkin, V. Petrov, S.K. Scott, and K. Showalter. Wave-induced chemical chaos. *Phys. Rev. Letters*, 76:546–549, 1996.

[426] S. Michelson and J. Leith. Autocrine and paracrine growth factors in tumor growth: a mathematical model. *Bull. Math. Biol.*, 53:639–656, 1991.

[427] S. Michelson, B.E. Miller, A.S. Glicksman, and J. Leith. Tumor micro-ecology and competitive interactions. *J. Theor. Biol.*, 128:233–246, 1987.

[428] A.S. Mikhailov and V.I. Krinsky. Rotating spiral waves in excitable media: The analytical results. *Physica D*, 9:346–371, 1983.

[429] M. Mimura and T. Tsujikawa. Aggregating pattern dynamics in a chemotaxis model including growth. *Physica A*, 230:499–543, 1996.

[430] M. Mimura and M. Yamaguti. Pattern formation in interacting and diffusing systems in population biology. *Adv. Biophys.*, 15:19–65, 1982.

[431] M. Mimura, H. Sakaguchi, and M. Matsushita. Reaction-diffusion modelling of bacterial colony patterns. *Physica A*, 282:283–303, 2000.

[432] M. Mina and E.J. Kollar. The induction of odontogenesis in non-dental mesenchyme combined with early murine mandibular arch epithelium. *Archs. Oral Biol.*, 32(2):123–127, 1987.

[433] J.E. Mittenthal and R.M. Mazo. A model for shape generation by strain and cell-cell adhesion in the epithelium of an arthropod leg segment. *J. Theor. Biol.*, 100:443–483, 1983.

[434] A. Mogilner and L. Edelstein-Keshet. Spatio-angular order in populations of self-aligning objects: formation of oriented patches. *Physica D*, 89:346–367, 1996.

[435] A. Mogilner and G.F. Oster. Cell motility driven by actin polimerization. *Biophys. J.*, 71:3030–3045, 1996.

[436] D. Mollison. Dependence of epidemic and population velocities on basic parameters. *Math. Biosci.*, 107:255–287, 1991.

[437] P.R. Montague and M.J. Friedlander. Morphogenesis and territorial coverage by isolated mammalian retinal ganglion cells. *J. Neurosci.*, 11:1440–1457, 1991.

[438] P. Moorcroft. *Territoriality and Carnivore Home Ranges*. PhD thesis, Princeton University, Princeton, NJ, 1997.

[439] P. Moorcroft and M.A. Lewis. *Home Range Patterns: Mechanistic Approaches to the Analysis of Animal Movement*. Princeton University Press, Princeton, NJ, 2001.

[440] P. Moorcroft, M.A. Lewis, and R. Crabtree. Analysis of coyote home range using a mechanistic home range model. *Ecology*, 80:1656–1665, 1999.

[441] L.H. Morgan. *Ancient Society*. Charles H. Kerr, Chicago, 1877.

[442] S.C. Müller, T. Plesser, and B. Hess. The structure of the core of the spiral wave in the Belousov–Zhabotinskii reaction. *Science*, 230:661–663, 1985.

[443] S.C. Müller, T. Plesser, and B. Hess. Two-dimensional spectrophotometry and pseudo-color representation of chemical patterns. *Naturwiss.*, 73:165–179, 1986.

[444] S.C. Müller, T. Plesser, and B. Hess. Two-dimensional spectrophotometry of spiral wave propagation in the Belousov–Zhabotinskii reaction. I. Experiments and digital representation. II. Geometric and kinematic parameters. *Physica D*, 24:71–96, 1987.

[445] J.D. Murray. Singular perturbations of a class of nonlinear hyperbolic and parabolic equations. *J. Maths. and Physics*, 47:111–133, 1968.

[446] J.D. Murray. Perturbation effects on the decay of discontinuous solutions of nonlinear first order wave equations. *SIAM J. Appl. Math.*, 19:273–298, 1970a.

[447] J.D. Murray. On the Gunn effect and other physical examples of perturbed conservation equations. *J. Fluid Mech.*, 44:315–346, 1970b.

[448] J.D. Murray. On Burgers' model equations for turbulence. *J. Fluid Mech.*, 59:263–279, 1973.

[449] J.D. Murray. Non-existence of wave solutions for a class of reaction diffusion equations given by the Volterra interacting-population equations with diffusion. *J. Theor. Biol.*, 52:459–469, 1975.

[450] J.D. Murray. On travelling wave solutions in a model for the Belousov–Zhabotinskii reaction. *J. Theor. Biol.*, 52:329–353, 1976.

[451] J.D. Murray. *Nonlinear Differential Equation Models in Biology*. Clarendon, Oxford, 1977.

[452] J.D. Murray. A pattern formation mechanism and its application to mammalian coat markings. In *'Vito Volterra' Symposium on Mathematical Models in Biology. Accademia dei Lincei, Rome, Dec. 1979*, volume 39 of *Lect. Notes in Biomathematics*, pages 360–399. Springer-Verlag, Berlin-Heidelberg-New York, 1980.

[453] J.D. Murray. On pattern formation mechanisms for Lepidopteran wing patterns and mammalian coat markings. *Phil. Trans. R. Soc. Lond. B*, 295:473–496, 1981a.

[454] J.D. Murray. A pre-pattern formation mechanism for animal coat markings. *J. Theor. Biol.*, 88:161–199, 1981b.

[455] J.D. Murray. Parameter space for Turing instability in reaction diffusion mechanisms: a comparison of models. *J. Theor. Biol.*, 98:143–163, 1982.

[456] J.D. Murray. On a mechanical model for morphogenesis: Mesenchymal patterns. In W. Jäger and J.D. Murray, editors, *Conference on: Modelling of Patterns in Space and Time, Heidelberg 1983*. Lecture Notes in Biomathematics series, 55:279–291, Springer-Verlag, Berlin-Heidelberg-New York, 1983.

[457] J.D. Murray. *Asymptotic Analysis*. Springer-Verlag, Berlin-Heidelberg-New York, second edition, 1984.

[458] J.D. Murray. How the leopard gets its spots. *Sci. Amer.*, 258(3):80–87, 1988.

[459] J.D. Murray. Modelling the pattern generating mechanism in the formation of stripes on alligators. In B. Simon, A. Truman, and I.M. Davies, editors, *IXth Intern. Congr. Mathematical Physics 1988*, pages 208–213, 1989. Adam Hilger, Bristol.

[460] J.D. Murray. Complex pattern formation and tissue interaction. In J. Demongeot and V. Capasso, editors, *1st Europ. Confer. on Applics. Maths. to Medic. and Biol. 1990*, pages 495–505, Winnipeg, Canada, 1993. Wuerz Publishing.

[461] J.D. Murray and G.C. Cruywagen. Threshold bifurcation in tissue interaction models for spatial pattern formation. *Proc. R. Soc. Lond. A*, 443:1–16, 1994.

[462] J.D. Murray and C.L. Frenzen. A cell justification for Gompertz' equation. *SIAM J. Appl. Math.*, 46:614–629, 1986.

[463] J.D. Murray and P.M. Kulesa. On a dynamic reaction-diffusion mechanism: the spatial patterning of teeth primordia in the alligator. *J. Chem. Soc., Faraday Trans.*, 92:2927–2932, 1996.

[464] J.D. Murray and P.K. Maini. A new approach to the generation of pattern and form in embryology. *Sci. Prog. (Oxf.)*, 70:539–553, 1986.

[465] J.D. Murray and M.R. Myerscough. Pigmentation pattern formation on snakes. *J. Theor. Biol.*, 149:339–360, 1991.

[466] J.D. Murray and G.F. Oster. Cell traction models for generating pattern and form in morphogenesis. *J. Math. Biol.*, 19:265–279, 1984a.

[467] J.D. Murray and G.F. Oster. Generation of biological pattern and form. *IMA J. Math. Appl. in Medic. and Biol.*, 1:51–75, 1984b.

[468] J.D. Murray and W.L. Seward. On the spatial spread of rabies among foxes with immunity. *J. Theor. Biol.*, 156:327–348, 1992.

[469] J.D. Murray, D.C. Deeming, and M.W.J. Ferguson. Size-dependent pigmentation-pattern formation in embryos of *Alligator mississippiensis*: Time of initiation of pattern generation mechanism. *Proc. R. Soc. Lond. B*, 239:279–293, 1990.

[470] J.D. Murray, P.K. Maini, and R.T. Tranquillo. Mechanical models for generating biological pattern and form in development. *Physics Reports*, 171:60–84, 1988.

[471] J.D. Murray, D. Manoussaki, S.R. Lubkin, and R.B. Vernon. A mechanical theory of *in vitro* vascular network formation. In C.D. Little, V. Mironov, and E.H. Sage, editors, *Vascular Morphogenesis: In Vivo, In Vitro and In Mente*, pages 173–188, Birkhaüser, Boston, 1998.

[472] J.D. Murray, G.F. Oster, and A.K. Harris. A mechanical model for mesenchymal morphogenesis. *J. Math. Biol.*, 17:125–129, 1983.

[473] J.D. Murray, E.A. Stanley, and D.L. Brown. On the spatial spread of rabies among foxes. *Proc. R. Soc. Lond. B*, 229:111–150, 1986.

[474] J.D. Murray and J.E.R. Cohen. On nonlinear convection dispersal effects in an interacting population model. *SIAM J. Appl. Math.*, 43:66–78, 1983.

[475] J.D. Murray and R.P. Sperb. Minimum domains for spatial patterns in a class of reaction diffusion equations. *J. Math. Biol.*, 18:169–184, 1983.

[476] J.D. Murray and K.R. Swanson. On the mechanical theory of biological pattern formation with applications to wound healing and angiogenesis. In M.A.J. Chaplain, G.D. Singh, and J.C. McLachlan, editors, *On Growth and Form: Spatio-Temporal Pattern Formation in Biology*, pages 251–285, John Wiley, Chichester, UK, 1999.

[477] J.D. Murray, J. Cook, S.R. Lubkin, and R.C. Tyson. Spatial pattern formation in biology: I. Dermal wound healing. II. Bacterial patterns. *J. Franklin Inst.*, 335:303–332, 1998.

[478] M.R. Myerscough and J.D. Murray. Analysis of propagating pattern in a chemotaxis system. *Bull. Math. Biol.*, 54:77–94, 1992.

[479] M.R. Myerscough, P.K. Maini, J.D. Murray, and K.H. Winters. Two-dimensional pattern formation in a chemotactic system. In T.L. Vincent, A.I. Mees, and L.S. Jennings, editors, *Dynamics of Complex Interconnected Biological Systems*, pages 65–83. Birkhauser, Boston, 1990.

[480] H. Nagawa and Y. Nakanishi. Mechanical aspects of the mesenchymal influence on epithelial branching morphogenesis of mouse salivary gland. *Development*, 101:491–500, 1987.

[481] B.N. Nagorcka. The role of a reaction-diffusion system in the initiation of skin organ primordia. I. The first wave of initiation. *J. Theor. Biol.*, 121:449–475, 1986.

[482] B.N. Nagorcka. A pattern formation mechanism to control spatial organisation in the embryo of *Drosophila melanogaster*. *J. Theor. Biol.*, 132:277–306, 1988.

[483] B.N. Nagorcka and D.A. Adelson. Pattern formation mechanisms in skin and hair: Some experimental tests. In M.A.J. Chaplain, G.D. Singh, and J.C. McLachlan, editors, *On Growth and Form: Spatio-Temporal Pattern Formation in Biology*, pages 89–110, John Wiley, Chichester, UK, 1999.

[484] B.N. Nagorcka and J.R. Mooney. The role of a reaction-diffusion system in the formation of hair fibres. *J. Theor. Biol.*, 98:575–607, 1982.

[485] B.N. Nagorcka and J.R. Mooney. The role of a reaction-diffusion system in the initiation of primary hair follicles. *J. Theor. Biol.*, 114:243–272, 1985.

[486] B.N. Nagorcka and J.R. Mooney. From stripes to spots: prepatterns which can be produced in the skin by reaction-diffusion systems. *IMA J. Math. Appl. Med. and Biol.*, 9:249–267, 1992.

[487] B.N. Nagorcka, V.S. Manoranjan, and J.D. Murray. Complex spatial patterns from tissue interactions—An illustrative model. *J. Theor. Biol.*, 128:359–374, 1987.

[488] J. M . Nazzaro and E. A. Neuwelt. The role of surgery in the management of supranterorial intermediate and high-grade astrocytomas in adults. *J. Neurosurg.*, 73:331–344, 1990.

[489] M.E. Nelson and L.D. Mech. Wolf predation risk associated with white-tailed deer movements. *Can. J. Zool.*, 69:2696–2699, 1991.

[490] P.C. Newell. Attraction and adhesion in the slime mold *Dictyostelium*. In J.E. Smith, editor, *Fungal Differentiation: A Contemporary Synthesis*, pages 43–71. Marcel Dekker, New York, 1983.

[491] H.F. Nijhout. Wing pattern formation in Lepidoptera: a model. *J. Exp. Zool.*, 206:119–136, 1978.

[492] H.F. Nijhout. Pattern formation in Lepidopteran wings: determination of an eyespot. *Devl. Biol.*, 80:267–274, 1980a.

[493] H.F. Nijhout. Ontogeny of the color pattern formation on the wings of Precis coenia (Lepidoptera: Nymphalidae). *Devl. Biol.*, 80:275–288, 1980b.

[494] H.F. Nijhout. Colour pattern modification by coldshock in Lepidoptera. *J. Embryol. Exp. Morph.*, 81:287–305, 1984.

[495] H.F. Nijhout. The developmental physiology of colour patterns in Lepidoptera. *Adv. Insect Physiol.*, 18:181–247, 1985a.

[496] H.F. Nijhout. Cautery-induced colour patterns in *Precis coenia* (Lepidoptera: Zool. Nymphalidae). *J. Embryol. Exp. Morph.*, 86:191–302, 1985b.

[497] H.F. Nijhout. *The Development and Evolution of Butterfly Wing Patterns*. Smithsonian Institution Press, Washington, DC, 1991.

[498] J.V. Noble. Geographic and temporal development of plagues. *Nature*, 250:726–728, 1974.

[499] B. Obrink. Epithelial cell adhesion molecules. *Exp. Cell Res.*, 163:1–21, 1986.

[500] G. Odell, G.F. Oster, B. Burnside, and P. Alberch. The mechanical basis for morphogenesis. *Dev. Biol.*, 85:446–462, 1981.

[501] M. Ohgiwara, M. Matsushita, and T. Matsuyama. Morphological changes in growth phenomena of bacterial colony patterns. *J. Phys. Soc. Japan*, 61:816–822, 1992.

[502] M. Okajima. A methodological approach to the development of epidermal ridges. *Prog. in Dermatol. Res.*, 20:175–188, 1982.

[503] M. Okajima and L. Newell-Morris. Development of dermal ridges in volar skin of fetal pigtailed macaques *Macaca nemestrina*. *Amer. J. Anat.*, 183:323–327, 1988.

[504] A. Okubo. *Diffusion and Ecological Problems: Mathematical Models*. Springer-Verlag, Berlin-Heidelberg-New York, 1980.

[505] A. Okubo. Dynamical aspects of animal grouping: swarms, schools, flocks and herds. *Adv. Biophys.*, 22:1–94, 1986.

[506] A. Okubo, P.K. Maini, M.H. Williamson, and J.D. Murray. On the spatial spread of the grey squirrel in Britain. *Proc. R. Soc. Lond. B*, 238:113–125, 1989.

[507] L. Olsen, J.A. Sherratt, and P.K. Maini. A mechanochemical model for adult dermal wound contraction and the permanence of the contracted tissue displacement profile. *J. Theor. Biol.*, 177:113–128, 1995.

[508] L. Olsen, J.A. Sherratt, and P.K. Maini. A mathematical model for fibro-proliferative wound healing disorders. *Bull. Math. Biol.*, 58:787–808, 1996.

[509] M.S. O'Reilly, T. Boehm, Y. Shing, N. Fukai, G. Vasios, W.S. Lane, E. Flynn, J.R. Birkhead, B.R. Olsen, and J. Folkman. Endostatin: an endogenous inhibitor of angiogenesis and tumor growth. *Cell*, 88:277–285, 1997.

[510] M.S. O'Reilly, L. Holmgren, C.C. Chen, and J. Folkman. Angiostatin induces and sustains dormancy of human primary tumors in mice. *Nature Medicine*, 2:689–692, 1996.

[511] P.J. Ortoleva and S.L. Schmidt. The structure and variety of chemical waves. In R.J. Field and M. Burger, editors, *Oscillations and Travelling Waves in Chemical Systems*, pages 333–418. John Wiley, New York, 1985.

[512] J.W. Osborn. New approach to Zahnreihen. *Nature, Lond.*, 225:343–346, 1970.

[513] J.W. Osborn. The ontogeny of tooth succession in *Lacerta vivpara Jacquin*. *Proc. R. Soc. Lond.*, B179:261–289, 1971.

[514] J.W. Osborn. Morphogenetic gradients: fields versus clones. In P.M. Butler and K.A. Joysey, editors, *Development, Function and Evolution of Teeth*, pages 171–201. Academic, London and New York, 1978.

[515] J.W. Osborn. A model simulating tooth morphogenesis without morphogenes. *J. Theor. Biol.*, 165:429–445, 1993.

[516] Gerold F. Oster. 'Phosphenes'—The patterns we see when we close our eyes are clues to how the eye works. *Sci. Amer.*, pages 82–88, February, 1970.

[517] G. F. Oster and G.M. Odell. The mechanochemistry of cytogels. *Physica D*, 12:333–350, 1984.

[518] G.F. Oster. On the crawling of cells. *J. Embryol. Exp. Morphol.*, 83:329–364, 1984. (Suppl.).

[519] G.F. Oster and P. Alberch. Evolution and bifurcation of developmental programs. *Evolution*, 36:444–459, 1982.

[520] G.F. Oster and J.D. Murray. Pattern formation models and developmental constraints. *J. Exp. Zool.*, 251:186–202, 1989.

[521] G.F. Oster, J.D. Murray, and A.K. Harris. Mechanical aspects of mesenchymal morphogenesis. *J. Embryol. Exp. Morphol.*, 78:83–125, 1983.

[522] G.F. Oster, J.D. Murray, and P.K. Maini. A model for chondrogenic condensations in the developing limb: the role of extracellular matrix and cell tractions. *J. Embryol. Exp. Morphol.*, 89:93–112, 1985a.

[523] G.F. Oster, J.D. Murray, and G.M. Odell. The formation of microvilli. In *Molecular Determinants of Animal Form*, pages 365–384. Alan R. Liss, New York, 1985b.

[524] G.F. Oster, N. Shubin, J.D. Murray, and P. Alberch. Evolution and morphogenetic rules. The shape of the vertebrate limb in ontogeny and phylogeny. *Evolution*, 42:862–884, 1988.

[525] H. Othmer. *Interactions of Reaction and Diffusion in Open Systems*. PhD thesis, Chemical Engineering Department and University of Minnesota, 1969.

[526] H.G. Othmer. Current problems in pattern formation. *Amer. Math. Assoc. Lects. on Math. in the Life Sciences*, 9:57–85, 1977.

[527] H.G. Othmer, P.K. Maini, and J.D. Murray, editors. *Mathematical Models for Biological Pattern Formation*. Plenum, New York, 1993.

[528] J.H. Ottaway. Normalization in the fitting of data by iterative methods: application to tracer kinetics and enzyme kinetics. *Biochem. J.*, 134:729–736, 1973.

[529] H. Otto. Die Beschuppung der Brevilinguir und Ascaleten. *Jenaische Zeit. Wiss.*, 44:193–252, 1908.

[530] Q. Ouyang and H.L. Swinney. Transition from a uniform state to hexagonal and striped Turing patterns. *Nature*, 352:610–612, 1991.

[531] Q. Ouyang, V. Castets, J. Boissonade, J.C. Roux, P. De Kepper, and H.L. Swinney. Sustained patterns in chlorite-iodide reactions in a one-dimensional reactor. *J. Chem. Phys.*, 95:352–360, 1990.

[532] Q. Ouyang, G.H. Gunaratne, and H.L. Swinney. Rhombic patterns: broken hexagonal symmetry. *Chaos*, 3:707–711, 1993.

[533] M.R. Owen and J.A. Sherratt. Pattern formation and spatiotemporal irregularity in a model for macrophage-tumour interactions. *J. Theor. Biol.*, 189:63–80, 1997.

[534] K.J. Painter. Models for pigment pattern formation in the skin of fishes. In P.K. Maini and H.G. Othmer, editors, *Mathematical Models for Biological Pattern Formation*, pages 59–82. Springer, New York, 2000.

[535] K.J. Painter, P.K. Maini, and H.G. Othmer. Stripe formation in juvenile *Pomacanthus* explained by a generalised Turing mechanism with chemotaxis. *Proc. Nat. Acad. Sci. (U.S.)*, 96:5549–5554, 1999.

[536] J.-L. Pallister. *Introduction to "On Monsters and Marvels" by Ambroise Paré*. University of Chicago Press, Chicago, 1982.

[537] A.M. Partanen, P. Ekblom, and I. Thesleff. Epidermal growth factor inhibits morphogenesis and cell differentiation in cultured mouse embryonic teeth. *Dev. Biol.*, 111:84–94, 1985.

[538] M. Pascual. Diffusion-induced chaos in a spatial predator-prey system. *Proc. R. Soc. Lond. B*, 251:1–7, 1993.

[539] P.P. Pastoret. Evolution of fox rabies in Belgium and the European Union. *Bull. et Mem. Acad. Roy. Medicine de Belgique*, 153:93–98, 1998.

[540] S. Patan, L.L. Munn, and R.K. Jain. Intussusceptive microvascular growth in a human colon adenocarcinoma xenograft: a novel mechanism of tumor angiogenesis. *Microvascular Research*, 51:229–249, 1996.

[541] E. Pate and H.G. Othmer. Applications of a model for scale-invariant pattern formation. *Differentiation*, 28:1–8, 1984.

[542] M. Patou. Analyse de la morphogenèse du pied des oiseaux a l'aise de melanges cellulaires interspecifiques. I. Étude morphologique. *J. Embryol. Exp. Morphol.*, 29:175–196, 1973.

[543] E.E. Peacock. *Wound Repair*. W.B. Saunders, Philapdelphia, 1984.

[544] R. Penrose. The topology of ridge systems. *Ann. Hum. Genet. Lond.*, 42:435–444, 1979.

[545] A.S. Perelson, P.K. Maini, J.D. Murray, J.M. Hyman, and G.F. Oster. Nonlinear pattern selection in a mechanical model for morphogenesis. *J. Math. Biol.*, 24:525–541, 1986.

[546] A.J. Perumpanani, D.L. Simmons, A.J.H. Gearing, K.M. Miller, G. Ward, J. Norbury, M. Scheemann, and J.A. Sherratt. Extracellular matrix-mediated chemotaxis can impede cell migration. *Proc. R. Soc. Lond. B*, 265:2347–2352, 1998.

[547] R. Peters and L.D. Mech. Scent-marking in wolves. *Am. Nat.*, 63:628–637, 1975.

[548] R.O. Peterson. Ecological studies of wolves on Isle Royale. Technical Report, School of Forestry, Michigan Technical University, 1999.

[549] W.M. Petroll, H.D. Cavanagh, P. Barry, P. Andrews, and J.V. Jester. Quantitative analysis of stress fibre orientation during corneal wound contraction. *J. Cell Sci.*, 104:353–363, 1993.

[550] V. Petrov, S.K. Scott, and K. Showalter. Excitability, wave reflection, and wave splitting in a cubic autocatalysis reaction-diffusion system. *Phil. Trans. R. Soc. A*, 347:631–642, 1994.

[551] B.R. Phillips, J.A. Quinn, and H. Goldfine. Random motility of swimming bacteria: Single cells compared to cell populations. *AIChE J.*, 40:334–348, 1994.

[552] G. J. Pilkington. Glioma heterogeneity *in vitro*: the significance of growth factors and gangliosides. *Neuropathol. Appl. Neurobiol.*, 18:434–442, 1992.

[553] G. J. Pilkington. The paradox of neoplastic glial cell invasion of the brain and apparent metastatic failure. *Anticancer Res.*, 17:4103–4106, 1997a.

[554] G. J. Pilkington. In vitro and in vivo models for the study of brain tumor invasion. *Anticancer Res.*, 17:4107–4110, 1997b.

[555] R.I. Pocock. Description of a new species of cheetah (*Acinonyx*). *Proc. R. Soc. Lond.*, 1927:245–252, 1927.

[556] T.D. Pollard. Actin. *Curr. Op. Cell Biol.*, 2:33–40, 1990.

[557] A. Portmann. *Animal Forms and Patterns. A Study of the Appearance of Animals. (English translation).* Faber and Faber, London, 1952.

[558] C.S. Potten, W.J. Hume, and E.K. Parkinson. Migration and mitosis in the epidermis. *Br. J. Dermatol.*, 111:695–699, 1984.

[559] R.J. Price and R. Skalak. Circumferential wall stress as a mechanism for arteriolar rarefaction and proliferation in a network model. *Microvascular Research*, 47:188–202, 1994.

[560] T. Price and M. Pavelka. Evolution of a colour pattern: history, development, and selection. *J. Evol. Biol.*, 9:451–470, 1996.

[561] G. Prota. *Melanins and Melanogenesis*. Academic, London, 1992.

[562] E. Purcell. Life at low Reynolds number. *Amer. J. Phys.*, 45:1–11, 1977.

[563] G. Radice. The spreading of epithelial cells during wound closure in *Xenopus* larvae. *Dev. Biol.*, 76:26–46, 1980.

[564] La Rage. *Centre Nationale d'Études sur la Rage*. CERN, Paris, 1977.

[565] G.F. Raggett. Modelling the Eyam plague. *Bull. Inst. Math. and its Applic.*, 18:221–226, 1982.

[566] R. Ramina, M. C. Neto, M. Meneses, W. O. Arruda, S. C. Hunhevicz, and A. A. Pedrozo. Management of deep-seated gliomas. *Critical Rev. Neurosurg.*, 9:34–40, 1999.

[567] M. Rawles. Tissue interactions in scale and feather development as studied in dermal-epidermal recombinations. *J. Embryol. Exp. Morph.*, 11:765–789, 1963.

[568] J.C. Reynolds. Details of the geographic replacement of the red squirrel (Sciurus vulgaris) by the grey squirrel (Sciurus carolinensis) in Eastern England. *J. Animal Ecology*, 54:149–162, 1985.

[569] M.K. Richardson, A. Hornbruch, and L. Wolpert. Pigment patterns in neural crest chimeras constructed from quail and guinea fowl embryos. *Dev. Biol.*, 143:309–319, 1991.

[570] M.K. Richardson and G. Keuck. A question of intent: when is a 'schematic' illustration or fraud? *Nature*, 410:144, 2001.

[571] R.D. Riddle and C.J. Tabin. How limbs develop. *Scientific American*, pages 54–59, February 1999.

[572] J. Rinzel. Models in neurobiology. In R.H. Enns, B.L. Jones, R.M. Miura, and S.S. Rangnekar, editors, *Nonlinear Phenomena in Physics and Biology*, pages 345–367. Plenum, New York, 1981.

[573] J. Rinzel and J.B. Keller. Traveling wave solutions of a nerve conduction equation. *Biophys. J.*, 13:1313–1337, 1973.

[574] J. Rinzel and D. Terman. Propagation phenomena in a bistable reaction-diffusion system. *SIAM J. Appl. Math.*, 42:1111–1137, 1982.

[575] G.B. Risse. A long pull, a strong pull and all together—San Francisco and bubonic plague 1907-1908. *Bull.Hist. Med.*, 66:260–286, 1992.

[576] H. Ritvo. *The Animal Estate. The English and Other Creatures in the Victorian Age.* Harvard University Press, Cambridge, 1987.

[577] E.K. Rodriguez, A. Hoger, and A.D. McCulloch. Stress-dependent finite growth in soft elastic tissues. *J. Biomech.*, 27:455–467, 1994.

[578] A.S. Romer. *Vertebrate Palaeontology.* University of Chicago Press, Chicago, 1977.

[579] E. Röse. Ueber die Zahnentwicklung der Crocodile. *Morph. Arbeit.*, 3:195–228, 1894.

[580] T. Rytömaa and K. Kiviniemi. Chloroma repression induced by the granulocytic chalone. *Nature*, 222:995–996, 1969.

[581] T. Rytömaa and K. Kiviniemi. Regression of generalised leukemia in rat induced by the granulocytic chalone. *Eur. J. Cancer*, 6:401–410, 1970.

[582] E.H. Sage. Pieces of eight: bioactive fragments of extracellular proteins as regulators of angiogenesis. *Trends in Cell Biol.*, 7:182–186, 1997a.

[583] E.H. Sage. Terms of attachment: SPARC and tumorigenesis. *Nature Medicine*, 3:171–176, 1997b.

[584] D. Savic. Models of pattern formation in animal coatings. *J. Theor. Biol.*, 172:299–303, 1995.

[585] B. Schaumann and M. Alter. *Dermatoglyphics in Medical Disorders.* Springer-Verlag, New York, 1976.

[586] G.W. Scherer, H. Hdach, and J. Phalippou. Thermal expansion of gels: A novel method for measuring permeability. *J. of Non-Crystalline Solids*, 130:157–170, 1991.

[587] S. Schmidt and P. Ortoleva. Asymptotic solutions of the FKN chemical wave equation. *J. Chem. Phys.*, 72:2733–2736, 1980.

[588] J. Schnackenberg. Simple chemical reaction systems with limit cycle behaviour. *J. Theor. Biol.*, 81:389–400, 1979.

[589] A.M. Schor, S.L. Schor, and R. Baillie. Angiogenesis: experimental data relevant to theoretical analysis. In M.A.J. Chaplain, G.D. Singh, and J.C. McLachlan, editors, *On Growth and Form. Spatio-Temporal Pattern Formation in Biology*, pages 201–224. John Wiley, New York, 1999.

[590] B.N. Schwanwitsch. On the ground-plan of wing-pattern in nymphalids and other families of rhopalocerous Lepidoptera. *Proc. Zool. Soc. (Lond.)*, 34:509–528, 1924.

[591] V. Schwartz. Neue Versuche zur Determination des zentralen Symmetriesystems bei Plodia interpunctella. *Biol. Zentr.*, 81:19–44, 1962.

[592] A.G. Searle. *Comparative Genetics of Coat Colour in Mammals.* Academic, London, 1968.

[593] L.A. Segel, editor. *Mathematics Applied to Continuum Mechanics.* Macmillan, New York, 1977.

[594] T. Sekimura, P.K. Maini, J.B. Nardi, M. Zhu, and J.D. Murray. Pattern formation in lepidopteran wings. *Comments in Theor. Biol.*, 5:69–87, 1998.

[595] T. Sekimura, M. Zhu, J. Cook, P.K. Maini, and J.D. Murray. Pattern formation of scale cells in lepidoptera differential origin-dependent cell adhesion. *Bull. Math. Biol.*, 61:807–827, 1999.

[596] P. Sengel. *Morphogenesis of Skin.* Cambridge University Press, Cambridge, UK, 1976.

[597] L.J. Shaw and J.D. Murray. Analysis of a model for complex skin patterns. *SIAM J. Appl. Math.*, 50:279–293, 1990.

[598] P.R. Sheldon. Parallel gradualistic evolution of Ordovician trilobites. *Nature*, 330:561–563, 1987.

[599] J.A. Sherratt. *Mathematical Models of Wound Healing.* PhD thesis, University of Oxford, 1991.

[600] J.A. Sherratt. Actin aggregation and embryonic epidermal wound healing. *J. Math. Biol.*, 31:703–716, 1993.

[601] J.A. Sherratt. Chemotaxis and chemokinesis in eukaryotic cells: the Keller–Segel equations as an approximation to a detailed model. *Bull. Math. Biol.*, 56:129–146, 1994.

[602] J.A. Sherratt and J. Lewis. Stress-induced alignment of actin filaments and the mechanics of cytogel. *Bull. Math. Biol.*, 55:637–654, 1993.

[603] J.A. Sherratt and J.D. Murray. Models of epidermal wound healing. *Proc. R. Soc. Lond. B*, 241:29–36, 1990.

[604] J.A. Sherratt and J.D. Murray. Mathematical analysis of a basic model for epidermal wound healing. *J. Math. Biol.*, 29:389–404, 1991.

[605] J.A. Sherratt and J.D. Murray. Epidermal wound healing: A theoretical approach. *Comm. Theor. Biol.*, 2:315–333, 1992a.

[606] J.A. Sherratt and J.D. Murray. Epidermal wound healing: the clinical implications of a simple mathematical model. *Cell Transplant*, 1:365–371, 1992b.

[607] J.A. Sherratt, M.A. Lewis, and A.C. Fowler. Ecological chaos in the wake of invasion. *Proc. Nat. Acad. Sci. USA*, 92:2524–2528, 1995.

[608] J.A. Sherratt, P. Martin, J.D. Murray, and J. Lewis. Mathematical models of wound healing in embryonic and adult epidermis. *IMA J. Math. Appl. Med. & Biol.*, 9:177–196, 1992.

[609] J.A. Sherratt, E.H. Sage, and J.D. Murray. Chemical control of eukaryotic cell movement: a new model. *J. Theor. Biol.*, 162:23–44, 1993b.

[610] M. Shibata and J. Bureš. Reverberation of cortical spreading depression along closed pathways in rat cerebral cortex. *J. Neurophysiol.*, 35:381–388, 1972.

[611] M. Shibata and J. Bureš. Optimum topographical conditions for reverberating cortical spreading depression in rats. *J. Neurobiol.*, 5:107–118, 1974.

[612] N. Shigesada and K. Kawasaki. *Biological Invasions: Theory and Practice*. Oxford University Press, Oxford, 1997.

[613] N. Shigesada, K. Kawasaki, and E. Teramoto. Travelling periodic waves in heterogeneous environments. *Theor. Popul. Biol.*, 30:143–160, 1986.

[614] E. Shochat, D. Hart, and Z. Agur. Using computer simulation for evaluating the efficacy of breast cancer chemotherapy protocols. *Math. Models and Methods in Appl. Sciences*, 9:599–615, 1999.

[615] N. Shubin and P. Alberch. A morphogenetic appraoch to the origin and basic organisation of the tetrapod limb. In M. Hecht, B. Wallace, and W. Steere, editors, *Evolutionary Biology*, volume 20, pages 319–387. Plenum, New York, 1986.

[616] S. Shuster. The cause of striae distensae. *Acta Dermato-Venereologica*, 59:161–169, 1979. (Supplement 85).

[617] A. Sibatani. Wing homeosis in Lepidoptera: a survey. *Devl. Biol.*, 79:1–18, 1981.

[618] D.L. Silbergeld and M.R. Chicoine. Isolation and characterization of human malignant glioma cells from histologically normal brain. *J. Neurosurg.*, 86:525–531, March 1997.

[619] D.L. Silbergeld, R.C. Rostomily, and E.C. Alvord, Jr. The cause of death in patients with glioblastoma is multifactorial: Clinical factors and autopsy findings in 117 cases of supratentorial glioblastoma in adults. *J. Neuro-Oncol.*, 10:179–185, 1991.

[620] R. Skalak and R.J. Price. The role of mechanical stresses in microvascular remodeling. *Microcirculation*, 3:143–165, 1996.

[621] R. Skalak, G. Dasgupta, M. Moss, E. Otten, P. Dullemeijer, and H. Vilmann. Analytical description of growth. *J. Theor. Biol.*, 94:555–577, 1982.

[622] J.M.W. Slack. *From Egg to Embryo. Determinative Events in Early Development*. Cambridge University Press, Cambridge, UK, 1983.

[623] J.L.R.M. Smeets, M.A. Allessie, W.J.E.P. Lammers, F.I.M. Bonke, and J. Hollen. The wavelength of the cardiac impulse and the reentrant arrhythmias in isolated rabbit atrium. The role of heart rate, autonomic transmitters, temperature and potassium. *Circ. Res.*, 73:96–108, 1986.

[624] J.C. Smith and L. Wolpert. Pattern formation along the anterioposterior axis of the chick wing: the increase in width following a polarizing region graft and the effect of X-irradiation. *J. Embryol. Exp. Morph.*, 63:127–144, 1981.

[625] J. Smoller. *Shock Waves and Reaction-Diffusion Equations*. Springer-Verlag, Berlin-Heidelberg-New York, 1983.

[626] J. Sneyd, A. Atri, M.W.J. Ferguson, M.A. Lewis, W. Seward, and J.D. Murray. A model for the spatial patterning of teeth primordia in the Alligator: Initiation of the dental determinant. *J. Theor. Biol.*, 165:633–658, 1993.

[627] J.M. Snowden. Wound closure: an analysis of the relative contributions of contraction and epithelialization. *J. Surg. Res.*, 37:453–463, 1984.

[628] M.K. Sparrow and P.J. Sparrow. *Topological Approach to the Matching of Single Fingerprints: Development of Algorithms for Use on Rolled Impressions*. US Govt. Printing Office (NBS/SP-50/124), Washington, DC, 1985.

[629] A. Spiros and L. Edelstein-Keshet. Testing a model for the dynamics of actin structures with biological parameter values. *Bull. Math. Biol.*, 60:275–305, 1998.

[630] F. Steck and A. Wandeler. The epidemiology of fox rabies in Europe. *Epidem. Rev.*, 2:71–96, 1980.

[631] O. Steinbock, P. Kuttunen, and K. Showalter. Chemical wave logic gates. *J. Chem. Phys.*, 100:18970–18975, 1996.

[632] M.G. Stern, M.T. Longaker, and R. Stern. Hyaluronic acid and its modulation in fetal and adult wound. In N.S. Adzick and M.T. Longaker, editors, *Fetal Wound Healing*, pages 189–198. Elsevier, New York, 1992.

[633] C.R. Stockard. The artificial production of one-eyed monsters and other defects, which occur in nature, by the use of chemicals. *Proc. Assoc. of Amer. Anatomists*, III(4):167–173, 1909.

[634] C.R. Stockard. Development rate and structural expression: an experimental study of twins, "double monsters" and single deformities, and the interaction among embryonic organs during their origin and development. *Amer. J. Anatomy*, 28:115–266, 1921.

[635] S.H. Strogatz. *Nonlinear Dynamics and Chaos: with Applications in Physics, Biology, Chemistry, and Engineering*. Addison-Wesley, Reading, MA, 1994.

[636] F.A. Stuart, K.H. Mahmood, J.L. Stanford, and D.G. Pritchard. Development of diagnostic test for, and vaccination against, tuberculosis in badgers. *Mammal Review*, 18:74–75, 1988.

[637] F. Suffert. Zur vergleichenden Analyse der Schmetterlingszeichnung. *Bull. Zentr.*, 47:385–413, 1927.

[638] G.W. Swan. Tumour growth models and cancer therapy. In J.R. Thomson and B.W. Brown, editors, *Cancer Modeling*, pages 91–104. Marcel Dekker, New York, 1987.

[639] K.R. Swanson. *Mathematical Modeling of the Growth and Control of Tumors*. PhD thesis, University of Washington, Seattle, WA, 1999.

[640] K.R. Swanson, E.C. Alvord, Jr, and J.D. Murray. A quantitative model for differential motility of gliomas in grey and white matter. *Cell Prolif.*, 33:317–329, 2000.

[641] N.V. Swindale. A model for the formation of ocular dominance stripes. *Proc. R. Soc. Lond. B*, 208:243–264, 1980.

[642] L.A. Taber. Biomechanics of growth, remodeling, and morphogenesis. *Appl. Mech. Rev.*, 48:487–545, 1995.

[643] P. Tass. Cortical pattern formation during visual hallucinations. *J. Biol. Phys.*, 21:177–210, 1995.

[644] O.R. Taylor. The past and possible future spread of Africanized honeybees in the Americas. *Bee World*, 58:19–30, 1977.

[645] L. Teulières and P. Saliou. Rabies in France, 100 years after Pasteur. *Presse Medicale*, 24:134–135, 1995.

[646] I. Thesleff and A.M. Partanen. Localization and quantitation of 125I-epidermal growth factor binding in mouse embryonic tooth and other embryonic tissues at different developmental stages. *Dev. Biol.*, 120:186–197, 1987.

[647] R. Thoma. *Untersuchungen ber die Histogenese und Histomechanik*. Enkeverlag, Stuttgart, 1893.

[648] D. Thomas. Artificial enzyme membranes, transport, memory, and oscillatory phenomena. In D. Thomas and J.-P. Kernevez, editors, *Analysis and Control of Immobilized Enzyme Systems*, pages 115–150. Springer-Verlag, Berlin-Heidelberg-New York, 1975.

[649] D'Arcy W. Thompson. *On Growth and Form*. Cambridge University Press, Cambridge, UK, 1917.

[650] P. Thorogood. Morphogenesis of cartilage. In B.K. Hall, editor, *Cartilage: Development, Differentiation and Growth*, volume 2. Academic, New York, 1983.

[651] C. Tickle. Development of the vertebrate limb: a model for growth and patterning. In M.A.J. Chaplain, G.D. Singh, and J.C McLachlan, editors, *On Growth and Form. Spatio-Temporal Pattern Formation in Biology*, pages 13–29. John Wiley, New York, 1999.

[652] C. Tickle, J. Lee, and G. Eichele. A quantitative analysis of the effect of all-trans-retinoic acid on the pattern of chick wing development. *Dev. Biol.*, 109:82–95, 1985.

[653] D. Tilman and P. Kareiva, editors. *Spatial Ecology. The Role of Space in Population Dynamics and Interspecific Interactions*. Princeton University Press, Princeton, NJ, 1997.

[654] E.J.F. Timmenga, T.T. Andreassen, H.J. Houthoff, and P.J. Klopper. The effect of mechanical stress on healing skin wounds: an experimental study in rabbits using tissue expansion. *Brit. J. Plastic Surg.*, 44:514–519, 1991.

[655] E.C. Titchmarsh. *Eigenfunctions Expansions Associated with Second-Order Differential Equations*. Clarendon, Oxford, 1964.

[656] A.W. Toga, E.M. Santori, R. Hazani, and K. Ambach. A 3D digital map of the rat brain. *Brain Res. Bull.*, 38:77–85, 1995.

[657] B. Toma and L. Andral. Epidemiology of fox rabies. *Adv. Vir. Res.*, 21:15, 1977.

[658] A. Tozeren and R. Skalak. Interaction of stress and growth in a fibrous tissue. *J. Theor. Biol.*, 130:337–350, 1988.

[659] P. Tracqui. From passive diffusion to active cellular migration in mathematical models of tumour invasion. *Acta Biotheoretica*, 43:443–464, 1995.

[660] P. Tracqui, G.C. Cruywagen, D.E. Woodward, G.T. Bartoo, J.D. Murray, and E.C. Alvord, Jr. A mathematical model of glioma growth: the effect of chemotherapy on spatio-temporal growth. *Cell Proliferation*, 28:17–31, 1995.

[661] P. Tracqui, D.E. Woodward, G.C. Cruywagen, and J.D. Murray. A mechanical model for fibroblast-driven wound healing. *J. Biol. Syst.*, 3:1075–1085, 1993.

[662] L. Tranqui and P. Tracqui. Mechanical signalling and angiogenesis. the integration of cell-extracellular matrix couplings. *C.R. Acad. Sci. Paris, Science de la Vie*, 323:31–47, 2000.

[663] R.T. Tranquillo and D.A. Lauffenburger. Stochastic model of leukocyte chemosensory movement. *J. Math. Biol.*, 25:229–262, 1987.

[664] R.T. Tranquillo and J.D. Murray. Continuum model of fibroblast-driven wound contraction: inflammation mediation. *J. Theor. Biol.*, 158:135–172, 1992.

[665] R.T. Tranquillo and J.D. Murray. Mechanistic model of wound contraction. *J. Surg. Research*, 55:233–247, 1993.

[666] R.W. Treadwell. Time and sequence of appearance of certain gross structures in *Pituophis melanoleucus sayi* embryos. *Herpetologica*, 18:120–124, 1962.

[667] J.P. Trinkaus. Formation of protrusions of the cell surface during tissue cell movement. In R.D. Hynes and C.E. Fox, editors, *Tumor Cell Surfaces and Malignancy*, pages 887–906. Alan R. Liss, New York, 1980.

[668] J.P. Trinkaus. *Cells into Organs. The Forces that Shape the Embryo*. Prentice-Hall, Englewood Cliffs, NJ, 1984.

[669] T. Tsujikawa, T. Nagai, M. Mimura, R. Kobayashi, and H. Ikeda. Stability properties of traveling pulse solutions of the higher dimensional FitzHugh–Nagumo equations. *Japan J. Appl. Math.*, 6:341–366, 1989.

[670] A.M. Turing. The chemical basis of morphogenesis. *Phil. Trans. R. Soc. Lond. B*, 237:37–72, 1952.

[671] W. Turner. *A Compleat History Of the Most Remarkable Providence, both of Judgement and Mercy, Which have hapened in this Present Age*. John Duynton, Raven, Jewet Street, London, 1697.

[672] J.J. Tyson and P.C. Fife. Target patterns in a realistic model of the Belousov–Zhabotinskii reaction. *J. Chem. Phys.*, 75:2224–2237, 1980.

[673] J.J. Tyson and J.P. Keener. Singular perturbation theory of travelling waves in excitable media (a review). *Physica D*, 32:327–361, 1988.

[674] J.J. Tyson, K.A. Alexander, V.S. Manoranjan, and J.D. Murray. Cyclic-AMP waves during aggregation of *Dictyostelium* amoebae. *Development*, 106:421–426, 1989a.

[675] J.J. Tyson, K.A. Alexander, V.S. Manoranjan, and J.D. Murray. Spiral waves of cyclic-AMP in a model of slime mold aggregation. *Physica D*, 34:193–207, 1989b.

[676] R.C. Tyson. *Pattern Formation by E. coli—Mathematical and Numerical Investigation of a Biological Phenomenon*. PhD thesis, Department of Applied Mathematics, University of Washington, Seattle, WA, 1996.

[677] R.C. Tyson, S.R. Lubkin, and J.D. Murray. A minimal mechanism for bacterial pattern formation. *Proc. R. Soc. Lond. B*, 266:299–304, 1998.

[678] R.C. Tyson, S.R. Lubkin, and J.D. Murray. Model and analysis of chemotactic bacterial patterns in liquid medium. *J. Math. Biol.*, 38:359–375, 1999.

[679] S. Vainio, I. Karanvanova, A. Jowett, and I. Thesleff. Identification of BMP-4 as a signal mediating secondary induction between epithelial and mesenchymal tissues during early tooth development. *Cell*, 75:45–58, 1993.

[680] E.S. Valenstein. History of psychosurgery. In S.H. Greenblatt, editor, *A History of Neurosurgery*, pages 499–516. American Society of Neurological Surgeons, Park Ridge, IL, 1997.

[681] V. van Ballenberghe, A.W. Erickson, and D. Byman. Ecology of the timber wolf in Northeastern Minnesota. *Wildl. Monogr.*, 43:1–43, 1975.

[682] H.A.S. Van den Brenk. Studies in restorative growth processes in mammalian wound healing. *Br. J. Surg.*, 43:525–550, 1956.

[683] J.W. Verano, L.S. Anderson, and R. Franco. Foot amputation by the Moche of ancient Peru: osteological evidence and archaeological context. *Int. J. Osteoarchaeol.*, 10:177–188, 2000.

[684] J.W. Verano, S. Uceda, C. Chapdelaine, R. Tello, M.I. Paredes, and V. Pimentel. Modified human skulls from the urban sector of the pyramids of Moche, northern Peru. *Latin American Antiquity*, 10:59–70, 1999.

[685] R.B. Vernon and E.H. Sage. Between molecules and morphology. extracellular matrix and creation of vascular form. *Amer. J. of Pathol.*, 147:873–883, 1995.

[686] R.B. Vernon, J.C. Angello, M.L. Iruela-Arispe, T.F. Lane, and E.H. Sage. Reorganization of basement membrane matrices by cellular traction promotes the formation of cellular networks *in vitro*. *Laboratory Investigation*, 66:536–547, 1992.

[687] R.B. Vernon, S.L. Lara, M.L. Drake, M.L. Iruela-Arispe, J.C. Angello, C.D. Little, T.N. Wight, and E.H. Sage. Organized type I collagen influences endothelial patterns during "spontaneous" angiogenesis *in vitro*: Planar cultures as models of vascular development. *In Vitro Vascular and Dev. Biol.*, 31:120–131, 1995.

[688] C.H. Waddington and J. Cowe. Computer simulations of a molluscan pigmentation pattern. *J. Theor. Biol.*, 25:219–225, 1969.

[689] V. Walbot and N. Holder. *Developmental Biology*. Random House, New York, 1987.

[690] H. Walker, editor. *A History of Neurological Surgery*. Williams and Wilkins, Baltimore, 1951.

[691] J. Walker. About phosphenes: luminous patterns that appear when the eyes are closed. *Scientific American*, 244:174–184, 1981.

[692] M. Walter, A. Fournier, and M. Reimers. Clonal mosaic model for the synthesis of mammalian coat patterns. *Graphics Interface '98 (Intern. Computer Graphics Conf. Vancouver)*, pages 82–91, 1998.

[693] A. Wandeler. Rabies virus. In M.J. Appel, editor, *Virus Infections of Carnivores*, pages 449–461. Elsevier Science, Amsterdam, 1987.

[694] D.A. Warrell. Rabies in man. In C. Kapalan, editor, *Rabies: the Facts*, pages 32–52. Oxford University Press, Oxford, 1977.

[695] F.M. Watt. The extracellular matrix and cell shape. *Trends in Biochem. Sci.*, 11:482–485, 1986.

[696] F.M. Watt. Proliferation and terminal differentiation of human epidermal keratinocytes in culture. *Biochem. Soc. Trans.*, 16:666–668, 1988a.

[697] F.M. Watt. The epidermal keratinocyte. *Bioessays*, 8:163–167, 1988b.

[698] G.T. Watt. Wound shape and tissue tension in healing. *Brit. J. Surg.*, 47:555–561, 1959.

[699] M.P. Welch, G.F. Odland, and A.F. Clark. Temporal relationships of F-actin bundle formation, collagen and fibronectin matrix assembly, and fibronectin receptor expression to wound contraction. *J. Cell Biol.*, 110:133–145, 1990.

[700] K.F. Wellmann. North American Indian rock art and hallucinogenic drugs. *J. Amer. Medical Assoc.*, 239:1524–1527, 1978.

[701] B.J. Welsh, J. Gomatam, and A.E. Burgess. Three-dimensional chemical waves in the Belousov–Zhabotinskii reaction. *Nature*, 304:611–614, 1983.

[702] S. Werner, H. Smola, X. Liao, M.T. Longaker, T. Krieg, P.H. Hofschneider, and L.T. Williams. The function of KGF in epithelial morphogenesis and wound reepithelialization. *Science*, 266:819–822, 1994.

[703] N. Wessells. *Tissue Interaction in Development*. W. A. Benjamin, Menlo Park, CA, 1977.

[704] B. Westergaard and M.W.J. Ferguson. Development of the dentition in *Alligator mississippiensis*. early embryonic development in the lower jaw. *J. Zool. Lond.*, 210:575–597, 1986.

[705] B. Westergaard and M.W.J. Ferguson. Development of the dentition in *Alligator mississippiensis*. later development in the lower jaws of embryos, hatchlings and young juveniles. *J. Zool. Lond.*, 212:191–222, 1987.

[706] B. Westergaard and M.W.J. Ferguson. Development of the dentition in *Alligator mississippiensis*. upper jaw dental and craniofacial development in embryos, hatchlings, and young juveniles. with a comparison to lower jaw development. *Amer. J. Anatomy*, 187:393–421, 1990.

[707] J.M. Westergaard. Measures applied in Denmark to control the rabies epizootic in 1977–1980. *Comp. Immunol. Microbiol. Infect. Dis.*, 5:383–387, 1982.

[708] D.J. Whitby and M.W.J. Ferguson. The extracellular matrix of lip wounds in fetal, neonatal and adult mice. *Development*, 112:651–668, 1991.

[709] K.A.J. White. *Territoriality and Survival in Wolf-Deer Interactions*. PhD thesis, Department of Applied Mathematics, University of Washington, Seattle, WA, 1995.

[710] K.A.J. White, M.A. Lewis, and J.D. Murray. A model for wolf-pack territory formation and maintenance. *J. Theor. Biol.*, 178:29–43, 1996a.

[711] K.A.J. White, M.A. Lewis, and J.D. Murray. Wolf-deer interactions: a mathematical model. *Proc. R. Soc. Lond. B*, 263:299–305, 1996b.

[712] K.A.J. White, M.A. Lewis, and J.D. Murray. On wolf territoriality and deer survival. In J. Bascompte and R.V. Sole, editors, *Modeling Spatiotemporal Dynamics in Ecology*, pages 105–126. Springer-Verlag, New York, 1998.

[713] M.H. Williamson. *Biological Invasions*. Chapman and Hall, London, 1996.

[714] M.H. Williamson and K.C. Brown. The analysis and modelling of British invasions. *Phil. Trans. R. Soc. Lond. B*, 314:505–522, 1986.

[715] P.G. Williamson. Palaeontological documentation of speciation in Cenozoic molluscs from Turkana Basan. *Nature*, 293:437–443, 1981a.

[716] P.G. Williamson. Morphological stasis and developmental constraints: real problems for neo-Darwinism. *Nature*, 294:214–215, 1981b.

[717] A.T. Winfree. Spiral waves of chemical activity. *Science*, 175:634–636, 1972.

[718] A.T. Winfree. Rotating chemical reactions. *Sci. Amer.*, 230(6):82–95, 1974.

[719] A.T. Winfree. The rotor in reaction-diffusion problems and in sudden cardiac death. In M. Cosnard and J. Demongeot, editors, *Lect. Notes in Biomathematics (Luminy Symposium on Oscillations, 1981)*, volume 49, pages 201–207, Berlin-Heidelberg-New York, 1983a. Springer-Verlag.

[720] A.T. Winfree. Sudden cardiac death: a problem in topology. *Sci. Amer.*, 248(5):144–161, 1983b.

[721] A.T. Winfree. *The Timing of Biological Clocks*. Scientific American Books, New York, 1987a.

[722] A.T. Winfree. *When Time Breaks Down: The Three-Dimensional Dynamics of Electrochemical Waves and Cardiac Arrhythmias*. Princeton University Press, Princeton, NJ, 1987b.

[723] A.T. Winfree. Electrical turbulence in three-dimensional heart muscle. *Science*, 266:1003–1006, 1994a.

[724] A.T. Winfree. Persistent tangled vortex rings in generic excitable media. *Nature*, 371:233–236, 1994b.

[725] A.T. Winfree. Mechanisms of cardiac fibrillation. *Science*, 270:1222–1225, 1995.

[726] A.T. Winfree. Heart muscle as a reaction diffusion medium: the roles of electrical potential diffusion, activation front curvature, and anisotropy. *Internat. J. Bifurc. and Chaos Appl. Sci. Engrg.*, 7:487–526, 1997.

[727] A.T. Winfree. *The Geometry of Biological Time*. Springer-Verlag, Berlin-Heidelberg-New York, 2nd edition, 2000.

[728] A.T. Winfree and S.H. Strogatz. Organising centres for three-dimensional chemical waves. *Nature*, 311:611–615, 1984.

[729] A.T. Winfree, G. Caudle, P. McGuire, and Z. Szilagyi. Quantitative optical tomography of chemical waves and their organising centers. *Chaos*, 6:617–626, 1996.

[730] K.H. Winters, M.R. Myerscough, P.K. Maini, and J.D. Murray. Tracking bifurcating solutions of a model biological pattern generator. *IMPACT of Computing in Sci. and Eng.*, 2:355–371, 1990.

[731] R. Wittenberg. *Models of Self-Organisation in Biological Development*. Master's thesis, University of Cape Town, 1993.

[732] M.W. Woerdeman. Beitrage zur Entwicklungsgeschichte von Zahnen und Gebiss der Reptilien. Beitrag I. Die Anlage und Entwicklung des embryonalen Gebisses als Ganzes. *Arch. mikrosk. Anat.*, 92:104–192, 1919.

[733] M.W. Woerdeman. Beitrage zur Entwicklungsgeschichte von Zahnen und Gebiss der Reptilien. Beitrag IV. Uber die Anlage und Entwicklung der Zahne. *Arch. mikrosk. Anat.*, 95:265–395, 1921.

[734] S. Wolfram. Cellular automata as models of complexity. *Nature*, 311:419–424, 1984.

[735] D.J. Wollkind and L.E. Stephenson. Chemical Turing pattern formation analyses: Comparison of theory and experiment. *SIAM J. Appl. Math.*, 61:387–431, 2000a.

[736] D.J. Wollkind and L.E. Stephenson. Chemical Turing patterns: a model system of a paradigm for morphogenesis. In P.K. Maini and H.G. Othmer, editors, *Mathematical Models for Biological Pattern Formation*, pages 113–142. Springer-Verlag, New York, 2000b.

[737] D.J. Wollkind, V.S. Manoranjan, and L. Zhang. Weakly nonlinear stability analyses of prototype reaction-diffusion model equations. *SIAM Rev.*, 36:176–214, 1994.

[738] L. Wolpert. Positional information and the spatial pattern of cellular differentiation. *J. Theor. Biol.*, 25:1–47, 1969.

[739] L. Wolpert. Positional information and pattern formation. *Curr. Top. Dev. Biol.*, 6:183–224, 1971.

[740] L. Wolpert. The development of pattern and form in animals. *Carolina Biol. Readers*, 1(5):1–16, 1977.

[741] L. Wolpert. Positional information and pattern formation. *Phil. Trans. R. Soc. Lond. B*, 295:441–450, 1981.

[742] L. Wolpert and A. Hornbruch. Positional signalling and the development of the humerus in the chick limb bud. *Development*, 100:333–338, 1987.

[743] L. Wolpert and W.D. Stein. Molecular aspects of early development. In G.M. Malacinski and S.V. Bryant, editors, *Proc. Symp. on Molecular Aspects of Early Development (Annual Meeting Amer. Soc. Zoologists, Louisville, 1982)*, pages 2–21, MacMillan, New York, 1984.

[744] D. E. Woodward, J. Cook, P. Tracqui, G.C. Cruywagen, J.D. Murray, and E.C. Alvord, Jr. A mathematical model of glioma growth: the effect of extent of surgical resection. *Cell Proliferation*, 29:269–288, 1996.

[745] D.E. Woodward, R.C. Tyson, J.D. Murray, E.O. Budrene, and H. Berg. Spatio-temporal patterns generated by *Salmonella typhimurium*. *Biophysical J.*, 68:2181–2189, 1995.

[746] N.A. Wright. Cell proliferation kinetics of the epidermis. In L.A. Goldsmith, editor, *Biochemistry and Physiology of the Skin*, pages 203–229. Oxford University Press, Oxford, 1983.

[747] N.A. Wright and M. Alison. *Biology of Epithelial Cell Populations*. Clarendon, Oxford, 1984.

[748] T. Wyatt. The biology of *Oikopleara dioica* and *Fritillaria borealis* in the Southern Bight. *Mar. Biol.*, 22:137–158, 1973.

[749] Y. Xu, C.M. Vest, and J.D. Murray. Holographic interferometry used to demonstrate a theory of pattern formation in animal coats. *Appl. Optics.*, 22:3479–3483, 1983.

[750] S. Yachi, K. Kawasaki, N. Shigesada, and E. Teramoto. Spatial patterns of propagating waves of fox rabies. *Forma*, 4:3–12, 1989.

[751] H. Yagisita, M. Mimura, and M. Yamada. Spiral wave behaviors in an excitable reaction-diffusion system on a sphere. *Physica D*, 124:126–136, 1998.

[752] T. Yamaguchi, T. Hirobe, Y. Kinjo, and K. Manaka. The effect of chalone on the cell cycle in the epidermis during wound healing. *Exp. Cell Res.*, 89:247–254, 1974.

[753] A. Yoshikawa and M. Yamaguti. On some further properties of solutions to a certain semi-linear system of partial differential equations. *Publ. RIMS, Kyoto Univ.*, 9:577–595, 1974.

[754] D.A. Young. A local activator-inhibitor model of vertebrate skin patterns. *Math. Biosciences*, 72:51–58, 1984.

[755] G.L. Yount, D.A. Haas-Kogan, K.S. Levine, K.D. Aldape, and M.A. Israel. Ionizing radiation inhibits chemotherapy-induced apoptosis in cultured glioma cells: implications for combined modality therapy. *Cancer Res.*, 58:3819–3825, 1998.

[756] A.N. Zaikin and A.M. Zhabotinskii. Concentration wave propagation in two-dimensional liquid-phase self-organising system. *Nature*, 225:535–537, 1970.

[757] L.R. Zehr. Stages in the development of the common garter snake *Thamnophis sirtalis sirtalis*. *Copeia*, 1962:322–329, 1962.

[758] M. Zhu and J.D. Murray. Parameter domains for generating spatial pattern: a comparison of reaction-diffusion and cell-chemotaxis models. *Intern. J. Bifurcation & Chaos*, 5:1503–1524, 1995.

[759] J.D. Zieske, S.C. Higashij, S.J. Spurrmic, and I.K. Gipson. Biosynthetic response of the rabbit cornea to a keratectomy wound. *Invest. Ophthalmol. Vis. Sci.*, 28:1668–1677, 1987.

[760] R.G. Zweifel. Genetics of color pattern polymorphism in the California king snake. *J. Heredity*, 72:238–244, 1981.

[761] V.S. Zykov. *Modelling of Wave Processes in Excitable Media*. Manchester University Press, Manchester, 1988.

Index

Interdisciplinary Applied Mathematics

1. *Gutzwiller:* Chaos in Classical and Quantum Mechanics
2. *Wiggins:* Chaotic Transport in Dynamical Systems
3. *Joseph/Renardy:* Fundamentals of Two-Fluid Dynamics:
   Part I: Mathematical Theory and Applications
4. *Joseph/Renardy:* Fundamentals of Two-Fluid Dynamics:
   Part II: Lubricated Transport, Drops and Miscible Liquids
5. *Seydel:* Practical Bifurcation and Stability Analysis:
   From Equilibrium to Chaos
6. *Hornung:* Homogenization and Porous Media
7. *Simo/Hughes:* Computational Inelasticity
8. *Keener/Sneyd:* Mathematical Physiology
9. *Han/Reddy:* Plasticity: Mathematical Theory and Numerical Analysis
10. *Sastry:* Nonlinear Systems: Analysis, Stability, and Control
11. *McCarthy:* Geometric Design of Linkages
12. *Winfree:* The Geometry of Biological Time (Second Edition)
13. *Bleistein/Cohen/Stockwell:* Mathematics of Multidimensional
    Seismic Imaging, Migration, and Inversion
14. *Okubo/Levin:* Diffusion and Ecological Problems: Modern Perspectives
    (Second Edition)
15. *Logan:* Transport Modeling in Hydrogeochemical Systems
16. *Torquato:* Random Heterogeneous Materials: Microstructure and
    Macroscopic Properties
17. *Murray:* Mathematical Biology I: An Introduction (Third Edition)
18. *Murray:* Mathematical Biology II: Spatial Models and Biomedical
    Applications (Third Edition)
19. *Kimmel/Axelrod:* Branching Processes in Biology
20. *Fall/Marland/Wagner/Tyson:* Computational Cell Biology
21. *Schlick:* Molecular Modeling and Simulation: An Interdisciplinary Guide

Interdisciplinary Applied Mathematics

1. *Gutzwiller:* Chaos in Classical and Quantum Mechanics
2. *Wiggins:* Chaotic Transport in Dynamical Systems
3. *Joseph/Renardy:* Fundamentals of Two-Fluid Dynamics:
   Part I: Mathematical Theory and Applications
4. *Joseph/Renardy:* Fundamentals of Two-Fluid Dynamics:
   Part II: Lubricated Transport, Drops and Miscible Liquids
5. *Seydel:* Practical Bifurcation and Stability Analysis:
   From Equilibrium to Chaos
6. *Hornung:* Homogenization and Porous Media
7. *Simo/Hughes:* Computational Inelasticity
8. *Keener/Sneyd:* Mathematical Physiology
9. *Han/Reddy:* Plasticity: Mathematical Theory and Numerical Analysis
10. *Sastry:* Nonlinear Systems: Analysis, Stability, and Control
11. *McCarthy:* Geometric Design of Linkages
12. *Winfree:* The Geometry of Biological Time (Second Edition)
13. *Bleistein/Cohen/Stockwell:* Mathematics of Multidimensional
    Seismic Imaging, Migration, and Inversion
14. *Okubo/Levin:* Diffusion and Ecological Problems: Modern Perspectives
    (Second Edition)
15. *Logan:* Transport Modeling in Hydrogeochemical Systems
16. *Torquato:* Random Heterogeneous Materials: Microstructure and
    Macroscopic Properties
17. *Murray:* Mathematical Biology I: An Introduction (Third Edition)
18. *Murray:* Mathematical Biology II: Spatial Models and Biomedical
    Applications (Third Edition)
19. *Kimmel/Axelrod:* Branching Processes in Biology
20. *Fall/Marland/Wagner/Tyson:* Computational Cell Biology
21. *Schlick:* Molecular Modeling and Simulation: An Interdisciplinary Guide